area A
perimeter P
length l

width w
surface area S
altitude (height) h

base b
circumference C
radius r

volume V
area of base B
slant height s

Rectangle

$$A = lw \qquad P = 2l + 2w$$

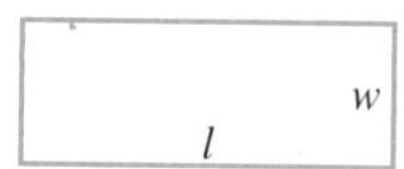

Triangle

$$A = \frac{1}{2}bh$$

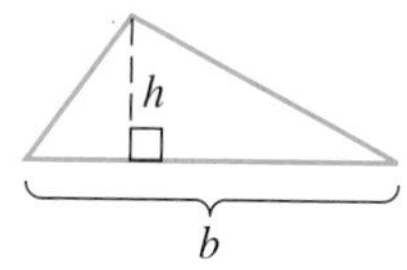

Square

$$A = s^2 \qquad P = 4s$$

Parallelogram

$$A = bh$$

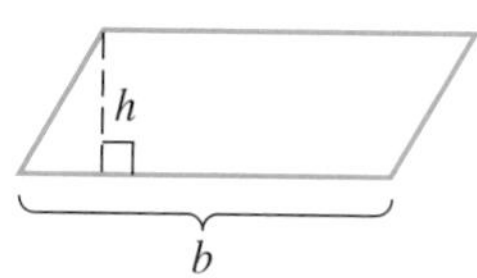

Trapezoid

$$A = \frac{1}{2}h(b_1 + b_2)$$

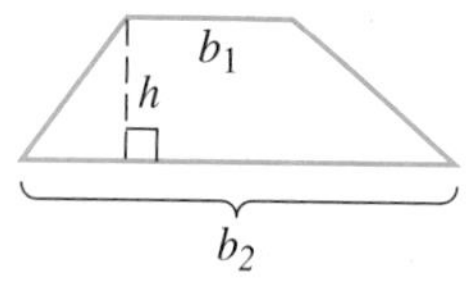

Circle

$$A = \pi r^2 \qquad C = 2\pi r$$

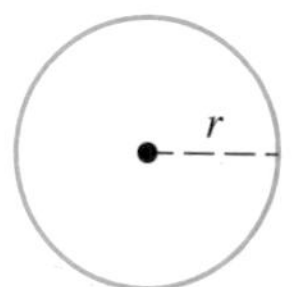

30°–60° Right Triangle

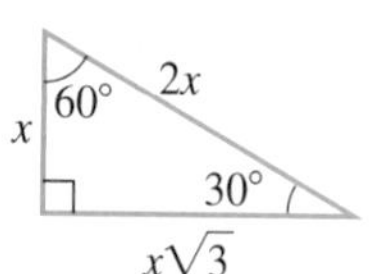

Right Triangle

$$a^2 + b^2 = c^2$$

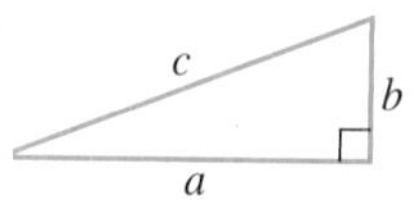

Isosceles Right Triangle

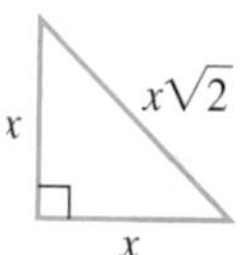

Right Circular Cylinder

$$V = \pi r^2 h \qquad S = 2\pi r^2 + 2\pi rh$$

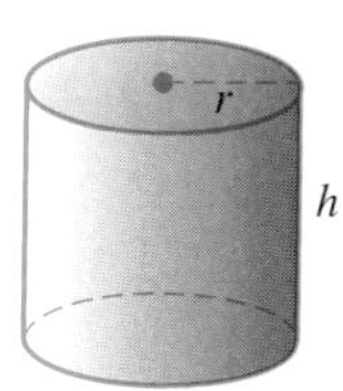

Sphere

$$S = 4\pi r^2 \qquad V = \frac{4}{3}\pi r^3$$

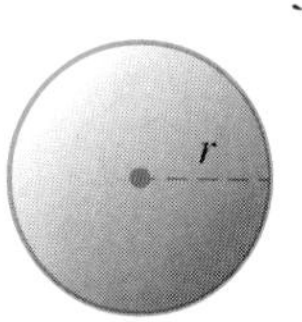

Right Circular Cone

$$V = \frac{1}{3}\pi r^2 h \qquad S = \pi r^2 + \pi rs$$

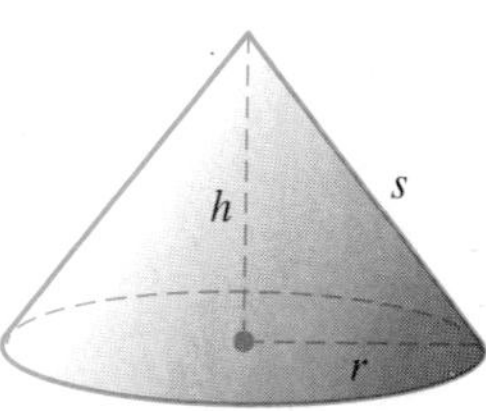

Pyramid

$$V = \frac{1}{3}Bh$$

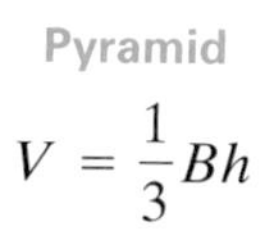

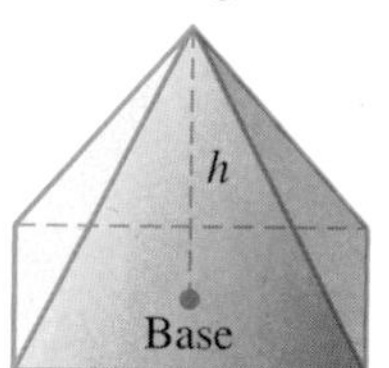

Prism

$$V = Bh$$

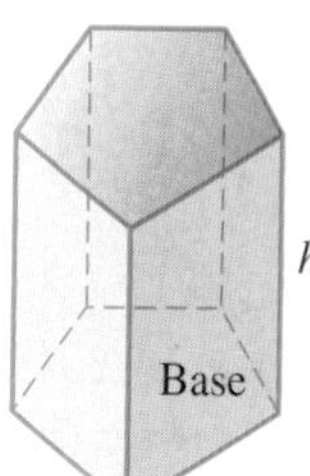

INTERMEDIATE ALGEBRA

Jerome E. Kaufmann
Karen L. Schwitters
SEMINOLE COMMUNITY COLLEGE

BROOKS/COLE
CENGAGE Learning
Australia • Brazil • Japan • Korea • Mexico • Singapore • Spain • United Kingdom • United States

Intermediate Algebra
Jerome E. Kaufmann
Karen L. Schwitters

Acquisitions Editor: Marc Bove
Publisher: Charles Van Wagner
Development Editor: Laura Localio
Assistant Editor: Stephanie Beeck
Editorial Assistant: Mary de la Cruz
Technology Project Manager: Lynh Pham
Marketing Manager: Joe Rogove
Marketing Assistant: Angela Kim
Marketing Communications Manager: Katherine Malatesta
Project Manager, Editorial Production: Hal Humphrey
Art Director: Vernon Boes
Print Buyer: Paula Vang
Permissions Editor: Bob Kauser
Production Service: Fran Andersen, Newgen North America
Text Designer: Diane Beasley
Photo Researcher: Pre-Press PMG
Copy Editor: Matt Darnell
Illustrators: Accurate Art, Precision Graphics, Rolin Graphics, Pre-Press PMG
Cover Designer: Diane Beasley
Cover Image: Don Farrall/Getty Images
Compositor: Newgen

Library of Congress Control Number: 2008934178

ISBN-13: 978-0-495-38798-5
ISBN-10: 0-495-38798-3

Brooks/Cole
10 Davis Drive
Belmont, CA 94002-3098
USA

Cengage Learning is a leading provider of customized learning solutions with office locations around the globe, including Singapore, the United Kingdom, Australia, Mexico, Brazil, and Japan. Locate your local office at **international.cengage.com/region**.

Cengage Learning products are represented in Canada by Nelson Education, Ltd.

For your course and learning solutions, visit **academic.cengage.com**.
Purchase any of our products at your local college store or at our preferred online store **www.ichapters.com**.

Printed in Canada
1 2 3 4 5 6 7 12 11 10 09 08

Contents

4 Systems of Equations 179

5 Polynomials 223

6 Rational Expressions 291

7 Exponents and Radicals 359

11 Exponential and Logarithmic Functions 607

Appendices 663

Preface

When preparing *Intermediate Algebra*, First Edition, we attempted to preserve the features that made the previous editions of our hardcover series successful. At the same time, we made a special effort to incorporate many changes and improvements, suggested by reviewers, to create a book that would serve the needs of students and instructors who prefer a paperback text. In our experience, instructors who prefer a paperback are more interested in reinforcing skills through practice. Hence, the text has a structured pedagogy that includes in-text practice exercises, detailed examples, learning objectives, an extensive selection of problem-set exercises, and well-organized end-of-chapter reviews and assessments.

This text was written for college students who need an algebra course that bridges the gap between beginning algebra and more advanced courses in precalculus mathematics. The basic concepts of intermediate algebra are presented in a simple, straightforward manner. The structure for explaining mathematical techniques and concepts has proven successful. Concepts are developed through examples, continuously reinforced through additional examples, and then applied in problem-solving situations.

A common thread runs throughout this book: **learn a skill, use the skills to help solve equations**, and then **use the equations to solve the application problems**. The examples are the "learning" portion, the Practice Your Skill in-text problems are the "using" portion, and the "Apply Your Skill" examples are the "apply" portion. This thread influences many of the decisions we made in preparing this text.

Early chapters are organized to start the book at the appropriate mathematical level with the right amount of review. We have added **Learning Objectives** to the section openers and repeat those learning objectives within the sections and exercise sets. In every example, we added a practice problem, **Practice Your Skill**, to provide an exercise that will immediately reinforce the skill presented in the example. We have added **Concept Quizzes** before each problem set to assess student's mastery of the mathematical ideas and vocabulary presented in the section. By broadening the topics in the problem sets, we show students that mathematics is part of everyday life. Problems and examples include references to career areas such as the electronics, mechanics, and healthcare fields. By strengthening the examples, we give students more support with problem solving in the main text.

Further, we have designed the structure of the problem sets to stress learning outcomes and easy student access to the objectives. The exercises in problem sets are grouped by learning objectives. To recap the chapter and its learning outcomes, the examples in the chapter summary are grouped by learning objective in a grid format. Using the learning objectives to organize the problem sets and the end-of-chapter summary grid gives students a strong sense of the objectives for the topics.

Key Features to the Series

- The table of contents is organized to present the standard intermediate algebra topics and provide review at the beginning of the book. Chapter 1 provides a firm foundation for algebra by presenting the real number system and its properties. For students needing a more thorough review, Appendix A covers operations with fractions in detail. Chapter 2 progresses to equation solving and problem solving. As an extension of equation solving, Chapter 3 covers solving linear equations in two variables. Graphing equations and inequalities is presented as a means to display the solution sets. Chapter 4 continues with equation solving by presenting systems of equations. All of the equation solving is followed by application problems.

Chapters 5 through 8 cover traditional polynomial algebra, leading up to and including solving quadratic equations. Chapters 9 through 11 are devoted to traditional intermediate algebra topics of conic sections, functions, and logarithms. For intermediate algebra courses that want to delve further into systems of equations, the appendices present sections on matrices and Cramer's rule.

- The book takes a practical approach to problem solving. Instructors will notice how much we stress a practical way to learn to solve problems in Chapter 2. We bring problem solving in early, and we stress problem solving often.
- The structure of the main text and exercise sets centers on learning objectives and learning outcomes. A list of learning objectives opens each section. The objectives are repeated as subheads within the section to organize the material, and the exercises are grouped by objective so that an instructor can easily see which concepts a student has mastered. At the end of the sections, a concept quiz reviews the "big ideas" of the section.
  - Expressly for this series, **Practice Your Skill** problems are added to the worked examples as a way of enhancing the material. On-the-spot problems help students to master the content of the examples more readily. Worked-out solutions to the practice problems are located at the back of the text.

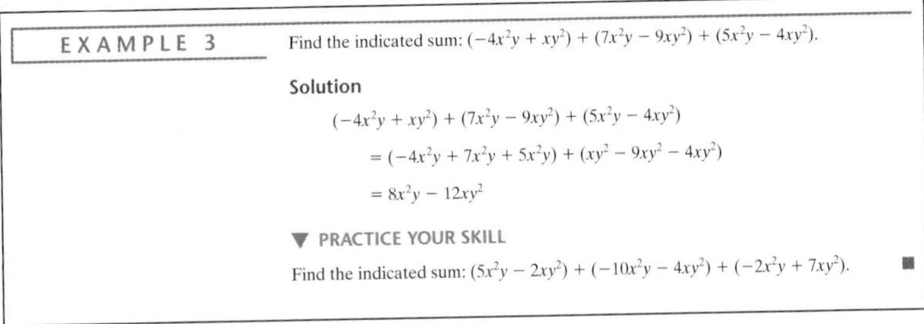

EXAMPLE 3

Find the indicated sum: $(-4x^2y + xy^2) + (7x^2y - 9xy^2) + (5x^2y - 4xy^2)$.

Solution

$$(-4x^2y + xy^2) + (7x^2y - 9xy^2) + (5x^2y - 4xy^2)$$
$$= (-4x^2y + 7x^2y + 5x^2y) + (xy^2 - 9xy^2 - 4xy^2)$$
$$= 8x^2y - 12xy^2$$

▼ PRACTICE YOUR SKILL

Find the indicated sum: $(5x^2y - 2xy^2) + (-10x^2y - 4xy^2) + (-2x^2y + 7xy^2)$. ■

- **Concept Quizzes** are included, immediately preceding each section problem set. These problems predominantly rely on the true/false format that allows students to check their understanding of the mathematical concepts introduced in the section. Users have reacted very favorably to concept quizzes, and they indicated that they used the problems for many different purposes.

CONCEPT QUIZ

For Problems 1–5, answer true or false.

1. Graphing a system of equations is the most accurate method to find the solution of a system.
2. To begin solving a system of equations by substitution, one of the equations is solved for one variable in terms of the other variable.
3. When solving a system of equations by substitution, deciding what variable to solve for may allow you to avoid working with fractions.
4. When finding the solution of the system $\begin{pmatrix} x = 2y + 4 \\ x = -y + 5 \end{pmatrix}$, you need only to find a value for x.
5. The ordered pairs (1, 3) and (5, 11) are both solutions of the system $\begin{pmatrix} y = 2x + 1 \\ 4x - 2y = -2 \end{pmatrix}$.

- We wanted an easy-to-use-and information-rich-chapter summary in grid format. This highly structured recap organizes the chapter content for easy accessibility. The summary grid includes **Objective** in the left-most column as the organizing information. To the right of the objective, a **Summary** column recaps the mathematical technique or concept in simple language. To reinforce the objective, we offer a new example in the **Example** column, and to reinforce the need to practice, we list the appropriate **Chapter Review Problems** for the objective in the rightmost column. We think that this new way of organizing the information will attract student interest and offer a valuable feature to instructors.

Chapter 2 Summary

OBJECTIVE	SUMMARY	EXAMPLE	CHAPTER REVIEW PROBLEMS
Solve first-degree equations. (Sec. 2.1, Obj. 1, p. 50)	Solving an algebraic equation refers to the process of finding the number (or numbers) that make(s) the algebraic equation a true numerical statement. Two properties of equality play an important role in solving equations. **Addition Property of Equality** $a = b$ if and only if $a + c = b + c$. **Multiplication Property of Equality** For $c \neq 0$, $a = b$ if and only if $ac = bc$.	Solve $3(2x - 1) = 2x + 6 - 5x$. Solution $3(2x - 1) = 2x + 6 - 5x$ $6x - 3 = -3x + 6$ $9x - 3 = 6$ $9x = 9$ $x = 1$ The solution set is $\{1\}$.	Problems 1–4
Solve equations involving fractions. (Sec. 2.2, Obj. 1, p. 58)	It is usually easiest to begin by multiplying both sides of the equation by the least common multiple of all the denominators in the equation. This process clears the equation of fractions.	Solve $\frac{x}{2} - \frac{x}{5} = \frac{7}{10}$. Solution $\frac{x}{2} - \frac{x}{5} = \frac{7}{10}$ $10\left(\frac{x}{2} - \frac{x}{5}\right) = 10\left(\frac{7}{10}\right)$ $10\left(\frac{x}{2}\right) - 10\left(\frac{x}{5}\right) = 7$ $5x - 2x = 7$ $3x = 7$ $x = \frac{7}{3}$ The solution set is $\left\{\frac{7}{3}\right\}$.	Problems 5–10

- Answer boxes for the Practice Your Skill in-text problems and Concept Quiz questions are conveniently placed at the end of the section problem sets. By doing so, we encourage students to study a worked example, practice an on-the-spot problem, and find the answer without having to search the appendices at the back of the book.

Answers to the Concept Quiz

1. False **2.** True **3.** True **4.** False **5.** True **6.** True **7.** False **8.** False **9.** False **10.** True

Answers to the Example Practice Skills

1. $\frac{1}{m^2(n + 4)}$ **2.** $\frac{y + 1}{y^2(y + 3)}$ **3.** $\frac{3x - 1}{x + 4}$ **4.** $\frac{4x}{y^2}$ **5.** $\frac{1}{y + 3}$ **6.** $\frac{y^2}{5y + 2}$ **7.** $\frac{x^2 + 4}{xy(x + 1)}$

Problem-Solving Approach

As mentioned, a common thread you will see throughout the text—and in all the texts we currently publish—is our well-known problem-solving approach. We keep students focused on problem solving by understanding and applying three easy steps: *learn a skill, use the skill to solve equations and inequalities, and then use equations and inequalities as problem-solving tools*. This straightforward approach has been the inspiration for many of the features in this text.

Learn a Skill. Algebraic skills are demonstrated in the many worked-out examples. Learning the skill is immediately reinforced with the Practice Your Skill problem within the example.

Use a Skill. Newly acquired skills are used as soon as possible to solve equations and inequalities. Therefore, equations and inequalities are introduced early in the text and then used throughout in a large variety of problem-solving situations.

Use equations and inequalities as problem-solving tools. Many word problems are scattered throughout the text. These problems deal with a large variety of applications and constantly show the connections between mathematics and the world around us. Problem-solving suggestions are offered throughout, with special discussion in several sections.

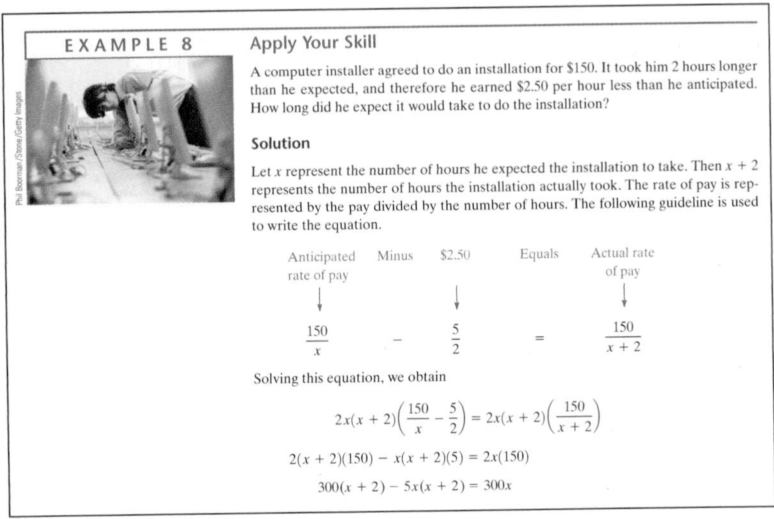

EXAMPLE 8

Phil Boorman/Stone/Getty Images

Apply Your Skill

A computer installer agreed to do an installation for \$150. It took him 2 hours longer than he expected, and therefore he earned \$2.50 per hour less than he anticipated. How long did he expect it would take to do the installation?

Solution

Let x represent the number of hours he expected the installation to take. Then $x + 2$ represents the number of hours the installation actually took. The rate of pay is represented by the pay divided by the number of hours. The following guideline is used to write the equation.

Anticipated rate of pay	Minus	\$2.50	Equals	Actual rate of pay
↓		↓		↓
$\frac{150}{x}$	$-$	$\frac{5}{2}$	$=$	$\frac{150}{x+2}$

Solving this equation, we obtain

$$2x(x+2)\left(\frac{150}{x} - \frac{5}{2}\right) = 2x(x+2)\left(\frac{150}{x+2}\right)$$

$$2(x+2)(150) - x(x+2)(5) = 2x(150)$$

$$300(x+2) - 5x(x+2) = 300x$$

Other Special Features

- Many examples contain helpful explanations in the form of line-by-line annotations indicated in blue and placed alongside the solution steps. Also, some examples contain remarks with added information below the example solution. Many examples contain check steps, which verify that an answer is correct and encourage students to check their work, a valuable habit. These checks are accompanied by a checkmark icon and are located at the end of an example.
- As recommended by the American Mathematical Association of Two-Year Colleges, many basic geometric concepts are integrated in problem-solving settings. The following geometric concepts are presented in problem-solving situations: complementary and supplementary angles, the sum of the measures of the angles of a triangle equals 180°, area and volume formulas, perimeter and circumference formulas, ratio, proportion, Pythagorean theorem, isosceles right triangle, and 30°– 60° right triangle relationships.
- Every chapter opener is followed by an **Internet Project**. These projects ask the student to conduct an Internet search on a topic that is relevant to the mathematics presented in that chapter. Many of the projects cover topics concerning the history of math or famous mathematicians.
- Problems called **Thoughts into Words** are included in every problem set except the review exercises. These problems are designed to encourage students to express in written form their thoughts about various mathematical ideas.
- Problems called **Further Investigations** appear in many of the problem sets. These are designed to lead into upcoming topics or to be slightly more complex. These problems encompass a variety of ideas: some exhibit different approaches to topics covered in the text, some are proofs, some bring in supplementary topics and relationships, and some are more challenging

problems. These problems add variety and flexibility to the problem sets and to the classroom experience, but they can be omitted entirely without disrupting the continuity pattern of the text.

- Every chapter includes a **Chapter Summary, Chapter Review Problem Set,** and **Chapter Test**. In addition, **Cumulative Review Problem Sets** are placed after Chapters 4, 6, 8, and 10. The cumulative reviews help students retain essential skills.
- All the answers for the Chapter Review Problem Sets, Chapter Tests, and Cumulative Review Problem Sets appear in the back of the text, along with answers to the odd-numbered problems.
- We think this text has exceptionally pleasing design features, including the functional use of color. The open format makes the flow of reading continuous and easy. In this design, we hope to capture the spirit of the way we present information: open, clean, friendly, and accessible.

Ancillaries

For the Instructor

***Annotated Instructor's Edition.* (0-495-38809-2)**
In the AIE, answers are printed next to all respective exercises. Graphs, tables, and other answers appear in an answer section at the back of the text. Problems that are available in electronic form in Enhanced WebAssign are identified by a bulleted problem number. To create an assessment, whether a quiz, homework assignment, or test, the instructor can select the problems by problem number from the identified problems in the text.

***Complete Solutions Manual.* (0-495-38801-7)**
Karen L. Schwitters and Laurel Fischer
The Complete Solutions Manual provides worked-out solutions to all of the problems in the text.

***Power Lecture CD-ROM with ExamView and JoinIn™.* (0-495-38799-1)**
New! This CD-ROM provides the instructor with dynamic media tools for teaching. Create, deliver, and customize tests (both print and online) in minutes with *ExamView® Computerized Testing Featuring Algorithmic Equations. JoinIn™ Student Response System* allows you to pose book-specific questions and display students' answers seamlessly within the Microsoft® PowerPoint® slides of your own lecture, in conjunction with the "clicker" hardware of your choice. Easily build solution sets for homework or exams using *Solution Builder*'s online solutions manual. Microsoft® PowerPoint® lecture slides, figures from the book, and Test Bank, in electronic format, are also included on this CD-ROM.

***Enhanced WebAssign.* (0-495-38804-1)**
WebAssign, the most widely used homework system in higher education, allows you to assign, collect, grade, and record homework assignments via the Web. Through a partnership between WebAssign and Cengage Learning Brooks/Cole, this proven homework system has been enhanced to include links to textbook sections, video examples, and problem-specific tutorials.

***Text-Specific DVDs.* (0-495-38808-4)**
Rena Petrello, Moorpark College
New! These highly praised videos feature valuable 10- to 20-minute demonstrations of nearly every learning objective lesson covered in the text. They may be used as a supplement to classroom learning or as the primary content for an online student. Videos will be available by DVD and online download.

For the Student

Student Solutions Manual **(0-495-38800-9)**
Karen L. Schwitters and Laurel Fischer
The Student Solutions Manual provides worked-out solutions to the odd-numbered problems in the text and worked-out solutions for the Chapter Review Problem Sets, Chapter Tests, and Cumulative Review Problem Sets.

Text-Specific Videos. **(0-495-38808-4)**
Rena Petrello, Moorpark College
New! These highly praised videos feature valuable 10- to 20-minute demonstrations of nearly every learning objective lesson covered in the text. They may be used as a supplement to classroom learning or as the primary content for an online student. Videos will be available by DVD and online download.

Acknowledgments

We would like to take this opportunity to thank the following people who served as reviewers for the first edition of this project:

Kochi Angar
Nash Community College

Amir Fazi Arabi
Central Virginia Community College

Sarah E. Baxter
Gloucester County College

Annette Benbow
Tarrant County College

A. Elena Bogardus
Camden County College

Dorothy Brown
Camden County College

Terry F. Clark
Keiser University

Linda P. Davis
Keiser University

Archie Earl
Norfolk State University

Arlene Eliason
Minnesota School of Business

Deborah D. Fries
Wor-Wic Community College

Nathaniel Gay
Keiser University

Margaret Hathaway
Kansas City Kansas Community College

Louis C. Henderson, Jr.
Coppin State University

Patricia Horacek
Pensacola Junior College

Kelly Jackson
Camden County College

Tom Johnson
University of Akron

Elias M. Jureidini
Lamar State College

Carolyn Krause
Delaware Tech and Community College

Patricia Labonne
Cumberland County College

Dottie Lapre
Nash Community College

Bruce H. Laster
Keiser University

Maria Luisa Mendez
Laredo Community College

Sunny Norfleet
St. Petersburg College

Ann Ostberg
Grace University

Armando I. Perez
Laredo Community College

Peter Peterson
John Tyler Community College

Vien Pham
Keiser University

Maria Pickle
St. Petersburg College

Sita Ramamurti
Trinity Washington University

Denver Riffe
National College-Bluefield

Daryl Schrader
St. Petersburg College

Lee Ann Spahr
Durham Technical College

Janet E. Thompson
University of Akron

Mary Lou Townsend
Wor-Wic Community College

Bonnie Filer-Tubaugh
University of Akron

Susan Twigg
Wor-Wic Community College

We are very grateful to the staff of Brooks/Cole, especially Gary Whalen, Kristin Marrs, Laura Localio, Greta Kleinert, and Lynh Pham, for their continuous cooperation and assistance throughout this project. We would also like to express our sincere gratitude to Fran Andersen and to Hal Humphrey. They continue to make life as an author so much easier by carrying out the details of production in a dedicated and caring way. Additional thanks are due to Arlene Kaufmann who spends numerous hours reading page proofs.

Jerome E. Kaufmann
Karen L. Schwitters

Basic Concepts and Properties

Numbers from the set of integers are used to express temperatures that are below 0°F.

The temperature at 6 p.m. was -3°F. By 11 p.m. the temperature had dropped another 5°F. We can use the **numerical expression** $-3 - 5$ to determine the temperature at 11 p.m.

Justin has p pennies, n nickels, and d dimes in his pocket. The **algebraic expression** $p + 5n + 10d$ represents that amount of money in cents.

Algebra is often described as a **generalized arithmetic**. That description may not tell the whole story, but it does convey an important idea: A good understanding of arithmetic provides a sound basis for the study of algebra. In this chapter we use the concepts of **numerical expression** and **algebraic expression** to review some ideas from arithmetic and to begin the transition to algebra. Be sure that you thoroughly understand the basic concepts we review in this first chapter.

Video tutorials for all section learning objectives are available in a variety of delivery modes.

INTERNET PROJECT

Symbols are used to indicate the arithmetic operations of addition, subtraction, multiplication, and division. Conduct an Internet search to determine the origin of the plus sign, +, that symbolizes addition. The use of symbols in the study of algebra necessitated an agreement on the order in which arithmetic operations should be performed. Search the Internet for an interactive site where you can practice order-of-operations problems, and share this site with other students.

1.1 Sets, Real Numbers, and Numerical Expressions

OBJECTIVES

1. Identify Certain Sets of Numbers
2. Apply the Properties of Equality
3. Simplify Numerical Expressions

1 Identify Certain Sets of Numbers

In arithmetic, we use symbols such as 6, $\frac{2}{3}$, 0.27, and π to represent numbers. The symbols $+$, $-$, $\cdot$, and $\div$ commonly indicate the basic operations of addition, subtraction, multiplication, and division, respectively. Thus we can form specific **numerical expressions**. For example, we can write the indicated sum of six and eight as $6 + 8$.

In algebra, the concept of a variable provides the basis for generalizing arithmetic ideas. For example, by using x and y to represent any numbers, we can use the expression $x + y$ to represent the indicated sum of any two numbers. The x and y in such an expression are called **variables**, and the phrase $x + y$ is called an **algebraic expression**.

We can extend to algebra many of the notational agreements we make in arithmetic, with a few modifications. The following chart summarizes the notational agreements that pertain to the four basic operations.

Operation	Arithmetic	Algebra	Vocabulary
Addition	$4 + 6$	$x + y$	The **sum** of x and y
Subtraction	$14 - 10$	$a - b$	The **difference** of a and b
Multiplication	$7 \cdot 5$ or 7×5	$a \cdot b$, $a(b)$, $(a)b$, $(a)(b)$, or ab	The **product** of a and b
Division	$8 \div 4$, $\frac{8}{4}$, or $4\overline{)8}$	$x \div y$, $\frac{x}{y}$, or $y\overline{)x}$	The **quotient** of x and y

Note the different ways to indicate a product, including the use of parentheses. The ab form is the simplest and probably the most widely used form. Expressions such as abc, $6xy$, and $14xyz$ all indicate multiplication. We also call your attention to the various forms that indicate division. In algebra, we usually use the fractional form, $\frac{x}{y}$, although the other forms do serve a purpose at times.

We can use some of the basic vocabulary and symbolism associated with the concept of sets in the study of algebra. A **set** is a collection of objects, and the objects are called **elements** or **members** of the set. In arithmetic and algebra the elements of a set are usually numbers.

The use of set braces, { }, to enclose the elements (or a description of the elements) and the use of capital letters to name sets provide a convenient way to communicate about sets. For example, we can represent a set A, which consists of the vowels of the English alphabet, in any of the following ways:

$A = \{\text{vowels of the English alphabet}\}$	Word description
$A = \{a, e, i, o, u\}$	List or roster description
$A = \{x \mid x \text{ is a vowel}\}$	Set builder notation

We can modify the listing approach if the number of elements is quite large. For example, all of the letters of the English alphabet can be listed as

$$\{a, b, c, \ldots, z\}$$

We simply begin by writing enough elements to establish a pattern; then the three dots indicate that the set continues in that pattern. The final entry indicates the last element of the pattern. If we write

$$\{1, 2, 3, \ldots\}$$

the set begins with the counting numbers 1, 2, and 3. The three dots indicate that it continues in a like manner forever; there is no last element. A set that consists of no elements is called the **null set** (written $\varnothing$).

Set builder notation combines the use of braces and the concept of a variable. For example, $\{x \mid x \text{ is a vowel}\}$ is read "the set of all x such that x is a vowel." Note that the vertical line is read "such that." We can use set builder notation to describe the set $\{1, 2, 3, \ldots\}$ as $\{x \mid x > 0 \text{ and } x \text{ is a whole number}\}$.

We use the symbol $\in$ to denote set membership. Thus if $A = \{a, e, i, o, u\}$, we can write $e \in A$, which we read as "e is an element of A." The slash symbol, /, is commonly used in mathematics as a negation symbol. For example, $m \notin A$ is read as "m is not an element of A."

Two sets are said to be *equal* if they contain exactly the same elements. For example,

$$\{1, 2, 3\} = \{2, 1, 3\}$$

because both sets contain the same elements; the order in which the elements are written doesn't matter. The slash mark through the equality symbol denotes "is not equal to." Thus if $A = \{1, 2, 3\}$ and $B = \{1, 2, 3, 4\}$, we can write $A \neq B$, which we read as "set A is not equal to set B."

We refer to most of the algebra that we will study in this text as the **algebra of real numbers**. This simply means that the variables represent real numbers. Therefore, it is necessary for us to be familiar with the various terms that are used to classify different types of real numbers.

$\{1, 2, 3, 4, \ldots\}$	Natural numbers, counting numbers, positive integers
$\{0, 1, 2, 3, \ldots\}$	Whole numbers, nonnegative integers
$\{\ldots, -3, -2, -1\}$	Negative integers
$\{\ldots, -3, -2, -1, 0\}$	Nonpositive integers
$\{\ldots, -3, -2, -1, 0, 1, 2, 3, \ldots\}$	Integers

We define a rational number as follows:

Definition 1.1 Rational Number

A rational number is any number that can be written in the form $\frac{a}{b}$, where a and b are integers and b does not equal zero.

We can easily recognize that each of the following numbers fits the definition of a rational number:

$$\frac{-3}{4} \quad \frac{2}{3} \quad \frac{15}{4} \quad \frac{1}{-5}$$

However, numbers such as -4, 0, 0.3, and $6\frac{1}{2}$ are also rational numbers. All of these numbers could be written in the form of $\frac{a}{b}$ as follows.

-4 can be written as $\frac{-4}{1}$ or $\frac{4}{-1}$ 0 can be written as $\frac{0}{1} = \frac{0}{2} = \frac{0}{3} = \dots$

0.3 can be written as $\frac{3}{10}$ $6\frac{1}{2}$ can be written as $\frac{13}{2}$

We can also define a rational number in terms of a decimal representation. We can classify decimals as terminating, repeating, or nonrepeating.

Type	Definition	Examples	Rational numbers
Terminating	A terminating decimal ends.	0.3, 0.46, 0.6234, 1.25	Yes
Repeating	A repeating decimal has a block of digits that repeats indefinitely.	0.66666 . . . 0.141414 . . . 0.694694694 . . . 0.23171717 . . .	Yes
Nonrepeating	A nonrepeating decimal does not terminate and does not have a block of digits that repeat indefinitely.	3.1415926535 . . . 1.414213562 . . . 0.276314583 . . .	No

A repeating decimal has a block of digits that repeats indefinitely. This repeating block of digits may be of any number of digits and may or may not begin immediately after the decimal point. A small horizontal bar (overbar) is commonly used to indicate the repeat block. Thus 0.6666 . . . is written as $0.\overline{6}$, and 0.2317171717 . . . is written as $0.23\overline{17}$.

In terms of decimals, we define a **rational number** as a number that has either a terminating or a repeating decimal representation. The following examples illustrate some rational numbers written in $\frac{a}{b}$ form and in decimal form.

$$\frac{3}{4} = 0.75 \qquad \frac{3}{11} = 0.\overline{27} \qquad \frac{1}{8} = 0.125 \qquad \frac{1}{7} = 0.\overline{142857} \qquad \frac{1}{3} = 0.\overline{3}$$

We define an **irrational number** as a number that *cannot* be expressed in $\frac{a}{b}$ form, where a and b are integers, and b is not zero. Furthermore, an irrational number has a nonrepeating and nonterminating decimal representation. Some examples of irrational numbers and a partial decimal representation for each follow.

$$\sqrt{2} = 1.414213562373095\dots \qquad \sqrt{3} = 1.73205080756887\dots$$

$$\pi = 3.14159265358979\dots$$

The entire set of **real numbers** is composed of the rational numbers along with the irrationals. Every real number is either a rational number or an irrational number. The following tree diagram summarizes the various classifications of the real number system.

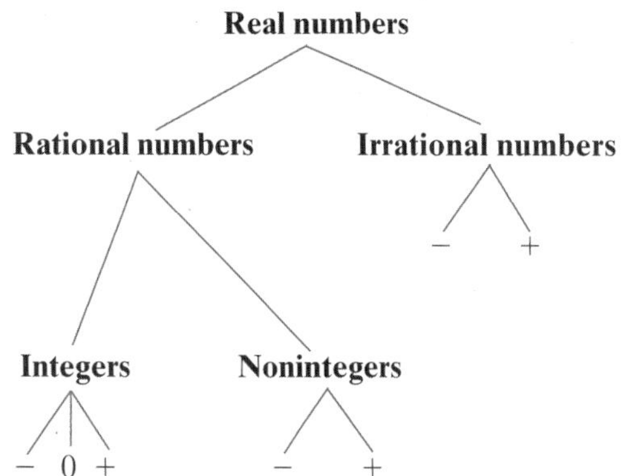

We can trace any real number down through the diagram as follows:

7 is real, rational, an integer, and positive.

$-\frac{2}{3}$ is real, rational, noninteger, and negative.

$\sqrt{7}$ is real, irrational, and positive.

0.38 is real, rational, noninteger, and positive.

Remark: We usually refer to the set of nonnegative integers, $\{0, 1, 2, 3, \ldots\}$, as the set of **whole numbers**, and we refer to the set of positive integers, $\{1, 2, 3, \ldots\}$, as the set of **natural numbers**. The set of whole numbers differs from the set of natural numbers by the inclusion of the number zero.

The concept of subset is convenient to use at this time. A set A is a **subset** of a set B if and only if every element of A is also an element of B. This is written as $A \subseteq B$ and read as "A is a subset of B." For example, if $A = \{1, 2, 3\}$ and $B = \{1, 2, 3, 5, 9\}$, then $A \subseteq B$ because every element of A is also an element of B. The slash mark again denotes negation, so if $A = \{1, 2, 5\}$ and $B = \{2, 4, 7\}$, we can say that A is not a subset of B by writing $A \nsubseteq B$. Figure 1.1 represents the subset

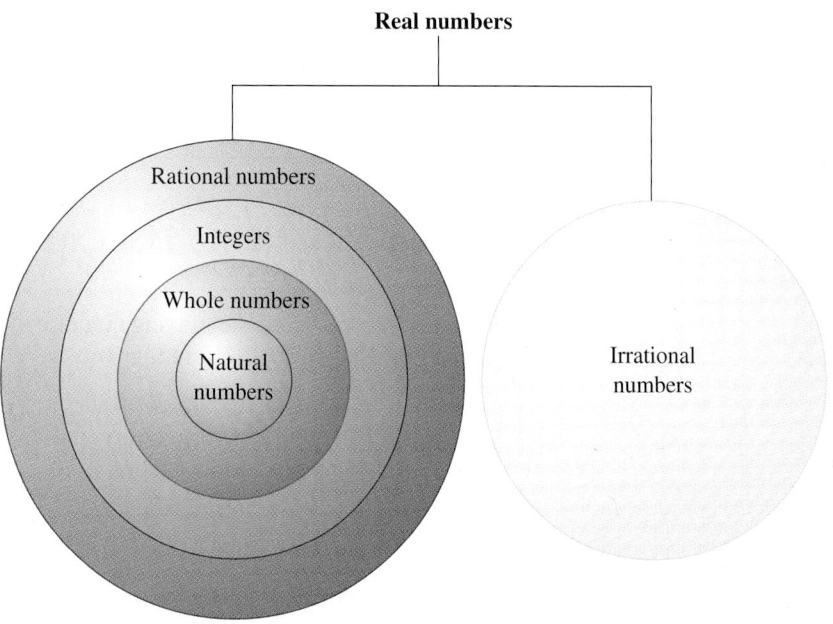

Figure 1.1

relationships for the set of real numbers. Refer to Figure 1.1 as you study the following statements that use subset vocabulary and subset symbolism.

1. The set of whole numbers is a subset of the set of integers.

$$\{0, 1, 2, 3, \ldots\} \subseteq \{\ldots, -2, -1, 0, 1, 2, \ldots\}$$

2. The set of integers is a subset of the set of rational numbers.

$$\{\ldots, -2, -1, 0, 1, 2, \ldots\} \subseteq \{x \mid x \text{ is a rational number}\}$$

3. The set of rational numbers is a subset of the set of real numbers.

$$\{x \mid x \text{ is a rational number}\} \subseteq \{y \mid y \text{ is a real number}\}$$

2 Apply the Properties of Equality

The relation **equality** plays an important role in mathematics—especially when we are manipulating real numbers and algebraic expressions that represent real numbers. An equality is a statement in which two symbols, or groups of symbols, are names for the same number. The symbol = is used to express an equality. Thus we can write

$$6 + 1 = 7 \qquad 18 - 2 = 16 \qquad 36 \div 4 = 9$$

(The symbol $\neq$ means *is not equal to*.) The following four basic properties of equality are self-evident, but we do need to keep them in mind. (We will expand this list in Chapter 2 when we work with solutions of equations.)

Properties of equality	Definition: For real numbers a, b, and c,	Example
Reflexive property	$a = a$.	$14 = 14$, $x = x$, $a + b = a + b$
Symmetric property	If $a = b$, then $b = a$.	If $3 + 1 = 4$, then $4 = 3 + 1$. If $x = 10$, then $10 = x$.
Transitive property	If $a = b$ and $b = c$, then $a = c$.	If $x = 7$ and $7 = y$, then $x = y$. If $x + 5 = y$ and $y = 8$, then $x + 5 = 8$.
Substitution property	If $a = b$, then a may be replaced by b, or b may be replaced by a, without changing the meaning of the statement.	If $x + y = 4$ and $x = 2$, then we can replace x in the first equation with the value 2, yielding $2 + y = 4$.

3 Simplify Numerical Expressions

Let's conclude this section by *simplifying some numerical expressions* that involve whole numbers. When simplifying numerical expressions, we perform the operations in the following order. Be sure that you agree with the result in each example.

1. Perform the operations inside the symbols of inclusion (parentheses, brackets, and braces) and above and below each fraction bar. Start with the innermost inclusion symbol.
2. Perform all multiplications and divisions in the order in which they appear from left to right.
3. Perform all additions and subtractions in the order in which they appear from left to right.

EXAMPLE 1 Simplify $20 + 60 \div 10 \cdot 2$

Solution

First do the division.

$$20 + 60 \div 10 \cdot 2 = 20 + 6 \cdot 2$$

Next do the multiplication.

$$20 + 6 \cdot 2 = 20 + 12$$

Then do the addition.

$$20 + 12 = 32$$

Thus $20 + 60 \div 10 \cdot 2$ simplifies to 32.

▼ **PRACTICE YOUR SKILL**

Simplify $15 + 45 \div 5 \cdot 3$. ■

EXAMPLE 2 Simplify $7 \cdot 4 \div 2 \cdot 3 \cdot 2 \div 4$.

Solution

The multiplications and divisions are to be done from left to right in the order in which they appear.

$$\begin{aligned} 7 \cdot 4 \div 2 \cdot 3 \cdot 2 \div 4 &= 28 \div 2 \cdot 3 \cdot 2 \div 4 \\ &= 14 \cdot 3 \cdot 2 \div 4 \\ &= 42 \cdot 2 \div 4 \\ &= 84 \div 4 \\ &= 21 \end{aligned}$$

Thus $7 \cdot 4 \div 2 \cdot 3 \cdot 2 \div 4$ simplifies to 21.

▼ **PRACTICE YOUR SKILL**

Simplify $5 \cdot 6 \div 3 \cdot 2$. ■

EXAMPLE 3 Simplify $5 \cdot 3 + 4 \div 2 - 2 \cdot 6 - 28 \div 7$.

Solution

First we do the multiplications and divisions in the order in which they appear. Then we do the additions and subtractions in the order in which they appear. Our work may take on the following format.

$$5 \cdot 3 + 4 \div 2 - 2 \cdot 6 - 28 \div 7 = 15 + 2 - 12 - 4 = 1$$

▼ **PRACTICE YOUR SKILL**

Simplify $2 \cdot 5 + 15 \div 5 - 2$. ■

EXAMPLE 4

Simplify $(4 + 6)(7 + 8)$.

Solution

We use the parentheses to indicate the *product* of the quantities $4 + 6$ and $7 + 8$. We perform the additions inside the parentheses first and then multiply.

$$(4 + 6)(7 + 8) = (10)(15) = 150$$

▼ **PRACTICE YOUR SKILL**

Simplify $(18 - 6)(2 + 3)$. ■

EXAMPLE 5

Simplify $(3 \cdot 2 + 4 \cdot 5)(6 \cdot 8 - 5 \cdot 7)$.

Solution

First we do the multiplications inside the parentheses.

$$(3 \cdot 2 + 4 \cdot 5)(6 \cdot 8 - 5 \cdot 7) = (6 + 20)(48 - 35)$$

Then we do the addition and subtraction inside the parentheses.

$$(6 + 20)(48 - 35) = (26)(13)$$

Then we find the final product.

$$(26)(13) = 338$$

▼ **PRACTICE YOUR SKILL**

Simplify $(3 \cdot 4 - 5 \cdot 2)(3 \cdot 6 - 2 \cdot 4)$. ■

EXAMPLE 6

Simplify $6 + 7[3(4 + 6)]$.

Solution

We use brackets for the same purposes as parentheses. In such a problem we need to simplify *from the inside out*; that is, we perform the operations in the innermost parentheses first. We thus obtain

$$\begin{aligned} 6 + 7[3(4 + 6)] &= 6 + 7[3(10)] \\ &= 6 + 7[30] \\ &= 6 + 210 \\ &= 216 \end{aligned}$$

▼ **PRACTICE YOUR SKILL**

Simplify $1 + 3[2(8 - 3)]$. ■

EXAMPLE 7

Simplify $\dfrac{6 \cdot 8 \div 4 - 2}{5 \cdot 4 - 9 \cdot 2}$.

Solution

First we perform the operations above and below the fraction bar. Then we find the final quotient.

$$\frac{6 \cdot 8 \div 4 - 2}{5 \cdot 4 - 9 \cdot 2} = \frac{48 \div 4 - 2}{20 - 18} = \frac{12 - 2}{2} = \frac{10}{2} = 5$$

▼ PRACTICE YOUR SKILL

Simplify $\frac{12 \cdot 6 \div 3 - 3}{7 \cdot 5 - 4 \cdot 8}$. ■

CONCEPT QUIZ

For Problems 1–10, answer true or false.

1. The expression ab indicates the sum of a and b.
2. The set $\{1, 2, 3, \ldots\}$ contains infinitely many elements.
3. The sets $A = \{1, 2, 4, 6\}$ and $B = \{6, 4, 1, 2\}$ are equal sets.
4. Every irrational number is also classified as a real number.
5. To evaluate $24 \div 6 \cdot 2$, the first operation that should be performed is to multiply 6 times 2.
6. To evaluate $6 + 8 \cdot 3$, the first operation that should be performed is to multiply 8 times 3.
7. The number 0.15 is real, irrational, and positive.
8. If $4 = x + 3$, then $x + 3 = 4$ is an example of the symmetric property of equality.
9. The numerical expression $6 \cdot 2 + 3 \cdot 5 - 6$ simplifies to 21.
10. The number represented by $0.\overline{12}$ is a rational number.

Problem Set 1.1

1 Identify Certain Sets of Numbers

For Problems 1–10, identify each statement as true or false.

1. Every irrational number is a real number.
2. Every rational number is a real number.
3. If a number is real, then it is irrational.
4. Every real number is a rational number.
5. All integers are rational numbers.
6. Some irrational numbers are also rational numbers.
7. Zero is a positive integer.
8. Zero is a rational number.
9. All whole numbers are integers.
10. Zero is a negative integer.

For Problems 11–18, from the list 0, 14, $\frac{2}{3}$, π, $\sqrt{7}$, $-\frac{11}{14}$, 2.34, $3.2\overline{1}$, $\frac{55}{8}$, $-\sqrt{17}$, -19, and -2.6, identify each of the following.

11. The whole numbers
12. The natural numbers
13. The rational numbers
14. The integers
15. The nonnegative integers
16. The irrational numbers
17. The real numbers
18. The nonpositive integers

For Problems 19–28, use the following set designations.

$N = \{x \mid x \text{ is a natural number}\}$
$Q = \{x \mid x \text{ is a rational number}\}$
$W = \{x \mid x \text{ is a whole number}\}$
$H = \{x \mid x \text{ is an irrational number}\}$
$I = \{x \mid x \text{ is an integer}\}$
$R = \{x \mid x \text{ is a real number}\}$

Place $\subseteq$ or $\nsubseteq$ in each blank to make a true statement.

19. R ______ N
20. N ______ R
21. I ______ Q
22. N ______ I
23. Q ______ H
24. H ______ Q
25. N ______ W
26. W ______ I
27. I ______ N
28. I ______ W

For Problems 29–32, classify the real number by tracing through the diagram in the text (see page 5).

29. -8
30. 0.9
31. $-\sqrt{2}$
32. $\frac{5}{6}$

For Problems 33–42, list the elements of each set. For example, the elements of $\{x \mid x \text{ is a natural number less than } 4\}$ can be listed as $\{1, 2, 3\}$.

33. $\{x \mid x \text{ is a natural number less than } 3\}$

34. $\{x \mid x \text{ is a natural number greater than } 3\}$

35. $\{n \mid n \text{ is a whole number less than } 6\}$

36. $\{y \mid y \text{ is an integer greater than } -4\}$

37. $\{y \mid y \text{ is an integer less than } 3\}$

38. $\{n \mid n \text{ is a positive integer greater than } -7\}$

39. $\{x \mid x \text{ is a whole number less than } 0\}$

40. $\{x \mid x \text{ is a negative integer greater than } -3\}$

41. $\{n \mid n \text{ is a nonnegative integer less than } 5\}$

42. $\{n \mid n \text{ is a nonpositive integer greater than } 3\}$

2 Apply the Properties of Equality

For Problems 43–50, replace each question mark to make the given statement an application of the indicated property of equality. For example, $16 = ?$ becomes $16 = 16$ because of the reflexive property of equality.

43. If $y = x$ and $x = -6$, then $y = ?$ (Transitive property of equality)

44. $5x + 7 = ?$ (Reflexive property of equality)

45. If $n = 2$ and $3n + 4 = 10$, then $3(?) + 4 = 10$ (Substitution property of equality)

46. If $y = x$ and $x = z + 2$, then $y = ?$ (Transitive property of equality)

47. If $4 = 3x + 1$, then $? = 4$ (Symmetric property of equality)

48. If $t = 4$ and $s + t = 9$, then $s + ? = 9$ (Substitution property of equality)

49. $5x = ?$ (Reflexive property of equality)

50. If $5 = n + 3$, then $n + 3 = ?$ (Symmetric property of equality)

3 Simplify Numerical Expressions

For Problems 51–74, simplify each of the numerical expressions.

51. $16 + 9 - 4 - 2 + 8 - 1$

52. $18 + 17 - 9 - 2 + 14 - 11$

53. $9 \div 3 \cdot 4 \div 2 \cdot 14$

54. $21 \div 7 \cdot 5 \cdot 2 \div 6$

55. $7 + 8 \cdot 2$

56. $21 - 4 \cdot 3 + 2$

57. $9 \cdot 7 - 4 \cdot 5 - 3 \cdot 2 + 4 \cdot 7$

58. $6 \cdot 3 + 5 \cdot 4 - 2 \cdot 8 + 3 \cdot 2$

59. $(17 - 12)(13 - 9)(7 - 4)$

60. $(14 - 12)(13 - 8)(9 - 6)$

61. $13 + (7 - 2)(5 - 1)$

62. $48 - (14 - 11)(10 - 6)$

63. $(5 \cdot 9 - 3 \cdot 4)(6 \cdot 9 - 2 \cdot 7)$

64. $(3 \cdot 4 + 2 \cdot 1)(5 \cdot 2 + 6 \cdot 7)$

65. $7[3(6 - 2)] - 64$

66. $12 + 5[3(7 - 4)]$

67. $[3 + 2(4 \cdot 1 - 2)][18 - (2 \cdot 4 - 7 \cdot 1)]$

68. $3[4(6 + 7)] + 2[3(4 - 2)]$

69. $14 + 4\left(\frac{8 - 2}{12 - 9}\right) - 2\left(\frac{9 - 1}{19 - 15}\right)$

70. $12 + 2\left(\frac{12 - 2}{7 - 2}\right) - 3\left(\frac{12 - 9}{17 - 14}\right)$

71. $[7 + 2 \cdot 3 \cdot 5 - 5] \div 8$

72. $[27 - (4 \cdot 2 + 5 \cdot 2)][(5 \cdot 6 - 4) - 20]$

73. $\frac{3 \cdot 8 - 4 \cdot 3}{5 \cdot 7 - 34} + 19$

74. $\frac{4 \cdot 9 - 3 \cdot 5 - 3}{18 - 12}$

75. You must, of course, be able to do calculations like those in Problems 51–74 both with and without a calculator. Furthermore, different types of calculators handle the priority-of-operations issue in different ways. Be sure you can do Problems 51–74 with *your* calculator.

THOUGHTS INTO WORDS

76. Explain in your own words the difference between the reflexive property of equality and the symmetric property of equality.

77. Your friend keeps getting an answer of 30 when simplifying $7 + 8(2)$. What mistake is he making and how would you help him?

78. Do you think $3\sqrt{2}$ is a rational or an irrational number? Defend your answer.

79. Explain why every integer is a rational number but not every rational number is an integer.

80. Explain the difference between $1.\overline{3}$ and 1.3.

Answers to the Concept Quiz

1. False **2.** True **3.** True **4.** True **5.** False **6.** True **7.** False **8.** True **9.** True **10.** True

Answers to the Example Practice Skills

1. 42 **2.** 20 **3.** 11 **4.** 60 **5.** 20 **6.** 31 **7.** 7

1.2 Operations with Real Numbers

OBJECTIVES

1 Review the Real Number Line

2 Find the Absolute Value of a Number

3 Add Real Numbers

4 Subtract Real Numbers

5 Multiply Real Numbers

6 Divide Real Numbers

7 Simplify Numerical Expressions

8 Use Real Numbers to Represent Problems

1 Review the Real Number Line

Before we review the four basic operations with real numbers, let's briefly discuss some concepts and terminology we commonly use with this material. It is often helpful to have a geometric representation of the set of real numbers, shown as in Figure 1.2. Such a representation, called the **real number line**, indicates a one-to-one correspondence between the set of real numbers and the points on a line. In other words, to each real number there corresponds one and only one point on the line, and to each point on the line there corresponds one and only one real number. The number associated with each point on the line is called the **coordinate** of the point.

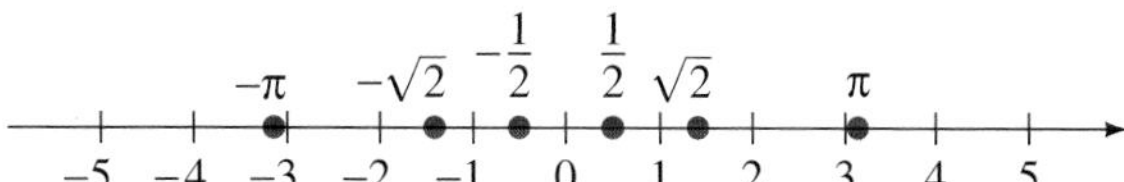

Figure 1.2

Many operations, relations, properties, and concepts pertaining to real numbers can be given a geometric interpretation on the real number line. For example, the addition problem $(-1) + (-2)$ can be depicted on the number line as in Figure 1.3.

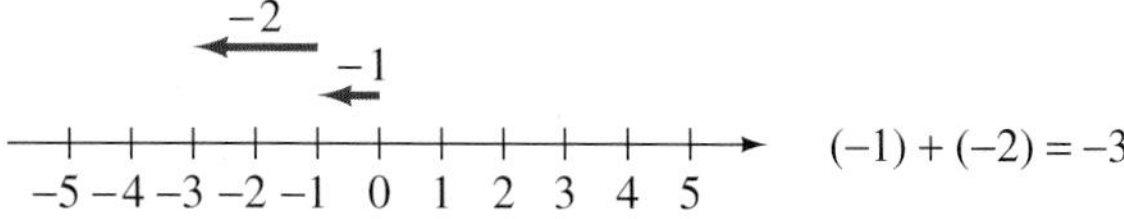

Figure 1.3

Figure 1.4

The inequality relations also have a geometric interpretation. The statement $a > b$ (which is read "a is greater than b") means that a is to the right of b, and the statement $c < d$ (which is read "c is less than d") means that c is to the left of d, as shown in Figure 1.4. The symbol $\leq$ means *is less than or equal to*, and the symbol $\geq$ means *is greater than or equal to*.

The property $-(-x) = x$ can be represented on the number line by following the sequence of steps shown in Figure 1.5.

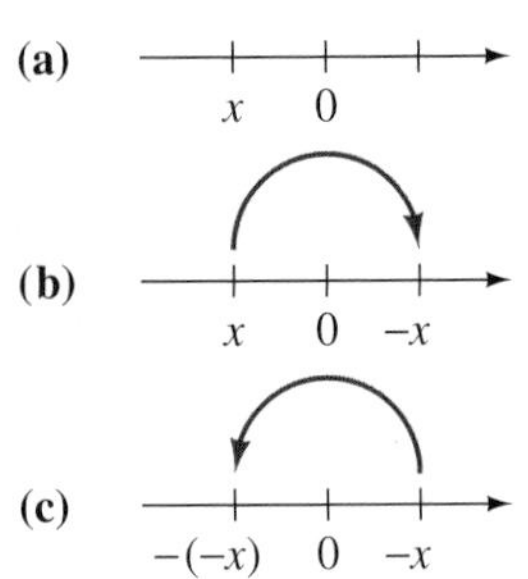

Figure 1.5

1. Choose a point having a coordinate of x.
2. Locate its opposite, written as $-x$, on the other side of zero.
3. Locate the opposite of $-x$, written as $-(-x)$, on the other side of zero.

Therefore, we conclude that **the opposite of the opposite of any real number is the number itself**, and we symbolically express this by $-(-x) = x$.

Remark: The symbol -1 can be read "negative one," "the negative of one," "the opposite of one," or "the additive inverse of one." The opposite-of and additive-inverse-of terminology is especially meaningful when working with variables. For example, the symbol $-x$, which is read "the opposite of x" or "the additive inverse of x," emphasizes an important issue. Because x can be any real number, $-x$ (the opposite of x) can be zero, positive, or negative. If x is positive, then $-x$ is negative. If x is negative, then $-x$ is positive. If x is zero, then $-x$ is zero.

2 Find the Absolute Value of a Number

We can use the concept of **absolute value** to describe precisely how to operate with positive and negative numbers. Geometrically, the absolute value of any number is the distance between the number and zero on the number line. For example, the absolute value of 2 is 2. The absolute value of -3 is 3. The absolute value of 0 is 0 (see Figure 1.6).

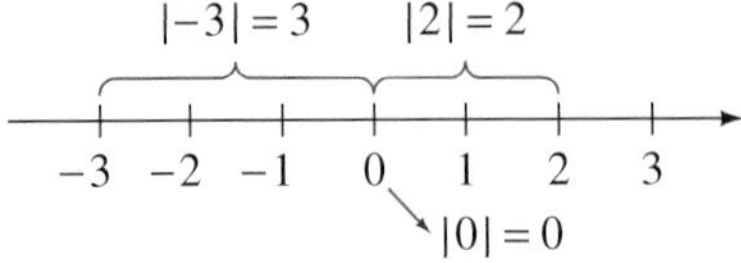

Figure 1.6

Symbolically, absolute value is denoted with vertical bars. Thus we write

$$|2| = 2 \qquad |-3| = 3 \qquad |0| = 0$$

More formally, we define the concept of absolute value as follows.

Definition 1.2

For all real numbers a,

1. If $a \geq 0$, then $|a| = a$.
2. If $a < 0$, then $|a| = -a$.

According to Definition 1.2, we obtain

$|6| = 6$ By applying part 1 of Definition 1.2

$|0| = 0$ By applying part 1 of Definition 1.2

$|-7| = -(-7) = 7$ By applying part 2 of Definition 1.2

Note that the absolute value of a positive number is the number itself, but the absolute value of a negative number is its opposite. Thus the absolute value of any number except zero is positive, and the absolute value of zero is zero. Together, these facts indicate that the absolute value of any real number is equal to the absolute value of its opposite. We summarize these ideas in the following properties.

Properties of Absolute Value

The variables a and b represent any real number.

1. $|a| \geq 0$

2. $|a| = |-a|$

3. $|a - b| = |b - a|$ $\quad$ $a - b$ and $b - a$ are opposites of each other.

3 Add Real Numbers

We can use various physical models to describe the addition of real numbers. For example, profits and losses pertaining to investments: A loss of \$25.75 (written as -25.75) on one investment, along with a profit of \$22.20 (written as 22.20) on a second investment, produces an overall loss of \$3.55. Thus $(-25.75) + 22.20 = -3.55$. Think in terms of profits and losses for each of the following examples.

$$50 + 75 = 125 \qquad 20 + (-30) = -10$$

$$-4.3 + (-6.2) = -10.5 \qquad -27 + 43 = 16$$

$$\frac{7}{8} + \left(-\frac{1}{4}\right) = \frac{5}{8} \qquad -3\frac{1}{2} + \left(-3\frac{1}{2}\right) = -7$$

Though all problems that involve addition of real numbers could be solved using the profit–loss interpretation, it is sometimes convenient to have a more precise description of the addition process. For this purpose we use the concept of absolute value.

Addition of Real Numbers

Two Positive Numbers The sum of two positive real numbers is the sum of their absolute values.

Two Negative Numbers The sum of two negative real numbers is the opposite of the sum of their absolute values.

One Positive and One Negative Number The sum of a positive real number and a negative real number can be found by subtracting the smaller absolute value from the larger absolute value and giving the result the sign of the original number that has the larger absolute value. If the two numbers have the same absolute value, then their sum is 0.

Zero and Another Number The sum of 0 and any real number is the real number itself.

Now consider the following examples in terms of the previous description of addition. These examples include operations with rational numbers in common fraction form. If you need a review on operations with fractions, see Appendix A.

EXAMPLE 1 Find the sum.

(a) $(-6) + (-8)$ **(b)** $6\frac{3}{4} + \left(-2\frac{1}{2}\right)$ **(c)** $14 + (-21)$ **(d)** $-72.4 + 72.4$

Solution

(a) $(-6) + (-8) = -(|-6| + |-8|) = -(6 + 8) = -14$

(b) $6\frac{3}{4} + \left(-2\frac{1}{2}\right) = \left(\left|6\frac{3}{4}\right| - \left|-2\frac{1}{2}\right|\right) = \left(6\frac{3}{4} - 2\frac{1}{2}\right) = \left(6\frac{3}{4} - 2\frac{2}{4}\right) = 4\frac{1}{4}$

(c) $14 + (-21) = -(|-21| - |14|) = -(21 - 14) = -7$

(d) $-72.4 + 72.4 = 0$

▼ **PRACTICE YOUR SKILL**

Find the sum.

(a) $-8.42 + 10.75$ **(b)** $\left(-\frac{2}{3}\right) + \left(-\frac{1}{4}\right)$ **(c)** $145 + (-213)$ ■

4 Subtract Real Numbers

We can describe the subtraction of real numbers in terms of addition.

Subtraction of Real Numbers

If a and b are real numbers, then

$$a - b = a + (-b)$$

It may be helpful for you to read $a - b = a + (-b)$ as "a minus b is equal to a plus the opposite of b." In other words, every subtraction problem can be changed to an equivalent addition problem. Consider the following examples.

EXAMPLE 2 Find the difference.

(a) $7 - 9$ **(b)** $-5 - (-13)$ **(c)** $6.1 - (-14.2)$ **(d)** $-\frac{7}{8} - \left(-\frac{1}{4}\right)$

Solution

(a) $7 - 9 = 7 + (-9) = -2$

(b) $-5 - (-13) = -5 + 13 = 8$

(c) $6.1 - (-14.2) = 6.1 + 14.2 = 20.3$

(d) $-\frac{7}{8} - \left(-\frac{1}{4}\right) = -\frac{7}{8} + \frac{1}{4} = -\frac{7}{8} + \frac{2}{8} = -\frac{5}{8}$

▼ **PRACTICE YOUR SKILL**

Find the difference.

(a) $-2 - 9$ **(b)** $6 - (-10)$ **(c)** $-3.2 - (-7.2)$ **(d)** $\frac{3}{4} - \left(-\frac{1}{2}\right)$ ■

It should be apparent that addition is a key operation. To simplify numerical expressions that involve addition and subtraction, we can first change all subtractions to additions and then perform the additions.

EXAMPLE 3 Simplify $7 - 9 - 14 + 12 - 6 + 4$.

Solution

$$\begin{aligned} 7 - 9 - 14 + 12 - 6 + 4 &= 7 + (-9) + (-14) + 12 + (-6) + 4 \\ &= -6 \end{aligned}$$

▼ **PRACTICE YOUR SKILL**

Simplify $4 - 10 - 3 + 12 - 2 - 8$. ■

EXAMPLE 4 Simplify $-2\frac{1}{8} + \frac{3}{4} - \left(-\frac{3}{8}\right) - \frac{1}{2}$.

Solution

$$\begin{aligned} -2\frac{1}{8} + \frac{3}{4} - \left(-\frac{3}{8}\right) - \frac{1}{2} &= -2\frac{1}{8} + \frac{3}{4} + \frac{3}{8} + \left(-\frac{1}{2}\right) \\ &= -\frac{17}{8} + \frac{6}{8} + \frac{3}{8} + \left(-\frac{4}{8}\right) \\ &= -\frac{12}{8} = -\frac{3}{2} \end{aligned}$$

Change to equivalent fractions with a common denominator.

▼ **PRACTICE YOUR SKILL**

Simplify $-1\frac{3}{4} - \left(-\frac{3}{8}\right) + \frac{1}{2}$. ■

It is often helpful to convert subtractions to additions *mentally*. In the next two examples, the work shown in the dashed boxes could be done in your head.

EXAMPLE 5 Simplify $4 - 9 - 18 + 13 - 10$.

Solution

$$\begin{aligned} 4 - 9 - 18 + 13 - 10 &= 4 + (-9) + (-18) + 13 + (-10) \\ &= -20 \end{aligned}$$

▼ **PRACTICE YOUR SKILL**

Simplify $8 - 4 - 3 + 6 - 1$. ■

EXAMPLE 6 Simplify $\left(\frac{2}{3}-\frac{1}{5}\right)-\left(\frac{1}{2}-\frac{7}{10}\right)$.

Solution

$$\left(\frac{2}{3}-\frac{1}{5}\right)-\left(\frac{1}{2}-\frac{7}{10}\right)=\left[\frac{2}{3}+\left(-\frac{1}{5}\right)\right]-\left[\frac{1}{2}+\left(-\frac{7}{10}\right)\right]$$

$$=\left[\frac{10}{15}+\left(-\frac{3}{15}\right)\right]-\left[\frac{5}{10}+\left(-\frac{7}{10}\right)\right]$$

Within the brackets, change to equivalent fractions with a common denominator.

$$=\left(\frac{7}{15}\right)-\left(-\frac{2}{10}\right)$$

$$=\left(\frac{7}{15}\right)+\left(+\frac{2}{10}\right)$$

$$=\frac{14}{30}+\left(+\frac{6}{30}\right)$$

Change to equivalent fractions with a common denominator.

$$=\frac{20}{30}=\frac{2}{3}$$

▼ **PRACTICE YOUR SKILL**

Simplify $\left(\frac{1}{3}-\frac{1}{2}\right)-\left(\frac{2}{5}-\frac{1}{10}\right)$. ■

5 Multiply Real Numbers

To determine the product of a positive number and a negative number, we can use the interpretation of multiplication of whole numbers as repeated addition. For example, $4 \cdot 2$ means four 2s; thus $4 \cdot 2 = 2 + 2 + 2 + 2 = 8$. Applying this concept to the product of 4 and -2 yields

$$4(-2) = -2 + (-2) + (-2) + (-2) = -8$$

Because the order in which we multiply two numbers does not change the product, we know that

$$4(-2) = -2(4) = -8$$

Therefore, the product of a positive real number and a negative real number, in either order, is a negative number.

Finally, let's consider the product of two negative integers. The following pattern using integers helps with the reasoning.

$$4(-2) = -8 \qquad 3(-2) = -6 \qquad 2(-2) = -4$$

$$1(-2) = -2 \qquad 0(-2) = 0 \qquad (-1)(-2) = ?$$

To continue this pattern, the product of -1 and -2 has to be 2. In general, this type of reasoning helps us realize that the product of any two negative real numbers is a positive real number. Using the concept of absolute value, we can describe the **multiplication of real numbers** as follows.

Multiplication of Real Numbers

1. The product of two positive or two negative real numbers is the product of their absolute values.
2. The product of a positive real number and a negative real number (either order) is the opposite of the product of their absolute values.
3. The product of zero and any real number is zero.

The following example illustrates this description of multiplication. Again, the steps shown in the dashed boxes are usually performed mentally.

EXAMPLE 7

Find the product for each of the following.

(a) $(-6)(-7)$ **(b)** $(8)(-9)$ **(c)** $\left(-\frac{3}{4}\right)\left(\frac{1}{3}\right)$

Solution

(a) $(-6)(-7) = |-6| \cdot |-7| = 6 \cdot 7 = 42$

(b) $(8)(-9) = -(|8| \cdot |-9|) = -(8 \cdot 9) = -72$

(c) $\left(-\frac{3}{4}\right)\left(\frac{1}{3}\right) = -\left(\left|-\frac{3}{4}\right| \cdot \left|\frac{1}{3}\right|\right) = -\left(\frac{3}{4} \cdot \frac{1}{3}\right) = -\frac{1}{4}$

▼ **PRACTICE YOUR SKILL**

Find the product for each of the following.

(a) $(-12)(3)$ **(b)** $(-1)(-9)$ **(c)** $\left(\frac{5}{8}\right)\left(-\frac{2}{5}\right)$ ■

The previous example illustrated a step-by-step process for multiplying real numbers. In practice, however, the key is to remember that the product of two positive or two negative numbers is positive and that the product of a positive number and a negative number (either order) is negative.

6 Divide Real Numbers

The relationship between multiplication and division provides the basis for dividing real numbers. For example, we know that $8 \div 2 = 4$ because $2 \cdot 4 = 8$. In other words, the quotient of two numbers can be found by looking at a related multiplication problem. In the following examples, we used this same type of reasoning to determine some quotients that involve integers.

$\frac{6}{-2} = -3$ because $(-2)(-3) = 6$

$\frac{-12}{3} = -4$ because $(3)(-4) = -12$

$\frac{-18}{-2} = 9$ because $(-2)(9) = -18$

$\frac{0}{-5} = 0$ because $(-5)(0) = 0$

$\frac{-8}{0}$ is undefined Remember that division by zero is undefined!

A precise description for **division of real numbers** follows.

Division of Real Numbers

1. The quotient of two positive or two negative real numbers is the quotient of their absolute values.
2. The quotient of a positive real number and a negative real number or of a negative real number and a positive real number is the opposite of the quotient of their absolute values.
3. The quotient of zero and any nonzero real number is zero.
4. The quotient of any nonzero real number and zero is undefined.

The following example illustrates this description of division. Again, for practical purposes, the key is to remember whether the quotient is positive or negative.

EXAMPLE 8

Find the quotient for each of the following.

(a) $\frac{-16}{-4}$ **(b)** $\frac{28}{-7}$ **(c)** $\frac{-3.6}{4}$ **(d)** $\frac{0}{\frac{7}{8}}$

Solution

(a) $\frac{-16}{-4} = \frac{|-16|}{|-4|} = \frac{16}{4} = 4$ **(b)** $\frac{28}{-7} = -\left(\frac{|28|}{|-7|}\right) = -\left(\frac{28}{7}\right) = -4$

(c) $\frac{-3.6}{4} = -\left(\frac{|-3.6|}{|4|}\right) = -\left(\frac{3.6}{4}\right) = -0.9$ **(d)** $\frac{0}{\frac{7}{8}} = 0$

▼ **PRACTICE YOUR SKILL**

Find the quotient for each of the following.

(a) $\frac{-24}{3}$ **(b)** $\frac{-4.8}{-0.8}$ **(c)** $\frac{0}{-7}$ ■

7 Simplify Numerical Expressions

Now let's simplify some numerical expressions that involve the four basic operations with real numbers. Remember that multiplications and divisions are done first, from left to right, before additions and subtractions are performed.

EXAMPLE 9

Simplify $-2\frac{1}{3} + 4\left(-\frac{2}{3}\right) - (-5)\left(-\frac{1}{3}\right)$.

Solution

$$-2\frac{1}{3} + 4\left(-\frac{2}{3}\right) - (-5)\left(-\frac{1}{3}\right) = -2\frac{1}{3} + \left(-\frac{8}{3}\right) + \left(-\frac{5}{3}\right)$$

$$= -\frac{7}{3} + \left(-\frac{8}{3}\right) + \left(-\frac{5}{3}\right) \quad \text{Change to improper fraction.}$$

$$= -\frac{20}{3}$$

▼ **PRACTICE YOUR SKILL**

Simplify $5\frac{3}{4} - \left(-1\frac{1}{4}\right) - (-2)\left(-\frac{1}{2}\right)$. ■

EXAMPLE 10 Simplify $-24 \div 4 + 8(-5) - (-5)(3)$.

Solution

$$\begin{aligned} -24 \div 4 + 8(-5) - (-5)(3) &= -6 + (-40) - (-15) \\ &= -6 + (-40) + 15 \\ &= -31 \end{aligned}$$

▼ **PRACTICE YOUR SKILL**

Simplify $12 - 8 \div (-2) + 6(-4)$. ■

EXAMPLE 11 Simplify $-7.3 - 2[-4.6(6 - 7)]$.

Solution

$$\begin{aligned} -7.3 - 2[-4.6(6 - 7)] = -7.3 - 2[-4.6(-1)] &= -7.3 - 2[4.6] \\ &= -7.3 - 9.2 \\ &= -7.3 + (-9.2) \\ &= -16.5 \end{aligned}$$

▼ **PRACTICE YOUR SKILL**

Simplify $-6.8 - 3[8 - (-2.1)(-5)]$. ■

EXAMPLE 12 Simplify $[3(-7) - 2(9)][5(-7) + 3(9)]$.

Solution

$$\begin{aligned} [3(-7) - 2(9)][5(-7) + 3(9)] &= [-21 - 18][-35 + 27] \\ &= [-39][-8] \\ &= 312 \end{aligned}$$

▼ **PRACTICE YOUR SKILL**

Simplify $[-2(-4) - 3(6)][4(-3) + 2(-1)]$. ■

8 Use Real Numbers to Represent Problems

EXAMPLE 13

Eray Haciosmanoglu/Used under license from Shutterstock

Apply Your Skill

On a flight from Orlando to Washington, D.C., the airline sold 52 economy seats, 25 business-class seats, and 12 first-class seats, and had 20 empty seats. The airline has determined that it makes a profit of \$550 per first-class seat and \$100 profit per business-class seat. However, the airline incurs a loss of \$20 per economy seat and a loss of \$75 per empty seat. Determine the profit (or loss) for the flight.

Solution

Let the profit be represented by positive numbers and the loss be represented by negative numbers. Then the following expression would represent the profit or loss for this flight.

$$52(-20) + 25(100) + 12(550) + 20(-75)$$

Simplify this expression as follows:

$$52(-20) + 25(100) + 12(550) + 20(-75)$$
$$= -1040 + 2500 + 6600 - 1500 = 6560$$

Therefore, the flight had a profit of \$6560.

▼ PRACTICE YOUR SKILL

The following scale is used by a human resource department to score a multiple-choice personality survey.

Answer	A	B	C	D	E
Points	5	3	−1	−2	−3

Determine John's score if he answered A ten times, B three times, C eight times, D four times, and E five times. ■

CONCEPT QUIZ

For Problems 1–10, answer true or false.

1. The product of two negative real numbers is a positive real number.
2. The quotient of two negative integers is a negative integer.
3. The quotient of any nonzero real number and zero is zero.
4. If x represents any real number, then $-x$ represents a negative real number.
5. The product of three negative real numbers is a negative real number.
6. The statement $|6 - 4| = |4 - 6|$ is a true statement.
7. The numerical expression $-\frac{3}{4} + \frac{2}{3} - \frac{1}{2}$ simplifies to $-\frac{7}{12}$.
8. The numerical expression $3\left(\frac{1}{5}\right) - 2\left(\frac{2}{3}\right) - 5\left(\frac{1}{2}\right)$ simplifies to $-\frac{31}{10}$.
9. The absolute value of every real number is a positive real number.
10. The numerical expression $0.3(2.4) - 0.4(1.6) + 0.2(5.3)$ simplifies to 1.14.

Problem Set 1.2

1 Review the Real Number Line

1. Graph the following points and their opposites on the real number line: 1, −2, and 4.

2. Graph the following points and their opposites on the real number line: −3, −1, and 5.

2 Find the Absolute Value of a Number

3. Find the following absolute values:
(a) $|-7|$ **(b)** $|0|$ **(c)** $|15|$

4. Find the following absolute values:
(a) $|2|$ **(b)** $|-1|$ **(c)** $|-10|$

3 Add Real Numbers

For Problems 5–14, find the sum.

5. $8 + (-15)$
6. $9 + (-18)$
7. $(-12) + (-7)$
8. $(-7) + (-14)$
9. $-2\frac{3}{8} + 5\frac{7}{8}$
10. $-1\frac{1}{5} + 3\frac{4}{5}$
11. $-17.3 + 12.5$
12. $-16.3 + 19.6$
13. $\left(-\frac{1}{3}\right) + \left(-\frac{3}{4}\right)$
14. $\left(-\frac{5}{6}\right) + \frac{3}{8}$

4 Subtract Real Numbers

For Problems 15–30, find the difference.

15. $-8 - 14$
16. $-17 - 9$
17. $9 - 16$
18. $8 - 22$
19. $4\frac{1}{3} - \left(-1\frac{1}{6}\right)$
20. $1\frac{1}{12} - \left(-5\frac{3}{4}\right)$
21. $-21 - 39$
22. $-23 - 38$

23. $21.42 - 7.29$

24. $2.73 - 8.14$

25. $-21.4 - (-14.9)$

26. $-32.6 - (-9.8)$

27. $-\frac{3}{2} - \left(-\frac{3}{4}\right)$

28. $\frac{5}{8} - \frac{11}{12}$

29. $-\frac{2}{3} - \frac{7}{9}$

30. $\frac{5}{6} - \left(-\frac{2}{9}\right)$

5 Multiply Real Numbers

For Problems 31–40, find the product.

31. $(-9)(-12)$

32. $(-6)(-13)$

33. $(5)(-14)$

34. $(-17)(4)$

35. $\left(-\frac{1}{3}\right)\left(\frac{2}{5}\right)$

36. $(-8)\left(\frac{1}{3}\right)$

37. $(5.4)(-7.2)$

38. $(-8.5)(-3.3)$

39. $\left(-\frac{3}{4}\right)\left(\frac{4}{5}\right)$

40. $\left(\frac{1}{2}\right)\left(-\frac{4}{5}\right)$

6 Divide Real Numbers

For Problems 41–54, find the quotient.

41. $(-56) \div (-4)$

42. $(-81) \div (-3)$

43. $\frac{-112}{16}$

44. $\frac{-75}{5}$

45. $\frac{1}{2} \div \left(-\frac{1}{8}\right)$

46. $\frac{2}{3} \div \left(-\frac{1}{6}\right)$

47. $0 \div (-14)$

48. $(-19) \div 0$

49. $(-21) \div 0$

50. $0 \div (-11)$

51. $\frac{-1.2}{-6}$

52. $\frac{-6.3}{0.7}$

53. $\frac{3}{4} \div \left(-\frac{1}{2}\right)$

54. $\left(-\frac{5}{6}\right) \div \left(-\frac{7}{8}\right)$

7 Simplify Numerical Expressions

For Problems 55–94, simplify each numerical expression.

55. $9 - 12 - 8 + 5 - 6$

56. $6 - 9 + 11 - 8 - 7 + 14$

57. $-21 + (-17) - 11 + 15 - (-10)$

58. $-16 - (-14) + 16 + 17 - 19$

59. $7\frac{1}{8} - \left(2\frac{1}{4} - 3\frac{7}{8}\right)$

60. $-4\frac{3}{5} - \left(1\frac{1}{5} - 2\frac{3}{10}\right)$

61. $16 - 18 + 19 - [14 - 22 - (31 - 41)]$

62. $-19 - [15 - 13 - (-12 + 8)]$

63. $[14 - (16 - 18)] - [32 - (8 - 9)]$

64. $[-17 - (14 - 18)] - [21 - (-6 - 5)]$

65. $4\frac{1}{12} - \frac{1}{2}\left(\frac{1}{3}\right)$

66. $-\frac{4}{5} - \frac{1}{2}\left(-\frac{3}{5}\right)$

67. $-5 + (-2)(7) - (-3)(8)$

68. $-9 - 4(-2) + (-7)(6)$

69. $\frac{2}{5}\left(-\frac{3}{4}\right) - \left(-\frac{1}{2}\right)\left(\frac{3}{5}\right)$

70. $-\frac{2}{3}\left(\frac{1}{4}\right) + \left(-\frac{1}{3}\right)\left(\frac{5}{4}\right)$

71. $(-6)(-9) + (-7)(4)$

72. $(-7)(-7) - (-6)(4)$

73. $3(5 - 9) - 3(-6)$

74. $7(8 - 9) + (-6)(4)$

75. $(6 - 11)(4 - 9)$

76. $(7 - 12)(-3 - 2)$

77. $-6(-3 - 9 - 1)$

78. $-8(-3 - 4 - 6)$

79. $56 \div (-8) - (-6) \div (-2)$

80. $-65 \div 5 - (-13)(-2) + (-36) \div 12$

81. $-3[5 - (-2)] - 2(-4 - 9)$

82. $-2(-7 + 13) + 6(-3 - 2)$

83. $\frac{-6 + 24}{-3} + \frac{-7}{-6 - 1}$

84. $\frac{-12 + 20}{-4} + \frac{-7 - 11}{-9}$

85. $14.1 - (17.2 - 13.6)$

86. $-9.3 - (10.4 + 12.8)$

87. $3(2.1) - 4(3.2) - 2(-1.6)$

88. $5(-1.6) - 3(2.7) + 5(6.6)$

89. $7(6.2 - 7.1) - 6(-1.4 - 2.9)$

90. $-3(2.2 - 4.5) - 2(1.9 + 4.5)$

91. $\frac{2}{3} - \left(\frac{3}{4} - \frac{5}{6}\right)$

92. $-\frac{1}{2} - \left(\frac{3}{8} + \frac{1}{4}\right)$

93. $3\left(\frac{1}{2}\right) + 4\left(\frac{2}{3}\right) - 2\left(\frac{5}{6}\right)$

94. $2\left(\frac{3}{8}\right) - 5\left(\frac{1}{2}\right) + 6\left(\frac{3}{4}\right)$

95. Use a calculator to check your answers for Problems 55–94.

8 Use Real Numbers to Represent Problems

96. A scuba diver was 32 feet below sea level when he noticed that his partner had his extra knife. He ascended 13 feet to meet his partner and then continued to dive down for another 50 feet. How far below sea level is the diver?

97. Jeff played 18 holes of golf on Saturday. On each of six holes he was 1 under par, on each of four holes he was 2 over par, on one hole he was 3 over par, on each of two holes he shot par, and on each of five holes he was 1 over par. How did he finish relative to par?

98. After dieting for 30 days, Ignacio has lost 18 pounds. What number describes his average weight change per day?

99. Michael bet \$5 on each of the nine races at the racetrack. His only winnings were \$28.50 on one race. How much did he win (or lose) for the day?

100. Max bought a piece of trim molding that measured $11\frac{3}{8}$ feet in length. Because of defects in the wood, he had to trim $1\frac{5}{8}$ feet off one end, and he also had to remove $\frac{3}{4}$ of a foot off the other end. How long was the piece of molding after he trimmed the ends?

101. Natasha recorded the daily gains or losses for her company stock for a week. On Monday it gained 1.25 dollars; on Tuesday it gained 0.88 dollars; on Wednesday it lost 0.50 dollars; on Thursday it lost 1.13 dollars; on Friday it gained 0.38 dollars. What was the net gain (or loss) for the week?

102. On a summer day in Florida, the afternoon temperature was 96°F. After a thunderstorm, the temperature dropped 8°F. What would be the temperature if the sun came back out and the temperature rose 5°F?

103. In an attempt to lighten a dragster, the racing team exchanged two rear wheels for wheels that each weighed 15.6 pounds less. They also exchanged the crankshaft for one that weighed 4.8 pounds less. They changed the rear axle for one that weighed 23.7 pounds less but had to add an additional roll bar that weighed 10.6 pounds. If they wanted to lighten the dragster by 50 pounds, did they meet their goal?

104. A large corporation has five divisions. Two of the divisions had earnings of \$2,300,000 each. The other three divisions had a loss of \$1,450,000, a loss of \$640,000, and a gain of \$1,850,000, respectively. What was the net gain (or loss) of the corporation for the year?

THOUGHTS INTO WORDS

105. Explain why $\frac{0}{8} = 0$, but $\frac{8}{0}$ is undefined.

106. The following simplification problem is incorrect. The answer should be -11. Find and correct the error.

$$\begin{aligned} 8 \div (-4)(2) - 3(4) \div 2 + (-1) &= (-2)(2) - 12 \div 1 \\ &= -4 - 12 \\ &= -16 \end{aligned}$$

Answers to the Concept Quiz

1. True **2.** False **3.** False **4.** False **5.** True **6.** True **7.** True **8.** False **9.** False **10.** True

Answers to the Example Practice Skills

1. (a) 2.33 **(b)** $-\frac{11}{12}$ **(c)** -68 **2. (a)** -11 **(b)** 16 **(c)** 4 **(d)** $\frac{5}{4}$ **3.** -7 **4.** $-\frac{7}{8}$ **5.** 6 **6.** $-\frac{7}{15}$

7. (a) -36 **(b)** 9 **(c)** $-\frac{1}{4}$ **8. (a)** -8 **(b)** 6 **(c)** 0 **9.** 6 **10.** -8 **11.** 0.7 **12.** 140 **13.** 28

1.3 Properties of Real Numbers and the Use of Exponents

OBJECTIVES

1 Review Real Number Properties

2 Apply Properties to Simplify Expressions

3 Evaluate Exponential Expressions

1 Review Real Number Properties

At the beginning of this section we will list and briefly discuss some of the basic properties of real numbers. Be sure that you understand these properties, for they not only facilitate manipulations with real numbers but also serve as the basis for many algebraic computations.

Closure Property for Addition

If a and b are real numbers, then $a + b$ is a unique real number.

Closure Property for Multiplication

If a and b are real numbers, then ab is a unique real number.

We say that the set of real numbers is *closed* with respect to addition and also with respect to multiplication. That is, the sum of two real numbers is a unique real number, and the product of two real numbers is a unique real number. We use the word *unique* to indicate *exactly one*.

Commutative Property of Addition

If a and b are real numbers, then

$$a + b = b + a$$

Commutative Property of Multiplication

If a and b are real numbers, then

$$ab = ba$$

We say that addition and multiplication are commutative operations. This means that the order in which we add or multiply two numbers does not affect the result. For example, $6 + (-8) = (-8) + 6$ and $(-4)(-3) = (-3)(-4)$. It is also important to realize that subtraction and division are *not* commutative operations; order *does* make a difference. For example, $3 - 4 = -1$ but $4 - 3 = 1$. Likewise, $2 \div 1 = 2$ but $1 \div 2 = \frac{1}{2}$.

Associative Property of Addition

If a, b, and c are real numbers, then

$$(a + b) + c = a + (b + c)$$

Associative Property of Multiplication

If a, b, and c are real numbers, then

$$(ab)c = a(bc)$$

Addition and multiplication are **binary operations**. That is, we add (or multiply) two numbers at a time. The associative properties apply if more than two numbers are to be added or multiplied; they are grouping properties. For example, $(-8 + 9) + 6 = -8 + (9 + 6)$; changing the grouping of the numbers does not affect the final sum. This is also true for multiplication, which is illustrated by $[(-4)(-3)](2) = (-4)[(-3)(2)]$. Subtraction and division are *not* associative operations. For example, $(8 - 6) - 10 = -8$, but $8 - (6 - 10) = 12$. An example showing that division is not associative is $(8 \div 4) \div 2 = 1$, but $8 \div (4 \div 2) = 4$.

Identity Property of Addition

If a is any real number, then

$$a + 0 = 0 + a = a$$

Zero is called the identity element for addition. This merely means that the sum of any real number and zero is identically the same real number. For example, $-87 + 0 = 0 + (-87) = -87$.

Identity Property of Multiplication

If a is any real number, then

$$a(1) = 1(a) = a$$

We call 1 the identity element for multiplication. The product of any real number and 1 is identically the same real number. For example, $(-119)(1) = (1)(-119) = -119$.

Additive Inverse Property

For every real number a, there exists a unique real number $-a$ such that

$$a + (-a) = -a + a = 0$$

The real number $-a$ is called the **additive inverse of a** or the **opposite of a**. For example, 16 and -16 are additive inverses, and their sum is 0. The additive inverse of 0 is 0.

Multiplication Property of Zero

If a is any real number, then

$$(a)(0) = (0)(a) = 0$$

The product of any real number and zero is zero. For example, $(-17)(0) = 0(-17) = 0$.

Multiplication Property of Negative One

If a is any real number, then

$$(a)(-1) = (-1)(a) = -a$$

The product of any real number and -1 is the opposite of the real number. For example, $(-1)(52) = (52)(-1) = -52$.

Multiplicative Inverse Property

For every nonzero real number a, there exists a unique real number $\frac{1}{a}$ such that

$$a\left(\frac{1}{a}\right) = \frac{1}{a}(a) = 1$$

The number $\frac{1}{a}$ is called the **multiplicative inverse of *a*** or the **reciprocal of *a*.** For example, the reciprocal of 2 is $\frac{1}{2}$ and $2\left(\frac{1}{2}\right) = \frac{1}{2}(2) = 1$. Likewise, the reciprocal of $\frac{1}{2}$ is $\frac{1}{\frac{1}{2}} = 2$. Therefore, 2 and $\frac{1}{2}$ are said to be reciprocals (or multiplicative inverses) of each other. Because division by zero is undefined, zero does not have a reciprocal.

Distributive Property

If a, b, and c are real numbers, then

$$a(b + c) = ab + ac$$

The distributive property ties together the operations of addition and multiplication. We say that **multiplication distributes over addition**. For example, $7(3 + 8) = 7(3) + 7(8)$. Because $b - c = b + (-c)$, it follows that **multiplication also distributes over subtraction**. This can be expressed symbolically as $a(b - c) = ab - ac$. For example, $6(8 - 10) = 6(8) - 6(10)$.

2 Apply Properties to Simplify Expressions

The following examples illustrate the use of the properties of real numbers to facilitate certain types of manipulations.

EXAMPLE 1 Simplify $[74 + (-36)] + 36$.

Solution

In such a problem, it is much more advantageous to group -36 and 36.

$$[74 + (-36)] + 36 = 74 + [(-36) + 36] \quad \text{By using the associative property for addition}$$
$$= 74 + 0 = 74$$

▼ **PRACTICE YOUR SKILL**

Simplify $25 + [(-25) + 119]$. ■

EXAMPLE 2 Simplify $[(-19)(25)](-4)$.

Solution

It is much easier to group 25 and -4. Thus

$$[(-19)(25)](-4) = (-19)[(25)(-4)] \quad \text{By using the associative property for multiplication}$$
$$= (-19)(-100)$$
$$= 1900$$

▼ **PRACTICE YOUR SKILL**

Simplify $4[(-25)(-57)]$. ■

EXAMPLE 3 Simplify $17 + (-14) + (-18) + 13 + (-21) + 15 + (-33)$.

Solution

We could add in the order in which the numbers appear. However, because addition is commutative and associative, we could change the order and group in any convenient way. For example, we could add all of the positive integers and add all of the negative integers, and then find the sum of these two results. It might be convenient to use the vertical format as follows:

$$\begin{array}{rrr} & -14 & \\ 17 & -18 & \\ 13 & -21 & -86 \\ \underline{15} & \underline{-33} & \underline{45} \\ 45 & -86 & -41 \end{array}$$

▼ **PRACTICE YOUR SKILL**

Simplify $22 + (-14) + (-42) + 12 + (-11) + 15$. ■

EXAMPLE 4 Simplify $-25(-2 + 100)$.

Solution

For this problem, it might be easiest to apply the distributive property first and then simplify.

$$\begin{aligned} -25(-2 + 100) &= (-25)(-2) + (-25)(100) \\ &= 50 + (-2500) \\ &= -2450 \end{aligned}$$

▼ **PRACTICE YOUR SKILL**

Simplify $-20(-5 + 150)$. ■

EXAMPLE 5 Simplify $(-87)(-26 + 25)$.

Solution

For this problem, it would be better not to apply the distributive property but instead to add the numbers inside the parentheses first and then find the indicated product.

$$\begin{aligned} (-87)(-26 + 25) &= (-87)(-1) \\ &= 87 \end{aligned}$$

▼ **PRACTICE YOUR SKILL**

Simplify $-15(47 - 44)$. ■

EXAMPLE 6 Simplify $3.7(104) + 3.7(-4)$.

Solution

Remember that the distributive property allows us to change from the form $a(b + c)$ to $ab + ac$ or from the form $ab + ac$ to $a(b + c)$. In this problem, we want to use the latter change. Thus

$$\begin{aligned} 3.7(104) + 3.7(-4) &= 3.7[104 + (-4)] \\ &= 3.7(100) \\ &= 370 \end{aligned}$$

▼ **PRACTICE YOUR SKILL**

Simplify $1.4(-5) + 1.4(15)$. ■

Examples 4, 5, and 6 illustrate an important issue. Sometimes the form $a(b + c)$ is more convenient, but at other times the form $ab + ac$ is better. In these cases, as well as in the cases of other properties, you should *think first* and decide whether or not the properties can be used to make the manipulations easier.

3 Evaluate Exponential Expressions

Exponents are used to indicate repeated multiplication. For example, we can write $4 \cdot 4 \cdot 4$ as 4^3, where the "raised 3" indicates that 4 is to be used as a factor 3 times. The following general definition is helpful.

Definition 1.3

If n is a positive integer and b is any real number, then

$$b^n = \underbrace{bbb \cdots b}_{n \text{ factors of } b}$$

We refer to b as the **base** and to n as the **exponent**. The expression b^n can be read "b to the nth power." We commonly associate the terms *squared* and *cubed* with exponents of 2 and 3, respectively. For example, b^2 is read "b squared" and b^3 as "b cubed." An exponent of 1 is usually not written, so b^1 is written as b. The following examples illustrate Definition 1.3.

$$2^3 = 2 \cdot 2 \cdot 2 = 8 \qquad \left(\frac{1}{2}\right)^5 = \frac{1}{2} \cdot \frac{1}{2} \cdot \frac{1}{2} \cdot \frac{1}{2} \cdot \frac{1}{2} = \frac{1}{32}$$

$$3^4 = 3 \cdot 3 \cdot 3 \cdot 3 = 81 \qquad (0.7)^2 = (0.7)(0.7) = 0.49$$

$$-5^2 = -(5 \cdot 5) = -25 \qquad (-5)^2 = (-5)(-5) = 25$$

Please take special note of the last two examples. Note that $(-5)^2$ means that -5 is the base and is to be used as a factor twice. However, -5^2 means that 5 is the base and that after it is squared, we take the opposite of that result.

Simplifying numerical expressions that contain exponents creates no trouble if we keep in mind that exponents are used to indicate repeated multiplication. Let's consider some examples.

EXAMPLE 7

Simplify $3(-4)^2 + 5(-3)^2$.

Solution

$$3(-4)^2 + 5(-3)^2 = 3(16) + 5(9) \qquad \text{Find the powers}$$
$$= 48 + 45$$
$$= 93$$

▼ PRACTICE YOUR SKILL

Simplify $5(-3)^2+6(-2)^2$. ■

EXAMPLE 8

Simplify $(2 + 3)^2$.

Solution

$$(2 + 3)^2 = (5)^2 \qquad \text{Add inside the parentheses before applying the exponent}$$
$$= 25 \qquad \text{Square the 5}$$

▼ PRACTICE YOUR SKILL

Simplify $(2 - 6)^2$. ■

EXAMPLE 9

Simplify $[3(-1) - 2(1)]^3$.

Solution

$$\begin{aligned}[3(-1) - 2(1)]^3 &= [-3 - 2]^3 \\ &= [-5]^3 \\ &= -125\end{aligned}$$

▼ **PRACTICE YOUR SKILL**

Simplify $[-5(-3) - 2(6)]^3$. ■

EXAMPLE 10

Simplify $4\left(\frac{1}{2}\right)^3 - 3\left(\frac{1}{2}\right)^2 + 6\left(\frac{1}{2}\right) + 2$.

Solution

$$\begin{aligned}4\left(\frac{1}{2}\right)^3 - 3\left(\frac{1}{2}\right)^2 + 6\left(\frac{1}{2}\right) + 2 &= 4\left(\frac{1}{8}\right) - 3\left(\frac{1}{4}\right) + 6\left(\frac{1}{2}\right) + 2 \\ &= \frac{1}{2} - \frac{3}{4} + 3 + 2 \\ &= \frac{19}{4}\end{aligned}$$

▼ **PRACTICE YOUR SKILL**

Simplify $2 + 8\left(\frac{1}{2}\right) - 3\left(\frac{1}{2}\right)^2 + 4\left(\frac{1}{2}\right)^3$. ■

CONCEPT QUIZ

For Problems 1–10, answer true or false.

1. Addition is a commutative operation.
2. Subtraction is a commutative operation.
3. Zero is the identity element for addition.
4. The multiplicative inverse of 0 is 0.
5. The numerical expression $(-25)(-16)(-4)$ simplifies to -1600.
6. The numerical expression $82(8) + 82(2)$ simplifies to 820.
7. Exponents are used to indicate repeated additions.
8. The numerical expression $65(7^2) + 35(7^2)$ simplifies to 4900.
9. In the expression $(-4)^3$, the base is 4.
10. In the expression -4^3, the base is 4.

Problem Set 1.3

1 Review Real Number Properties

For Problems 1–14, state the property that justifies each of the statements. For example, $3 + (-4) = (-4) + 3$ because of the commutative property of addition.

1. $[6 + (-2)] + 4 = 6 + [(-2) + 4]$
2. $x(3) = 3(x)$
3. $42 + (-17) = -17 + 42$
4. $1(x) = x$
5. $-114 + 114 = 0$
6. $(-1)(48) = -48$
7. $-1(x + y) = -(x + y)$
8. $-3(2 + 4) = -3(2) + (-3)(4)$
9. $12yx = 12xy$

10. $[(-7)(4)](-25) = (-7)[4(-25)]$

11. $7(4) + 9(4) = (7 + 9)4$

12. $(x + 3) + (-3) = x + [3 + (-3)]$

13. $[(-14)(8)](25) = (-14)[8(25)]$

14. $\left(\frac{3}{4}\right)\left(\frac{4}{3}\right) = 1$

2 Apply Properties to Simplify Expressions

For Problems 15–26, simplify each numerical expression. Be sure to take advantage of the properties whenever they can be used to make the computations easier.

15. $36 + (-14) + (-12) + 21 + (-9) - 4$

16. $-37 + 42 + 18 + 37 + (-42) - 6$

17. $[83 + (-99)] + 18$

18. $[63 + (-87)] + (-64)$

19. $(25)(-13)(4)$

20. $(14)(25)(-13)(4)$

21. $17(97) + 17(3)$

22. $-86[49 + (-48)]$

23. $14 - 12 - 21 - 14 + 17 - 18 + 19 - 32$

24. $16 - 14 - 13 - 18 + 19 + 14 - 17 + 21$

25. $(-50)(15)(-2) - (-4)(17)(25)$

26. $(2)(17)(-5) - (4)(13)(-25)$

3 Evaluate Exponential Expressions

For Problems 27–54, simplify each of the numerical expressions.

27. $2^3 - 3^3$

28. $3^2 - 2^4$

29. $-5^2 - 4^2$

30. $-7^2 + 5^2$

31. $(-2)^3 - 3^2$

32. $(-3)^3 + 3^2$

33. $3(-1)^3 - 4(3)^2$

34. $4(-2)^3 - 3(-1)^4$

35. $7(2)^3 + 4(-2)^3$

36. $-4(-1)^2 - 3(2)^3$

37. $-3(-2)^3 + 4(-1)^5$

38. $5(-1)^3 - (-3)^3$

39. $(-3)^2 - 3(-2)(5) + 4^2$

40. $(-2)^2 - 3(-2)(6) - (-5)^2$

41. $2^3 + 3(-1)^3(-2)^2 - 5(-1)(2)^2$

42. $-2(3)^2 - 2(-2)^3 - 6(-1)^5$

43. $(3 + 4)^2$

44. $(4 - 9)^2$

45. $[3(-2)^2 - 2(-3)^2]^3$

46. $[-3(-1)^3 - 4(-2)^2]^2$

47. $2(-1)^3 - 3(-1)^2 + 4(-1) - 5$

48. $(-2)^3 + 2(-2)^2 - 3(-2) - 1$

49. $2^4 - 2(2)^3 - 3(2)^2 + 7(2) - 10$

50. $3(-3)^3 + 4(-3)^2 - 5(-3) + 7$

51. $3\left(\frac{1}{2}\right)^4 - 2\left(\frac{1}{2}\right)^3 + 5\left(\frac{1}{2}\right)^2 - 4\left(\frac{1}{2}\right) + 1$

52. $4(0.1)^2 - 6(0.1) + 0.7$

53. $-\left(\frac{2}{3}\right)^2 + 5\left(\frac{2}{3}\right) - 4$

54. $4\left(\frac{1}{3}\right)^3 + 3\left(\frac{1}{3}\right)^2 + 2\left(\frac{1}{3}\right) + 6$

55. Use your calculator to check your answers for Problems 27–52.

For Problems 56–64, use your calculator to evaluate each numerical expression.

56. 2^{10}

57. 3^7

58. $(-2)^8$

59. $(-2)^{11}$

60. -4^9

61. -5^6

62. $(3.14)^3$

63. $(1.41)^4$

64. $(1.73)^5$

THOUGHTS INTO WORDS

65. State, in your own words, the multiplication property of negative one.

66. Explain how the associative and commutative properties can help simplify $[(25)(97)](-4)$.

67. Your friend keeps getting an answer of 64 when simplifying -2^6. What mistake is he making, and how would you help him?

68. Write a sentence explaining in your own words how to evaluate the expression $(-8)^2$. Also write a sentence explaining how to evaluate -8^2.

69. For what natural numbers n does $(-1)^n = -1$? For what natural numbers n does $(-1)^n = 1$? Explain your answers.

70. Is the set $\{0, 1\}$ closed with respect to addition? Is the set $\{0, 1\}$ closed with respect to multiplication? Explain your answers.

Answers to the Concept Quiz

1. True **2.** False **3.** True **4.** False **5.** True **6.** True **7.** False **8.** True **9.** False **10.** True

Answers to the Example Practice Skills

1. 119 **2.** 5700 **3.** -18 **4.** -2900 **5.** -45 **6.** 14 **7.** 69 **8.** 16 **9.** 27 **10.** $\frac{23}{4}$

1.4 Algebraic Expressions

OBJECTIVES

1. Simplify Algebraic Expressions
2. Evaluate Algebraic Expressions
3. Translate from English to Algebra

1 Simplify Algebraic Expressions

Algebraic expressions such as

$$2x, \quad 8xy, \quad 3xy^2, \quad -4a^2b^3c, \quad \text{and} \quad z$$

are called **terms**. A term is an indicated product that may have any number of factors. The variables involved in a term are called **literal factors**, and the numerical factor is called the **numerical coefficient**. Thus in $8xy$, the x and y are literal factors and 8 is the numerical coefficient. The numerical coefficient of the term $-4a^2bc$ is -4. Because $1(z) = z$, the numerical coefficient of the term z is understood to be 1. Terms that have the same literal factors are called **similar terms** or **like terms**. Some examples of similar terms are

$$3x \quad \text{and} \quad 14x \qquad\qquad 5x^2 \quad \text{and} \quad 18x^2$$

$$7xy \quad \text{and} \quad -9xy \qquad\qquad 9x^2y \quad \text{and} \quad -14x^2y$$

$$2x^3y^2, \quad 3x^3y^2, \quad \text{and} \quad -7x^3y^2$$

By the symmetric property of equality, we can write the distributive property as

$$ab + ac = a(b + c)$$

Then the commutative property of multiplication can be applied to change the form to

$$ba + ca = (b + c)a$$

This latter form provides the basis for simplifying algebraic expressions by **combining similar terms**. Consider the following examples.

$$3x + 5x = (3 + 5)x$$
$$= 8x$$

$$-6xy + 4xy = (-6 + 4)xy$$
$$= -2xy$$

$$5x^2 + 7x^2 + 9x^2 = (5 + 7 + 9)x^2$$
$$= 21x^2$$

$$4x - x = 4x - 1x$$
$$= (4 - 1)x = 3x$$

More complicated expressions might require that we first rearrange the terms by applying the commutative property of addition.

$$
\begin{aligned}
7x + 2y + 9x + 6y &= 7x + 9x + 2y + 6y \\
&= (7 + 9)x + (2 + 6)y && \text{Distributive property} \\
&= 16x + 8y
\end{aligned}
$$

$$
\begin{aligned}
6a - 5 - 11a + 9 &= 6a + (-5) + (-11a) + 9 \\
&= 6a + (-11a) + (-5) + 9 && \text{Commutative property} \\
&= (6 + (-11))a + 4 && \text{Distributive property} \\
&= -5a + 4
\end{aligned}
$$

As soon as you thoroughly understand the various simplifying steps, you may want to do the steps mentally. Then you could go directly from the given expression to the simplified form, as follows:

$$14x + 13y - 9x + 2y = 5x + 15y$$

$$3x^2y - 2y + 5x^2y + 8y = 8x^2y + 6y$$

$$-4x^2 + 5y^2 - x^2 - 7y^2 = -5x^2 - 2y^2$$

Applying the distributive property to remove parentheses and then to combine similar terms sometimes simplifies an algebraic expression, as the next example illustrates.

EXAMPLE 1

Simplify the following.

(a) $4(x + 2) + 3(x + 6)$ **(b)** $-5(y + 3) - 2(y - 8)$
(c) $5(x - y) - (x + y)$

Solution

(a)
$$
\begin{aligned}
4(x + 2) + 3(x + 6) &= 4(x) + 4(2) + 3(x) + 3(6) \\
&= 4x + 8 + 3x + 18 \\
&= 4x + 3x + 8 + 18 \\
&= (4 + 3)x + 26 \\
&= 7x + 26
\end{aligned}
$$

(b)
$$
\begin{aligned}
-5(y + 3) - 2(y - 8) &= -5(y) - 5(3) - 2(y) - 2(-8) \\
&= -5y - 15 - 2y + 16 \\
&= -5y - 2y - 15 + 16 \\
&= -7y + 1
\end{aligned}
$$

(c)
$$
\begin{aligned}
5(x - y) - (x + y) &= 5(x - y) - 1(x + y) && \text{Remember, } -a = -1(a). \\
&= 5(x) - 5(y) - 1(x) - 1(y) \\
&= 5x - 5y - 1x - 1y \\
&= 4x - 6y
\end{aligned}
$$

▼ PRACTICE YOUR SKILL

Simplify the following.

(a) $3(x - 4) + 5(x + 2)$ **(b)** $-3(b + 8) - 5(b - 1)$
(c) $-2(a + b) - (a + b)$ ■

When we are multiplying two terms such as 3 and $2x$, the associative property of multiplication provides the basis for simplifying the product.

$$3(2x) = (3 \cdot 2)x = 6x$$

This idea is put to use in the following example.

EXAMPLE 2

Simplify $3(2x + 5y) + 4(3x + 2y)$.

Solution

$$\begin{aligned} 3(2x + 5y) + 4(3x + 2y) &= 3(2x) + 3(5y) + 4(3x) + 4(2y) \\ &= 6x + 15y + 12x + 8y \\ &= 6x + 12x + 15y + 8y \\ &= 18x + 23y \end{aligned}$$

▼ **PRACTICE YOUR SKILL**

Simplify $-4(3x + 7y) + 2(-5x + 3y)$. ■

After you are sure of each step, a more simplified format may be used, as the following examples illustrate.

$$\begin{aligned} 5(a + 4) - 7(a + 3) &= 5a + 20 - 7a - 21 \quad \text{Be careful with this sign.} \\ &= -2a - 1 \end{aligned}$$

$$\begin{aligned} 3(x^2 + 2) + 4(x^2 - 6) &= 3x^2 + 6 + 4x^2 - 24 \\ &= 7x^2 - 18 \end{aligned}$$

$$\begin{aligned} 2(3x - 4y) - 5(2x - 6y) &= 6x - 8y - 10x + 30y \\ &= -4x + 22y \end{aligned}$$

2 Evaluate Algebraic Expressions

An algebraic expression takes on a numerical value whenever each variable in the expression is replaced by a real number. For example, if x is replaced by 5 and y by 9, the algebraic expression $x + y$ becomes the numerical expression $5 + 9$, which simplifies to 14. We say that $x + y$ has a value of 14 when x equals 5 and y equals 9. If $x = -3$ and $y = 7$, then $x + y$ has a value of $-3 + 7 = 4$. The following examples illustrate the process of finding a value of an algebraic expression. We commonly refer to the process as **evaluating algebraic expressions**.

EXAMPLE 3

Find the value of $3x - 4y$ when $x = 2$ and $y = -3$.

Solution

$$\begin{aligned} 3x - 4y &= 3(2) - 4(-3) \quad \text{when } x = 2 \text{ and } y = -3 \\ &= 6 + 12 \\ &= 18 \end{aligned}$$

▼ **PRACTICE YOUR SKILL**

Find the value of $5a - 3b$ when $a = -8$ and $b = 4$. ■

EXAMPLE 4 Evaluate $x^2 - 2xy + y^2$ for $x = -2$ and $y = -5$.

Solution

$$\begin{aligned} x^2 - 2xy + y^2 &= (-2)^2 - 2(-2)(-5) + (-5)^2 \quad \text{when } x = -2 \text{ and } y = -5 \\ &= 4 - 20 + 25 \\ &= 9 \end{aligned}$$

▼ **PRACTICE YOUR SKILL**

Evaluate $x^2 + 3xy - y^2$ for $x = -1$ and $y = 3$. ■

EXAMPLE 5 Evaluate $(a + b)^2$ for $a = 6$ and $b = -2$.

Solution

$$\begin{aligned} (a + b)^2 &= [6 + (-2)]^2 \quad \text{when } a = 6 \text{ and } b = -2 \\ &= (4)^2 \\ &= 16 \end{aligned}$$

▼ **PRACTICE YOUR SKILL**

Evaluate $(x + y)^3$ for $x = -6$ and $y = 2$. ■

EXAMPLE 6 Evaluate $(3x + 2y)(2x - y)$ for $x = 4$ and $y = -1$.

Solution

$$\begin{aligned} (3x + 2y)(2x - y) &= [3(4) + 2(-1)][2(4) - (-1)] \quad \text{when } x = 4 \text{ and } y = -1 \\ &= (12 - 2)(8 + 1) \\ &= (10)(9) \\ &= 90 \end{aligned}$$

▼ **PRACTICE YOUR SKILL**

Evaluate $(2a + 5b)(a - 2b)$ for $a = -3$ and $b = -1$. ■

EXAMPLE 7 Evaluate $7x - 2y + 4x - 3y$ for $x = -\frac{1}{2}$ and $y = \frac{2}{3}$.

Solution

Let's first simplify the given expression.

$$7x - 2y + 4x - 3y = 11x - 5y$$

Now we can substitute $-\frac{1}{2}$ for x and $\frac{2}{3}$ for y.

$$\begin{aligned} 11x - 5y &= 11\left(-\frac{1}{2}\right) - 5\left(\frac{2}{3}\right) \\ &= -\frac{11}{2} - \frac{10}{3} \end{aligned}$$

$$= -\frac{33}{6} - \frac{20}{6}$$ Change to equivalent fractions with a common denominator.

$$= -\frac{53}{6}$$

▼ **PRACTICE YOUR SKILL**

Evaluate $4a - 3b - 6a + 2b$ for $a = -\frac{3}{4}$ and $b = \frac{1}{3}$. ■

EXAMPLE 8 Evaluate $2(3x + 1) - 3(4x - 3)$ for $x = -6.2$.

Solution

Let's first simplify the given expression.

$$2(3x + 1) - 3(4x - 3) = 6x + 2 - 12x + 9$$
$$= -6x + 11$$

Now we can substitute -6.2 for x.

$$-6x + 11 = -6(-6.2) + 11$$
$$= 37.2 + 11$$
$$= 48.2$$

▼ **PRACTICE YOUR SKILL**

Evaluate $-4(2y + 3) + 5(3y - 1)$ for $y = -3.1$. ■

EXAMPLE 9 Evaluate $2(a^2 + 1) - 3(a^2 + 5) + 4(a^2 - 1)$ for $a = 10$.

Solution

Let's first simplify the given expression.

$$2(a^2 + 1) - 3(a^2 + 5) + 4(a^2 - 1) = 2a^2 + 2 - 3a^2 - 15 + 4a^2 - 4$$
$$= 3a^2 - 17$$

Substituting $a = 10$, we obtain

$$3a^2 - 17 = 3(10)^2 - 17$$
$$= 3(100) - 17$$
$$= 300 - 17$$
$$= 283$$

▼ **PRACTICE YOUR SKILL**

Evaluate $3(x^2 + 2) - 2(x^2 - 1) - 5(x^2 - 3)$ for $x = 4$. ■

3 Translate from English to Algebra

To use the tools of algebra to solve problems, we must be able to translate from English to algebra. This translation process requires that we recognize key phrases in the English language that translate into algebraic expressions (which involve the operations of addition, subtraction, multiplication, and division). Some of these key phrases and their algebraic counterparts are listed in the following table. The variable n represents the number being referred to in each phrase. When translating, remember that the commutative property holds only for the operations of addition and multiplication. Therefore, order will be crucial to algebraic expressions that involve subtraction and division.

English phrase	Algebraic expression
Addition	
The sum of a number and 4	$n + 4$
7 more than a number	$n + 7$
A number plus 10	$n + 10$
A number increased by 6	$n + 6$
8 added to a number	$n + 8$
Subtraction	
14 minus a number	$14 - n$
12 less than a number	$n - 12$
A number decreased by 10	$n - 10$
The difference between a number and 2	$n - 2$
5 subtracted from a number	$n - 5$
Multiplication	
14 times a number	$14n$
The product of 4 and a number	$4n$
$\frac{3}{4}$ of a number	$\frac{3}{4}n$
Twice a number	$2n$
Multiply a number by 12	$12n$
Division	
The quotient of 6 and a number	$\frac{6}{n}$
The quotient of a number and 6	$\frac{n}{6}$
A number divided by 9	$\frac{n}{9}$
The ratio of a number and 4	$\frac{n}{4}$
Mixture of operations	
4 more than three times a number	$3n + 4$
5 less than twice a number	$2n - 5$
3 times the sum of a number and 2	$3(n + 2)$
2 more than the quotient of a number and 12	$\frac{n}{12} + 2$
7 times the difference of 6 and a number	$7(6 - n)$

An English statement may not always contain a key word such as *sum, difference, product,* or *quotient.* Instead, the statement may describe a physical situation, and from this description we must deduce the operations involved. Some suggestions for handling such situations are given in the following examples.

EXAMPLE 10

Sonya can keyboard 65 words per minute. How many words will she keyboard in m minutes?

Solution

The total number of words keyboarded equals the product of the rate per minute and the number of minutes. Therefore, Sonya should be able to keyboard $65m$ words in m minutes.

▼ PRACTICE YOUR SKILL

A machine can paint eight automobile parts per hour. How many parts will be painted in h hours? ■

EXAMPLE 11

Russ has n nickels and d dimes. Express this amount of money in cents.

Solution

Each nickel is worth 5 cents and each dime is worth 10 cents. We represent the amount in cents by $5n + 10d$.

▼ PRACTICE YOUR SKILL

Michelle has q quarters and d dimes. Express this amount of money in cents. ■

EXAMPLE 12

The cost of a 50-pound sack of fertilizer is d dollars. What is the cost per pound for the fertilizer?

Solution

We calculate the cost per pound by dividing the total cost by the number of pounds. We represent the cost per pound by $\frac{d}{50}$.

▼ PRACTICE YOUR SKILL

Bart paid d dollars for a 25-pound bag of dog food. What is the cost per pound for the dog food? ■

The English statement we want to translate into algebra may contain some geometric ideas. Tables 1.1 and 1.2 contain some of the basic relationships that pertain to linear measurement in the English and metric systems, respectively.

Table 1.1 English system

12 inches	=	1 foot
3 feet	=	1 yard
1760 yards	=	1 mile
5280 feet	=	1 mile

Table 1.2 Metric system

1 kilometer	=	1000 meters
1 hectometer	=	100 meters
1 dekameter	=	10 meters
1 decimeter	=	0.1 meter
1 centimeter	=	0.01 meter
1 millimeter	=	0.001 meter

EXAMPLE 13

The distance between two cities is k kilometers. Express this distance in meters.

Solution

Because 1 kilometer equals 1000 meters, the distance in meters is represented by $1000k$.

▼ PRACTICE YOUR SKILL

The distance between two concession stands in a theater is y yards. Express this distance in feet. ■

EXAMPLE 14

The length of a rope is y yards and f feet. Express this length in inches.

Solution

Because 1 foot equals 12 inches and 1 yard equals 36 inches, the length of the rope in inches can be represented by $36y + 12f$.

▼ PRACTICE YOUR SKILL

The height of a hybrid corn plant is m meters and c centimeters. Express this height in millimeters. ■

EXAMPLE 15

The length of a rectangle is l centimeters and the width is w centimeters. Express the perimeter of the rectangle in meters.

Solution

A sketch of the rectangle may be helpful (Figure 1.7).

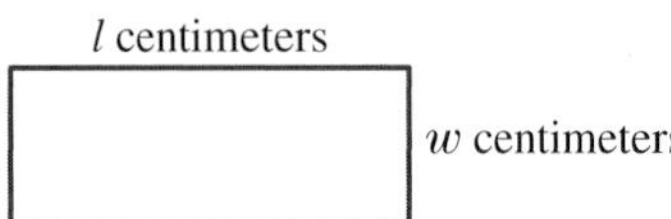

Figure 1.7

The perimeter of a rectangle is the sum of the lengths of the four sides. Thus the perimeter in centimeters is $l + w + l + w$, which simplifies to $2l + 2w$. Now, because 1 centimeter equals 0.01 meter, the perimeter, in meters, is $0.01(2l + 2w)$. This could also be written as $\frac{2l + 2w}{100} = \frac{2(l + w)}{100} = \frac{l + w}{50}$.

▼ PRACTICE YOUR SKILL

The length of a rectangle is l inches and the width is w inches. Express the perimeter in feet. ■

CONCEPT QUIZ

For Problems 1–10, answer true or false.

1. The numerical coefficient of the term xy is 1.
2. The terms $5x^2y$ and $6xy^2$ are similar terms.
3. The algebraic expression $-3(2x - y)$ simplifies to 9 if x is replaced by -4 and y is replaced by -5.

4. The algebraic expression $xy - 2x + 3y - xy + y$ simplifies to 4 if x is replaced by $\frac{1}{2}$ and y is replaced by $-\frac{3}{4}$.
5. The algebraic expression $-2(x - y) + 3(3x + 2y) - (x - y)$ simplifies to $6x + 9y$.
6. The value of $-3(2x + 4) + 4(2x - 1)$ is -9.72 when $x = 3.14$.
7. The algebraic expression $(x - y) - (x - y)$ simplifies to $2x - 2y$.
8. In the metric system, 1 centimeter = 10 millimeters.
9. The English phrase "4 less than twice the number n" translates into the algebraic expression $2n - 4$.
10. If the length of a rectangle is l inches and its width is w inches, then the perimeter in feet can be represented by $24(l + w)$.

Problem Set 1.4

1 Simplify Algebraic Expressions

Simplify the algebraic expressions in Problems 1–14 by combining similar terms.

1. $-7x + 11x$
2. $5x - 8x + x$
3. $5a^2 - 6a^2$
4. $12b^3 - 17b^3$
5. $4n - 9n - n$
6. $6n + 13n - 15n$
7. $4x - 9x + 2y$
8. $7x - 9y - 10x - 13y$
9. $-3a^2 + 7b^2 + 9a^2 - 2b^2$
10. $-xy + z - 8xy - 7z$
11. $15x - 4 + 6x - 9$
12. $5x - 2 - 7x + 4 - x - 1$
13. $5a^2b - ab^2 - 7a^2b$
14. $8xy^2 - 5x^2y + 2xy^2 + 7x^2y$

Simplify the algebraic expressions in Problems 15–34 by removing parentheses and combining similar terms.

15. $3(x + 2) + 5(x + 3)$
16. $5(x - 1) + 7(x + 4)$
17. $-2(a - 4) - 3(a + 2)$
18. $-7(a + 1) - 9(a + 4)$
19. $3(n^2 + 1) - 8(n^2 - 1)$
20. $4(n^2 + 3) + (n^2 - 7)$
21. $-6(x^2 - 5) - (x^2 - 2)$
22. $3(x + y) - 2(x - y)$
23. $5(2x + 1) + 4(3x - 2)$
24. $5(3x - 1) + 6(2x + 3)$
25. $3(2x - 5) - 4(5x - 2)$
26. $3(2x - 3) - 7(3x - 1)$
27. $-2(n^2 - 4) - 4(2n^2 + 1)$
28. $-4(n^2 + 3) - (2n^2 - 7)$
29. $3(2x - 4y) - 2(x + 9y)$
30. $-7(2x - 3y) + 9(3x + y)$
31. $3(2x - 1) - 4(x + 2) - 5(3x + 4)$
32. $-2(x - 1) - 5(2x + 1) + 4(2x - 7)$
33. $-(3x - 1) - 2(5x - 1) + 4(-2x - 3)$
34. $4(-x - 1) + 3(-2x - 5) - 2(x + 1)$

2 Evaluate Algebraic Expressions

Evaluate the algebraic expressions in Problems 35–57 for the given values of the variables.

35. $3x + 7y$, $x = -1$ and $y = -2$
36. $5x - 9y$, $x = -2$ and $y = 5$
37. $4x^2 - y^2$, $x = 2$ and $y = -2$
38. $3a^2 + 2b^2$, $a = 2$ and $b = 5$
39. $2a^2 - ab + b^2$, $a = -1$ and $b = -2$
40. $-x^2 + 2xy + 3y^2$, $x = -3$ and $y = 3$
41. $2x^2 - 4xy - 3y^2$, $x = 1$ and $y = -1$
42. $4x^2 + xy - y^2$, $x = 3$ and $y = -2$
43. $3xy - x^2y^2 + 2y^2$, $x = 5$ and $y = -1$
44. $x^2y^3 - 2xy + x^2y^2$, $x = -1$ and $y = -3$
45. $7a - 2b - 9a + 3b$, $a = 4$ and $b = -6$
46. $-4x + 9y - 3x - y$, $x = -4$ and $y = 7$
47. $(x - y)^2$, $x = 5$ and $y = -3$
48. $2(a + b)^2$, $a = 6$ and $b = -1$
49. $-2a - 3a + 7b - b$, $a = -10$ and $b = 9$
50. $3(x - 2) - 4(x + 3)$, $x = -2$
51. $-2(x + 4) - (2x - 1)$, $x = -3$
52. $-4(2x - 1) + 7(3x + 4)$, $x = 4$
53. $2(x - 1) - (x + 2) - 3(2x - 1)$, $x = -1$
54. $-3(x + 1) + 4(-x - 2) - 3(-x + 4)$, $x = -\frac{1}{2}$
55. $3(x^2 - 1) - 4(x^2 + 1) - (2x^2 - 1)$, $x = \frac{2}{3}$
56. $2(n^2 + 1) - 3(n^2 - 3) + 3(5n^2 - 2)$, $n = \frac{1}{4}$
57. $5(x - 2y) - 3(2x + y) - 2(x - y)$, $x = \frac{1}{3}$ and $y = -\frac{3}{4}$

For Problems 58–63, use your calculator and evaluate each of the algebraic expressions for the indicated values. Express the final answers to the nearest tenth.

58. πr^2, $\pi = 3.14$ and $r = 2.1$

59. πr^2, $\pi = 3.14$ and $r = 8.4$

60. $\pi r^2 h$, $\pi = 3.14$, $r = 1.6$, and $h = 11.2$

61. $\pi r^2 h$, $\pi = 3.14$, $r = 4.8$, and $h = 15.1$

62. $2\pi r^2 + 2\pi rh$, $\pi = 3.14$, $r = 3.9$, and $h = 17.6$

63. $2\pi r^2 + 2\pi rh$, $\pi = 3.14$, $r = 7.8$, and $h = 21.2$

3 Translate from English to Algebra

For Problems 64–78, translate each English phrase into an algebraic expression and use n to represent the unknown number.

64. The sum of a number and 4

65. A number increased by 12

66. A number decreased by 7

67. Five less than a number

68. A number subtracted from 75

69. The product of a number and 50

70. One-third of a number

71. Four less than one-half of a number

72. Seven more than three times a number

73. The quotient of a number and 8

74. The quotient of 50 and a number

75. Nine less than twice a number

76. Six more than one-third of a number

77. Ten times the difference of a number and 6

78. Twelve times the sum of a number and 7

For Problems 79–99, answer the question with an algebraic expression.

79. Brian is n years old. How old will he be in 20 years?

80. Crystal is n years old. How old was she 5 years ago?

81. Pam is t years old, and her mother is 3 less than twice as old as Pam. What is the age of Pam's mother?

82. The sum of two numbers is 65, and one of the numbers is x. What is the other number?

83. The difference of two numbers is 47, and the smaller number is n. What is the other number?

84. The product of two numbers is 98, and one of the numbers is n. What is the other number?

85. The quotient of two numbers is 8, and the smaller number is y. What is the other number?

86. The perimeter of a square is c centimeters. How long is each side of the square?

87. The perimeter of a square is m meters. How long, in centimeters, is each side of the square?

88. Jesse has n nickels, d dimes, and q quarters in his bank. How much money, in cents, does he have in his bank?

89. Tina has c cents, which is all in quarters. How many quarters does she have?

90. If n represents a whole number, what represents the next larger whole number?

91. If n represents an odd integer, what represents the next larger odd integer?

92. If n represents an even integer, what represents the next larger even integer?

93. The cost of a 5-pound box of candy is c cents. What is the price per pound?

94. Larry's annual salary is d dollars. What is his monthly salary?

95. Mila's monthly salary is d dollars. What is her annual salary?

96. The perimeter of a square is i inches. What is the perimeter expressed in feet?

97. The perimeter of a rectangle is y yards and f feet. What is the perimeter expressed in feet?

98. The length of a line segment is d decimeters. How long is the line segment expressed in meters?

99. The distance between two cities is m miles. How far is this, expressed in feet?

100. Use your calculator to check your answers for Problems 35–57.

THOUGHTS INTO WORDS

101. Explain the difference between simplifying a numerical expression and evaluating an algebraic expression.

102. How would you help someone who is having difficulty expressing n nickels and d dimes in terms of cents?

103. When asked to write an algebraic expression for "8 more than a number," you wrote $x + 8$ and another student wrote $8 + x$. Are both expressions correct? Explain your answer.

104. When asked to write an algebraic expression for "6 less than a number," you wrote $x - 6$ and another student wrote $6 - x$. Are both expressions correct? Explain your answer.

Answers to the Concept Quiz

1. True **2.** False **3.** True **4.** False **5.** True **6.** True **7.** False **8.** False **9.** True **10.** False

Answers to the Example Practice Skills

1. (a) $8x - 2$ **(b)** $-8b - 19$ **(c)** $-3a - 3b$ **2.** $-22x - 22y$ **3.** -52 **4.** -17 **5.** -64 **6.** 11 **7.** $\frac{7}{6}$ **8.** -38.7 **9.** -41 **10.** $8h$ **11.** $25q + 10d$ **12.** $\frac{d}{25}$ **13.** $3y$ **14.** $1000m + 100c$ **15.** $\frac{l + w}{6}$

Chapter 1 Summary

OBJECTIVE	SUMMARY	EXAMPLE	CHAPTER REVIEW PROBLEMS
Identify certain sets of numbers (Sec. 1.1, Obj. 1, p. 2)	A set is a collection of objects. The objects are called elements or members of the set. The sets of natural numbers, whole numbers, integers, rational numbers, and irrational numbers are all subsets of the set of real numbers.	From the list $-4, \frac{7}{5}, 0.35, \sqrt{2},$ and 0, identify the integers. **Solution** The integers are -4 and 0.	Problem 1
Apply the properties of equality and the properties of real numbers (Sec. 1.1, Obj. 2, p. 6; Sec. 1.3, Obj. 1, p. 23)	The properties of real numbers help with numerical manipulations and serve as a basis for algebraic computation. The properties of equality are listed on page 6 and the properties of real numbers are listed on pages 23–25.	State the property that justifies the statement If $x = y$ and $y = 7$, then $x = 7$. **Solution** The statement is justified by the transitive property of equality.	Problems 2–10
Find the absolute value of a number (Sec. 1.2, Obj. 2, p. 12)	Geometrically, the absolute value of any number is the distance between the number and zero on the number line. More formally, the absolute value of a real number a is defined as follows: 1. If $a \geq 0$, then $\lvert a\rvert = a$. 2. If $a < 0$, then $\lvert a\rvert = -a$.	Find the absolute value of the following. **(a)** $\lvert -2\rvert$ **(b)** $\left\lvert \frac{15}{4}\right\rvert$ **(c)** $\lvert -\sqrt{3}\rvert$ **Solutions** **(a)** $\lvert -2\rvert = -(-2) = 2$ **(b)** $\left\lvert \frac{15}{4}\right\rvert = \frac{15}{4}$ **(c)** $\lvert -\sqrt{3}\rvert = -(-\sqrt{3}) = \sqrt{3}$	Problems 11–14
Simplify numerical expressions *Addition* *Subtraction* *Multiplication and Division* (Sec. 1.1, Obj. 3, p. 6; Sec. 1.2, Obj. 7, p. 18)	Remember that multiplications and divisions are done first, from left to right, before additions and subtractions are done. The rules for addition of real numbers are on page 13. Applying the principle $a - b = a + (-b)$ changes every subtraction to an equivalent addition problem. 1. The product (or quotient) of two positive numbers or two negative numbers is the product (or quotient) of their absolute values. 2. The product (or quotient) of one positive and one negative number is the opposite of the product (or quotient) of their absolute values.	Simplify $30 + 50 \div 5 \cdot (-2) - 15$. **Solution** $30 + 50 \div 5 \cdot (-2) - 15$ $= 30 + 10 \cdot (-2) - 15$ $= 30 + (-20) - 15$ $= 10 - 15$ $= -5$	Problems 15–22 *(continued)*

OBJECTIVE	SUMMARY	EXAMPLE	CHAPTER REVIEW PROBLEMS
Evaluate exponential expressions (Sec. 1.3, Obj. 3, p. 27)	Exponents are used to indicate repeated multiplications. The expression b^n can be read "b to the nth power". We refer to b as the base and n as the exponent.	Simplify $2(-5)^3 + 3(-2)^2$. **Solution** $2(-5)^3 + 3(-2)^2$ $= 2(-125) + 3(4)$ $= -250 + 12$ $= -238$	Problems 23–26
Simplify algebraic expressions (Sec. 1.4, Obj. 1, p. 31)	Algebraic expressions such as $2x$, $3xy^2$, and $-4a^2b^3c$ are called terms. We call the variables in a term the literal factors and we call the numerical factor the numerical coefficient. Terms that have the same literal factors are called similar or like terms. The distributive property in the form $ba + ca = (b + c)a$ serves as a basis for combining like terms.	Simplify $5x^2 + 3x - 2x^2 - 7x$. **Solution** $5x^2 + 3x - 2x^2 - 7x$ $= 5x^2 - 2x^2 + 3x - 7x$ $= (5 - 2)x^2 + (3 - 7)x$ $= 3x^2 + (-4)x$ $= 3x^2 - 4x$	Problems 27–36
Evaluate algebraic expressions (Sec. 1.4, Obj. 2, p. 33)	An algebraic expression takes on a numerical value whenever each variable in the expression is replaced by a real number. The process of finding a value of an algebraic expression is referred to as evaluating the algebraic expression.	Evaluate $x^2 - 2xy + y^2$ when $x = 3$ and $y = -4$. **Solution** $x^2 - 2xy + y^2$ $= (3)^2 - 2(3)(-4) + (-4)^2$ when $x = 3$ and $y = -4$; $(3)^2 - 2(3)(-4) + (-4)^2$ $= 9 + 24 + 16 = 49$.	Problems 37–46
Translate from English to algebra (Sec. 1.4, Obj. 3, p. 35)	To translate English phrases into algebraic expressions, you must be familiar with key phrases that signal whether we are to find a sum, difference, product, or quotient.	Translate the English phrase *six less than twice a number* into an algebraic expression. **Solution** Let n represent the number. Six less than means that 6 will be subtracted from twice the number. Twice the number means that the number will be multiplied by 2. The phrase *six less than twice a number* translates into $2n - 6$.	Problems 47–64
Use real numbers to represent problems (Sec. 1.2, Obj. 8, p. 19)	Real numbers can be used to represent many situations in the real world.	A patient in the hospital had a body temperature of 106.7°. Over the next three hours his temperature fell 1.2° per hour. What was his temperature after the three hours? **Solution** $106.7 - 3(1.2)$ $= 106.7 - 3.6$ $= 103.1$; his temperature was 103.1°.	Problems 64–68

Chapter 1 Review Problem Set

1. From the list 0, $\sqrt{2}$, $\frac{3}{4}$, $-\frac{5}{6}$, $\frac{25}{3}$, $-\sqrt{3}$, -8, 0.34, $0.2\overline{3}$, 67, and $\frac{9}{7}$, identify each of the following.

a. The natural numbers

b. The integers

c. The nonnegative integers

d. The rational numbers

e. The irrational numbers

For Problems 2–10, state the property of equality or the property of real numbers that justifies each of the statements. For example, $6(-7) = -7(6)$ because of the commutative property of multiplication; and if $2 = x + 3$, then $x + 3 = 2$ is true because of the symmetric property of equality.

2. $7 + (3 + (-8)) = (7 + 3) + (-8)$

3. If $x = 2$ and $x + y = 9$, then $2 + y = 9$.

4. $-1(x + 2) = -(x + 2)$

5. $3(x + 4) = 3(x) + 3(4)$

6. $[(17)(4)](25) = (17)[(4)(25)]$

7. $x + 3 = 3 + x$

8. $3(98) + 3(2) = 3(98 + 2)$

9. $\left(\frac{3}{4}\right)\left(\frac{4}{3}\right) = 1$

10. If $4 = 3x - 1$, then $3x - 1 = 4$.

For Problems 11–14, find the absolute value.

11. $|-6.2|$

12. $\left|\frac{7}{3}\right|$

13. $|-\sqrt{15}|$

14. $|-8|$

For Problems 15–26, simplify each of the numerical expressions.

15. $-8\frac{1}{4} + \left(-4\frac{5}{8}\right) - \left(-6\frac{3}{8}\right)$

16. $9\frac{1}{3} - 12\frac{1}{2} + \left(-4\frac{1}{6}\right) - \left(-1\frac{1}{6}\right)$

17. $-8(2) - 16 \div (-4) + (-2)(-2)$

18. $4(-3) - 12 \div (-4) + (-2)(-1) - 8$

19. $-3(2 - 4) - 4(7 - 9) + 6$

20. $[48 + (-73)] + 74$

21. $[5(-2) - 3(-1)][-2(-1) + 3(2)]$

22. $3 - [-2(3 - 4)] + 7$

23. $-4^2 - 2^3$

24. $(-2)^4 + (-1)^3 - 3^2$

25. $2(-1)^2 - 3(-1)(2) - 2^2$

26. $[4(-1) - 2(3)]^2$

For Problems 27–36, simplify each of the algebraic expressions by combining similar terms.

27. $3a^2 - 2b^2 - 7a^2 - 3b^2$

28. $4x - 6 - 2x - 8 + x + 12$

29. $\frac{1}{5}ab^2 - \frac{3}{10}ab^2 + \frac{2}{5}ab^2 + \frac{7}{10}ab^2$

30. $-\frac{2}{3}x^2y - \left(-\frac{3}{4}x^2y\right) - \frac{5}{12}x^2y - 2x^2y$

31. $3(2n^2 + 1) + 4(n^2 - 5)$

32. $-2(3a - 1) + 4(2a + 3) - 5(3a + 2)$

33. $-(n - 1) - (n + 2) + 3$

34. $3(2x - 3y) - 4(3x + 5y) - x$

35. $4(a - 6) - (3a - 1) - 2(4a - 7)$

36. $-5(x^2 - 4) - 2(3x^2 + 6) + (2x^2 - 1)$

For Problems 37–46, evaluate each of the algebraic expressions for the given values of the variables.

37. $-5x + 4y$ for $x = \frac{1}{2}$ and $y = -1$

38. $3x^2 - 2y^2$ for $x = \frac{1}{4}$ and $y = -\frac{1}{2}$

39. $-5(2x - 3y)$ for $x = 1$ and $y = -3$

40. $(3a - 2b)^2$ for $a = -2$ and $b = 3$

41. $a^2 + 3ab - 2b^2$ for $a = 2$ and $b = -2$

42. $3n^2 - 4 - 4n^2 + 9$ for $n = 7$

43. $3(2x - 1) + 2(3x + 4)$ for $x = 1.2$

44. $-4(3x - 1) - 5(2x - 1)$ for $x = -2.3$

45. $2(n^2 + 3) - 3(n^2 + 1) + 4(n^2 - 6)$ for $n = -\frac{2}{3}$

46. $5(3n - 1) - 7(-2n + 1) + 4(3n - 1)$ for $n = \frac{1}{2}$

For Problems 47–54, translate each English phrase into an algebraic expression and use n to represent the unknown number.

47. Four increased by twice a number

48. Fifty subtracted from three times a number

49. Six less than two-thirds of a number

50. Ten times the difference of a number and 14

51. Eight subtracted from five times a number

52. The quotient of a number and three less than the number

53. Three less than five times the sum of a number and 2

54. Three-fourths of the sum of a number and 12

For Problems 55–64, answer the question with an algebraic expression.

55. The sum of two numbers is 37 and one of the numbers is n. What is the other number?

56. Yuriko can type w words in an hour. What is her typing rate per minute?

57. Harry is y years old. His brother is 7 years less than twice as old as Harry. How old is Harry's brother?

58. If n represents a multiple of 3, what represents the next largest multiple of 3?

59. Celia has p pennies, n nickels, and q quarters. How much, in cents, does Celia have?

60. The perimeter of a square is i inches. How long, in feet, is each side of the square?

61. The length of a rectangle is y yards and the width is f feet. What is the perimeter of the rectangle expressed in inches?

62. The length of a piece of wire is d decimeters. What is the length expressed in centimeters?

63. Joan is f feet and i inches tall. How tall is she in inches?

64. The perimeter of a rectangle is 50 centimeters. If the rectangle is c centimeters long, how wide is it?

65. Kayla has the capacity to record 4 minutes of video on her cellular phone. She currently has $3\frac{1}{2}$ minutes of video clips. How much recording capacity will she have left if she deletes $2\frac{1}{4}$ minutes of clips and adds $1\frac{3}{4}$ minutes of recording?

66. During the week, the price of a stock recorded the following gains and losses: Monday lost \$1.25, Tuesday lost \$0.45, Wednesday gained \$0.67, Thursday gained \$1.10, and Friday lost \$0.22. What is the average daily gain or loss for the week?

67. A crime-scene investigator has 3.4 ounces of a sample. He needs to conduct four tests that each require 0.6 ounces of the sample and one test that requires 0.8 ounces of the sample. How much of the sample remains after he uses it for the five tests?

68. For week 1 of a weight loss competition, Team A had three members lose 8 pounds each, two members lose 5 pounds each, one member loses 4 pounds, and two members gain 3 pounds. What was the total weight loss for Team A in the first week of the competition?

Chapter 1 Test

1. ______
2. ______
3. ______
4. ______
5. ______
6. ______
7. ______
8. ______
9. ______
10. ______
11. ______
12. ______
13. ______
14. ______
15. ______
16. ______
17. ______
18. ______
19. ______
20. ______
21. ______
22. ______

1. State the property of equality that justifies writing $x + 4 = 6$ for $6 = x + 4$.

2. State the property of real numbers that justifies writing $5(10 + 2)$ as $5(10) + 5(2)$.

For Problems 3–11, simplify each numerical expression.

3. $-4 - (-3) + (-5) - 7 + 10$

4. $7 - 8 - 3 + 4 - 9 - 4 + 2 - 12$

5. $5\left(-\frac{1}{3}\right) - 3\left(-\frac{1}{2}\right) + 7\left(-\frac{2}{3}\right) + 1$

6. $(-6) \cdot 3 \div (-2) - 8 \div (-4)$

7. $-\frac{1}{2}(3 - 7) - \frac{2}{5}(2 - 17)$

8. $[48 + (-93)] + (-49)$

9. $3(-2)^3 + 4(-2)^2 - 9(-2) - 14$

10. $[2(-6) + 5(-4)][-3(-4) - 7(6)]$

11. $[-2(-3) - 4(2)]^5$

12. Simplify $6x^2 - 3x - 7x^2 - 5x - 2$ by combining similar terms.

13. Simplify $3(3n - 1) - 4(2n + 3) + 5(-4n - 1)$ by removing parentheses and combining similar terms.

For Problems 14–20, evaluate each algebraic expression for the given values of the variables.

14. $-7x - 3y$ for $x = -6$ and $y = 5$

15. $3a^2 - 4b^2$ for $a = -\frac{3}{4}$ and $b = \frac{1}{2}$

16. $6x - 9y - 8x + 4y$ for $x = \frac{1}{2}$ and $y = -\frac{1}{3}$

17. $-5n^2 - 6n + 7n^2 + 5n - 1$ for $n = -6$

18. $-7(x - 2) + 6(x - 1) - 4(x + 3)$ for $x = 3.7$

19. $-2xy - x + 4y$ for $x = -3$ and $y = 9$

20. $4(n^2 + 1) - (2n^2 + 3) - 2(n^2 + 3)$ for $n = -4$

For Problems 21 and 22, translate the English phrase into an algebraic expression using n to represent the unknown number.

21. Thirty subtracted from six times a number

22. Four more than three times the sum of a number and 8

For Problems 23–25, answer each question with an algebraic expression.

23. The product of two numbers is 72 and one of the numbers is n. What is the other number? **23.** ________________

24. Tao has n nickels, d dimes, and q quarters. How much money, in cents, does she have? **24.** ________________

25. The length of a rectangle is x yards and the width is y feet. What is the perimeter of the rectangle expressed in feet? **25.** ________________

Equations, Inequalities, and Problem Solving

© Jimin Lai/AFP/Getty Images

■ *Most shoppers take advantage of the discounts offered by retailers. When making decisions about purchases, it is beneficial to be able to compute the sale prices.*

A retailer of sporting goods bought a putter for \$18. He wants to price the putter to make a profit of 40% of the selling price. What price should he mark on the putter? The equation $s = 18 + 0.4s$ can be used to determine that the putter should be sold for \$30.

Throughout this text, we develop algebraic skills, use these skills to help solve equations and inequalities, and then use equations and inequalities to solve applied problems. In this chapter, we review and expand concepts that are important to the development of problem-solving skills.

Video tutorials for all section learning objectives are available in a variety of delivery modes.

INTERNET PROJECT

Many students study algebra but are unaware of why the subject is called "algebra." Conduct an Internet search to find the origin of the term and find two variations of the term *algebra*. Then do another search to determine who is considered the "father" of algebra.

2.1 Solving First-Degree Equations

OBJECTIVES

1 Solve First-Degree Equations

2 Use Equations to Solve Word Problems

1 Solve First-Degree Equations

In Section 1.1, we stated that an equality (equation) is a statement where two symbols, or groups of symbols, are names for the same number. It should be further stated that an equation may be true or false. For example, the equation $3 + (-8) = -5$ is true, but the equation $-7 + 4 = 2$ is false.

Algebraic equations contain one or more variables. The following are examples of algebraic equations.

$$3x + 5 = 8 \qquad 4y - 6 = -7y + 9 \qquad x^2 - 5x - 8 = 0$$

$$3x + 5y = 4 \qquad x^3 + 6x^2 - 7x - 2 = 0$$

An algebraic equation such as $3x + 5 = 8$ is neither true nor false as it stands, and we often refer to it as an "open sentence." Each time that a number is substituted for x, the algebraic equation $3x + 5 = 8$ becomes a numerical statement that is true or false. For example, if $x = 0$, then $3x + 5 = 8$ becomes $3(0) + 5 = 8$, which is a false statement. If $x = 1$, then $3x + 5 = 8$ becomes $3(1) + 5 = 8$, which is a true statement. **Solving an equation** refers to the process of finding the number (or numbers) that make(s) an algebraic equation a true numerical statement. We call such numbers the **solutions** or **roots** of the equation, and we say that they **satisfy** the equation. We call the set of all solutions of an equation its **solution set**. Thus $\{1\}$ is the solution set of $3x + 5 = 8$.

In this chapter, we will consider techniques for solving **first-degree equations in one variable**. This means that the equations contain only one variable and that this variable has an exponent of 1. The following are examples of first-degree equations in one variable.

$$3x + 5 = 8 \qquad \frac{2}{3}y + 7 = 9$$

$$7a - 6 = 3a + 4 \qquad \frac{x-2}{4} = \frac{x-3}{5}$$

Equivalent equations are equations that have the same solution set. For example,

1. $3x + 5 = 8$
2. $3x = 3$
3. $x = 1$

are all equivalent equations because $\{1\}$ is the solution set of each.

The general procedure for solving an equation is to continue replacing the given equation with equivalent but simpler equations until we obtain an equation of the form *variable* = *constant* or *constant* = *variable*. Thus in the example above, $3x + 5 = 8$ was simplified to $3x = 3$, which was further simplified to $x = 1$, from which the solution set {1} is obvious.

To solve equations we need to use the various properties of equality. In addition to the reflexive, symmetric, transitive, and substitution properties we listed in Section 1.1, the following properties of equality play an important role.

Addition Property of Equality

For all real numbers a, b, and c,

$$a = b \quad \text{if and only if } a + c = b + c$$

Multiplication Property of Equality

For all real numbers a, b, and c, where $c \neq 0$,

$$a = b \quad \text{if and only if } ac = bc$$

The addition property of equality states that when the same number is added to both sides of an equation, an equivalent equation is produced. The multiplication property of equality states that we obtain an equivalent equation whenever we multiply both sides of an equation by the same *nonzero* real number. The following examples demonstrate the use of these properties to solve equations.

EXAMPLE 1 Solve $2x - 1 = 13$.

Solution

$$2x - 1 = 13$$

$$2x - 1 + 1 = 13 + 1 \qquad \text{Add 1 to both sides}$$

$$2x = 14$$

$$\frac{1}{2}(2x) = \frac{1}{2}(14) \qquad \text{Multiply both sides by } \frac{1}{2}$$

$$x = 7$$

The solution set is {7}.

▼ PRACTICE YOUR SKILL

Solve $-4x + 3 = -41$. ■

To check an apparent solution, we can substitute it into the original equation and see if we obtain a true numerical statement.

✔ Check

$$2x - 1 = 13$$

$$2(7) - 1 \stackrel{?}{=} 13$$

$$14 - 1 \stackrel{?}{=} 13$$

$$13 = 13$$

Now we know that {7} is the solution set of $2x - 1 = 13$. We will not show our checks for every example in this text, but do remember that checking is a way to detect arithmetic errors.

EXAMPLE 2 Solve $-7 = -5a + 9$.

Solution

$$-7 = -5a + 9$$

$$-7 + (-9) = 5a + 9 + (-9) \quad \text{Add } -9 \text{ to both sides}$$

$$-16 = -5a$$

$$-\frac{1}{5}(-16) = -\frac{1}{5}(-5a) \quad \text{Multiply both sides by } -\frac{1}{5}$$

$$\frac{16}{5} = a$$

The solution set is $\left\{\frac{16}{5}\right\}$.

▼ PRACTICE YOUR SKILL

Solve $15 = -2x + 38$. ■

Note that in Example 2 the final equation is $\frac{16}{5} = a$ instead of $a = \frac{16}{5}$. Technically, the symmetric property of equality (if $a = b$, then $b = a$) would permit us to change from $\frac{16}{5} = a$ to $a = \frac{16}{5}$, but such a change is not necessary to determine that the solution is $\frac{16}{5}$. Note that we could use the symmetric property at the very beginning to change $-7 = -5a + 9$ to $-5a + 9 = -7$; some people prefer having the variable on the left side of the equation.

Let's clarify another point. We stated the properties of equality in terms of only two operations, addition and multiplication. We could also include the operations of subtraction and division in the statements of the properties. That is, we could think in terms of subtracting the same number from both sides of an equation and also in terms of dividing both sides of an equation by the same nonzero number. For example, in the solution of Example 2, we could subtract 9 from both sides rather than adding -9 to both sides. Likewise, we could divide both sides by -5 instead of multiplying both sides by $-\frac{1}{5}$.

EXAMPLE 3 Solve $7x - 3 = 5x + 9$.

Solution

$$7x - 3 = 5x + 9$$

$$7x - 3 + (-5x) = 5x + 9 + (-5x) \quad \text{Add } -5x \text{ to both sides}$$

$$2x - 3 = 9$$

$$2x - 3 + 3 = 9 + 3 \quad \text{Add 3 to both sides}$$

$$2x = 12$$

$$\frac{1}{2}(2x) = \frac{1}{2}(12) \qquad \text{Multiply both sides by } \frac{1}{2}$$

$$x = 6$$

The solution set is {6}.

▼ PRACTICE YOUR SKILL

Solve $3y + 4 = 8y - 26$. ■

EXAMPLE 4 Solve $4(y - 1) + 5(y + 2) = 3(y - 8)$.

Solution

$$4(y - 1) + 5(y + 2) = 3(y - 8)$$

$$4y - 4 + 5y + 10 = 3y - 24 \qquad \text{Remove parentheses by applying the distributive property}$$

$$9y + 6 = 3y - 24 \qquad \text{Simplify the left side by combining similar terms}$$

$$9y + 6 + (-3y) = 3y - 24 + (-3y) \qquad \text{Add } -3y \text{ to both sides}$$

$$6y + 6 = -24$$

$$6y + 6 + (-6) = -24 + (-6) \qquad \text{Add } -6 \text{ to both sides}$$

$$6y = -30$$

$$\frac{1}{6}(6y) = \frac{1}{6}(-30) \qquad \text{Multiply both sides by } \frac{1}{6}$$

$$y = -5$$

The solution set is {−5}.

▼ PRACTICE YOUR SKILL

Solve $5(x - 4) + 3(x + 7) = 2(x - 1)$. ■

We can summarize the process of solving first-degree equations in one variable as follows.

Step 1 Simplify both sides of the equation as much as possible.

Step 2 Use the addition property of equality to isolate a term that contains the variable on one side of the equation and a constant on the other side.

Step 3 Use the multiplication property of equality to make the coefficient of the variable 1; that is, multiply both sides of the equation by the reciprocal of the numerical coefficient of the variable. The solution set should now be obvious.

Step 4 Check each solution by substituting it in the original equation and verifying that the resulting numerical statement is true.

2 Use Equations to Solve Word Problems

To use the tools of algebra to solve problems, we must be able to translate back and forth between the English language and the language of algebra. More specifically, we need to translate English sentences into algebraic equations. Such translations allow us to use our knowledge of equation solving to solve word problems. Let's consider an example.

EXAMPLE 5

Apply Your Skill

If we subtract 27 from three times a certain number, the result is 18. Find the number.

Solution

Let n represent the number to be found. The sentence "If we subtract 27 from three times a certain number, the result is 18" translates into the equation $3n - 27 = 18$. Solving this equation, we obtain

$$3n - 27 = 18$$

$$3n = 45 \quad \text{Add 27 to both sides}$$

$$n = 15 \quad \text{Multiply both sides by } \frac{1}{3}$$

The number to be found is 15.

▼ PRACTICE YOUR SKILL

If we add 43 to twice a number, the result is -19. Find the number. ■

We often refer to the statement "Let n represent the number to be found" as **declaring the variable**. We need to choose a letter to use as a variable and indicate what it represents for a specific problem. This may seem like an insignificant idea, but as the problems become more complex, the process of declaring the variable becomes even more important. Furthermore, it is true that you could probably solve a problem such as Example 5 without setting up an algebraic equation. However, as problems increase in difficulty, the translation from English to algebra becomes a key issue. Therefore, even with these relatively easy problems, we suggest that you concentrate on the translation process.

The next example involves the use of integers. Remember that the set of integers consists of $\{\ldots, -2, -1, 0, 1, 2, \ldots\}$. Furthermore, the integers can be classified as even, $\{\ldots, -4, -2, 0, 2, 4, \ldots\}$, or odd, $\{\ldots, -3, -1, 1, 3, \ldots\}$.

EXAMPLE 6

Apply Your Skill

The sum of three consecutive integers is 13 greater than twice the smallest of the three integers. Find the integers.

Solution

Because consecutive integers differ by 1, we will represent them as follows: Let n represent the smallest of the three consecutive integers; then $n + 1$ represents the second largest and $n + 2$ represents the largest.

$$\underbrace{n + (n + 1) + (n + 2)}_{\text{The sum of the three consecutive integers}} = \underbrace{2n + 13}_{\text{13 greater than twice the smallest}}$$

$$3n + 3 = 2n + 13$$

$$n = 10$$

The three consecutive integers are 10, 11, and 12.

▼ PRACTICE YOUR SKILL

For three consecutive integers, the sum of the first two integers is 14 more than the third integer. Find the integers. ■

To check our answers for Example 6, we must determine whether or not they satisfy the conditions stated in the original problem. Because 10, 11, and 12 are consecutive integers whose sum is 33, and because twice the smallest plus 13 is also 33 ($2(10) + 13 = 33$), we know that our answers are correct. (Remember, in checking a result for a word problem, it is *not* sufficient to check the result in the equation set up to solve the problem; the equation itself may be in error!)

In the two previous examples, the equation formed was almost a direct translation of a sentence in the statement of the problem. Now let's consider a situation where we need to think in terms of a guideline not explicitly stated in the problem.

EXAMPLE 7

Dynamic Graphics/Jupiter Images

Apply Your Skill

Khoa received a car repair bill for \$106. This included \$23 for parts, \$22 per hour for each hour of labor, and \$6 for taxes. Find the number of hours of labor.

Solution

See Figure 2.1. Let h represent the number of hours of labor. Then $22h$ represents the total charge for labor.

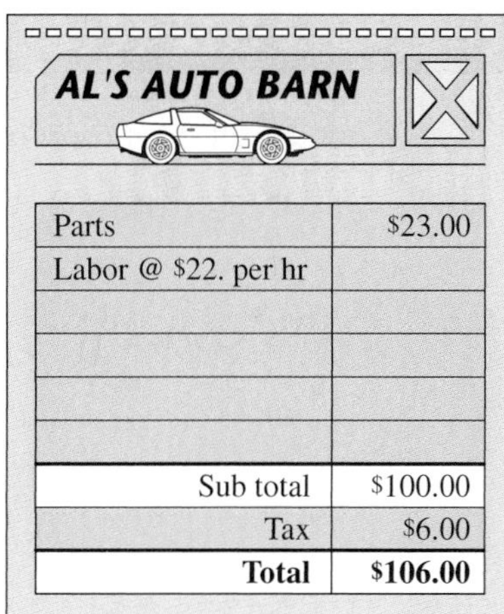

AL'S AUTO BARN

Parts	\$23.00
Labor @ \$22. per hr	
Sub total	\$100.00
Tax	\$6.00
Total	**\$106.00**

Figure 2.1

We can use a guideline of *charge for parts plus charge for labor plus tax equals the total bill* to set up the following equation.

$$\underset{\text{Parts}}{23} + \underset{\text{Labor}}{22h} + \underset{\text{Tax}}{6} = \underset{\text{Total bill}}{106}$$

Solving this equation, we obtain

$$22h + 29 = 106$$
$$22h = 77$$
$$h = 3\frac{1}{2}$$

Khoa was charged for $3\frac{1}{2}$ hours of labor.

▼ **PRACTICE YOUR SKILL**

Wallace received a cell-phone bill for \$89.00. This included \$49.00 for the monthly service charge, \$21.00 for taxes, and \$0.05 per minute for each minute of cell-phone use. Find the number of minutes the phone was used. ■

CONCEPT QUIZ

For Problems 1–10, answer true or false.

1. Equivalent equations have the same solution set.
2. $x^2 = 9$ is a first-degree equation.
3. The set of all solutions is called a solution set.
4. If the solution set is the null set, then the equation has at least one solution.
5. Solving an equation refers to obtaining any other equivalent equation.
6. If 5 is a solution, then a true numerical statement is formed when 5 is substituted for the variable in the equation.
7. Any number can be subtracted from both sides of an equation, and the result is an equivalent equation.
8. Any number can divide both sides of an equation to obtain an equivalent equation.
9. The solution set for the equation $3(x - 2) - (x - 3) = -2$ is $\left\{\frac{1}{2}\right\}$.
10. The solution set for the equation $-3(2x - 3) = -2(2x - 3)$ is $\left\{-\frac{3}{2}\right\}$.

Problem Set 2.1

1 Solve First-Degree Equations

For problems 1–50, solve each equation.

1. $3x + 4 = 16$
2. $4x + 2 = 22$
3. $5x + 1 = -14$
4. $7x + 4 = -31$
5. $-x - 6 = 8$
6. $8 - x = -2$
7. $4y - 3 = 21$
8. $6y - 7 = 41$
9. $3x - 4 = 15$
10. $5x + 1 = 12$
11. $-4 = 2x - 6$
12. $-14 = 3a - 2$
13. $-6y - 4 = 16$
14. $-8y - 2 = 18$
15. $4x - 1 = 2x + 7$
16. $9x - 3 = 6x + 18$
17. $5y + 2 = 2y - 11$
18. $9y + 3 = 4y - 10$
19. $3x + 4 = 5x - 2$
20. $2x - 1 = 6x + 15$
21. $-7a + 6 = -8a + 14$
22. $-6a - 4 = -7a + 11$
23. $5x + 3 - 2x = x - 15$
24. $4x - 2 - x = 5x + 10$
25. $6y + 18 + y = 2y + 3$
26. $5y + 14 + y = 3y - 7$
27. $4x - 3 + 2x = 8x - 3 - x$
28. $x - 4 - 4x = 6x + 9 - 8x$
29. $6n - 4 - 3n = 3n + 10 + 4n$
30. $2n - 1 - 3n = 5n - 7 - 3n$
31. $4(x - 3) = -20$
32. $3(x + 2) = -15$
33. $-3(x - 2) = 11$
34. $-5(x - 1) = 12$
35. $5(2x + 1) = 4(3x - 7)$
36. $3(2x - 1) = 2(4x + 7)$
37. $5x - 4(x - 6) = -11$
38. $3x - 5(2x + 1) = 13$
39. $-2(3x - 1) - 3 = -4$
40. $-6(x - 4) - 10 = -12$
41. $-2(3x + 5) = -3(4x + 3)$
42. $-(2x - 1) = -5(2x + 9)$
43. $3(x - 4) - 7(x + 2) = -2(x + 18)$
44. $4(x - 2) - 3(x - 1) = 2(x + 6)$
45. $-2(3n - 1) + 3(n + 5) = -4(n - 4)$
46. $-3(4n + 2) + 2(n - 6) = -2(n + 1)$
47. $3(2a - 1) - 2(5a + 1) = 4(3a + 4)$
48. $4(2a + 3) - 3(4a - 2) = 5(4a - 7)$
49. $-2(n - 4) - (3n - 1) = -2 + (2n - 1)$
50. $-(2n - 1) + 6(n + 3) = -4 - (7n - 11)$

2 Use Equations to Solve Word Problems

For Problems 51–66, use an algebraic approach to solve each problem.

51. If 15 is subtracted from three times a certain number, the result is 27. Find the number.
52. If 1 is subtracted from seven times a certain number, the result is the same as if 31 is added to three times the number. Find the number.

53. Find three consecutive integers whose sum is 42.

54. Find four consecutive integers whose sum is -118.

55. Find three consecutive odd integers such that three times the second minus the third is 11 more than the first.

56. Find three consecutive even integers such that four times the first minus the third is 6 more than twice the second.

57. The difference of two numbers is 67. The larger number is 3 less than six times the smaller number. Find the numbers.

58. The sum of two numbers is 103. The larger number is 1 more than five times the smaller number. Find the numbers.

59. Angelo is paid double time for each hour he works over 40 hours in a week. Last week he worked 46 hours and earned $572. What is his normal hourly rate?

60. Suppose that a plumbing repair bill, not including tax, was $130. This included $25 for parts and an amount for 5 hours of labor. Find the hourly rate that was charged for labor.

61. Suppose that Maria has 150 coins consisting of pennies, nickels, and dimes. The number of nickels she has is 10 less than twice the number of pennies; the number of dimes she has is 20 less than three times the number of pennies. How many coins of each kind does she have?

62. Hector has a collection of nickels, dimes, and quarters totaling 122 coins. The number of dimes he has is 3 more than four times the number of nickels, and the number of quarters he has is 19 less than the number of dimes. How many coins of each kind does he have?

63. The selling price of a ring is $750. This represents $150 less than three times the cost of the ring. Find the cost of the ring.

64. In a class of 62 students, the number of females is 1 less than twice the number of males. How many females and how many males are there in the class?

65. An apartment complex contains 230 apartments each having one, two, or three bedrooms. The number of two-bedroom apartments is 10 more than three times the number of three-bedroom apartments. The number of one-bedroom apartments is twice the number of two-bedroom apartments. How many apartments of each kind are in the complex?

66. Barry sells bicycles on a salary-plus-commission basis. He receives a monthly salary of $300 and a commission of $15 for each bicycle that he sells. How many bicycles must he sell in a month to have a total monthly income of $750?

THOUGHTS INTO WORDS

67. Explain the difference between a numerical statement and an algebraic equation.

68. Are the equations $7 = 9x - 4$ and $9x - 4 = 7$ equivalent equations? Defend your answer.

69. Suppose that your friend shows you the following solution to an equation.

$$17 = 4 - 2x$$
$$17 + 2x = 4 - 2x + 2x$$
$$17 + 2x = 4$$
$$17 + 2x - 17 = 4 - 17$$
$$2x = -13$$
$$x = \frac{-13}{2}$$

Is this a correct solution? What suggestions would you have in terms of the method used to solve the equation?

70. Explain in your own words what it means to declare a variable when solving a word problem.

71. Make up an equation whose solution set is the null set and explain why this is the solution set.

72. Make up an equation whose solution set is the set of all real numbers and explain why this is the solution set.

FURTHER INVESTIGATIONS

73. Solve each of the following equations.

(a) $5x + 7 = 5x - 4$

(b) $4(x - 1) = 4x - 4$

(c) $3(x - 4) = 2(x - 6)$

(d) $7x - 2 = -7x + 4$

(e) $2(x - 1) + 3(x + 2) = 5(x - 7)$

(f) $-4(x - 7) = -2(2x + 1)$

74. Verify that for any three consecutive integers, the sum of the smallest and largest is equal to twice the middle integer. [*Hint:* Use n, $n + 1$, and $n + 2$ to represent the three consecutive integers.]

Answers to the Concept Quiz

1. True **2.** False **3.** True **4.** False **5.** False **6.** True **7.** True **8.** False **9.** True **10.** False

Answers to the Example Practice Skills

1. $\{11\}$ **2.** $\left\{\frac{23}{2}\right\}$ **3.** $\{6\}$ **4.** $\left\{-\frac{1}{2}\right\}$ **5.** -31 **6.** 15, 16, and 17 **7.** 380 minutes

2.2 Equations Involving Fractional Forms

OBJECTIVES

1. Solve Equations Involving Fractions
2. Solve Word Problems

1 Solve Equations Involving Fractions

To solve equations that involve fractions, it is usually easiest to begin by **clearing the equation of all fractions**. This can be accomplished by multiplying both sides of the equation by the least common multiple of all the denominators in the equation. Remember that the least common multiple of a set of whole numbers is the smallest nonzero whole number that is divisible by each of the numbers. For example, the least common multiple of 2, 3, and 6 is 12. When working with fractions, we refer to the least common multiple of a set of denominators as the **least common denominator** (LCD). Let's consider some equations involving fractions.

EXAMPLE 1 Solve $\frac{1}{2}x + \frac{2}{3} = \frac{3}{4}$.

Solution

$$\frac{1}{2}x + \frac{2}{3} = \frac{3}{4}$$

$$12\left(\frac{1}{2}x + \frac{2}{3}\right) = 12\left(\frac{3}{4}\right)$$ Multiply both sides by 12, which is the LCD of 2, 3, and 4

$$12\left(\frac{1}{2}x\right) + 12\left(\frac{2}{3}\right) = 12\left(\frac{3}{4}\right)$$ Apply the distributive property to the left side

$$6x + 8 = 9$$

$$6x = 1$$

$$x = \frac{1}{6}$$

The solution set is $\left\{\frac{1}{6}\right\}$.

✔ **Check**

$$\frac{1}{2}x + \frac{2}{3} = \frac{3}{4}$$

$$\frac{1}{2}\left(\frac{1}{6}\right) + \frac{2}{3} \stackrel{?}{=} \frac{3}{4}$$

$$\frac{1}{12} + \frac{2}{3} \stackrel{?}{=} \frac{3}{4}$$

$$\frac{1}{12} + \frac{8}{12} \stackrel{?}{=} \frac{3}{4}$$

$$\frac{9}{12} \stackrel{?}{=} \frac{3}{4}$$

$$\frac{3}{4} = \frac{3}{4}$$

▼ **PRACTICE YOUR SKILL**

Solve $\frac{2}{5}a - \frac{3}{2} = \frac{1}{3}$. ■

EXAMPLE 2 Solve $\frac{x}{2} + \frac{x}{3} = 10$.

Solution

$$\frac{x}{2} + \frac{x}{3} = 10 \qquad \text{Recall that } \frac{x}{2} = \frac{1}{2}x$$

$$6\left(\frac{x}{2} + \frac{x}{3}\right) = 6(10) \qquad \text{Multiply both sides by the LCD}$$

$$6\left(\frac{x}{2}\right) + 6\left(\frac{x}{3}\right) = 6(10) \qquad \text{Apply the distributive property to the left side}$$

$$3x + 2x = 60$$

$$5x = 60$$

$$x = 12$$

The solution set is $\{12\}$.

▼ **PRACTICE YOUR SKILL**

Solve $\frac{y}{6} + \frac{y}{4} = 8$. ■

As you study the examples in this section, pay special attention to the steps shown in the solutions. There are no hard-and-fast rules as to which steps should be performed mentally; this is an individual decision. When you solve problems, show enough steps to allow the flow of the process to be understood and to minimize the chances of making careless computational errors.

EXAMPLE 3 Solve $\frac{x-2}{3} + \frac{x+1}{8} = \frac{5}{6}$.

Solution

$$\frac{x-2}{3} + \frac{x+1}{8} = \frac{5}{6}$$

$$24\left(\frac{x-2}{3} + \frac{x+1}{8}\right) = 24\left(\frac{5}{6}\right) \qquad \text{Multiply both sides by the LCD}$$

$$24\left(\frac{x-2}{3}\right) + 24\left(\frac{x+1}{8}\right) = 24\left(\frac{5}{6}\right)$$ Apply the distributive property to the left side

$$8(x-2) + 3(x+1) = 20$$

$$8x - 16 + 3x + 3 = 20$$

$$11x - 13 = 20$$

$$11x = 33$$

$$x = 3$$

The solution set is $\{3\}$.

▼ PRACTICE YOUR SKILL

Solve $\frac{y+4}{4} + \frac{y-1}{3} = \frac{5}{2}$. ■

EXAMPLE 4 Solve $\frac{3t-1}{5} - \frac{t-4}{3} = 1$.

Solution

$$\frac{3t-1}{5} - \frac{t-4}{3} = 1$$

$$15\left(\frac{3t-1}{5} - \frac{t-4}{3}\right) = 15(1)$$ Multiply both sides by the LCD

$$15\left(\frac{3t-1}{5}\right) - 15\left(\frac{t-4}{3}\right) = 15(1)$$ Apply the distributive property to the left side

$$3(3t-1) - 5(t-4) = 15$$

$$9t - 3 - 5t + 20 = 15$$ Be careful with this sign!

$$4t + 17 = 15$$

$$4t = -2$$

$$t = -\frac{2}{4} = -\frac{1}{2}$$ Reduce!

The solution set is $\left\{-\frac{1}{2}\right\}$.

▼ PRACTICE YOUR SKILL

Solve $\frac{2a+5}{3} - \frac{a-6}{2} = 1$. ■

2 Solve Word Problems

As we expand our skills for solving equations, we also expand our capabilities for solving word problems. There is no definitive procedure that will ensure success at solving word problems, but the following suggestions can be helpful.

Suggestions for Solving Word Problems

1. Read the problem carefully and make certain that you understand the meanings of all of the words. Be especially alert for any technical terms used in the statement of the problem.
2. Read the problem a second time (perhaps even a third time) to get an overview of the situation being described. Determine the known facts as well as what is to be found.
3. Sketch any figure, diagram, or chart that might be helpful in analyzing the problem.
4. Choose a meaningful variable to represent an unknown quantity in the problem (perhaps t, if time is an unknown quantity) and represent any other unknowns in terms of that variable.
5. Look for a guideline that you can use to set up an equation. A guideline might be a formula, such as *distance equals rate times time*, or a statement of a relationship, such as "The sum of the two numbers is 28."
6. Form an equation that contains the variable and that translates the conditions of the guideline from English to algebra.
7. Solve the equation, and use the solution to determine all facts requested in the problem.
8. Check all answers back into the **original statement of the problem.**

Keep these suggestions in mind as we continue to solve problems. We will elaborate on some of these suggestions at different times throughout the text. Now let's consider some problems.

EXAMPLE 5

Apply Your Skill

Find a number such that three-eighths of the number minus one-half of it is 14 less than three-fourths of the number.

Solution

Let n represent the number to be found.

$$\frac{3}{8}n - \frac{1}{2}n = \frac{3}{4}n - 14$$

$$8\left(\frac{3}{8}n - \frac{1}{2}n\right) = 8\left(\frac{3}{4}n - 14\right)$$

$$8\left(\frac{3}{8}n\right) - 8\left(\frac{1}{2}n\right) = 8\left(\frac{3}{4}n\right) - 8(14)$$

$$3n - 4n = 6n - 112$$

$$-n = 6n - 112$$

$$-7n = -112$$

$$n = 16$$

The number is 16. Check it!

▼ PRACTICE YOUR SKILL

Find a number such that three-fourths of the number plus one-third of the number is 2 more than the number.

■

EXAMPLE 6 **Apply Your Skill**

The width of a rectangular parking lot is 8 feet less than three-fifths of the length. The perimeter of the lot is 400 feet. Find the length and width of the lot.

Solution

Let l represent the length of the lot. Then $\frac{3}{5}l - 8$ represents the width (Figure 2.2).

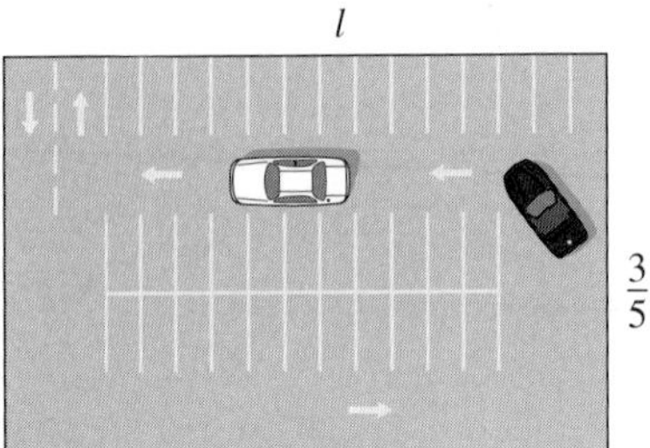

Figure 2.2

A guideline for this problem is the formula, *the perimeter of a rectangle equals twice the length plus twice the width* $(P = 2l + 2w)$. Use this formula to form the following equation.

$$\begin{array}{ccccc} P & = & 2l + & & 2w \\ \downarrow & & \downarrow & & \downarrow \\ 400 & = & 2l + & 2 & \left(\frac{3}{5}l - 8\right) \end{array}$$

Solving this equation, we obtain

$$400 = 2l + \frac{6l}{5} - 16$$

$$5(400) = 5\left(2l + \frac{6l}{5} - 16\right)$$

$$2000 = 10l + 6l - 80$$

$$2000 = 16l - 80$$

$$2080 = 16l$$

$$130 = l$$

The length of the lot is 130 feet, and the width is $\frac{3}{5}(130) - 8 = 70$ feet.

▼ PRACTICE YOUR SKILL

The width of a sports field on campus is 20 feet less than three-fourths of the length of the field. The perimeter of the field is 1080 feet. Find the length and width of the field. ■

In Examples 5 and 6, note the use of different letters as variables. It is helpful to choose a variable that has significance for the problem you are working on. For example, in Example 6 the choice of l to represent the length seems natural and meaningful. (Certainly this is another matter of personal preference, but you might consider it.)

In Example 6 a geometric relationship, $P = 2l + 2w$, serves as a guideline for setting up the equation. The following geometric relationships pertaining to angle measure may also serve as guidelines.

1. Complementary angles are two angles the sum of whose measures is 90°.
2. Supplementary angles are two angles the sum of whose measures is 180°.
3. The sum of the measures of the three angles of a triangle is 180°.

EXAMPLE 7

Apply Your Skill

One of two complementary angles is 6° larger than one-half of the other angle. Find the measure of each of the angles.

Solution

Let a represent the measure of one of the angles. Then $\frac{1}{2}a + 6$ represents the measure of the other angle. Because they are complementary angles, the sum of their measures is 90°.

$$a + \left(\frac{1}{2}a + 6\right) = 90$$

$$2a + a + 12 = 180$$

$$3a + 12 = 180$$

$$3a = 168$$

$$a = 56$$

If $a = 56$, then $\frac{1}{2}a + 6$ becomes $\frac{1}{2}(56) + 6 = 34$. The angles have measures of 34° and 56°.

▼ PRACTICE YOUR SKILL

One of two supplementary angles is 4° larger than three-fifths of the other angle. Find the measure of each angle. ■

EXAMPLE 8

Andersen Ross/Iconica/Getty Images

Apply Your Skill

Dominic's present age is 10 years more than Michele's present age. In 5 years, Michele's age will be three-fifths of Dominic's age. What are their present ages?

Solution

Let x represent Michele's present age. Then Dominic's age will be represented by $x + 10$. In 5 years, everyone's age is increased by 5 years, so we need to add 5 to Michele's present age and 5 to Dominic's present age to represent their ages in 5 years. Therefore, in 5 years Michele's age will be represented by $x + 5$, and Dominic's age will be represented by $x + 15$. Thus we can set up the equation reflecting the fact that in 5 years, Michele's age will be three-fifths of Dominic's age.

$$x + 5 = \frac{3}{5}(x + 15)$$

$$5(x + 5) = 5\left[\frac{3}{5}(x + 15)\right]$$

$$5x + 25 = 3(x + 15)$$

$$5x + 25 = 3x + 45$$

$$2x + 25 = 45$$

$$2x = 20$$
$$x = 10$$

Because x represents Michele's present age, we know her age is 10. Dominic's present age is represented by $x + 10$, so his age is 20.

▼ **PRACTICE YOUR SKILL**

Raymond's present age is 6 years less than Kay's present age. In 4 years Raymond's age will be five-eighths of Kay's age. What are their present ages? ■

Keep in mind that the problem-solving suggestions offered in this section simply outline a general algebraic approach to solving problems. You will add to this list throughout this course and in any subsequent mathematics courses that you take. Furthermore, you will be able to pick up additional problem-solving ideas from your instructor and from fellow classmates as you discuss problems in class. Always be on the alert for any ideas that might help you become a better problem solver.

CONCEPT QUIZ

For Problems 1–10, answer true or false.

1. When solving an equation that involves fractions, the equation can be cleared of all the fractions by multiplying both sides of the equation by the least common multiple of all the denominators in the problem.
2. The least common multiple of a set of denominators is referred to as the lowest common denominator.
3. The least common multiple of 4, 6, and 9 is 36.
4. The least common multiple of 3, 9, and 18 is 36.
5. Answers for word problems need to be checked back into the original statement of the problem.
6. In a right triangle, the two acute angles are complementary angles.
7. A triangle can have two supplementary angles.
8. The sum of the measure of the three angles in a triangle is 100°.
9. If x represents Eric's present age, then $5x$ represents his age in 5 years.
10. If x represents Joni's present age, then $x - 4$ represents her age in 4 years.

Problem Set 2.2

1 Solve Equations Involving Fractions

For Problems 1–40, solve each equation.

1. $\frac{3}{4}x = 9$
2. $\frac{2}{3}x = -14$
3. $\frac{-2x}{3} = \frac{2}{5}$
4. $\frac{-5x}{4} = \frac{7}{2}$
5. $\frac{n}{2} - \frac{2}{3} = \frac{5}{6}$
6. $\frac{n}{4} - \frac{5}{6} = \frac{5}{12}$
7. $\frac{5n}{6} - \frac{n}{8} = \frac{-17}{12}$
8. $\frac{2n}{5} - \frac{n}{6} = \frac{-7}{10}$
9. $\frac{a}{4} - 1 = \frac{a}{3} + 2$
10. $\frac{3a}{7} - 1 = \frac{a}{3}$
11. $\frac{h}{4} + \frac{h}{5} = 1$
12. $\frac{h}{6} + \frac{3h}{8} = 1$
13. $\frac{h}{2} - \frac{h}{3} + \frac{h}{6} = 1$
14. $\frac{3h}{4} + \frac{2h}{5} = 1$
15. $\frac{x-2}{3} + \frac{x+3}{4} = \frac{11}{6}$
16. $\frac{x+4}{5} + \frac{x-1}{4} = \frac{37}{10}$
17. $\frac{x+2}{2} - \frac{x-1}{5} = \frac{3}{5}$
18. $\frac{2x+1}{3} - \frac{x+1}{7} = -\frac{1}{3}$
19. $\frac{n+2}{4} - \frac{2n-1}{3} = \frac{1}{6}$
20. $\frac{n-1}{9} - \frac{n+2}{6} = \frac{3}{4}$

21. $\frac{y}{3} + \frac{y-5}{10} = \frac{4y+3}{5}$

22. $\frac{y}{3} + \frac{y-2}{8} = \frac{6y-1}{12}$

23. $\frac{4x-1}{10} - \frac{5x+2}{4} = -3$

24. $\frac{2x-1}{2} - \frac{3x+1}{4} = \frac{3}{10}$

25. $\frac{2x-1}{8} - 1 = \frac{x+5}{7}$

26. $\frac{3x+1}{9} + 2 = \frac{x-1}{4}$

27. $\frac{2a-3}{6} + \frac{3a-2}{4} + \frac{5a+6}{12} = 4$

28. $\frac{3a-1}{4} + \frac{a-2}{3} - \frac{a-1}{5} = \frac{21}{20}$

29. $x + \frac{3x-1}{9} - 4 = \frac{3x+1}{3}$

30. $\frac{2x+7}{8} + x - 2 = \frac{x-1}{2}$

31. $\frac{x+3}{2} + \frac{x+4}{5} = \frac{3}{10}$

32. $\frac{x-2}{5} - \frac{x-3}{4} = -\frac{1}{20}$

33. $n + \frac{2n-3}{9} - 2 = \frac{2n+1}{3}$

34. $n - \frac{3n+1}{6} - 1 = \frac{2n+4}{12}$

35. $\frac{3}{4}(t-2) - \frac{2}{5}(2t-3) = \frac{1}{5}$

36. $\frac{2}{3}(2t+1) - \frac{1}{2}(3t-2) = 2$

37. $\frac{1}{2}(2x-1) - \frac{1}{3}(5x+2) = 3$

38. $\frac{2}{5}(4x-1) + \frac{1}{4}(5x+2) = -1$

39. $3x - 1 + \frac{2}{7}(7x-2) = -\frac{11}{7}$

40. $2x + 5 + \frac{1}{2}(6x-1) = -\frac{1}{2}$

2 Solve Word Problems

For Problems 41–58, use an algebraic approach to solve each problem.

41. Find a number such that one-half of the number is 3 less than two-thirds of the number.

42. One-half of a number plus three-fourths of the number is 2 more than four-thirds of the number. Find the number.

43. Suppose that the width of a certain rectangle is 1 inch more than one-fourth of its length. The perimeter of the rectangle is 42 inches. Find the length and width of the rectangle.

44. Suppose that the width of a rectangle is 3 centimeters less than two-thirds of its length. The perimeter of the rectangle is 114 centimeters. Find the length and width of the rectangle.

45. Find three consecutive integers such that the sum of the first plus one-third of the second plus three-eighths of the third is 25.

46. Lou is paid $1\frac{1}{2}$ times his normal hourly rate for each hour he works over 40 hours in a week. Last week he worked 44 hours and earned \$276. What is his normal hourly rate?

47. A board 20 feet long is cut into two pieces such that the length of one piece is two-thirds of the length of the other piece. Find the length of the shorter piece of board.

48. Jody has a collection of 116 coins consisting of dimes, quarters, and silver dollars. The number of quarters is 5 less than three-fourths of the number of dimes. The number of silver dollars is 7 more than five-eighths of the number of dimes. How many coins of each kind are in her collection?

49. The sum of the present ages of Angie and her mother is 64 years. In eight years Angie will be three-fifths as old as her mother at that time. Find the present ages of Angie and her mother.

50. Annilee's present age is two-thirds of Jessie's present age. In 12 years the sum of their ages will be 54 years. Find their present ages.

51. Sydney's present age is one-half of Marcus's present age. In 12 years, Sydney's age will be five-eighths of Marcus's age. Find their present ages.

52. The sum of the present ages of Ian and his brother is 45. In 5 years, Ian's age will be five-sixths of his brother's age. Find their present ages.

53. Aura took three biology exams and has an average score of 88. Her second exam score was 10 points better than her first, and her third exam score was 4 points better than her second exam. What were her three exam scores?

54. The average of the salaries of Tim, Maida, and Aaron is \$24,000 per year. Maida earns \$10,000 more than Tim, and Aaron's salary is \$2000 more than twice Tim's salary. Find the salary of each person.

55. One of two supplementary angles is 4° more than one-third of the other angle. Find the measure of each of the angles.

56. If one-half of the complement of an angle plus three-fourths of the supplement of the angle equals 110°, find the measure of the angle.

57. If the complement of an angle is 5° less than one-sixth of its supplement, find the measure of the angle.

58. In $\triangle ABC$, angle B is 8° less than one-half of angle A and angle C is 28° larger than angle A. Find the measures of the three angles of the triangle.

THOUGHTS INTO WORDS

59. Explain why the solution set of the equation $x + 3 = x + 4$ is the null set.

60. Explain why the solution set of the equation $\frac{x}{3} + \frac{x}{2} = \frac{5x}{6}$ is the entire set of real numbers.

61. Why must potential answers to word problems be checked back into the original statement of the problem?

62. Suppose your friend solved the problem, *find two consecutive odd integers whose sum is 28*, like this:

$$x + x + 1 = 28$$
$$2x = 27$$
$$x = \frac{27}{2} = 13\frac{1}{2}$$

She claims that $13\frac{1}{2}$ will check in the equation. Where has she gone wrong and how would you help her?

Answers to the Concept Quiz

1. True **2.** True **3.** True **4.** False **5.** True **6.** True **7.** False **8.** False **9.** False **10.** False

Answers to the Example Practice Skills

1. $\left\{\frac{55}{12}\right\}$ **2.** $\left\{\frac{96}{5}\right\}$ **3.** $\left\{\frac{22}{7}\right\}$ **4.** $\{-22\}$ **5.** 24 **6.** Width is 220 ft; length is 320 ft **7.** 70°, 110° **8.** Raymond is 6 years old and Kay is 12 years old

2.3 Equations Involving Decimals and Problem Solving

OBJECTIVES

1 Solve Equations Involving Decimals

2 Solve Word Problems Including Discount and Selling Price

1 Solve Equations Involving Decimals

In solving equations that involve fractions, usually the procedure is to clear the equation of all fractions. For solving equations that involve decimals, there are two commonly used procedures. One procedure is to keep the numbers in decimal form and solve the equation by applying the properties. Another procedure is to multiply both sides of the equation by an appropriate power of 10 to clear the equation of all decimals. Which technique to use depends on your personal preference and on the complexity of the equation. The following examples demonstrate both techniques.

EXAMPLE 1

Solve $0.2x + 0.24 = 0.08x + 0.72$.

Solution

Let's clear the decimals by multiplying both sides of the equation by 100.

$$0.2x + 0.24 = 0.08x + 0.72$$
$$100(0.2x + 0.24) = 100(0.08x + 0.72)$$

$$100(0.2x) + 100(0.24) = 100(0.08x) + 100(0.72)$$
$$20x + 24 = 8x + 72$$
$$12x + 24 = 72$$
$$12x = 48$$
$$x = 4$$

✔ Check

$$0.2x + 0.24 = 0.08x + 0.72$$
$$0.2(4) + 0.24 \stackrel{?}{=} 0.08(4) + 0.72$$
$$0.8 + 0.24 \stackrel{?}{=} 0.32 + 0.72$$
$$1.04 = 1.04$$

The solution set is {4}.

▼ **PRACTICE YOUR SKILL**

Solve $0.14a - 0.8 = 0.07a + 3.4$. ■

EXAMPLE 2 Solve $0.07x + 0.11x = 3.6$.

Solution

Let's keep this problem in decimal form.

$$0.07x + 0.11x = 3.6$$
$$0.18x = 3.6$$
$$x = \frac{3.6}{0.18}$$
$$x = 20$$

✔ Check

$$0.07x + 0.11x = 3.6$$
$$0.07(20) + 0.11(20) \stackrel{?}{=} 3.6$$
$$1.4 + 2.2 \stackrel{?}{=} 3.6$$
$$3.6 = 3.6$$

The solution set is {20}.

▼ **PRACTICE YOUR SKILL**

Solve $0.4y + 1.1y = 3.15$. ■

EXAMPLE 3 Solve $s = 1.95 + 0.35s$.

Solution

Let's keep this problem in decimal form.

$$s = 1.95 + 0.35s$$
$$s + (-0.35s) = 1.95 + 0.35s + (-0.35s)$$

$$0.65s = 1.95 \qquad \text{Remember, } s = 1.00s$$

$$s = \frac{1.95}{0.65}$$

$$s = 3$$

The solution set is {3}. Check it!

▼ PRACTICE YOUR SKILL

Solve $x = 4.5 + 0.25x$. ■

EXAMPLE 4

Solve $0.12x + 0.11(7000 - x) = 790$.

Solution

Let's clear the decimals by multiplying both sides of the equation by 100.

$$0.12x + 0.11(7000 - x) = 790$$

$$100[0.12x + 0.11(7000 - x)] = 100(790) \qquad \text{Multiply both sides by 100}$$

$$100(0.12x) + 100[0.11(7000 - x)] = 100(790)$$

$$12x + 11(7000 - x) = 79{,}000$$

$$12x + 77{,}000 - 11x = 79{,}000$$

$$x + 77{,}000 = 79{,}000$$

$$x = 2000$$

The solution set is {2000}.

▼ PRACTICE YOUR SKILL

Solve $0.06n + 0.05(3000 - n) = 167$. ■

2 Solve Word Problems Including Discount and Selling Price

We can solve many consumer problems with an algebraic approach. For example, let's consider some discount sale problems involving the relationship, *original selling price minus discount equals discount sale price.*

Original selling price − Discount = Discount sale price

EXAMPLE 5

Apply Your Skill

Karyl bought a dress at a 35% discount sale for $32.50. What was the original price of the dress?

Solution

Let p represent the original price of the dress. Using the discount sale relationship as a guideline, we find that the problem translates into an equation as follows:

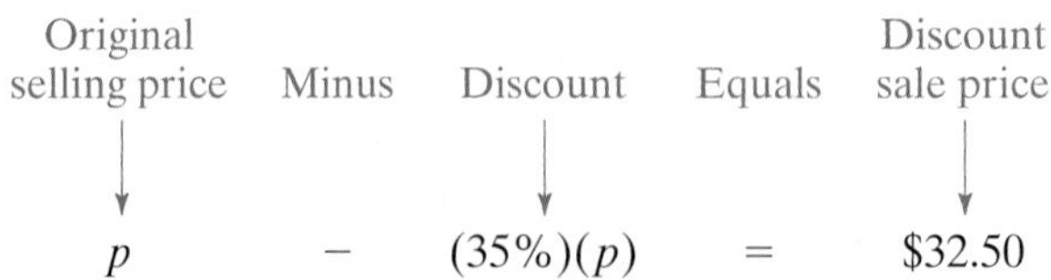

Switching this equation to decimal form and solving the equation, we obtain

$$p - (35\%)(p) = 32.50$$
$$(65\%)(p) = 32.50$$
$$0.65p = 32.50$$
$$p = 50$$

The original price of the dress was \$50.

▼ PRACTICE YOUR SKILL

Lucas paid \$45.00 for a pair of jeans that were on sale for a 40% discount. What was the original price of the jeans? ■

EXAMPLE 6

Dave Porter/Alamy Limited

Apply Your Skill

A pair of jogging shoes that was originally priced at \$50 is on sale for 20% off. Find the discount sale price of the shoes.

Solution

Let s represent the discount sale price.

Original price	Minus	Discount	Equals	Sale price
↓		↓		↓
\$50	−	(20%)(\$50)	=	s

Solving this equation we obtain

$$50 - (20\%)(50) = s$$
$$50 - (0.2)(50) = s$$
$$50 - 10 = s$$
$$40 = s$$

The shoes are on sale for \$40.

▼ PRACTICE YOUR SKILL

Jason received a private mailing coupon from an electronics store that offered 12% off any item. If he uses the coupon, how much will he have to pay for a laptop computer that is priced at \$980? ■

Remark: Keep in mind that if an item is on sale for 35% off, then the purchaser will pay 100% − 35% = 65% of the original price. Thus in Example 5 you could begin with the equation $0.65p = 32.50$. Likewise in Example 6 you could start with the equation $s = 0.8(50)$.

Another basic relationship that pertains to consumer problems is *selling price equals cost plus profit*. We can state profit (also called markup, markon, and margin of profit) in different ways. Profit may be stated as a percent of the selling price, as a percent of the cost, or simply in terms of dollars and cents. We shall consider some problems for which the profit is calculated either as a percent of the cost or as a percent of the selling price.

$$\text{Selling price} = \text{Cost} + \text{Profit}$$

EXAMPLE 7

Mikael Andersson/Nordic Photos/PhotoLibrary

Apply Your Skill

A retailer has some shirts that cost \$20 each. She wants to sell them at a profit of 60% of the cost. What selling price should be marked on the shirts?

Solution

Let s represent the selling price. Use the relationship *selling price equals cost plus profit* as a guideline.

Selling price	Equals	Cost	Plus	Profit
↓		↓		↓
s	$=$	\$20	$+$	(60%)(\$20)

Solving this equation yields

$$s = 20 + (60\%)(20)$$

$$s = 20 + (0.6)(20)$$

$$s = 20 + 12$$

$$s = 32$$

The selling price should be \$32.

▼ PRACTICE YOUR SKILL

Heather bought some artwork at an online auction for \$400. She wants to resell the artwork online and make a profit of 40% of the cost. What price should Heather list online to make her profit? ■

Remark: A profit of 60% of the cost means that the selling price is 100% of the cost plus 60% of the cost, or 160% of the cost. Thus in Example 7 we could solve the equation $s = 1.6(20)$.

EXAMPLE 8

Frances M. Roberts/Alamy Limited

Apply Your Skill

A retailer of sporting goods bought a putter for \$18. He wants to price the putter such that he will make a profit of 40% of the selling price. What price should he mark on the putter?

Solution

Let s represent the selling price.

Selling price	Equals	Cost	Plus	Profit
↓		↓		↓
s	$=$	\$18	$+$	(40%)(s)

Solving this equation yields

$$s = 18 + (40\%)(s)$$

$$s = 18 + 0.4s$$

$$0.6s = 18$$

$$s = 30$$

The selling price should be \$30.

▼ PRACTICE YOUR SKILL

A college bookstore purchased math textbooks for \$54 each. At what price should the bookstore sell the books if it wants to make a profit of 60% of the selling price? ■

EXAMPLE 9

Janusz Wrobel/Alamy Limited

Apply Your Skill

If a maple tree costs a landscaper \$55.00 and he sells it for \$80.00, what is his rate of profit based on the cost? Round the rate to the nearest tenth of a percent.

Solution

Let r represent the rate of profit, and use the following guideline.

Selling price	Equals	Cost	Plus	Profit
↓		↓		↓
80.00	$=$	55.00	$+$	$r(55.00)$
25.00	$=$	$r(55.00)$		
$\frac{25.00}{55.00}$	$=$	r		
0.455	$\approx$	r		

To change the answer to a percent, multiply 0.455 by 100. Thus his rate of profit is 45.5%.

▼ PRACTICE YOUR SKILL

If a bicycle cost a bike dealer \$200 and he sells it for \$300, what is his rate of profit based on the cost? ■

We can solve certain types of investment and money problems by using an algebraic approach. Consider the following examples.

EXAMPLE 10

Burke/Triolo Productions/Brand X Pictures/Jupiter Images

Apply Your Skill

Erick has 40 coins, consisting only of dimes and nickels, worth \$3.35. How many dimes and how many nickels does he have?

Solution

Let x represent the number of dimes. Then the number of nickels can be represented by the total number of coins minus the number of dimes. Hence $40 - x$ represents the number of nickels. Because we know the amount of money Erick has, we need to multiply the number of each coin by its value. Use the following guideline.

Money from the dimes	Plus	Money from the nickels	Equals	Total money	
↓		↓		↓	
$0.10x$	$+$	$0.05(40 - x)$	$=$	3.35	
$10x$	$+$	$5(40 - x)$	$=$	335	Multiply both sides by 100
$10x$	$+$	$200 - 5x$	$=$	335	
		$5x + 200$	$=$	335	
		$5x$	$=$	135	
		x	$=$	27	

The number of dimes is 27, and the number of nickels is $40 - x = 13$. So Erick has 27 dimes and 13 nickels.

▼ **PRACTICE YOUR SKILL**

Lane has 20 coins, consisting only of quarters and dimes, worth \$3.95. How many quarters and how many dimes does he have? ■

EXAMPLE 11

Steve Allen/Brand X Pictures/Jupiter Images

Apply Your Skill

A man invests \$8000, part of it at 6% and the remainder at 8%. His total yearly interest from the two investments is \$580. How much did he invest at each rate?

Solution

Let x represent the amount he invested at 6%. Then $8000 - x$ represents the amount he invested at 8%. Use the following guideline.

Interest earned from 6% investment	+	Interest earned from 8% investment	=	Total amount of interest earned
↓		↓		↓
$(6\%)(x)$	+	$(8\%)(8000 - x)$	=	\$580

Solving this equation yields

$$(6\%)(x) + (8\%)(8000 - x) = 580$$
$$0.06x + 0.08(8000 - x) = 580$$
$$6x + 8(8000 - x) = 58{,}000 \quad \text{Multiply both sides by 100}$$
$$6x + 64{,}000 - 8x = 58{,}000$$
$$-2x + 64{,}000 = 58{,}000$$
$$-2x = -6000$$
$$x = 3000$$

Therefore, \$3000 was invested at 6%, and \$8000 − \$3000 = \$5000 was invested at 8%.

Don't forget to check word problems; determine whether the answers satisfy the conditions stated in the *original* problem. A check for Example 11 follows.

✔ **Check**

We claim that \$3000 is invested at 6% and \$5000 at 8%, and this satisfies the condition that \$8000 is invested. The \$3000 at 6% produces \$180 of interest, and the \$5000 at 8% produces \$400. Therefore, the interest from the investments is \$580. The conditions of the problem are satisfied, and our answers are correct.

▼ **PRACTICE YOUR SKILL**

A person invested \$10,000, part of it at 7% and the remainder at 5%. The total yearly interest from the two investments is \$630. How much is invested at each rate? ■

As you tackle word problems throughout this text, keep in mind that our primary objective is to expand your repertoire of problem-solving techniques. We have chosen problems that provide you with the opportunity to use a variety of approaches to solving problems. Don't fall into the trap of thinking "I will never be faced with this kind of problem." That is not the issue; the goal is to develop problem-solving techniques. In the examples we are sharing some of our ideas for solving problems, but don't hesitate to use your own ingenuity. Furthermore, don't become discouraged—all of us have difficulty with some problems. Give each your best shot!

CONCEPT QUIZ

For Problems 1–10, answer true or false.

1. To solve an equation involving decimals, you must first multiply both sides of the equation by a power of 10.
2. When using the formula "selling price = cost + profit" the profit is always a percentage of the cost.
3. If Kim bought a putter for \$50 and then sold it to a friend for \$60, her rate of profit based on the cost was 10%.
4. To determine the selling price when the profit is a percent of the selling price, you can subtract the percent of profit from 100% and then divide the cost by that result.
5. If an item is bought for \$30, then it should be sold for \$37.50 in order to obtain a profit of 20% based on the selling price.
6. A discount of 10% followed by a discount of 20% is the same as a discount of 30%.
7. If an item is bought for \$25, then it should be sold for \$30 in order to obtain a profit of 20% based on the cost.
8. To solve the equation $0.4x + 0.15 = 0.06x + 0.71$ one could start by multiplying both sides of the equation by 100.
9. A 10% discount followed by a 40% discount is the same as a 40% discount followed by a 10% discount.
10. Multiplying both sides of the equation $0.4(x - 1.2) = 0.6$ by 10 produces the equivalent equation $4(x - 12) = 6$.

Problem Set 2.3

1 Solve Equations Involving Decimals

For Problems 1–28, solve each equation.

1. $0.14x = 2.8$
2. $1.6x = 8$
3. $0.09y = 4.5$
4. $0.07y = 0.42$
5. $n + 0.4n = 56$
6. $n - 0.5n = 12$
7. $s = 9 + 0.25s$
8. $s = 15 + 0.4s$
9. $s = 3.3 + 0.45s$
10. $s = 2.1 + 0.6s$
11. $0.11x + 0.12(900 - x) = 104$
12. $0.09x + 0.11(500 - x) = 51$
13. $0.08(x + 200) = 0.07x + 20$
14. $0.07x = 152 - 0.08(2000 - x)$
15. $0.12t - 2.1 = 0.07t - 0.2$
16. $0.13t - 3.4 = 0.08t - 0.4$
17. $0.92 + 0.9(x - 0.3) = 2x - 5.95$
18. $0.3(2n - 5) = 11 - 0.65n$
19. $0.1d + 0.11(d + 1500) = 795$
20. $0.8x + 0.9(850 - x) = 715$
21. $0.12x + 0.1(5000 - x) = 560$
22. $0.10t + 0.12(t + 1000) = 560$
23. $0.09(x + 200) = 0.08x + 22$
24. $0.09x = 1650 - 0.12(x + 5000)$
25. $0.3(2t + 0.1) = 8.43$
26. $0.5(3t + 0.7) = 20.6$
27. $0.1(x - 0.1) - 0.4(x + 2) = -5.31$
28. $0.2(x + 0.2) + 0.5(x - 0.4) = 5.44$

2 Solve Word Problems Including Discount and Selling Price

For Problems 29–50, use an algebraic approach to solve each problem.

29. Judy bought a coat at a 20% discount sale for \$72. What was the original price of the coat?
30. Jim bought a pair of slacks at a 25% discount sale for \$24. What was the original price of the slacks?

31. Find the discount sale price of a \$64 item that is on sale for 15% off.

32. Find the discount sale price of a \$72 item that is on sale for 35% off.

33. A retailer has some skirts that cost \$30 each. She wants to sell them at a profit of 60% of the cost. What price should she charge for the skirts?

34. The owner of a pizza parlor wants to make a profit of 70% of the cost for each pizza sold. If it costs \$2.50 to make a pizza, at what price should each pizza be sold?

35. If a ring costs a jeweler \$200, at what price should it be sold to yield a profit of 50% on the selling price?

36. If a head of lettuce costs a retailer \$0.32, at what price should it be sold to yield a profit of 60% on the selling price?

37. If a pair of shoes costs a retailer \$24 and he sells them for \$39.60, what is his rate of profit based on the cost?

38. A retailer has some skirts that cost her \$45 each. If she sells them for \$83.25 per skirt, find her rate of profit based on the cost.

39. If a computer costs an electronics dealer \$300 and she sells them for \$800, what is her rate of profit based on the selling price?

40. A textbook costs a bookstore \$45, and the store sells it for \$60. Find the rate of profit based on the selling price.

41. Mitsuko's salary for next year is \$34,775. This represents a 7% increase over this year's salary. Find Mitsuko's present salary.

42. Don bought a used car for \$15,794, with 6% tax included. What was the price of the car without the tax?

43. Eva invested a certain amount of money at 10% interest and \$1500 more than that amount at 11%. Her total yearly interest was \$795. How much did she invest at each rate?

44. A total of \$4000 was invested, part of it at 8% interest and the remainder at 9%. If the total yearly interest amounted to \$350, how much was invested at each rate?

45. A sum of \$95,000 is split between two investments, one paying 6% and the other 9%. If the total yearly interest amounted to \$7290, how much was invested at 9%?

46. If \$1500 is invested at 6% interest, how much money must be invested at 9% so that the total return for both investments is \$301.50?

47. Suppose that Javier has a handful of coins, consisting of pennies, nickels, and dimes, worth \$2.63. The number of nickels is 1 less than twice the number of pennies, and the number of dimes is 3 more than the number of nickels. How many coins of each kind does he have?

48. Sarah has a collection of nickels, dimes, and quarters worth \$15.75. She has 10 more dimes than nickels and twice as many quarters as dimes. How many coins of each kind does she have?

49. A collection of 70 coins consisting of dimes, quarters, and half-dollars has a value of \$17.75. There are three times as many quarters as dimes. Find the number of each kind of coin.

50. Abby has 37 coins, consisting only of dimes and quarters, worth \$7.45. How many dimes and how many quarters does she have?

THOUGHTS INTO WORDS

51. Return to Problem 39 and calculate the rate of profit based on cost. Compare the rate of profit based on cost to the rate of profit based on selling price. From a consumer's viewpoint, would you prefer that a retailer figure its profit on the basis of the cost of an item or on the basis of its selling price? Explain your answer.

52. Is a 10% discount followed by a 30% discount the same as a 30% discount followed by a 10% discount? Justify your answer.

53. What is wrong with the following solution, and how should it be done?

$$\begin{aligned} 1.2x + 2 &= 3.8 \\ 10(1.2x) + 2 &= 10(3.8) \\ 12x + 2 &= 38 \\ 12x &= 36 \\ x &= 3 \end{aligned}$$

FURTHER INVESTIGATIONS

For Problems 54–63, solve each equation and express the solutions in decimal form. Be sure to check your solutions. Use your calculator whenever it seems helpful.

54. $1.2x + 3.4 = 5.2$

55. $0.12x - 0.24 = 0.66$

56. $0.12x + 0.14(550 - x) = 72.5$

57. $0.14t + 0.13(890 - t) = 67.95$

58. $0.7n + 1.4 = 3.92$

59. $0.14n - 0.26 = 0.958$

60. $0.3(d + 1.8) = 4.86$

61. $0.6(d - 4.8) = 7.38$

62. $0.8(2x - 1.4) = 19.52$

63. $0.5(3x + 0.7) = 20.6$

64. The following formula can be used to determine the selling price of an item when the profit is based on a percent of the selling price.

$$\text{Selling price} = \frac{\text{Cost}}{100\% - \text{Percent of profit}}$$

Show how this formula is developed.

65. A retailer buys an item for \$90, resells it for \$100, and claims that she is making only a 10% profit. Is this claim correct?

66. Is a 10% discount followed by a 20% discount equal to a 30% discount? Defend your answer.

Answers to the Concept Quiz

1. False **2.** False **3.** False **4.** True **5.** True **6.** False **7.** True **8.** True **9.** True **10.** False

Answers to the Example Practice Skills

1. {60} **2.** {2.1} **3.** {6} **4.** {1700} **5.** \$75.00 **6.** \$862.40 **7.** \$560 **8.** \$135 **9.** 50% **10.** 13 quarters and 7 dimes **11.** \$6500 at 7% and \$3500 at 5%

2.4 Formulas

OBJECTIVES

1. Evaluate Formulas for Given Values
2. Solve Formulas for a Specified Variable
3. Use Formulas to Solve Problems

1 Evaluate Formulas for Given Values

To find the distance traveled in 4 hours at a rate of 55 miles per hour, we multiply the rate times the time; thus the distance is $55(4) = 220$ miles. We can state the rule *distance equals rate times time* as a formula: $d = rt$. Formulas are rules we state in symbolic form, usually as equations.

Formulas are typically used in two different ways. At times a formula is solved for a specific variable when we are given the numerical values for the other variables. This is much like evaluating an algebraic expression. At other times we need to change the form of an equation by solving for one variable in terms of the other variables. Throughout our work on formulas, we will use the properties of equality and the techniques we have previously learned for solving equations. Let's consider some examples.

EXAMPLE 1

If we invest P dollars at r percent for t years, then the amount of simple interest i is given by the formula $i = Prt$. Find the amount of interest earned by \$500 at 7% for 2 years.

Solution

By substituting \$500 for P, 7% for r, and 2 for t, we obtain

$$i = Prt$$
$$i = (500)(7\%)(2)$$
$$i = (500)(0.07)(2)$$
$$i = 70$$

Thus we earn \$70 in interest.

▼ PRACTICE YOUR SKILL

Use the formula $i = Prt$ to find the amount of interest earned by $2500 invested at 6% for 3 years. ■

EXAMPLE 2

If we invest P dollars at a simple rate of r percent, then the amount A accumulated after t years is given by the formula $A = P + Prt$. If we invest $500 at 8%, how many years will it take to accumulate $600?

Solution

Substituting $500 for P, 8% for r, and $600 for A, we obtain

$$A = P + Prt$$

$$600 = 500 + 500(8\%)(t)$$

Solving this equation for t yields

$$600 = 500 + 500(0.08)(t)$$

$$600 = 500 + 40t$$

$$100 = 40t$$

$$2\frac{1}{2} = t$$

It will take $2\frac{1}{2}$ years to accumulate $600.

▼ PRACTICE YOUR SKILL

Use the formula $A = P + Prt$ to determine how many years it will take $1000 invested at 5% to accumulate to $1800. ■

When we are using a formula, it is sometimes convenient first to change its form. For example, suppose we are to use the *perimeter* formula for a rectangle ($P = 2l + 2w$) to complete the following chart.

Perimeter (P)	32	24	36	18	56	80
Length (l)	10	7	14	5	15	22
Width (w)	?	?	?	?	?	?

All in centimeters

Because w is the unknown quantity, it would simplify the computational work if we first solved the formula for w in terms of the other variables as follows:

$$P = 2l + 2w$$

$$P - 2l = 2w \qquad \text{Add } -2l \text{ to both sides}$$

$$\frac{P - 2l}{2} = w \qquad \text{Multiply both sides by } \frac{1}{2}$$

$$w = \frac{P - 2l}{2} \qquad \text{Apply the symmetric property of equality}$$

Now, for each value for P and l, we can easily determine the corresponding value for w. Be sure you agree with the following values for w: 6, 5, 4, 4, 13, and 18. Likewise, we can also solve the formula $P = 2l + 2w$ for l in terms of P and w. The result would be $l = \frac{P - 2w}{2}$.

2 Solve Formulas for a Specified Variable

Let's consider some other often-used formulas and see how we can use the properties of equality to alter their forms. Here we will be solving a formula for a specified variable in terms of the other variables. The key is to isolate the term that contains the variable being solved for. Then, by appropriately applying the multiplication property of equality, we will solve the formula for the specified variable. Throughout this section, we will identify formulas when we first use them. (Some geometric formulas are also given on the endsheets.)

EXAMPLE 3

Solve $A = \frac{1}{2}bh$ for h (area of a triangle).

Solution

$$A = \frac{1}{2}bh$$

$$2A = bh \quad \text{Multiply both sides by 2}$$

$$\frac{2A}{b} = h \quad \text{Multiply both sides by } \frac{1}{b}$$

$$h = \frac{2A}{b} \quad \text{Apply the symmetric property of equality}$$

▼ PRACTICE YOUR SKILL

Solve $V = \frac{1}{3}Bh$ for B (volume of a pyramid). ■

EXAMPLE 4

Solve $A = P + Prt$ for t.

Solution

$$A = P + Prt$$

$$A - P = Prt \quad \text{Add } -P \text{ to both sides}$$

$$\frac{A - P}{Pr} = t \quad \text{Multiply both sides by } \frac{1}{Pr}$$

$$t = \frac{A - P}{Pr} \quad \text{Apply the symmetric property of equality}$$

▼ PRACTICE YOUR SKILL

Solve $S = 4lw + 2lh$ for h. ■

EXAMPLE 5

Solve $A = P + Prt$ for P.

Solution

$$A = P + Prt$$

$$A = P(1 + rt) \quad \text{Apply the distributive property to the right side}$$

$$\frac{A}{1 + rt} = P \quad \text{Multiply both sides by } \frac{1}{1 + rt}$$

$$P = \frac{A}{1 + rt} \quad \text{Apply the symmetric property of equality}$$

▼ **PRACTICE YOUR SKILL**

Solve $S = ad + an$ for a. ■

EXAMPLE 6 Solve $A = \frac{1}{2}h(b_1 + b_2)$ for b_1 (area of a trapezoid).

Solution

$$A = \frac{1}{2}h(b_1 + b_2)$$

$$2A = h(b_1 + b_2) \quad \text{Multiply both sides by 2}$$

$$2A = hb_1 + hb_2 \quad \text{Apply the distributive property to right side}$$

$$2A - hb_2 = hb_1 \quad \text{Add } -hb_2 \text{ to both sides}$$

$$\frac{2A - hb_2}{h} = b_1 \quad \text{Multiply both sides by } \frac{1}{h}$$

$$b_1 = \frac{2A - hb_2}{h} \quad \text{Apply the symmetric property of equality}$$

▼ **PRACTICE YOUR SKILL**

Solve $P = 2(l + w)$ for w. ■

In order to isolate the term containing the variable being solved for, we will apply the distributive property in different ways. In Example 5 you *must* use the distributive property to change from the form $P + Prt$ to $P(1 + rt)$. However, in Example 6 we used the distributive property to change $h(b_1 + b_2)$ to $hb_1 + hb_2$. In both problems the key is to isolate the term that contains the variable being solved for, so that an appropriate application of the multiplication property of equality will produce the desired result. Also note the use of subscripts to identify the two bases of a trapezoid. Subscripts enable us to use the same letter b to identify the bases, but b_1 represents one base and b_2 the other.

Sometimes we are faced with equations such as $ax + b = c$, where x is the variable and a, b, and c are referred to as *arbitrary constants*. Again we can use the properties of equality to solve the equation for x as follows:

$$ax + b = c$$

$$ax = c - b \quad \text{Add } -b \text{ to both sides}$$

$$x = \frac{c - b}{a} \quad \text{Multiply both sides by } \frac{1}{a}$$

In Chapter 3, we will be working with equations such as $2x - 5y = 7$, which are called equations of two variables in x and y. Often we need to change the form of such equations by solving for one variable in terms of the other variable. The properties of equality provide the basis for doing this.

EXAMPLE 7 Solve $2x - 5y = 7$ for y in terms of x.

Solution

$$2x - 5y = 7$$

$$-5y = 7 - 2x \quad \text{Add } -2x \text{ to both sides}$$

$$y = \frac{7 - 2x}{-5} \qquad \text{Multiply both sides by } -\frac{1}{5}$$

$$y = \frac{2x - 7}{5}$$ Multiply the numerator and denominator of the fraction on the right by -1. (This final step is not absolutely necessary, but usually we prefer to have a positive number as a denominator.)

▼ PRACTICE YOUR SKILL

Solve $3x - 4y = 5$ for y. ■

Equations of two variables may also contain arbitrary constants. For example, the equation $\frac{x}{a} + \frac{y}{b} = 1$ contains the variables x and y and the arbitrary constants a and b.

EXAMPLE 8

Solve the equation $\frac{x}{a} + \frac{y}{b} = 1$ for x.

Solution

$$\frac{x}{a} + \frac{y}{b} = 1$$

$$ab\left(\frac{x}{a} + \frac{y}{b}\right) = ab(1) \qquad \text{Multiply both sides by } ab$$

$$bx + ay = ab$$

$$bx = ab - ay \qquad \text{Add } -ay \text{ to both sides}$$

$$x = \frac{ab - ay}{b} \qquad \text{Multiply both sides by } \frac{1}{b}$$

▼ PRACTICE YOUR SKILL

Solve $\frac{x}{c} + \frac{y}{d} = 1$ for x. ■

Remark: Traditionally, equations that contain more than one variable, such as those in Examples 3–8, are called **literal equations**. As illustrated, it is sometimes necessary to solve a literal equation for one variable in terms of the other variable(s).

3 Use Formulas to Solve Problems

We often use formulas as guidelines for setting up an appropriate algebraic equation when solving a word problem. Let's consider an example to illustrate this point.

EXAMPLE 9

Harnett/Hanzon/PhotoLibrary

Apply Your Skill

How long will it take \$500 to double itself if we invest it at 8% simple interest?

Solution

For \$500 to grow into \$1000 (double itself), it must earn \$500 in interest. Thus we let t represent the number of years it will take \$500 to earn \$500 in interest. Now we can use the formula $i = Prt$ as a guideline.

$$i = Prt$$

$$500 = 500(8\%)(t)$$

Solving this equation, we obtain

$$500 = 500(0.08)(t)$$

$$1 = 0.08t$$

$$100 = 8t$$

$$12\frac{1}{2} = t$$

It will take $12\frac{1}{2}$ years.

▼ **PRACTICE YOUR SKILL**

How long will it take \$3000 to grow into \$4200 if it is invested at 5% simple interest? ■

Sometimes we use formulas in the analysis of a problem but not as the main guideline for setting up the equation. For example, uniform motion problems involve the formula $d = rt$, but the main guideline for setting up an equation for such problems is usually a statement about times, rates, or distances. Let's consider an example to demonstrate.

EXAMPLE 10

Apply Your Skill

Mercedes starts jogging at 5 miles per hour. One-half hour later, Karen starts jogging on the same route at 7 miles per hour. How long will it take Karen to catch Mercedes?

Solution

First, let's sketch a diagram and record some information (Figure 2.3).

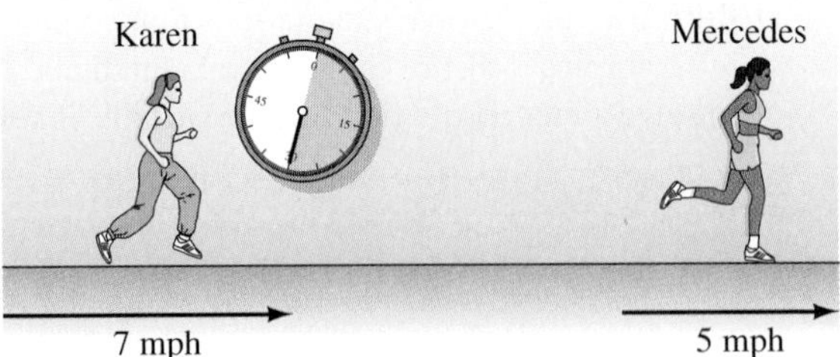

Figure 2.3

If we let t represent Karen's time, then $t + \frac{1}{2}$ represents Mercedes' time. We can use the statement *Karen's distance equals Mercedes' distance* as a guideline.

$$\underbrace{7t}_{\text{Karen's distance}} = \underbrace{5\left(t + \frac{1}{2}\right)}_{\text{Mercedes' distance}}$$

Solving this equation, we obtain

$$7t = 5t + \frac{5}{2}$$

$$2t = \frac{5}{2}$$

$$t = \frac{5}{4}$$

Karen should catch Mercedes in $1\frac{1}{4}$ hours.

▼ **PRACTICE YOUR SKILL**

Brittany starts out bicycling at 8 miles per hour. An hour later, Franco starts bicycling on the same route at 12 miles per hour. How long will it take Franco to catch up with Brittany? ■

Remark: An important part of problem solving is the ability to sketch a meaningful figure that can be used to record the given information and help in the analysis of the problem. Our sketches were done by professional artists for aesthetic purposes. Your sketches can be very roughly drawn as long as they depict the situation in a way that helps you analyze the problem.

Note that in the solution of Example 10 we used a figure and a simple arrow diagram to record and organize the information pertinent to the problem. Some people find it helpful to use a chart for that purpose. We shall use a chart in Example 11. Keep in mind that we are not trying to dictate a particular approach; you decide what works best for you.

EXAMPLE 11

GeoStock/Photodisc/Getty Images

Apply Your Skill

Two trains leave a city at the same time, one traveling east and the other traveling west. At the end of $9\frac{1}{2}$ hours, they are 1292 miles apart. If the rate of the train traveling east is 8 miles per hour faster than the rate of the other train, find their rates.

Solution

If we let r represent the rate of the westbound train, then $r + 8$ represents the rate of the eastbound train. Now we can record the times and rates in a chart and then use the distance formula ($d = rt$) to represent the distances.

	Rate	Time	Distance ($d = rt$)
Westbound train	r	$9\frac{1}{2}$	$\frac{19}{2}r$
Eastbound train	$r + 8$	$9\frac{1}{2}$	$\frac{19}{2}(r + 8)$

Because the distance that the westbound train travels plus the distance that the eastbound train travels equals 1292 miles, we can set up and solve the following equation.

$$\begin{array}{c} \text{Eastbound} \\ \text{distance} \end{array} + \begin{array}{c} \text{Westbound} \\ \text{distance} \end{array} = \begin{array}{c} \text{Miles} \\ \text{apart} \end{array}$$

$$\frac{19r}{2} + \frac{19(r+8)}{2} = 1292$$

$$19r + 19(r+8) = 2584$$

$$19r + 19r + 152 = 2584$$

$$38r = 2432$$

$$r = 64$$

The westbound train travels at a rate of 64 miles per hour, and the eastbound train travels at a rate of $64 + 8 = 72$ miles per hour.

▼ PRACTICE YOUR SKILL

Two trucks leave the warehouse at the same time, one traveling south and the other traveling north. At the end of $1\frac{1}{2}$ hours, the trucks are 159 miles apart. If the rate of the truck traveling south is 6 miles per hour less than the rate of the truck traveling north, find their rates. ■

Now let's consider a problem that is often referred to as a mixture problem. There is no basic formula that applies to all of these problems, but we suggest that you think in terms of a pure substance, which is often helpful in setting up a guideline. Also keep in mind that the phrase "a 40% solution of some substance" means that the solution contains 40% of that particular substance and 60% of something else mixed with it. For example, a 40% salt solution contains 40% salt, and the other 60% is something else, probably water. Now let's illustrate what we mean by suggesting that you think in terms of a pure substance.

EXAMPLE 12

Don Mason/Brand X Pictures/Jupiter Images

Apply Your Skill

Bryan's Pest Control stocks a 7% solution of insecticide for lawns and also a 15% solution. How many gallons of each should be mixed to produce 40 gallons that is 12% insecticide?

Solution

The key idea in solving such a problem is to recognize the following guideline.

$$\begin{pmatrix} \text{Amount of insecticide} \\ \text{in the 7\% solution} \end{pmatrix} + \begin{pmatrix} \text{Amount of insecticide} \\ \text{in the 15\% solution} \end{pmatrix} = \begin{pmatrix} \text{Amount of insecticide in} \\ \text{40 gallons of 15\% solution} \end{pmatrix}$$

Let x represent the gallons of 7% solution. Then $40 - x$ represents the gallons of 15% solution. The guideline translates into the following equation.

$$(7\%)(x) + (15\%)(40 - x) = (12\%)(40)$$

Solving this equation yields

$$0.07x + 0.15(40 - x) = 0.12(40)$$

$$0.07x + 6 - 0.15x = 4.8$$

$$-0.08x + 6 = 4.8$$

$$-0.08x = -1.2$$

$$x = 15$$

Thus 15 gallons of 7% solution and $40 - x = 25$ gallons of 15% solution need to be mixed to obtain 40 gallons of 12% solution.

▼ **PRACTICE YOUR SKILL**

A pharmacist has a 6% solution of cough syrup and a 14% solution of the same cough syrup. How many ounces of each must be mixed to make 16 ounces of a 10% solution of cough syrup? ■

EXAMPLE 13

Apply Your Skill

How many liters of pure alcohol must we add to 20 liters of a 40% solution to obtain a 60% solution?

Solution

The key idea in solving such a problem is to recognize the following guideline.

$$\begin{pmatrix}\text{Amount of pure}\\ \text{alcohol in the}\\ \text{original solution}\end{pmatrix} + \begin{pmatrix}\text{Amount of}\\ \text{pure alcohol}\\ \text{to be added}\end{pmatrix} = \begin{pmatrix}\text{Amount of pure}\\ \text{alcohol in the}\\ \text{final solution}\end{pmatrix}$$

Let l represent the number of liters of pure alcohol to be added; then the guideline translates into the following equation.

$$(40\%)(20) + l = 60\%(20 + l)$$

Solving this equation yields

$$\begin{aligned} 0.4(20) + l &= 0.6(20 + l) \\ 8 + l &= 12 + 0.6l \\ 0.4l &= 4 \\ l &= 10 \end{aligned}$$

We need to add 10 liters of pure alcohol. (Remember to check this answer back into the original statement of the problem.)

▼ **PRACTICE YOUR SKILL**

How many quarts of pure antifreeze must be added to 12 quarts of a 30% antifreeze solution to obtain a 40% antifreeze solution? ■

CONCEPT QUIZ

For Problems 1–10, answer true or false.

1. Formulas are rules stated in symbolic form, usually as algebraic expressions.
2. The properties of equality that apply to solving equations also apply to solving formulas.
3. The formula $A = P + Prt$ can be solved for r or t but not for P.
4. The formula $i = Prt$ is equivalent to $P = \frac{i}{rt}$.
5. The equation $y = mx + b$ is equivalent to $x = \frac{y - b}{m}$.
6. The formula $\text{F} = \frac{9}{5}\text{C} + 32$ is equivalent to $\text{C} = \frac{5}{9}(\text{F} - 32)$.
7. The formula $\text{F} = \frac{9}{5}\text{C} + 32$ means that a temperature of 30° Celsius is equal to 86° Fahrenheit.
8. The formula $\text{C} = \frac{5}{9}(\text{F} - 32)$ means that a temperature of 32° Fahrenheit is equal to 0° Celsius.
9. The amount of pure acid in 30 ounces of a 20% acid solution is 10 ounces.
10. For an equation such as $ax + b = c$ in which x is the variable, a, b, and c are referred to as arbitrary constants.

Problem Set 2.4

1 Evaluate Formulas for Given Values

1. Solve $i = Prt$ for i, given that $P = \$300$, $r = 8\%$, and $t = 5$ years.

2. Solve $i = Prt$ for i, given that $P = \$500$, $r = 9\%$, and $t = 3\frac{1}{2}$ years.

3. Solve $i = Prt$ for t, given that $P = \$400$, $r = 11\%$, and $i = \$132$.

4. Solve $i = Prt$ for t, given that $P = \$250$, $r = 12\%$, and $i = \$120$.

5. Solve $i = Prt$ for r, given that $P = \$600$, $t = 2\frac{1}{2}$ years, and $i = \$90$. Express r as a percent.

6. Solve $i = Prt$ for r, given that $P = \$700$, $t = 2$ years, and $i = \$126$. Express r as a percent.

7. Solve $i = Prt$ for P, given that $r = 9\%$, $t = 3$ years, and $i = \$216$.

8. Solve $i = Prt$ for P, given that $r = 8\frac{1}{2}\%$, $t = 2$ years, and $i = \$204$.

9. Solve $A = P + Prt$ for A, given that $P = \$1000$, $r = 12\%$, and $t = 5$ years.

10. Solve $A = P + Prt$ for A, given that $P = \$850$, $r = 9\frac{1}{2}\%$, and $t = 10$ years.

11. Solve $A = P + Prt$ for r, given that $A = \$1372$, $P = \$700$, and $t = 12$ years. Express r as a percent.

12. Solve $A = P + Prt$ for r, given that $A = \$516$, $P = \$300$, and $t = 8$ years. Express r as a percent.

13. Solve $A = P + Prt$ for P, given that $A = \$326$, $r = 7\%$, and $t = 9$ years.

14. Solve $A = P + Prt$ for P, given that $A = \$720$, $r = 8\%$, and $t = 10$ years.

15. Use the formula $A = \frac{1}{2}h(b_1 + b_2)$ and complete the following chart.

A	98	104	49	162	$16\frac{1}{2}$	$38\frac{1}{2}$	square feet
h	14	8	7	9	3	11	feet
b_1	8	12	4	16	4	5	feet
b_2	?	?	?	?	?	?	feet

A = area, h = height, b_1 = one base, b_2 = other base

16. Use the formula $P = 2l + 2w$ and complete the following chart. (You may want to change the form of the formula.)

P	28	18	12	34	68	centimeters
w	6	3	2	7	14	centimeters
l	?	?	?	?	?	centimeters

P = perimeter, w = width, l = length

2 Solve Formulas for a Specified Variable

Solve each of the following for the indicated variable.

17. $V = Bh$ for h (volume of a prism)

18. $A = lw$ for l (area of a rectangle)

19. $V = \pi r^2 h$ for h (volume of a circular cylinder)

20. $V = \frac{1}{3}Bh$ for B (volume of a pyramid)

21. $C = 2\pi r$ for r (circumference of a circle)

22. $A = 2\pi r^2 + 2\pi rh$ for h (surface area of a circular cylinder)

23. $I = \frac{100M}{C}$ for C (intelligence quotient)

24. $A = \frac{1}{2}h(b_1 + b_2)$ for h (area of a trapezoid)

25. $\text{F} = \frac{9}{5}\text{C} + 32$ for C (Celsius to Fahrenheit)

26. $\text{C} = \frac{5}{9}(\text{F} - 32)$ for F (Fahrenheit to Celsius)

For Problems 27–36, solve each equation for x.

27. $y = mx + b$

28. $\frac{x}{a} + \frac{y}{b} = 1$

29. $y - y_1 = m(x - x_1)$

30. $a(x + b) = c$

31. $a(x + b) = b(x - c)$

32. $x(a - b) = m(x - c)$

33. $\frac{x - a}{b} = c$

34. $\frac{x}{a} - 1 = b$

35. $\frac{1}{3}x + a = \frac{1}{2}b$

36. $\frac{2}{3}x - \frac{1}{4}a = b$

For Problems 37–46, solve each equation for the indicated variable.

37. $2x - 5y = 7$ for x

38. $5x - 6y = 12$ for x

39. $-7x - y = 4$ for y

40. $3x - 2y = -1$ for y

41. $3(x - 2y) = 4$ for x

42. $7(2x + 5y) = 6$ for y

43. $\frac{y - a}{b} = \frac{x + b}{c}$ for x

44. $\frac{x - a}{b} = \frac{y - a}{c}$ for y

45. $(y + 1)(a - 3) = x - 2$ for y

46. $(y - 2)(a + 1) = x$ for y

3 Use Formulas to Solve Problems

Solve each of Problems 47–62 by setting up and solving an appropriate algebraic equation.

47. Suppose that the length of a certain rectangle is 2 meters less than four times its width. The perimeter of the rectangle is 56 meters. Find the length and width of the rectangle.

48. The perimeter of a triangle is 42 inches. The second side is 1 inch more than twice the first side, and the third side is 1 inch less than three times the first side. Find the lengths of the three sides of the triangle.

49. How long will it take \$500 to double itself at 9% simple interest?

50. How long will it take \$700 to triple itself at 10% simple interest?

51. How long will it take P dollars to double itself at 9% simple interest?

52. How long will it take P dollars to triple itself at 10% simple interest?

53. Two airplanes leave Chicago at the same time and fly in opposite directions. If one travels at 450 miles per hour and the other at 550 miles per hour, how long will it take for them to be 4000 miles apart?

54. Look at Figure 2.4. Tyrone leaves city A on a moped traveling toward city B at 18 miles per hour. At the same time, Tina leaves city B on a bicycle traveling toward city A at 14 miles per hour. The distance between the two cities is 112 miles. How long will it take before Tyrone and Tina meet?

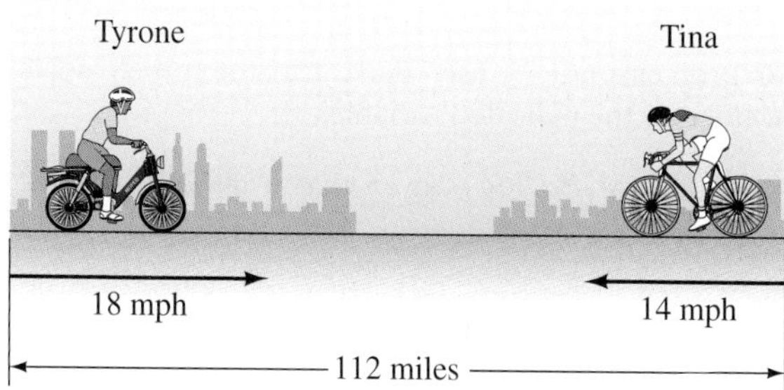

Figure 2.4

55. Juan starts walking at 4 miles per hour. An hour and a half later, Cathy starts jogging along the same route at 6 miles per hour. How long will it take Cathy to catch up with Juan?

56. A car leaves a town at 60 kilometers per hour. How long will it take a second car, traveling at 75 kilometers per hour, to catch the first car if the second car leaves 1 hour later?

57. Bret started on a 70-mile bicycle ride at 20 miles per hour. After a time he became a little tired and slowed down to 12 miles per hour for the rest of the trip. The entire trip of 70 miles took $4\frac{1}{2}$ hours. How far had Bret ridden when he reduced his speed to 12 miles per hour?

58. How many gallons of a 12% salt solution must be mixed with 6 gallons of a 20% salt solution to obtain a 15% salt solution?

59. Suppose that you have a supply of a 30% solution of alcohol and a 70% solution of alcohol. How many quarts of each should be mixed to produce 20 quarts that is 40% alcohol?

60. How many cups of grapefruit juice must be added to 40 cups of punch that is 5% grapefruit juice to obtain a punch that is 10% grapefruit juice?

61. How many milliliters of pure acid must be added to 150 milliliters of a 30% solution of acid to obtain a 40% solution?

62. A 16-quart radiator contains a 50% solution of antifreeze. How much needs to be drained out and replaced with pure antifreeze to obtain a 60% antifreeze solution?

THOUGHTS INTO WORDS

63. Some people subtract 32 and then divide by 2 to estimate the change from a Fahrenheit reading to a Celsius reading. Why does this give an estimate and how good is the estimate?

64. One of your classmates analyzes Problem 56 as follows: "The first car has traveled 60 kilometers before the second car starts. Because the second car travels 15 kilometers per hour faster, it will take $\frac{60}{15} = 4$ hours for the second car to overtake the first car." How would you react to this analysis of the problem?

65. Summarize the new ideas relative to problem solving that you have acquired thus far in this course.

FURTHER INVESTIGATIONS

For Problems 66–73, use your calculator to help solve each formula for the indicated variable.

66. Solve $i = Prt$ for i, given that $P = \$875$, $r = 12\frac{1}{2}\%$, and $t = 4$ years.

67. Solve $i = Prt$ for i, given that $P = \$1125$, $r = 13\frac{1}{4}\%$, and $t = 4$ years.

68. Solve $i = Prt$ for t, given that $i = \$453.25$, $P = \$925$, and $r = 14\%$.

69. Solve $i = Prt$ for t, given that $i = \$243.75$, $P = \$1250$, and $r = 13\%$.

70. Solve $i = Prt$ for r, given that $i = \$356.50$, $P = \$1550$, and $t = 2$ years. Express r as a percent.

71. Solve $i = Prt$ for r, given that $i = \$159.50$, $P = \$2200$, and $t = 0.5$ of a year. Express r as a percent.

72. Solve $A = P + Prt$ for P, given that $A = \$1423.50$, $r = 9\frac{1}{2}\%$, and $t = 1$ year.

73. Solve $A = P + Prt$ for P, given that $A = \$2173.75$, $r = 8\frac{3}{4}\%$, and $t = 2$ years.

74. If you have access to computer software that includes spreadsheets, return to Problems 15 and 16. You should be able to enter the given information in rows. Then, when you enter a formula in a cell below the information and drag that formula across the columns, the software should produce all the answers.

Answers to the Concept Quiz

1. False **2.** True **3.** False **4.** True **5.** True **6.** True **7.** True **8.** True **9.** False **10.** True

Answers to the Example Practice Skills

1. \$450 **2.** 16 yr **3.** $B = \frac{3V}{h}$ **4.** $h = \frac{S - 4lw}{2l}$ **5.** $a = \frac{S}{d + n}$ **6.** $w = \frac{P - 2l}{2}$ **7.** $y = \frac{3x - 5}{4}$ **8.** $x = \frac{cd - cy}{d}$ **9.** 8 yr **10.** 2 hr **11.** Southbound 50 mph, northbound 56 mph **12.** 8 oz of the 6% solution and 8 oz of the 14% solution **13.** 2 qt

2.5 Inequalities

OBJECTIVES

1. Write Solution Sets in Interval Notation
2. Solve Inequalities

1 Write Solution Sets in Interval Notation

We listed the basic inequality symbols in Section 1.2. With these symbols we can make various **statements of inequality**:

$a < b$ means a is less than b.

$a \leq b$ means a is less than or equal to b.

$a > b$ means a is greater than b.

$a \geq b$ means a is greater than or equal to b.

Here are some examples of **numerical statements of inequality**:

$7 + 8 > 10$	$-4 + (-6) \geq -10$
$-4 > -6$	$7 - 9 \leq -2$
$7 - 1 < 20$	$3 + 4 > 12$
$8(-3) < 5(-3)$	$7 - 1 < 0$

Note that only $3 + 4 > 12$ and $7 - 1 < 0$ are *false*; the other six are *true* numerical statements.

Algebraic inequalities contain one or more variables. The following are examples of algebraic inequalities.

$$x + 4 > 8 \qquad 3x + 2y \leq 4$$

$$3x - 1 < 15 \qquad x^2 + y^2 + z^2 \geq 7$$

$$y^2 + 2y - 4 \geq 0$$

An algebraic inequality such as $x + 4 > 8$ is neither true nor false as it stands, and we call it an **open sentence**. For each numerical value we substitute for x, the algebraic inequality $x + 4 > 8$ becomes a numerical statement of inequality that is true or false. For example, if $x = -3$, then $x + 4 > 8$ becomes $-3 + 4 > 8$, which is false. If $x = 5$, then $x + 4 > 8$ becomes $5 + 4 > 8$, which is true. **Solving an inequality** is the process of finding the numbers that make an algebraic inequality a true numerical statement. We call such numbers the *solutions* of the inequality; the solutions *satisfy* the inequality.

There are various ways to display the solution set of an inequality. The three most common ways to show the solution set are set builder notation, a line graph of the solution, or interval notation. The examples in Figure 2.5 contain some simple algebraic inequalities, their solution sets, graphs of the solution sets, and the solution sets written in interval notation. Look them over carefully to be sure you understand the symbols.

Algebraic inequality	Solution set	Graph of solution set	Interval notation
$x < 2$	$\{x \mid x < 2\}$	−5 −4 −3 −2 −1 0 1 2 3 4 5	$(-\infty, 2)$
$x > -1$	$\{x \mid x > -1\}$	−5 −4 −3 −2 −1 0 1 2 3 4 5	$(-1, \infty)$
$3 < x$	$\{x \mid x > 3\}$	−5 −4 −3 −2 −1 0 1 2 3 4 5	$(3, \infty)$
$x \geq 1$ ($\geq$ is read "greater than or equal to")	$\{x \mid x \geq 1\}$	−5 −4 −3 −2 −1 0 1 2 3 4 5	$[1, \infty)$
$x \leq 2$ ($\leq$ is read "less than or equal to")	$\{x \mid x \leq 2\}$	−5 −4 −3 −2 −1 0 1 2 3 4 5	$(-\infty, 2]$
$1 \geq x$	$\{x \mid x \leq 1\}$	−5 −4 −3 −2 −1 0 1 2 3 4 5	$(-\infty, 1]$

Figure 2.5

EXAMPLE 1

Express the given inequalities in interval notation and graph the interval on a number line.

(a) $x > -2$ **(b)** $x \leq -1$ **(c)** $x < 3$ **(d)** $x \geq 2$

Solution

(a) For the solution set of the inequality $x > -2$, we want all the numbers greater than -2 but not including -2. In interval notation, the solution set is written as $(-2, \infty)$, where parentheses are used to indicate exclusion of the endpoint. The use of a parenthesis carries over to the graph of the solution set. On the graph, the left-hand parenthesis at -2 indicates that -2 is *not* a solution, and the red part of the line to the right of -2 indicates that all real numbers greater than -2 are solutions. We refer to the red portion of the number line as the *graph* of the solution set.

Inequality	Interval notation	Graph
$x > -2$	$(-2, \infty)$	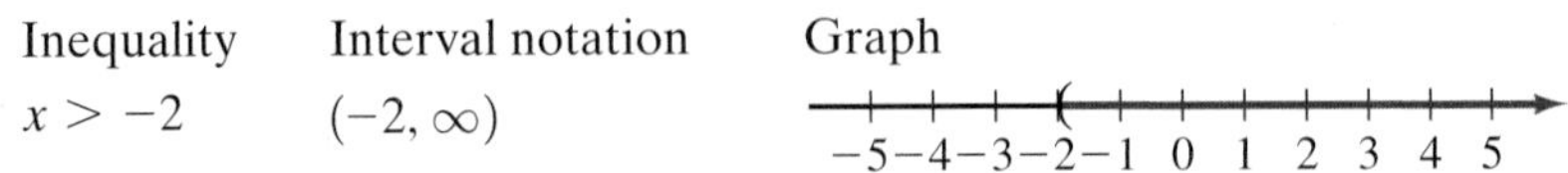

(b) For the solution set of the inequality $x \leq -1$, we want all the numbers less than or equal to -1. In interval notation, the solution set is written as $(-\infty, -1]$, where a square bracket is used to indicate inclusion of the endpoint. The use of a square bracket carries over to the graph of the solution set. On the graph, the right-hand square bracket at -1 indicates that -1 is part of the solution, and the red part of the line to the left of -1 indicates that all real numbers less than -1 are solutions.

Inequality	Interval notation	Graph
$x \leq -1$	$(-\infty, -1]$	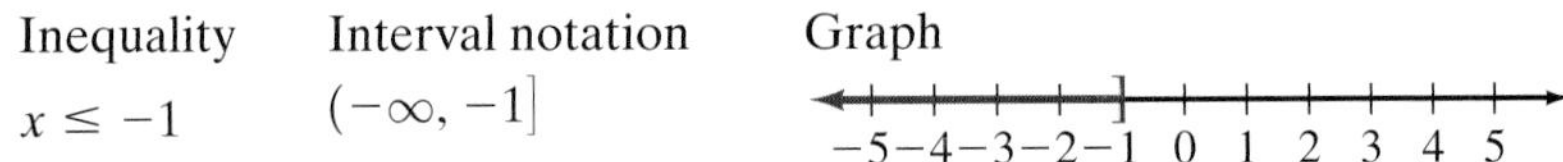

(c) For the solution set of the inequality $x < 3$, we want all the numbers less than 3 but not including 3. In interval notation, the solution set is written as $(-\infty, 3)$.

Inequality	Interval notation	Graph
$x < 3$	$(-\infty, 3)$	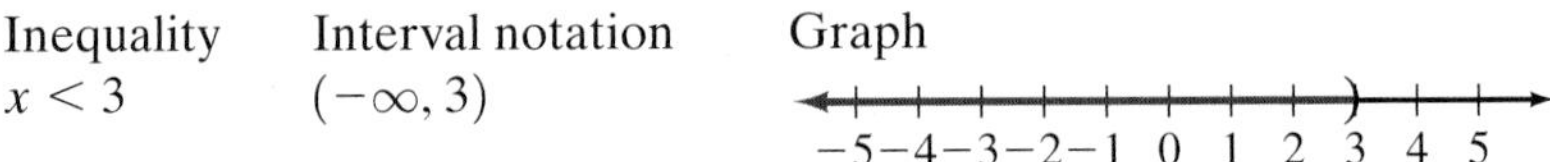

(d) For the solution set of the inequality $x \geq 2$, we want all the numbers greater than or equal to 2. In interval notation, the solution set is written as $[2, \infty)$.

Inequality	Interval notation	Graph
$x \geq 2$	$[2, \infty)$	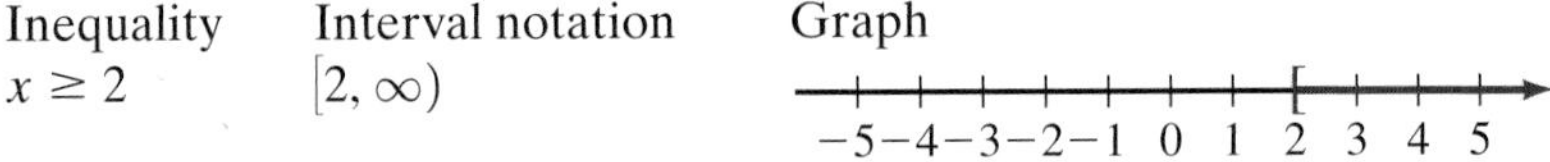

Remark: Note that the infinity symbol always has a parenthesis next to it because no actual endpoint could be included.

▼ PRACTICE YOUR SKILL

Express the given inequality in interval notation and graph the interval on a number line.

(a) $x < 4$ **(b)** $x \geq 3$ **(c)** $x < -4$ **(d)** $x \leq 0$ ■

2 Solve Inequalities

The general process for solving inequalities closely parallels the process for solving equations. We continue to replace the given inequality with equivalent, but simpler, inequalities. For example,

$$3x + 4 > 10 \quad \textbf{(1)}$$

$$3x > 6 \quad \textbf{(2)}$$

$$x > 2 \quad \textbf{(3)}$$

are all equivalent inequalities; that is, they all have the same solutions. By inspection we see that the solutions for (3) are all numbers greater than 2. Thus (1) has the same solutions.

The exact procedure for simplifying inequalities so that we can determine the solutions is based primarily on two properties. The first of these is the addition property of inequality.

Addition Property of Inequality

For all real numbers a, b, and c,

$$a > b \quad \text{if and only if } a + c > b + c$$

The addition property of inequality states that we can add any number to both sides of an inequality to produce an equivalent inequality. We have stated the property in terms of $>$, but analogous properties exist for $<$, $\geq$, and $\leq$.

Before we state the multiplication property of inequality, let's look at some numerical examples.

$2 < 5$	Multiply both sides by 4:	$4(2) < 4(5)$	$8 < 20$
$-3 > -7$	Multiply both sides by 2:	$2(-3) > 2(-7)$	$-6 > -14$
$-4 < 6$	Multiply both sides by 10:	$10(-4) < 10(6)$	$-40 < 60$
$4 < 8$	Multiply both sides by -3:	$-3(4) > -3(8)$	$-12 > -24$
$3 > -2$	Multiply both sides by -4:	$-4(3) < -4(-2)$	$-12 < 8$
$-4 < -1$	Multiply both sides by -2:	$-2(-4) > -2(-1)$	$8 > 2$

Notice in the first three examples that, when we multiply both sides of an inequality by a *positive number*, we obtain an inequality of the *same sense*. That means that if the original inequality is *less than*, then the new inequality is *less than*; and if the original inequality is *greater than*, then the new inequality is *greater than*. The last three examples illustrate that when we multiply both sides of an inequality by a *negative number* we get an inequality of the *opposite sense*.

We can state the multiplication property of inequality as follows.

Multiplication Property of Inequality

(a) For all real numbers a, b, and c, with $c > 0$,

$$a > b \quad \text{if and only if } ac > bc$$

(b) For all real numbers a, b, and c, with $c < 0$,

$$a > b \quad \text{if and only if } ac < bc$$

Similar properties hold if we reverse each inequality or if we replace $>$ with $\geq$ and $<$ with $\leq$. For example, if $a \leq b$ and $c < 0$, then $ac \geq bc$.

Now let's use the addition and multiplication properties of inequality to help solve some inequalities.

EXAMPLE 2 Solve $3x - 4 > 8$ and graph the solutions.

Solution

$$3x - 4 > 8$$

$$3x - 4 + 4 > 8 + 4 \qquad \text{Add 4 to both sides}$$

$$3x > 12$$

$$\frac{1}{3}(3x) > \frac{1}{3}(12) \quad \text{Multiply both sides by } \frac{1}{3}$$

$$x > 4$$

The solution set is $(4, \infty)$. Figure 2.6 shows the graph of the solution set.

Figure 2.6

▼ **PRACTICE YOUR SKILL**

Solve $5x + 7 < 22$ and graph the solution set. ■

EXAMPLE 3 Solve $-2x + 1 > 5$ and graph the solutions.

Solution

$$-2x + 1 > 5$$

$$-2x + 1 + (-1) > 5 + (-1) \quad \text{Add } -1 \text{ to both sides}$$

$$-2x > 4$$

$$-\frac{1}{2}(-2x) < -\frac{1}{2}(4) \quad \text{Multiply both sides by } -\frac{1}{2}$$

Note that the sense of the inequality has been reversed

$$x < -2$$

The solution set is $(-\infty, -2)$, which can be illustrated on a number line as in Figure 2.7.

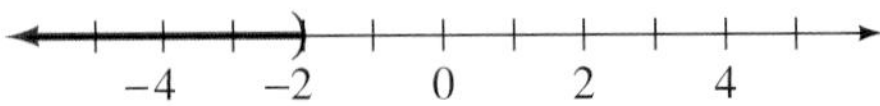

Figure 2.7

▼ **PRACTICE YOUR SKILL**

Solve $-3x - 4 < 47$ and graph the solution set. ■

Checking solutions for an inequality presents a problem. Obviously, we cannot check all of the infinitely many solutions for a particular inequality. However, by checking at least one solution, especially when the multiplication property has been used, we might catch the common mistake of forgetting to change the sense of an inequality. In Example 3 we are claiming that all numbers less than -2 will satisfy the original inequality. Let's check one such number, say -4.

$$-2x + 1 > 5$$

$$-2(-4) + 1 \overset{?}{>} 5 \quad \text{when } x = -4$$

$$8 + 1 \overset{?}{>} 5$$

$$9 > 5$$

Thus -4 satisfies the original inequality. Had we forgotten to switch the sense of the inequality when both sides were multiplied by $-\frac{1}{2}$, our answer would have been $x > -2$, and we would have detected such an error by the check.

Many of the same techniques used to solve equations, such as removing parentheses and combining similar terms, may be used to solve inequalities. However, we must be extremely careful when using the multiplication property of inequality. Study each of the following examples carefully. The format we used highlights the major steps of a solution.

EXAMPLE 4 Solve $-3x + 5x - 2 \geq 8x - 7 - 9x$.

Solution

$$-3x + 5x - 2 \geq 8x - 7 - 9x$$

$$2x - 2 \geq -x - 7 \qquad \text{Combine similar terms on both sides}$$

$$3x - 2 \geq -7 \qquad \text{Add } x \text{ to both sides}$$

$$3x \geq -5 \qquad \text{Add 2 to both sides}$$

$$\frac{1}{3}(3x) \geq \frac{1}{3}(-5) \qquad \text{Multiply both sides by } \frac{1}{3}$$

$$x \geq -\frac{5}{3}$$

The solution set is $\left[-\frac{5}{3}, \infty\right)$.

▼ **PRACTICE YOUR SKILL**

Solve $-x + 4x + 8 \geq 6x - 5 - 2x$. ■

EXAMPLE 5 Solve $-5(x - 1) \leq 10$ and graph the solutions.

Solution

$$-5(x - 1) \leq 10$$

$$-5x + 5 \leq 10 \qquad \text{Apply the distributive property on the left}$$

$$-5x \leq 5 \qquad \text{Add } -5 \text{ to both sides}$$

$$-\frac{1}{5}(-5x) \geq -\frac{1}{5}(5) \qquad \text{Multiply both sides by } -\frac{1}{5}\text{, which reverses the inequality}$$

$$x \geq -1$$

The solution set is $[-1, \infty)$, and it can be graphed as in Figure 2.8.

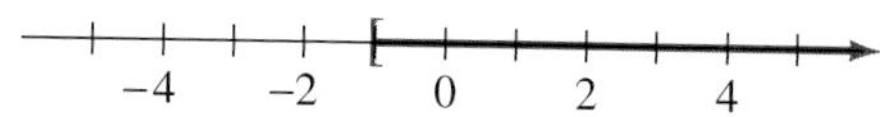

Figure 2.8

▼ **PRACTICE YOUR SKILL**

Solve $-4(x + 3) \geq 28$. ■

EXAMPLE 6 Solve $4(x - 3) > 9(x + 1)$.

Solution

$$
\begin{aligned}
4(x - 3) &> 9(x + 1) \\
4x - 12 &> 9x + 9 && \text{Apply the distributive property} \\
-5x - 12 &> 9 && \text{Add } -9x \text{ to both sides} \\
-5x &> 21 && \text{Add 12 to both sides} \\
-\frac{1}{5}(-5x) &< -\frac{1}{5}(21) && \text{Multiply both sides by } -\frac{1}{5}\text{, which reverses the inequality} \\
x &< -\frac{21}{5}
\end{aligned}
$$

The solution set is $\left(-\infty, -\frac{21}{5}\right)$.

▼ **PRACTICE YOUR SKILL**

Solve $2(x - 1) \geq 5(x + 3)$. ■

The next example will solve the inequality without indicating the justification for each step. Be sure that you can supply the reasons for the steps.

EXAMPLE 7 Solve $3(2x + 1) - 2(2x + 5) < 5(3x - 2)$.

Solution

$$
\begin{aligned}
3(2x + 1) - 2(2x + 5) &< 5(3x - 2) \\
6x + 3 - 4x - 10 &< 15x - 10 \\
2x - 7 &< 15x - 10 \\
-13x - 7 &< -10 \\
-13x &< -3 \\
-\frac{1}{13}(-13x) &> -\frac{1}{13}(-3) \\
x &> \frac{3}{13}
\end{aligned}
$$

The solution set is $\left(\frac{3}{13}, \infty\right)$.

▼ **PRACTICE YOUR SKILL**

Solve $2(3x - 4) - 5(x + 1) < 3(2x + 5)$. ■

CONCEPT QUIZ For Problems 1–10, answer true or false.

1. Numerical statements of inequality are always true.
2. The algebraic statement $x + 4 > 6$ is called an open sentence.
3. The algebraic inequality $2x > 10$ has one solution.
4. The algebraic inequality $x < 3$ has an infinite number of solutions.
5. The solution set for the inequality $-3x - 1 > 2$ is $(-1, \infty)$.

6. When graphing the solution set of an inequality, a square bracket is used to include the endpoint.
7. The solution set of the inequality $x \geq 4$ is written $(4, \infty)$.
8. The solution set of the inequality $x < -5$ is written $(-\infty, -5)$.
9. When multiplying both sides of an inequality by a negative number, the sense of the inequality stays the same.
10. When adding a negative number to both sides of an inequality, the sense of the inequality stays the same.

Problem Set 2.5

1 Write Solution Sets in Interval Notation

For Problems 1–8, express the given inequality in interval notation and sketch a graph of the interval.

1. $x > 1$
2. $x > -2$
3. $x \geq -1$
4. $x \geq 3$
5. $x < -2$
6. $x < 1$
7. $x \leq 2$
8. $x \leq 0$

For Problems 9–16, express each interval as an inequality using the variable x. For example, we can express the interval $[5, \infty)$ as $x \geq 5$.

9. $(-\infty, 4)$
10. $(-\infty, -2)$
11. $(-\infty, -7]$
12. $(-\infty, 9]$
13. $(8, \infty)$
14. $(-5, \infty)$
15. $[-7, \infty)$
16. $[10, \infty)$

2 Solve Inequalities

For Problems 17–40, solve each of the inequalities and graph the solution set on a number line.

17. $x - 3 > -2$
18. $x + 2 < 1$
19. $-2x \geq 8$
20. $-3x \leq -9$
21. $5x \leq -10$
22. $4x \geq -4$
23. $2x + 1 < 5$
24. $2x + 2 > 4$
25. $3x - 2 > -5$
26. $5x - 3 < -3$
27. $-7x - 3 \leq 4$
28. $-3x - 1 \geq 8$
29. $2 + 6x > -10$
30. $1 + 6x > -17$
31. $5 - 3x < 11$
32. $4 - 2x < 12$
33. $15 < 1 - 7x$
34. $12 < 2 - 5x$
35. $-10 \leq 2 + 4x$
36. $-9 \leq 1 + 2x$
37. $3(x + 2) > 6$
38. $2(x - 1) < -4$
39. $5x + 2 \geq 4x + 6$
40. $6x - 4 \leq 5x - 4$

For Problems 41–70, solve each inequality and express the solution set using interval notation.

41. $2x - 1 > 6$
42. $3x - 2 < 12$
43. $-5x - 2 < -14$
44. $5 - 4x > -2$
45. $-3(2x + 1) \geq 12$
46. $-2(3x + 2) \leq 18$
47. $4(3x - 2) \geq -3$
48. $3(4x - 3) \leq -11$
49. $6x - 2 > 4x - 14$
50. $9x + 5 < 6x - 10$
51. $2x - 7 < 6x + 13$
52. $2x - 3 > 7x + 22$
53. $4(x - 3) \leq -2(x + 1)$
54. $3(x - 1) \geq -(x + 4)$
55. $5(x - 4) - 6(x + 2) < 4$
56. $3(x + 2) - 4(x - 1) < 6$
57. $-3(3x + 2) - 2(4x + 1) \geq 0$
58. $-4(2x - 1) - 3(x + 2) \geq 0$
59. $-(x - 3) + 2(x - 1) < 3(x + 4)$
60. $3(x - 1) - (x - 2) > -2(x + 4)$
61. $7(x + 1) - 8(x - 2) < 0$
62. $5(x - 6) - 6(x + 2) < 0$
63. $-5(x - 1) + 3 > 3x - 4 - 4x$
64. $3(x + 2) + 4 < -2x + 14 + x$
65. $3(x - 2) - 5(2x - 1) \geq 0$
66. $4(2x - 1) - 3(3x + 4) \geq 0$
67. $-5(3x + 4) < -2(7x - 1)$
68. $-3(2x + 1) > -2(x + 4)$
69. $-3(x + 2) > 2(x - 6)$
70. $-2(x - 4) < 5(x - 1)$

THOUGHTS INTO WORDS

71. Do the *less than* and *greater than* relations possess a symmetric property similar to the symmetric property of equality? Defend your answer.

72. Give a step-by-step description of how you would solve the inequality $-3 > 5 - 2x$.

73. How would you explain to someone why it is necessary to reverse the inequality symbol when multiplying both sides of an inequality by a negative number?

FURTHER INVESTIGATIONS

74. Solve each of the following inequalities.

(a) $5x - 2 > 5x + 3$
(b) $3x - 4 < 3x + 7$
(c) $4(x + 1) < 2(2x + 5)$
(d) $-2(x - 1) > 2(x + 7)$
(e) $3(x - 2) < -3(x + 1)$
(f) $2(x + 1) + 3(x + 2) < 5(x - 3)$

Answers to the Concept Quiz

1. False **2.** True **3.** False **4.** True **5.** False **6.** True **7.** False **8.** True **9.** False **10.** True

Answers to the Example Practice Skills

1. (a) $(-\infty, 4)$ 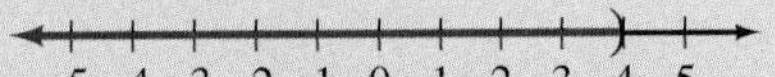

(b) $[3, \infty)$ 

(c) $(-\infty, -4)$ 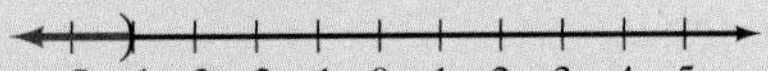

(d) $(-\infty, 0]$ 

2. $(-\infty, 3)$ 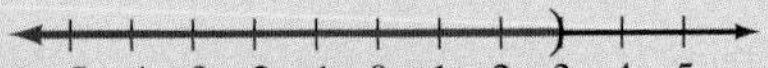

3. $(-17, \infty)$ 

4. $(-\infty, 13]$ **5.** $(-\infty, 10]$ 

6. $\left(-\infty, -\frac{17}{3}\right]$ **7.** $\left(-\frac{28}{5}, \infty\right)$

2.6 More on Inequalities and Problem Solving

OBJECTIVES

1 Solve Inequalities Involving Fractions or Decimals

2 Solve Inequalities That Are Compound Statements

3 Use Inequalities to Solve Word Problems

1 Solve Inequalities Involving Fractions or Decimals

When we discussed solving equations that involve fractions, we found that **clearing the equation of all fractions** is frequently an effective technique. To accomplish this, we multiply both sides of the equation by the least common denominator of all the denominators in the equation. This same basic approach also works very well with inequalities that involve fractions, as the next examples demonstrate.

EXAMPLE 1

Solve $\frac{2}{3}x - \frac{1}{2}x > \frac{3}{4}$.

Solution

$$\frac{2}{3}x - \frac{1}{2}x > \frac{3}{4}$$

$$12\left(\frac{2}{3}x - \frac{1}{2}x\right) > 12\left(\frac{3}{4}\right) \quad \text{Multiply both sides by 12, which is the LCD of 3, 2, and 4}$$

$$12\left(\frac{2}{3}x\right) - 12\left(\frac{1}{2}x\right) > 12\left(\frac{3}{4}\right) \quad \text{Apply the distributive property}$$

$$8x - 6x > 9$$

$$2x > 9$$

$$x > \frac{9}{2}$$

The solution set is $\left(\frac{9}{2}, \infty\right)$.

▼ PRACTICE YOUR SKILL

Solve $\frac{2}{5}x - \frac{1}{3}x > -\frac{4}{5}$. ■

EXAMPLE 2

Solve $\frac{x+2}{4} + \frac{x-3}{8} < 1$.

Solution

$$\frac{x+2}{4} + \frac{x-3}{8} < 1$$

$$8\left(\frac{x+2}{4} + \frac{x-3}{8}\right) < 8(1) \quad \text{Multiply both sides by 8, which is the LCD of 4 and 8}$$

$$8\left(\frac{x+2}{4}\right) + 8\left(\frac{x-3}{8}\right) < 8(1)$$

$$2(x+2) + (x-3) < 8$$

$$2x + 4 + x - 3 < 8$$

$$3x + 1 < 8$$

$$3x < 7$$

$$x < \frac{7}{3}$$

The solution set is $\left(-\infty, \frac{7}{3}\right)$.

▼ PRACTICE YOUR SKILL

Solve $\frac{x-1}{2} + \frac{x+4}{5} \leq 3$. ■

EXAMPLE 3

Solve $\frac{x}{2} - \frac{x-1}{5} \geq \frac{x+2}{10} - 4$.

Solution

$$\frac{x}{2} - \frac{x-1}{5} \geq \frac{x+2}{10} - 4$$

$$10\left(\frac{x}{2} - \frac{x-1}{5}\right) \geq 10\left(\frac{x+2}{10} - 4\right)$$

$$10\left(\frac{x}{2}\right) - 10\left(\frac{x-1}{5}\right) \geq 10\left(\frac{x+2}{10}\right) - 10(4)$$

$$5x - 2(x - 1) \geq x + 2 - 40$$
$$5x - 2x + 2 \geq x - 38$$
$$3x + 2 \geq x - 38$$
$$2x + 2 \geq -38$$
$$2x \geq -40$$
$$x \geq -20$$

The solution set is $[-20, \infty)$.

▼ PRACTICE YOUR SKILL

Solve $\frac{y}{3} - \frac{y-2}{5} > \frac{y+1}{15} - 1$. ■

The idea of **clearing all decimals** works with inequalities in much the same way as it does with equations. We can multiply both sides of an inequality by an appropriate power of 10 and then proceed to solve in the usual way. The next two examples illustrate this procedure.

EXAMPLE 4

Solve $x \geq 1.6 + 0.2x$.

Solution

$$x \geq 1.6 + 0.2x$$
$$10(x) \geq 10(1.6 + 0.2x) \quad \text{Multiply both sides by 10}$$
$$10x \geq 16 + 2x$$
$$8x \geq 16$$
$$x \geq 2$$

The solution set is $[2, \infty)$.

▼ PRACTICE YOUR SKILL

Solve $0.12x + 0.6 \leq 0.48$. ■

EXAMPLE 5

Solve $0.08x + 0.09(x + 100) \geq 43$.

Solution

$$0.08x + 0.09(x + 100) \geq 43$$
$$100(0.08x + 0.09(x + 100)) \geq 100(43) \quad \text{Multiply both sides by 100}$$
$$8x + 9(x + 100) \geq 4300$$
$$8x + 9x + 900 \geq 4300$$
$$17x + 900 \geq 4300$$
$$17x \geq 3400$$
$$x \geq 200$$

The solution set is $[200, \infty)$.

▼ PRACTICE YOUR SKILL

Solve $0.05x + 0.07(x + 500) \geq 287$. ■

2 Solve Inequalities That Are Compound Statements

We use the words "and" and "or" in mathematics to form **compound statements.** The following are examples of compound numerical statements that use "and." We call such statements **conjunctions**. We agree to call a conjunction true only if all of its component parts are true. Statements 1 and 2 below are true, but statements 3, 4, and 5 are false.

1. $3 + 4 = 7$ and $-4 < -3$. True

2. $-3 < -2$ and $-6 > -10$. True

3. $6 > 5$ and $-4 < -8$. False

4. $4 < 2$ and $0 < 10$. False

5. $-3 + 2 = 1$ and $5 + 4 = 8$. False

We call compound statements that use "or" **disjunctions**. The following are examples of disjunctions that involve numerical statements.

6. $0.14 > 0.13$ or $0.235 < 0.237$. True

7. $\frac{3}{4} > \frac{1}{2}$ or $-4 + (-3) = 10$. True

8. $-\frac{2}{3} > \frac{1}{3}$ or $(0.4)(0.3) = 0.12$. True

9. $\frac{2}{5} < -\frac{2}{5}$ or $7 + (-9) = 16$. False

A disjunction is true if at least one of its component parts is true. In other words, disjunctions are false only if all of the component parts are false. Thus statements 6, 7, and 8 are true, but statement 9 is false.

Now let's consider finding solutions for some compound statements that involve algebraic inequalities. Keep in mind that our previous agreements for labeling conjunctions and disjunctions true or false form the basis for our reasoning.

EXAMPLE 6

Graph the solution set for the conjunction $x > -1$ and $x < 3$.

Solution

The key word is "and," so we need to satisfy both inequalities. Thus all numbers between -1 and 3 are solutions, and we can indicate this on a number line as in Figure 2.9.

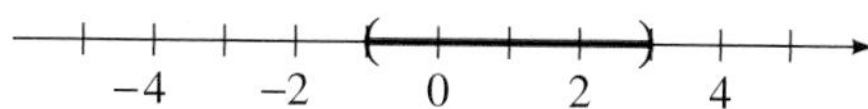

Figure 2.9

Using interval notation, we can represent the interval enclosed in parentheses in Figure 2.9 by $(-1, 3)$. Using set builder notation, we can express the same interval as $\{x \mid -1 < x < 3\}$, where the statement $-1 < x < 3$ is read "Negative one is less than x, and x is less than three." In other words, x is between -1 and 3.

▼ **PRACTICE YOUR SKILL**

Graph the solution set for the conjunction $x \geq 1$ and $x \leq 6$. ■

Example 6 represents another concept that pertains to sets. The set of all elements common to two sets is called the **intersection** of the two sets. Thus in Example 6, we found the intersection of the two sets $\{x \mid x > -1\}$ and $\{x \mid x < 3\}$ to be the set $\{x \mid -1 < x < 3\}$. In general, we define the intersection of two sets as follows.

Definition 2.1

The **intersection** of two sets A and B (written $A \cap B$) is the set of all elements that are in both A and in B. Using set builder notation, we can write

$$A \cap B = \{x \mid x \in A \text{ and } x \in B\}$$

EXAMPLE 7

Solve the conjunction $3x + 1 > -5$ and $2x + 5 > 7$, and graph its solution set on a number line.

Solution

First, let's simplify both inequalities.

$$\begin{aligned} 3x + 1 &> -5 \quad \text{and} \quad & 2x + 5 &> 7 \\ 3x &> -6 \quad \text{and} \quad & 2x &> 2 \\ x &> -2 \quad \text{and} \quad & x &> 1 \end{aligned}$$

Because this is a conjunction, we must satisfy both inequalities. Thus all numbers greater than 1 are solutions, and the solution set is $(1, \infty)$. We show the graph of the solution set in Figure 2.10.

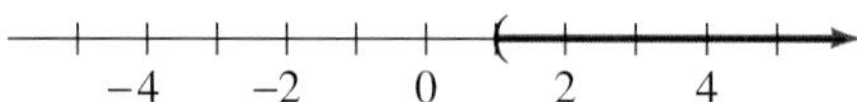

Figure 2.10

▼ **PRACTICE YOUR SKILL**

Solve the conjunction $2x - 4 > -6$ and $3x + 5 > 14$, and graph its solution set on a number line. ■

We can solve a conjunction such as $3x + 1 > -3$ and $3x + 1 < 7$, in which the same algebraic expression (in this case $3x + 1$) is contained in both inequalities, by using the **compact form** $-3 < 3x + 1 < 7$ as follows:

$$-3 < 3x + 1 < 7$$

$$-4 < 3x < 6 \qquad \text{Add } -1 \text{ to the left side, middle, and right side}$$

$$-\frac{4}{3} < x < 2 \qquad \text{Multiply through by } \frac{1}{3}$$

The solution set is $\left(-\frac{4}{3}, 2\right)$.

The word *and* ties the concept of a conjunction to the set concept of intersection. In a like manner, the word *or* links the idea of a disjunction to the set concept of **union**. We define the union of two sets as follows.

Definition 2.2

The **union** of two sets A and B (written $A \cup B$) is the set of all elements that are in A or in B, or in both. Using set builder notation, we can write

$$A \cup B = \{x \mid x \in A \text{ or } x \in B\}$$

EXAMPLE 8

Graph the solution set for the disjunction $x < -1$ or $x > 2$, and express it using interval notation.

Solution

The key word is "or," so all numbers that satisfy either inequality (or both) are solutions. Thus all numbers less than -1, along with all numbers greater than 2, are the solutions. The graph of the solution set is shown in Figure 2.11.

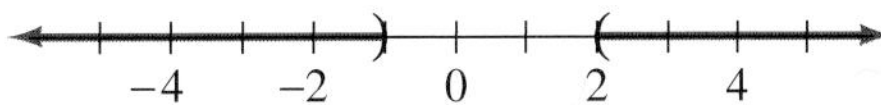

Figure 2.11

Using interval notation and the set concept of union, we can express the solution set as $(-\infty, -1) \cup (2, \infty)$.

▼ **PRACTICE YOUR SKILL**

Graph the solution set for the disjunction $x \le 0$ or $x \ge 5$, and express it using interval notation. ■

Example 8 illustrates that in terms of set vocabulary, the solution set of a disjunction is the union of the solution sets of the component parts of the disjunction. Note that there is no compact form for writing $x < -1$ *or* $x > 2$ or for any disjunction.

EXAMPLE 9

Solve the disjunction $2x - 5 < -11$ or $5x + 1 \ge 6$, and graph its solution set on a number line.

Solution

First, let's simplify both inequalities.

$$\begin{aligned} 2x - 5 &< -11 \quad \text{or} \quad & 5x + 1 &\ge 6 \\ 2x &< -6 \quad \text{or} \quad & 5x &\ge 5 \\ x &< -3 \quad \text{or} \quad & x &\ge 1 \end{aligned}$$

This is a disjunction, and all numbers less than -3, along with all numbers greater than or equal to 1, will satisfy it. Thus the solution set is $(-\infty, -3) \cup [1, \infty)$. Its graph is shown in Figure 2.12.

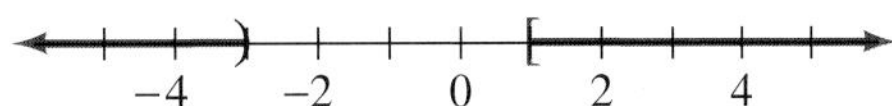

Figure 2.12

▼ **PRACTICE YOUR SKILL**

Solve the disjunction $3x - 1 < 5$ or $2x + 5 > 15$, and graph its solution set on a number line. ■

In summary, to solve a compound sentence involving an inequality, proceed as follows.

1. Solve separately each inequality in the compound sentence.
2. If it is a conjunction, the solution set is the intersection of the solution sets of each inequality.
3. If it is a disjunction, the solution set is the union of the solution sets of each inequality.

Figure 2.13 shows some conventions associated with interval notation. These are in addition to the previous list in Figure 2.5.

Set	Graph	Interval notation
$\{x \mid 2 < x < 4\}$	−4 −2 0 2 4	$(2, 4)$
$\{x \mid 2 \leq x < 4\}$	−4 −2 0 2 4	$[2, 4)$
$\{x \mid 2 < x \leq 4\}$	−4 −2 0 2 4	$(2, 4]$
$\{x \mid 2 \leq x \leq 4\}$	−4 −2 0 2 4	$[2, 4]$

Figure 2.13

3 Use Inequalities to Solve Word Problems

We will conclude this section with some word problems that contain inequality statements.

EXAMPLE 10

Dave & Les Jacobs/Jupiter Images

Apply Your Skill

Sari had scores of 94, 84, 86, and 88 on her first four exams of the semester. What score must she obtain on the fifth exam to have an average of 90 or better for the five exams?

Solution

Let s represent the score Sari needs on the fifth exam. Because the average is computed by adding all scores and dividing by the number of scores, we have the following inequality to solve.

$$\frac{94 + 84 + 86 + 88 + s}{5} \geq 90$$

Solving this inequality, we obtain

$$\frac{352 + s}{5} \geq 90$$

$$5\left(\frac{352 + s}{5}\right) \geq 5(90) \quad \text{Multiply both sides by 5}$$

$$352 + s \geq 450$$

$$s \geq 98$$

Sari must receive a score of 98 or better.

▼ PRACTICE YOUR SKILL

Matt scored 86, 75, 71, and 80 on his first four exams. To keep his scholarship he must have at least an 80 average. What must he score on the fifth exam to have an average of 80 or better? ■

EXAMPLE 11

George Diebold/Riser/Getty Images

Apply Your Skill

An investor has $1000 to invest. Suppose she invests $500 at 8% interest. At what rate must she invest the other $500 so that the two investments together yield more than $100 of yearly interest?

Solution

Let r represent the unknown rate of interest. We can use the following guideline to set up an inequality.

Interest from 8% investment	+	Interest from r percent investment	>	$100
↓		↓		↓
(8%)($500)	+	r($500)	>	$100

Solving this inequality yields

$$40 + 500r > 100$$
$$500r > 60$$
$$r > \frac{60}{500}$$
$$r > 0.12 \qquad \text{Change to a decimal}$$

She must invest the other $500 at a rate greater than 12%.

▼ PRACTICE YOUR SKILL

Mary has $5000 to invest. If she invests $1500 at 6% interest, then at what rate must she invest the other $3500 so that the two investments yield more than $335? ■

EXAMPLE 12

Creatas Images/Jupiter Images

Apply Your Skill

If the temperature for a 24-hour period ranged between 41°F and 59°F, inclusive (that is, $41 \leq F \leq 59$), what was the range in Celsius degrees?

Solution

Use the formula $F = \frac{9}{5}C + 32$, to solve the following compound inequality.

$$41 \leq \frac{9}{5}C + 32 \leq 59$$

Solving this yields

$$9 \leq \frac{9}{5}C \leq 27 \qquad \text{Add } -32$$
$$\frac{5}{9}(9) \leq \frac{5}{9}\left(\frac{9}{5}C\right) \leq \frac{5}{9}(27) \qquad \text{Multiply by } \frac{5}{9}$$
$$5 \leq C \leq 15$$

The range was between 5°C and 15°C, inclusive.

▼ PRACTICE YOUR SKILL

A nursery advertises that a particular plant only thrives between the temperatures of 50°F and 86°F, inclusive. The nursery wants to display this information in both Fahrenheit and Celsius scales on an international website. What temperature range in Celsius should the nursery display for this particular plant? ■

CONCEPT QUIZ

For Problems 1–5, answer true or false.

1. The solution set of a compound inequality formed by the word "and" is an intersection of the solution sets of the two inequalities.
2. The solution set of any compound inequality is the union of the solution sets of the two inequalities.
3. The intersection of two sets contains the elements that are common to both sets.
4. The union of two sets contains all the elements in both sets.
5. The intersection of set A and set B is denoted by $A \cap B$.

For Problems 6–10, match the compound statement with the graph of its solution set.

6. $x > 4$ or $x < -1$ — A. −5 −4 −3 −2 −1 0 1 2 3 4 5
7. $x > 4$ and $x > -1$ — B. −5 −4 −3 −2 −1 0 1 2 3 4 5
8. $x > 4$ or $x > -1$ — C. −5 −4 −3 −2 −1 0 1 2 3 4 5
9. $x \leq 4$ and $x \geq -1$ — D. −5 −4 −3 −2 −1 0 1 2 3 4 5
10. $x > 4$ or $x \geq -1$ — E. −5 −4 −3 −2 −1 0 1 2 3 4 5

Problem Set 2.6

1 Solve Inequalities Involving Fractions or Decimals

For Problems 1–18, solve each of the inequalities and express the solution sets in interval notation.

1. $\frac{2}{5}x + \frac{1}{3}x > \frac{44}{15}$
2. $\frac{1}{4}x - \frac{4}{3}x < -13$
3. $x - \frac{5}{6} < \frac{x}{2} + 3$
4. $x + \frac{2}{7} > \frac{x}{2} - 5$
5. $\frac{x-2}{3} + \frac{x+1}{4} \geq \frac{5}{2}$
6. $\frac{x-1}{3} + \frac{x+2}{5} \leq \frac{3}{5}$
7. $\frac{3-x}{6} + \frac{x+2}{7} \leq 1$
8. $\frac{4-x}{5} + \frac{x+1}{6} \geq 2$
9. $\frac{x+3}{8} - \frac{x+5}{5} \geq \frac{3}{10}$
10. $\frac{x-4}{6} - \frac{x-2}{9} \leq \frac{5}{18}$
11. $\frac{4x-3}{6} - \frac{2x-1}{12} < -2$
12. $\frac{3x+2}{9} - \frac{2x+1}{3} > -1$
13. $0.06x + 0.08(250 - x) \geq 19$
14. $0.08x + 0.09(2x) \geq 130$
15. $0.09x + 0.1(x + 200) > 77$
16. $0.07x + 0.08(x + 100) > 38$
17. $x \geq 3.4 + 0.15x$
18. $x \geq 2.1 + 0.3x$

2 Solve Inequalities That Are Compound Statements

For Problems 19–34, graph the solution set for each compound inequality, and express the solution sets in interval notation.

19. $x > -1$ and $x < 2$
20. $x > 1$ and $x < 4$
21. $x \leq 2$ and $x > -1$
22. $x \leq 4$ and $x \geq -2$

23. $x > 2$ or $x < -1$

24. $x > 1$ or $x < -4$

25. $x \leq 1$ or $x > 3$

26. $x < -2$ or $x \geq 1$

27. $x > 0$ and $x > -1$

28. $x > -2$ and $x > 2$

29. $x < 0$ and $x > 4$

30. $x > 1$ or $x < 2$

31. $x > -2$ or $x < 3$

32. $x > 3$ and $x < -1$

33. $x > -1$ or $x > 2$

34. $x < -2$ or $x < 1$

For Problems 35–44, solve each compound inequality and graph the solution sets. Express the solution sets in interval notation.

35. $x - 2 > -1$ and $x - 2 < 1$

36. $x + 3 > -2$ and $x + 3 < 2$

37. $x + 2 < -3$ or $x + 2 > 3$

38. $x - 4 < -2$ or $x - 4 > 2$

39. $2x - 1 \geq 5$ and $x > 0$

40. $3x + 2 > 17$ and $x \geq 0$

41. $5x - 2 < 0$ and $3x - 1 > 0$

42. $x + 1 > 0$ and $3x - 4 < 0$

43. $3x + 2 < -1$ or $3x + 2 > 1$

44. $5x - 2 < -2$ or $5x - 2 > 2$

For Problems 45–56, solve each compound inequality using the compact form. Express the solution sets in interval notation.

45. $-3 < 2x + 1 < 5$

46. $-7 < 3x - 1 < 8$

47. $-17 \leq 3x - 2 \leq 10$

48. $-25 \leq 4x + 3 \leq 19$

49. $1 < 4x + 3 < 9$

50. $0 < 2x + 5 < 12$

51. $-6 < 4x - 5 < 6$

52. $-2 < 3x + 4 < 2$

53. $-4 \leq \frac{x - 1}{3} \leq 4$

54. $-1 \leq \frac{x + 2}{4} \leq 1$

55. $-3 < 2 - x < 3$

56. $-4 < 3 - x < 4$

3 Use Inequalities to Solve Word Problems

For Problems 57–67, solve each problem by setting up and solving an appropriate inequality.

57. Suppose that Lance has \$500 to invest. If he invests \$300 at 9% interest, at what rate must he invest the remaining \$200 so that the two investments yield more than \$47 in yearly interest?

58. Mona invests \$100 at 8% yearly interest. How much does she have to invest at 9% so that the total yearly interest from the two investments exceeds \$26?

59. The average height of the two forwards and the center of a basketball team is 6 feet and 8 inches. What must the average height of the two guards be so that the team average is at least 6 feet and 4 inches?

60. Thanh has scores of 52, 84, 65, and 74 on his first four math exams. What score must he make on the fifth exam to have an average of 70 or better for the five exams?

61. Marsha bowled 142 and 170 in her first two games. What must she bowl in the third game to have an average of at least 160 for the three games?

62. Candace had scores of 95, 82, 93, and 84 on her first four exams of the semester. What score must she obtain on the fifth exam to have an average of 90 or better for the five exams?

63. Suppose that Derwin shot rounds of 82, 84, 78, and 79 on the first four days of a golf tournament. What must he shoot on the fifth day of the tournament to average 80 or less for the five days?

64. The temperatures for a 24-hour period ranged between $-4°F$ and $23°F$, inclusive. What was the range in Celsius degrees? $\left(\text{Use } F = \frac{9}{5}C + 32.\right)$

65. Oven temperatures for baking various foods usually range between 325°F and 425°F, inclusive. Express this range in Celsius degrees. (Round answers to the nearest degree.)

66. A person's intelligence quotient (I) is found by dividing mental age (M), as indicated by standard tests, by chronological age (C) and then multiplying this ratio by 100. The formula $I = \frac{100M}{C}$ can be used. If the I range of a group of 11-year-olds is given by $80 \leq I \leq 140$, find the range of the mental age of this group.

67. Repeat Problem 66 for an I range of 70 to 125, inclusive, for a group of 9-year-olds.

THOUGHTS INTO WORDS

68. Explain the difference between a conjunction and a disjunction. Give an example of each (outside the field of mathematics).

69. How do you know by inspection that the solution set of the inequality $x + 3 > x + 2$ is the entire set of real numbers?

70. Find the solution set for each of the following compound statements, and in each case explain your reasoning.

(a) $x < 3$ and $5 > 2$

(b) $x < 3$ or $5 > 2$

(c) $x < 3$ and $6 < 4$

(d) $x < 3$ or $6 < 4$

Answers to the Concept Quiz

1. True **2.** False **3.** True **4.** True **5.** True **6.** B **7.** E **8.** A **9.** D **10.** C

Answers to the Example Practice Skills

1. $(-12, \infty)$ **2.** $\left(-\infty, \frac{27}{7}\right]$ **3.** $(-20, \infty)$ **4.** $(-\infty, -1]$ **5.** $[2100, \infty)$ **6.** 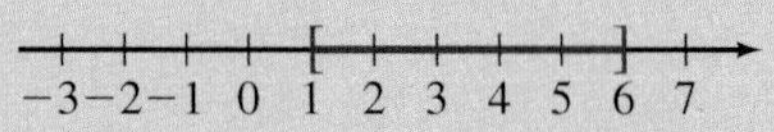

7. $(3, \infty)$ **8.** $(-\infty, 0] \cup [5, \infty)$ 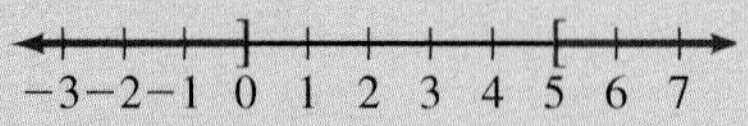 **9.** $(-\infty, 2) \cup (5, \infty)$

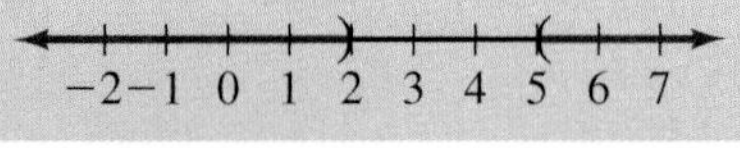

10. 88 or better **11.** Greater than 7% **12.** Between 10°C and 30°C inclusive

2.7 Equations and Inequalities Involving Absolute Value

OBJECTIVES

1. Solve Absolute Value Equations
2. Solve Absolute Value Inequalities

1 Solve Absolute Value Equations

In Section 1.2, we defined the absolute value of a real number by

$$|a| = \begin{cases} a & \text{if } a \geq 0 \\ -a & \text{if } a < 0 \end{cases}$$

We also interpreted the absolute value of any real number to be the distance between the number and zero on a number line. For example, $|6| = 6$ translates to 6 units between 6 and 0. Likewise, $|-8| = 8$ translates to 8 units between -8 and 0.

The interpretation of absolute value as distance on a number line provides a straightforward approach to solving a variety of equations and inequalities involving absolute value. First, let's consider some equations.

EXAMPLE 1

Solve $|x| = 2$.

Solution

Think in terms of distance between the number and zero, and you will see that x must be 2 or -2. That is, the equation $|x| = 2$ is equivalent to

$$x = -2 \quad \text{or} \quad x = 2$$

The solution set is $\{-2, 2\}$.

▼ PRACTICE YOUR SKILL

Solve $|x| = 7$. ■

EXAMPLE 2

Solve $|x + 2| = 5$.

Solution

The number represented by $x + 2$ must be equal to -5 or 5. Thus $|x + 2| = 5$ is equivalent to

$$x + 2 = -5 \quad \text{or} \quad x + 2 = 5$$

Solving each equation of the disjunction yields

$$x + 2 = -5 \quad \text{or} \quad x + 2 = 5$$

$$x = -7 \quad \text{or} \quad x = 3$$

The solution set is $\{-7, 3\}$.

✔ **Check**

$$|x + 2| = 5 \qquad |x + 2| = 5$$

$$|-7 + 2| \stackrel{?}{=} 5 \qquad |3 + 2| \stackrel{?}{=} 5$$

$$|-5| \stackrel{?}{=} 5 \qquad |5| \stackrel{?}{=} 5$$

$$5 = 5 \qquad 5 = 5$$

▼ **PRACTICE YOUR SKILL**

Solve $|x - 9| = 4$. ■

The following general property should seem reasonable given the distance interpretation of absolute value.

Property 2.1

$|x| = k$ is equivalent to $x = -k$ or $x = k$, where k is a positive number.

Example 3 demonstrates our format for solving equations of the form $|x| = k$.

EXAMPLE 3

Solve $|5x + 3| = 7$.

Solution

$$|5x + 3| = 7$$

$$5x + 3 = -7 \quad \text{or} \quad 5x + 3 = 7$$

$$5x = -10 \quad \text{or} \quad 5x = 4$$

$$x = -2 \quad \text{or} \quad x = \frac{4}{5}$$

The solution set is $\left\{-2, \frac{4}{5}\right\}$. Check these solutions!

▼ **PRACTICE YOUR SKILL**

Solve $|4x - 6| = 10$. ■

EXAMPLE 4

Solve $|2x + 5| - 3 = 8$.

Solution

First isolate the absolute value expression by adding 3 to both sides of the equation.

$$|2x + 5| - 3 = 8$$

$$|2x + 5| - 3 + 3 = 8 + 3$$

$$|2x + 5| = 11$$

$$2x + 5 = 11 \quad \text{or} \quad 2x + 5 = -11$$
$$2x = 6 \quad \text{or} \quad 2x = -16$$
$$x = 3 \quad \text{or} \quad x = -8$$

The solution set is $\{-8, 3\}$.

▼ PRACTICE YOUR SKILL

Solve $|x - 1| + 7 = 15$. ■

2 Solve Absolute Value Inequalities

The distance interpretation for absolute value also provides a good basis for solving some inequalities that involve absolute value. Consider the following examples.

EXAMPLE 5

Solve $|x| < 2$ and graph the solution set.

Solution

The number represented by x must be less than two units away from zero. Thus $|x| < 2$ is equivalent to

$$x > -2 \quad \text{and} \quad x < 2$$

The solution set is $(-2, 2)$, and its graph is shown in Figure 2.14.

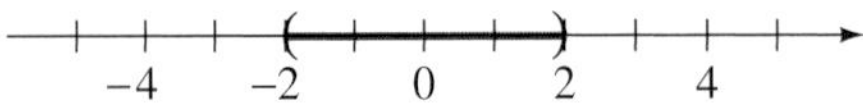

Figure 2.14

▼ PRACTICE YOUR SKILL

Solve $|x| < 4$ and graph the solution. ■

EXAMPLE 6

Solve $|x + 3| < 1$ and graph the solutions.

Solution

Let's continue to think in terms of distance on a number line. The number represented by $x + 3$ must be less than one unit away from zero. Thus $|x + 3| < 1$ is equivalent to

$$x + 3 > -1 \quad \text{and} \quad x + 3 < 1$$

Solving this conjunction yields

$$x + 3 > -1 \quad \text{and} \quad x + 3 < 1$$
$$x > -4 \quad \text{and} \quad x < -2$$

The solution set is $(-4, -2)$, and its graph is shown in Figure 2.15.

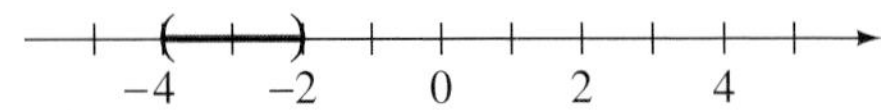

Figure 2.15

▼ PRACTICE YOUR SKILL

Solve $|x - 2| \leq 1$ and graph the solution. ■

Take another look at Examples 5 and 6. The following general property should seem reasonable.

Property 2.2

$|x| < k$ is equivalent to $x > -k$ and $x < k$, where k is a positive number.

Remember that we can write a conjunction such as $x > -k$ and $x < k$ in the compact form $-k < x < k$. The compact form provides a convenient format for solving inequalities such as $|3x - 1| < 8$, as Example 7 illustrates.

EXAMPLE 7

Solve $|3x - 1| < 8$ and graph the solutions.

Solution

$$|3x - 1| < 8$$

$$-8 < 3x - 1 < 8$$

$$-7 < 3x < 9 \quad \text{Add 1 to left side, middle, and right side}$$

$$\frac{1}{3}(-7) < \frac{1}{3}(3x) < \frac{1}{3}(9) \quad \text{Multiply through by } \frac{1}{3}$$

$$-\frac{7}{3} < x < 3$$

The solution set is $\left(-\frac{7}{3}, 3\right)$, and its graph is shown in Figure 2.16.

Figure 2.16

▼ **PRACTICE YOUR SKILL**

Solve $|2x - 3| < 9$ and graph the solution. ■

The distance interpretation also clarifies a property that pertains to *greater than* situations involving absolute value. Consider the following examples.

EXAMPLE 8

Solve $|x| > 1$ and graph the solutions.

Solution

The number represented by x must be more than one unit away from zero. Thus $|x| > 1$ is equivalent to

$$x < -1 \quad \text{or} \quad x > 1$$

The solution set is $(-\infty, -1) \cup (1, \infty)$, and its graph is shown in Figure 2.17.

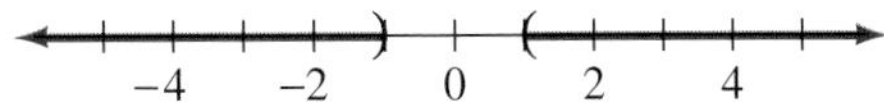

Figure 2.17

▼ **PRACTICE YOUR SKILL**

Solve $|x| > 4$ and graph the solution. ■

EXAMPLE 9 Solve $|x - 1| > 3$ and graph the solutions.

Solution

The number represented by $x - 1$ must be more than three units away from zero. Thus $|x - 1| > 3$ is equivalent to

$$x - 1 < -3 \quad \text{or} \quad x - 1 > 3$$

Solving this disjunction yields

$$x - 1 < -3 \quad \text{or} \quad x - 1 > 3$$

$$x < -2 \quad \text{or} \quad x > 4$$

The solution set is $(-\infty, -2) \cup (4, \infty)$, and its graph is shown in Figure 2.18.

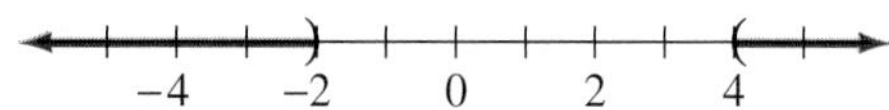

Figure 2.18

▼ **PRACTICE YOUR SKILL**

Solve $|x + 2| \geq 4$ and graph the solution. ■

Examples 8 and 9 illustrate the following general property.

Property 2.3

$|x| > k$ is equivalent to $x < -k$ or $x > k$, where k is a positive number.

Therefore, solving inequalities of the form $|x| > k$ can take the format shown in Example 10.

EXAMPLE 10 Solve $|3x - 1| + 4 > 6$ and graph the solution.

Solution

First isolate the absolute value expression by subtracting 4 from both sides of the equation.

$$|3x - 1| + 4 > 6$$

$$|3x - 1| + 4 - 4 > 6 - 4 \qquad \text{Subtract 4 from both sides}$$

$$|3x - 1| > 2$$

$$3x - 1 < -2 \quad \text{or} \quad 3x - 1 > 2$$

$$3x < -1 \quad \text{or} \quad 3x > 3$$

$$x < -\frac{1}{3} \quad \text{or} \quad x > 1$$

The solution set is $\left(-\infty, -\frac{1}{3}\right) \cup (1, \infty)$, and its graph is shown in Figure 2.19.

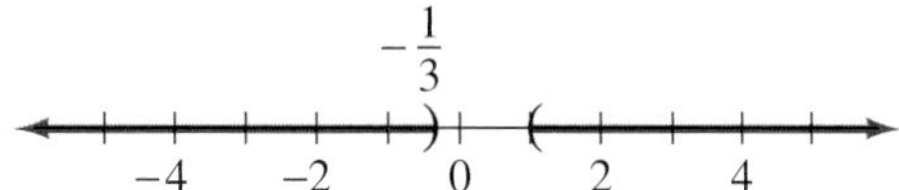

Figure 2.19

▼ **PRACTICE YOUR SKILL**

Solve $|2x - 5| - 3 > 1$ and graph the solution. ■

Properties 2.1, 2.2, and 2.3 provide the basis for solving a variety of equations and inequalities that involve absolute value. However, if at any time you become doubtful about what property applies, don't forget the distance interpretation. Furthermore, note that in each of the properties, k is a positive number. If k is a nonpositive number, then we can determine the solution sets by inspection, as indicated by the following examples.

$|x + 3| = 0$ has a solution of $x = -3$, because the number $x + 3$ has to be 0. The solution set of $|x + 3| = 0$ is $\{-3\}$.

$|2x - 5| = -3$ has no solutions, because the absolute value (distance) cannot be negative. The solution set is $\varnothing$, the null set.

$|x - 7| < -4$ has no solutions, because we cannot obtain an absolute value less than -4. The solution set is $\varnothing$.

$|2x - 1| > -1$ is satisfied by all real numbers because the absolute value of $(2x - 1)$, regardless of what number is substituted for x, will always be greater than -1. The solution set is the set of all real numbers, which we can express in interval notation as $(-\infty, \infty)$.

CONCEPT QUIZ

For Problems 1–10, answer true or false.

1. The absolute value of a negative number is the opposite of the number.
2. The absolute value of a number is always positive or zero.
3. The absolute value of a number is equal to the absolute value of its opposite.
4. The compound statement $x < 1$ or $x > 3$ can be written in compact form $3 < x < 1$.
5. The solution set for the equation $|x + 5| = 0$ is the null set, $\varnothing$
6. The solution set for $|x - 2| \geq -6$ is all real numbers.
7. The solution set for $|x + 1| < -3$ is all real numbers.
8. The solution set for $|x - 4| \leq 0$ is $|4|$.
9. If a solution set in interval notation is $(-4, -2)$, then it can be expressed as $\{x \mid -4 < x < -2\}$ in set builder notation.
10. If a solution set in interval notation is $(-\infty, -2) \cup (4, \infty)$, then it can be expressed as $\{x \mid x < -2 \text{ or } x > 4\}$ in set builder notation.

Problem Set 2.7

1 Solve Absolute Value Equations

For Problems 1–16, solve each equation.

1. $|x - 1| = 8$
2. $|x + 2| = 9$
3. $|2x - 4| = 6$
4. $|3x - 4| = 14$
5. $|3x + 4| = 11$
6. $|5x - 7| = 14$
7. $|4 - 2x| = 6$
8. $|3 - 4x| = 8$
9. $\left|x - \frac{3}{4}\right| = \frac{2}{3}$
10. $\left|x + \frac{1}{2}\right| = \frac{3}{5}$
11. $|2x - 3| + 2 = 5$
12. $|3x - 1| - 1 = 9$
13. $|x + 2| - 6 = -2$
14. $|x - 3| - 4 = -1$
15. $|4x - 3| + 2 = 2$
16. $|5x + 1| + 4 = 4$

2 Solve Absolute Value Inequalities

For Problems 17–30, solve each inequality and graph the solutions.

17. $|x| < 5$

18. $|x| < 1$

19. $|x| \leq 2$

20 $|x| \leq 4$

21. $|x| > 2$

22. $|x| > 3$

23. $|x - 1| < 2$

24. $|x - 2| < 4$

25. $|x + 2| \leq 4$

26. $|x + 1| \leq 1$

27. $|x + 2| > 1$

28. $|x + 1| > 3$

29. $|x - 3| \geq 2$

30. $|x - 2| \geq 1$

For Problems 31–54, solve each inequality.

31. $|x - 2| > 6$

32. $|x - 3| > 9$

33. $|x + 3| < 5$

34. $|x + 1| < 8$

35. $|2x - 1| \leq 9$

36. $|3x + 1| \leq 13$

37. $|4x + 2| \geq 12$

38. $|5x - 2| \geq 10$

39. $|2 - x| > 4$

40. $|4 - x| > 3$

41. $|1 - 2x| < 2$

42. $|2 - 3x| < 5$

43. $|5x + 9| \leq 16$

44. $|7x - 6| \geq 22$

45. $|-2x + 7| \leq 13$

46. $|-3x - 4| \leq 15$

47. $\left|\frac{x - 3}{4}\right| < 2$

48. $\left|\frac{x + 2}{3}\right| < 1$

49. $\left|\frac{2x + 1}{2}\right| > 1$

50. $\left|\frac{3x - 1}{4}\right| > 3$

51. $|x + 7| - 3 \geq 4$

52. $|x - 2| + 4 \geq 10$

53. $|2x - 1| + 1 \leq 6$

54. $|4x + 3| - 2 \leq 5$

For Problems 55–64, solve each equation and inequality *by inspection*.

55. $|2x + 1| = -4$

56. $|5x - 1| = -2$

57. $|3x - 1| > -2$

58. $|4x + 3| < -4$

59. $|5x - 2| = 0$

60. $|3x - 1| = 0$

61. $|4x - 6| < -1$

62. $|x + 9| > -6$

63. $|x + 4| < 0$

64. $|x + 6| > 0$

THOUGHTS INTO WORDS

65. Explain how you would solve the inequality $|2x + 5| > -3$.

66. Why is 2 the only solution for $|x - 2| \leq 0$?

67. Explain how you would solve the equation $|2x - 3| = 0$.

FURTHER INVESTIGATIONS

Consider the equation $|x| = |y|$. This equation will be a true statement if x is equal to y or if x is equal to the opposite of y. Use the following format, $x = y$ or $x = -y$, to solve the equations in Problems 68–73.

For Problems 68–73, solve each equation.

68. $|3x + 1| = |2x + 3|$

69. $|-2x - 3| = |x + 1|$

70. $|2x - 1| = |x - 3|$

71. $|x - 2| = |x + 6|$

72. $|x + 1| = |x - 4|$

73. $|x + 1| = |x - 1|$

74. Use the definition of absolute value to help prove Property 2.1.

75. Use the definition of absolute value to help prove Property 2.2.

76. Use the definition of absolute value to help prove Property 2.3.

Answers to the Concept Quiz

1. True **2.** True **3.** True **4.** False **5.** False **6.** True **7.** False **8.** True **9.** True **10.** True

Answers to the Example Practice Skills

1. $\{-7, 7\}$ **2.** $\{5, 13\}$ **3.** $\{-1, 4\}$ **4.** $\{-7, 9\}$ **5.** $(-4, 4)$

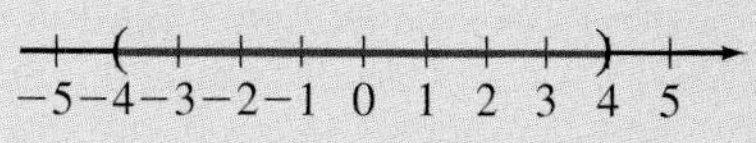

−5 −4 −3 −2 −1 0 1 2 3 4 5

6. $[1, 3]$

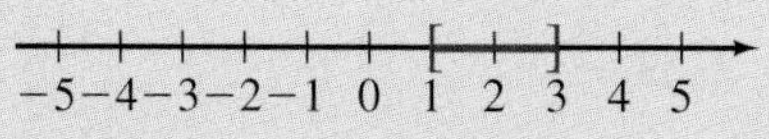

−5 −4 −3 −2 −1 0 1 2 3 4 5

7. $(-3, 6)$

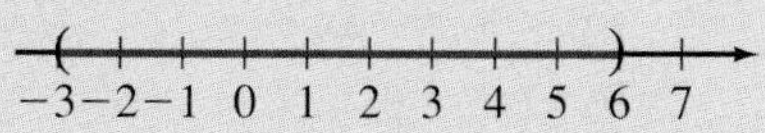

−3 −2 −1 0 1 2 3 4 5 6 7

8. $(-\infty, -4) \cup (4, \infty)$

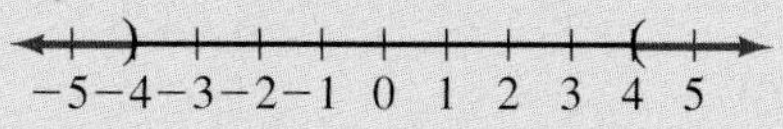

−5 −4 −3 −2 −1 0 1 2 3 4 5

9. $(-\infty, -6] \cup [2, \infty)$

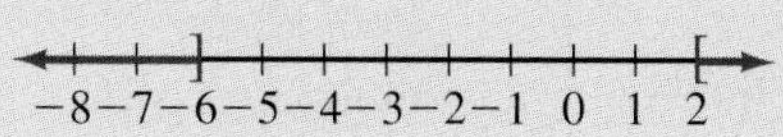

−8 −7 −6 −5 −4 −3 −2 −1 0 1 2

10. $\left(-\infty, \frac{1}{2}\right) \cup \left(\frac{9}{2}, \infty\right)$

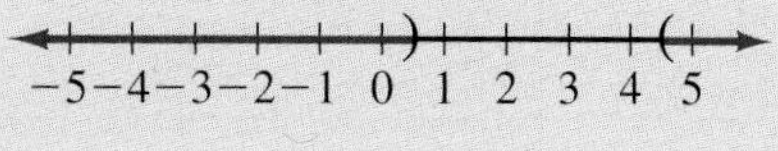

−5 −4 −3 −2 −1 0 1 2 3 4 5

Chapter 2 Summary

OBJECTIVE	SUMMARY	EXAMPLE	CHAPTER REVIEW PROBLEMS
Solve first-degree equations. (Sec. 2.1, Obj. 1, p. 50)	Solving an algebraic equation refers to the process of finding the number (or numbers) that make(s) the algebraic equation a true numerical statement. Two properties of equality play an important role in solving equations. **Addition Property of Equality** $a = b$ if and only if $a + c = b + c$. **Multiplication Property of Equality** For $c \neq 0$, $a = b$ if and only if $ac = bc$.	Solve $3(2x - 1) = 2x + 6 - 5x$. **Solution** $3(2x - 1) = 2x + 6 - 5x$ $6x - 3 = -3x + 6$ $9x - 3 = 6$ $9x = 9$ $x = 1$ The solution set is $\{1\}$.	Problems 1–4
Solve equations involving fractions. (Sec. 2.2, Obj. 1, p. 58)	It is usually easiest to begin by multiplying both sides of the equation by the least common multiple of all the denominators in the equation. This process clears the equation of fractions.	Solve $\frac{x}{2} - \frac{x}{5} = \frac{7}{10}$. **Solution** $\frac{x}{2} - \frac{x}{5} = \frac{7}{10}$ $10\left(\frac{x}{2} - \frac{x}{5}\right) = 10\left(\frac{7}{10}\right)$ $10\left(\frac{x}{2}\right) - 10\left(\frac{x}{5}\right) = 7$ $5x - 2x = 7$ $3x = 7$ $x = \frac{7}{3}$ The solution set is $\left\{\frac{7}{3}\right\}$.	Problems 5–10
Solve equations involving decimals. (Sec. 2.3, Obj. 1, p. 66)	To solve equations that contain decimals, you can clear the equation of the decimals by multiplying both sides by an appropriate power of 10, or you can keep the problem in decimal form and perform the calculations with decimals.	Solve $0.04x + 0.07(2x) = 90$. **Solution** $0.04x + 0.07(2x) = 90$ $100[0.04x + 0.07(2x)] = 100(90)$ $4x + 7(2x) = 9000$ $4x + 14x = 9000$ $18x = 9000$ $x = 500$ The solution set is $\{500\}$.	Problems 11–14 *(continued)*

OBJECTIVE	SUMMARY	EXAMPLE	CHAPTER REVIEW PROBLEMS
Use equations to solve word problems. (Sec. 2.1, Obj. 2, p. 53; Sec. 2.2, Obj. 2, p. 60)	Keep the following suggestions in mind as you solve word problems. 1. Read the problem carefully. 2. Sketch any figure, diagram, or chart that might be helpful. 3. Choose a meaningful variable. 4. Look for a guideline. 5. Form an equation. 6. Solve the equation. 7. Check your answers.	The length of a rectangle is 4 feet less than twice the width. The perimeter of the rectangle is 34 feet. Find the length and width. **Solution** Let w represent the width; then $2w - 4$ represents the length. Use the formula $P = 2w + 2l$. $34 = 2w + 2(2w - 4)$ $34 = 2w + 4w - 8$ $42 = 6w$ $7 = w$ So the width is 7 feet and the length is $2(7) - 4 = 10$ feet.	Problems 15–20
Solve word problems involving discount and selling price. (Sec. 2.3, Obj. 2, p. 68)	Discount sale problems involve the relationship *original selling price minus discount equals sale price.* Another basic relationship is *selling price equals cost plus profit.* Profit may be stated as a percent of the selling price, as a percent of the cost, or as an amount.	A car repair shop has some brake pads that cost \$30 each. He wants to sell them at a profit of 70% of the cost. What selling price will be charged to the customer? **Solution** Selling price = Cost + Profit $s = 30 + (60\%)(30)$ $s = 30 + (0.60)(30)$ $s = 30 + 18 = 48$ The selling price would be \$48.00.	Problems 21–24
Evaluate formulas for given values. (Sec. 2.4, Obj. 1, p. 75)	A formula can be solved for a specific variable when we are given the numerical values for the other variables.	Solve $i = Prt$ for r, given that $P = \$1200$, $t = 4$ years, and $i = \$360$. **Solution** $i = Prt$ $360 = (1200)(r)(4)$ $360 = 4800r$ $r = \frac{360}{4800} = 0.075$ The rate would be 7.5%.	Problems 25–28
Solve formulas for a specified variable. (Sec. 2.4, Obj. 2, p. 77)	We can change the form of an equation by solving for one variable in terms of the other variables.	Solve $A = \frac{1}{2}bh$ for b. **Solution** $A = \frac{1}{2}bh$ $2A = 2\left(\frac{1}{2}bh\right)$ $2A = bh$ $\frac{2A}{h} = b$	Problems 29–38 *(continued)*

OBJECTIVE	SUMMARY	EXAMPLE	CHAPTER REVIEW PROBLEMS
Use formulas to solve problems. (Sec. 2.4, Obj. 3, p. 79)	Formulas are often used as guidelines for setting up an algebraic equation when solving a word problem. Sometimes formulas are used in the analysis of a problem but not as the main guideline. For example, uniform motion problems use the formula $d = rt$, but the guideline is usually a statement about times, rates, or distances.	How long will it take \$400 to triple if it is invested at 8% simple interest? **Solution** Use the formula $i = Prt$. For \$400 to triple (be worth \$1200), it must earn \$800 in interest. $800 = 400(8\%)(t)$ $800 = 400(0.08)(t)$ $2 = 0.08t$ $t = \frac{2}{0.08} = 25$ It will take 25 years to triple.	Problems 39–42
Write solution sets in interval notation. (Sec. 2.5, Obj. 1, p. 87)	The solution set for an algebraic inequality can be written in interval notation. See Figure 2.20 below for examples of various algebraic inequalities and how their solution sets would be written in interval notation.	Express the solution set for $x \leq 4$ in interval notation. **Solution** For the solution set we want all numbers less than or equal to 4. In interval notation, the solution set is written $(-\infty, 4]$.	Problems 43–46

Solution set	Graph	Interval notation
$\{x \mid x > 1\}$	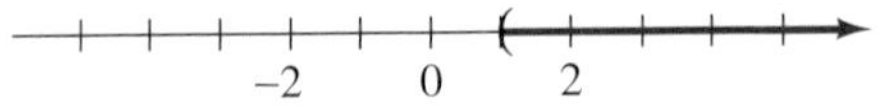 -2 0 2	$(1, \infty)$
$\{x \mid x \geq 2\}$	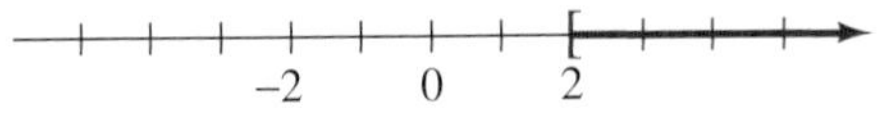 -2 0 2	$[2, \infty)$
$\{x \mid x < 0\}$	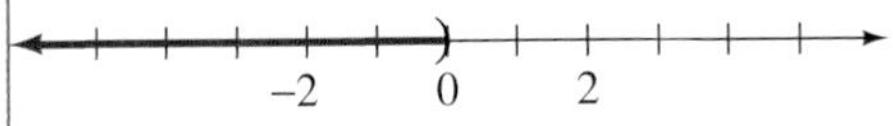 -2 0 2	$(-\infty, 0)$
$\{x \mid x \leq -1\}$	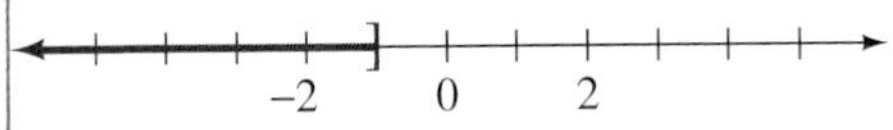 -2 0 2	$(-\infty, -1]$
$\{x \mid -2 < x \leq 2\}$	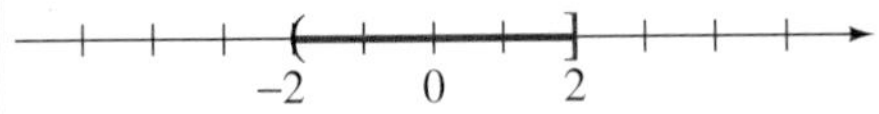 -2 0 2	$(-2, 2]$
$\{x \mid x \leq -1 \text{ or } x > 1\}$	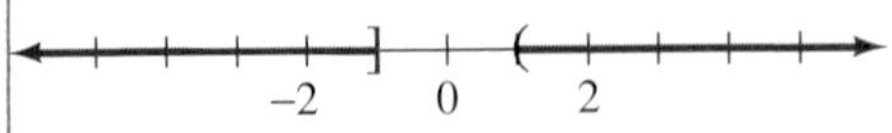 -2 0 2	$(-\infty, -1] \cup (1, \infty)$

Figure 2.20

(continued)

OBJECTIVE	SUMMARY	EXAMPLE	CHAPTER REVIEW PROBLEMS
Solve inequalities. (Sec. 2.5, Obj. 2, p. 88)	The addition property of equality states that any number can be added to each side of an inequality to produce an equivalent inequality. The multiplication property of equality states that both sides of an inequality can be multiplied by a positive number to produce an equivalent inequality. If both sides of an inequality are multiplied by a negative number, then an inequality of the *opposite sense* is produced. When multiplying or dividing both sides of an inequality by a negative number, be sure to reverse the inequality symbol.	Solve $-8x + 2(x - 7) < 40$. **Solution** $-8x + 2(x - 7) < 40$ $-8x + 2x - 14 < 40$ $-6x - 14 < 40$ $-6x < 54$ $\frac{-6x}{-6} > \frac{54}{-6}$ $x > -9$ The solution set is $(-9, \infty)$.	Problems 47–51
Solve inequalities involving fractions or decimals. (Sec. 2.6, Obj. 1, p. 94)	When solving inequalities that involve fractions, usually the inequality is multiplied by the least common multiple of all the denominators to clear the equation of fractions. The same technique can be used for inequalities involving decimals.	Solve $\frac{x+5}{3} - \frac{x+1}{2} < \frac{5}{6}$. **Solution** Multiply both sides of the inequality by 6. $6\left(\frac{x+5}{3} - \frac{x+1}{2}\right) < 6\left(\frac{5}{6}\right)$ $2(x + 5) - 3(x + 1) < 5$ $2x + 10 - 3x - 3 < 5$ $-x + 7 < 5$ $-x < -2$ $-1(-x) > -1(-2)$ $x > 2$ The solution set is $(2, \infty)$.	Problems 52–56
Solve inequalities that are compound statements. (Sec. 2.6, Obj. 2, p. 97)	Inequalities connected with the word "and" form a compound statement called a conjunction. A conjunction is true only if all of its component parts are true. Inequalities connected with the word "or" form a compound statement called a disjunction. A disjunction is true if at least one of its component parts is true.	Solve the compound statement $x + 4 \leq -10$ or $x - 2 \geq 1$. **Solution** Simplify each inequality. $x + 4 \leq -10$ or $x - 2 \geq 1$ $x \leq -14$ or $x \geq 3$ The solution set is $(-\infty, -14] \cup [3, \infty)$.	Problems 57–64 *(continued)*

OBJECTIVE	SUMMARY	EXAMPLE	CHAPTER REVIEW PROBLEMS
Use inequalities to solve word problems. (Sec. 2.6, Obj. 3, p. 100)	To solve word problems involving inequalities, use the same suggestions given for solving word problems; however, the guideline will translate into an inequality rather than an equation.	Cheryl bowled 156 and 180 in her first two games. What must she bowl in the third game to have an average of at least 170 for the three games? **Solution** Let s represent the score in the third game. $\frac{156 + 180 + s}{3} \geq 170$ $156 + 180 + s \geq 510$ $336 + s \geq 510$ $s \geq 174$ She must bowl 174 or greater.	Problems 65–66
Solve absolute value equations. (Sec. 2.7, Obj. 1, p. 104)	Property 2.1 states that $\|x\| = k$ is equivalent to $x = k$ or $x = -k$, where k is a positive number. This property is applied to solve absolute value equations.	Solve $\|2x - 5\| = 9$. **Solution** $\|2x - 5\| = 9$ $2x - 5 = 9$ or $2x - 5 = -9$ $2x = 14$ or $2x = -4$ $x = 7$ or $x = -2$ The solution set is $\{-2, 7\}$.	Problems 67–70
Solve absolute value inequalities. (Sec. 2.7, Obj. 2, p. 106)	Property 2.2 states that $\|x\| < k$ is equivalent to $x > -k$ and $x < k$, where k is a positive number. This conjunction can be written in compact form as $-k < x < k$. For example, $\|x + 3\| < 7$ can be written as $-7 < x + 3 < 7$ to begin the process of solving the inequality. Property 2.3 states that $\|x\| > k$ is equivalent to $x < -k$ or $x > k$, where k is a positive number. This disjunction cannot be written in a compact form.	Solve $\|x + 5\| > 8$. **Solution** $\|x + 5\| > 8$ $x + 5 < -8$ or $x + 5 > 8$ $x < -13$ or $x > 3$ The solution set is $(-\infty, -13) \cup (3, \infty)$.	Problems 71–74

Chapter 2 Review Problem Set

For Problems 1–15, solve each of the equations.

1. $5(x - 6) = 3(x + 2)$
2. $2(2x + 1) - (x - 4) = 4(x + 5)$
3. $-(2n - 1) + 3(n + 2) = 7$
4. $2(3n - 4) + 3(2n - 3) = -2(n + 5)$
5. $\frac{3t - 2}{4} = \frac{2t + 1}{3}$
6. $\frac{x + 6}{5} + \frac{x - 1}{4} = 2$
7. $1 - \frac{2x - 1}{6} = \frac{3x}{8}$
8. $\frac{2x + 1}{3} + \frac{3x - 1}{5} = \frac{1}{10}$
9. $\frac{3n - 1}{2} - \frac{2n + 3}{7} = 1$
10. $\frac{5x + 6}{2} - \frac{x - 4}{3} = \frac{5}{6}$
11. $0.06x + 0.08(x + 100) = 15$
12. $0.4(t - 6) = 0.3(2t + 5)$
13. $0.1(n + 300) = 0.09n + 32$
14. $0.2(x - 0.5) - 0.3(x + 1) = 0.4$

Solve each of Problems 15–24 by setting up and solving an appropriate equation.

15. The width of a rectangle is 2 meters more than one-third of the length. The perimeter of the rectangle is 44 meters. Find the length and width of the rectangle.
16. Find three consecutive integers such that the sum of one-half of the smallest and one-third of the largest is 1 less than the other integer.
17. Pat is paid time-and-a-half for each hour he works over 36 hours in a week. Last week he worked 42 hours for a total of \$472.50. What is his normal hourly rate?
18. Marcela has a collection of nickels, dimes, and quarters worth \$24.75. The number of dimes is 10 more than twice the number of nickels, and the number of quarters is 25 more than the number of dimes. How many coins of each kind does she have?
19. If the complement of an angle is one-tenth of the supplement of the angle, find the measure of the angle.
20. A total of \$500 was invested, part of it at 7% interest and the remainder at 8%. If the total yearly interest from both investments amounted to \$38, how much was invested at each rate?
21. A retailer has some sweaters that cost her \$38 each. She wants to sell them at a profit of 20% of her cost. What price should she charge for each sweater?
22. If a necklace cost a jeweler \$60, at what price should it be sold to yield a profit of 80% based on the selling price?
23. If a DVD player costs a retailer \$40 and they sell for \$100, what is the rate of profit based on the selling price?
24. Yuri bought a pair of running shoes at a 25% discount sale for \$48. What was the original price of the running shoes?
25. Solve $i = Prt$ for P, given that $r = 6\%$, $t = 3$ years, and $i = \$1440$.
26. Solve $A = P + Prt$ for r, given that $A = \$3706$, $P = \$3400$, and $t = 2$ years. Express r as a percent.
27. Solve $P = 2w + 2l$ for w, given that $P = 86$ meters and $l = 32$ meters.
28. Solve $C = \frac{5}{9}(F - 32)$ for C, given that $F = -4°$.

For Problems 29–33, solve each equation for x.

29. $ax - b = b + 2$
30. $ax = bx + c$
31. $m(x + a) = p(x + b)$
32. $5x - 7y = 11$
33. $\frac{x - a}{b} = \frac{y + 1}{c}$

For Problems 34–38, solve each of the formulas for the indicated variable.

34. $A = \pi r^2 + \pi rs$ for s
35. $A = \frac{1}{2}h(b_1 + b_2)$ for b_2
36. $S_n = \frac{n(a_1 + a_2)}{2}$ for n
37. $\frac{1}{R} = \frac{1}{R_1} + \frac{1}{R_2}$ for R
38. $ax + by = c$ for y
39. How many pints of a 1% hydrogen peroxide solution should be mixed with a 4% hydrogen peroxide solution to obtain 10 pints of a 2% hydrogen peroxide solution?

40. Gladys leaves a town driving at a rate of 40 miles per hour. Two hours later, Reena leaves from the same place traveling the same route. She catches Gladys in 5 hours and 20 minutes. How fast was Reena traveling?

41. In $1\frac{1}{4}$ hours more time, Rita, riding her bicycle at 12 miles per hour rode 2 miles farther than Sonya, who was riding her bicycle at 16 miles per hour. How long did each girl ride?

42. How many cups of orange juice must be added to 50 cups of a punch that is 10% orange juice to obtain a punch that is 20% orange juice?

For Problems 43–46, express the given inequality in interval notation.

43. $x \geq -2$

44. $x > 6$

45. $x < -1$

46. $x \leq 0$

For Problems 47–56, solve each of the inequalities.

47. $5x - 2 \geq 4x - 7$

48. $3 - 2x < -5$

49. $2(3x - 1) - 3(x - 3) > 0$

50. $3(x + 4) \leq 5(x - 1)$

51. $-3(2t - 1) - (t + 2) > -6(t - 3)$

52. $\frac{5}{6}n - \frac{1}{3}n < \frac{1}{6}$

53. $\frac{n-4}{5} + \frac{n-3}{6} > \frac{7}{15}$

54. $\frac{2}{3}(x - 1) + \frac{1}{4}(2x + 1) < \frac{5}{6}(x - 2)$

55. $s \geq 4.5 + 0.25s$

56. $0.07x + 0.09(500 - x) \geq 43$

For Problems 57–64, graph the solutions of each compound inequality.

57. $x > -1$ and $x < 1$

58. $x > 2$ or $x \leq -3$

59. $x > 2$ and $x > 3$

60. $x < 2$ or $x > -1$

61. $2x + 1 > 3$ or $2x + 1 < -3$

62. $2 \leq x + 4 \leq 5$

63. $-1 < 4x - 3 \leq 9$

64. $x + 1 > 3$ and $x - 3 < -5$

65. Susan's average score for her first three psychology exams is 84. What must she get on the fourth exam so that her average for the four exams is 85 or better?

66. Marci invests $3000 at 6% yearly interest. How much does she have to invest at 8% so that the yearly interest from the two investments exceeds $500?

For Problems 67–70, solve each of the equations.

67. $|3x - 1| = 11$

68. $|2n + 3| = 4$

69. $|3x + 1| - 8 = 2$

70. $\left|\frac{1}{2}x + 3\right| - 1 = 5$

For Problems 71–74, solve each of the inequalities.

71. $|2x - 1| < 11$

72. $|3x + 1| > 10$

73. $|5x - 4| \geq 8$

74. $\left|\frac{1}{4}x + 1\right| \leq 6$

Chapter 2 Test

For Problems 1–10, solve each equation.

1. $5x - 2 = 2x - 11$ — 1. ______

2. $6(n - 2) - 4(n + 3) = -14$ — 2. ______

3. $-3(x + 4) = 3(x - 5)$ — 3. ______

4. $3(2x - 1) - 2(x + 5) = -(x - 3)$ — 4. ______

5. $\frac{3t - 2}{4} = \frac{5t + 1}{5}$ — 5. ______

6. $\frac{5x + 2}{3} - \frac{2x + 4}{6} = -\frac{4}{3}$ — 6. ______

7. $|4x - 3| = 9$ — 7. ______

8. $\frac{1 - 3x}{4} + \frac{2x + 3}{3} = 1$ — 8. ______

9. $2 - \frac{3x - 1}{5} = -4$ — 9. ______

10. $0.05x + 0.06(1500 - x) = 83.5$ — 10. ______

11. Solve $\frac{2}{3}x - \frac{3}{4}y = 2$ for y — 11. ______

12. Solve $S = 2\pi r(r + h)$ for h — 12. ______

For Problems 13–20, solve each inequality and express the solution set using interval notation.

13. $7x - 4 > 5x - 8$ — 13. ______

14. $-3x - 4 \leq x + 12$ — 14. ______

15. $2(x - 1) - 3(3x + 1) \geq -6(x - 5)$ — 15. ______

16. $\frac{3}{5}x - \frac{1}{2}x < 1$ — 16. ______

17. $\frac{x - 2}{6} - \frac{x + 3}{9} > -\frac{1}{2}$ — 17. ______

18. $0.05x + 0.07(800 - x) \geq 52$ — 18. ______

19. $|6x - 4| < 10$ — 19. ______

20. $|4x + 5| \geq 6$ — 20. ______

For Problems 21–25, solve each problem by setting up and solving an appropriate equation or inequality.

21. Dela bought a dress at a 20% discount sale for \$57.60. Find the original price of the dress. — 21. ______

22. The length of a rectangle is 1 centimeter more than three times its width. If the perimeter of the rectangle is 50 centimeters, find the length of the rectangle. — 22. ______

23. ______________________

24. ______________________

25. ______________________

23. How many cups of grapefruit juice must be added to 30 cups of a punch that is 8% grapefruit juice to obtain a punch that is 10% grapefruit juice?

24. Rex has scores of 85, 92, 87, 88, and 91 on the first five exams. What score must he make on the sixth exam to have an average of 90 or better for all six exams?

25. If the complement of an angle is $\frac{2}{11}$ of the supplement of the angle, find the measure of the angle.

Linear Equations and Inequalities in Two Variables

3

René Descartes, a philosopher and mathematician, developed a system for locating a point on a plane. This system is our current rectangular coordinate grid used for graphing; it is named the Cartesian coordinate system.

René Descartes, a French mathematician of the 17th century, was able to transform geometric problems into an algebraic setting so that he could use the tools of algebra to solve the problems. This connecting of algebraic and geometric ideas is the foundation of a branch of mathematics called **analytic geometry**, today more commonly called **coordinate geometry**. Basically, there are two kinds of problems in coordinate geometry: Given an algebraic equation, find its geometric graph; and given a set of conditions pertaining to a geometric graph, find its algebraic equation. We discuss problems of both types in this chapter.

Video tutorials for all section learning objectives are available in a variety of delivery modes.

INTERNET PROJECT

In this chapter the rectangular coordinate system is used for graphing. Another two-dimensional coordinate system is the polar coordinate system. Conduct an Internet search to see an example of the polar coordinate system. How are the coordinates of a point determined in the polar coordinate system?

3.1 Rectangular Coordinate System and Linear Equations

OBJECTIVES

1. Find Solutions for Linear Equations in Two Variables
2. Review of the Rectangular Coordinate System
3. Graph the Solutions for Linear Equations
4. Graph Linear Equations by Finding the x and y Intercepts
5. Graph Lines Passing through the Origin, Vertical Lines, and Horizontal Lines
6. Apply Graphing to Linear Relationships
7. Introduce Graphing Utilities (Optional Exercises)

1 Find Solutions for Linear Equations in Two Variables

In this chapter we want to consider solving equations in two variables. Let's begin by considering the solutions for the equation $y = 3x + 2$. A **solution** of an equation in two variables is an ordered pair of real numbers that satisfies the equation. When using the variables x and y, we agree that the first number of an ordered pair is a value of x and the second number is a value of y. We see that (1, 5) is a solution for $y = 3x + 2$ because if x is replaced by 1 and y by 5, the result is the true numerical statement $5 = 3(1) + 2$. Likewise, (2, 8) is a solution because $8 = 3(2) + 2$ is a true numerical statement. We can find infinitely many pairs of real numbers that satisfy $y = 3x + 2$ by arbitrarily choosing values for x and then, for each chosen value of x, determining a corresponding value for y. Let's use a table to record some of the solutions for $y = 3x + 2$.

x value	y value determined from $y = 3x + 2$	Ordered pairs
−3	−7	(−3, −7)
−1	−1	(−1, −1)
0	2	(0, 2)
1	5	(1, 5)
2	8	(2, 8)
4	14	(4, 14)

EXAMPLE 1

Determine some ordered-pair solutions for the equation $y = 2x - 5$ and record the values in a table.

Solution

We can start by arbitrarily choosing values for x and then determine the corresponding y value. Even though you can arbitrarily choose values for x, it is good practice to choose some negative values, zero, and some positive values.

Let $x = -4$; then, according to our equation, $y = 2(-4) - 5 = -13$.

Let $x = -1$; then, according to our equation, $y = 2(-1) - 5 = -7$.

Let $x = 0$; then, according to our equation, $y = 2(0) - 5 = -5$.

Let $x = 2$; then, according to our equation, $y = 2(2) - 5 = -1$.

Let $x = 4$; then, according to our equation, $y = 2(4) - 5 = 3$.

Organizing this information in a chart gives the following table.

x value	y value determined from $y = 2x - 5$	Ordered pair
-4	-13	$(-4, -13)$
-1	-7	$(-1, -7)$
0	-5	$(0, -5)$
2	-1	$(2, -1)$
4	3	$(4, 3)$

▼ **PRACTICE YOUR SKILL**

Determine the ordered-pair solutions for the equation $y = -2x + 4$ for the x values of $-4, -2, 0, 1$, and 3. Organize the information into a table. ■

A table can show some of the infinite number of solutions for a linear equation in two variables, but for a visual display, solutions are plotted on a coordinate system. Let's review the rectangular coordinate system and then we can use a graph to display the solutions of an equation in two variables.

2 Review of the Rectangular Coordinate System

Consider two number lines, one vertical and one horizontal, perpendicular to each other at the point we associate with zero on both lines (Figure 3.1). We refer to these number lines as the **horizontal and vertical axes** or, together, as the **coordinate axes**. They partition the plane into four regions called **quadrants**. The quadrants are numbered counterclockwise from I through IV as indicated in Figure 3.1. The point of intersection of the two axes is called the **origin**.

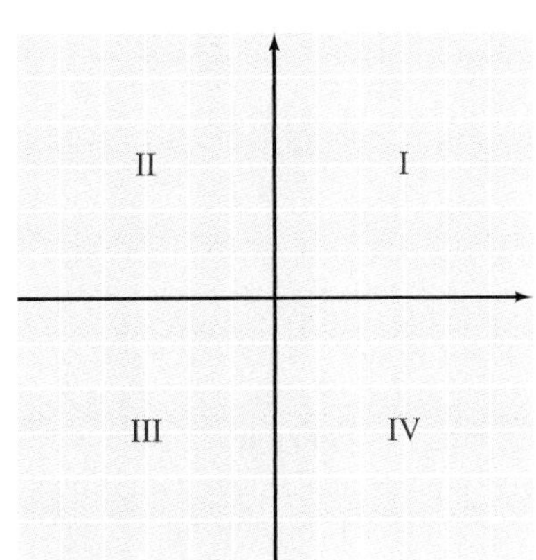

Figure 3.1

It is now possible to set up a one-to-one correspondence between **ordered pairs** of real numbers and the points in a plane. To each ordered pair of real numbers there corresponds a unique point in the plane, and to each point in the plane there corresponds a unique ordered pair of real numbers. A part of this correspondence is illustrated in Figure 3.2. The ordered pair (3, 2) means that the point A is located three units to the right of, and two units up from, the origin. (The ordered pair (0, 0) is associated with the origin O.) The ordered pair $(-3, -5)$ means that the point D is located three units to the left and five units down from the origin.

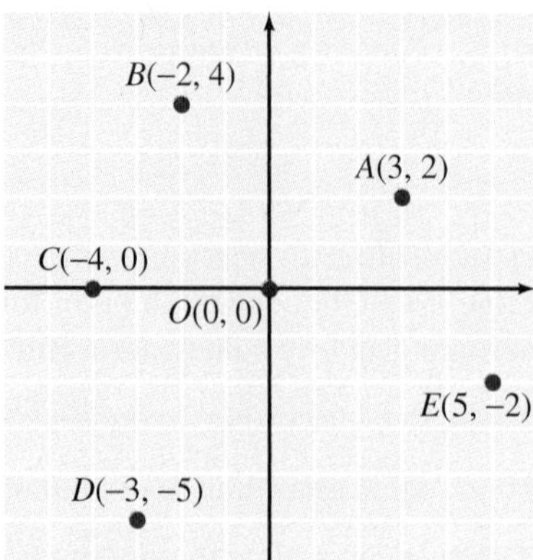

Figure 3.2

Remark: The notation $(-2, 4)$ was used earlier in this text to indicate an interval of the real number line. Now we are using the same notation to indicate an ordered pair of real numbers. This double meaning should not be confusing because the context of the material will always indicate which meaning of the notation is being used. Throughout this chapter, we will be using the ordered-pair interpretation.

In general we refer to the real numbers a and b in an ordered pair (a, b) associated with a point as the **coordinates of the point**. The first number, a, called the **abscissa**, is the directed distance of the point from the vertical axis measured parallel to the horizontal axis. The second number, b, called the **ordinate**, is the directed distance of the point from the horizontal axis measured parallel to the vertical axis (Figure 3.3a). Thus in the first quadrant all points have a positive abscissa and a positive ordinate. In the second quadrant all points have a negative abscissa and a positive ordinate. We have indicated the sign situations for all four quadrants in Figure 3.3(b). This system of associating points in a plane with pairs of real numbers is called the **rectangular coordinate system** or the **Cartesian coordinate system**.

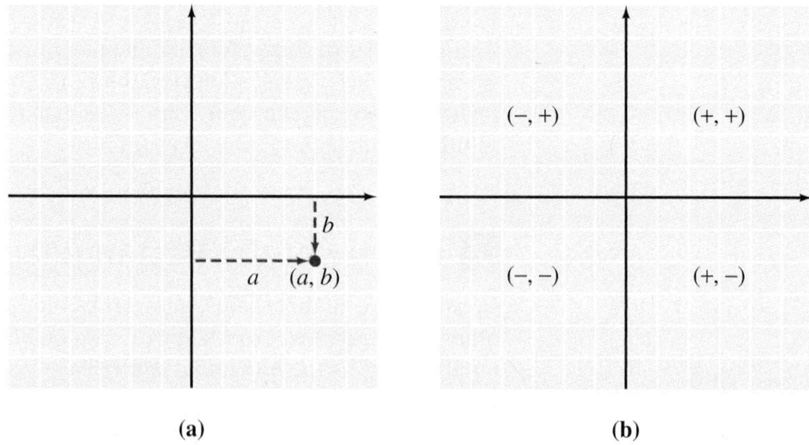

Figure 3.3

Historically, the rectangular coordinate system provided the basis for the development of the branch of mathematics called **analytic geometry**, or what we presently refer to as **coordinate geometry**. In this discipline, René Descartes, a French 17th-century mathematician, was able to transform geometric problems into an algebraic setting and then use the tools of algebra to solve the problems. Basically, there are two kinds of problems to solve in coordinate geometry:

1. Given an algebraic equation, find its geometric graph.
2. Given a set of conditions pertaining to a geometric figure, find its algebraic equation.

In this chapter we will discuss problems of both types. Let's start by finding the graph of an algebraic equation.

3 Graph the Solutions for Linear Equations

Let's begin by determining some solutions for the equation $y = x + 2$ and then plot the solutions on a rectangular coordinate system to produce a graph of the equation. Let's use a table to record some of the solutions.

Choose x	Determine y from $y = x + 2$	Solutions for $y = x + 2$
0	2	(0, 2)
1	3	(1, 3)
3	5	(3, 5)
5	7	(5, 7)
−2	0	(−2, 0)
−4	−2	(−4, −2)
−6	−4	(−6, −4)

We can plot the ordered pairs as points in a coordinate system and use the horizontal axis as the x axis and the vertical axis as the y axis, as in Figure 3.4(a). Connecting the points with a straight line as in Figure 3.4(b) produces a **graph of the equation** $y = x + 2$. Every point on the line has coordinates that are solutions of the equation $y = x + 2$. The graph provides a visual display of all the infinite solutions for the equation.

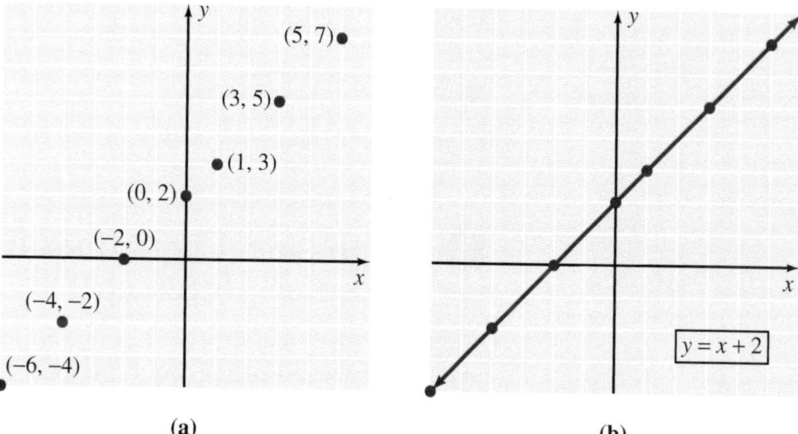

Figure 3.4

EXAMPLE 2 Graph the equation $y = -x + 4$.

Solution

Let's begin by determining some solutions for the equation $y = -x + 4$ and then plot the solutions on a rectangular coordinate system to produce a graph of the equation. Let's use a table to record some of the solutions.

x value	y value determined from $y = -x + 4$	Ordered pairs
-3	7	$(-3, 7)$
-1	5	$(-1, 5)$
0	4	$(0, 4)$
2	2	$(2, 2)$
4	0	$(4, 0)$
6	-2	$(6, -2)$

We can plot the ordered pairs on a coordinate system as shown in Figure 3.5(a). The graph of the equation is produced by drawing a straight line through the plotted points as in Figure 3.5(b).

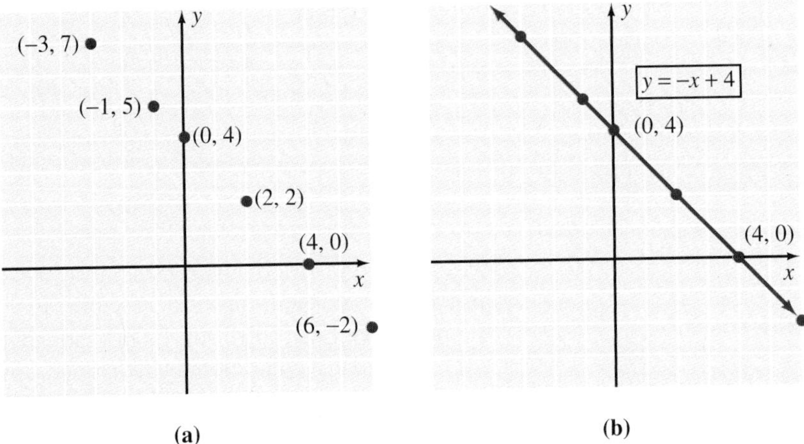

Figure 3.5

▼ **PRACTICE YOUR SKILL**

Graph the equation $y = 2x - 2$. ■

4 Graph Linear Equations by Finding the x and y Intercepts

The points (4, 0) and (0, 4) in Figure 3.5(b) are special points. They are the points of the graph that are on the coordinate axes. That is, they yield the x intercept and the y intercept of the graph. Let's define in general the *intercepts* of a graph.

The x coordinates of the points that a graph has in common with the x axis are called the ***x* intercepts** of the graph. (To compute the x intercepts, let $y = 0$ and solve for x.)

The y coordinates of the points that a graph has in common with the y axis are called the ***y* intercepts** of the graph. (To compute the y intercepts, let $x = 0$ and solve for y.)

It is advantageous to be able to recognize the kind of graph that a certain type of equation produces. For example, if we recognize that the graph of $3x + 2y = 12$ is a straight line, then it becomes a simple matter to find two points and sketch the line. Let's pursue the graphing of straight lines in a little more detail.

In general, any equation of the form $Ax + By = C$, where A, B, and C are constants (A and B not both zero) and x and y are variables, is a **linear equation**, and its graph is a straight line. Two points of clarification about this description of a linear equation should be made. First, the choice of x and y for variables is arbitrary. Any two letters could be used to represent the variables. For example, an equation such as $3r + 2s = 9$ can be considered a linear equation in two variables. So that we are not constantly changing the labeling of the coordinate axes when graphing equations, however, it is much easier to use the same two variables in all equations. Thus we will go along with convention and use x and y as variables. Second, the phrase "any equation of the form $Ax + By = C$" technically means "any equation of the form $Ax + By = C$ or equivalent to that form." For example, the equation $y = 2x - 1$ is equivalent to $-2x + y = -1$ and thus is linear and produces a straight-line graph.

The knowledge that any equation of the form $Ax + By = C$ produces a straight-line graph, along with the fact that two points determine a straight line, makes graphing linear equations a simple process. We merely find two solutions (such as the intercepts), plot the corresponding points, and connect the points with a straight line. It is usually wise to find a third point as a check point. Let's consider an example.

EXAMPLE 3

Graph $3x - 2y = 12$.

Solution

First, let's find the intercepts. Let $x = 0$; then

$$3(0) - 2y = 12$$
$$-2y = 12$$
$$y = -6$$

Thus $(0, -6)$ is a solution. Let $y = 0$; then

$$3x - 2(0) = 12$$
$$3x = 12$$
$$x = 4$$

Thus $(4, 0)$ is a solution. Now let's find a third point to serve as a check point. Let $x = 2$; then

$$3(2) - 2y = 12$$
$$6 - 2y = 12$$
$$-2y = 6$$
$$y = -3$$

Thus $(2, -3)$ is a solution. Plot the points associated with these three solutions and connect them with a straight line to produce the graph of $3x - 2y = 12$ in Figure 3.6.

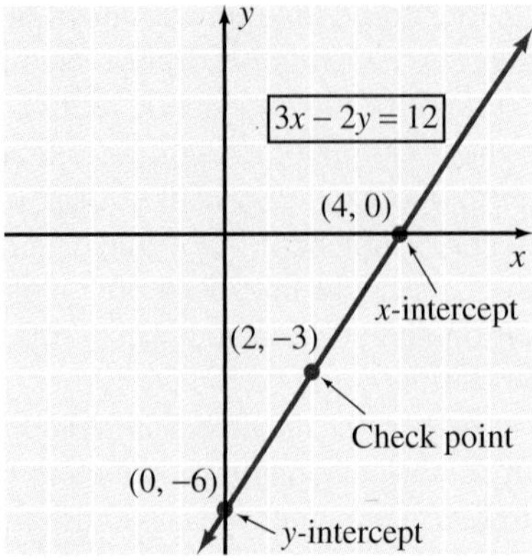

Figure 3.6

▼ PRACTICE YOUR SKILL

Graph $3x - y = 6$. ■

Let's review our approach to Example 3. Note that we did not solve the equation for y in terms of x or for x in terms of y. Because we know the graph is a straight line, there is no need for any extensive table of values. Furthermore, the solution $(2, -3)$ served as a check point. If it had not been on the line determined by the two intercepts, then we would have known that an error had been made.

EXAMPLE 4

Graph $2x + 3y = 7$.

Solution

Without showing all of our work, the following table indicates the intercepts and a check point. The points from the table are plotted, and the graph of $2x + 3y = 7$ is shown in Figure 3.7.

x	y	
0	$\frac{7}{3}$	
$\frac{7}{2}$	0	Intercepts
2	1	Check point

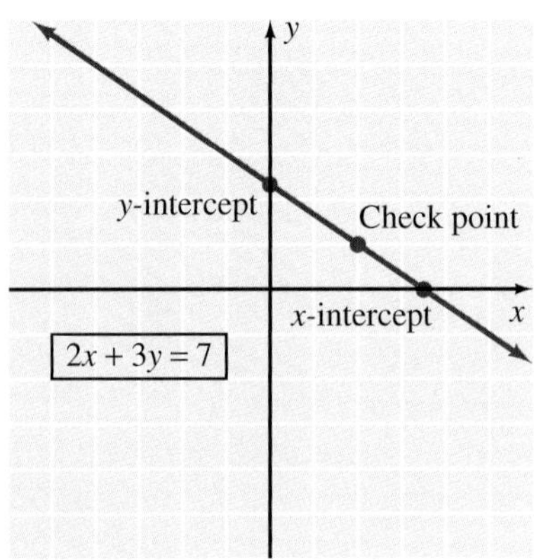

Figure 3.7

▼ PRACTICE YOUR SKILL

Graph $x + 2y = 3$. ■

5 Graph Lines Passing through the Origin, Vertical Lines, and Horizontal Lines

It is helpful to recognize some *special* straight lines. For example, the graph of any equation of the form $Ax + By = C$, where $C = 0$ (the constant term is zero), is a straight line that contains the origin. Let's consider an example.

EXAMPLE 5

Graph $y = 2x$.

Solution

Obviously (0, 0) is a solution. (Also, notice that $y = 2x$ is equivalent to $-2x + y = 0$; thus it fits the condition $Ax + By = C$, where $C = 0$.) Because both the x intercept and the y intercept are determined by the point (0, 0), another point is necessary to determine the line. Then a third point should be found as a check point. The graph of $y = 2x$ is shown in Figure 3.8.

x	y	
0	0	Intercepts
2	4	Additional point
−1	−2	Check point

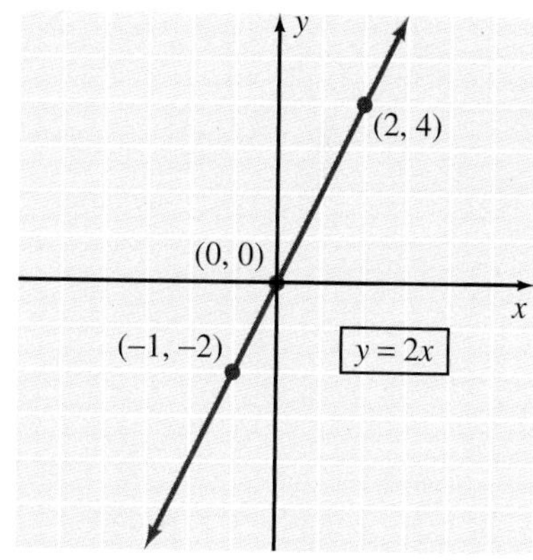

Figure 3.8

▼ **PRACTICE YOUR SKILL**

Graph $y = -3x$. ■

EXAMPLE 6

Graph $x = 2$.

Solution

Because we are considering linear equations in *two variables*, the equation $x = 2$ is equivalent to $x + 0(y) = 2$. Now we can see that any value of y can be used, but the x value must always be 2. Therefore, some of the solutions are $(2, 0), (2, 1), (2, 2), (2, -1)$, and $(2, -2)$. The graph of all solutions of $x = 2$ is the vertical line in Figure 3.9.

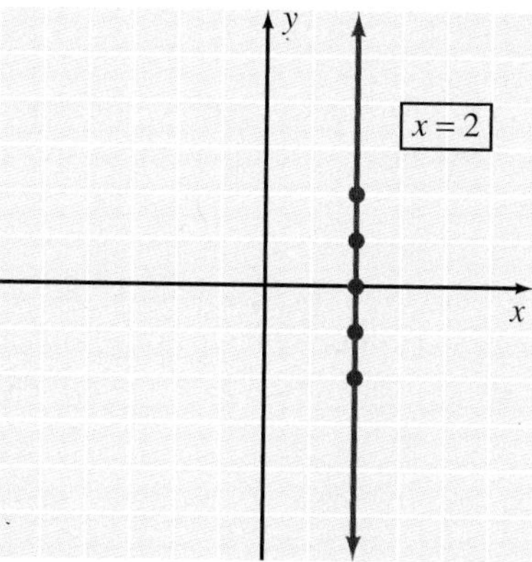

Figure 3.9

▼ **PRACTICE YOUR SKILL**

Graph $x = -3$. ■

EXAMPLE 7 Graph $y = -3$.

Solution

The equation $y = -3$ is equivalent to $0(x) + y = -3$. Thus any value of x can be used, but the value of y must be -3. Some solutions are $(0, -3)$, $(1, -3)$, $(2, -3)$, $(-1, -3)$, and $(-2, -3)$. The graph of $y = -3$ is the horizontal line in Figure 3.10.

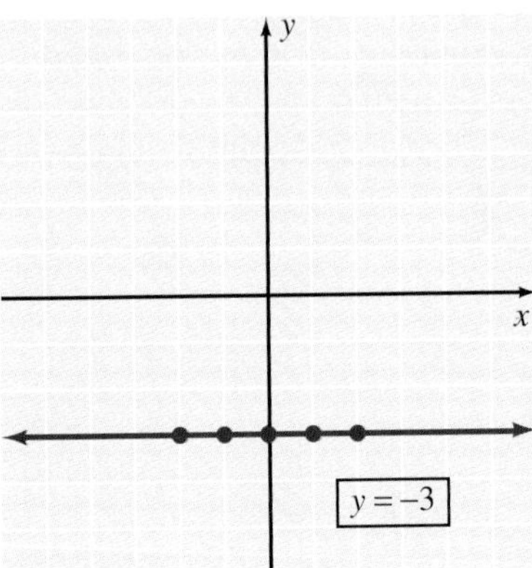

Figure 3.10

▼ **PRACTICE YOUR SKILL**

Graph $y = 4$. ■

In general, the graph of any equation of the form $Ax + By = C$, where $A = 0$ or $B = 0$ (not both), is a line parallel to one of the axes. More specifically, any equation of the form $x = a$, where a is a constant, is a line parallel to the y axis that has an x intercept of a. Any equation of the form $y = b$, where b is a constant, is a line parallel to the x axis that has a y intercept of b.

6 Apply Graphing to Linear Relationships

There are numerous applications of linear relationships. For example, suppose that a retailer has a number of items that she wants to sell at a profit of 30% of the cost of each item. If we let s represent the selling price and c the cost of each item, then the equation

$$s = c + 0.3c = 1.3c$$

can be used to determine the selling price of each item based on the cost of the item. In other words, if the cost of an item is \$4.50, then it should be sold for $s = (1.3)(4.5) = \$5.85$.

The equation $s = 1.3c$ can be used to determine the following table of values. Reading from the table, we see that if the cost of an item is \$15, then it should be sold for \$19.50 in order to yield a profit of 30% of the cost. Furthermore, because this is a linear relationship, we can obtain exact values between values given in the table.

c	1	5	10	15	20
s	1.3	6.5	13	19.5	26

For example, a c value of 12.5 is halfway between c values of 10 and 15, so the corresponding s value is halfway between the s values of 13 and 19.5. Therefore, a c value of 12.5 produces an s value of

$$s = 13 + \frac{1}{2}(19.5 - 13) = 16.25$$

Thus, if the cost of an item is \$12.50, it should be sold for \$16.25.

Now let's graph this linear relationship. We can label the horizontal axis c, label the vertical axis s, and use the origin along with one ordered pair from the table to produce the straight-line graph in Figure 3.11. (Because of the type of application, we use only nonnegative values for c and s.)

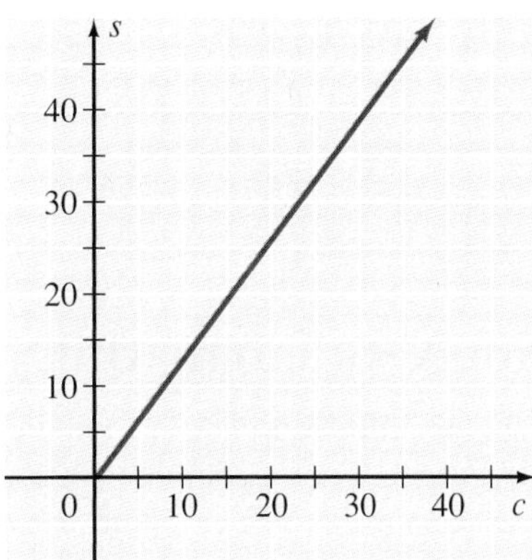

Figure 3.11

From the graph we can approximate s values on the basis of given c values. For example, if $c = 30$, then by reading up from 30 on the c axis to the line and then across to the s axis, we see that s is a little less than 40. (An exact s value of 39 is obtained by using the equation $s = 1.3c$.)

Many formulas that are used in various applications are linear equations in two variables. For example, the formula $C = \frac{5}{9}(F - 32)$, which is used to convert temperatures from the Fahrenheit scale to the Celsius scale, is a linear relationship. Using this equation, we can determine that 14°F is equivalent to $C = \frac{5}{9}(14 - 32) = \frac{5}{9}(-18) = -10$°C. Let's use the equation $C = \frac{5}{9}(F - 32)$ to complete the following table.

F	−22	−13	5	32	50	68	86
C	−30	−25	−15	0	10	20	30

Reading from the table, we see, for example, that −13°F = −25°C and 68°F = 20°C.

To graph the equation $C = \frac{5}{9}(F - 32)$ we can label the horizontal axis F, label the vertical axis C, and plot two ordered pairs (F, C) from the table. Figure 3.12 shows the graph of the equation.

From the graph we can approximate C values on the basis of given F values. For example, if F = 80°, then by reading up from 80 on the F axis to the line and then across to the C axis, we see that C is approximately 25°. Likewise, we can obtain approximate F values on the basis of given C values. For example, if C = −25°, then by reading across from -25 on the C axis to the line and then up to the F axis, we see that F is approximately $-15°$.

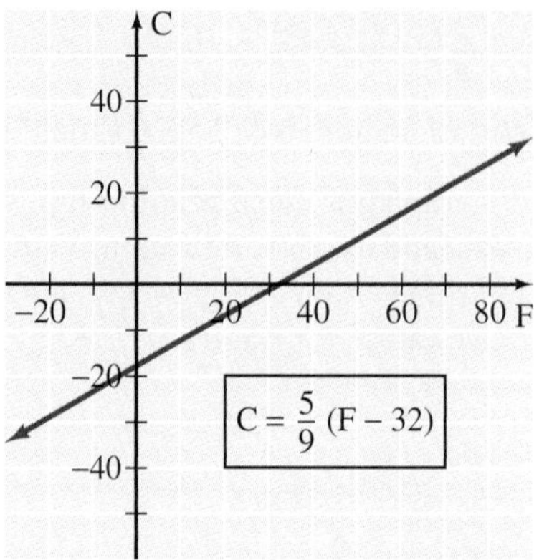

Figure 3.12

7 Introduce Graphing Utilities

The term **graphing utility** is used in current literature to refer to either a graphing calculator (see Figure 3.13) or a computer with a graphing software package. (We will frequently use the phrase *use a graphing calculator* to mean "use a graphing calculator or a computer with the appropriate software.")

These devices have a large range of capabilities that enable the user not only to obtain a quick sketch of a graph but also to study various characteristics of it, such as the x intercepts, y intercepts, and turning points of a curve. We will introduce some of these features of graphing utilities as we need them in the text. Because there are so many different types of graphing utilities available, we will use mostly generic terminology and let you consult your user's manual for specific key-punching instructions. We urge you to study the graphing utility examples in this text even if you do not have access to a graphing calculator or a computer. The examples were chosen to reinforce concepts under discussion.

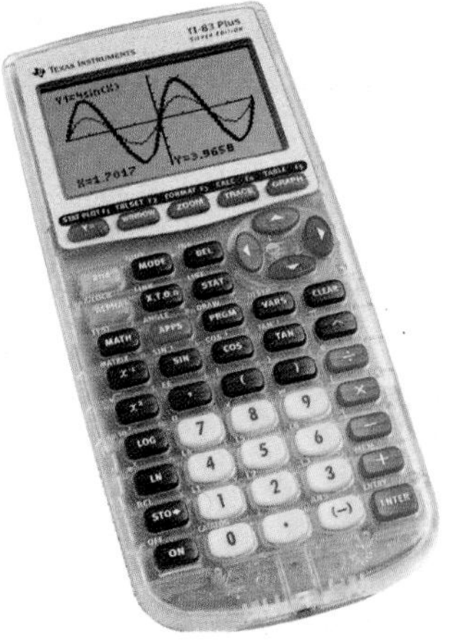

Courtesy Texas Instruments

Figure 3.13

EXAMPLE 8

Use a graphing utility to obtain a graph of the line $2.1x + 5.3y = 7.9$.

Solution

First, let's solve the equation for y in terms of x.

$$2.1x + 5.3y = 7.9$$

$$5.3y = 7.9 - 2.1x$$

$$y = \frac{7.9 - 2.1x}{5.3}$$

Now we can enter the expression $\frac{7.9 - 2.1x}{5.3}$ for Y_1 and obtain the graph as shown in Figure 3.14.

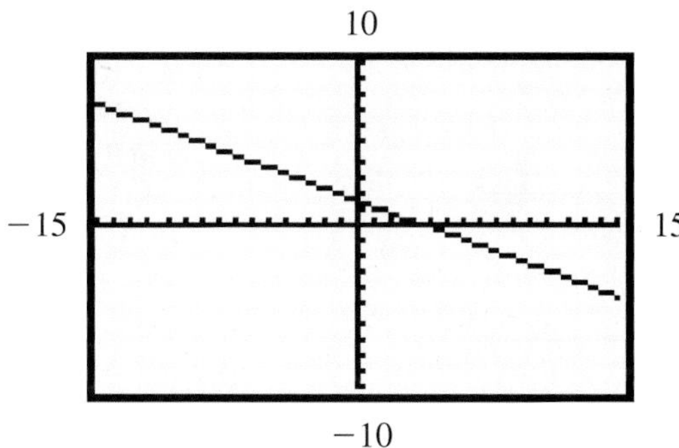

Figure 3.14

▼ PRACTICE YOUR SKILL

Use a graphing utility to obtain a graph of the line $3.4x - 2.5y = 6.8$. ■

CONCEPT QUIZ

For Problems 1–10, answer true or false.

1. In a rectangular coordinate system, the coordinate axes partition the plane into four parts called quadrants.
2. Quadrants are named with Roman numerals and are numbered clockwise.
3. The real numbers in an ordered pair are referred to as the coordinates of the point.
4. If the abscissa of an ordered pair is negative, then the point is in either the 3rd or 4th quadrant.
5. The equation $y = x + 3$ has an infinite number of ordered pairs that satisfy the equation.
6. The graph of $y = x^2$ is a straight line.
7. The y intercept of the graph of $3x + 4y = -4$ is -4.
8. The graph of $y = 4$ is a vertical line.
9. The graph of $x = 4$ has an x intercept of 4.
10. The graph of every linear equation has a y intercept.

Problem Set 3.1

1 Find Solutions for Linear Equations in Two Variables

For Problems 1–4, determine which of the ordered pairs are solutions to the given equation.

1. $y = 3x - 2$ $(2, 4), (-1, -5), (0, 1)$
2. $y = 2x + 3$ $(2, 5), (1, 5), (-1, 1)$
3. $2x + y = 6$ $(-2, 10), (-1, 5), (3, 0)$
4. $-3x + 2y = 2$ $\left(3, \frac{11}{2}\right), (-2, -2) \left(-1, -\frac{1}{2}\right)$

3 Graph the Solutions for Linear Equations

For Problems 5–8, complete the table of values for the equation and graph the equation.

5. $y = -x + 3$

x	−2	−1	0	4
y				

6. $y = 2x - 1$

x	−3	−1	0	2
y				

7. $2x - y = 6$

x	−2	0	2	4
y				

8. $2x - 3y = -6$

x	−3	0	2	3
y				

4 Graph Linear Equations by Finding the x and y Intercepts

For Problems 9–28, graph each of the linear equations by finding the x and y intercepts.

9. $x + 2y = 4$
10. $2x + y = 6$
11. $2x - y = 2$
12. $3x - y = 3$
13. $3x + 2y = 6$
14. $2x + 3y = 6$
15. $5x - 4y = 20$
16. $4x - 3y = -12$
17. $x + 4y = -6$
18. $5x + y = -2$
19. $-x - 2y = 3$
20. $-3x - 2y = 12$
21. $y = x + 3$
22. $y = x - 1$
23. $y = -2x - 1$
24. $y = 4x + 3$
25. $y = \frac{1}{2}x + \frac{2}{3}$
26. $y = \frac{2}{3}x - \frac{3}{4}$
27. $-3y = -x + 3$
28. $2y = x - 2$

5 Graph Lines Passing through the Origin, Vertical Lines, and Horizontal Lines

For Problems 29–40, graph each of the linear equations.

29. $y = -x$
30. $y = x$
31. $y = 3x$
32. $y = -4x$
33. $2x - 3y = 0$
34. $3x + 4y = 0$
35. $x = 0$
36. $y = 0$
37. $y = 2$
38. $x = -3$
39. $x = -4$
40. $y = -1$

6 Apply Graphing to Linear Relationships

41. **(a)** Digital Solutions charges for help-desk services according to the equation $c = 0.25m + 10$, where c represents the cost in dollars and m represents the minutes of service. Complete the following table.

m	5	10	15	20	30	60
c						

(b) Label the horizontal axis m and the vertical axis c, and graph the equation $c = 0.25m + 10$ for nonnegative values of m.

(c) Use the graph from part (b) to approximate values for c when $m = 25, 40$, and 45.

(d) Check the accuracy of your readings from the graph in part (c) by using the equation $c = 0.25m + 10$.

42. **(a)** The equation $F = \frac{9}{5}C + 32$ can be used to convert from degrees Celsius to degrees Fahrenheit. Complete the following table.

C	0	5	10	15	20	−5	−10	−15	−20	−25
F										

(b) Graph the equation $F = \frac{9}{5}C + 32$.

(c) Use your graph from part (b) to approximate values for F when C = 25°, 30°, −30°, and −40°.

(d) Check the accuracy of your readings from the graph in part (c) by using the equation $F = \frac{9}{5}C + 32$.

43. **(a)** A doctor's office wants to chart and graph the linear relationship between the hemoglobin A1c reading and the average blood glucose level. The equation $G = 30h - 60$ describes the relationship, where h is the hemoglobin A1c reading and G is the average blood glucose reading. Complete this chart of values:

Hemoglobin A1c, h	6.0	6.5	7.0	8.0	8.5	9.0	10.0
Blood glucose, G							

(b) Label the horizontal axis h and the vertical axis G, then graph the equation $G = 30h - 60$ for h values between 4.0 and 12.0.

(c) Use the graph from part (b) to approximate values for G when $h = 5.5$ and 7.5.

(d) Check the accuracy of your readings from the graph in part (c) by using the equation $G = 30h - 60$.

44. Suppose that the daily profit from an ice cream stand is given by the equation $p = 2n - 4$, where n represents the gallons of ice cream mix used in a day and p represents the dollars of profit. Label the horizontal axis n and the vertical axis p, and graph the equation $p = 2n - 4$ for nonnegative values of n.

45. The cost (c) of playing an online computer game for a time (t) in hours is given by the equation $c = 3t + 5$. Label the horizontal axis t and the vertical axis c, and graph the equation for nonnegative values of t.

46. The area of a sidewalk whose width is fixed at 3 feet can be given by the equation $A = 3l$, where A represents the area in square feet and l represents the length in feet. Label the horizontal axis l and the vertical axis A, and graph the equation $A = 3l$ for nonnegative values of l.

47. An online grocery store charges for delivery based on the equation $C = 0.30p$, where C represents the cost in dollars and p represents the weight of the groceries in pounds. Label the horizontal axis p and the vertical axis C, and graph the equation $C = 0.30p$ for nonnegative values of p.

THOUGHTS INTO WORDS

48. How do we know that the graph of $y = -3x$ is a straight line that contains the origin?

49. How do we know that the graphs of $2x - 3y = 6$ and $-2x + 3y = -6$ are the same line?

50. What is the graph of the conjunction $x = 2$ and $y = 4$? What is the graph of the disjunction $x = 2$ or $y = 4$? Explain your answers.

51. Your friend claims that the graph of the equation $x = 2$ is the point (2, 0). How do you react to this claim?

FURTHER INVESTIGATIONS

From our work with absolute value, we know that $|x + y| = 1$ is equivalent to $x + y = 1$ or $x + y = -1$. Therefore, the graph of $|x + y| = 1$ consists of the two lines $x + y = 1$ and $x + y = -1$. Graph each of the following.

52. $|x + y| = 1$

53. $|x - y| = 4$

54. $|2x - y| = 4$

55. $|3x + 2y| = 6$

GRAPHING CALCULATOR ACTIVITIES

This is the first of many appearances of a group of problems called graphing calculator activities. These problems are specifically designed for those of you who have access to a graphing calculator or a computer with an appropriate software package. Within the framework of these problems, you will be given the opportunity to reinforce concepts we discussed in the text; lay groundwork for concepts we will introduce later in the text; predict shapes and locations of graphs on the basis of your previous graphing experiences; solve problems that are unreasonable or perhaps impossible to solve without a graphing utility; and in general become familiar with the capabilities and limitations of your graphing utility.

56. (a) Graph $y = 3x + 4$, $y = 2x + 4$, $y = -4x + 4$, and $y = -2x + 4$ on the same set of axes.

(b) Graph $y = \frac{1}{2}x - 3$, $y = 5x - 3$, $y = 0.1x - 3$, and $y = -7x - 3$ on the same set of axes.

(c) What characteristic do all lines of the form $y = ax + 2$ (where a is any real number) share?

57. (a) Graph $y = 2x - 3$, $y = 2x + 3$, $y = 2x - 6$, and $y = 2x + 5$ on the same set of axes.

(b) Graph $y = -3x + 1$, $y = -3x + 4$, $y = -3x - 2$, and $y = -3x - 5$ on the same set of axes.

(c) Graph $y = \frac{1}{2}x + 3$, $y = \frac{1}{2}x - 4$, $y = \frac{1}{2}x + 5$, and $y = \frac{1}{2}x - 2$ on the same set of axes.

(d) What relationship exists among all lines of the form $y = 3x + b$, where b is any real number?

58. (a) Graph $2x + 3y = 4$, $2x + 3y = -6$, $4x - 6y = 7$, and $8x + 12y = -1$ on the same set of axes.

(b) Graph $5x - 2y = 4$, $5x - 2y = -3$, $10x - 4y = 3$, and $15x - 6y = 30$ on the same set of axes.

(c) Graph $x + 4y = 8$, $2x + 8y = 3$, $x - 4y = 6$, and $3x + 12y = 10$ on the same set of axes.

(d) Graph $3x - 4y = 6$, $3x + 4y = 10$, $6x - 8y = 20$, and $6x - 8y = 24$ on the same set of axes.

(e) For each of the following pairs of lines, (a) predict whether they are parallel lines, and (b) graph each pair of lines to check your prediction.

(1) $5x - 2y = 10$ and $5x - 2y = -4$

(2) $x + y = 6$ and $x - y = 4$

(3) $2x + y = 8$ and $4x + 2y = 2$

(4) $y = 0.2x + 1$ and $y = 0.2x - 4$

(5) $3x - 2y = 4$ and $3x + 2y = 4$

(6) $4x - 3y = 8$ and $8x - 6y = 3$

(7) $2x - y = 10$ and $6x - 3y = 6$

(8) $x + 2y = 6$ and $3x - 6y = 6$

59. Now let's use a graphing calculator to get a graph of $C = \frac{5}{9}(F - 32)$. By letting $F = x$ and $C = y$, we obtain Figure 3.15.

Pay special attention to the boundaries on x. These values were chosen so that the fraction

$$\frac{(\text{Maximum value of } x) \text{ minus } (\text{Minimum value of } x)}{95}$$

would be equal to 1. The viewing window of the graphing calculator used to produce Figure 3.15 is 95 pixels (dots) wide. Therefore, we use 95 as the denominator of the fraction. We chose the boundaries for y to make sure that the cursor would be visible on the screen when we looked for certain values.

Now let's use the TRACE feature of the graphing calculator to complete the following table. Note that the cursor moves in increments of 1 as we trace along the graph.

F	−5	5	9	11	12	20	30	45	60
C									

(This was accomplished by setting the aforementioned fraction equal to 1.) By moving the cursor to each of the F values, we can complete the table as follows.

F	−5	5	9	11	12	20	30	45	60
C	−21	−15	−13	−12	−11	−7	−1	7	16

The C values are expressed to the nearest degree. Use your calculator and check the values in the table by using the equation $C = \frac{5}{9}(F - 32)$.

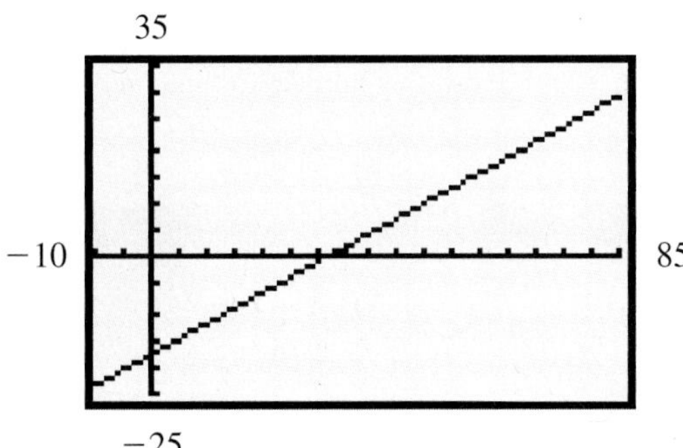

Figure 3.15

60. **(a)** Use your graphing calculator to graph $F = \frac{9}{5}C + 32$. Be sure to set boundaries on the horizontal axis so that when you are using the trace feature, the cursor will move in increments of 1.

(b) Use the TRACE feature and check your answers for part (a) of Problem 42.

Answers to the Concept Quiz

1. True **2.** False **3.** True **4.** False **5.** True **6.** False **7.** False **8.** False **9.** True **10.** False

Answers to the Example Practice Skills

1. $(-4, 12), (-2, 8), (0, 4), (1, 2), (3, -2)$ **2.**

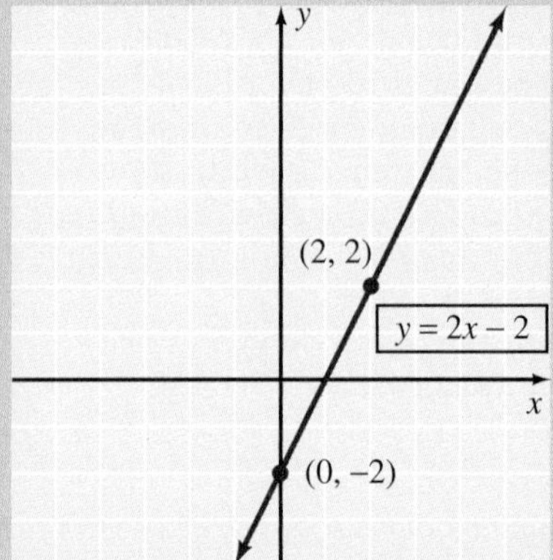

3.

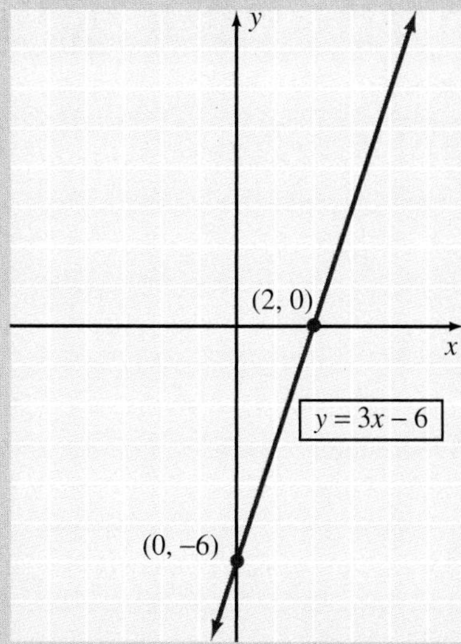
y
x
(2, 0)
y = 3x − 6
(0, −6)

4.

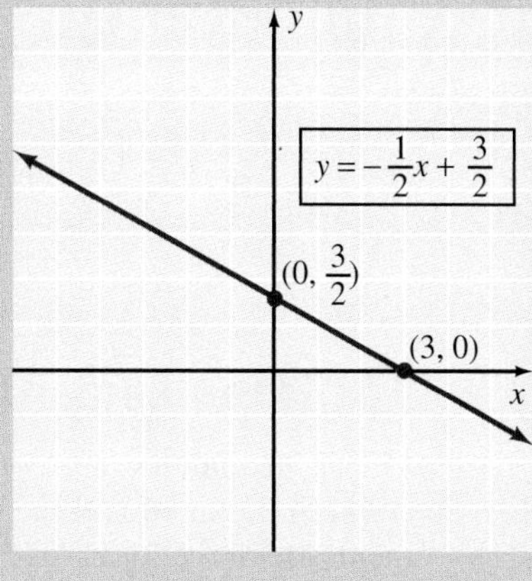
y
x
$y = -\frac{1}{2}x + \frac{3}{2}$
$(0, \frac{3}{2})$
(3, 0)

5.

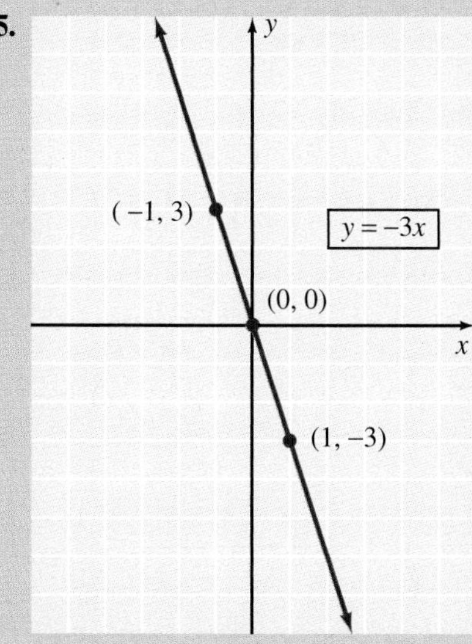
y
x
(−1, 3)
y = −3x
(0, 0)
(1, −3)

6.

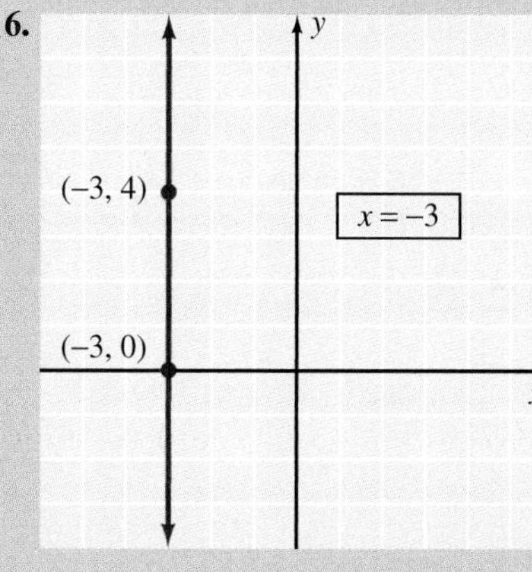
y
x
(−3, 4)
x = −3
(−3, 0)

7.

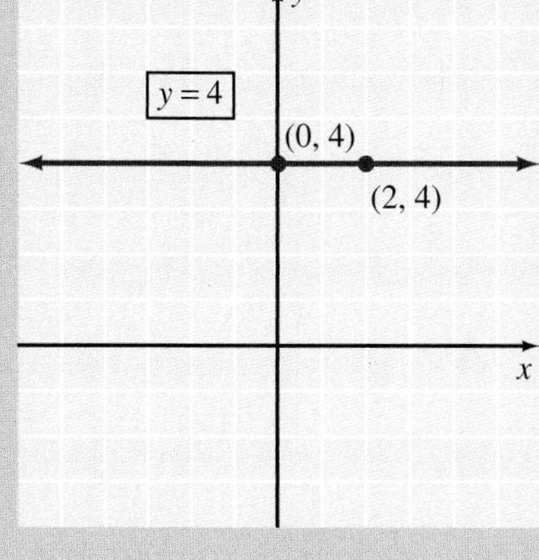
y
x
y = 4
(0, 4)
(2, 4)

8.

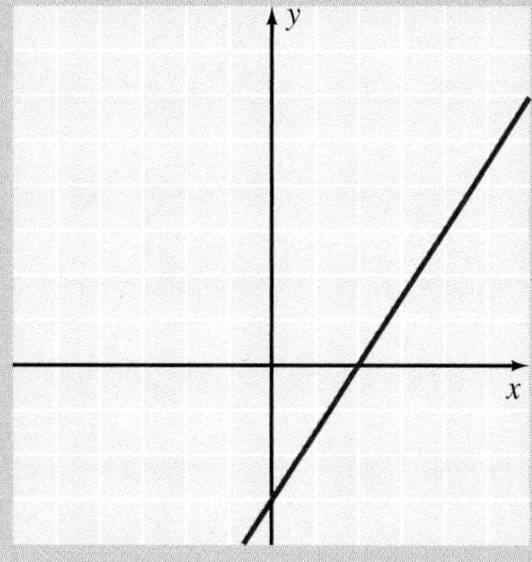
y
x

3.2 Linear Inequalities in Two Variables

OBJECTIVES

1 Graph Linear Inequalities

1 Graph Linear Inequalities

Linear inequalities in two variables are of the form $Ax + By > C$ or $Ax + By < C$, where A, B, and C are real numbers. (Combined linear equality and inequality statements are of the form $Ax + By \geq C$ or $Ax + By \leq C$.)

Graphing linear inequalities is almost as easy as graphing linear equations. The following discussion leads into a simple, step-by-step process. Let's consider the following equation and related inequalities.

$$x + y = 2 \qquad x + y > 2 \qquad x + y < 2$$

The graph of $x + y = 2$ is shown in Figure 3.16. The line divides the plane into two half planes, one above the line and one below the line. In Figure 3.17(a) we have indicated

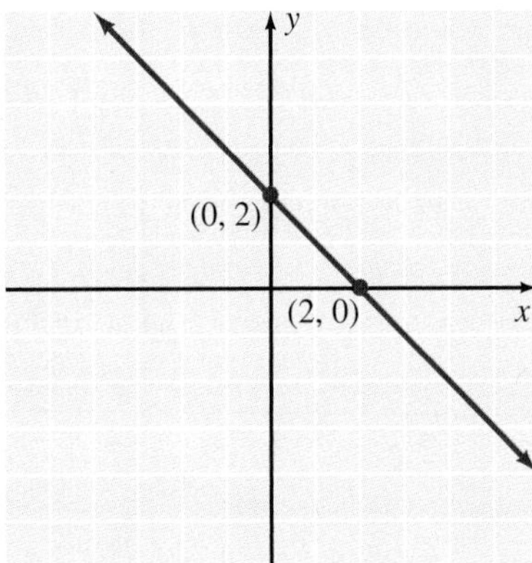

Figure 3.16

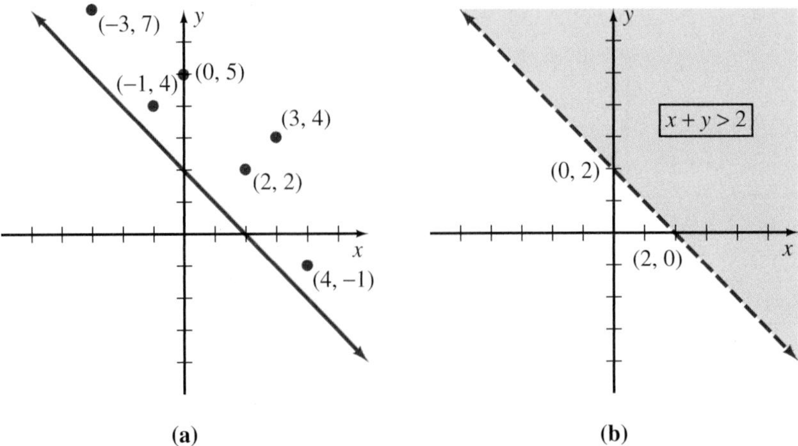

Figure 3.17

several points in the half-plane above the line. Note that for each point, the ordered pair of real numbers satisfies the inequality $x + y > 2$. This is true for *all points* in the half-plane above the line. Therefore, the graph of $x + y > 2$ is the half-plane above the line, as indicated by the shaded portion in Figure 3.17(b). We use a dashed line to indicate that points on the line do *not* satisfy $x + y > 2$. We would use a solid line if we were graphing $x + y \geq 2$.

In Figure 3.18(a), several points were indicated in the half-plane below the line $x + y = 2$. Note that for each point, the ordered pair of real numbers satisfies the inequality $x + y < 2$. This is true for *all points* in the half-plane below the line. Thus the graph of $x + y < 2$ is the half-plane below the line, as indicated in Figure 3.18(b).

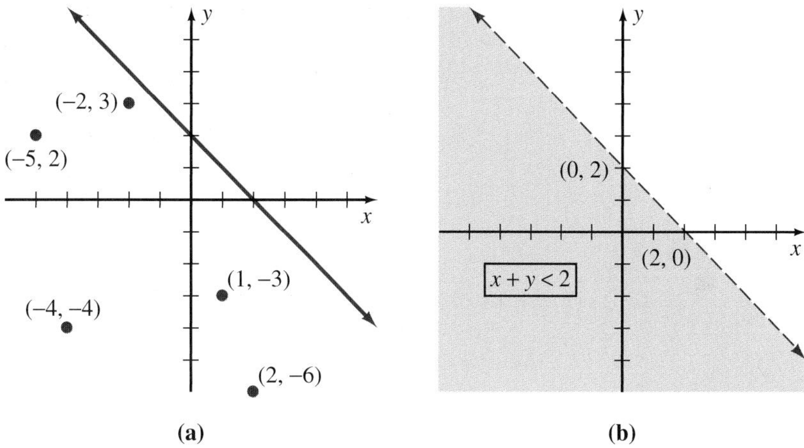

Figure 3.18

To graph a linear inequality, we suggest the following steps.

1. First, graph the corresponding equality. Use a solid line if equality is included in the original statement; use a dashed line if equality is not included.
2. Choose a "test point" not on the line and substitute its coordinates into the inequality. (The origin is a convenient point to use if it is not on the line.)
3. The graph of the original inequality is
   (a) the half-plane that contains the test point if the inequality is satisfied by that point, or
   (b) the half-plane that does not contain the test point if the inequality is not satisfied by the point.

Let's apply these steps to some examples.

EXAMPLE 1

Graph $x - 2y > 4$.

Solution

Step 1 Graph $x - 2y = 4$ as a dashed line because equality is not included in $x - 2y > 4$ (Figure 3.19).

Step 2 Choose the origin as a test point and substitute its coordinates into the inequality.

$$x - 2y > 4 \qquad \text{becomes } 0 - 2(0) > 4, \text{ which is false.}$$

Step 3 Because the test point did not satisfy the given inequality, the graph is the half-plane that does not contain the test point. Thus the graph of $x - 2y > 4$ is the half-plane below the line, as indicated in Figure 3.19.

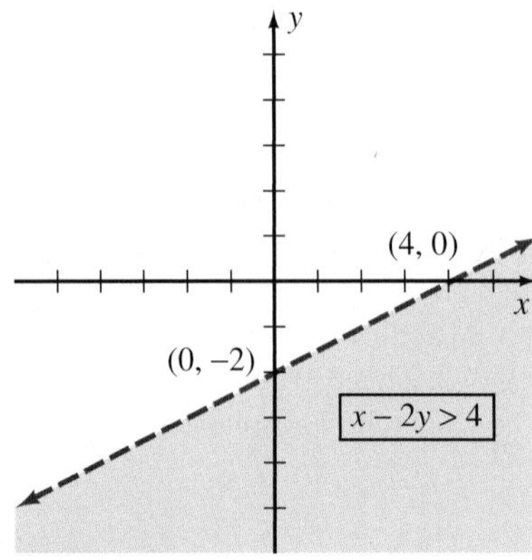

Figure 3.19

▼ PRACTICE YOUR SKILL

Graph $3x + y < 3$. ■

EXAMPLE 2 Graph $3x + 2y \leq 6$.

Solution

Step 1 Graph $3x + 2y = 6$ as a solid line because equality is included in $3x + 2y \leq 6$ (Figure 3.20).

Step 2 Choose the origin as a test point and substitute its coordinates into the given statement.

$3x + 2y \leq 6$ becomes $3(0) + 2(0) \leq 6$, which is true.

Step 3 Because the test point satisfies the given statement, all points in the same half-plane as the test point satisfy the statement. Thus the graph of $3x + 2y \leq 6$ consists of the line and the half-plane below the line (Figure 3.20).

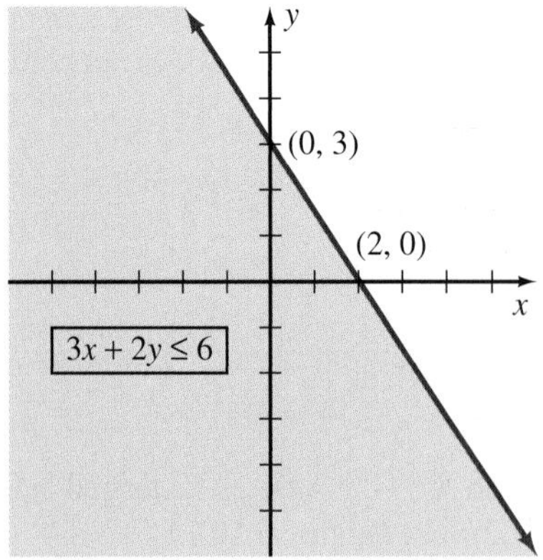

Figure 3.20

▼ PRACTICE YOUR SKILL

Graph $x - 4y \geq 4$. ■

EXAMPLE 3 Graph $y \leq 3x$.

Solution

Step 1 Graph $y = 3x$ as a solid line because equality is included in the statement $y \leq 3x$ (Figure 3.21).

Step 2 The origin is on the line, so we must choose some other point as a test point. Let's try (2, 1).

$$y \leq 3x \qquad \text{becomes } 1 \leq 3(2), \text{ which is a true statement.}$$

Step 3 Because the test point satisfies the given inequality, the graph is the half-plane that contains the test point. Thus the graph of $y \leq 3x$ consists of the line and the half-plane below the line, as indicated in Figure 3.21.

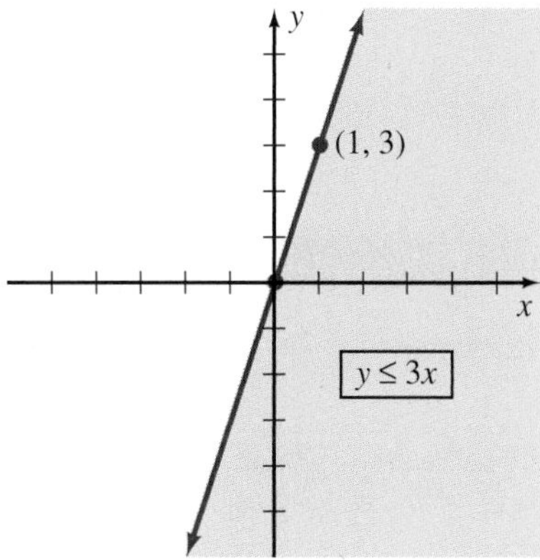

Figure 3.21

▼ PRACTICE YOUR SKILL

Graph $y > 2x$. ■

CONCEPT QUIZ

For Problems 1–10, answer true or false.

1. The ordered pair $(2, -3)$ satisfies the inequality $2x + y > 1$.
2. A dashed line on the graph indicates that the points on the line do not satisfy the inequality.
3. Any point can be used as a test point to determine the half-plane that is the solution of the inequality.
4. The ordered pair $(3, -2)$ satisfies the inequality $5x - 2y \geq 19$.
5. The ordered pair $(1, -3)$ satisfies the inequality $-2x - 3y < 4$.
6. The graph of $x > 0$ is the half-plane above the x axis.
7. The graph of $y < 0$ is the half-plane below the x axis.
8. The graph of $-x + y > 4$ is the half-plane above the line $-x + y = 4$.
9. The origin can serve as a test point to determine the half-plane that satisfies the inequality $3y > 2x$.
10. The ordered pair $(-2, -1)$ can be used as a test point to determine the half-plane that satisfies the inequality $y < -3x - 7$.

Problem Set 3.2

1 Graph Linear Inequalities

For Problems 1–24, graph each of the inequalities.

1. $x - y > 2$
2. $x + y > 4$
3. $x + 3y < 3$
4. $2x - y > 6$
5. $2x + 5y \geq 10$
6. $3x + 2y \leq 4$
7. $y \leq -x + 2$
8. $y \geq -2x - 1$
9. $y > -x$
10. $y < x$
11. $2x - y \geq 0$
12. $x + 2y \geq 0$
13. $-x + 4y - 4 \leq 0$
14. $-2x + y - 3 \leq 0$
15. $y > -\frac{3}{2}x - 3$
16. $2x + 5y > -4$
17. $y < -\frac{1}{2}x + 2$
18. $y < -\frac{1}{3}x + 1$
19. $x \leq 3$
20. $y \geq -2$
21. $x > 1$ and $y < 3$
22. $x > -2$ and $y > -1$
23. $x \leq -1$ and $y < 1$
24. $x < 2$ and $y \geq -2$

THOUGHTS INTO WORDS

25. Why is the point $(-4, 1)$ not a good test point to use when graphing $5x - 2y > -22$?
26. Explain how you would graph the inequality $-3 > x - 3y$.

FURTHER INVESTIGATIONS

27. Graph $|x| < 2$. [*Hint:* Remember that $|x| < 2$ is equivalent to $-2 < x < 2$.]
28. Graph $|y| > 1$.
29. Graph $|x + y| < 1$.
30. Graph $|x - y| > 2$.

GRAPHING CALCULATOR ACTIVITIES

31. This is a good time for you to become acquainted with the DRAW features of your graphing calculator. Again, you may need to consult your user's manual for specific key-punching instructions. Return to Examples 1, 2, and 3 of this section, and use your graphing calculator to graph the inequalities.
32. Use a graphing calculator to check your graphs for Problems 1–24.
33. Use the DRAW feature of your graphing calculator to draw each of the following.
    (a) A line segment between $(-2, -4)$ and $(-2, 5)$
    (b) A line segment between $(2, 2)$ and $(5, 2)$
    (c) A line segment between $(2, 3)$ and $(5, 7)$
    (d) A triangle with vertices at $(1, -2)$, $(3, 4)$, and $(-3, 6)$

Answers to the Concept Quiz

1. False **2.** True **3.** False **4.** True **5.** False **6.** False **7.** True **8.** True **9.** False **10.** False

Answers to the Example Practice Skills

1.

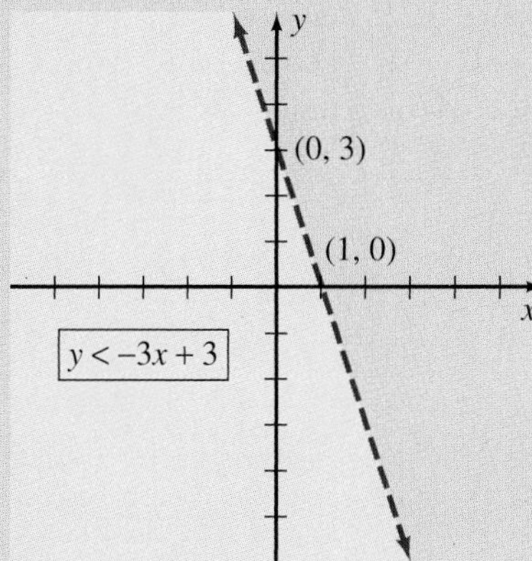

2.

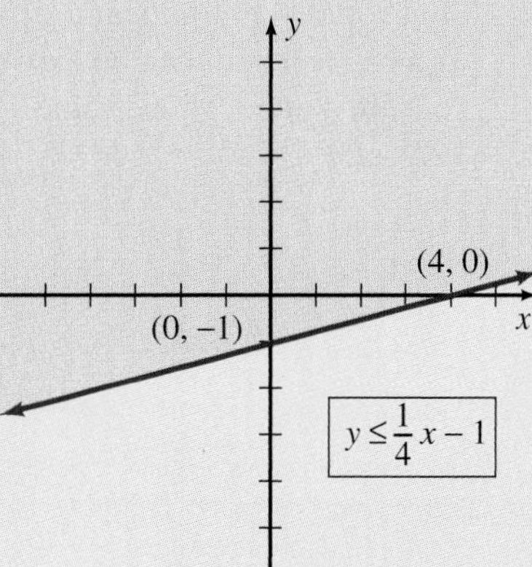

3.

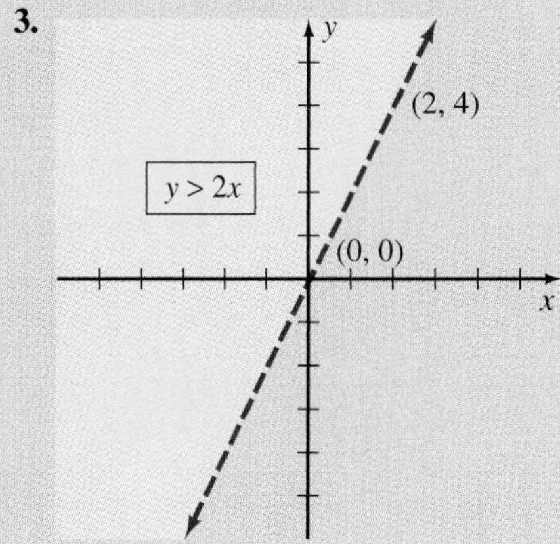

3.3 Distance and Slope

OBJECTIVES

1. Find the Distance between Two Points
2. Find the Slope of a Line
3. Use Slope to Graph Lines
4. Apply Slope to Solve Problems

1 Find the Distance between Two Points

As we work with the rectangular coordinate system, it is sometimes necessary to express the length of certain line segments. In other words, we need to be able to find the distance between two points. Let's first consider two specific examples and then develop the general distance formula.

EXAMPLE 1

Find the distance between the points $A(2, 2)$ and $B(5, 2)$ and also between the points $C(-2, 5)$ and $D(-2, -4)$.

Solution

Let's plot the points and draw $\overline{AB}$ as in Figure 3.22. Because $\overline{AB}$ is parallel to the x axis, its length can be expressed as $|5 - 2|$ or $|2 - 5|$. (The absolute value is used to ensure a nonnegative value.) Thus the length of $\overline{AB}$ is 3 units. Likewise, the length of $\overline{CD}$ is $|5 - (-4)| = |-4 - 5| = 9$ units.

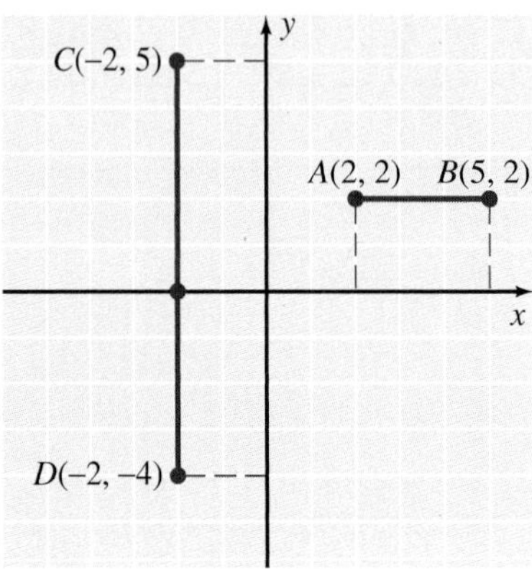

Figure 3.22

▼ **PRACTICE YOUR SKILL**

Find the distance between the points $A(3, 6)$ and $B(3, -2)$. ■

EXAMPLE 2

Find the distance between the points $A(2, 3)$ and $B(5, 7)$.

Solution

Let's plot the points and form a right triangle as indicated in Figure 3.23. Note that the coordinates of point C are $(5, 3)$. Because $\overline{AC}$ is parallel to the horizontal axis, its length is easily determined to be 3 units. Likewise, $\overline{CB}$ is parallel to the vertical axis and its length is 4 units. Let d represent the length of $\overline{AB}$, and apply the Pythagorean theorem to obtain

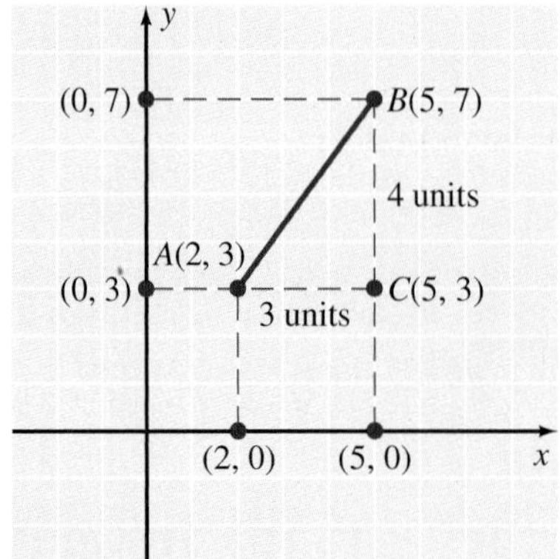

Figure 3.23

$$d^2 = 3^2 + 4^2$$
$$d^2 = 9 + 16$$
$$d^2 = 25$$
$$d = \pm\sqrt{25} = \pm 5$$

"Distance between" is a nonnegative value, so the length of $\overline{AB}$ is 5 units.

▼ **PRACTICE YOUR SKILL**

Find the distance between the points $A(-4, 1)$ and $B(8, 6)$. ■

We can use the approach we used in Example 2 to develop a general distance formula for finding the distance between any two points in a coordinate plane. The development proceeds as follows:

1. Let $P_1(x_1, y_1)$ and $P_2(x_2, y_2)$ represent any two points in a coordinate plane.
2. Form a right triangle as indicated in Figure 3.24. The coordinates of the vertex of the right angle, point R, are (x_2, y_1).

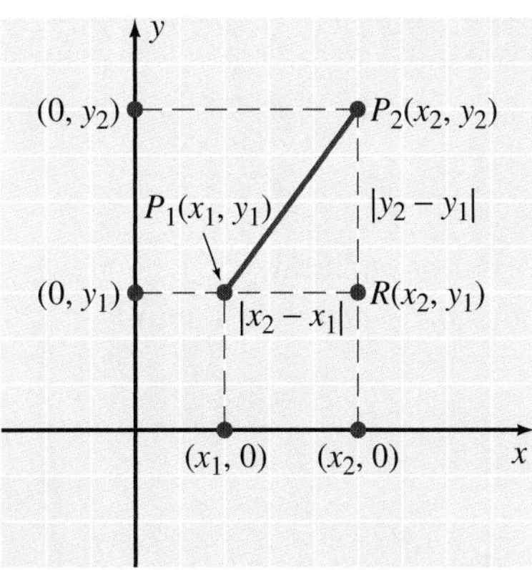

Figure 3.24

The length of $\overline{P_1R}$ is $|x_2 - x_1|$ and the length of $\overline{RP_2}$ is $|y_2 - y_1|$. (Again, the absolute value is used to ensure a nonnegative value.) Let d represent the length of $\overline{P_1P_2}$ and apply the Pythagorean theorem to obtain

$$d^2 = |x_2 - x_1|^2 + |y_2 - y_1|^2$$

Because $|a|^2 = a^2$, the **distance formula** can be stated as

$$d = \sqrt{(x_2 - x_1)^2 + (y_2 - y_1)^2}$$

It makes no difference which point you call P_1 or P_2 when using the distance formula. If you forget the formula, don't panic. Just form a right triangle and apply the Pythagorean theorem as we did in Example 2. Let's consider an example that demonstrates the use of the distance formula.

Answers to the distance problems can be left in square-root form or approximated using a calculator. Radical answers in this chapter will be restricted to radicals that are perfect squares or radicals that do not need to be simplified. The skill of simplifying radicals is covered in Chapter 7, after which you will be able to simplify the answers for distance problems.

EXAMPLE 3

Find the distance between $(-1, 5)$ and $(1, 2)$.

Solution

Let $(-1, 5)$ be P_1 and $(1, 2)$ be P_2. Using the distance formula, we obtain

$$\begin{aligned} d &= \sqrt{[(1 - (-1))]^2 + (2 - 5)^2} \\ &= \sqrt{2^2 + (-3)^2} \\ &= \sqrt{4 + 9} \\ &= \sqrt{13} \end{aligned}$$

The distance between the two points is $\sqrt{13}$ units.

▼ **PRACTICE YOUR SKILL**

Find the distance between the points $A(-1, 3)$ and $B(4, 5)$. ■

In Example 3, we did not sketch a figure because of the simplicity of the problem. However, sometimes it is helpful to use a figure to organize the given information and aid in analyzing the problem, as we see in the next example.

EXAMPLE 4

Verify that the points (2, 2), (11, 7), and (4, 9) are vertices of an isosceles triangle. (An isosceles triangle has two sides of the same length.)

Solution

Let's plot the points and draw the triangle (Figure 3.25). Use the distance formula to find the lengths d_1, d_2, and d_3, as follows:

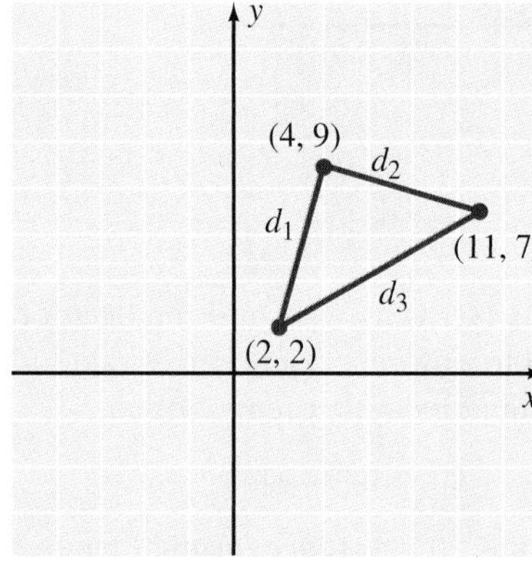

Figure 3.25

$$d_1 = \sqrt{(4-2)^2 + (9-2)^2} = \sqrt{2^2 + 7^2} = \sqrt{4 + 49} = \sqrt{53}$$

$$d_2 = \sqrt{(11-4)^2 + (7-9)^2} = \sqrt{7^2 + (-2)^2} = \sqrt{49 + 4} = \sqrt{53}$$

$$d_3 = \sqrt{(11-2)^2 + (7-2)^2} = \sqrt{9^2 + 5^2} = \sqrt{81 + 25} = \sqrt{106}$$

Because $d_1 = d_2$, we know that it is an isosceles triangle.

▼ **PRACTICE YOUR SKILL**

Verify that the points (−2, −2), (7, −1), and (2, 3) are vertices of an isosceles triangle. ■

2 Find the Slope of a Line

In coordinate geometry, the concept of **slope** is used to describe the "steepness" of lines. The slope of a line is the ratio of the vertical change to the horizontal change as we move from one point on a line to another point. This is illustrated in Figure 3.26 with points P_1 and P_2.

A precise definition for slope can be given by considering the coordinates of the points P_1, P_2, and R as indicated in Figure 3.27. The horizontal change as we move from P_1 to P_2 is $x_2 - x_1$ and the vertical change is $y_2 - y_1$. Thus the following definition for slope is given.

Definition 3.1

If points P_1 and P_2 with coordinates (x_1, y_1) and (x_2, y_2), respectively, are any two different points on a line, then the slope of the line (denoted by m) is

$$m = \frac{y_2 - y_1}{x_2 - x_1}, \qquad x_2 \neq x_1$$

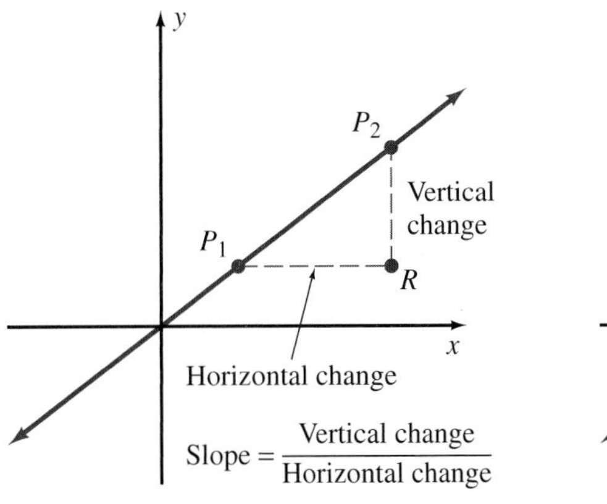

Figure 3.26

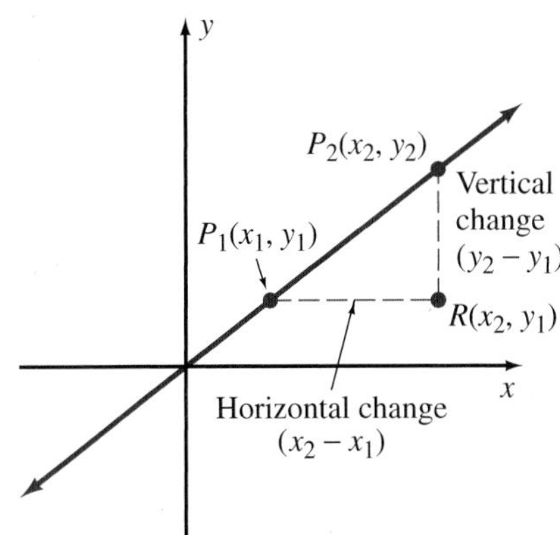

Figure 3.27

Because $\frac{y_2 - y_1}{x_2 - x_1} = \frac{y_1 - y_2}{x_1 - x_2}$, how we designate P_1 and P_2 is not important. Let's use Definition 3.1 to find the slopes of some lines.

EXAMPLE 5

Find the slope of the line determined by each of the following pairs of points, and graph the lines.

(a) $(-1, 1)$ and $(3, 2)$ **(b)** $(4, -2)$ and $(-1, 5)$

(c) $(2, -3)$ and $(-3, -3)$

Solution

(a) Let $(-1, 1)$ be P_1 and $(3, 2)$ be P_2 (Figure 3.28).

$$m = \frac{y_2 - y_1}{x_2 - x_1} = \frac{2 - 1}{3 - (-1)} = \frac{1}{4}$$

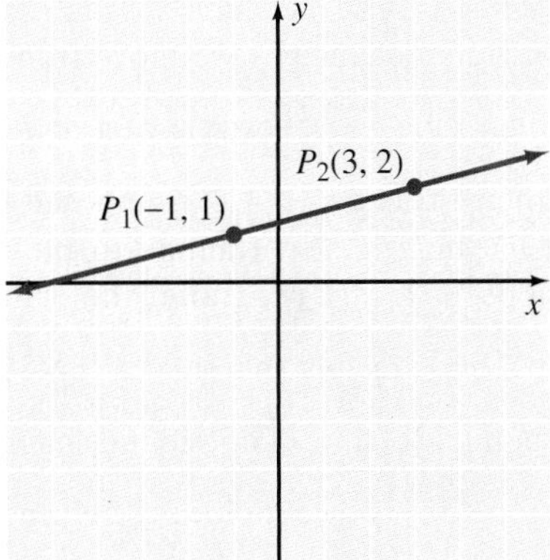

Figure 3.28

(b) Let $(4, -2)$ be P_1 and $(-1, 5)$ be P_2 (Figure 3.29).

$$m = \frac{y_2 - y_1}{x_2 - x_1} = \frac{5 - (-2)}{-1 - 4} = \frac{7}{-5} = -\frac{7}{5}$$

(c) Let $(2, -3)$ be P_1 and $(-3, -3)$ be P_2 (Figure 3.30).

$$m = \frac{y_2 - y_1}{x_2 - x_1}$$

$$= \frac{-3 - (-3)}{-3 - 2}$$

$$= \frac{0}{-5} = 0$$

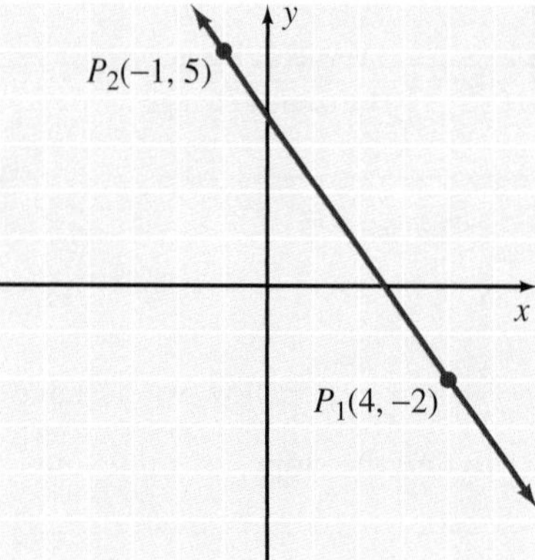

Figure 3.29

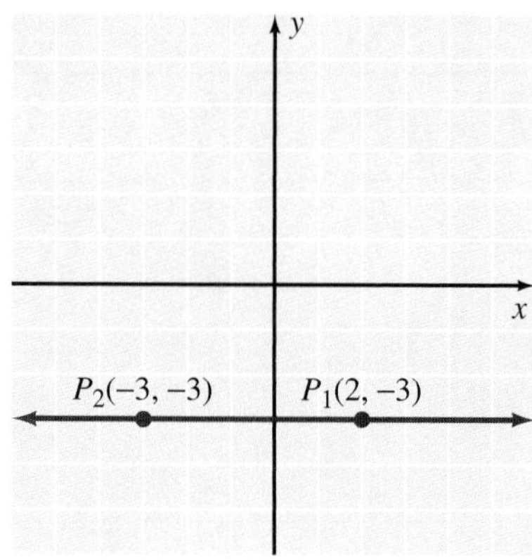

Figure 3.30

▼ PRACTICE YOUR SKILL

Find the slope of the line determined by each of the following pairs of points, and graph the lines.

(a) $(-4, 2)$ and $(2, 5)$ **(b)** $(-3, 4)$ and $(1, 4)$
(c) $(3, -2)$ and $(0, 2)$ ■

The three parts of Example 5 represent the three basic possibilities for slope; that is, the slope of a line can be positive, negative, or zero. A line that has a positive slope rises as we move from left to right, as in Figure 3.28. A line that has a negative slope falls as we move from left to right, as in Figure 3.29. A horizontal line, as in Figure 3.30, has a slope of zero. Finally, we need to realize that *the concept of slope is undefined for vertical lines.* This is due to the fact that for any vertical line, the horizontal change as we move from one point on the line to another is zero. Thus the ratio $\frac{y_2 - y_1}{x_2 - x_1}$ will have a denominator of zero and be undefined. Accordingly, the restriction $x_2 \neq x_1$ is imposed in Definition 3.1.

One final idea pertaining to the concept of slope needs to be emphasized. The slope of a line is a **ratio**, the ratio of vertical change to horizontal change. A slope of $\frac{2}{3}$ means that for every 2 units of vertical change there must be a corresponding 3 units of horizontal change. Thus, starting at some point on a line that has a slope of $\frac{2}{3}$, we could locate other points on the line as follows:

$\frac{2}{3} = \frac{4}{6}$ → by moving 4 units *up* and 6 units to the *right*

$\frac{2}{3} = \frac{8}{12}$ → by moving 8 units *up* and 12 units to the *right*

$\frac{2}{3} = \frac{-2}{-3}$ → by moving 2 units *down* and 3 units to the *left*

Likewise, if a line has a slope of $-\frac{3}{4}$, then by starting at some point on the line we could locate other points on the line as follows:

$-\frac{3}{4} = \frac{-3}{4}$ → by moving 3 units *down* and 4 units to the *right*

$-\frac{3}{4} = \frac{3}{-4}$ → by moving 3 units *up* and 4 units to the *left*

$-\frac{3}{4} = \frac{-9}{12}$ → by moving 9 units *down* and 12 units to the *right*

$-\frac{3}{4} = \frac{15}{-20}$ → by moving 15 units *up* and 20 units to the *left*

3 Use Slope to Graph Lines

EXAMPLE 6

Graph the line that passes through the point (0, −2) and has a slope of $\frac{1}{3}$.

Solution

To graph, plot the point (0, −2). Furthermore, because the slope is equal to $\frac{\text{vertical change}}{\text{horizontal change}} = \frac{1}{3}$, we can locate another point on the line by starting from the point (0, −2) and moving 1 unit up and 3 units to the right to obtain the point (3, −1). Because two points determine a line, we can draw the line (Figure 3.31).

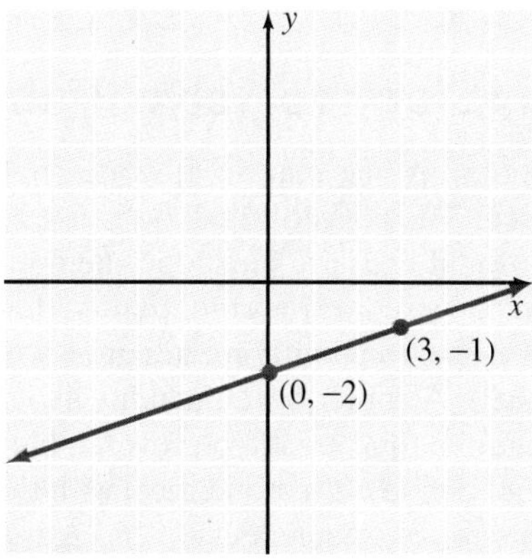

Figure 3.31

Remark: Because $m = \frac{1}{3} = \frac{-1}{-3}$, we can locate another point by moving 1 unit down and 3 units to the left from the point (0, −2).

▼ PRACTICE YOUR SKILL

Graph the line that passes through the point (−3, 2) and has a slope of $\frac{2}{5}$. ■

EXAMPLE 7 Graph the line that passes through the point (1, 3) and has a slope of −2.

Solution

To graph the line, plot the point (1, 3). We know that $m = -2 = \frac{-2}{1}$. Furthermore, because the slope $= \frac{\text{vertical change}}{\text{horizontal change}} = \frac{-2}{1}$, we can locate another point on the line by starting from the point (1, 3) and moving 2 units down and 1 unit to the right to obtain the point (2, 1). Because two points determine a line, we can draw the line (Figure 3.32).

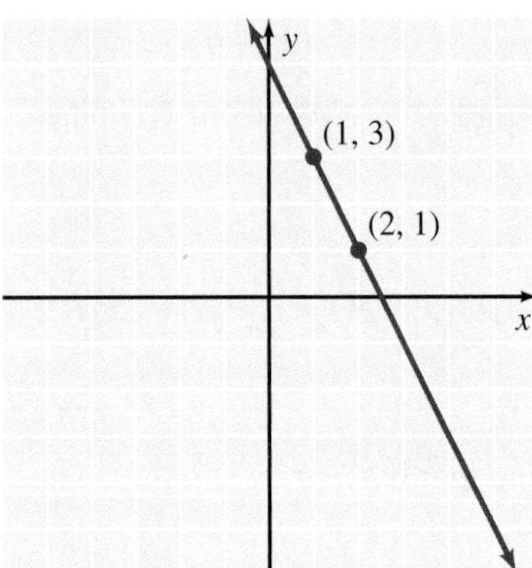

Figure 3.32

Remark: Because $m = -2 = \frac{-2}{1} = \frac{2}{-1}$ we can locate another point by moving 2 units up and 1 unit to the left from the point (1, 3).

▼ **PRACTICE YOUR SKILL**

Graph the line that passes through the point (2, 0) and has a slope of $m = -\frac{1}{3}$. ■

4 Apply Slope to Solve Problems

The concept of slope has many real-world applications even though the word *slope* is often not used. The concept of slope is used in most situations where an incline is involved. Hospital beds are hinged in the middle so that both the head end and the foot end can be raised or lowered; that is, the slope of either end of the bed can be changed. Likewise, treadmills are designed so that the incline (slope) of the platform can be adjusted. A roofer, when making an estimate to replace a roof, is concerned not only about the total area to be covered but also about the pitch of the roof. (Contractors do not define *pitch* as identical with the mathematical definition of slope, but both concepts refer to "steepness.") In Figure 3.33, the two roofs might require the same amount of shingles, but the roof on the left will take longer to complete because the pitch is so great that scaffolding will be required.

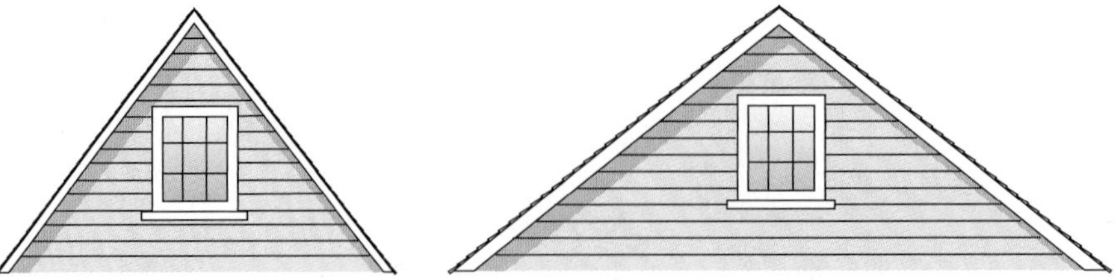

Figure 3.33

The concept of slope is also used in the construction of flights of stairs (Figure 3.34). The terms *rise* and *run* are commonly used, and the steepness (slope) of the stairs can be expressed as the ratio of rise to run. In Figure 3.34, the stairs on the left, where the ratio of rise to run is $\frac{10}{11}$, are steeper than the stairs on the right, which have a ratio of $\frac{7}{11}$.

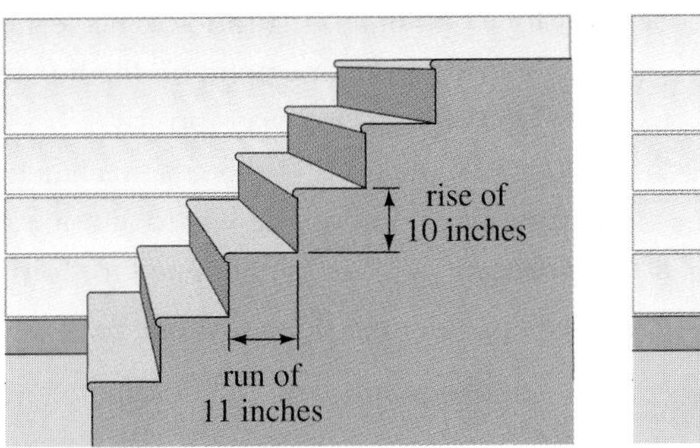

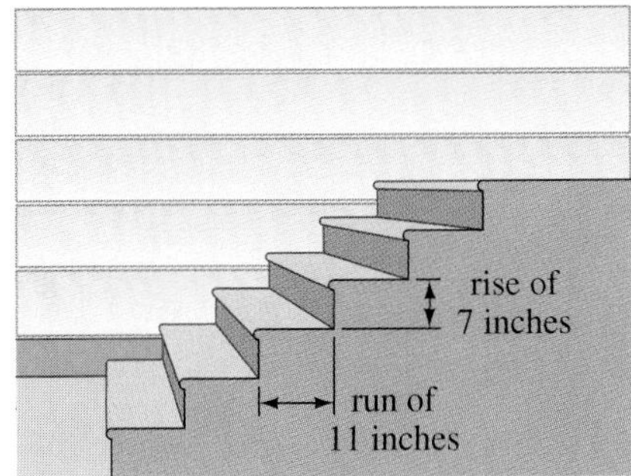

Figure 3.34

In highway construction, the word *grade* is used for the concept of slope. For example, in Figure 3.35 the highway is said to have a grade of 17%. This means that for every horizontal distance of 100 feet, the highway rises or drops 17 feet. In other words, the slope of the highway is $\frac{17}{100}$.

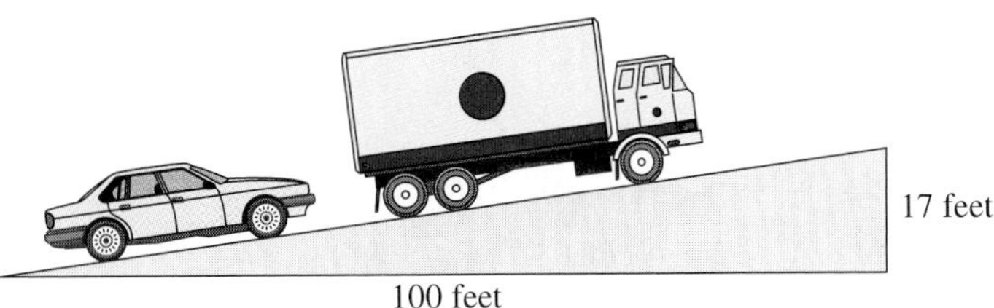

Figure 3.35

EXAMPLE 8

A certain highway has a 3% grade. How many feet does it rise in a horizontal distance of 1 mile?

Solution

A 3% grade means a slope of $\frac{3}{100}$. Therefore, if we let y represent the unknown vertical distance and use the fact that 1 mile = 5280 feet, we can set up and solve the following proportion.

$$\frac{3}{100} = \frac{y}{5280}$$
$$100y = 3(5280) = 15{,}840$$
$$y = 158.4$$

The highway rises 158.4 feet in a horizontal distance of 1 mile.

▼ PRACTICE YOUR SKILL

A certain highway has a 2.5% grade. How many feet does it rise in a horizontal distance of 2000 feet? ■

CONCEPT QUIZ

For Problems 1–10, answer true or false.

1. When applying the distance formula $\sqrt{(x_2 - x_1)^2 + (y_2 - y_1)^2}$ to find the distance between two points, you can designate either of the two points as P_1.
2. An isosceles triangle has two sides of the same length.
3. The distance between the points $(-1, 4)$ and $(-1, -2)$ is 2 units.
4. The distance between the points $(3, -4)$ and $(3, 2)$ is undefined.
5. The slope of a line is the ratio of the vertical change to the horizontal change when moving from one point on the line to another point on the line.
6. The slope of a line is always positive.
7. A slope of 0 means that there is no change in the vertical direction when moving from one point on the line to another point on the line.
8. The concept of slope is undefined for horizontal lines.
9. When applying the slope formula $m = \frac{y_2 - y_1}{x_2 - x_1}$ to find the slope of a line between two points, you can designate either of the two points as P_2.
10. If the ratio of the rise to the run for some steps is $\frac{3}{4}$ and the rise is 9 inches, then the run is $6\frac{3}{4}$ inches.

Problem Set 3.3

1 Find the Distance between Two Points

For Problems 1–12, find the distance between each of the pairs of points. Express answers in radical form.

1. $(-2, -1), (7, 11)$
2. $(2, 1), (10, 7)$
3. $(1, -1), (3, -4)$
4. $(-1, 3), (2, -2)$
5. $(6, -5), (9, -7)$
6. $(-4, 2), (-1, 6)$
7. $(-3, 3), (0, -2)$
8. $(-1, -4), (4, 0)$
9. $(1, -6), (-5, -6)$
10. $(-2, 3), (-2, -7)$
11. $(1, 7), (4, -1)$
12. $(6, 4), (3, -8)$

13. Verify that the points $(0, 2)$, $(0, 7)$, and $(12, 7)$ are vertices of a right triangle. [*Hint:* If $a^2 + b^2 = c^2$, then it is a right triangle with the right angle opposite side c.]

14. Verify that the points $(0, 4)$, $(-3, 0)$, and $(3, 0)$ are vertices of an isosceles triangle.

15. Verify that the points $(3, 5)$ and $(5, 8)$ divide the line segment joining $(1, 2)$ and $(7, 11)$ into three segments of equal length.

16. Verify that $(5, 1)$ is the midpoint of the line segment joining $(2, 6)$ and $(8, -4)$.

2 Find the Slope of a Line

For Problems 17–28, graph the line determined by the two points and find the slope of the line.

17. $(1, 2), (4, 6)$
18. $(3, 1), (-2, -2)$
19. $(-4, 5), (-1, -2)$
20. $(-2, 5), (3, -1)$
21. $(2, 6), (6, -2)$
22. $(-2, -1), (2, -5)$
23. $(-6, 1), (-1, 4)$
24. $(-3, 3), (2, 3)$
25. $(-2, -4), (2, -4)$
26. $(1, -5), (4, -1)$
27. $(0, -2), (4, 0)$
28. $(-4, 0), (0, -6)$

29. Find x if the line through $(-2, 4)$ and $(x, 6)$ has a slope of $\frac{2}{9}$.

30. Find y if the line through $(1, y)$ and $(4, 2)$ has a slope of $\frac{5}{3}$.

31. Find x if the line through $(x, 4)$ and $(2, -5)$ has a slope of $-\frac{9}{4}$.

32. Find y if the line through $(5, 2)$ and $(-3, y)$ has a slope of $-\frac{7}{8}$.

For Problems 33–40, you are given one point on a line and the slope of the line. Find the coordinates of three other points on the line.

33. $(2, 5)$, $m = \frac{1}{2}$
34. $(3, 4)$, $m = \frac{5}{6}$
35. $(-3, 4)$, $m = 3$
36. $(-3, -6)$, $m = 1$
37. $(5, -2)$, $m = -\frac{2}{3}$
38. $(4, -1)$, $m = -\frac{3}{4}$
39. $(-2, -4)$, $m = -2$
40. $(-5, 3)$, $m = -3$

For Problems 41–50, find the coordinates of two points on the given line, and then use those coordinates to find the slope of the line.

41. $2x + 3y = 6$

42. $4x + 5y = 20$

43. $x - 2y = 4$

44. $3x - y = 12$

45. $4x - 7y = 12$

46. $2x + 7y = 11$

47. $y = 4$

48. $x = 3$

49. $y = -5x$

50. $y - 6x = 0$

3 Use Slope to Graph Lines

For Problems 51–58, graph the line that passes through the given point and has the given slope.

51. $(3, 1)\quad m = \frac{2}{3}$

52. $(-1, 0)\quad m = \frac{3}{4}$

53. $(-2, 3)\quad m = -1$

54. $(1, -4)\quad m = -3$

55. $(0, 5)\quad m = \frac{-1}{4}$

56. $(-3, 4)\quad m = \frac{-3}{2}$

57. $(2, -2)\quad m = \frac{3}{2}$

58. $(3, -4)\quad m = \frac{5}{2}$

4 Apply Slope to Solve Problems

59. A certain highway has a 2% grade. How many feet does it rise in a horizontal distance of 1 mile? (1 mile = 5280 feet)

60. The grade of a highway up a hill is 30%. How much change in horizontal distance is there if the vertical height of the hill is 75 feet?

61. Suppose that a highway rises a distance of 215 feet in a horizontal distance of 2640 feet. Express the grade of the highway to the nearest tenth of a percent.

62. If the ratio of rise to run is to be $\frac{3}{5}$ for some steps and the rise is 19 centimeters, find the run to the nearest centimeter.

63. If the ratio of rise to run is to be $\frac{2}{3}$ for some steps and the run is 28 centimeters, find the rise to the nearest centimeter.

64. Suppose that a county ordinance requires a $2\frac{1}{4}$% "fall" for a sewage pipe from the house to the main pipe at the street. How much vertical drop must there be for a horizontal distance of 45 feet? Express the answer to the nearest tenth of a foot.

THOUGHTS INTO WORDS

65. How would you explain the concept of slope to someone who was absent from class the day it was discussed?

66. If one line has a slope of $\frac{2}{5}$ and another line has a slope of $\frac{3}{7}$, which line is steeper? Explain your answer.

67. Suppose that a line has a slope of $\frac{2}{3}$ and contains the point (4, 7). Are the points (7, 9) and (1, 3) also on the line? Explain your answer.

FURTHER INVESTIGATIONS

68. Sometimes it is necessary to find the coordinate of a point on a number line that is located somewhere between two given points. For example, suppose that we want to find the coordinate (x) of the point located two-thirds of the distance from 2 to 8. Because the total distance from 2 to 8 is $8 - 2 = 6$ units, we can start at 2 and move $\frac{2}{3}(6) = 4$ units toward 8. Thus $x = 2 + \frac{2}{3}(6) = 2 + 4 = 6$.

For each of the following, find the coordinate of the indicated point on a number line.

(a) Two-thirds of the distance from 1 to 10
(b) Three-fourths of the distance from −2 to 14
(c) One-third of the distance from −3 to 7
(d) Two-fifths of the distance from −5 to 6
(e) Three-fifths of the distance from −1 to −11
(f) Five-sixths of the distance from 3 to −7

69. Now suppose that we want to find the coordinates of point P, which is located two-thirds of the distance from $A(1, 2)$ to $B(7, 5)$ in a coordinate plane. We have plotted the given points A and B in Figure 3.36 to help with the analysis of this problem. Point D is two-thirds of the distance from A to C because parallel lines cut off proportional segments on every transversal that intersects the lines. Thus $\overline{AC}$ can be treated as a segment of a number line, as shown in Figure 3.37.

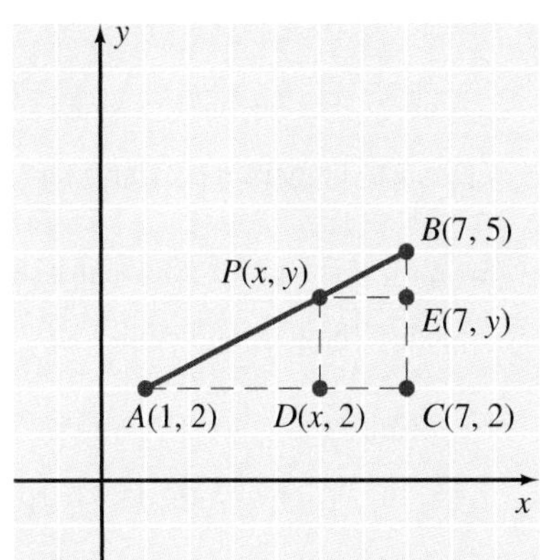

Figure 3.36

Figure 3.37

Therefore,

$$x = 1 + \frac{2}{3}(7 - 1) = 1 + \frac{2}{3}(6) = 5$$

Similarly, $\overline{CB}$ can be treated as a segment of a number line, as shown in Figure 3.38. Therefore,

B 5
E y
C 2

$$y = 2 + \frac{2}{3}(5 - 2) = 2 + \frac{2}{3}(3) = 4$$

The coordinates of point P are (5, 4).

Figure 3.38

For each of the following, find the coordinates of the indicated point in the xy plane.

(a) One-third of the distance from (2, 3) to (5, 9)
(b) Two-thirds of the distance from (1, 4) to (7, 13)
(c) Two-fifths of the distance from (−2, 1) to (8, 11)
(d) Three-fifths of the distance from (2, −3) to (−3, 8)
(e) Five-eighths of the distance from (−1, −2) to (4, −10)
(f) Seven-eighths of the distance from (−2, 3) to (−1, −9)

70. Suppose we want to find the coordinates of the midpoint of a line segment. Let $P(x, y)$ represent the midpoint of the line segment from $A(x_1, y_1)$ to $B(x_2, y_2)$. Using the method in Problem 68, the formula for the x coordinate of the midpoint is $x = x_1 + \frac{1}{2}(x_2 - x_1)$. This formula can be manipulated algebraically to produce a simpler formula:

$$x = x_1 + \frac{1}{2}(x_2 - x_1)$$

$$x = x_1 + \frac{1}{2}x_2 - \frac{1}{2}x_1$$

$$x = \frac{1}{2}x_1 + \frac{1}{2}x_2$$

$$x = \frac{x_1 + x_2}{2}$$

Hence the x coordinate of the midpoint can be interpreted as the average of the x coordinates of the endpoints of the line segment. A similar argument for the y coordinate of the midpoint gives the following formula:

$$y = \frac{y_1 + y_2}{2}$$

For each of the pairs of points, use the formula to find the midpoint of the line segment between the points.

(a) (3, 1) and (7, 5)
(b) (−2, 8) and (6, 4)
(c) (−3, 2) and (5, 8)
(d) (4, 10) and (9, 25)
(e) (−4, −1) and (−10, 5)
(f) (5, 8) and (−1, 7)

GRAPHING CALCULATOR ACTIVITIES

71. Remember that we did some work with parallel lines back in the graphing calculator activities in Problem Set 3.1. Now let's do some work with perpendicular lines. Be sure to set your boundaries so that the distance between tick marks is the same on both axes.

(a) Graph $y = 4x$ and $y = -\frac{1}{4}x$ on the same set of axes. Do they appear to be perpendicular lines?

(b) Graph $y = 3x$ and $y = \frac{1}{3}x$ on the same set of axes. Do they appear to be perpendicular lines?

(c) Graph $y = \frac{2}{5}x - 1$ and $y = -\frac{5}{2}x + 2$ on the same set of axes. Do they appear to be perpendicular lines?

(d) Graph $y = \frac{3}{4}x - 3$, $y = \frac{4}{3}x + 2$, and $y = -\frac{4}{3}x + 2$ on the same set of axes. Does there appear to be a pair of perpendicular lines?

(e) On the basis of your results in parts (a) through (d), make a statement about how we can recognize perpendicular lines from their equations.

72. For each of the following pairs of equations: (1) predict whether they represent parallel lines, perpendicular lines, or lines that intersect but are not perpendicular; and (2) graph each pair of lines to check your prediction.

(a) $5.2x + 3.3y = 9.4$ and $5.2x + 3.3y = 12.6$
(b) $1.3x - 4.7y = 3.4$ and $1.3x - 4.7y = 11.6$
(c) $2.7x + 3.9y = 1.4$ and $2.7x - 3.9y = 8.2$
(d) $5x - 7y = 17$ and $7x + 5y = 19$
(e) $9x + 2y = 14$ and $2x + 9y = 17$
(f) $2.1x + 3.4y = 11.7$ and $3.4x - 2.1y = 17.3$

Answers to the Concept Quiz

1. True **2.** True **3.** False **4.** False **5.** True **6.** False **7.** True **8.** False **9.** True **10.** False

Answers to the Example Practice Skills

1. 8 units **2.** 13 units **3.** $\sqrt{29}$ units **4.** $d_1 = d_2 = \sqrt{41}, d_3 = \sqrt{82}$

5. (a) $m = \frac{1}{2}$

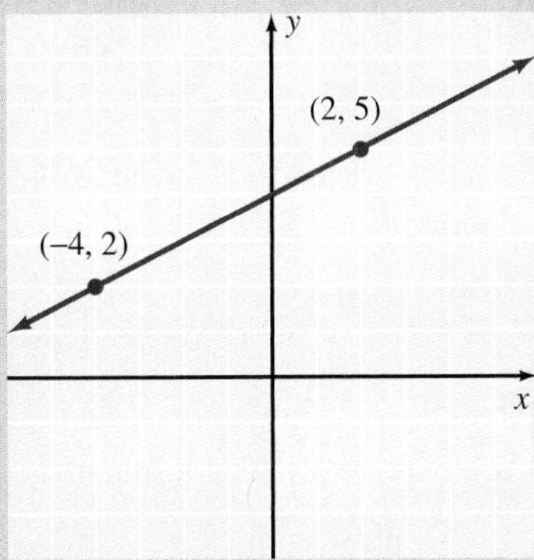

(b) $m = 0$

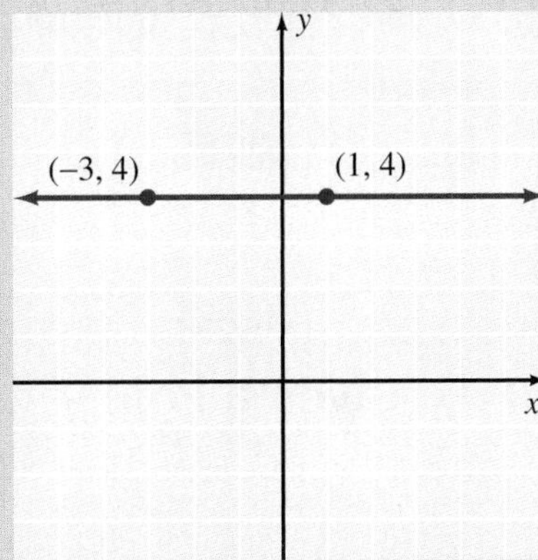

(c) $m = -\frac{4}{3}$

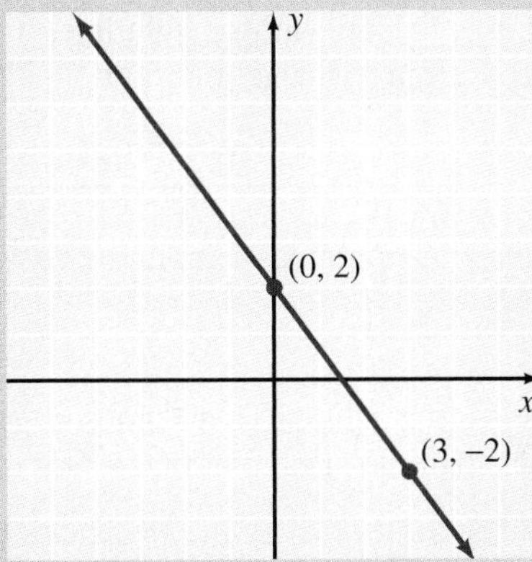

6.

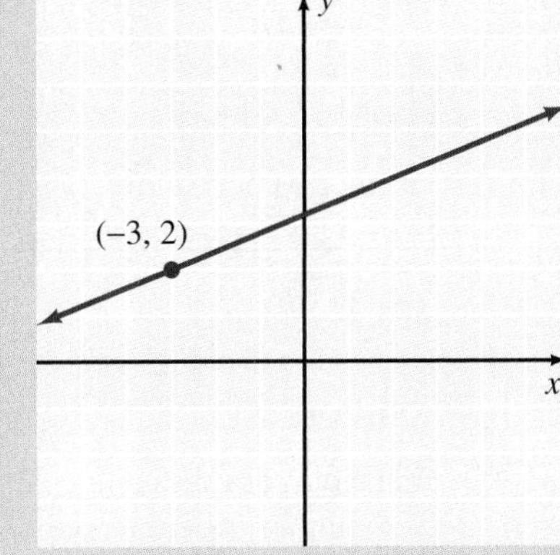

7.

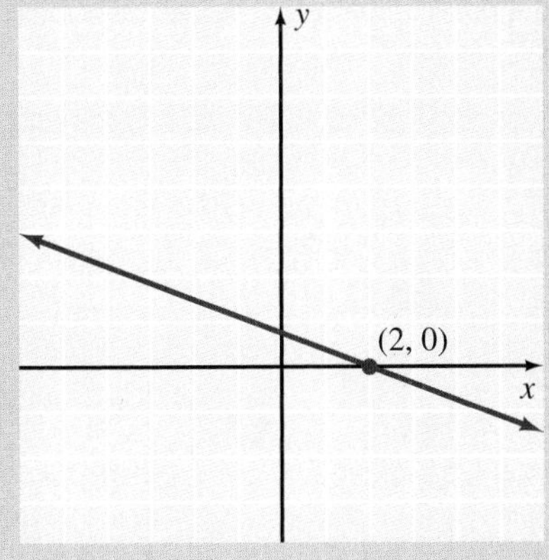

8. 50 ft

3.4 Determining the Equation of a Line

OBJECTIVES

1. Find the Equation of a Line Given a Point and a Slope
2. Find the Equation of a Line Given Two Points
3. Find the Equation of a Line Given the Slope and y Intercept
4. Use the Point-Slope Form to Write Equations of Lines
5. Apply the Slope-Intercept Form of an Equation
6. Find the Equations for Parallel or Perpendicular Lines

1 Find the Equation of a Line Given a Point and a Slope

To review, there are basically two types of problems to solve in coordinate geometry:

1. Given an algebraic equation, find its geometric graph.
2. Given a set of conditions pertaining to a geometric figure, find its algebraic equation.

Problems of type 1 have been our primary concern thus far in this chapter. Now let's analyze some problems of type 2 that deal specifically with straight lines. Given certain facts about a line, we need to be able to determine its algebraic equation. Let's consider some examples.

EXAMPLE 1

Find the equation of the line that has a slope of $\frac{2}{3}$ and contains the point (1, 2).

Solution

First, let's draw the line and record the given information. Then choose a point (x, y) that represents any point on the line other than the given point (1, 2). (See Figure 3.39.)

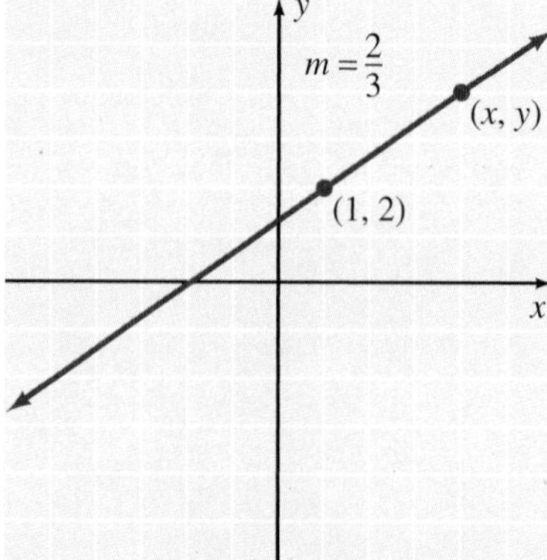

Figure 3.39

The slope determined by (1, 2) and (x, y) is $\frac{2}{3}$. Thus

$$\frac{y-2}{x-1} = \frac{2}{3}$$

$$2(x-1) = 3(y-2)$$

$$2x - 2 = 3y - 6$$

$$2x - 3y = -4$$

▼ **PRACTICE YOUR SKILL**

Find the equation of the line that has a slope of $\frac{3}{4}$ and contains the point $(3, -1)$. ■

2 Find the Equation of a Line Given Two Points

EXAMPLE 2 Find the equation of the line that contains (3, 2) and (−2, 5).

Solution

First, let's draw the line determined by the given points (Figure 3.40); if we know two points, we can find the slope.

$$m = \frac{y_2 - y_1}{x_2 - x_1} = \frac{3}{-5} = -\frac{3}{5}$$

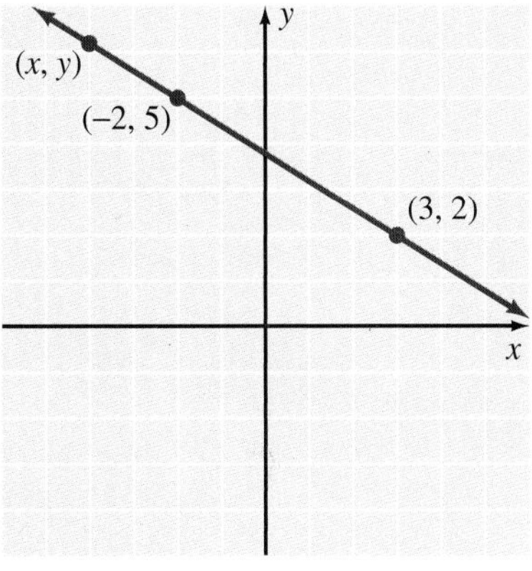

Figure 3.40

Now we can use the same approach as in Example 1.

Form an equation using a variable point (x, y), one of the two given points, and the slope of $-\frac{3}{5}$.

$$\frac{y - 5}{x + 2} = \frac{3}{-5} \qquad \left(-\frac{3}{5} = \frac{3}{-5}\right)$$

$$3(x + 2) = -5(y - 5)$$

$$3x + 6 = -5y + 25$$

$$3x + 5y = 19$$

▼ **PRACTICE YOUR SKILL**

Find the equation of the line that contains (2, 5) and (−4, 10). ■

3 Find the Equation of a Line Given the Slope and y Intercept

EXAMPLE 3 Find the equation of the line that has a slope of $\frac{1}{4}$ and a y intercept of 2.

Solution

A y intercept of 2 means that the point (0, 2) is on the line (Figure 3.41).

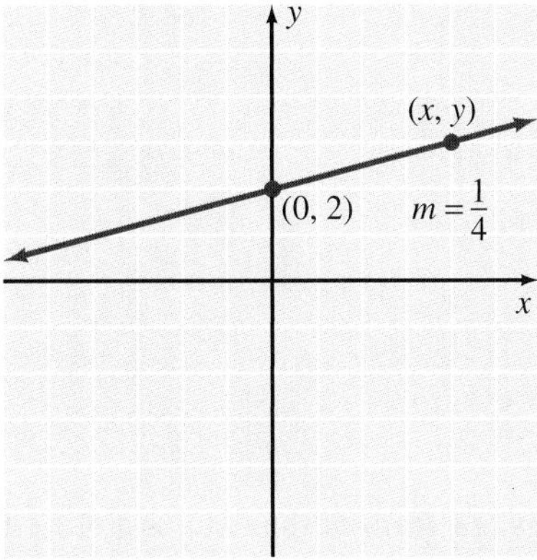

Figure 3.41

Choose a variable point (x, y) and proceed as in the previous examples.

$$\frac{y-2}{x-0} = \frac{1}{4}$$

$$1(x-0) = 4(y-2)$$

$$x = 4y - 8$$

$$x - 4y = -8$$

▼ **PRACTICE YOUR SKILL**

Find the equation of the line that has a slope of $\frac{3}{2}$ and a y intercept of -4. ■

Perhaps it would be helpful to pause a moment and look back over Examples 1, 2, and 3. Note that we used the same basic approach in all three situations. We chose a variable point (x, y) and used it to determine the equation that satisfies the conditions given in the problem. The approach we took in the previous examples can be generalized to produce some special forms of equations of straight lines.

4 Use the Point-Slope Form to Write Equations of Lines

Generalizing from the previous examples, let's find the equation of a line that has a slope of m and contains the point (x_1, y_1). To use the slope formula we will need two points. Choosing a point (x, y) to represent any other point on the line (Figure 3.42) and using the given point (x_1, y_1), we can determine slope to be

$$m = \frac{y - y_1}{x - x_1}, \quad \text{where } x \neq x_1$$

Simplifying gives us the equation $y - y_1 = m(x - x_1)$.

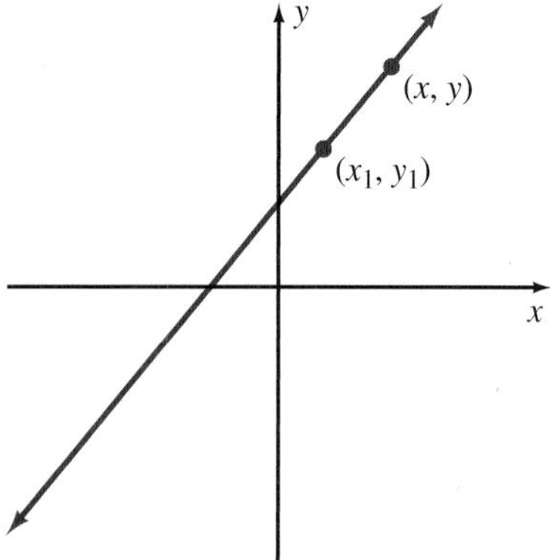

Figure 3.42

We refer to the equation

$$y - y_1 = m(x - x_1)$$

as the **point-slope form** of the equation of a straight line. Instead of the approach we used in Example 1, we could use the point-slope form to write the equation of a line with a given slope that contains a given point.

EXAMPLE 4

Use the point-slope form to find the equation of a line that has a slope of $\frac{3}{5}$ and contains the point (2, 4).

Solution

We can determine the equation of the line by substituting $\frac{3}{5}$ for m and (2, 4) for (x_1, y_1) in the point-slope form.

$$y - y_1 = m(x - x_1)$$

$$y - 4 = \frac{3}{5}(x - 2)$$

$$5(y - 4) = 3(x - 2)$$

$$5y - 20 = 3x - 6$$

$$-14 = 3x - 5y$$

Thus the equation of the line is $3x - 5y = -14$.

▼ **PRACTICE YOUR SKILL**

Use the point-slope form to find the equation of a line that has a slope of $\frac{4}{3}$ and contains the point $(-2, 5)$. ■

5 Apply the Slope-Intercept Form of an Equation

Another special form of the equation of a line is the slope-intercept form. Let's use the point-slope form to find the equation of a line that has a slope of m and a y intercept of b. A y intercept of b means that the line contains the point $(0, b)$, as in Figure 3.43. Therefore, we can use the point-slope form as follows:

$$y - y_1 = m(x - x_1)$$

$$y - b = m(x - 0)$$

$$y - b = mx$$

$$y = mx + b$$

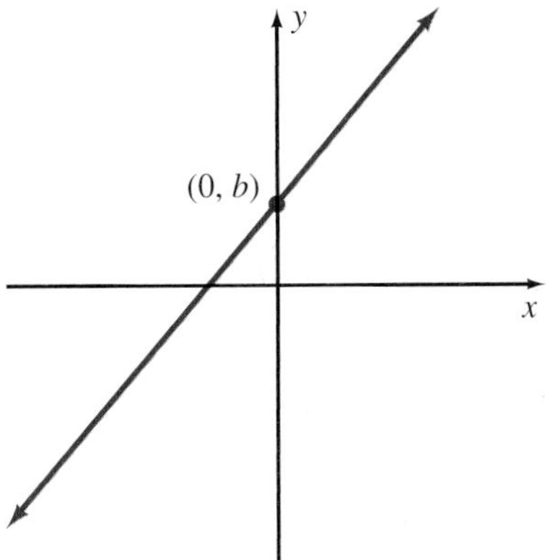

Figure 3.43

We refer to the equation

$$y = mx + b$$

as the **slope-intercept form** of the equation of a straight line. We use it for three primary purposes, as the next three examples illustrate.

EXAMPLE 5 Find the equation of the line that has a slope of $\frac{1}{4}$ and a y intercept of 2.

Solution

This is a restatement of Example 3, but this time we will use the slope-intercept form ($y = mx + b$) of a line to write its equation. Because $m = \frac{1}{4}$ and $b = 2$, we can substitute these values into $y = mx + b$.

$$y = mx + b$$

$$y = \frac{1}{4}x + 2$$

$$4y = x + 8 \quad \text{Multiply both sides by 4}$$

$$x - 4y = -8 \quad \text{Same result as in Example 3}$$

▼ **PRACTICE YOUR SKILL**

Find the equation of the line that has a slope of -3 and a y intercept of 8. ■

EXAMPLE 6 Find the slope of the line when the equation is $3x + 2y = 6$.

Solution

We can solve the equation for y in terms of x and then compare it to the slope-intercept form to determine its slope. Thus

$$3x + 2y = 6$$

$$2y = -3x + 6$$

$$y = -\frac{3}{2}x + 3$$

$$y = -\frac{3}{2}x + 3 \qquad y = mx + b$$

The slope of the line is $-\frac{3}{2}$. Furthermore, the y intercept is 3.

▼ **PRACTICE YOUR SKILL**

Find the slope of the line when the equation is $4x - 5y = 10$. ■

EXAMPLE 7

Graph the line determined by the equation $y = \frac{2}{3}x - 1$.

Solution

Comparing the given equation to the general slope-intercept form, we see that the slope of the line is $\frac{2}{3}$ and the y intercept is -1. Because the y intercept is -1, we can plot the point $(0, -1)$. Then, because the slope is $\frac{2}{3}$, let's move 3 units to the right and 2 units up from $(0, -1)$ to locate the point $(3, 1)$. The two points $(0, -1)$ and $(3, 1)$ determine the line in Figure 3.44. (Again, you should determine a third point as a check point.)

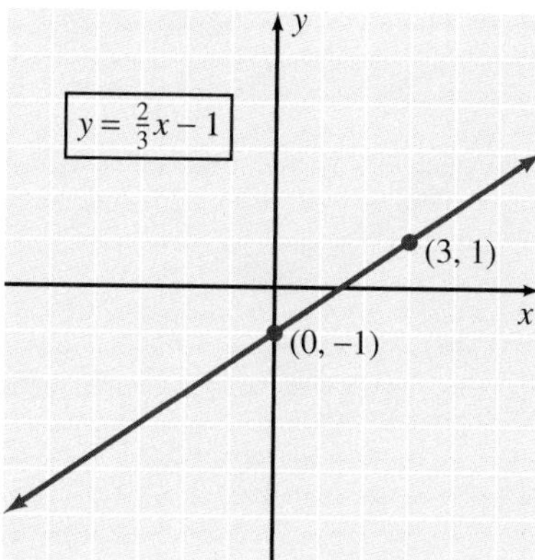

Figure 3.44

▼ **PRACTICE YOUR SKILL**

Graph the line determined by the equation $y = \frac{1}{4}x + 2$. ■

In general, if the equation of a nonvertical line is written in slope-intercept form ($y = mx + b$), then the coefficient of x is the slope of the line and the constant term is the y intercept. (Remember that the concept of slope is not defined for a vertical line.)

We use two forms of equations of straight lines extensively. They are the **standard form** and the **slope-intercept form**, and we describe them as follows.

Standard Form. $Ax + By = C$, where B and C are integers and A is a nonnegative integer (A and B not both zero).

Slope-Intercept Form. $y = mx + b$, where m is a real number representing the slope and b is a real number representing the y intercept.

6 Find the Equations for Parallel or Perpendicular Lines

We can use two important relationships between lines and their slopes to solve certain kinds of problems. It can be shown that nonvertical parallel lines have the same slope and that two nonvertical lines are perpendicular if the product of their slopes is -1. (Details for verifying these facts are left to another course.) In other words, if two lines have slopes m_1 and m_2, respectively, then

1. The two lines are parallel if and only if $m_1 = m_2$.
2. The two lines are perpendicular if and only if $(m_1)(m_2) = -1$.

The following examples demonstrate the use of these properties.

EXAMPLE 8

(a) Verify that the graphs of $2x + 3y = 7$ and $4x + 6y = 11$ are parallel lines.
(b) Verify that the graphs of $8x - 12y = 3$ and $3x + 2y = 2$ are perpendicular lines.

Solution

(a) Let's change each equation to slope-intercept form.

$$2x + 3y = 7 \quad \rightarrow \quad 3y = -2x + 7$$
$$y = -\frac{2}{3}x + \frac{7}{3}$$

$$4x + 6y = 11 \quad \rightarrow \quad 6y = -4x + 11$$
$$y = -\frac{4}{6}x + \frac{11}{6}$$
$$y = -\frac{2}{3}x + \frac{11}{6}$$

Both lines have a slope of $-\frac{2}{3}$, but they have different y intercepts. Therefore, the two lines are parallel.

(b) Solving each equation for y in terms of x, we obtain

$$8x - 12y = 3 \quad \rightarrow \quad -12y = -8x + 3$$
$$y = \frac{8}{12}x - \frac{3}{12}$$
$$y = \frac{2}{3}x - \frac{1}{4}$$

$$3x + 2y = 2 \quad \rightarrow \quad 2y = -3x + 2$$
$$y = -\frac{3}{2}x + 1$$

Because $\left(\frac{2}{3}\right)\left(-\frac{3}{2}\right) = -1$ (the product of the two slopes is -1), the lines are perpendicular.

▼ PRACTICE YOUR SKILL

(a) Verify that the graphs of $x - 3y = 2$ and $2x - 6y = 7$ are parallel lines.
(b) Verify that the graphs of $2x - 5y = 3$ and $5x + 2y = 8$ are perpendicular lines. ■

Remark: The statement "the product of two slopes is -1" is the same as saying that the two slopes are negative reciprocals of each other; that is, $m_1 = -\frac{1}{m_2}$.

EXAMPLE 9

Find the equation of the line that contains the point (1, 4) and is parallel to the line determined by $x + 2y = 5$.

Solution

First, let's draw a figure to help in our analysis of the problem (Figure 3.45). Because the line through (1, 4) is to be parallel to the line determined by $x + 2y = 5$, it must have the same slope. Let's find the slope by changing $x + 2y = 5$ to the slope-intercept form:

$$x + 2y = 5$$

$$2y = -x + 5$$

$$y = -\frac{1}{2}x + \frac{5}{2}$$

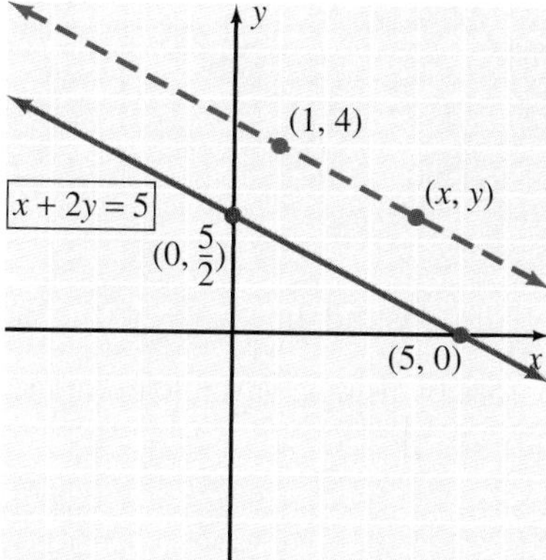

Figure 3.45

The slope of both lines is $-\frac{1}{2}$. Now we can choose a variable point (x, y) on the line through (1, 4) and proceed as we did in earlier examples.

$$\frac{y - 4}{x - 1} = \frac{1}{-2}$$

$$1(x - 1) = -2(y - 4)$$

$$x - 1 = -2y + 8$$

$$x + 2y = 9$$

▼ PRACTICE YOUR SKILL

Find the equation of the line that contains the point (−2, 7) and is parallel to the line determined by $3x + y = 4$. ■

EXAMPLE 10

Find the equation of the line that contains the point (−1, −2) and is perpendicular to the line determined by $2x - y = 6$.

Solution

First, let's draw a figure to help in our analysis of the problem (Figure 3.46). Because the line through (−1, −2) is to be perpendicular to the line determined by $2x - y = 6$, its slope must be the negative reciprocal of the slope of $2x - y = 6$. Let's find the slope of $2x - y = 6$ by changing it to the slope-intercept form.

$$2x - y = 6$$
$$-y = -2x + 6$$
$$y = 2x - 6 \quad \text{The slope is 2.}$$

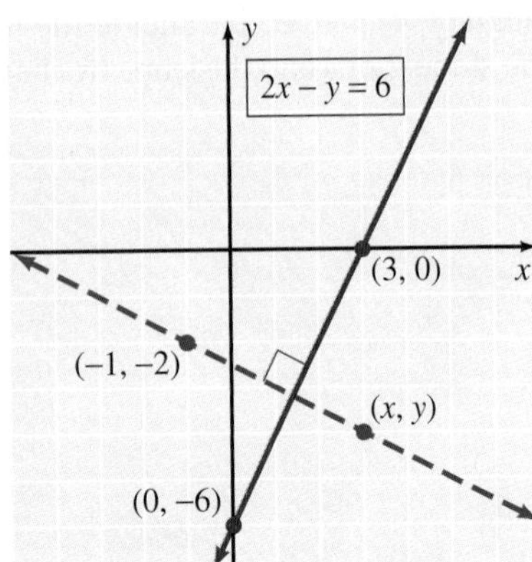

Figure 3.46

The slope of the desired line is $-\frac{1}{2}$ (the negative reciprocal of 2), and we can proceed as before by using a variable point (x, y).

$$\frac{y + 2}{x + 1} = \frac{1}{-2}$$
$$1(x + 1) = -2(y + 2)$$
$$x + 1 = -2y - 4$$
$$x + 2y = -5$$

▼ PRACTICE YOUR SKILL

Find the equation of the line that contains the point $(3, -1)$ and is perpendicular to the line determined by $5x + 2y = -10$. ■

CONCEPT QUIZ

For Problems 1–10, answer true or false.

1. If two lines have the same slope, then the lines are parallel.
2. If the slopes of two lines are reciprocals, then the lines are perpendicular.
3. In the standard form of the equation of a line $Ax + By = C$, A can be a rational number in fractional form.
4. In the slope-intercept form of an equation of a line $y = mx + b$, m is the slope.
5. In the standard form of the equation of a line $Ax + By = C$, A is the slope.
6. The slope of the line determined by the equation $3x - 2y = -4$ is $\frac{3}{2}$.
7. The concept of a slope is not defined for the line $y = 2$.
8. The concept of slope is not defined for the line $x = 2$.
9. The lines determined by the equations $x - 3y = 4$ and $2x - 6y = 11$ are parallel lines.
10. The lines determined by the equations $x - 3y = 4$ and $x + 3y = 4$ are perpendicular lines.

Problem Set 3.4

1 Find the Equation of a Line Given a Point and a Slope

For Problems 1–8, write the equation of the line that has the indicated slope and contains the indicated point. Express final equations in standard form.

1. $m = \frac{1}{2}$, $(3, 5)$
2. $m = \frac{1}{3}$, $(2, 3)$
3. $m = 3$, $(-2, 4)$
4. $m = -2$, $(-1, 6)$
5. $m = -\frac{3}{4}$, $(-1, -3)$
6. $m = -\frac{3}{5}$, $(-2, -4)$
7. $m = \frac{5}{4}$, $(4, -2)$
8. $m = \frac{3}{2}$, $(8, -2)$
9. x intercept of -3 and slope of $-\frac{5}{8}$
10. x intercept of 5 and slope of $-\frac{3}{10}$

2 Find the Equation of a Line Given Two Points

For Problems 11–22, write the equation of the line that contains the indicated pair of points. Express final equations in standard form.

11. $(2, 1), (6, 5)$
12. $(-1, 2), (2, 5)$
13. $(-2, -3), (2, 7)$
14. $(-3, -4), (1, 2)$
15. $(-3, 2), (4, 1)$
16. $(-2, 5), (3, -3)$
17. $(-1, -4), (3, -6)$
18. $(3, 8), (7, 2)$
19. $(0, 0), (5, 7)$
20. $(0, 0), (-5, 9)$
21. x intercept of 2 and y intercept of -4
22. x intercept of -1 and y intercept of -3

For Problems 23–28, the situations can be described by the use of linear equations in two variables. If two pairs of values are known, then we can determine the equation by using the approach used in Example 2 of this section. For each of the following, assume that the relationship can be expressed as a linear equation in two variables, and use the given information to determine the equation. Express the equation in slope-intercept form.

23. A company uses 7 pounds of fertilizer for a lawn that measures 5000 square feet and 12 pounds for a lawn that measures 10,000 square feet. Let y represent the pounds of fertilizer and x the square footage of the lawn.

24. A new diet fad claims that a person weighing 140 pounds should consume 1490 daily calories and that a 200-pound person should consume 1700 calories. Let y represent the calories and x the weight of the person in pounds.

25. Two banks on opposite corners of a town square had signs that displayed the current temperature. One bank displayed the temperature in degrees Celsius and the other in degrees Fahrenheit. A temperature of 10°C was displayed at the same time as a temperature of 50°F. On another day, a temperature of −5°C was displayed at the same time as a temperature of 23°F. Let y represent the temperature in degrees Fahrenheit and x the temperature in degrees Celsius.

26. An accountant has a schedule of depreciation for some business equipment. The schedule shows that after 12 months the equipment is worth \$7600 and that after 20 months it is worth \$6000. Let y represent the worth and x represent the time in months.

27. A diabetic patient was told on a doctor visit that her HA1c reading of 6.5 corresponds to an average blood glucose level of 135. At the next checkup three months later, the patient had an HA1c reading of 6.0 and was told that it corresponds to an average blood glucose level of 120. Let y represent the HA1c reading and let x represent the average blood glucose level.

28. Hal purchased a 500-minute calling card for \$17.50. After he used all the minutes on that card he purchased another card from the same company at a price of \$26.25 for 750 minutes. Let y represent the cost of the card in dollars and let x represent the number of minutes.

3 Find the Equation of a Line Given the Slope and y Intercept

For Problems 29–36, write the equation of the line that has the indicated slope (m) and y intercept (b). Express final equations in slope-intercept form.

29. $m = \frac{3}{7}$, $b = 4$
30. $m = \frac{2}{9}$, $b = 6$
31. $m = 2$, $b = -3$
32. $m = -3$, $b = -1$
33. $m = -\frac{2}{5}$, $b = 1$
34. $m = -\frac{3}{7}$, $b = 4$
35. $m = 0$, $b = -4$
36. $m = \frac{1}{5}$, $b = 0$

4 Use the Point-Slope Form to Write Equations of Lines

For Problems 37–42, use the point-slope form to write the equation of the line that has the indicated slope and contains the indicated point. Express the final answer in standard form.

37. $m = \frac{5}{2}$, $(-3, 4)$

38. $m = \frac{2}{3}$, $(1, -4)$

39. $m = -2$, $(5, 8)$

40. $m = -1$, $(-6, 2)$

41. $m = -\frac{1}{3}$, $(5, 0)$

42. $m = -\frac{3}{4}$, $(0, 1)$

5 Apply the Slope-Intercept Form of an Equation

For Problems 43–48, change the equation to slope-intercept form and determine the slope and y intercept of the line.

43. $3x + y = 7$

44. $5x - y = 9$

45. $3x + 2y = 9$

46. $x - 4y = 3$

47. $x = 5y + 12$

48. $-4x - 7y = 14$

For Problems 49–56, use the slope-intercept form to graph the following lines.

49. $y = \frac{2}{3}x - 4$

50. $y = \frac{1}{4}x + 2$

51. $y = 2x + 1$

52. $y = 3x - 1$

53. $y = -\frac{3}{2}x + 4$

54. $y = -\frac{5}{3}x + 3$

55. $y = -x + 2$

56. $y = -2x + 4$

For Problems 57–66, graph the following lines using the technique that seems most appropriate.

57. $y = -\frac{2}{5}x - 1$

58. $y = -\frac{1}{2}x + 3$

59. $x + 2y = 5$

60. $2x - y = 7$

61. $-y = -4x + 7$

62. $3x = 2y$

63. $7y = -2x$

64. $y = -3$

65. $x = 2$

66. $y = -x$

6 Find the Equations for Parallel or Perpendicular Lines

For Problems 67–78, write the equation of the line that satisfies the given conditions. Express final equations in standard form.

67. Contains the point $(2, -4)$ and is parallel to the y axis

68. Contains the point $(-3, -7)$ and is parallel to the x axis

69. Contains the point $(5, 6)$ and is perpendicular to the y axis

70. Contains the point $(-4, 7)$ and is perpendicular to the x axis

71. Contains the point $(1, 3)$ and is parallel to the line $x + 5y = 9$

72. Contains the point $(-1, 4)$ and is parallel to the line $x - 2y = 6$

73. Contains the origin and is parallel to the line $4x - 7y = 3$

74. Contains the origin and is parallel to the line $-2x - 9y = 4$

75. Contains the point $(-1, 3)$ and is perpendicular to the line $2x - y = 4$

76. Contains the point $(-2, -3)$ and is perpendicular to the line $x + 4y = 6$

77. Contains the origin and is perpendicular to the line $-2x + 3y = 8$

78. Contains the origin and is perpendicular to the line $y = -5x$

THOUGHTS INTO WORDS

79. What does it mean to say that two points determine a line?

80. How would you help a friend determine the equation of the line that is perpendicular to $x - 5y = 7$ and contains the point $(5, 4)$?

81. Explain how you would find the slope of the line $y = 4$.

FURTHER INVESTIGATIONS

82. The equation of a line that contains the two points (x_1, y_1) and (x_2, y_2) is $\frac{y - y_1}{x - x_1} = \frac{y_2 - y_1}{x_2 - x_1}$. We often refer to this as the **two-point form** of the equation of a straight line. Use the two-point form and write the equation of the line that contains each of the indicated pairs of points. Express final equations in standard form.

(a) $(1, 1)$ and $(5, 2)$
(b) $(2, 4)$ and $(-2, -1)$
(c) $(-3, 5)$ and $(3, 1)$
(d) $(-5, 1)$ and $(2, -7)$

83. Let $Ax + By = C$ and $A'x + B'y = C'$ represent two lines. Change both of these equations to slope-intercept form, and then verify each of the following properties.

(a) If $\frac{A}{A'} = \frac{B}{B'} \neq \frac{C}{C'}$, then the lines are parallel.
(b) If $AA' = -BB'$, then the lines are perpendicular.

84. The properties in Problem 83 provide us with another way to write the equation of a line parallel or perpendicular to a given line that contains a given point not on the line. For example, suppose that we want the equation of the line perpendicular to $3x + 4y = 6$ that contains the point $(1, 2)$. The form $4x - 3y = k$, where k is a constant, represents a family of lines perpendicular to $3x + 4y = 6$ because we have satisfied the condition $AA' = -BB'$. Therefore, to find what specific line of the family contains $(1, 2)$, we substitute 1 for x and 2 for y to determine k.

$$4x - 3y = k$$
$$4(1) - 3(2) = k$$
$$-2 = k$$

Thus the equation of the desired line is $4x - 3y = -2$.

Use the properties from Problem 83 to help write the equation of each of the following lines.

(a) Contains $(1, 8)$ and is parallel to $2x + 3y = 6$
(b) Contains $(-1, 4)$ and is parallel to $x - 2y = 4$
(c) Contains $(2, -7)$ and is perpendicular to $3x - 5y = 10$
(d) Contains $(-1, -4)$ and is perpendicular to $2x + 5y = 12$

85. The problem of finding the perpendicular bisector of a line segment presents itself often in the study of analytic geometry. As with any problem of writing the equation of a line, you must determine the slope of the line and a point that the line passes through. A perpendicular bisector passes through the midpoint of the line segment and has a slope that is the negative reciprocal of the slope of the line segment. The problem can be solved as follows:

Find the perpendicular bisector of the line segment between the points $(1, -2)$ and $(7, 8)$.

The midpoint of the line segment is $\left(\frac{1+7}{2}, \frac{-2+8}{2}\right) = (4, 3)$.

The slope of the line segment is $m = \frac{8 - (-2)}{7 - 1} = \frac{10}{6} = \frac{5}{3}$.

Hence the perpendicular bisector will pass through the point $(4, 3)$ and have a slope of $m = -\frac{3}{5}$.

$$y - 3 = -\frac{3}{5}(x - 4)$$
$$5(y - 3) = -3(x - 4)$$
$$5y - 15 = -3x + 12$$
$$3x + 5y = 27$$

Thus the equation of the perpendicular bisector of the line segment between the points $(1, -2)$ and $(7, 8)$ is $3x + 5y = 27$.

Find the perpendicular bisector of the line segment between the points for the following. Write the equation in standard form.

(a) $(-1, 2)$ and $(3, 0)$
(b) $(6, -10)$ and $(-4, 2)$
(c) $(-7, -3)$ and $(5, 9)$
(d) $(0, 4)$ and $(12, -4)$

GRAPHING CALCULATOR ACTIVITIES

86. Predict whether each of the following pairs of equations represents parallel lines, perpendicular lines, or lines that intersect but are not perpendicular. Then graph each pair of lines to check your predictions. (The properties presented in Problem 83 should be very helpful.)

(a) $5.2x + 3.3y = 9.4$ and $5.2x + 3.3y = 12.6$
(b) $1.3x - 4.7y = 3.4$ and $1.3x - 4.7y = 11.6$
(c) $2.7x + 3.9y = 1.4$ and $2.7x - 3.9y = 8.2$
(d) $5x - 7y = 17$ and $7x + 5y = 19$
(e) $9x + 2y = 14$ and $2x + 9y = 17$
(f) $2.1x + 3.4y = 11.7$ and $3.4x - 2.1y = 17.3$
(g) $7.1x - 2.3y = 6.2$ and $2.3x + 7.1y = 9.9$
(h) $-3x + 9y = 12$ and $9x - 3y = 14$
(i) $2.6x - 5.3y = 3.4$ and $5.2x - 10.6y = 19.2$
(j) $4.8x - 5.6y = 3.4$ and $6.1x + 7.6y = 12.3$

Answers to the Concept Quiz

1. True **2.** False **3.** False **4.** True **5.** False **6.** True **7.** False **8.** True **9.** True **10.** False

Answers to the Example Practice Skills

1. $3x - 4y = 13$ **2.** $5x + 6y = 40$ **3.** $3x - 2y = 8$ **4.** $4x - 3y = -23$ **5.** $3x + y = 8$ **6.** $m = \frac{4}{5}$

7.

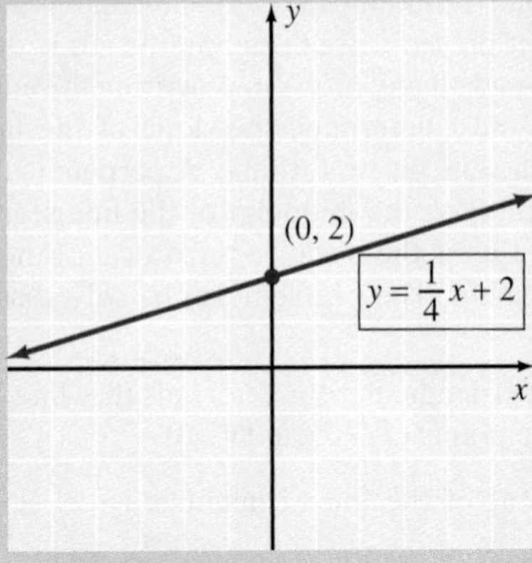

8. **(a)** $m_1 = m_2 = \frac{1}{3}$ **(b)** $m_1 = \frac{2}{5}, m_2 = -\frac{5}{2}$ **9.** $3x + y = 1$ **10.** $2x - 5y = 11$

Chapter 3 Summary

OBJECTIVE	SUMMARY	EXAMPLE	CHAPTER REVIEW PROBLEMS
Find solutions for linear equations in two variables. (Sec. 3.1, Obj. 1, p. 122)	A solution of an equation in two variables is an ordered pair of real numbers that satisfies the equation.	Find a solution for the equation $2x - 3y = -6$. **Solution** Choose an arbitrary value for x and determine the corresponding y value. Let $x = 3$; then substitute 3 for x in the equation. $2(3) - 3y = -6$ $6 - 3y = -6$ $-3y = -12$ $y = 4$ Therefore, the ordered pair (3, 4) is a solution.	Problems 1–4
Graph the solutions for linear equations. (Sec. 3.1, Obj. 3, p. 125)	A graph provides a visual display of all the infinite solutions of an equation in two variables. The ordered pair solutions for a linear equation can be plotted as points on a rectangular coordinate system. Connecting the points with a straight line produces a graph of the equation.	Graph $y = 2x - 3$. **Solution** Find at least three ordered-pair solutions for the equation. We can determine that $(-1, -5)$, $(0, -3)$, and $(1, -1)$ are solutions. The graph is shown below. 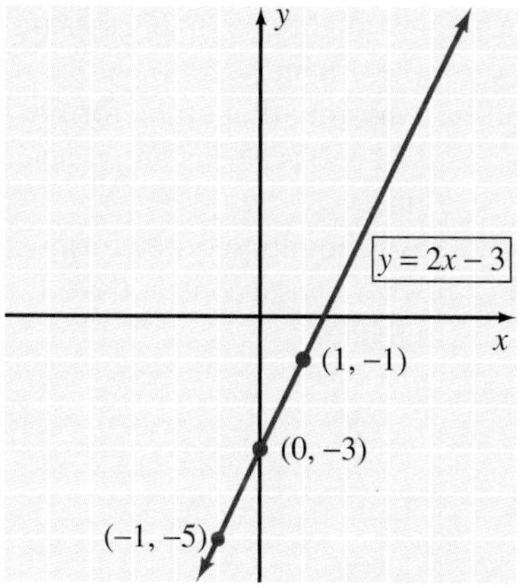 y y = 2x – 3 x (1, –1) (0, –3) (–1, –5)	Problems 5–8 *(continued)*

OBJECTIVE	SUMMARY	EXAMPLE	CHAPTER REVIEW PROBLEMS
Graph linear equations by finding the x and y intercepts. (Sec. 3.1, Obj. 4, p. 126)	The x intercept is the x coordinate of the point where the graph intersects the x axis. The y intercept is the y coordinate of the point where the graph intersects the y axis. To find the x intercept, substitute 0 for y in the equation and then solve for x. To find the y intercept, substitute 0 for x in the equation and then solve for y. Plot the intercepts and connect them with a straight line to produce the graph.	Graph $x - 2y = 4$. **Solution** Let $y = 0$. $x - 2(0) = 4$ $x = 4$ Let $x = 0$. $0 - 2y = 4$ $y = -2$ 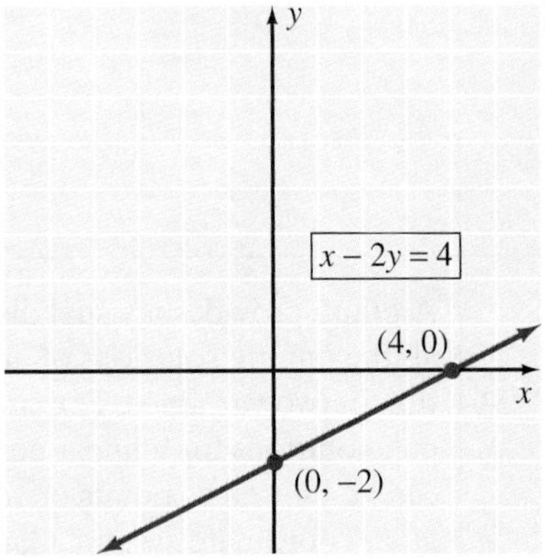	Problems 9–12
Graph lines passing through the origin, vertical lines, and horizontal lines. (Sec. 3.1, Obj. 5, p. 129)	The graph of any equation of the form $Ax + By = C$, where $C = 0$, is a straight line that passes through the origin. Any equation of the form $x = a$, where a is a constant, is a vertical line. Any equation of the form $y = b$, where b is a constant, is a horizontal line.	Graph $3x + 2y = 0$. **Solution** The equation indicates that the graph will be a line passing through the origin. Solving the equation for y gives us $y = -\frac{3}{2}x$. Find at least three ordered-pair solutions for the equation. We can determine that $(-2, 3)$, $(0, 0)$, and $(2, -3)$ are solutions. The graph is shown below. 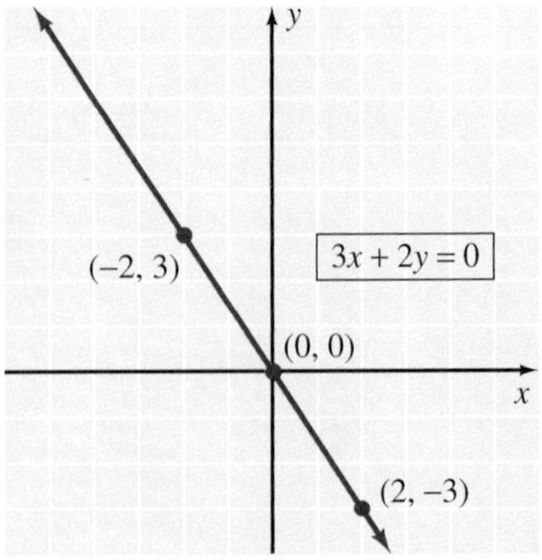	Problems 13–18

(continued)

OBJECTIVE	SUMMARY	EXAMPLE	CHAPTER REVIEW PROBLEMS
Apply graphing to linear relationships. (Sec. 3.1, Obj. 6, p. 130)	Many relationships between two quantities are linear relationships. Graphs of these relationships can be used to present information about the relationship.	Let c represent the cost in dollars and let w represent the gallons of water used; then the equation $c = 0.004w + 20$ can be used to determine the cost of a water bill for a household. Graph the relationship. **Solution** Label the vertical axis c and the horizontal axis w. Because of the type of application, we use only nonnegative values for w. 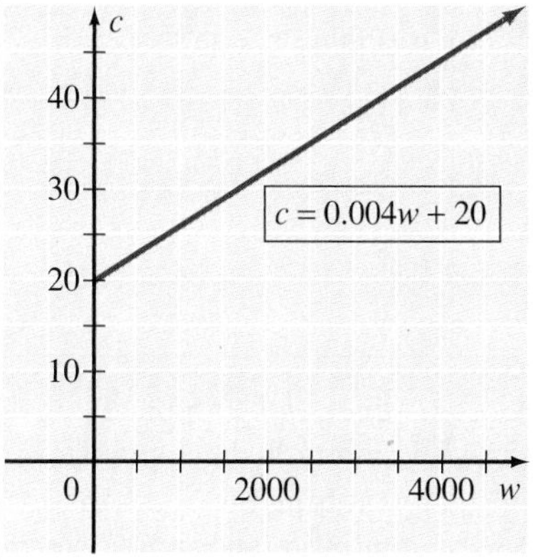 $c = 0.004w + 20$	Problems 19–20
Graph linear inequalities. (Sec. 3.2, Obj. 1, p. 138)	To graph a linear inequality, first graph the line for the corresponding equality. Use a solid line if the equality is included in the given statement or a dashed line if the equality is not included. Then a test point is used to determine which half-plane is included in the solution set. See page 139 for the detailed steps.	Graph $x - 2y \leq -4$. **Solution** First graph $x - 2y = -4$. Choose $(0, 0)$ as a test point. Substituting $(0, 0)$ into the inequality yields $0 \leq -4$. Because the test point $(0, 0)$ makes the inequality a false statement, the half-plane not containing the point $(0, 0)$ is in the solution. 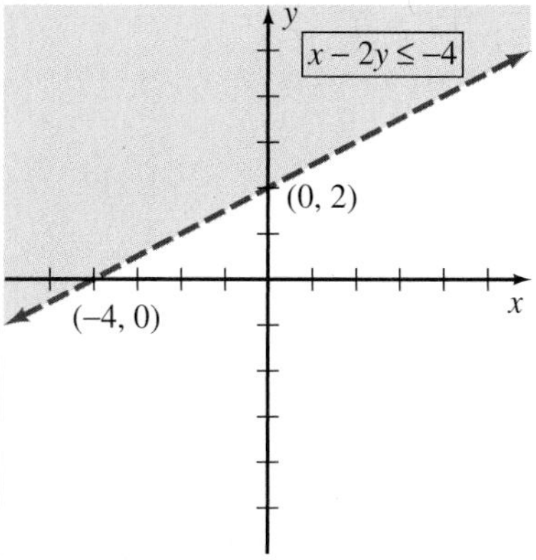 $x - 2y \leq -4$; $(0, 2)$; $(-4, 0)$	Problems 21–26

(continued)

OBJECTIVE	SUMMARY	EXAMPLE	CHAPTER REVIEW PROBLEMS
Find the distance between two points. (Sec. 3.3, Obj. 1, p. 143)	The distance between any two points (x_1, y_1) and (x_2, y_2) is given by the distance formula $d = \sqrt{(x_2 - x_1)^2 + (y_2 - y_1)^2}$	Find the distance between $(1, -5)$ and $(4, 2)$. **Solution** $d = \sqrt{(x_2 - x_1)^2 + (y_2 - y_1)^2}$ $d = \sqrt{(4 - 1)^2 + (2 - (-5))^2}$ $d = \sqrt{(3)^2 + (7)^2}$ $d = \sqrt{9 + 49} = \sqrt{58}$	Problems 27–29
Find the slope of a line. (Sec. 3.3, Obj. 2, p. 146)	The slope (denoted by m) of a line determined by the points (x_1, y_1) and (x_2, y_2) is given by the slope formula $m = \frac{y_2 - y_1}{x_2 - x_1}$ where $x_2 \neq x_1$	Find the slope of a line that contains the points $(-1, 2)$ and $(7, 8)$. **Solution** Use the slope formula: $m = \frac{8 - 2}{7 - (-1)} = \frac{6}{8} = \frac{3}{4}$ Thus the slope of the line is $\frac{3}{4}$.	Problems 30–32
Use slope to graph lines. (Sec. 3.3, Obj. 3, p. 149)	A line can be graphed knowing a point on the line and the slope by plotting the point and from that point using the slope to locate another point on the line. Then those two points can be connected with a straight line to produce the graph.	Graph the line that contains the point $(-3, -2)$ and has a slope of $\frac{5}{2}$. **Solution** From the point $(-3, -2)$, locate another point by moving up 5 units and to the right 2 units to obtain the point $(-1, 3)$. Then draw the line. 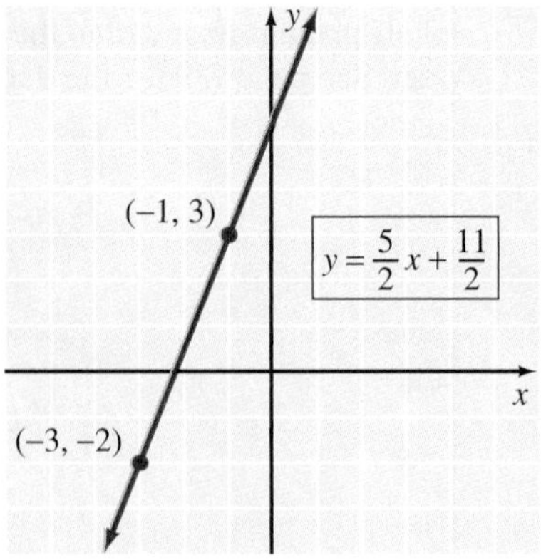 y (−1, 3) $y = \frac{5}{2}x + \frac{11}{2}$ x (−3, −2)	Problems 33–36 *(continued)*

OBJECTIVE	SUMMARY	EXAMPLE	CHAPTER REVIEW PROBLEMS
Apply slope to solve problems. (Sec. 3.3, Obj. 4, p. 150)	The concept of slope is used in most situations where an incline is involved. In highway construction the word grade is used for slope.	A certain highway has a grade of 2%. How many feet does it rise in a horizontal distance of one-third of a mile (1760 feet)? **Solution** A 2% grade is equivalent to a slope of $\frac{2}{100}$. We can set up the proportion $\frac{2}{100} = \frac{y}{1760}$; then solving for y gives us $y = 35.2$. So the highway rises 35.2 feet in one-third of a mile.	Problems 37–38
Apply the slope intercept form of an equation of a line. (Sec. 3.4, Obj. 5, p. 159)	The equation $y = mx + b$ is referred to as the slope-intercept form of the equation of a line. If the equation of a nonvertical line is written in this form, then the coefficient of x is the slope and the constant term is the y intercept.	Change the equation $2x + 7y = -21$ to slope-intercept form and determine the slope and y intercept. **Solution** Solve the equation $2x + 7y = -21$ for y. $2x + 7y = -21$ $7y = -2x - 21$ $y = \frac{-2}{7}x - 3$ The slope is $-\frac{2}{7}$ and the y intercept is -3.	Problems 39–41
Find the equation of a line given the slope and a point contained in the line. (Sec. 3.4, Obj. 1, p. 156)	To determine the equation of a straight line given a set of conditions, we can use the point-slope form $y - y_1 = m(x - x_1)$, or $m = \frac{y - y_1}{x - x_1}$. The result can be expressed in standard form or slope-intercept form.	Find the equation of a line that contains the point $(1, -4)$ and has a slope of $\frac{3}{2}$. **Solution** Substitute $\frac{3}{2}$ for m and $(1, -4)$ for (x_1, y_1) into the formula $m = \frac{y - y_1}{x - x_1}$: $\frac{3}{2} = \frac{y - (-4)}{x - 1}$ Simplifying this equation yields $3x - 2y = 11$.	Problems 42–44 *(continued)*

OBJECTIVE	SUMMARY	EXAMPLE	CHAPTER REVIEW PROBLEMS
Find the equation of a line given two points contained in the line. (Sec. 3.4, Obj. 2, p. 157)	First calculate the slope of the line. Substitute the slope and the coordinates of one of the points into $y - y_1 = m(x - x_1)$ or $m = \dfrac{y - y_1}{x - x_1}$.	Find the equation of a line that contains the points $(-3, 4)$ and $(-6, 10)$. **Solution** First calculate the slope. $m = \dfrac{10 - 4}{-6 - (-3)} = \dfrac{6}{-3} = -2$ Now substitute -2 for m and $(-3, 4)$ for (x_1, y_1) in the formula $y - y_1 = m(x - x_1)$. $y - 4 = -2(x - (-3))$ Simplifying this equation yields $2x + y = -2$.	Problems 45–46, 50–53
Find the equations for parallel and perpendicular lines. (Sec. 3.4, Obj. 6, p. 161)	If two lines have slopes m_1 and m_2, respectively, then: 1. The two lines are parallel if and only if $m_1 = m_2$. 2. The two lines are perpendicular if and only if $(m_1)(m_2) = -1$.	Find the equation of a line that contains the point $(2, 1)$ and is parallel to the line $y = 3x + 4$. **Solution** The slope of the parallel line is 3. Therefore, use this slope and the point $(2, 1)$ to determine the equation: $y - 1 = 3(x - 2)$ Simplifying this equation yields $y = 3x - 5$.	Problems 47–49

Chapter 3 Review Problem Set

For Problems 1–4, determine which of the ordered pairs are solutions of the given equation.

1. $4x + y = 6$; $(1, 2), (6, 0), (-1, 10)$

2. $-x + 2y = 4$; $(-4, 1), (-4, -1), (0, 2)$

3. $3x + 2y = 12$; $(2, 3), (-2, 9), (3, 2)$

4. $2x - 3y = -6$; $(0, -2), (-3, 0), (1, 2)$

For Problems 5–8, complete the table of values for the equation and graph the equation.

5. $y = 2x - 5$

x	-1	0	1	4
y				

6. $y = -2x - 1$

x	-3	-1	0	2
y				

7. $y = \dfrac{3x - 4}{2}$

x	-2	0	2	4
y				

8. $2x - 3y = 3$

x	-3	0	3
y			

For Problems 9–12, graph each equation by finding the x and y intercepts.

9. $2x - y = 6$

10. $-3x - 2y = 6$

11. $x - 2y = 4$

12. $5x - y = -5$

For Problems 13–18, graph each equation.

13. $y = -4x$

14. $2x + 3y = 0$

15. $x = 1$

16. $y = -2$

17. $y = 4$

18. $x = -3$

19. **(a)** An apartment moving company charges according to the equation $c = 75h + 150$, where c represents the charge in dollars and h represents the number of hours for the move. Complete the following table.

h	1	2	3	4
c				

(b) Labeling the horizontal axis h and the vertical axis c, graph the equation $c = 75h + 150$ for nonnegative values of h.

(c) Use the graph from part (b) to approximate values of c when $h = 1.5$ and 3.5.

(d) Check the accuracy of your reading from the graph in part (c) by using the equation $c = 75h + 150$.

20. **(a)** The value added tax is computed by the equation $t = 0.15v$, where t represents the tax and v represents the value of the goods. Complete the following table.

v	100	200	350	400
t				

(b) Labeling the horizontal axis v and the vertical axis t, graph the equation $t = 0.15v$ for nonnegative values of v.

(c) Use the graph from part (b) to approximate values of t when $v = 250$ and $v = 300$.

(d) Check the accuracy of your reading from the graph in part (c) by using the equation $t = 0.15v$.

For Problems 21–26, graph each inequality.

21. $-x + 3y < -6$

22. $x + 2y \geq 4$

23. $2x - 3y \leq 6$

24. $y > -\frac{1}{2}x + 3$

25. $y < 2x - 5$

26. $y \geq \frac{2}{3}x$

27. Find the distance between each of the pairs of points.
(a) $(-1, 5)$ and $(1, -2)$
(b) $(5, 0)$ and $(2, 7)$

28. Find the lengths of the sides of a triangle whose vertices are at $(2, 3)$, $(5, -1)$, and $(-4, -5)$.

29. Verify that $(1, 2)$ is the midpoint of the line segment joining $(-3, -1)$ and $(5, 5)$.

30. Find the slope of the line determined by each pair of points.

(a) $(3, 4), (-2, -2)$ **(b)** $(-2, 3), (4, -1)$

31. Find y if the line through $(-4, 3)$ and $(12, y)$ has a slope of $\frac{1}{8}$.

32. Find x if the line through $(x, 5)$ and $(3, -1)$ has a slope of $-\frac{3}{2}$.

For Problems 33–36, graph the line that has the indicated slope and contains the indicated point.

33. $m = -\frac{1}{2}$, $(0, 3)$

34. $m = \frac{3}{5}$, $(0, -4)$

35. $m = 3$, $(-1, 2)$

36. $m = -2$, $(1, 4)$

37. A certain highway has a 6% grade. How many feet does it rise in a horizontal distance of 1 mile (5280 feet)?

38. If the ratio of rise to run is to be $\frac{2}{3}$ for the steps of a staircase and the run is 12 inches, find the rise.

39. Find the slope of each of the following lines.
(a) $4x + y = 7$ **(b)** $2x - 7y = 3$

40. Find the slope of any line that is perpendicular to the line $-3x + 5y = 7$.

41. Find the slope of any line that is parallel to the line $4x + 5y = 10$.

For Problems 42–49, write the equation of the line that satisfies the stated conditions. Express final equations in standard form.

42. Having a slope of $-\frac{3}{7}$ and a y intercept of 4

43. Containing the point $(-1, -6)$ and having a slope of $\frac{2}{3}$

44. Containing the point $(3, -5)$ and having a slope of -1

45. Containing the points $(-1, 2)$ and $(3, -5)$

46. Containing the points $(0, 4)$ and $(2, 6)$

47. Containing the point $(2, 5)$ and parallel to the line $x - 2y = 4$

48. Containing the point $(-2, -6)$ and perpendicular to the line $3x + 2y = 12$

49. Containing the point $(-8, 3)$ and parallel to the line $4x + y = 7$

50. The taxes for a primary residence can be described by a linear relationship. Find the equation for the relationship if the taxes for a home valued at \$200,000 are \$2400, and the taxes are \$3150 when the home is valued at \$250,000. Let y be the taxes and x the value of the home. Write the equation in slope-intercept form.

51. The freight charged by a trucking firm for a parcel under 200 pounds depends on the miles it is being shipped. To ship a 150-pound parcel 300 miles, it costs \$40. If the same parcel is shipped 1000 miles, the cost is \$180. Assume the relationship between the cost and miles is linear. Find the equation for the relationship. Let y be the cost and x be the miles. Write the equation in slope-intercept form.

52. On a final exam in math class, the number of points earned has a linear relationship with the number of correct answers. John got 96 points when he answered 12 questions correctly. Kimberly got 144 points when she answered 18 questions correctly. Find the equation for the relationship. Let y be the number of points and x be the number of correct answers. Write the equation in slope-intercept form.

53. The time needed to install computer cables has a linear relationship with the number of feet of cable being installed. It takes $1\frac{1}{2}$ hours to install 300 feet, and 1050 feet can be installed in 4 hours. Find the equation for the relationship. Let y be the feet of cable installed and x be the time in hours. Write the equation in slope-intercept form.

Chapter 3 Test

1. Determine which of the ordered pairs are solutions of the equation $-2x + y = 6$: $(-1, 4)$, $(2, 2)$, $(4, -2)$, $(-3, 0)$, $(10, 26)$. 1. ________

2. Find the slope of the line determined by the points $(-2, 4)$ and $(3, -2)$. 2. ________

3. Find the slope of the line determined by the equation $3x - 7y = 12$. 3. ________

4. Find the length of the line segment whose endpoints are $(4, 2)$ and $(-3, -1)$. 4. ________

5. What is the slope of all lines that are parallel to the line $7x - 2y = 9$? 5. ________

6. What is the slope of all lines that are perpendicular to the line $4x + 9y = -6$? 6. ________

7. The grade of a highway up a hill is 25%. How much change in horizontal distance is there if the vertical height of the hill is 120 feet? 7. ________

8. Suppose that a highway rises 200 feet in a horizontal distance of 3000 feet. Express the grade of the highway to the nearest tenth of a percent. 8. ________

9. If the ratio of rise to run is to be $\frac{3}{4}$ for the steps of a staircase and the rise is 32 centimeters, find the run to the nearest centimeter. 9. ________

10. Find the x intercept of the line $3x - y = -6$. 10. ________

11. Find the y intercept of the line $y = \frac{3}{5}x - \frac{2}{3}$. 11. ________

12. Graph the line that contains the point $(-2, 3)$ and has a slope of $-\frac{1}{4}$. 12. ________

13. Find the x and y intercepts for the line $x - 4y = -4$ and graph the line. 13. ________

For Problems 14–18, graph each equation.

14. $y = -x - 3$ 14. ________

15. $-3x + y = 5$ 15. ________

16. $3y = 2x$ 16. ________

17. $y = 3$ 17. ________

18. $y = \dfrac{-x - 1}{4}$ 18. ________

For Problems 19 and 20, graph each inequality.

19. $2x - y < 4$ 19. ________

20. $3x + 2y \geq 6$ 20. ________

21. Find the equation of the line that has a slope of $-\frac{3}{2}$ and contains the point $(4, -5)$. Express the equation in standard form. 21. ________

22. Find the equation of the line that contains the points $(-4, 2)$ and $(2, 1)$. Express the equation in slope-intercept form. 22. ________

23. ____________________

24. ____________________

25. ____________________

23. Find the equation of the line that is parallel to the line $5x + 2y = 7$ and contains the point $(-2, -4)$. Express the equation in standard form.

24. Find the equation of the line that is perpendicular to the line $x - 6y = 9$ and contains the point $(4, 7)$. Express the equation in standard form.

25. The monthly bill for a cellular phone can be described by a linear relationship. Find the equation for this relationship if the bill for 750 minutes used is \$35.00 and the bill for 550 minutes used is \$31.00. Let y represent the amount of the bill and let x represent the number of minutes used. Write the equation in slope-intercept form.

Systems of Equations

4

Michael Stevens/FogStock/Alamy Limited

■ *When mixing different solutions, a chemist could use a system of equations to determine how much of each solution is needed to produce a specific concentration.*

A 10% salt solution is to be mixed with a 20% salt solution to produce 20 gallons of a 17.5% salt solution. How many gallons of the 10% solution and how many gallons of the 20% solution will be needed? The two equations $x + y = 20$ and $0.10x + 0.20y = 0.175(20)$, where x represents the number of gallons of the 10% solution and y represents the number of gallons of the 20% solution, algebraically represent the conditions of the problem. The two equations considered together form a **system of linear equations**, and the problem can be solved by solving this system of equations.

Throughout most of this chapter, we will consider systems of linear equations and their applications. We will discuss various techniques for solving systems of linear equations.

Video tutorials for all section learning objectives are available in a variety of delivery modes.

INTERNET PROJECT

Three methods for solving systems of linear equations are presented in this chapter. Another method called "Cramer's rule" is presented in Appendix C. Conduct an Internet search to find the year that Cramer published his rule. Would you consider Cramer's rule a recent development in mathematics?

4.1 Systems of Two Linear Equations and Linear Inequalities in Two Variables

OBJECTIVES

1. Solve Systems of Linear Equations by Graphing
2. Solve Systems of Linear Inequalities

1 Solve Systems of Linear Equations by Graphing

In Chapter 3, we stated that any equation of the form $Ax + By = C$, where A, B, and C are real numbers (A and B not both zero), is a *linear equation* in the two variables x and y, and its graph is a straight line. Two linear equations in two variables considered together form a **system of two linear equations in two variables**. Here are a few examples:

$$\begin{pmatrix} x + y = 6 \\ x - y = 2 \end{pmatrix} \qquad \begin{pmatrix} 3x + 2y = 1 \\ 5x - 2y = 23 \end{pmatrix} \qquad \begin{pmatrix} 4x - 5y = 21 \\ 3x + 7y = -38 \end{pmatrix}$$

To **solve a system**, such as one of the above, means to find all of the ordered pairs that satisfy both equations in the system. For example, if we graph the two equations $x + y = 6$ and $x - y = 2$ on the same set of axes, as in Figure 4.1, then the ordered pair associated with the point of intersection of the two lines is the solution of the system. Thus we say that $\{(4, 2)\}$ is the solution set of the system

$$\begin{pmatrix} x + y = 6 \\ x - y = 2 \end{pmatrix}$$

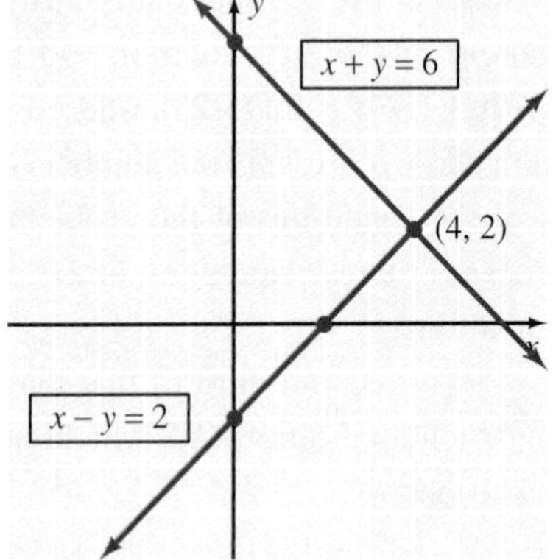

Figure 4.1

To check, substitute 4 for x and 2 for y in the two equations, which yields

$x + y$ becomes $4 + 2 = 6$ A true statement

$x - y$ becomes $4 - 2 = 2$ A true statement

Because the graph of a linear equation in two variables is a straight line, there are three possible situations that can occur when we solve a system of two linear equations in two variables. We illustrate these cases in Figure 4.2.

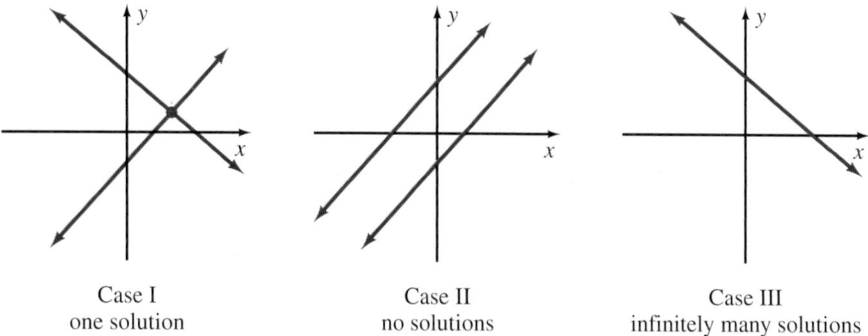

Figure 4.2

Case I The graphs of the two equations are two lines intersecting in *one* point. There is *one solution*, and the system is called a **consistent system**.

Case II The graphs of the two equations are parallel lines. There is *no solution*, and the system is called an **inconsistent system**.

Case III The graphs of the two equations are the same line, and there are *infinitely many solutions* to the system. Any pair of real numbers that satisfies one of the equations will also satisfy the other equation, and we say that the equations are **dependent**.

Thus as we solve a system of two linear equations in two variables, we know what to expect. The system will have no solutions, one ordered pair as a solution, or infinitely many ordered pairs as solutions.

EXAMPLE 1

Solve the system $\begin{pmatrix} 2x - y = -2 \\ 4x + y = 8 \end{pmatrix}$.

Solution

Graph both lines on the same coordinate system. Let's graph the lines by determining intercepts and a check point for each of the lines.

$2x - y = -2$

x	y
0	2
−1	0
2	6

$4x + y = 8$

x	y
0	8
2	0
1	4

Figure 4.3 shows the graphs of the two equations. It appears that (1, 4) is the solution of the system.

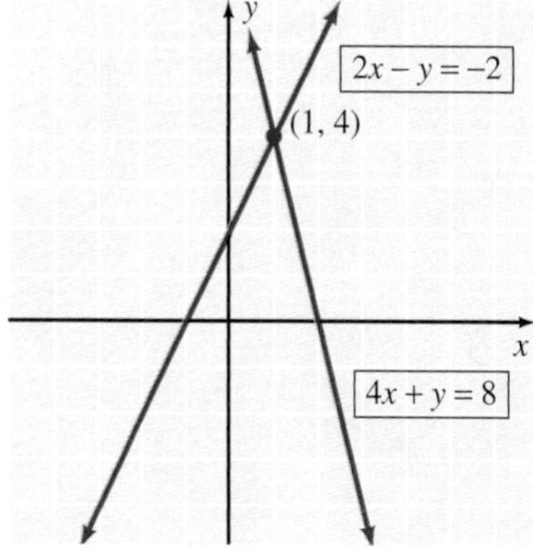

Figure 4.3

To check it, we can substitute 1 for x and 4 for y in both equations.

$2x - y = -2$ becomes $2(1) - 4 = -2$ A true statement

$4x + y = 8$ becomes $4(1) + 4 = 8$ A true statement

Therefore, $\{(1, 4)\}$ is the solution set.

▼ PRACTICE YOUR SKILL

Solve the system of equations $\begin{pmatrix} y = -x - 2 \\ x + 2y = -4 \end{pmatrix}$. ■

EXAMPLE 2 Solve the system $\begin{pmatrix} x - 3y = 3 \\ 2x - 6y = 6 \end{pmatrix}$.

Solution

Graph both lines on the same coordinate system. Let's graph the lines by determining intercepts and a check point for each of the lines.

$x - 3y = 3$

x	y
0	−1
3	0
−3	−2

$2x - 6y = 6$

x	y
0	−1
3	0
1	$-\frac{2}{3}$

Figure 4.4 shows the graph of this system. Since the graphs of both equations are the same line, the coordinates of any point on the line satisfy both equations. Hence the system has infinitely many solutions. Informally, the solution is stated as infinitely many solutions. In set notation the solution would be written as $\{(x, y) \mid x - 3y = 3\}$. This is read as the set of ordered pairs (x, y) such that $x - 3y = 3$ and means the coordinates of every point on the line $x - 3y = 3$ are solutions to the system.

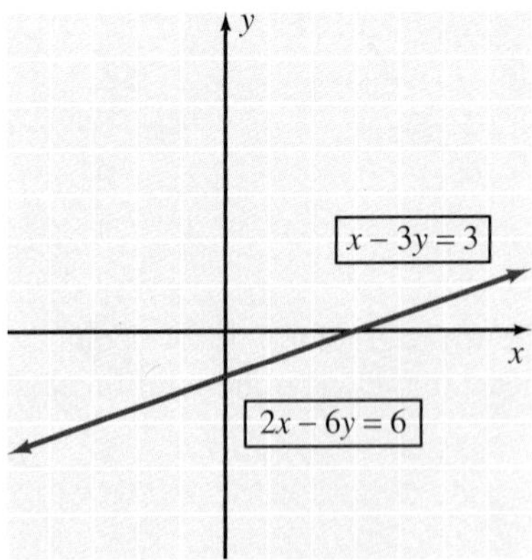

Figure 4.4

▼ PRACTICE YOUR SKILL

Solve the system $\begin{pmatrix} 4x + 2y = -2 \\ 2x + y = -1 \end{pmatrix}$. ■

EXAMPLE 3

Solve the system $\begin{pmatrix} y = 2x + 3 \\ 2x - y = 8 \end{pmatrix}$.

Solution

Graph both lines on the same coordinate system. Let's graph the lines by determining intercepts and a check point for each of the lines.

$y = 2x + 3$

x	y
0	3
$-\frac{3}{2}$	0
1	5

$2x - y = 8$

x	y
0	−8
4	0
2	−4

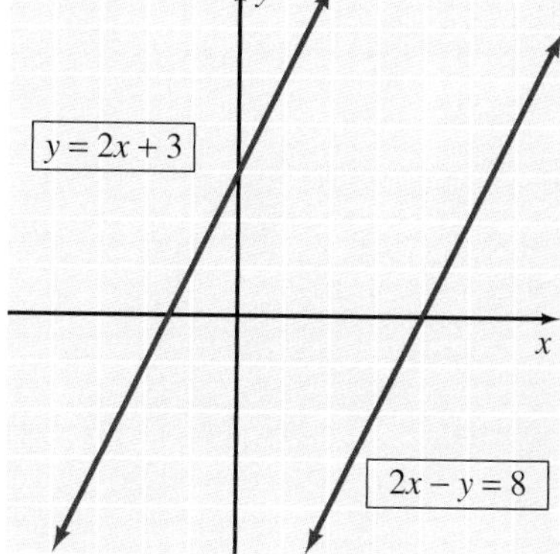

Figure 4.5

Figure 4.5 shows the graph of this system. Since the lines are parallel, there is no solution to the system. The solution set is ∅.

▼ PRACTICE YOUR SKILL

Solve the system $\begin{pmatrix} x + y = 5 \\ y = -x + 2 \end{pmatrix}$.

■

2 Solve Systems of Linear Inequalities

Finding solution sets for systems of linear inequalities relies heavily on the graphing approach. The solution set of a system of linear inequalities, such as

$$\begin{pmatrix} x + y > 2 \\ x - y < 2 \end{pmatrix}$$

is the intersection of the solution sets of the individual inequalities. In Figure 4.6(a) we indicated the solution set for $x + y > 2$, and in Figure 4.6(b) we indicated the solution set for $x - y < 2$. Then, in Figure 4.6(c), we shaded the region that represents the intersection of the two solution sets from parts (a) and (b); thus it is the graph of the system. Remember that dashed lines are used to indicate that the points on the lines are not included in the solution set.

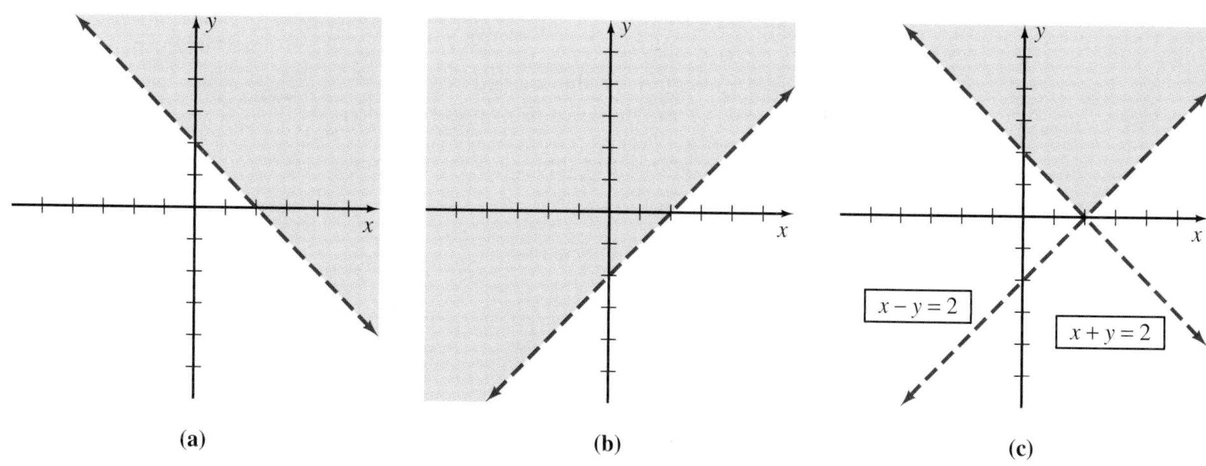

Figure 4.6

In the following examples, we indicated only the final solution set for the system.

EXAMPLE 4

Solve the following system by graphing.

$$\begin{pmatrix} 2x - y \geq 4 \\ x + 2y < 2 \end{pmatrix}$$

Solution

The graph of $2x - y \geq 4$ consists of all points *on or below* the line $2x - y = 4$. The graph of $x + 2y < 2$ consists of all points *below* the line $x + 2y = 2$. The graph of the system is indicated by the shaded region in Figure 4.7. Note that all points in the shaded region are on or below the line $2x - y = 4$ *and* below the line $x + 2y = 2$.

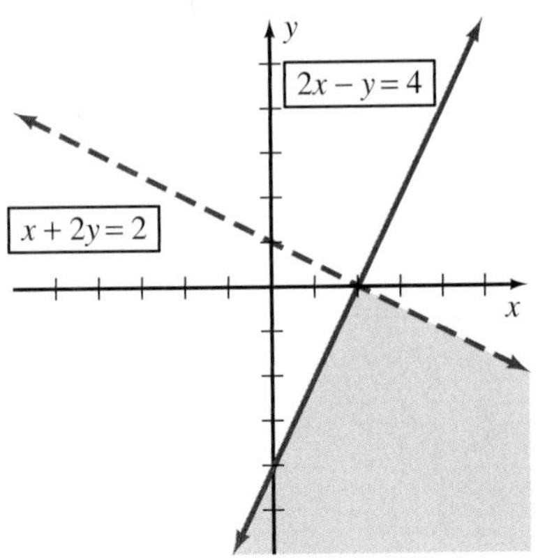

Figure 4.7

▼ PRACTICE YOUR SKILL

Solve the following system by graphing. $\begin{pmatrix} 3x + y > -3 \\ 2x - y \geq 0 \end{pmatrix}$ ■

EXAMPLE 5

Solve the following system by graphing. $\begin{pmatrix} y > \frac{2}{5}x - 4 \\ y < -\frac{1}{3}x - 1 \end{pmatrix}$

Solution

The graph of $y > \frac{2}{5}x - 4$ consists of all points *above* the line $y = \frac{2}{5}x - 4$. The graph of $y < -\frac{1}{3}x - 1$ consists of all points *below* the line $y = -\frac{1}{3}x - 1$. The graph of the system is indicated by the shaded regions in Figure 4.8. Note that all points in the shaded region are *above* the line $y = \frac{2}{5}x - 4$ and *below* the line $y = -\frac{1}{3}x - 1$.

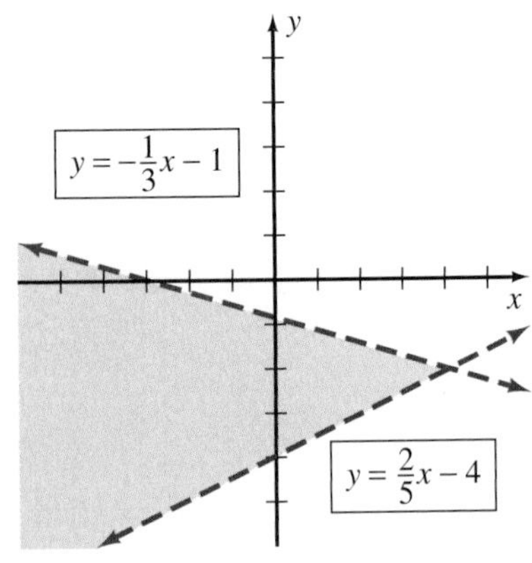

Figure 4.8

▼ PRACTICE YOUR SKILL

Solve the following system by graphing. $\begin{pmatrix} y < -\frac{4}{3}x + 5 \\ y < x - 2 \end{pmatrix}$ ■

EXAMPLE 6

Solve the following system by graphing.

$$\begin{pmatrix} x \le 2 \\ y \ge -1 \end{pmatrix}$$

Solution

Remember that even though each inequality contains only one variable, we are working in a rectangular coordinate system that involves ordered pairs. That is, the system could be written as

$$\begin{pmatrix} x + 0(y) \le 2 \\ 0(x) + y \ge -1 \end{pmatrix}$$

The graph of the system is the shaded region in Figure 4.9. Note that all points in the shaded region are on or to the left of the line $x = 2$ and on or above the line $y = -1$.

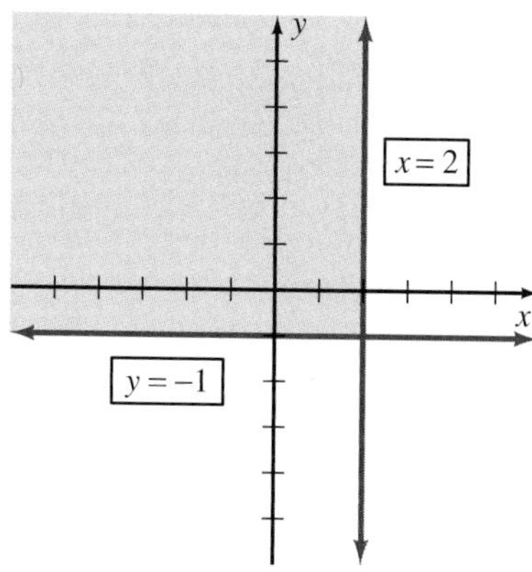

Figure 4.9

▼ **PRACTICE YOUR SKILL**

Solve the following system by graphing. $\begin{pmatrix} y < 4 \\ x > -1 \end{pmatrix}$ ■

In our final example of this section, we will use a graphing utility to help solve a system of equations.

EXAMPLE 7

Solve the system $\begin{pmatrix} 1.14x + 2.35y = -7.12 \\ 3.26x - 5.05y = 26.72 \end{pmatrix}$.

Solution

First, we need to solve each equation for y in terms of x. Thus the system becomes

$$\begin{pmatrix} y = \dfrac{-7.12 - 1.14x}{2.35} \\ y = \dfrac{3.26x - 26.72}{5.05} \end{pmatrix}$$

Now we can enter both of these equations into a graphing utility and obtain Figure 4.10. In this figure it appears that the point of intersection is at approximately $x = 2$ and $y = -4$. By direct substitution into the given equations, we can verify that the point of intersection is exactly $(2, -4)$.

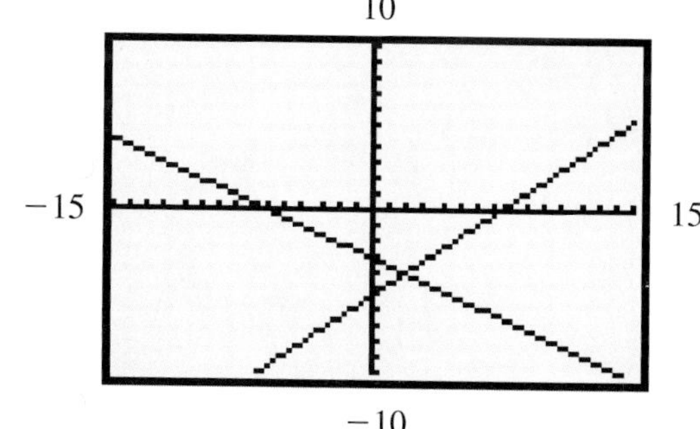

Figure 4.10

▼ **PRACTICE YOUR SKILL**

Solve the following system by graphing. $\begin{pmatrix} 2.35x + 4.16y = 10.13 \\ 5.18x - 1.17y = -8.69 \end{pmatrix}$ ■

CONCEPT QUIZ

For Problems 1–10, answer true or false.

1. To solve a system of equations means to find all the ordered pairs that satisfy all the equations in the system.
2. A consistent system of linear equations will have more than one solution.
3. If the graph of a system of two linear equations results in two distinct parallel lines, then the system has no solutions.
4. Every system of equations has a solution.
5. If the graphs of the two equations in a system are the same line, then the equations in the system are dependent.
6. The solution of a system of linear inequalities is the intersection of the solution sets of the individual inequalities.
7. For the system of inequalities $\begin{pmatrix} 2x + y > 4 \\ x - 3y < 6 \end{pmatrix}$, the points on the line $2x + y = 4$ are included in the solution.
8. The solution set of the system of inequalities $\begin{pmatrix} y > 2x + 5 \\ y < 2x + 1 \end{pmatrix}$ is the null set.
9. The ordered pair (1, 4) satisfies the system of inequalities $\begin{pmatrix} x + y > 2 \\ -2x + y < 3 \end{pmatrix}$.
10. The ordered pair $(-4, -1)$ satisfies the system of inequalities $\begin{pmatrix} x + y < 5 \\ 2x - 3y > 6 \end{pmatrix}$.

Problem Set 4.1

1 Solve Systems of Linear Equations by Graphing

For Problems 1–16, use the graphing approach to determine whether the system is consistent, the system is inconsistent, or the equations are dependent. If the system is consistent, find the solution set from the graph and check it.

1. $\begin{pmatrix} x - y = 1 \\ 2x + y = 8 \end{pmatrix}$

2. $\begin{pmatrix} 3x + y = 0 \\ x - 2y = -7 \end{pmatrix}$

3. $\begin{pmatrix} 4x + 3y = -5 \\ 2x - 3y = -7 \end{pmatrix}$

4. $\begin{pmatrix} 2x - y = 9 \\ 4x - 2y = 11 \end{pmatrix}$

5. $\begin{pmatrix} \frac{1}{2}x + \frac{1}{4}y = 9 \\ 4x + 2y = 72 \end{pmatrix}$

6. $\begin{pmatrix} 5x + 2y = -9 \\ 4x - 3y = 2 \end{pmatrix}$

7. $\begin{pmatrix} \frac{1}{2}x - \frac{1}{3}y = 3 \\ x + 4y = -8 \end{pmatrix}$

8. $\begin{pmatrix} 4x - 9y = -60 \\ \frac{1}{3}x - \frac{3}{4}y = -5 \end{pmatrix}$

9. $\begin{pmatrix} x - \frac{y}{2} = -4 \\ 8x - 4y = -1 \end{pmatrix}$

10. $\begin{pmatrix} 3x - 2y = 7 \\ 6x + 5y = -4 \end{pmatrix}$

11. $\begin{pmatrix} x + 2y = 4 \\ 2x - y = 3 \end{pmatrix}$

12. $\begin{pmatrix} 2x - y = -8 \\ x + y = 2 \end{pmatrix}$

13. $\begin{pmatrix} y = 2x + 5 \\ x + 3y = -6 \end{pmatrix}$

14. $\begin{pmatrix} y = 4 - 2x \\ y = 7 - 3x \end{pmatrix}$

15. $\begin{pmatrix} y = 2x \\ 3x - 2y = -2 \end{pmatrix}$

16. $\begin{pmatrix} y = -2x \\ 3x - y = 0 \end{pmatrix}$

2 Solve Systems of Linear Inequalities

For Problems 17–32, indicate the solution set for each system of inequalities by shading the appropriate region.

17. $\begin{pmatrix} 3x - 4y \geq 0 \\ 2x + 3y \leq 0 \end{pmatrix}$

18. $\begin{pmatrix} 3x + 2y \leq 6 \\ 2x - 3y \geq 6 \end{pmatrix}$

19. $\begin{pmatrix} x - 3y < 6 \\ x + 2y \geq 4 \end{pmatrix}$

20. $\begin{pmatrix} 2x - y \leq 4 \\ 2x + y > 4 \end{pmatrix}$

21. $\begin{pmatrix} x + y < 4 \\ x - y > 2 \end{pmatrix}$

22. $\begin{pmatrix} x + y > 1 \\ x - y < 1 \end{pmatrix}$

23. $\begin{pmatrix} y < x + 1 \\ y \geq x \end{pmatrix}$

24. $\begin{pmatrix} y > x - 3 \\ y < x \end{pmatrix}$

25. $\begin{pmatrix} y > x \\ y > 2 \end{pmatrix}$

26. $\begin{pmatrix} 2x + y > 6 \\ 2x + y < 2 \end{pmatrix}$

27. $\begin{pmatrix} x \geq -1 \\ y < 4 \end{pmatrix}$

28. $\begin{pmatrix} x < 3 \\ y > 2 \end{pmatrix}$

29. $\begin{pmatrix} 2x - y > 4 \\ 2x - y > 0 \end{pmatrix}$

30. $\begin{pmatrix} x + y > 4 \\ x + y > 6 \end{pmatrix}$

31. $\begin{pmatrix} 3x - 2y < 6 \\ 2x - 3y < 6 \end{pmatrix}$

32. $\begin{pmatrix} 2x + 5y > 10 \\ 5x + 2y > 10 \end{pmatrix}$

THOUGHTS INTO WORDS

33. How do you know by inspection, without graphing, that the solution set of the system $\begin{pmatrix} 3x - 2y > 5 \\ 3x - 2y < 2 \end{pmatrix}$ is the null set?

34. Is it possible for a system of two linear equations in two variables to have exactly two solutions? Defend your answer.

GRAPHING CALCULATOR ACTIVITIES

35. Use your graphing calculator to help determine whether, in Problems 1–16, the system is consistent, the system is inconsistent, or the equations are dependent.

36. Use your graphing calculator to help determine the solution set for each of the following systems. Be sure to check your answers.

(a) $\begin{pmatrix} 3x - y = 30 \\ 5x - y = 46 \end{pmatrix}$

(b) $\begin{pmatrix} 1.2x + 3.4y = 25.4 \\ 3.7x - 2.3y = 14.4 \end{pmatrix}$

(c) $\begin{pmatrix} 1.98x + 2.49y = 13.92 \\ 1.19x + 3.45y = 16.18 \end{pmatrix}$

(d) $\begin{pmatrix} 2x - 3y = 10 \\ 3x + 5y = 53 \end{pmatrix}$

(e) $\begin{pmatrix} 4x - 7y = -49 \\ 6x + 9y = 219 \end{pmatrix}$

(f) $\begin{pmatrix} 3.7x - 2.9y = -14.3 \\ 1.6x + 4.7y = -30 \end{pmatrix}$

Answers to the Concept Quiz

1. True **2.** False **3.** True **4.** False **5.** True **6.** True **7.** False **8.** True **9.** True **10.** False

Answers to the Example Practice Skills

1. $\{(0, -2)\}$ **2.** $\{(x, y) \mid 2x + y = -1\}$ **3.** $\varnothing$

4.

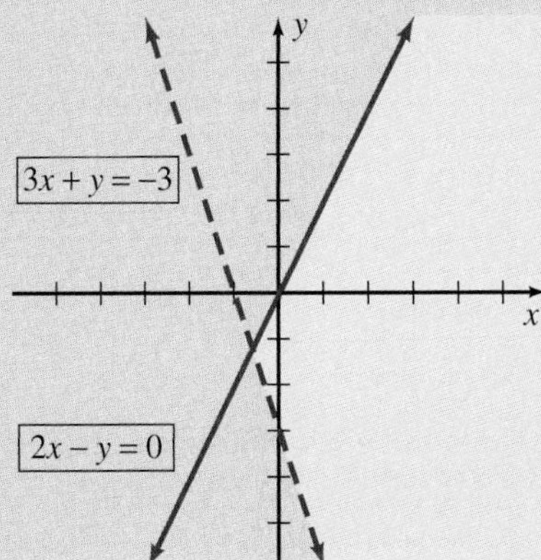

5.

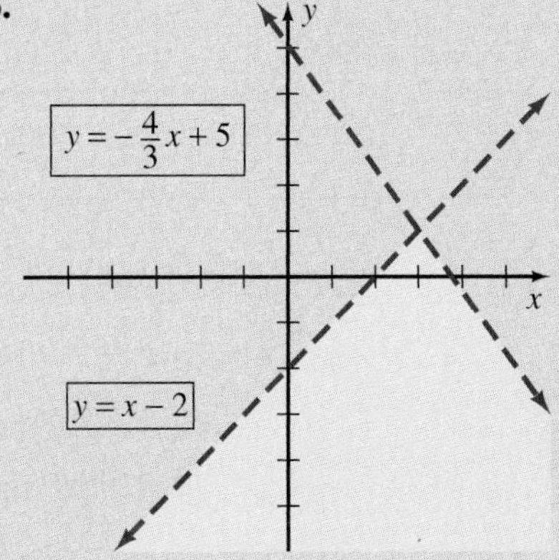

6.

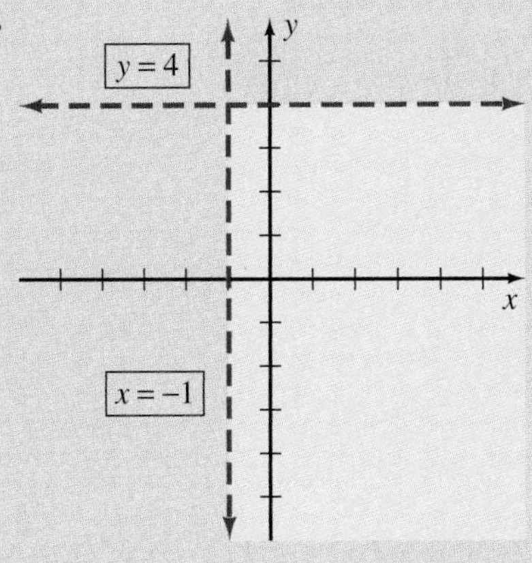

7. $\{(-1, 3)\}$

4.2 Substitution Method

OBJECTIVES

1 Solve Systems of Linear Equations by Substitution

2 Use Systems of Equations to Solve Problems

1 Solve Systems of Linear Equations by Substitution

It should be evident that solving systems of equations by graphing requires accurate graphs. In fact, unless the solutions are integers, it is quite difficult to obtain exact solutions from a graph. Thus we will consider some other methods for solving systems of equations.

We describe the **substitution method**, which works quite well with systems of two linear equations in two unknowns, as follows.

Step 1 Solve one of the equations for one variable in terms of the other variable if neither equation is in such a form. (If possible, make a choice that will avoid fractions.)

Step 2 Substitute the expression obtained in Step 1 into the other equation to produce an equation with one variable.

Step 3 Solve the equation obtained in Step 2.

Step 4 Use the solution obtained in Step 3, along with the expression obtained in Step 1, to determine the solution of the system.

Now let's look at some examples that illustrate the substitution method.

EXAMPLE 1

Solve the system $\begin{pmatrix} x + y = 16 \\ y = x + 2 \end{pmatrix}$.

Solution

Because the second equation states that y equals $x + 2$, we can substitute $x + 2$ for y in the first equation.

$$x + y = 16 \xrightarrow{\text{Substitute } x + 2 \text{ for } y} x + (x + 2) = 16$$

Now we have an equation with one variable that we can solve in the usual way.

$$x + (x + 2) = 16$$
$$2x + 2 = 16$$
$$2x = 14$$
$$x = 7$$

Substituting 7 for x in one of the two original equations (let's use the second one) yields

$$y = 7 + 2 = 9$$

To check, we can substitute 7 for x and 9 for y in both of the original equations.

$7 + 9 = 16$ A true statement

$9 = 7 + 2$ A true statement

The solution set is $\{(7, 9)\}$.

▼ PRACTICE YOUR SKILL

Solve the system $\begin{pmatrix} 3x + y = 5 \\ y = x - 11 \end{pmatrix}$. ■

EXAMPLE 2

Solve the system $\begin{pmatrix} x = 3y - 25 \\ 4x + 5y = 19 \end{pmatrix}$.

Solution

In this case the first equation states that x equals $3y - 25$. Therefore, we can substitute $3y - 25$ for x in the second equation.

$$4x + 5y = 19 \xrightarrow{\text{Substitute } 3y - 25 \text{ for } x} 4(3y - 25) + 5y = 19$$

Solving this equation yields

$$4(3y - 25) + 5y = 19$$
$$12y - 100 + 5y = 19$$
$$17y = 119$$
$$y = 7$$

Substituting 7 for y in the first equation produces

$$x = 3(7) - 25$$
$$= 21 - 25 = -4$$

The solution set is $\{(-4, 7)\}$; check it.

▼ PRACTICE YOUR SKILL

Solve the system $\begin{pmatrix} x = 2y - 13 \\ 2x + 3y = 30 \end{pmatrix}$. ■

EXAMPLE 3

Solve the system $\begin{pmatrix} 3x - 7y = 2 \\ x + 4y = 1 \end{pmatrix}$.

Solution

Let's solve the second equation for x in terms of y.

$$x + 4y = 1$$
$$x = 1 - 4y$$

Now we can substitute $1 - 4y$ for x in the first equation.

$$3x - 7y = 2 \xrightarrow{\text{Substitute } 1 - 4y \text{ for } x} 3(1 - 4y) - 7y = 2$$

Let's solve this equation for y.

$$3(1 - 4y) - 7y = 2$$
$$3 - 12y - 7y = 2$$

$$-19y = -1$$
$$y = \frac{1}{19}$$

Finally, we can substitute $\frac{1}{19}$ for y in the equation $x = 1 - 4y$.

$$x = 1 - 4\left(\frac{1}{19}\right)$$
$$= 1 - \frac{4}{19}$$
$$= \frac{15}{19}$$

The solution set is $\left\{\left(\frac{15}{19}, \frac{1}{19}\right)\right\}$.

▼ PRACTICE YOUR SKILL

Solve the system $\begin{pmatrix} x + 3y = 5 \\ 2x - 6y = -8 \end{pmatrix}$. ■

EXAMPLE 4

Solve the system $\begin{pmatrix} 5x - 6y = -4 \\ 3x + 2y = -8 \end{pmatrix}$.

Solution

Note that solving either equation for either variable will produce a fractional form. Let's solve the second equation for y in terms of x.

$$3x + 2y = -8$$
$$2y = -8 - 3x$$
$$y = \frac{-8 - 3x}{2}$$

Now we can substitute $\frac{-8 - 3x}{2}$ for y in the first equation.

$$5x - 6y = -4 \xrightarrow{\text{Substitute } \frac{-8-3x}{2} \text{ for } y} 5x - 6\left(\frac{-8 - 3x}{2}\right) = -4$$

Solving this equation yields

$$5x - 6\left(\frac{-8 - 3x}{2}\right) = -4$$
$$5x - 3(-8 - 3x) = -4$$
$$5x + 24 + 9x = -4$$
$$14x = -28$$
$$x = -2$$

Substituting -2 for x in $y = \frac{-8 - 3x}{2}$ yields

$$\begin{aligned} y &= \frac{-8 - 3(-2)}{2} \\ &= \frac{-8 + 6}{2} \\ &= \frac{-2}{2} \\ &= -1 \end{aligned}$$

The solution set is $\{(-2, -1)\}$.

▼ PRACTICE YOUR SKILL

Solve the system $\begin{pmatrix} 3x - 4y = 7 \\ 2x + 3y = 16 \end{pmatrix}$. ■

2 Use Systems of Equations to Solve Problems

Many word problems that we solved earlier in this text using one variable and one equation can also be solved using a system of two linear equations in two variables. In fact, in many of these problems you may find it more natural to use two variables. Let's consider some examples.

EXAMPLE 5

Mariano N. Ruiz/Used under license from Shutterstock

Apply Your Skill

Anita invested some money at 8% and \$400 more than that amount at 9%. The yearly interest from the two investments was \$87. How much did Anita invest at each rate?

Solution

Let x represent the amount invested at 8% and let y represent the amount invested at 9%. The problem translates into the following system.

Amount invested at 9% was \$400 more than at 8%. ⟶
Yearly interest from the two investments was \$87. ⟶
$$\begin{pmatrix} y = x + 400 \\ 0.08x + 0.09y = 87 \end{pmatrix}$$

From the first equation we can substitute $x + 400$ for y in the second equation and then solve for x.

$$\begin{aligned} 0.08x + 0.09(x + 400) &= 87 \\ 0.08x + 0.09x + 36 &= 87 \\ 0.17x &= 51 \\ x &= 300 \end{aligned}$$

Therefore, \$300 is invested at 8% and \$300 + \$400 = \$700 is invested at 9%.

▼ PRACTICE YOUR SKILL

Chris invested some money at 6% and \$600 more than that amount at 7%. The yearly interest from the two investments was \$484. How much did Chris invest at each rate? ■

EXAMPLE 6

Apply Your Skill

The perimeter of a rectangle is 66 inches. The width of the rectangle is 7 inches less than the length of the rectangle. Find the dimensions of the rectangle.

Solution

Let l represent the length of the rectangle and w the width of the rectangle. The problem translates into the following system.

$$\begin{pmatrix} 2w + 2l = 66 \\ w = l - 7 \end{pmatrix}$$

From the second equation, we can substitute $l - 7$ for w in the first equation and solve:

$$\begin{aligned} 2w + 2l &= 66 \\ 2(l - 7) + 2l &= 66 \\ 2l - 14 + 2l &= 66 \\ 4l &= 80 \\ l &= 20 \end{aligned}$$

Substitute 20 for l in $w = l - 7$ to obtain

$$w = 20 - 7 = 13$$

Therefore, the dimensions of the rectangle are 13 inches by 20 inches.

▼ **PRACTICE YOUR SKILL**

The perimeter of a rectangle is 50 inches. The length of the rectangle is 9 inches more than the width of the rectangle. Find the dimensions of the rectangle. ■

CONCEPT QUIZ

For Problems 1–5, answer true or false.

1. Graphing a system of equations is the most accurate method to find the solution of a system.
2. To begin solving a system of equations by substitution, one of the equations is solved for one variable in terms of the other variable.
3. When solving a system of equations by substitution, deciding what variable to solve for may allow you to avoid working with fractions.
4. When finding the solution of the system $\begin{pmatrix} x = 2y + 4 \\ x = -y + 5 \end{pmatrix}$, you need only to find a value for x.
5. The ordered pairs (1, 3) and (5, 11) are both solutions of the system $\begin{pmatrix} y = 2x + 1 \\ 4x - 2y = -2 \end{pmatrix}$.

Problem Set 4.2

1 Solve Systems of Linear Equations by Substitution

For Problems 1–26, solve each system by using the substitution method.

1. $\begin{pmatrix} x + y = 20 \\ x = y - 4 \end{pmatrix}$
2. $\begin{pmatrix} x + y = 23 \\ y = x - 5 \end{pmatrix}$
3. $\begin{pmatrix} y = -3x - 18 \\ 5x - 2y = -8 \end{pmatrix}$
4. $\begin{pmatrix} 4x - 3y = 33 \\ x = -4y - 25 \end{pmatrix}$
5. $\begin{pmatrix} x = -3y \\ 7x - 2y = -69 \end{pmatrix}$
6. $\begin{pmatrix} 9x - 2y = -38 \\ y = -5x \end{pmatrix}$
7. $\begin{pmatrix} 2x + 3y = 11 \\ 3x - 2y = -3 \end{pmatrix}$
8. $\begin{pmatrix} 3x - 4y = -14 \\ 4x + 3y = 23 \end{pmatrix}$
9. $\begin{pmatrix} 3x - 4y = 9 \\ x = 4y - 1 \end{pmatrix}$
10. $\begin{pmatrix} y = 3x - 5 \\ 2x + 3y = 6 \end{pmatrix}$

11. $\begin{pmatrix} y = \frac{2}{5}x - 1 \\ 3x + 5y = 4 \end{pmatrix}$

12. $\begin{pmatrix} y = \frac{3}{4}x - 5 \\ 5x - 4y = 9 \end{pmatrix}$

13. $\begin{pmatrix} 7x - 3y = -2 \\ x = \frac{3}{4}y + 1 \end{pmatrix}$

14. $\begin{pmatrix} 5x - y = 9 \\ x = \frac{1}{2}y - 3 \end{pmatrix}$

15. $\begin{pmatrix} 2x + y = 12 \\ 3x - y = 13 \end{pmatrix}$

16. $\begin{pmatrix} -x + 4y = -22 \\ x - 7y = 34 \end{pmatrix}$

17. $\begin{pmatrix} 4x + 3y = -40 \\ 5x - y = -12 \end{pmatrix}$

18. $\begin{pmatrix} x - 5y = 33 \\ -4x + 7y = -41 \end{pmatrix}$

19. $\begin{pmatrix} 3x + y = 2 \\ 11x - 3y = 5 \end{pmatrix}$

20. $\begin{pmatrix} 2x - y = 9 \\ 7x + 4y = 1 \end{pmatrix}$

21. $\begin{pmatrix} 3x + 5y = 22 \\ 4x - 7y = -39 \end{pmatrix}$

22. $\begin{pmatrix} 2x - 3y = -16 \\ 6x + 7y = 16 \end{pmatrix}$

23. $\begin{pmatrix} 4x - 5y = 3 \\ 8x + 15y = -24 \end{pmatrix}$

24. $\begin{pmatrix} 2x + 3y = 3 \\ 4x - 9y = -4 \end{pmatrix}$

25. $\begin{pmatrix} 6x - 3y = 4 \\ 5x + 2y = -1 \end{pmatrix}$

26. $\begin{pmatrix} 7x - 2y = 1 \\ 4x + 5y = 2 \end{pmatrix}$

2 Use Systems of Equations to Solve Problems

For Problems 27–40, solve each problem by setting up and solving an appropriate system of equations.

27. Doris invested some money at 7% and some money at 8%. She invested $6000 more at 8% than she did at 7%. Her total yearly interest from the two investments was $780. How much did Doris invest at each rate?

28. Suppose that Gus invested a total of $8000, part of it at 8% and the remainder at 9%. His yearly income from the two investments was $690. How much did he invest at each rate?

29. Find two numbers whose sum is 131 such that one number is 5 less than three times the other.

30. The length of a rectangle is twice the width of the rectangle. Given that the perimeter of the rectangle is 72 centimeters, find the dimensions.

31. Two angles are complementary, and the measure of one of the angles is 10° less than four times the measure of the other angle. Find the measure of each angle.

32. The difference of two numbers is 75. The larger number is 3 less than four times the smaller number. Find the numbers.

33. In a class of 50 students, the number of females is 2 more than five times the number of males. How many females are there in the class?

34. In a recent survey, one thousand registered voters were asked about their political preferences. The number of males in the survey was five less than one-half of the number of females. Find the number of males in the survey.

35. The perimeter of a rectangle is 94 inches. The length of the rectangle is 7 inches more than the width. Find the dimensions of the rectangle.

36. Two angles are supplementary, and the measure of one of them is 20° less than three times the measure of the other angle. Find the measure of each angle.

37. A deposit slip listed $700 in cash to be deposited. There were 100 bills, some of them five-dollar bills and the remainder ten-dollar bills. How many bills of each denomination were deposited?

38. Cindy has 30 coins, consisting of dimes and quarters, that total $5.10. How many coins of each kind does she have?

39. The income from a student production was $47,500. The price of a student ticket was $15, and nonstudent tickets were sold at $20 each. Three thousand tickets were sold. How many tickets of each kind were sold?

40. Sue bought three packages of cookies and two sacks of potato chips for $3.65. Later she bought two more packages of cookies and five additional sacks of potato chips for $4.23. Find the price of a package of cookies.

THOUGHTS INTO WORDS

41. Give a general description of how to use the substitution method to solve a system of two linear equations in two variables.

42. Explain how you would solve the system $\begin{pmatrix} 2x + 5y = 5 \\ 5x - y = 9 \end{pmatrix}$ using the substitution method.

GRAPHING CALCULATOR ACTIVITIES

43. Use your graphing calculator to help determine whether, in Problems 1–10, the system is consistent, the system is inconsistent, or the equations are dependent.

44. Use your graphing calculator to help determine the solution set for each of the following systems. Be sure to check your answers.

(a) $\begin{pmatrix} 3x - y = 30 \\ 5x - y = 46 \end{pmatrix}$

(b) $\begin{pmatrix} 1.2x + 3.4y = 25.4 \\ 3.7x - 2.3y = 14.4 \end{pmatrix}$

(c) $\begin{pmatrix} 1.98x + 2.49y = 13.92 \\ 1.19x + 3.45y = 16.18 \end{pmatrix}$

(d) $\begin{pmatrix} 2x - 3y = 10 \\ 3x + 5y = 53 \end{pmatrix}$

(e) $\begin{pmatrix} 4x - 7y = -49 \\ 6x + 9y = 219 \end{pmatrix}$

(f) $\begin{pmatrix} 3.7x - 2.9y = -14.3 \\ 1.6x + 4.7y = -30 \end{pmatrix}$

Answers to the Concept Quiz

1. False **2.** True **3.** True **4.** False **5.** True

Answers to the Example Practice Skills

1. $\{(4, -7)\}$ **2.** $\{(3, 8)\}$ **3.** $\left\{\left(\frac{1}{2}, \frac{3}{2}\right)\right\}$ **4.** $\{(5, 2)\}$ **5.** \$3400 at 6%, \$4000 at 7% **6.** 8 in. by 17 in.

4.3 Elimination-by-Addition Method

OBJECTIVES

1. Solve Systems of Equations Using the Elimination-by-Addition Method
2. Determine Which Method to Use to Solve a System of Equations
3. Use Systems of Equations to Solve Problems

1 Solve Systems of Equations Using the Elimination-by-Addition Method

We found in the previous section that the substitution method for solving a system of two equations and two unknowns works rather well. However, as the number of equations and unknowns increases, the substitution method becomes quite unwieldy. In this section we are going to introduce another method, called the **elimination-by-addition** method. We shall introduce it here using systems of two linear equations in two unknowns and then, in the next section, extend its use to three linear equations in three unknowns.

The elimination-by-addition method involves replacing systems of equations with simpler, equivalent systems until we obtain a system whereby we can easily extract the solutions. **Equivalent systems of equations are systems that have exactly the same solution set.** We can apply the following operations or transformations to a system of equations to produce an equivalent system.

1. Any two equations of the system can be interchanged.
2. Both sides of an equation of the system can be multiplied by any nonzero real number.
3. Any equation of the system can be replaced by the *sum* of that equation and a nonzero multiple of another equation.

Now let's see how to apply these operations to solve a system of two linear equations in two unknowns.

EXAMPLE 1

Solve the system $\begin{pmatrix} 3x + 2y = 1 \\ 5x - 2y = 23 \end{pmatrix}$. (1) (2)

Solution

Let's replace equation (2) with an equation we form by multiplying equation (1) by 1 and then adding that result to equation (2).

$$\begin{pmatrix} 3x + 2y = 1 \\ 8x \qquad = 24 \end{pmatrix} \quad \begin{matrix} (3) \\ (4) \end{matrix}$$

From equation (4) we can easily obtain the value of x.

$$8x = 24$$
$$x = 3$$

Then we can substitute 3 for x in equation (3).

$$3x + 2y = 1$$
$$3(3) + 2y = 1$$
$$2y = -8$$
$$y = -4$$

The solution set is $\{(3, -4)\}$. Check it!

▼ PRACTICE YOUR SKILL

Solve the system $\begin{pmatrix} 3x + 5y = 14 \\ -3x + 4y = 22 \end{pmatrix}$. ■

EXAMPLE 2

Solve the system $\begin{pmatrix} x + 5y = -2 \\ 3x - 4y = -25 \end{pmatrix}$. (1) (2)

Solution

Let's replace equation (2) with an equation we form by multiplying equation (1) by -3 and then adding that result to equation (2).

$$\begin{pmatrix} x + 5y = -2 \\ -19y = -19 \end{pmatrix} \quad \begin{matrix} (3) \\ (4) \end{matrix}$$

From equation (4) we can obtain the value of y.

$$-19y = -19$$
$$y = 1$$

Now we can substitute 1 for y in equation (3).

$$x + 5y = -2$$
$$x + 5(1) = -2$$
$$x = -7$$

The solution set is $\{(-7, 1)\}$.

▼ PRACTICE YOUR SKILL

Solve the system $\begin{pmatrix} 2x + y = -4 \\ 5x + 3y = -9 \end{pmatrix}$. ■

Note that our objective has been to produce an equivalent system of equations whereby one of the variables can be *eliminated* from one equation. We accomplish this by multiplying one equation of the system by an appropriate number and then *adding* that result to the other equation. Thus the method is called **elimination by addition**. Let's look at another example.

EXAMPLE 3

Solve the system $\begin{pmatrix} 2x + 5y = 4 \\ 5x - 7y = -29 \end{pmatrix}$. (1) (2)

Solution

Let's form an equivalent system where the second equation has no x term. First, we can multiply equation (2) by -2.

$$\begin{pmatrix} 2x + 5y = 4 \\ -10x + 14y = 58 \end{pmatrix} \quad \begin{matrix} (3) \\ (4) \end{matrix}$$

Now we can replace equation (4) with an equation that we form by multiplying equation (3) by 5 and then adding that result to equation (4).

$$\begin{pmatrix} 2x + 5y = 4 \\ 39y = 78 \end{pmatrix} \quad \begin{matrix} (5) \\ (6) \end{matrix}$$

From equation (6) we can find the value of y.

$$39y = 78$$
$$y = 2$$

Now we can substitute 2 for y in equation (5).

$$2x + 5y = 4$$
$$2x + 5(2) = 4$$
$$2x = -6$$
$$x = -3$$

The solution set is $\{(-3, 2)\}$.

▼ **PRACTICE YOUR SKILL**

Solve the system $\begin{pmatrix} 3x - 5y = 23 \\ 5x + 4y = 26 \end{pmatrix}$. ■

EXAMPLE 4

Solve the system $\begin{pmatrix} 3x - 2y = 5 \\ 2x + 7y = 9 \end{pmatrix}$. (1) (2)

Solution

We can start by multiplying equation (2) by -3.

$$\begin{pmatrix} 3x - 2y = 5 \\ -6x - 21y = -27 \end{pmatrix} \quad \begin{matrix} (3) \\ (4) \end{matrix}$$

Now we can replace equation (4) with an equation we form by multiplying equation (3) by 2 and then adding that result to equation (4).

$$\begin{pmatrix} 3x - 2y = 5 \\ -25y = -17 \end{pmatrix} \quad \begin{matrix} (5) \\ (6) \end{matrix}$$

From equation (6) we can find the value of y.

$$-25y = -17$$

$$y = \frac{17}{25}$$

Now we can substitute $\frac{17}{25}$ for y in equation (5).

$$3x - 2y = 5$$

$$3x - 2\left(\frac{17}{25}\right) = 5$$

$$3x - \frac{34}{25} = 5$$

$$3x = 5 + \frac{34}{25}$$

$$3x = \frac{125}{25} + \frac{34}{25}$$

$$3x = \frac{159}{25}$$

$$x = \left(\frac{159}{25}\right)\left(\frac{1}{3}\right) = \frac{53}{25}$$

The solution set is $\left\{\left(\frac{53}{25}, \frac{17}{25}\right)\right\}$. (Perhaps you should check this result!)

▼ **PRACTICE YOUR SKILL**

Solve the system $\begin{pmatrix} 6x + 5y = 5 \\ 5x - 2y = 6 \end{pmatrix}$. ■

2 Determine Which Method to Use to Solve a System of Equations

Both the elimination-by-addition and the substitution methods can be used to obtain exact solutions for any system of two linear equations in two unknowns. Sometimes the issue is that of deciding which method to use on a particular system. As we have seen with the examples thus far in this section and those of the previous section, many systems lend themselves to one or the other method by virtue of the original format of the equations. Let's emphasize that point with some more examples.

EXAMPLE 5

Solve the system $\begin{pmatrix} 4x - 3y = 4 \\ 10x + 9y = -1 \end{pmatrix}$. (1) (2)

Solution

Because changing the form of either equation in preparation for the substitution method would produce a fractional form, we are probably better off using the elimination-by-addition method. Let's replace equation (2) with an equation we form by multiplying equation (1) by 3 and then adding that result to equation (2).

$$\begin{pmatrix} 4x - 3y = 4 \\ 22x \quad\quad = 11 \end{pmatrix} \quad \begin{matrix} (3) \\ (4) \end{matrix}$$

From equation (4) we can determine the value of x.

$$22x = 11$$
$$x = \frac{11}{22} = \frac{1}{2}$$

Now we can substitute $\frac{1}{2}$ for x in equation (3).

$$4x - 3y = 4$$
$$4\left(\frac{1}{2}\right) - 3y = 4$$
$$2 - 3y = 4$$
$$-3y = 2$$
$$y = -\frac{2}{3}$$

The solution set is $\left\{\left(\frac{1}{2}, -\frac{2}{3}\right)\right\}$.

▼ PRACTICE YOUR SKILL

Solve the system $\begin{pmatrix} 2x + 6y = -2 \\ 4x + 9y = -3 \end{pmatrix}$. ■

EXAMPLE 6 Solve the system $\begin{pmatrix} 6x + 5y = -3 \\ y = -2x - 7 \end{pmatrix}$. (1) (2)

Solution

Because the second equation is of the form "y equals," let's use the substitution method. From the second equation we can substitute $-2x - 7$ for y in the first equation.

$$6x + 5y = -3 \xrightarrow{\text{Substitute } -2x - 7 \text{ for } y} 6x + 5(-2x - 7) = -3$$

Solving this equation yields

$$6x + 5(-2x - 7) = -3$$
$$6x - 10x - 35 = -3$$
$$-4x - 35 = -3$$
$$-4x = 32$$
$$x = -8$$

Substitute -8 for x in the second equation to obtain

$$y = -2(-8) - 7 = 16 - 7 = 9$$

The solution set is $\{(-8, 9)\}$.

▼ PRACTICE YOUR SKILL

Solve the system $\begin{pmatrix} x = 3y + 14 \\ 2x + y = 7 \end{pmatrix}$. ■

Sometimes we need to simplify the equations of a system before we can decide which method to use for solving the system. Let's consider an example of that type.

EXAMPLE 7

Solve the system $\begin{pmatrix} \dfrac{x-2}{4} + \dfrac{y+1}{3} = 2 \\ \dfrac{x+1}{7} + \dfrac{y-3}{2} = \dfrac{1}{2} \end{pmatrix}$. (1) (2)

Solution

First, we need to simplify the two equations. Let's multiply both sides of equation (1) by 12 and simplify.

$$12\left(\frac{x-2}{4} + \frac{y+1}{3}\right) = 12(2)$$

$$3(x-2) + 4(y+1) = 24$$

$$3x - 6 + 4y + 4 = 24$$

$$3x + 4y - 2 = 24$$

$$3x + 4y = 26$$

Let's multiply both sides of equation (2) by 14.

$$14\left(\frac{x+1}{7} + \frac{y-3}{2}\right) = 14\left(\frac{1}{2}\right)$$

$$2(x+1) + 7(y-3) = 7$$

$$2x + 2 + 7y - 21 = 7$$

$$2x + 7y - 19 = 7$$

$$2x + 7y = 26$$

Now we have the following system to solve.

$$\begin{pmatrix} 3x + 4y = 26 \\ 2x + 7y = 26 \end{pmatrix} \quad \begin{matrix} (3) \\ (4) \end{matrix}$$

Probably the easiest approach is to use the elimination-by-addition method. We can start by multiplying equation (4) by -3.

$$\begin{pmatrix} 3x + 4y = 26 \\ -6x - 21y = -78 \end{pmatrix} \quad \begin{matrix} (5) \\ (6) \end{matrix}$$

Now we can replace equation (6) with an equation we form by multiplying equation (5) by 2 and then adding that result to equation (6).

$$\begin{pmatrix} 3x + 4y = 26 \\ -13y = -26 \end{pmatrix} \quad \begin{matrix} (7) \\ (8) \end{matrix}$$

From equation (8) we can find the value of y.

$$-13y = -26$$

$$y = 2$$

Now we can substitute 2 for y in equation (7).

$$3x + 4y = 26$$

$$3x + 4(2) = 26$$

$$3x = 18$$

$$x = 6$$

The solution set is $\{(6, 2)\}$.

▼ **PRACTICE YOUR SKILL**

Solve the system $\begin{pmatrix} \dfrac{x+1}{5} + \dfrac{y+1}{2} = 2 \\ \dfrac{x+6}{2} + \dfrac{y-3}{3} = \dfrac{5}{2} \end{pmatrix}$. ■

Remark: Don't forget that to check a problem like Example 7 you must check the potential solutions back in the original equations.

In Section 4.1, we discussed the fact that you can tell whether a system of two linear equations in two unknowns has no solution, one solution, or infinitely many solutions by graphing the equations of the system. That is, the two lines may be parallel (no solution), or they may intersect in one point (one solution), or they may coincide (infinitely many solutions).

From a practical viewpoint, the systems that have one solution deserve most of our attention. However, we need to be able to deal with the other situations; they do occur occasionally. Let's use two examples to illustrate the type of thing that happens when we encounter *no solution* or *infinitely many solutions* when using either the elimination-by-addition method or the substitution method.

EXAMPLE 8

Solve the system $\begin{pmatrix} y = 3x - 1 \\ -9x + 3y = 4 \end{pmatrix}$. (1) (2)

Solution

Using the substitution method, we can proceed as follows:

$$-9x + 3y = 4 \xrightarrow{\text{Substitute } 3x - 1 \text{ for } y} -9x + 3(3x - 1) = 4$$

Solving this equation yields

$$-9x + 3(3x - 1) = 4$$
$$-9x + 9x - 3 = 4$$
$$-3 = 4$$

The *false numerical statement*, $-3 = 4$, implies that the system has *no solution*. (You may want to graph the two lines to verify this conclusion!)

▼ **PRACTICE YOUR SKILL**

Solve the system $\begin{pmatrix} y = -2x + 1 \\ 4x + 2y = 3 \end{pmatrix}$. ■

EXAMPLE 9

Solve the system $\begin{pmatrix} 5x + y = 2 \\ 10x + 2y = 4 \end{pmatrix}$. (1) (2)

Solution

Use the elimination-by-addition method and proceed as follows: Let's replace equation (2) with an equation we form by multiplying equation (1) by -2 and then adding that result to equation (2).

$$\begin{pmatrix} 5x + y = 2 \\ 0 + 0 = 0 \end{pmatrix} \qquad \begin{matrix} (3) \\ (4) \end{matrix}$$

The *true numerical statement*, $0 + 0 = 0$, implies that the system has *infinitely many solutions*. Any ordered pair that satisfies one of the equations will also satisfy the other equation. Thus the solution set can be expressed as

$$\{(x, y) \mid 5x + y = 2\}$$

▼ **PRACTICE YOUR SKILL**

Solve the system $\begin{pmatrix} 4x - 6y = 2 \\ 2x - 3y = 1 \end{pmatrix}$. ■

CONCEPT QUIZ

For Problems 1–10, answer true or false.

1. The elimination-by-addition method involves replacing systems of equations with simpler, equivalent systems until the solution can easily be determined.
2. Equivalent systems of equations are systems that have exactly the same solution set.
3. Any two equations of a system can be interchanged to obtain an equivalent system.
4. Any equation of a system can be multiplied on both sides by zero to obtain an equivalent system.
5. Any equation of the system can be replaced by the difference of that equation and a nonzero multiple of another equation.
6. The objective of the elimination-by-addition method is to produce an equivalent system with an equation where one of the variables has been eliminated.
7. The elimination-by-addition method is used for solving a system of equations only if the substitution method cannot be used.
8. If an equivalent system for an original system is $\begin{pmatrix} 3x - 5y = 7 \\ 0 + 0 = 0 \end{pmatrix}$, then the original system is inconsistent and has no solution.
9. The system $\begin{pmatrix} 5x - 2y = 3 \\ 5x - 2y = 9 \end{pmatrix}$ has infinitely many solutions.
10. The solution set of the system $\begin{pmatrix} x = 3y + 7 \\ 2x - 6y = 9 \end{pmatrix}$ is the null set.

Problem Set 4.3

1 Solve Systems of Equations Using the Elimination-by-Addition Method

For Problems 1–16, use the elimination-by-addition method to solve each system.

1. $\begin{pmatrix} 2x + 3y = -1 \\ 5x - 3y = 29 \end{pmatrix}$
2. $\begin{pmatrix} 3x - 4y = -30 \\ 7x + 4y = 10 \end{pmatrix}$
3. $\begin{pmatrix} 6x - 7y = 15 \\ 6x + 5y = -21 \end{pmatrix}$
4. $\begin{pmatrix} 5x + 2y = -4 \\ 5x - 3y = 6 \end{pmatrix}$
5. $\begin{pmatrix} x - 2y = -12 \\ 2x + 9y = 2 \end{pmatrix}$
6. $\begin{pmatrix} x - 4y = 29 \\ 3x + 2y = -11 \end{pmatrix}$
7. $\begin{pmatrix} 4x + 7y = -16 \\ 6x - y = -24 \end{pmatrix}$
8. $\begin{pmatrix} 6x + 7y = 17 \\ 3x + y = -4 \end{pmatrix}$
9. $\begin{pmatrix} 10x - 8y = -11 \\ 8x + 4y = -1 \end{pmatrix}$
10. $\begin{pmatrix} 4x + 3y = -4 \\ 3x - 7y = 34 \end{pmatrix}$
11. $\begin{pmatrix} 7x - 2y = 4 \\ 7x - 2y = 9 \end{pmatrix}$
12. $\begin{pmatrix} 8x - 3y = 13 \\ 4x + 9y = 3 \end{pmatrix}$
13. $\begin{pmatrix} 5x + 4y = 1 \\ 3x - 2y = -1 \end{pmatrix}$
14. $\begin{pmatrix} 2x - 7y = -2 \\ 3x + y = 1 \end{pmatrix}$
15. $\begin{pmatrix} 5x - y = 6 \\ 10x - 2y = 12 \end{pmatrix}$
16. $\begin{pmatrix} 3x - 2y = 5 \\ 2x + 5y = -3 \end{pmatrix}$

2 Determine Which Method to Use to Solve a System of Equations

For Problems 17–44, solve each system by using either the substitution or the elimination-by-addition method, whichever seems more appropriate.

17. $\begin{pmatrix} 5x + 3y = -7 \\ 7x - 3y = 55 \end{pmatrix}$
18. $\begin{pmatrix} 4x - 7y = 21 \\ -4x + 3y = -9 \end{pmatrix}$

19. $\begin{pmatrix} x = 5y + 7 \\ 4x + 9y = 28 \end{pmatrix}$

20. $\begin{pmatrix} 11x - 3y = -60 \\ y = -38 - 6x \end{pmatrix}$

21. $\begin{pmatrix} x = -6y + 79 \\ x = 4y - 41 \end{pmatrix}$

22. $\begin{pmatrix} y = 3x + 34 \\ y = -8x - 54 \end{pmatrix}$

23. $\begin{pmatrix} 4x - 3y = 2 \\ 5x - y = 3 \end{pmatrix}$

24. $\begin{pmatrix} 3x - y = 9 \\ 5x + 7y = 1 \end{pmatrix}$

25. $\begin{pmatrix} 5x - 2y = 1 \\ 10x - 4y = 7 \end{pmatrix}$

26. $\begin{pmatrix} 4x + 7y = 2 \\ 9x - 2y = 1 \end{pmatrix}$

27. $\begin{pmatrix} 3x - 2y = 7 \\ 5x + 7y = 1 \end{pmatrix}$

28. $\begin{pmatrix} 2x - 3y = 4 \\ y = \frac{2}{3}x - \frac{4}{3} \end{pmatrix}$

29. $\begin{pmatrix} -2x + 5y = -16 \\ x = \frac{3}{4}y + 1 \end{pmatrix}$

30. $\begin{pmatrix} y = \frac{2}{3}x - \frac{3}{4} \\ 2x + 3y = 11 \end{pmatrix}$

31. $\begin{pmatrix} y = \frac{2}{3}x - 4 \\ 5x - 3y = 9 \end{pmatrix}$

32. $\begin{pmatrix} 5x - 3y = 7 \\ x = \frac{3y}{4} - \frac{1}{3} \end{pmatrix}$

33. $\begin{pmatrix} \frac{x}{6} + \frac{y}{3} = 3 \\ \frac{5x}{2} - \frac{y}{6} = -17 \end{pmatrix}$

34. $\begin{pmatrix} \frac{3x}{4} - \frac{2y}{3} = 31 \\ \frac{7x}{5} + \frac{y}{4} = 22 \end{pmatrix}$

35. $\begin{pmatrix} -(x - 6) + 6(y + 1) = 58 \\ 3(x + 1) - 4(y - 2) = -15 \end{pmatrix}$

36. $\begin{pmatrix} -2(x + 2) + 4(y - 3) = -34 \\ 3(x + 4) - 5(y + 2) = 23 \end{pmatrix}$

37. $\begin{pmatrix} 5(x + 1) - (y + 3) = -6 \\ 2(x - 2) + 3(y - 1) = 0 \end{pmatrix}$

38. $\begin{pmatrix} 2(x - 1) - 3(y + 2) = 30 \\ 3(x + 2) + 2(y - 1) = -4 \end{pmatrix}$

39. $\begin{pmatrix} \frac{1}{2}x - \frac{1}{3}y = 12 \\ \frac{3}{4}x + \frac{2}{3}y = 4 \end{pmatrix}$

40. $\begin{pmatrix} \frac{2}{3}x + \frac{1}{5}y = 0 \\ \frac{3}{2}x - \frac{3}{10}y = -15 \end{pmatrix}$

41. $\begin{pmatrix} \frac{2x}{3} - \frac{y}{2} = -\frac{5}{4} \\ \frac{x}{4} + \frac{5y}{6} = \frac{17}{16} \end{pmatrix}$

42. $\begin{pmatrix} \frac{x}{2} + \frac{y}{3} = \frac{5}{72} \\ \frac{x}{4} + \frac{5y}{2} = -\frac{17}{48} \end{pmatrix}$

43. $\begin{pmatrix} \frac{3x + y}{2} + \frac{x - 2y}{5} = 8 \\ \frac{x - y}{3} - \frac{x + y}{6} = \frac{10}{3} \end{pmatrix}$

44. $\begin{pmatrix} \frac{x - y}{4} - \frac{2x - y}{3} = -\frac{1}{4} \\ \frac{2x + y}{3} + \frac{x + y}{2} = \frac{17}{6} \end{pmatrix}$

3 Use Systems of Equations to Solve Problems

For Problems 45–56, solve each problem by setting up and solving an appropriate system of equations.

45. A 10% salt solution is to be mixed with a 20% salt solution to produce 20 gallons of a 17.5% salt solution. How many gallons of the 10% solution and how many gallons of the 20% solution will be needed?

46. A small-town library buys a total of 35 books that cost \$462. Some of the books cost \$12 each, and the remainder cost \$14 each. How many books of each price did the library buy?

47. Suppose that on a particular day the cost of three tennis balls and two golf balls is \$7. The cost of six tennis balls and three golf balls is \$12. Find the cost of one tennis ball and the cost of one golf ball.

48. For moving purposes, the Hendersons bought 25 cardboard boxes for \$97.50. There were two kinds of boxes; the large ones cost \$7.50 per box and the small ones were \$3 per box. How many boxes of each kind did they buy?

49. A motel in a suburb of Chicago rents double rooms for \$120 per day and single rooms for \$90 per day. If a total of 55 rooms were rented for \$6150, how many of each kind were rented?

50. Suppose that one solution contains 50% alcohol and another solution contains 80% alcohol. How many liters of each solution should be mixed to make 10.5 liters of a 70% alcohol solution?

51. A college fraternity house spent \$670 for an order of 85 pizzas. The order consisted of cheese pizzas costing \$5 each and Supreme pizzas costing \$12 each. Find the number of each kind of pizza ordered.

52. Part of \$8400 is invested at 5% and the remainder is invested at 8%. The total yearly interest from the two investments is \$576. Determine how much is invested at each rate.

53. If the numerator of a certain fraction is increased by 5 and the denominator is decreased by 1, the resulting fraction is $\frac{8}{3}$. However, if the numerator of the original fraction is doubled and the denominator is increased by 7, then the resulting fraction is $\frac{6}{11}$. Find the original fraction.

54. A man bought 2 pounds of coffee and 1 pound of butter for a total of \$9.25. A month later, the prices had not changed (this makes it a fictitious problem), and he bought 3 pounds of coffee and 2 pounds of butter for \$15.50. Find the price per pound of both the coffee and the butter.

55. Suppose that we have a rectangular-shaped book cover. If the width is increased by 2 centimeters and the length is decreased by 1 centimeter, the area is increased by 28 square centimeters. However, if the width is decreased by 1 centimeter and the length is increased by

2 centimeters, then the area is increased by 10 square centimeters. Find the dimensions of the book cover.

56. A blueprint indicates a master bedroom in the shape of a rectangle. If the width is increased by 2 feet and the length remains the same, then the area is increased by 36 square feet. However, if the width is increased by 1 foot and the length is increased by 2 feet, then the area is increased by 48 square feet. Find the dimensions of the room as indicated on the blueprint.

THOUGHTS INTO WORDS

57. Give a general description of how to use the elimination-by-addition method to solve a system of two linear equations in two variables.

58. Explain how you would solve the system

$$\begin{pmatrix} 3x - 4y = -1 \\ 2x - 5y = 9 \end{pmatrix}$$

using the elimination-by-addition method.

59. How do you decide whether to solve a system of linear equations in two variables by using the substitution method or by using the elimination-by-addition method?

FURTHER INVESTIGATIONS

60. There is another way of telling whether a system of two linear equations in two unknowns is consistent or inconsistent, or whether the equations are dependent, without taking the time to graph each equation. It can be shown that any system of the form

$$a_1x + b_1y = c_1$$
$$a_2x + b_2y = c_2$$

has one and only one solution if

$$\frac{a_1}{a_2} \neq \frac{b_1}{b_2}$$

that it has no solution if

$$\frac{a_1}{a_2} = \frac{b_1}{b_2} \neq \frac{c_1}{c_2}$$

and that it has infinitely many solutions if

$$\frac{a_1}{a_2} = \frac{b_1}{b_2} = \frac{c_1}{c_2}$$

For each of the following systems, determine whether the system is consistent, the system is inconsistent, or the equations are dependent.

(a) $\begin{pmatrix} 4x - 3y = 7 \\ 9x + 2y = 5 \end{pmatrix}$ **(b)** $\begin{pmatrix} 5x - y = 6 \\ 10x - 2y = 19 \end{pmatrix}$

(c) $\begin{pmatrix} 5x - 4y = 11 \\ 4x + 5y = 12 \end{pmatrix}$ **(d)** $\begin{pmatrix} x + 2y = 5 \\ x - 2y = 9 \end{pmatrix}$

(e) $\begin{pmatrix} x - 3y = 5 \\ 3x - 9y = 15 \end{pmatrix}$ **(f)** $\begin{pmatrix} 4x + 3y = 7 \\ 2x - y = 10 \end{pmatrix}$

(g) $\begin{pmatrix} 3x + 2y = 4 \\ y = -\frac{3}{2}x - 1 \end{pmatrix}$ **(h)** $\begin{pmatrix} y = \frac{4}{3}x - 2 \\ 4x - 3y = 6 \end{pmatrix}$

61. A system such as

$$\begin{pmatrix} \frac{3}{x} + \frac{2}{y} = 2 \\ \frac{2}{x} - \frac{3}{y} = \frac{1}{4} \end{pmatrix}$$

is not a system of linear equations but can be transformed into a linear system by changing variables. For example, when we substitute u for $\frac{1}{x}$ and v for $\frac{1}{y}$, the system cited becomes

$$\begin{pmatrix} 3u + 2v = 2 \\ 2u - 3v = \frac{1}{4} \end{pmatrix}$$

We can solve this "new" system either by elimination by addition or by substitution (we will leave the details for you) to produce $u = \frac{1}{2}$ and $v = \frac{1}{4}$. Therefore, because $u = \frac{1}{x}$ and $v = \frac{1}{y}$, we have

$$\frac{1}{x} = \frac{1}{2} \quad \text{and} \quad \frac{1}{y} = \frac{1}{4}$$

Solving these equations yields

$$x = 2 \quad \text{and} \quad y = 4$$

The solution set of the original system is {(2, 4)}. Solve each of the following systems.

(a) $\begin{pmatrix} \frac{1}{x} + \frac{2}{y} = \frac{7}{12} \\ \frac{3}{x} - \frac{2}{y} = \frac{5}{12} \end{pmatrix}$ **(b)** $\begin{pmatrix} \frac{2}{x} + \frac{3}{y} = \frac{19}{15} \\ -\frac{2}{x} + \frac{1}{y} = -\frac{7}{15} \end{pmatrix}$

(c) $\begin{pmatrix} \frac{3}{x} - \frac{2}{y} = \frac{13}{6} \\ \frac{2}{x} + \frac{3}{y} = 0 \end{pmatrix}$ **(d)** $\begin{pmatrix} \frac{4}{x} + \frac{1}{y} = 11 \\ \frac{3}{x} - \frac{5}{y} = -9 \end{pmatrix}$

(e) $\begin{pmatrix} \frac{5}{x} - \frac{2}{y} = 23 \\ \frac{4}{x} + \frac{3}{y} = \frac{23}{2} \end{pmatrix}$ **(f)** $\begin{pmatrix} \frac{2}{x} - \frac{7}{y} = \frac{9}{10} \\ \frac{5}{x} + \frac{4}{y} = -\frac{41}{20} \end{pmatrix}$

62. Solve the following system for x and y.

$$\begin{pmatrix} a_1x + b_1y = c_1 \\ a_2x + b_2y = c_2 \end{pmatrix}$$

GRAPHING CALCULATOR ACTIVITIES

63. Use a graphing calculator to check your answers for Problem 60.

64. Use a graphing calculator to check your answers for Problem 61.

Answers to the Concept Quiz

1. True **2.** True **3.** True **4.** False **5.** True **6.** True **7.** False **8.** False **9.** False **10.** True

Answers to the Example Practice Skills

1. $\{(-2, 4)\}$ **2.** $\{(-3, 2)\}$ **3.** $\{(6, -1)\}$ **4.** $\left\{\left(\frac{40}{37}, -\frac{11}{37}\right)\right\}$ **5.** $\left\{\left(0, -\frac{1}{3}\right)\right\}$ **6.** $\{(5, -3)\}$ **7.** $\{(-1, 3)\}$ **8.** $\varnothing$ **9.** $\{(x, y)|2x - 3y = 1\}$

4.4 Systems of Three Linear Equations in Three Variables

OBJECTIVES

1 Solve Systems of Three Linear Equations

2 Use Systems of Three Linear Equations to Solve Problems

1 Solve Systems of Three Linear Equations

Consider a linear equation in three variables x, y, and z, such as $3x - 2y + z = 7$. Any **ordered triple** (x, y, z) that makes the equation a true numerical statement is said to be a solution of the equation. For example, the ordered triple $(2, 1, 3)$ is a solution because $3(2) - 2(1) + 3 = 7$. However, the ordered triple $(5, 2, 4)$ is not a solution because $3(5) - 2(2) + 4 \neq 7$. There are infinitely many solutions in the solution set.

Remark: The concept of a *linear* equation is generalized to include equations of more than two variables. Thus an equation such as $5x - 2y + 9z = 8$ is called a linear equation in three variables; the equation $5x - 7y + 2z - 11w = 1$ is called a linear equation in four variables; and so on.

To *solve* a system of three linear equations in three variables, such as

$$\begin{pmatrix} 3x - y + 2z = 13 \\ 4x + 2y + 5z = 30 \\ 5x - 3y - z = 3 \end{pmatrix}$$

means to find all of the ordered triples that satisfy all three equations. In other words, the solution set of the system is the intersection of the solution sets of all three equations in the system.

The graph of a linear equation in three variables is a **plane**, not a line. In fact, graphing equations in three variables requires the use of a three-dimensional coordinate system. Thus using a graphing approach to solve systems of three linear equations in three variables is not at all practical. However, a simple graphical analysis does give us some idea of what we can expect as we begin solving such systems.

In general, because each linear equation in three variables produces a plane, a system of three such equations produces three planes. There are various ways in which three planes can be related. For example, they may be mutually parallel, or two of the planes may be parallel and the third one intersect each of the two. (You may want to analyze all of the other possibilities for the three planes!) However, for our purposes at this time, we need to realize that from a solution set viewpoint, a system of three linear equations in three variables produces one of the following possibilities.

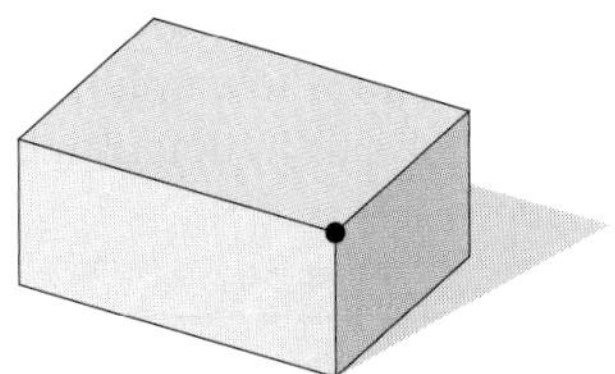

Figure 4.11

1. There is *one ordered triple* that satisfies all three equations. The three planes have a common point of intersection, as indicated in Figure 4.11.
2. There are *infinitely many* ordered triples in the solution set, all of which are coordinates of points on a line common to the planes. This can happen if three planes have a common line of intersection (Figure 4.12a) or if two of the planes coincide, and the third plane intersects them (Figure 4.12b).

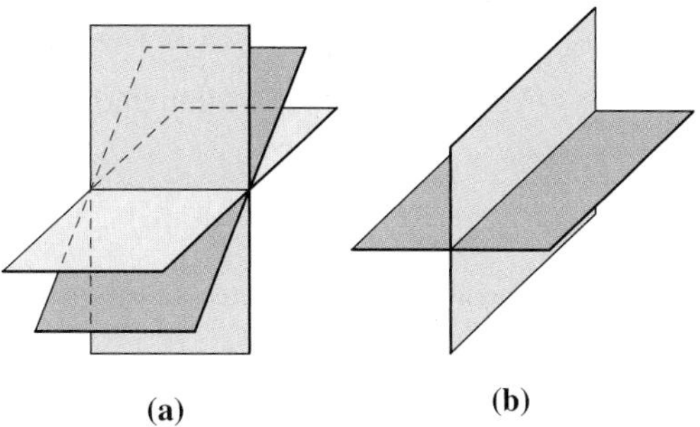

Figure 4.12

3. There are *infinitely many* ordered triples in the solution set, all of which are coordinates of points on a plane. This happens if the three planes coincide, as illustrated in Figure 4.13.

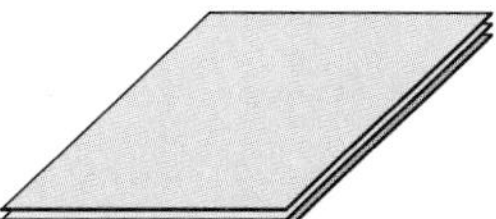

Figure 4.13

4. The solution set is *empty*; it is ∅. This can happen in various ways, as we see in Figure 4.14. Note that in each situation there are no points common to all three planes.

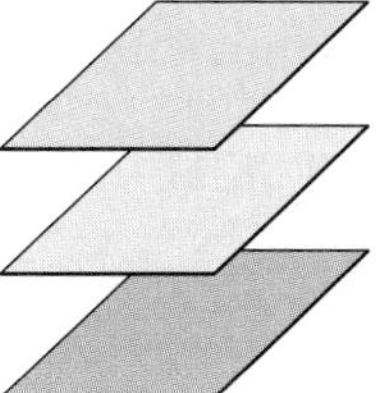

(a) Three parallel planes

(b) Two planes coincide and the third one is parallel to the coinciding planes.

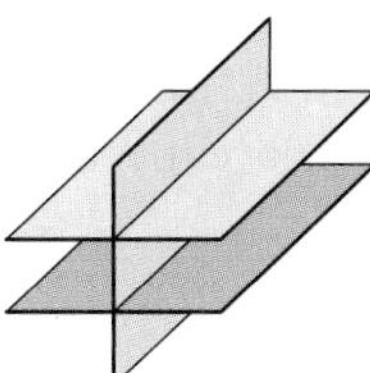

(c) Two planes are parallel and the third intersects them in parallel lines.

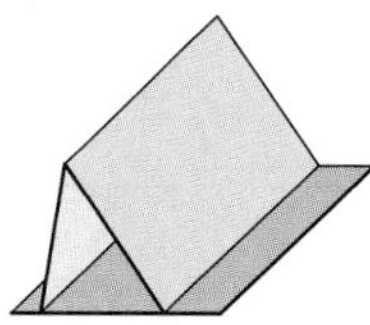

(d) No two planes are parallel, but two of them intersect in a line that is parallel to the third plane.

Figure 4.14

Now that we know what possibilities exist, let's consider finding the solution sets for some systems. Our approach will be the elimination-by-addition method, whereby we replace systems with equivalent systems until we obtain a system whose solution set can be easily determined. Let's start with an example that allows us to determine the solution set without changing to another, equivalent system.

EXAMPLE 1

Solve the system $\begin{pmatrix} 2x - 3y + 5z = -5 \\ 2y - 3z = 4 \\ 4z = -8 \end{pmatrix}$. (1) (2) (3)

Solution

From equation (3) we can find the value of z.

$$4z = -8$$
$$z = -2$$

Now we can substitute -2 for z in equation (2).

$$2y - 3z = 4$$
$$2y - 3(-2) = 4$$
$$2y + 6 = 4$$
$$2y = -2$$
$$y = -1$$

Finally, we can substitute -2 for z and -1 for y in equation (1).

$$2x - 3y + 5z = -5$$
$$2x - 3(-1) + 5(-2) = -5$$
$$2x + 3 - 10 = -5$$
$$2x - 7 = -5$$
$$2x = 2$$
$$x = 1$$

The solution set is $\{(1, -1, -2)\}$.

▼ **PRACTICE YOUR SKILL**

Solve the system $\begin{pmatrix} 2x + 2y + z = 5 \\ 3y + z = -9 \\ 2z = 6 \end{pmatrix}$. ■

Note the format of the equations in the system of Example 1. The first equation contains all three variables, the second equation has only two variables, and the third equation has only one variable. This allowed us to solve the third equation and then to use "back-substitution" to find the values of the other variables. Now let's consider an example where we have to make one replacement of an equivalent system.

EXAMPLE 2

Solve the system $\begin{pmatrix} 3x + 2y - 7z = -34 \\ y + 5z = 21 \\ 3y - 2z = -22 \end{pmatrix}$. (1) (2) (3)

Solution

Let's replace equation (3) with an equation we form by multiplying equation (2) by -3 and then adding that result to equation (3).

$$\begin{pmatrix} 3x + 2y - 7z = -34 \\ y + 5z = 21 \\ -17z = -85 \end{pmatrix}. \quad \begin{matrix} (4) \\ (5) \\ (6) \end{matrix}$$

From equation (6), we can find the value of z.

$$-17z = -85$$
$$z = 5$$

Now we can substitute 5 for z in equation (5).

$$y + 5z = 21$$
$$y + 5(5) = 21$$
$$y = -4$$

Finally, we can substitute 5 for z and -4 for y in equation (4).

$$3x + 2y - 7z = -34$$
$$3x + 2(-4) - 7(5) = -34$$
$$3x - 8 - 35 = -34$$
$$3x - 43 = -34$$
$$3x = 9$$
$$x = 3$$

The solution set is $\{(3, -4, 5)\}$.

▼ PRACTICE YOUR SKILL

Solve the system $\begin{pmatrix} 3x - 2y + 4z = -16 \\ 5y + 2z = 13 \\ y - 4z = 7 \end{pmatrix}$. ■

Now let's consider some examples where we have to make more than one replacement of equivalent systems.

EXAMPLE 3

Solve the system $\begin{pmatrix} x - y + 4z = -29 \\ 3x - 2y - z = -6 \\ 2x - 5y + 6z = -55 \end{pmatrix}$. $\begin{matrix} (1) \\ (2) \\ (3) \end{matrix}$

Solution

Let's replace equation (2) with an equation we form by multiplying equation (1) by -3 and then adding that result to equation (2). Let's also replace equation (3) with an equation we form by multiplying equation (1) by -2 and then adding that result to equation (3).

$$\begin{pmatrix} x - y + 4z = -29 \\ y - 13z = 81 \\ -3y - 2z = 3 \end{pmatrix} \quad \begin{matrix} (4) \\ (5) \\ (6) \end{matrix}$$

Now let's replace equation (6) with an equation we form by multiplying equation (5) by 3 and then adding that result to equation (6).

$$\begin{pmatrix} x - y + 4z = -29 \\ y - 13z = 81 \\ -41z = 246 \end{pmatrix} \quad \begin{matrix} (7) \\ (8) \\ (9) \end{matrix}$$

From equation (9) we can determine the value of z.

$$-41z = 246$$
$$z = -6$$

Now we can substitute -6 for z in equation (8).

$$y - 13z = 81$$
$$y - 13(-6) = 81$$
$$y + 78 = 81$$
$$y = 3$$

Finally, we can substitute -6 for z and 3 for y in equation (7).

$$x - y + 4z = -29$$
$$x - 3 + 4(-6) = -29$$
$$x - 3 - 24 = -29$$
$$x - 27 = -29$$
$$x = -2$$

The solution set is $\{(-2, 3, -6)\}$.

▼ PRACTICE YOUR SKILL

Solve the system $\begin{pmatrix} x - 3y + 2z = 3 \\ 2x + 7y - 3z = -3 \\ 3x - 2y + 2z = 3 \end{pmatrix}$. ■

EXAMPLE 4

Solve the system $\begin{pmatrix} 3x - 4y + z = 14 \\ 5x + 3y - 2z = 27 \\ 7x - 9y + 4z = 31 \end{pmatrix}$. (1) (2) (3)

Solution

A glance at the coefficients in the system indicates that eliminating the z terms from equations (2) and (3) would be easy. Let's replace equation (2) with an equation we form by multiplying equation (1) by 2 and then adding that result to equation (2). Let's also replace equation (3) with an equation we form by multiplying equation (1) by -4 and then adding that result to equation (3).

$$\begin{pmatrix} 3x - 4y + z = 14 \\ 11x - 5y = 55 \\ -5x + 7y = -25 \end{pmatrix} \quad \begin{matrix} (4) \\ (5) \\ (6) \end{matrix}$$

Now let's eliminate the y terms from equations (5) and (6). First let's multiply equation (6) by 5.

$$\begin{pmatrix} 3x - 4y + z = 14 \\ 11x - 5y = 55 \\ -25x + 35y = -125 \end{pmatrix} \quad \begin{matrix} (7) \\ (8) \\ (9) \end{matrix}$$

Now we can replace equation (9) with an equation we form by multiplying equation (8) by 7 and then adding that result to equation (9).

$$\begin{pmatrix} 3x - 4y + z = 14 \\ 11x - 5y \quad = 55 \\ 52x \quad = 260 \end{pmatrix} \begin{matrix} (10) \\ (11) \\ (12) \end{matrix}$$

From equation (12), we can determine the value of x.

$$52x = 260$$

$$x = 5$$

Now we can substitute 5 for x in equation (11).

$$11x - 5y = 55$$

$$11(5) - 5y = 55$$

$$-5y = 0$$

$$y = 0$$

Finally, we can substitute 5 for x and 0 for y in equation (10).

$$3x - 4y + z = 14$$

$$3(5) - 4(0) + z = 14$$

$$15 - 0 + z = 14$$

$$z = -1$$

The solution set is $\{(5, 0, -1)\}$.

▼ **PRACTICE YOUR SKILL**

Solve the system $\begin{pmatrix} 3x + y + 2z = 10 \\ 4x + 2y + z = 11 \\ 2x - 3y + 3z = 12 \end{pmatrix}$. ■

EXAMPLE 5

Solve the system $\begin{pmatrix} x - 2y + 3z = 1 \\ 3x - 5y - 2z = 4 \\ 2x - 4y + 6z = 7 \end{pmatrix}$. (1) (2) (3)

Solution

A glance at the coefficients indicates that it should be easy to eliminate the x terms from equations (2) and (3). We can replace equation (2) with an equation we form by multiplying equation (1) by -3 and then adding that result to equation (2). Likewise, we can replace equation (3) with an equation we form by multiplying equation (1) by -2 and then adding that result to equation (3).

$$\begin{pmatrix} x - 2y + 3z = 1 \\ y - 11z = 1 \\ 0 + 0 + 0 = 5 \end{pmatrix} \begin{matrix} (4) \\ (5) \\ (6) \end{matrix}$$

The false statement, $0 = 5$, indicates that the system is inconsistent and that the solution set is therefore $\varnothing$. (If you were to graph this system, equations (1) and (3) would produce parallel planes, which is the situation depicted in Figure 4.14c.)

▼ **PRACTICE YOUR SKILL**

Solve the system $\begin{pmatrix} x + 2y + 5z = 8 \\ 3x - 5y + 2z = 12 \\ 2x + 4y + 10z = 14 \end{pmatrix}$. ■

EXAMPLE 6

Solve the system $\begin{pmatrix} 2x - y + 4z = 1 \\ 3x + 2y - z = 5 \\ 5x - 6y + 17z = -1 \end{pmatrix}$.

(1)
(2)
(3)

Solution

A glance at the coefficients indicates that it is easy to eliminate the y terms from equations (2) and (3). We can replace equation (2) with an equation we form by multiplying equation (1) by 2 and then adding that result to equation (2). Likewise, we can replace equation (3) with an equation we form by multiplying equation (1) by -6 and then adding that result to equation (3).

$$\begin{pmatrix} 2x - y + 4z = 1 \\ 7x + 7z = 7 \\ -7x - 7z = -7 \end{pmatrix} \quad \begin{matrix} (4) \\ (5) \\ (6) \end{matrix}$$

Now let's replace equation (6) with an equation we form by multiplying equation (5) by 1 and then adding that result to equation (6).

$$\begin{pmatrix} 2x - y + 4z = 1 \\ 7x + 7z = 7 \\ 0 + 0 = 0 \end{pmatrix} \quad \begin{matrix} (7) \\ (8) \\ (9) \end{matrix}$$

The true numerical statement, $0 + 0 = 0$, indicates that the system has infinitely many solutions. (The graph of this system is shown in Figure 4.12a.)

Remark: It can be shown that the solutions for the system in Example 6 are of the form $(t, 3 - 2t, 1 - t)$, where t is any real number. For example, if we let $t = 2$ then we get the ordered triple $(2, -1, -1)$, and this triple will satisfy all three of the original equations. For our purposes in this text, we shall simply indicate that such a system has infinitely many solutions.

▼ PRACTICE YOUR SKILL

Solve the system $\begin{pmatrix} 3x + y + 2z = 3 \\ x + 2y + 5z = 8 \\ -2x + y + 3z = 5 \end{pmatrix}$. ■

2 Use Systems of Three Linear Equations to Solve Problems

When using a system of equations to solve a problem that involves three variables, it will be necessary to write a system of equations with three equations. In the next example, a system of three will be set up and we will omit the details in solving the system.

EXAMPLE 7

Part of \$50,000 is invested at 4%, another part at 6%, and the remainder at 7%. The total yearly income from the three investments is \$3050. The sum of the amounts invested at 4% and 6% equals the amount invested at 7%. Determine how much is invested at each rate.

Solution

Let x represent the amount invested at 4%, let y represent the amount invested at 6%, and let z represent the amount invested at 7%. Knowing that all three parts equal the total amount invested, \$50,000, we can form the equation $x + y + z = 50{,}000$. We can determine the yearly interest from each part by multiplying the amount invested times the interest rate. Hence, the next equation is $0.04x + 0.06y + 0.07z = 3050$. We obtain the third equation from the information that the sum of the amounts invested at 4% and 6% equals the amount invested at 7%. So the third equation is $x + y = z$. These equations form a system of equations as follows.

$$\begin{pmatrix} x + y + z = 50{,}000 \\ 0.04x + 0.06y + 0.07z = 3050 \\ x + y - z = 0 \end{pmatrix}$$

Solving this system, it can be determined that \$10,000 is invested at 4%, \$15,000 is invested at 6%, and \$25,000 is invested at 7%.

▼ PRACTICE YOUR SKILL

Part of \$30,000 is invested at 6%, another part at 8%, and the remainder at 9%. The total yearly income from the three investments is \$2340. The sum of the amounts invested at 6% and 8% equals \$10,000 more than the amount invested at 9%. Determine how much is invested at each rate.

CONCEPT QUIZ

For Problems 1–10, answer true or false.

1. The graph of a linear equation in three variables is a line.
2. A system of three linear equations in three variables produces three planes when graphed.
3. Three planes can be related by intersecting in exactly two points.
4. One way three planes can be related is if two of the planes are parallel and the third plane intersects them in parallel lines.
5. A system of three linear equations in three variables always has an infinite number of solutions.
6. A system of three linear equations in three variables can have one ordered triple as a solution.
7. The solution set of the system $\begin{pmatrix} 2x - y + 3z = 4 \\ y - z = 12 \\ 2z = 6 \end{pmatrix}$ is $\{(5, 15, 3)\}$.
8. The solution set of the system $\begin{pmatrix} x - y + z = 4 \\ x - y + z = 6 \\ 3y - 2z = 9 \end{pmatrix}$ is $\{(3, 1, 2)\}$.
9. It is possible for a system of three linear equations in three variables to have a solution set consisting of $\{(0, 0, 0)\}$.
10. The solution set of the system $\begin{pmatrix} x - 3z = 4 \\ 3x - 2y + 7z = -1 \\ 2x + z = 9 \end{pmatrix}$ is $\left\{\left(\frac{31}{7}, \frac{107}{14}, \frac{1}{7}\right)\right\}$.

Problem Set 4.4

1 Solve Systems of Three Linear Equations

Solve each of the following systems. If the solution set is ∅ or if it contains infinitely many solutions, then so indicate.

1. $\begin{pmatrix} x + 2y - 3z = 2 \\ 3y - z = 13 \\ 3y + 5z = 25 \end{pmatrix}$

2. $\begin{pmatrix} 2x + 3y - 4z = -10 \\ 2y + 3z = 16 \\ 2y - 5z = -16 \end{pmatrix}$

3. $\begin{pmatrix} 3x + 2y - 2z = 14 \\ x - 6z = 16 \\ 2x + 5z = -2 \end{pmatrix}$

4. $\begin{pmatrix} 3x + 2y - z = -11 \\ 2x - 3y = -1 \\ 4x + 5y = -13 \end{pmatrix}$

5. $\begin{pmatrix} 2x - y + z = 0 \\ 3x - 2y + 4z = 11 \\ 5x + y - 6z = -32 \end{pmatrix}$

6. $\begin{pmatrix} x - 2y + 3z = 7 \\ 2x + y + 5z = 17 \\ 3x - 4y - 2z = 1 \end{pmatrix}$

7. $\begin{pmatrix} 4x - y + z = 5 \\ 3x + y + 2z = 4 \\ x - 2y - z = 1 \end{pmatrix}$

8. $\begin{pmatrix} 2x - y + 3z = -14 \\ 4x + 2y - z = 12 \\ 6x - 3y + 4z = -22 \end{pmatrix}$

9. $\begin{pmatrix} x - y + 2z = 4 \\ 2x - 2y + 4z = 7 \\ 3x - 3y + 6z = 1 \end{pmatrix}$

10. $\begin{pmatrix} x + y - z = 2 \\ 3x - 4y + 2z = 5 \\ 2x + 2y - 2z = 7 \end{pmatrix}$

11. $\begin{pmatrix} x - 2y + z = -4 \\ 2x + 4y - 3z = -1 \\ -3x - 6y + 7z = 4 \end{pmatrix}$

12. $\begin{pmatrix} 2x - y + 3z = 1 \\ 4x + 7y - z = 7 \\ x + 4y - 2z = 3 \end{pmatrix}$

13. $\begin{pmatrix} 3x - 2y + 4z = 6 \\ 9x + 4y - z = 0 \\ 6x - 8y - 3z = 3 \end{pmatrix}$

14. $\begin{pmatrix} 2x - y + 3z = 0 \\ 3x + 2y - 4z = 0 \\ 5x - 3y + 2z = 0 \end{pmatrix}$

15. $\begin{pmatrix} 3x - y + 4z = 9 \\ 3x + 2y - 8z = -12 \\ 9x + 5y - 12z = -23 \end{pmatrix}$

16. $\begin{pmatrix} 5x - 3y + z = 1 \\ 2x - 5y = -2 \\ 3x - 2y - 4z = -27 \end{pmatrix}$

17. $\begin{pmatrix} 4x - y + 3z = -12 \\ 2x + 3y - z = 8 \\ 6x + y + 2z = -8 \end{pmatrix}$

18. $\begin{pmatrix} x + 3y - 2z = 19 \\ 3x - y - z = 7 \\ -2x + 5y + z = 2 \end{pmatrix}$

19. $\begin{pmatrix} x + y + z = 1 \\ 2x - 3y + 6z = 1 \\ -x + y + z = 0 \end{pmatrix}$

20. $\begin{pmatrix} 3x + 2y - 2z = -2 \\ x - 3y + 4z = -13 \\ -2x + 5y + 6z = 29 \end{pmatrix}$

2 Use Systems of Three Linear Equations to Solve Problems

Solve each of the following problems by setting up and solving a system of three linear equations in three variables.

21. The sum of the digits of a three-digit number is 14. The number is 14 larger than 20 times the tens digit. The sum of the tens digit and the units digit is 12 larger than the hundreds digit. Find the number.

22. The sum of the digits of a three-digit number is 13. The sum of the hundreds digit and the tens digit is 1 less than the units digit. The sum of three times the hundreds digit and four times the units digit is 26 more than twice the tens digit. Find the number.

23. Two bottles of catsup, two jars of peanut butter, and one jar of pickles cost \$7.78. Three bottles of catsup, four jars of peanut butter, and two jars of pickles cost \$14.34. Four bottles of catsup, three jars of peanut butter, and five jars of pickles cost \$19.19. Find the cost per bottle of catsup, the cost per jar of peanut butter, and the cost per jar of pickles.

24. Five pounds of potatoes, 1 pound of onions, and 2 pounds of apples cost \$3.80. Two pounds of potatoes, 3 pounds of onions, and 4 pounds of apples cost \$5.78. Three pounds of potatoes, 4 pounds of onions, and 1 pound of apples cost \$4.08. Find the price per pound for each item.

25. The sum of three numbers is 20. The sum of the first and third numbers is 2 more than twice the second number. The third number minus the first yields three times the second number. Find the numbers.

26. The sum of three numbers is 40. The third number is 10 less than the sum of the first two numbers. The second number is 1 larger than the first. Find the numbers.

27. The sum of the measures of the angles of a triangle is 180°. The largest angle is twice the smallest angle. The sum of the smallest and the largest angle is twice the other angle. Find the measure of each angle.

28. A box contains \$2 in nickels, dimes, and quarters. There are 19 coins in all, and there are twice as many nickels as dimes. How many coins of each kind are there?

29. Part of \$3000 is invested at 12%, another part at 13%, and the remainder at 14%. The total yearly income from the three investments is \$400. The sum of the amounts invested at 12% and 13% equals the amount invested at 14%. Determine how much is invested at each rate.

30. The perimeter of a triangle is 45 centimeters. The longest side is 4 centimeters less than twice the shortest side. The sum of the lengths of the shortest and longest sides is 7 centimeters less than three times the length of the remaining side. Find the lengths of all three sides of the triangle.

THOUGHTS INTO WORDS

31. Give a step-by-step description of how to solve the system

$$\begin{pmatrix} x - 2y + 3z = -23 \\ 5y - 2z = 32 \\ 4z = -24 \end{pmatrix}$$

32. Describe how you would solve the system

$$\begin{pmatrix} x - 3z = 4 \\ 3x - 2y + 7z = -1 \\ 2x + z = 9 \end{pmatrix}$$

Answers to the Concept Quiz

1. False **2.** True **3.** False **4.** True **5.** False **6.** True **7.** True **8.** False **9.** True **10.** True

Answers to the Example Practice Skills

1. $\{(5, -4, 3)\}$ **2.** $\{(-2, 3, -1)\}$ **3.** $\{(-1, 2, 5)\}$ **4.** $\{(3, -1, 1)\}$ **5.** $\varnothing$ **6.** Infinitely many solutions **7.** \$8000 at 6%, \$12,000 at 8%, \$10,000 at 9%

Chapter 4 Summary

<table>
<tr><th>OBJECTIVE</th><th>SUMMARY</th><th>EXAMPLE</th><th>CHAPTER REVIEW PROBLEMS</th></tr>
<tr>
<td>Solve systems of two linear equations by graphing.
(Sec. 4.1, Obj. 1, p. 180)</td>
<td>Graphing a system of two linear equations in two variables produces one of the following results.

1. The graphs of the two equations are two intersecting lines, which indicates that there is one unique solution of the system. Such a system is called a consistent system.
2. The graphs of the two equations are two parallel lines, which indicates that there is no solution for the system. It is called an inconsistent system.
3. The graphs of the two equations are the same line, which indicates infinitely many solutions for the system. The equations are called dependent equations.</td>
<td>Solve $\begin{pmatrix} x - 3y = 6 \\ 2x + 3y = 3 \end{pmatrix}$ by graphing.

Solution

Graph the lines by determining the x and y intercepts and a check point.

$x - 3y = 6$
<table>
<tr><td>x</td><td>0</td><td>6</td><td>−3</td></tr>
<tr><td>y</td><td>−2</td><td>0</td><td>−3</td></tr>
</table>
$2x + 3y = 3$
<table>
<tr><td>x</td><td>0</td><td>$\frac{3}{2}$</td><td>−1</td></tr>
<tr><td>y</td><td>1</td><td>0</td><td>$\frac{5}{3}$</td></tr>
</table>

It appears that (3, −1) is the solution. Checking these values in the equations, we can determine that the solution set is {(3, −1)}.</td>
<td>Problems 1–3

(continued)</td>
</tr>
</table>

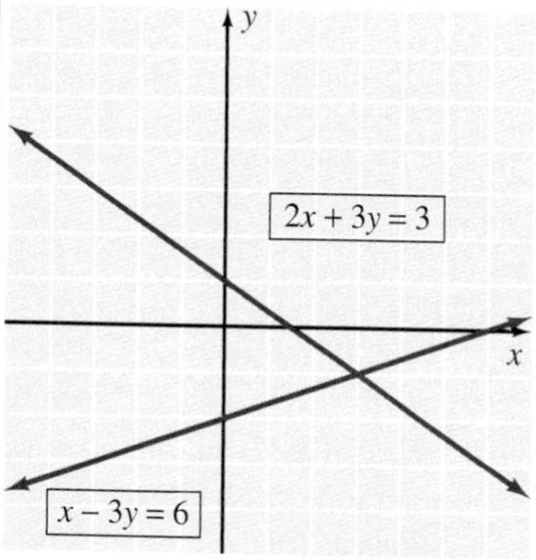

OBJECTIVE	SUMMARY	EXAMPLE	CHAPTER REVIEW PROBLEMS
Solve systems of linear inequalities. (Sec. 4.1, Obj. 2, p. 183)	The solution set of a system of linear inequalities is the intersection of the solution sets of the individual inequalities.	Solve the system $\begin{pmatrix} y < -x + 2 \\ y > \frac{1}{2}x + 1 \end{pmatrix}$. **Solution** The graph of $y < -x + 2$ consists of all the points below the line $y = -x + 2$. The graph of $y > \frac{1}{2}x + 1$ consists of all the points above the line $y = \frac{1}{2}x + 1$. [Graph: $y = \frac{1}{2}x + 1$; $y = -x + 2$] The graph of the system is indicated by the shaded region.	Problems 4–6
Solve systems of linear equations using substitution. (Sec. 4.2, Obj. 1, p. 188)	We can describe the substitution method of solving a system of equations as follows. Step 1: Solve one of the equations for one variable in terms of the other variable if neither equation is in such a form. (If possible, make a choice that will avoid fractions.) Step 2: Substitute the expression obtained in Step 1 into the other equation to produce an equation with one variable. Step 3: Solve the equation obtained in Step 2. Step 4: Use the solution obtained in Step 3, along with the expression obtained in Step 1, to determine the solution of the system.	Solve the system $\begin{pmatrix} 3x + y = -9 \\ 2x + 3y = 8 \end{pmatrix}$. **Solution** Solving the first equation for y gives the equation $y = -3x - 9$. In the second equation, substitute $-3x - 9$ for y and solve. $2x + 3(-3x - 9) = 8$ $2x - 9x - 27 = 8$ $-7x = 35$ $x = -5$ Now, to find the value of y, substitute 5 for x in the equation $y = -3x - 9$. $y = -3(-5) - 9 = 6$ The solution set of the system is $\{(-5, 6)\}$.	Problems 7–10

(continued)

OBJECTIVE	SUMMARY	EXAMPLE	CHAPTER REVIEW PROBLEMS
Solve systems of equations using the elimination-by-addition method. (Sec. 4.3, Obj. 1, p. 194)	The elimination-by-addition method involves the replacement of a system of equations with equivalent systems until a system is obtained whereby the solutions can be easily determined. The following operations or transformations can be performed on a system to produce an equivalent system. **1.** Any two equations of the system can be interchanged. **2.** Both sides of any equation of the system can be multiplied by any nonzero real number. **3.** Any equation of the system can be replaced by the *sum* of that equation and a nonzero multiple of another equation.	Solve the system $\begin{pmatrix} 2x - 5y = 31 \\ 4x + 3y = 23 \end{pmatrix}$. **Solution** Let's multiply the first equation by -2 and add the result to the second equation to eliminate the x variable. Then the equivalent system is $\begin{pmatrix} 2x - 5y = 31 \\ 13y = -39 \end{pmatrix}$. Now, solving the second equation for y, we obtain $y = -3$. Substitute -3 for y in either of the original equations and solve for x. $2x - 5(-3) = 31$ $2x + 15 = 31$ $2x = 16$ $x = 8$ The solution set of the system is $\{(8, -3)\}$.	Problems 7–10
Determine which method to use to solve a system of equations. (Sec. 4.3, Obj. 2, p. 197)	Graphing a system provides visual support for the solution, but it may be impossible to get exact solutions from a graph. Both the substitution method and the elimination-by-addition method provide exact solutions. Many systems lend themselves to one or the other method. Substitution is usually the preferred method if one of the equations in the system is already solved for a variable.	Solve $\begin{pmatrix} x - 3y = -13 \\ 2x + 5y = -18 \end{pmatrix}$. **Solution** Because the first equation can be solved for x without involving any fractions, the system is a good candidate for solving by the substitution method. The system could also be solved very easily using the elimination-by-addition method by multiplying the first equation by -2 and adding the result to the second equation. Either method will produce the solution set of $\{(-1, 4)\}$.	Problems 11–22 *(continued)*

OBJECTIVE	SUMMARY	EXAMPLE	CHAPTER REVIEW PROBLEMS
Use systems of equations to solve problems. (Sec. 4.2, Obj. 2, p. 191; Sec. 4.3, Obj. 3, p. 203)	Many problems that were solved earlier using only one variable may seem easier to solve by using two variables and a system of equations.	A car dealership has 220 vehicles on the lot. The number of cars on the lot is 5 less than twice the number of trucks. Find the number of cars and trucks on the lot. **Solution** Letting x represent the number of cars and y the number of trucks, we obtain the following system. $\begin{pmatrix} x + y = 220 \\ x = 2y - 5 \end{pmatrix}$ Solving the system, we can determine that the dealership has 145 cars and 75 trucks on the lot.	Problems 23–30
Solve systems of three linear equations. (Sec. 4.4, Obj. 1, p. 204)	Solving a system of three linear equations in three variables produces one of the following results. **1.** There is one ordered triple that satisfies all three equations. **2.** There are infinitely many ordered triples in the solution set, all of which are coordinates of points on a line common to the planes. **3.** There are infinitely many ordered triples in the solution set, all of which are coordinates of points on a plane. **4.** The solution set is empty; it is ∅.	Solve $\begin{pmatrix} 4x + 3y - 2z = -5 \\ 2y + 3z = -7 \\ y - 3z = -8 \end{pmatrix}$ **Solution** Replacing the third equation with the sum of the second equation and the third equation yields $3y = -15$. Therefore we can determine that $y = -5$. Substituting -5 for y in the third equation gives $-5 - 3z = -8$. Solving this equation yields $z = 1$. Substituting -5 for y and 1 for z in the first equation gives $4x + 3(-5) - 2(1) = -5$. Solving this equation gives $x = 3$. The solution set for the system is $\{(3, -5, 1)\}$.	Problems 31–36 *(continued)*

OBJECTIVE	SUMMARY	EXAMPLE	CHAPTER REVIEW PROBLEMS
Use systems of three linear equations to solve problems. (Sec. 4.4, Obj. 2, p. 210)	Many word problems involving three variables can be solved using a system of three linear equations.	The sum of the measures of the angles in a triangle is 180°. The largest angle is eight times the smallest angle. The sum of the smallest and the largest angle is three times the other angle. Find the measure of each angle. **Solution** Let x represent the measure of the largest angle, let y represent the measure of the middle angle, and let z represent the measure of the smallest angle. From the information in the problem, we can write the following system of equations: $\begin{pmatrix} x + y + z = 180 \\ x = 8z \\ x + z = 3y \end{pmatrix}$ By solving this system, we can determine that the measures of the angles of the triangle are 15°, 45°, and 120°.	Problems 37–40

Chapter 4 Review Problem Set

For Problems 1–3, solve by graphing.

1. $\begin{pmatrix} x - 2y = -4 \\ x + y = 5 \end{pmatrix}$

2. $\begin{pmatrix} y = \frac{1}{3}x + 2 \\ y = -\frac{1}{3}x \end{pmatrix}$

3. $\begin{pmatrix} 3x + 2y = 6 \\ y = -\frac{3}{2}x + 1 \end{pmatrix}$

For Problems 4–6, graph the solution set for the system.

4. $\begin{pmatrix} 3x + y > 6 \\ x - 2y \leq 4 \end{pmatrix}$

5. $\begin{pmatrix} y < -\frac{2}{3}x + 4 \\ y > \frac{1}{2}x + 3 \end{pmatrix}$

6. $\begin{pmatrix} y < 1 \\ x \leq -2 \end{pmatrix}$

For Problems 7–10, solve each system of equations using (a) the substitution method and (b) the elimination method.

7. $\begin{pmatrix} 3x - 2y = -6 \\ 2x + 5y = 34 \end{pmatrix}$

8. $\begin{pmatrix} x + 4y = 25 \\ y = -3x - 2 \end{pmatrix}$

9. $\begin{pmatrix} x = 5y - 49 \\ 4x + 3y = -12 \end{pmatrix}$

10. $\begin{pmatrix} x - 6y = 7 \\ 3x + 5y = 9 \end{pmatrix}$

For Problems 11–22, solve each system using the method that seems most appropriate to you.

11. $\begin{pmatrix} x - 3y = 25 \\ -3x + 2y = -26 \end{pmatrix}$

12. $\begin{pmatrix} 5x - 7y = -66 \\ x + 4y = 30 \end{pmatrix}$

13. $\begin{pmatrix} 4x + 3y = -9 \\ 3x - 5y = 15 \end{pmatrix}$

14. $\begin{pmatrix} 2x + 5y = 47 \\ 4x - 7y = -25 \end{pmatrix}$

15. $\begin{pmatrix} 7x - 3y = 25 \\ y = 3x - 9 \end{pmatrix}$

16. $\begin{pmatrix} x = -4 - 5y \\ y = 4x + 16 \end{pmatrix}$

17. $\begin{pmatrix} \frac{1}{2}x + \frac{2}{3}y = 6 \\ \frac{3}{4}x - \frac{5}{6}y = -24 \end{pmatrix}$

18. $\begin{pmatrix} \frac{3}{4}x - \frac{1}{2}y = 14 \\ \frac{5}{12}x + \frac{3}{4}y = 16 \end{pmatrix}$

19. $\begin{pmatrix} 6x - 4y = 7 \\ 9x + 8y = 0 \end{pmatrix}$

20. $\begin{pmatrix} 4x - 5y = -5 \\ 6x - 10y = -9 \end{pmatrix}$

21. $\begin{pmatrix} 2x + 3y = 9 \\ y = -\frac{2}{3}x + 2 \end{pmatrix}$

22. $\begin{pmatrix} 2x + \frac{1}{2}y = -2 \\ y = -4x - 4 \end{pmatrix}$

For Problems 23–30, solve each problem by setting up and solving a system of two equations and two unknowns.

23. At a local confectionery, 7 pounds of cashews and 5 pounds of Spanish peanuts cost \$88, and 3 pounds of cashews and 2 pounds of Spanish peanuts cost \$37. Find the price per pound for cashews and for Spanish peanuts.

24. We bought two cartons of pop and 4 pounds of candy for \$12. The next day we bought three cartons of pop and 2 pounds of candy for \$9. Find the price of a carton of pop and also the price of a pound of candy.

25. Suppose that a mail-order company charges a fixed fee for shipping merchandise that weighs 1 pound or less, plus an additional fee for each pound over 1 pound. If the shipping charge for 5 pounds is \$2.40 and for 12 pounds is \$3.10, find the fixed fee and the additional fee.

26. How many quarts of milk that is 1% fat must be mixed with milk that is 4% fat to obtain 10 quarts of milk that is 2% fat?

27. The perimeter of a rectangle is 56 centimeters. The length of the rectangle is three times the width. Find the dimensions of the rectangle.

28. Antonio had a total of \$4200 debt on two credit cards. One of the cards charged 1% interest per month and the other card charged 1.5% interest per month. Find the amount of debt on each card if he paid \$57 in interest charges for the month.

29. After working her shift as a waitress, Kelly had collected 30 bills as tips consisting of one-dollar bills and five-dollar bills. If her tips amounted to \$50, how many of each type of bill did she have?

30. In an ideal textbook, every problem set had a fixed number of review problems and a fixed number of problems on the new material. Professor Kelly always assigned 80% of the review problems and 40% of the problems on the new material, which amounted to 56 problems. Professor Edward always assigned 100% of the review problems and 60% of the problems on the new material, which amounted to 78 problems. How many problems of each type are in the problem sets?

For Problems 31–36, solve each system of equations.

31. $\begin{pmatrix} x - 2y + 4z = -14 \\ 3x - 5y + z = 20 \\ -2x + y - 5z = 22 \end{pmatrix}$

32. $\begin{pmatrix} x + 3y - 2z = 28 \\ 2x - 8y + 3z = -63 \\ 3x + 8y - 5z = 72 \end{pmatrix}$

33. $\begin{pmatrix} x + y - z = -2 \\ 2x - 3y + 4z = 17 \\ -3x + 2y + 5z = -7 \end{pmatrix}$

34. $\begin{pmatrix} -x - y + z = -3 \\ 3x + 2y - 4z = 12 \\ 5x + y + 2z = 5 \end{pmatrix}$

35. $\begin{pmatrix} 3x + y - z = -6 \\ 3x + 2y + 3z = 9 \\ 6x - 2y + 2z = 8 \end{pmatrix}$

36. $\begin{pmatrix} x - 3y + z = 2 \\ 2x - 5y - 3z = 22 \\ -4x + 3y + 5z = -26 \end{pmatrix}$

37. The perimeter of a triangle is 33 inches. The longest side is 3 inches more than twice the shortest side. The sum of the lengths of the shortest side and the longest side is 9 more than the remaining side. Find the length of all three sides of the triangle.

38. Kenisha has a Bank of US credit card that charges 1% interest per month, a Community Bank credit card that charges 1.5% interest per month, and a First National credit card that charges 2% interest per month. In total she has \$6400 charged between the three credit cards. The total interest for the month for all three cards is \$99. The amount charged on the Community Bank card is \$500 less than the amount charged on the Bank of US card. Find the amount charged on each card.

39. The measure of the largest angle of a triangle is twice the measure of the smallest angle of the triangle. The sum of the measures of the largest angle and the smallest angle of a triangle is twice the measure of the remaining angle of the triangle. Find the measure of all three angles of the triangle.

40. At the end of an evening selling flowers, a vendor had collected 64 bills consisting of five-dollar bills, ten-dollar bills, and twenty-dollar bills that amounted to \$620. The number of ten-dollar bills was three times the number of twenty-dollar bills. Find the number of each kind of bill.

Chapter 4 Test

For Problems 1–4, refer to the following systems of equations:

I. $\begin{pmatrix} 5x - 2y = 12 \\ 2x + 5y = 7 \end{pmatrix}$ **II.** $\begin{pmatrix} x - 4y = 1 \\ 2x - 8y = 2 \end{pmatrix}$

III. $\begin{pmatrix} 4x - 5y = 6 \\ 4x - 5y = 1 \end{pmatrix}$ **IV.** $\begin{pmatrix} 2x + 3y = 9 \\ 7x - 4y = 9 \end{pmatrix}$

1. For which of these systems are the equations said to be dependent? **1.** ______

2. For which of these systems does the solution set consist of a single ordered pair? **2.** ______

3. For which of these systems are the graphs parallel lines? **3.** ______

4. For which of these systems are the graphs perpendicular lines? **4.** ______

For Problems 5–6, solve the system by graphing.

5. $\begin{pmatrix} y = -2x - 1 \\ y = -\frac{1}{2}x + 2 \end{pmatrix}$ **5.** ______

6. $\begin{pmatrix} y = \frac{1}{3}x - 2 \\ x - 3y = 6 \end{pmatrix}$ **6.** ______

7. Use the elimination-by-addition method to solve the system $\begin{pmatrix} 2x - 3y = -17 \\ 5x + y = 17 \end{pmatrix}$. **7.** ______

8. Use the substitution method to solve the system $\begin{pmatrix} -5x + 4y = 35 \\ x - 3y = -18 \end{pmatrix}$. **8.** ______

For Problems 9–14, solve each of the systems using the method that seems most appropriate to you.

9. $\begin{pmatrix} 2x - 7y = -8 \\ 4x + 5y = 3 \end{pmatrix}$ **9.** ______

10. $\begin{pmatrix} \frac{2}{3}x - \frac{1}{2}y = 7 \\ \frac{1}{4}x + \frac{1}{3}y = 12 \end{pmatrix}$ **10.** ______

11. $\begin{pmatrix} 3x + 5y = -18 \\ 2x + y = 2 \end{pmatrix}$ **11.** ______

12. $\begin{pmatrix} y = \frac{3}{4}x - 3 \\ 3x - 4y = 6 \end{pmatrix}$ **12.** ______

13. $\begin{pmatrix} 2x + 5y = -6 \\ 3x - 4y = -9 \end{pmatrix}$ **13.** ______

14. $\begin{pmatrix} x + 3y = 24 \\ 4x + y = 19 \end{pmatrix}$ **14.** ______

15. ______

16. ______

17. ______

18. ______

19. ______

20. ______

21. ______

22. ______

23. ______

24. ______

25. ______

15. Find the value of x in the solution for the system $\begin{pmatrix} x = -2y + 5 \\ 7x + 3y = 46 \end{pmatrix}$.

For Problems 16–17, solve the system of equations.

16. $\begin{pmatrix} 4x + y + 3z = 5 \\ 3y + 2z = -7 \\ 4z = -8 \end{pmatrix}$

17. $\begin{pmatrix} x + 6y + 4z = 17 \\ 5y - 2z = -6 \\ 2y + 3z = 9 \end{pmatrix}$

For Problems 18–21, graph the solution for the system.

18. $\begin{pmatrix} x + 3y < 3 \\ 2x - y > 2 \end{pmatrix}$

19. $\begin{pmatrix} x - 3y > 3 \\ x - 3y < -3 \end{pmatrix}$

20. $\begin{pmatrix} y < 2x \\ y > -\frac{3}{5}x + 4 \end{pmatrix}$

21. $\begin{pmatrix} x \geq -4 \\ y < 3 \end{pmatrix}$

For Problems 22–25, set up and solve a system of equations to help solve each problem.

22. The perimeter of a rectangle is 82 inches. The length of the rectangle is 4 inches less than twice the width of the rectangle. Find the dimensions of the rectangle.

23. Allison distributed \$4000 between two investments. One investment paid 7% annual interest rate and the other paid 8% annual interest rate. How much was invested at each rate if she received \$306 in interest for the year?

24. One solution contains 30% alcohol and another solution contains 80% alcohol. Some of each of the two solutions is mixed to produce 5 liters of a 60% alcohol solution. How many liters of the 80% alcohol solution are used?

25. A box contains \$7.80 in nickels, dimes, and quarters. There are 6 more dimes than nickels and three times as many quarters as nickels. Find the number of quarters.

Chapters 1–4 Cumulative Review Problem Set

For Problems 1–10, evaluate each algebraic expression for the given values of the variables. Don't forget that in some cases it may be helpful to simplify the algebraic expression before evaluating it.

1. $x^2 - 2xy + y^2$ for $x = -2$ and $y = -4$
2. $-n^2 + 2n - 4$ for $n = -3$
3. $2x^2 - 5x + 6$ for $x = 3$
4. $3(2x - 1) - 2(x + 4) - 4(2x - 7)$ for $x = -1$
5. $-(2n - 1) + 5(2n - 3) - 6(3n + 4)$ for $n = 4$
6. $2(a - 4) - (a - 1) + (3a - 6)$ for $a = -5$
7. $(3x^2 - 4x - 7) - (4x^2 - 7x + 8)$ for $x = -4$
8. $-2(3x - 5y) - 4(x + 2y) + 3(-2x - 3y)$ for $x = 2$ and $y = -3$
9. $5(-x^2 - x + 3) - (2x^2 - x + 6) - 2(x^2 + 4x - 6)$ for $x = 2$
10. $3(x^2 - 4xy + 2y^2) - 2(x^2 - 6xy - y^2)$ for $x = -5$ and $y = -2$

For Problems 11–17, solve each of the equations.

11. $-2(n - 1) + 3(2n + 1) = -11$
12. $\frac{3}{4}(x - 2) - \frac{2}{5}(2x - 3) = \frac{1}{5}$
13. $0.1(x - 0.1) - 0.4(x + 2) = -5.31$
14. $\frac{2x - 1}{2} - \frac{5x + 2}{3} = 3$
15. $|3n - 2| = 7$
16. $|2x - 1| = |x + 4|$
17. $0.08(x + 200) = 0.07x + 20$

For Problems 18–23, solve each equation for the indicated variable.

18. $5x - 2y = 6$ for x
19. $3x + 4y = 12$ for y
20. $V = 2\pi rh + 2\pi r^2$ for h
21. $\frac{1}{R} = \frac{1}{R_1} + \frac{1}{R_2}$ for R_1
22. Solve $A = P + Prt$ for r, given that $A = \$4997$, $P = \$3800$, and $t = 3$ years.
23. Solve $C = \frac{5}{9}(F - 32)$ for C, given that $F = 5°$.

For Problems 24–31, solve each of the inequalities.

24. $-5(3n + 4) < -2(7n - 1)$
25. $7(x + 1) - 8(x - 2) < 0$
26. $|2x - 1| > 7$
27. $|3x + 7| < 14$
28. $0.09x + 0.1(x + 200) > 77$
29. $\frac{2x - 1}{4} - \frac{x - 2}{6} \leq \frac{3}{8}$
30. $-(x - 1) + 2(3x - 1) \geq 2(x + 4) - (x - 1)$
31. $\frac{1}{4}(x - 2) + \frac{3}{7}(2x - 1) < \frac{3}{14}$
32. Determine which of the ordered pairs are solutions of the equation. $4x - y = -4$; $(0, 4)$, $(1, 0)$, $(-2, -4)$, $\left(-\frac{1}{2}, 2\right)$

For Problems 33–36, find the x and y intercepts and graph the equation.

33. $3x - 4y = -12$
34. $x + 2y = 4$
35. $x - y = 5$
36. $3x - 2y = -6$
37. Graph the line that has a slope of $\frac{3}{4}$ and contains the point $(-3, 0)$.
38. Graph the line that has a slope of -3 and contains the point $(1, 5)$.
39. Graph the line whose equation is $y = 2$.
40. Graph the line whose equation is $x = -4$.

For Problems 41–46, graph the solution set.

41. $y < -2x + 4$
42. $2x - y \leq -2$
43. $y > -3x$
44. $\begin{pmatrix} y < \frac{1}{2}x - 1 \\ y < -\frac{3}{2}x + 3 \end{pmatrix}$
45. $\begin{pmatrix} y > -1 \\ x < 3 \end{pmatrix}$
46. $\begin{pmatrix} x - 2y > -2 \\ 3x + y \leq 3 \end{pmatrix}$
47. Find the distance between the points $(-4, -1)$ and $(1, 11)$.
48. Find the distance between the points $(-2, 3)$ and $(7, 1)$.
49. Find the slope of the line that contains the points $(4, -6)$ and $(7, 6)$.
50. Find the slope of the line that contains the points $(-2, 3)$ and $(1, 4)$.
51. Change the equation $2x - 5y = 10$ to slope-intercept form and determine the slope and y intercept.
52. Change the equation $3x + 4y = 0$ to slope-intercept form and determine the slope and y intercept.

For Problems 53–56, write the equation of the line that satisfies the given conditions. Express final equations in standard form.

53. Contains the point (1, 6) and has slope of -2

54. x intercept of 3 and y intercept of -2

55. Contains the point $(-4, 2)$ and is parallel to the line $3x + y = 4$

56. Contains the point $(5, -1)$ and is perpendicular to the line $2x + 5y = -6$

57. Assuming that the following situation can be expressed as a linear relationship between two variables, write a linear equation in two variables that describes the relationship. Use slope-intercept form to express the final equation. For infants weighing more than 117 ounces, the dose of medicine for an infant depends upon the weight of the infant. An infant that weighs 131 ounces should receive a dose of 2 grams and an infant that weighs 166 ounces should receive a dose of 7 grams. Let y represent the dose of medicine in grams and let x represent the weight of the infant in ounces.

58. Solve the system of equations $\begin{pmatrix} y = 3x + 6 \\ y = x + 4 \end{pmatrix}$ by graphing.

For Problems 59–62, solve each system.

59. $\begin{pmatrix} 2x - y = -19 \\ 2x + 5y = -1 \end{pmatrix}$

60. $\begin{pmatrix} 2x + 5y = 48 \\ 3x - 4y = -20 \end{pmatrix}$

61. $\begin{pmatrix} y = \frac{1}{2}x - 7 \\ 3x + 4y = 2 \end{pmatrix}$

62. $\begin{pmatrix} 2x + y + z = 3 \\ 3x - 2y + z = 5 \\ x + 4y + 3z = -1 \end{pmatrix}$

For Problems 63–67, solve each problem by setting up and solving an appropriate equation or inequality.

63. Find three consecutive odd integers such that three times the first minus the second is 1 more than the third.

64. The sum of the present ages of Joey and his mother is 46 years. In 4 years, Joey will be 3 years less than one-half as old as his mother at that time. Find the present ages of Joey and his mother.

65. Sandy starts off with her bicycle at 8 miles per hour. Fifty minutes later, Billie starts riding along the same route at 12 miles per hour. How long will it take Billie to overtake Sandy?

66. A retailer has some carpet that cost him \$18.00 a square yard. If he sells it for \$30 a square yard, what is his rate of profit based on the selling price?

67. Brad had scores of 88, 92, 93, and 89 on his first four algebra tests. What score must he obtain on the fifth test to have an average of 90 or better for the five tests?

For Problems 68–72, solve by setting up and solving a system of equations.

68. Inez has a collection of 48 coins consisting of nickels, dimes, and quarters. The number of dimes is 1 less than twice the number of nickels, and the number of quarters is 10 greater than the number of dimes. How many coins of each denomination are there in the collection?

69. The difference of the measures of two supplementary angles is 56°. Find the measure of each angle.

70. Norm invested a certain amount of money at 8% interest and \$200 more than that amount at 9%. His total yearly interest was \$86. How much did he invest at each rate?

71. Sanchez has a collection of pennies, nickels, and dimes worth \$9.35. He has 5 more nickels than pennies and twice as many dimes as pennies. How may coins of each kind does he have?

72. How many milliliters of pure acid must be added to 150 milliliters of a 30% solution of acid to obtain a 40% solution?

Polynomials

5

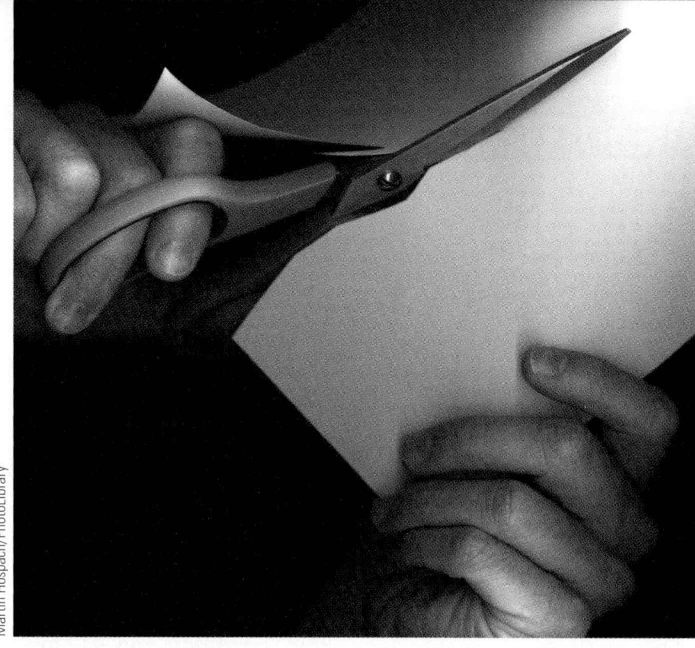

Martin Hospach/PhotoLibrary

A quadratic equation can be solved to determine the width of a uniform strip trimmed off both sides and ends of a sheet of paper to obtain a specified area for the sheet of paper.

A strip of uniform width cut off of both sides and both ends of an 8-inch by 11-inch sheet of paper must reduce the size of the paper to an area of 40 square inches. Find the width of the strip. With the equation $(11 - 2x)(8 - 2x) = 40$, you can determine that the strip should be 1.5 inches wide.

The main object of this text is to help you develop algebraic skills, use these skills to solve equations and inequalities, and use equations and inequalities to solve word problems. The work in this chapter will focus on a class of algebraic expressions called **polynomials**.

Video tutorials for all section learning objectives are available in a variety of delivery modes.

INTERNET PROJECT

Multiplying binomials is one of the topics of this chapter. When we look for a product such as $(x + 2)^5$, we call the process "binomial expansion." Pascal's triangle can be used to find the coefficients in a binomial expansion. Conduct an Internet search for an example of Pascal's triangle and construct a triangle with eight rows. Apply your results to Problem 93 on page 247 in Section 5.3.

5.1 Polynomials: Sums and Differences

OBJECTIVES

1. Find the Degree of a Polynomial
2. Add Polynomials
3. Subtract Polynomials
4. Simplify Polynomial Expressions
5. Use Polynomials in Geometry Problems

1 Find the Degree of a Polynomial

Recall that algebraic expressions such as $5x$, $-6y^2$, $7xy$, $14a^2b$, and $-17ab^2c^3$ are called terms. A **term** is an indicated product and may contain any number of factors. The variables in a term are called **literal factors**, and the numerical factor is called the **numerical coefficient**. Thus in $7xy$, the x and y are literal factors, 7 is the numerical coefficient, and the term is in two variables (x and y).

Terms that contain variables with only whole numbers as exponents are called **monomials**. The previously listed terms, $5x$, $-6y^2$, $7xy$, $14a^2b$, and $-17ab^2c^3$, are all monomials. (We shall work later with some algebraic expressions, such as $7x^{-1}y^{-1}$ and $6a^{-2}b^{-3}$, that are not monomials.)

The **degree** of a monomial is the sum of the exponents of the literal factors.

$7xy$ is of degree 2.

$14a^2b$ is of degree 3.

$-17ab^2c^3$ is of degree 6.

$5x$ is of degree 1.

$-6y^2$ is of degree 2.

If the monomial contains only one variable, then the exponent of the variable is the degree of the monomial. The last two examples illustrate this point. We say that any nonzero constant term is of degree zero.

A **polynomial** is a monomial or a finite sum (or difference) of monomials. Thus

$$4x^2, \quad 3x^2 - 2x - 4, \quad 7x^4 - 6x^3 + 4x^2 + x - 1,$$
$$3x^2y - 2xy^2, \quad \frac{1}{5}a^2 - \frac{2}{3}b^2, \quad \text{and} \quad 14$$

are examples of polynomials. In addition to calling a polynomial with one term a **monomial**, we also classify polynomials with two terms as **binomials** and those with three terms as **trinomials**.

The **degree of a polynomial** is the degree of the term with the highest degree in the polynomial. The following examples illustrate some of this terminology.

The polynomial $4x^3y^4$ is a monomial in two variables of degree 7.

The polynomial $4x^2y - 2xy$ is a binomial in two variables of degree 3.

The polynomial $9x^2 - 7x + 1$ is a trinomial in one variable of degree 2.

2 Add Polynomials

Remember that *similar terms*, or *like terms*, are terms that have the same literal factors. In the preceding chapters, we have frequently simplified algebraic expressions by combining similar terms, as the next examples illustrate.

$$\begin{aligned} 2x + 3y + 7x + 8y &= 2x + 7x + 3y + 8y \\ &= (2 + 7)x + (3 + 8)y \\ &= 9x + 11y \end{aligned}$$

Steps in dashed boxes are usually done mentally

$$\begin{aligned} 4a - 7 - 9a + 10 &= 4a + (-7) + (-9a) + 10 \\ &= 4a + (-9a) + (-7) + 10 \\ &= (4 + (-9))a + (-7) + 10 \\ &= -5a + 3 \end{aligned}$$

Both addition and subtraction of polynomials rely on basically the same ideas. The commutative, associative, and distributive properties provide the basis for rearranging, regrouping, and combining similar terms. Let's consider some examples.

EXAMPLE 1

Add $4x^2 + 5x + 1$ and $7x^2 - 9x + 4$.

Solution

We generally use the horizontal format for such work. Thus

$$\begin{aligned} (4x^2 + 5x + 1) + (7x^2 - 9x + 4) &= (4x^2 + 7x^2) + (5x - 9x) + (1 + 4) \\ &= 11x^2 - 4x + 5 \end{aligned}$$

▼ PRACTICE YOUR SKILL

Add $3x^2 - 7x + 3$ and $5x^2 + 11x - 7$. ■

$8x^2+4x-4$

EXAMPLE 2

Add $5x - 3$, $3x + 2$, and $8x + 6$.

Solution

$$\begin{aligned} (5x - 3) + (3x + 2) + (8x + 6) &= (5x + 3x + 8x) + (-3 + 2 + 6) \\ &= 16x + 5 \end{aligned}$$

▼ PRACTICE YOUR SKILL

Add $7x + 2$, $2x + 6$, and $6x - 1$. ■

$15x+7$

EXAMPLE 3

Find the indicated sum: $(-4x^2y + xy^2) + (7x^2y - 9xy^2) + (5x^2y - 4xy^2)$.

Solution

$$\begin{aligned}(-4x^2y + xy^2) &+ (7x^2y - 9xy^2) + (5x^2y - 4xy^2)\\ &= (-4x^2y + 7x^2y + 5x^2y) + (xy^2 - 9xy^2 - 4xy^2)\\ &= 8x^2y - 12xy^2\end{aligned}$$

▼ PRACTICE YOUR SKILL

Find the indicated sum: $(5x^2y - 2xy^2) + (-10x^2y - 4xy^2) + (-2x^2y + 7xy^2)$. ■

3 Subtract Polynomials

The idea of subtraction as adding the opposite extends to polynomials in general. Hence the expression $a - b$ is equivalent to $a + (-b)$. We can form the opposite of a polynomial by taking the opposite of each term. For example, the opposite of $3x^2 - 7x + 1$ is $-3x^2 + 7x - 1$. We express this in symbols as

$$-(3x^2 - 7x + 1) = -3x^2 + 7x - 1$$

Now consider the following subtraction problems.

EXAMPLE 4

Subtract $3x^2 + 7x - 1$ from $7x^2 - 2x - 4$.

Solution

Use the horizontal format to obtain

$$\begin{aligned}(7x^2 - 2x - 4) - (3x^2 + 7x - 1) &= (7x^2 - 2x - 4) + (-3x^2 - 7x + 1)\\ &= (7x^2 - 3x^2) + (-2x - 7x) + (-4 + 1)\\ &= 4x^2 - 9x - 3\end{aligned}$$

▼ PRACTICE YOUR SKILL

Subtract $2x^2 + 4x - 3$ from $5x^2 - 6x + 8$. ■

EXAMPLE 5

Subtract $-3y^2 + y - 2$ from $4y^2 + 7$.

Solution

Because subtraction is not a commutative operation, be sure to perform the subtraction in the correct order.

$$\begin{aligned}(4y^2 + 7) - (-3y^2 + y - 2) &= (4y^2 + 7) + (3y^2 - y + 2)\\ &= (4y^2 + 3y^2) + (-y) + (7 + 2)\\ &= 7y^2 - y + 9\end{aligned}$$

▼ PRACTICE YOUR SKILL

Subtract $-5y^2 + 3y - 6$ from $2y^2 + 10$. ■

The next example demonstrates the use of the vertical format for this work.

EXAMPLE 6 Subtract $4x^2 - 7xy + 5y^2$ from $3x^2 - 2xy + y^2$.

Solution

$$\begin{array}{r} 3x^2 - 2xy + y^2 \\ \underline{4x^2 - 7xy + 5y^2} \end{array}$$

Note which polynomial goes on the bottom and how the similar terms are aligned

Now we can mentally form the opposite of the bottom polynomial and add.

$$\begin{array}{r} 3x^2 - 2xy + y^2 \\ \underline{4x^2 - 7xy + 5y^2} \\ -x^2 + 5xy - 4y^2 \end{array}$$

The opposite of $4x^2 - 7xy + 5y^2$ is $-4x^2 + 7xy - 5y^2$

▼ **PRACTICE YOUR SKILL**

Subtract $x^2 + 4xy + 6y^2$ from $7x^2 - 3xy + y^2$. ■

4 Simplify Polynomial Expressions

We can also use the distributive property and the properties $a = 1(a)$ and $-a = -1(a)$ when adding and subtracting polynomials. The next examples illustrate this approach.

EXAMPLE 7 Perform the indicated operations: $(5x - 2) + (2x - 1) - (3x + 4)$.

Solution

$$\begin{aligned} (5x - 2) + (2x - 1) - (3x + 4) &= 1(5x - 2) + 1(2x - 1) - 1(3x + 4) \\ &= 1(5x) - 1(2) + 1(2x) - 1(1) - 1(3x) - 1(4) \\ &= 5x - 2 + 2x - 1 - 3x - 4 \\ &= 5x + 2x - 3x - 2 - 1 - 4 \\ &= 4x - 7 \end{aligned}$$

▼ **PRACTICE YOUR SKILL**

Perform the indicated operations: $(3x - 5) + (4x - 6) - (3x + 4)$. ■

We can do some of the steps mentally and simplify our format, as shown in the next two examples.

EXAMPLE 8 Perform the indicated operations: $(5a^2 - 2b) - (2a^2 + 4) + (-7b - 3)$.

Solution

$$\begin{aligned} (5a^2 - 2b) - (2a^2 + 4) + (-7b - 3) &= 5a^2 - 2b - 2a^2 - 4 - 7b - 3 \\ &= 3a^2 - 9b - 7 \end{aligned}$$

▼ **PRACTICE YOUR SKILL**

Perform the indicated operations: $(8a^2 - 3b) - (a^2 + 5) + (-6b - 8)$. ■

EXAMPLE 9 Simplify $(4t^2 - 7t - 1) - (t^2 + 2t - 6)$.

Solution

$$\begin{aligned}(4t^2 - 7t - 1) - (t^2 + 2t - 6) &= 4t^2 - 7t - 1 - t^2 - 2t + 6 \\ &= 3t^2 - 9t + 5\end{aligned}$$

▼ **PRACTICE YOUR SKILL**

Perform the indicated operations: $(6a^2 - 2a - 3) - (8a^2 + 4a - 5)$. ■

Remember that a polynomial in parentheses preceded by a negative sign can be written without the parentheses by replacing each term with its opposite. Thus in Example 9, $-(t^2 + 2t - 6) = -t^2 - 2t + 6$. Finally, let's consider a simplification problem that contains grouping symbols within grouping symbols.

EXAMPLE 10 Simplify $7x + [3x - (2x + 7)]$.

Solution

$$\begin{aligned}7x + [3x - (2x + 7)] &= 7x + [3x - 2x - 7] && \text{Remove the innermost parentheses first} \\ &= 7x + [x - 7] \\ &= 7x + x - 7 \\ &= 8x - 7\end{aligned}$$

▼ **PRACTICE YOUR SKILL**

Simplify $10x + [4x - (x + 5)]$. ■

5 Use Polynomials in Geometry Problems

Sometimes we encounter polynomials in a geometric setting. The next example shows that a polynomial can represent the total surface area of a rectangular solid.

EXAMPLE 11 Find a polynomial that represents the total surface area of the rectangular solid with the dimensions shown in Figure 5.1. Use the polynomial to determine the surface area for some specific solids.

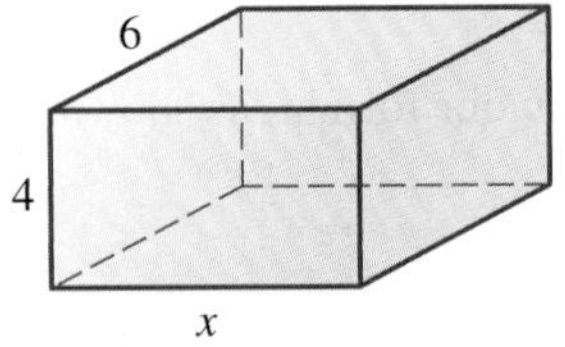

Figure 5.1

Solution

The total surface area would be the sum of the areas of all six sides of the solid. Using the dimensions in Figure 5.1, the sum of the areas of the sides is as follows:

$4x$	+	$4x$	+	$6x$	+	$6x$	+	24	+	24
↑		↑		↑		↑		↑		↑
Area of front		Area of back		Area of top		Area of bottom		Area of left side		Area of right side

Simplifying $4x + 4x + 6x + 6x + 24 + 24$, we obtain the polynomial $20x + 48$, which represents the total surface area of the rectangular solid. Furthermore, by evaluating the polynomial $20x + 48$ for different positive values of x, we can determine the total surface area of any rectangular solid for which two dimensions are 4 and 6. The following chart contains some specific rectangular solids.

x	4 by 6 by x rectangular solid	Total surface area $(20x + 48)$
2	4 by 6 by 2	$20(2) + 48 = 88$
4	4 by 6 by 4	$20(4) + 48 = 128$
5	4 by 6 by 5	$20(5) + 48 = 148$
7	4 by 6 by 7	$20(7) + 48 = 188$
12	4 by 6 by 12	$20(12) + 48 = 288$

▼ **PRACTICE YOUR SKILL**

Find a polynomial that represents the total surface area of the rectangular solid that has a width of 5, a length of 8, and a height of x. Use the polynomial to determine the total surface when the height is 4, 6, or 12. ■

CONCEPT QUIZ

For Problems 1–10, answer true or false.

1. The degree of the monomial $4x^2y$ is 3.
2. The degree of the polynomial $2x^4 - 5x^3 + 7x^2 - 4x + 6$ is 10.
3. A three-term polynomial is called a binomial.
4. A polynomial is a monomial or a finite sum of monomials.
5. Monomial terms must have whole number exponents for each variable.
6. The sum of $-2x - 1$, $-x + 4$, and $5x - 7$ is $8x - 4$.
7. If $-2x^2 + 3x - 4$ is subtracted from $-3x^2 - 7x + 2$, the result is $-x^2 - 10x + 6$.
8. Polynomials must be of the same degree if they are to be added.
9. If $-x - 1$ is subtracted from the sum of $2x - 1$ and $-4x - 6$, the result is $-x - 6$.
10. If the sum of $2x^2 - 4x - 8$ and $-2x + 6$ is subtracted from $-3x - 6$, the result is $-2x^2 + 3x - 4$.

Problem Set 5.1

1 Find the Degree of a Polynomial

For Problems 1–10, determine the degree of the given polynomials.

1. $7xy + 6y$
2. $-5x^2y^2 - 6xy^2 + x$
3. $-x^2y + 2xy^2 - xy$
4. $5x^3y^2 - 6x^3y^3$
5. $5x^2 - 7x - 2$
6. $7x^3 - 2x + 4$
7. $8x^6 + 9$
8. $5y^6 + y^4 - 2y^2 - 8$
9. -12
10. $7x - 2y$

2 Add Polynomials

For Problems 11–20, add the given polynomials.

11. $3x - 7$ and $7x + 4$
12. $9x + 6$ and $5x - 3$
13. $-5t - 4$ and $-6t + 9$
14. $-7t + 14$ and $-3t - 6$
15. $3x^2 - 5x - 1$ and $-4x^2 + 7x - 1$
16. $6x^2 + 8x + 4$ and $-7x^2 - 7x - 10$
17. $12a^2b^2 - 9ab$ and $5a^2b^2 + 4ab$
18. $15a^2b^2 - ab$ and $-20a^2b^2 - 6ab$
19. $2x - 4$, $-7x + 2$, and $-4x + 9$
20. $-x^2 - x - 4$, $2x^2 - 7x + 9$, and $-3x^2 + 6x - 10$

3 Subtract Polynomials

For Problems 21–30, subtract the polynomials using the horizontal format.

21. $5x - 2$ from $3x + 4$
22. $7x + 5$ from $2x - 1$
23. $-4a - 5$ from $6a + 2$
24. $5a + 7$ from $-a - 4$

25. $3x^2 - x + 2$ from $7x^2 + 9x + 8$

26. $5x^2 + 4x - 7$ from $3x^2 + 2x - 9$

27. $2a^2 - 6a - 4$ from $-4a^2 + 6a + 10$

28. $-3a^2 - 6a + 3$ from $3a^2 + 6a - 11$

29. $2x^3 + x^2 - 7x - 2$ from $5x^3 + 2x^2 + 6x - 13$

30. $6x^3 + x^2 + 4$ from $9x^3 - x - 2$

For Problems 31–40, subtract the polynomials using the vertical format.

31. $5x - 2$ from $12x + 6$

32. $3x - 7$ from $2x + 1$

33. $-4x + 7$ from $-7x - 9$

34. $-6x - 2$ from $5x + 6$

35. $2x^2 + x + 6$ from $4x^2 - x - 2$

36. $4x^2 - 3x - 7$ from $-x^2 - 6x + 9$

37. $x^3 + x^2 - x - 1$ from $-2x^3 + 6x^2 - 3x + 8$

38. $2x^3 - x + 6$ from $x^3 + 4x^2 + 1$

39. $-5x^2 + 6x - 12$ from $2x - 1$

40. $2x^2 - 7x - 10$ from $-x^3 - 12$

4 Simplify Polynomial Expressions

For Problems 41–46, perform the operations as described.

41. Subtract $2x^2 - 7x - 1$ from the sum of $x^2 + 9x - 4$ and $-5x^2 - 7x + 10$.

42. Subtract $4x^2 + 6x + 9$ from the sum of $-3x^2 - 9x + 6$ and $-2x^2 + 6x - 4$.

43. Subtract $-x^2 - 7x - 1$ from the sum of $4x^2 + 3$ and $-7x^2 + 2x$.

44. Subtract $-4x^2 + 6x - 3$ from the sum of $-3x + 4$ and $9x^2 - 6$.

45. Subtract the sum of $5n^2 - 3n - 2$ and $-7n^2 + n + 2$ from $-12n^2 - n + 9$.

46. Subtract the sum of $-6n^2 + 2n - 4$ and $4n^2 - 2n + 4$ from $-n^2 - n + 1$.

For Problems 47–56, perform the indicated operations.

47. $(5x + 2) + (7x - 1) + (-4x - 3)$

48. $(-3x + 1) + (6x - 2) + (9x - 4)$

49. $(12x - 9) - (-3x + 4) - (7x + 1)$

50. $(6x + 4) - (4x - 2) - (-x - 1)$

51. $(2x^2 - 7x - 1) + (-4x^2 - x + 6) + (-7x^2 - 4x - 1)$

52. $(5x^2 + x + 4) + (-x^2 + 2x + 4) + (-14x^2 - x + 6)$

53. $(7x^2 - x - 4) - (9x^2 - 10x + 8) + (12x^2 + 4x - 6)$

54. $(-6x^2 + 2x + 5) - (4x^2 + 4x - 1) + (7x^2 + 4)$

55. $(n^2 - 7n - 9) - (-3n + 4) - (2n^2 - 9)$

56. $(6n^2 - 4) - (5n^2 + 9) - (6n + 4)$

For Problems 57–70, simplify by removing the inner parentheses first and working outward.

57. $3x - [5x - (x + 6)]$

58. $7x - [2x - (-x - 4)]$

59. $2x^2 - [-3x^2 - (x^2 - 4)]$

60. $4x^2 - [-x^2 - (5x^2 - 6)]$

61. $-2n^2 - [n^2 - (-4n^2 + n + 6)]$

62. $-7n^2 - [3n^2 - (-n^2 - n + 4)]$

63. $[4t^2 - (2t + 1) + 3] - [3t^2 + (2t - 1) - 5]$

64. $-(3n^2 - 2n + 4) - [2n^2 - (n^2 + n + 3)]$

65. $[2n^2 - (2n^2 - n + 5)] + [3n^2 + (n^2 - 2n - 7)]$

66. $3x^2 - [4x^2 - 2x - (x^2 - 2x + 6)]$

67. $[7xy - (2x - 3xy + y)] - [3x - (x - 10xy - y)]$

68. $[9xy - (4x + xy - y)] - [4y - (2x - xy + 6y)]$

69. $[4x^3 - (2x^2 - x - 1)] - [5x^3 - (x^2 + 2x - 1)]$

70. $[x^3 - (x^2 - x + 1)] - [-x^3 + (7x^2 - x + 10)]$

5 Use Polynomials in Geometry Problems

71. Find a polynomial that represents the perimeter of each of the following figures (Figures 5.2, 5.3, and 5.4).

(a)

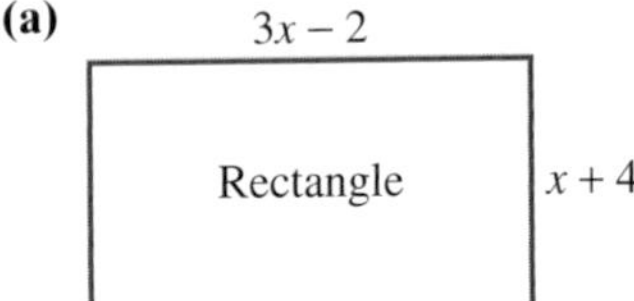

Figure 5.2

(b)

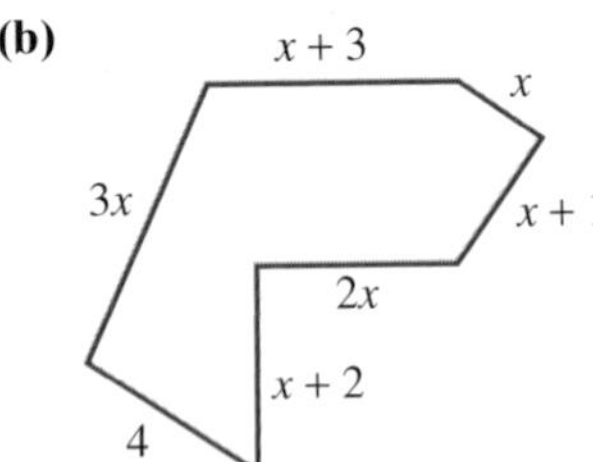

Figure 5.3

(c)

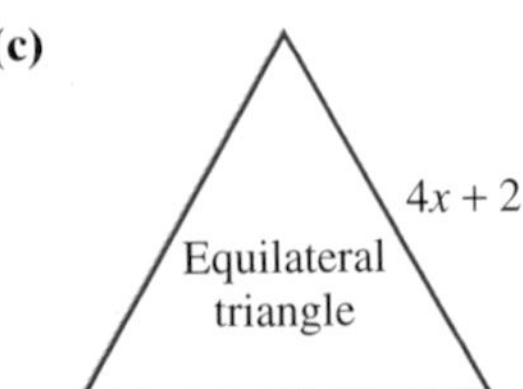

Figure 5.4

72. Find a polynomial that represents the total surface area of the rectangular solid in Figure 5.5.

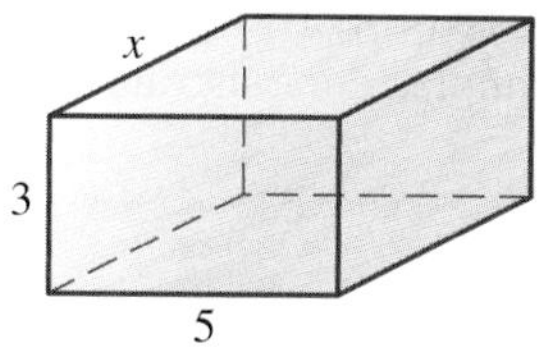

Figure 5.5

Now use that polynomial to determine the total surface area of each of the following rectangular solids.

(a) 3 by 5 by 4 **(b)** 3 by 5 by 7

(c) 3 by 5 by 11 **(d)** 3 by 5 by 13

73. Find a polynomial that represents the total surface area of the right circular cylinder in Figure 5.6. Now use that polynomial to determine the total surface area of each of the following right circular cylinders that have a base with a radius of 4. Use 3.14 for π, and express the answers to the nearest tenth.

(a) $h = 5$ **(b)** $h = 7$

(c) $h = 14$ **(d)** $h = 18$

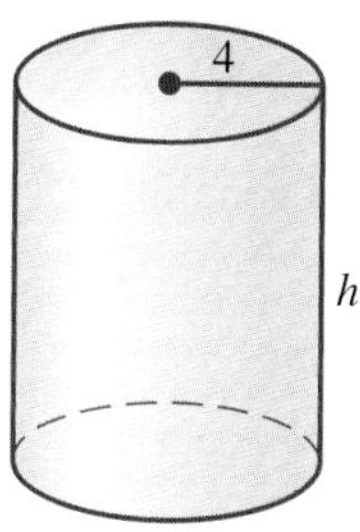

Figure 5.6

THOUGHTS INTO WORDS

74. Explain how to subtract the polynomial $-3x^2 + 2x - 4$ from $4x^2 + 6$.

75. Is the sum of two binomials always another binomial? Defend your answer.

76. Explain how to simplify the expression

$$7x - [3x - (2x - 4) + 2] - x$$

Answers to the Concept Quiz

1. True **2.** False **3.** False **4.** True **5.** True **6.** False **7.** True **8.** False **9.** True **10.** True

Answers to the Example Practice Skills

1. $8x^2 + 4x - 4$ **2.** $15x + 7$ **3.** $-7x^2y + xy^2$ **4.** $3x^2 - 10x + 11$ **5.** $7y^2 - 3y + 16$ **6.** $6x^2 - 7xy - 5y^2$ **7.** $4x - 15$ **8.** $7a^2 - 9b - 13$ **9.** $-2a^2 - 6a + 2$ **10.** $13x - 5$ **11.** $26x + 80$; 184; 236; 392

5.2 Products and Quotients of Monomials

OBJECTIVES

1. Multiply Monomials
2. Raise a Monomial to an Exponent
3. Divide Monomials
4. Use Polynomials in Geometry Problems

1 Multiply Monomials

Suppose that we want to find the product of two monomials such as $3x^2y$ and $4x^3y^2$. To proceed, use the properties of real numbers, and keep in mind that exponents indicate repeated multiplication.

$$\begin{aligned}(3x^2y)(4x^3y^2) &= (3 \cdot x \cdot x \cdot y)(4 \cdot x \cdot x \cdot x \cdot y \cdot y) \\ &= 3 \cdot 4 \cdot x \cdot x \cdot x \cdot x \cdot x \cdot y \cdot y \cdot y \\ &= 12x^5y^3\end{aligned}$$

You can use such an approach to find the product of any two monomials. However, there are some basic properties of exponents that make the process of multiplying monomials a much easier task. Let's consider each of these properties and illustrate its use when multiplying monomials. The following examples demonstrate the first property.

$$x^2 \cdot x^3 = (x \cdot x)(x \cdot x \cdot x) = x^5$$

$$a^4 \cdot a^2 = (a \cdot a \cdot a \cdot a)(a \cdot a) = a^6$$

$$b^3 \cdot b^4 = (b \cdot b \cdot b)(b \cdot b \cdot b \cdot b) = b^7$$

In general,

$$\begin{aligned} b^n \cdot b^m &= \underbrace{(b \cdot b \cdot b \cdot \ldots b)}_{\substack{n \text{ factors} \\ \text{of } b}}\underbrace{(b \cdot b \cdot b \cdot \ldots b)}_{\substack{m \text{ factors} \\ \text{of } b}} \\ &= \underbrace{b \cdot b \cdot b \cdot \ldots b}_{(n+m) \text{ factors of } b} \\ &= b^{n+m} \end{aligned}$$

We can state the first property as follows.

Property 5.1

If b is any real number and n and m are positive integers, then

$$b^n \cdot b^m = b^{n+m}$$

Property 5.1 says that to find the product of two positive integral powers of the same base, we add the exponents and use this sum as the exponent of the common base.

$$x^7 \cdot x^8 = x^{7+8} = x^{15} \qquad y^6 \cdot y^4 = y^{6+4} = y^{10}$$

$$2^3 \cdot 2^8 = 2^{3+8} = 2^{11} \qquad (-3)^4 \cdot (-3)^5 = (-3)^{4+5} = (-3)^9$$

$$\left(\frac{2}{3}\right)^7 \cdot \left(\frac{2}{3}\right)^5 = \left(\frac{2}{3}\right)^{5+7} = \left(\frac{2}{3}\right)^{12}$$

The following examples illustrate the use of Property 5.1, along with the commutative and associative properties of multiplication, to form the basis for multiplying monomials. The steps enclosed in the dashed boxes could be performed mentally.

EXAMPLE 1

$$\begin{aligned} (3x^2y)(4x^3y^2) &= 3 \cdot 4 \cdot x^2 \cdot x^3 \cdot y \cdot y^2 \\ &= 12x^{2+3}y^{1+2} \\ &= 12x^5y^3 \end{aligned}$$

▼ PRACTICE YOUR SKILL

Find the product $(5xy^2)(2x^4y^2)$. ■

EXAMPLE 2

$$\begin{aligned} (-5a^3b^4)(7a^2b^5) &= -5 \cdot 7 \cdot a^3 \cdot a^2 \cdot b^4 \cdot b^5 \\ &= -35a^{3+2}b^{4+5} \\ &= -35a^5b^9 \end{aligned}$$

▼ PRACTICE YOUR SKILL

Find the product $(-2a^2b^3)(6a^4b^5)$. ■

EXAMPLE 3

$$\left(\frac{3}{4}xy\right)\left(\frac{1}{2}x^5y^6\right) = \frac{3}{4} \cdot \frac{1}{2} \cdot x \cdot x^5 \cdot y \cdot y^6$$

$$= \frac{3}{8}x^{1+5}y^{1+6}$$

$$= \frac{3}{8}x^6y^7$$

▼ **PRACTICE YOUR SKILL**

Find the product $\left(\frac{2}{3}x^2y\right)\left(\frac{1}{5}x^3y^5\right)$. ■

EXAMPLE 4

$$(-ab^2)(-5a^2b) = (-1)(-5)(a)(a^2)(b^2)(b)$$

$$= 5a^{1+2}b^{2+1}$$

$$= 5a^3b^3$$

▼ **PRACTICE YOUR SKILL**

Find the product $(-3a^2b)(-a^3b^2)$. ■

EXAMPLE 5

$$(2x^2y^2)(3x^2y)(4y^3) = 2 \cdot 3 \cdot 4 \cdot x^2 \cdot x^2 \cdot y^2 \cdot y \cdot y^3$$

$$= 24x^{2+2}y^{2+1+3}$$

$$= 24x^4y^6$$

▼ **PRACTICE YOUR SKILL**

Find the product $(5x^3y)(2xy^3)(3x^2)$. ■

2 Raise a Monomial to an Exponent

The following examples demonstrate another useful property of exponents.

$$(x^2)^3 = x^2 \cdot x^2 \cdot x^2 = x^{2+2+2} = x^6$$

$$(a^3)^2 = a^3 \cdot a^3 = a^{3+3} = a^6$$

$$(b^4)^3 = b^4 \cdot b^4 \cdot b^4 = b^{4+4+4} = b^{12}$$

In general,

$$(b^n)^m = \underbrace{b^n \cdot b^n \cdot b^n \cdot \ldots b^n}_{m \text{ factors of } b^n}$$

$$= b^{\overbrace{n+n+n+\cdots+n}^{\text{Adding } m \text{ of these}}}$$

$$= b^{mn}$$

We can state this property as follows.

Property 5.2

If b is any real number, and m and n are positive integers, then

$$(b^n)^m = b^{mn}$$

The following examples show how Property 5.2 is used to find "the power of a power."

$$(x^4)^5 = x^{5(4)} = x^{20} \qquad (y^6)^3 = y^{3(6)} = y^{18} \qquad (2^3)^7 = 2^{7(3)} = 2^{21}$$

A third property of exponents pertains to raising a monomial to a power. Consider the following examples, which we use to introduce the property.

$$(3x)^2 = (3x)(3x) = 3 \cdot 3 \cdot x \cdot x = 3^2 \cdot x^2$$

$$(4y^2)^3 = (4y^2)(4y^2)(4y^2) = 4 \cdot 4 \cdot 4 \cdot y^2 \cdot y^2 \cdot y^2 = (4)^3(y^2)^3$$

$$(-2a^3b^4)^2 = (-2a^3b^4)(-2a^3b^4) = (-2)(-2)(a^3)(a^3)(b^4)(b^4)$$
$$= (-2)^2(a^3)^2(b^4)^2$$

In general,

$$(ab)^n = \underbrace{(ab)(ab)(ab) \cdot \ldots (ab)}_{n \text{ factors of } ab}$$

$$= \underbrace{(a \cdot a \cdot a \cdot a \cdot \ldots a)}_{n \text{ factors of } a}\underbrace{(b \cdot b \cdot b \cdot \ldots b)}_{n \text{ factors of } b}$$

$$= a^n b^n$$

We can formally state Property 5.3 as follows.

Property 5.3

If a and b are real numbers, and n is a positive integer, then

$$(ab)^n = a^n b^n$$

Property 5.3 and Property 5.2 form the basis for raising a monomial to a power, as in the next examples.

EXAMPLE 6

$(x^2y^3)^4 = (x^2)^4(y^3)^4$ Use $(ab)^n = a^nb^n$

$= x^8y^{12}$ Use $(b^n)^m = b^{mn}$

▼ PRACTICE YOUR SKILL

$(x^3y^4)^3$ ■

EXAMPLE 7

$(3a^5)^3 = (3)^3(a^5)^3$

$= 27a^{15}$

▼ PRACTICE YOUR SKILL

$(2a^3)^4$ ■

EXAMPLE 8

$$(-2xy^4)^5 = (-2)^5(x)^5(y^4)^5$$
$$= -32x^5y^{20}$$

▼ PRACTICE YOUR SKILL

$(-3x^2y^5)^3$

■

3 Divide Monomials

To develop an effective process for dividing by a monomial, we need yet another property of exponents. This property is a direct consequence of the definition of an exponent. Study the following examples.

$$\frac{x^4}{x^3} = \frac{x \cdot x \cdot x \cdot x}{x \cdot x \cdot x} = x \qquad \frac{x^3}{x^3} = \frac{x \cdot x \cdot x}{x \cdot x \cdot x} = 1$$

$$\frac{a^5}{a^2} = \frac{a \cdot a \cdot a \cdot a \cdot a}{a \cdot a} = a^3 \qquad \frac{y^5}{y^5} = \frac{y \cdot y \cdot y \cdot y \cdot y}{y \cdot y \cdot y \cdot y \cdot y} = 1$$

$$\frac{y^8}{y^4} = \frac{y \cdot y \cdot y \cdot y \cdot y \cdot y \cdot y \cdot y}{y \cdot y \cdot y \cdot y} = y^4$$

We can state the general property as follows:

Property 5.4

If b is any nonzero real number, and m and n are positive integers, then

1. $\dfrac{b^n}{b^m} = b^{n-m}$, when $n > m$

2. $\dfrac{b^n}{b^m} = 1$, when $n = m$

Applying Property 5.4 to the previous examples yields

$$\frac{x^4}{x^3} = x^{4-3} = x^1 = x \qquad \frac{x^3}{x^3} = 1$$

$$\frac{a^5}{a^2} = a^{5-2} = a^3 \qquad \frac{y^5}{y^5} = 1$$

$$\frac{y^8}{y^4} = y^{8-4} = y^4$$

(We will discuss the situation when $n < m$ in a later chapter.)

Property 5.4, along with our knowledge of dividing integers, provides the basis for dividing monomials. The following example demonstrates the process.

EXAMPLE 9

Simplify the following.

(a) $\dfrac{24x^5}{3x^2}$ **(b)** $\dfrac{-36a^{13}}{-12a^5}$ **(c)** $\dfrac{-56x^9}{7x^4}$

(d) $\dfrac{72b^5}{8b^5}$ **(e)** $\dfrac{48y^7}{-12y}$ **(f)** $\dfrac{12x^4y^7}{2x^2y^4}$

Solution

(a) $\dfrac{24x^5}{3x^2} = 8x^{5-2} = 8x^3$ **(b)** $\dfrac{-36a^{13}}{-12a^5} = 3a^{13-5} = 3a^8$

(c) $\dfrac{-56x^9}{7x^4} = -8x^{9-4} = -8x^5$ **(d)** $\dfrac{72b^5}{8b^5} = 9 \quad \left(\dfrac{b^5}{b^5} = 1\right)$

(e) $\dfrac{48y^7}{-12y} = -4y^{7-1} = -4y^6$ **(f)** $\dfrac{12x^4y^7}{2x^2y^4} = 6x^{4-2}y^{7-4} = 6x^2y^3$

▼ **PRACTICE YOUR SKILL**

Simplify the following.

(a) $\dfrac{72a^4b}{-8ab}$ **(b)** $\dfrac{-5m^5n^4}{m^3n^2}$ **(c)** $\dfrac{-16x^3}{2x^2}$ ■

CONCEPT QUIZ

For Problems 1–10, answer true or false.

1. When multiplying factors with the same base, add the exponents.
2. $3^2 \cdot 3^2 = 9^4$
3. $2x^2 \cdot 3x^3 = 6x^6$
4. $(x^2)^3 = x^5$
5. $(-4x^3)^2 = -4x^6$
6. To simplify $(3x^2y)(2x^3y^2)^4$ according to the order of operations, first raise $2x^3y^2$ to the fourth power and then multiply the monomials.
7. $\dfrac{-8x^6}{2x^2} = -4x^3$
8. $\dfrac{24x^3y^2}{-xy} = -24x^2y$
9. $\dfrac{-14xy^3}{-7xy^3} = 2$
10. $\dfrac{36a^2b^3c}{-18ab^2} = -2abc$

Problem Set 5.2

1 Multiply Monomials

For Problems 1–36, find each product.

1. $(4x^3)(9x)$
2. $(6x^3)(7x^2)$
3. $(-2x^2)(6x^3)$
4. $(2xy)(-4x^2y)$
5. $(-a^2b)(-4ab^3)$
6. $(-8a^2b^2)(-3ab^3)$
7. $(x^2yz^2)(-3xyz^4)$
8. $(-2xy^2z^2)(-x^2y^3z)$
9. $(5xy)(-6y^3)$
10. $(-7xy)(4x^4)$
11. $(3a^2b)(9a^2b^4)$
12. $(-8a^2b^2)(-12ab^5)$
13. $(m^2n)(-mn^2)$
14. $(-x^3y^2)(xy^3)$
15. $\left(\frac{2}{5}xy^2\right)\left(\frac{3}{4}x^2y^4\right)$
16. $\left(\frac{1}{2}x^2y^6\right)\left(\frac{2}{3}xy\right)$
17. $\left(-\frac{3}{4}ab\right)\left(\frac{1}{5}a^2b^3\right)$
18. $\left(-\frac{2}{7}a^2\right)\left(\frac{3}{5}ab^3\right)$
19. $\left(-\frac{1}{2}xy\right)\left(\frac{1}{3}x^2y^3\right)$
20. $\left(\frac{3}{4}x^4y^5\right)(-x^2y)$
21. $(3x)(-2x^2)(-5x^3)$
22. $(-2x)(-6x^3)(x^2)$
23. $(-6x^2)(3x^3)(x^4)$
24. $(-7x^2)(3x)(4x^3)$
25. $(x^2y)(-3xy^2)(x^3y^3)$
26. $(xy^2)(-5xy)(x^2y^4)$
27. $(-3y^2)(-2y^2)(-4y^5)$
28. $(-y^3)(-6y)(-8y^4)$
29. $(4ab)(-2a^2b)(7a)$
30. $(3b)(-2ab^2)(7a)$
31. $(-ab)(-3ab)(-6ab)$
32. $(-3a^2b)(-ab^2)(-7a)$
33. $\left(\frac{2}{3}xy\right)(-3x^2y)(5x^4y^5)$
34. $\left(\frac{3}{4}x\right)(-4x^2y^2)(9y^3)$
35. $(12y)(-5x)\left(-\frac{5}{6}x^4y\right)$
36. $(-12x)(3y)\left(-\frac{3}{4}xy^6\right)$

For Problems 37–52, find each product. Assume that the variables in the exponents represent positive integers. For example,

$$(x^{2n})(x^{3n}) = x^{2n+3n} = x^{5n}$$

37. $(2x^n)(3x^{2n})$

38. $(3x^{2n})(x^{3n-1})$

39. $(a^{2n-1})(a^{3n+4})$

40. $(a^{5n-1})(a^{5n+1})$

41. $(x^{3n-2})(x^{n+2})$

42. $(x^{n-1})(x^{4n+3})$

43. $(a^{5n-2})(a^3)$

44. $(x^{3n-4})(x^4)$

45. $(2x^n)(-5x^n)$

46. $(4x^{2n-1})(-3x^{n+1})$

47. $(-3a^2)(-4a^{n+2})$

48. $(-5x^{n-1})(-6x^{2n+4})$

49. $(x^n)(2x^{2n})(3x^2)$

50. $(2x^n)(3x^{3n-1})(-4x^{2n+5})$

51. $(3x^{n-1})(x^{n+1})(4x^{2-n})$

52. $(-5x^{n+2})(x^{n-2})(4x^{3-2n})$

2 Raise a Monomial to an Exponent

For Problems 53–74, raise each monomial to the indicated power.

53. $(3xy^2)^3$

54. $(4x^2y^3)^3$

55. $(-2x^2y)^5$

56. $(-3xy^4)^3$

57. $(-x^4y^5)^4$

58. $(-x^5y^2)^4$

59. $(ab^2c^3)^6$

60. $(a^2b^3c^5)^5$

61. $(2a^2b^3)^6$

62. $(2a^3b^2)^6$

63. $(9xy^4)^2$

64. $(8x^2y^5)^2$

65. $(-3ab^3)^4$

66. $(-2a^2b^4)^4$

67. $-(2ab)^4$

68. $-(3ab)^4$

69. $-(xy^2z^3)^6$

70. $-(xy^2z^3)^8$

71. $(-5a^2b^2c)^3$

72. $(-4abc^4)^3$

73. $(-xy^4z^2)^7$

74. $(-x^2y^4z^5)^5$

3 Divide Monomials

For Problems 75–90, find each quotient.

75. $\dfrac{9x^4y^5}{3xy^2}$

76. $\dfrac{12x^2y^7}{6x^2y^3}$

77. $\dfrac{25x^5y^6}{-5x^2y^4}$

78. $\dfrac{56x^6y^4}{-7x^2y^3}$

79. $\dfrac{-54ab^2c^3}{-6abc}$

80. $\dfrac{-48a^3bc^5}{-6a^2c^4}$

81. $\dfrac{-18x^2y^2z^6}{xyz^2}$

82. $\dfrac{-32x^4y^5z^8}{x^2yz^3}$

83. $\dfrac{a^3b^4c^7}{-abc^5}$

84. $\dfrac{-a^4b^5c}{a^2b^4c}$

85. $\dfrac{-72x^2y^4}{-8x^2y^4}$

86. $\dfrac{-96x^4y^5}{12x^4y^4}$

87. $\dfrac{14ab^3}{-14ab}$

88. $\dfrac{-12abc^2}{12bc}$

89. $\dfrac{-36x^3y^5}{2y^5}$

90. $\dfrac{-48xyz^2}{2xz}$

4 Use Polynomials in Geometry Problems

91. Find a polynomial that represents the total surface area of the rectangular solid in Figure 5.7. Also find a polynomial that represents the volume.

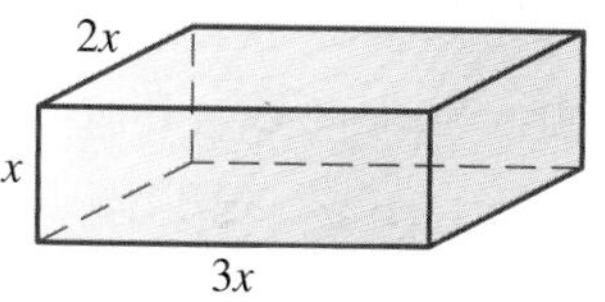

Figure 5.7

92. Find a polynomial that represents the total surface area of the rectangular solid in Figure 5.8. Also find a polynomial that represents the volume.

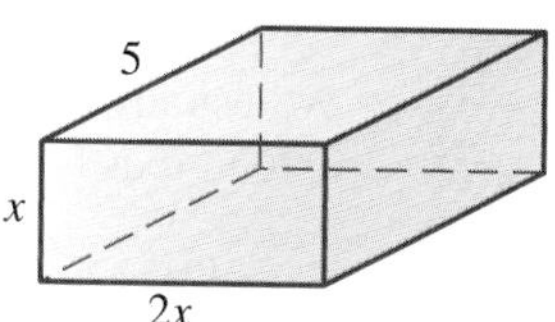

Figure 5.8

93. Find a polynomial that represents the area of the shaded region in Figure 5.9. The length of a radius of the larger circle is r units, and the length of a radius of the smaller circle is 6 units.

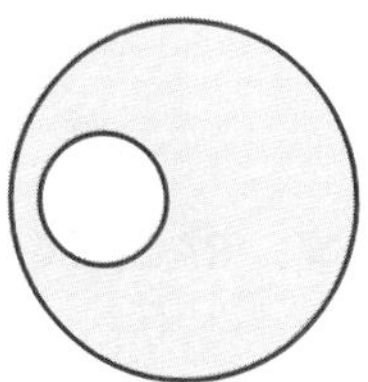

Figure 5.9

THOUGHTS INTO WORDS

94. How would you convince someone that $\dfrac{x^6}{x^2}$ is x^4 and not x^3?

95. Your friend simplifies $2^3 \cdot 2^2$ as follows:

$$2^3 \cdot 2^2 = 4^{3+2} = 4^5 = 1024$$

What has she done incorrectly, and how would you help her?

Answers to the Concept Quiz

1. True **2.** False **3.** False **4.** False **5.** False **6.** True **7.** False **8.** True **9.** True **10.** True

Answers to the Example Practice Skills

1. $10x^5y^4$ **2.** $-12a^6b^8$ **3.** $\frac{2}{15}x^5y^6$ **4.** $3a^5b^3$ **5.** $30x^6y^4$ **6.** x^9y^{12} **7.** $16a^{12}$ **8.** $-27x^6y^{15}$

9. (a) $-9a^3$ **(b)** $-5m^2n^2$ **(c)** $-8x$

5.3 Multiplying Polynomials

OBJECTIVES

1. Multiply Polynomials
2. Multiply Two Binomials Using a Shortcut Pattern
3. Find the Square of a Binomial Using a Shortcut Pattern
4. Use a Pattern to Find the Product of $(a + b)(a - b)$
5. Find the Cube of a Binomial
6. Use Polynomials in Geometry Problems

1 Multiply Polynomials

We usually state the distributive property as $a(b + c) = ab + ac$; however, we can extend it as follows:

$$a(b + c + d) = ab + ac + ad$$

$$a(b + c + d + e) = ab + ac + ad + ae \qquad \text{etc.}$$

We apply the commutative and associative properties, the properties of exponents, and the distributive property together to find the product of a monomial and a polynomial. The following examples illustrate this idea.

EXAMPLE 1

$$\begin{aligned} 3x^2(2x^2 + 5x + 3) &= 3x^2(2x^2) + 3x^2(5x) + 3x^2(3) \\ &= 6x^4 + 15x^3 + 9x^2 \end{aligned}$$

▼ **PRACTICE YOUR SKILL**

$4x^3(3x^2 + 2x - 5)$ ■

EXAMPLE 2

$$\begin{aligned} -2xy(3x^3 - 4x^2y - 5xy^2 + y^3) &= -2xy(3x^3) - (-2xy)(4x^2y) \\ &\quad -(-2xy)(5xy^2) + (-2xy)(y^3) \\ &= -6x^4y + 8x^3y^2 + 10x^2y^3 - 2xy^4 \end{aligned}$$

▼ **PRACTICE YOUR SKILL**

$-4ab^2(2a^2 + ab^2 - 3ab + b^2)$ ■

Now let's consider the product of two polynomials neither of which is a monomial. Consider the following examples.

EXAMPLE 3

$$(x + 2)(y + 5) = x(y + 5) + 2(y + 5)$$
$$= x(y) + x(5) + 2(y) + 2(5)$$
$$= xy + 5x + 2y + 10$$

▼ PRACTICE YOUR SKILL

$(a + 3)(b + 4)$ ■

Note that each term of the first polynomial is multiplied by each term of the second polynomial.

EXAMPLE 4

$$(x - 3)(y + z + 3) = x(y + z + 3) - 3(y + z + 3)$$
$$= xy + xz + 3x - 3y - 3z - 9$$

▼ PRACTICE YOUR SKILL

$(a - 4)(a + b + 5)$ ■

Multiplying polynomials often produces similar terms that can be combined to simplify the resulting polynomial.

EXAMPLE 5

$$(x + 5)(x + 7) = x(x + 7) + 5(x + 7)$$
$$= x^2 + 7x + 5x + 35$$
$$= x^2 + 12x + 35$$

▼ PRACTICE YOUR SKILL

$(a + 8)(a + 4)$ ■

EXAMPLE 6

$$(x - 2)(x^2 - 3x + 4) = x(x^2 - 3x + 4) - 2(x^2 - 3x + 4)$$
$$= x^3 - 3x^2 + 4x - 2x^2 + 6x - 8$$
$$= x^3 - 5x^2 + 10x - 8$$

▼ PRACTICE YOUR SKILL

$(a - 3)(a^2 + 2a - 5)$ ■

In Example 6, we are claiming that

$$(x - 2)(x^2 - 3x + 4) = x^3 - 5x^2 + 10x - 8$$

for all real numbers. In addition to going back over our work, how can we verify such a claim? Obviously, we cannot try all real numbers, but trying at least one number gives us a partial check. Let's try the number 4.

$$(x - 2)(x^2 - 3x + 4) = (4 - 2)(4^2 - 3(4) + 4) \qquad \text{When } x = 4$$
$$= 2(16 - 12 + 4)$$
$$= 2(8)$$
$$= 16$$

$$\begin{aligned} x^3 - 5x^2 + 10x - 8 &= 4^3 - 5(4)^2 + 10(4) - 8 \quad \text{When } x = 4 \\ &= 64 - 80 + 40 - 8 \\ &= 16 \end{aligned}$$

EXAMPLE 7

$$\begin{aligned} (3x - 2y)(x^2 + xy - y^2) &= 3x(x^2 + xy - y^2) - 2y(x^2 + xy - y^2) \\ &= 3x^3 + 3x^2y - 3xy^2 - 2x^2y - 2xy^2 + 2y^3 \\ &= 3x^3 + x^2y - 5xy^2 + 2y^3 \end{aligned}$$

▼ **PRACTICE YOUR SKILL**

$(4a - b)(3a^2 - ab - b^2)$ ■

2 Multiply Two Binomials Using a Shortcut Pattern

It helps to be able to find the product of two binomials without showing all of the intermediate steps. This is quite easy to do with the *three-step shortcut pattern* demonstrated by Figures 5.10 and 5.11 in the following examples.

EXAMPLE 8

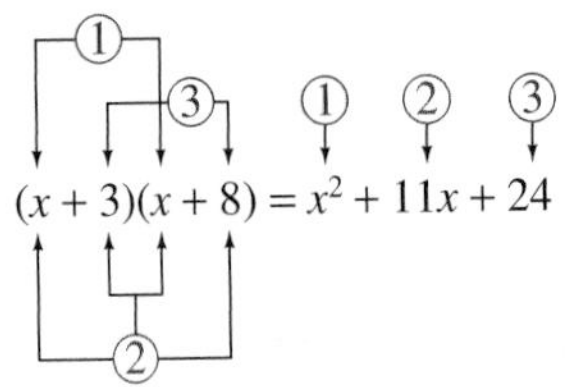

Figure 5.10

Step ①. Multiply $x \cdot x$.
Step ②. Multiply $3 \cdot x$ and $8 \cdot x$ and combine.
Step ③. Multiply $3 \cdot 8$.

▼ **PRACTICE YOUR SKILL**

$(a + 5)(a + 2)$ ■

EXAMPLE 9

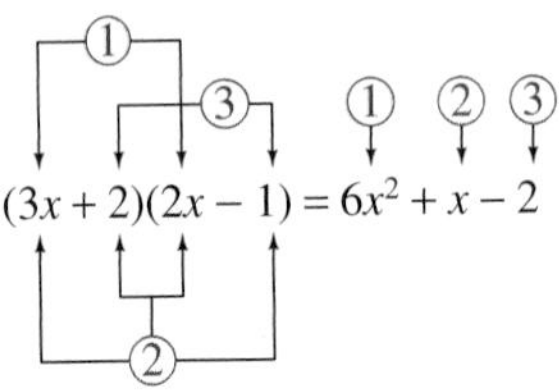

Figure 5.11

▼ **PRACTICE YOUR SKILL**

$(5a - 1)(3a + 2)$ ■

The mnemonic device FOIL is often used to remember the pattern for multiplying binomials. The letters in FOIL represent, First, Outside, Inside, and Last. If you look back at Examples 8 and 9, step 1 is to find the product of the first terms in the binomial;

step 2 is to find the sum of the product of the outside terms and the product of the inside terms; and step 3 is to find the product of the last terms in each binomial.

Now see if you can use the pattern to find the following products.

$$(x + 2)(x + 6) = ?$$

$$(x - 3)(x + 5) = ?$$

$$(2x + 5)(3x + 7) = ?$$

$$(3x - 1)(4x - 3) = ?$$

Your answers should be $x^2 + 8x + 12$, $x^2 + 2x - 15$, $6x^2 + 29x + 35$, and $12x^2 - 13x + 3$. Keep in mind that this shortcut pattern applies only to finding the product of two binomials.

3 Find the Square of a Binomial Using a Shortcut Pattern

We can use exponents to indicate repeated multiplication of polynomials. For example, $(x + 3)^2$ means $(x + 3)(x + 3)$ and $(x + 4)^3$ means $(x + 4)(x + 4) \cdot (x + 4)$. To square a binomial, we can simply write it as the product of two equal binomials and apply the shortcut pattern. Thus

$$(x + 3)^2 = (x + 3)(x + 3) = x^2 + 6x + 9$$

$$(x - 6)^2 = (x - 6)(x - 6) = x^2 - 12x + 36 \qquad \text{and}$$

$$(3x - 4)^2 = (3x - 4)(3x - 4) = 9x^2 - 24x + 16$$

When squaring binomials, be careful not to forget the middle term. That is to say, $(x + 3)^2 \neq x^2 + 3^2$; instead, $(x + 3)^2 = x^2 + 6x + 9$.

When multiplying binomials, there are some special patterns that you should recognize. We can use these patterns to find products, and later we will use some of them when factoring polynomials.

PATTERN 1

$$(a + b)^2 = (a + b)(a + b) = a^2 + 2ab + b^2$$

Square of first term of binomial + Twice the product of the two terms of binomial + Square of second term of binomial

EXAMPLE 10

Expand the following squares of binomials.

(a) $(x + 4)^2$ **(b)** $(2x + 3y)^2$ **(c)** $(5a + 7b)^2$

Solution

Square of the first term of binomial + Twice the product of the terms of binomial + Square of second term of binomial

(a) $(x + 4)^2 = x^2 + 8x + 16$

(b) $(2x + 3y)^2 = 4x^2 + 12xy + 9y^2$

(c) $(5a + 7b)^2 = 25a^2 + 70ab + 49b^2$

▼ **PRACTICE YOUR SKILL**

Expand the following squares of binomials.

(a) $(x + 3)^2$ **(b)** $(3x + y)^2$ **(c)** $(3a + 5b)^2$ ■

PATTERN 2

$$(a - b)^2 = (a - b)(a - b) = a^2 - 2ab + b^2$$

Square of first term of binomial − Twice the product of the two terms of binomial + Square of second term of binomial

EXAMPLE 11

Expand the following squares of binomials.

(a) $(x - 8)^2$ **(b)** $(3x - 4y)^2$ **(c)** $(4a - 9b)^2$

Solution

Square of the first term of binomial − Twice the product of the terms of binomial + Square of second term of binomial

(a) $(x - 8)^2 = x^2 - 16x + 64$

(b) $(3x - 4y)^2 = 9x^2 - 24xy + 16y^2$

(c) $(4a - 9b)^2 = 16a^2 - 72ab + 81b^2$

▼ **PRACTICE YOUR SKILL**

Expand the following squares of binomials.

(a) $(x - 5)^2$ **(b)** $(x - 2y)^2$ **(c)** $(2a - 3b)^2$ ■

4 Use a Pattern to Find the Product of $(a + b)(a - b)$

PATTERN 3

$$(a + b)(a - b) = a^2 - b^2$$

Square of first term of binomials − Square of second term of binomials

EXAMPLE 12

Find the product for the following.

(a) $(x + 7)(x - 7)$ **(b)** $(2x + y)(2x - y)$ **(c)** $(3a - 2b)(3a + 2b)$

Solution

Square of the first term of binomial − Square of second term of binomial

(a) $(x + 7)(x - 7) = x^2 - 49$

(b) $(2x + y)(2x - y) = 4x^2 - y^2$

(c) $(3a - 2b)(3a + 2b) = 9a^2 - 4b^2$

▼ **PRACTICE YOUR SKILL**

Find the product for the following.

(a) $(x + 6)(x - 6)$ **(b)** $(x + 9y)(x - 9y)$ **(c)** $(7a - 5b)(7a + 5b)$

5 Find the Cube of a Binomial

Now suppose that we want to cube a binomial. One approach is as follows:

$$\begin{aligned}(x + 4)^3 &= (x + 4)(x + 4)(x + 4)\\ &= (x + 4)(x^2 + 8x + 16)\\ &= x(x^2 + 8x + 16) + 4(x^2 + 8x + 16)\\ &= x^3 + 8x^2 + 16x + 4x^2 + 32x + 64\\ &= x^3 + 12x^2 + 48x + 64\end{aligned}$$

Another approach is to cube a *general* binomial and then use the resulting pattern, as follows.

PATTERN 4

$$\begin{aligned}(a + b)^3 &= (a + b)(a + b)(a + b)\\ &= (a + b)(a^2 + 2ab + b^2)\\ &= a(a^2 + 2ab + b^2) + b(a^2 + 2ab + b^2)\\ &= a^3 + 2a^2b + ab^2 + a^2b + 2ab^2 + b^3\\ &= a^3 + 3a^2b + 3ab^2 + b^3\end{aligned}$$

EXAMPLE 13

Expand $(x + 4)^3$.

Solution

Let's use the pattern $(a + b)^3 = a^3 + 3a^2b + 3ab^2 + b^3$ to cube the binomial $x + 4$.

$$\begin{aligned}(x + 4)^3 &= x^3 + 3x^2(4) + 3x(4)^2 + 4^3\\ &= x^3 + 12x^2 + 48x + 64\end{aligned}$$

▼ **PRACTICE YOUR SKILL**

Expand $(x + 5)^3$. ■

Because $a - b = a + (-b)$, we can easily develop a pattern for cubing $a - b$.

PATTERN 5

$$\begin{aligned}(a - b)^3 &= [a + (-b)]^3\\ &= a^3 + 3a^2(-b) + 3a(-b)^2 + (-b)^3\\ &= a^3 - 3a^2b + 3ab^2 - b^3\end{aligned}$$

EXAMPLE 14

Expand $(3x - 2y)^3$.

Solution

Now let's use the pattern $(a - b)^3 = a^3 - 3a^2b + 3ab^2 - b^3$ to cube the binomial $3x - 2y$.

$$\begin{aligned}(3x - 2y)^3 &= (3x)^3 - 3(3x)^2(2y) + 3(3x)(2y)^2 - (2y)^3\\ &= 27x^3 - 54x^2y + 36xy^2 - 8y^3\end{aligned}$$

▼ **PRACTICE YOUR SKILL**

Expand $(4x - 3y)^3$. ■

Finally, we need to realize that if the patterns are forgotten or do not apply, then we can revert to applying the distributive property.

$$\begin{aligned}(2x - 1)(x^2 - 4x + 6) &= 2x(x^2 - 4x + 6) - 1(x^2 - 4x + 6)\\ &= 2x^3 - 8x^2 + 12x - x^2 + 4x - 6\\ &= 2x^3 - 9x^2 + 16x - 6\end{aligned}$$

6 Use Polynomials in Geometry Problems

As you might expect, there are geometric interpretations for many of the algebraic concepts we present in this section. We will give you the opportunity to make some of these connections between algebra and geometry in the next problem set. Let's conclude this section with a problem that allows us to use some algebra and geometry.

EXAMPLE 15

A rectangular piece of tin is 16 inches long and 12 inches wide, as shown in Figure 5.12. From each corner a square piece x inches on a side is cut out. The flaps are then turned up to form an open box. Find polynomials that represent the volume and outside surface area of the box.

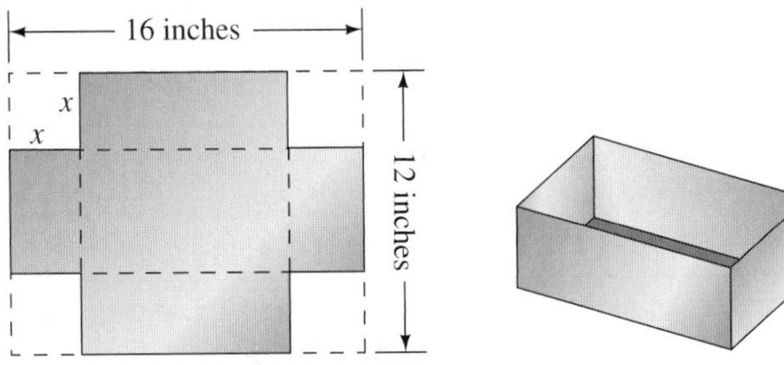

Figure 5.12

Solution

The length of the box will be $16 - 2x$, the width $12 - 2x$, and the height x. With the volume formula $V = lwh$, the polynomial $(16 - 2x)(12 - 2x)(x)$, which simplifies to $4x^3 - 56x^2 + 192x$, represents the volume.

The outside surface area of the box is the area of the original piece of tin minus the four corners that were cut off. Therefore, the polynomial $16(12) - 4x^2$, or $192 - 4x^2$, represents the outside surface area of the box.

▼ **PRACTICE YOUR SKILL**

A square piece of cardboard has sides that measure 8 inches. From each corner, a square piece x inches on a side is cut out. The flaps are then turned up to form an open box. Find polynomials that represent the volume and outside surface area of the box. ■

Remark: Recall that in Section 5.1 we found the total surface area of a rectangular solid by adding the areas of the sides, top, and bottom. Use this approach for the open box in Example 15 to check our answer of $192 - 4x^2$. Keep in mind that the box has no top.

CONCEPT QUIZ

For Problems 1–10, answer true or false.

1. The algebraic expression $(x + y)^2$ is called the square of a binomial.
2. The algebraic expression $(x + y)(x + 2xy + y)$ is called the product of two binomials.
3. The mnemonic device FOIL stands for first, outside, inside, and last.
4. Although the distributive property is usually stated as $a(b + c) = ab + ac$, it can be extended, as in $a(b + c + d + e) = ab + ac + ad + ae$, when multiplying polynomials.
5. Multiplying polynomials often produces similar terms that can be combined to simplify the resulting product.
6. The pattern for $(a + b)^2$ is $a^2 + b^2$.
7. The pattern for $(a - b)^2$ is $a^2 - 2ab - b^2$.
8. The pattern for $(a + b)(a - b)$ is $a^2 - b^2$.
9. The pattern for $(a + b)^3$ is $a^3 + 3ab + b^3$.
10. The pattern for $(a - b)^3$ is $a^3 + 3a^2b - 3ab^2 - b^3$.

Problem Set 5.3

1 Multiply Polynomials

For Problems 1–24, find each indicated product.

1. $2xy(5xy^2 + 3x^2y^3)$
2. $3x^2y(6y^2 - 5x^2y^4)$
3. $-3a^2b(4ab^2 - 5a^3)$
4. $-7ab^2(2b^3 - 3a^2)$
5. $8a^3b^4(3ab - 2ab^2 + 4a^2b^2)$
6. $9a^3b(2a - 3b + 7ab)$
7. $-x^2y(6xy^2 + 3x^2y^3 - x^3y)$
8. $-ab^2(5a + 3b - 6a^2b^3)$
9. $(a + 2b)(x + y)$
10. $(t - s)(x + y)$
11. $(a - 3b)(c + 4d)$
12. $(a - 4b)(c - d)$
13. $(t + 3)(t^2 - 3t - 5)$
14. $(t - 2)(t^2 + 7t + 2)$
15. $(x - 4)(x^2 + 5x - 4)$
16. $(x + 6)(2x^2 - x - 7)$
17. $(2x - 3)(x^2 + 6x + 10)$
18. $(3x + 4)(2x^2 - 2x - 6)$
19. $(4x - 1)(3x^2 - x + 6)$
20. $(5x - 2)(6x^2 + 2x - 1)$
21. $(x^2 + 2x + 1)(x^2 + 3x + 4)$
22. $(x^2 - x + 6)(x^2 - 5x - 8)$
23. $(2x^2 + 3x - 4)(x^2 - 2x - 1)$
24. $(3x^2 - 2x + 1)(2x^2 + x - 2)$

2 Multiply Two Binomials Using a Shortcut Pattern

For Problems 25–42, find the indicated product using the shortcut pattern for multiplying binomials.

25. $(x + 6)(x + 10)$
26. $(x + 2)(x + 10)$
27. $(y - 5)(y + 11)$
28. $(y - 3)(y + 9)$
29. $(n + 2)(n - 7)$
30. $(n + 3)(n - 12)$
31. $(x - 6)(x - 8)$
32. $(x - 3)(x - 13)$
33. $(4x + 5)(x + 7)$
34. $(6x + 5)(x + 3)$
35. $(7x - 2)(2x + 1)$
36. $(6x - 1)(3x + 2)$
37. $(1 + t)(5 - 2t)$
38. $(3 - t)(2 + 4t)$
39. $(6x + 7)(3x - 10)$
40. $(4x - 7)(7x + 4)$
41. $(2x - 5y)(x + 3y)$
42. $(x - 4y)(3x + 7y)$

For Problems 43–46, find the indicated product. Use the shortcut pattern for multiplying two binomials; then use the distributive property to determine the final product.

43. $(x + 1)(x - 2)(x - 3)$

44. $(x - 1)(x + 4)(x - 6)$

45. $(x - 3)(x + 3)(x - 1)$

46. $(x - 5)(x + 5)(x - 8)$

For Problems 47–56, find the indicated product. Assume all variables that appear as exponents represent positive integers.

47. $(x^n - 4)(x^n + 4)$

48. $(x^{3a} - 1)(x^{3a} + 1)$

49. $(x^a + 6)(x^a - 2)$

50. $(x^a + 4)(x^a - 9)$

51. $(2x^n + 5)(3x^n - 7)$

52. $(3x^n + 5)(4x^n - 9)$

53. $(x^{2a} - 7)(x^{2a} - 3)$

54. $(x^{2a} + 6)(x^{2a} - 4)$

55. $(2x^n + 5)^2$

56. $(3x^n - 7)^2$

3 Find the Square of a Binomial Using a Shortcut Pattern

For Problems 57–66, find the indicated product using the shortcut pattern.

57. $(x - 6)^2$

58. $(x - 2)^2$

59. $(t + 9)^2$

60. $(t + 13)^2$

61. $(y - 7)^2$

62. $(y - 4)^2$

63. $(3t + 7)^2$

64. $(4t + 6)^2$

65. $(7x - 4)^2$

66. $(5x - 7)^2$

4 Use a Pattern to Find the Product of $(a + b)(a - b)$

For Problems 67–74, find the indicated product using the shortcut pattern.

67. $(x + 6)(x - 6)$

68. $(t + 8)(t - 8)$

69. $(3y - 1)(3y + 1)$

70. $(5y - 2)(5y + 2)$

71. $(2 - 5x)(2 + 5x)$

72. $(6 - 3x)(6 + 3x)$

73. $(5x - 2a)(5x + 2a)$

74. $(9x - 2y)(9x + 2y)$

5 Find the Cube of a Binomial

For Problems 75–84, find the indicated product using the shortcut pattern for the cube of a binomial.

75. $(x + 2)^3$

76. $(x + 1)^3$

77. $(x - 4)^3$

78. $(x - 5)^3$

79. $(2x + 3)^3$

80. $(3x + 1)^3$

81. $(4x - 1)^3$

82. $(3x - 2)^3$

83. $(5x + 2)^3$

84. $(4x - 5)^3$

6 Use Polynomials in Geometry Problems

85. Explain how Figure 5.13 can be used to demonstrate geometrically that $(x + 2)(x + 6) = x^2 + 8x + 12$.

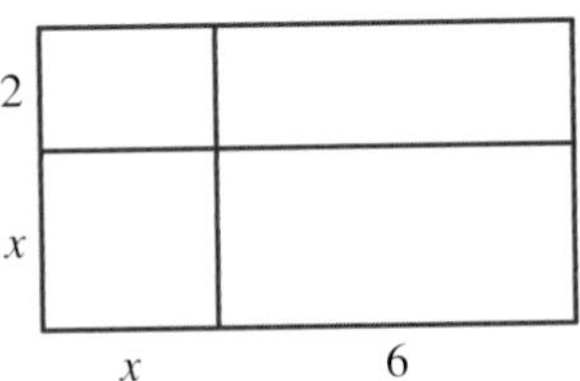

Figure 5.13

86. Find a polynomial that represents the sum of the areas of the two rectangles shown in Figure 5.14.

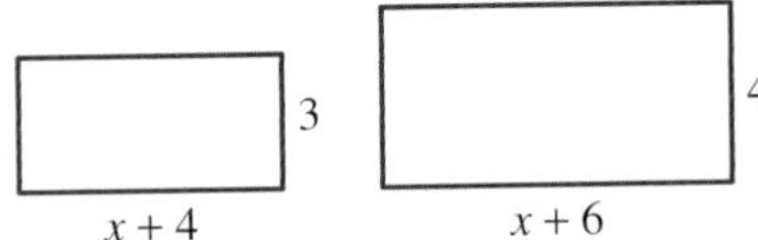

Figure 5.14

87. Find a polynomial that represents the area of the shaded region in Figure 5.15.

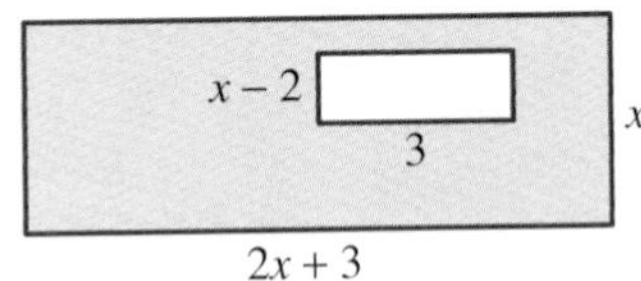

Figure 5.15

88. Explain how Figure 5.16 can be used to demonstrate geometrically that $(x + 7)(x - 3) = x^2 + 4x - 21$.

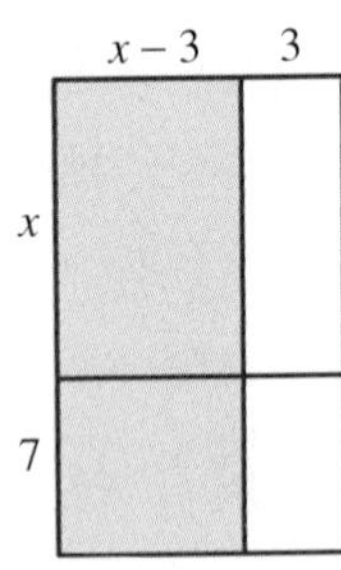

Figure 5.16

89. A square piece of cardboard is 16 inches on a side. A square piece x inches on a side is cut out from each corner. The flaps are then turned up to form an open box. Find polynomials that represent the volume and outside surface area of the box.

THOUGHTS INTO WORDS

90. How would you simplify $(2^3 + 2^2)^2$? Explain your reasoning.

91. Describe the process of multiplying two polynomials.

92. Determine the number of terms in the product of $(x + y)$ and $(a + b + c + d)$ without doing the multiplication. Explain how you arrived at your answer.

FURTHER INVESTIGATIONS

93. We have used the following two multiplication patterns.

$$(a + b)^2 = a^2 + 2ab + b^2$$

$$(a + b)^3 = a^3 + 3a^2b + 3ab^2 + b^3$$

By multiplying, we can extend these patterns as follows:

$$(a + b)^4 = a^4 + 4a^3b + 6a^2b^2 + 4ab^3 + b^4$$

$$(a + b)^5 = a^5 + 5a^4b + 10a^3b^2 + 10a^2b^3 + 5a^4 + b^5$$

On the basis of these results, see if you can determine a pattern that will enable you to complete each of the following without using the long multiplication process.

(a) $(a + b)^6$ **(b)** $(a + b)^7$

(c) $(a + b)^8$ **(d)** $(a + b)^9$

94. Find each of the following indicated products. These patterns will be used again in Section 5.5.

(a) $(x - 1)(x^2 + x + 1)$ **(b)** $(x + 1)(x^2 - x + 1)$

(c) $(x + 3)(x^2 - 3x + 9)$ **(d)** $(x - 4)(x^2 + 4x + 16)$

(e) $(2x - 3)(4x^2 + 6x + 9)$

(f) $(3x + 5)(9x^2 - 15x + 25)$

95. Some of the product patterns can be used to do arithmetic computations mentally. For example, let's use the pattern $(a + b)^2 = a^2 + 2ab + b^2$ to compute 31^2 mentally. Your thought process should be "$31^2 = (30 + 1)^2 = 30^2 + 2(30)(1) + 1^2 = 961$." Compute each of the following numbers mentally, and then check your answers.

(a) 21^2 **(b)** 41^2 **(c)** 71^2

(d) 32^2 **(e)** 52^2 **(f)** 82^2

96. Use the pattern $(a - b)^2 = a^2 - 2ab + b^2$ to compute each of the following numbers mentally, and then check your answers.

(a) 19^2 **(b)** 29^2 **(c)** 49^2

(d) 79^2 **(e)** 38^2 **(f)** 58^2

97. Every whole number with a units digit of 5 can be represented by the expression $10x + 5$, where x is a whole number. For example, $35 = 10(3) + 5$ and $145 = 10(14) + 5$. Now let's observe the following pattern when squaring such a number.

$$(10x + 5)^2 = 100x^2 + 100x + 25$$

$$= 100x(x + 1) + 25$$

The pattern inside the dashed box can be stated as "add 25 to the product of x, $x + 1$, and 100." Thus, to compute 35^2 mentally, we can think "$35^2 = 3(4)(100) + 25 = 1225$." Compute each of the following numbers mentally, and then check your answers.

(a) 15^2 **(b)** 25^2 **(c)** 45^2

(d) 55^2 **(e)** 65^2 **(f)** 75^2

(g) 85^2 **(h)** 95^2 **(i)** 105^2

Answers to the Concept Quiz

1. True **2.** False **3.** True **4.** True **5.** True **6.** False **7.** False **8.** True **9.** False **10.** False

Answers to the Example Practice Skills

1. $12x^5 + 8x^4 - 20x^3$ **2.** $-8a^3b^2 - 4a^2b^4 + 12a^2b^3 - 4ab^4$ **3.** $ab + 4a + 3b + 12$ **4.** $a^2 + ab + a - 4b - 20$ **5.** $a^2 + 12a + 32$ **6.** $a^3 - a^2 - 11a + 15$ **7.** $12a^3 - 7a^2b - 3ab^2 + b^3$ **8.** $a^2 + 7a + 10$ **9.** $15a^2 + 7a - 2$ **10.** **(a)** $x^2 + 6x + 9$ **(b)** $9x^2 + 6xy + y^2$ **(c)** $9a^2 + 30ab + 25b^2$ **11.** **(a)** $x^2 - 10x + 25$ **(b)** $x^2 - 4xy + 4y^2$ **(c)** $4a^2 - 12ab + 9b^2$ **12.** **(a)** $x^2 - 36$ **(b)** $x^2 - 81y^2$ **(c)** $49a^2 - 25b^2$ **13.** $x^3 + 15x^2 + 75x + 125$ **14.** $64x^3 - 144x^2y + 108xy^2 - 27y^3$ **15.** Area is $64 - 4x^2$; volume is $4x^3 - 32x^2 + 64x$

5.4 Factoring: Use of the Distributive Property

OBJECTIVES

1 Classify Numbers as Prime or Composite
2 Factor Composite Numbers into a Product of Prime Numbers
3 Understand the Rules about Completely Factored Form
4 Factor Out the Highest Common Monomial Factor
5 Factor Out a Common Binomial Factor
6 Factor by Grouping
7 Use Factoring to Solve Equations
8 Solve Word Problems That Involve Factoring

1 Classify Numbers as Prime or Composite

Recall that 2 and 3 are said to be *factors* of 6 because the product of 2 and 3 is 6. Likewise, in an indicated product such as $7ab$, the 7, a, and b are called factors of the product. If a positive integer greater than 1 has no factors that are positive integers other than itself and 1, then it is called a **prime number**. Thus the prime numbers less than 20 are 2, 3, 5, 7, 11, 13, 17, and 19. A positive integer greater than 1 that is not a prime number is called a **composite number**. The composite numbers less than 20 are 4, 6, 8, 9, 10, 12, 14, 15, 16, and 18.

2 Factor Composite Numbers into a Product of Prime Numbers

Every composite number is the product of prime numbers. Consider the following examples.

$4 = 2 \cdot 2$ $\quad$ $63 = 3 \cdot 3 \cdot 7$

$12 = 2 \cdot 2 \cdot 3$ $\quad$ $121 = 11 \cdot 11$

$35 = 5 \cdot 7$

The indicated product form that contains only prime factors is called the **prime factorization form** of a number. Thus the prime factorization form of 63 is $3 \cdot 3 \cdot 7$. We also say that the number has been **completely factored** when it is in the prime factorization form.

3 Understand the Rules about Completely Factored Form

In general, factoring is the reverse of multiplication. Previously, we have used the distributive property to find the product of a monomial and a polynomial, as shown in the table.

Use the Distributive Property to Find a Product

Expression	Rewrite by applying the distributive property	Product
$3(x + 2)$	$3(x) + 3(2)$	$3x + 6$
$5(2x - 1)$	$5(2x) + 5(-1)$	$10x - 5$
$x(x^2 + 6x - 4)$	$x(x^2) + x(6x) + x(-4)$	$x^3 + 6x^2 - 4x$

We shall also use the distributive property [in the form $ab + ac = a(b + c)$] to reverse the process—that is, to factor a given polynomial. Consider the examples in the following table.

Use the Distributive Property to Factor

Expression	Rewrite the expression	Factored form by applying the distributive property
$3x + 6$	$3(x) + 3(2)$	$3(x + 2)$
$10x - 5$	$5(2x) + 5(-1)$	$5(2x - 1)$
$x^3 + 6x^2 - 4x$	$x(x^2) + x(6x) + x(-4)$	$x(x^2 + 6x - 4)$

Note that in each example a given polynomial has been factored into the product of a monomial and a polynomial. Obviously, polynomials could be factored in a variety of ways. Consider some factorizations of $3x^2 + 12x$.

$$3x^2 + 12x = 3x(x + 4) \quad \text{or} \quad 3x^2 + 12x = 3(x^2 + 4x) \quad \text{or}$$

$$3x^2 + 12x = x(3x + 12) \quad \text{or} \quad 3x^2 + 12x = \frac{1}{2}(6x^2 + 24x)$$

We are, however, primarily interested in the first of the previous factorization forms, which we refer to as the **completely factored form**. A polynomial with integral coefficients is in completely factored form if:

1. it is expressed as a product of polynomials with *integral coefficients* and
2. no polynomial, other than a monomial, within the factored form can be further factored into polynomials with integral coefficients.

Do you see why only the first of the preceding factored forms of $3x^2 + 12x$ is said to be in completely factored form? In each of the other three forms, the polynomial inside the parentheses can be factored further. Moreover, in the last form, $\frac{1}{2}(6x^2 + 24x)$, the condition of using only integral coefficients is violated.

EXAMPLE 1

For each of the following, determine if the factorization is in completely factored form. If it is not in completely factored form, state which rule is violated.

(a) $4m^3 + 8m^4n = 4m^2(m + 2m^2n)$ **(b)** $32p^2q^4 + 8pq = 8pq(4pq^3 + 1)$
(c) $8x^2y^5 + 4x^3y^2 = 8x^2y^2(y^3 + 0.5x)$ **(d)** $10ab^3 + 20a^4b = 2ab(5b^2 + 10a^3)$

Solution

(a) No, it is not completely factored. The polynomial inside the parentheses can be factored further.
(b) Yes, it is completely factored.
(c) No, it is not completely factored. The coefficient of 0.5 is not an integer.
(d) No, it is not completely factored. The polynomial inside the parentheses can be factored further.

▼ PRACTICE YOUR SKILL

For each of the following, determine if the factorization is in completely factored form. If it is not in completely factored form, state which rule is violated.

(a) $9x^3y^2 + 3x^2y^2 = 6x^2y^2\left(\frac{3}{2}x + \frac{1}{2}\right)$

(b) $x^2y^4 + 8xy = y(x^2y^3 + 8xy)$

(c) $6x^3y^5 + 4x^3y^2 = 2x^3y^2(3y^3 + 2)$ ■

4 Factor Out the Highest Common Monomial Factor

The factoring process that we discuss in this section, $ab + ac = a(b + c)$, is often referred to as **factoring out the highest common monomial factor**. The key idea in this process is to recognize the monomial factor that is common to all terms. For example, we observe that each term of the polynomial $2x^3 + 4x^2 + 6x$ has a factor of $2x$. Thus we write

$$2x^3 + 4x^2 + 6x = 2x(\qquad)$$

and insert within the parentheses the appropriate polynomial factor. We determine the terms of this polynomial factor by dividing each term of the original polynomial by the factor of $2x$. The final, completely factored form is

$$2x^3 + 4x^2 + 6x = 2x(x^2 + 2x + 3)$$

The following examples further demonstrate this process of factoring out the highest common monomial factor.

$$12x^3 + 16x^2 = 4x^2(3x + 4) \qquad 6x^2y^3 + 27xy^4 = 3xy^3(2x + 9y)$$
$$8ab - 18b = 2b(4a - 9) \qquad 8y^3 + 4y^2 = 4y^2(2y + 1)$$
$$30x^3 + 42x^4 - 24x^5 = 6x^3(5 + 7x - 4x^2)$$

Note that in each example, the common monomial factor itself is not in a completely factored form. For example, $4x^2(3x + 4)$ is not written as $2 \cdot 2 \cdot x \cdot x \cdot (3x + 4)$.

EXAMPLE 2

Factor out the highest common factor for each of the following.

(a) $3x^4 + 15x^3 - 21x^2$ **(b)** $8x^3y^2 - 2x^4y - 12xy^2$

Solution

(a) Each term of the polynomial has a common factor of $3x^2$.

$$3x^4 + 15x^3 - 21x^2 = 3x^2(x^2 + 5x - 7)$$

(b) Each term of the polynomial has a common factor of $2xy$.

$$8x^3y^2 - 2x^4y - 12xy^2 = 2xy(4x^2y - x^3 - 6y)$$

▼ PRACTICE YOUR SKILL

Factor out the highest common factor for each of the following.

(a) $10a^2 - 15a^3 + 35a^4$ **(b)** $2mn - 8m^3$ ■

5 Factor Out a Common Binomial Factor

Sometimes there may be a common binomial factor rather than a common monomial factor. For example, each of the two terms of the expression $x(y + 2) + z(y + 2)$ has a binomial factor of $(y + 2)$. Thus we can factor $(y + 2)$ from each term, and our result is

$$x(y + 2) + z(y + 2) = (y + 2)(x + z)$$

Consider a few more examples that involve a common binomial factor.

EXAMPLE 3

For each of the following, factor out the common binomial factor.

(a) $a^2(b + 1) + 2(b + 1)$ **(b)** $x(2y - 1) - y(2y - 1)$
(c) $x(x + 2) + 3(x + 2)$

Solution

(a) $a^2(b + 1) + 2(b + 1) = (b + 1)(a^2 + 2)$
(b) $x(2y - 1) - y(2y - 1) = (2y - 1)(x - y)$
(c) $x(x + 2) + 3(x + 2) = (x + 2)(x + 3)$

▼ **PRACTICE YOUR SKILL**

For each of the following, factor out the common binomial factor.

(a) $6(xy + 8) + z(xy + 8)$ **(b)** $x^2(x + y) - y^3(x + y)$
(c) $x(2x + y) + y(2x + y) + z(2x + y)$ ■

6 Factor by Grouping

It may be that the original polynomial exhibits no apparent common monomial or binomial factor, which is the case with $ab + 3a + bc + 3c$. However, by factoring a from the first two terms and c from the last two terms, we get

$$ab + 3a + bc + 3c = a(b + 3) + c(b + 3)$$

Now a common binomial factor of $(b + 3)$ is obvious, and we can proceed as before:

$$a(b + 3) + c(b + 3) = (b + 3)(a + c)$$

We refer to this factoring process as **factoring by grouping**. Let's consider a few more examples of this type.

EXAMPLE 4

Factor the following using factoring by grouping.

(a) $ab^2 - 4b^2 + 3a - 12$ **(b)** $x^2 - x + 5x - 5$ **(c)** $x^2 + 2x - 3x - 6$

Solution

(a) $ab^2 - 4b^2 + 3a - 12 = b^2(a - 4) + 3(a - 4)$ Factor b^2 from the first two terms and 3 from the last two terms

$= (a - 4)(b^2 + 3)$ Factor common binomial from both terms

(b) $x^2 - x + 5x - 5 = x(x - 1) + 5(x - 1)$ Factor x from the first two terms and 5 from the last two terms

$= (x - 1)(x + 5)$ Factor common binomial from both terms

(c) $x^2 + 2x - 3x - 6 = x(x + 2) - 3(x + 2)$ Factor x from the first two terms and -3 from the last two terms

$= (x + 2)(x - 3)$ Factor common binomial factor from both terms

▼ **PRACTICE YOUR SKILL**

Factor the following using factoring by grouping.

(a) $4x^3y + 8xy + 3x^2 + 6$ **(b)** $7y^2 - 14y + 5y - 10$
(c) $2x^2 + 3xy - 4xy - 6y^2$ ■

It may be necessary to rearrange some terms before applying the distributive property. Terms that contain common factors need to be grouped together, and this may be done in more than one way. The next example illustrates this idea.

Method 1 $4a^2 - bc^2 - a^2b + 4c^2 = 4a^2 - a^2b + 4c^2 - bc^2$

$= a^2(4 - b) + c^2(4 - b)$

$= (4 - b)(a^2 + c^2)$ or

Method 2 $4a^2 - bc^2 - a^2b + 4c^2 = 4a^2 + 4c^2 - bc^2 - a^2b$

$$= 4(a^2 + c^2) - b(c^2 + a^2)$$
$$= 4(a^2 + c^2) - b(a^2 + c^2)$$
$$= (a^2 + c^2)(4 - b)$$

7 Use Factoring to Solve Equations

One reason why factoring is an important algebraic skill is that it extends our techniques for solving equations. Each time we examine a factoring technique, we will then use it to help solve certain types of equations.

We need another property of equality before we consider some equations where the highest-common-factor technique is useful. Suppose that the product of two numbers is zero. Can we conclude that at least one of these numbers must itself be zero? Yes. Let's state a property that formalizes this idea. Property 5.5, along with the highest-common-factor pattern, provides us with another technique for solving equations.

Property 5.5

Let a and b be real numbers. Then

$ab = 0$ if and only if $a = 0$ or $b = 0$

EXAMPLE 5

Solve $x^2 + 6x = 0$.

Solution

$$x^2 + 6x = 0$$
$$x(x + 6) = 0 \quad \text{Factor the left side}$$
$$x = 0 \quad \text{or} \quad x + 6 = 0 \quad ab = 0 \text{ if and only if } a = 0 \text{ or } b = 0$$
$$x = 0 \quad \text{or} \quad x = -6$$

Thus both 0 and -6 will satisfy the original equation, and the solution set is $\{-6, 0\}$.

▼ **PRACTICE YOUR SKILL**

Solve $y^2 - 4y = 0$. ■

EXAMPLE 6

Solve $a^2 = 11a$.

Solution

$$a^2 = 11a$$
$$a^2 - 11a = 0 \quad \text{Add } -11a \text{ to both sides}$$
$$a(a - 11) = 0 \quad \text{Factor the left side}$$
$$a = 0 \quad \text{or} \quad a - 11 = 0 \quad ab = 0 \text{ if and only if } a = 0 \text{ or } b = 0$$
$$a = 0 \quad \text{or} \quad a = 11$$

The solution set is $\{0, 11\}$.

▼ **PRACTICE YOUR SKILL**

Solve $x^2 = -12x$. ■

Remark: Note that in Example 6 we did *not* divide both sides of the equation by a. This would cause us to lose the solution of 0.

EXAMPLE 7

Solve $3n^2 - 5n = 0$.

Solution

$$3n^2 - 5n = 0$$

$$n(3n - 5) = 0$$

$$n = 0 \quad \text{or} \quad 3n - 5 = 0$$

$$n = 0 \quad \text{or} \quad 3n = 5$$

$$n = 0 \quad \text{or} \quad n = \frac{5}{3}$$

The solution set is $\left\{0, \frac{5}{3}\right\}$.

▼ **PRACTICE YOUR SKILL**

Solve $7y^2 + 2y = 0$. ■

EXAMPLE 8

Solve $3ax^2 + bx = 0$ for x.

Solution

$$3ax^2 + bx = 0$$

$$x(3ax + b) = 0$$

$$x = 0 \quad \text{or} \quad 3ax + b = 0$$

$$x = 0 \quad \text{or} \quad 3ax = -b$$

$$x = 0 \quad \text{or} \quad x = -\frac{b}{3a}$$

The solution set is $\left\{0, -\frac{b}{3a}\right\}$.

▼ **PRACTICE YOUR SKILL**

Solve $8cy^2 - dy = 0$ for y. ■

8 Solve Word Problems That Involve Factoring

Many of the problems that we solve in the next few sections have a geometric setting. Some basic geometric figures, along with appropriate formulas, are listed in the inside front cover of this text. You may need to refer to them to refresh your memory.

EXAMPLE 9

Apply Your Skill

The area of a square is three times its perimeter. Find the length of a side of the square.

Solution

Let s represent the length of a side of the square (Figure 5.17). The area is represented by s^2 and the perimeter by $4s$. Thus

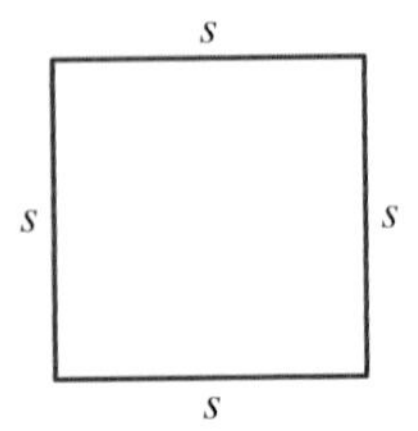

$$s^2 = 3(4s) \quad \text{The area is to be three times the perimeter}$$

$$s^2 = 12s$$

$$s^2 - 12s = 0$$

$$s(s - 12) = 0$$

$$s = 0 \quad \text{or} \quad s = 12$$

s s s s

Figure 5.17

Because 0 is not a reasonable solution, it must be a 12-by-12 square. (Be sure to check this answer in the original statement of the problem!)

▼ PRACTICE YOUR SKILL

The area of a square is twice its perimeter. Find the length of a side of the square. ■

EXAMPLE 10

Apply Your Skill

Suppose that the volume of a right circular cylinder is numerically equal to the total surface area of the cylinder. If the height of the cylinder is equal to the length of a radius of the base, find the height.

Solution

Because $r = h$, the formula for volume $V = \pi r^2 h$ becomes $V = \pi r^3$, and the formula for the total surface area $S = 2\pi r^2 + 2\pi rh$ becomes $S = 2\pi r^2 + 2\pi r^2$ or $S = 4\pi r^2$. Therefore, we can set up and solve the following equation.

$$\pi r^3 = 4\pi r^2 \quad \text{Volume is equal to the surface area}$$

$$\pi r^3 - 4\pi r^2 = 0$$

$$\pi r^2(r - 4) = 0$$

$$\pi r^2 = 0 \quad \text{or} \quad r - 4 = 0$$

$$r = 0 \quad \text{or} \quad r = 4$$

Zero is not a reasonable answer; therefore, the height must be 4 units.

▼ PRACTICE YOUR SKILL

Suppose that the volume of a cube is numerically equal to the total surface area of the cube. Find the length of an edge of the cube. ■

CONCEPT QUIZ

For Problems 1–10, answer true or false.

1. The greatest common factor of $6x^2y^3 - 12x^3y^2 + 18x^4y$ is $2x^2y$.
2. If the factored form of a polynomial can be factored further, then it has not met the conditions to be considered "factored completely."
3. Common factors are always monomials.
4. If the product of x and y is zero, then x is zero or y is zero.
5. The factored form, $3a(2a^2 + 4)$, is factored completely.
6. The solutions for the equation $x(x + 2) = 7$ are 7 and 5.
7. The solution set for $x^2 = 7x$ is $\{7\}$.
8. The solution set for $x(x - 2) - 3(x - 2) = 0$ is $\{2, 3\}$.
9. The solution set for $-3x = x^2$ is $\{-3, 0\}$.
10. The solution set for $x(x + 6) = 2(x + 6)$ is $\{-6\}$.

Problem Set 5.4

1 Classify Numbers as Prime or Composite

For Problems 1–10, classify each number as prime or composite.

1. 63
2. 81
3. 59
4. 83
5. 51
6. 69
7. 91
8. 119
9. 71
10. 101

2 Factor Composite Numbers into a Product of Prime Numbers

For Problems 11–20, factor each of the composite numbers into the product of prime numbers. For example, $30 = 2 \cdot 3 \cdot 5$.

11. 28
12. 39
13. 44
14. 49
15. 56
16. 64
17. 72
18. 84
19. 87
20. 91

3 Understand the Rules about Completely Factored Form

For Problems 21–24, state if the polynomial is factored completely.

21. $6x^2y + 12xy^2 = 2xy(3x + 6y)$
22. $2a^3b^2 + 4a^2b^2 = 4a^2b^2\left(\frac{1}{2}a + 1\right)$
23. $10m^2n^3 + 15m^4n^2 = 5m^2n(2n^2 + 3m^2n)$
24. $24ab + 12bc - 18bd = 6b(4a + 2c - 3d)$

4 Factor Out the Highest Common Monomial Factor

For Problems 25–40, factor completely.

25. $28y^2 - 4y$
26. $42y^2 - 6y$
27. $20xy - 15x$
28. $27xy - 36y$
29. $7x^3 + 10x^2$
30. $12x^3 - 10x^2$
31. $18a^2b + 27ab^2$
32. $24a^3b^2 + 36a^2b$
33. $12x^3y^4 - 39x^4y^3$
34. $15x^4y^2 - 45x^5y^4$
35. $8x^4 + 12x^3 - 24x^2$
36. $6x^5 - 18x^3 + 24x$
37. $5x + 7x^2 + 9x^4$
38. $9x^2 - 17x^4 + 21x^5$
39. $15x^2y^3 + 20xy^2 + 35x^3y^4$
40. $8x^5y^3 - 6x^4y^5 + 12x^2y^3$

5 Factor Out a Common Binomial Factor

For Problems 41–46, factor completely.

41. $x(y + 2) + 3(y + 2)$
42. $x(y - 1) + 5(y - 1)$
43. $3x(2a + b) - 2y(2a + b)$
44. $5x(a - b) + y(a - b)$
45. $x(x + 2) + 5(x + 2)$
46. $x(x - 1) - 3(x - 1)$

6 Factor by Grouping

For Problems 47–64, factor by grouping.

47. $ax + 4x + ay + 4y$
48. $ax - 2x + ay - 2y$
49. $ax - 2bx + ay - 2by$
50. $2ax - bx + 2ay - by$
51. $3ax - 3bx - ay + by$
52. $5ax - 5bx - 2ay + 2by$
53. $2ax + 2x + ay + y$
54. $3bx + 3x + by + y$
55. $ax^2 - x^2 + 2a - 2$
56. $ax^2 - 2x^2 + 3a - 6$
57. $2ac + 3bd + 2bc + 3ad$
58. $2bx + cy + cx + 2by$

59. $ax - by + bx - ay$

60. $2a^2 - 3bc - 2ab + 3ac$

61. $x^2 + 9x + 6x + 54$

62. $x^2 - 2x + 5x - 10$

63. $2x^2 + 8x + x + 4$

64. $3x^2 + 18x - 2x - 12$

7 Use Factoring to Solve Equations

For Problems 65–80, solve each of the equations.

65. $x^2 + 7x = 0$

66. $x^2 + 9x = 0$

67. $x^2 - x = 0$

68. $x^2 - 14x = 0$

69. $a^2 = 5a$

70. $b^2 = -7b$

71. $-2y = 4y^2$

72. $-6x = 2x^2$

73. $3x^2 + 7x = 0$

74. $-4x^2 + 9x = 0$

75. $4x^2 = 5x$

76. $3x = 11x^2$

77. $x - 4x^2 = 0$

78. $x - 6x^2 = 0$

79. $12a = -a^2$

80. $-5a = -a^2$

For Problems 81–86, solve each equation for the indicated variable.

81. $5bx^2 - 3ax = 0$ for x

82. $ax^2 + bx = 0$ for x

83. $2by^2 = -3ay$ for y

84. $3ay^2 = by$ for y

85. $y^2 - ay + 2by - 2ab = 0$ for y

86. $x^2 + ax + bx + ab = 0$ for x

8 Solve Word Problems That Involve Factoring

For Problems 87–96, set up an equation and solve each of the following problems.

87. The square of a number equals seven times the number. Find the number.

88. Suppose that the area of a square is six times its perimeter. Find the length of a side of the square.

89. The area of a circular region is numerically equal to three times the circumference of the circle. Find the length of a radius of the circle.

90. Find the length of a radius of a circle such that the circumference of the circle is numerically equal to the area of the circle.

91. Suppose that the area of a circle is numerically equal to the perimeter of a square and that the length of a radius of the circle is equal to the length of a side of the square. Find the length of a side of the square. Express your answer in terms of π.

92. Find the length of a radius of a sphere such that the surface area of the sphere is numerically equal to the volume of the sphere.

93. Suppose that the area of a square lot is twice the area of an adjoining rectangular plot of ground. If the rectangular plot is 50 feet wide and its length is the same as the length of a side of the square lot, find the dimensions of both the square and the rectangle.

94. The area of a square is one-fourth as large as the area of a triangle. One side of the triangle is 16 inches long, and the altitude to that side is the same length as a side of the square. Find the length of a side of the square.

95. Suppose that the volume of a sphere is numerically equal to twice the surface area of the sphere. Find the length of a radius of the sphere.

96. Suppose that a radius of a sphere is equal in length to a radius of a circle. If the volume of the sphere is numerically equal to four times the area of the circle, find the length of a radius for both the sphere and the circle.

THOUGHTS INTO WORDS

97. Is $2 \cdot 3 \cdot 5 \cdot 7 \cdot 11 + 7$ a prime or a composite number? Defend your answer.

98. Suppose that your friend factors $36x^2y + 48xy^2$ as follows:

$$\begin{aligned} 36x^2y + 48xy^2 &= (4xy)(9x + 12y) \\ &= (4xy)(3)(3x + 4y) \\ &= 12xy(3x + 4y) \end{aligned}$$

Is this a correct approach? Would you have any suggestion to offer your friend?

99. Your classmate solves the equation $3ax + bx = 0$ for x as follows:

$$\begin{aligned} 3ax + bx &= 0 \\ 3ax &= -bx \\ x &= \frac{-bx}{3a} \end{aligned}$$

How should he know that the solution is incorrect? How would you help him obtain the correct solution?

FURTHER INVESTIGATIONS

100. The total surface area of a right circular cylinder is given by the formula $A = 2\pi r^2 + 2\pi rh$, where r represents the radius of a base and h represents the height of the cylinder. For computational purposes, it may be more convenient to change the form of the right side of the formula by factoring it.

$$A = 2\pi r^2 + 2\pi rh$$
$$= 2\pi r(r + h)$$

Use $A = 2\pi r(r + h)$ to find the total surface area of each of the following cylinders. Also, use $\frac{22}{7}$ as an approximation for π.

(a) $r = 7$ centimeters and $h = 12$ centimeters

(b) $r = 14$ meters and $h = 20$ meters

(c) $r = 3$ feet and $h = 4$ feet

(d) $r = 5$ yards and $h = 9$ yards

For Problems 101–106, factor each expression. Assume that all variables that appear as exponents represent positive integers.

101. $2x^{2a} - 3x^a$

102. $6x^{2a} + 8x^a$

103. $y^{3m} + 5y^{2m}$

104. $3y^{5m} - y^{4m} - y^{3m}$

105. $2x^{6a} - 3x^{5a} + 7x^{4a}$

106. $6x^{3a} - 10x^{2a}$

Answers to the Concept Quiz

1. False **2.** True **3.** False **4.** True **5.** False **6.** False **7.** False **8.** True **9.** True **10.** False

Answers to the Example Practice Skills

1. (a) No, it is not completely factored. There are coefficients that are not integers. **(b)** No, it is not completely factored. The polynomial inside the parentheses can be factored further. **(c)** Yes, it is completely factored. **2. (a)** $5a^2(2 - 3a + 7a^2)$ **(b)** $2m(n - 4m^2)$ **3. (a)** $(xy + 8)(6 + z)$ **(b)** $(x + y)(x^2 - y^3)$ **(c)** $(2x + y)(x + y + z)$ **4. (a)** $(4xy + 3)(x^2 + 2)$ **(b)** $(7y + 5)(y - 2)$ **(c)** $(2x + 3y)(x - 2y)$ **5.** $\{0, 4\}$ **6.** $\{-12, 0\}$ **7.** $\left\{-\frac{2}{7}, 0\right\}$ **8.** $\left\{0, \frac{d}{8c}\right\}$ **9.** 8-by-8 square **10.** 6 units

5.5 Factoring: Difference of Two Squares and Sum or Difference of Two Cubes

OBJECTIVES

1. Factor the Difference of Two Squares
2. Factor the Sum or Difference of Two Cubes
3. Use Factoring to Solve Equations
4. Solve Word Problems That Involve Factoring

1 Factor the Difference of Two Squares

In Section 5.3, we examined some special multiplication patterns. One of these patterns was

$$(a + b)(a - b) = a^2 - b^2$$

This same pattern, viewed as a factoring pattern, is referred to as the difference of two squares.

Difference of Two Squares

$$a^2 - b^2 = (a + b)(a - b)$$

Applying the pattern is fairly simple, as the next example demonstrates.

EXAMPLE 1

Factor each of the following.

(a) $x^2 - 16$ **(b)** $4x^2 - 25$ **(c)** $16x^2 - 9y^2$ **(d)** $1 - a^2$

Solution

(a) $x^2 - 16 = (x)^2 - (4)^2 = (x + 4)(x - 4)$

(b) $4x^2 - 25 = (2x)^2 - (5)^2 = (2x + 5)(2x - 5)$

(c) $16x^2 - 9y^2 = (4x)^2 - (3y)^2 = (4x + 3y)(4x - 3y)$

(d) $1 - a^2 = (1)^2 - (a)^2 = (1 + a)(1 - a)$

▼ **PRACTICE YOUR SKILL**

Factor each of the following.

(a) $m^2 - 36$ **(b)** $9y^2 - 49$ **(c)** $64 - 25b^2$ ■

Multiplication is commutative, so the order of writing the factors is not important. For example, $(x + 4)(x - 4)$ can also be written as $(x - 4)(x + 4)$.

You must be careful not to assume an analogous factoring pattern for the *sum* of two squares; *it does not exist.* For example, $x^2 + 4 \neq (x + 2)(x + 2)$ because $(x + 2)(x + 2) = x^2 + 4x + 4$. We say that a polynomial such as $x^2 + 4$ is a **prime polynomial** or that it is not factorable using integers.

Sometimes the difference-of-two-squares pattern can be applied more than once, as the next example illustrates.

EXAMPLE 2

Completely factor each of the following.

(a) $x^4 - y^4$ **(b)** $16x^4 - 81y^4$

Solution

(a) $x^4 - y^4 = (x^2 + y^2)(x^2 - y^2) = (x^2 + y^2)(x + y)(x - y)$

(b) $16x^4 - 81y^4 = (4x^2 + 9y^2)(4x^2 - 9y^2) = (4x^2 + 9y^2)(2x + 3y)(2x - 3y)$

▼ **PRACTICE YOUR SKILL**

Factor completely $256a^4 - b^4$. ■

It may also be that the squares are other than simple monomial squares, as in the next example.

EXAMPLE 3

Completely factor each of the following.

(a) $(x + 3)^2 - y^2$ **(b)** $4x^2 - (2y + 1)^2$ **(c)** $(x - 1)^2 - (x + 4)^2$

Solution

(a) $(x + 3)^2 - y^2 = [(x + 3) + y][(x + 3) - y] = (x + 3 + y)(x + 3 - y)$

(b) $4x^2 - (2y + 1)^2 = [2x + (2y + 1)][2x - (2y + 1)]$

$= (2x + 2y + 1)(2x - 2y - 1)$

(c) $(x - 1)^2 - (x + 4)^2 = [(x - 1) + (x + 4)][(x - 1) - (x + 4)]$

$= (x - 1 + x + 4)(x - 1 - x - 4)$

$= (2x + 3)(-5)$

▼ **PRACTICE YOUR SKILL**

Factor completely $(2x + y)^2 - 9$. ■

It is possible to apply both the technique of factoring out a common monomial factor and the pattern of the difference of two squares to the same problem. In general, it is best to look first for a common monomial factor. Consider the following example.

EXAMPLE 4

Completely factor each of the following.

(a) $2x^2 - 50$ **(b)** $9x^2 - 36$ **(c)** $48y^3 - 27y$

Solution

(a) $2x^2 - 50 = 2(x^2 - 25) = 2(x + 5)(x - 5)$

(b) $9x^2 - 36 = 9(x^2 - 4) = 9(x + 2)(x - 2)$

(c) $48y^3 - 27y = 3y(16y^2 - 9) = 3y(4y + 3)(4y - 3)$

▼ **PRACTICE YOUR SKILL**

Factor completely $18a^2 - 50$. ■

Word of Caution The polynomial $9x^2 - 36$ can be factored as follows:

$9x^2 - 36 = (3x + 6)(3x - 6)$

$= 3(x + 2)(3)(x - 2)$

$= 9(x + 2)(x - 2)$

However, when one takes this approach, there seems to be a tendency to stop at the step $(3x + 6)(3x - 6)$. Therefore, remember the suggestion to *look first for a common monomial factor*.

The following examples should help you summarize all of the factoring techniques we have considered thus far.

$7x^2 + 28 = 7(x^2 + 4)$

$4x^2y - 14xy^2 = 2xy(2x - 7y)$

$x^2 - 4 = (x + 2)(x - 2)$

$18 - 2x^2 = 2(9 - x^2) = 2(3 + x)(3 - x)$

$y^2 + 9$ is not factorable using integers

$5x + 13y$ is not factorable using integers

$x^4 - 16 = (x^2 + 4)(x^2 - 4) = (x^2 + 4)(x + 2)(x - 2)$

2 Factor the Sum or Difference of Two Cubes

As we pointed out before, there exists no sum-of-squares pattern analogous to the difference-of-squares factoring pattern. That is, a polynomial such as $x^2 + 9$ is not factorable using integers. However, patterns do exist for both the sum and the difference of two cubes. These patterns are as follows.

Sum and Difference of Two Cubes

$$a^3 + b^3 = (a + b)(a^2 - ab + b^2)$$
$$a^3 - b^3 = (a - b)(a^2 + ab + b^2)$$

Note how we apply these patterns in the next example.

EXAMPLE 5

Factor each of the following.

(a) $x^3 + 27$ **(b)** $8a^3 + 125b^3$ **(c)** $x^3 - 1$ **(d)** $27y^3 - 64x^3$

Solution

(a) $x^3 + 27 = (x)^3 + (3)^3 = (x + 3)(x^2 - 3x + 9)$

(b) $8a^3 + 125b^3 = (2a)^3 + (5b)^3 = (2a + 5b)(4a^2 - 10ab + 25b^2)$

(c) $x^3 - 1 = (x)^3 - (1)^3 = (x - 1)(x^2 + x + 1)$

(d) $27y^3 - 64x^3 = (3y)^3 - (4x)^3 = (3y - 4x)(9y^2 + 12xy + 16x^2)$

▼ **PRACTICE YOUR SKILL**

Factor completely $x^3 + 8y^3$. ■

3 Use Factoring to Solve Equations

Remember that each time we pick up a new factoring technique we also develop more power for solving equations. Let's consider how we can use the difference-of-two-squares factoring pattern to help solve certain types of equations.

EXAMPLE 6

Solve $x^2 = 16$.

Solution

$$x^2 = 16$$
$$x^2 - 16 = 0$$
$$(x + 4)(x - 4) = 0$$

$$x + 4 = 0 \quad \text{or} \quad x - 4 = 0$$
$$x = -4 \quad \text{or} \quad x = 4$$

The solution set is $\{-4, 4\}$. (Be sure to check these solutions in the original equation!)

▼ **PRACTICE YOUR SKILL**

Solve $m^2 = 81$. ■

EXAMPLE 7 Solve $9x^2 = 64$.

Solution

$$9x^2 = 64$$
$$9x^2 - 64 = 0$$
$$(3x + 8)(3x - 8) = 0$$
$$3x + 8 = 0 \quad \text{or} \quad 3x - 8 = 0$$
$$3x = -8 \quad \text{or} \quad 3x = 8$$
$$x = -\frac{8}{3} \quad \text{or} \quad x = \frac{8}{3}$$

The solution set is $\left\{-\frac{8}{3}, \frac{8}{3}\right\}$.

▼ **PRACTICE YOUR SKILL**

Solve $25a^2 = 36$. ■

EXAMPLE 8 Solve $7x^2 - 7 = 0$.

Solution

$$7x^2 - 7 = 0$$
$$7(x^2 - 1) = 0$$
$$x^2 - 1 = 0 \quad \text{Multiply both sides by } \frac{1}{7}$$
$$(x + 1)(x - 1) = 0$$
$$x + 1 = 0 \quad \text{or} \quad x - 1 = 0$$
$$x = -1 \quad \text{or} \quad x = 1$$

The solution set is $\{-1, 1\}$.

▼ **PRACTICE YOUR SKILL**

Solve $3 - 12x^2 = 0$. ■

In the previous examples we have been using the property $ab = 0$ if and only if $a = 0$ or $b = 0$. This property can be extended to any number of factors whose product is zero. Thus for three factors, the property could be stated $abc = 0$ if and only if $a = 0$ or $b = 0$ or $c = 0$. The next two examples illustrate this idea.

EXAMPLE 9 Solve $x^4 - 16 = 0$.

Solution

$$x^4 - 16 = 0$$
$$(x^2 + 4)(x^2 - 4) = 0$$
$$(x^2 + 4)(x + 2)(x - 2) = 0$$
$$x^2 + 4 = 0 \quad \text{or} \quad x + 2 = 0 \quad \text{or} \quad x - 2 = 0$$
$$x^2 = -4 \quad \text{or} \quad x = -2 \quad \text{or} \quad x = 2$$

The solution set is $\{-2, 2\}$. (Because no real numbers, when squared, will produce -4, the equation $x^2 = -4$ yields no additional real number solutions.)

▼ **PRACTICE YOUR SKILL**

Solve $x^4 - 81 = 0$. ■

EXAMPLE 10 Solve $x^3 - 49x = 0$.

Solution

$$x^3 - 49x = 0$$
$$x(x^2 - 49) = 0$$
$$x(x + 7)(x - 7) = 0$$
$$x = 0 \quad \text{or} \quad x + 7 = 0 \quad \text{or} \quad x - 7 = 0$$
$$x = 0 \quad \text{or} \quad x = -7 \quad \text{or} \quad x = 7$$

The solution set is $\{-7, 0, 7\}$.

▼ **PRACTICE YOUR SKILL**

Solve $y^3 - 16y = 0$. ■

4 Solve Word Problems That Involve Factoring

The more we know about solving equations, the more capability we have for solving word problems.

EXAMPLE 11 **Apply Your Skill**

The combined area of two squares is 40 square centimeters. Each side of one square is three times as long as a side of the other square. Find the dimensions of each of the squares.

Solution

Let s represent the length of a side of the smaller square. Then $3s$ represents the length of a side of the larger square (Figure 5.18).

$$s^2 + (3s)^2 = 40$$
$$s^2 + 9s^2 = 40$$
$$10s^2 = 40$$
$$s^2 = 4$$

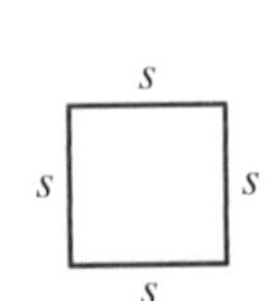

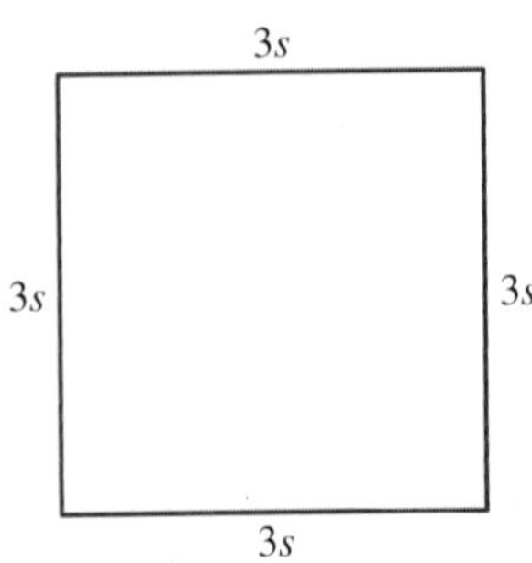

Figure 5.18

$$s^2 - 4 = 0$$

$$(s + 2)(s - 2) = 0$$

$$s + 2 = 0 \quad \text{or} \quad s - 2 = 0$$

$$s = -2 \quad \text{or} \quad s = 2$$

Because s represents the length of a side of a square, the solution -2 must be disregarded. Thus the length of a side of the small square is 2 centimeters, and the large square has sides of length $3(2) = 6$ centimeters.

▼ **PRACTICE YOUR SKILL**

The combined area of two squares is 125 square inches. Each side of one square is twice as long as a side of the other square. Find the dimensions of each square. ■

CONCEPT QUIZ

For Problems 1–10, answer true or false.

1. A binomial that has two perfect square terms that are subtracted is called the difference of two squares.
2. The sum of two squares is factorable using integers.
3. When factoring it is usually best to look for a common factor first.
4. The polynomial $4x^2 + y^2$ factors into $(2x + y)(2x + y)$.
5. The completely factored form of $y^4 - 81$ is $(y^2 + 9)(y^2 - 9)$.
6. The solution set for $x^2 = -16$ is $\{-4\}$.
7. The solution set for $5x^3 - 5x = 0$ is $\{-1, 0, 1\}$.
8. The solution set for $x^4 - 9x^2 = 0$ is $\{-3, 0, 3\}$.
9. $1 - x^3 = (1 - x)(1 + x - x^2)$
10. $8 + x^3 = (2 + x)(4 - 2x + x^2)$

Problem Set 5.5

1 Factor the Difference of Two Squares

For Problems 1–20, use the difference-of-squares pattern to factor each of the following.

1. $x^2 - 1$ **2.** $x^2 - 9$

3. $16x^2 - 25$ **4.** $4x^2 - 49$

5. $9x^2 - 25y^2$ **6.** $x^2 - 64y^2$

7. $25x^2y^2 - 36$ **8.** $x^2y^2 - a^2b^2$

9. $4x^2 - y^4$ **10.** $x^6 - 9y^2$

11. $1 - 144n^2$ **12.** $25 - 49n^2$

13. $(x + 2)^2 - y^2$ **14.** $(3x + 5)^2 - y^2$

15. $4x^2 - (y + 1)^2$ **16.** $x^2 - (y - 5)^2$

17. $9a^2 - (2b + 3)^2$ **18.** $16s^2 - (3t + 1)^2$

19. $(x + 2)^2 - (x + 7)^2$ **20.** $(x - 1)^2 - (x - 8)^2$

For Problems 21–44, factor each of the following polynomials completely. Indicate any that are not factorable using integers. Don't forget to look first for a common monomial factor.

21. $9x^2 - 36$ **22.** $8x^2 - 72$

23. $5x^2 + 5$ **24.** $7x^2 + 28$

25. $8y^2 - 32$ **26.** $5y^2 - 80$

27. $a^3b - 9ab$ **28.** $x^3y^2 - xy^2$

29. $16x^2 + 25$ **30.** $x^4 - 16$

31. $n^4 - 81$ **32.** $4x^2 + 9$

33. $3x^3 + 27x$ **34.** $20x^3 + 45x$

35. $4x^3y - 64xy^3$ **36.** $12x^3 - 27xy^2$

37. $6x - 6x^3$ **38.** $1 - 16x^4$

39. $1 - x^4y^4$ **40.** $20x - 5x^3$

41. $4x^2 - 64y^2$ **42.** $9x^2 - 81y^2$

43. $3x^4 - 48$ **44.** $2x^5 - 162x$

2 Factor the Sum or Difference of Two Cubes

For Problems 45–56, use the sum-of-two-cubes or the difference-of-two-cubes pattern to factor each of the following.

45. $a^3 - 64$

46. $a^3 - 27$

47. $x^3 + 1$

48. $x^3 + 8$

49. $27x^3 + 64y^3$

50. $8x^3 + 27y^3$

51. $1 - 27a^3$

52. $1 - 8x^3$

53. $x^3y^3 - 1$

54. $125x^3 + 27y^3$

55. $x^6 - y^6$

56. $x^6 + y^6$

3 Use Factoring to Solve Equations

For Problems 57–70, find all real number solutions for each equation.

57. $x^2 - 25 = 0$

58. $x^2 - 1 = 0$

59. $9x^2 - 49 = 0$

60. $4y^2 = 25$

61. $8x^2 - 32 = 0$

62. $3x^2 - 108 = 0$

63. $3x^3 = 3x$

64. $4x^3 = 64x$

65. $20 - 5x^2 = 0$

66. $54 - 6x^2 = 0$

67. $x^4 - 81 = 0$

68. $x^5 - x = 0$

69. $6x^3 + 24x = 0$

70. $4x^3 + 12x = 0$

4 Solve Word Problems That Involve Factoring

For Problems 71–80, set up an equation and solve each of the following problems.

71. The cube of a number equals nine times the same number. Find the number.

72. The cube of a number equals the square of the same number. Find the number.

73. The combined area of two circles is 80π square centimeters. The length of a radius of one circle is twice the length of a radius of the other circle. Find the length of the radius of each circle.

74. The combined area of two squares is 26 square meters. The sides of the larger square are five times as long as the sides of the smaller square. Find the dimensions of each of the squares.

75. A rectangle is twice as long as it is wide, and its area is 50 square meters. Find the length and the width of the rectangle.

76. Suppose that the length of a rectangle is one and one-third times as long as its width. The area of the rectangle is 48 square centimeters. Find the length and width of the rectangle.

77. The total surface area of a right circular cylinder is 54π square inches. If the altitude of the cylinder is twice the length of a radius, find the altitude of the cylinder.

78. The total surface area of a right circular cone is 108π square feet. If the slant height of the cone is twice the length of a radius of the base, find the length of a radius.

79. The sum of the areas of a circle and a square is $(16\pi + 64)$ square yards. If a side of the square is twice the length of a radius of the circle, find the length of a side of the square.

80. The length of an altitude of a triangle is one-third the length of the side to which it is drawn. If the area of the triangle is 6 square centimeters, find the length of that altitude.

THOUGHTS INTO WORDS

81. Explain how you would solve the equation $4x^3 = 64x$.

82. What is wrong with the following factoring process?

$$25x^2 - 100 = (5x + 10)(5x - 10)$$

How would you correct the error?

83. Consider the following solution:

$$6x^2 - 24 = 0$$

$$6(x^2 - 4) = 0$$

$$6(x + 2)(x - 2) = 0$$

$$6 = 0 \quad \text{or} \quad x + 2 = 0 \quad \text{or} \quad x - 2 = 0$$

$$6 = 0 \quad \text{or} \quad x = -2 \quad \text{or} \quad x = 2$$

The solution set is $\{-2, 2\}$.

Is this a correct solution? Would you have any suggestion to offer the person who used this approach?

Answers to the Concept Quiz

1. True **2.** False **3.** True **4.** False **5.** False **6.** False **7.** True **8.** True **9.** False **10.** True

Answers to the Example Practice Skills

1. (a) $(m+6)(m-6)$ **(b)** $(3y+7)(3y-7)$ **(c)** $(8-5b)(8+5b)$ **2.** $(4a-b)(4a+b)(16a^2+b^2)$
3. $(2x+y+3)(2x+y-3)$ **4.** $2(3a-5)(3a+5)$ **5.** $(x+2y)(x^2-2xy+4y^2)$ **6.** $\{-9, 9\}$ **7.** $\left\{-\frac{6}{5}, \frac{6}{5}\right\}$
8. $\left\{-\frac{1}{2}, \frac{1}{2}\right\}$ **9.** $\{-3, 3\}$ **10.** $\{-4, 0, 4\}$ **11.** Side of small square is 5 inches; Side of large square is 10 inches

5.6 Factoring Trinomials

OBJECTIVES

1. Factor Trinomials of the Form $x^2 + bx + c$
2. Factor Trinomials of the Form $ax^2 + bx + c$
3. Factor Perfect-Square Trinomials
4. Summary of Factoring Techniques

1 Factor Trinomials of the Form $x^2 + bx + c$

One of the most common types of factoring used in algebra is expressing a trinomial as the product of two binomials. To develop a factoring technique, we first look at some multiplication ideas. Let's consider the product $(x + a)(x + b)$ and use the distributive property to show how each term of the resulting trinomial is formed.

$$\begin{aligned}(x+a)(x+b) &= x(x+b) + a(x+b)\\ &= x(x) + x(b) + a(x) + a(b)\\ &= x^2 + (a+b)x + ab\end{aligned}$$

Note that the coefficient of the middle term is the sum of a and b and that the last term is the product of a and b. These two relationships can be used to factor trinomials. Let's consider some examples.

EXAMPLE 1

Factor $x^2 + 8x + 12$.

Solution

We need to complete the following with two integers whose sum is 8 and whose product is 12.

$$x^2 + 8x + 12 = (x + \_\_\_\_)(x + \_\_\_\_)$$

The possible pairs of factors of 12 are 1(12), 2(6), and 3(4). Because $6 + 2 = 8$, we can complete the factoring as follows:

$$x^2 + 8x + 12 = (x+6)(x+2)$$

To check our answer, we find the product of $(x + 6)$ and $(x + 2)$.

▼ PRACTICE YOUR SKILL

Factor $y^2 + 11y + 24$. ■

EXAMPLE 2

Factor $x^2 - 10x + 24$.

Solution

We need two integers whose product is 24 and whose sum is -10. Let's use a small table to organize our thinking.

Factors	Product of the factors	Sum of the factors
$(-1)(-24)$	24	-25
$(-2)(-12)$	24	-14
$(-3)(-8)$	24	-11
$(-4)(-6)$	24	-10

The bottom line contains the numbers that we need. Thus

$$x^2 - 10x + 24 = (x - 4)(x - 6)$$

▼ **PRACTICE YOUR SKILL**

Factor $a^2 - 18a + 32$. ■

EXAMPLE 3

Factor $x^2 + 7x - 30$.

Solution

We need two integers whose product is -30 and whose sum is 7.

Factors	Product of the factors	Sum of the factors
$(-1)(30)$	-30	29
$(1)(-30)$	-30	-29
$(2)(-15)$	-30	-13
$(-2)(15)$	-30	13
$(-3)(10)$	-30	7

No need to search any further

The numbers that we need are -3 and 10, and we can complete the factoring.

$$x^2 + 7x - 30 = (x + 10)(x - 3)$$

▼ **PRACTICE YOUR SKILL**

Factor $y^2 - 8y - 20$. ■

EXAMPLE 4

Factor $x^2 + 7x + 16$.

Solution

We need two integers whose product is 16 and whose sum is 7.

Factors	Product of the factors	Sum of the factors
$(1)(16)$	16	17
$(2)(8)$	16	10
$(4)(4)$	16	8

We have exhausted all possible pairs of factors of 16 and no two factors have a sum of 7, so we conclude that $x^2 + 7x + 16$ *is not factorable using integers.*

▼ **PRACTICE YOUR SKILL**

Factor $m^2 + 8m + 24$. ■

The tables in Examples 2, 3, and 4 were used to illustrate one way of organizing your thoughts for such problems. Normally you would probably factor such problems mentally without taking the time to formulate a table. Note, however, that in Example 4 the table helped us to be absolutely sure that we tried all the possibilities. Whether or not you use the table, keep in mind that the key ideas are the product and sum relationships.

EXAMPLE 5

Factor $n^2 - n - 72$.

Solution

Note that the coefficient of the middle term is -1. Hence we are looking for two integers whose product is -72, and because their sum is -1, the absolute value of the negative number must be 1 larger than the positive number. The numbers are -9 and 8, and we can complete the factoring.

$$n^2 - n - 72 = (n - 9)(n + 8)$$

▼ **PRACTICE YOUR SKILL**

Factor $a^2 + a - 30$. ■

EXAMPLE 6

Factor $t^2 + 2t - 168$.

Solution

We need two integers whose product is -168 and whose sum is 2. Because the absolute value of the constant term is rather large, it might help to look at it in prime factored form.

$$168 = 2 \cdot 2 \cdot 2 \cdot 3 \cdot 7$$

Now we can mentally form two numbers by using all of these factors in different combinations. Using two 2s and a 3 in one number and the other 2 and the 7 in the second number produces $2 \cdot 2 \cdot 3 = 12$ and $2 \cdot 7 = 14$. The coefficient of the middle term of the trinomial is 2, so we know that we must use 14 and -12. Thus we obtain

$$t^2 + 2t - 168 = (t + 14)(t - 12)$$

▼ **PRACTICE YOUR SKILL**

Factor $y^2 - 6y - 216$. ■

2 Factor Trinomials of the Form $ax^2 + bx + c$

We have been factoring trinomials of the form $x^2 + bx + c$—that is, trinomials where the coefficient of the squared term is 1. Now let's consider factoring trinomials where the coefficient of the squared term is not 1. First, let's illustrate an informal trial-and-error technique that works quite well for certain types of trinomials. This technique is based on our knowledge of multiplication of binomials.

EXAMPLE 7

Factor $2x^2 + 11x + 5$.

Solution

By looking at the first term, $2x^2$, and the positive signs of the other two terms, we know that the binomials are of the form

$$(x + \_\_\_)(2x + \_\_\_)$$

Because the factors of the last term, 5, are 1 and 5, we have only the following two possibilities to try.

$$(x + 1)(2x + 5) \qquad \text{or} \qquad (x + 5)(2x + 1)$$

By checking the middle term formed in each of these products, we find that the second possibility yields the correct middle term of $11x$. Therefore,

$$2x^2 + 11x + 5 = (x + 5)(2x + 1)$$

▼ PRACTICE YOUR SKILL

Factor $3m^2 + 11m + 10$. ■

EXAMPLE 8

Factor $10x^2 - 17x + 3$.

Solution

First, observe that $10x^2$ can be written as $x \cdot 10x$ or $2x \cdot 5x$. Second, because the middle term of the trinomial is negative and the last term is positive, we know that the binomials are of the form

$$(x - \_\_\_)(10x - \_\_\_) \qquad \text{or} \qquad (2x - \_\_\_)(5x - \_\_\_)$$

The factors of the last term, 3, are 1 and 3, so the following possibilities exist:

$$(x - 1)(10x - 3) \qquad (2x - 1)(5x - 3)$$

$$(x - 3)(10x - 1) \qquad (2x - 3)(5x - 1)$$

By checking the middle term formed in each of these products, we find that the product $(2x - 3)(5x - 1)$ yields the desired middle term of $-17x$. Therefore,

$$10x^2 - 17x + 3 = (2x - 3)(5x - 1)$$

▼ PRACTICE YOUR SKILL

Factor $14a^2 - 37a + 5$. ■

EXAMPLE 9

Factor $4x^2 + 6x + 9$.

Solution

The first term, $4x^2$, and the positive signs of the middle and last terms indicate that the binomials are of the form

$$(x + \_\_\_)(4x + \_\_\_) \qquad \text{or} \qquad (2x + \_\_\_)(2x + \_\_\_)$$

Because the factors of 9 are 1 and 9 or 3 and 3, we have the following five possibilities to try.

$$(x + 1)(4x + 9) \qquad (2x + 1)(2x + 9)$$

$$(x + 9)(4x + 1) \qquad (2x + 3)(2x + 3)$$

$$(x + 3)(4x + 3)$$

When we try all of these possibilities we find that none of them yields a middle term of $6x$. Therefore, $4x^2 + 6x + 9$ is not factorable using integers.

▼ PRACTICE YOUR SKILL

Factor $6b^2 + 10b + 3$. ■

By now it is obvious that factoring trinomials of the form $ax^2 + bx + c$ can be tedious. The key idea is to organize your work so that you consider all possibilities. We suggested one possible format in the previous three examples. As you practice such problems, you may come across a format of your own. Whatever works best for you is the right approach.

There is another, more systematic technique that you may wish to use with some trinomials. It is an extension of the technique we used at the beginning of this section. To see the basis of this technique, let's look at the following product.

$$\begin{aligned}(px + r)(qx + s) &= px(qx) + px(s) + r(qx) + r(s)\\ &= (pq)x^2 + (ps + rq)x + rs\end{aligned}$$

Note that the product of the coefficient of the x^2 term and the constant term is $pqrs$. Likewise, the product of the two coefficients of x, ps and rq, is also $pqrs$. Therefore, when we are factoring the trinomial $(pq)x^2 + (ps + rq)x + rs$, the two coefficients of x must have a sum of $(ps) + (rq)$ and a product of $pqrs$. Let's see how this works in some examples.

EXAMPLE 10

Factor $6x^2 - 11x - 10$

Solution

Step 1 Multiply the coefficient of the x^2 term, 6, and the constant term, -10.

$$(6)(-10) = -60$$

Step 2 Find two integers whose sum is -11 and whose product is -60. It will be helpful to make a listing of the factor pairs for 60.

(1)(60)	(4)(15)
(2)(30)	(5)(12)
(3)(20)	(6)(10)

Because our product from Step 1 is -60, we want a pair of factors for which the absolute value of their difference is 11. These factors are 4 and 15. To make the sum be -11 and the product -60, assign the signs so that we have $+4$ and -15.

Step 3 Rewrite the original problem, expressing the middle term as a sum of terms using the factors found in Step 2 as the coefficients of the terms.

Original problem	Problem rewritten
$6x^2 - 11x - 10$	$6x^2 - 15x + 4x - 10$

Step 4 Now use factoring by grouping to factor the rewritten problem:

$$\begin{aligned}6x^2 - 15x + 4x - 10 &= 3x(2x - 5) + 2(3x - 5)\\ &= (2x - 5)(3x + 2)\end{aligned}$$

Thus $6x^2 - 11x - 10 = (2x - 5)(3x + 2)$.

▼ PRACTICE YOUR SKILL

Factor $4a^2 - 3a - 10$. ■

EXAMPLE 11 Factor $4x^2 - 29x + 30$

Solution

Step 1 Multiply the coefficient of the x^2 term, 4, and the constant term, 30:

$$(4)(30) = 120$$

Step 2 Find two integers whose sum is -29 and whose product is 120. It will be helpful to make a listing of the factor pairs for 120.

(1)(120)	(5)(24)
(2)(60)	(6)(20)
(3)(40)	(8)(15)
(4)(30)	(10)(12)

Because our product from Step 1 is 120, we want a pair of factors for which the absolute value of their sum is 29. These factors are 5 and 24. To make the sum be -29 and the product 120, assign the signs so that we have -5 and -24.

Step 3 Rewrite the original problem, expressing the middle term as a sum of terms using the factors found in Step 2 as the coefficients of the terms.

Original problem	Problem rewritten
$4x^2 - 29x + 30$	$4x^2 - 5x - 24x + 30$

Step 4 Now use factoring by grouping to factor the rewritten problem:

$$\begin{aligned} 4x^2 - 5x - 24x + 30 &= x(4x - 5) - 6(4x - 5) \\ &= (4x - 5)(x - 6) \end{aligned}$$

Thus $4x^2 - 29x + 30 = (4x - 5)(x - 6)$.

▼ **PRACTICE YOUR SKILL**

Factor $12a^2 + a - 6$. ■

The technique presented in Examples 10 and 11 has concrete steps to follow. Examples 7 through 9 were factored by trial-and-error technique. Both of the techniques we used have their strengths and weaknesses. Which technique to use depends on the complexity of the problem and on your personal preference. The more that you work with both techniques, the more comfortable you will feel using them.

3 Factor Perfect-Square Trinomials

Before we summarize our work with factoring techniques, let's look at two more special factoring patterns. In Section 5.3 we used the following two patterns to square binomials.

$$(a + b)^2 = a^2 + 2ab + b^2 \quad \text{and} \quad (a - b)^2 = a^2 - 2ab + b^2$$

These patterns can also be used for factoring purposes.

$$a^2 + 2ab + b^2 = (a + b)^2 \quad \text{and} \quad a^2 - 2ab + b^2 = (a - b)^2$$

The trinomials on the left sides are called **perfect-square trinomials**; they are the result of squaring a binomial. We can always factor perfect-square trinomials using the usual techniques for factoring trinomials. However, they are easily recognized by the nature of their terms. For example, $4x^2 + 12x + 9$ is a perfect-square trinomial because

1. The first term is a perfect square.
2. The last term is a perfect square.
3. The middle term is twice the product of the quantities being squared in the first and last terms.

Likewise, $9x^2 - 30x + 25$ is a perfect-square trinomial because

1. The first term is a perfect square.
2. The last term is a perfect square.
3. The middle term is the negative of twice the product of the quantities being squared in the first and last terms.

Once we know that we have a perfect-square trinomial, the factors follow immediately from the two basic patterns. Thus

$$4x^2 + 12x + 9 = (2x + 3)^2 \qquad 9x^2 - 30x + 25 = (3x - 5)^2$$

The next example illustrates perfect-square trinomials and their factored forms.

EXAMPLE 12

Factor each of the following.

(a) $x^2 + 14x + 49$ **(b)** $n^2 - 16n + 64$ **(c)** $36a^2 + 60ab + 25b^2$
(d) $16x^2 - 8xy + y^2$

Solution

(a) $x^2 + 14x + 49 = (x)^2 + 2(x)(7) + (7) = (x + 7)^2$

(b) $n^2 - 16n + 64 = (n)^2 - 2(n)(8) + (8)^2 = (n - 8)^2$

(c) $36a^2 + 60ab + 25b^2 = (6a)^2 + 2(6a)(5b) + (5b)^2 = (6a + 5b)^2$

(d) $16x^2 - 8xy + y^2 = (4x)^2 - 2(4x)(y) + (y)^2 = (4x - y)^2$

▼ **PRACTICE YOUR SKILL**

Factor each of the following.

(a) $a^2 + 22x + 121$ **(b)** $25x^2 - 60xy + 36y^2$ ■

4 Summary of Factoring Techniques

As we have indicated, factoring is an important algebraic skill. We learned some basic factoring techniques one at a time, but you must be able to apply whichever is (or are) appropriate to the situation. Let's review the techniques and consider examples that demonstrate their use.

1. As a general guideline, always look for a common factor first. The common factor could be a binomial term.

$$3x^2y^3 + 27xy = 3xy(x^2y^2 + 9) \qquad x(y + 2) + 5(y + 2) = (y + 2)(x + 5)$$

2. If the polynomial has two terms, then its pattern could be the difference of the squares or the sum or difference of two cubes.

$$9a^2 - 25 = (3a + 5)(3a - 5) \qquad 8x^3 + 125 = (2x + 5)(4x^2 - 10x + 25)$$

3. If the polynomial has three terms, then the polynomial may factor into the product of two binomials. Examples 10 and 11 presented concrete steps for factoring trinomials. Examples 7 through 9 were factored by trial-and-error. The perfect-square trinomial pattern is a special case of the technique.

$$30n^2 - 31n + 5 = (5n - 1)(6n - 5) \qquad t^4 + 3t^2 + 2 = (t^2 + 2)(t^2 + 1)$$

4. If the polynomial has four or more terms, then factoring by grouping may apply. It may be necessary to rearrange the terms before factoring.

$$ab + ac + 4b + 4c = a(b + c) + 4(b + c) = (b + c)(a + 4)$$

5. If none of the mentioned patterns or techniques work, then the polynomial may not be factorable using integers.

$x^2 + 5x + 12$ Not factorable using integers

CONCEPT QUIZ

For Problems 1–10, answer true or false.

1. To factor $x^2 - 4x - 60$ we look for two numbers whose product is -60 and whose sum is -4.
2. To factor $2x^2 - x - 3$ we look for two numbers whose product is -3 and whose sum is -1.
3. A trinomial of the form $x^2 + bx + c$ will never have a common factor other than 1.
4. A trinomial of the form $ax^2 + bx + c$ will never have a common factor other than 1.
5. The polynomial $x^2 + 25x + 72$ is not factorable using integers.
6. The polynomial $x^2 + 27x + 72$ is not factorable using integers.
7. The polynomial $2x^2 + 5x - 3$ is not factorable using integers.
8. The trinomial $49x^2 - 42x + 9$ is a perfect-square trinomial.
9. The trinomial $25x^2 + 80x - 64$ is a perfect-square trinomial.
10. The completely factored form of $12x^2 - 38x + 30$ is $2(2x - 3)(3x - 5)$.

Problem Set 5.6

1 Factor Trinomials of the Form $x^2 + bx + c$

For Problems 1–30, factor completely each of the polynomials and indicate any that are not factorable using integers.

1. $x^2 + 9x + 20$
2. $x^2 + 11x + 24$
3. $x^2 - 11x + 28$
4. $x^2 - 8x + 12$
5. $a^2 + 5a - 36$
6. $a^2 + 6a - 40$
7. $y^2 + 20y + 84$
8. $y^2 + 21y + 98$
9. $x^2 - 5x - 14$
10. $x^2 - 3x - 54$
11. $x^2 + 9x + 12$
12. $35 - 2x - x^2$
13. $6 + 5x - x^2$
14. $x^2 + 8x - 24$
15. $x^2 + 15xy + 36y^2$
16. $x^2 - 14xy + 40y^2$
17. $a^2 - ab - 56b^2$
18. $a^2 + 2ab - 63b^2$
19. $x^2 + 25x + 150$
20. $x^2 + 21x + 108$
21. $n^2 - 36n + 320$
22. $n^2 - 26n + 168$
23. $t^2 + 3t - 180$
24. $t^2 - 2t - 143$
25. $t^4 - 5t^2 + 6$
26. $t^4 + 10t^2 + 24$
27. $x^4 - 9x^2 + 8$
28 $x^4 - x^2 - 12$
29. $x^4 - 17x^2 + 16$
30. $x^4 - 13x^2 + 36$

2 Factor Trinomials of the Form $ax^2 + bx + c$

For Problems 31–56, factor completely each of the polynomials and indicate any that are not factorable using integers.

31. $15x^2 + 23x + 6$
32. $9x^2 + 30x + 16$
33. $12x^2 - x - 6$
34. $20x^2 - 11x - 3$
35. $4a^2 + 3a - 27$
36. $12a^2 + 4a - 5$
37. $3n^2 - 7n - 20$
38. $4n^2 + 7n - 15$
39. $3x^2 + 10x + 4$
40. $4n^2 - 19n + 21$
41. $10n^2 - 29n - 21$
42. $4x^2 - x + 6$

43. $8x^2 + 26x - 45$

44. $6x^2 + 13x - 33$

45. $6 - 35x - 6x^2$

46. $4 - 4x - 15x^2$

47. $20y^2 + 31y - 9$

48. $8y^2 + 22y - 21$

49. $24n^2 - 2n - 5$

50. $3n^2 - 16n - 35$

51. $5n^2 + 33n + 18$

52. $7n^2 + 31n + 12$

53. $10x^4 + 3x^2 - 4$

54. $3x^4 + 7x^2 - 6$

55. $18n^4 + 25n^2 - 3$

56. $4n^4 + 3n^2 - 27$

3 Factor Perfect-Square Trinomials

For Problems 57–62, factor completely each of the polynomials.

57. $y^2 - 16y + 64$

58. $a^2 + 30a + 225$

59. $4x^2 + 12xy + 9y^2$

60. $25x^2 - 60xy + 36y^2$

61. $8y^2 - 8y + 2$

62. $12x^2 + 36x + 27$

4 Summary of Factoring Techniques

Problems 63–100 should help you pull together all of the factoring techniques of this chapter. Factor completely each polynomial, and indicate any that are not factorable using integers.

63. $2t^2 - 8$

64. $14w^2 - 29w - 15$

65. $12x^2 + 7xy - 10y^2$

66. $8x^2 + 2xy - y^2$

67. $18n^3 + 39n^2 - 15n$

68. $n^2 + 18n + 77$

69. $n^2 - 17n + 60$

70. $(x + 5)^2 - y^2$

71. $36a^2 - 12a + 1$

72. $2n^2 - n - 5$

73. $6x^2 + 54$

74. $x^5 - x$

75. $3x^2 + x - 5$

76. $5x^2 + 42x - 27$

77. $x^2 - (y - 7)^2$

78. $2n^3 + 6n^2 + 10n$

79. $1 - 16x^4$

80. $9a^2 - 30a + 25$

81. $4n^2 + 25n + 36$

82. $x^3 - 9x$

83. $n^3 - 49n$

84. $4x^2 + 16$

85. $x^2 - 7x - 8$

86. $x^2 + 3x - 54$

87. $3x^4 - 81x$

88. $x^3 + 125$

89. $x^4 + 6x^2 + 9$

90. $18x^2 - 12x + 2$

91. $x^4 - 5x^2 - 36$

92. $6x^4 - 5x^2 - 21$

93. $6w^2 - 11w - 35$

94. $10x^3 + 15x^2 + 20x$

95. $25n^2 + 64$

96. $4x^2 - 37x + 40$

97. $2n^3 + 14n^2 - 20n$

98. $25t^2 - 100$

99. $2xy + 6x + y + 3$

100. $3xy + 15x - 2y - 10$

THOUGHTS INTO WORDS

101. How can you determine that $x^2 + 5x + 12$ is not factorable using integers?

102. Explain your thought process when factoring $30x^2 + 13x - 56$.

103. Consider the following approach to factoring $12x^2 + 54x + 60$:

$$\begin{aligned} 12x^2 + 54x + 60 &= (3x + 6)(4x + 10) \\ &= 3(x + 2)(2)(2x + 5) \\ &= 6(x + 2)(2x + 5) \end{aligned}$$

Is this a correct factoring process? Do you have any suggestion for the person using this approach?

FURTHER INVESTIGATIONS

For Problems 104–109, factor each trinomial and assume that all variables that appear as exponents represent positive integers.

104. $x^{2a} + 2x^a - 24$

105. $x^{2a} + 10x^a + 21$

106. $6x^{2a} - 7x^a + 2$

107. $4x^{2a} + 20x^a + 25$

108. $12x^{2n} + 7x^n - 12$

109. $20x^{2n} + 21x^n - 5$

Consider the following approach to factoring $(x - 2)^2 + 3(x - 2) - 10$:

$$\begin{aligned} &(x - 2)^2 + 3(x - 2) - 10 \\ &= y^2 + 3y - 10 \\ &= (y + 5)(y - 2) \\ &= (x - 2 + 5)(x - 2 - 2) \\ &= (x + 3)(x - 4) \end{aligned}$$

Use this approach to factor Problems 110–115.

110. $(x - 3)^2 + 10(x - 3) + 24$

111. $(x + 1)^2 - 8(x + 1) + 15$

112. $(2x + 1)^2 + 3(2x + 1) - 28$

113. $(3x - 2)^2 - 5(3x - 2) - 36$

114. $6(x - 4)^2 + 7(x - 4) - 3$

115. $15(x + 2)^2 - 13(x + 2) + 2$

Answers to the Concept Quiz

1. True **2.** False **3.** True **4.** False **5.** True **6.** False **7.** False **8.** True **9.** False **10.** True

Answers to the Example Practice Skills

1. $(y + 3)(y + 8)$ **2.** $(a - 2)(a - 16)$ **3.** $(y - 10)(y + 2)$ **4.** Not factorable **5.** $(a + 6)(a - 5)$ **6.** $(y - 18)(y + 12)$ **7.** $(3m + 5)(m + 2)$ **8.** $(2a - 5)(7a - 1)$ **9.** Not factorable **10.** $(4a + 5)(a - 2)$ **11.** $(4a + 3)(3a - 2)$ **12. (a)** $(a + 11)^2$ **(b)** $(5x - 6y)^2$

5.7 Equations and Problem Solving

OBJECTIVES

1 Solve Equations

2 Solve Word Problems

1 Solve Equations

The techniques for factoring trinomials that were presented in the previous section provide us with more power to solve equations. That is, the property "$ab = 0$ if and only if $a = 0$ or $b = 0$" continues to play an important role as we solve equations that contain factorable trinomials. Let's consider some examples.

EXAMPLE 1

Solve $x^2 - 11x - 12 = 0$.

Solution

$$x^2 - 11x - 12 = 0$$

$$(x - 12)(x + 1) = 0$$

$$x - 12 = 0 \quad \text{or} \quad x + 1 = 0$$

$$x = 12 \quad \text{or} \quad x = -1$$

The solution set is $\{-1, 12\}$.

▼ **PRACTICE YOUR SKILL**

Solve $y^2 - 5y - 24 = 0$. ■

EXAMPLE 2 Solve $20x^2 + 7x - 3 = 0$.

Solution

$$20x^2 + 7x - 3 = 0$$
$$(4x - 1)(5x + 3) = 0$$
$$4x - 1 = 0 \quad \text{or} \quad 5x + 3 = 0$$
$$4x = 1 \quad \text{or} \quad 5x = -3$$
$$x = \frac{1}{4} \quad \text{or} \quad x = -\frac{3}{5}$$

The solution set is $\left\{-\frac{3}{5}, \frac{1}{4}\right\}$.

▼ PRACTICE YOUR SKILL

Solve $6a^2 + a - 5 = 0$. ■

EXAMPLE 3 Solve $-2n^2 - 10n + 12 = 0$.

Solution

$$-2n^2 - 10n + 12 = 0$$
$$-2(n^2 + 5n - 6) = 0$$
$$n^2 + 5n - 6 = 0 \qquad \text{Multiply both sides by } -\frac{1}{2}$$
$$(n + 6)(n - 1) = 0$$
$$n + 6 = 0 \quad \text{or} \quad n - 1 = 0$$
$$n = -6 \quad \text{or} \quad n = 1$$

The solution set is $\{-6, 1\}$.

▼ PRACTICE YOUR SKILL

Solve $3m^2 - 6m - 24 = 0$. ■

EXAMPLE 4 Solve $16x^2 - 56x + 49 = 0$.

Solution

$$16x^2 - 56x + 49 = 0$$
$$(4x - 7)^2 = 0$$
$$(4x - 7)(4x - 7) = 0$$
$$4x - 7 = 0 \quad \text{or} \quad 4x - 7 = 0$$
$$4x = 7 \quad \text{or} \quad 4x = 7$$
$$x = \frac{7}{4} \quad \text{or} \quad x = \frac{7}{4}$$

The only solution is $\frac{7}{4}$; thus the solution set is $\left\{\frac{7}{4}\right\}$.

▼ PRACTICE YOUR SKILL

Solve $9y^2 + 48y + 64 = 0$. ■

EXAMPLE 5

Solve $9a(a + 1) = 4$.

Solution

$$9a(a + 1) = 4$$
$$9a^2 + 9a = 4$$
$$9a^2 + 9a - 4 = 0$$
$$(3a + 4)(3a - 1) = 0$$
$$3a + 4 = 0 \quad \text{or} \quad 3a - 1 = 0$$
$$3a = -4 \quad \text{or} \quad 3a = 1$$
$$a = -\frac{4}{3} \quad \text{or} \quad a = \frac{1}{3}$$

The solution set is $\left\{-\frac{4}{3}, \frac{1}{3}\right\}$.

▼ PRACTICE YOUR SKILL

Solve $x(2x - 1) = 10$. ■

EXAMPLE 6

Solve $(x - 1)(x + 9) = 11$.

Solution

$$(x - 1)(x + 9) = 11$$
$$x^2 + 8x - 9 = 11$$
$$x^2 + 8x - 20 = 0$$
$$(x + 10)(x - 2) = 0$$
$$x + 10 = 0 \quad \text{or} \quad x - 2 = 0$$
$$x = -10 \quad \text{or} \quad x = 2$$

The solution set is $\{-10, 2\}$.

▼ PRACTICE YOUR SKILL

Solve $(x + 7)(x - 5) = 13$. ■

2 Solve Word Problems

As you might expect, the increase in our power to solve equations broadens our base for solving problems. Now we are ready to tackle some problems using equations of the types presented in this section.

EXAMPLE 7

Image Source/Jupiter Images

Apply Your Skill

A room contains 78 chairs. The number of chairs per row is one more than twice the number of rows. Find the number of rows and the number of chairs per row.

Solution

Let r represent the number of rows. Then $2r + 1$ represents the number of chairs per row.

$$r(2r + 1) = 78 \quad \text{The number of rows times the number of chairs per row yields the total number of chairs}$$

$$2r^2 + r = 78$$

$$2r^2 + r - 78 = 0$$

$$(2r + 13)(r - 6) = 0$$

$$2r + 13 = 0 \quad \text{or} \quad r - 6 = 0$$

$$2r = -13 \quad \text{or} \quad r = 6$$

$$r = -\frac{13}{2} \quad \text{or} \quad r = 6$$

The solution $-\frac{13}{2}$ must be disregarded, so there are 6 rows and $2r + 1$ or $2(6) + 1 = 13$ chairs per row.

▼ PRACTICE YOUR SKILL

A cryptographer needs to arrange 60 numbers in a rectangular array where the number of columns is 2 more than twice the number of rows. Find the number of rows and the number of columns. ■

EXAMPLE 8 Apply Your Skill

A strip of uniform width cut from both sides and both ends of an 8-inch by 11-inch sheet of paper reduces the size of the paper to an area of 40 square inches. Find the width of the strip.

Solution

Let x represent the width of the strip, as indicated in Figure 5.19.

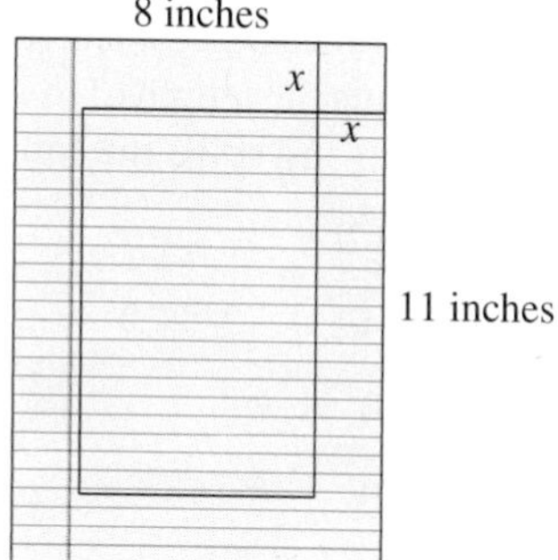

Figure 5.19

The length of the paper after the strips of width x are cut from both ends and both sides will be $11 - 2x$, and the width of the newly formed rectangle will be $8 - 2x$. Because the area ($A = lw$) is to be 40 square inches, we can set up and solve the following equation.

$$(11 - 2x)(8 - 2x) = 40$$

$$88 - 38x + 4x^2 = 40$$

$$4x^2 - 38x + 48 = 0$$

$$2x^2 - 19x + 24 = 0$$
$$(2x - 3)(x - 8) = 0$$
$$2x - 3 = 0 \quad \text{or} \quad x - 8 = 0$$
$$2x = 3 \quad \text{or} \quad x = 8$$
$$x = \frac{3}{2} \quad \text{or} \quad x = 8$$

The solution of 8 must be discarded because the width of the original sheet is only 8 inches. Therefore, the strip to be cut from all four sides must be $1\frac{1}{2}$ inches wide. (Check this answer!)

▼ **PRACTICE YOUR SKILL**

A rectangular digital image that is 5 inches by 7 inches needs to have a uniform amount cropped from both ends and both sides to reduce the area to 15 square inches. Find the width of the amount to be cropped. ■

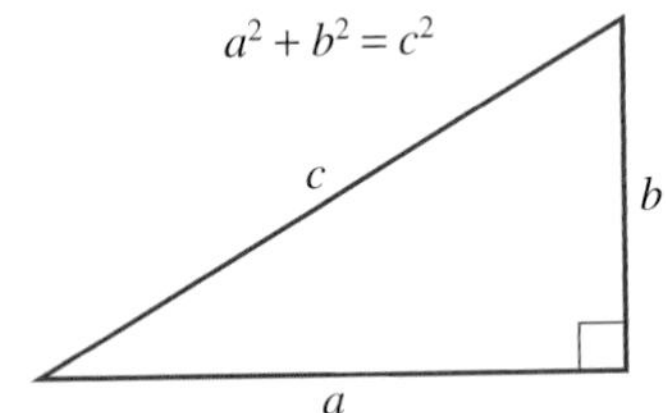

Figure 5.20

The Pythagorean theorem, an important theorem pertaining to right triangles, can sometimes serve as a guideline for solving problems that deal with right triangles (see Figure 5.20). The Pythagorean theorem states that "in any right triangle, the square of the longest side (called the hypotenuse) is equal to the sum of the squares of the other two sides (called legs)." Let's use this relationship to help solve a problem.

EXAMPLE 9

Apply Your Skill

One leg of a right triangle is 2 centimeters more than twice as long as the other leg. The hypotenuse is 1 centimeter longer than the longer of the two legs. Find the lengths of the three sides of the right triangle.

Solution

Let l represent the length of the shortest leg. Then $2l + 2$ represents the length of the other leg and $2l + 3$ represents the length of the hypotenuse. Use the Pythagorean theorem as a guideline to set up and solve the following equation.

$$l^2 + (2l + 2)^2 = (2l + 3)^2$$
$$l^2 + 4l^2 + 8l + 4 = 4l^2 + 12l + 9$$
$$l^2 - 4l - 5 = 0$$
$$(l - 5)(l + 1) = 0$$
$$l - 5 = 0 \quad \text{or} \quad l + 1 = 0$$
$$l = 5 \quad \text{or} \quad l = -1$$

The negative solution must be discarded, so the length of one leg is 5 centimeters, the other leg is $2(5) + 2 = 12$ centimeters long, and the hypotenuse is $2(5) + 3 = 13$ centimeters long.

▼ **PRACTICE YOUR SKILL**

One leg of a right triangle is 1 inch more than the other leg. The hypotenuse is 2 inches more than the shorter of the two legs. Find the lengths of all three sides. ■

CONCEPT QUIZ

For Problems 1–5, answer true or false.

1. If $xy = 0$, then $x = 0$ or $y = 0$.
2. If the product of three numbers is zero, then at least one of the numbers must be zero.
3. The Pythagorean theorem is true for all triangles.
4. The longest side of a right triangle is called the hypotenuse.
5. If we know the length of any two sides of a right triangle, then the third side can be determined by using the Pythagorean theorem.

Problem Set 5.7

1 Solve Equations

For Problems 1–54, solve each equation. You will need to use the factoring techniques that we discussed throughout this chapter.

1. $x^2 + 4x + 3 = 0$
2. $x^2 + 7x + 10 = 0$
3. $x^2 + 18x + 72 = 0$
4. $n^2 + 20n + 91 = 0$
5. $n^2 - 13n + 36 = 0$
6. $n^2 - 10n + 16 = 0$
7. $x^2 + 4x - 12 = 0$
8. $x^2 + 7x - 30 = 0$
9. $w^2 - 4w = 5$
10. $s^2 - 4s = 21$
11. $n^2 + 25n + 156 = 0$
12. $n(n - 24) = -128$
13. $3t^2 + 14t - 5 = 0$
14. $4t^2 - 19t - 30 = 0$
15. $6x^2 + 25x + 14 = 0$
16. $25x^2 + 30x + 8 = 0$
17. $3t(t - 4) = 0$
18. $1 - x^2 = 0$
19. $-6n^2 + 13n - 2 = 0$
20. $(x + 1)^2 - 4 = 0$
21. $2n^3 = 72n$
22. $a(a - 1) = 2$
23. $(x - 5)(x + 3) = 9$
24. $3w^3 - 24w^2 + 36w = 0$
25. $16 - x^2 = 0$
26. $16t^2 - 72t + 81 = 0$
27. $n^2 + 7n - 44 = 0$
28. $2x^3 = 50x$
29. $3x^2 = 75$
30. $x^2 + x - 2 = 0$
31. $15x^2 + 34x + 15 = 0$
32. $20x^2 + 41x + 20 = 0$
33. $8n^2 - 47n - 6 = 0$
34. $7x^2 + 62x - 9 = 0$
35. $28n^2 - 47n + 15 = 0$
36. $24n^2 - 38n + 15 = 0$
37. $35n^2 - 18n - 8 = 0$
38. $8n^2 - 6n - 5 = 0$
39. $-3x^2 - 19x + 14 = 0$
40. $5x^2 = 43x - 24$
41. $n(n + 2) = 360$
42. $n(n + 1) = 182$
43. $9x^4 - 37x^2 + 4 = 0$
44. $4x^4 - 13x^2 + 9 = 0$
45. $3x^2 - 46x - 32 = 0$
46. $x^4 - 9x^2 = 0$
47. $2x^2 + x - 3 = 0$
48. $x^3 + 5x^2 - 36x = 0$
49. $12x^3 + 46x^2 + 40x = 0$
50. $5x(3x - 2) = 0$
51. $(3x - 1)^2 - 16 = 0$
52. $(x + 8)(x - 6) = -24$
53. $4a(a + 1) = 3$
54. $-18n^2 - 15n + 7 = 0$

2 Solve Word Problems

For Problems 55–70, set up an equation and solve each problem.

55. Find two consecutive integers whose product is 72.

56. Find two consecutive even whole numbers whose product is 224.

57. Find two integers whose product is 105 such that one of the integers is one more than twice the other integer.

58. Find two integers whose product is 104 such that one of the integers is three less than twice the other integer.

59. The perimeter of a rectangle is 32 inches, and the area is 60 square inches. Find the length and width of the rectangle.

60. Suppose that the length of a certain rectangle is two centimeters more than three times its width. If the area of the rectangle is 56 square centimeters, find its length and width.

61. The sum of the squares of two consecutive integers is 85. Find the integers.

62. The sum of the areas of two circles is 65π square feet. The length of a radius of the larger circle is 1 foot less than twice the length of a radius of the smaller circle. Find the length of a radius of each circle.

63. The combined area of a square and a rectangle is 64 square centimeters. The width of the rectangle is 2 centimeters more than the length of a side of the square, and the length of the rectangle is 2 centimeters more than its width. Find the dimensions of the square and the rectangle.

64. The Ortegas have an apple orchard that contains 90 trees. The number of trees in each row is 3 more than twice the number of rows. Find the number of rows and the number of trees per row.

65. The lengths of the three sides of a right triangle are represented by consecutive whole numbers. Find the lengths of the three sides.

66. The area of the floor of the rectangular room shown in Figure 5.21 is 175 square feet. The length of the room is $1\frac{1}{2}$ feet longer than the width. Find the length of the room.

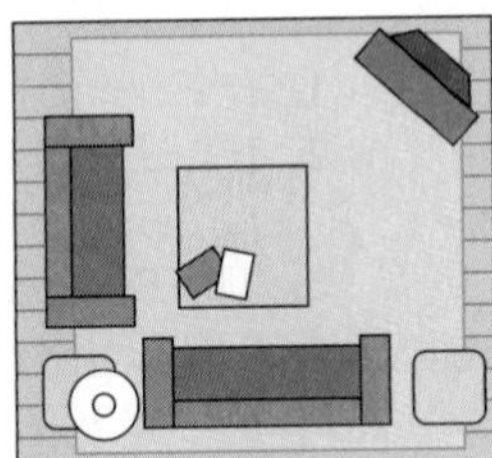

Figure 5.21

67. Suppose that the length of one leg of a right triangle is 3 inches more than the length of the other leg. If the length of the hypotenuse is 15 inches, find the lengths of the two legs.

68. The lengths of the three sides of a right triangle are represented by consecutive even whole numbers. Find the lengths of the three sides.

69. The area of a triangular sheet of paper is 28 square inches. One side of the triangle is 2 inches more than three times the length of the altitude to that side. Find the length of that side and the altitude to the side.

70. A strip of uniform width is shaded along both sides and both ends of a rectangular poster that measures 12 inches by 16 inches (see Figure 5.22). How wide is the shaded strip if one-half of the poster is shaded?

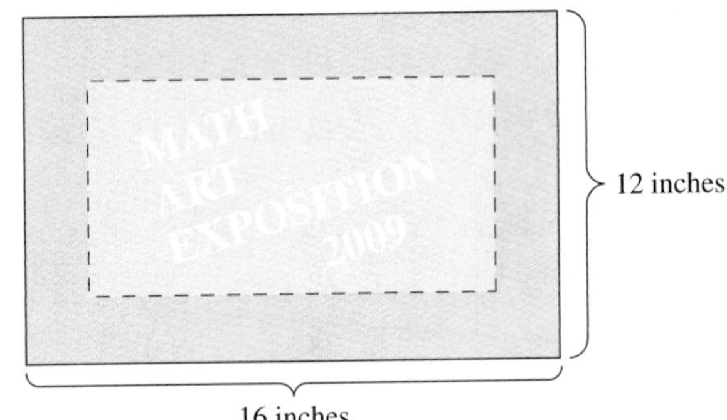

Figure 5.22

THOUGHTS INTO WORDS

71. Discuss the role that factoring plays in solving equations.

72. Explain how you would solve the equation $(x + 6)(x - 4) = 0$ and also how you would solve $(x + 6)(x - 4) = -16$.

73. Explain how you would solve the equation $3(x - 1)(x + 2) = 0$ and also how you would solve the equation $x(x - 1)(x + 2) = 0$.

74. Consider the following two solutions for the equation $(x + 3)(x - 4) = (x + 3)(2x - 1)$.

Solution A

$$(x + 3)(x - 4) = (x + 3)(2x - 1)$$
$$(x + 3)(x - 4) - (x + 3)(2x - 1) = 0$$
$$(x + 3)[x - 4 - (2x - 1)] = 0$$
$$(x + 3)(x - 4 - 2x + 1) = 0$$
$$(x + 3)(-x - 3) = 0$$
$$x + 3 = 0 \quad \text{or} \quad -x - 3 = 0$$
$$x = -3 \quad \text{or} \quad -x = 3$$
$$x = -3 \quad \text{or} \quad x = -3$$

The solution set is $\{-3\}$.

Solution B

$$(x + 3)(x - 4) = (x + 3)(2x - 1)$$
$$x^2 - x - 12 = 2x^2 + 5x - 3$$
$$0 = x^2 + 6x + 9$$
$$0 = (x + 3)^2$$
$$x + 3 = 0$$
$$x = -3$$

The solution set is $\{-3\}$.

Are both approaches correct? Which approach would you use, and why?

Answers to the Concept Quiz

1. True **2.** True **3.** False **4.** True **5.** True

Answers to the Example Practice Skills

1. $\{-3, 8\}$ **2.** $\left\{-1, \frac{5}{6}\right\}$ **3.** $\{-2, 4\}$ **4.** $\left\{-\frac{8}{3}\right\}$ **5.** $\left\{-2, \frac{5}{2}\right\}$ **6.** $\{-8, 6\}$ **7.** 5 rows and 12 columns **8.** 1 in. **9.** 3 in., 4 in., and 5 in.

Chapter 5 Summary

OBJECTIVE	SUMMARY	EXAMPLE	CHAPTER REVIEW PROBLEMS
Find the degree of a polynomial. (Sec. 5.1, Obj. 1, p. 224)	The degree of a monomial is the sum of the exponents of the literal factors. The degree of a polynomial is the degree of the term with the highest degree in the polynomial.	Find the degree of the given polynomial. $6x^4 - 7x^3 + 8x^2 + 2x - 10$ **Solution** The degree of the polynomial is 4, because the term with the highest degree, $6x^4$, has degree of 4.	Problems 1–4
Add, subtract, and simplify polynomial expressions. (Sec. 5.1, Obj. 2, p. 225; Sec. 5.1, Obj. 3, p. 226; Sec. 5.1, Obj. 4, p. 227)	Similar (or like) terms have the same literal factors. The commutative, associative, and distributive properties provide the basis for rearranging, regrouping, and combining similar terms.	Perform the indicated operations. $4x-[9x^2 - 2(7x - 3x^2)]$ **Solution** $4x-[9x^2 - 2(7x - 3x^2)]$ $= 4x - [9x^2 - 14x + 6x^2]$ $= 4x - [15x^2 - 14x]$ $= 4x - 15x^2 + 14x$ $= -15x^2 + 18x$	Problems 5–10
Multiply monomials and raising a monomial to an exponent. (Sec. 5.2, Obj. 1, p. 231; Sec. 5.2, Obj. 2, p. 233)	The following properties provide the basis for multiplying monomials. 1. $b^n \cdot b^m = b^{n+m}$ 2. $(b^n)^m = b^{mn}$ 3. $(ab)^n = a^n b^n$	Simplify each of the following. **(a)** $(-5a^4b)(2a^2b^3)$ **(b)** $(-3x^3y)^2$ **Solution** **(a)** $(-5a^4b)(2a^2b^3) = -10a^6b^4$ **(b)** $(-3x^3y)^2 = (-3)^2(x^3)^2(y)^2$ $= 9x^6y^2$	Problems 11–18
Divide monomials. (Sec. 5.2, Obj. 3, p. 235)	The following properties provide the basis for dividing monomials. 1. $\dfrac{b^n}{b^m} = b^{n-m}$ if $n > m$ 2. $\dfrac{b^n}{b^m} = 1$ if $n = m$	Find the quotient. $\dfrac{8x^5y^2}{-8xy^4}$ **Solution** $\dfrac{8x^5y^2}{-8xy^4} = -\dfrac{x^4}{y^2}$	Problems 19–22
Multiply polynomials. (Sec. 5.3, Obj. 1, p. 238)	To multiply two polynomials, every term of the first polynomial is multiplied by each term of the second polynomial. Multiplying polynomials often produces similar terms that can be combined to simplify the resulting polynomial.	Find the indicated product. $(3x + 4)(x^2 + 6x - 5)$ **Solution** $(3x + 4)(x^2 + 6x - 5)$ $= 3x(x^2 + 6x - 5)$ $+ 4(x^2 + 6x - 5)$ $= 3x^3 + 18x^2 - 15x + 4x^2$ $+ 24x - 20$ $= 3x^3 + 22x^2 + 9x - 20$	Problems 23–28 *(continued)*

OBJECTIVE	SUMMARY	EXAMPLE	CHAPTER REVIEW PROBLEMS
Multiply two binomials using a shortcut pattern. (Sec. 5.3, Obj. 2, p. 240)	A three-step shortcut pattern, often referred to as FOIL, is used to find the product of two binomials.	Find the indicated product. $(3x + 5)(x - 4)$ **Solution** $(3x + 5)(x - 4)$ $= 3x^2 + (-12x + 5x) - 20$ $= 3x^2 - 7x - 20$	Problems 29–32
Find the square of a binomial using a shortcut pattern. (Sec. 5.3, Obj. 3, p. 241)	The patterns for squaring a binomial are $(a + b)^2 = a^2 + 2ab + b^2$ and $(a - b)^2 = a^2 - 2ab + b^2$.	Expand $(4x - 3)^2$. **Solution** $(4x - 3)^2 = (4x)^2 + 2(4x)(-3) + (-3)^2$ $= 16x^2 - 24x + 9$	Problems 33–36
Use a pattern to find the product of $(a+ b)(a - b)$. (Sec. 5.3, Obj. 4, p. 242)	The pattern is $(a + b)(a - b) = a^2 - b^2$.	Find the product. $(x - 3y)(x + 3y)$ **Solution** $(x - 3y)(x + 3y) = (x)^2 - (3y)^2$ $= x^2 - 9y^2$	Problems 37–38
Find the cube of a binomial. (Sec. 5.3, Obj. 5, p. 243)	The patterns for cubing a binomial are $(a + b)^3 = a^3 + 3a^2b + 3ab^2 + b^3$ and $(a - b)^3 = a^3 - 3a^2b + 3ab^2 - b^3$.	Expand $(2a + 5)^3$. **Solution** $(2a + 5)^3$ $= (2a)^3 + 3(2a)^2(5) + 3(2a)(5)^2 + (5)^3$ $= 8a^3 + 60a^2 + 150a + 125$	Problems 39–40
Use polynomials in geometry problems. (Sec. 5.1, Obj. 5, p. 228; Sec. 5.2, Obj. 4, p. 237; Sec. 5.3, Obj. 6, p. 244)	Sometimes polynomials are encountered in a geometric setting. A polynomial may be used to represent area or volume.	A rectangular piece of cardboard is 20 inches long and 10 inches wide. From each corner a square piece x inches on a side is cut out. The flaps are turned up to form an open box. Find a polynomial that represents the volume. **Solution** The length of the box will be $20 - 2x$, the width of the box will be $10 - 2x$, and the height will be x, so $V = (20 - 2x)(10 - 2x)(x)$. Simplifying the polynomial gives $V = x^3 - 30x^2 + 200x$.	Problems 41–42 *(continued)*

OBJECTIVE	SUMMARY	EXAMPLE	CHAPTER REVIEW PROBLEMS
Understand the rules about completely factored form. (Sec. 5.4, Obj. 3, p. 248)	A polynomial with integral coefficients is completely factored if: 1. it is expressed as a product of polynomials with integral coefficients; and 2. no polynomial, other than a monomial, within the factored form can be further factored into polynomials with integral coefficients.	Which of the following is the completely factored form of $2x^3y + 6x^2y^2$? **(a)** $2x^3y + 6x^2y^2 = x^2y(2x + 6y)$ **(b)** $2x^3y + 6x^2y^2 = 6x^2y\left(\frac{1}{3}x + y\right)$ **(c)** $2x^3y + 6x^2y^2 = 2x^2y(x + 6y)$ **(d)** $2x^3y + 6x^2y^2 = 2xy(x^2 + 6xy)$ **Solution** Only (c) is completely factored. For parts (a) and (c), the polynomial inside the parentheses can be factored further. For part (b), the coefficients are not integers.	
Factor out the highest common monomial factor. (Sec. 5.4, Obj. 4, p. 250)	The distributive property in the form $ab + ac = a(b + c)$ is the basis for factoring out the highest common monomial factor.	Factor $-4x^3y^4 - 2x^4y^3 - 6x^5y^2$. **Solution** $-4x^3y^4 - 2x^4y^3 - 6x^5y^2 = -2x^3y^2(2y^2 + xy + 3x^2)$	Problems 43–44
Factor out a common binomial factor. (Sec. 5.4, Obj. 5, p. 250)	The common factor can be a binomial factor.	Factor $y(x - 4) + 6(x - 4)$. **Solution** $y(x - 4) + 6(x - 4) = (x - 4)(y + 6)$	Problems 45–46
Factor by grouping. (Sec. 5.4, Obj. 6, p. 251)	It may be that the polynomial exhibits no common monomial or binomial factor. However, after factoring common factors from groups of terms, a common factor may be evident.	Factor $2xz + 6x + yz + 3y$. **Solution** $2xz + 6x + yz + 3y = 2x(z + 3) + y(z + 3) = (z + 3)(2x + y)$	Problems 47–48
Factor the difference of two squares. (Sec. 5.5, Obj. 1, p. 257)	The factoring pattern $a^2 - b^2 = (a + b)(a - b)$ is called the difference of two squares.	Factor $36a^2 - 25b^2$. **Solution** $36a^2 - 25b^2 = (6a - 5b)(6a + 5b)$	Problems 49–50
Factor the sum or difference of two cubes. (Sec. 5.5, Obj. 2, p. 260)	The factoring patterns $a^3 + b^3 = (a + b)(a^2 - ab + b^2)$ and $a^3 - b^3 = (a - b)(a^2 + ab + b^2)$ are called the sum of two cubes and the difference of two cubes, respectively.	Factor $8x^3 + 27y^3$. **Solution** $8x^3 + 27y^3 = (2x + 3y)(4x^2 - 6xy + 9y^2)$	Problems 51–52 *(continued)*

OBJECTIVE	SUMMARY	EXAMPLE	CHAPTER REVIEW PROBLEMS
Factor trinomials of the form $x^2 + bx + c$. (Sec. 5.6, Obj. 1, p. 265)	Expressing a trinomial (for which the coefficient of the squared term is 1) as a product of two binomials is based on the relationship $(x + a)(x + b) = x^2 + (a + b)x + ab$. The coefficient of the middle term is the sum of a and b, and the last term is the product of a and b.	Factor $x^2 - 2x - 35$. **Solution** $x^2 - 2x - 35 = (x - 7)(x + 5)$	Problems 53–56
Factor trinomials of the form $ax^2 + bx + c$. (Sec. 5.6, Obj. 2, p. 267)	Two methods were presented for factoring trinomials of the form $ax^2 + bx + c$. One technique is to try the various possibilities of factors and check by multiplying. This method is referred to as trial-and-error. The other method is a structured technique that is shown in Examples 10 and 11 of Section 5.6.	Factor $4x^2 + 16x + 15$. **Solution** Multiply 4 times 15 to get 60. The factors of 60 that add to 16 are 6 and 10. Rewrite the problem and factor by grouping: $4x^2 + 16x + 15$ $= 4x^2 + 10x + 6x + 15$ $= 2x(2x + 5) + 3(2x + 5)$ $= (2x + 5)(2x + 3)$	Problems 57–60
Factor perfect-square trinomials. (Sec. 5.6, Obj. 3, p. 270)	A perfect-square trinomial is the result of squaring a binomial. There are two basic perfect-square trinomial factoring patterns, $a^2 + 2ab + b^2 = (a + b)^2$ and $a^2 - 2ab + b^2 = (a - b)^2$.	Factor $16x^2 + 40x + 25$. **Solution** $16x^2 + 40x + 25 = (4x + 5)^2$	Problems 61–62
Summary of factoring techniques. (Sec. 5.6, Obj. 4, p. 271)	1. As a general guideline, always look for a common factor first. The common factor could be a binomial term. 2. If the polynomial has two terms, then its pattern could be the difference of squares or the sum or difference of two cubes. 3. If the polynomial has three terms, then the polynomial may factor into the product of two binomials. 4. If the polynomial has four or more terms, then factoring by grouping may apply. It may be necessary to rearrange the terms before factoring. 5. If none of the mentioned patterns or techniques work, then the polynomial may not be factorable using integers.	Factor $18x^2 - 50$. **Solution** First factor out a common factor of 2: $18x^2 - 50 = 2(9x^2 - 25)$ Now factor the difference of squares: $18x^2 - 50$ $= 2(9x^2 - 25)$ $= 2(3x - 5)(3x + 5)$	Problems 63–84 *(continued)*

OBJECTIVE	SUMMARY	EXAMPLE	CHAPTER REVIEW PROBLEMS
Solve equations. (Sec. 5.4, Obj. 7, p. 252; Sec. 5.5, Obj. 3, p. 260; Sec. 5.7, Obj. 1, p. 274)	The factoring techniques in this chapter, along with the property $ab = 0$, provide the basis for some additional equation-solving skills.	Solve $x^2 - 11x + 28 = 0$. **Solution** $x^2 - 11x + 28 = 0$ $(x - 7)(x - 4) = 0$ $x - 7 = 0$ or $x - 4 = 0$ $x = 7$ or $x = 4$ The solution set is $\{4, 7\}$.	Problems 85–104
Solve word problems. (Sec. 5.4, Obj. 8, p. 253; Sec. 5.5, Obj. 4, p. 262; Sec. 5.7, Obj. 2, p. 276)	The ability to solve more types of equations increased our capabilities to solve word problems.	Suppose that the area of a square is numerically equal to three times its perimeter. Find the length of a side of the square. **Solution** Let x represent the length of a side of the square. The area is x^2 and the perimeter is $4x$. Because the area is numerically equal to three times the perimeter, we have the equation $x^2 = 3(4x)$. By solving this equation, we can determine that the length of a side of the square is 12 units.	Problems 105–114

Chapter 5 Review Problem Set

For Problems 1–4, find the degree of the polynomial.

1. $-2x^3 + 4x^2 - 8x + 10$
2. $x^4 + 11x^2 - 15$
3. $5x^3y + 4x^4y^2 - 3x^3y^2$
4. $5xy^3 + 2x^2y^2 - 3x^3y^2$

For Problems 5–40, perform the indicated operations and then simplify.

5. $(3x - 2) + (4x - 6) + (-2x + 5)$
6. $(8x^2 + 9x - 3) - (5x^2 - 3x - 1)$
7. $(6x^2 - 2x - 1) + (4x^2 + 2x + 5) - (-2x^2 + x - 1)$
8. $(-3x^2 - 4x + 8) + (5x^2 + 7x + 2) - (-9x^2 + x + 6)$
9. $[3x - (2x - 3y + 1)] - [2y - (x - 1)]$
10. $[8x - (5x - y + 3)] - [-4y - (2x + 1)]$
11. $(-5x^2y^3)(4x^3y^4)$
12. $(-2a^2)(3ab^2)(a^2b^3)$
13. $\left(\frac{1}{2}ab\right)(8a^3b^2)(-2a^3)$
14. $\left(\frac{3}{4}x^2y^3\right)(12x^3y^2)(3y^3)$
15. $(4x^2y^3)^4$
16. $(-2x^2y^3z)^3$
17. $-(3ab)(2a^2b^3)^2$
18. $(3x^{n+1})(2x^{3n-1})$
19. $\dfrac{-39x^3y^4}{3xy^3}$
20. $\dfrac{30x^5y^4}{15x^2y}$
21. $\dfrac{12a^2b^5}{-3a^2b^3}$
22. $\dfrac{20a^4b^6}{5ab^3}$
23. $5a^2(3a^2 - 2a - 1)$
24. $-2x^3(4x^2 - 3x - 5)$
25. $(x + 4)(3x^2 - 5x - 1)$
26. $(3x + 2)(2x^2 - 5x + 1)$

27. $(x^2 - 2x - 5)(x^2 + 3x - 7)$

28. $(3x^2 - x - 4)(x^2 + 2x - 5)$

29. $(4x - 3y)(6x + 5y)$

30. $(7x - 9)(x + 4)$

31. $(7 - 3x)(3 + 5x)$

32. $(x^2 - 3)(x^2 + 8)$

33. $(2x - 3)^2$

34. $(5x - 1)^2$

35. $(4x + 3y)^2$

36. $(2x + 5y)^2$

37. $(2x - 7)(2x + 7)$

38. $(3x - 1)(3x + 1)$

39. $(x - 2)^3$

40. $(2x + 5)^3$

41. Find a polynomial that represents the area of the shaded region in Figure 5.23.

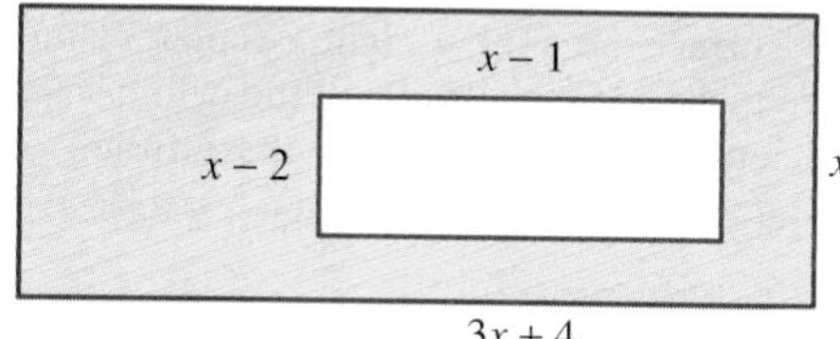

Figure 5.23

42. Find a polynomial that represents the volume of the rectangular solid in Figure 5.24.

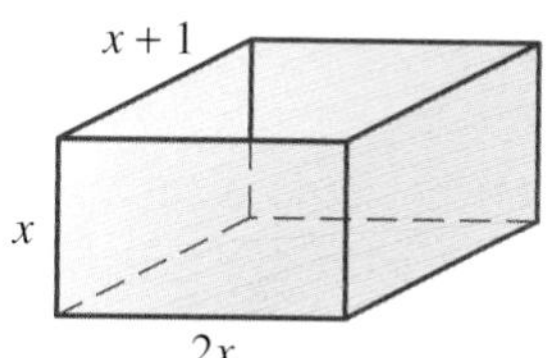

Figure 5.24

For Problems 43–62, factor each polynomial.

43. $10a^2b - 5ab^3 - 15a^3b^2$

44. $3xy - 5x^2y^2 - 15x^3y^3$

45. $a(x + 4) + b(x + 4)$

46. $y(3x - 1) + 7(3x - 1)$

47. $6x^3 + 3x^2y + 2xz^2 + yz^2$

48. $mn + 5n^2 - 4m - 20n$

49. $49a^2 - 25b^2$

50. $36x^2 - y^2$

51. $125a^3 - 8$

52. $27x^3 + 64y^3$

53. $x^2 - 9x + 18$

54. $x^2 + 11x + 28$

55. $x^2 - 4x - 21$

56. $x^2 + 6x - 16$

57. $2x^2 + 9x + 4$

58. $6x^2 - 11x + 4$

59. $12x^2 - 5x - 2$

60. $8x^2 - 10x - 3$

61. $4x^2 - 12xy + 9y^2$

62. $x^2 + 16xy + 64y^2$

For Problems 63–84, factor each polynomial completely. Indicate any that are not factorable using integers.

63. $x^2 + 3x - 28$

64. $2t^2 - 18$

65. $4n^2 + 9$

66. $12n^2 - 7n + 1$

67. $x^6 - x^2$

68. $x^3 - 6x^2 - 72x$

69. $6a^3b + 4a^2b^2 - 2a^2bc$

70. $x^2 - (y - 1)^2$

71. $8x^2 + 12$

72. $12x^2 + x - 35$

73. $16n^2 - 40n + 25$

74. $4n^2 - 8n$

75. $3w^3 + 18w^2 - 24w$

76. $20x^2 + 3xy - 2y^2$

77. $16a^2 - 64a$

78. $3x^3 - 15x^2 - 18x$

79. $n^2 - 8n - 128$

80. $t^4 - 22t^2 - 75$

81. $35x^2 - 11x - 6$

82. $15 - 14x + 3x^2$

83. $64n^3 - 27$

84. $16x^3 + 250$

For Problems 85–104, solve each equation.

85. $4x^2 - 36 = 0$

86. $x^2 + 5x - 6 = 0$

87. $49n^2 - 28n + 4 = 0$

88. $(3x - 1)(5x + 2) = 0$

89. $(3x - 4)^2 - 25 = 0$

90. $6a^3 = 54a$

91. $x^5 = x$

92. $-n^2 + 2n + 63 = 0$

93. $7n(7n + 2) = 8$

94. $30w^2 - w - 20 = 0$

95. $5x^4 - 19x^2 - 4 = 0$

96. $9n^2 - 30n + 25 = 0$

97. $n(2n + 4) = 96$

98. $7x^2 + 33x - 10 = 0$

99. $(x + 1)(x + 2) = 42$

100. $x^2 + 12x - x - 12 = 0$

101. $2x^4 + 9x^2 + 4 = 0$

102. $30 - 19x - 5x^2 = 0$

103. $3t^3 - 27t^2 + 24t = 0$

104. $-4n^2 - 39n + 10 = 0$

For Problems 105–114, set up an equation and solve each problem.

105. Find three consecutive integers such that the product of the smallest and the largest is one less than 9 times the middle integer.

106. Find two integers whose sum is 2 and whose product is -48.

107. Find two consecutive odd whole numbers whose product is 195.

108. Two cars leave an intersection at the same time, one traveling north and the other traveling east. Some time later, they are 20 miles apart, and the car going east has traveled 4 miles farther than the other car. How far has each car traveled?

109. The perimeter of a rectangle is 32 meters, and its area is 48 square meters. Find the length and width of the rectangle.

110. A room contains 144 chairs. The number of chairs per row is two less than twice the number of rows. Find the number of rows and the number of chairs per row.

111. The area of a triangle is 39 square feet. The length of one side is 1 foot more than twice the altitude to that side. Find the length of that side and the altitude to the side.

112. A rectangular-shaped pool 20 feet by 30 feet has a sidewalk of uniform width around the pool (see Figure 5.25). The area of the sidewalk is 336 square feet. Find the width of the sidewalk.

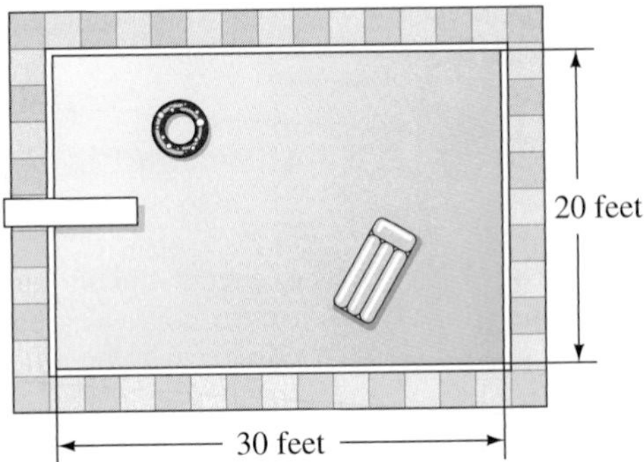

Figure 5.25

113. The sum of the areas of two squares is 89 square centimeters. The length of a side of the larger square is 3 centimeters more than the length of a side of the smaller square. Find the dimensions of each square.

114. The total surface area of a right circular cylinder is 32π square inches. If the altitude of the cylinder is three times the length of a radius, find the altitude of the cylinder.

Chapter 5 Test

For Problems 1–8, perform the indicated operations and simplify each expression.

1. $(-3x - 1) + (9x - 2) - (4x + 8)$ 1. ______
2. $(-6xy^2)(8x^3y^2)$ 2. ______
3. $(-3x^2y^4)^3$ 3. ______
4. $(5x - 7)(4x + 9)$ 4. ______
5. $(3n - 2)(2n - 3)$ 5. ______
6. $(x - 4y)^3$ 6. ______
7. $(x + 6)(2x^2 - x - 5)$ 7. ______
8. $\dfrac{-70x^4y^3}{5xy^2}$ 8. ______

For Problems 9–14, factor each expression completely.

9. $6x^2 + 19x - 20$ 9. ______
10. $12x^2 - 3$ 10. ______
11. $64 + t^3$ 11. ______
12. $30x + 4x^2 - 16x^3$ 12. ______
13. $x^2 - xy + 4x - 4y$ 13. ______
14. $24n^2 + 55n - 24$ 14. ______

For Problems 15–22, solve each equation.

15. $x^2 + 8x - 48 = 0$ 15. ______
16. $4n^2 = n$ 16. ______
17. $4x^2 - 12x + 9 = 0$ 17. ______
18. $(n - 2)(n + 7) = -18$ 18. ______
19. $3x^3 + 21x^2 - 54x = 0$ 19. ______
20. $12 + 13x - 35x^2 = 0$ 20. ______
21. $n(3n - 5) = 2$ 21. ______
22. $9x^2 - 36 = 0$ 22. ______

For Problems 23–25, set up an equation and solve each problem.

23. The perimeter of a rectangle is 30 inches, and its area is 54 square inches. Find the length of the longest side of the rectangle. 23. ______
24. A room contains 105 chairs arranged in rows. The number of rows is one more than twice the number of chairs per row. Find the number of rows. 24. ______
25. The combined area of a square and a rectangle is 57 square feet. The width of the rectangle is 3 feet more than the length of a side of the square, and the length of the rectangle is 5 feet more than the length of a side of the square. Find the length of the rectangle. 25. ______

Rational Expressions

Colorblind/Corbis Premium RF/Alamy Limited

■ *Computers often work together to compile large processing jobs. Rational numbers are used to express the rate of the processing speed of a computer.*

It takes Pat 12 hours to complete a task. After he had been working on this task for 3 hours, he was joined by his brother, Liam, and together they finished the job in 5 hours. How long would it take Liam to do the job by himself? We can use the **fractional equation** $\frac{5}{12} + \frac{5}{h} = \frac{3}{4}$ to determine that Liam could do the entire job by himself in 15 hours.

Rational expressions are to algebra what rational numbers are to arithmetic. Most of the work we will do with rational expressions in this chapter parallels the work you have previously done with arithmetic fractions. The same basic properties we use to explain reducing, adding, subtracting, multiplying, and dividing arithmetic fractions will serve as a basis for our work with rational expressions. The techniques of factoring that we studied in Chapter 5 will also play an important role in our discussions. At the end of this chapter, we will work with some fractional equations that contain rational expressions.

Video tutorials for all section learning objectives are available in a variety of delivery modes.

INTERNET PROJECT

The term "rational" in mathematics is derived from the word "ratio." One of the most commonly used ratios in art and architecture is the golden ratio. Do an Internet search to determine the approximate value of the golden ratio. Many Renaissance artists and architects used the golden ratio to proportion their work in the form of the golden rectangle; in the golden rectangle, the ratio of the longer side to the shorter side is the golden ratio. Is the rectangular cover of this text, which measures 11 inches by 8.5 inches, a golden rectangle?

6.1 Simplifying Rational Expressions

OBJECTIVES

1 Reduce Rational Numbers

2 Simplify Rational Expressions

1 Reduce Rational Numbers

We reviewed the basic operations with rational numbers in an informal setting in Chapter 1. In this review, we relied primarily on your knowledge of arithmetic. At this time, we want to become a little more formal with our review so that we can use the work with rational numbers as a basis for operating with rational expressions. We will define a rational expression shortly.

You will recall that any number that can be written in the form $\frac{a}{b}$, where a and b are integers and $b \neq 0$, is called a rational number. The following are examples of rational numbers:

$$\frac{1}{2} \qquad \frac{3}{4} \qquad \frac{15}{7} \qquad \frac{-5}{6} \qquad \frac{7}{-8} \qquad \frac{-12}{-17}$$

Numbers such as 6, -4, 0, $4\frac{1}{2}$, 0.7, and 0.21 are also rational, because we can express them as the indicated quotient of two integers. For example,

$$6 = \frac{6}{1} = \frac{12}{2} = \frac{18}{3} \quad \text{and so on} \qquad 4\frac{1}{2} = \frac{9}{2}$$

$$-4 = \frac{4}{-1} = \frac{-4}{1} = \frac{8}{-2} \quad \text{and so on} \qquad 0.7 = \frac{7}{10}$$

$$0 = \frac{0}{1} = \frac{0}{2} = \frac{0}{3} \quad \text{and so on} \qquad 0.21 = \frac{21}{100}$$

Because a rational number is the quotient of two integers, our previous work with division of integers can help us understand the various forms of rational numbers. If the signs of the numerator and denominator are different, then the rational number is negative. If the signs of the numerator and denominator are the same, then the rational number is positive. The next examples and Property 6.1 show the equivalent forms of rational numbers. Generally, it is preferable to express the denominator of a rational number as a positive integer.

$$\frac{8}{-2} = \frac{-8}{2} = -\frac{8}{2} = -4 \qquad \frac{12}{3} = \frac{-12}{-3} = 4$$

Observe the following general properties.

Property 6.1

1. $\dfrac{-a}{b} = \dfrac{a}{-b} = -\dfrac{a}{b}$, where $b \neq 0$

2. $\dfrac{-a}{-b} = \dfrac{a}{b}$, where $b \neq 0$

Therefore, a rational number such as $\frac{-2}{5}$ can also be written as $\frac{2}{-5}$ or $-\frac{2}{5}$.

We use the following property, often referred to as the **fundamental principle of fractions**, to reduce fractions to lowest terms or express fractions in simplest or reduced form.

Property 6.2 Fundamental Principle of Fractions

If b and k are nonzero integers and a is any integer, then

$$\frac{a \cdot k}{b \cdot k} = \frac{a}{b}$$

Let's apply Properties 6.1 and 6.2 to the following examples.

EXAMPLE 1

Reduce $\frac{18}{24}$ to lowest terms.

Solution

$$\frac{18}{24} = \frac{3 \cdot 6}{4 \cdot 6} = \frac{3}{4}$$

▼ PRACTICE YOUR SKILL

Reduce $\frac{12}{20}$ to lowest terms. ■

EXAMPLE 2

Change $\frac{40}{48}$ to simplest form.

Solution

$$\frac{\overset{5}{\cancel{40}}}{\underset{6}{\cancel{48}}} = \frac{5}{6}$$ A common factor of 8 was divided out of both numerator and denominator

▼ PRACTICE YOUR SKILL

Reduce $\frac{45}{65}$ to lowest terms. ■

EXAMPLE 3

Express $\frac{-36}{63}$ in reduced form.

Solution

$$\frac{-36}{63} = -\frac{36}{63} = -\frac{4 \cdot 9}{7 \cdot 9} = -\frac{4}{7}$$

▼ **PRACTICE YOUR SKILL**

Reduce $\frac{-20}{68}$ to lowest terms. ■

EXAMPLE 4

Reduce $\frac{72}{-90}$ to simplest form.

Solution

$$\frac{72}{-90} = -\frac{72}{90} = -\frac{\not{2} \cdot 2 \cdot 2 \cdot \not{3} \cdot \not{3}}{\not{2} \cdot \not{3} \cdot \not{3} \cdot 5} = -\frac{4}{5}$$

▼ **PRACTICE YOUR SKILL**

Reduce $\frac{84}{-120}$ to lowest terms. ■

Note the different terminology used in Examples 1–4. Regardless of the terminology, keep in mind that the number is not being changed; rather, the form of the numeral representing the number is being changed. In Example 1, $\frac{18}{24}$ and $\frac{3}{4}$ are equivalent fractions: they name the same number. Also note the use of prime factors in Example 4.

2 Simplify Rational Expressions

A **rational expression** is the indicated quotient of two polynomials. The following are examples of rational expressions.

$$\frac{3x^2}{5} \qquad \frac{x-2}{x+3} \qquad \frac{x^2+5x-1}{x^2-9} \qquad \frac{xy^2+x^2y}{xy} \qquad \frac{a^3-3a^2-5a-1}{a^4+a^3+6}$$

Because we must avoid division by zero, no values that create a denominator of zero can be assigned to variables. Thus the rational expression $\frac{x-2}{x+3}$ is meaningful for all values of x except $x = -3$. Rather than making restrictions for each individual expression, we will merely assume that all denominators represent nonzero real numbers.

Property 6.2 $\left(\frac{a \cdot k}{b \cdot k} = \frac{a}{b}\right)$ serves as the basis for simplifying rational expressions, as the next examples illustrate.

EXAMPLE 5

Simplify $\frac{15xy}{25y}$.

Solution

$$\frac{15xy}{25y} = \frac{3 \cdot \not{5} \cdot x \cdot \not{y}}{\not{5} \cdot 5 \cdot \not{y}} = \frac{3x}{5}$$

▼ **PRACTICE YOUR SKILL**

Simplify $\frac{18ab}{4a}$. ■

EXAMPLE 6

Simplify $\frac{-9}{18x^2y}$.

Solution

$$\frac{-9}{18x^2y} = -\frac{\overset{1}{\cancel{9}}}{\underset{2}{\cancel{18}}x^2y} = -\frac{1}{2x^2y}$$ A common factor of 9 was divided out of numerator and denominator

▼ **PRACTICE YOUR SKILL**

Simplify $\frac{-6}{24ab^3}$. ■

EXAMPLE 7

Simplify $\frac{-28a^2b^2}{-63a^2b^3}$.

Solution

$$\frac{-28a^2b^2}{-63a^2b^3} = \frac{4 \cdot \cancel{7} \cdot \cancel{a^2} \cdot \cancel{b^2}}{9 \cdot \cancel{7} \cdot \cancel{a^2} \cdot \underset{b}{\cancel{b^3}}} = \frac{4}{9b}$$

▼ **PRACTICE YOUR SKILL**

Simplify $\frac{-42x^3y^4}{-60x^5y^4}$. ■

The factoring techniques from Chapter 5 can be used to factor numerators and/or denominators so that we can apply the property $\frac{a \cdot k}{b \cdot k} = \frac{a}{b}$. Examples 8–12 should clarify this process.

EXAMPLE 8

Simplify $\frac{x^2 + 4x}{x^2 - 16}$.

Solution

$$\frac{x^2 + 4x}{x^2 - 16} = \frac{x\cancel{(x + 4)}}{(x - 4)\cancel{(x + 4)}} = \frac{x}{x - 4}$$

▼ **PRACTICE YOUR SKILL**

Simplify $\frac{y^2 - 5y}{y^2 - 25}$. ■

EXAMPLE 9

Simplify $\frac{4a^2 + 12a + 9}{2a + 3}$.

Solution

$$\frac{4a^2 + 12a + 9}{2a + 3} = \frac{\cancel{(2a + 3)}(2a + 3)}{1\cancel{(2a + 3)}} = \frac{2a + 3}{1} = 2a + 3$$

▼ **PRACTICE YOUR SKILL**

Simplify $\frac{9x^2 + 24x + 16}{3x + 4}$. ■

EXAMPLE 10 Simplify $\frac{5n^2 + 6n - 8}{10n^2 - 3n - 4}$.

Solution

$$\frac{5n^2 + 6n - 8}{10n^2 - 3n - 4} = \frac{(5n - 4)(n + 2)}{(5n - 4)(2n + 1)} = \frac{n + 2}{2n + 1}$$

▼ **PRACTICE YOUR SKILL**

Simplify $\frac{2x^2 - 7x - 15}{4x^2 + 12x + 9}$. ■

EXAMPLE 11 Simplify $\frac{6x^3y - 6xy}{x^3 + 5x^2 + 4x}$.

Solution

$$\frac{6x^3y - 6xy}{x^3 + 5x^2 + 4x} = \frac{6xy(x^2 - 1)}{x(x^2 + 5x + 4)} = \frac{6xy(x + 1)(x - 1)}{x(x + 1)(x + 4)} = \frac{6y(x - 1)}{x + 4}$$

▼ **PRACTICE YOUR SKILL**

Simplify $\frac{3x^3y - 12xy}{x^3 + 3x^2 - 10x}$. ■

Note that in Example 11 we left the numerator of the final fraction in factored form. This is often done if expressions other than monomials are involved. Both $\frac{6y(x - 1)}{x + 4}$ and $\frac{6xy - 6y}{x + 4}$ are acceptable answers.

Remember that the quotient of any nonzero real number and its opposite is -1. For example, $\frac{6}{-6} = -1$ and $\frac{-8}{8} = -1$. Likewise, the indicated quotient of any polynomial and its opposite is equal to -1; that is,

$\frac{a}{-a} = -1$ because a and $-a$ are opposites

$\frac{a - b}{b - a} = -1$ because $a - b$ and $b - a$ are opposites

$\frac{x^2 - 4}{4 - x^2} = -1$ because $x^2 - 4$ and $4 - x^2$ are opposites

Example 12 shows how we use this idea when simplifying rational expressions.

EXAMPLE 12 Simplify $\frac{6a^2 - 7a + 2}{10a - 15a^2}$.

Solution

$$\frac{6a^2 - 7a + 2}{10a - 15a^2} = \frac{(2a - 1)}{5a}\frac{(3a - 2)}{(2 - 3a)} \qquad \frac{3a - 2}{2 - 3a} = -1$$

$$= (-1)\left(\frac{2a - 1}{5a}\right)$$

$$= -\frac{2a - 1}{5a} \quad \text{or} \quad \frac{1 - 2a}{5a}$$

▼ **PRACTICE YOUR SKILL**

Simplify $\frac{x^2 - 9}{6x - 2x^2}$. ■

CONCEPT QUIZ

For Problems 1–10, answer true or false.

1. When a rational number is being reduced, the form of the numeral is being changed but not the number it represents.
2. A rational number is the ratio of two integers where the denominator is not zero.
3. -3 is a rational number.
4. The rational expression $\frac{x + 2}{x + 3}$ is meaningful for all values of x except when $x = -2$ and $x = 3$.
5. The binomials $x - y$ and $y - x$ are opposites.
6. The binomials $x + 3$ and $x - 3$ are opposites.
7. The rational expression $\frac{2 - x}{x + 2}$ reduces to -1.
8. The rational expression $\frac{x - y}{y - x}$ reduces to -1.
9. $\frac{x^2 + 5x - 14}{x^2 + 2x + 1} = \frac{5x - 14}{2x + 1}$
10. The rational expression $\frac{2x - x^2}{x^2 - 4}$ reduces to $\frac{x}{x + 2}$.

Problem Set 6.1

1 Reduce Rational Numbers

For Problems 1–8, express each rational number in reduced form.

1. $\frac{27}{36}$
2. $\frac{14}{21}$
3. $\frac{45}{54}$
4. $\frac{-14}{42}$
5. $\frac{24}{-60}$
6. $\frac{45}{-75}$
7. $\frac{-16}{-56}$
8. $\frac{-30}{-42}$

2 Simplify Rational Expressions

For Problems 9–50, simplify each rational expression.

9. $\frac{12xy}{42y}$
10. $\frac{21xy}{35x}$
11. $\frac{18a^2}{45ab}$
12. $\frac{48ab}{84b^2}$
13. $\frac{-14y^3}{56xy^2}$
14. $\frac{-14x^2y^3}{63xy^2}$
15. $\frac{54c^2d}{-78cd^2}$
16. $\frac{60x^3z}{-64xyz^2}$
17. $\frac{-40x^3y}{-24xy^4}$
18. $\frac{-30x^2y^2z^2}{-35xz^3}$
19. $\frac{x^2 - 4}{x^2 + 2x}$
20. $\frac{xy + y^2}{x^2 - y^2}$
21. $\frac{18x + 12}{12x - 6}$
22. $\frac{20x + 50}{15x - 30}$
23. $\frac{a^2 + 7a + 10}{a^2 - 7a - 18}$
24. $\frac{a^2 + 4a - 32}{3a^2 + 26a + 16}$
25. $\frac{2n^2 + n - 21}{10n^2 + 33n - 7}$
26. $\frac{4n^2 - 15n - 4}{7n^2 - 30n + 8}$
27. $\frac{5x^2 + 7}{10x}$
28. $\frac{12x^2 + 11x - 15}{20x^2 - 23x + 6}$
29. $\frac{6x^2 + x - 15}{8x^2 - 10x - 3}$
30. $\frac{4x^2 + 8x}{x^3 + 8}$
31. $\frac{3x^2 - 12x}{x^3 - 64}$
32. $\frac{x^2 - 14x + 49}{6x^2 - 37x - 35}$
33. $\frac{3x^2 + 17x - 6}{9x^2 - 6x + 1}$
34. $\frac{9y^2 - 1}{3y^2 + 11y - 4}$
35. $\frac{2x^3 + 3x^2 - 14x}{x^2y + 7xy - 18y}$
36. $\frac{3x^3 + 12x}{9x^2 + 18x}$

37. $\dfrac{5y^2 + 22y + 8}{25y^2 - 4}$

38. $\dfrac{16x^3y + 24x^2y^2 - 16xy^3}{24x^2y + 12xy^2 - 12y^3}$

39. $\dfrac{15x^3 - 15x^2}{5x^3 + 5x}$

40. $\dfrac{5n^2 + 18n - 8}{3n^2 + 13n + 4}$

41. $\dfrac{4x^2y + 8xy^2 - 12y^3}{18x^3y - 12x^2y^2 - 6xy^3}$

42. $\dfrac{3 + x - 2x^2}{2 + x - x^2}$

43. $\dfrac{3n^2 + 16n - 12}{7n^2 + 44n + 12}$

44. $\dfrac{x^4 - 2x^2 - 15}{2x^4 + 9x^2 + 9}$

45. $\dfrac{8 + 18x - 5x^2}{10 + 31x + 15x^2}$

46. $\dfrac{6x^4 - 11x^2 + 4}{2x^4 + 17x^2 - 9}$

47. $\dfrac{27x^4 - x}{6x^3 + 10x^2 - 4x}$

48. $\dfrac{64x^4 + 27x}{12x^3 - 27x^2 - 27x}$

49. $\dfrac{-40x^3 + 24x^2 + 16x}{20x^3 + 28x^2 + 8x}$

50. $\dfrac{-6x^3 - 21x^2 + 12x}{-18x^3 - 42x^2 + 120x}$

For Problems 51–58, simplify each rational expression. You will need to use factoring by grouping.

51. $\dfrac{xy + ay + bx + ab}{xy + ay + cx + ac}$

52. $\dfrac{xy + 2y + 3x + 6}{xy + 2y + 4x + 8}$

53. $\dfrac{ax - 3x + 2ay - 6y}{2ax - 6x + ay - 3y}$

54. $\dfrac{x^2 - 2x + ax - 2a}{x^2 - 2x + 3ax - 6a}$

55. $\dfrac{5x^2 + 5x + 3x + 3}{5x^2 + 3x - 30x - 18}$

56. $\dfrac{x^2 + 3x + 4x + 12}{2x^2 + 6x - x - 3}$

57. $\dfrac{2st - 30 - 12s + 5t}{3st - 6 - 18s + t}$

58. $\dfrac{nr - 6 - 3n + 2r}{nr + 10 + 2r + 5n}$

For Problems 59–68, simplify each rational expression. You may want to refer to Example 12 of this section.

59. $\dfrac{5x - 7}{7 - 5x}$

60. $\dfrac{4a - 9}{9 - 4a}$

61. $\dfrac{n^2 - 49}{7 - n}$

62. $\dfrac{9 - y}{y^2 - 81}$

63. $\dfrac{2y - 2xy}{x^2y - y}$

64. $\dfrac{3x - x^2}{x^2 - 9}$

65. $\dfrac{2x^3 - 8x}{4x - x^3}$

66. $\dfrac{x^2 - (y - 1)^2}{(y - 1)^2 - x^2}$

67. $\dfrac{n^2 - 5n - 24}{40 + 3n - n^2}$

68. $\dfrac{x^2 + 2x - 24}{20 - x - x^2}$

THOUGHTS INTO WORDS

69. Compare the concept of a rational number in arithmetic to the concept of a rational expression in algebra.

70. What role does factoring play in the simplifying of rational expressions?

71. Why is the rational expression $\dfrac{x + 3}{x^2 - 4}$ undefined for $x = 2$ and $x = -2$ but defined for $x = -3$?

72. How would you convince someone that $\dfrac{x - 4}{4 - x} = -1$ for all real numbers except 4?

Answers to the Concept Quiz

1. True 2. True 3. True 4. False 5. True 6. False 7. False 8. True 9. False 10. False

Answers to the Example Practice Skills

1. $\dfrac{3}{5}$ 2. $\dfrac{9}{13}$ 3. $-\dfrac{5}{17}$ 4. $-\dfrac{7}{10}$ 5. $\dfrac{9b}{2}$ 6. $-\dfrac{1}{4ab^3}$ 7. $\dfrac{7}{10x^2}$ 8. $\dfrac{y}{y + 5}$ 9. $3x + 4$ 10. $\dfrac{x - 5}{2x + 3}$ 11. $\dfrac{3y(x + 2)}{x + 5}$ 12. $-\dfrac{x + 3}{2x}$

6.2 Multiplying and Dividing Rational Expressions

OBJECTIVES

1. Multiply Rational Numbers
2. Multiply Rational Expressions
3. Divide Rational Numbers
4. Divide Rational Expressions
5. Simplify Problems That Involve Both Multiplication and Division

1 Multiply Rational Numbers

We define multiplication of rational numbers in common fraction form as follows:

Definition 6.1

If a, b, c, and d are integers, and b and d are not equal to zero, then

$$\frac{a}{b} \cdot \frac{c}{d} = \frac{a \cdot c}{b \cdot d} = \frac{ac}{bd}$$

To multiply rational numbers in common fraction form, we merely **multiply numerators and multiply denominators**, as the following examples demonstrate. (The steps in the dashed boxes are usually done mentally.)

$$\frac{2}{3} \cdot \frac{4}{5} = \frac{2 \cdot 4}{3 \cdot 5} = \frac{8}{15}$$

$$\frac{-3}{4} \cdot \frac{5}{7} = \frac{-3 \cdot 5}{4 \cdot 7} = \frac{-15}{28} = -\frac{15}{28}$$

$$-\frac{5}{6} \cdot \frac{13}{3} = \frac{-5}{6} \cdot \frac{13}{3} = \frac{-5 \cdot 13}{6 \cdot 3} = \frac{-65}{18} = -\frac{65}{18}$$

We also agree, when multiplying rational numbers, to express the final product in reduced form. The following examples show three different formats used to multiply and simplify rational numbers.

$$\frac{3}{4} \cdot \frac{4}{7} = \frac{3 \cdot \cancel{4}}{\cancel{4} \cdot 7} = \frac{3}{7}$$

$$\frac{\overset{1}{\cancel{8}}}{\underset{1}{\cancel{9}}} \cdot \frac{\overset{3}{\cancel{27}}}{\underset{4}{\cancel{32}}} = \frac{3}{4}$$

A common factor of 9 was divided out of 9 and 27, and a common factor of 8 was divided out of 8 and 32

$$\left(-\frac{28}{25}\right)\left(-\frac{65}{78}\right) = \frac{2 \cdot 2 \cdot 7 \cdot \cancel{5} \cdot \cancel{13}}{\cancel{5} \cdot 5 \cdot \cancel{2} \cdot 3 \cdot \cancel{13}} = \frac{14}{15}.$$

We should recognize that a negative times a negative is positive. Also, note the use of prime factors to help us recognize common factors.

2 Multiply Rational Expressions

Multiplication of rational expressions follows the same basic pattern as multiplication of rational numbers in common fraction form. That is to say, we multiply numerators and multiply denominators and express the final product in simplified or reduced form. Let's consider some examples.

$$\frac{3x}{4y}\cdot\frac{8y^2}{9x}=\frac{3\cdot 8\cdot x\cdot y^2}{4\cdot 9\cdot x\cdot y}=\frac{2y}{3}$$

Note that we use the commutative property of multiplication to rearrange the factors in a form that allows us to identify common factors of the numerator and denominator

$$\frac{-4a}{6a^2b^2}\cdot\frac{9ab}{12a^2}=-\frac{4\cdot 9\cdot a^2\cdot b}{6\cdot 12\cdot a^4\cdot b^2}=-\frac{1}{2a^2b}$$

$$\frac{12x^2y}{-18xy}\cdot\frac{-24xy^2}{56y^3}=\frac{12\cdot 24\cdot x^3\cdot y^3}{18\cdot 56\cdot x\cdot y^4}=\frac{2x^2}{7y}$$

You should recognize that the first fraction is equivalent to $-\frac{12x^2y}{18xy}$ and the second to $-\frac{24xy^2}{56y^3}$; thus the product is positive

If the rational expressions contain polynomials (other than monomials) that are factorable, then our work may take on the following format.

EXAMPLE 1

Multiply and simplify $\frac{y}{x^2-4}\cdot\frac{x+2}{y^2}$.

Solution

$$\frac{y}{x^2-4}\cdot\frac{x+2}{y^2}=\frac{y(x+2)}{y^2(x+2)(x-2)}=\frac{1}{y(x-2)}$$

▼ PRACTICE YOUR SKILL

Multiply and simplify $\frac{m}{n^2-16}\cdot\frac{n-4}{m^3}$. ■

In Example 1, note that we combined the steps of multiplying numerators and denominators and factoring the polynomials. Also note that we left the final answer in factored form. Either $\frac{1}{y(x-2)}$ or $\frac{1}{xy-2y}$ would be an acceptable answer.

EXAMPLE 2

Multiply and simplify $\frac{x^2-x}{x+5}\cdot\frac{x^2+5x+4}{x^4-x^2}$.

Solution

$$\frac{x^2-x}{x+5}\cdot\frac{x^2+5x+4}{x^4-x^2}=\frac{x(x-1)}{x+5}\cdot\frac{(x+1)(x+4)}{x^2(x-1)(x+1)}$$

$$=\frac{x(x-1)(x+1)(x+4)}{(x+5)(x^2)(x-1)(x+1)}=\frac{x+4}{x(x+5)}$$

▼ **PRACTICE YOUR SKILL**

Multiply and simplify $\frac{y^2 + 2y}{y + 3} \cdot \frac{y^2 - 4y - 5}{y^5 - 3y^4 - 10y^3}$. ■

EXAMPLE 3

Multiply and simplify $\frac{6n^2 + 7n - 5}{n^2 + 2n - 24} \cdot \frac{4n^2 + 21n - 18}{12n^2 + 11n - 15}$.

Solution

$$\frac{6n^2 + 7n - 5}{n^2 + 2n - 24} \cdot \frac{4n^2 + 21n - 18}{12n^2 + 11n - 15}$$

$$= \frac{\cancel{(3n + 5)}(2n - 1)\cancel{(4n - 3)}\cancel{(n + 6)}}{\cancel{(n + 6)}(n - 4)\cancel{(3n + 5)}\cancel{(4n - 3)}} = \frac{2n - 1}{n - 4}$$

▼ **PRACTICE YOUR SKILL**

Multiply and simplify $\frac{12x^2 - x - 1}{x^2 + 2x - 8} \cdot \frac{3x^2 - 4x - 4}{12x^2 + 11x + 2}$. ■

3 Divide Rational Numbers

We define division of rational numbers in common fraction form as follows.

> **Definition 6.2**
>
> If a, b, c, and d are integers and b, c, and d are not equal to zero, then
>
> $$\frac{a}{b} \div \frac{c}{d} = \frac{a}{b} \cdot \frac{d}{c} = \frac{ad}{bc}$$

Definition 6.2 states that to divide two rational numbers in fraction form, we **invert the divisor and multiply**. We call the numbers $\frac{c}{d}$ and $\frac{d}{c}$ "reciprocals" or "multiplicative inverses" of each other because their product is 1. Thus we can describe division by saying "to divide by a fraction, multiply by its reciprocal." The following examples demonstrate the use of Definition 6.2.

$$\frac{7}{8} \div \frac{5}{6} = \frac{7}{\underset{4}{\cancel{8}}} \cdot \frac{\overset{3}{\cancel{6}}}{5} = \frac{21}{20} \qquad \frac{-5}{9} \div \frac{15}{18} = -\frac{\cancel{5}}{\cancel{9}} \cdot \frac{\overset{2}{\cancel{18}}}{\underset{3}{\cancel{15}}} = -\frac{2}{3}$$

$$\frac{14}{-19} \div \frac{21}{-38} = \left(-\frac{14}{19}\right) \div \left(-\frac{21}{38}\right) = \left(-\frac{\overset{2}{\cancel{14}}}{\cancel{19}}\right)\left(-\frac{\overset{2}{\cancel{38}}}{\underset{3}{\cancel{21}}}\right) = \frac{4}{3}$$

4 Divide Rational Expressions

We define division of algebraic rational expressions in the same way that we define division of rational numbers. That is, the quotient of two rational expressions is the product we obtain when we multiply the first expression by the reciprocal of the second. Consider the following examples.

EXAMPLE 4

Divide and simplify $\frac{16x^2y}{24xy^3} \div \frac{9xy}{8x^2y^2}$.

Solution

$$\frac{16x^2y}{24xy^3} \div \frac{9xy}{8x^2y^2} = \frac{16x^2y}{24xy^3} \cdot \frac{8x^2y^2}{9xy} = \frac{16 \cdot \cancel{8} \cdot \cancel{x^4}^{x^2} \cdot \cancel{y^3}}{\cancel{24}_3 \cdot 9 \cdot \cancel{x^2} \cdot \cancel{y^4}_y} = \frac{16x^2}{27y}$$

▼ PRACTICE YOUR SKILL

Divide and simplify $\frac{20xy^3}{15x^2y} \div \frac{4y^5}{12x^2y}$. ■

EXAMPLE 5

Divide and simplify $\frac{3a^2 + 12}{3a^2 - 15a} \div \frac{a^4 - 16}{a^2 - 3a - 10}$.

Solution

$$\frac{3a^2 + 12}{3a^2 - 15a} \div \frac{a^4 - 16}{a^2 - 3a - 10} = \frac{3a^2 + 12}{3a^2 - 15a} \cdot \frac{a^2 - 3a - 10}{a^4 - 16}$$

$$= \frac{3(a^2 + 4)}{3a(a - 5)} \cdot \frac{(a - 5)(a + 2)}{(a^2 + 4)(a + 2)(a - 2)}$$

$$= \frac{\cancel{3}^1\cancel{(a^2 + 4)}\cancel{(a - 5)}\cancel{(a + 2)}}{\cancel{3}_1a\cancel{(a - 5)}\cancel{(a^2 + 4)}\cancel{(a + 2)}(a - 2)}$$

$$= \frac{1}{a(a - 2)}$$

▼ PRACTICE YOUR SKILL

Divide and simplify $\frac{2y^2 + 18}{2y + 12} \div \frac{y^4 - 81}{y^2 + 3y - 18}$. ■

EXAMPLE 6

Divide and simplify $\frac{28t^3 - 51t^2 - 27t}{49t^2 + 42t + 9} \div (4t - 9)$.

Solution

$$\frac{28t^3 - 51t^2 - 27t}{49t^2 + 42t + 9} \div \frac{4t - 9}{1} = \frac{28t^3 - 51t^2 - 27t}{49t^2 + 42t + 9} \cdot \frac{1}{4t - 9}$$

$$= \frac{t(7t + 3)(4t - 9)}{(7t + 3)(7t + 3)} \cdot \frac{1}{(4t - 9)}$$

$$= \frac{t\cancel{(7t + 3)}\cancel{(4t - 9)}}{\cancel{(7t + 3)}(7t + 3)\cancel{(4t - 9)}}$$

$$= \frac{t}{7t + 3}$$

▼ PRACTICE YOUR SKILL

Divide and simplify $\frac{10y^4 - y^3 - 2y^2}{25y^2 + 20y + 4} \div (2y - 1)$. ■

In a problem such as Example 6, it may be helpful to write the divisor with a denominator of 1. Thus we write $4t - 9$ as $\frac{4t-9}{1}$; its reciprocal is obviously $\frac{1}{4t-9}$.

5 Simplify Problems That Involve Both Multiplication and Division

Let's consider one final example that involves both multiplication and division.

EXAMPLE 7

Perform the indicated operations and simplify.

$$\frac{x^2+5x}{3x^2-4x-20} \cdot \frac{x^2y+y}{2x^2+11x+5} \div \frac{xy^2}{6x^2-17x-10}$$

Solution

$$\frac{x^2+5x}{3x^2-4x-20} \cdot \frac{x^2y+y}{2x^2+11x+5} \div \frac{xy^2}{6x^2-17x-10}$$

$$= \frac{x^2+5x}{3x^2-4x-20} \cdot \frac{x^2y+y}{2x^2+11x+5} \cdot \frac{6x^2-17x-10}{xy^2}$$

$$= \frac{x(x+5)}{(3x-10)(x+2)} \cdot \frac{y(x^2+1)}{(2x+1)(x+5)} \cdot \frac{(2x+1)(3x-10)}{xy^2}$$

$$= \frac{\cancel{x}\cancel{(x+5)}(\cancel{y})(x^2+1)\cancel{(2x+1)}\cancel{(3x-10)}}{\cancel{(3x-10)}(x+2)\cancel{(2x+1)}\cancel{(x+5)}(\cancel{x})(\cancel{y^2})_{y}} = \frac{x^2+1}{y(x+2)}$$

▼ **PRACTICE YOUR SKILL**

Simplify $\frac{3x+6}{2x^2+5x+3} \cdot \frac{x^2+4}{x^2+x-2} \div \frac{3xy}{2x^2+x-3}$. ■

CONCEPT QUIZ

For Problems 1–10, answer true or false.

1. To multiply two rational numbers in fraction form, we need to change to equivalent fractions with a common denominator.
2. When multiplying rational expressions that contain polynomials, the polynomials are factored so that common factors can be divided out.
3. In the division problem $\frac{2x^2y}{3z} \div \frac{4x^3}{5y^2}$, the fraction $\frac{4x^3}{5y^2}$ is the divisor.
4. The numbers $-\frac{2}{3}$ and $\frac{3}{2}$ are multiplicative inverses.
5. To divide two numbers in fraction form, we invert the divisor and multiply.
6. If $x \neq 0$, then $\left(\frac{4xy}{x}\right)\left(\frac{3y}{2x}\right) = \frac{6y^2}{x}$.
7. $\frac{3}{4} \div \frac{4}{3} = 1$.
8. If $x \neq 0$ and $y \neq 0$, then $\frac{5x^2y}{2y} \div \frac{10x^2}{3y} = \frac{3}{4}$.
9. If $x \neq 0$ and $y \neq 0$, then $\frac{1}{x} \div \frac{1}{y} = xy$.
10. If $x \neq y$, then $\frac{1}{x-y} \div \frac{1}{y-x} = -1$.

Problem Set 6.2

1 Multiply Rational Numbers

For Problems 1–6, multiply. Express final answers in reduced form.

1. $\frac{7}{12} \cdot \frac{6}{35}$

2. $\frac{5}{8} \cdot \frac{12}{20}$

3. $\frac{-4}{9} \cdot \frac{18}{30}$

4. $\frac{-6}{9} \cdot \frac{36}{48}$

5. $\frac{3}{-8} \cdot \frac{-6}{12}$

6. $\frac{-12}{16} \cdot \frac{18}{-32}$

2 Multiply Rational Expressions

For Problems 7–50, perform the indicated operations involving rational expressions. Express final answers in simplest form.

7. $\frac{6xy}{9y^4} \cdot \frac{30x^3y}{-48x}$

8. $\frac{-14xy^4}{18y^2} \cdot \frac{24x^2y^3}{35y^2}$

9. $\frac{5a^2b^2}{11ab} \cdot \frac{22a^3}{15ab^2}$

10. $\frac{10a^2}{5b^2} \cdot \frac{15b^3}{2a^4}$

11. $\frac{5xy}{8y^2} \cdot \frac{18x^2y}{15}$

12. $\frac{4x^2}{5y^2} \cdot \frac{15xy}{24x^2y^2}$

13. $\frac{9x^2y^3}{14x} \cdot \frac{21y}{15xy^2} \cdot \frac{10x}{12y^3}$

14. $\frac{5xy}{7a} \cdot \frac{14a^2}{15x} \cdot \frac{3a}{8y}$

15. $\frac{3x+6}{5y} \cdot \frac{x^2+4}{x^2+10x+16}$

16. $\frac{5xy}{x+6} \cdot \frac{x^2-36}{x^2-6x}$

17. $\frac{5a^2+20a}{a^3-2a^2} \cdot \frac{a^2-a-12}{a^2-16}$

18. $\frac{2a^2+6}{a^2-a} \cdot \frac{a^3-a^2}{8a-4}$

19. $\frac{3n^2+15n-18}{3n^2+10n-48} \cdot \frac{6n^2-n-40}{4n^2+6n-10}$

20. $\frac{6n^2+11n-10}{3n^2+19n-14} \cdot \frac{2n^2+6n-56}{2n^2-3n-20}$

21. $\frac{5-14n-3n^2}{1-2n-3n^2} \cdot \frac{9+7n-2n^2}{27-15n+2n^2}$

22. $\frac{6-n-2n^2}{12-11n+2n^2} \cdot \frac{24-26n+5n^2}{2+3n+n^2}$

23. $\frac{3x^4+2x^2-1}{3x^4+14x^2-5} \cdot \frac{x^4-2x^2-35}{x^4-17x^2+70}$

24. $\frac{2x^4+x^2-3}{2x^4+5x^2+2} \cdot \frac{3x^4+10x^2+8}{3x^4+x^2-4}$

25. $\frac{10t^3+25t}{20t+10} \cdot \frac{2t^2-t-1}{t^5-t}$

26. $\frac{t^4-81}{t^2-6t+9} \cdot \frac{6t^2-11t-21}{5t^2+8t-21}$

27. $\frac{4t^2+t-5}{t^3-t^2} \cdot \frac{t^4+6t^3}{16t^2+40t+25}$

28. $\frac{9n^2-12n+4}{n^2-4n-32} \cdot \frac{n^2+4n}{3n^3-2n^2}$

29. $\frac{nr+3n+2r+6}{nr+3n-3r-9} \cdot \frac{n^2-9}{n^3-4n}$

30. $\frac{xy+xc+ay+ac}{xy-2xc+ay-2ac} \cdot \frac{2x^3-8x}{12x^3+20x^2-8x}$

3 Divide Rational Numbers

31. $\left(-\frac{5}{7}\right) \div \frac{6}{7}$

32. $\left(-\frac{5}{9}\right) \div \frac{10}{3}$

33. $\frac{-9}{5} \div \frac{27}{10}$

34. $\frac{4}{7} \div \frac{16}{-21}$

35. $\frac{4}{9} \cdot \frac{6}{11} \div \frac{4}{15}$

36. $\frac{2}{3} \cdot \frac{6}{7} \div \frac{8}{3}$

4 Divide Rational Expressions

37. $\frac{5x^4}{12x^2y^3} \div \frac{9}{5xy}$

38. $\frac{7x^2y}{9xy^3} \div \frac{3x^4}{2x^2y^2}$

39. $\frac{9a^2c}{12bc^2} \div \frac{21ab}{14c^3}$

40. $\frac{3ab^3}{4c} \div \frac{21ac}{12bc^3}$

41. $\frac{9y^2}{x^2+12x+36} \div \frac{12y}{x^2+6x}$

42. $\frac{7xy}{x^2-4x+4} \div \frac{14y}{x^2-4}$

43. $\frac{x^2-4xy+4y^2}{7xy^2} \div \frac{4x^2-3xy-10y^2}{20x^2y+25xy^2}$

44. $\frac{x^2+5xy-6y^2}{xy^2-y^3} \div \frac{xy+4y^2}{2x^2+15xy+18y^2}$

45. $\frac{3x^2-20x+25}{2x^2-7x-15} \div \frac{9x^2-3x-20}{12x^2+28x+15}$

46. $\frac{21t^2+t-2}{2t^2-17t-9} \div \frac{12t^2-5t-3}{8t^2-2t-3}$

5 Simplify Problems That Involve Both Multiplication and Division

47. $\frac{x^2-x}{4y} \cdot \frac{10xy^2}{2x-2} \div \frac{3x^2+3x}{15x^2y^2}$

48. $\frac{4xy^2}{7x} \cdot \frac{14x^3y}{12y} \div \frac{7y}{9x^3}$

49. $\frac{a^2-4ab+4b^2}{6a^2-4ab} \cdot \frac{3a^2+5ab-2b^2}{6a^2+ab-b^2} \div \frac{a^2-4b^2}{8a+4b}$

50. $\frac{2x^2+3x}{2x^3-10x^2} \cdot \frac{x^2-8x+15}{3x^3-27x} \div \frac{14x+21}{x^2-6x-27}$

THOUGHTS INTO WORDS

51. Explain in your own words how to divide two rational expressions.

52. Suppose that your friend missed class the day the material in this section was discussed. How could you draw on her background in arithmetic to explain to her how to multiply and divide rational expressions?

53. Give a step-by-step description of how to do the following multiplication problem.

$$\frac{x^2 + 5x + 6}{x^2 - 2x - 8} \cdot \frac{x^2 - 16}{16 - x^2}$$

Answers to the Concept Quiz

1. False **2.** True **3.** True **4.** False **5.** True **6.** True **7.** False **8.** False **9.** False **10.** True

Answers to the Example Practice Skills

1. $\frac{1}{m^2(n + 4)}$ **2.** $\frac{y + 1}{y^2(y + 3)}$ **3.** $\frac{3x - 1}{x + 4}$ **4.** $\frac{4x}{y^2}$ **5.** $\frac{1}{y + 3}$ **6.** $\frac{y^2}{5y + 2}$ **7.** $\frac{x^2 + 4}{xy(x + 1)}$

6.3 Adding and Subtracting Rational Expressions

OBJECTIVES

1 Add and Subtract Rational Numbers

2 Add and Subtract Rational Expressions

1 Add and Subtract Rational Numbers

We can define addition and subtraction of rational numbers as follows:

Definition 6.3

If a, b, and c are integers and b is not zero, then

$$\frac{a}{b} + \frac{c}{b} = \frac{a + c}{b} \quad \text{Addition}$$

$$\frac{a}{b} - \frac{c}{b} = \frac{a - c}{b} \quad \text{Subtraction}$$

We can add or subtract rational numbers with a common denominator by adding or subtracting the numerators and placing the result over the common denominator. The following examples illustrate Definition 6.3.

$$\frac{2}{9} + \frac{3}{9} = \frac{2 + 3}{9} = \frac{5}{9}$$

$$\frac{7}{8} - \frac{3}{8} = \frac{7 - 3}{8} = \frac{4}{8} = \frac{1}{2} \quad \text{Don't forget to reduce!}$$

$$\frac{4}{6} + \frac{-5}{6} = \frac{4 + (-5)}{6} = \frac{-1}{6} = -\frac{1}{6}$$

$$\frac{7}{10} + \frac{4}{-10} = \frac{7}{10} + \frac{-4}{10} = \frac{7 + (-4)}{10} = \frac{3}{10}$$

If rational numbers that do not have a common denominator are to be added or subtracted, then we apply the fundamental principle of fractions $\left(\frac{a}{b} = \frac{ak}{bk}\right)$ **to obtain equivalent fractions with a common denominator.** Equivalent fractions are fractions such as $\frac{1}{2}$ and $\frac{2}{4}$ that name the same number. Consider the following example.

$$\frac{1}{2} + \frac{1}{3} = \frac{3}{6} + \frac{2}{6} = \frac{3+2}{6} = \frac{5}{6}$$

$\frac{1}{2}$ and $\frac{3}{6}$ are equivalent fractions

$\frac{1}{3}$ and $\frac{2}{6}$ are equivalent fractions

Note that we chose 6 as our common denominator and that 6 is the **least common multiple** of the original denominators 2 and 3. (The least common multiple of a set of whole numbers is the smallest nonzero whole number divisible by each of the numbers.) In general, we use the least common multiple of the denominators of the fractions to be added or subtracted as a **least common denominator** (LCD).

A least common denominator may be found by inspection or by using the prime-factored forms of the numbers. Let's consider some examples and use each of these techniques.

EXAMPLE 1

Subtract $\frac{5}{6} - \frac{3}{8}$.

Solution

By inspection, we can see that the LCD is 24. Thus both fractions can be changed to equivalent fractions, each with a denominator of 24.

$$\frac{5}{6} - \frac{3}{8} = \left(\frac{5}{6}\right)\left(\frac{4}{4}\right) - \left(\frac{3}{8}\right)\left(\frac{3}{3}\right) = \frac{20}{24} - \frac{9}{24} = \frac{11}{24}$$

Form of 1 Form of 1

▼ PRACTICE YOUR SKILL

Subtract $\frac{13}{15} - \frac{3}{10}$. ■

In Example 1, note that the fundamental principle of fractions, $\frac{a}{b} = \frac{a \cdot k}{b \cdot k}$, can be written as $\frac{a}{b} = \left(\frac{a}{b}\right)\left(\frac{k}{k}\right)$. This latter form emphasizes the fact that 1 is the multiplication identity element.

EXAMPLE 2

Perform the indicated operations: $\frac{3}{5} + \frac{1}{6} - \frac{13}{15}$.

Solution

Again by inspection, we can determine that the LCD is 30. Thus we can proceed as follows:

$$\frac{3}{5} + \frac{1}{6} - \frac{13}{15} = \left(\frac{3}{5}\right)\left(\frac{6}{6}\right) + \left(\frac{1}{6}\right)\left(\frac{5}{5}\right) - \left(\frac{13}{15}\right)\left(\frac{2}{2}\right)$$

$$= \frac{18}{30} + \frac{5}{30} - \frac{26}{30} = \frac{18 + 5 - 26}{30}$$

$$= \frac{-3}{30} = -\frac{1}{10} \quad \text{Don't forget to reduce!}$$

▼ **PRACTICE YOUR SKILL**

Perform the indicated operations: $\frac{3}{8} + \frac{1}{6} - \frac{3}{4}$. ■

EXAMPLE 3

Add $\frac{7}{18} + \frac{11}{24}$.

Solution

Let's use the prime-factored forms of the denominators to help find the LCD.

$$18 = 2 \cdot 3 \cdot 3 \qquad 24 = 2 \cdot 2 \cdot 2 \cdot 3$$

The LCD must contain three factors of 2 because 24 contains three 2s. The LCD must also contain two factors of 3 because 18 has two 3s. Thus the LCD $= 2 \cdot 2 \cdot 2 \cdot 3 \cdot 3 = 72$. Now we can proceed as usual.

$$\frac{7}{18} + \frac{11}{24} = \left(\frac{7}{18}\right)\left(\frac{4}{4}\right) + \left(\frac{11}{24}\right)\left(\frac{3}{3}\right) = \frac{28}{72} + \frac{33}{72} = \frac{61}{72}$$

▼ **PRACTICE YOUR SKILL**

Add $\frac{7}{24} + \frac{3}{20}$. ■

2 Add and Subtract Rational Expressions

We use the same *common denominator* approach when adding or subtracting rational expressions, as in these next examples.

$$\frac{3}{x} + \frac{9}{x} = \frac{3 + 9}{x} = \frac{12}{x}$$

$$\frac{8}{x - 2} - \frac{3}{x - 2} = \frac{8 - 3}{x - 2} = \frac{5}{x - 2}$$

$$\frac{9}{4y} + \frac{5}{4y} = \frac{9+5}{4y} = \frac{14}{4y} = \frac{7}{2y} \quad \text{Don't forget to simplify the final answer!}$$

$$\frac{n^2}{n-1} - \frac{1}{n-1} = \frac{n^2-1}{n-1} = \frac{(n+1)\cancel{(n-1)}}{\cancel{n-1}} = n+1$$

$$\frac{6a^2}{2a+1} + \frac{13a+5}{2a+1} = \frac{6a^2+13a+5}{2a+1} = \frac{\cancel{(2a+1)}(3a+5)}{\cancel{2a+1}} = 3a+5$$

In each of the previous examples that involve rational expressions, we should technically restrict the variables to exclude division by zero. For example, $\frac{3}{x} + \frac{9}{x} = \frac{12}{x}$ is true for all real number values for x, except $x = 0$. Likewise, $\frac{8}{x-2} - \frac{3}{x-2} = \frac{5}{x-2}$ as long as x does not equal 2. Rather than taking the time and space to write down restrictions for each problem, we will merely assume that such restrictions exist.

To add and subtract rational expressions with different denominators, follow the same basic routine that you follow when you add or subtract rational numbers with different denominators. Study the following examples carefully and note the similarity to our previous work with rational numbers.

EXAMPLE 4

Add $\frac{x+2}{4} + \frac{3x+1}{3}$.

Solution

By inspection, we see that the LCD is 12.

$$\begin{aligned}\frac{x+2}{4} + \frac{3x+1}{3} &= \left(\frac{x+2}{4}\right)\left(\frac{3}{3}\right) + \left(\frac{3x+1}{3}\right)\left(\frac{4}{4}\right)\\ &= \frac{3(x+2)}{12} + \frac{4(3x+1)}{12}\\ &= \frac{3(x+2)+4(3x+1)}{12}\\ &= \frac{3x+6+12x+4}{12}\\ &= \frac{15x+10}{12}\end{aligned}$$

▼ PRACTICE YOUR SKILL

Subtract $\frac{x-4}{3} - \frac{x-2}{9}$. ■

Note the final result in Example 4. The numerator, $15x + 10$, could be factored as $5(3x + 2)$. However, because this produces no common factors with the denominator, the fraction cannot be simplified. Thus the final answer can be left as $\frac{15x+10}{12}$. It would also be acceptable to express it as $\frac{5(3x+2)}{12}$.

EXAMPLE 5

Subtract $\frac{a-2}{2} - \frac{a-6}{6}$.

Solution

By inspection, we see that the LCD is 6.

$$\frac{a-2}{2} - \frac{a-6}{6} = \left(\frac{a-2}{2}\right)\left(\frac{3}{3}\right) - \frac{a-6}{6}$$

$$= \frac{3(a-2)}{6} - \frac{a-6}{6}$$

$$= \frac{3(a-2)-(a-6)}{6}$$ Be careful with this sign as you move to the next step!

$$= \frac{3a-6-a+6}{6}$$

$$= \frac{2a}{6} = \frac{a}{3}$$ Don't forget to simplify

▼ **PRACTICE YOUR SKILL**

Perform the indicated operations: $\frac{y-5}{12} + \frac{y+3}{18} - \frac{y-5}{9}$. ■

EXAMPLE 6

Perform the indicated operations: $\frac{x+3}{10} + \frac{2x+1}{15} - \frac{x-2}{18}$.

Solution

If you cannot determine the LCD by inspection, then use the prime-factored forms of the denominators.

$$10 = 2 \cdot 5 \qquad 15 = 3 \cdot 5 \qquad 18 = 2 \cdot 3 \cdot 3$$

The LCD must contain one factor of 2, two factors of 3, and one factor of 5. Thus the LCD is $2 \cdot 3 \cdot 3 \cdot 5 = 90$.

$$\frac{x+3}{10} + \frac{2x+1}{15} - \frac{x-2}{18} = \left(\frac{x+3}{10}\right)\left(\frac{9}{9}\right) + \left(\frac{2x+1}{15}\right)\left(\frac{6}{6}\right) - \left(\frac{x-2}{18}\right)\left(\frac{5}{5}\right)$$

$$= \frac{9(x+3)}{90} + \frac{6(2x+1)}{90} - \frac{5(x-2)}{90}$$

$$= \frac{9(x+3)+6(2x+1)-5(x-2)}{90}$$

$$= \frac{9x+27+12x+6-5x+10}{90}$$

$$= \frac{16x+43}{90}$$

▼ **PRACTICE YOUR SKILL**

Perform the indicated operations: $\frac{y+2}{4} + \frac{3y-1}{6} - \frac{y-4}{18}$. ■

A denominator that contains variables does not create any serious difficulties; our approach remains basically the same.

EXAMPLE 7

Add $\frac{3}{2x} + \frac{5}{3y}$.

Solution

Using an LCD of $6xy$, we can proceed as follows:

$$\frac{3}{2x} + \frac{5}{3y} = \left(\frac{3}{2x}\right)\left(\frac{3y}{3y}\right) + \left(\frac{5}{3y}\right)\left(\frac{2x}{2x}\right)$$

$$= \frac{9y}{6xy} + \frac{10x}{6xy}$$

$$= \frac{9y + 10x}{6xy}$$

▼ **PRACTICE YOUR SKILL**

Add $\frac{7}{3a} + \frac{4}{5b}$. ■

EXAMPLE 8

Subtract $\frac{7}{12ab} - \frac{11}{15a^2}$.

Solution

We can prime-factor the numerical coefficients of the denominators to help find the LCD.

$$\left.\begin{aligned} 12ab &= 2 \cdot 2 \cdot 3 \cdot a \cdot b \\ 15a^2 &= 3 \cdot 5 \cdot a^2 \end{aligned}\right\} \longrightarrow \text{LCD} = 2 \cdot 2 \cdot 3 \cdot 5 \cdot a^2 \cdot b = 60a^2b$$

$$\frac{7}{12ab} - \frac{11}{15a^2} = \left(\frac{7}{12ab}\right)\left(\frac{5a}{5a}\right) - \left(\frac{11}{15a^2}\right)\left(\frac{4b}{4b}\right)$$

$$= \frac{35a}{60a^2b} - \frac{44b}{60a^2b}$$

$$= \frac{35a - 44b}{60a^2b}$$

▼ **PRACTICE YOUR SKILL**

Subtract $\frac{5}{6xy} - \frac{3}{10y^2}$. ■

EXAMPLE 9

Add $\frac{x}{x - 3} + \frac{4}{x}$.

Solution

By inspection, the LCD is $x(x - 3)$.

$$\frac{x}{x - 3} + \frac{4}{x} = \left(\frac{x}{x - 3}\right)\left(\frac{x}{x}\right) + \left(\frac{4}{x}\right)\left(\frac{x - 3}{x - 3}\right)$$

$$= \frac{x^2}{x(x - 3)} + \frac{4(x - 3)}{x(x - 3)}$$

$$= \frac{x^2 + 4(x-3)}{x(x-3)}$$

$$= \frac{x^2 + 4x - 12}{x(x-3)} \quad \text{or} \quad \frac{(x+6)(x-2)}{x(x-3)}$$

▼ **PRACTICE YOUR SKILL**

Add $\frac{8}{y} + \frac{2y}{y+1}$. ■

EXAMPLE 10 Subtract $\frac{2x}{x+1} - 3$.

Solution

$$\frac{2x}{x+1} - 3 = \frac{2x}{x+1} - 3\left(\frac{x+1}{x+1}\right)$$

$$= \frac{2x}{x+1} - \frac{3(x+1)}{x+1}$$

$$= \frac{2x - 3(x+1)}{x+1}$$

$$= \frac{2x - 3x - 3}{x+1}$$

$$= \frac{-x-3}{x+1}$$

▼ **PRACTICE YOUR SKILL**

Subtract $\frac{4y}{y+5} - 7$. ■

CONCEPT QUIZ

For Problems 1–10, answer true or false.

1. The addition problem $\frac{2x}{x+4} + \frac{1}{x+4}$ is equal to $\frac{2x+1}{x+4}$ for all values of x except $x = -\frac{1}{2}$ and $x = -4$.
2. Any common denominator can be used to add rational expressions, but typically we can use the least common denominator.
3. The fractions $\frac{2x^2}{3y}$ and $\frac{10x^2z}{15yz}$ are equivalent fractions.
4. The least common multiple of the denominators is always the lowest common denominator.
5. To simplify the expression $\frac{5}{2x-1} + \frac{3}{1-2x}$, we could use $2x - 1$ for the common denominator.
6. If $x \neq \frac{1}{2}$, then $\frac{5}{2x-1} + \frac{3}{1-2x} = \frac{2}{2x-1}$.
7. $\frac{3}{-4} - \frac{-2}{3} = \frac{17}{12}$

8. $\frac{4x-1}{5}+\frac{2x+1}{6}=\frac{x}{5}$

9. $\frac{x}{4}-\frac{3x}{2}+\frac{5x}{3}=\frac{5x}{12}$

10. If $x \neq 0$, then $\frac{2}{3x}-\frac{3}{2x}-1=\frac{-5-6x}{6x}$

Problem Set 6.3

1 Add and Subtract Rational Numbers

For Problems 1–12, perform the indicated operations involving rational numbers. Be sure to express your answers in reduced form.

1. $\frac{1}{4}+\frac{5}{6}$

2. $\frac{3}{5}+\frac{1}{6}$

3. $\frac{7}{8}-\frac{3}{5}$

4. $\frac{7}{9}-\frac{1}{6}$

5. $\frac{6}{5}+\frac{1}{-4}$

6. $\frac{7}{8}+\frac{5}{-12}$

7. $\frac{8}{15}+\frac{3}{25}$

8. $\frac{5}{9}-\frac{11}{12}$

9. $\frac{1}{5}+\frac{5}{6}-\frac{7}{15}$

10. $\frac{2}{3}-\frac{7}{8}+\frac{1}{4}$

11. $\frac{1}{3}-\frac{1}{4}-\frac{3}{14}$

12. $\frac{5}{6}-\frac{7}{9}-\frac{3}{10}$

2 Add and Subtract Rational Expressions

For Problems 13–66, add or subtract the rational expressions as indicated. Be sure to express your answers in simplest form.

13. $\frac{2x}{x-1}+\frac{4}{x-1}$

14. $\frac{3x}{2x+1}-\frac{5}{2x+1}$

15. $\frac{4a}{a+2}+\frac{8}{a+2}$

16. $\frac{6a}{a-3}-\frac{18}{a-3}$

17. $\frac{3(y-2)}{7y}+\frac{4(y-1)}{7y}$

18. $\frac{2x-1}{4x^2}+\frac{3(x-2)}{4x^2}$

19. $\frac{x-1}{2}+\frac{x+3}{3}$

20. $\frac{x-2}{4}+\frac{x+6}{5}$

21. $\frac{2a-1}{4}+\frac{3a+2}{6}$

22. $\frac{a-4}{6}+\frac{4a-1}{8}$

23. $\frac{n+2}{6}-\frac{n-4}{9}$

24. $\frac{2n+1}{9}-\frac{n+3}{12}$

25. $\frac{3x-1}{3}-\frac{5x+2}{5}$

26. $\frac{4x-3}{6}-\frac{8x-2}{12}$

27. $\frac{x-2}{5}-\frac{x+3}{6}+\frac{x+1}{15}$

28. $\frac{x+1}{4}+\frac{x-3}{6}-\frac{x-2}{8}$

29. $\frac{3}{8x}+\frac{7}{10x}$

30. $\frac{5}{6x}-\frac{3}{10x}$

31. $\frac{5}{7x}-\frac{11}{4y}$

32. $\frac{5}{12x}-\frac{9}{8y}$

33. $\frac{4}{3x}+\frac{5}{4y}-1$

34. $\frac{7}{3x}-\frac{8}{7y}-2$

35. $\frac{7}{10x^2}+\frac{11}{15x}$

36. $\frac{7}{12a^2}-\frac{5}{16a}$

37. $\frac{10}{7n}-\frac{12}{4n^2}$

38. $\frac{6}{8n^2}-\frac{3}{5n}$

39. $\frac{3}{n^2}-\frac{2}{5n}+\frac{4}{3}$

40. $\frac{1}{n^2}+\frac{3}{4n}-\frac{5}{6}$

41. $\frac{3}{x}-\frac{5}{3x^2}-\frac{7}{6x}$

42. $\frac{7}{3x^2}-\frac{9}{4x}-\frac{5}{2x}$

43. $\frac{6}{5t^2}-\frac{4}{7t^3}+\frac{9}{5t^3}$

44. $\frac{5}{7t}+\frac{3}{4t^2}+\frac{1}{14t}$

45. $\frac{5b}{24a^2}-\frac{11a}{32b}$

46. $\frac{9}{14x^2y}-\frac{4x}{7y^2}$

47. $\frac{7}{9xy^3}-\frac{4}{3x}+\frac{5}{2y^2}$

48. $\frac{7}{16a^2b}+\frac{3a}{20b^2}$

49. $\frac{2x}{x-1}+\frac{3}{x}$

50. $\frac{3x}{x-4}-\frac{2}{x}$

51. $\frac{a-2}{a}-\frac{3}{a+4}$

52. $\frac{a+1}{a}-\frac{2}{a+1}$

53. $\frac{-3}{4n+5}-\frac{8}{3n+5}$

54. $\frac{-2}{n-6}-\frac{6}{2n+3}$

55. $\frac{-1}{x+4}+\frac{4}{7x-1}$

56. $\frac{-3}{4x+3}+\frac{5}{2x-5}$

57. $\frac{7}{3x-5} - \frac{5}{2x+7}$

58. $\frac{5}{x-1} - \frac{3}{2x-3}$

59. $\frac{5}{3x-2} + \frac{6}{4x+5}$

60. $\frac{3}{2x+1} + \frac{2}{3x+4}$

61. $\frac{3x}{2x+5} + 1$

62. $2 + \frac{4x}{3x-1}$

63. $\frac{4x}{x-5} - 3$

64. $\frac{7x}{x+4} - 2$

65. $-1 - \frac{3}{2x+1}$

66. $-2 - \frac{5}{4x-3}$

67. Recall that the indicated quotient of a polynomial and its opposite is -1. For example, $\frac{x-2}{2-x}$ simplifies to -1. Keep this idea in mind as you add or subtract the following rational expressions.

(a) $\frac{1}{x-1} - \frac{x}{x-1}$

(b) $\frac{3}{2x-3} - \frac{2x}{2x-3}$

(c) $\frac{4}{x-4} - \frac{x}{x-4} + 1$

(d) $-1 + \frac{2}{x-2} - \frac{x}{x-2}$

68. Consider the addition problem $\frac{8}{x-2} + \frac{5}{2-x}$. Note that the denominators are opposites of each other. If the property $\frac{a}{-b} = -\frac{a}{b}$ is applied to the second fraction, we have $\frac{5}{2-x} = -\frac{5}{x-2}$. Thus we proceed as follows:

$$\frac{8}{x-2} + \frac{5}{2-x} = \frac{8}{x-2} - \frac{5}{x-2} = \frac{8-5}{x-2} = \frac{3}{x-2}$$

Use this approach to do the following problems.

(a) $\frac{7}{x-1} + \frac{2}{1-x}$

(b) $\frac{5}{2x-1} + \frac{8}{1-2x}$

(c) $\frac{4}{a-3} - \frac{1}{3-a}$

(d) $\frac{10}{a-9} - \frac{5}{9-a}$

(e) $\frac{x^2}{x-1} - \frac{2x-3}{1-x}$

(f) $\frac{x^2}{x-4} - \frac{3x-28}{4-x}$

THOUGHTS INTO WORDS

69. What is the difference between the concept of least common multiple and the concept of least common denominator?

70. A classmate tells you that she finds the least common multiple of two counting numbers by listing the multiples of each number and then choosing the smallest number that appears in both lists. Is this a correct procedure? What is the weakness of this procedure?

71. For which real numbers does $\frac{x}{x-3} + \frac{4}{x}$ equal $\frac{(x+6)(x-2)}{x(x-3)}$? Explain your answer.

72. Suppose that your friend does an addition problem as follows:

$$\frac{5}{8} + \frac{7}{12} = \frac{5(12) + 8(7)}{8(12)} = \frac{60+56}{96} = \frac{116}{96} = \frac{29}{24}$$

Is this answer correct? If not, what advice would you offer your friend?

Answers to the Concept Quiz

1. False **2.** True **3.** True **4.** True **5.** True **6.** True **7.** False **8.** False **9.** True **10.** True

Answers to the Example Practice Skills

1. $\frac{17}{30}$ **2.** $-\frac{5}{24}$ **3.** $\frac{53}{120}$ **4.** $\frac{2x-10}{9}$ **5.** $\frac{y+11}{36}$ **6.** $\frac{25y+20}{36}$ **7.** $\frac{35b+12a}{15ab}$ **8.** $\frac{25y-9x}{30xy^2}$ **9.** $\frac{2(y+2)^2}{y(y+1)}$ or $\frac{2y^2+8y+8}{y(y+1)}$ **10.** $\frac{-3y-35}{y+5}$

6.4 More on Rational Expressions and Complex Fractions

OBJECTIVES

1 Add and Subtract Rational Expressions

2 Simplify Complex Fractions

1 Add and Subtract Rational Expressions

In this section, we expand our work with adding and subtracting rational expressions and discuss the process of simplifying complex fractions. Before we begin, however, this seems like an appropriate time to offer a bit of advice regarding your study of algebra. Success in algebra depends on having a good understanding of the concepts as well as on being able to perform the various computations. As for the computational work, you should adopt a carefully organized format that shows as many steps as you need in order to minimize the chances of making careless errors. Don't be eager to find shortcuts for certain computations before you have a thorough understanding of the steps involved in the process. This advice is especially appropriate at the beginning of this section.

Study Examples 1–4 very carefully. Note that the same basic procedure is followed in solving each problem:

Step 1 Factor the denominators.

Step 2 Find the LCD.

Step 3 Change each fraction to an equivalent fraction that has the LCD as its denominator.

Step 4 Combine the numerators and place over the LCD.

Step 5 Simplify by performing the addition or subtraction.

Step 6 Look for ways to reduce the resulting fraction.

EXAMPLE 1

Add $\frac{8}{x^2 - 4x} + \frac{2}{x}$.

Solution

$$\frac{8}{x^2 - 4x} + \frac{2}{x} = \frac{8}{x(x - 4)} + \frac{2}{x}$$ Factor the denominators

The LCD is $x(x - 4)$. Find the LCD

$$= \frac{8}{x(x - 4)} + \left(\frac{2}{x}\right)\left(\frac{x - 4}{x - 4}\right)$$ Change each fraction to an equivalent fraction that has the LCD as its denominator

$$= \frac{8 + 2(x - 4)}{x(x - 4)}$$ Combine numerators and place over the LCD

$$= \frac{8 + 2x - 8}{x(x - 4)}$$ Simplify by performing addition or subtraction

$$= \frac{2x}{x(x-4)}$$

$$= \frac{2}{x-4} \qquad \text{Reduce}$$

▼ **PRACTICE YOUR SKILL**

Add $\dfrac{8}{y^2 - 3y} + \dfrac{2}{y}$. ■

EXAMPLE 2

Subtract $\dfrac{a}{a^2 - 4} - \dfrac{3}{a + 2}$.

Solution

$$\frac{a}{a^2-4} - \frac{3}{a+2} = \frac{a}{(a+2)(a-2)} - \frac{3}{a+2} \qquad \text{Factor the denominators}$$

The LCD is $(a + 2)(a - 2)$. Find the LCD

$$= \frac{a}{(a+2)(a-2)} - \left(\frac{3}{a+2}\right)\left(\frac{a-2}{a-2}\right) \qquad \text{Change each fraction to an equivalent fraction that has the LCD as its denominator}$$

$$= \frac{a - 3(a-2)}{(a+2)(a-2)} \qquad \text{Combine numerators and place over the LCD}$$

$$= \frac{a - 3a + 6}{(a+2)(a-2)} \qquad \text{Simplify by performing addition or subtraction}$$

$$= \frac{-2a + 6}{(a+2)(a-2)} \quad \text{or} \quad \frac{-2(a-3)}{(a+2)(a-2)}$$

▼ **PRACTICE YOUR SKILL**

Subtract $\dfrac{2x}{x^2 - 16} - \dfrac{3}{x - 4}$. ■

EXAMPLE 3

Add $\dfrac{3n}{n^2 + 6n + 5} + \dfrac{4}{n^2 - 7n - 8}$.

Solution

$$\frac{3n}{n^2 + 6n + 5} + \frac{4}{n^2 - 7n - 8}$$

$$= \frac{3n}{(n+5)(n+1)} + \frac{4}{(n-8)(n+1)} \qquad \text{Factor the denominators}$$

The LCD is $(n + 5)(n + 1)(n - 8)$. Find the LCD

$$= \left(\frac{3n}{(n+5)(n+1)}\right)\left(\frac{n-8}{n-8}\right)$$

$$+ \left(\frac{4}{(n-8)(n+1)}\right)\left(\frac{n+5}{n+5}\right) \qquad \text{Change each fraction to an equivalent fraction that has the LCD as its denominator}$$

$$= \frac{3n(n-8) + 4(n+5)}{(n+5)(n+1)(n-8)} \qquad \text{Combine numerators and place over the LCD}$$

$$= \frac{3n^2 - 24n + 4n + 20}{(n + 5)(n + 1)(n - 8)}$$ Simplify by performing addition or subtraction

$$= \frac{3n^2 - 20n + 20}{(n + 5)(n + 1)(n - 8)}$$

▼ **PRACTICE YOUR SKILL**

Add $\frac{x}{x^2 + 6x + 8} + \frac{6}{x^2 + x - 12}$. ■

EXAMPLE 4

Perform the indicated operations.

$$\frac{2x^2}{x^4 - 1} + \frac{x}{x^2 - 1} - \frac{1}{x - 1}$$

Solution

$$\frac{2x^2}{x^4 - 1} + \frac{x}{x^2 - 1} - \frac{1}{x - 1}$$

$$= \frac{2x^2}{(x^2 + 1)(x + 1)(x - 1)} + \frac{x}{(x + 1)(x - 1)} - \frac{1}{x - 1}$$ Factor the denominators

The LCD is $(x^2 + 1)(x + 1)(x - 1)$. Find the LCD

$$= \frac{2x^2}{(x^2 + 1)(x + 1)(x - 1)} + \left(\frac{x}{(x + 1)(x - 1)}\right)\left(\frac{x^2 + 1}{x^2 + 1}\right) - \left(\frac{1}{x - 1}\right)\frac{(x^2 + 1)(x + 1)}{(x^2 + 1)(x + 1)}$$ Change each fraction to an equivalent fraction that has the LCD as its denominator

$$= \frac{2x^2 + x(x^2 + 1) - (x^2 + 1)(x + 1)}{(x^2 + 1)(x + 1)(x - 1)}$$ Combine numerators and place over the LCD

$$= \frac{2x^2 + x^3 + x - x^3 - x^2 - x - 1}{(x^2 + 1)(x + 1)(x - 1)}$$ Simplify by performing addition or subtraction

$$= \frac{x^2 - 1}{(x^2 + 1)(x + 1)(x - 1)}$$

$$= \frac{(x + 1)(x - 1)}{(x^2 + 1)(x + 1)(x - 1)}$$

$$= \frac{1}{x^2 + 1}$$ Reduce

▼ **PRACTICE YOUR SKILL**

Perform the indicated operations: $\frac{3}{y - 2} - \frac{y}{y^2 - 4} + \frac{y^2}{y^4 - 16}$. ■

2 Simplify Complex Fractions

Complex fractions are fractional forms that contain rational numbers or rational expressions in the numerators and/or denominators. The following are examples of complex fractions.

$$= \frac{3y + 2x}{xy} \div \frac{5y^2 - 6x}{xy^2}$$

$$= \frac{3y + 2x}{\cancel{xy}} \cdot \frac{\cancel{xy^2}^{\,y}}{5y^2 - 6x}$$

$$= \frac{y(3y + 2x)}{5y^2 - 6x}$$

Solution B

Here we find the LCD of all four denominators (x, y, x, and y^2). The LCD is xy^2. Use this LCD to multiply the entire complex fraction by a form of 1, specifically $\frac{xy^2}{xy^2}$.

$$\frac{\frac{3}{x} + \frac{2}{y}}{\frac{5}{x} - \frac{6}{y^2}} = \left(\frac{xy^2}{xy^2}\right)\left(\frac{\frac{3}{x} + \frac{2}{y}}{\frac{5}{x} - \frac{6}{y^2}}\right)$$

$$= \frac{xy^2\left(\frac{3}{x} + \frac{2}{y}\right)}{xy^2\left(\frac{5}{x} - \frac{6}{y^2}\right)}$$

$$= \frac{xy^2\left(\frac{3}{x}\right) + xy^2\left(\frac{2}{y}\right)}{xy^2\left(\frac{5}{x}\right) - xy^2\left(\frac{6}{y^2}\right)}$$

$$= \frac{3y^2 + 2xy}{5y^2 - 6x} \quad \text{or} \quad \frac{y(3y + 2x)}{5y^2 - 6x}$$

▼ **PRACTICE YOUR SKILL**

Simplify $\frac{\frac{4}{a^2} + \frac{3}{b^2}}{\frac{2}{a} - \frac{5}{b}}$. ■

Certainly either approach (Solution A or Solution B) will work with problems such as Examples 6 and 7. Examine Solution B in both examples carefully. This approach works effectively with complex fractions where the LCD of all the denominators is easy to find. (Don't be misled by the length of Solution B for Example 6; we were especially careful to show every step.)

EXAMPLE 8

Simplify $\frac{\frac{1}{x} + \frac{1}{y}}{2}$.

Solution

The number 2 can be written as $\frac{2}{1}$; thus the LCD of all three denominators (x, y, and 1) is xy. Therefore, let's multiply the entire complex fraction by a form of 1, specifically $\frac{xy}{xy}$.

$$\left(\frac{\frac{1}{x}+\frac{1}{y}}{\frac{2}{1}}\right)\left(\frac{xy}{xy}\right) = \frac{xy\left(\frac{1}{x}\right)+xy\left(\frac{1}{y}\right)}{2xy}$$

$$= \frac{y+x}{2xy}$$

▼ **PRACTICE YOUR SKILL**

Simplify $\dfrac{\frac{1}{a}+\frac{1}{b}}{5}$. ■

EXAMPLE 9

Simplify $\dfrac{-3}{\frac{2}{x}-\frac{3}{y}}$.

Solution

$$\left(\frac{\frac{-3}{1}}{\frac{2}{x}-\frac{3}{y}}\right)\left(\frac{xy}{xy}\right) = \frac{-3(xy)}{xy\left(\frac{2}{x}\right)-xy\left(\frac{3}{y}\right)}$$

$$= \frac{-3xy}{2y-3x}$$

▼ **PRACTICE YOUR SKILL**

Simplify $\dfrac{-4}{\frac{2}{a}-\frac{5}{b}}$. ■

Let's conclude this section with an example that has a complex fraction as part of an algebraic expression.

EXAMPLE 10

Simplify $1-\dfrac{n}{1-\frac{1}{n}}$.

Solution

First simplify the complex fraction $\dfrac{n}{1-\frac{1}{n}}$ by multiplying by $\dfrac{n}{n}$.

$$\left(\frac{n}{1-\frac{1}{n}}\right)\left(\frac{n}{n}\right) = \frac{n^2}{n-1}$$

Now we can perform the subtraction.

$$1-\frac{n^2}{n-1} = \left(\frac{n-1}{n-1}\right)\left(\frac{1}{1}\right)-\frac{n^2}{n-1}$$

$$= \frac{n-1}{n-1}-\frac{n^2}{n-1}$$

$$= \frac{n-1-n^2}{n-1} \quad \text{or} \quad \frac{-n^2+n-1}{n-1}$$

▼ **PRACTICE YOUR SKILL**

Simplify $2 - \dfrac{x}{1 + \dfrac{3}{x}}$.

CONCEPT QUIZ

For Problems 1–7, answer true or false.

1. A complex fraction can be described as a fraction within a fraction.
2. Division can simplify the complex fraction $\dfrac{\dfrac{2y}{x}}{\dfrac{6}{x^2}}$.
3. The complex fraction $\dfrac{\dfrac{3}{x-2} + \dfrac{2}{x+2}}{\dfrac{7x}{(x+2)(x-2)}}$ simplifies to $\dfrac{5x+2}{7x}$ for all values of x except $x = 0$.
4. The complex fraction $\dfrac{\dfrac{1}{3} - \dfrac{5}{6}}{\dfrac{1}{6} + \dfrac{5}{9}}$ simplifies to $-\dfrac{9}{13}$.
5. One method for simplifying a complex fraction is to multiply the entire fraction by a form of 1.
6. The complex fraction $\dfrac{\dfrac{3}{4} - \dfrac{1}{2}}{\dfrac{2}{3}}$ simplifies to $\dfrac{3}{8}$.
7. The complex fraction $\dfrac{\dfrac{7}{8} - \dfrac{1}{18}}{\dfrac{5}{6} + \dfrac{4}{15}}$ simplifies to $\dfrac{59}{33}$.
8. Arrange in order the following steps for adding rational expressions.
   A. Combine numerators and place over the LCD.
   B. Find the LCD.
   C. Reduce.
   D. Factor the denominators.
   E. Simplify by performing addition or subtraction.
   F. Change each fraction to an equivalent fraction that has the LCD as its denominator.

Problem Set 6.4

1 Add and Subtract Rational Expressions

For Problems 1–40, perform the indicated operations and express your answers in simplest form.

1. $\dfrac{2x}{x^2 + 4x} + \dfrac{5}{x}$
2. $\dfrac{3x}{x^2 - 6x} + \dfrac{4}{x}$
3. $\dfrac{4}{x^2 + 7x} - \dfrac{1}{x}$
4. $\dfrac{-10}{x^2 - 9x} - \dfrac{2}{x}$
5. $\dfrac{x}{x^2 - 1} + \dfrac{5}{x + 1}$
6. $\dfrac{2x}{x^2 - 16} + \dfrac{7}{x - 4}$
7. $\dfrac{6a + 4}{a^2 - 1} - \dfrac{5}{a - 1}$
8. $\dfrac{4a - 4}{a^2 - 4} - \dfrac{3}{a + 2}$
9. $\dfrac{2n}{n^2 - 25} - \dfrac{3}{4n + 20}$
10. $\dfrac{3n}{n^2 - 36} - \dfrac{2}{5n + 30}$
11. $\dfrac{5}{x} - \dfrac{5x - 30}{x^2 + 6x} + \dfrac{x}{x + 6}$

12. $\dfrac{3}{x+1} + \dfrac{x+5}{x^2-1} - \dfrac{3}{x-1}$

13. $\dfrac{3}{x^2+9x+14} + \dfrac{5}{2x^2+15x+7}$

14. $\dfrac{6}{x^2+11x+24} + \dfrac{4}{3x^2+13x+12}$

15. $\dfrac{1}{a^2-3a-10} - \dfrac{4}{a^2+4a-45}$

16. $\dfrac{6}{a^2-3a-54} - \dfrac{10}{a^2+5a-6}$

17. $\dfrac{3a}{8a^2-2a-3} + \dfrac{1}{4a^2+13a-12}$

18. $\dfrac{2a}{6a^2+13a-5} + \dfrac{a}{2a^2+a-10}$

19. $\dfrac{5}{x^2+3} - \dfrac{2}{x^2+4x-21}$

20. $\dfrac{7}{x^2+1} - \dfrac{3}{x^2+7x-60}$

21. $\dfrac{3x}{x^2-6x+9} - \dfrac{2}{x-3}$

22. $\dfrac{3}{x+4} + \dfrac{2x}{x^2+8x+16}$

23. $\dfrac{5}{x^2-1} + \dfrac{9}{x^2+2x+1}$

24. $\dfrac{6}{x^2-9} - \dfrac{9}{x^2-6x+9}$

25. $\dfrac{2}{y^2+6y-16} - \dfrac{4}{y+8} - \dfrac{3}{y-2}$

26. $\dfrac{7}{y-6} - \dfrac{10}{y+12} + \dfrac{4}{y^2+6y-72}$

27. $x - \dfrac{x^2}{x-2} + \dfrac{3}{x^2-4}$

28. $x + \dfrac{5}{x^2-25} - \dfrac{x^2}{x+5}$

29. $\dfrac{x+3}{x+10} + \dfrac{4x-3}{x^2+8x-20} + \dfrac{x-1}{x-2}$

30. $\dfrac{2x-1}{x+3} + \dfrac{x+4}{x-6} + \dfrac{3x-1}{x^2-3x-18}$

31. $\dfrac{n}{n-6} + \dfrac{n+3}{n+8} + \dfrac{12n+26}{n^2+2n-48}$

32. $\dfrac{n-1}{n+4} + \dfrac{n}{n+6} + \dfrac{2n+18}{n^2+10n+24}$

33. $\dfrac{4x-3}{2x^2+x-1} - \dfrac{2x+7}{3x^2+x-2} - \dfrac{3}{3x-2}$

34. $\dfrac{2x+5}{x^2+3x-18} - \dfrac{3x-1}{x^2+4x-12} + \dfrac{5}{x-2}$

35. $\dfrac{n}{n^2+1} + \dfrac{n^2+3n}{n^4-1} - \dfrac{1}{n-1}$

36. $\dfrac{2n^2}{n^4-16} - \dfrac{n}{n^2-4} + \dfrac{1}{n+2}$

37. $\dfrac{15x^2-10}{5x^2-7x+2} - \dfrac{3x+4}{x-1} - \dfrac{2}{5x-2}$

38. $\dfrac{32x+9}{12x^2+x-6} - \dfrac{3}{4x+3} - \dfrac{x+5}{3x-2}$

39. $\dfrac{t+3}{3t-1} + \dfrac{8t^2+8t+2}{3t^2-7t+2} - \dfrac{2t+3}{t-2}$

40. $\dfrac{t-3}{2t+1} + \dfrac{2t^2+19t-46}{2t^2-9t-5} - \dfrac{t+4}{t-5}$

2 Simplify Complex Fractions

For Problems 41–64, simplify each complex fraction.

41. $\dfrac{\frac{1}{2}-\frac{1}{4}}{\frac{5}{8}+\frac{3}{4}}$

42. $\dfrac{\frac{3}{8}+\frac{3}{4}}{\frac{5}{8}-\frac{7}{12}}$

43. $\dfrac{\frac{3}{28}-\frac{5}{14}}{\frac{5}{7}+\frac{1}{4}}$

44. $\dfrac{\frac{5}{9}+\frac{7}{36}}{\frac{3}{18}-\frac{5}{12}}$

45. $\dfrac{\frac{5}{6y}}{\frac{10}{3xy}}$

46. $\dfrac{\frac{9}{8xy^2}}{\frac{5}{4x^2}}$

47. $\dfrac{\frac{3}{x}-\frac{2}{y}}{\frac{4}{y}-\frac{7}{xy}}$

48. $\dfrac{\frac{9}{x}+\frac{7}{x^2}}{\frac{5}{y}+\frac{3}{y^2}}$

49. $\dfrac{\frac{6}{a}-\frac{5}{b^2}}{\frac{12}{a^2}+\frac{2}{b}}$

50. $\dfrac{\frac{4}{ab}-\frac{3}{b^2}}{\frac{1}{a}+\frac{3}{b}}$

51. $\dfrac{\frac{2}{x}-3}{\frac{3}{y}+4}$

52. $\dfrac{1+\frac{3}{x}}{1-\frac{6}{x}}$

53. $\dfrac{3+\frac{2}{n+4}}{5-\frac{1}{n+4}}$

54. $\dfrac{4+\frac{6}{n-1}}{7-\frac{4}{n-1}}$

55. $\dfrac{5-\frac{2}{n-3}}{4-\frac{1}{n-3}}$

56. $\dfrac{\frac{3}{n-5}-2}{1-\frac{4}{n-5}}$

57. $\dfrac{\dfrac{-1}{y-2}+\dfrac{5}{x}}{\dfrac{3}{x}-\dfrac{4}{xy-2x}}$

58. $\dfrac{\dfrac{-2}{x}-\dfrac{4}{x+2}}{\dfrac{3}{x^2+2x}+\dfrac{3}{x}}$

59. $\dfrac{\dfrac{2}{x-3}-\dfrac{3}{x+3}}{\dfrac{5}{x^2-9}-\dfrac{2}{x-3}}$

60. $\dfrac{\dfrac{2}{x-y}+\dfrac{3}{x+y}}{\dfrac{5}{x+y}-\dfrac{1}{x^2-y^2}}$

61. $\dfrac{3a}{2-\dfrac{1}{a}}-1$

62. $\dfrac{a}{\dfrac{1}{a}+4}+1$

63. $2-\dfrac{x}{3-\dfrac{2}{x}}$

64. $1+\dfrac{x}{1+\dfrac{1}{x}}$

THOUGHTS INTO WORDS

65. Which of the two techniques presented in the text would you use to simplify $\dfrac{\dfrac{1}{4}+\dfrac{1}{3}}{\dfrac{3}{4}-\dfrac{1}{6}}$? Which technique would you use to simplify $\dfrac{\dfrac{3}{8}-\dfrac{5}{7}}{\dfrac{7}{9}+\dfrac{6}{25}}$? Explain your choice for each problem.

66. Give a step-by-step description of how to do the following addition problem.

$$\frac{3x+4}{8}+\frac{5x-2}{12}$$

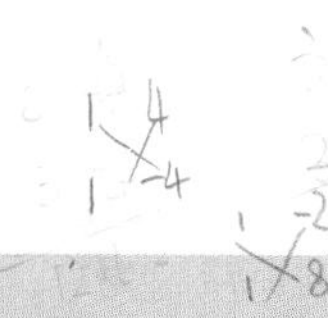

Answers to the Concept Quiz

1. True **2.** True **3.** False **4.** True **5.** True **6.** True **7.** False **8.** D, B, F, A, E, C

Answers to the Example Practice Skills

1. $\dfrac{2y+2}{y(y-3)}$ **2.** $\dfrac{-x-12}{(x+4)(x-4)}$ **3.** $\dfrac{x+1}{(x+4)(x-3)}$ **4.** $\dfrac{2y^3+7y^2-8y+24}{(y+2)(y-2)(y^2+4)}$ **5.** $\dfrac{2}{a}$ **6.** $\dfrac{4}{3}$ **7.** $\dfrac{4b^2+3a^2}{ab(2b-5a)}$ **8.** $\dfrac{b+a}{5ab}$ **9.** $\dfrac{-5ab}{2b-5a}$ **10.** $\dfrac{2x+6-x^2}{x+3}$

6.5 Dividing Polynomials

OBJECTIVES

1. Divide Polynomials
2. Use Synthetic Division to Divide Polynomials

1 Divide Polynomials

In Chapter 5, we saw how the property $\dfrac{b^n}{b^m}=b^{n-m}$, along with our knowledge of dividing integers, is used to divide monomials. For example,

$$\frac{12x^3}{3x}=4x^2 \qquad \frac{-36x^4y^5}{4xy^2}=-9x^3y^3$$

In Section 6.3, we used $\dfrac{a}{b}+\dfrac{c}{b}=\dfrac{a+c}{b}$ and $\dfrac{a}{b}-\dfrac{c}{b}=\dfrac{a-c}{b}$ as the basis for adding and subtracting rational expressions. These same equalities, viewed as

$\frac{a+b}{c} = \frac{a}{c} + \frac{b}{c}$ and $\frac{a-c}{b} = \frac{a}{b} - \frac{c}{b}$, along with our knowledge of dividing monomials, provide the basis for dividing polynomials by monomials. Consider the following examples:

$$\frac{18x^3 + 24x^2}{6x} = \frac{18x^3}{6x} + \frac{24x^2}{6x} = 3x^2 + 4x$$

$$\frac{35x^2y^3 - 55x^3y^4}{5xy^2} = \frac{35x^2y^3}{5xy^2} - \frac{55x^3y^4}{5xy^2} = 7xy - 11x^2y^2$$

To divide a polynomial by a monomial, we divide each term of the polynomial by the monomial. As with many skills, once you feel comfortable with the process, you may then want to perform some of the steps mentally. Your work could take on the following format.

$$\frac{40x^4y^5 + 72x^5y^7}{8x^2y} = 5x^2y^4 + 9x^3y^6 \qquad \frac{36a^3b^4 - 45a^4b^6}{-9a^2b^3} = -4ab + 5a^2b^3$$

In Section 6.1, we saw that a fraction like $\frac{3x^2 + 11x - 4}{x+4}$ can be simplified as follows:

$$\frac{3x^2 + 11x - 4}{x+4} = \frac{(3x-1)\cancel{(x+4)}}{\cancel{x+4}} = 3x - 1$$

We can obtain the same result by using a dividing process similar to long division in arithmetic.

Step 1 Use the conventional long-division format, and arrange both the dividend and the divisor in descending powers of the variable.

$$x + 4\overline{)3x^2 + 11x - 4}$$

Step 2 Find the first term of the quotient by dividing the first term of the dividend by the first term of the divisor.

$$\begin{array}{r} 3x \\ x + 4\overline{)3x^2 + 11x - 4} \end{array}$$

Step 3 Multiply the entire divisor by the term of the quotient found in Step 2, and position the product to be subtracted from the dividend.

$$\begin{array}{r} 3x \\ x + 4\overline{)3x^2 + 11x - 4} \\ \underline{3x^2 + 12x} \end{array}$$

Step 4 Subtract.

$$\begin{array}{r} 3x \\ x + 4\overline{)3x^2 + 11x - 4} \\ \underline{3x^2 + 12x} \\ -x - 4 \end{array}$$

Remember to add the opposite!
$(3x^2 + 11x - 4) - (3x^2 + 12x) = -x - 4$

Step 5 Repeat the process beginning with Step 2; use the polynomial that resulted from the subtraction in Step 4 as a new dividend.

$$\begin{array}{r} 3x \; -1 \\ x + 4\overline{)3x^2 + 11x - 4} \\ \underline{3x^2 + 12x} \\ -x - 4 \\ \underline{-x - 4} \end{array}$$

In the next example, let's *think* in terms of the previous step-by-step procedure but arrange our work in a more compact form.

EXAMPLE 1

Divide $5x^2 + 6x - 8$ by $x + 2$.

Solution

$$\begin{array}{r} 5x \; - 4 \\ x + 2\overline{)5x^2 + 6x - 8} \\ \underline{5x^2 + 10x} \\ -4x - 8 \\ \underline{-4x - 8} \\ 0 \end{array}$$

Think Steps

1. $\frac{5x^2}{x} = 5x.$
2. $5x(x + 2) = 5x^2 + 10x.$
3. $(5x^2 + 6x - 8) - (5x^2 + 10x) = -4x - 8.$
4. $\frac{-4x}{x} = -4.$ **5.** $-4(x + 2) = -4x - 8.$

▼ PRACTICE YOUR SKILL

Divide $2x^2 + 7x + 3$ by $x + 3$. ■

Recall that to check a division problem, we can multiply the divisor by the quotient and then add the remainder. In other words,

$$\text{Dividend} = (\text{Divisor})(\text{Quotient}) + (\text{Remainder})$$

Sometimes the remainder is expressed as a fractional part of the divisor. The relationship then becomes

$$\frac{\text{Dividend}}{\text{Divisor}} = \text{Quotient} + \frac{\text{Remainder}}{\text{Divisor}}$$

EXAMPLE 2 Divide $2x^2 - 3x + 1$ by $x - 5$.

Solution

$$\begin{array}{r}
2x + 7 \\
x - 5\overline{)2x^2 - 3x + 1} \\
\underline{2x^2 - 10x} \\
7x + 1 \\
\underline{7x - 35} \\
36
\end{array}$$

36 ← Remainder

Thus

$$\frac{2x^2 - 3x + 1}{x - 5} = 2x + 7 + \frac{36}{x - 5} \qquad x \neq 5$$

✔ **Check**

$$(x - 5)(2x + 7) + 36 \stackrel{?}{=} 2x^2 - 3x + 1$$
$$2x^2 - 3x - 35 + 36 \stackrel{?}{=} 2x^2 - 3x + 1$$
$$2x^2 - 3x + 1 = 2x^2 - 3x + 1$$

▼ PRACTICE YOUR SKILL

Divide $3x^2 - 10x - 4$ by $x - 4$. ■

Each of the next two examples illustrates another point regarding the division process. Study them carefully, and then you should be ready to work the exercises in the next problem set.

EXAMPLE 3 Divide $t^3 - 8$ by $t - 2$.

Solution

$$\begin{array}{r}
t^2 + 2t + 4 \\
t - 2\overline{)t^3 + 0t^2 + 0t - 8} \\
\underline{t^3 - 2t^2} \\
2t^2 + 0t - 8 \\
\underline{2t^2 - 4t} \\
4t - 8 \\
\underline{4t - 8} \\
0
\end{array}$$

← Note the insertion of a "t^2 term" and a "t term" with zero coefficients

Check this result!

▼ PRACTICE YOUR SKILL

Divide $y^3 + 27$ by $y + 3$. ■

EXAMPLE 4

Divide $y^3 + 3y^2 - 2y - 1$ by $y^2 + 2y$.

Solution

$$
\begin{array}{r}
y + 1 \\
y^2 + 2y \overline{)\, y^3 + 3y^2 - 2y - 1} \\
\underline{y^3 + 2y^2} \\
y^2 - 2y - 1 \\
\underline{y^2 + 2y} \\
-4y - 1
\end{array}
\longleftarrow \text{Remainder of } -4y - 1
$$

(The division process is complete when the degree of the remainder is less than the degree of the divisor.) Thus

$$\frac{y^3 + 3y^2 - 2y - 1}{y^2 + 2y} = y + 1 + \frac{-4y - 1}{y^2 + 2y}$$

▼ **PRACTICE YOUR SKILL**

Divide $x^3 - x^2 - 9x + 2$ by $x^2 + 3x$. ■

2 Use Synthetic Division to Divide Polynomials

If the divisor is of the form $x - k$, where the coefficient of the x term is 1, then the format of the division process described in this section can be simplified by a procedure called **synthetic division**. The procedure is a shortcut for this type of polynomial division. If you are continuing on to study college algebra, then you will want to know synthetic division. If you are not continuing on to college algebra, then you probably will not need a shortcut and the long-division process will be sufficient.

First, let's consider an example and use the usual division process. Then, in step-by-step fashion, we can observe some shortcuts that will lead us into the synthetic-division procedure. Consider the division problem $(2x^4 + x^3 - 17x^2 + 13x + 2) \div (x - 2)$:

$$
\begin{array}{r}
2x^3 + 5x^2 - 7x - 1 \\
x - 2 \overline{)\, 2x^4 + x^3 - 17x^2 + 13x + 2} \\
\underline{2x^4 - 4x^3} \\
5x^3 - 17x^2 \\
\underline{5x^3 - 10x^2} \\
-7x^2 + 13x \\
\underline{-7x^2 + 14x} \\
-x + 2 \\
\underline{-x + 2}
\end{array}
$$

Note that, because the dividend $(2x^4 + x^3 - 17x^2 + 13x + 2)$ is written in descending powers of x, the quotient $(2x^3 + 5x^2 - 7x - 1)$ is produced, also in descending powers of x. In other words, the numerical coefficients are the important numbers. Thus let's rewrite this problem in terms of its coefficients.

$$
\begin{array}{r}
2 + 5 - 7 - 1 \\
1 - 2 \overline{)\, 2 + 1 - 17 + 13 + 2} \\
\underline{\textcircled{2} - 4} \\
5 \textcircled{-17} \\
\underline{\textcircled{5} - 10} \\
-7 + \textcircled{13} \\
\underline{\textcircled{-7} + 14} \\
-1 + \textcircled{2} \\
\underline{\textcircled{-1} + 2}
\end{array}
$$

Now observe that the numbers that are circled are simply repetitions of the numbers directly above them in the format. Therefore, by removing the circled numbers, we can write the process in a more compact form as

$$\begin{array}{rrrrrrr}
 & 2 & 5 & -7 & -1 & & (1)\\
-2) & 2 & 1 & -17 & -13 & 2 & (2)\\
 & & -4 & -10 & 14 & 2 & (3)\\ \hline
 & & 5 & -7 & -1 & 0 & (4)
\end{array}$$

where the repetitions are omitted and where 1, the coefficient of x in the divisor, is omitted.

Note that line (4) reveals all of the coefficients of the quotient, line (1), except for the first coefficient of 2. Thus we can begin line (4) with the first coefficient and then use the following form.

$$\begin{array}{rrrrrrr}
-2) & 2 & 1 & -17 & 13 & 2 & (5)\\
 & & -4 & -10 & 14 & 2 & (6)\\ \hline
 & 2 & 5 & -7 & -1 & 0 & (7)
\end{array}$$

Line (7) contains the coefficients of the quotient, where the 0 indicates the remainder.

Finally, by changing the constant in the divisor to 2 (instead of -2), we can add the corresponding entries in lines (5) and (6) rather than subtract. Hence the final synthetic division form for this problem is

$$\begin{array}{rrrrrr}
2) & 2 & 1 & -17 & 13 & 2\\
 & & 4 & 10 & -14 & -2\\ \hline
 & 2 & 5 & -7 & -1 & 0
\end{array}$$

Now let's consider another problem that illustrates a step-by-step procedure for carrying out the synthetic-division process. Suppose that we want to divide $3x^3 - 2x^2 + 6x - 5$ by $x + 4$.

Step 1 Write the coefficients of the dividend as

$$\begin{array}{rrrrr}
) & 3 & -2 & 6 & -5
\end{array}$$

Step 2 In the divisor, $(x + 4)$, use -4 instead of 4 so that later we can add rather than subtract.

$$\begin{array}{rrrrr}
-4) & 3 & -2 & 6 & -5
\end{array}$$

Step 3 Bring down the first coeffecient of the dividend (3).

$$\begin{array}{rrrrr}
-4) & 3 & -2 & 6 & -5\\
 & & & & \\ \hline
 & 3 & & &
\end{array}$$

Step 4 Multiply(3)(-4), which yields -12; this result is to be added to the second coefficient of the dividend (-2).

$$\begin{array}{rrrrr}
-4) & 3 & -2 & 6 & -5\\
 & & -12 & & \\ \hline
 & 3 & -14 & &
\end{array}$$

Step 5 Multiply (-14)(-4), which yields 56; this result is to be added to the third coefficient of the dividend (6).

$$\begin{array}{rrrrr}
-4) & 3 & -2 & 6 & -5\\
 & & -12 & 56 & \\ \hline
 & 3 & -14 & 62 &
\end{array}$$

Step 6 Multiply $(62)(-4)$, which yields -248; this result is added to the last term of the dividend (-5).

$$\begin{array}{r|rrrr} -4 & 3 & -2 & 6 & -5 \\ & & -12 & 56 & -248 \\ \hline & 3 & -14 & 62 & -253 \end{array}$$

The last row indicates a quotient of $3x^2 - 14x + 62$ and a remainder of -253. Thus we have

$$\frac{3x^3 - 2x^2 + 6x - 5}{x + 4} = 3x^2 - 14x + 62 - \frac{253}{x + 4}$$

We will consider one more example, which shows only the final, compact form for synthetic division.

EXAMPLE 5

Find the quotient and remainder for $(4x^4 - 2x^3 + 6x - 1) \div (x - 1)$.

Solution

$$\begin{array}{r|rrrrr} 1 & 4 & -2 & 0 & 6 & -1 \\ & & 4 & 2 & 2 & 8 \\ \hline & 4 & 2 & 2 & 8 & 7 \end{array}$$

Note that a zero has been inserted as the coefficient of the missing x^2 term

Therefore,

$$\frac{4x^4 - 2x^3 + 6x - 1}{x - 1} = 4x^3 + 2x^2 + 2x + 8 + \frac{7}{x - 1}$$

▼ **PRACTICE YOUR SKILL**

Divide $3x^4 - 7x^3 + 6x^2 - 3x - 10$ by $x - 2$. ■

CONCEPT QUIZ

For Problems 1–10, answer true or false.

1. A division problem written as $(x^2 - x - 6) \div (x - 1)$ could also be written as $\frac{x^2 - x - 6}{x - 1}$.
2. The division of $\frac{x^2 + 7x + 12}{x + 3} = x + 4$ could be checked by multiplying $(x + 4)$ by $(x + 3)$.
3. For the division problem $(2x^2 + 5x + 9) \div (2x + 1)$, the remainder is 7. The remainder for the division problem can be expressed as $\frac{7}{2x + 1}$.
4. In general, to check a division problem we can multiply the divisor by the quotient and subtract the remainder.
5. If a term is inserted to act as a placeholder, then the coefficient of the term must be zero.
6. When performing division, the process ends when the degree of the remainder is less than the degree of the divisor.
7. The remainder is 0 when $x^3 - 1$ is divided by $x - 1$.
8. The remainder is 0 when $x^3 + 1$ is divided by $x + 1$.
9. The remainder is 0 when $x^3 - 1$ is divided by $x + 1$.
10. The remainder is 0 when $x^3 + 1$ is divided by $x - 1$.

Problem Set 6.5

1 Divide Polynomials

For Problems 1–10, perform the indicated divisions of polynomials by monomials.

1. $\frac{9x^4 + 18x^3}{3x}$

2. $\frac{12x^3 - 24x^2}{6x^2}$

3. $\frac{-24x^6 + 36x^8}{4x^2}$

4. $\frac{-35x^5 - 42x^3}{-7x^2}$

5. $\frac{15a^3 - 25a^2 - 40a}{5a}$

6. $\frac{-16a^4 + 32a^3 - 56a^2}{-8a}$

7. $\frac{13x^3 - 17x^2 + 28x}{-x}$

8. $\frac{14xy - 16x^2y^2 - 20x^3y^4}{-xy}$

9. $\frac{-18x^2y^2 + 24x^3y^2 - 48x^2y^3}{6xy}$

10. $\frac{-27a^3b^4 - 36a^2b^3 + 72a^2b^5}{9a^2b^2}$

For Problems 11–52, perform the indicated divisions.

11. $\frac{x^2 - 7x - 78}{x + 6}$

12. $\frac{x^2 + 11x - 60}{x - 4}$

13. $(x^2 + 12x - 160) \div (x - 8)$

14. $(x^2 - 18x - 175) \div (x + 7)$

15. $\frac{2x^2 - x - 4}{x - 1}$

16. $\frac{3x^2 - 2x - 7}{x + 2}$

17. $\frac{15x^2 + 22x - 5}{3x + 5}$

18. $\frac{12x^2 - 32x - 35}{2x - 7}$

19. $\frac{3x^3 + 7x^2 - 13x - 21}{x + 3}$

20. $\frac{4x^3 - 21x^2 + 3x + 10}{x - 5}$

21. $(2x^3 + 9x^2 - 17x + 6) \div (2x - 1)$

22. $(3x^3 - 5x^2 - 23x - 7) \div (3x + 1)$

23. $(4x^3 - x^2 - 2x + 6) \div (x - 2)$

24. $(6x^3 - 2x^2 + 4x - 3) \div (x + 1)$

25. $(x^4 - 10x^3 + 19x^2 + 33x - 18) \div (x - 6)$

26. $(x^4 + 2x^3 - 16x^2 + x + 6) \div (x - 3)$

27. $\frac{x^3 - 125}{x - 5}$

28. $\frac{x^3 + 64}{x + 4}$

29. $(x^3 + 64) \div (x + 1)$

30. $(x^3 - 8) \div (x - 4)$

31. $(2x^3 - x - 6) \div (x + 2)$

32. $(5x^3 + 2x - 3) \div (x - 2)$

33. $\frac{4a^2 - 8ab + 4b^2}{a - b}$

34. $\frac{3x^2 - 2xy - 8y^2}{x - 2y}$

35. $\frac{4x^3 - 5x^2 + 2x - 6}{x^2 - 3x}$

36. $\frac{3x^3 + 2x^2 - 5x - 1}{x^2 + 2x}$

37. $\frac{8y^3 - y^2 - y + 5}{y^2 + y}$

38. $\frac{5y^3 - 6y^2 - 7y - 2}{y^2 - y}$

39. $(2x^3 + x^2 - 3x + 1) \div (x^2 + x - 1)$

40. $(3x^3 - 4x^2 + 8x + 8) \div (x^2 - 2x + 4)$

41. $(4x^3 - 13x^2 + 8x - 15) \div (4x^2 - x + 5)$

42. $(5x^3 + 8x^2 - 5x - 2) \div (5x^2 - 2x - 1)$

43. $(5a^3 + 7a^2 - 2a - 9) \div (a^2 + 3a - 4)$

44. $(4a^3 - 2a^2 + 7a - 1) \div (a^2 - 2a + 3)$

45. $(2n^4 + 3n^3 - 2n^2 + 3n - 4) \div (n^2 + 1)$

46. $(3n^4 + n^3 - 7n^2 - 2n + 2) \div (n^2 - 2)$

47. $(x^5 - 1) \div (x - 1)$

48. $(x^5 + 1) \div (x + 1)$

49. $(x^4 - 1) \div (x + 1)$

50. $(x^4 - 1) \div (x - 1)$

51. $(3x^4 + x^3 - 2x^2 - x + 6) \div (x^2 - 1)$

52. $(4x^3 - 2x^2 + 7x - 5) \div (x^2 + 2)$

2 Use Synthetic Division to Divide Polynomials

For problems 53–64, use synthetic division to determine the quotient and remainder.

53. $(x^2 - 8x + 12) \div (x - 2)$

54. $(x^2 + 9x + 18) \div (x + 3)$

55. $(x^2 + 2x - 10) \div (x - 4)$

56. $(x^2 - 10x + 15) \div (x - 8)$

57. $(x^3 - 2x^2 - x + 2) \div (x - 2)$

58. $(x^3 - 5x^2 + 2x + 8) \div (x + 1)$

59. $(x^3 - 7x - 6) \div (x + 2)$

60. $(x^3 + 6x^2 - 5x - 1) \div (x - 1)$

61. $(2x^3 - 5x^2 - 4x + 6) \div (x - 2)$

62. $(3x^4 - x^3 + 2x^2 - 7x - 1) \div (x + 1)$

63. $(x^4 + 4x^3 - 7x - 1) \div (x - 3)$

64. $(2x^4 + 3x^2 + 3) \div (x + 2)$

THOUGHTS INTO WORDS

65. Describe the process of long division of polynomials.

66. Give a step-by-step description of how you would do the following division problem.

$$(4 - 3x - 7x^3) \div (x + 6)$$

67. How do you know by inspection that $3x^2 + 5x + 1$ cannot be the correct answer for the division problem $(3x^3 - 7x^2 - 22x + 8) \div (x - 4)$?

Answers to the Concept Quiz

1. True **2.** True **3.** True **4.** False **5.** True **6.** True **7.** True **8.** True **9.** False **10.** False

Answers to the Example Practice Skills

1. $2x + 1$ **2.** $3x + 2 + \frac{4}{x - 4}$ **3.** $y^2 - 3y + 9$ **4.** $x - 4 + \frac{3x + 2}{x^2 + 3x}$ **5.** $3x^3 - x^2 + 4x + 5$

6.6 Fractional Equations

OBJECTIVES

1. Solve Rational Equations
2. Solve Proportions
3. Solve Word Problems Involving Ratios

1 Solve Rational Equations

The fractional equations used in this text are of two basic types. One type has only constants as denominators, and the other type contains variables in the denominators.

In Chapter 2, we considered fractional equations that involve only constants in the denominators. Let's briefly review our approach to solving such equations, because we will be using that same basic technique to solve any type of fractional equation.

EXAMPLE 1

Solve $\frac{x - 2}{3} + \frac{x + 1}{4} = \frac{1}{6}$.

Solution

$$\frac{x - 2}{3} + \frac{x + 1}{4} = \frac{1}{6}$$

$$12\left(\frac{x - 2}{3} + \frac{x + 1}{4}\right) = 12\left(\frac{1}{6}\right) \quad \text{Multiply both sides by 12, which is the LCD of all of the denominators}$$

$$4(x - 2) + 3(x + 1) = 2$$

$$4x - 8 + 3x + 3 = 2$$

$$7x - 5 = 2$$

$$7x = 7$$

$$x = 1$$

The solution set is $\{1\}$. Check it!

▼ **PRACTICE YOUR SKILL**

Solve $\frac{x-2}{2} + \frac{x-2}{4} = \frac{x+3}{8}$. ■

If an equation contains a variable (or variables) in one or more denominators, then we proceed in essentially the same way as in Example 1 **except that we must avoid any value of the variable that makes a denominator zero**. Consider the following examples.

EXAMPLE 2

Solve $\frac{5}{n} + \frac{1}{2} = \frac{9}{n}$.

Solution

First, we need to realize that n cannot equal zero. (Let's indicate this restriction so that it is not forgotten!) Then we can proceed.

$$\frac{5}{n} + \frac{1}{2} = \frac{9}{n}, \qquad n \neq 0$$

$$2n\left(\frac{5}{n} + \frac{1}{2}\right) = 2n\left(\frac{9}{n}\right) \qquad \text{Multiply both sides by the LCD, which is } 2n$$

$$10 + n = 18$$

$$n = 8$$

The solution set is {8}. Check it!

▼ **PRACTICE YOUR SKILL**

Solve $\frac{5}{n} + \frac{1}{3} = \frac{7}{n}$. ■

EXAMPLE 3

Solve $\frac{35 - x}{x} = 7 + \frac{3}{x}$.

Solution

$$\frac{35 - x}{x} = 7 + \frac{3}{x}, \qquad x \neq 0$$

$$x\left(\frac{35 - x}{x}\right) = x\left(7 + \frac{3}{x}\right) \qquad \text{Multiply both sides by } x$$

$$35 - x = 7x + 3$$

$$32 = 8x$$

$$4 = x$$

The solution set is {4}.

▼ **PRACTICE YOUR SKILL**

Solve $\frac{30 - x}{x} = 3 + \frac{2}{x}$. ■

EXAMPLE 4 Solve $\frac{3}{a-2} = \frac{4}{a+1}$.

Solution

$$\frac{3}{a-2} = \frac{4}{a+1}, \qquad a \neq 2 \text{ and } a \neq -1$$

$$(a-2)(a+1)\left(\frac{3}{a-2}\right) = (a-2)(a+1)\left(\frac{4}{a+1}\right) \qquad \text{Multiply both sides by } (a-2)(a+1)$$

$$3(a+1) = 4(a-2)$$

$$3a + 3 = 4a - 8$$

$$11 = a$$

The solution set is $\{11\}$.

▼ **PRACTICE YOUR SKILL**

Solve $\frac{3}{x-2} = \frac{6}{x+2}$. ■

Keep in mind that listing the restrictions at the beginning of a problem does not replace checking the potential solutions. In Example 4, the answer 11 needs to be checked in the original equation.

EXAMPLE 5 Solve $\frac{a}{a-2} + \frac{2}{3} = \frac{2}{a-2}$.

Solution

$$\frac{a}{a-2} + \frac{2}{3} = \frac{2}{a-2}, \qquad a \neq 2$$

$$3(a-2)\left(\frac{a}{a-2} + \frac{2}{3}\right) = 3(a-2)\left(\frac{2}{a-2}\right) \qquad \text{Multiply both sides by } 3(a-2)$$

$$3a + 2(a-2) = 6$$

$$3a + 2a - 4 = 6$$

$$5a = 10$$

$$a = 2$$

Because our initial restriction was $a \neq 2$, we conclude that this equation has no solution. Thus the solution set is $\varnothing$.

▼ **PRACTICE YOUR SKILL**

Solve $\frac{x}{x-4} + \frac{1}{2} = \frac{4}{x-4}$. ■

2 Solve Proportions

A **ratio** is the comparison of two numbers by division. We often use the fractional form to express ratios. For example, we can write the ratio of a to b as $\frac{a}{b}$. A statement of equality between two ratios is called a **proportion**. Thus, if $\frac{a}{b}$ and $\frac{c}{d}$ are two equal ratios then we can form the proportion $\frac{a}{b} = \frac{c}{d}$ ($b \neq 0$ and $d \neq 0$). We deduce an important property of proportions as follows:

$$\frac{a}{b} = \frac{c}{d}, \quad b \neq 0 \text{ and } d \neq 0$$

$$bd\left(\frac{a}{b}\right) = bd\left(\frac{c}{d}\right) \quad \text{Multiply both sides by } bd$$

$$ad = bc$$

Cross-Multiplication Property of Proportions

If $\frac{a}{b} = \frac{c}{d}$ ($b \neq 0$ and $d \neq 0$), then $ad = bc$.

We can treat some fractional equations as proportions and solve them by using the cross-multiplication idea, as in the next examples.

EXAMPLE 6

Solve $\frac{5}{x+6} = \frac{7}{x-5}$.

Solution

$$\frac{5}{x+6} = \frac{7}{x-5}, \quad x \neq -6 \text{ and } x \neq 5$$

$$5(x-5) = 7(x+6) \quad \text{Apply the cross-multiplication property}$$

$$5x - 25 = 7x + 42$$

$$-67 = 2x$$

$$-\frac{67}{2} = x$$

The solution set is $\left\{-\frac{67}{2}\right\}$.

▼ **PRACTICE YOUR SKILL**

Solve $\frac{7}{x+3} = \frac{2}{x-8}$. ■

EXAMPLE 7

Solve $\frac{x}{7} = \frac{4}{x+3}$.

Solution

$$\frac{x}{7} = \frac{4}{x+3}, \quad x \neq -3$$

$$x(x+3) = 7(4) \quad \text{Cross-multiplication property}$$

$$x^2 + 3x = 28$$

$$x^2 + 3x - 28 = 0$$
$$(x + 7)(x - 4) = 0$$
$$x + 7 = 0 \quad \text{or} \quad x - 4 = 0$$
$$x = -7 \quad \text{or} \quad x = 4$$

The solution set is $\{-7, 4\}$. Check these solutions in the original equation.

▼ **PRACTICE YOUR SKILL**

Solve $\frac{y}{5} = \frac{3}{y - 2}$. ■

3 Solve Word Problems Involving Ratios

The ability to solve fractional equations broadens our base for solving word problems. We are now ready to tackle some word problems that translate into rational equations.

EXAMPLE 8

Apply Your Skill

On a certain map, $1\frac{1}{2}$ inches represents 25 miles. If two cities are $5\frac{1}{4}$ inches apart on the map, find the number of miles between the cities (see Figure 6.1).

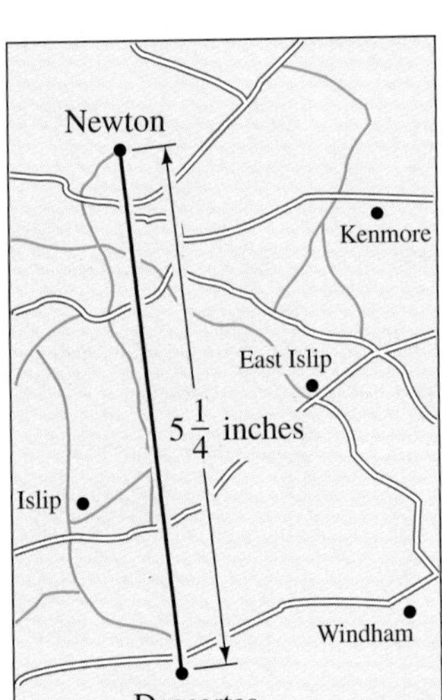

Figure 6.1

Solution

Let m represent the number of miles between the two cities. To set up the proportion, we will use a ratio of inches on the map to miles. Be sure to keep the ratio "inches on the map to miles" the same for both sides of the proportion.

$$\frac{1\frac{1}{2}}{25} = \frac{5\frac{1}{4}}{m}, \quad m \neq 0$$

$$\frac{\frac{3}{2}}{25} = \frac{\frac{21}{4}}{m}$$

$$\frac{3}{2}m = 25\left(\frac{21}{4}\right) \qquad \text{Cross-multiplication property}$$

$$\frac{2}{3}\left(\frac{3}{2}m\right) = \frac{2}{3}(25)\left(\frac{21}{4}\right) \qquad \text{Multiply both sides by } \frac{2}{3}$$

$$m = \frac{175}{2}$$

$$= 87\frac{1}{2}$$

The distance between the two cities is $87\frac{1}{2}$ miles.

▼ **PRACTICE YOUR SKILL**

On a scaled drawing of anatomy 2 centimeters represents 1.5 millimeters. If two blood vessels are 8 centimeters apart on the drawing, find the actual number of millimeters between the blood vessels. ■

EXAMPLE 9 Apply Your Skill

A sum of \$750 is to be divided between two people in the ratio of 2 to 3. How much does each person receive?

Solution

Let d represent the amount of money that one person receives. Then $750 - d$ represents the amount for the other person.

$$\frac{d}{750 - d} = \frac{2}{3}, \qquad d \neq 750$$

$$3d = 2(750 - d)$$

$$3d = 1500 - 2d$$

$$5d = 1500$$

$$d = 300$$

If $d = 300$, then $750 - d$ equals 450. Therefore, one person receives \$300 and the other person receives \$450.

▼ PRACTICE YOUR SKILL

A sum of \$2000 is to be divided between two people in the ratio of 1 to 4. How much does each person receive? ■

CONCEPT QUIZ

For Problems 1–3, answer true or false.

1. In solving rational equations, any value of the variable that makes a denominator zero cannot be a solution of the equation.
2. One method of solving rational equations is to multiply both sides of the equation by the lowest common denominator of the fractions in the equation.
3. In solving a rational equation that is a proportion, cross products can be set equal to each other.

For Problems 4–8, match each equation with its solution set.

4. $\frac{x+2}{5} = \frac{2x-3}{3}$	**A.** $\emptyset$
5. $\frac{7}{x-1} = \frac{4}{x+2}$	**B.** $\left\{\frac{43}{9}\right\}$
6. $\frac{46-n}{n} = 8 + \frac{3}{n}$	**C.** $\{3\}$
7. $\frac{x}{x-3} - \frac{3}{2} = \frac{3}{x-3}$	**D.** $\{-5, 0\}$
8. $-1 + \frac{2x}{x+3} = \frac{-4}{x+4}$	**E.** $\{-6\}$

9. Identify the following equations as a proportion or not a proportion.

   (a) $\frac{2x}{x+1} + x = \frac{7}{x+1}$ **(b)** $\frac{x-8}{2x+5} = \frac{7}{9}$ **(c)** $5 + \frac{2x}{x+6} = \frac{x-3}{x+4}$

10. Select all the equations that could represent the following problem. John bought three bottles of energy drink for \$5.07. If the price remains the same, what will eight bottles of the energy drink cost?

    (a) $\frac{3}{5.07} = \frac{x}{8}$ **(b)** $\frac{5.07}{8} = \frac{x}{3}$ **(c)** $\frac{3}{8} = \frac{5.07}{x}$ **(d)** $\frac{5.07}{3} = \frac{x}{8}$

Problem Set 6.6

1 Solve Rational Equations

For Problems 1–32, solve each equation.

1. $\frac{x+1}{4} + \frac{x-2}{6} = \frac{3}{4}$

2. $\frac{x+2}{5} + \frac{x-1}{6} = \frac{3}{5}$

3. $\frac{x+3}{2} - \frac{x-4}{7} = 1$

4. $\frac{x+4}{3} - \frac{x-5}{9} = 1$

5. $\frac{5}{n} + \frac{1}{3} = \frac{7}{n}$

6. $\frac{3}{n} + \frac{1}{6} = \frac{11}{3n}$

7. $\frac{7}{2x} + \frac{3}{5} = \frac{2}{3x}$

8. $\frac{9}{4x} + \frac{1}{3} = \frac{5}{2x}$

9. $\frac{3}{4x} + \frac{5}{6} = \frac{4}{3x}$

10. $\frac{5}{7x} - \frac{5}{6} = \frac{1}{6x}$

11. $\frac{47-n}{n} = 8 + \frac{2}{n}$

12. $\frac{45-n}{n} = 6 + \frac{3}{n}$

13. $\frac{n}{65-n} = 8 + \frac{2}{65-n}$

14. $\frac{n}{70-n} = 7 + \frac{6}{70-n}$

15. $n + \frac{1}{n} = \frac{17}{4}$

16. $n + \frac{1}{n} = \frac{37}{6}$

17. $n - \frac{2}{n} = \frac{23}{5}$

18. $n - \frac{3}{n} = \frac{26}{3}$

19. $\frac{x}{x+1} - 2 = \frac{3}{x-3}$

20. $\frac{x}{x-2} + 1 = \frac{8}{x-1}$

21. $\frac{a}{a+5} - 2 = \frac{3a}{a+5}$

22. $\frac{a}{a-3} - \frac{3}{2} = \frac{3}{a-3}$

23. $\frac{x}{x-6} - 3 = \frac{6}{x-6}$

24. $\frac{x}{x+1} + 3 = \frac{4}{x+1}$

25. $\frac{3s}{s+2} + 1 = \frac{35}{2(3s+1)}$

26. $\frac{s}{2s-1} - 3 = \frac{-32}{3(s+5)}$

27. $2 - \frac{3x}{x-4} = \frac{14}{x+7}$

28. $-1 + \frac{2x}{x+3} = \frac{-4}{x+4}$

29. $\frac{3n}{n-1} - \frac{1}{3} = \frac{-40}{3n-18}$

30. $\frac{n}{n+1} + \frac{1}{2} = \frac{-2}{n+2}$

31. $\frac{2x}{x-2} + \frac{15}{x^2-7x+10} = \frac{3}{x-5}$

32. $\frac{x}{x-4} - \frac{2}{x+3} = \frac{20}{x^2-x-12}$

2 Solve Proportions

For Problems 33–44, solve each proportion.

33. $\frac{5}{7x-3} = \frac{3}{4x-5}$

34. $\frac{3}{2x-1} = \frac{5}{3x+2}$

35. $\frac{-2}{x-5} = \frac{1}{x+9}$

36. $\frac{5}{2a-1} = \frac{-6}{3a+2}$

37. $\frac{5}{x+6} = \frac{6}{x-3}$

38. $\frac{3}{x-1} = \frac{4}{x+2}$

39. $\frac{3x-7}{10} = \frac{2}{x}$

40. $\frac{x}{-4} = \frac{3}{12x-25}$

41. $\frac{n+6}{27} = \frac{1}{n}$

42. $\frac{n}{5} = \frac{10}{n-5}$

43. $\frac{-3}{4x+5} = \frac{2}{5x-7}$

44. $\frac{7}{x+4} = \frac{3}{x-8}$

3 Solve Word Problems Involving Ratios

For Problems 45–56, set up an algebraic equation and solve each problem.

45. A sum of \$1750 is to be divided between two people in the ratio of 3 to 4. How much does each person receive?

46. A blueprint has a scale where 1 inch represents 5 feet. Find the dimensions of a rectangular room that measures $3\frac{1}{2}$ inches by $5\frac{3}{4}$ inches on the blueprint.

47. One angle of a triangle has a measure of 60° and the measures of the other two angles are in the ratio of 2 to 3. Find the measures of the other two angles.

48. The ratio of the complement of an angle to its supplement is 1 to 4. Find the measure of the angle.

49. If a home valued at \$150,000 is assessed \$2500 in real estate taxes, then how much, at the same rate, are the taxes on a home valued at \$210,000?

50. The ratio of male students to female students at a certain university is 5 to 7. If there is a total of 16,200 students, find the number of male students and the number of female students.

51. Suppose that Laura and Tammy together sold \$120.75 worth of candy for the annual school fair. If the ratio of Tammy's sales to Laura's sales was 4 to 3, how much did each sell?

52. The total value of a house and a lot is \$168,000. If the ratio of the value of the house to the value of the lot is 7 to 1, find the value of the house.

53. A 20-foot board is to be cut into two pieces whose lengths are in the ratio of 7 to 3. Find the lengths of the two pieces.

54. An inheritance of \$300,000 is to be divided between a son and the local heart fund in the ratio of 3 to 1. How much money will the son receive?

55. Suppose that, in a certain precinct, 1150 people voted in the last presidential election. If the ratio of female voters to male voters was 3 to 2, how many females and how many males voted?

56. The perimeter of a rectangle is 114 centimeters. If the ratio of its width to its length is 7 to 12, find the dimensions of the rectangle.

THOUGHTS INTO WORDS

57. How could you do Problem 53 without using algebra?

58. Now do Problem 55 using the same approach that you used in Problem 57. What difficulties do you encounter?

59. How can you tell by inspection that the equation $\frac{x}{x+2} = \frac{-2}{x+2}$ has no solution?

60. How would you help someone solve the equation $\frac{3}{x} - \frac{4}{x} = \frac{-1}{x}$?

Answers to the Concept Quiz

1. True **2.** True **3.** True **4.** C **5.** E **6.** B **7.** A **8.** D **9.** **(a)** Not a proportion **(b)** Proportion **(c)** Not a proportion **10.** **c** and **d**

Answers to the Example Practice Skills

1. $\{3\}$ **2.** $\{6\}$ **3.** $\{7\}$ **4.** $\{6\}$ **5.** $\varnothing$ **6.** $\left\{\frac{62}{5}\right\}$ **7.** $\{-3, 5\}$ **8.** 6 mm **9.** \$400, \$1600

6.7 More Rational Equations and Applications

OBJECTIVES

1. Solve Rational Equations Where Denominators Require Factoring
2. Solve Formulas That Are in Fractional Form
3. Solve Rate–Time Word Problems

1 Solve Rational Equations Where Denominators Require Factoring

Let's begin this section by considering a few more rational equations. We will continue to solve them using the same basic techniques as in the previous section. That is, we will multiply both sides of the equation by the least common denominator of all of the denominators in the equation, with the necessary restrictions to avoid division by zero. Some of the denominators in these problems will require factoring before we can determine a least common denominator.

EXAMPLE 1

Solve $\frac{x}{2x-8} + \frac{16}{x^2-16} = \frac{1}{2}$.

Solution

$$\frac{x}{2x-8} + \frac{16}{x^2-16} = \frac{1}{2}$$

$$\frac{x}{2(x-4)} + \frac{16}{(x+4)(x-4)} = \frac{1}{2}, \quad x \neq 4 \text{ and } x \neq -4$$

$$2(x-4)(x+4)\left(\frac{x}{2(x-4)} + \frac{16}{(x+4)(x-4)}\right) = 2(x+4)(x-4)\left(\frac{1}{2}\right)$$

Multiply both sides by the LCD, $2(x-4)(x+4)$

$$x(x+4) + 2(16) = (x+4)(x-4)$$

$$x^2 + 4x + 32 = x^2 - 16$$

$$4x = -48$$

$$x = -12$$

The solution set is $\{-12\}$. Perhaps you should check it!

▼ **PRACTICE YOUR SKILL**

Solve $\frac{x}{3x+6} + \frac{8}{x^2-4} = \frac{1}{3}$. ■

In Example 1, note that the restrictions were not indicated until the denominators were expressed in factored form. It is usually easier to determine the necessary restrictions at this step.

EXAMPLE 2

Solve $\frac{3}{n-5} - \frac{2}{2n+1} = \frac{n+3}{2n^2-9n-5}$.

Solution

$$\frac{3}{n-5} - \frac{2}{2n+1} = \frac{n+3}{2n^2-9n-5}$$

$$\frac{3}{n-5} - \frac{2}{2n+1} = \frac{n+3}{(2n+1)(n-5)}, \quad n \neq -\frac{1}{2} \text{ and } n \neq 5$$

$$(2n+1)(n-5)\left(\frac{3}{n-5} - \frac{2}{2n+1}\right) = (2n+1)(n-5)\left(\frac{n+3}{(2n+1)(n-5)}\right)$$

Multiply both sides by the LCD, $(2n+1)(n-5)$

$$3(2n+1) - 2(n-5) = n+3$$

$$6n + 3 - 2n + 10 = n + 3$$

$$4n + 13 = n + 3$$

$$3n = -10$$

$$n = -\frac{10}{3}$$

The solution set is $\left\{-\frac{10}{3}\right\}$.

▼ **PRACTICE YOUR SKILL**

Solve $\frac{5}{x-2} - \frac{2}{3x+1} = \frac{x+1}{3x^2-5x-2}$. ■

EXAMPLE 3

Solve $2 + \frac{4}{x-2} = \frac{8}{x^2 - 2x}$.

Solution

$$2 + \frac{4}{x-2} = \frac{8}{x^2 - 2x}$$

$$2 + \frac{4}{x-2} = \frac{8}{x(x-2)}, \qquad x \neq 0 \text{ and } x \neq 2$$

$$x(x-2)\left(2 + \frac{4}{x-2}\right) = x(x-2)\left(\frac{8}{x(x-2)}\right) \qquad \text{Multiply both sides by the LCD, } x(x-2)$$

$$2x(x-2) + 4x = 8$$

$$2x^2 - 4x + 4x = 8$$

$$2x^2 = 8$$

$$x^2 = 4$$

$$x^2 - 4 = 0$$

$$(x+2)(x-2) = 0$$

$$x + 2 = 0 \quad \text{or} \quad x - 2 = 0$$

$$x = -2 \quad \text{or} \quad x = 2$$

Because our initial restriction indicated that $x \neq 2$, the only solution is -2. Thus the solution set is $\{-2\}$.

▼ PRACTICE YOUR SKILL

Solve $2 - \frac{6}{x+3} = \frac{18}{x^2 + 3x}$. ■

2 Solve Formulas That Are in Fractional Form

In Section 2.4 we discussed using the properties of equality to change the form of various formulas. For example, we considered the simple interest formula $A = P + Prt$ and changed its form by solving for P as follows:

$$A = P + Prt$$

$$A = P(1 + rt)$$

$$\frac{A}{1 + rt} = P \qquad \text{Multiply both sides by } \frac{1}{1 + rt}$$

If the formula is in the form of a rational equation, then the techniques of these last two sections are applicable. Consider the following example.

EXAMPLE 4

If the original cost of some business property is C dollars and it is depreciated linearly over N years, then its value V at the end of T years is given by

$$V = C\left(1 - \frac{T}{N}\right)$$

Solve this formula for N in terms of V, C, and T.

Solution

$$V = C\left(1 - \frac{T}{N}\right)$$

$$V = C - \frac{CT}{N}$$

$$N(V) = N\left(C - \frac{CT}{N}\right) \quad \text{Multiply both sides by } N$$

$$NV = NC - CT$$

$$NV - NC = -CT$$

$$N(V - C) = -CT$$

$$N = \frac{-CT}{V - C}$$

$$N = -\frac{CT}{V - C}$$

▼ **PRACTICE YOUR SKILL**

Solve $S = \dfrac{a}{1 - r}$ for r. ■

3 Solve Rate–Time Word Problems

In Section 2.4 we solved some uniform motion problems. The formula $d = rt$ was used in the analysis of the problems, and we used guidelines that involve distance relationships. Now let's consider some uniform motion problems where guidelines that involve either times or rates are appropriate. These problems will generate fractional equations to solve.

EXAMPLE 5

Apply Your Skill

An airplane travels 2050 miles in the same time that a car travels 260 miles. If the rate of the plane is 358 miles per hour greater than the rate of the car, find the rate of each.

Solution

Let r represent the rate of the car. Then $r + 358$ represents the rate of the plane. The fact that the times are equal can be a guideline. Remember from the basic formula, $d = rt$, that $t = \dfrac{d}{r}$.

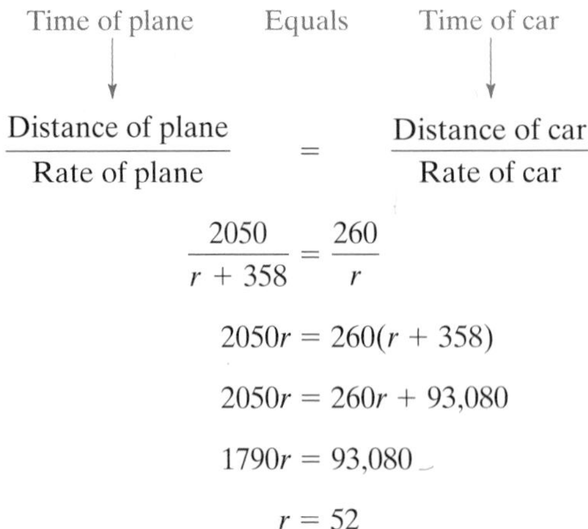

Time of plane ↓ Equals Time of car ↓

$$\frac{\text{Distance of plane}}{\text{Rate of plane}} = \frac{\text{Distance of car}}{\text{Rate of car}}$$

$$\frac{2050}{r + 358} = \frac{260}{r}$$

$$2050r = 260(r + 358)$$

$$2050r = 260r + 93{,}080$$

$$1790r = 93{,}080$$

$$r = 52$$

If $r = 52$, then $r + 358$ equals 410. Thus the rate of the car is 52 miles per hour and the rate of the plane is 410 miles per hour.

▼ **PRACTICE YOUR SKILL**

An airplane travels 2200 miles in the same time that a car travels 280 miles. If the rate of the plane is 480 miles per hour greater than the rate of the car, find the rate of each. ■

EXAMPLE 6

Apply Your Skill

Darren Hedges/Used under license from Shutterstock

It takes a freight train 2 hours longer to travel 300 miles than it takes an express train to travel 280 miles. The rate of the express train is 20 miles per hour greater than the rate of the freight train. Find the times and rates of both trains.

Solution

Let t represent the time of the express train. Then $t + 2$ represents the time of the freight train. Let's record the information of this problem in a table.

	Distance	Time	Rate $= \dfrac{\text{Distance}}{\text{Time}}$
Express train	280	t	$\dfrac{280}{t}$
Freight train	300	$t + 2$	$\dfrac{300}{t+2}$

The fact that the rate of the express train is 20 miles per hour greater than the rate of the freight train can be a guideline.

Rate of express ↓ Equals Rate of freight train plus 20 ↓

$$\frac{280}{t} = \frac{300}{t+2} + 20$$

$$t(t+2)\left(\frac{280}{t}\right) = t(t+2)\left(\frac{300}{t+2} + 20\right)$$

$$280(t+2) = 300t + 20t(t+2)$$

$$280t + 560 = 300t + 20t^2 + 40t$$

$$280t + 560 = 340t + 20t^2$$

$$0 = 20t^2 + 60t - 560$$

$$0 = t^2 + 3t - 28$$

$$0 = (t+7)(t-4)$$

$$t + 7 = 0 \quad \text{or} \quad t - 4 = 0$$

$$t = -7 \quad \text{or} \quad t = 4$$

The negative solution must be discarded, so the time of the express train (t) is 4 hours, and the time of the freight train ($t + 2$) is 6 hours. The rate of the express train $\left(\frac{280}{t}\right)$ is $\frac{280}{4} = 70$ miles per hour, and the rate of the freight train $\left(\frac{300}{t+2}\right)$ is $\frac{300}{6} = 50$ miles per hour.

▼ PRACTICE YOUR SKILL

It takes a tour bus 1 hour longer to travel 220 miles than it takes a car to travel 210 miles. The rate of the car is 15 miles per hour greater than the rate of the tour bus. Find the times and rates of both the tour bus and the car. ■

Remark: Note that to solve Example 5 we went directly to a guideline without the use of a table, but for Example 6 we used a table. Again, remember that this is a personal preference; we are merely acquainting you with a variety of techniques.

Uniform motion problems are a special case of a larger group of problems we refer to as **rate–time problems**. For example, if a certain machine can produce 150 items in 10 minutes, then we say that the machine is producing at a rate of $\frac{150}{10} = 15$ items per minute. Likewise, if a person can do a certain job in 3 hours then, assuming a constant rate of work, we say that the person is working at a rate of $\frac{1}{3}$ of the job per hour. In general, if Q is the quantity of something done in t units of time then the rate, r, is given by $r = \frac{Q}{t}$. We state the rate in terms of *so much quantity per unit of time.* (In uniform motion problems, the "quantity" is distance.) Let's consider some examples of rate–time problems.

EXAMPLE 7

David R. Frazier Photolibrary, Inc./Alamy Limited

Apply Your Skill

If Jim can mow a lawn in 50 minutes and if his son, Todd, can mow the same lawn in 40 minutes, then how long will it take them to mow the lawn if they work together?

Solution

Jim's rate is $\frac{1}{50}$ of the lawn per minute and Todd's rate is $\frac{1}{40}$ of the lawn per minute. If we let m represent the number of minutes that they work together, then $\frac{1}{m}$ represents their rate when working together. Therefore, because the sum of the individual rates must equal the rate working together, we can set up and solve the following equation.

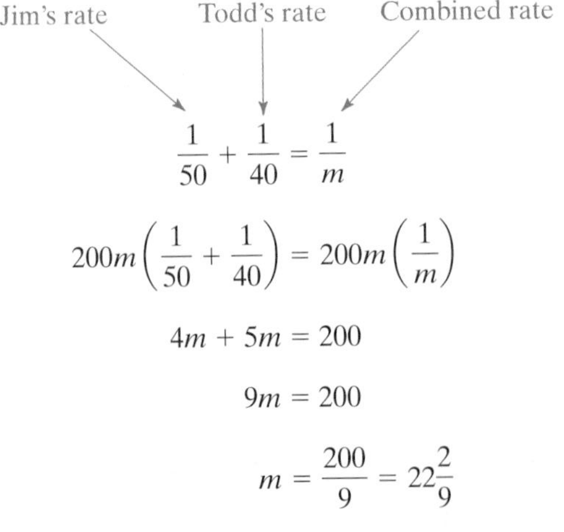

$$\frac{1}{50} + \frac{1}{40} = \frac{1}{m}$$

$$200m\left(\frac{1}{50} + \frac{1}{40}\right) = 200m\left(\frac{1}{m}\right)$$

$$4m + 5m = 200$$

$$9m = 200$$

$$m = \frac{200}{9} = 22\frac{2}{9}$$

It should take them $22\frac{2}{9}$ minutes.

▼ PRACTICE YOUR SKILL

If Maria can clean a house in 120 minutes and Stephanie can clean the same house in 200 minutes, how long will it take them to clean the house if they work together? ■

EXAMPLE 8

Creatas/Superstock

Apply Your Skill

Working together, Linda and Kathy can type a term paper in $3\frac{3}{5}$ hours. Linda can type the paper by herself in 6 hours. How long would it take Kathy to type the paper by herself?

Solution

Their rate working together is $\dfrac{1}{3\frac{3}{5}} = \dfrac{1}{\frac{18}{5}} = \dfrac{5}{18}$ of the job per hour, and Linda's rate is $\dfrac{1}{6}$ of the job per hour. If we let h represent the number of hours that it would take Kathy to do the job by herself, then her rate is $\dfrac{1}{h}$ of the job per hour. Thus we have

$$\underset{\text{Linda's rate}}{\frac{1}{6}} + \underset{\text{Kathy's rate}}{\frac{1}{h}} = \underset{\text{Combined rate}}{\frac{5}{18}}$$

Solving this equation yields

$$18h\left(\frac{1}{6} + \frac{1}{h}\right) = 18h\left(\frac{5}{18}\right)$$

$$3h + 18 = 5h$$

$$18 = 2h$$

$$9 = h$$

It would take Kathy 9 hours to type the paper by herself.

▼ PRACTICE YOUR SKILL

Working together, Josh and Mayra can print and fold the school newspaper in $2\frac{2}{3}$ hours. Josh can print and fold the paper by himself in 4 hours. How long would it take Mayra to print and fold the paper by herself? ■

Our final example of this section illustrates another approach that some people find meaningful for rate–time problems. For this approach, think in terms of fractional parts of the job. For example, if a person can do a certain job in 5 hours, then at the end of 2 hours, he or she has done $\frac{2}{5}$ of the job. (Again, assume a constant rate of work.) At the end of 4 hours, he or she has finished $\frac{4}{5}$ of the job; and, in general, at the end of h hours, he or she has done $\frac{h}{5}$ of the job. Then, just as in the motion problems where distance equals rate multiplied by the time, here the fractional part done equals the working rate multiplied by the time. Let's see how this works in a problem.

EXAMPLE 9

Jim West/Alamy Limited

Apply Your Skill

It takes Pat 12 hours to complete a task. After he had been working for 3 hours, he was joined by his brother Mike, and together they finished the task in 5 hours. How long would it take Mike to do the job by himself?

Solution

Let h represent the number of hours that it would take Mike to do the job by himself. The fractional part of the job that Pat does equals his working rate multiplied by his time. Because it takes Pat 12 hours to do the entire job, his working rate is $\frac{1}{12}$. He works for 8 hours (3 hours before Mike and then 5 hours with Mike). Therefore, Pat's part of the job is $\frac{1}{12}(8) = \frac{8}{12}$. The fractional part of the job that Mike does equals his working rate multiplied by his time. Because h represents Mike's time to do the entire job, his working rate is $\frac{1}{h}$; he works for 5 hours. Therefore, Mike's part of the job is $\frac{1}{h}(5) = \frac{5}{h}$. Adding the two fractional parts together results in 1 entire job being done. Let's also show this information in chart form and set up our guideline. Then we can set up and solve the equation.

	Time to do entire job	Working rate	Time working	Fractional part of the job done
Pat	12	$\frac{1}{12}$	8	$\frac{8}{12}$
Mike	h	$\frac{1}{h}$	5	$\frac{5}{h}$

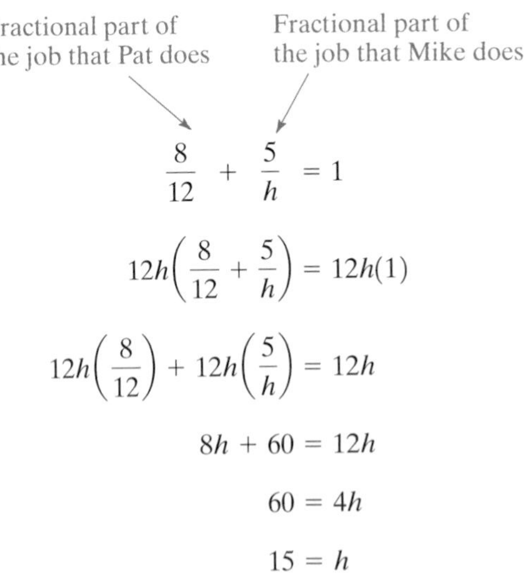

$$\frac{8}{12} + \frac{5}{h} = 1$$

$$12h\left(\frac{8}{12} + \frac{5}{h}\right) = 12h(1)$$

$$12h\left(\frac{8}{12}\right) + 12h\left(\frac{5}{h}\right) = 12h$$

$$8h + 60 = 12h$$

$$60 = 4h$$

$$15 = h$$

It would take Mike 15 hours to do the entire job by himself.

▼ PRACTICE YOUR SKILL

It takes John 8 hours to detail a boat. After working for 2 hours he was joined by Franco, and together they finished the boat in 3 hours. How long would it take Franco to detail the boat by himself? ■

CONCEPT QUIZ

For Problems 1–10, answer true or false.

1. Assuming uniform motion, the rate at which a car travels is equal to the time traveled divided by the distance traveled.
2. If a worker can lay 640 square feet of tile in 8 hours, we can say his rate of work is 80 square feet per hour.
3. If a person can complete two jobs in 5 hours, then the person is working at the rate of $\frac{5}{2}$ of the job per hour.
4. In a time–rate problem involving two workers, the sum of their individual rates must equal the rate working together.
5. If a person works at the rate of $\frac{2}{15}$ of the job per hour, then at the end of 3 hours the job would be $\frac{6}{15}$ completed.
6. If a person can do a job in 7 hours, then at the end of 5 hours he or she will have completed $\frac{5}{7}$ of the job.
7. Solving the equation $y = \frac{a}{b}x + \frac{c}{d}$ for x yields $x = \frac{y - bc}{ad}$.
8. Solving the equation $y = \frac{3}{4}(x - 4)$ for x yields $x = \frac{4y + 12}{3}$.
9. If Zorka can complete a certain task in 5 hours and Mitzie can complete the same task in 9 hours, then working together they should be able to complete the task in 7 hours.
10. The solution set for the equation $\frac{7x + 2}{12x^2 + 11x - 15} - \frac{1}{3x + 5} = \frac{2}{4x - 3}$ is $\varnothing$.

Problem Set 6.7

1 Solve Rational Equations Where Denominators Require Factoring

For Problems 1–30, solve each equation.

1. $\frac{x}{4x - 4} + \frac{5}{x^2 - 1} = \frac{1}{4}$
2. $\frac{x}{3x - 6} + \frac{4}{x^2 - 4} = \frac{1}{3}$
3. $3 + \frac{6}{t - 3} = \frac{6}{t^2 - 3t}$
4. $2 + \frac{4}{t - 1} = \frac{4}{t^2 - t}$
5. $\frac{3}{n - 5} + \frac{4}{n + 7} = \frac{2n + 11}{n^2 + 2n - 35}$
6. $\frac{2}{n + 3} + \frac{3}{n - 4} = \frac{2n - 1}{n^2 - n - 12}$
7. $\frac{5x}{2x + 6} - \frac{4}{x^2 - 9} = \frac{5}{2}$
8. $\frac{3x}{5x + 5} - \frac{2}{x^2 - 1} = \frac{3}{5}$
9. $1 + \frac{1}{n - 1} = \frac{1}{n^2 - n}$
10. $3 + \frac{9}{n - 3} = \frac{27}{n^2 - 3n}$
11. $\frac{2}{n - 2} - \frac{n}{n + 5} = \frac{10n + 15}{n^2 + 3n - 10}$
12. $\frac{n}{n + 3} + \frac{1}{n - 4} = \frac{11 - n}{n^2 - n - 12}$
13. $\frac{2}{2x - 3} - \frac{2}{10x^2 - 13x - 3} = \frac{x}{5x + 1}$
14. $\frac{1}{3x + 4} + \frac{6}{6x^2 + 5x - 4} = \frac{x}{2x - 1}$
15. $\frac{2x}{x + 3} - \frac{3}{x - 6} = \frac{29}{x^2 - 3x - 18}$
16. $\frac{x}{x - 4} - \frac{2}{x + 8} = \frac{63}{x^2 + 4x - 32}$
17. $\frac{a}{a - 5} + \frac{2}{a - 6} = \frac{2}{a^2 - 11a + 30}$
18. $\frac{a}{a + 2} + \frac{3}{a + 4} = \frac{14}{a^2 + 6a + 8}$
19. $\frac{-1}{2x - 5} + \frac{2x - 4}{4x^2 - 25} = \frac{5}{6x + 15}$
20. $\frac{-2}{3x + 2} + \frac{x - 1}{9x^2 - 4} = \frac{3}{12x - 8}$

21. $\frac{7y+2}{12y^2+11y-15} - \frac{1}{3y+5} = \frac{2}{4y-3}$

22. $\frac{5y-4}{6y^2+y-12} - \frac{2}{2y+3} = \frac{5}{3y-4}$

23. $\frac{2n}{6n^2+7n-3} - \frac{n-3}{3n^2+11n-4} = \frac{5}{2n^2+11n+12}$

24. $\frac{x+1}{2x^2+7x-4} - \frac{x}{2x^2-7x+3} = \frac{1}{x^2+x-12}$

25. $\frac{1}{2x^2-x-1} + \frac{3}{2x^2+x} = \frac{2}{x^2-1}$

26. $\frac{2}{n^2+4n} + \frac{3}{n^2-3n-28} = \frac{5}{n^2-6n-7}$

27. $\frac{x+1}{x^3-9x} - \frac{1}{2x^2+x-21} = \frac{1}{2x^2+13x+21}$

28. $\frac{x}{2x^2+5x} - \frac{x}{2x^2+7x+5} = \frac{2}{x^2+x}$

29. $\frac{4t}{4t^2-t-3} + \frac{2-3t}{3t^2-t-2} = \frac{1}{12t^2+17t+6}$

30. $\frac{2t}{2t^2+9t+10} + \frac{1-3t}{3t^2+4t-4} = \frac{4}{6t^2+11t-10}$

2 Solve Formulas That Are in Fractional Form

For Problems 31–44, solve each equation for the indicated variable.

31. $y = \frac{5}{6}x + \frac{2}{9}$ for x

32. $y = \frac{3}{4}x - \frac{2}{3}$ for x

33. $\frac{-2}{x-4} = \frac{5}{y-1}$ for y

34. $\frac{7}{y-3} = \frac{3}{x+1}$ for y

35. $I = \frac{100M}{C}$ for M

36. $V = C\left(1 - \frac{T}{N}\right)$ for T

37. $\frac{R}{S} = \frac{T}{S+T}$ for R

38. $\frac{1}{R} = \frac{1}{S} + \frac{1}{T}$ for R

39. $\frac{y-1}{x-3} = \frac{b-1}{a-3}$ for y

40. $y = -\frac{a}{b}x + \frac{c}{d}$ for x

41. $\frac{x}{a} + \frac{y}{b} = 1$ for y

42. $\frac{y-b}{x} = m$ for y

43. $\frac{y-1}{x+6} = \frac{-2}{3}$ for y

44. $\frac{y+5}{x-2} = \frac{3}{7}$ for y

3 Solve Rate–Time Word Problems

Set up an equation and solve each of the following problems.

45. Kent drives his Mazda 270 miles in the same time that it takes Dave to drive his Nissan 250 miles. If Kent averages 4 miles per hour faster than Dave, find their rates.

46. Suppose that Wendy rides her bicycle 30 miles in the same time that it takes Kim to ride her bicycle 20 miles. If Wendy rides 5 miles per hour faster than Kim, find the rate of each.

47. An inlet pipe can fill a tank (see Figure 6.2) in 10 minutes. A drain can empty the tank in 12 minutes. If the tank is empty and both the pipe and drain are open, how long will it take before the tank overflows?

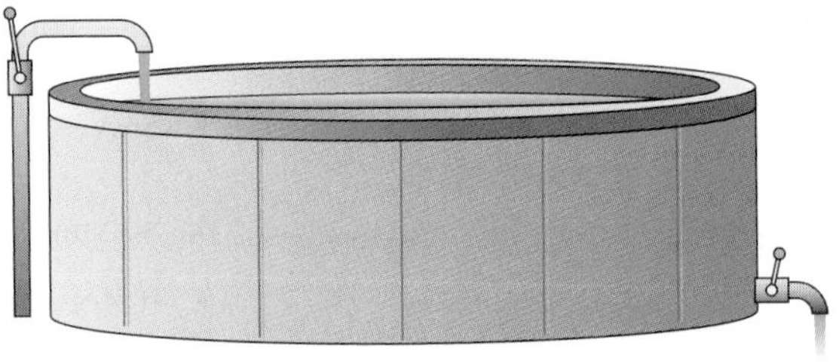

Figure 6.2

48. Barry can do a certain job in 3 hours, whereas it takes Sanchez 5 hours to do the same job. How long would it take them to do the job working together?

49. Connie can type 600 words in 5 minutes less than it takes Katie to type 600 words. If Connie types at a rate of 20 words per minute faster than Katie types, find the typing rate of each woman.

50. Walt can mow a lawn in 1 hour and his son, Malik, can mow the same lawn in 50 minutes. One day Malik started mowing the lawn by himself and worked for 30 minutes. Then Walt joined him and they finished the lawn. How long did it take them to finish mowing the lawn after Walt started to help?

51. Plane A can travel 1400 miles in 1 hour less time than it takes plane B to travel 2000 miles. The rate of plane B is 50 miles per hour greater than the rate of plane A. Find the times and rates of both planes.

52. To travel 60 miles, it takes Sue, riding a moped, 2 hours less time than it takes Doreen to travel 50 miles riding a bicycle. Sue travels 10 miles per hour faster than Doreen. Find the times and rates of both girls.

53. It takes Amy twice as long to deliver papers as it does Nancy. How long would it take each girl to deliver the papers by herself if they can deliver the papers together in 40 minutes?

54. If two inlet pipes are both open, they can fill a pool in 1 hour and 12 minutes. One of the pipes can fill the pool by itself in 2 hours. How long would it take the other pipe to fill the pool by itself?

55. Rod agreed to mow a vacant lot for \$12. It took him an hour longer than he had anticipated, so he earned \$1 per hour less than he had originally calculated. How long had he anticipated that it would take him to mow the lot?

56. Last week Al bought some golf balls for \$20. The next day they were on sale for \$0.50 per ball less, and he bought \$22.50 worth of balls. If he purchased 5 more balls on the second day than on the first day, how many did he buy each day and at what price per ball?

57. Debbie rode her bicycle out into the country for a distance of 24 miles. On the way back, she took a much shorter route of 12 miles and made the return trip in one-half hour less time. If her rate out into the country was 4 miles per hour greater than her rate on the return trip, find both rates.

58. Felipe jogs for 10 miles and then walks another 10 miles. He jogs $2\frac{1}{2}$ miles per hour faster than he walks, and the entire distance of 20 miles takes 6 hours. Find the rate at which he walks and the rate at which he jogs.

THOUGHTS INTO WORDS

59. Why is it important to consider more than one way to solve a problem?

60. Write a paragraph or two summarizing the new ideas about problem solving you have acquired so far in this course.

Answers to the Concept Quiz

1. False **2.** True **3.** False **4.** True **5.** True **6.** True **7.** False **8.** True **9.** False **10.** True

Answers to the Example Practice Skills

1. $\{14\}$ **2.** $\left\{-\frac{2}{3}\right\}$ **3.** $\{3\}$ **4.** $r = \frac{S - a}{S}$ **5.** Plane, 550 mph; car, 70 mph **6.** Tour bus, 55 mph for 4 hr; car, 70 mph for 3 hr **7.** 75 min **8.** 8 hr **9.** 8 hr

Chapter 6 Summary

OBJECTIVE	SUMMARY	EXAMPLE	CHAPTER REVIEW PROBLEMS
Reduce rational numbers and rational expressions. (Sec. 6.1, Obj. 1, p. 292; Sec. 6.1, Obj. 2, p. 294)	Any number that can be written in the form $\frac{a}{b}$, where a and b are integers and $b \neq 0$, is a rational number. A rational expression is defined as the indicated quotient of two polynomials. The fundamental principle of fractions, $\frac{a \cdot k}{b \cdot k} = \frac{a}{b}$, is used when reducing rational numbers or rational expressions.	Simplify $\frac{x^2 - 2x - 15}{x^2 + x - 6}$. **Solution** $\frac{x^2 - 2x - 15}{x^2 + x - 6}$ $= \frac{(x + 3)(x - 5)}{(x + 3)(x - 2)} = \frac{x - 5}{x - 2}$	Problems 1–6
Multiply rational numbers and rational expressions. (Sec. 6.2, Obj. 1, p. 299; Sec. 6.2, Obj. 2, p. 300)	Multiplication of rational expressions is based on the following definition: $\frac{a}{b} \cdot \frac{c}{d} = \frac{ac}{bd}$.	Find the product $\frac{3y^2 + 12y}{y^3 - 2y^2} \cdot \frac{y^2 - 3y + 2}{y^2 + 7y + 12}$. **Solution** $\frac{3y^2 + 12y}{y^3 - 2y^2} \cdot \frac{y^2 - 3y + 2}{y^2 + 7y + 12}$ $= \frac{3y(y + 4)}{y^2(y - 2)} \cdot \frac{(y - 2)(y - 1)}{(y + 3)(y + 4)}$ $= \frac{3y(y + 4)}{y^2(y - 2)} \cdot \frac{(y - 2)(y - 1)}{(y + 3)(y + 4)}$ $= \frac{3(y - 1)}{y(y + 3)}$	Problems 7–8
Divide rational numbers and rational expressions. (Sec. 6.2, Obj. 3, p. 301; Sec. 6.2, Obj. 4, p. 301)	Division of rational expressions is based on the following definition: $\frac{a}{b} \div \frac{c}{d} = \frac{a}{b} \cdot \frac{d}{c} = \frac{ad}{bc}$.	Find the quotient $\frac{6xy}{x^2 - 6x + 9} \div \frac{18x}{x^2 - 9}$. **Solution** $\frac{6xy}{x^2 - 6x + 9} \div \frac{18x}{x^2 - 9}$ $= \frac{6xy}{x^2 - 6x + 9} \cdot \frac{x^2 - 9}{18x}$ $= \frac{6xy}{(x - 3)(x - 3)} \cdot \frac{(x + 3)(x - 3)}{18x}$ $= \frac{6xy}{(x - 3)(x - 3)} \cdot \frac{(x + 3)(x - 3)}{18x}$ $= \frac{y(x + 3)}{3(x - 3)}$	Problems 9–10 *(continued)*

<table>
<tr><th>OBJECTIVE</th><th>SUMMARY</th><th>EXAMPLE</th><th>CHAPTER REVIEW PROBLEMS</th></tr>
<tr><td>Simplify problems that involve both multiplication and division. (Sec. 6.2, Obj. 5, p. 303)</td><td>Change the divisions to multiplying by the reciprocal and then find the product.</td><td>Perform the indicated operations:
$\frac{6xy^3}{5x} \div \frac{3xy}{10} \cdot \frac{y}{7x^2}$
Solution
$\frac{6xy^3}{5x} \div \frac{3xy}{10} \cdot \frac{y}{7x^2}$
$= \frac{6xy^3}{5x} \cdot \frac{10}{3xy} \cdot \frac{y}{7x^2}$
$= \frac{6xy^3}{5x} \cdot \frac{10}{3xy} \cdot \frac{y}{7x^2}$
$= \frac{4y^3}{7x^3}$</td><td>Problems 11–14</td></tr>
<tr><td>Add and subtract rational numbers or rational expressions. (Sec. 6.3, Obj. 1, p. 305; Sec. 6.3, Obj. 2, p. 307; Sec. 6.4, Obj. 1, p. 314)</td><td>Addition and subtraction of rational expressions are based on the following definitions.
$\frac{a}{b} + \frac{c}{b} = \frac{a+c}{b}$ Addition
$\frac{a}{b} - \frac{c}{b} = \frac{a-c}{b}$ Subtraction
The following basic procedure is used to add or subtract rational expressions.
1. Factor the denominators.
2. Find the LCD.
3. Change each fraction to an equivalent fraction that has the LCD as the denominator.
4. Combine the numerators and place over the LCD.
5. Simplify by performing the addition or subtraction in the numerator.
6. If possible, reduce the resulting fraction.</td><td>Subtract
$\frac{2}{x^2 - 2x - 3} - \frac{5}{x^2 + 5x + 4}$.
Solution
$\frac{2}{x^2 - 2x - 3} - \frac{5}{x^2 + 5x + 4}$
$= \frac{2}{(x-3)(x+1)}$
$- \frac{5}{(x+1)(x+4)}$
The LCD is $(x-3)(x+1)(x+4)$.
$= \frac{2(x+4)}{(x-3)(x+1)(x+4)}$
$- \frac{5(x-3)}{(x+1)(x+4)(x-3)}$
$= \frac{2(x+4) - 5(x-3)}{(x-3)(x+1)(x+4)}$
$= \frac{2x + 8 - 5x + 15}{(x-3)(x+1)(x+4)}$
$= \frac{-3x + 23}{(x-3)(x+1)(x+4)}$</td><td>Problems 15–20

(continued)</td></tr>
</table>

OBJECTIVE	SUMMARY	EXAMPLE	CHAPTER REVIEW PROBLEMS
Simplify complex fractions. (Sec. 6.4, Obj. 2, p. 316)	Fractions that contain rational numbers or rational expressions in the numerators or denominators are called complex fractions. In Section 6.4, two methods were shown for simplifying complex fractions.	Simplify $\dfrac{\frac{2}{x}-\frac{3}{y}}{\frac{4}{x^2}+\frac{5}{y}}$. **Solution** $\dfrac{\frac{2}{x}-\frac{3}{y}}{\frac{4}{x^2}+\frac{5}{y}}$ Multiply the numerator and denominator by x^2y. $\dfrac{x^2y\left(\frac{2}{x}-\frac{3}{y}\right)}{x^2y\left(\frac{4}{x^2}+\frac{5}{y}\right)}$ $=\dfrac{x^2y\left(\frac{2}{x}\right)+x^2y\left(-\frac{3}{y}\right)}{x^2y\left(\frac{4}{x^2}\right)+x^2y\left(\frac{5}{y}\right)}$ $=\dfrac{2xy-3x^2}{4y+5x^2}$	Problems 21–24
Divide polynomials. (Sec. 6.5, Obj. 1, p. 323)	1. To divide a polynomial by a monomial, divide each term of the polynomial by the monomial. 2. The procedure for dividing a polynomial by a polynomial resembles the long-division process.	Divide $2x^2+11x+19$ by $x+3$. **Solution** $\begin{array}{r} 2x+5 \\ x+3\overline{)2x^2+11x+19} \\ \underline{2x^2+6x} \\ 5x+19 \\ \underline{5x+15} \\ 4 \end{array}$ Thus $\dfrac{2x^2+11x+19}{x+3}$ $=2x+5+\dfrac{4}{x+3}$.	Problems 25–26
Use synthetic division to divide polynomials. (Sec. 6.5, Obj. 2, p. 326)	Synthetic division is a shortcut to the long-division process when the divisor is of the form $x-k$.	Divide x^4-3x^2+5x+6 by $x+2$. **Solution** $\begin{array}{r\|rrrrr} -2 & 1 & 0 & -3 & 5 & 6 \\ & & -2 & 4 & -2 & -6 \\ \hline & 1 & -2 & 1 & 3 & 0 \end{array}$ Thus $\dfrac{x^4-3x^2+5x+6}{x+2}$ $=x^3-2x^2+x+3$.	*(continued)*

OBJECTIVE	SUMMARY	EXAMPLE	CHAPTER REVIEW PROBLEMS
Solve rational equations. (Sec. 6.6, Obj. 1, p. 330)	To solve a rational equation, it is often easiest to begin by multiplying both sides of the equation by the LCD of all the denominators in the equation. Recall that any value of the variable that makes the denominator zero cannot be a solution to the equation.	Solve $\frac{2}{3x} + \frac{5}{12} = \frac{1}{4x}$. **Solution** $\frac{2}{3x} + \frac{5}{12} = \frac{1}{4x}$ Multiply both sides by $12x$. $12x\left(\frac{2}{3x} + \frac{5}{12}\right) = 12x\left(\frac{1}{4x}\right)$ $12x\left(\frac{2}{3x}\right) + 12x\left(\frac{5}{12}\right) = 12x\left(\frac{1}{4x}\right)$ $8 + 5x = 3$ $5x = -5$ $x = -1$ The solution set is $\{-1\}$.	Problems 29–33
Solve proportions. (Sec. 6.6, Obj. 2, p. 332)	A ratio is the comparison of two numbers by division. A proportion is a statement of equality between two ratios. Proportions can be solved using the cross-multiplication property of proportions.	Solve $\frac{5}{2x - 1} = \frac{3}{x + 4}$. **Solution** $\frac{5}{2x - 1} = \frac{3}{x + 4}$ $3(2x - 1) = 5(x + 4)$ $6x - 3 = 5x + 20$ $x = 23$ The solution set is $\{23\}$.	Problems 34–35
Solve rational equations where the denominators require factoring. (Sec. 6.7, Obj. 1, p. 337)	It may be necessary to factor the denominators in a rational equation in order to determine the LCD of all the denominators.	Solve $\frac{7x}{3x + 12} - \frac{2}{x^2 - 16} = \frac{7}{3}$. **Solution** $\frac{7x}{3x + 12} - \frac{2}{x^2 - 16} = \frac{7}{3}$ $\frac{7x}{3(x + 4)} - \frac{2}{(x - 4)(x + 4)} = \frac{7}{3}$ Multiply both sides by $3(x + 4)(x - 4)$. $7x(x - 4) - 2(3) = 7(x + 4)(x - 4)$ $7x^2 - 28x - 6 = 7x^2 - 112$ $-28x = -106$ $x = \frac{-106}{-28} = \frac{53}{14}$ The solution set is $\left\{\frac{53}{14}\right\}$.	Problems 36–38

(continued)

OBJECTIVE	SUMMARY	EXAMPLE	CHAPTER REVIEW PROBLEMS
Solve formulas that are in fractional form. (Sec. 6.7, Obj. 2, p. 339)	The techniques that are used for solving rational equations can also be used to change the form of formulas.	Solve $\frac{x}{2a} - \frac{y}{2b} = 1$ for y. **Solution** $\frac{x}{2a} - \frac{y}{2b} = 1$ Multiply both sides by $2ab$. $2ab\left(\frac{x}{2a} - \frac{y}{2b}\right) = 2ab(1)$ $bx - ay = 2ab$ $-ay = 2ab - bx$ $y = \frac{2ab - bx}{-a}$ $y = \frac{-2ab + bx}{a}$	Problems 39–40
Solve word problems involving ratios. (Sec. 6.6, Obj. 3, p. 334)	Many real-world situations can be solved by using ratios and setting up a proportion to be solved.	At a law firm, the ratio of female attorneys to male attorneys is 1 to 15. If the firm has a total of 125 attorneys, find the number of female attorneys. **Solution** Let x represent the number of female attorneys. Then $125 - x$ represents the numbers of male attorneys. The following proportion can be set up. $\frac{x}{125 - x} = \frac{1}{4}$ Solve by cross-multiplication. $\frac{x}{125 - x} = \frac{1}{4}$ $4x = 1(125 - x)$ $4x = 125 - x$ $5x = 125$ $x = 25$ There are 25 female attorneys.	Problems 41–42 *(continued)*

OBJECTIVE	SUMMARY	EXAMPLE	CHAPTER REVIEW PROBLEMS
Solve rate–time word problems. (Sec. 6.7, Obj. 3, p. 340)	Uniform motion problems are a special case of rate–time problems. In general, if Q is the quantity of something done in t time units, then the rate, r, is given by $r = \frac{Q}{t}$.	At a veterinarian clinic, it takes Laurie twice as long to feed the animals as it does Janet. How long would it take each person to feed the animals by herself if they can feed the animals together in 60 minutes? **Solution** Let t represent the time it takes Janet to feed the animals. Then $2t$ represents the time it would take Laurie to feed the animals. Laurie's rate plus Janet's rate equals the rate working together. $\frac{1}{2t} + \frac{1}{t} = \frac{1}{60}$ Multiply both sides by $60t$. $60t\left(\frac{1}{2t} + \frac{1}{t}\right) = 60t\left(\frac{1}{60}\right)$ $30 + 60 = t$ $90 = t$ It would take Janet 90 minutes working alone to feed the animals, and it would take Laurie 180 minutes working alone to feed the animals.	Problems 43–47

Chapter 6 Review Problem Set

For Problems 1–6, simplify each rational expression.

1. $\frac{26x^2y^3}{39x^4y^2}$

2. $\frac{a^2 - 9}{a^2 + 3a}$

3. $\frac{n^2 - 3n - 10}{n^2 + n - 2}$

4. $\frac{x^4 - 1}{x^3 - x}$

5. $\frac{8x^3 - 2x^2 - 3x}{12x^2 - 9x}$

6. $\frac{x^4 - 7x^2 - 30}{2x^4 + 7x^2 + 3}$

For Problems 7–20, perform the indicated operations and express your answers in simplest form.

7. $\frac{9ab}{3a + 6} \cdot \frac{a^2 - 4a - 12}{a^2 - 6a}$

8. $\frac{n^2 + 10n + 25}{n^2 - n} \cdot \frac{5n^3 - 3n^2}{5n^2 + 22n - 15}$

9. $\frac{6xy^2}{7y^3} \div \frac{15x^2y}{5x^2}$

10. $\frac{x^2 - 2xy - 3y^2}{x^2 + 9y^2} \div \frac{2x^2 + xy - y^2}{2x^2 - xy}$

11. $\frac{2x^2y}{3x} \cdot \frac{xy^2}{6} \div \frac{x}{9y}$

12. $\frac{10x^4y^3}{8x^2y} \div \frac{5}{xy^2} \cdot \frac{3y}{x}$

13. $\frac{8}{2x - 6} \div \frac{2x - 1}{x^2 - 9} \div \frac{2x^2 + x - 1}{x^2 + 7x + 12}$

14. $\frac{2 - x}{6} \cdot \frac{x + 1}{x^2 - 4} \div \frac{x^2 + 2x + 1}{10}$

15. $\frac{2x + 1}{5} + \frac{3x - 2}{4}$

16. $\frac{3}{2n} + \frac{5}{3n} - \frac{1}{9}$

17. $\frac{3x}{x + 7} - \frac{2}{x}$

18. $\frac{10}{x^2 - 5x} + \frac{2}{x}$

19. $\dfrac{3}{n^2 - 5n - 36} + \dfrac{2}{n^2 + 3n - 4}$

20. $\dfrac{3}{2y + 3} + \dfrac{5y - 2}{2y^2 - 9y - 18} - \dfrac{1}{y - 6}$

For Problems 21–24, simplify each complex fraction.

21. $\dfrac{\dfrac{5}{8} - \dfrac{1}{2}}{\dfrac{1}{6} + \dfrac{3}{4}}$

22. $\dfrac{\dfrac{3}{2x} + \dfrac{5}{3y}}{\dfrac{4}{x} - \dfrac{3}{4y}}$

23. $\dfrac{\dfrac{3}{x - 2} - \dfrac{4}{x^2 - 4}}{\dfrac{2}{x + 2} + \dfrac{1}{x - 2}}$

24. $1 - \dfrac{1}{2 - \dfrac{1}{x}}$

For Problems 25 and 26, perform the long division.

25. $(18x^2 + 9x - 2) \div (3x + 2)$

26. $(3x^3 + 5x^2 - 6x - 2) \div (x + 4)$

For Problems 27 and 28, divide using synthetic division.

27. Divide $3x^4 - 14x^3 + 7x^2 + 6x - 8$ by $x - 4$.

28. Divide $2x^4 + x^2 - x + 3$ by $x + 1$.

For Problems 29–40, solve each equation.

29. $\dfrac{4x + 5}{3} + \dfrac{2x - 1}{5} = 2$

30. $\dfrac{3}{4x} + \dfrac{4}{5} = \dfrac{9}{10x}$

31. $\dfrac{a}{a - 2} - \dfrac{3}{2} = \dfrac{2}{a - 2}$

32. $n + \dfrac{1}{n} = \dfrac{53}{14}$

33. $\dfrac{x}{2x + 1} - 1 = \dfrac{-4}{7(x - 2)}$

34. $\dfrac{4}{5y - 3} = \dfrac{2}{3y + 7}$

35. $\dfrac{2x}{-5} = \dfrac{3}{4x - 13}$

36. $\dfrac{1}{2x - 7} + \dfrac{x - 5}{4x^2 - 49} = \dfrac{4}{6x - 21}$

37. $\dfrac{2n}{2n^2 + 11n - 21} - \dfrac{n}{n^2 + 5n - 14} = \dfrac{3}{n^2 + 5n - 14}$

38. $\dfrac{2}{t^2 - t - 6} + \dfrac{t + 1}{t^2 + t - 12} = \dfrac{t}{t^2 + 6t + 8}$

39. Solve $\dfrac{y - 6}{x + 1} = \dfrac{3}{4}$ for y.

40. Solve $\dfrac{x}{a} - \dfrac{y}{b} = 1$ for y.

For Problems 41–47, set up an equation, and solve the problem.

41. A sum of $1400 is to be divided between two people in the ratio of $\frac{3}{5}$. How much does each person receive?

42. At a restaurant the tips are split between the busboy and the waiter in the ratio of 2 to 7. Find the amount each received in tips if there was a total of $162 in tips.

43. Working together, Dan and Julio can mow a lawn in 12 minutes. Julio can mow the lawn by himself in 10 minutes less time than it takes Dan by himself. How long does it take each of them to mow the lawn alone?

44. Suppose that car A can travel 250 miles in 3 hours less time than it takes car B to travel 440 miles. The rate of car B is 5 miles per hour faster than that of car A. Find the rates of both cars.

45. Mark can overhaul an engine in 20 hours, and Phil can do the same job by himself in 30 hours. If they both work together for a time and then Mark finishes the job by himself in 5 hours, how long did they work together?

46. Kelly contracted to paint a house for $640. It took him 20 hours longer than he had anticipated, so he earned $1.60 per hour less than he had calculated. How long had he anticipated that it would take him to paint the house?

47. Nasser rode his bicycle 66 miles in $4\frac{1}{2}$ hours. For the first 40 miles he averaged a certain rate, and then for the last 26 miles he reduced his rate by 3 miles per hour. Find his rate for the last 26 miles.

Chapter 6 Test

For Problems 1–4, simplify each rational expression.

1. $\dfrac{39x^2y^3}{72x^3y}$ 1. ________

2. $\dfrac{3x^2 + 17x - 6}{x^3 - 36x}$ 2. ________

3. $\dfrac{6n^2 - 5n - 6}{3n^2 + 14n + 8}$ 3. ________

4. $\dfrac{2x - 2x^2}{x^2 - 1}$ 4. ________

For Problems 5–13, perform the indicated operations and express your answers in simplest form.

5. $\dfrac{5x^2y}{8x} \cdot \dfrac{12y^2}{20xy}$ 5. ________

6. $\dfrac{5a + 5b}{20a + 10b} \cdot \dfrac{a^2 - ab}{2a^2 + 2ab}$ 6. ________

7. $\dfrac{3x^2 + 10x - 8}{5x^2 + 19x - 4} \div \dfrac{3x^2 - 23x + 14}{x^2 - 3x - 28}$ 7. ________

8. $\dfrac{3x - 1}{4} + \dfrac{2x + 5}{6}$ 8. ________

9. $\dfrac{5x - 6}{3} - \dfrac{x - 12}{6}$ 9. ________

10. $\dfrac{3}{5n} + \dfrac{2}{3} - \dfrac{7}{3n}$ 10. ________

11. $\dfrac{3x}{x - 6} + \dfrac{2}{x}$ 11. ________

12. $\dfrac{9}{x^2 - x} - \dfrac{2}{x}$ 12. ________

13. $\dfrac{3}{2n^2 + n - 10} + \dfrac{5}{n^2 + 5n - 14}$ 13. ________

14. Divide $3x^3 + 10x^2 - 9x - 4$ by $x + 4$. 14. ________

15. Simplify the complex fraction $\dfrac{\dfrac{3}{2x} - \dfrac{1}{6}}{\dfrac{2}{3x} + \dfrac{3}{4}}$. 15. ________

16. Solve $\dfrac{x + 2}{y - 4} = \dfrac{3}{4}$ for y. 16. ________

For Problems 17–22, solve each equation.

17. $\dfrac{x - 1}{2} - \dfrac{x + 2}{5} = -\dfrac{3}{5}$ 17. ________

18. $\dfrac{5}{4x} + \dfrac{3}{2} = \dfrac{7}{5x}$ 18. ________

19. ______

20. ______

21. ______

22. ______

23. ______

24. ______

25. ______

19. $\dfrac{-3}{4n - 1} = \dfrac{-2}{3n + 11}$

20. $n - \dfrac{5}{n} = 4$

21. $\dfrac{6}{x - 4} - \dfrac{4}{x + 3} = \dfrac{8}{x - 4}$

22. $\dfrac{1}{3x - 1} + \dfrac{x - 2}{9x^2 - 1} = \dfrac{7}{6x - 2}$

For Problems 23–25, set up an equation and solve the problem.

23. The denominator of a rational number is 9 less than three times the numerator. The number in simplest form is $\frac{3}{8}$. Find the number.

24. It takes Jodi three times as long to deliver papers as it does Jannie. Together they can deliver the papers in 15 minutes. How long would it take Jodi by herself?

25. René can ride her bike 60 miles in 1 hour less time than it takes Sue to ride 60 miles. René's rate is 3 miles per hour faster than Sue's rate. Find René's rate.

Chapters 1–6 Cumulative Review Problem Set

1. Simplify the numerical expression $16 \div 4(2) + 8$.

2. Simplify the numerical expression $(-2)^2 + (-2)^3 - 3^2$.

3. Evaluate $-2xy + 5y^2$ for $x = -3$ and $y = 4$.

4. Evaluate $3(n - 2) + 4(n - 4) - 8(n - 3)$ for $n = -\frac{1}{2}$.

For Problems 5–14, perform the indicated operations and then simplify.

5. $(6a^2 + 3a - 4) + (8a + 6) + (a^2 - 1)$

6. $(x^2 + 5x + 2) - (3x^2 - 4x + 6)$

7. $(2x^2y)(-xy^4)$

8. $(4xy^3)^2$

9. $(-3a^3)^2(4ab^2)$

10. $(4a^2b)(-3a^3b^2)(2ab)$

11. $-3x^2(6x^2 - x + 4)$

12. $(5x + 3y)(2x - y)$

13. $(x + 4y)^2$

14. $(a + 3b)(a^2 - 4ab + b^2)$

For Problems 15–20, factor each polynomial completely.

15. $x^2 - 5x + 6$

16. $6x^2 - 5x - 4$

17. $2x^2 - 8x + 6$

18. $3x^2 + 18x - 48$

19. $9m^2 - 16n^2$

20. $27a^3 + 8$

21. Simplify $\dfrac{-28x^2y^5}{4x^4y}$

22. Simplify $\dfrac{4x - x^2}{x - 4}$

For Problems 23–28, perform the indicated operations and express the answer in simplest form.

23. $\dfrac{6xy}{2x + 4} \cdot \dfrac{x^2 - 3x - 10}{3xy - 3y}$

24. $\dfrac{x^2 - 3x - 4}{x^2 - 1} \div \dfrac{x^2 - x - 12}{x^2 + 6x - 7}$

25. $\dfrac{7n - 3}{5} - \dfrac{n + 4}{2}$

26. $\dfrac{3}{x^2 + x - 6} + \dfrac{5}{x^2 - 9}$

27. $\dfrac{\frac{2}{x} + \frac{3}{y}}{6}$

28. $\dfrac{\frac{1}{n^2} - \frac{1}{m^2}}{\frac{1}{m} + \frac{1}{n}}$

29. Divide $(6x^3 + 7x^2 + 5x + 12)$ by $(2x + 3)$.

30. Use synthetic division to divide $(2x^3 - 3x^2 - 23x + 14)$ by $(x - 4)$.

31. Find the slope between the points $(4, -3)$ and $(-2, 6)$.

For Problems 32–35, graph the equation.

32. $x + 3y = -3$

33. $2x - 5y = 10$

34. $y = -2x$

35. $y = 3$

For Problems 36–37, graph the solution set of the inequality.

36. $y > -x + 3$

37. $x + 2y < 4$

38. Write the equation of a line that has a slope of $\frac{2}{3}$ and contains the point $(-1, 4)$. Express the answer in standard form.

39. Write the equation of a line that contains the points $(0, 4)$ and $(3, -2)$. Express the answer in slope-intercept form.

40. Write the equation of a line that is parallel to the line $x - 2y = 3$ and contains the point $(1, 5)$. Express the answer in standard form.

For Problems 41–50, solve the equation.

41. $8n - 3(n + 2) = 2n + 12$

42. $0.2(y - 6) = 0.02y + 3.12$

43. $\dfrac{x + 1}{4} + \dfrac{3x + 2}{2} = 5$

44. $\dfrac{5}{8}(x + 2) - \dfrac{1}{2}x = 2$

45. $|3x - 2| = 8$

46. $|x + 8| - 4 = 16$

47. $x^2 + 7x - 8 = 0$

48. $2x^2 + 13x + 15 = 0$

49. $n - \frac{3}{n} = \frac{26}{3}$

50. $\frac{3}{n-7} + \frac{4}{n+2} = \frac{27}{n^2 - 5n - 14}$

For Problems 51–54, solve the system of equations.

51. $\begin{pmatrix} 3x + y = -9 \\ 4x + 3y = -2 \end{pmatrix}$

52. $\begin{pmatrix} x + 2y = 16 \\ 4x - y = -8 \end{pmatrix}$

53. $\begin{pmatrix} 2x + 5y = 3 \\ 3x + 2y = 10 \end{pmatrix}$

54. $\begin{pmatrix} x + 2y + z = -3 \\ 2x + y + 3z = -6 \\ 4x + 3y + z = 2 \end{pmatrix}$

55. Solve the formula $A = P + Prt$ for P.

For Problems 56–60, solve the inequality and express the solution in interval notation.

56. $-3x + 2(x - 4) \geq -10$

57. $-10 < 3x + 2 < 8$

58. $|4x + 3| < 15$

59. $|2x + 6| \geq 20$

60. $|x + 4| - 6 > 0$

61. The owner of a local café wants to make a profit of 80% of the cost for each Caesar salad sold. If it costs $3.20 to make a Caesar salad, at what price should each salad be sold.

62. Find the discount sale price of a $920 television that is on sale for 25% off.

63. Suppose that the length of a rectangle is 8 inches less than twice the width. The perimeter of the rectangle is 122 inches. Find the length and width of the rectangle.

64. Two planes leave Kansas City at the same time and fly in opposite directions. If one travels at 450 miles per hour and the other travels at 400 miles per hour, how long will it take for them to be 3400 miles apart?

65. A sum of $68,000 is to be divided between two partners in the ratio of $\frac{1}{4}$. How much does each person receive?

66. Victor can rake the lawn in 20 minutes, and his sister Lucia can rake the same lawn in 30 minutes. How long will it take them to rake the lawn if they work together?

67. One leg of a right triangle is 7 inches less than the other leg. The hypotenuse is 1 inch longer than the longer of the two legs. Find the length of the three sides of the right triangle.

68. How long will it take $1500 to double itself at 6% simple interest?

69. A collection of 40 coins consisting of dimes and quarters has a value of $5.95. Find the number of each kind of coin.

70. Suppose that you have a supply of 10% saline solution and 40% saline solution. How many liters of each should be mixed to produce 30 liters of a 28% saline solution?

Exponents and Radicals

7

■ *By knowing the time it takes for the pendulum to swing from one side to the other side and back, the formula* $T = 2\pi\sqrt{\frac{L}{32}}$ *can be solved to find the length of the pendulum.*

How long will it take a pendulum that is 1.5 feet long to swing from one side to the other side and back? The formula $T = 2\pi\sqrt{\frac{L}{32}}$ can be used to determine that it will take approximately 1.4 seconds.

It is not uncommon in mathematics to find two separately developed concepts that are closely related to each other. In this chapter, we will first develop the concepts of exponent and root individually and then show how they merge to become even more functional as a unified idea.

Video tutorials for all section learning objectives are available in a variety of delivery modes.

INTERNET PROJECT

If the lengths of the three sides of the triangle are known, the area can be calculated with Heron's formula. In ancient Greece, Heron of Alexandria was a mathematician as well as an engineer. Among his inventions are the first recorded steam engine and the first vending machine. Do an Internet search to determine how the first vending machine operated and what it dispensed.

7.1 Using Integers as Exponents

OBJECTIVES

1. Simplify Numerical Expressions That Have Positive and Negative Exponents
2. Simplify Algebraic Expressions That Have Positive and Negative Exponents
3. Multiply and Divide Algebraic Expressions That Have Positive and Negative Exponents
4. Simplify Sums and Differences of Expressions Involving Positive and Negative Exponents

1 Simplify Numerical Expressions That Have Positive and Negative Exponents

Thus far in the text we have used only positive integers as exponents. In Chapter 1 the expression b^n, where b is any real number and n is a positive integer, was defined by

$$b^n = b \cdot b \cdot b \cdot \ldots \cdot b \qquad n \text{ factors of } b$$

Then, in Chapter 5, some of the parts of the following property served as a basis for manipulation with polynomials.

Property 7.1

If m and n are positive integers and a and b are real numbers (and $b \neq 0$ whenever it appears in a denominator), then

1. $b^n \cdot b^m = b^{n+m}$

2. $(b^n)^m = b^{mn}$

3. $(ab)^n = a^n b^n$

4. $\left(\frac{a}{b}\right)^n = \frac{a^n}{b^n}$

5. $\frac{b^n}{b^m} = b^{n-m}$ when $n > m$

$\frac{b^n}{b^m} = 1$ when $n = m$

$\frac{b^n}{b^m} = \frac{1}{b^{m-n}}$ when $n < m$

We are now ready to extend the concept of an exponent to include the use of zero and the negative integers as exponents.

First, let's consider the use of zero as an exponent. We want to use zero in such a way that the previously listed properties continue to hold. If $b^n \cdot b^m = b^{n+m}$ is to hold, then $x^4 \cdot x^0 = x^{4+0} = x^4$. In other words, x^0 *acts like* 1 because $x^4 \cdot x^0 = x^4$. This line of reasoning suggests the following definition.

Definition 7.1

If b is a nonzero real number, then

$$b^0 = 1$$

According to Definition 7.1, the following statements are all true.

$$5^0 = 1 \qquad (-413)^0 = 1$$

$$\left(\frac{3}{11}\right)^0 = 1 \qquad n^0 = 1, \quad n \neq 0$$

$$(x^3y^4)^0 = 1, \qquad x \neq 0, y \neq 0$$

We can use a similar line of reasoning to motivate a definition for the use of negative integers as exponents. Consider the example $x^4 \cdot x^{-4}$. If $b^n \cdot b^m = b^{n+m}$ is to hold, then $x^4 \cdot x^{-4} = x^{4+(-4)} = x^0 = 1$. Thus x^{-4} must be the reciprocal of x^4, because their product is 1. That is,

$$x^{-4} = \frac{1}{x^4}$$

This suggests the following general definition.

Definition 7.2

If n is a positive integer and b is a nonzero real number, then

$$b^{-n} = \frac{1}{b^n}$$

According to Definition 7.2, the following statements are all true.

$$x^{-5} = \frac{1}{x^5} \qquad 2^{-4} = \frac{1}{2^4} = \frac{1}{16}$$

$$10^{-2} = \frac{1}{10^2} = \frac{1}{100} \quad \text{or} \quad 0.01 \qquad \frac{2}{x^{-3}} = \frac{2}{\frac{1}{x^3}} = (2)\left(\frac{x^3}{1}\right) = 2x^3$$

$$\left(\frac{3}{4}\right)^{-2} = \frac{1}{\left(\frac{3}{4}\right)^2} = \frac{1}{\frac{9}{16}} = \frac{16}{9}$$

It can be verified (although it is beyond the scope of this text) that all of the parts of Property 7.1 hold for *all integers*. In fact, the following equality can replace the three separate statements for part (5).

$$\frac{b^n}{b^m} = b^{n-m} \quad \text{for all integers } n \text{ and } m$$

Let's restate Property 7.1 as it holds for all integers and include, at the right, a "name tag" for easy reference.

Property 7.2

If m and n are integers and a and b are real numbers (and $b \neq 0$ whenever it appears in a denominator), then

1. $b^n \cdot b^m = b^{n+m}$ — Product of two powers
2. $(b^n)^m = b^{mn}$ — Power of a power
3. $(ab)^n = a^n b^n$ — Power of a product
4. $\left(\dfrac{a}{b}\right)^n = \dfrac{a^n}{b^n}$ — Power of a quotient
5. $\dfrac{b^n}{b^m} = b^{n-m}$ — Quotient of two powers

Having the use of all integers as exponents enables us to work with a large variety of numerical and algebraic expressions. Let's consider some examples that illustrate the use of the various parts of Property 7.2.

EXAMPLE 1

Simplify each of the following numerical expressions.

(a) $10^{-3} \cdot 10^2$ **(b)** $(2^{-3})^{-2}$ **(c)** $(2^{-1} \cdot 3^2)^{-1}$

(d) $\left(\dfrac{2^{-3}}{3^{-2}}\right)^{-1}$ **(e)** $\dfrac{10^{-2}}{10^{-4}}$

Solution

(a) $10^{-3} \cdot 10^2 = 10^{-3+2}$ — Product of two powers

$$= 10^{-1}$$

$$= \frac{1}{10^1} = \frac{1}{10}$$

(b) $(2^{-3})^{-2} = 2^{(-2)(-3)}$ — Power of a power

$$= 2^6 = 64$$

(c) $(2^{-1} \cdot 3^2)^{-1} = (2^{-1})^{-1}(3^2)^{-1}$ — Power of a product

$$= 2^1 \cdot 3^{-2}$$

$$= \frac{2^1}{3^2} = \frac{2}{9}$$

(d) $\left(\dfrac{2^{-3}}{3^{-2}}\right)^{-1} = \dfrac{(2^{-3})^{-1}}{(3^{-2})^{-1}}$ — Power of a quotient

$$= \frac{2^3}{3^2} = \frac{8}{9}$$

(e) $\dfrac{10^{-2}}{10^{-4}} = 10^{-2-(-4)}$ — Quotient of two powers

$$= 10^2 = 100$$

▼ PRACTICE YOUR SKILL

Simplify each of the following.

(a) $6^{-2} \cdot 6^5$ **(b)** $(2^{-2})^{-2}$ **(c)** $(2^{-2} \cdot 3^{-1})^{-1}$ **(d)** $\left(\dfrac{3^{-3}}{2^{-2}}\right)^{-1}$ **(e)** $\dfrac{4^{-3}}{4^{-6}}$ ■

2 Simplify Algebraic Expressions That Have Positive and Negative Exponents

EXAMPLE 2

Simplify each of the following; express final results without using zero or negative integers as exponents.

(a) $x^2 \cdot x^{-5}$ **(b)** $(x^{-2})^4$ **(c)** $(x^2y^{-3})^{-4}$ **(d)** $\left(\frac{a^3}{b^{-5}}\right)^{-2}$ **(e)** $\frac{x^{-4}}{x^{-2}}$

Solution

(a) $x^2 \cdot x^{-5} = x^{2+(-5)}$ Product of two powers

$= x^{-3}$

$= \frac{1}{x^3}$

(b) $(x^{-2})^4 = x^{4(-2)}$ Power of a power

$= x^{-8}$

$= \frac{1}{x^8}$

(c) $(x^2y^{-3})^{-4} = (x^2)^{-4}(y^{-3})^{-4}$ Power of a product

$= x^{-4(2)}y^{-4(-3)}$

$= x^{-8}y^{12}$

$= \frac{y^{12}}{x^8}$

(d) $\left(\frac{a^3}{b^{-5}}\right)^{-2} = \frac{(a^3)^{-2}}{(b^{-5})^{-2}}$ Power of a quotient

$= \frac{a^{-6}}{b^{10}}$

$= \frac{1}{a^6b^{10}}$

(e) $\frac{x^{-4}}{x^{-2}} = x^{-4-(-2)}$ Quotient of two powers

$= x^{-2}$

$= \frac{1}{x^2}$

▼ PRACTICE YOUR SKILL

Simplify each of the following; express final results without using zero or negative integers as exponents.

(a) $y^4 \cdot y^{-1}$ **(b)** $(x^3)^{-2}$ **(c)** $(a^{-2}b^3)^{-3}$ **(d)** $\left(\frac{x^2}{y^3}\right)^{-3}$ **(e)** $\frac{y^{-2}}{y^{-5}}$ ■

3 Multiply and Divide Algebraic Expressions That Have Positive and Negative Exponents

EXAMPLE 3

Find the indicated products and quotients; express your results using positive integral exponents only.

(a) $(3x^2y^{-4})(4x^{-3}y)$ **(b)** $\frac{12a^3b^2}{-3a^{-1}b^5}$ **(c)** $\left(\frac{15x^{-1}y^2}{5xy^{-4}}\right)^{-1}$

Solution

(a) $(3x^2y^{-4})(4x^{-3}y) = 12x^{2+(-3)}y^{-4+1}$

$= 12x^{-1}y^{-3}$

$= \dfrac{12}{xy^3}$

(b) $\dfrac{12a^3b^2}{-3a^{-1}b^5} = -4a^{3-(-1)}b^{2-5}$

$= -4a^4b^{-3}$

$= -\dfrac{4a^4}{b^3}$

(c) $\left(\dfrac{15x^{-1}y^2}{5xy^{-4}}\right)^{-1} = (3x^{-1-1}y^{2-(-4)})^{-1}$ Note that we are first simplifying inside the parentheses

$= (3x^{-2}y^6)^{-1}$

$= 3^{-1}x^2y^{-6}$

$= \dfrac{x^2}{3y^6}$

▼ **PRACTICE YOUR SKILL**

Find the indicated products and quotients; express the results using positive integral exponents only.

(a) $(2xy^{-3})(5x^{-2}y^5)$ **(b)** $\dfrac{8x^2y}{-2x^{-3}y^4}$ **(c)** $\left(\dfrac{12a^{-2}b^3}{3a^3b^{-4}}\right)^{-1}$ ■

4 Simplify Sums and Differences of Expressions Involving Positive and Negative Exponents

The final examples of this section show the simplification of numerical and algebraic expressions that involve sums and differences. In such cases, we use Definition 7.2 to change from negative to positive exponents so that we can proceed in the usual way.

EXAMPLE 4 Simplify $2^{-3} + 3^{-1}$.

Solution

$2^{-3} + 3^{-1} = \dfrac{1}{2^3} + \dfrac{1}{3^1}$

$= \dfrac{1}{8} + \dfrac{1}{3}$

$= \dfrac{3}{24} + \dfrac{8}{24}$ Use 24 as the LCD

$= \dfrac{11}{24}$

▼ **PRACTICE YOUR SKILL**

Simplify $3^{-2} + 4^{-1}$. ■

EXAMPLE 5

Simplify $(4^{-1} - 3^{-2})^{-1}$.

Solution

$$(4^{-1} - 3^{-2})^{-1} = \left(\frac{1}{4^1} - \frac{1}{3^2}\right)^{-1} \quad \text{Apply } b^{-n} = \frac{1}{b^n} \text{ to } 4^{-1} \text{ and to } 3^{-2}$$

$$= \left(\frac{1}{4} - \frac{1}{9}\right)^{-1}$$

$$= \left(\frac{9}{36} - \frac{4}{36}\right)^{-1} \quad \text{Use 36 as the LCD}$$

$$= \left(\frac{5}{36}\right)^{-1}$$

$$= \frac{1}{\left(\frac{5}{36}\right)^1} \quad \text{Apply } b^{-n} = \frac{1}{b^n}$$

$$= \frac{1}{\frac{5}{36}} = \frac{36}{5}$$

▼ PRACTICE YOUR SKILL

Simplify $(2^{-3} - 3^{-1})^{-1}$. ■

EXAMPLE 6

Express $a^{-1} + b^{-2}$ as a single fraction involving positive exponents only.

Solution

$$a^{-1} + b^{-2} = \frac{1}{a^1} + \frac{1}{b^2} \quad \text{Use } ab^2 \text{ as the LCD}$$

$$= \left(\frac{1}{a}\right)\left(\frac{b^2}{b^2}\right) + \left(\frac{1}{b^2}\right)\left(\frac{a}{a}\right) \quad \text{Change to equivalent fractions with } ab^2 \text{ as the LCD}$$

$$= \frac{b^2}{ab^2} + \frac{a}{ab^2}$$

$$= \frac{b^2 + a}{ab^2}$$

▼ PRACTICE YOUR SKILL

Simplify $x^{-3} + y^{-2}$. ■

CONCEPT QUIZ

For Problems 1–10, answer true or false.

1. $\left(\frac{2}{5}\right)^{-2} = \left(\frac{5}{2}\right)^2$
2. $(3)^0(3)^2 = 9^2$
3. $(2)^{-4}(2)^4 = 2$
4. $(4^{-2})^{-1} = 16$
5. $(2^{-2} \cdot 2^{-3})^{-1} = \frac{1}{16}$

6. $\left(\frac{3^{-2}}{3^{-1}}\right)^2 = \frac{1}{9}$

7. $\frac{1}{\left(\frac{2}{3}\right)^{-3}} = \frac{8}{27}$

8. $(10^4)(10^{-6}) = \frac{1}{100}$

9. $\frac{x^{-6}}{x^{-3}} = x^2$

10. $x^{-1} - x^{-2} = \frac{x-1}{x^2}$

Problem Set 7.1

1 Simplify Numerical Expressions That Have Positive and Negative Exponents

For Problems 1–42, simplify each numerical expression.

1. 3^{-3}
2. 2^{-4}
3. -10^{-2}
4. 10^{-3}
5. $\frac{1}{3^{-4}}$
6. $\frac{1}{2^{-6}}$
7. $-\left(\frac{1}{3}\right)^{-3}$
8. $\left(\frac{1}{2}\right)^{-3}$
9. $\left(-\frac{1}{2}\right)^{-3}$
10. $\left(\frac{2}{7}\right)^{-2}$
11. $\left(-\frac{3}{4}\right)^{0}$
12. $\frac{1}{\left(\frac{4}{5}\right)^{-2}}$
13. $\frac{1}{\left(\frac{3}{7}\right)^{-2}}$
14. $-\left(\frac{5}{6}\right)^{0}$
15. $2^7 \cdot 2^{-3}$
16. $3^{-4} \cdot 3^6$
17. $10^{-5} \cdot 10^2$
18. $10^4 \cdot 10^{-6}$
19. $10^{-1} \cdot 10^{-2}$
20. $10^{-2} \cdot 10^{-2}$
21. $(3^{-1})^{-3}$
22. $(2^{-2})^{-4}$
23. $(5^3)^{-1}$
24. $(3^{-1})^3$
25. $(2^3 \cdot 3^{-2})^{-1}$
26. $(2^{-2} \cdot 3^{-1})^{-3}$
27. $(4^2 \cdot 5^{-1})^2$
28. $(2^{-3} \cdot 4^{-1})^{-1}$
29. $\left(\frac{2^{-1}}{5^{-2}}\right)^{-1}$
30. $\left(\frac{2^{-4}}{3^{-2}}\right)^{-2}$
31. $\left(\frac{2^{-1}}{3^{-2}}\right)^{2}$
32. $\left(\frac{3^{2}}{5^{-1}}\right)^{-1}$
33. $\frac{3^3}{3^{-1}}$
34. $\frac{2^{-2}}{2^3}$
35. $\frac{10^{-2}}{10^2}$
36. $\frac{10^{-2}}{10^{-5}}$
37. $2^{-2} + 3^{-2}$
38. $2^{-4} + 5^{-1}$
39. $\left(\frac{1}{3}\right)^{-1} - \left(\frac{2}{5}\right)^{-1}$
40. $\left(\frac{3}{2}\right)^{-1} - \left(\frac{1}{4}\right)^{-1}$
41. $(2^{-3} + 3^{-2})^{-1}$
42. $(5^{-1} - 2^{-3})^{-1}$

2 Simplify Algebraic Expressions That Have Positive and Negative Exponents

For Problems 43–62, simplify each expression. Express final results without using zero or negative integers as exponents.

43. $x^2 \cdot x^{-8}$
44. $x^{-3} \cdot x^{-4}$
45. $a^3 \cdot a^{-5} \cdot a^{-1}$
46. $b^{-2} \cdot b^3 \cdot b^{-6}$
47. $(a^{-4})^2$
48. $(b^4)^{-3}$
49. $(x^2y^{-6})^{-1}$
50. $(x^5y^{-1})^{-3}$
51. $(ab^3c^{-2})^{-4}$
52. $(a^3b^{-3}c^{-2})^{-5}$
53. $(2x^3y^{-4})^{-3}$
54. $(4x^5y^{-2})^{-2}$
55. $\left(\frac{x^{-1}}{y^{-4}}\right)^{-3}$
56. $\left(\frac{y^{3}}{x^{-4}}\right)^{-2}$
57. $\left(\frac{3a^{-2}}{2b^{-1}}\right)^{-2}$
58. $\left(\frac{2xy^{2}}{5a^{-1}b^{-2}}\right)^{-1}$
59. $\frac{x^{-6}}{x^{-4}}$
60. $\frac{a^{-2}}{a^2}$
61. $\frac{a^3b^{-2}}{a^{-2}b^{-4}}$
62. $\frac{x^{-3}y^{-4}}{x^2y^{-1}}$

3 Multiply and Divide Algebraic Expressions That Have Positive and Negative Exponents

For Problems 63–74, find the indicated products and quotients. Express final results using positive integral exponents only.

63. $(2xy^{-1})(3x^{-2}y^4)$

64. $(-4x^{-1}y^2)(6x^3y^{-4})$

65. $(-7a^2b^{-5})(-a^{-2}b^7)$

66. $(-9a^{-3}b^{-6})(-12a^{-1}b^4)$

67. $\dfrac{28x^{-2}y^{-3}}{4x^{-3}y^{-1}}$

68. $\dfrac{63x^2y^{-4}}{7xy^{-4}}$

69. $\dfrac{-72a^2b^{-4}}{6a^3b^{-7}}$

70. $\dfrac{108a^{-5}b^{-4}}{9a^{-2}b}$

71. $\left(\dfrac{35x^{-1}y^{-2}}{7x^4y^3}\right)^{-1}$

72. $\left(\dfrac{-48ab^2}{-6a^3b^5}\right)^{-2}$

73. $\left(\dfrac{-36a^{-1}b^{-6}}{4a^{-1}b^4}\right)^{-2}$

74. $\left(\dfrac{8xy^3}{-4x^4y}\right)^{-3}$

4 Simplify Sums and Differences of Expressions Involving Positive and Negative Exponents

For Problems 75–84, express each of the following as a single fraction involving positive exponents only.

75. $x^{-2} + x^{-3}$

76. $x^{-1} + x^{-5}$

77. $x^{-3} - y^{-1}$

78. $2x^{-1} - 3y^{-2}$

79. $3a^{-2} + 4b^{-1}$

80. $a^{-1} + a^{-1}b^{-3}$

81. $x^{-1}y^{-2} - xy^{-1}$

82. $x^2y^{-2} - x^{-1}y^{-3}$

83. $2x^{-1} - 3x^{-2}$

84. $5x^{-2}y + 6x^{-1}y^{-2}$

THOUGHTS INTO WORDS

85. Is the following simplification process correct?

$$(3^{-2})^{-1} = \left(\frac{1}{3^2}\right)^{-1} = \left(\frac{1}{9}\right)^{-1} = \frac{1}{\left(\frac{1}{9}\right)^1} = 9$$

Could you suggest a better way to do the problem?

86. Explain how to simplify $(2^{-1} \cdot 3^{-2})^{-1}$ and also how to simplify $(2^{-1} + 3^{-2})^{-1}$.

FURTHER INVESTIGATIONS

87. Use a calculator to check your answers for Problems 1–42.

88. Use a calculator to simplify each of the following numerical expressions. Express your answers to the nearest hundredth.

(a) $(2^{-3} + 3^{-3})^{-2}$

(b) $(4^{-3} - 2^{-1})^{-2}$

(c) $(5^{-3} - 3^{-5})^{-1}$

(d) $(6^{-2} + 7^{-4})^{-2}$

(e) $(7^{-3} - 2^{-4})^{-2}$

(f) $(3^{-4} + 2^{-3})^{-3}$

Answers to the Concept Quiz

1. True **2.** False **3.** False **4.** True **5.** False **6.** True **7.** True **8.** True **9.** False **10.** True

Answers to the Example Practice Skills

1. (a) 216 **(b)** 16 **(c)** 12 **(d)** $\frac{27}{4}$ **(e)** 64 **2. (a)** y^3 **(b)** $\frac{1}{x^6}$ **(c)** $\frac{a^6}{b^9}$ **(d)** $\frac{y^9}{x^6}$ **(e)** y^3

3. (a) $\frac{10y^2}{x}$ **(b)** $-\frac{4x^5}{y^3}$ **(c)** $\frac{a^5}{4b^7}$ **4.** $\frac{13}{36}$ **5.** $-\frac{24}{5}$ **6.** $\frac{y^2 + x^3}{x^3y^2}$

7.2 Roots and Radicals

OBJECTIVES

1. Evaluate Roots of Numbers
2. Express a Radical in Simplest Radical Form
3. Rationalize the Denominator to Simplify Radicals
4. Applications of Radicals

1 Evaluate Roots of Numbers

To **square a number** means to raise it to the second power — that is, to use the number as a factor twice.

$4^2 = 4 \cdot 4 = 16$ Read "four squared equals sixteen"

$10^2 = 10 \cdot 10 = 100$

$\left(\frac{1}{2}\right)^2 = \frac{1}{2} \cdot \frac{1}{2} = \frac{1}{4}$

$(-3)^2 = (-3)(-3) = 9$

A **square root of a number** is one of its two equal factors. Thus 4 is a square root of 16 because $4 \cdot 4 = 16$. Likewise, -4 is also a square root of 16 because $(-4)(-4) = 16$. In general, a is a square root of b if $a^2 = b$. The following generalizations are a direct consequence of the previous statement.

1. Every positive real number has two square roots; one is positive and the other is negative. They are opposites of each other.
2. Negative real numbers have no real number square roots because any real number except zero is positive when squared.
3. The square root of 0 is 0.

The symbol $\sqrt{\ }$, called a **radical sign**, is used to designate the nonnegative square root. The number under the radical sign is called the **radicand**. The entire expression, such as $\sqrt{16}$, is called a **radical**.

$\sqrt{16} = 4$ $\sqrt{16}$ indicates the nonnegative or **principal square root** of 16

$-\sqrt{16} = -4$ $-\sqrt{16}$ indicates the negative square root of 16

$\sqrt{0} = 0$ Zero has only one square root. Technically, we could write $-\sqrt{0} = -0 = 0$.

$\sqrt{-4}$ is not a real number.

$-\sqrt{-4}$ is not a real number.

In general, the following definition is useful.

Definition 7.3

If $a \geq 0$ and $b \geq 0$, then $\sqrt{b} = a$ if and only if $a^2 = b$; a is called the **principal square root** of b.

To **cube a number** means to raise it to the third power — that is, to use the number as a factor three times.

$$2^3 = 2 \cdot 2 \cdot 2 = 8 \quad \text{Read "two cubed equals eight"}$$

$$4^3 = 4 \cdot 4 \cdot 4 = 64$$

$$\left(\frac{2}{3}\right)^3 = \frac{2}{3} \cdot \frac{2}{3} \cdot \frac{2}{3} = \frac{8}{27}$$

$$(-2)^3 = (-2)(-2)(-2) = -8$$

A **cube root of a number** is one of its three equal factors. Thus 2 is a cube root of 8 because $2 \cdot 2 \cdot 2 = 8$. (In fact, 2 is the only real number that is a cube root of 8.) Furthermore, -2 is a cube root of -8 because $(-2)(-2)(-2) = -8$. (In fact, -2 is the only real number that is a cube root of -8.)

In general, a is a cube root of b if $a^3 = b$. The following generalizations are a direct consequence of the previous statement.

1. Every positive real number has one positive real number cube root.
2. Every negative real number has one negative real number cube root.
3. The cube root of 0 is 0.

Remark: Technically, every nonzero real number has three cube roots, but only one of them is a real number. The other two roots are classified as complex numbers. We are restricting our work at this time to the set of real numbers.

The symbol $\sqrt[3]{}$ designates the cube root of a number. Thus we can write

$$\sqrt[3]{8} = 2 \qquad \sqrt[3]{\frac{1}{27}} = \frac{1}{3}$$

$$\sqrt[3]{-8} = -2 \qquad \sqrt[3]{-\frac{1}{27}} = -\frac{1}{3}$$

In general, the following definition is useful.

Definition 7.4

$\sqrt[3]{b} = a$ if and only if $a^3 = b$.

In Definition 7.4, if b is a positive number, then a, the cube root, is a positive number; whereas if b is a negative number, then a, the cube root, is a negative number. The number a is called the principal cube root of b or simply the cube root of b.

The concept of root can be extended to fourth roots, fifth roots, sixth roots, and, in general, nth roots.

Definition 7.5

The nth root of b is a if and only if $a^n = b$.

We can make the following generalizations.

If n is an even positive integer, then the following statements are true.

1. Every positive real number has exactly two real nth roots — one positive and one negative. For example, the real fourth roots of 16 are 2 and -2.
2. Negative real numbers do not have real nth roots. For example, there are no real fourth roots of -16.

If n is an odd positive integer greater than 1, then the following statements are true.

1. Every real number has exactly one real nth root.
2. The real nth root of a positive number is positive. For example, the fifth root of 32 is 2.
3. The real nth root of a negative number is negative. For example, the fifth root of -32 is -2.

The symbol $\sqrt[n]{\ }$ designates the principal nth root. To complete our terminology, the n in the radical $\sqrt[n]{b}$ is called the *index* of the radical. If $n = 2$, we commonly write $\sqrt{b}$ instead of $\sqrt[2]{b}$.

The following chart can help summarize this information with respect to $\sqrt[n]{b}$, where n is a positive integer greater than 1.

	If b is		
	Positive	**Zero**	**Negative**
n is even	$\sqrt[n]{b}$ is a positive real number	$\sqrt[n]{b} = 0$	$\sqrt[n]{b}$ is not a real number
n is odd	$\sqrt[n]{b}$ is a positive real number	$\sqrt[n]{b} = 0$	$\sqrt[n]{b}$ is a negative real number

Consider the following examples.

$\sqrt[4]{81} = 3$ because $3^4 = 81$

$\sqrt[5]{32} = 2$ because $2^5 = 32$

$\sqrt[5]{-32} = -2$ because $(-2)^5 = -32$

$\sqrt[4]{-16}$ is not a real number because any real number, except zero, is positive when raised to the fourth power

The following property is a direct consequence of Definition 7.5.

Property 7.3

1. $(\sqrt[n]{b})^n = b$ n is any positive integer greater than 1
2. $\sqrt[n]{b^n} = b$ n is any positive integer greater than 1 if $b \geq 0$; n is an odd positive integer greater than 1 if $b < 0$

Because the radical expressions in parts (1) and (2) of Property 7.3 are both equal to b, by the transitive property they are equal to each other. Hence $\sqrt[n]{b^n} = (\sqrt[n]{b})^n$. The arithmetic is usually easier to simplify when we use the form $(\sqrt[n]{b})^n$. The following examples demonstrate the use of Property 7.3.

$$\sqrt{144^2} = (\sqrt{144})^2 = 12^2 = 144$$

$$\sqrt[3]{64^3} = (\sqrt[3]{64})^3 = 4^3 = 64$$

$$\sqrt[3]{(-8)^3} = (\sqrt[3]{-8})^3 = (-2)^3 = -8$$

$$\sqrt[4]{16^4} = (\sqrt[4]{16})^4 = 2^4 = 16$$

Let's use some examples to lead into the next very useful property of radicals.

$\sqrt{4 \cdot 9} = \sqrt{36} = 6$ and $\sqrt{4} \cdot \sqrt{9} = 2 \cdot 3 = 6$

$\sqrt{16 \cdot 25} = \sqrt{400} = 20$ and $\sqrt{16} \cdot \sqrt{25} = 4 \cdot 5 = 20$

$\sqrt[3]{8 \cdot 27} = \sqrt[3]{216} = 6$ and $\sqrt[3]{8} \cdot \sqrt[3]{27} = 2 \cdot 3 = 6$

$\sqrt[3]{(-8)(27)} = \sqrt[3]{-216} = -6$ and $\sqrt[3]{-8} \cdot \sqrt[3]{27} = (-2)(3) = -6$

In general, we can state the following property.

Property 7.4

$\sqrt[n]{bc} = \sqrt[n]{b}\sqrt[n]{c}$ $\quad \sqrt[n]{b}$ and $\sqrt[n]{c}$ are real numbers

Property 7.4 states that **the *n*th root of a product is equal to the product of the *n*th roots.**

2 Express a Radical in Simplest Radical Form

The definition of *n*th root, along with Property 7.4, provides the basis for changing radicals to simplest radical form. The concept of **simplest radical form** takes on additional meaning as we encounter more complicated expressions, but for now it simply means that the radicand does not contain any perfect powers of the index. Let's consider some examples to clarify this idea.

EXAMPLE 1

Express each of the following in simplest radical form.

(a) $\sqrt{8}$ **(b)** $\sqrt{45}$ **(c)** $\sqrt[3]{24}$ **(d)** $\sqrt[3]{54}$

Solution

(a) $\sqrt{8} = \sqrt{4 \cdot 2} = \sqrt{4}\sqrt{2} = 2\sqrt{2}$

↑ 4 is a perfect square

(b) $\sqrt{45} = \sqrt{9 \cdot 5} = \sqrt{9}\sqrt{5} = 3\sqrt{5}$

↑ 9 is a perfect square

(c) $\sqrt[3]{24} = \sqrt[3]{8 \cdot 3} = \sqrt[3]{8}\sqrt[3]{3} = 2\sqrt[3]{3}$

↑ 8 is a perfect cube

(d) $\sqrt[3]{54} = \sqrt[3]{27 \cdot 2} = \sqrt[3]{27}\sqrt[3]{2} = 3\sqrt[3]{2}$

↑ 27 is a perfect cube

▼ **PRACTICE YOUR SKILL**

Express each of the following in simplest radical form.

(a) $\sqrt{20}$ **(b)** $\sqrt{18}$ **(c)** $\sqrt[3]{32}$ **(d)** $\sqrt[3]{128}$ ■

The first step in each example is to express the radicand of the given radical as the product of two factors, one of which must be a perfect nth power other than 1. Also, observe the radicands of the final radicals. In each case, the radicand cannot have a factor that is a perfect nth power other than 1. We say that the final radicals $2\sqrt{2}$, $3\sqrt{5}$, $2\sqrt[3]{3}$, and $3\sqrt[3]{2}$ are in **simplest radical form**.

You may vary the steps somewhat in changing to simplest radical form, but the final result should be the same. Consider some different approaches to changing $\sqrt{72}$ to simplest form:

$$\sqrt{72} = \sqrt{9}\sqrt{8} = 3\sqrt{8} = 3\sqrt{4}\sqrt{2} = 3 \cdot 2\sqrt{2} = 6\sqrt{2} \quad \text{or}$$

$$\sqrt{72} = \sqrt{4}\sqrt{18} = 2\sqrt{18} = 2\sqrt{9}\sqrt{2} = 2 \cdot 3\sqrt{2} = 6\sqrt{2} \quad \text{or}$$

$$\sqrt{72} = \sqrt{36}\sqrt{2} = 6\sqrt{2}$$

Another variation of the technique for changing radicals to simplest form is to prime factor the radicand and then to look for perfect nth powers in exponential form. The following example illustrates the use of this technique.

EXAMPLE 2

Express each of the following in simplest radical form.

(a) $\sqrt{50}$ **(b)** $3\sqrt{80}$ **(c)** $\sqrt[3]{108}$

Solution

(a) $\sqrt{50} = \sqrt{2 \cdot 5 \cdot 5} = \sqrt{5^2}\sqrt{2} = 5\sqrt{2}$

(b) $3\sqrt{80} = 3\sqrt{2 \cdot 2 \cdot 2 \cdot 2 \cdot 5} = 3\sqrt{2^4}\sqrt{5} = 3 \cdot 2^2\sqrt{5} = 12\sqrt{5}$

(c) $\sqrt[3]{108} = \sqrt[3]{2 \cdot 2 \cdot 3 \cdot 3 \cdot 3} = \sqrt[3]{3^3}\sqrt[3]{4} = 3\sqrt[3]{4}$

▼ PRACTICE YOUR SKILL

Express each of the following in simplest radical form.

(a) $\sqrt{48}$ **(b)** $5\sqrt{32}$ **(c)** $\sqrt[3]{375}$ ■

Another property of nth roots is demonstrated by the following examples.

$$\sqrt{\frac{36}{9}} = \sqrt{4} = 2 \quad \text{and} \quad \frac{\sqrt{36}}{\sqrt{9}} = \frac{6}{3} = 2$$

$$\sqrt[3]{\frac{64}{8}} = \sqrt[3]{8} = 2 \quad \text{and} \quad \frac{\sqrt[3]{64}}{\sqrt[3]{8}} = \frac{4}{2} = 2$$

$$\sqrt[3]{\frac{-8}{64}} = \sqrt[3]{-\frac{1}{8}} = -\frac{1}{2} \quad \text{and} \quad \frac{\sqrt[3]{-8}}{\sqrt[3]{64}} = \frac{-2}{4} = -\frac{1}{2}$$

In general, we can state the following property.

Property 7.5

$$\sqrt[n]{\frac{b}{c}} = \frac{\sqrt[n]{b}}{\sqrt[n]{c}} \qquad \sqrt[n]{b} \text{ and } \sqrt[n]{c} \text{ are real numbers, and } c \neq 0$$

Property 7.5 states that **the nth root of a quotient is equal to the quotient of the nth roots**.

To evaluate radicals such as $\sqrt{\frac{4}{25}}$ and $\sqrt[3]{\frac{27}{8}}$, for which the numerator and denominator of the fractional radicand are perfect nth powers, you may use Property 7.5 or merely rely on the definition of nth root.

$$\sqrt{\frac{4}{25}} = \frac{\sqrt{4}}{\sqrt{25}} = \frac{2}{5} \quad \text{or} \quad \sqrt{\frac{4}{25}} = \frac{2}{5} \quad \text{because } \frac{2}{5} \cdot \frac{2}{5} = \frac{4}{25}$$

Property 7.5   Definition of nth root

$$\sqrt[3]{\frac{27}{8}} = \frac{\sqrt[3]{27}}{\sqrt[3]{8}} = \frac{3}{2} \quad \text{or} \quad \sqrt[3]{\frac{27}{8}} = \frac{3}{2} \quad \text{because } \frac{3}{2} \cdot \frac{3}{2} \cdot \frac{3}{2} = \frac{27}{8}$$

Radicals such as $\sqrt{\frac{28}{9}}$ and $\sqrt[3]{\frac{24}{27}}$, in which only the denominators of the radicand are perfect nth powers, can be simplified as follows:

$$\sqrt{\frac{28}{9}} = \frac{\sqrt{28}}{\sqrt{9}} = \frac{\sqrt{28}}{3} = \frac{\sqrt{4}\sqrt{7}}{3} = \frac{2\sqrt{7}}{3}$$

$$\sqrt[3]{\frac{24}{27}} = \frac{\sqrt[3]{24}}{\sqrt[3]{27}} = \frac{\sqrt[3]{24}}{3} = \frac{\sqrt[3]{8}\sqrt[3]{3}}{3} = \frac{2\sqrt[3]{3}}{3}$$

Before we consider more examples, let's summarize some ideas that pertain to the simplifying of radicals. A radical is said to be in **simplest radical form** if the following conditions are satisfied.

1. No fraction appears with a radical sign. $\sqrt{\frac{3}{4}}$ violates this condition.
2. No radical appears in the denominator. $\frac{\sqrt{2}}{\sqrt{3}}$ violates this condition.
3. No radicand, when expressed in prime-factored form, contains a factor raised to a power equal to or greater than the index.
   $\sqrt{2^3 \cdot 5}$ violates this condition.

3 Rationalize the Denominator to Simplify Radicals

Now let's consider an example in which neither the numerator nor the denominator of the radicand is a perfect nth power.

EXAMPLE 3 Simplify $\sqrt{\frac{2}{3}}$.

Solution

$$\sqrt{\frac{2}{3}} = \frac{\sqrt{2}}{\sqrt{3}} = \frac{\sqrt{2}}{\sqrt{3}} \cdot \frac{\sqrt{3}}{\sqrt{3}} = \frac{\sqrt{6}}{3}$$

Form of 1

▼ **PRACTICE YOUR SKILL**

Simplify $\sqrt{\frac{3}{5}}$. ■

We refer to the process we used to simplify the radical in Example 3 as **rationalizing the denominator**. Note that the denominator becomes a rational number. The process of rationalizing the denominator can often be accomplished in more than one way, as we will see in the next example.

EXAMPLE 4

Simplify $\frac{\sqrt{5}}{\sqrt{8}}$.

Solution A

$$\frac{\sqrt{5}}{\sqrt{8}} = \frac{\sqrt{5}}{\sqrt{8}} \cdot \frac{\sqrt{8}}{\sqrt{8}} = \frac{\sqrt{40}}{8} = \frac{\sqrt{4}\sqrt{10}}{8} = \frac{2\sqrt{10}}{8} = \frac{\sqrt{10}}{4}$$

Solution B

$$\frac{\sqrt{5}}{\sqrt{8}} = \frac{\sqrt{5}}{\sqrt{8}} \cdot \frac{\sqrt{2}}{\sqrt{2}} = \frac{\sqrt{10}}{\sqrt{16}} = \frac{\sqrt{10}}{4}$$

Solution C

$$\frac{\sqrt{5}}{\sqrt{8}} = \frac{\sqrt{5}}{\sqrt{4}\sqrt{2}} = \frac{\sqrt{5}}{2\sqrt{2}} = \frac{\sqrt{5}}{2\sqrt{2}} \cdot \frac{\sqrt{2}}{\sqrt{2}} = \frac{\sqrt{10}}{2\sqrt{4}} = \frac{\sqrt{10}}{2(2)} = \frac{\sqrt{10}}{4}$$

▼ **PRACTICE YOUR SKILL**

Simplify $\frac{\sqrt{7}}{\sqrt{12}}$. ■

The three approaches to Example 4 again illustrate the need to think first and only then push the pencil. You may find one approach easier than another. To conclude this section, study the following examples and check the final radicals against the three conditions previously listed for **simplest radical form**.

EXAMPLE 5

Simplify each of the following.

(a) $\frac{3\sqrt{2}}{5\sqrt{3}}$ **(b)** $\frac{3\sqrt{7}}{2\sqrt{18}}$ **(c)** $\sqrt[3]{\frac{5}{9}}$ **(d)** $\frac{\sqrt[3]{5}}{\sqrt[3]{16}}$

Solution

(a) $$\frac{3\sqrt{2}}{5\sqrt{3}} = \frac{3\sqrt{2}}{5\sqrt{3}} \cdot \underbrace{\frac{\sqrt{3}}{\sqrt{3}}}_{\text{Form of 1}} = \frac{3\sqrt{6}}{5\sqrt{9}} = \frac{3\sqrt{6}}{15} = \frac{\sqrt{6}}{5}$$

(b) $$\frac{3\sqrt{7}}{2\sqrt{18}} = \frac{3\sqrt{7}}{2\sqrt{18}} \cdot \underbrace{\frac{\sqrt{2}}{\sqrt{2}}}_{\text{Form of 1}} = \frac{3\sqrt{14}}{2\sqrt{36}} = \frac{3\sqrt{14}}{12} = \frac{\sqrt{14}}{4}$$

(c) $$\sqrt[3]{\frac{5}{9}} = \frac{\sqrt[3]{5}}{\sqrt[3]{9}} = \frac{\sqrt[3]{5}}{\sqrt[3]{9}} \cdot \underbrace{\frac{\sqrt[3]{3}}{\sqrt[3]{3}}}_{\text{Form of 1}} = \frac{\sqrt[3]{15}}{\sqrt[3]{27}} = \frac{\sqrt[3]{15}}{3}$$

(d) $\dfrac{\sqrt[3]{5}}{\sqrt[3]{16}} = \dfrac{\sqrt[3]{5}}{\sqrt[3]{16}} \cdot \dfrac{\sqrt[3]{4}}{\sqrt[3]{4}} = \dfrac{\sqrt[3]{20}}{\sqrt[3]{64}} = \dfrac{\sqrt[3]{20}}{4}$

Form of 1 (pointing to $\dfrac{\sqrt[3]{4}}{\sqrt[3]{4}}$)

▼ PRACTICE YOUR SKILL

Simplify each of the following.

(a) $\dfrac{5\sqrt{2}}{3\sqrt{5}}$ **(b)** $\dfrac{2\sqrt{5}}{7\sqrt{12}}$ **(c)** $\sqrt[3]{\dfrac{3}{4}}$ **(d)** $\dfrac{\sqrt[3]{7}}{\sqrt[3]{32}}$ ■

4 Applications of Radicals

Many real-world applications involve radical expressions. For example, police often use the formula $S = \sqrt{30Df}$ to estimate the speed of a car on the basis of the length of the skid marks at the scene of an accident. In this formula, S represents the speed of the car in miles per hour, D represents the length of the skid marks in feet, and f represents a coefficient of friction. For a particular situation, the coefficient of friction is a constant that depends on the type and condition of the road surface.

EXAMPLE 6

Using 0.35 as a coefficient of friction, determine how fast a car was traveling if it skidded 325 feet.

Solution

Substitute 0.35 for f and 325 for D in the formula.

$$S = \sqrt{30Df} = \sqrt{30(325)(0.35)} = 58, \quad \text{to the nearest whole number}$$

The car was traveling at approximately 58 miles per hour.

▼ PRACTICE YOUR SKILL

Using 0.25 as a coefficient of friction, determine how fast a car was traveling, to the nearest whole number, if it skidded 290 feet. ■

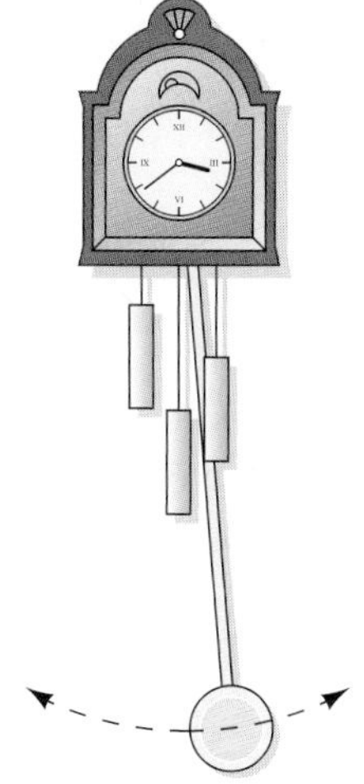

Figure 7.1

The **period** of a pendulum is the time it takes to swing from one side to the other side and back. The formula

$$T = 2\pi\sqrt{\frac{L}{32}}$$

where T represents the time in seconds and L the length in feet, can be used to determine the period of a pendulum (see Figure 7.1).

EXAMPLE 7

Find, to the nearest tenth of a second, the period of a pendulum of length 3.5 feet.

Solution

Let's use 3.14 as an approximation for π and substitute 3.5 for L in the formula.

$$T = 2\pi\sqrt{\frac{L}{32}} = 2(3.14)\sqrt{\frac{3.5}{32}} = 2.1, \quad \text{to the nearest tenth}$$

The period is approximately 2.1 seconds.

▼ **PRACTICE YOUR SKILL**

Find, to the nearest tenth, the period of a pendulum of length 4.5 feet. ■

Radical expressions are also used in some geometric applications. For example, the area of a triangle can be found by using a formula that involves a square root. If a, b, and c represent the lengths of the three sides of a triangle, the formula $K = \sqrt{s(s-a)(s-b)(s-c)}$, known as Heron's formula, can be used to determine the area (K) of the triangle. The letter s represents the semiperimeter of the triangle; that is, $s = \frac{a+b+c}{2}$.

EXAMPLE 8

Find the area of a triangular piece of sheet metal that has sides of length 17 inches, 19 inches, and 26 inches.

Solution

First, let's find the value of s, the semiperimeter of the triangle.

$$s = \frac{17+19+26}{2} = 31$$

Now we can use Heron's formula.

$$\begin{aligned} K = \sqrt{s(s-a)(s-b)(s-c)} &= \sqrt{31(31-17)(31-19)(31-26)} \\ &= \sqrt{31(14)(12)(5)} \\ &= \sqrt{26{,}040} \\ &= 161.4, \quad \text{to the nearest tenth} \end{aligned}$$

Thus the area of the piece of sheet metal is approximately 161.4 square inches.

▼ **PRACTICE YOUR SKILL**

Find the area of a triangle, to the nearest tenth, that has sides of lengths 14 inches, 18 inches, and 20 inches. ■

Remark: Note that in Examples 6–8 we did not simplify the radicals. When one is using a calculator to approximate the square roots, there is no need to simplify first.

CONCEPT QUIZ

For Problems 1–10, answer true or false.

1. The cube root of a number is one of its three equal factors.
2. Every positive real number has one positive real number square root.
3. The principal square root of a number is the positive square root of the number.
4. The symbol $\sqrt{\ }$ is called a radical.
5. The square root of 0 is not a real number.
6. The number under the radical sign is called the radicand.
7. Every positive real number has two square roots.
8. The n in the radical $\sqrt[n]{a}$ is called the index of the radical.
9. If n is an odd integer greater than 1 and b is a negative real number, then $\sqrt[n]{b}$ is a negative real number.
10. $\frac{3\sqrt{24}}{8}$ is in simplest radical form.

Problem Set 7.2

1 Evaluate Roots of Numbers

For Problems 1–20, evaluate each of the following. For example, $\sqrt{25} = 5$.

1. $\sqrt{64}$
2. $\sqrt{49}$
3. $-\sqrt{100}$
4. $-\sqrt{81}$
5. $\sqrt[3]{27}$
6. $\sqrt[3]{216}$
7. $\sqrt[3]{-64}$
8. $\sqrt[3]{-125}$
9. $\sqrt[4]{81}$
10. $-\sqrt[4]{16}$
11. $\sqrt{\frac{16}{25}}$
12. $\sqrt{\frac{25}{64}}$
13. $-\sqrt{\frac{36}{49}}$
14. $\sqrt{\frac{16}{64}}$
15. $\sqrt{\frac{9}{36}}$
16. $\sqrt{\frac{144}{36}}$
17. $\sqrt[3]{\frac{27}{64}}$
18. $\sqrt[3]{-\frac{8}{27}}$
19. $\sqrt[3]{8^3}$
20. $\sqrt[4]{16^4}$

2 Express a Radical in Simplest Radical Form

For Problems 21–48, change each radical to simplest radical form.

21. $\sqrt{27}$
22. $\sqrt{48}$
23. $\sqrt{32}$
24. $\sqrt{98}$
25. $\sqrt{80}$
26. $\sqrt{125}$
27. $\sqrt{160}$
28. $\sqrt{112}$
29. $4\sqrt{18}$
30. $5\sqrt{32}$
31. $-6\sqrt{20}$
32. $-4\sqrt{54}$
33. $\frac{2}{5}\sqrt{75}$
34. $\frac{1}{3}\sqrt{90}$
35. $\frac{3}{2}\sqrt{24}$
36. $\frac{3}{4}\sqrt{45}$
37. $-\frac{5}{6}\sqrt{28}$
38. $-\frac{2}{3}\sqrt{96}$
39. $\sqrt{\frac{19}{4}}$
40. $\sqrt{\frac{22}{9}}$
41. $\sqrt{\frac{27}{16}}$
42. $\sqrt{\frac{8}{25}}$
43. $\sqrt{\frac{75}{81}}$
44. $\sqrt{\frac{24}{49}}$
45. $\sqrt[3]{16}$
46. $\sqrt[3]{40}$
47. $2\sqrt[3]{81}$
48. $-3\sqrt[3]{54}$

3 Rationalize the Denominator to Simplify Radicals

For Problems 49–74, rationalize the denominator and express each radical in simplest form.

49. $\sqrt{\frac{2}{7}}$
50. $\sqrt{\frac{3}{8}}$
51. $\sqrt{\frac{2}{3}}$
52. $\sqrt{\frac{7}{12}}$
53. $\frac{\sqrt{5}}{\sqrt{12}}$
54. $\frac{\sqrt{3}}{\sqrt{7}}$
55. $\frac{\sqrt{11}}{\sqrt{24}}$
56. $\frac{\sqrt{5}}{\sqrt{48}}$
57. $\frac{\sqrt{18}}{\sqrt{27}}$
58. $\frac{\sqrt{10}}{\sqrt{20}}$
59. $\frac{\sqrt{35}}{\sqrt{7}}$
60. $\frac{\sqrt{42}}{\sqrt{6}}$
61. $\frac{2\sqrt{3}}{\sqrt{7}}$
62. $\frac{3\sqrt{2}}{\sqrt{6}}$
63. $-\frac{4\sqrt{12}}{\sqrt{5}}$
64. $\frac{-6\sqrt{5}}{\sqrt{18}}$
65. $\frac{3\sqrt{2}}{4\sqrt{3}}$
66. $\frac{6\sqrt{5}}{5\sqrt{12}}$
67. $\frac{-8\sqrt{18}}{10\sqrt{50}}$
68. $\frac{4\sqrt{45}}{-6\sqrt{20}}$
69. $\frac{2}{\sqrt[3]{9}}$
70. $\frac{3}{\sqrt[3]{3}}$
71. $\frac{\sqrt[3]{27}}{\sqrt[3]{4}}$
72. $\frac{\sqrt[3]{8}}{\sqrt[3]{16}}$
73. $\frac{\sqrt[3]{6}}{\sqrt[3]{4}}$
74. $\frac{\sqrt[3]{4}}{\sqrt[3]{2}}$

4 Applications of Radicals

75. Use a coefficient of friction of 0.4 in the formula from Example 6 to find the speeds of cars that left skid marks of lengths 150 feet, 200 feet, and 350 feet. Express your answers to the nearest mile per hour.

76. Use the formula from Example 7 to find the periods of pendulums of lengths 2 feet, 3 feet, and 5.5 feet. Express your answers to the nearest tenth of a second.

77. Find, to the nearest square centimeter, the area of a triangle that measures 14 centimeters by 16 centimeters by 18 centimeters.

78. Find, to the nearest square yard, the area of a triangular plot of ground that measures 45 yards by 60 yards by 75 yards.

79. Find the area of an equilateral triangle, each of whose sides is 18 inches long. Express the area to the nearest square inch.

80. Find, to the nearest square inch, the area of the quadrilateral in Figure 7.2.

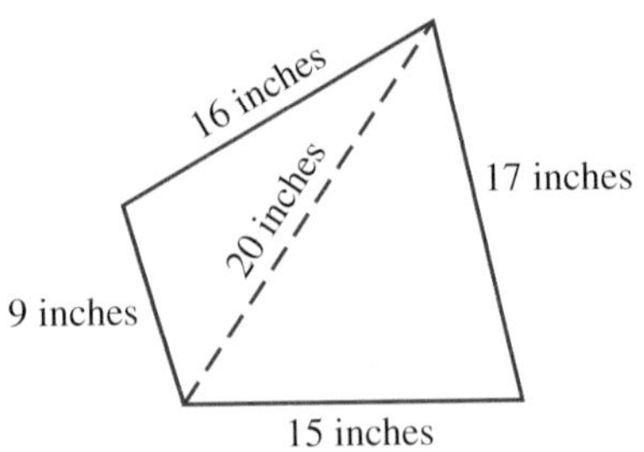

Figure 7.2

THOUGHTS INTO WORDS

81. Why is $\sqrt{-9}$ not a real number?

82. Why is it that we say 25 has two square roots (5 and -5) but we write $\sqrt{25} = 5$?

83. How is the multiplication property of 1 used when simplifying radicals?

84. How could you find a whole number approximation for $\sqrt{2750}$ if you did not have a calculator or table available?

FURTHER INVESTIGATIONS

85. Use your calculator to find a rational approximation, to the nearest thousandth, for (a) through (i).

(a) $\sqrt{2}$ **(b)** $\sqrt{75}$ **(c)** $\sqrt{156}$

(d) $\sqrt{691}$ **(e)** $\sqrt{3249}$ **(f)** $\sqrt{45{,}123}$

(g) $\sqrt{0.14}$ **(h)** $\sqrt{0.023}$ **(i)** $\sqrt{0.8649}$

86. Sometimes a fairly good estimate can be made of a radical expression by using whole number approximations. For example, $5\sqrt{35} + 7\sqrt{50}$ is approximately $5(6) + 7(7) = 79$. Using a calculator, we find that $5\sqrt{35} + 7\sqrt{50} = 79.1$, to the nearest tenth. In this case our whole number estimate is very good. For (a) through (f), first make a whole number estimate and then use your calculator to see how well you estimated.

(a) $3\sqrt{10} - 4\sqrt{24} + 6\sqrt{65}$

(b) $9\sqrt{27} + 5\sqrt{37} - 3\sqrt{80}$

(c) $12\sqrt{5} + 13\sqrt{18} + 9\sqrt{47}$

(d) $3\sqrt{98} - 4\sqrt{83} - 7\sqrt{120}$

(e) $4\sqrt{170} + 2\sqrt{198} + 5\sqrt{227}$

(f) $-3\sqrt{256} - 6\sqrt{287} + 11\sqrt{321}$

Answers to the Concept Quiz

1. True **2.** True **3.** True **4.** False **5.** False **6.** True **7.** True **8.** True **9.** True **10.** False

Answers to the Example Practice Skills

1. (a) $2\sqrt{5}$ **(b)** $3\sqrt{2}$ **(c)** $2\sqrt[3]{4}$ **(d)** $4\sqrt[3]{2}$ **2. (a)** $4\sqrt{3}$ **(b)** $20\sqrt{2}$ **(c)** $5\sqrt[3]{3}$ **3.** $\frac{\sqrt{15}}{5}$ **4.** $\frac{\sqrt{21}}{6}$

5. (a) $\frac{\sqrt{10}}{3}$ **(b)** $\frac{\sqrt{15}}{21}$ **(c)** $\frac{\sqrt[3]{6}}{2}$ **(d)** $\frac{\sqrt[3]{14}}{4}$ **6.** 47 mph **7.** 2.4 sec **8.** 122.4 in.^{2}

7.3 Combining Radicals and Simplifying Radicals That Contain Variables

OBJECTIVES

1. Simplify Expressions by Combining Radicals
2. Simplify Radicals That Contain Variables

1 Simplify Expressions by Combining Radicals

Recall our use of the distributive property as the basis for combining similar terms. For example,

$$3x + 2x = (3 + 2)x = 5x$$

$$8y - 5y = (8 - 5)y = 3y$$

$$\frac{2}{3}a^2 + \frac{3}{4}a^2 = \left(\frac{2}{3} + \frac{3}{4}\right)a^2 = \left(\frac{8}{12} + \frac{9}{12}\right)a^2 = \frac{17}{12}a^2$$

In a like manner, expressions that contain radicals can often be simplified by using the distributive property, as follows:

$$3\sqrt{2} + 5\sqrt{2} = (3 + 5)\sqrt{2} = 8\sqrt{2}$$

$$7\sqrt[3]{5} - 3\sqrt[3]{5} = (7 - 3)\sqrt[3]{5} = 4\sqrt[3]{5}$$

$$4\sqrt{7} + 5\sqrt{7} + 6\sqrt{11} - 2\sqrt{11} = (4 + 5)\sqrt{7} + (6 - 2)\sqrt{11}$$
$$= 9\sqrt{7} + 4\sqrt{11}$$

Note that *in order to be added or subtracted, radicals must have the same index and the same radicand*. Thus we cannot simplify an expression such as $5\sqrt{2} + 7\sqrt{11}$.

Simplifying by combining radicals sometimes requires that you first express the given radicals in simplest form and then apply the distributive property. The following examples illustrate this idea.

EXAMPLE 1

Simplify $3\sqrt{8} + 2\sqrt{18} - 4\sqrt{2}$.

Solution

$$3\sqrt{8} + 2\sqrt{18} - 4\sqrt{2} = 3\sqrt{4}\sqrt{2} + 2\sqrt{9}\sqrt{2} - 4\sqrt{2}$$
$$= 3 \cdot 2 \cdot \sqrt{2} + 2 \cdot 3 \cdot \sqrt{2} - 4\sqrt{2}$$
$$= 6\sqrt{2} + 6\sqrt{2} - 4\sqrt{2}$$
$$= (6 + 6 - 4)\sqrt{2} = 8\sqrt{2}$$

▼ **PRACTICE YOUR SKILL**

Simplify $5\sqrt{12} - 3\sqrt{75} + 6\sqrt{27}$. ■

EXAMPLE 2

Simplify $\frac{1}{4}\sqrt{45} + \frac{1}{3}\sqrt{20}$.

Solution

$$\begin{aligned}\frac{1}{4}\sqrt{45} + \frac{1}{3}\sqrt{20} &= \frac{1}{4}\sqrt{9}\sqrt{5} + \frac{1}{3}\sqrt{4}\sqrt{5} \\ &= \frac{1}{4}\cdot 3\cdot\sqrt{5} + \frac{1}{3}\cdot 2\cdot\sqrt{5} \\ &= \frac{3}{4}\sqrt{5} + \frac{2}{3}\sqrt{5} = \left(\frac{3}{4}+\frac{2}{3}\right)\sqrt{5} \\ &= \left(\frac{9}{12}+\frac{8}{12}\right)\sqrt{5} = \frac{17}{12}\sqrt{5}\end{aligned}$$

▼ PRACTICE YOUR SKILL

Simplify $\frac{1}{5}\sqrt{8} + \frac{1}{3}\sqrt{18}$. ■

EXAMPLE 3

Simplify $5\sqrt[3]{2} - 2\sqrt[3]{16} - 6\sqrt[3]{54}$.

Solution

$$\begin{aligned}5\sqrt[3]{2} - 2\sqrt[3]{16} - 6\sqrt[3]{54} &= 5\sqrt[3]{2} - 2\sqrt[3]{8}\sqrt[3]{2} - 6\sqrt[3]{27}\sqrt[3]{2} \\ &= 5\sqrt[3]{2} - 2\cdot 2\cdot\sqrt[3]{2} - 6\cdot 3\cdot\sqrt[3]{2} \\ &= 5\sqrt[3]{2} - 4\sqrt[3]{2} - 18\sqrt[3]{2} \\ &= (5-4-18)\sqrt[3]{2} \\ &= -17\sqrt[3]{2}\end{aligned}$$

▼ PRACTICE YOUR SKILL

Simplify $\sqrt[3]{2} - 4\sqrt[3]{54} + 3\sqrt[3]{16}$. ■

2 Simplify Radicals That Contain Variables

Before we discuss the process of simplifying radicals that contain variables, there is one technicality that we should call to your attention. Let's look at some examples to clarify the point. Consider the radical $\sqrt{x^2}$.

Let $x = 3$; then $\sqrt{x^2} = \sqrt{3^2} = \sqrt{9} = 3$.

Let $x = -3$; then $\sqrt{x^2} = \sqrt{(-3)^2} = \sqrt{9} = 3$.

Thus if $x \geq 0$ then $\sqrt{x^2} = x$, *but* if $x < 0$ then $\sqrt{x^2} = -x$. Using the concept of absolute value, we can state that for all real numbers, $\sqrt{x^2} = |x|$.

Now consider the radical $\sqrt{x^3}$. Because x^3 is negative when x is negative, we need to restrict x to the nonnegative reals when working with $\sqrt{x^3}$. Thus we can write, "if $x \geq 0$, then $\sqrt{x^3} = \sqrt{x^2}\sqrt{x} = x\sqrt{x}$," and no absolute value sign is necessary. Finally, let's consider the radical $\sqrt[3]{x^3}$.

Let $x = 2$; then $\sqrt[3]{x^3} = \sqrt[3]{2^3} = \sqrt[3]{8} = 2$.

Let $x = -2$; then $\sqrt[3]{x^3} = \sqrt[3]{(-2)^3} = \sqrt[3]{-8} = -2$.

Thus it is correct to write, "$\sqrt[3]{x^3} = x$ for all real numbers," and again no absolute value sign is necessary.

The previous discussion indicates that technically, every radical expression involving variables in the radicand needs to be analyzed individually in terms of any necessary restrictions imposed on the variables. To help you gain experience with this skill, examples and problems are discussed under Further Investigations in the problem set. For now, however, to avoid considering such restrictions on a problem-to-problem basis, we shall merely assume that all variables represent positive real numbers. Let's consider the process of simplifying radicals that contain variables in the radicand. Study the following examples, and note that the same basic approach we used in Section 7.2 is applied here.

EXAMPLE 4

Simplify each of the following.

(a) $\sqrt{8x^3}$ **(b)** $\sqrt{45x^3y^7}$ **(c)** $\sqrt{180a^4b^3}$ **(d)** $\sqrt[3]{40x^4y^8}$

Solution

(a) $\sqrt{8x^3} = \sqrt{4x^2}\sqrt{2x} = 2x\sqrt{2x}$

↑ $4x^2$ is a perfect square

(b) $\sqrt{45x^3y^7} = \sqrt{9x^2y^6}\sqrt{5xy} = 3xy^3\sqrt{5xy}$

↑ $9x^2y^6$ is a perfect square

(c) If the numerical coefficient of the radicand is quite large, then you may want to look at it in the prime-factored form.

$$\begin{aligned}\sqrt{180a^4b^3} &= \sqrt{2 \cdot 2 \cdot 3 \cdot 3 \cdot 5 \cdot a^4 \cdot b^3}\\ &= \sqrt{36 \cdot 5 \cdot a^4 \cdot b^3}\\ &= \sqrt{36a^4b^2}\sqrt{5b}\\ &= 6a^2b\sqrt{5b}\end{aligned}$$

(d) $\sqrt[3]{40x^4y^8} = \sqrt[3]{8x^3y^6}\sqrt[3]{5xy^2} = 2xy^2\sqrt[3]{5xy^2}$

↑ $8x^3y^6$ is a perfect cube

▼ PRACTICE YOUR SKILL

Simplify each of the following.

(a) $\sqrt{98y^3}$ **(b)** $\sqrt{28a^5b^3}$ **(c)** $\sqrt{240x^3y^4}$ **(d)** $\sqrt[3]{54a^2b^5}$ ■

Before we consider more examples, let's restate (in such a way as to include radicands containing variables) the conditions necessary for a radical to be in simplest radical form.

1. A radicand contains no polynomial factor raised to a power equal to or greater than the index of the radical. $\sqrt{x^3}$ violates this condition
2. No fraction appears within a radical sign. $\sqrt{\frac{2x}{3y}}$ violates this condition
3. No radical appears in the denominator. $\frac{3}{\sqrt[3]{4x}}$ violates this condition

EXAMPLE 5

Express each of the following in simplest radical form.

(a) $\sqrt{\frac{2x}{3y}}$ (b) $\frac{\sqrt{5}}{\sqrt{12a^3}}$ (c) $\frac{\sqrt{8x^2}}{\sqrt{27y^5}}$

(d) $\frac{3}{\sqrt[3]{4x}}$ (e) $\frac{\sqrt[3]{16x^2}}{\sqrt[3]{9y^5}}$

Solution

(a) $\sqrt{\frac{2x}{3y}} = \frac{\sqrt{2x}}{\sqrt{3y}} = \frac{\sqrt{2x}}{\sqrt{3y}} \cdot \frac{\sqrt{3y}}{\sqrt{3y}} = \frac{\sqrt{6xy}}{3y}$

Form of 1 (pointing to $\frac{\sqrt{3y}}{\sqrt{3y}}$)

(b) $\frac{\sqrt{5}}{\sqrt{12a^3}} = \frac{\sqrt{5}}{\sqrt{12a^3}} \cdot \frac{\sqrt{3a}}{\sqrt{3a}} = \frac{\sqrt{15a}}{\sqrt{36a^4}} = \frac{\sqrt{15a}}{6a^2}$

Form of 1 (pointing to $\frac{\sqrt{3a}}{\sqrt{3a}}$)

(c) $\frac{\sqrt{8x^2}}{\sqrt{27y^5}} = \frac{\sqrt{4x^2}\sqrt{2}}{\sqrt{9y^4}\sqrt{3y}} = \frac{2x\sqrt{2}}{3y^2\sqrt{3y}} = \frac{2x\sqrt{2}}{3y^2\sqrt{3y}} \cdot \frac{\sqrt{3y}}{\sqrt{3y}}$

$= \frac{2x\sqrt{6y}}{(3y^2)(3y)} = \frac{2x\sqrt{6y}}{9y^3}$

(d) $\frac{3}{\sqrt[3]{4x}} = \frac{3}{\sqrt[3]{4x}} \cdot \frac{\sqrt[3]{2x^2}}{\sqrt[3]{2x^2}} = \frac{3\sqrt[3]{2x^2}}{\sqrt[3]{8x^3}} = \frac{3\sqrt[3]{2x^2}}{2x}$

(e) $\frac{\sqrt[3]{16x^2}}{\sqrt[3]{9y^5}} = \frac{\sqrt[3]{16x^2}}{\sqrt[3]{9y^5}} \cdot \frac{\sqrt[3]{3y}}{\sqrt[3]{3y}} = \frac{\sqrt[3]{48x^2y}}{\sqrt[3]{27y^6}} = \frac{\sqrt[3]{8}\sqrt[3]{6x^2y}}{3y^2} = \frac{2\sqrt[3]{6x^2y}}{3y^2}$

▼ PRACTICE YOUR SKILL

Express each of the following in simplest radical form.

(a) $\sqrt{\frac{3a}{5b}}$ (b) $\frac{\sqrt{7}}{\sqrt{18x^3}}$ (c) $\frac{\sqrt{75a^3}}{\sqrt{8b}}$ (d) $\frac{2}{\sqrt[3]{3y^2}}$ (e) $\frac{\sqrt[3]{81a^2}}{\sqrt[3]{2b}}$ ■

Note that in part (c) we did some simplifying first before rationalizing the denominator, whereas in part (b) we proceeded immediately to rationalize the denominator. This is an individual choice, and you should probably do it both ways a few times to decide which you prefer.

CONCEPT QUIZ

For Problems 1–10, answer true or false.

1. In order to be combined when adding, radicals must have the same index and the same radicand.
2. If $x \geq 0$, then $\sqrt{x^2} = x$.
3. For all real numbers, $\sqrt{x^2} = x$.
4. For all real numbers, $\sqrt[3]{x^3} = x$.
5. A radical is not in simplest radical form if it has a fraction within the radical sign.
6. If a radical contains a factor raised to a power that is equal to the index of the radical, then the radical is not in simplest radical form.
7. The radical $\frac{1}{\sqrt{x}}$ is in simplest radical form.
8. $3\sqrt{2} + 4\sqrt{3} = 7\sqrt{5}$.
9. If $x > 0$, then $\sqrt{45x^3} = 3x^2\sqrt{5x}$.
10. If $x > 0$, then $\frac{4\sqrt{x^5}}{3\sqrt{4x^2}} = \frac{2x\sqrt{x}}{3}$.

Problem Set 7.3

1 Simplify Expressions by Combining Radicals

For Problems 1–20, use the distributive property to help simplify each of the following. For example,

$$\begin{aligned}3\sqrt{8} - \sqrt{32} &= 3\sqrt{4}\sqrt{2} - \sqrt{16}\sqrt{2}\\ &= 3(2)\sqrt{2} - 4\sqrt{2}\\ &= 6\sqrt{2} - 4\sqrt{2}\\ &= (6-4)\sqrt{2} = 2\sqrt{2}\end{aligned}$$

1. $5\sqrt{18} - 2\sqrt{2}$
2. $7\sqrt{12} + 4\sqrt{3}$
3. $7\sqrt{12} + 10\sqrt{48}$
4. $6\sqrt{8} - 5\sqrt{18}$
5. $-2\sqrt{50} - 5\sqrt{32}$
6. $-2\sqrt{20} - 7\sqrt{45}$
7. $3\sqrt{20} - \sqrt{5} - 2\sqrt{45}$
8. $6\sqrt{12} + \sqrt{3} - 2\sqrt{48}$
9. $-9\sqrt{24} + 3\sqrt{54} - 12\sqrt{6}$
10. $13\sqrt{28} - 2\sqrt{63} - 7\sqrt{7}$
11. $\frac{3}{4}\sqrt{7} - \frac{2}{3}\sqrt{28}$
12. $\frac{3}{5}\sqrt{5} - \frac{1}{4}\sqrt{80}$
13. $\frac{3}{5}\sqrt{40} + \frac{5}{6}\sqrt{90}$
14. $\frac{3}{8}\sqrt{96} - \frac{2}{3}\sqrt{54}$
15. $\frac{3\sqrt{18}}{5} - \frac{5\sqrt{72}}{6} + \frac{3\sqrt{98}}{4}$
16. $\frac{-2\sqrt{20}}{3} + \frac{3\sqrt{45}}{4} - \frac{5\sqrt{80}}{6}$
17. $5\sqrt[3]{3} + 2\sqrt[3]{24} - 6\sqrt[3]{81}$
18. $-3\sqrt[3]{2} - 2\sqrt[3]{16} + \sqrt[3]{54}$
19. $-\sqrt[3]{16} + 7\sqrt[3]{54} - 9\sqrt[3]{2}$
20. $4\sqrt[3]{24} - 6\sqrt[3]{3} + 13\sqrt[3]{81}$

2 Simplify Radicals That Contain Variables

For Problems 21–64, express each of the following in simplest radical form. All variables represent positive real numbers.

21. $\sqrt{32x}$
22. $\sqrt{50y}$
23. $\sqrt{75x^2}$
24. $\sqrt{108y^2}$
25. $\sqrt{20x^2y}$
26. $\sqrt{80xy^2}$
27. $\sqrt{64x^3y^7}$
28. $\sqrt{36x^5y^6}$
29. $\sqrt{54a^4b^3}$
30. $\sqrt{96a^7b^8}$
31. $\sqrt{63x^6y^8}$
32. $\sqrt{28x^4y^{12}}$
33. $2\sqrt{40a^3}$
34. $4\sqrt{90a^5}$
35. $\frac{2}{3}\sqrt{96xy^3}$
36. $\frac{4}{5}\sqrt{125x^4y}$
37. $\sqrt{\frac{2x}{5y}}$
38. $\sqrt{\frac{3x}{2y}}$
39. $\sqrt{\frac{5}{12x^4}}$
40. $\sqrt{\frac{7}{8x^2}}$
41. $\frac{5}{\sqrt{18y}}$
42. $\frac{3}{\sqrt{12x}}$
43. $\frac{\sqrt{7x}}{\sqrt{8y^5}}$
44. $\frac{\sqrt{5y}}{\sqrt{18x^3}}$

45. $\dfrac{\sqrt{18y^3}}{\sqrt{16x}}$

46. $\dfrac{\sqrt{2x^3}}{\sqrt{9y}}$

47. $\dfrac{\sqrt{24a^2b^3}}{\sqrt{7ab^6}}$

48. $\dfrac{\sqrt{12a^2b}}{\sqrt{5a^3b^3}}$

49. $\sqrt[3]{24y}$

50. $\sqrt[3]{16x^2}$

51. $\sqrt[3]{16x^4}$

52. $\sqrt[3]{54x^3}$

53. $\sqrt[3]{56x^6y^8}$

54. $\sqrt[3]{81x^5y^6}$

55. $\sqrt[3]{\dfrac{7}{9x^2}}$

56. $\sqrt[3]{\dfrac{5}{2x}}$

57. $\dfrac{\sqrt[3]{3y}}{\sqrt[3]{16x^4}}$

58. $\dfrac{\sqrt[3]{2y}}{\sqrt[3]{3x}}$

59. $\dfrac{\sqrt[3]{12xy}}{\sqrt[3]{3x^2y^5}}$

60. $\dfrac{5}{\sqrt[3]{9xy^2}}$

61. $\sqrt{8x + 12y}$ [*Hint*: $\sqrt{8x + 12y} = \sqrt{4(2x + 3y)}$]

62. $\sqrt{4x + 4y}$

63. $\sqrt{16x + 48y}$

64. $\sqrt{27x + 18y}$

For Problems 65–74, use the distributive property to help simplify each of the following. All variables represent positive real numbers.

65. $-3\sqrt{4x} + 5\sqrt{9x} + 6\sqrt{16x}$

66. $-2\sqrt{25x} - 4\sqrt{36x} + 7\sqrt{64x}$

67. $2\sqrt{18x} - 3\sqrt{8x} - 6\sqrt{50x}$

68. $4\sqrt{20x} + 5\sqrt{45x} - 10\sqrt{80x}$

69. $5\sqrt{27n} - \sqrt{12n} - 6\sqrt{3n}$

70. $4\sqrt{8n} + 3\sqrt{18n} - 2\sqrt{72n}$

71. $7\sqrt{4ab} - \sqrt{16ab} - 10\sqrt{25ab}$

72. $4\sqrt{ab} - 9\sqrt{36ab} + 6\sqrt{49ab}$

73. $-3\sqrt{2x^3} + 4\sqrt{8x^3} - 3\sqrt{32x^3}$

74. $2\sqrt{40x^5} - 3\sqrt{90x^5} + 5\sqrt{160x^5}$

THOUGHTS INTO WORDS

75. Is the expression $3\sqrt{2} + \sqrt{50}$ in simplest radical form? Defend your answer.

76. Your friend simplified $\dfrac{\sqrt{6}}{\sqrt{8}}$ as follows:

$$\frac{\sqrt{6}}{\sqrt{8}} \cdot \frac{\sqrt{8}}{\sqrt{8}} = \frac{\sqrt{48}}{8} = \frac{\sqrt{16}\sqrt{3}}{8} = \frac{4\sqrt{3}}{8} = \frac{\sqrt{3}}{2}$$

Is this a correct procedure? Can you show her a better way to do this problem?

77. Does $\sqrt{x + y}$ equal $\sqrt{x} + \sqrt{y}$? Defend your answer.

FURTHER INVESTIGATIONS

78. Use your calculator and evaluate each expression in Problems 1–16. Then evaluate the simplified expression that you obtained when doing these problems. Your two results for each problem should be the same.

Consider these problems, where the variables could represent any real number. However, we would still have the restriction that the radical would represent a real number. In other words, the radicand must be nonnegative.

$\sqrt{98x^2} = \sqrt{49x^2}\,\sqrt{2} = 7|x|\sqrt{2}$ An absolute value sign is necessary to ensure that the principal root is nonnegative.

$\sqrt{24x^4} = \sqrt{4x^4}\sqrt{6} = 2x^2\sqrt{6}$ Because x^2 is nonnegative, there is no need for an absolute value sign to ensure that the principal root is nonnegative.

$\sqrt{25x^3} = \sqrt{25x^2}\sqrt{x} = 5x\sqrt{x}$ Because the radicand is defined to be nonnegative, x must be nonnegative, and there is no need for an absolute value sign to ensure that the principal root is nonnegative.

$\sqrt{18b^5} = \sqrt{9b^4}\sqrt{2b} = 3b^2\sqrt{2b}$ An absolute value sign is not necessary to ensure that the principal root is nonnegative.

$\sqrt{12y^6} = \sqrt{4y^6}\sqrt{3} = 2|y^3|\sqrt{3}$ An absolute value sign is necessary to ensure that the principal root is nonnegative.

79. Do the following problems, where the variable could be any real number as long as the radical represents a real number. Use absolute value signs in the answers as necessary.

(a) $\sqrt{125x^2}$ **(b)** $\sqrt{16x^4}$

(c) $\sqrt{8b^3}$ **(d)** $\sqrt{3y^5}$

(e) $\sqrt{288x^6}$ **(f)** $\sqrt{28m^8}$

(g) $\sqrt{128c^{10}}$ **(h)** $\sqrt{18d^7}$

(i) $\sqrt{49x^2}$ **(j)** $\sqrt{80n^{20}}$

(k) $\sqrt{81h^3}$

Answers to the Concept Quiz

1. True **2.** True **3.** False **4.** True **5.** True **6.** True **7.** False **8.** False **9.** False **10.** True

Answers to the Example Practice Skills

1. $13\sqrt{3}$ **2.** $\frac{7}{5}\sqrt{2}$ **3.** $-5\sqrt[3]{2}$ **4. (a)** $7y\sqrt{2y}$ **(b)** $2a^2b\sqrt{7ab}$ **(c)** $4xy^2\sqrt{15x}$ **(d)** $3b\sqrt[3]{2a^2b^2}$

5. (a) $\frac{\sqrt{15ab}}{5b}$ **(b)** $\frac{\sqrt{14x}}{6x^2}$ **(c)** $\frac{5a\sqrt{6ab}}{4b}$ **(d)** $\frac{2\sqrt[3]{9y}}{3y}$ **(e)** $\frac{3\sqrt[3]{12a^2b^2}}{2b}$

7.4 Products and Quotients Involving Radicals

OBJECTIVES

1. Multiply Two Radicals
2. Use the Distributive Property to Multiply Radical Expressions
3. Rationalize Binomial Denominators

1 Multiply Two Radicals

As we have seen, Property 7.4 ($\sqrt[n]{bc} = \sqrt[n]{b}\sqrt[n]{c}$) is used to express one radical as the product of two radicals and also to express the product of two radicals as one radical. In fact, we have used the property for both purposes within the framework of simplifying radicals. For example,

$$\frac{\sqrt{3}}{\sqrt{32}} = \frac{\sqrt{3}}{\sqrt{16}\sqrt{2}} = \frac{\sqrt{3}}{4\sqrt{2}} = \frac{\sqrt{3}}{4\sqrt{2}} \cdot \frac{\sqrt{2}}{\sqrt{2}} = \frac{\sqrt{6}}{8}$$

$\sqrt[n]{bc} = \sqrt[n]{b}\sqrt[n]{c}$ $\sqrt[n]{b}\sqrt[n]{c} = \sqrt[n]{bc}$

The following examples demonstrate the use of Property 7.4 to multiply radicals and to express the product in simplest form.

EXAMPLE 1

Multiply and simplify where possible.

(a) $(2\sqrt{3})(3\sqrt{5})$ **(b)** $(3\sqrt{8})(5\sqrt{2})$

(c) $(7\sqrt{6})(3\sqrt{8})$ **(d)** $(2\sqrt[3]{6})(5\sqrt[3]{4})$

Solution

(a) $(2\sqrt{3})(3\sqrt{5}) = 2 \cdot 3 \cdot \sqrt{3} \cdot \sqrt{5} = 6\sqrt{15}$

(b) $(3\sqrt{8})(5\sqrt{2}) = 3 \cdot 5 \cdot \sqrt{8} \cdot \sqrt{2} = 15\sqrt{16} = 15 \cdot 4 = 60$

(c) $(7\sqrt{6})(3\sqrt{8}) = 7 \cdot 3 \cdot \sqrt{6} \cdot \sqrt{8} = 21\sqrt{48} = 21\sqrt{16}\sqrt{3}$

$= 21 \cdot 4 \cdot \sqrt{3} = 84\sqrt{3}$

(d) $(2\sqrt[3]{6})(5\sqrt[3]{4}) = 2 \cdot 5 \cdot \sqrt[3]{6} \cdot \sqrt[3]{4} = 10\sqrt[3]{24}$

$= 10\sqrt[3]{8}\sqrt[3]{3}$

$= 10 \cdot 2 \cdot \sqrt[3]{3}$

$= 20\sqrt[3]{3}$

▼ **PRACTICE YOUR SKILL**

Multiply and simplify where possible.

(a) $(2\sqrt{6})(2\sqrt{5})$ **(b)** $(5\sqrt{18})(2\sqrt{2})$ **(c)** $(2\sqrt{21})(4\sqrt{3})$

(d) $(4\sqrt[3]{6})(5\sqrt[3]{9})$

■

2 Use the Distributive Property to Multiply Radical Expressions

Recall the use of the distributive property when finding the product of a monomial and a polynomial. For example, $3x^2(2x + 7) = 3x^2(2x) + 3x^2(7) = 6x^3 + 21x^2$. In a similar manner, the distributive property and Property 7.4 provide the basis for finding certain special products that involve radicals. The following examples illustrate this idea.

EXAMPLE 2

Multiply and simplify where possible.

(a) $\sqrt{3}(\sqrt{6} + \sqrt{12})$ **(b)** $2\sqrt{2}(4\sqrt{3} - 5\sqrt{6})$

(c) $\sqrt{6x}(\sqrt{8x} + \sqrt{12xy})$ **(d)** $\sqrt[3]{2}(5\sqrt[3]{4} - 3\sqrt[3]{16})$

Solution

(a) $\sqrt{3}(\sqrt{6} + \sqrt{12}) = \sqrt{3}\sqrt{6} + \sqrt{3}\sqrt{12}$

$= \sqrt{18} + \sqrt{36}$

$= \sqrt{9}\sqrt{2} + 6$

$= 3\sqrt{2} + 6$

(b) $2\sqrt{2}(4\sqrt{3} - 5\sqrt{6}) = (2\sqrt{2})(4\sqrt{3}) - (2\sqrt{2})(5\sqrt{6})$

$= 8\sqrt{6} - 10\sqrt{12}$

$= 8\sqrt{6} - 10\sqrt{4}\sqrt{3}$

$= 8\sqrt{6} - 20\sqrt{3}$

(c) $\sqrt{6x}(\sqrt{8x} + \sqrt{12xy}) = (\sqrt{6x})(\sqrt{8x}) + (\sqrt{6x})(\sqrt{12xy})$

$= \sqrt{48x^2} + \sqrt{72x^2y}$

$= \sqrt{16x^2}\sqrt{3} + \sqrt{36x^2}\sqrt{2y}$

$= 4x\sqrt{3} + 6x\sqrt{2y}$

(d) $\sqrt[3]{2}(5\sqrt[3]{4} - 3\sqrt[3]{16}) = (\sqrt[3]{2})(5\sqrt[3]{4}) - (\sqrt[3]{2})(3\sqrt[3]{16})$

$= 5\sqrt[3]{8} - 3\sqrt[3]{32}$

$$= 5 \cdot 2 - 3\sqrt[3]{8}\sqrt[3]{4}$$
$$= 10 - 6\sqrt[3]{4}$$

▼ PRACTICE YOUR SKILL

Multiply and simplify where possible.

(a) $\sqrt{2}(\sqrt{10} + \sqrt{8})$ **(b)** $4\sqrt{3}(5\sqrt{2} - 3\sqrt{6})$

(c) $\sqrt{2a}(\sqrt{14a} + \sqrt{12ab})$ **(d)** $\sqrt[3]{3}(4\sqrt[3]{9} + \sqrt[3]{81})$ ■

The distributive property also plays a central role in determining the product of two binomials. For example, $(x + 2)(x + 3) = x(x + 3) + 2(x + 3) = x^2 + 3x + 2x + 6 = x^2 + 5x + 6$. Finding the product of two binomial expressions that involve radicals can be handled in a similar fashion, as in the next examples.

EXAMPLE 3

Find the following products and simplify.

(a) $(\sqrt{3} + \sqrt{5})(\sqrt{2} + \sqrt{6})$ **(b)** $(2\sqrt{2} - \sqrt{7})(3\sqrt{2} + 5\sqrt{7})$

(c) $(\sqrt{8} + \sqrt{6})(\sqrt{8} - \sqrt{6})$ **(d)** $(\sqrt{x} + \sqrt{y})(\sqrt{x} - \sqrt{y})$

Solution

(a) $(\sqrt{3} + \sqrt{5})(\sqrt{2} + \sqrt{6}) = \sqrt{3}(\sqrt{2} + \sqrt{6}) + \sqrt{5}(\sqrt{2} + \sqrt{6})$
$$= \sqrt{3}\sqrt{2} + \sqrt{3}\sqrt{6} + \sqrt{5}\sqrt{2} + \sqrt{5}\sqrt{6}$$
$$= \sqrt{6} + \sqrt{18} + \sqrt{10} + \sqrt{30}$$
$$= \sqrt{6} + 3\sqrt{2} + \sqrt{10} + \sqrt{30}$$

(b) $(2\sqrt{2} - \sqrt{7})(3\sqrt{2} + 5\sqrt{7}) = 2\sqrt{2}(3\sqrt{2} + 5\sqrt{7})$
$$-\sqrt{7}(3\sqrt{2} + 5\sqrt{7})$$
$$= (2\sqrt{2})(3\sqrt{2}) + (2\sqrt{2})(5\sqrt{7})$$
$$- (\sqrt{7})(3\sqrt{2}) - (\sqrt{7})(5\sqrt{7})$$
$$= 12 + 10\sqrt{14} - 3\sqrt{14} - 35$$
$$= -23 + 7\sqrt{14}$$

(c) $(\sqrt{8} + \sqrt{6})(\sqrt{8} - \sqrt{6}) = \sqrt{8}(\sqrt{8} - \sqrt{6}) + \sqrt{6}(\sqrt{8} - \sqrt{6})$
$$= \sqrt{8}\sqrt{8} - \sqrt{8}\sqrt{6} + \sqrt{6}\sqrt{8} - \sqrt{6}\sqrt{6}$$
$$= 8 - \sqrt{48} + \sqrt{48} - 6$$
$$= 2$$

(d) $(\sqrt{x} + \sqrt{y})(\sqrt{x} - \sqrt{y}) = \sqrt{x}(\sqrt{x} - \sqrt{y}) + \sqrt{y}(\sqrt{x} - \sqrt{y})$
$$= \sqrt{x}\sqrt{x} - \sqrt{x}\sqrt{y} + \sqrt{y}\sqrt{x} - \sqrt{y}\sqrt{y}$$
$$= x - \sqrt{xy} + \sqrt{xy} - y$$
$$= x - y$$

▼ PRACTICE YOUR SKILL

Multiply and simplify where possible.

(a) $(\sqrt{6} + \sqrt{3})(\sqrt{5} + \sqrt{14})$ **(b)** $(5\sqrt{2} + \sqrt{3})(4\sqrt{2} - 2\sqrt{3})$

(c) $(\sqrt{5} + \sqrt{3})(\sqrt{5} - \sqrt{3})$ **(d)** $(\sqrt{a} - \sqrt{b})(\sqrt{a} + \sqrt{b})$ ■

3 Rationalize Binomial Denominators

Note parts (c) and (d) of Example 3; they fit the special-product pattern $(a + b)(a - b) = a^2 - b^2$. Furthermore, in each case the final product is in rational form. The factors $a + b$ and $a - b$ are called **conjugates**. This suggests a way of rationalizing the denominator in an expression that contains a binomial denominator with radicals. We will multiply by the conjugate of the binomial denominator. Consider the following example.

EXAMPLE 4

Simplify $\dfrac{4}{\sqrt{5} + \sqrt{2}}$ by rationalizing the denominator.

Solution

Form of 1

$$\frac{4}{\sqrt{5} + \sqrt{2}} = \frac{4}{\sqrt{5} + \sqrt{2}} \cdot \left(\frac{\sqrt{5} - \sqrt{2}}{\sqrt{5} - \sqrt{2}}\right)$$

$$= \frac{4(\sqrt{5} - \sqrt{2})}{(\sqrt{5} + \sqrt{2})(\sqrt{5} - \sqrt{2})} = \frac{4(\sqrt{5} - \sqrt{2})}{5 - 2}$$

$$= \frac{4(\sqrt{5} - \sqrt{2})}{3} \quad \text{or} \quad \frac{4\sqrt{5} - 4\sqrt{2}}{3}$$

Either answer is acceptable

▼ PRACTICE YOUR SKILL

Simplify $\dfrac{5}{\sqrt{7} + \sqrt{3}}$ by rationalizing the denominator. ■

The next examples further illustrate the process of rationalizing and simplifying expressions that contain binomial denominators.

EXAMPLE 5

For each of the following, rationalize the denominator and simplify.

(a) $\dfrac{\sqrt{3}}{\sqrt{6} - 9}$ **(b)** $\dfrac{7}{3\sqrt{5} + 2\sqrt{3}}$

(c) $\dfrac{\sqrt{x} + 2}{\sqrt{x} - 3}$ **(d)** $\dfrac{2\sqrt{x} - 3\sqrt{y}}{\sqrt{x} + \sqrt{y}}$

Solution

(a)
$$\frac{\sqrt{3}}{\sqrt{6} - 9} = \frac{\sqrt{3}}{\sqrt{6} - 9} \cdot \frac{\sqrt{6} + 9}{\sqrt{6} + 9}$$

$$= \frac{\sqrt{3}(\sqrt{6} + 9)}{(\sqrt{6} - 9)(\sqrt{6} + 9)}$$

$$= \frac{\sqrt{18} + 9\sqrt{3}}{6 - 81}$$

$$= \frac{3\sqrt{2} + 9\sqrt{3}}{-75}$$

$$= \frac{3(\sqrt{2} + 3\sqrt{3})}{(-3)(25)}$$

$$= -\frac{\sqrt{2} + 3\sqrt{3}}{25} \quad \text{or} \quad \frac{-\sqrt{2} - 3\sqrt{3}}{25}$$

(b) $$\frac{7}{3\sqrt{5} + 2\sqrt{3}} = \frac{7}{3\sqrt{5} + 2\sqrt{3}} \cdot \frac{3\sqrt{5} - 2\sqrt{3}}{3\sqrt{5} - 2\sqrt{3}}$$

$$= \frac{7(3\sqrt{5} - 2\sqrt{3})}{(3\sqrt{5} + 2\sqrt{3})(3\sqrt{5} - 2\sqrt{3})}$$

$$= \frac{7(3\sqrt{5} - 2\sqrt{3})}{45 - 12}$$

$$= \frac{7(3\sqrt{5} - 2\sqrt{3})}{33} \quad \text{or} \quad \frac{21\sqrt{5} - 14\sqrt{3}}{33}$$

(c) $$\frac{\sqrt{x} + 2}{\sqrt{x} - 3} = \frac{\sqrt{x} + 2}{\sqrt{x} - 3} \cdot \frac{\sqrt{x} + 3}{\sqrt{x} + 3} = \frac{(\sqrt{x} + 2)(\sqrt{x} + 3)}{(\sqrt{x} - 3)(\sqrt{x} + 3)}$$

$$= \frac{x + 3\sqrt{x} + 2\sqrt{x} + 6}{x - 9}$$

$$= \frac{x + 5\sqrt{x} + 6}{x - 9}$$

(d) $$\frac{2\sqrt{x} - 3\sqrt{y}}{\sqrt{x} + \sqrt{y}} = \frac{2\sqrt{x} - 3\sqrt{y}}{\sqrt{x} + \sqrt{y}} \cdot \frac{\sqrt{x} - \sqrt{y}}{\sqrt{x} - \sqrt{y}}$$

$$= \frac{(2\sqrt{x} - 3\sqrt{y})(\sqrt{x} - \sqrt{y})}{(\sqrt{x} + \sqrt{y})(\sqrt{x} - \sqrt{y})}$$

$$= \frac{2x - 2\sqrt{xy} - 3\sqrt{xy} + 3y}{x - y}$$

$$= \frac{2x - 5\sqrt{xy} + 3y}{x - y}$$

▼ PRACTICE YOUR SKILL

For each of the following, rationalize the denominator and simplify.

(a) $\frac{\sqrt{2}}{\sqrt{6} - 4}$ **(b)** $\frac{4}{3\sqrt{2} - 5\sqrt{3}}$ **(c)** $\frac{\sqrt{a} - 5}{\sqrt{a} + 4}$ **(d)** $\frac{\sqrt{a} - 2\sqrt{b}}{\sqrt{a} - \sqrt{b}}$ ■

CONCEPT QUIZ

For Problems 1–10, answer true or false.

1. The property $\sqrt[n]{x}\sqrt[n]{y} = \sqrt[n]{xy}$ can be used to express the product of two radicals as one radical.
2. The product of two radicals always results in an expression that has a radical even after simplifying.
3. The conjugate of $5 + \sqrt{3}$ is $-5 - \sqrt{3}$.
4. The product of $2 - \sqrt{7}$ and $2 + \sqrt{7}$ is a rational number.
5. To rationalize the denominator for the expression $\frac{2\sqrt{5}}{4 - \sqrt{5}}$, we would multiply by $\frac{\sqrt{5}}{\sqrt{5}}$.
6. $\frac{\sqrt{8} + \sqrt{12}}{\sqrt{2}} = 2 + \sqrt{6}$.

7. $\dfrac{\sqrt{2}}{\sqrt{8}+\sqrt{12}} = \dfrac{1}{2+\sqrt{6}}$.

8. The product of $5 + \sqrt{3}$ and $-5 - \sqrt{3}$ is -28.

9. $\dfrac{\sqrt{2}}{3+\sqrt{6}} = \dfrac{3\sqrt{2}-2\sqrt{3}}{3}$

10. $(-\sqrt{2}-\sqrt{3})(\sqrt{2}+3\sqrt{3}) = -11 - 4\sqrt{6}$

Problem Set 7.4

1 Multiply Two Radicals

For Problems 1–14, multiply and simplify where possible.

1. $\sqrt{6}\sqrt{12}$
2. $\sqrt{8}\sqrt{6}$
3. $(3\sqrt{3})(2\sqrt{6})$
4. $(5\sqrt{2})(3\sqrt{12})$
5. $(4\sqrt{2})(-6\sqrt{5})$
6. $(-7\sqrt{3})(2\sqrt{5})$
7. $(-3\sqrt{3})(-4\sqrt{8})$
8. $(-5\sqrt{8})(-6\sqrt{7})$
9. $(5\sqrt{6})(4\sqrt{6})$
10. $(3\sqrt{7})(2\sqrt{7})$
11. $(2\sqrt[3]{4})(6\sqrt[3]{2})$
12. $(4\sqrt[3]{3})(5\sqrt[3]{9})$
13. $(4\sqrt[3]{6})(7\sqrt[3]{4})$
14. $(9\sqrt[3]{6})(2\sqrt[3]{9})$

2 Use the Distributive Property to Multiply Radical Expressions

For Problems 15–52, find the following products and express answers in simplest radical form. All variables represent nonnegative real numbers.

15. $\sqrt{2}(\sqrt{3}+\sqrt{5})$
16. $\sqrt{3}(\sqrt{7}+\sqrt{10})$
17. $3\sqrt{5}(2\sqrt{2}-\sqrt{7})$
18. $5\sqrt{6}(2\sqrt{5}-3\sqrt{11})$
19. $2\sqrt{6}(3\sqrt{8}-5\sqrt{12})$
20. $4\sqrt{2}(3\sqrt{12}+7\sqrt{6})$
21. $-4\sqrt{5}(2\sqrt{5}+4\sqrt{12})$
22. $-5\sqrt{3}(3\sqrt{12}-9\sqrt{8})$
23. $3\sqrt{x}(5\sqrt{2}+\sqrt{y})$
24. $\sqrt{2x}(3\sqrt{y}-7\sqrt{5})$
25. $\sqrt{xy}(5\sqrt{xy}-6\sqrt{x})$
26. $4\sqrt{x}(2\sqrt{xy}+2\sqrt{x})$
27. $\sqrt{5y}(\sqrt{8x}+\sqrt{12y^2})$
28. $\sqrt{2x}(\sqrt{12xy}-\sqrt{8y})$
29. $5\sqrt{3}(2\sqrt{8}-3\sqrt{18})$
30. $2\sqrt{2}(3\sqrt{12}-\sqrt{27})$
31. $(\sqrt{3}+4)(\sqrt{3}-7)$
32. $(\sqrt{2}+6)(\sqrt{2}-2)$
33. $(\sqrt{5}-6)(\sqrt{5}-3)$
34. $(\sqrt{7}-2)(\sqrt{7}-8)$
35. $(3\sqrt{5}-2\sqrt{3})(2\sqrt{7}+\sqrt{2})$
36. $(\sqrt{2}+\sqrt{3})(\sqrt{5}-\sqrt{7})$
37. $(2\sqrt{6}+3\sqrt{5})(\sqrt{8}-3\sqrt{12})$
38. $(5\sqrt{2}-4\sqrt{6})(2\sqrt{8}+\sqrt{6})$
39. $(2\sqrt{6}+5\sqrt{5})(3\sqrt{6}-\sqrt{5})$
40. $(7\sqrt{3}-\sqrt{7})(2\sqrt{3}+4\sqrt{7})$
41. $(3\sqrt{2}-5\sqrt{3})(6\sqrt{2}-7\sqrt{3})$
42. $(\sqrt{8}-3\sqrt{10})(2\sqrt{8}-6\sqrt{10})$
43. $(\sqrt{6}+4)(\sqrt{6}-4)$
44. $(\sqrt{7}-2)(\sqrt{7}+2)$
45. $(\sqrt{2}+\sqrt{10})(\sqrt{2}-\sqrt{10})$
46. $(2\sqrt{3}+\sqrt{11})(2\sqrt{3}-\sqrt{11})$
47. $(\sqrt{2x}+\sqrt{3y})(\sqrt{2x}-\sqrt{3y})$
48. $(2\sqrt{x}-5\sqrt{y})(2\sqrt{x}+5\sqrt{y})$
49. $2\sqrt[3]{3}(5\sqrt[3]{4}+\sqrt[3]{6})$
50. $2\sqrt[3]{2}(3\sqrt[3]{6}-4\sqrt[3]{5})$
51. $3\sqrt[3]{4}(2\sqrt[3]{2}-6\sqrt[3]{4})$
52. $3\sqrt[3]{3}(4\sqrt[3]{9}+5\sqrt[3]{7})$

3 Rationalize Binomial Denominators

For Problems 53–76, rationalize the denominator and simplify. All variables represent positive real numbers.

53. $\dfrac{2}{\sqrt{7}+1}$
54. $\dfrac{6}{\sqrt{5}+2}$
55. $\dfrac{3}{\sqrt{2}-5}$
56. $\dfrac{-4}{\sqrt{6}-3}$
57. $\dfrac{1}{\sqrt{2}+\sqrt{7}}$
58. $\dfrac{3}{\sqrt{3}+\sqrt{10}}$
59. $\dfrac{\sqrt{2}}{\sqrt{10}-\sqrt{3}}$
60. $\dfrac{\sqrt{3}}{\sqrt{7}-\sqrt{2}}$
61. $\dfrac{\sqrt{3}}{2\sqrt{5}+4}$
62. $\dfrac{\sqrt{7}}{3\sqrt{2}-5}$
63. $\dfrac{6}{3\sqrt{7}-2\sqrt{6}}$
64. $\dfrac{5}{2\sqrt{5}+3\sqrt{7}}$
65. $\dfrac{\sqrt{6}}{3\sqrt{2}+2\sqrt{3}}$
66. $\dfrac{3\sqrt{6}}{5\sqrt{3}-4\sqrt{2}}$
67. $\dfrac{2}{\sqrt{x}+4}$
68. $\dfrac{3}{\sqrt{x}+7}$

69. $\dfrac{\sqrt{x}}{\sqrt{x}-5}$

70. $\dfrac{\sqrt{x}}{\sqrt{x}-1}$

71. $\dfrac{\sqrt{x}-2}{\sqrt{x}+6}$

72. $\dfrac{\sqrt{x}+1}{\sqrt{x}-10}$

73. $\dfrac{\sqrt{x}}{\sqrt{x}+2\sqrt{y}}$

74. $\dfrac{\sqrt{y}}{2\sqrt{x}-\sqrt{y}}$

75. $\dfrac{3\sqrt{y}}{2\sqrt{x}-3\sqrt{y}}$

76. $\dfrac{2\sqrt{x}}{3\sqrt{x}+5\sqrt{y}}$

THOUGHTS INTO WORDS

77. How would you help someone rationalize the denominator and simplify $\dfrac{4}{\sqrt{8}+\sqrt{12}}$?

78. Discuss how the distributive property has been used so far in this chapter.

79. How would you simplify the expression $\dfrac{\sqrt{8}+\sqrt{12}}{\sqrt{2}}$?

FURTHER INVESTIGATIONS

80. Use your calculator to evaluate each expression in Problems 53–66, and compare the results you obtained when you did the problems.

Answers to the Concept Quiz

1. True **2.** False **3.** False **4.** True **5.** False **6.** True **7.** False **8.** False **9.** True **10.** True

Answers to the Example Practice Skills

1. (a) $4\sqrt{30}$ **(b)** 60 **(c)** $24\sqrt{7}$ **(d)** $60\sqrt[3]{2}$ **2. (a)** $2\sqrt{5}+4$ **(b)** $20\sqrt{6}-36\sqrt{2}$ **(c)** $2a\sqrt{7}+2a\sqrt{6b}$ **(d)** $12+3\sqrt[3]{9}$ **3. (a)** $\sqrt{30}+2\sqrt{21}+\sqrt{15}+\sqrt{42}$ **(b)** $34-6\sqrt{6}$ **(c)** 2 **(d)** $a-b$ **4.** $\dfrac{5(\sqrt{7}-\sqrt{3})}{4}$ **5. (a)** $-\dfrac{\sqrt{3}+2\sqrt{2}}{5}$ **(b)** $-\dfrac{12\sqrt{2}+20\sqrt{3}}{57}$ **(c)** $\dfrac{a-9\sqrt{a}+20}{a-16}$ **(d)** $\dfrac{a-\sqrt{ab}-2b}{a-b}$

7.5 Equations Involving Radicals

OBJECTIVES

1 Solve Radical Equations

2 Apply Solving Radical Equations to Problems

1 Solve Radical Equations

We often refer to equations that contain radicals with variables in a radicand as **radical equations**. In this section we discuss techniques for solving such equations that contain one or more radicals. To solve radical equations, we need the following property of equality.

Property 7.6

Let a and b be real numbers and let n be a positive integer.

If $a=b$, then $a^n=b^n$.

Property 7.6 states that we can raise both sides of an equation to a positive integral power. However, raising both sides of an equation to a positive integral power sometimes produces results that do not satisfy the original equation. Let's consider two examples to illustrate this point.

EXAMPLE 1

Solve $\sqrt{2x - 5} = 7$.

Solution

$$\sqrt{2x - 5} = 7$$

$$(\sqrt{2x - 5})^2 = 7^2 \quad \text{Square both sides}$$

$$2x - 5 = 49$$

$$2x = 54$$

$$x = 27$$

✔ **Check**

$$\sqrt{2x - 5} = 7$$

$$\sqrt{2(27) - 5} \stackrel{?}{=} 7$$

$$\sqrt{49} \stackrel{?}{=} 7$$

$$7 = 7$$

The solution set for $\sqrt{2x - 5} = 7$ is $\{27\}$.

▼ **PRACTICE YOUR SKILL**

Solve $\sqrt{3x + 1} = 8$. ■

EXAMPLE 2

Solve $\sqrt{3a + 4} = -4$.

Solution

$$\sqrt{3a + 4} = -4$$

$$(\sqrt{3a + 4})^2 = (-4)^2 \quad \text{Square both sides}$$

$$3a + 4 = 16$$

$$3a = 12$$

$$a = 4$$

✔ **Check**

$$\sqrt{3a + 4} = -4$$

$$\sqrt{3(4) + 4} \stackrel{?}{=} -4$$

$$\sqrt{16} \stackrel{?}{=} -4$$

$$4 \neq -4$$

Because 4 does not check, the original equation has no real number solution. Thus the solution set is $\varnothing$.

▼ **PRACTICE YOUR SKILL**

Solve $\sqrt{5x - 1} = -3$. ■

In general, raising both sides of an equation to a positive integral power produces an equation that has all of the solutions of the original equation, but it may also have some extra solutions that do not satisfy the original equation. Such extra solutions are called **extraneous solutions**. Therefore, when using Property 7.6, you *must* check each potential solution in the original equation.

Let's consider some examples to illustrate different situations that arise when we are solving radical equations.

EXAMPLE 3

Solve $\sqrt{2t - 4} = t - 2$.

Solution

$$\sqrt{2t - 4} = t - 2$$

$$(\sqrt{2t - 4})^2 = (t - 2)^2 \qquad \text{Square both sides}$$

$$2t - 4 = t^2 - 4t + 4$$

$$0 = t^2 - 6t + 8$$

$$0 = (t - 2)(t - 4) \qquad \text{Factor the right side}$$

$$t - 2 = 0 \quad \text{or} \quad t - 4 = 0 \qquad \text{Apply: } ab = 0 \text{ if and only if } a = 0 \text{ or } b = 0$$

$$t = 0 \quad \text{or} \quad t = 4$$

✔ **Check**

$$\sqrt{2t - 4} = t - 2 \qquad\qquad \sqrt{2t - 4} = t - 2$$

$$\sqrt{2(2) - 4} \stackrel{?}{=} 2 - 2, \quad \text{when } t = 2 \qquad \text{or} \qquad \sqrt{2(4) - 4} \stackrel{?}{=} 4 - 2, \quad \text{when } t = 4$$

$$\sqrt{0} \stackrel{?}{=} 0 \qquad\qquad \sqrt{4} \stackrel{?}{=} 2$$

$$0 = 0 \qquad\qquad 2 = 2$$

The solution set is $\{2, 4\}$.

▼ **PRACTICE YOUR SKILL**

Solve $\sqrt{2x - 6} = x - 3$. ■

EXAMPLE 4

Solve $\sqrt{y} + 6 = y$.

Solution

$$\sqrt{y} + 6 = y$$

$$\sqrt{y} = y - 6$$

$$(\sqrt{y})^2 = (y - 6)^2 \qquad \text{Square both sides}$$

$$y = y^2 - 12y + 36$$

$$0 = y^2 - 13y + 36$$

$$0 = (y - 4)(y - 9) \qquad \text{Factor the right side}$$

$$y - 4 = 0 \quad \text{or} \quad y - 9 = 0 \qquad \text{Apply: } ab = 0 \text{ if and only if } a = 0 \text{ or } b = 0$$

$$y = 4 \quad \text{or} \quad y = 9$$

✔ **Check**

$$\sqrt{y} + 6 = y \qquad\qquad \sqrt{y} + 6 = y$$

$$\sqrt{4} + 6 \stackrel{?}{=} 4, \quad \text{when } y = 4 \qquad \text{or} \qquad \sqrt{9} + 6 \stackrel{?}{=} 9, \quad \text{when } y = 9$$

$$2 + 6 \stackrel{?}{=} 4 \qquad\qquad 3 + 6 \stackrel{?}{=} 9$$
$$8 \neq 4 \qquad\qquad 9 = 9$$

The only solution is 9, so the solution set is $\{9\}$.

▼ **PRACTICE YOUR SKILL**

Solve $\sqrt{x} + 12 = x$. ■

In Example 4, note that we changed the form of the original equation $\sqrt{y} + 6 = y$ to $\sqrt{y} = y - 6$ before we squared both sides. Squaring both sides of $\sqrt{y} + 6 = y$ produces $y + 12\sqrt{y} + 36 = y^2$, which is a much more complex equation that still contains a radical. Here again, it pays to think ahead before carrying out all the steps. Now let's consider an example involving a cube root.

EXAMPLE 5

Solve $\sqrt[3]{n^2 - 1} = 2$.

Solution

$$\sqrt[3]{n^2 - 1} = 2$$
$$(\sqrt[3]{n^2 - 1})^3 = 2^3 \quad \text{Cube both sides}$$
$$n^2 - 1 = 8$$
$$n^2 - 9 = 0$$
$$(n + 3)(n - 3) = 0$$
$$n + 3 = 0 \quad \text{or} \quad n - 3 = 0$$
$$n = -3 \quad \text{or} \quad n = 3$$

✔ **Check**

$$\sqrt[3]{n^2 - 1} = 2 \qquad\qquad \sqrt[3]{n^2 - 1} = 2$$
$$\sqrt[3]{(-3)^2 - 1} \stackrel{?}{=} 2, \text{ when } n = -3 \quad \text{or} \quad \sqrt[3]{3^2 - 1} \stackrel{?}{=} 2, \text{ when } n = 3$$
$$\sqrt[3]{8} \stackrel{?}{=} 2 \qquad\qquad \sqrt[3]{8} \stackrel{?}{=} 2$$
$$2 = 2 \qquad\qquad 2 = 2$$

The solution set is $\{-3, 3\}$.

▼ **PRACTICE YOUR SKILL**

Solve $\sqrt[3]{y^2 + 2} = 3$. ■

It may be necessary to square both sides of an equation, simplify the resulting equation, and then square both sides again. The next example illustrates this type of problem.

EXAMPLE 6

Solve $\sqrt{x + 2} = 7 - \sqrt{x + 9}$.

Solution

$$\sqrt{x + 2} = 7 - \sqrt{x + 9}$$
$$(\sqrt{x + 2})^2 = (7 - \sqrt{x + 9})^2 \quad \text{Square both sides}$$
$$x + 2 = 49 - 14\sqrt{x + 9} + x + 9$$

$$x + 2 = x + 58 - 14\sqrt{x + 9}$$

$$-56 = -14\sqrt{x + 9}$$

$$4 = \sqrt{x + 9}$$

$$(4)^2 = (\sqrt{x + 9})^2 \quad \text{Square both sides}$$

$$16 = x + 9$$

$$7 = x$$

✔ **Check**

$$\sqrt{x + 2} = 7 - \sqrt{x + 9}$$

$$\sqrt{7 + 2} \stackrel{?}{=} 7 - \sqrt{7 + 9}$$

$$\sqrt{9} \stackrel{?}{=} 7 - \sqrt{16}$$

$$3 \stackrel{?}{=} 7 - 4$$

$$3 = 3$$

The solution set is $\{7\}$.

▼ **PRACTICE YOUR SKILL**

Solve $\sqrt{x + 8} = 1 + \sqrt{x + 1}$. ■

2 Apply Solving Radical Equations to Problems

In Section 7.1 we used the formula $S = \sqrt{30Df}$ to approximate how fast a car was traveling on the basis of the length of skid marks. (Remember that S represents the speed of the car in miles per hour, D represents the length of the skid marks in feet, and f represents a coefficient of friction.) This same formula can be used to estimate the length of skid marks that are produced by cars traveling at different rates on various types of road surfaces. To use the formula for this purpose, let's change the form of the equation by solving for D.

$$\sqrt{30Df} = S$$

$$30Df = S^2 \quad \text{The result of squaring both sides of the original equation}$$

$$D = \frac{S^2}{30f} \quad \text{D, S, and f are positive numbers, so this final equation and the original one are equivalent}$$

EXAMPLE 7

Suppose that, for a particular road surface, the coefficient of friction is 0.35. How far will a car skid when the brakes are applied at 60 miles per hour?

Solution

We can substitute 0.35 for f and 60 for S in the formula $D = \dfrac{S^2}{30f}$.

$$D = \frac{60^2}{30(0.35)} = 343, \quad \text{to the nearest whole number}$$

The car will skid approximately 343 feet.

▼ **PRACTICE YOUR SKILL**

Suppose that, for a particular road surface, the coefficient of friction is 0.45. How far will a car skid when the brakes are applied at 70 miles per hour? ■

Remark: Pause for a moment and think about the result in Example 7. The coefficient of friction 0.35 refers to a wet concrete road surface. Note that a car traveling at 60 miles per hour on such a surface will skid more than the length of a football field.

CONCEPT QUIZ

For Problems 1–10, answer true or false.

1. To solve a radical equation, we can raise each side of the equation to a positive integer power.
2. Solving the equation that results from squaring each side of an original equation may not give all the solutions of the original equation.
3. The equation $\sqrt[3]{x-1} = -2$ has a solution.
4. Potential solutions that do not satisfy the original equation are called extraneous solutions.
5. The equation $\sqrt{x+1} = -2$ has no solutions.
6. The solution set for $\sqrt{x} + 2 = x$ is $\{1, 4\}$.
7. The solution set for $\sqrt{x+1} + \sqrt{x-2} = -3$ is the null set.
8. The solution set for $\sqrt[3]{x+2} = -2$ is the null set.
9. The solution set for the equation $\sqrt{x^2 - 2x + 1} = x - 3$ is $\{2\}$.
10. The solution set for the equation $\sqrt{5x+1} + \sqrt{x+4} = 3$ is $\{0\}$.

Problem Set 7.5

1 Solve Radical Equations

For Problems 1–56, solve each equation. Don't forget to check each of your potential solutions.

1. $\sqrt{5x} = 10$
2. $\sqrt{3x} = 9$
3. $\sqrt{2x} + 4 = 0$
4. $\sqrt{4x} + 5 = 0$
5. $2\sqrt{n} = 5$
6. $5\sqrt{n} = 3$
7. $3\sqrt{n} - 2 = 0$
8. $2\sqrt{n} - 7 = 0$
9. $\sqrt{3y+1} = 4$
10. $\sqrt{2y-3} = 5$
11. $\sqrt{4y-3} - 6 = 0$
12. $\sqrt{3y+5} - 2 = 0$
13. $\sqrt{3x-1} + 1 = 4$
14. $\sqrt{4x-1} - 3 = 2$
15. $\sqrt{2n+3} - 2 = -1$
16. $\sqrt{5n+1} - 6 = -4$
17. $\sqrt{2x-5} = -1$
18. $\sqrt{4x-3} = -4$
19. $\sqrt{5x+2} = \sqrt{6x+1}$
20. $\sqrt{4x+2} = \sqrt{3x+4}$
21. $\sqrt{3x+1} = \sqrt{7x-5}$
22. $\sqrt{6x+5} = \sqrt{2x+10}$
23. $\sqrt{3x-2} - \sqrt{x+4} = 0$
24. $\sqrt{7x-6} - \sqrt{5x+2} = 0$
25. $5\sqrt{t} - 1 = 6$
26. $4\sqrt{t} + 3 = 6$
27. $\sqrt{x^2+7} = 4$
28. $\sqrt{x^2+3} - 2 = 0$
29. $\sqrt{x^2 + 13x + 37} = 1$
30. $\sqrt{x^2 + 5x - 20} = 2$
31. $\sqrt{x^2 - x + 1} = x + 1$
32. $\sqrt{n^2 - 2n - 4} = n$
33. $\sqrt{x^2 + 3x + 7} = x + 2$
34. $\sqrt{x^2 + 2x + 1} = x + 3$
35. $\sqrt{-4x + 17} = x - 3$
36. $\sqrt{2x-1} = x - 2$
37. $\sqrt{n+4} = n + 4$
38. $\sqrt{n+6} = n + 6$
39. $\sqrt{3y} = y - 6$
40. $2\sqrt{n} = n - 3$
41. $4\sqrt{x} + 5 = x$
42. $\sqrt{-x} - 6 = x$
43. $\sqrt[3]{x-2} = 3$
44. $\sqrt[3]{x+1} = 4$
45. $\sqrt[3]{2x+3} = -3$
46. $\sqrt[3]{3x-1} = -4$
47. $\sqrt[3]{2x+5} = \sqrt[3]{4-x}$
48. $\sqrt[3]{3x-1} = \sqrt[3]{2-5x}$
49. $\sqrt{x+19} - \sqrt{x+28} = -1$
50. $\sqrt{x+4} = \sqrt{x-1} + 1$
51. $\sqrt{3x+1} + \sqrt{2x+4} = 3$
52. $\sqrt{2x-1} - \sqrt{x+3} = 1$
53. $\sqrt{n-4} + \sqrt{n+4} = 2\sqrt{n-1}$
54. $\sqrt{n-3} + \sqrt{n+5} = 2\sqrt{n}$
55. $\sqrt{t+3} - \sqrt{t-2} = \sqrt{7-t}$
56. $\sqrt{t+7} - 2\sqrt{t-8} = \sqrt{t-5}$

2 Apply Solving Radical Equations to Problems

57. Use the formula given in Example 7 with a coefficient of friction of 0.95. How far will a car skid at 40 miles per hour? At 55 miles per hour? At 65 miles per hour? Express the answers to the nearest foot.

58. Solve the formula $T = 2\pi\sqrt{\frac{L}{32}}$ for L. (Recall that in this formula, which was used in Section 7.2, T represents the period of a pendulum expressed in seconds and L represents the length of the pendulum in feet.)

59. In Problem 58, you should have obtained the equation $L = \frac{8T^2}{\pi^2}$. What is the length of a pendulum that has a period of 2 seconds? Of 2.5 seconds? Of 3 seconds? Express your answers to the nearest tenth of a foot.

THOUGHTS INTO WORDS

60. Explain the concept of extraneous solutions.

61. Explain why possible solutions for radical equations *must* be checked.

62. Your friend makes an effort to solve the equation $3 + 2\sqrt{x} = x$ as follows:

$$(3 + 2\sqrt{x})^2 = x^2$$

$$9 + 12\sqrt{x} + 4x = x^2$$

At this step he stops and doesn't know how to proceed. What help would you give him?

Answers to the Concept Quiz

1. True **2.** False **3.** True **4.** True **5.** True **6.** False **7.** True **8.** False **9.** False **10.** True

Answers to the Example Practice Skills

1. {21} **2.** ∅ **3.** {3, 5} **4.** {16} **5.** {−5, 5} **6.** {8} **7.** Approximately 363 ft

7.6 Merging Exponents and Roots

OBJECTIVES

1. Evaluate a Number Raised to a Rational Exponent
2. Write an Expression with Rational Exponents as a Radical
3. Write Radical Expressions as Expressions with Rational Exponents
4. Simplify Algebraic Expressions That Have Rational Exponents
5. Multiply and Divide Radicals with Different Indexes

1 Evaluate a Number Raised to a Rational Exponent

Recall that the basic properties of positive integral exponents led to a definition for the use of negative integers as exponents. In this section, the properties of integral exponents are used to form definitions for the use of rational numbers as exponents. These definitions will tie together the concepts of exponent and root.

Let's consider the following comparisons.

From our study of radicals, we know that	If $(b^n)^m = b^{mn}$ is to hold when n equals a rational number of the form $\frac{1}{p}$, where p is a positive integer greater than 1, then
↓	↓
$(\sqrt{5})^2 = 5$	$(5^{\frac{1}{2}})^2 = 5^{2(\frac{1}{2})} = 5^1 = 5$
$(\sqrt[3]{8})^3 = 8$	$(8^{\frac{1}{3}})^3 = 8^{3(\frac{1}{3})} = 8^1 = 8$
$(\sqrt[4]{21})^4 = 21$	$(21^{\frac{1}{4}})^4 = 21^{4(\frac{1}{4})} = 21^1 = 21$

It would seem reasonable to make the following definition.

Definition 7.6

If b is a real number, n is a positive integer greater than 1, and $\sqrt[n]{b}$ exists, then

$$b^{\frac{1}{n}} = \sqrt[n]{b}$$

Definition 7.6 states that $b^{\frac{1}{n}}$ means the nth root of b. We shall assume that b and n are chosen so that $\sqrt[n]{b}$ exists. For example, $(-25)^{\frac{1}{2}}$ is not meaningful at this time because $\sqrt{-25}$ is not a real number. Consider the following examples, which demonstrate the use of Definition 7.6.

$$25^{\frac{1}{2}} = \sqrt{25} = 5 \qquad 16^{\frac{1}{4}} = \sqrt[4]{16} = 2$$

$$8^{\frac{1}{3}} = \sqrt[3]{8} = 2 \qquad \left(\frac{36}{49}\right)^{\frac{1}{2}} = \sqrt{\frac{36}{49}} = \frac{6}{7}$$

$$(-27)^{\frac{1}{3}} = \sqrt[3]{-27} = -3$$

The following definition provides the basis for the use of *all* rational numbers as exponents.

Definition 7.7

If $\frac{m}{n}$ is a rational number, where n is a positive integer greater than 1 and b is a real number such that $\sqrt[n]{b}$ exists, then

$$b^{\frac{m}{n}} = \sqrt[n]{b^m} = (\sqrt[n]{b})^m$$

In Definition 7.7, note that the denominator of the exponent is the index of the radical and that the numerator of the exponent is either the exponent of the radicand or the exponent of the root.

Whether we use the form $\sqrt[n]{b^m}$ or the form $(\sqrt[n]{b})^m$ for computational purposes depends somewhat on the magnitude of the problem. Let's use both forms on two problems to illustrate this point.

$$8^{\frac{2}{3}} = \sqrt[3]{8^2} = \sqrt[3]{64} = 4 \qquad \text{or} \qquad 8^{\frac{2}{3}} = (\sqrt[3]{8})^2 = 2^2 = 4$$

$$27^{\frac{2}{3}} = \sqrt[3]{27^2} = \sqrt[3]{729} = 9 \qquad \text{or} \qquad 27^{\frac{2}{3}} = (\sqrt[3]{27})^2 = 3^2 = 9$$

To compute $8^{\frac{2}{3}}$, either form seems to work about as well as the other. However, to compute $27^{\frac{2}{3}}$, it should be obvious that $(\sqrt[3]{27})^2$ is much easier to handle than $\sqrt[3]{27^2}$.

EXAMPLE 1

Simplify each of the following numerical expressions.

(a) $25^{\frac{3}{2}}$ **(b)** $16^{\frac{3}{4}}$ **(c)** $(32)^{-\frac{2}{5}}$

(d) $(-64)^{\frac{2}{3}}$ **(e)** $-8^{\frac{1}{3}}$

Solution

(a) $25^{\frac{3}{2}} = (\sqrt{25})^3 = 5^3 = 125$

(b) $16^{\frac{3}{4}} = (\sqrt[4]{16})^3 = 2^3 = 8$

(c) $(32)^{-\frac{2}{5}} = \dfrac{1}{(32)^{\frac{2}{5}}} = \dfrac{1}{(\sqrt[5]{32})^2} = \dfrac{1}{2^2} = \dfrac{1}{4}$

(d) $(-64)^{\frac{2}{3}} = (\sqrt[3]{-64})^2 = (-4)^2 = 16$

(e) $-8^{\frac{1}{3}} = -\sqrt[3]{8} = -2$

▼ PRACTICE YOUR SKILL

Simplify each of the following numerical expressions.

(a) $36^{\frac{1}{2}}$ **(b)** $81^{\frac{3}{4}}$ **(c)** $(9)^{-\frac{3}{2}}$ **(d)** $(-125)^{\frac{2}{3}}$ **(e)** $-16^{\frac{1}{2}}$ ■

2 Write an Expression with Rational Exponents as a Radical

The basic laws of exponents that we stated in Property 7.2 are true for all rational exponents. Therefore, from now on we will use Property 7.2 for rational as well as integral exponents.

Some problems can be handled better in exponential form and others in radical form. Thus we must be able to switch forms with a certain amount of ease. Let's consider some examples where we switch from one form to the other.

EXAMPLE 2

Write each of the following expressions in radical form.

(a) $x^{\frac{3}{4}}$ **(b)** $3y^{\frac{2}{5}}$ **(c)** $x^{\frac{1}{4}}y^{\frac{3}{4}}$ **(d)** $(x + y)^{\frac{2}{3}}$

Solution

(a) $x^{\frac{3}{4}} = \sqrt[4]{x^3}$ **(b)** $3y^{\frac{2}{5}} = 3\sqrt[5]{y^2}$

(c) $x^{\frac{1}{4}}y^{\frac{3}{4}} = (xy^3)^{\frac{1}{4}} = \sqrt[4]{xy^3}$ **(d)** $(x + y)^{\frac{2}{3}} = \sqrt[3]{(x + y)^2}$

▼ PRACTICE YOUR SKILL

Write each of the following expressions in radical form.

(a) $y^{\frac{5}{6}}$ **(b)** $4a^{\frac{3}{5}}$ **(c)** $a^{\frac{3}{5}}b^{\frac{4}{5}}$ **(d)** $(a - b)^{\frac{3}{4}}$ ■

3 Write Radical Expressions as Expressions with Rational Exponents

EXAMPLE 3

Write each of the following using positive rational exponents.

(a) $\sqrt{xy}$ **(b)** $\sqrt[4]{a^3b}$ **(c)** $4\sqrt[3]{x^2}$ **(d)** $\sqrt[5]{(x + y)^4}$

Solution

(a) $\sqrt{xy} = (xy)^{\frac{1}{2}} = x^{\frac{1}{2}}y^{\frac{1}{2}}$ **(b)** $\sqrt[4]{a^3b} = (a^3b)^{\frac{1}{4}} = a^{\frac{3}{4}}b^{\frac{1}{4}}$

(c) $4\sqrt[3]{x^2} = 4x^{\frac{2}{3}}$ **(d)** $\sqrt[5]{(x + y)^4} = (x + y)^{\frac{4}{5}}$

▼ PRACTICE YOUR SKILL

Write each of the following using positive rational exponents.

(a) $\sqrt{ab}$ **(b)** $\sqrt[5]{a^2b^4}$ **(c)** $7\sqrt[3]{x}$ **(d)** $\sqrt[5]{(a - b)^2}$ ■

4 Simplify Algebraic Expressions That Have Rational Exponents

The properties of exponents provide the basis for simplifying algebraic expressions that contain rational exponents, as these next examples illustrate.

EXAMPLE 4

Simplify each of the following. Express final results using positive exponents only.

(a) $(3x^{\frac{1}{2}})(4x^{\frac{2}{3}})$ **(b)** $(5a^{\frac{1}{3}}b^{\frac{1}{2}})^2$ **(c)** $\dfrac{12y^{\frac{1}{3}}}{6y^{\frac{1}{2}}}$ **(d)** $\left(\dfrac{3x^{\frac{2}{5}}}{2y^{\frac{2}{3}}}\right)^4$

Solution

(a) $(3x^{\frac{1}{2}})(4x^{\frac{2}{3}}) = 3 \cdot 4 \cdot x^{\frac{1}{2}} \cdot x^{\frac{2}{3}}$

$= 12x^{\frac{1}{2}+\frac{2}{3}}$ $\quad b^n \cdot b^m = b^{n+m}$

$= 12x^{\frac{3}{6}+\frac{4}{6}}$ $\quad$ Use 6 as LCD

$= 12x^{\frac{7}{6}}$

(b) $(5a^{\frac{1}{3}}b^{\frac{1}{2}})^2 = 5^2 \cdot (a^{\frac{1}{3}})^2 \cdot (b^{\frac{1}{2}})^2$ $\quad (ab)^n = a^nb^n$

$= 25a^{\frac{2}{3}}b$ $\quad (b^n)^m = b^{mn}$

(c) $\dfrac{12y^{\frac{1}{3}}}{6y^{\frac{1}{2}}} = 2y^{\frac{1}{3}-\frac{1}{2}}$ $\quad \dfrac{b^n}{b^m} = b^{n-m}$

$= 2y^{\frac{2}{6}-\frac{3}{6}}$

$= 2y^{-\frac{1}{6}}$

$= \dfrac{2}{y^{\frac{1}{6}}}$

(d) $\left(\dfrac{3x^{\frac{2}{5}}}{2y^{\frac{2}{3}}}\right)^4 = \dfrac{(3x^{\frac{2}{5}})^4}{(2y^{\frac{2}{3}})^4}$ $\quad \left(\dfrac{a}{b}\right)^n = \dfrac{a^n}{b^n}$

$= \dfrac{3^4 \cdot (x^{\frac{2}{5}})^4}{2^4 \cdot (y^{\frac{2}{3}})^4}$ $\quad (ab)^n = a^nb^n$

$= \dfrac{81x^{\frac{8}{5}}}{16y^{\frac{8}{3}}}$ $\quad (b^n)^m = b^{mn}$

▼ **PRACTICE YOUR SKILL**

Simplify each of the following. Express final results using positive exponents only.

(a) $(6a^{\frac{3}{4}})(3a^{\frac{5}{8}})$ **(b)** $(8x^{\frac{1}{5}}y^{\frac{1}{4}})^2$ **(c)** $\dfrac{15a^{\frac{1}{4}}}{5a^{\frac{2}{3}}}$ **(d)** $\left(\dfrac{3a^{\frac{1}{3}}}{4b^{\frac{2}{5}}}\right)^2$

■

5 Multiply and Divide Radicals with Different Indexes

The link between exponents and roots also provides a basis for multiplying and dividing some radicals even if they have different indexes. The general procedure is as follows:

1. Change from radical form to exponential form.
2. Apply the properties of exponents.
3. Then change back to radical form.

The three parts of Example 5 illustrate this process.

EXAMPLE 5 Perform the indicated operations and express the answers in simplest radical form.

(a) $\sqrt{2}\sqrt[3]{2}$ **(b)** $\dfrac{\sqrt{5}}{\sqrt[3]{5}}$ **(c)** $\dfrac{\sqrt{4}}{\sqrt[3]{2}}$

Solution

(a) $\sqrt{2}\sqrt[3]{2} = 2^{\frac{1}{2}} \cdot 2^{\frac{1}{3}}$

$= 2^{\frac{1}{2}+\frac{1}{3}}$

$= 2^{\frac{3}{6}+\frac{2}{6}}$ Use 6 as LCD

$= 2^{\frac{5}{6}}$

$= \sqrt[6]{2^5} = \sqrt[6]{32}$

(b) $\dfrac{\sqrt{5}}{\sqrt[3]{5}} = \dfrac{5^{\frac{1}{2}}}{5^{\frac{1}{3}}}$

$= 5^{\frac{1}{2}-\frac{1}{3}}$

$= 5^{\frac{3}{6}-\frac{2}{6}}$ Use 6 as LCD

$= 5^{\frac{1}{6}} = \sqrt[6]{5}$

(c) $\dfrac{\sqrt{4}}{\sqrt[3]{2}} = \dfrac{4^{\frac{1}{2}}}{2^{\frac{1}{3}}}$

$= \dfrac{(2^2)^{\frac{1}{2}}}{2^{\frac{1}{3}}}$

$= \dfrac{2^1}{2^{\frac{1}{3}}}$

$= 2^{1-\frac{1}{3}}$

$= 2^{\frac{2}{3}} = \sqrt[3]{2^2} = \sqrt[3]{4}$

▼ **PRACTICE YOUR SKILL**

Perform the indicated operation and express the answer in simplest form.

(a) $\sqrt[3]{5}\sqrt[4]{5}$ **(b)** $\dfrac{\sqrt[3]{6}}{\sqrt[4]{6}}$ **(c)** $\dfrac{\sqrt[3]{9}}{\sqrt[4]{3}}$ ■

CONCEPT QUIZ

For Problems 1–10, answer true or false.

1. Assuming the nth root of x exists, $\sqrt[n]{x}$ can be written as $x^{\frac{1}{n}}$.
2. An exponent of $\frac{1}{3}$ means that we need to find the cube root of the number.
3. To evaluate $16^{\frac{2}{3}}$ we would find the square root of 16 and then cube the result.
4. When an expression with a rational exponent is written as a radical expression, the denominator of the rational exponent is the index of the radical.
5. The expression $\sqrt[n]{x^m}$ is equivalent to $(\sqrt[n]{x})^m$.
6. $-16^{-3} = \dfrac{1}{64}$
7. $\dfrac{\sqrt{7}}{\sqrt[3]{7}} = \sqrt[6]{7}$
8. $(16)^{-\frac{3}{4}} = \dfrac{1}{8}$
9. $\dfrac{\sqrt[3]{16}}{\sqrt{2}} = 2\sqrt{2}$
10. $\sqrt[3]{64^2} = 16$

Problem Set 7.6

1 Evaluate a Number Raised to a Rational Exponent

For Problems 1–30, evaluate each numerical expression.

1. $81^{\frac{1}{2}}$
2. $64^{\frac{1}{2}}$
3. $27^{\frac{1}{3}}$
4. $(-32)^{\frac{1}{5}}$
5. $(-8)^{\frac{1}{3}}$
6. $\left(-\frac{27}{8}\right)^{\frac{1}{3}}$
7. $-25^{\frac{1}{2}}$
8. $-64^{\frac{1}{3}}$
9. $36^{-\frac{1}{2}}$
10. $81^{-\frac{1}{2}}$
11. $\left(\frac{1}{27}\right)^{-\frac{1}{3}}$
12. $\left(-\frac{8}{27}\right)^{-\frac{1}{3}}$
13. $4^{\frac{3}{2}}$
14. $64^{\frac{2}{3}}$
15. $27^{\frac{4}{3}}$
16. $4^{\frac{7}{2}}$
17. $(-1)^{\frac{7}{3}}$
18. $(-8)^{\frac{4}{3}}$
19. $-4^{\frac{5}{2}}$
20. $-16^{\frac{3}{2}}$
21. $\left(\frac{27}{8}\right)^{\frac{4}{3}}$
22. $\left(\frac{8}{125}\right)^{\frac{2}{3}}$
23. $\left(\frac{1}{8}\right)^{-\frac{2}{3}}$
24. $\left(-\frac{1}{27}\right)^{-\frac{2}{3}}$
25. $64^{-\frac{7}{6}}$
26. $32^{-\frac{4}{5}}$
27. $-25^{\frac{3}{2}}$
28. $-16^{\frac{3}{4}}$
29. $125^{\frac{4}{3}}$
30. $81^{\frac{5}{4}}$

2 Write an Expression with Rational Exponents as a Radical

For Problems 31–44, write each of the following in radical form. For example,

$$3x^{\frac{2}{3}} = 3\sqrt[3]{x^2}$$

31. $x^{\frac{4}{3}}$
32. $x^{\frac{2}{5}}$
33. $3x^{\frac{1}{2}}$
34. $5x^{\frac{1}{4}}$
35. $(2y)^{\frac{1}{3}}$
36. $(3xy)^{\frac{1}{2}}$
37. $(2x - 3y)^{\frac{1}{2}}$
38. $(5x + y)^{\frac{1}{3}}$
39. $(2a - 3b)^{\frac{2}{3}}$
40. $(5a + 7b)^{\frac{3}{5}}$
41. $x^{\frac{2}{3}}y^{\frac{1}{3}}$
42. $x^{\frac{3}{7}}y^{\frac{5}{7}}$
43. $-3x^{\frac{1}{5}}y^{\frac{2}{5}}$
44. $-4x^{\frac{3}{4}}y^{\frac{1}{4}}$

3 Write Radical Expressions as Expressions with Rational Exponents

For Problems 45–58, write each of the following using positive rational exponents. For example,

$$\sqrt{ab} = (ab)^{\frac{1}{2}} = a^{\frac{1}{2}}b^{\frac{1}{2}}$$

45. $\sqrt{5y}$
46. $\sqrt{2xy}$
47. $3\sqrt{y}$
48. $5\sqrt{ab}$
49. $\sqrt[3]{xy^2}$
50. $\sqrt[5]{x^2y^4}$
51. $\sqrt[4]{a^2b^3}$
52. $\sqrt[6]{ab^5}$
53. $\sqrt[5]{(2x - y)^3}$
54. $\sqrt[7]{(3x - y)^4}$
55. $5x\sqrt{y}$
56. $4y\sqrt[3]{x}$
57. $-\sqrt[3]{x + y}$
58. $-\sqrt[5]{(x - y)^2}$

4 Simplify Algebraic Expressions That Have Rational Exponents

For Problems 59–80, simplify each of the following. Express final results using positive exponents only. For example,

$$(2x^{\frac{1}{2}})(3x^{\frac{1}{3}}) = 6x^{\frac{5}{6}}$$

59. $(2x^{\frac{2}{5}})(6x^{\frac{1}{4}})$
60. $(3x^{\frac{1}{4}})(5x^{\frac{1}{3}})$
61. $(y^{\frac{2}{3}})(y^{-\frac{1}{4}})$
62. $(y^{\frac{3}{4}})(y^{-\frac{1}{2}})$
63. $(x^{\frac{2}{5}})(4x^{-\frac{1}{2}})$
64. $(2x^{\frac{1}{3}})(x^{-\frac{1}{2}})$
65. $(4x^{\frac{1}{2}}y)^2$
66. $(3x^{\frac{1}{4}}y^{\frac{1}{5}})^3$
67. $(8x^6y^3)^{\frac{1}{3}}$
68. $(9x^2y^4)^{\frac{1}{2}}$
69. $\frac{24x^{\frac{3}{5}}}{6x^{\frac{1}{3}}}$
70. $\frac{18x^{\frac{1}{2}}}{9x^{\frac{1}{3}}}$
71. $\frac{48b^{\frac{1}{3}}}{12b^{\frac{3}{4}}}$
72. $\frac{56a^{\frac{1}{6}}}{8a^{\frac{1}{4}}}$
73. $\left(\frac{6x^{\frac{2}{5}}}{7y^{\frac{2}{3}}}\right)^2$
74. $\left(\frac{2x^{\frac{1}{3}}}{3y^{\frac{1}{4}}}\right)^4$
75. $\left(\frac{x^2}{y^3}\right)^{-\frac{1}{2}}$
76. $\left(\frac{a^3}{b^{-2}}\right)^{-\frac{1}{3}}$
77. $\left(\frac{18x^{\frac{1}{3}}}{9x^{\frac{1}{4}}}\right)^2$
78. $\left(\frac{72x^{\frac{3}{4}}}{6x^{\frac{1}{2}}}\right)^2$
79. $\left(\frac{60a^{\frac{1}{5}}}{15a^{\frac{3}{4}}}\right)^2$
80. $\left(\frac{64a^{\frac{1}{3}}}{16a^{\frac{5}{9}}}\right)^3$

5 Multiply and Divide Radicals with Different Indexes

For Problems 81–90, perform the indicated operations and express answers in simplest radical form. (See Example 5.)

81. $\sqrt[3]{3}\sqrt{3}$ **82.** $\sqrt{2}\sqrt[4]{2}$

83. $\sqrt[4]{6}\sqrt{6}$ **84.** $\sqrt[3]{5}\sqrt{5}$

85. $\dfrac{\sqrt[3]{3}}{\sqrt[4]{3}}$ **86.** $\dfrac{\sqrt{2}}{\sqrt[3]{2}}$

87. $\dfrac{\sqrt[3]{8}}{\sqrt[4]{4}}$ **88.** $\dfrac{\sqrt{9}}{\sqrt[3]{3}}$

89. $\dfrac{\sqrt[4]{27}}{\sqrt{3}}$ **90.** $\dfrac{\sqrt[3]{16}}{\sqrt[6]{4}}$

THOUGHTS INTO WORDS

91. Your friend keeps getting an error message when evaluating $-4^{\frac{5}{2}}$ on his calculator. What error is he probably making?

92. Explain how you would evaluate $27^{\frac{2}{3}}$ without a calculator.

FURTHER INVESTIGATIONS

93. Use your calculator to evaluate each of the following.

(a) $\sqrt[3]{1728}$ **(b)** $\sqrt[3]{5832}$

(c) $\sqrt[4]{2401}$ **(d)** $\sqrt[4]{65{,}536}$

(e) $\sqrt[5]{161{,}051}$ **(f)** $\sqrt[5]{6{,}436{,}343}$

94. Definition 7.7 states that

$$b^{\frac{m}{n}} = \sqrt[n]{b^m} = (\sqrt[n]{b})^m$$

Use your calculator to verify each of the following.

(a) $\sqrt[3]{27^2} = (\sqrt[3]{27})^2$ **(b)** $\sqrt[3]{8^5} = (\sqrt[3]{8})^5$

(c) $\sqrt[4]{16^3} = (\sqrt[4]{16})^3$ **(d)** $\sqrt[3]{16^2} = (\sqrt[3]{16})^2$

(e) $\sqrt[5]{9^4} = (\sqrt[5]{9})^4$ **(f)** $\sqrt[3]{12^4} = (\sqrt[3]{12})^4$

95. Use your calculator to evaluate each of the following.

(a) $16^{\frac{5}{2}}$ **(b)** $25^{\frac{7}{2}}$

(c) $16^{\frac{9}{4}}$ **(d)** $27^{\frac{5}{3}}$

(e) $343^{\frac{2}{3}}$ **(f)** $512^{\frac{4}{3}}$

96. Use your calculator to estimate each of the following to the nearest one-thousandth.

(a) $7^{\frac{4}{3}}$ **(b)** $10^{\frac{4}{5}}$

(c) $12^{\frac{3}{5}}$ **(d)** $19^{\frac{2}{5}}$

(e) $7^{\frac{3}{4}}$ **(f)** $10^{\frac{5}{4}}$

97. (a) Because $\frac{4}{5} = 0.8$, we can evaluate $10^{\frac{4}{5}}$ by evaluating $10^{0.8}$, which involves a shorter sequence of "calculator steps." Evaluate parts (b), (c), (d), (e), and (f) of Problem 96 and take advantage of decimal exponents.

(b) What problem is created when we try to evaluate $7^{\frac{4}{3}}$ by changing the exponent to decimal form?

Answers to the Concept Quiz

1. True **2.** True **3.** False **4.** True **5.** True **6.** False **7.** True **8.** True **9.** False **10.** True

Answers to the Example Practice Skills

1. (a) 6 **(b)** 27 **(c)** $\frac{1}{27}$ **(d)** 25 **(e)** -4 **2. (a)** $\sqrt[6]{y^5}$ **(b)** $4\sqrt[5]{a^3}$ **(c)** $\sqrt[5]{a^3b^4}$ **(d)** $\sqrt[4]{(a-b)^3}$ **3. (a)** $a^{\frac{1}{2}}b^{\frac{1}{2}}$ **(b)** $a^{\frac{2}{5}}b^{\frac{4}{5}}$ **(c)** $7x^{\frac{1}{3}}$ **(d)** $(a-b)^{\frac{2}{5}}$ **4. (a)** $18a^{\frac{11}{8}}$ **(b)** $64x^{\frac{2}{5}}y^{\frac{1}{2}}$ **(c)** $\dfrac{3}{a^{\frac{5}{12}}}$ **(d)** $\dfrac{9a^{\frac{2}{3}}}{16b^{\frac{4}{5}}}$ **5. (a)** $\sqrt[12]{5^7}$ **(b)** $\sqrt[12]{6}$ **(c)** $\sqrt[12]{3^5}$

7.7 Scientific Notation

OBJECTIVES

1 Write Numbers in Scientific Notation

2 Convert Numbers from Scientific Notation to Ordinary Decimal Notation

3 Perform Calculations with Numbers Using Scientific Notation

1 Write Numbers in Scientific Notation

Many applications of mathematics involve the use of very large or very small numbers.

1. The speed of light is approximately 29,979,200,000 centimeters per second.
2. A light year—the distance that light travels in 1 year—is approximately 5,865,696,000,000 miles.
3. A millimicron equals 0.000000001 of a meter.

Working with numbers of this type in standard decimal form is quite cumbersome. It is much more convenient to represent very small and very large numbers in **scientific notation**. Although negative numbers can be written in scientific form, we will restrict our discussion to positive numbers. The expression $(N)(10)^k$, where N is a number greater than or equal to 1 and less than 10, written in decimal form, and k is any integer, is commonly called scientific notation or the scientific form of a number. Consider the following examples, which show a comparison between ordinary decimal notation and scientific notation.

Ordinary notation	Scientific notation
2.14	$(2.14)(10^0)$
31.78	$(3.178)(10^1)$
412.9	$(4.129)(10^2)$
8,000,000	$(8)(10^6)$
0.14	$(1.4)(10^{-1})$
0.0379	$(3.79)(10^{-2})$
0.00000049	$(4.9)(10^{-7})$

To switch from ordinary notation to scientific notation, you can use the following procedure.

> Write the given number as the product of a power of 10 and a number greater than or equal to 1 and less than 10. The exponent of 10 is determined by counting the number of places that the decimal point was moved when going from the original number to the number greater than or equal to 1 and less than 10. This exponent is (a) negative if the original number is less than 1, (b) positive if the original number is greater than 10, or (c) 0 if the original number itself is between 1 and 10.

Thus we can write

$$0.00467 = (4.67)(10^{-3})$$

$$87{,}000 = (8.7)(10^4)$$

$$3.1416 = (3.1416)(10^0)$$

We can express the constants given earlier in scientific notation as follows:

Speed of light 29,979,200,000 = $(2.99792)(10^{10})$ centimeters per second.

Light year 5,865,696,000,000 = $(5.865696)(10^{12})$ miles.

Metric units A millimicron is 0.000000001 = $(1)(10^{-9})$ meter.

2 Convert Numbers from Scientific Notation to Ordinary Decimal Notation

To switch from scientific notation to ordinary decimal notation, you can use the following procedure.

> Move the decimal point the number of places indicated by the exponent of 10. The decimal point is moved to the right if the exponent is positive and to the left if the exponent is negative.

Thus we can write

$$(4.78)(10^4) = 47{,}800$$

$$(8.4)(10^{-3}) = 0.0084$$

3 Perform Calculations with Numbers Using Scientific Notation

Scientific notation can frequently be used to simplify numerical calculations. We merely change the numbers to scientific notation and use the appropriate properties of exponents. Consider the following examples.

EXAMPLE 1

Convert each number to scientific notation and perform the indicated operation. Express the result in ordinary decimal form.

(a) $(0.00024)(20{,}000)$ **(b)** $\dfrac{7{,}800{,}000}{0.0039}$

(c) $\dfrac{(0.00069)(0.0034)}{(0.0000017)(0.023)}$ **(d)** $\sqrt{0.000004}$

Solution

(a)
$$\begin{aligned}(0.00024)(20{,}000) &= (2.4)(10^{-4})(2)(10^4)\\ &= (2.4)(2)(10^{-4})(10^4)\\ &= (4.8)(10^0)\\ &= (4.8)(1)\\ &= 4.8\end{aligned}$$

(b)
$$\begin{aligned}\frac{7{,}800{,}000}{0.0039} &= \frac{(7.8)(10^6)}{(3.9)(10^{-3})}\\ &= (2)(10^9)\\ &= 2{,}000{,}000{,}000\end{aligned}$$

(c) $\dfrac{(0.00069)(0.0034)}{(0.0000017)(0.023)} = \dfrac{(6.9)(10^{-4})(3.4)(10^{-3})}{(1.7)(10^{-6})(2.3)(10^{-2})}$

$$= \frac{(\overset{3}{\cancel{6.9}})(\overset{2}{\cancel{3.4}})(10^{-7})}{(\cancel{1.7})(\cancel{2.3})(10^{-8})}$$

$$= (6)(10^{1})$$

$$= 60$$

(d) $\sqrt{0.00004} = \sqrt{(4)(10^{-6})}$

$$= ((4)(10^{-6}))^{\frac{1}{2}}$$

$$= 4^{\frac{1}{2}}(10^{-6})^{\frac{1}{2}}$$

$$= (2)(10^{-3})$$

$$= 0.002$$

▼ PRACTICE YOUR SKILL

Convert each number to scientific notation and perform the indicated operation. Express the result in ordinary decimal form.

(a) $(0.000031)(3000)$ **(b)** $\dfrac{4{,}500{,}000}{0.15}$ **(c)** $\dfrac{(0.00000036)(5400)}{(270{,}000)(0.00012)}$ ■

(d) $\sqrt{0.00000016}$

EXAMPLE 2

The speed of light is approximately $(1.86)(10^5)$ miles per second. When the earth is $(9.3)(10^7)$ miles away from the sun, how long does it take light from the sun to reach the earth?

Solution

We will use the formula $t = \dfrac{d}{r}$.

$$t = \frac{(9.3)(10^7)}{(1.86)(10^5)}$$

$$t = \frac{(9.3)}{(1.86)}(10^2) \quad \text{Subtract exponents}$$

$$t = (5)(10^2) = 500 \text{ seconds}$$

At this distance it takes light about 500 seconds to travel from the sun to the earth. To find the answer in minutes, divide 500 seconds by 60 seconds/minute. That gives a result of approximately 8.33 minutes.

▼ PRACTICE YOUR SKILL

A large virus has a diameter of length 100 nanometers or 0.0000001 meters. An *E. coli* cell has a diameter of 2 micrometers or 0.000002 meters. How many times larger in length is the *E. coli* cell than the virus? ■

Many calculators are equipped to display numbers in scientific notation. The display panel shows the number between 1 and 10 and the appropriate exponent of 10. For example, evaluating $(3{,}800{,}000)^2$ yields

`1.444E13`

Thus $(3{,}800{,}000)^2 = (1.444)(10^{13}) = 14{,}440{,}000{,}000{,}000$.

Similarly, the answer for $(0.000168)^2$ is displayed as

2.8224E-8

Thus $(0.000168)^2 = (2.8224)(10^{-8}) = 0.000000028224$.

Calculators vary as to the number of digits displayed in the number between 1 and 10 when scientific notation is used. For example, we used two different calculators to estimate $(6729)^6$ and obtained the following results.

9.2833E22

9.283316768E22

Obviously, you need to know the capabilities of your calculator when working with problems in scientific notation. Many calculators also allow the entry of a number in scientific notation. Such calculators are equipped with an enter-the-exponent key (often labeled as [EE] or [EEX]). Thus a number such as $(3.14)(10^8)$ might be entered as follows:

Enter	Press	Display
3.14	[EE]	3.14E
8		3.14E8

or

Enter	Press	Display
3.14	[EE]	$3.14^{\ 00}$
8		$3.14^{\ 08}$

A [MODE] key is often used on calculators to let you choose normal decimal notation, scientific notation, or engineering notation. (The abbreviations Norm, Sci, and Eng are commonly used.) If the calculator is in scientific mode, then a number can be entered and changed to scientific form by pressing the [ENTER] key. For example, when we enter 589 and press the [ENTER] key, the display will show 5.89E2. Likewise, when the calculator is in scientific mode, the answers to computational problems are given in scientific form. For example, the answer for (76)(533) is given as 4.0508E4.

It should be evident from this brief discussion that you need to have a thorough understanding of scientific notation even when you are using a calculator.

CONCEPT QUIZ

For Problems 1–10, answer true or false.

1. A positive number written in scientific notation has the form $(N)(10^k)$, where $1 \le N < 10$ and k is an integer.
2. A number is less than zero if the exponent is negative when the number is written in scientific notation.
3. $(3.11)(10^{-2}) = 311$
4. $(5.24)(10^{-1}) = 0.524$
5. $(8.91)(10^2) = 89.1$
6. $(4.163)(10^{-5}) = 0.00004163$
7. $0.00715 = (7.15)(10^{-3})$
8. Scientific notation provides a way of working with numbers that are very large or very small in magnitude.
9. $(0.0012)(5000) = 60$
10. $\dfrac{6{,}200{,}000}{0.0031} = 2{,}000{,}000{,}000$

Problem Set 7.7

1 Write Numbers in Scientific Notation

For Problems 1–18, write each of the following in scientific notation. For example,

$$27{,}800 = (2.78)(10^4)$$

1. 89
2. 117
3. 4290
4. 812,000
5. 6,120,000
6. 72,400,000
7. 40,000,000
8. 500,000,000
9. 376.4
10. 9126.21
11. 0.347
12. 0.2165
13. 0.0214
14. 0.0037
15. 0.00005
16. 0.00000082
17. 0.00000000194
18. 0.00000000003

2 Convert Numbers from Scientific Notation to Ordinary Decimal Notation

For Problems 19–32, write each of the following in ordinary decimal notation. For example,

$$(3.18)(10^2) = 318$$

19. $(2.3)(10^1)$
20. $(1.62)(10^2)$
21. $(4.19)(10^3)$
22. $(7.631)(10^4)$
23. $(5)(10^8)$
24. $(7)(10^9)$
25. $(3.14)(10^{10})$
26. $(2.04)(10^{12})$
27. $(4.3)(10^{-1})$
28. $(5.2)(10^{-2})$
29. $(9.14)(10^{-4})$
30. $(8.76)(10^{-5})$
31. $(5.123)(10^{-8})$
32. $(6)(10^{-9})$

3 Perform Calculations with Numbers Using Scientific Notation

For Problems 33–50, use scientific notation and the properties of exponents to help you perform the following operations.

33. $(0.0037)(0.00002)$
34. $(0.00003)(0.00025)$
35. $(0.00007)(11{,}000)$
36. $(0.000004)(120{,}000)$
37. $\dfrac{360{,}000{,}000}{0.0012}$
38. $\dfrac{66{,}000{,}000{,}000}{0.022}$
39. $\dfrac{0.000064}{16{,}000}$
40. $\dfrac{0.00072}{0.0000024}$
41. $\dfrac{(60{,}000)(0.006)}{(0.0009)(400)}$
42. $\dfrac{(0.00063)(960{,}000)}{(3{,}200)(0.0000021)}$
43. $\dfrac{(0.0045)(60{,}000)}{(1800)(0.00015)}$
44. $\dfrac{(0.00016)(300)(0.028)}{0.064}$
45. $\sqrt{9{,}000{,}000}$
46. $\sqrt{0.00000009}$
47. $\sqrt[3]{8000}$
48. $\sqrt[3]{0.001}$
49. $(90{,}000)^{\frac{3}{2}}$
50. $(8000)^{\frac{2}{3}}$

51. Avogadro's number, 602,000,000,000,000,000,000,000, is the number of atoms in 1 mole of a substance. Express this number in scientific notation.

52. The Social Security program paid out approximately $44,000,000,000 in benefits in May 2005. Express this number in scientific notation.

53. Carlos's first computer had a processing speed of $(1.6)(10^6)$ hertz. He recently purchased a laptop computer with a processing speed of $(1.33)(10^9)$ hertz. Approximately how many times faster is the processing speed of his laptop than that of his first computer? Express the result in decimal form.

54. Alaska has an area of approximately $(6.15)(10^5)$ square miles. In 2006 the state had a population of approximately 670,000 people. Compute the population density to the nearest hundredth. (Population density is the number of people per square mile.) Express the result in decimal form rounded to the nearest hundredth.

55. In the year 2007 the public debt of the United States was approximately $9,000,000,000,000. For July 2007, the census reported that approximately 300,000,000 people lived in the United States. Convert these figures to scientific notation, and compute the average debt per person. Express the result in scientific notation.

56. The space shuttle can travel at approximately 410,000 miles per day. If the shuttle could travel to Mars, and Mars was 140,000,000 miles away, how many days would it take the shuttle to travel to Mars? Express the result in decimal form.

57. Atomic masses are measured in atomic mass units (amu). The amu, $(1.66)(10^{-27})$ kilograms, is defined as $\frac{1}{12}$ the mass of a common carbon atom. Find the mass of a carbon atom in kilograms. Express the result in scientific notation.

58. The field of view of a microscope is $(4)(10^{-4})$ meters. If a single-cell organism occupies $\frac{1}{5}$ of the field of view, find the length of the organism in meters. Express the result in scientific notation.

59. The mass of an electron is $(9.11)(10^{-31})$ kilogram, and the mass of a proton is $(1.67)(10^{-27})$ kilogram. Approximately how many times more is the weight of a proton than the weight of an electron? Express the result in decimal form.

60. A square pixel on a computer screen has a side of length $(1.17)(10^{-2})$ inches. Find the approximate area of the pixel in square inches. Express the result in decimal form.

THOUGHTS INTO WORDS

61. Explain the importance of scientific notation.

62. Why do we need scientific notation even when using calculators and computers?

FURTHER INVESTIGATIONS

63. Sometimes it is more convenient to express a number as a product of a power of 10 and a number that is *not* between 1 and 10. For example, suppose that we want to calculate $\sqrt{640{,}000}$. We can proceed as follows:

$$\begin{aligned}\sqrt{640{,}000} &= \sqrt{(64)(10^4)}\\ &= ((64)(10^4))^{\frac{1}{2}}\\ &= (64)^{\frac{1}{2}}(10^4)^{\frac{1}{2}}\\ &= (8)(10^2)\\ &= 8(100) = 800\end{aligned}$$

Compute each of the following without a calculator, and then use a calculator to check your answers.

(a) $\sqrt{49{,}000{,}000}$ **(b)** $\sqrt{0.0025}$

(c) $\sqrt{14{,}400}$ **(d)** $\sqrt{0.000121}$

(e) $\sqrt[3]{27{,}000}$ **(f)** $\sqrt[3]{0.000064}$

64. Use your calculator to evaluate each of the following. Express final answers in ordinary notation.

(a) $(27{,}000)^2$ **(b)** $(450{,}000)^2$

(c) $(14{,}800)^2$ **(d)** $(1700)^3$

(e) $(900)^4$ **(f)** $(60)^5$

(g) $(0.0213)^2$ **(h)** $(0.000213)^2$

(i) $(0.000198)^2$ **(j)** $(0.000009)^3$

65. Use your calculator to estimate each of the following. Express final answers in scientific notation with the number between 1 and 10 rounded to the nearest one-thousandth.

(a) $(4576)^4$ **(b)** $(719)^{10}$

(c) $(28)^{12}$ **(d)** $(8619)^6$

(e) $(314)^5$ **(f)** $(145{,}723)^2$

66. Use your calculator to estimate each of the following. Express final answers in ordinary notation rounded to the nearest one-thousandth.

(a) $(1.09)^5$ **(b)** $(1.08)^{10}$

(c) $(1.14)^7$ **(d)** $(1.12)^{20}$

(e) $(0.785)^4$ **(f)** $(0.492)^5$

Answers to the Concept Quiz

1. True **2.** False **3.** False **4.** True **5.** False **6.** True **7.** True **8.** True **9.** False **10.** True

Answers to the Example Practice Skills

1. (a) 0.093 **(b)** 30,000,000 **(c)** 0.00006 **(d)** 0.0004 **2.** 20 times

Chapter 7 Summary

OBJECTIVE	SUMMARY	EXAMPLE	CHAPTER REVIEW PROBLEMS
Simplify numerical expressions that have positive and negative exponents. (Sec. 7.1, Obj. 1, p. 360)	The concept of exponent is expanded to include negative exponents and exponents of zero. If b is a nonzero number, then $b^0 = 1$. If n is a positive integer and b is a nonzero number, then $b^{-n} = \frac{1}{b^n}$.	Simplify $\left(\frac{2}{5}\right)^{-2}$. **Solution** $\left(\frac{2}{5}\right)^{-2} = \frac{2^{-2}}{5^{-2}} = \frac{5^2}{2^2} = \frac{25}{4}$	Problems 1–6
Simplify algebraic expressions that have positive and negative exponents. (Sec. 7.1, Obj. 2, p. 363)	The properties for integer exponents listed on page 362 form the basis for manipulating with integer exponents. These properties, along with knowing from Definition 7.2 that $b^{-n} = \frac{1}{b^n}$, enable us to simplify algebraic expressions and express the results with positive exponents.	Simplify $(2x^{-3}y)^{-2}$ and express the final result using positive exponents. **Solution** $(2x^{-3}y)^{-2} = 2^{-2}x^6y^{-2} = \frac{x^6}{2^2y^2} = \frac{x^6}{4y^2}$	Problems 7–10
Multiply and divide algebraic expressions that have positive and negative exponents. (Sec. 7.1, Obj. 3, p. 363)	The previous remark also applies to simplifying multiplication and division problems that involve integer exponents.	Simplify $(-3x^5y^{-2})(4x^{-1}y^{-1})$ and express the final result using positive exponents. **Solution** $(-3x^5y^{-2})(4x^{-1}y^{-1}) = -12x^4y^{-3} = -\frac{12x^4}{y^3}$	Problems 11–18
Simplify sums and differences of expressions involving positive and negative exponents. (Sec. 7.1, Obj. 4, p. 364)	Change from negative exponents to positive and perform the indicated operation. It may be necessary to find a common denominator.	Simplify $5x^{-2} + 6y^{-1}$ and express the result as a single fraction involving positive exponents only. **Solution** $5x^{-2} + 6y^{-1} = \frac{5}{x^2} + \frac{6}{y} = \frac{5}{x^2} \cdot \frac{y}{y} + \frac{6}{y} \cdot \frac{x^2}{x^2} = \frac{5y + 6x^2}{x^2y}$	Problems 19–22

(continued)

OBJECTIVE	SUMMARY	EXAMPLE	CHAPTER REVIEW PROBLEMS
Express a radical in simplest radical form. (Sec. 7.2, Obj. 2, p. 372)	A radical expression is in simplest form if: 1. a radicand contains no polynomial factor raised to a power equal to or greater than the index of the radical; 2. no fraction appears within a radical sign; and 3. no radical appears in the denominator. The following properties are used to express radicals in simplest form: $\sqrt[n]{bc} = \sqrt[n]{b}\sqrt[n]{c}$ $\sqrt[n]{\frac{b}{c}} = \frac{\sqrt[n]{b}}{\sqrt[n]{c}}$	Simplify $\sqrt{150a^3b^2}$. Assume all variables represent nonnegative values. **Solution** $\sqrt{150a^3b^2} = \sqrt{25a^2b^2}\sqrt{6b}$ $= 5ab\sqrt{6b}$	Problems 23–28
Rationalize the denominator to simplify radicals. (Sec. 7.2, Obj. 3, p. 373)	If a radical appears in the denominator, then it will be necessary to rationalize the denominator for the expression to be in simplest form.	Simplify $\frac{2\sqrt{18}}{\sqrt{5}}$. **Solution** $\frac{2\sqrt{18}}{\sqrt{5}} = \frac{2\sqrt{9}\sqrt{2}}{\sqrt{5}}$ $= \frac{2(3)\sqrt{2}}{\sqrt{5}} = \frac{6\sqrt{2}}{\sqrt{5}}$ $= \frac{6\sqrt{2}}{\sqrt{5}} \cdot \frac{\sqrt{5}}{\sqrt{5}} = \frac{6\sqrt{10}}{\sqrt{25}}$ $= \frac{6\sqrt{10}}{5}$	Problems 29–34
Simplify expressions by combining radicals. (Sec. 7.3, Obj. 1, p. 379)	Simplifying by combining radicals sometimes requires that we first express the given radicals in simplest form.	Simplify $\sqrt{24} - \sqrt{54} + 8\sqrt{6}$. **Solution** $\sqrt{24} - \sqrt{54} + 8\sqrt{6}$ $= \sqrt{4}\sqrt{6} - \sqrt{9}\sqrt{6} + 8\sqrt{6}$ $= 2\sqrt{6} - 3\sqrt{6} + 8\sqrt{6}$ $= 7\sqrt{6}$	Problems 35–38
Multiply two radicals. (Sec. 7.4, Obj. 1, p. 385)	The property $\sqrt[n]{b}\sqrt[n]{c} = \sqrt[n]{bc}$ is used to find the product of two radicals.	Multiply $\sqrt[3]{4x^2y}\sqrt[3]{6x^2y^2}$. **Solution** $\sqrt[3]{4x^2y}\sqrt[3]{6x^2y^2} = \sqrt[3]{24x^4y^3}$ $= \sqrt[3]{8x^3y^3}\sqrt[3]{3x}$ $= 2xy\sqrt[3]{3x}$	Problems 39–42
Use the distributive property to multiply radical expressions. (Sec. 7.4, Obj. 2, p. 386)	The distributive property and the property $\sqrt[n]{b}\sqrt[n]{c} = \sqrt[n]{bc}$ are used to find products of radical expressions.	Multiply $\sqrt{2x}(\sqrt{6x} + \sqrt{18xy})$ and simplify where possible. **Solution** $\sqrt{2x}(\sqrt{6x} + \sqrt{18xy})$ $= \sqrt{12x^2} + \sqrt{36x^2y}$ $= \sqrt{4x^2}\sqrt{3} + \sqrt{36x^2}\sqrt{y}$ $= 2x\sqrt{3} + 6x\sqrt{y}$	Problems 43–48 *(continued)*

OBJECTIVE	SUMMARY	EXAMPLE	CHAPTER REVIEW PROBLEMS
Rationalize binomial denominators. (Sec. 7.4, Obj. 3, p. 388)	The factors $a - b$ and $a + b$ are called conjugates. To rationalize a binomial denominator, multiply by its conjugate.	Simplify $\frac{3}{\sqrt{7} - \sqrt{5}}$ by rationalizing the denominator. **Solution** $\frac{3}{\sqrt{7} - \sqrt{5}}$ $= \frac{3}{(\sqrt{7} - \sqrt{5})} \cdot \frac{(\sqrt{7} + \sqrt{5})}{(\sqrt{7} + \sqrt{5})}$ $= \frac{3(\sqrt{7} + \sqrt{5})}{\sqrt{49} - \sqrt{25}} = \frac{3(\sqrt{7} + \sqrt{5})}{7 - 5}$ $= \frac{3(\sqrt{7} + \sqrt{5})}{2}$	Problems 49–52
Solve radical equations. (Sec. 7.5, Obj. 1, p. 391)	Equations with variables in a radicand are called radical equations. Radical equations are solved by raising each side of the equation to the appropriate power. However, raising both sides of the equation to a power may produce extraneous roots. Therefore, you must check each potential solution.	Solve $\sqrt{x} + 20 = x$. **Solution** $\sqrt{x} + 20 = x$ $\sqrt{x} = x - 20$ Isolate the radical. $(\sqrt{x})^2 = (x - 20)^2$ $x = x^2 - 40x + 400$ $0 = x^2 - 41x + 400$ $0 = (x - 25)(x - 16)$ $x = 25$ or $x = 16$ Check $\sqrt{x} + 20 = x$ If $x = 25$: $\sqrt{25} + 20 = 25$, $25 = 25$ If $x = 16$: $\sqrt{16} + 20 = 25$, $24 \neq 25$ The solution set is $\{25\}$.	Problems 53–60
Apply solving radical equations to problems. (Sec. 7.5, Obj. 2, p. 395)	Various formulas involve radical equations. These formulas are solved in the same manner as radical equations.	Use the formula $\sqrt{30Df} = S$ (given in Section 7.5) to determine the coefficient of friction, to the nearest hundredth, if a car traveling at 50 miles per hour skidded 300 feet. **Solution** Solve $\sqrt{30Df} = S$ for f. $(\sqrt{30Df})^2 = S^2$ $30Df = S^2$ $f = \frac{S^2}{30D}$ Substituting the values for S and D gives $f = \frac{50^2}{30(300)} = 0.28$	Problems 61–62 *(continued)*

OBJECTIVE	SUMMARY	EXAMPLE	CHAPTER REVIEW PROBLEMS
Evaluate a number raised to a rational exponent. (Sec. 7.6, Obj. 1, p. 397)	If b is a real number, n is a positive integer greater than 1, and $\sqrt[n]{b}$ exists, then $b^{\frac{1}{n}} = \sqrt[n]{b}$. Thus $b^{\frac{1}{n}}$ denotes the nth root of b.	Simplify $16^{\frac{3}{2}}$. **Solution** $16^{\frac{3}{2}} = (16^{\frac{1}{2}})^3 = 4^3 = 64$	Problems 63–70
Write an expression with rational exponents as a radical. (Sec. 7.6, Obj. 2, p. 399)	If $\frac{m}{n}$ is a rational number, n is a positive integer greater than 1, and b is a real number such that $\sqrt[n]{b}$ exists, then $b^{\frac{m}{n}} = \sqrt[n]{b^m} = (\sqrt[n]{b})^m$.	Write $x^{\frac{3}{5}}$ in radical form. **Solution** $x^{\frac{3}{5}} = \sqrt[5]{x^3}$	Problems 71–74
Write radical expressions as expressions with rational exponents. (Sec. 7.6, Obj. 3, p. 399)	The index of the radical will be the denominator of the rational exponent.	Write $\sqrt[4]{x^3y}$ using positive rational exponents. **Solution** $\sqrt[4]{x^3y} = x^{\frac{3}{4}}y^{\frac{1}{4}}$	Problems 75–78
Simplify algebraic expressions that have rational exponents. (Sec. 7.6, Obj. 4, p. 400)	Properties of exponents are used to simplify products and quotients involving rational exponents.	Simplify $(4x^{\frac{1}{3}})(-3x^{-\frac{3}{4}})$ and express the result with positive exponents only. **Solution** $(4x^{\frac{1}{3}})(-3x^{-\frac{3}{4}}) = -12x^{\frac{1}{3}-\frac{3}{4}}$ $= -12x^{-\frac{5}{12}}$ $= \frac{-12}{x^{\frac{5}{12}}}$	Problems 79–84
Multiply and divide radicals with different indexes. (Sec. 7.6, Obj. 5, p. 400)	The link between rational exponents and roots provides the basis for multiplying and dividing radicals with different indexes.	Multiply $\sqrt[3]{y^2}\sqrt{y}$ and express in simplest radical form. **Solution** $\sqrt[3]{y^2}\sqrt{y} = y^{\frac{2}{3}}y^{\frac{1}{2}}$ $= y^{\frac{2}{3}+\frac{1}{2}} = y^{\frac{7}{6}}$ $= \sqrt[6]{y^7} = y\sqrt[6]{y}$	Problems 85–88
Write numbers in scientific notation. (Sec. 7.7, Obj. 1, p. 404)	Scientific notation is often used to write numbers that are very small or very large in magnitude. The scientific form of a number is expressed as $(N)(10)^k$, where the absolute value of N is a number greater than or equal to 1 and less than 10, written in decimal form, and k is an integer.	Write each of the following in scientific notation. **(a)** 0.000000843 **(b)** 456,000,000,000 **Solution** **(a)** $0.000000843 = (8.43)(10^{-7})$ **(b)** 456,000,000,000 $= (4.56)(10^{11})$	Problems 89–92 *(continued)*

OBJECTIVE	SUMMARY	EXAMPLE	CHAPTER REVIEW PROBLEMS
Convert numbers from scientific notation to ordinary decimal notation. (Sec. 7.7, Obj. 2, p. 405)	To switch from scientific notation to ordinary notation, move the decimal point the number of places indicated by the exponent of 10. The decimal point is moved to the right if the exponent is positive and to the left if the exponent is negative.	Write each of the following in ordinary decimal notation. **(a)** $(8.5)(10^{-5})$ **(b)** $(3.4)(10^{6})$ **Solution** **(a)** $(8.5)(10^{-5}) = 0.000085$ **(b)** $(3.4)(10^{6}) = 3{,}400{,}000$	Problems 93–96
Perform calculations with numbers using scientific notation. (Sec. 7.7, Obj. 3, p. 405)	Scientific notation can often be used to simplify numerical calculations.	Use scientific notation and the properties of exponents to simplify $\frac{0.0000084}{0.002}$. **Solution** Change the numbers to scientific notation and use the appropriate properties of exponents. Express the result in standard decimal notation. $\frac{0.0000084}{0.002} = \frac{(8.4)(10^{-6})}{(2)(10^{-3})}$ $= (4.2)(10^{-3})$ $= 0.0042$	Problems 97–104

Chapter 7 Review Problem Set

For Problems 1–6, evaluate the numerical expression.

1. 4^{-3}

2. $\left(\frac{2}{3}\right)^{-2}$

3. $(3^2 \cdot 3^{-3})^{-1}$

4. $(4^{-2} \cdot 4^2)^{-1}$

5. $\left(\frac{3^{-1}}{3^2}\right)^{-1}$

6. $\left(\frac{5^2}{5^{-1}}\right)^{-1}$

For Problems 7–18, simplify and express the final result using positive exponents.

7. $(x^{-3}y^4)^{-2}$

8. $\left(\frac{2a^{-1}}{3b^4}\right)^{-3}$

9. $\left(\frac{4a^{-2}}{3b^{-2}}\right)^{-2}$

10. $(5x^3y^{-2})^{-3}$

11. $\left(\frac{6x^{-2}}{2x^4}\right)^{-2}$

12. $\left(\frac{8y^2}{2y^{-1}}\right)^{-1}$

13. $(-5x^{-3})(2x^6)$

14. $(a^{-4}b^3)(3ab^2)$

15. $\frac{a^{-1}b^{-2}}{a^4b^{-5}}$

16. $\frac{x^3y^5}{x^{-1}y^6}$

17. $\frac{-12x^3}{6x^5}$

18. $\frac{10a^2b^3}{-5ab^4}$

For Problems 19–22, express as a single fraction involving positive exponents only.

19. $x^{-2} + y^{-1}$

20. $a^{-2} - 2a^{-1}b^{-1}$

21. $2x^{-1} + 3y^{-2}$

22. $(2x)^{-1} + 3y^{-2}$

For Problems 23–34, express the radical in simplest radical form. Assume the variables represent positive real numbers.

23. $\sqrt{54}$

24. $\sqrt{48x^3y}$

25. $\sqrt[3]{56}$

26. $\sqrt[3]{108x^4y^8}$

27. $\frac{3}{4}\sqrt{150}$

28. $\frac{2}{3}\sqrt{45xy^3}$

29. $\frac{4\sqrt{3}}{\sqrt{6}}$

30. $\sqrt{\frac{5}{12x^3}}$

31. $\frac{\sqrt[3]{2}}{\sqrt[3]{9}}$

32. $\sqrt{\frac{9}{5}}$

33. $\sqrt{\frac{3x^3}{7}}$

34. $\frac{\sqrt{8x^2}}{\sqrt{2x}}$

For Problems 35–38, use the distributive property to help simplify the expression.

35. $3\sqrt{45} - 2\sqrt{20} - \sqrt{80}$

36. $4\sqrt[3]{24} + 3\sqrt[3]{3} - 2\sqrt[3]{81}$

37. $3\sqrt{24} - \frac{2\sqrt{54}}{5} + \frac{\sqrt{96}}{4}$

38. $-2\sqrt{12x} + 3\sqrt{27x} - 5\sqrt{48x}$

For Problems 39–48, multiply and simplify. Assume the variables represent nonnegative real numbers.

39. $(3\sqrt{8})(4\sqrt{5})$

40. $(5\sqrt[3]{2})(6\sqrt[3]{4})$

41. $(\sqrt{6xy})(\sqrt{10x})$

42. $(-3\sqrt{6xy^3})(\sqrt{6y})$

43. $3\sqrt{2}(4\sqrt{6} - 2\sqrt{7})$

44. $(\sqrt{x} + 3)(\sqrt{x} - 5)$

45. $(2\sqrt{5} - \sqrt{3})(2\sqrt{5} + \sqrt{3})$

46. $(3\sqrt{2} + \sqrt{6})(5\sqrt{2} - 3\sqrt{6})$

47. $(2\sqrt{a} + \sqrt{b})(3\sqrt{a} - 4\sqrt{b})$

48. $(4\sqrt{8} - \sqrt{2})(\sqrt{8} + 3\sqrt{2})$

For Problems 49–52, rationalize the denominator and simplify.

49. $\frac{4}{\sqrt{7} - 1}$ **50.** $\frac{\sqrt{3}}{\sqrt{8} + \sqrt{5}}$

51. $\frac{3}{2\sqrt{3} + 3\sqrt{5}}$ **52.** $\frac{3\sqrt{2}}{2\sqrt{6} - \sqrt{10}}$

For Problems 53–60, solve the equation.

53. $\sqrt{7x - 3} = 4$ **54.** $\sqrt{2y + 1} = \sqrt{5y - 11}$

55. $\sqrt{2x} = x - 4$ **56.** $\sqrt{n^2 - 4n - 4} = n$

57. $\sqrt[3]{2x - 1} = 3$ **58.** $\sqrt{t^2 + 9t - 1} = 3$

59. $\sqrt{x^2 + 3x - 6} = x$ **60.** $\sqrt{x + 1} - \sqrt{2x} = -1$

61. The formula $S = \sqrt{30Df}$ is used to approximate the speed S, where D represents the length of the skid marks in feet and f represents the coefficient of friction for the road surface. Suppose that the coefficient of friction is 0.38. How far will a car skid, to the nearest foot, when the brakes are applied at 75 miles per hour?

62. The formula $T = 2\pi\sqrt{\frac{L}{32}}$ is used for pendulum motion, where T represents the period of the pendulum in seconds and L represents the length of the pendulum in feet. Find the length of a pendulum, to the nearest tenth of a foot, if the period is 2.4 seconds.

For Problems 63–70, simplify.

63. $4^{\frac{5}{2}}$ **64.** $(-1)^{\frac{2}{3}}$

65. $\left(\frac{8}{27}\right)^{\frac{2}{3}}$ **66.** $-16^{\frac{3}{2}}$

67. $(27)^{-\frac{2}{3}}$ **68.** $(32)^{-\frac{2}{5}}$

69. $9^{\frac{3}{2}}$ **70.** $16^{\frac{3}{4}}$

For Problems 71–74, write the expression in radical form.

71. $x^{\frac{1}{3}}y^{\frac{2}{3}}$ **72.** $a^{\frac{3}{4}}$

73. $4y^{\frac{1}{2}}$ **74.** $(x + 5y)^{\frac{2}{3}}$

For Problems 75–78, write the expression using positive rational exponents.

75. $\sqrt[5]{x^3y}$ **76.** $\sqrt[3]{4a^2}$

77. $6\sqrt[4]{y^2}$ **78.** $\sqrt[3]{(3a + b)^5}$

For Problems 79–84, simplify and express the final result using positive exponents.

79. $(4x^{\frac{1}{2}})(5x^{\frac{1}{5}})$ **80.** $\frac{42a^{\frac{3}{4}}}{6a^{\frac{1}{3}}}$

81. $\left(\frac{x^3}{y^4}\right)^{-\frac{1}{3}}$ **82.** $(-3a^{\frac{1}{4}})(2a^{-\frac{1}{2}})$

83. $(x^{\frac{4}{5}})^{-\frac{1}{2}}$ **84.** $\frac{-24y^{\frac{2}{3}}}{4y^{\frac{1}{4}}}$

For Problems 85–88, perform the indicated operation and express the answer in simplest radical form.

85. $\sqrt[4]{3}\sqrt{3}$ **86.** $\sqrt[3]{9}\sqrt{3}$

87. $\frac{\sqrt[3]{5}}{\sqrt[4]{5}}$ **88.** $\frac{\sqrt[3]{16}}{\sqrt{2}}$

For Problems 89–92, write the number in scientific notation.

89. 540,000,000 **90.** 84,000

91. 0.000000032 **92.** 0.000768

For Problems 93–96, write the number in ordinary decimal notation.

93. $(1.4)(10^{-6})$ **94.** $(6.38)(10^{-4})$

95. $(4.12)(10^{7})$ **96.** $(1.25)(10^{5})$

For Problems 97–104, use scientific notation and the properties of exponents to help perform the calculation.

97. (0.00002)(0.0003) **98.** (120,000)(300,000)

99. (0.000015)(400,000) **100.** $\frac{0.000045}{0.0003}$

101. $\frac{(0.00042)(0.0004)}{0.006}$ **102.** $\sqrt{0.000004}$

103. $\sqrt[3]{0.000000008}$ **104.** $(4{,}000{,}000)^{\frac{3}{2}}$

Chapter 7 Test

1. ______
2. ______
3. ______
4. ______

For Problems 1–4, simplify each of the numerical expressions.

1. $(4)^{-\frac{5}{2}}$
2. $-16^{\frac{5}{4}}$
3. $\left(\frac{2}{3}\right)^{-4}$
4. $\left(\frac{2^{-1}}{2^{-2}}\right)^{-2}$

5. ______
6. ______
7. ______
8. ______
9. ______

For Problems 5–9, express in simplest radical form. Assume the variables represent positive real numbers.

5. $\sqrt{63}$
6. $\sqrt[3]{108}$
7. $\sqrt{52x^4y^3}$
8. $\frac{5\sqrt{18}}{3\sqrt{12}}$
9. $\sqrt{\frac{7}{24x^3}}$

10. ______
11. ______
12. ______
13. ______
14. ______
15. ______
16. ______
17. ______
18. ______

10. Multiply and simplify: $(4\sqrt{6})(3\sqrt{12})$
11. Multiply and simplify: $(3\sqrt{2} + \sqrt{3})(\sqrt{2} - 2\sqrt{3})$
12. Simplify by combining similar radicals: $2\sqrt{50} - 4\sqrt{18} - 9\sqrt{32}$
13. Rationalize the denominator and simplify: $\frac{3\sqrt{2}}{4\sqrt{3} - \sqrt{8}}$
14. Simplify and express the answer using positive exponents: $\left(\frac{2x^{-1}}{3y}\right)^{-2}$
15. Simplify and express the answer using positive exponents: $\frac{-84a^{\frac{1}{2}}}{7a^{\frac{4}{5}}}$
16. Express $x^{-1} + y^{-3}$ as a single fraction involving positive exponents.
17. Multiply and express the answer using positive exponents: $(3x^{-\frac{1}{2}})(4x^{\frac{3}{4}})$
18. Multiply and simplify: $(3\sqrt{5} - 2\sqrt{3})(3\sqrt{5} + 2\sqrt{3})$

19. ______
20. ______

For Problems 19 and 20, use scientific notation and the properties of exponents to help with the calculations.

19. $\frac{(0.00004)(300)}{0.00002}$
20. $\sqrt{0.000009}$

21. ______
22. ______
23. ______
24. ______
25. ______

For Problems 21–25, solve the equation.

21. $\sqrt{3x + 1} = 3$
22. $\sqrt[3]{3x + 2} = 2$
23. $\sqrt{x} = x - 2$
24. $\sqrt{5x - 2} = \sqrt{3x + 8}$
25. $\sqrt{x^2 - 10x + 28} = 2$

Quadratic Equations and Inequalities

Jeff Greenberg/PhotoEdit

■ *The Pythagorean theorem is applied throughout the construction industry when right angles are involved.*

A page for a magazine contains 70 square inches of type. The height of the page is twice the width. If the margin around the type is 2 inches uniformly, what are the dimensions of a page? We can use the quadratic equation $(x - 4)(2x - 4) = 70$ to determine that the page measures 9 inches by 18 inches.

Solving equations is one of the central themes of this text. Let's pause for a moment and reflect on the different types of equations that we have solved in the previous seven chapters.

As the chart on the next page shows, we have solved second-degree equations in one variable, but only those for which the polynomial is factorable. In this chapter we will expand our work to include more general types of second-degree equations as well as inequalities in one variable.

Video tutorials for all section learning objectives are available in a variety of delivery modes.

INTERNET PROJECT

For the Internet Project in Chapter 6, you determined the approximate value of the golden ratio. Two quantities are in the golden ratio if the ratio between the sum of the two quantities and the larger quantity is the same as the ratio between the larger quantity and the smaller quantity. Conduct an Internet search to find the quadratic equation whose solution yields the exact value of the golden ratio. To find the exact value of the golden ratio, solve the quadratic equation using one of the methods we present in this chapter.

Type of equation	Examples
First-degree equations in one variable	$3x + 2x = x - 4$; $5(x + 4) = 12$; $\frac{x + 2}{3} + \frac{x - 1}{4} = 2$
Second-degree equations in one variable *that are factorable*	$x^2 + 5x = 0$; $x^2 + 5x + 6 = 0$; $x^2 - 9 = 0$; $x^2 - 10x + 25 = 0$
Fractional equations	$\frac{2}{x} + \frac{3}{x} = 4$; $\frac{5}{a - 1} = \frac{6}{a - 2}$; $\frac{2}{x^2 - 9} + \frac{3}{x + 3} = \frac{4}{x - 3}$
Radical equations	$\sqrt{x} = 2$; $\sqrt{3x - 2} = 5$; $\sqrt{5y + 1} = \sqrt{3y + 4}$

8.1 Complex Numbers

OBJECTIVES

1. Know About the Set of Complex Numbers
2. Add and Subtract Complex Numbers
3. Simplify Radicals Involving Negative Numbers
4. Perform Operations on Radicals Involving Negative Numbers
5. Multiply Complex Numbers
6. Divide Complex Numbers

1 Know About the Set of Complex Numbers

Because the square of any real number is nonnegative, a simple equation such as $x^2 = -4$ has no solutions in the set of real numbers. To handle this situation, we can expand the set of real numbers into a larger set called the **complex numbers**. In this section we will instruct you on how to manipulate complex numbers.

To provide a solution for the equation $x^2 + 1 = 0$, we use the number i, where

$$i^2 = -1$$

The number i is not a real number and is often called the **imaginary unit**, but the number i^2 is the real number -1. The imaginary unit i is used to define a complex number as follows:

Definition 8.1

A **complex number** is any number that can be expressed in the form

$$a + bi$$

where a and b are real numbers.

The form $a + bi$ is called the **standard form** of a complex number. The real number a is called the **real part** of the complex number, and b is called the **imaginary part**. (Note that b is a real number even though it is called the imaginary part.) The following list exemplifies this terminology.

1. The number $7 + 5i$ is a complex number that has a real part of 7 and an imaginary part of 5.
2. The number $\frac{2}{3} + i\sqrt{2}$ is a complex number that has a real part of $\frac{2}{3}$ and an imaginary part of $\sqrt{2}$. (It is easy to mistake $\sqrt{2i}$ for $\sqrt{2}i$. Thus we customarily write $i\sqrt{2}$ instead of $\sqrt{2}i$ to avoid any difficulties with the radical sign.)
3. The number $-4 - 3i$ can be written in the standard form $-4 + (-3i)$ and therefore is a complex number that has a real part of -4 and an imaginary part of -3. [The form $-4 - 3i$ is often used, but we know that it means $-4 + (-3i)$.]
4. The number $-9i$ can be written as $0 + (-9i)$; thus it is a complex number that has a real part of 0 and an imaginary part of -9. (Complex numbers, such as $-9i$, for which $a = 0$ and $b \neq 0$ are called **pure imaginary numbers**.)
5. The real number 4 can be written as $4 + 0i$ and is thus a complex number that has a real part of 4 and an imaginary part of 0.

Look at item 5 in this list. We see that the set of real numbers is a subset of the set of complex numbers. The following diagram indicates the organizational format of the complex numbers.

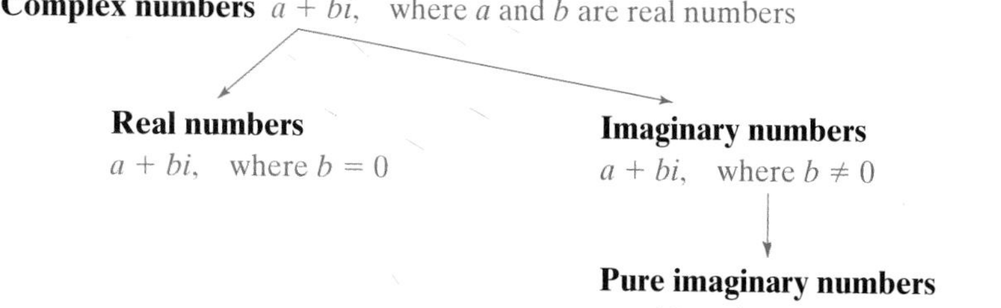

Two complex numbers $a + bi$ and $c + di$ are said to be **equal** if and only if $a = c$ and $b = d$.

2 Add and Subtract Complex Numbers

To **add complex numbers**, we simply add their real parts and add their imaginary parts. Thus

$$(a + bi) + (c + di) = (a + c) + (b + d)i$$

The following example shows addition of two complex numbers.

EXAMPLE 1

Add the complex numbers.

(a) $(4 + 3i) + (5 + 9i)$ **(b)** $(-6 + 4i) + (8 - 7i)$

(c) $\left(\frac{1}{2} + \frac{3}{4}i\right) + \left(\frac{2}{3} + \frac{1}{5}i\right)$

Solution

(a) $(4 + 3i) + (5 + 9i) = (4 + 5) + (3 + 9)i = 9 + 12i$

(b) $(-6 + 4i) + (8 - 7i) = (-6 + 8) + (4 - 7)i = 2 - 3i$

(c) $$\left(\frac{1}{2} + \frac{3}{4}i\right) + \left(\frac{2}{3} + \frac{1}{5}i\right) = \left(\frac{1}{2} + \frac{2}{3}\right) + \left(\frac{3}{4} + \frac{1}{5}\right)i$$
$$= \left(\frac{3}{6} + \frac{4}{6}\right) + \left(\frac{15}{20} + \frac{4}{20}\right)i$$
$$= \frac{7}{6} + \frac{19}{20}i$$

▼ **PRACTICE YOUR SKILL**

Add the complex numbers.

(a) $(6 + 2i) + (4 + 5i)$ **(b)** $(5 - 3i) + (-9 - 2i)$

(c) $\left(\frac{1}{4} + \frac{2}{3}i\right) + \left(\frac{1}{2} + \frac{3}{4}i\right)$ ■

The set of complex numbers is *closed* with respect to addition; that is, the sum of two complex numbers is a complex number. Furthermore, the commutative and associative properties of addition hold for all complex numbers. The addition identity element is $0 + 0i$ (or simply the real number 0). The additive inverse of $a + bi$ is $-a - bi$, because

$$(a + bi) + (-a - bi) = 0$$

To **subtract complex numbers**, for example, $c + di$ from $a + bi$, add the additive inverse of $c + di$. Thus

$$(a + bi) - (c + di) = (a + bi) + (-c - di)$$
$$= (a - c) + (b - d)i$$

In other words, we subtract the real parts and subtract the imaginary parts, as in the next examples.

1. $(9 + 8i) - (5 + 3i) = (9 - 5) + (8 - 3)i$
$= 4 + 5i$

2. $(3 - 2i) - (4 - 10i) = (3 - 4) + (-2 - (-10))i$
$= -1 + 8i$

3 Simplify Radicals Involving Negative Numbers

Because $i^2 = -1$, i is a square root of -1, so we let $i = \sqrt{-1}$. It should also be evident that $-i$ is a square root of -1, because

$$(-i)^2 = (-i)(-i) = i^2 = -1$$

Thus, in the set of complex numbers, -1 has two square roots, i and $-i$. We express these symbolically as

$$\sqrt{-1} = i \qquad \text{and} \qquad -\sqrt{-1} = -i$$

Let us extend our definition so that in the set of complex numbers every negative real number has two square roots. We simply define $\sqrt{-b}$, where b is a positive real number, to be the number whose square is $-b$. Thus

$$(\sqrt{-b})^2 = -b \quad \text{for } b > 0$$

Furthermore, because $(i\sqrt{b})(i\sqrt{b}) = i^2(b) = -1(b) = -b$, we see that

$$\sqrt{-b} = i\sqrt{b}$$

In other words, a square root of any negative real number can be represented as the product of a real number and the imaginary unit i. Consider the following example.

EXAMPLE 2

Simplify each of the following.

(a) $\sqrt{-4}$ **(b)** $\sqrt{-17}$ **(c)** $\sqrt{-24}$

Solution

(a) $\sqrt{-4} = i\sqrt{4} = 2i$

(b) $\sqrt{-17} = i\sqrt{17}$

(c) $\sqrt{-24} = i\sqrt{24} = i\sqrt{4}\sqrt{6} = 2i\sqrt{6}$ Note that we simplified the radical $\sqrt{24}$ to $2\sqrt{6}$

▼ **PRACTICE YOUR SKILL**

Simplify each of the following.

(a) $\sqrt{-25}$ **(b)** $\sqrt{-13}$ **(c)** $\sqrt{-18}$ ■

We should also observe that $-\sqrt{-b}$, where $b > 0$, is a square root of $-b$ because

$$(-\sqrt{-b})^2 = (-i\sqrt{b})^2 = i^2(b) = -1(b) = -b$$

Thus in the set of complex numbers, $-b$ (where $b > 0$) has two square roots, $i\sqrt{b}$ and $-i\sqrt{b}$. We express these symbolically as

$$\sqrt{-b} = i\sqrt{b} \qquad \text{and} \qquad -\sqrt{-b} = -i\sqrt{b}$$

We must be very careful with the use of the symbol $\sqrt{-b}$, where $b > 0$. Some real number properties that involve the square root symbol do not hold if the square root symbol does not represent a real number. For example, $\sqrt{a}\sqrt{b} = \sqrt{ab}$ does not hold if a and b are both negative numbers.

Correct $\sqrt{-4}\sqrt{-9} = (2i)(3i) = 6i^2 = 6(-1) = -6$

Incorrect $\sqrt{-4}\sqrt{-9} = \sqrt{(-4)(-9)} = \sqrt{36} = 6$

4 Perform Operations on Radicals Involving Negative Numbers

To avoid difficulty with this idea, you should rewrite all expressions of the form $\sqrt{-b}$, where $b > 0$, in the form $i\sqrt{b}$ before doing any computations. The following example further demonstrates this point.

EXAMPLE 3

Simplify each of the following.

(a) $\sqrt{-6}\sqrt{-8}$ **(b)** $\sqrt{-2}\sqrt{-8}$ **(c)** $\dfrac{\sqrt{-75}}{\sqrt{-3}}$ **(d)** $\dfrac{\sqrt{-48}}{\sqrt{12}}$

Solution

(a) $\sqrt{-6}\sqrt{-8} = (i\sqrt{6})(i\sqrt{8}) = i^2\sqrt{48} = (-1)\sqrt{16}\sqrt{3} = -4\sqrt{3}$

(b) $\sqrt{-2}\sqrt{-8} = (i\sqrt{2})(i\sqrt{8}) = i^2\sqrt{16} = (-1)(4) = -4$

(c) $\dfrac{\sqrt{-75}}{\sqrt{-3}} = \dfrac{i\sqrt{75}}{i\sqrt{3}} = \dfrac{\sqrt{75}}{\sqrt{3}} = \sqrt{\dfrac{75}{3}} = \sqrt{25} = 5$

(d) $\dfrac{\sqrt{-48}}{\sqrt{12}} = \dfrac{i\sqrt{48}}{\sqrt{12}} = i\sqrt{\dfrac{48}{12}} = i\sqrt{4} = 2i$

▼ **PRACTICE YOUR SKILL**

Simplify each of the following.

(a) $\sqrt{-3}\sqrt{-12}$ **(b)** $\sqrt{-10}\sqrt{-2}$ **(c)** $\dfrac{\sqrt{-98}}{\sqrt{-2}}$ **(d)** $\dfrac{\sqrt{-108}}{\sqrt{3}}$ ■

5 Multiply Complex Numbers

Complex numbers have a binomial form, so we find the product of two complex numbers in the same way that we find the product of two binomials. Then, by replacing i^2 with -1, we are able to simplify and express the final result in standard form. Consider the following example.

EXAMPLE 4

Find the product of each of the following.

(a) $(2 + 3i)(4 + 5i)$ **(b)** $(-3 + 6i)(2 - 4i)$ **(c)** $(1 - 7i)^2$

(d) $(2 + 3i)(2 - 3i)$

Solution

(a) $(2 + 3i)(4 + 5i) = 2(4 + 5i) + 3i(4 + 5i)$

$= 8 + 10i + 12i + 15i^2$

$= 8 + 22i + 15i^2$

$= 8 + 22i + 15(-1) = -7 + 22i$

(b) $(-3 + 6i)(2 - 4i) = -3(2 - 4i) + 6i(2 - 4i)$

$= -6 + 12i + 12i - 24i^2$

$= -6 + 24i - 24(-1)$

$= -6 + 24i + 24 = 18 + 24i$

(c) $(1 - 7i)^2 = (1 - 7i)(1 - 7i)$

$= 1(1 - 7i) - 7i(1 - 7i)$

$= 1 - 7i - 7i + 49i^2$

$$= 1 - 14i + 49(-1)$$
$$= 1 - 14i - 49$$
$$= -48 - 14i$$

(d) $(2 + 3i)(2 - 3i) = 2(2 - 3i) + 3i(2 - 3i)$
$$= 4 - 6i + 6i - 9i^2$$
$$= 4 - 9(-1)$$
$$= 4 + 9$$
$$= 13$$

▼ PRACTICE YOUR SKILL

Find the product of each of the following.

(a) $(4 + 3i)(6 + i)$ **(b)** $(-2 + 5i)(1 - 3i)$ **(c)** $(6 - 2i)^2$

(d) $(5 + 2i)(5 - 2i)$ ■

Example 4(d) illustrates an important situation: The complex numbers $2 + 3i$ and $2 - 3i$ are conjugates of each other. In general, two complex numbers $a + bi$ and $a - bi$ are called **conjugates** of each other. *The product of a complex number and its conjugate is always a real number*, which can be shown as follows:

$$(a + bi)(a - bi) = a(a - bi) + bi(a - bi)$$
$$= a^2 - abi + abi - b^2i^2$$
$$= a^2 - b^2(-1)$$
$$= a^2 + b^2$$

6 Divide Complex Numbers

We use conjugates to simplify expressions, such as $\frac{3i}{5 + 2i}$, that indicate the quotient of two complex numbers. To eliminate i in the denominator and change the indicated quotient to the standard form of a complex number, we can multiply both the numerator and the denominator by the conjugate of the denominator as follows:

$$\frac{3i}{5 + 2i} = \frac{3i(5 - 2i)}{(5 + 2i)(5 - 2i)}$$
$$= \frac{15i - 6i^2}{25 - 4i^2}$$
$$= \frac{15i - 6(-1)}{25 - 4(-1)}$$
$$= \frac{15i + 6}{29}$$
$$= \frac{6}{29} + \frac{15}{29}i$$

The following example further clarifies the process of dividing complex numbers.

EXAMPLE 5

Find the quotient of each of the following.

(a) $\dfrac{2-3i}{4-7i}$ **(b)** $\dfrac{4-5i}{2i}$

Solution

(a)
$$\frac{2-3i}{4-7i} = \frac{(2-3i)(4+7i)}{(4-7i)(4+7i)} \qquad 4+7i \text{ is the conjugate of } 4-7i$$
$$= \frac{8+14i-12i-21i^2}{16-49i^2}$$
$$= \frac{8+2i-21(-1)}{16-49(-1)}$$
$$= \frac{8+2i+21}{16+49}$$
$$= \frac{29+2i}{65}$$
$$= \frac{29}{65} + \frac{2}{65}i$$

(b)
$$\frac{4-5i}{2i} = \frac{(4-5i)(-2i)}{(2i)(-2i)} \qquad -2i \text{ is the conjugate of } 2i$$
$$= \frac{-8i+10i^2}{-4i^2}$$
$$= \frac{-8i+10(-1)}{-4(-1)}$$
$$= \frac{-8i-10}{4}$$
$$= -\frac{5}{2} - 2i$$

▼ **PRACTICE YOUR SKILL**

Find the quotient of each of the following.

(a) $\dfrac{3+5i}{6+i}$ **(b)** $\dfrac{5-8i}{3i}$ ■

In Example 5(b), where the denominator is a pure imaginary number, we can change to standard form by choosing a multiplier other than the conjugate. Consider the following alternative approach for Example 5(b).

$$\frac{4-5i}{2i} = \frac{(4-5i)(i)}{(2i)(i)}$$
$$= \frac{4i-5i^2}{2i^2}$$
$$= \frac{4i-5(-1)}{2(-1)}$$
$$= \frac{4i+5}{-2}$$
$$= -\frac{5}{2} - 2i$$

CONCEPT QUIZ

For Problems 1–10, answer true or false.

1. The number i is a real number and is called the imaginary unit.
2. The number $4 + 2i$ is a complex number that has a real part of 4.
3. The number $-3 - 5i$ is a complex number that has an imaginary part of 5.
4. Complex numbers that have a real part of 0 are called pure imaginary numbers.
5. The set of real numbers is a subset of the set of complex numbers.
6. Any real number x can be written as the complex number $x + 0i$.
7. By definition, i^2 is equal to -1.
8. The complex numbers $-2 + 5i$ and $2 - 5i$ are conjugates.
9. The product of two complex numbers is never a real number.
10. In the set of complex numbers, -16 has two square roots.

Problem Set 8.1

1 Know About the Set of Complex Numbers

For Problems 1–8, label each statement true or false.

1. Every complex number is a real number.
2. Every real number is a complex number.
3. The real part of the complex number $6i$ is 0.
4. Every complex number is a pure imaginary number.
5. The sum of two complex numbers is always a complex number.
6. The imaginary part of the complex number 7 is 0.
7. The sum of two complex numbers is sometimes a real number.
8. The sum of two pure imaginary numbers is always a pure imaginary number.

2 Add and Subtract Complex Numbers

For Problems 9–26, add or subtract as indicated.

9. $(6 + 3i) + (4 + 5i)$
10. $(5 + 2i) + (7 + 10i)$
11. $(-8 + 4i) + (2 + 6i)$
12. $(5 - 8i) + (-7 + 2i)$
13. $(3 + 2i) - (5 + 7i)$
14. $(1 + 3i) - (4 + 9i)$
15. $(-7 + 3i) - (5 - 2i)$
16. $(-8 + 4i) - (9 - 4i)$
17. $(-3 - 10i) + (2 - 13i)$
18. $(-4 - 12i) + (-3 + 16i)$
19. $(4 - 8i) - (8 - 3i)$
20. $(12 - 9i) - (14 - 6i)$
21. $(-1 - i) - (-2 - 4i)$
22. $(-2 - 3i) - (-4 - 14i)$
23. $\left(\frac{3}{2} + \frac{1}{3}i\right) + \left(\frac{1}{6} - \frac{3}{4}i\right)$
24. $\left(\frac{2}{3} - \frac{1}{5}i\right) + \left(\frac{3}{5} - \frac{3}{4}i\right)$
25. $\left(-\frac{5}{9} + \frac{3}{5}i\right) - \left(\frac{4}{3} - \frac{1}{6}i\right)$
26. $\left(\frac{3}{8} - \frac{5}{2}i\right) - \left(\frac{5}{6} + \frac{1}{7}i\right)$

3 Simplify Radicals Involving Negative Numbers

For Problems 27–42, write each of the following in terms of i and simplify. For example,

$$\sqrt{-20} = i\sqrt{20} = i\sqrt{4}\sqrt{5} = 2i\sqrt{5}$$

27. $\sqrt{-81}$
28. $\sqrt{-49}$
29. $\sqrt{-14}$
30. $\sqrt{-33}$
31. $\sqrt{-\frac{16}{25}}$
32. $\sqrt{-\frac{64}{36}}$
33. $\sqrt{-18}$
34. $\sqrt{-84}$
35. $\sqrt{-75}$
36. $\sqrt{-63}$
37. $3\sqrt{-28}$
38. $5\sqrt{-72}$
39. $-2\sqrt{-80}$
40. $-6\sqrt{-27}$
41. $12\sqrt{-90}$
42. $9\sqrt{-40}$

4 Perform Operations on Radicals Involving Negative Numbers

For Problems 43–60, write each of the following in terms of i, perform the indicated operations, and simplify. For example,

$$\begin{aligned}\sqrt{-3}\sqrt{-8} &= (i\sqrt{3})(i\sqrt{8})\\ &= i^2\sqrt{24}\\ &= (-1)\sqrt{4}\sqrt{6}\\ &= -2\sqrt{6}\end{aligned}$$

43. $\sqrt{-4}\sqrt{-16}$
44. $\sqrt{-81}\sqrt{-25}$
45. $\sqrt{-3}\sqrt{-5}$
46. $\sqrt{-7}\sqrt{-10}$
47. $\sqrt{-9}\sqrt{-6}$
48. $\sqrt{-8}\sqrt{-16}$
49. $\sqrt{-15}\sqrt{-5}$
50. $\sqrt{-2}\sqrt{-20}$
51. $\sqrt{-2}\sqrt{-27}$
52. $\sqrt{-3}\sqrt{-15}$

53. $\sqrt{6}\sqrt{-8}$

54. $\sqrt{-75}\sqrt{3}$

55. $\frac{\sqrt{-25}}{\sqrt{-4}}$

56. $\frac{\sqrt{-81}}{\sqrt{-9}}$

57. $\frac{\sqrt{-56}}{\sqrt{-7}}$

58. $\frac{\sqrt{-72}}{\sqrt{-6}}$

59. $\frac{\sqrt{-24}}{\sqrt{6}}$

60. $\frac{\sqrt{-96}}{\sqrt{2}}$

5 Multiply Complex Numbers

For Problems 61–84, find each of the products and express the answers in the standard form of a complex number.

61. $(5i)(4i)$

62. $(-6i)(9i)$

63. $(7i)(-6i)$

64. $(-5i)(-12i)$

65. $(3i)(2 - 5i)$

66. $(7i)(-9 + 3i)$

67. $(-6i)(-2 - 7i)$

68. $(-9i)(-4 - 5i)$

69. $(3 + 2i)(5 + 4i)$

70. $(4 + 3i)(6 + i)$

71. $(6 - 2i)(7 - i)$

72. $(8 - 4i)(7 - 2i)$

73. $(-3 - 2i)(5 + 6i)$

74. $(-5 - 3i)(2 - 4i)$

75. $(9 + 6i)(-1 - i)$

76. $(10 + 2i)(-2 - i)$

77. $(4 + 5i)^2$

78. $(5 - 3i)^2$

79. $(-2 - 4i)^2$

80. $(-3 - 6i)^2$

81. $(6 + 7i)(6 - 7i)$

82. $(5 - 7i)(5 + 7i)$

83. $(-1 + 2i)(-1 - 2i)$

84. $(-2 - 4i)(-2 + 4i)$

6 Divide Complex Numbers

For Problems 85–100, find each of the following quotients and express the answers in the standard form of a complex number.

85. $\frac{3i}{2 + 4i}$

86. $\frac{4i}{5 + 2i}$

87. $\frac{-2i}{3 - 5i}$

88. $\frac{-5i}{2 - 4i}$

89. $\frac{-2 + 6i}{3i}$

90. $\frac{-4 - 7i}{6i}$

91. $\frac{2}{7i}$

92. $\frac{3}{10i}$

93. $\frac{2 + 6i}{1 + 7i}$

94. $\frac{5 + i}{2 + 9i}$

95. $\frac{3 + 6i}{4 - 5i}$

96. $\frac{7 - 3i}{4 - 3i}$

97. $\frac{-2 + 7i}{-1 + i}$

98. $\frac{-3 + 8i}{-2 + i}$

99. $\frac{-1 - 3i}{-2 - 10i}$

100. $\frac{-3 - 4i}{-4 - 11i}$

101. Some of the solution sets for quadratic equations in the next sections will contain complex numbers such as $(-4 + \sqrt{-12})/2$ and $(-4 - \sqrt{-12})/2$. We can simplify the first number as follows.

$$\frac{-4 + \sqrt{-12}}{2} = \frac{-4 + i\sqrt{12}}{2}$$
$$= \frac{-4 + 2i\sqrt{3}}{2} = \frac{\cancel{2}(-2 + i\sqrt{3})}{\cancel{2}}$$
$$= -2 + i\sqrt{3}$$

Simplify each of the following complex numbers.

(a) $\frac{-4 - \sqrt{-12}}{2}$

(b) $\frac{6 + \sqrt{-24}}{4}$

(c) $\frac{-1 - \sqrt{-18}}{2}$

(d) $\frac{-6 + \sqrt{-27}}{3}$

(e) $\frac{10 + \sqrt{-45}}{4}$

(f) $\frac{4 - \sqrt{-48}}{2}$

THOUGHTS INTO WORDS

102. Why is the set of real numbers a subset of the set of complex numbers?

103. Can the sum of two nonreal complex numbers be a real number? Defend your answer.

104. Can the product of two nonreal complex numbers be a real number? Defend your answer.

Answers to the Concept Quiz

1. False **2.** True **3.** False **4.** True **5.** True **6.** True **7.** True **8.** False **9.** False **10.** True

Answers to the Example Practice Skills

1. (a) $10 + 7i$ **(b)** $-4 - 5i$ **(c)** $\frac{3}{4} + \frac{17}{12}i$ **2. (a)** $5i$ **(b)** $i\sqrt{13}$ **(c)** $3i\sqrt{2}$ **3. (a)** -6 **(b)** $-2\sqrt{5}$ **(c)** 7 **(d)** $6i$ **4. (a)** $21 + 22i$ **(b)** $13 + 11i$ **(c)** $32 - 24i$ **(d)** 29 **5. (a)** $\frac{23 + 27i}{37}$ **(b)** $-\frac{8}{3} - \frac{5}{3}i$

8.2 Quadratic Equations

OBJECTIVES

1. Solve Quadratic Equations by Factoring
2. Solve Quadratic Equations of the Form $x^2 = a$
3. Solve Problems Involving Right Triangles and 30°-60° Triangles

1 Solve Quadratic Equations by Factoring

A second-degree equation in one variable contains the variable with an exponent of 2 but no higher power. Such equations are also called **quadratic equations**. The following are examples of quadratic equations.

$$x^2 = 36 \qquad y^2 + 4y = 0 \qquad x^2 + 5x - 2 = 0$$
$$3n^2 + 2n - 1 = 0 \qquad 5x^2 + x + 2 = 3x^2 - 2x - 1$$

A quadratic equation in the variable x can also be defined as any equation that can be written in the form

$$ax^2 + bx + c = 0$$

where a, b, and c are real numbers and $a \neq 0$. The form $ax^2 + bx + c = 0$ is called the **standard form** of a quadratic equation.

In previous chapters you solved quadratic equations (the term *quadratic* was not used at that time) by factoring and applying the following property: $ab = 0$ if and only if $a = 0$ or $b = 0$. Let's review a few such examples.

EXAMPLE 1

Solve $3n^2 + 14n - 5 = 0$.

Solution

$$3n^2 + 14n - 5 = 0$$
$$(3n - 1)(n + 5) = 0 \qquad \text{Factor the left side}$$
$$3n - 1 = 0 \quad \text{or} \quad n + 5 = 0 \qquad \text{Apply: } ab = 0 \text{ if and only if } a = 0 \text{ or } b = 0$$
$$3n = 1 \quad \text{or} \quad n = -5$$
$$n = \frac{1}{3} \quad \text{or} \quad n = -5$$

The solution set is $\left\{-5, \frac{1}{3}\right\}$.

▼ **PRACTICE YOUR SKILL**

Solve $2x^2 - 7x - 15 = 0$. ■

EXAMPLE 2 Solve $2\sqrt{x} = x - 8$.

Solution

$$2\sqrt{x} = x - 8$$

$$(2\sqrt{x})^2 = (x - 8)^2 \qquad \text{Square both sides}$$

$$4x = x^2 - 16x + 64$$

$$0 = x^2 - 20x + 64$$

$$0 = (x - 16)(x - 4) \qquad \text{Factor the right side}$$

$$x - 16 = 0 \quad \text{or} \quad x - 4 = 0 \qquad \text{Apply: } ab = 0 \text{ if and only if } a = 0 \text{ or } b = 0$$

$$x = 16 \quad \text{or} \quad x = 4$$

✔ **Check**

$$2\sqrt{x} = x - 8 \qquad\qquad 2\sqrt{x} = x - 8$$

$$2\sqrt{16} \stackrel{?}{=} 16 - 8 \quad \text{or} \quad 2\sqrt{4} \stackrel{?}{=} 4 - 8$$

$$2(4) \stackrel{?}{=} 8 \qquad\qquad 2(2) \stackrel{?}{=} -4$$

$$8 = 8 \qquad\qquad 4 \neq -4$$

The solution set is $\{16\}$.

▼ **PRACTICE YOUR SKILL**

Solve $\sqrt{x} = x - 6$. ■

We should make two comments about Example 2. First, remember that applying the property "if $a = b$, then $a^n = b^n$" might produce extraneous solutions. Therefore, we *must* check all potential solutions. Second, the equation $2\sqrt{x} = x - 8$ is said to be of **quadratic form** because it can be written as $2x^{\frac{1}{2}} = \left(x^{\frac{1}{2}}\right)^2 - 8$. More will be said about the phrase *quadratic form* later.

2 Solve Quadratic Equations of the Form $x^2 = a$

Let's consider quadratic equations of the form $x^2 = a$, where x is the variable and a is any real number. We can solve $x^2 = a$ as follows:

$$x^2 = a$$

$$x^2 - a = 0$$

$$x^2 - (\sqrt{a})^2 = 0 \qquad a = (\sqrt{a})^2$$

$$(x - \sqrt{a})(x + \sqrt{a}) = 0 \qquad \text{Factor the left side}$$

$$x - \sqrt{a} = 0 \quad \text{or} \quad x + \sqrt{a} = 0 \qquad \text{Apply: } ab = 0 \text{ if and only if } a = 0 \text{ or } b = 0$$

$$x = \sqrt{a} \quad \text{or} \quad x = -\sqrt{a}$$

The solutions are $\sqrt{a}$ and $-\sqrt{a}$. We can state this result as a general property and use it to solve certain types of quadratic equations.

Property 8.1

For any real number a,

$$x^2 = a \quad \text{if and only if } x = \sqrt{a} \text{ or } x = -\sqrt{a}$$

(The statement $x = \sqrt{a}$ or $x = -\sqrt{a}$ can be written as $x = \pm\sqrt{a}$.)

Property 8.1, along with our knowledge of square roots, makes it easy to solve quadratic equations of the form $x^2 = a$.

EXAMPLE 3

Solve $x^2 = 45$.

Solution

$$x^2 = 45$$

$$x = \pm\sqrt{45}$$

$$x = \pm 3\sqrt{5} \qquad \sqrt{45} = \sqrt{9}\sqrt{5} = 3\sqrt{5}$$

The solution set is $\{\pm 3\sqrt{5}\}$.

▼ PRACTICE YOUR SKILL

Solve $y^2 = 75$. ■

EXAMPLE 4

Solve $x^2 = -9$.

Solution

$$x^2 = -9$$

$$x = \pm\sqrt{-9}$$

$$x = \pm 3i \qquad \sqrt{-9} = i\sqrt{9} = 3i$$

Thus the solution set is $\{\pm 3i\}$.

▼ PRACTICE YOUR SKILL

Solve $a^2 = -36$. ■

EXAMPLE 5

Solve $7n^2 = 12$.

Solution

$$7n^2 = 12$$

$$n^2 = \frac{12}{7}$$

$$n = \pm\sqrt{\frac{12}{7}}$$

$$n = \pm\frac{2\sqrt{21}}{7} \qquad \sqrt{\frac{12}{7}} = \frac{\sqrt{12}}{\sqrt{7}} \cdot \frac{\sqrt{7}}{\sqrt{7}} = \frac{\sqrt{84}}{7} = \frac{\sqrt{4}\sqrt{21}}{7} = \frac{2\sqrt{21}}{7}$$

The solution set is $\left\{\pm\frac{2\sqrt{21}}{7}\right\}$.

▼ **PRACTICE YOUR SKILL**

Solve $5x^2 = 48$. ■

EXAMPLE 6 Solve $(3n + 1)^2 = 25$.

Solution

$$(3n + 1)^2 = 25$$
$$(3n + 1) = \pm\sqrt{25}$$
$$3n + 1 = \pm 5$$
$$3n + 1 = 5 \quad \text{or} \quad 3n + 1 = -5$$
$$3n = 4 \quad \text{or} \quad 3n = -6$$
$$n = \frac{4}{3} \quad \text{or} \quad n = -2$$

The solution set is $\left\{-2, \frac{4}{3}\right\}$.

▼ **PRACTICE YOUR SKILL**

Solve $(4x - 3)^2 = 81$. ■

EXAMPLE 7 Solve $(x - 3)^2 = -10$.

Solution

$$(x - 3)^2 = -10$$
$$x - 3 = \pm\sqrt{-10}$$
$$x - 3 = \pm i\sqrt{10}$$
$$x = 3 \pm i\sqrt{10}$$

Thus the solution set is $\{3 \pm i\sqrt{10}\}$.

▼ **PRACTICE YOUR SKILL**

Solve $(x + 8)^2 = -15$. ■

Remark: Take another look at the equations in Examples 4 and 7. We should immediately realize that the solution sets will consist only of nonreal complex numbers, because the square of any nonzero real number is positive.

Sometimes it may be necessary to change the form before we can apply Property 8.1. Let's consider one example to illustrate this idea.

EXAMPLE 8 Solve $3(2x - 3)^2 + 8 = 44$.

Solution

$$3(2x - 3)^2 + 8 = 44$$
$$3(2x - 3)^2 = 36$$
$$(2x - 3)^2 = 12$$

$$2x - 3 = \pm\sqrt{12}$$

$$2x - 3 = \pm 2\sqrt{3}$$

$$2x = 3 \pm 2\sqrt{3}$$

$$x = \frac{3 \pm 2\sqrt{3}}{2}$$

The solution set is $\left\{\frac{3 \pm 2\sqrt{3}}{2}\right\}$.

▼ PRACTICE YOUR SKILL

Solve $2(5x + 1)^2 + 6 = 62$. ■

3 Solve Problems Involving Right Triangles and 30°-60° Triangles

Our work with radicals, Property 8.1, and the Pythagorean theorem form a basis for solving a variety of problems that pertain to right triangles.

EXAMPLE 9

Apply Your Skill

A 50-foot rope hangs from the top of a flagpole. When pulled taut to its full length, the rope reaches a point on the ground 18 feet from the base of the pole. Find the height of the pole to the nearest tenth of a foot.

Solution

Let's make a sketch (Figure 8.1) and record the given information.

Use the Pythagorean theorem to solve for p as follows:

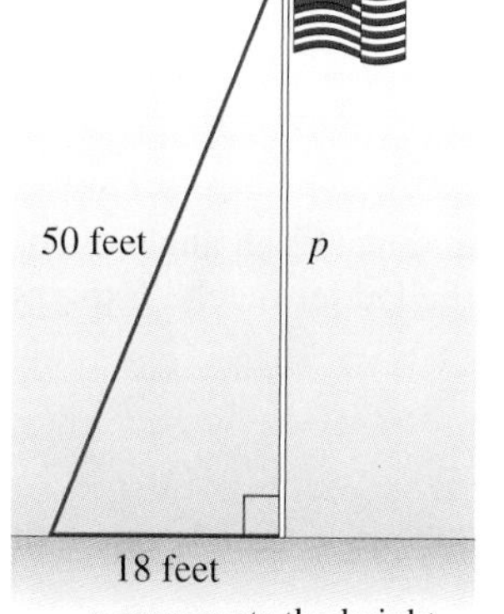

p represents the height of the flagpole.

Figure 8.1

$$p^2 + 18^2 = 50^2$$

$$p^2 + 324 = 2500$$

$$p^2 = 2176$$

$$p = \sqrt{2176} = 46.6 \quad \text{to the nearest tenth}$$

The height of the flagpole is approximately 46.6 feet.

▼ PRACTICE YOUR SKILL

A 120-foot guy-wire hangs from the top of a cell phone tower. When pulled taut the guy-wire reaches a point on the ground 90 feet from the base of the tower. Find the height of the tower to the nearest tenth of a foot. ■

There are two special kinds of right triangles that we use extensively in later mathematics courses. The first is the **isosceles right triangle**, which is a right triangle with both legs of the same length. Let's consider a problem that involves an isosceles right triangle.

EXAMPLE 10

Apply Your Skill

Find the length of each leg of an isosceles right triangle that has a hypotenuse of length 5 meters.

Solution

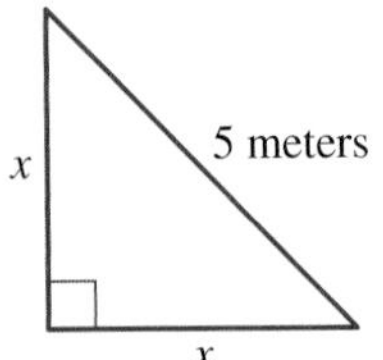

Figure 8.2

Let's sketch an isosceles right triangle and let x represent the length of each leg (Figure 8.2). Then we can apply the Pythagorean theorem.

$$x^2 + x^2 = 5^2$$

$$2x^2 = 25$$

$$x^2 = \frac{25}{2}$$

$$x = \pm\sqrt{\frac{25}{2}} = \pm\frac{5}{\sqrt{2}} = \pm\frac{5\sqrt{2}}{2}$$

Each leg is $\frac{5\sqrt{2}}{2}$ meters long.

▼ PRACTICE YOUR SKILL

Find the length of each leg of an isosceles right triangle that has a hypotenuse of length 8 inches. ■

Remark: In Example 9 we made no attempt to express $\sqrt{2176}$ in simplest radical form because the answer was to be given as a rational approximation to the nearest tenth. However, in Example 10 we left the final answer in radical form and therefore expressed it in simplest radical form.

The second special kind of right triangle that we use frequently is one that contains acute angles of 30° and 60°. In such a right triangle, which we refer to as a **30°-60° right triangle**, the side opposite the 30° angle is equal in length to one-half of the length of the hypotenuse. This relationship, along with the Pythagorean theorem, provides us with another problem-solving technique.

EXAMPLE 11

Apply Your Skill

Suppose that a 20-foot ladder is leaning against a building and makes an angle of 60° with the ground. How far up the building does the top of the ladder reach? Express your answer to the nearest tenth of a foot.

Solution

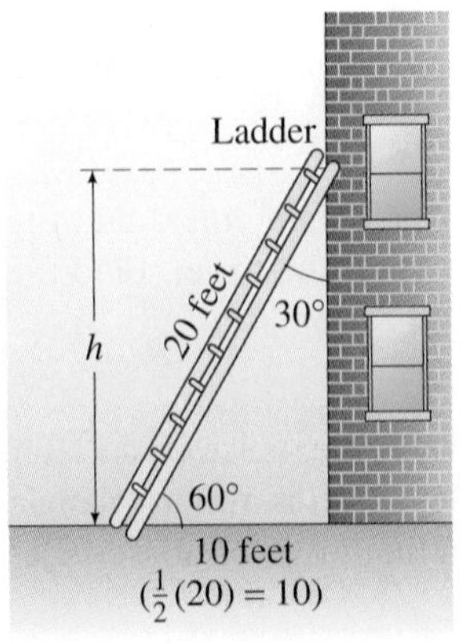

Figure 8.3

Figure 8.3 depicts this situation. The side opposite the 30° angle equals one-half of the hypotenuse, so it is of length $\frac{1}{2}(20) = 10$ feet. Now we can apply the Pythagorean theorem.

$$h^2 + 10^2 = 20^2$$

$$h^2 + 100 = 400$$

$$h^2 = 300$$

$$h = \sqrt{300} = 17.3 \quad \text{to the nearest tenth}$$

The top of the ladder touches the building at a point approximately 17.3 feet from the ground.

▼ PRACTICE YOUR SKILL

Suppose that a 16-foot ladder is leaning against a building and makes an angle of 60° with the ground. Will the ladder reach above a windowsill that is 13 feet above the ground? ■

CONCEPT QUIZ

For Problems 1–10, answer true or false.

1. The quadratic equation $-3x^2 + 5x - 8 = 0$ is in standard form.
2. The solution set of the equation $(x + 1)^2 = -25$ will consist only of nonreal complex numbers.
3. An isosceles right triangle is a right triangle that has a hypotenuse of the same length as one of the legs.
4. In a 30°-60° right triangle, the hypotenuse is equal in length to twice the length of the side opposite the 30° angle.
5. The equation $2x^2 + x^3 - x + 4 = 0$ is a quadratic equation.
6. The solution set for $4x^2 = 8x$ is $\{2\}$.
7. The solution set for $3x^2 = 8x$ is $\left\{0, \frac{8}{3}\right\}$.
8. The solution set for $x^2 - 8x - 48 = 0$ is $\{-12, 4\}$.
9. If the length of each leg of an isosceles right triangle is 4 inches, then the hypotenuse is of length $4\sqrt{2}$ inches.
10. If the length of the leg opposite the 30° angle in a right triangle is 6 centimeters, then the length of the other leg is 12 centimeters.

Problem Set 8.2

1 Solve Quadratic Equations by Factoring

For Problems 1–20, solve each of the quadratic equations by factoring and applying the property, $ab = 0$ if and only if $a = 0$ or $b = 0$. If necessary, return to Chapter 5 and review the factoring techniques presented there.

1. $x^2 - 9x = 0$
2. $x^2 + 5x = 0$
3. $x^2 = -3x$
4. $x^2 = 15x$
5. $3y^2 + 12y = 0$
6. $6y^2 - 24y = 0$
7. $5n^2 - 9n = 0$
8. $4n^2 + 13n = 0$
9. $x^2 + x - 30 = 0$
10. $x^2 - 8x - 48 = 0$
11. $x^2 - 19x + 84 = 0$
12. $x^2 - 21x + 104 = 0$
13. $2x^2 + 19x + 24 = 0$
14. $4x^2 + 29x + 30 = 0$
15. $15x^2 + 29x - 14 = 0$
16. $24x^2 + x - 10 = 0$
17. $25x^2 - 30x + 9 = 0$
18. $16x^2 - 8x + 1 = 0$
19. $6x^2 - 5x - 21 = 0$
20. $12x^2 - 4x - 5 = 0$

For Problems 21–26, solve each radical equation. Don't forget, you *must* check potential solutions.

21. $3\sqrt{x} = x + 2$
22. $3\sqrt{2x} = x + 4$
23. $\sqrt{2x} = x - 4$
24. $\sqrt{x} = x - 2$
25. $\sqrt{3x} + 6 = x$
26. $\sqrt{5x} + 10 = x$

2 Solve Quadratic Equations of the Form $x^2 = a$

For Problems 27–62, use Property 8.1 to help solve each quadratic equation.

27. $x^2 = 1$
28. $x^2 = 81$
29. $x^2 = -36$
30. $x^2 = -49$
31. $x^2 = 14$
32. $x^2 = 22$
33. $n^2 - 28 = 0$
34. $n^2 - 54 = 0$
35. $3t^2 = 54$
36. $4t^2 = 108$
37. $2t^2 = 7$
38. $3t^2 = 8$
39. $15y^2 = 20$
40. $14y^2 = 80$
41. $10x^2 + 48 = 0$
42. $12x^2 + 50 = 0$
43. $24x^2 = 36$
44. $12x^2 = 49$
45. $(x - 2)^2 = 9$
46. $(x + 1)^2 = 16$
47. $(x + 3)^2 = 25$
48. $(x - 2)^2 = 49$
49. $(x + 6)^2 = -4$
50. $(3x + 1)^2 = 9$
51. $(2x - 3)^2 = 1$
52. $(2x + 5)^2 = -4$
53. $(n - 4)^2 = 5$
54. $(n - 7)^2 = 6$

55. $(t + 5)^2 = 12$

56. $(t - 1)^2 = 18$

57. $(3y - 2)^2 = -27$

58. $(4y + 5)^2 = 80$

59. $3(x + 7)^2 + 4 = 79$

60. $2(x + 6)^2 - 9 = 63$

61. $2(5x - 2)^2 + 5 = 25$

62. $3(4x - 1)^2 + 1 = -17$

3 Solve Problems Involving Right Triangles and 30°-60° Triangles

For Problems 63–68, a and b represent the lengths of the legs of a right triangle and c represents the length of the hypotenuse. Express answers in simplest radical form.

63. Find c if $a = 4$ centimeters and $b = 6$ centimeters.

64. Find c if $a = 3$ meters and $b = 7$ meters.

65. Find a if $c = 12$ inches and $b = 8$ inches.

66. Find a if $c = 8$ feet and $b = 6$ feet.

67. Find b if $c = 17$ yards and $a = 15$ yards.

68. Find b if $c = 14$ meters and $a = 12$ meters.

For Problems 69–72, use the isosceles right triangle in Figure 8.4. Express your answers in simplest radical form.

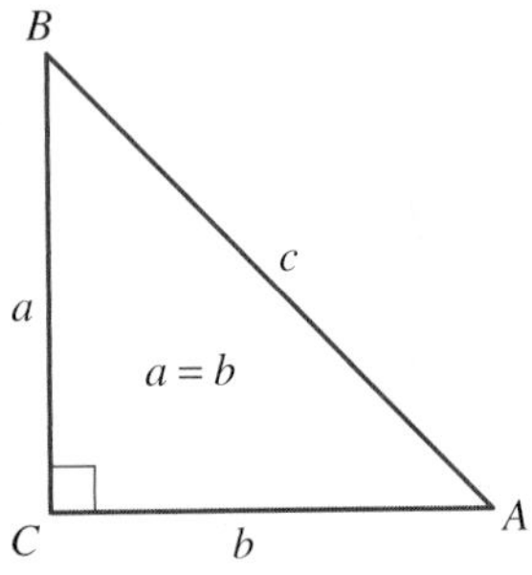

Figure 8.4

69. If $b = 6$ inches, find c.

70. If $a = 7$ centimeters, find c.

71. If $c = 8$ meters, find a and b.

72. If $c = 9$ feet, find a and b.

For Problems 73–78, use the triangle in Figure 8.5. Express your answers in simplest radical form.

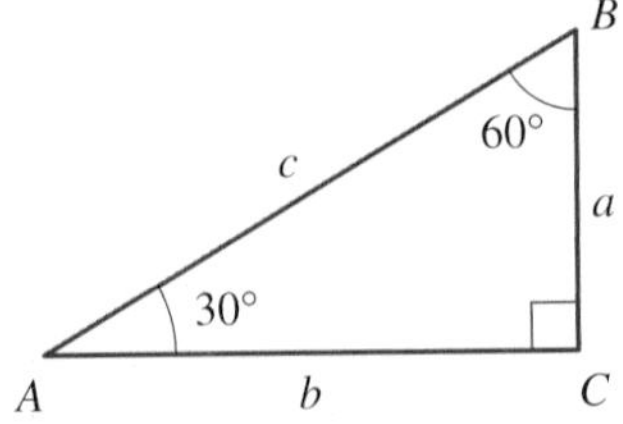

Figure 8.5

73. If $a = 3$ inches, find b and c.

74. If $a = 6$ feet, find b and c.

75. If $c = 14$ centimeters, find a and b.

76. If $c = 9$ centimeters, find a and b.

77. If $b = 10$ feet, find a and c.

78. If $b = 8$ meters, find a and c.

79. A 24-foot ladder resting against a house reaches a windowsill 16 feet above the ground. How far is the foot of the ladder from the foundation of the house? Express your answer to the nearest tenth of a foot.

80. A 62-foot guy-wire makes an angle of 60° with the ground and is attached to a telephone pole (see Figure 8.6). Find the distance from the base of the pole to the point on the pole where the wire is attached. Express your answer to the nearest tenth of a foot.

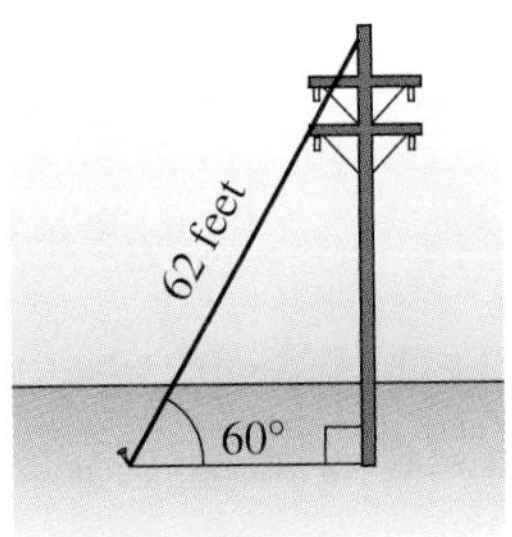

Figure 8.6

81. A rectangular plot measures 16 meters by 34 meters. Find, to the nearest meter, the distance from one corner of the plot to the corner diagonally opposite.

82. Consecutive bases of a square-shaped baseball diamond are 90 feet apart (see Figure 8.7). Find, to the nearest tenth of a foot, the distance from first base diagonally across the diamond to third base.

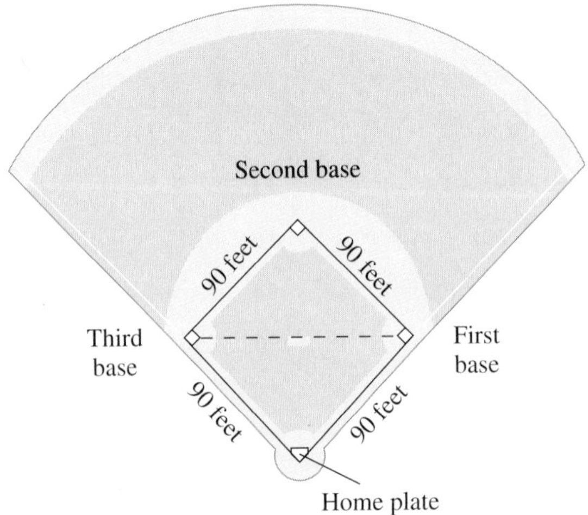

Figure 8.7

83. A diagonal of a square parking lot is 75 meters. Find, to the nearest meter, the length of a side of the lot.

THOUGHTS INTO WORDS

84. Explain why the equation $(x + 2)^2 + 5 = 1$ has no real number solutions.

85. Suppose that your friend solved the equation $(x + 3)^2 = 25$ as follows:

$$(x + 3)^2 = 25$$
$$x^2 + 6x + 9 = 25$$
$$x^2 + 6x - 16 = 0$$
$$(x + 8)(x - 2) = 0$$
$$x + 8 = 0 \quad \text{or} \quad x - 2 = 0$$
$$x = -8 \quad \text{or} \quad x = 2$$

Is this a correct approach to the problem? Would you offer any suggestion about an easier approach to the problem?

FURTHER INVESTIGATIONS

86. Suppose that we are given a cube with edges 12 centimeters in length. Find the length of a diagonal from a lower corner to the diagonally opposite upper corner. Express your answer to the nearest tenth of a centimeter.

87. Suppose that we are given a rectangular box with a length of 8 centimeters, a width of 6 centimeters, and a height of 4 centimeters. Find the length of a diagonal from a lower corner to the upper corner diagonally opposite. Express your answer to the nearest tenth of a centimeter.

88. The converse of the Pythagorean theorem is also true. It states: "If the measures a, b, and c of the sides of a triangle are such that $a^2 + b^2 = c^2$, then the triangle is a right triangle with a and b the measures of the legs and c the measure of the hypotenuse." Use the converse of the Pythagorean theorem to determine which of the triangles with sides of the following measures are right triangles.

(a) 9, 40, 41 **(b)** 20, 48, 52

(c) 19, 21, 26 **(d)** 32, 37, 49

(e) 65, 156, 169 **(f)** 21, 72, 75

89. Find the length of the hypotenuse (h) of an isosceles right triangle if each leg is s units long. Then use this relationship to redo Problems 69–72.

90. Suppose that the side opposite the 30° angle in a 30°-60° right triangle is s units long. Express the length of the hypotenuse and the length of the other leg in terms of s. Then use these relationships and redo Problems 73–78.

Answers to the Concept Quiz

1. True **2.** True **3.** False **4.** True **5.** False **6.** False **7.** True **8.** False **9.** True **10.** False

Answers to the Example Practice Skills

1. $\left\{-\frac{3}{2}, 5\right\}$ **2.** $\{9\}$ **3.** $\{\pm 5\sqrt{3}\}$ **4.** $\{\pm 6i\}$ **5.** $\left\{\pm\frac{4\sqrt{15}}{5}\right\}$ **6.** $\left\{-\frac{3}{2}, 3\right\}$ **7.** $\{-8 \pm i\sqrt{15}\}$ **8.** $\left\{\frac{-1 \pm 2\sqrt{7}}{5}\right\}$ **9.** 79.4 ft **10.** $4\sqrt{2}$ in. **11.** Yes, the ladder reaches 13.9 ft (to the nearest tenth of a foot)

8.3 Completing the Square

OBJECTIVE

1 Solve Quadratic Equations by Completing the Square

1 Solve Quadratic Equations by Completing the Square

Thus far we have solved quadratic equations by factoring and applying the property "$ab = 0$ if and only if $a = 0$ or $b = 0$" or by applying the property "$x^2 = a$ if and only if $x = \pm\sqrt{a}$." In this section we examine another method, called **completing the square**, which will give us the power to solve any quadratic equation.

A factoring technique we studied in Chapter 5 relied on recognizing **perfect-square trinomials**. In each of the following, the perfect-square trinomial on the right side is the result of squaring the binomial on the left side of the equation.

$$(x + 4)^2 = x^2 + 8x + 16 \qquad (x - 6)^2 = x^2 - 12x + 36$$

$$(x + 7)^2 = x^2 + 14x + 49 \qquad (x - 9)^2 = x^2 - 18x + 81$$

$$(x + a)^2 = x^2 + 2ax + a^2$$

Note that in each of the square trinomials, the constant term is equal to the square of one-half of the coefficient of the x term. This relationship enables us to form a perfect-square trinomial by adding a proper constant term. To find the constant term, take one-half of the coefficient of the x term and then square the result. For example, suppose that we want to form a perfect-square trinomial from $x^2 + 10x$. The coefficient of the x term is 10. Because $\frac{1}{2}(10) = 5$, and $5^2 = 25$, the constant term should be 25. Hence the perfect-square trinomial that can be formed is $x^2 + 10x + 25$. This perfect-square trinomial can be factored and expressed as $(x + 5)^2$. Let's use these ideas to help solve some quadratic equations.

EXAMPLE 1

Solve $x^2 + 10x - 2 = 0$.

Solution

$x^2 + 10x - 2 = 0$

$x^2 + 10x = 2$ — Isolate the x^2 and x terms

$\frac{1}{2}(10) = 5$ and $5^2 = 25$ — Take $\frac{1}{2}$ of the coefficient of the x term and then square the result

$x^2 + 10x + 25 = 2 + 25$ — Add 25 to *both* sides of the equation

$(x + 5)^2 = 27$ — Factor the perfect-square trinomial

$x + 5 = \pm\sqrt{27}$ — Now solve by applying Property 8.1

$x + 5 = \pm 3\sqrt{3}$

$x = -5 \pm 3\sqrt{3}$

The solution set is $\{-5 \pm 3\sqrt{3}\}$.

▼ **PRACTICE YOUR SKILL**

Solve $y^2 + 6y - 4 = 0$. ■

Note from Example 1 that the method of completing the square to solve a quadratic equation is just what the name implies. A perfect-square trinomial is formed, and then the equation can be changed to the necessary form for applying the property "$x^2 = a$ if and only if $x = \pm\sqrt{a}$." Let's consider another example.

EXAMPLE 2

Solve $x(x + 8) = -23$.

Solution

$x(x + 8) = -23$

$x^2 + 8x = -23$ — Apply the distributive property

$\frac{1}{2}(8) = 4$ and $4^2 = 16$ — Take $\frac{1}{2}$ of the coefficient of the x term and then square the result

$$x^2 + 8x + 16 = -23 + 16 \quad \text{Add 16 to } \textit{both} \text{ sides of the equation}$$

$$(x + 4)^2 = -7 \quad \text{Factor the perfect-square trinomial}$$

$$x + 4 = \pm\sqrt{-7} \quad \text{Now solve by applying Property 8.1}$$

$$x + 4 = \pm i\sqrt{7}$$

$$x = -4 \pm i\sqrt{7}$$

The solution set is $\{-4 \pm i\sqrt{7}\}$.

▼ **PRACTICE YOUR SKILL**

Solve $y(y + 4) = -10$. ■

EXAMPLE 3 Solve $x^2 - 3x + 1 = 0$.

Solution

$$x^2 - 3x + 1 = 0$$

$$x^2 - 3x = -1$$

$$x^2 - 3x + \frac{9}{4} = -1 + \frac{9}{4} \quad \frac{1}{2}(3) = \frac{3}{2} \text{ and } \left(\frac{3}{2}\right)^2 = \frac{9}{4}$$

$$\left(x - \frac{3}{2}\right)^2 = \frac{5}{4}$$

$$x - \frac{3}{2} = \pm\sqrt{\frac{5}{4}}$$

$$x - \frac{3}{2} = \pm\frac{\sqrt{5}}{2}$$

$$x = \frac{3}{2} \pm \frac{\sqrt{5}}{2}$$

$$x = \frac{3 \pm \sqrt{5}}{2}$$

The solution set is $\left\{\frac{3 \pm \sqrt{5}}{2}\right\}$.

▼ **PRACTICE YOUR SKILL**

Solve $y^2 + y + 3 = 0$. ■

In Example 3, note that because the coefficient of the x term is odd, we are forced into the realm of fractions. Using common fractions rather than decimals enables us to apply our previous work with radicals.

The relationship for a perfect-square trinomial that states that the constant term is equal to the square of one-half of the coefficient of the x term holds only if the coefficient of x^2 is 1. Thus we must make an adjustment when solving quadratic equations that have a coefficient of x^2 other than 1. We will need to apply the multiplication property of equality so that the coefficient of the x^2 term becomes 1. The next example shows how to make this adjustment.

EXAMPLE 4 Solve $2x^2 + 12x - 5 = 0$.

Solution

$$2x^2 + 12x - 5 = 0$$

$$2x^2 + 12x = 5$$

$$x^2 + 6x = \frac{5}{2} \qquad \text{Multiply both sides by } \frac{1}{2}$$

$$x^2 + 6x + 9 = \frac{5}{2} + 9 \qquad \frac{1}{2}(6) = 3 \text{ and } 3^2 = 9$$

$$x^2 + 6x + 9 = \frac{23}{2}$$

$$(x + 3)^2 = \frac{23}{2}$$

$$x + 3 = \pm\sqrt{\frac{23}{2}}$$

$$x + 3 = \pm\frac{\sqrt{46}}{2} \qquad \sqrt{\frac{23}{2}} = \frac{\sqrt{23}}{\sqrt{2}}\cdot\frac{\sqrt{2}}{\sqrt{2}} = \frac{\sqrt{46}}{2}$$

$$x = -3 \pm \frac{\sqrt{46}}{2}$$

$$x = \frac{-6}{2} \pm \frac{\sqrt{46}}{2} \qquad \text{Common denominator of 2}$$

$$x = \frac{-6 \pm \sqrt{46}}{2}$$

The solution set is $\left\{\frac{-6 \pm \sqrt{46}}{2}\right\}$.

▼ **PRACTICE YOUR SKILL**

Solve $3y^2 + 24y - 7 = 0$. ■

As mentioned earlier, we can use the method of completing the square to solve *any* quadratic equation. To illustrate, let's use it to solve an equation that could also be solved by factoring.

EXAMPLE 5 Solve $x^2 - 2x - 8 = 0$ by completing the square.

Solution

$$x^2 - 2x - 8 = 0$$

$$x^2 - 2x = 8$$

$$x^2 - 2x + 1 = 8 + 1 \qquad \frac{1}{2}(-2) = -1 \text{ and } (-1)^2 = 1$$

$$(x - 1)^2 = 9$$

$$x - 1 = \pm 3$$

$$x - 1 = 3 \quad \text{or} \quad x - 1 = -3$$

$$x = 4 \quad \text{or} \quad x = -2$$

The solution set is $\{-2, 4\}$.

▼ **PRACTICE YOUR SKILL**

Solve $y^2 - 6y - 72 = 0$. ■

Solving the equation in Example 5 by factoring would be easier than completing the square. Remember, however, that the method of completing the square will work with any quadratic equation.

CONCEPT QUIZ

For Problems 1–10, answer true or false.

1. In a perfect-square trinomial, the constant term is equal to one-half the coefficient of the x term.
2. The method of completing the square will solve any quadratic equation.
3. Every quadratic equation solved by completing the square will have real number solutions.
4. The completing-the-square method cannot be used if factoring could solve the quadratic equation.
5. To use the completing-the-square method for solving the equation $3x^2 + 2x = 5$, we would first divide both sides of the equation by 3.
6. The equation $x^2 + 2x = 0$ cannot be solved by using the method of completing the square.
7. To solve the equation $x^2 - 5x = 1$ by completing the square, we would start by adding $\frac{25}{4}$ to both sides of the equation.
8. To solve the equation $x^2 - 2x = 14$ by completing the square, we must first change the form of the equation to $x^2 - 2x - 14 = 0$.
9. The solution set of the equation $x^2 - 2x = 14$ is $\{1 \pm \sqrt{15}\}$.
10. The solution set of the equation $x^2 - 5x - 1 = 0$ is $\left\{\frac{5 \pm \sqrt{29}}{2}\right\}$.

Problem Set 8.3

1 Solve Quadratic Equations by Completing the Square

For Problems 1–14, solve each quadratic equation by using (a) the factoring method and (b) the method of completing the square.

1. $x^2 - 4x - 60 = 0$

2. $x^2 + 6x - 16 = 0$

3. $x^2 - 14x = -40$

4. $x^2 - 18x = -72$

5. $x^2 - 5x - 50 = 0$

6. $x^2 + 3x - 18 = 0$

7. $x(x + 7) = 8$

8. $x(x - 1) = 30$

9. $2n^2 - n - 15 = 0$

10. $3n^2 + n - 14 = 0$

11. $3n^2 + 7n - 6 = 0$

12. $2n^2 + 7n - 4 = 0$

13. $n(n + 6) = 160$

14. $n(n - 6) = 216$

For Problems 15–38, use the method of completing the square to solve each quadratic equation.

15. $x^2 + 4x - 2 = 0$

16. $x^2 + 2x - 1 = 0$

17. $x^2 + 6x - 3 = 0$

18. $x^2 + 8x - 4 = 0$

19. $y^2 - 10y = 1$

20. $y^2 - 6y = -10$

21. $n^2 - 8n + 17 = 0$

22. $n^2 - 4n + 2 = 0$

23. $n(n + 12) = -9$

24. $n(n + 14) = -4$

25. $n^2 + 2n + 6 = 0$

26. $n^2 + n - 1 = 0$

27. $x^2 + 3x - 2 = 0$

28. $x^2 + 5x - 3 = 0$

29. $x^2 + 5x + 1 = 0$

30. $x^2 + 7x + 2 = 0$

31. $y^2 - 7y + 3 = 0$

32. $y^2 - 9y + 30 = 0$

33. $2x^2 + 4x - 3 = 0$

34. $2t^2 - 4t + 1 = 0$

35. $3n^2 - 6n + 5 = 0$

36. $3x^2 + 12x - 2 = 0$

37. $3x^2 + 5x - 1 = 0$

38. $2x^2 + 7x - 3 = 0$

For Problems 39–60, solve each quadratic equation using the method that seems most appropriate.

39. $x^2 + 8x - 48 = 0$

40. $x^2 + 5x - 14 = 0$

41. $2n^2 - 8n = -3$

42. $3x^2 + 6x = 1$

43. $(3x - 1)(2x + 9) = 0$

44. $(5x + 2)(x - 4) = 0$

45. $(x + 2)(x - 7) = 10$

46. $(x - 3)(x + 5) = -7$

47. $(x - 3)^2 = 12$

48. $x^2 = 16x$

49. $3n^2 - 6n + 4 = 0$

50. $2n^2 - 2n - 1 = 0$

51. $n(n + 8) = 240$

52. $t(t - 26) = -160$

53. $3x^2 + 5x = -2$

54. $2x^2 - 7x = -5$

55. $4x^2 - 8x + 3 = 0$

56. $9x^2 + 18x + 5 = 0$

57. $x^2 + 12x = 4$

58. $x^2 + 6x = -11$

59. $4(2x + 1)^2 - 1 = 11$

60. $5(x + 2)^2 + 1 = 16$

61. Use the method of completing the square to solve $ax^2 + bx + c = 0$ for x, where a, b, and c are real numbers and $a \neq 0$.

THOUGHTS INTO WORDS

62. Explain the process of completing the square to solve a quadratic equation.

63. Give a step-by-step description of how to solve $3x^2 + 9x - 4 = 0$ by completing the square.

FURTHER INVESTIGATIONS

Solve Problems 64–67 for the indicated variable. Assume that all letters represent positive numbers.

64. $\frac{x^2}{a^2} - \frac{y^2}{b^2} = 1$ for y

65. $\frac{x^2}{a^2} + \frac{y^2}{b^2} = 1$ for x

66. $s = \frac{1}{2}gt^2$ for t

67. $A = \pi r^2$ for r

Solve each of the following equations for x.

68. $x^2 + 8ax + 15a^2 = 0$

69. $x^2 - 5ax + 6a^2 = 0$

70. $10x^2 - 31ax - 14a^2 = 0$

71. $6x^2 + ax - 2a^2 = 0$

72. $4x^2 + 4bx + b^2 = 0$

73. $9x^2 - 12bx + 4b^2 = 0$

Answers to the Concept Quiz

1. False **2.** True **3.** False **4.** False **5.** True **6.** False **7.** True **8.** False **9.** True **10.** True

Answers to the Example Practice Skills

1. $\{-3 \pm \sqrt{13}\}$ **2.** $\{-2 \pm i\sqrt{6}\}$ **3.** $\left\{\frac{-1 \pm i\sqrt{11}}{2}\right\}$ **4.** $\left\{\frac{-12 \pm \sqrt{165}}{3}\right\}$ **5.** $\{-6, 12\}$

8.4 Quadratic Formula

OBJECTIVES

1 Use the Quadratic Formula to Solve Quadratic Equations

2 Determine the Nature of Roots to Quadratic Equations

1 Use the Quadratic Formula to Solve Quadratic Equations

As we saw in the previous section, the method of completing the square can be used to solve any quadratic equation. Therefore, if we apply the method of completing the square to the equation $ax^2 + bx + c = 0$, where a, b, and c are real numbers and

$a \neq 0$, we can produce a formula for solving quadratic equations. This formula can then be used to solve any quadratic equation. Let's solve $ax^2 + bx + c = 0$ by completing the square.

$$ax^2 + bx + c = 0$$

$$ax^2 + bx = -c \qquad \text{Isolate the } x^2 \text{ and } x \text{ terms}$$

$$x^2 + \frac{b}{a}x = -\frac{c}{a} \qquad \text{Multiply both sides by } \frac{1}{a}$$

$$x^2 + \frac{b}{a}x + \frac{b^2}{4a^2} = -\frac{c}{a} + \frac{b^2}{4a^2} \qquad \frac{1}{2}\left(\frac{b}{a}\right) = \frac{b}{2a} \text{ and } \left(\frac{b}{2a}\right)^2 = \frac{b^2}{4a^2}$$

Complete the square by adding $\frac{b^2}{4a^2}$ to both sides

$$x^2 + \frac{b}{a}x + \frac{b^2}{4a^2} = -\frac{4ac}{4a^2} + \frac{b^2}{4a^2} \qquad \text{Common denominator of } 4a^2 \text{ on right side}$$

$$x^2 + \frac{b}{a}x + \frac{b^2}{4a^2} = \frac{b^2}{4a^2} - \frac{4ac}{4a^2} \qquad \text{Commutative property}$$

$$\left(x + \frac{b}{2a}\right)^2 = \frac{b^2 - 4ac}{4a^2}$$

$$x + \frac{b}{2a} = \pm\sqrt{\frac{b^2 - 4ac}{4a^2}}$$

The right side is combined into a single fraction

$$x + \frac{b}{2a} = \pm\frac{\sqrt{b^2 - 4ac}}{\sqrt{4a^2}}$$

$$x + \frac{b}{2a} = \pm\frac{\sqrt{b^2 - 4ac}}{2a} \qquad \sqrt{4a^2} = |2a|, \text{ but } 2a \text{ can be used because of the use of } \pm$$

$$x + \frac{b}{2a} = \frac{\sqrt{b^2 - 4ac}}{2a} \quad \text{or} \quad x + \frac{b}{2a} = -\frac{\sqrt{b^2 - 4ac}}{2a}$$

$$x = -\frac{b}{2a} + \frac{\sqrt{b^2 - 4ac}}{2a} \quad \text{or} \quad x = -\frac{b}{2a} - \frac{\sqrt{b^2 - 4ac}}{2a}$$

$$x = \frac{-b + \sqrt{b^2 - 4ac}}{2a} \quad \text{or} \quad x = \frac{-b - \sqrt{b^2 - 4ac}}{2a}$$

The quadratic formula is usually stated as follows.

Quadratic Formula

$$x = \frac{-b \pm \sqrt{b^2 - 4ac}}{2a}, \qquad a \neq 0$$

We can use the quadratic formula to solve *any* quadratic equation by expressing the equation in the standard form $ax^2 + bx + c = 0$ and then substituting the values for a, b, and c into the formula. Let's consider some examples.

EXAMPLE 1 Solve $x^2 + 5x + 2 = 0$.

Solution

$$x^2 + 5x + 2 = 0$$

The given equation is in standard form with $a = 1$, $b = 5$, and $c = 2$. Let's substitute these values into the formula and simplify.

$$x = \frac{-b \pm \sqrt{b^2 - 4ac}}{2a}$$

$$x = \frac{-5 \pm \sqrt{5^2 - 4(1)(2)}}{2(1)}$$

$$x = \frac{-5 \pm \sqrt{25 - 8}}{2}$$

$$x = \frac{-5 \pm \sqrt{17}}{2}$$

The solution set is $\left\{\frac{-5 \pm \sqrt{17}}{2}\right\}$.

▼ **PRACTICE YOUR SKILL**

Solve $x^2 + 7x + 5 = 0$. ■

EXAMPLE 2 Solve $x^2 - 2x - 4 = 0$.

Solution

$$x^2 - 2x - 4 = 0$$

We need to think of $x^2 - 2x - 4 = 0$ as $x^2 + (-2)x + (-4) = 0$ in order to determine the values $a = 1$, $b = -2$, and $c = -4$. Let's substitute these values into the quadratic formula and simplify.

$$x = \frac{-b \pm \sqrt{b^2 - 4ac}}{2a}$$

$$x = \frac{-(-2) \pm \sqrt{(-2)^2 - 4(1)(-4)}}{2(1)}$$

$$x = \frac{2 \pm \sqrt{4 + 16}}{2}$$

$$x = \frac{2 \pm \sqrt{20}}{2}$$

$$x = \frac{2 \pm 2\sqrt{5}}{2}$$

$$x = \frac{\cancel{2}(1 \pm \sqrt{5})}{\cancel{2}} = (1 \pm \sqrt{5})$$

The solution set is $\{1 \pm \sqrt{5}\}$.

▼ **PRACTICE YOUR SKILL**

Solve $x^2 - 6x - 4 = 0$. ■

EXAMPLE 3 Solve $x^2 - 2x + 19 = 0$.

Solution

$$x^2 - 2x + 19 = 0$$

We can substitute $a = 1$, $b = -2$, and $c = 19$.

$$x = \frac{-b \pm \sqrt{b^2 - 4ac}}{2a}$$

$$x = \frac{-(-2) \pm \sqrt{(-2)^2 - 4(1)(19)}}{2(1)}$$

$$x = \frac{2 \pm \sqrt{4 - 76}}{2}$$

$$x = \frac{2 \pm \sqrt{-72}}{2}$$

$$x = \frac{2 \pm 6i\sqrt{2}}{2} \qquad \sqrt{-72} = i\sqrt{72} = i\sqrt{36}\sqrt{2} = 6i\sqrt{2}$$

$$x = \frac{\cancel{2}(1 \pm 3i\sqrt{2})}{\cancel{2}} = 1 \pm 3i\sqrt{2}$$

The solution set is $\{1 \pm 3i\sqrt{2}\}$.

▼ **PRACTICE YOUR SKILL**

Solve $x^2 - 2x + 8 = 0$. ■

EXAMPLE 4 Solve $2x^2 + 4x - 3 = 0$.

Solution

$$2x^2 + 4x - 3 = 0$$

Here $a = 2$, $b = 4$, and $c = -3$. Solving by using the quadratic formula is unlike solving by completing the square in that there is no need to make the coefficient of x^2 equal to 1.

$$x = \frac{-b \pm \sqrt{b^2 - 4ac}}{2a}$$

$$x = \frac{-4 \pm \sqrt{4^2 - 4(2)(-3)}}{2(2)}$$

$$x = \frac{-4 \pm \sqrt{16 + 24}}{4}$$

$$x = \frac{-4 \pm \sqrt{40}}{4}$$

$$x = \frac{-4 \pm 2\sqrt{10}}{4}$$

$$x = \frac{2(-2 \pm \sqrt{10})}{4}$$

$$x = \frac{-2 \pm \sqrt{10}}{2}$$

The solution set is $\left\{\frac{-2 \pm \sqrt{10}}{2}\right\}$.

▼ PRACTICE YOUR SKILL

Solve $5x^2 + 3x - 4 = 0$.

EXAMPLE 5

Solve $n(3n - 10) = 25$.

Solution

$$n(3n - 10) = 25$$

First, we need to change the equation to the standard form $an^2 + bn + c = 0$.

$$n(3n - 10) = 25$$
$$3n^2 - 10n = 25$$
$$3n^2 - 10n - 25 = 0$$

Now we can substitute $a = 3$, $b = -10$, and $c = -25$ into the quadratic formula.

$$n = \frac{-b \pm \sqrt{b^2 - 4ac}}{2a}$$

$$n = \frac{-(-10) \pm \sqrt{(-10)^2 - 4(3)(-25)}}{2(3)}$$

$$n = \frac{10 \pm \sqrt{100 + 300}}{2(3)}$$

$$n = \frac{10 \pm \sqrt{400}}{6}$$

$$n = \frac{10 \pm 20}{6}$$

$$n = \frac{10 + 20}{6} \quad \text{or} \quad n = \frac{10 - 20}{6}$$

$$n = 5 \quad \text{or} \quad n = -\frac{5}{3}$$

The solution set is $\left\{-\frac{5}{3}, 5\right\}$.

▼ PRACTICE YOUR SKILL

Solve $n(2n + 5) = 12$.

In Example 5, note that we used the variable n. The quadratic formula is usually stated in terms of x, but it certainly can be applied to quadratic equations in other variables. Also note in Example 5 that the polynomial $3n^2 - 10n - 25$ can be factored as $(3n + 5)(n - 5)$. Therefore, we could also solve the equation $3n^2 - 10n - 25 = 0$ by using the factoring approach. Section 8.5 will offer some guidance in deciding which approach to use for a particular equation.

2 Determine the Nature of Roots to Quadratic Equations

The quadratic formula makes it easy to determine the nature of the roots of a quadratic equation without completely solving the equation. The number

$$b^2 - 4ac$$

which appears under the radical sign in the quadratic formula, is called the **discriminant** of the quadratic equation. The discriminant is the indicator of the kind of roots the equation has. For example, suppose that you start to solve the equation $x^2 - 4x + 7 = 0$ as follows:

$$x = \frac{-b \pm \sqrt{b^2 - 4ac}}{2a}$$

$$x = \frac{-(-4) \pm \sqrt{(-4)^2 - 4(1)(7)}}{2(1)}$$

$$x = \frac{4 \pm \sqrt{16 - 28}}{2}$$

$$x = \frac{4 \pm \sqrt{-12}}{2}$$

At this stage you should be able to look ahead and realize that you will obtain two complex solutions for the equation. (Note, by the way, that these solutions are complex conjugates.) In other words, the discriminant, -12, indicates what type of roots you will obtain.

We make the following general statements relative to the roots of a quadratic equation of the form $ax^2 + bx + c = 0$.

1. If $b^2 - 4ac < 0$, then the equation has two nonreal complex solutions.
2. If $b^2 - 4ac = 0$, then the equation has one real solution.
3. If $b^2 - 4ac > 0$, then the equation has two real solutions.

The following examples illustrate each of these situations. (You may want to solve the equations completely to verify the conclusions.)

Equation	Discriminant	Nature of roots
$x^2 - 3x + 7 = 0$	$b^2 - 4ac = (-3)^2 - 4(1)(7)$ $= 9 - 28$ $= -19$	Two nonreal complex solutions
$9x^2 - 12x + 4 = 0$	$b^2 - 4ac = (-12)^2 - 4(9)(4)$ $= 144 - 144$ $= 0$	One real solution
$2x^2 + 5x - 3 = 0$	$b^2 - 4ac = (5)^2 - 4(2)(-3)$ $= 25 + 24$ $= 49$	Two real solutions

Remark: A clarification is called for at this time. Previously, we made the statement that if $b^2 - 4ac = 0$, then the equation has one real solution. Technically, such an equation has two solutions, but they are equal. For example, each factor of $(x - 7)(x - 7) = 0$ produces a solution, but both solutions are the number 7. We sometimes refer to this as one real solution with a *multiplicity of two*. Using the idea of multiplicity of roots, we can say that every quadratic equation has two roots.

EXAMPLE 6

Use the discriminant to determine whether the equation $5x^2 + 2x + 7 = 0$ has two nonreal complex solutions, one real solution with a multiplicity of 2, or two real solutions.

Solution

For the equation $5x^2 + 2x + 7 = 0$, we have $a = 5$, $b = 2$, and $c = 7$.

$$\begin{aligned} b^2 - 4ac &= (2)^2 - 4(5)(7) \\ &= 4 - 140 \\ &= -136 \end{aligned}$$

Because the discriminant is negative, the solutions will be two nonreal complex numbers.

▼ PRACTICE YOUR SKILL

Use the discriminant to determine whether the equation $4x^2 + 12x + 9 = 0$ has two nonreal complex solutions, one real solution with a multiplicity of 2, or two real solutions. ■

CONCEPT QUIZ

For Problems 1–10, answer true or false.

1. The quadratic formula can be used to solve any quadratic equation.
2. The number $\sqrt{b^2 - 4ac}$ is called the discriminant of the quadratic equation.
3. Every quadratic equation will have two solutions.
4. The quadratic formula cannot be used if the quadratic equation can be solved by factoring.
5. To use the quadratic formula for solving the equation $3x^2 + 2x - 5 = 0$, you must first divide both sides of the equation by 3.
6. The equation $9x^2 + 30x + 25 = 0$ has one real solution with a multiplicity of 2.
7. The equation $2x^2 + 3x + 4 = 0$ has two nonreal complex solutions.
8. The equation $x^2 + 9 = 0$ has two real solutions.
9. The solution set for the equation $x^2 - 3x + 4 = 0$ is $\left\{\frac{3 \pm i\sqrt{7}}{2}\right\}$.
10. The solution set for the equation $x^2 - 10x + 24 = 0$ is $\{-4, 6\}$.

Problem Set 8.4

1 Use the Quadratic Formula to Solve Quadratic Equations

For Problems 1–40, use the quadratic formula to solve each of the quadratic equations. Check your solutions by using the sum and product relationships.

1. $x^2 + 2x - 1 = 0$

2. $x^2 + 4x - 1 = 0$

3. $n^2 + 5n - 3 = 0$

4. $n^2 + 3n - 2 = 0$

5. $a^2 - 8a = 4$

6. $a^2 - 6a = 2$

7. $n^2 + 5n + 8 = 0$

8. $2n^2 - 3n + 5 = 0$

9. $x^2 - 18x + 80 = 0$

10. $x^2 + 19x + 70 = 0$

11. $-y^2 = -9y + 5$

12. $-y^2 + 7y = 4$

13. $2x^2 + x - 4 = 0$

14. $2x^2 + 5x - 2 = 0$

15. $4x^2 + 2x + 1 = 0$

16. $3x^2 - 2x + 5 = 0$

17. $3a^2 - 8a + 2 = 0$

18. $2a^2 - 6a + 1 = 0$

19. $-2n^2 + 3n + 5 = 0$

20. $-3n^2 - 11n + 4 = 0$

21. $3x^2 + 19x + 20 = 0$

22. $2x^2 - 17x + 30 = 0$

23. $36n^2 - 60n + 25 = 0$

24. $9n^2 + 42n + 49 = 0$

25. $4x^2 - 2x = 3$

26. $6x^2 - 4x = 3$

27. $5x^2 - 13x = 0$

28. $7x^2 + 12x = 0$

29. $3x^2 = 5$

30. $4x^2 = 3$

31. $6t^2 + t - 3 = 0$

32. $2t^2 + 6t - 3 = 0$

33. $n^2 + 32n + 252 = 0$

34. $n^2 - 4n - 192 = 0$

35. $12x^2 - 73x + 110 = 0$

36. $6x^2 + 11x - 255 = 0$

37. $-2x^2 + 4x - 3 = 0$

38. $-2x^2 + 6x - 5 = 0$

39. $-6x^2 + 2x + 1 = 0$

40. $-2x^2 + 4x + 1 = 0$

2 Determine the Nature of Roots to Quadratic Equations

For each quadratic equation in Problems 41–50, first use the discriminant to determine whether the equation has two nonreal complex solutions, one real solution with a multiplicity of 2, or two real solutions. Then solve the equation.

41. $x^2 + 4x - 21 = 0$

42. $x^2 - 3x - 54 = 0$

43. $9x^2 - 6x + 1 = 0$

44. $4x^2 + 20x + 25 = 0$

45. $x^2 - 7x + 13 = 0$

46. $2x^2 - x + 5 = 0$

47. $15x^2 + 17x - 4 = 0$

48. $8x^2 + 18x - 5 = 0$

49. $3x^2 + 4x = 2$

50. $2x^2 - 6x = -1$

THOUGHTS INTO WORDS

51. Your friend states that the equation $-2x^2 + 4x - 1 = 0$ must be changed to $2x^2 - 4x + 1 = 0$ (by multiplying both sides by -1) before the quadratic formula can be applied. Is she right about this? If not, how would you convince her she is wrong?

52. Another of your friends claims that the quadratic formula can be used to solve the equation $x^2 - 9 = 0$. How would you react to this claim?

53. Why must we change the equation $3x^2 - 2x = 4$ to $3x^2 - 2x - 4 = 0$ before applying the quadratic formula?

FURTHER INVESTIGATIONS

The solution set for $x^2 - 4x - 37 = 0$ is $\{2 \pm \sqrt{41}\}$. With a calculator, we found a rational approximation, to the nearest one-thousandth, for each of these solutions.

$$2 - \sqrt{41} = -4.403 \quad \text{and} \quad 2 + \sqrt{41} = 8.403$$

Thus the solution set is $\{-4.403, 8.403\}$, with the answers rounded to the nearest one-thousandth.

Solve each of the equations in Problems 54–63, expressing solutions to the nearest one-thousandth.

54. $x^2 - 6x - 10 = 0$

55. $x^2 - 16x - 24 = 0$

56. $x^2 + 6x - 44 = 0$

57. $x^2 + 10x - 46 = 0$

58. $x^2 + 8x + 2 = 0$

59. $x^2 + 9x + 3 = 0$

60. $4x^2 - 6x + 1 = 0$

61. $5x^2 - 9x + 1 = 0$

62. $2x^2 - 11x - 5 = 0$

63. $3x^2 - 12x - 10 = 0$

For Problems 64–66, use the discriminant to help solve each problem.

64. Determine k so that the solutions of $x^2 - 2x + k = 0$ are complex but nonreal.

65. Determine k so that $4x^2 - kx + 1 = 0$ has two equal real solutions.

66. Determine k so that $3x^2 - kx - 2 = 0$ has real solutions.

Answers to the Concept Quiz

1. True **2.** False **3.** True **4.** False **5.** False **6.** True **7.** True **8.** False **9.** True **10.** False

Answers to the Example Practice Skills

1. $\left\{\frac{-7 \pm \sqrt{29}}{2}\right\}$ **2.** $\{3 \pm \sqrt{13}\}$ **3.** $\{1 \pm i\sqrt{7}\}$ **4.** $\left\{\frac{-3 \pm \sqrt{89}}{10}\right\}$ **5.** $\left\{-4, \frac{3}{2}\right\}$
6. One real solution with a multiplicity of 2

8.5 More Quadratic Equations and Applications

OBJECTIVES

1. Solve Quadratic Equations Selecting the Most Appropriate Method
2. Solve Word Problems Involving Quadratic Equations

1 Solve Quadratic Equations Selecting the Most Appropriate Method

Which method should be used to solve a particular quadratic equation? There is no hard-and-fast answer to that question; it depends on the type of equation and on your personal preference. In the following examples we will state reasons for choosing a specific technique. However, keep in mind that usually this is a decision you must make as the need arises. That's why you need to be familiar with the strengths and weaknesses of each method.

EXAMPLE 1

Solve $2x^2 - 3x - 1 = 0$.

Solution

Because of the leading coefficient of 2 and the constant term of -1, there are very few factoring possibilities to consider. Therefore, with such problems, first try the factoring approach. Unfortunately, this particular polynomial is not factorable using integers. Let's use the quadratic formula to solve the equation.

$$x = \frac{-b \pm \sqrt{b^2 - 4ac}}{2a}$$

$$x = \frac{-(-3) \pm \sqrt{(-3)^2 - 4(2)(-1)}}{2(2)}$$

$$x = \frac{3 \pm \sqrt{9 + 8}}{4}$$

$$x = \frac{3 \pm \sqrt{17}}{4}$$

The solution set is $\left\{\frac{3 \pm \sqrt{17}}{4}\right\}$.

▼ **PRACTICE YOUR SKILL**

Solve $3x^2 - 5x + 1 = 0$. ■

EXAMPLE 2

Solve $\frac{3}{n} + \frac{10}{n+6} = 1$.

Solution

$$\frac{3}{n} + \frac{10}{n+6} = 1, \qquad n \neq 0 \text{ and } n \neq -6$$

$$n(n+6)\left(\frac{3}{n} + \frac{10}{n+6}\right) = 1(n)(n+6) \qquad \text{Multiply both sides by } n(n+6), \text{ which is the LCD}$$

$$3(n+6) + 10n = n(n+6)$$

$$3n + 18 + 10n = n^2 + 6n$$

$$13n + 18 = n^2 + 6n$$

$$0 = n^2 - 7n - 18$$

This equation is an easy one to consider for possible factoring, and it factors as follows:

$$0 = (n-9)(n+2)$$

$$n - 9 = 0 \quad \text{or} \quad n + 2 = 0$$

$$n = 9 \quad \text{or} \quad n = -2$$

✔ Check

Substituting 9 and −2 back into the original equation, we obtain

$$\frac{3}{n} + \frac{10}{n+6} = 1 \qquad\qquad \frac{3}{n} + \frac{10}{n+6} = 1$$

$$\frac{3}{9} + \frac{10}{9+6} \stackrel{?}{=} 1 \qquad\qquad \frac{3}{-2} + \frac{10}{-2+6} \stackrel{?}{=} 1$$

$$\frac{1}{3} + \frac{10}{15} \stackrel{?}{=} 1 \qquad \text{or} \qquad -\frac{3}{2} + \frac{10}{4} \stackrel{?}{=} 1$$

$$\frac{1}{3} + \frac{2}{3} \stackrel{?}{=} 1 \qquad\qquad -\frac{3}{2} + \frac{5}{2} \stackrel{?}{=} 1$$

$$1 = 1 \qquad\qquad \frac{2}{2} = 1$$

The solution set is $\{-2, 9\}$.

▼ PRACTICE YOUR SKILL

Solve $\frac{6}{x} + \frac{3}{x+4} = 1.$ ■

In Example 2, note the indication of the initial restrictions $n \neq 0$ and $n \neq -6$. Remember that we need to do this when solving fractional equations.

EXAMPLE 3

Solve $x^2 + 22x + 112 = 0$.

Solution

The size of the constant term makes the factoring approach a little cumbersome for this problem. Furthermore, because the leading coefficient is 1 and the coefficient of the x term is even, the method of completing the square will work effectively.

$$x^2 + 22x + 112 = 0$$

$$x^2 + 22x = -112$$

$$x^2 + 22x + 121 = -112 + 121$$

$$(x + 11)^2 = 9$$

$$x + 11 = \pm\sqrt{9}$$

$$x + 11 = \pm 3$$

$$x + 11 = 3 \quad \text{or} \quad x + 11 = -3$$

$$x = -8 \quad \text{or} \quad x = -14$$

The solution set is $\{-14, -8\}$.

▼ PRACTICE YOUR SKILL

Solve $y^2 + 28y + 192 = 0$. ■

EXAMPLE 4 Solve $x^4 - 4x^2 - 96 = 0$.

Solution

An equation such as $x^4 - 4x^2 - 96 = 0$ is not a quadratic equation, but we can solve it using the techniques that we use on quadratic equations. That is, we can factor the polynomial and apply the property "$ab = 0$ if and only if $a = 0$ or $b = 0$" as follows:

$$x^4 - 4x^2 - 96 = 0$$

$$(x^2 - 12)(x^2 + 8) = 0$$

$$x^2 - 12 = 0 \quad \text{or} \quad x^2 + 8 = 0$$

$$x^2 = 12 \quad \text{or} \quad x^2 = -8$$

$$x = \pm\sqrt{12} \quad \text{or} \quad x = \pm\sqrt{-8}$$

$$x = \pm 2\sqrt{3} \quad \text{or} \quad x = \pm 2i\sqrt{2}$$

The solution set is $\{\pm 2\sqrt{3}, \pm 2i\sqrt{2}\}$.

▼ **PRACTICE YOUR SKILL**

Solve $y^4 - 17y^2 - 60 = 0$. ■

Remark: Another approach to Example 4 would be to substitute y for x^2 and y^2 for x^4. Then the equation $x^4 - 4x^2 - 96 = 0$ becomes the quadratic equation $y^2 - 4y - 96 = 0$. Thus we say that $x^4 - 4x^2 - 96 = 0$ is of *quadratic form*. Then we could solve the quadratic equation $y^2 - 4y - 96 = 0$ and use the equation $y = x^2$ to determine the solutions for x.

2 Solve Word Problems Involving Quadratic Equations

Before we conclude this section with some word problems that can be solved using quadratic equations, let's restate the suggestions we made in an earlier chapter for solving word problems.

Suggestions for Solving Word Problems

1. Read the problem carefully, and make certain that you understand the meanings of all the words. Be especially alert for any technical terms used in the statement of the problem.
2. Read the problem a second time (perhaps even a third time) to get an overview of the situation being described and to determine the known facts as well as what is to be found.
3. Sketch any figure, diagram, or chart that might be helpful in analyzing the problem.
4. Choose a meaningful variable to represent an unknown quantity in the problem (perhaps l, if the length of a rectangle is an unknown quantity), and represent any other unknowns in terms of that variable.
5. Look for a guideline that you can use to set up an equation. A guideline might be a formula such as $A = lw$ or a relationship such as "the fractional part of a job done by Bill plus the fractional part of the job done by Mary equals the total job."

6. Form an equation that contains the variable and that translates the conditions of the guideline from English into algebra.
7. Solve the equation and use the solutions to determine all facts requested in the problem.
8. **Check all answers back into the original statement of the problem.**

Keep these suggestions in mind as we now consider some word problems.

EXAMPLE 5

Helene Rogers/Alamy Limited

Apply Your Skill

A page for a magazine contains 70 square inches of type. The height of a page is twice the width. If the margin around the type is to be 2 inches uniformly, what are the dimensions of a page?

Solution

Let x represent the width of a page; then $2x$ represents the height of a page. Now let's draw and label a model of a page (Figure 8.8).

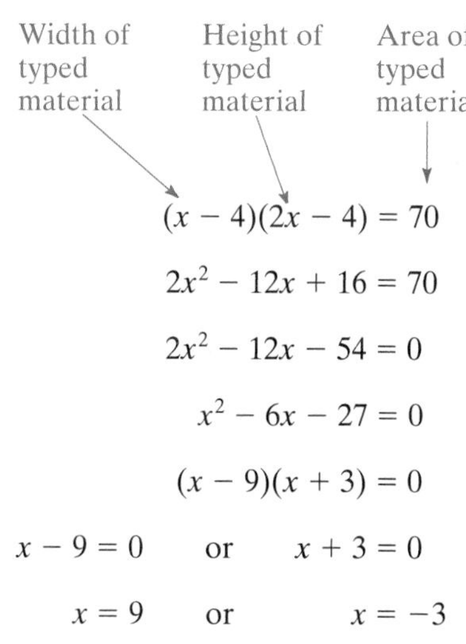

$$(x - 4)(2x - 4) = 70$$

$$2x^2 - 12x + 16 = 70$$

$$2x^2 - 12x - 54 = 0$$

$$x^2 - 6x - 27 = 0$$

$$(x - 9)(x + 3) = 0$$

$$x - 9 = 0 \quad \text{or} \quad x + 3 = 0$$

$$x = 9 \quad \text{or} \quad x = -3$$

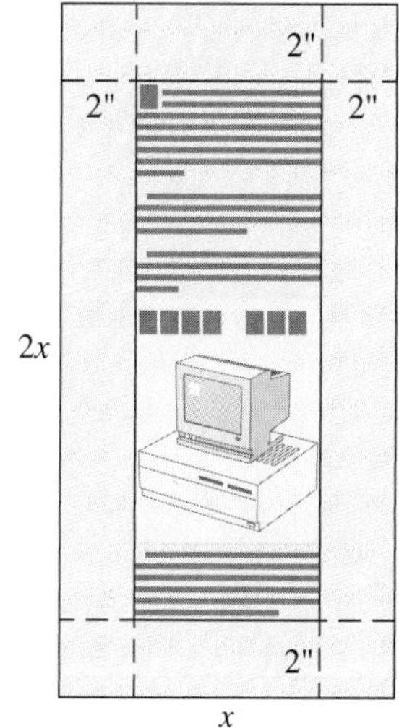

Figure 8.8

Disregard the negative solution; the page must be 9 inches wide, and its height is $2(9) = 18$ inches.

▼ PRACTICE YOUR SKILL

A rectangular digital image has a width that measures three times the length. If one centimeter is uniformly cropped from the image, then the area of the image is 64 square centimeters. What are the dimensions of the cropped image? ■

Let's use our knowledge of quadratic equations to analyze some applications in the business world. For example, if P dollars are invested at r rate of interest compounded annually for t years, then the amount of money, A, accumulated at the end of t years is given by the formula

$$A = P(1 + r)^t$$

This compound interest formula serves as a guideline for the next problem.

EXAMPLE 6

Apply Your Skill

Suppose that \$1000 is invested at a certain rate of interest compounded annually for 2 years. If the accumulated value at the end of 2 years is \$1200, find the rate of interest, rounded to the nearest tenth.

Solution

Let r represent the rate of interest. Substitute the known values into the compound interest formula to yield

$$A = P(1 + r)^t$$

$$1200 = 1000(1 + r)^2$$

Solving this equation, we obtain

$$\frac{1200}{1000} = (1 + r)^2$$

$$1.2 = (1 + r)^2$$

$$\pm\sqrt{1.2} = 1 + r$$

$$r = -1 \pm \sqrt{1.2}$$

$$r = -1 + \sqrt{1.2} \quad \text{or} \quad r = -1 - \sqrt{1.2}$$

We must disregard the negative solution, so that $r = -1 + \sqrt{1.2}$ is the only solution. Change $-1 + \sqrt{1.2}$ to a percent (rounded to the nearest tenth). The rate of interest is 9.5%.

▼ PRACTICE YOUR SKILL

Find the interest rate, rounded to the nearest tenth, that is necessary for an investment of \$3000 compounded annually to have an accumulated value of \$3500 at the end of two years. ■

EXAMPLE 7

Craig Lenihan/AP Photos

Apply Your Skill

On a 130-mile trip from Orlando to Sarasota, Roberto encountered a heavy thunderstorm for the last 40 miles of the trip. During the thunderstorm he averaged 20 miles per hour slower than before the storm. The entire trip took $2\frac{1}{2}$ hours. How fast did he travel before the storm?

Solution

Let x represent Roberto's rate before the thunderstorm; then $x - 20$ represents his speed during the thunderstorm. Because $t = \frac{d}{r}$, it follows that $\frac{90}{x}$ represents the time traveling before the storm and $\frac{40}{x - 20}$ represents the time traveling during the storm. The following guideline sums up the situation.

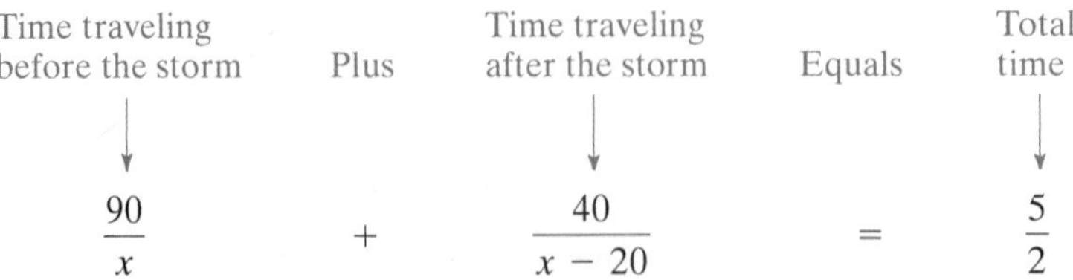

Solving this equation, we obtain

$$2x(x-20)\left(\frac{90}{x}+\frac{40}{x-20}\right)=2x(x-20)\left(\frac{5}{2}\right)$$

$$2x(x-20)\left(\frac{90}{x}\right)+2x(x-20)\left(\frac{40}{x-20}\right)=2x(x-20)\left(\frac{5}{2}\right)$$

$$180(x-20)+2x(40)=5x(x-20)$$

$$180x-3600+80x=5x^2-100x$$

$$0=5x^2-360x+3600$$

$$0=5(x^2-72x+720)$$

$$0=5(x-60)(x-12)$$

$$x-60=0 \quad \text{or} \quad x-12=0$$

$$x=60 \quad \text{or} \quad x=12$$

We discard the solution of 12 because it would be impossible to drive 20 miles per hour slower than 12 miles per hour; thus Roberto's rate before the thunderstorm was 60 miles per hour.

▼ PRACTICE YOUR SKILL

After 15 miles of a 20-mile bicycle trip, Pete had a flat tire and had to walk for the rest of the trip. While walking he averaged 6 miles per hour less than when he was bicycling. The entire trip took $2\frac{3}{4}$ hours. How fast did he bicycle? ■

EXAMPLE 8 Apply Your Skill

Phil Boorman/Stone/Getty Images

A computer installer agreed to do an installation for \$150. It took him 2 hours longer than he expected, and therefore he earned \$2.50 per hour less than he anticipated. How long did he expect it would take to do the installation?

Solution

Let x represent the number of hours he expected the installation to take. Then $x + 2$ represents the number of hours the installation actually took. The rate of pay is represented by the pay divided by the number of hours. The following guideline is used to write the equation.

Anticipated rate of pay	Minus	\$2.50	Equals	Actual rate of pay
↓		↓		↓
$\frac{150}{x}$	$-$	$\frac{5}{2}$	$=$	$\frac{150}{x+2}$

Solving this equation, we obtain

$$2x(x+2)\left(\frac{150}{x}-\frac{5}{2}\right)=2x(x+2)\left(\frac{150}{x+2}\right)$$

$$2(x+2)(150)-x(x+2)(5)=2x(150)$$

$$300(x+2)-5x(x+2)=300x$$

$$300x + 600 - 5x^2 - 10x = 300x$$
$$-5x^2 - 10x + 600 = 0$$
$$-5(x^2 + 2x - 120) = 0$$
$$-5(x + 12)(x - 10) = 0$$
$$x = -12 \quad \text{or} \quad x = 10$$

Disregard the negative answer. Therefore he anticipated that the installation would take 10 hours.

▼ **PRACTICE YOUR SKILL**

A tutor agreed to proofread a term paper for $24. It took her half an hour less than she expected, and therefore she earned $4 per hour more than she anticipated. How long did she expect it would take to proofread the term paper? ■

This next problem set contains a large variety of word problems. Not only are there some business applications similar to those we discussed in this section, but there are also more problems of the types discussed in Chapters 5 and 6. Try to give them your best shot without referring to the examples in earlier chapters.

CONCEPT QUIZ

For Problems 1–5, choose the method that you think is most appropriate for solving the given equation.

1. $2x^2 + 6x - 3 = 0$
2. $(x + 1)^2 = 36$
3. $x^2 - 3x + 2 = 0$
4. $x^2 + 6x = 19$
5. $4x^2 + 2x - 5 = 0$

A. Factoring
B. Square-root property (Property 8.1)
C. Completing the square
D. Quadratic formula

For Problems 6–10, match each question with its correct solution set.

6. $x^2 - 5x - 24 = 0$
7. $8x^2 + 31x - 4 = 0$
8. $3x^2 - x = -4$
9. $x^2 + 5x - 24 = 0$
10. $3x^2 + x = -4$

A. $\{-8, 3\}$
B. $\left\{\frac{1 \pm i\sqrt{47}}{6}\right\}$
C. $\left\{\frac{-1 \pm i\sqrt{47}}{6}\right\}$
D. $\{-3, 8\}$
E. $\left\{-4, \frac{1}{8}\right\}$

Problem Set 8.5

1 Solve Quadratic Equations Selecting the Most Appropriate Method

For Problems 1–20, solve each quadratic equation using the method that seems most appropriate to you.

1. $x^2 - 4x - 6 = 0$
2. $x^2 - 8x - 4 = 0$
3. $3x^2 + 23x - 36 = 0$
4. $n^2 + 22n + 105 = 0$
5. $x^2 - 18x = 9$
6. $x^2 + 20x = 25$
7. $2x^2 - 3x + 4 = 0$
8. $3y^2 - 2y + 1 = 0$
9. $135 + 24n + n^2 = 0$
10. $28 - x - 2x^2 = 0$
11. $(x - 2)(x + 9) = -10$
12. $(x + 3)(2x + 1) = -3$
13. $2x^2 - 4x + 7 = 0$
14. $3x^2 - 2x + 8 = 0$
15. $x^2 - 18x + 15 = 0$
16. $x^2 - 16x + 14 = 0$
17. $20y^2 + 17y - 10 = 0$
18. $12x^2 + 23x - 9 = 0$
19. $4t^2 + 4t - 1 = 0$
20. $5t^2 + 5t - 1 = 0$

For Problems 21–40, solve each equation.

21. $n + \frac{3}{n} = \frac{19}{4}$

22. $n - \frac{2}{n} = -\frac{7}{3}$

23. $\frac{3}{x} + \frac{7}{x-1} = 1$

24. $\frac{2}{x} + \frac{5}{x+2} = 1$

25. $\frac{12}{x-3} + \frac{8}{x} = 14$

26. $\frac{16}{x+5} - \frac{12}{x} = -2$

27. $\frac{3}{x-1} - \frac{2}{x} = \frac{5}{2}$

28. $\frac{4}{x+1} + \frac{2}{x} = \frac{5}{3}$

29. $\frac{6}{x} + \frac{40}{x+5} = 7$

30. $\frac{12}{t} + \frac{18}{t+8} = \frac{9}{2}$

31. $\frac{5}{n-3} - \frac{3}{n+3} = 1$

32. $\frac{3}{t+2} + \frac{4}{t-2} = 2$

33. $x^4 - 18x^2 + 72 = 0$

34. $x^4 - 21x^2 + 54 = 0$

35. $3x^4 - 35x^2 + 72 = 0$

36. $5x^4 - 32x^2 + 48 = 0$

37. $3x^4 + 17x^2 + 20 = 0$

38. $4x^4 + 11x^2 - 45 = 0$

39. $6x^4 - 29x^2 + 28 = 0$

40. $6x^4 - 31x^2 + 18 = 0$

2 Solve Word Problems Involving Quadratic Equations

For Problems 41–68, set up an equation and solve each problem.

41. Find two consecutive whole numbers such that the sum of their squares is 145.

42. Find two consecutive odd whole numbers such that the sum of their squares is 74.

43. Two positive integers differ by 3, and their product is 108. Find the numbers.

44. Suppose that the sum of two numbers is 20 and that the sum of their squares is 232. Find the numbers.

45. Find two numbers such that their sum is 10 and their product is 22.

46. Find two numbers such that their sum is 6 and their product is 7.

47. Suppose that the sum of two whole numbers is 9 and that the sum of their reciprocals is $\frac{1}{2}$. Find the numbers.

48. The difference between two whole numbers is 8, and the difference between their reciprocals is $\frac{1}{6}$. Find the two numbers.

49. The sum of the lengths of the two legs of a right triangle is 21 inches. If the length of the hypotenuse is 15 inches, find the length of each leg.

50. The length of a rectangular floor is 1 meter less than twice its width. If a diagonal of the rectangle is 17 meters, find the length and width of the floor.

51. A rectangular plot of ground measuring 12 meters by 20 meters is surrounded by a sidewalk of a uniform width (see Figure 8.9). The area of the sidewalk is 68 square meters. Find the width of the walk.

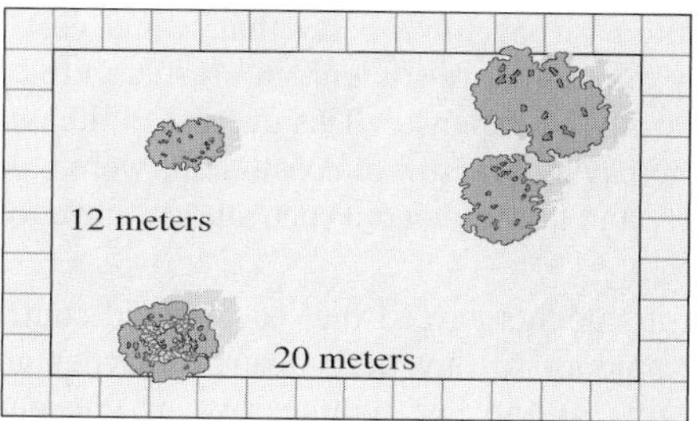

Figure 8.9

52. A 5-inch by 7-inch picture is surrounded by a frame of uniform width. The area of the picture and frame together is 80 square inches. Find the width of the frame.

53. The perimeter of a rectangle is 44 inches, and its area is 112 square inches. Find the length and width of the rectangle.

54. A rectangular piece of cardboard is 2 units longer than it is wide. From each of its corners a square piece 2 units on a side is cut out. The flaps are then turned up to form an open box that has a volume of 70 cubic units. Find the length and width of the original piece of cardboard.

55. Charlotte's time to travel 250 miles is 1 hour more than Lorraine's time to travel 180 miles. Charlotte drove 5 miles per hour faster than Lorraine. How fast did each one travel?

56. Larry's time to travel 156 miles is 1 hour more than Terrell's time to travel 108 miles. Terrell drove 2 miles per hour faster than Larry. How fast did each one travel?

57. On a 570-mile trip, Andy averaged 5 miles per hour faster for the last 240 miles than he did for the first 330 miles. The entire trip took 10 hours. How fast did he travel for the first 330 miles?

58. On a 135-mile bicycle excursion, Maria averaged 5 miles per hour faster for the first 60 miles than she did for the last 75 miles. The entire trip took 8 hours. Find her rate for the first 60 miles.

59. It takes Terry 2 hours longer to do a certain job than it takes Tom. They worked together for 3 hours; then Tom left and Terry finished the job in 1 hour. How long would it take each of them to do the job alone?

60. Suppose that Arlene can mow the entire lawn in 40 minutes less time with the power mower than she can with the push mower. One day the power mower broke down after she had been mowing for 30 minutes. She finished the lawn with the push mower in 20 minutes. How long does it take Arlene to mow the entire lawn with the power mower?

61. A student did a word processing job for \$24. It took him 1 hour longer than he expected, and therefore he earned \$4 per hour less than he anticipated. How long did he expect that it would take to do the job?

62. A group of students agreed that each would chip in the same amount to pay for a party that would cost \$100. Then they found 5 more students interested in the party and in sharing the expenses. This decreased the amount each had to pay by \$1. How many students were involved in the party and how much did each student have to pay?

63. A group of students agreed that each would contribute the same amount to buy their favorite teacher an \$80 birthday gift. At the last minute, two of the students decided not to chip in. This increased the amount that the remaining students had to pay by \$2 per student. How many students actually contributed to the gift?

64. The formula $D = \dfrac{n(n-3)}{2}$ yields the number of diagonals, D, in a polygon of n sides. Find the number of sides of a polygon that has 54 diagonals.

65. The formula $S = \dfrac{n(n+1)}{2}$ yields the sum, S, of the first n natural numbers 1, 2, 3, 4, ... How many consecutive natural numbers starting with 1 will give a sum of 1275?

66. At a point 16 yards from the base of a tower, the distance to the top of the tower is 4 yards more than the height of the tower (see Figure 8.10). Find the height of the tower.

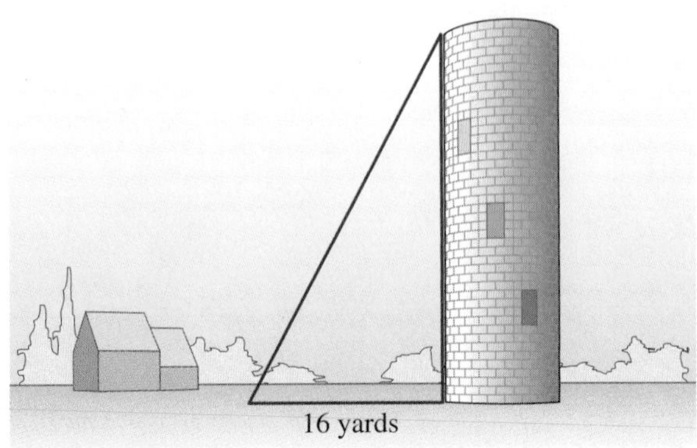

Figure 8.10

67. Suppose that \$500 is invested at a certain rate of interest compounded annually for 2 years. If the accumulated value at the end of 2 years is \$594.05, find the rate of interest.

68. Suppose that \$10,000 is invested at a certain rate of interest compounded annually for 2 years. If the accumulated value at the end of 2 years is \$12,544, find the rate of interest.

THOUGHTS INTO WORDS

69. How would you solve the equation $x^2 - 4x = 252$? Explain your choice of the method that you would use.

70. Explain how you would solve $(x-2)(x-7) = 0$ and also how you would solve $(x-2)(x-7) = 4$.

71. One of our problem-solving suggestions is to look for a guideline that can be used to help determine an equation. What does this suggestion mean to you?

72. Can a quadratic equation with integral coefficients have exactly one nonreal complex solution? Explain your answer.

FURTHER INVESTIGATIONS

For Problems 73–79, solve each equation.

73. $x - 9\sqrt{x} + 18 = 0$ [*Hint:* Let $y = \sqrt{x}$.]

74. $x - 4\sqrt{x} + 3 = 0$

75. $x + \sqrt{x} - 2 = 0$

76. $x^{\frac{2}{3}} + x^{\frac{1}{3}} - 6 = 0$ [*Hint:* Let $y = x^{\frac{1}{3}}$.]

77. $6x^{\frac{2}{3}} - 5x^{\frac{1}{3}} - 6 = 0$

78. $x^{-2} + 4x^{-1} - 12 = 0$

79. $12x^{-2} - 17x^{-1} - 5 = 0$

The following equations are also quadratic in form. To solve, begin by raising each side of the equation to the appropriate power so that the exponent will become an integer. Then, to solve the resulting quadratic equation, you may use the square-root property, factoring, or the quadratic formula, whichever is most appropriate. Be aware that raising each side of the equation to a power may introduce extraneous roots; therefore, be sure to check your solutions. Study the following example before you begin the problems.

Solve

$$(x+3)^{\frac{2}{3}} = 1$$

$$\left[(x+3)^{\frac{2}{3}}\right]^3 = 1^3 \quad \text{Raise both sides to the third power}$$

$$(x+3)^2 = 1$$

$$x^2 + 6x + 9 = 1$$
$$x^2 + 6x + 8 = 0$$
$$(x + 4)(x + 2) = 0$$
$$x + 4 = 0 \quad \text{or} \quad x + 2 = 0$$
$$x = -4 \quad \text{or} \quad x = -2$$

Both solutions do check. The solution set is $\{-4, -2\}$.

For problems 80–88, solve each equation.

80. $(5x + 6)^{\frac{1}{2}} = x$

81. $(3x + 4)^{\frac{1}{2}} = x$

82. $x^{\frac{2}{3}} = 2$

83. $x^{\frac{2}{5}} = 2$

84. $(2x + 6)^{\frac{1}{2}} = x$

85. $(2x - 4)^{\frac{2}{3}} = 1$

86. $(4x + 5)^{\frac{2}{3}} = 2$

87. $(6x + 7)^{\frac{1}{2}} = x + 2$

88. $(5x + 21)^{\frac{1}{2}} = x + 3$

Answers to the Concept Quiz

Answers for Problems 1–5 may vary. **1.** D **2.** B **3.** A **4.** C **5.** D **6.** D **7.** E **8.** B **9.** A **10.** C

Answers to the Example Practice Skills

1. $\left\{\frac{5 \pm \sqrt{13}}{6}\right\}$ **2.** $\{-3, 8\}$ **3.** $\{-16, -12\}$ **4.** $\{\pm 2\sqrt{5}, \pm i\sqrt{3}\}$ **5.** 4 cm by 16 cm **6.** 8.0% **7.** 10 mph **8.** 2 hr

8.6 Quadratic and Other Nonlinear Inequalities

OBJECTIVES

1. Solve Quadratic Inequalities
2. Solve Inequalities of Quotients

1 Solve Quadratic Inequalities

We refer to the equation $ax^2 + bx + c = 0$ as the standard form of a quadratic equation in one variable. Similarly, the following forms express **quadratic inequalities** in one variable.

$$ax^2 + bx + c > 0 \qquad ax^2 + bx + c < 0$$
$$ax^2 + bx + c \geq 0 \qquad ax^2 + bx + c \leq 0$$

We can use the number line very effectively to help solve quadratic inequalities where the quadratic polynomial is factorable. Let's consider some examples to illustrate the procedure.

EXAMPLE 1 Solve and graph the solutions for $x^2 + 2x - 8 > 0$.

Solution

First, let's factor the polynomial.

$$x^2 + 2x - 8 > 0$$
$$(x + 4)(x - 2) > 0$$

On a number line (Figure 8.11) we indicate that, at $x = 2$ and $x = -4$, the product $(x + 4)(x - 2)$ equals zero. The numbers -4 and 2 divide the number line into three intervals: (1) the numbers less than -4, (2) the numbers between -4 and 2, and (3) the numbers greater than 2. We can choose a **test number** from each of these intervals and

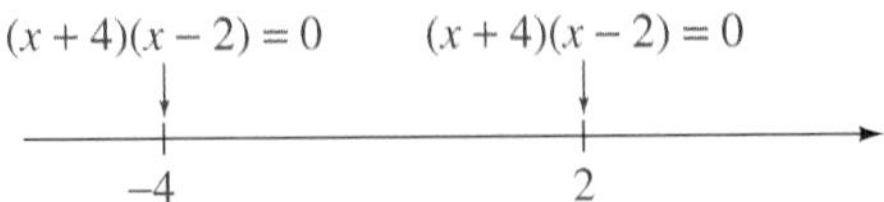

Figure 8.11

see how it affects the signs of the factors $x + 4$ and $x - 2$ and, consequently, the sign of the product of these factors. For example, if $x < -4$ (try $x = -5$), then $x + 4$ is negative and $x - 2$ is negative, so their product is positive. If $-4 < x < 2$ (try $x = 0$), then $x + 4$ is positive and $x - 2$ is negative, so their product is negative. If $x > 2$ (try $x = 3$), then $x + 4$ is positive and $x - 2$ is positive, so their product is positive. This information can be conveniently arranged using a number line, as shown in Figure 8.12. Note the open circles at -4 and 2, which indicate that they are not included in the solution set.

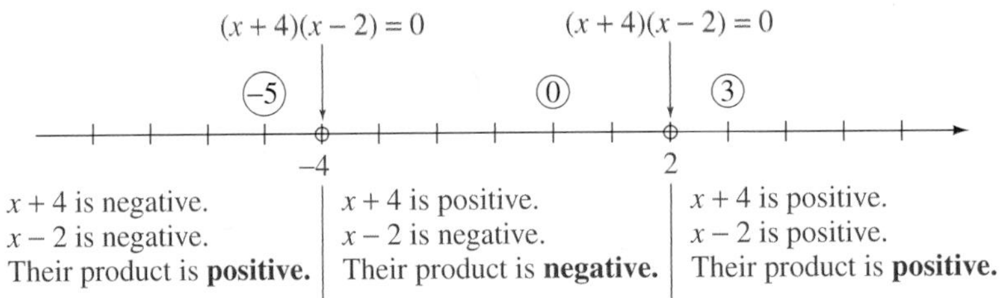

Figure 8.12

Thus the given inequality, $x^2 + 2x - 8 > 0$, is satisfied by numbers less than -4 along with numbers greater than 2. Using interval notation, the solution set is $(-\infty, -4) \cup (2, \infty)$. These solutions can be shown on a number line (Figure 8.13).

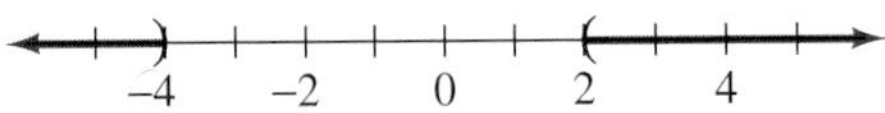

Figure 8.13

▼ PRACTICE YOUR SKILL

Solve and graph the solution for $y^2 - y - 30 > 0$. ■

We refer to numbers such as -4 and 2 in the preceding example (where the given polynomial or algebraic expression equals zero or is undefined) as **critical numbers**. Let's consider some additional examples that make use of critical numbers and test numbers.

EXAMPLE 2

Solve and graph the solutions for $x^2 + 2x - 3 \leq 0$.

Solution

First, factor the polynomial.

$$x^2 + 2x - 3 \leq 0$$

$$(x + 3)(x - 1) \leq 0$$

Second, locate the values for which $(x + 3)(x - 1)$ equals zero. We put solid dots at -3 and 1 to remind ourselves that these two numbers are to be included in the solution set because the given statement includes equality. Now let's choose a test number from each of the three intervals and record the sign behavior of the factors $(x + 3)$ and $(x - 1)$ (Figure 8.14).

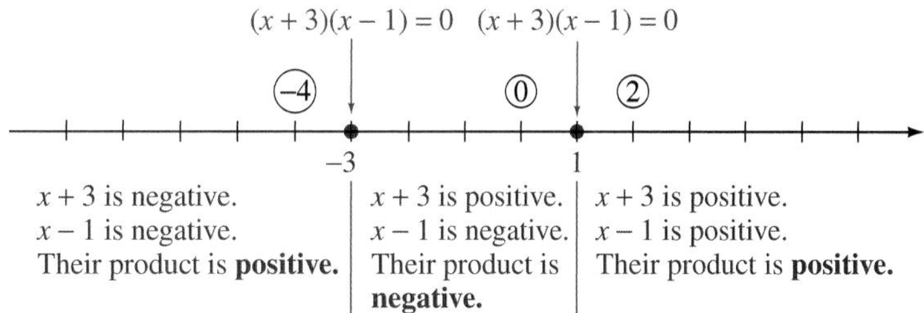

Figure 8.14

Therefore, the solution set is $[-3, 1]$, and it can be graphed as in Figure 8.15.

Figure 8.15

▼ PRACTICE YOUR SKILL

Solve and graph the solution for $y^2 - 7y + 10 \leq 0$. ■

2 Solve Inequalities of Quotients

Examples 1 and 2 have indicated a systematic approach for solving quadratic inequalities where the polynomial is factorable. This same type of number line analysis can also be used to solve indicated quotients such as $\frac{x+1}{x-5} > 0$.

EXAMPLE 3

Solve and graph the solutions for $\frac{x+1}{x-5} > 0$.

Solution

First, indicate that at $x = -1$ the given quotient equals zero and at $x = 5$ the quotient is undefined. Second, choose test numbers from each of the three intervals and record the sign behavior of $(x + 1)$ and $(x - 5)$ as in Figure 8.16.

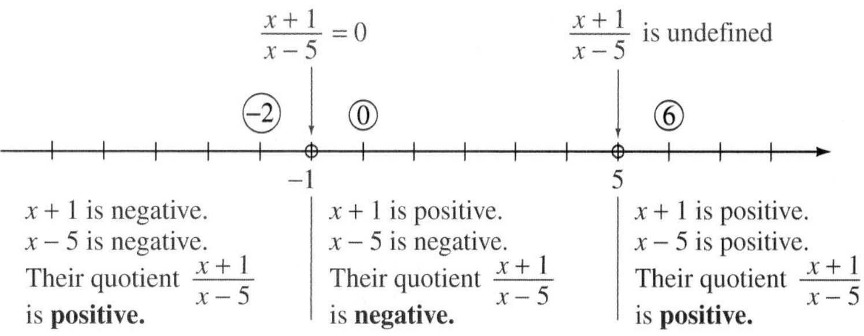

Figure 8.16

Therefore, the solution set is $(-\infty, -1) \cup (5, \infty)$, and its graph is shown in Figure 8.17.

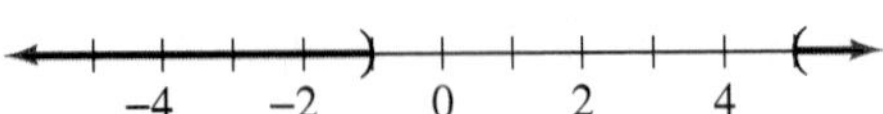

Figure 8.17

▼ **PRACTICE YOUR SKILL**

Solve and graph the solution for $\frac{x-4}{x+3} > 0$. ■

EXAMPLE 4

Solve $\frac{x+2}{x+4} \leq 0$.

Solution

The indicated quotient equals zero at $x = -2$ and is undefined at $x = -4$. (Note that -2 is to be included in the solution set but -4 is not to be included.) Now let's choose some test numbers and record the sign behavior of $(x + 2)$ and $(x + 4)$ as in Figure 8.18.

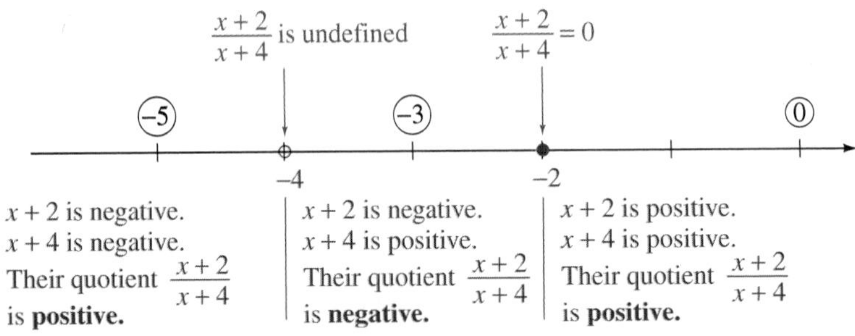

Figure 8.18

Therefore, the solution set is $(-4, -2]$.

▼ **PRACTICE YOUR SKILL**

Solve $\frac{x-6}{x-2} \leq 0$. ■

The final example illustrates that sometimes we need to change the form of the given inequality before we use the number-line analysis.

EXAMPLE 5

Solve $\frac{x}{x+2} \geq 3$.

Solution

First, let's change the form of the given inequality as follows:

$$\frac{x}{x+2} \geq 3$$

$$\frac{x}{x+2} - 3 \geq 0 \qquad \text{Add } -3 \text{ to both sides}$$

$$\frac{x - 3(x+2)}{x+2} \geq 0 \qquad \text{Express the left side over a common denominator}$$

$$\frac{x - 3x - 6}{x+2} \geq 0$$

$$\frac{-2x - 6}{x+2} \geq 0$$

Now we can proceed as we did with the previous examples. If $x = -3$, then $\frac{-2x - 6}{x + 2}$ equals zero; if $x = -2$, then $\frac{-2x - 6}{x + 2}$ is undefined. Then, choosing test numbers, we can record the sign behavior of $(-2x - 6)$ and $(x + 2)$ as in Figure 8.19.

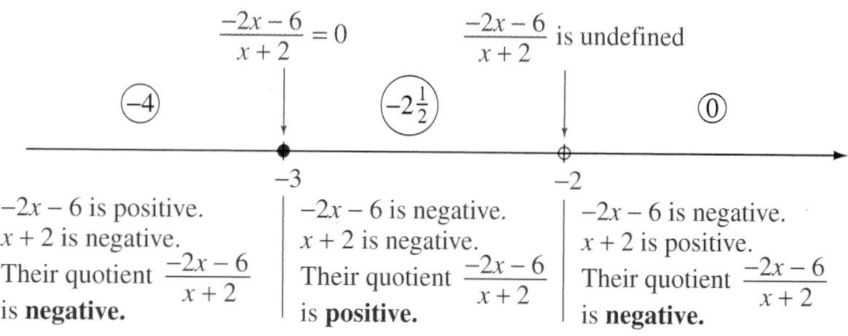

$-2x - 6$ is positive. $x + 2$ is negative. Their quotient $\frac{-2x-6}{x+2}$ is **negative.**	$-2x - 6$ is negative. $x + 2$ is negative. Their quotient $\frac{-2x-6}{x+2}$ is **positive.**	$-2x - 6$ is negative. $x + 2$ is positive. Their quotient $\frac{-2x-6}{x+2}$ is **negative.**

Figure 8.19

Therefore, the solution set is $[-3, -2)$. Perhaps you should check a few numbers from this solution set back into the original inequality!

▼ **PRACTICE YOUR SKILL**

Solve $\frac{5x}{x + 3} \geq 2$. ■

CONCEPT QUIZ

For Problems 1–10, answer true or false.

1. When solving the inequality $(x + 3)(x - 2) > 0$, we are finding values of x that make the product of $(x + 3)$ and $(x - 2)$ a positive number.
2. The solution set of the inequality $x^2 + 4 > 0$ is all real numbers.
3. The solution set of the inequality $x^2 \leq 0$ is the null set.
4. The critical numbers for the inequality $(x + 4)(x - 1) \leq 0$ are -4 and -1.
5. The number 2 is included in the solution set of the inequality $\frac{x + 4}{x - 2} \geq 0$.
6. The solution set of $(x - 2)^2 \geq 0$ is the set of all real numbers.
7. The solution set of $\frac{x + 2}{x - 3} \leq 0$ is $(-2, 3)$.
8. The solution set of $\frac{x - 1}{x} > 2$ is $(-1, 0)$.
9. The solution set of the inequality $(x - 2)^2(x + 1)^2 < 0$ is $\varnothing$.
10. The solution set of the inequality $(x - 4)(x + 3)^2 \leq 0$ is $(-\infty, 4]$.

Problem Set 8.6

1 Solve Quadratic Inequalities

For Problems 1–12, solve each inequality and graph its solution set on a number line.

1. $(x + 2)(x - 1) > 0$
2. $(x - 2)(x + 3) > 0$
3. $(x + 1)(x + 4) < 0$
4. $(x - 3)(x - 1) < 0$
5. $(2x - 1)(3x + 7) \geq 0$
6. $(3x + 2)(2x - 3) \geq 0$
7. $(x + 2)(4x - 3) \leq 0$
8. $(x - 1)(2x - 7) \leq 0$
9. $(x + 1)(x - 1)(x - 3) > 0$
10. $(x + 2)(x + 1)(x - 2) > 0$
11. $x(x + 2)(x - 4) \leq 0$
12. $x(x + 3)(x - 3) \leq 0$

For Problems 13–38, solve each inequality.

13. $x^2 + 2x - 35 < 0$

14. $x^2 + 3x - 54 < 0$

15. $x^2 - 11x + 28 > 0$

16. $x^2 + 11x + 18 > 0$

17. $3x^2 + 13x - 10 \le 0$

18. $4x^2 - x - 14 \le 0$

19. $8x^2 + 22x + 5 \ge 0$

20. $12x^2 - 20x + 3 \ge 0$

21. $x(5x - 36) > 32$

22. $x(7x + 40) < 12$

23. $x^2 - 14x + 49 \ge 0$

24. $(x + 9)^2 \ge 0$

25. $4x^2 + 20x + 25 \le 0$

26. $9x^2 - 6x + 1 \le 0$

27. $(x + 1)(x - 3)^2 > 0$

28. $(x - 4)^2(x - 1) \le 0$

29. $4 - x^2 < 0$

30. $2x^2 - 18 \ge 0$

31. $4(x^2 - 36) < 0$

32. $-4(x^2 - 36) \ge 0$

33. $5x^2 + 20 > 0$

34. $-3x^2 - 27 \ge 0$

35. $x^2 - 2x \ge 0$

36. $2x^2 + 6x < 0$

37. $3x^3 + 12x^2 > 0$

38. $2x^3 + 4x^2 \le 0$

2 Solve Inequalities of Quotients

For Problems 39–56, solve each inequality.

39. $\frac{x + 1}{x - 2} > 0$

40. $\frac{x - 1}{x + 2} > 0$

41. $\frac{x - 3}{x + 2} < 0$

42. $\frac{x + 2}{x - 4} < 0$

43. $\frac{2x - 1}{x} \ge 0$

44. $\frac{x}{3x + 7} \ge 0$

45. $\frac{-x + 2}{x - 1} \le 0$

46. $\frac{3 - x}{x + 4} \le 0$

47. $\frac{2x}{x + 3} > 4$

48. $\frac{x}{x - 1} > 2$

49. $\frac{x - 1}{x - 5} \le 2$

50. $\frac{x + 2}{x + 4} \le 3$

51. $\frac{x + 2}{x - 3} > -2$

52. $\frac{x - 1}{x - 2} < -1$

53. $\frac{3x + 2}{x + 4} \le 2$

54. $\frac{2x - 1}{x + 2} \ge -1$

55. $\frac{x + 1}{x - 2} < 1$

56. $\frac{x + 3}{x - 4} \ge 1$

THOUGHTS INTO WORDS

57. Explain how to solve the inequality $(x + 1)(x - 2)(x - 3) > 0$.

58. Explain how to solve the inequality $(x - 2)^2 > 0$ by inspection.

59. Your friend looks at the inequality $1 + \frac{1}{x} > 2$ and without any computation states that the solution set is all real numbers between 0 and 1. How can she do that?

60. Why is the solution set for $(x - 2)^2 \ge 0$ the set of all real numbers?

61. Why is the solution set for $(x - 2)^2 \le 0$ the set $\{2\}$?

FURTHER INVESTIGATIONS

62. The product $(x - 2)(x + 3)$ is positive if both factors are negative *or* if both factors are positive. Therefore, we can solve $(x - 2)(x + 3) > 0$ as follows:

$$(x - 2 < 0 \text{ and } x + 3 < 0) \text{ or } (x - 2 > 0 \text{ and } x + 3 > 0)$$

$$(x < 2 \text{ and } x < -3) \text{ or } (x > 2 \text{ and } x > -3)$$

$$x < -3 \text{ or } x > 2$$

The solution set is $(-\infty, -3) \cup (2, \infty)$. Use this type of analysis to solve each of the following.

(a) $(x - 2)(x + 7) > 0$

(b) $(x - 3)(x + 9) \ge 0$

(c) $(x + 1)(x - 6) \le 0$

(d) $(x + 4)(x - 8) < 0$

(e) $\frac{x + 4}{x - 7} > 0$

(f) $\frac{x - 5}{x + 8} \le 0$

Answers to the Concept Quiz

1. True **2.** True **3.** False **4.** False **5.** False **6.** True **7.** False **8.** True **9.** True **10.** True

Answers to the Example Practice Skills

1. $(-\infty, -5) \cup (6, \infty)$ **2.** $[2, 5]$ **3.** $(-\infty, -3) \cup (4, \infty)$ **4.** $(2, 6]$ **5.** $(-\infty, -3) \cup [2, \infty)$

Chapter 8 Summary

OBJECTIVE	SUMMARY	EXAMPLE	CHAPTER REVIEW PROBLEMS
Know about the set of complex numbers. (Sec. 8.1, Obj. 1, p. 418)	A number of the form $a + bi$, where a and b are real numbers and i is the imaginary unit defined by $i = \sqrt{-1}$, is a complex number. Two complex numbers are said to be equal if and only if $a = c$ and $b = d$.		
Add and subtract complex numbers. (Sec. 8.1, Obj. 2, p. 419)	We describe the addition and subtraction of complex numbers as follows: $(a + bi) + (c + di) = (a + c) + (b + d)i$ $(a + bi) - (c + di) = (a - c) + (b - d)i$	Add the complex numbers $(3 - 6i) + (-7 - 3i)$. **Solution** $(3 - 6i) + (-7 - 3i) = (3 - 7) + (-6 - 3)i = -4 - 9i$	Problems 1–4
Simplify radicals involving negative numbers. (Sec. 8.1, Obj. 3, p. 420)	We can represent a square root of any negative real number as the product of a real number and the imaginary unit i. That is, $\sqrt{-b} = i\sqrt{b}$, where b is a positive real number.	Write $\sqrt{-48}$ in terms of i and simplify. **Solution** $\sqrt{-48} = \sqrt{-1}\sqrt{48} = i\sqrt{16}\sqrt{3} = 4i\sqrt{3}$	Problems 5–8
Perform operations on radicals involving negative numbers. (Sec. 8.1, Obj. 4, p. 421)	Before performing any operations, represent a square root of any negative real number as the product of a real number and the imaginary unit i.	Perform the indicated operation and simplify. $\dfrac{\sqrt{-28}}{\sqrt{-4}}$ **Solution** $\dfrac{\sqrt{-28}}{\sqrt{-4}} = \dfrac{i\sqrt{28}}{i\sqrt{4}} = \dfrac{\sqrt{28}}{\sqrt{4}} = \sqrt{7}$	Problems 9–12
Multiply complex numbers. (Sec. 8.1, Obj. 5, p. 422)	The product of two complex numbers follows the same pattern as the product of two binomials. The conjugate of $a + bi$ is $a - bi$. The product of a complex number and its conjugate is a real number. When simplifying, replace any i^2 with -1.	Find the product $(2 + 3i)(4 - 5i)$ and express the answer in standard form of a complex number. **Solution** $(2 + 3i)(4 - 5i) = 8 + 2i - 15i^2 = 8 + 2i - 15(-1) = 23 + 2i$	Problems 13–16 *(continued)*

OBJECTIVE	SUMMARY	EXAMPLE	CHAPTER REVIEW PROBLEMS
Divide complex numbers. (Sec. 8.1, Obj. 6, p. 423)	To simplify expressions that indicate the quotient of complex numbers such as $\frac{4+3i}{5-2i}$, multiply the numerator and denominator by the conjugate of the denominator.	Find the quotient $\frac{2+3i}{4-i}$ and express the answer in standard form of a complex number. **Solution** Multiply the numerator and denominator by $4+i$, the conjugate of the denominator. $\frac{2+3i}{4-i} = \frac{(2+3i)}{(4-i)} \cdot \frac{(4+i)}{(4+i)}$ $= \frac{8+14i+3i^2}{16-i^2}$ $= \frac{8+14i+3(-1)}{16-(-1)}$ $= \frac{5+14i}{17} = \frac{5}{17} + \frac{14}{17}i$	Problems 17–20
Solve quadratic equations by factoring. (Sec. 8.2, Obj. 1, p. 427)	The standard form for a quadratic equation in one variable is $ax^2 + bx + c = 0$, where a, b, and c are real numbers and $a \neq 0$. Some quadratics can be solved by factoring and applying the property, $ab = 0$ if and only if $a = 0$ or $b = 0$.	Solve $2x^2 + x - 3 = 0$. **Solution** $2x^2 + x - 3 = 0$ $(2x+3)(x-1) = 0$ $2x + 3 = 0$ or $x - 1 = 0$ $x = -\frac{3}{2}$ or $x = 1$ The solution set is $\left\{-\frac{3}{2}, 1\right\}$.	Problems 21–24
Solve quadratic equations of the form $x^2 = a$. (Sec. 8.1, Obj. 2, p. 428)	We can solve some quadratic equations by applying the property, $x^2 = a$ if and only if $x = \pm\sqrt{a}$.	Solve $3(x+7)^2 = 24$. **Solution** $3(x+7)^2 = 24$ First divide both sides of the equation by 3. $(x+7)^2 = 8$ $x + 7 = \pm\sqrt{8}$ $x + 7 = \pm 2\sqrt{2}$ $x = -7 \pm 2\sqrt{2}$ The solution set is $\{-7 \pm 2\sqrt{2}\}$.	Problems 25–28 *(continued)*

<table>
<tr><th>OBJECTIVE</th><th>SUMMARY</th><th>EXAMPLE</th><th>CHAPTER REVIEW PROBLEMS</th></tr>
<tr><td>Solve quadratic equations by completing the square.
(Sec. 8.3, Obj. 1, p. 435)</td><td>To solve a quadratic equation by completing the square, first put the equation in the form $x^2 + bx = k$. Then (1) take one-half of b, square that result, and add to each side of the equation; (2) factor the left side; and (3) apply the property, $x^2 = a$ if and only if $x = \pm\sqrt{a}$.</td><td>Solve $x^2 + 12x - 2 = 0$.
Solution
$x^2 + 12x - 2 = 0$
$x^2 + 12x = 2$
$x^2 + 12x + 36 = 2 + 36$
$(x + 6)^2 = 38$
$x + 6 = \pm\sqrt{38}$
$x = -6 \pm \sqrt{38}$
The solution set is $\{-6 \pm \sqrt{38}\}$.</td><td>Problems 29–32</td></tr>
<tr><td>Use the quadratic formula to solve quadratic equations.
(Sec. 8.4, Obj. 1, p. 440)</td><td>Any quadratic equation of the form $ax^2 + bx + c = 0$ can be solved by the quadratic formula, which is usually stated as
$x = \dfrac{-b \pm \sqrt{b^2 - 4ac}}{2a}$.</td><td>Solve $3x^2 - 5x - 6 = 0$.
Solution
$3x^2 - 5x - 6 = 0$
$a = 3, b = -5,$ and $c = -6$
$x = \dfrac{-(-5) \pm \sqrt{(-5)^2 - 4(3)(-6)}}{2(3)}$
$x = \dfrac{5 \pm \sqrt{97}}{6}$
The solution set is $\left\{\dfrac{5 \pm \sqrt{97}}{6}\right\}$.</td><td>Problems 33–36</td></tr>
<tr><td>Determine the nature of roots to quadratic equations.
(Sec. 8.4, Obj. 2, p. 444)</td><td>The discriminant, $b^2 - 4ac$, can be used to determine the nature of the roots of a quadratic equation.
1. If $b^2 - 4ac$ is less than zero, then the equation has two nonreal complex solutions.
2. If $b^2 - 4ac$ is equal to zero, then the equation has two equal real solutions.
3. If $b^2 - 4ac$ is greater than zero, then the equation has two unequal real solutions.</td><td>Use the discriminant to determine the nature of the solutions for the equation $2x^2 + 3x + 5 = 0$.
Solution
$2x^2 + 3x + 5 = 0$
For $a = 2$, $b = 3$, and $c = 5$,
$b^2 - 4ac = (3)^2 - 4(2)(5) = -31$.
Because the discriminant is less than zero, the equation has two nonreal complex solutions.</td><td>Problems 37–40

(continued)</td></tr>
</table>

OBJECTIVE	SUMMARY	EXAMPLE	CHAPTER REVIEW PROBLEMS
Solve quadratic equations selecting the most appropriate method. (Sec. 8.5, Obj. 1, p. 447)	There are three major methods for solving a quadratic equation. 1. Factoring 2. Completing the square 3. Quadratic formula Consider which method is most appropriate before you begin solving the equation.	Solve $x^2 - 4x + 9 = 0$. **Solution** This equation does not factor. Because $a = 1$ and b is an even number, this equation can easily be solved by completing the square. $x^2 - 4x + 9 = 0$ $x^2 - 4x = -9$ $x^2 - 4x + 4 = -9 + 4$ $(x + 4)^2 = -5$ $x + 4 = \pm\sqrt{-5}$ $x = -4 \pm i\sqrt{5}$ The solution set is $\{-4 \pm i\sqrt{5}\}$.	Problems 41–59
Solve problems pertaining to right triangles and 30º-60º triangles. (Sec. 8.2, Obj. 3, p. 431)	There are two special kinds of right triangles that are used in later mathematics courses. The **isosceles right triangle** is a right triangle that has both legs of the same length. In a **30º-60º right triangle**, the side opposite the 30º angle is equal in length to one-half the length of the hypotenuse.	Find the length of each leg of an isosceles right triangle that has a hypotenuse of length 6 inches. **Solution** Let x represent the length of each leg. $x^2 + x^2 = 6^2$ $2x^2 = 36$ $x^2 = 18$ $x = \pm\sqrt{18} = \pm 3\sqrt{2}$ Disregard the negative solution. The length of each leg is $3\sqrt{2}$.	Problems 60–62
Solve word problems involving quadratic equations. (Sec. 8.5, Obj. 2, p. 450)	Keep the following suggestions in mind as you solve word problems. 1. Read the problem carefully. 2. Sketch any figure, diagram, or chart that might help you organize and analyze the problem. 3. Choose a meaningful variable. 4. Look for a guideline that can be used to set up an equation. 5. Form an equation that translates the guideline from English into algebra. 6. Solve the equation and answer the question posed in the problem. 7. Check all answers back into the original statement of the problem.	Find two consecutive odd whole numbers such that the sum of their squares is 290. **Solution** Let x represent the first whole number. Then $x + 2$ would represent the next consecutive odd whole number. $x^2 + (x + 2)^2 = 290$ $x^2 + x^2 + 4x + 4 = 290$ $2x^2 + 4x - 286 = 0$ $2(x^2 + 2x - 143) = 0$ $2(x + 13)(x - 11) = 0$ $x = -13$ or $x = 11$ Disregard the solution of −13 because it is not a whole number. The integers are 11 and 13.	Problems 63–70 *(continued)*

OBJECTIVE	SUMMARY	EXAMPLE	CHAPTER REVIEW PROBLEMS
Solve quadratic inequalities. (Sec. 8.6, Obj. 1, p. 457)	To solve quadratic inequalities that are factorable polynomials, the critical numbers are found by factoring the polynomial. The critical numbers partition the number line into regions. A test point from each region is used to determine if the values in that region make the inequality a true statement. The answer is usually expressed in interval notation.	Solve $x^2 + x - 6 \le 0$. **Solution** Solve the equation $x^2 + x - 6 = 0$ to find the critical numbers. $x^2 + x - 6 = 0$ $(x + 3)(x - 2) = 0$ $x = -3$ or $x = 2$ The critical numbers are -3 and 2. Choose a test point from each of the intervals $(-\infty, -3)$, $(-3, 2)$, and $(2, \infty)$. Evaluating the inequality $x^2 + x - 6 \le 0$ for each of the test points shows that $(-3, 2)$ is the only interval of values that makes the inequality a true statement. Because the inequality includes the endpoints of the interval, the solution is $[-3, 2]$.	Problems 71–74
Solve inequalities of quotients. (Sec. 8.6, Obj. 2, p. 459)	To solve inequalities involving quotients, use the same basic approach as for solving quadratic equations. Be careful to avoid any values that make the denominator zero.	Solve $\frac{x + 1}{2x - 3} \ge 0$. **Solution** Set the numerator equal to zero and then set the denominator equal to zero to find the critical numbers. $x + 1 = 0$ and $2x - 3 = 0$ $x = -1$ and $x = \frac{3}{2}$ The critical numbers are -1 and $\frac{3}{2}$. Evaluate the inequality with a test point from each of the intervals $(-\infty, -1)$, $\left(-1, \frac{3}{2}\right)$, and $\left(\frac{3}{2}, \infty\right)$; this shows that the values in the intervals $(-\infty, -1)$ and $\left(\frac{3}{2}, \infty\right)$ make the inequality a true statement. Because the inequality includes the "equal to" statement, the solution should include -1 but not $\frac{3}{2}$, because $\frac{3}{2}$ would make the quotient undefined. The solution set is $(-\infty, -1] \cup \left(\frac{3}{2}, \infty\right)$.	Problems 75–78

Chapter 8 Review Problem Set

For Problems 1–4, perform the indicated operations and express the answers in the standard form of a complex number.

1. $(-7 + 3i) + (9 - 5i)$

2. $(4 - 10i) - (7 - 9i)$

3. $(6 - 3i) - (-2 + 5i)$

4. $(-4 + i) + (2 + 3i)$

For Problems 5–8, write each expression in terms of i and simplify.

5. $\sqrt{-8}$

6. $\sqrt{-25}$

7. $3\sqrt{-16}$

8. $2\sqrt{-18}$

For Problems 9–18, perform the indicated operation and simplify.

9. $\sqrt{-2}\sqrt{-6}$

10. $\sqrt{-2}\sqrt{18}$

11. $\dfrac{\sqrt{-42}}{\sqrt{-6}}$

12. $\dfrac{\sqrt{-6}}{\sqrt{2}}$

13. $5i(3 - 6i)$

14. $(5 - 7i)(6 + 8i)$

15. $(-2 - 3i)(4 - 8i)$

16. $(4 - 3i)(4 + 3i)$

17. $\dfrac{4 + 3i}{6 - 2i}$

18. $\dfrac{-1 - i}{-2 + 5i}$

For Problems 19 and 20, perform the indicated operations and express the answer in the standard form of a complex number.

19. $\dfrac{3 + 4i}{2i}$

20. $\dfrac{-6 + 5i}{-i}$

For Problems 21–24, solve each of the quadratic equations by factoring.

21. $x^2 + 8x = 0$

22. $x^2 = 6x$

23. $x^2 - 3x - 28 = 0$

24. $2x^2 + x - 3 = 0$

For Problems 25–28, use Property 8.1 to help solve each quadratic equation.

25. $2x^2 = 90$

26. $(y - 3)^2 = -18$

27. $(2x + 3)^2 = 24$

28. $a^2 - 27 = 0$

For Problems 29–32, use the method of completing the square to solve the quadratic equation.

29. $y^2 + 18y - 10 = 0$

30. $n^2 + 6n + 20 = 0$

31. $x^2 - 10x + 1 = 0$

32. $x^2 + 5x - 2 = 0$

For Problems 33–36, use the quadratic formula to solve the equation.

33. $x^2 + 6x + 4 = 0$

34. $x^2 + 4x + 6 = 0$

35. $3x^2 - 2x + 4 = 0$

36. $5x^2 - x - 3 = 0$

For Problems 37–40, find the discriminant of each equation and determine whether the equation has (1) two nonreal complex solutions, (2) one real solution with a multiplicity of 2, or (3) two real solutions. Do not solve the equations.

37. $4x^2 - 20x + 25 = 0$

38. $5x^2 - 7x + 31 = 0$

39. $7x^2 - 2x - 14 = 0$

40. $5x^2 - 2x = 4$

For Problems 41–59, solve each equation.

41. $x^2 - 17x = 0$

42. $(x - 2)^2 = 36$

43. $(2x - 1)^2 = -64$

44. $x^2 - 4x - 21 = 0$

45. $x^2 + 2x - 9 = 0$

46. $x^2 - 6x = -34$

47. $4\sqrt{x} = x - 5$

48. $3n^2 + 10n - 8 = 0$

49. $n^2 - 10n = 200$

50. $3a^2 + a - 5 = 0$

51. $x^2 - x + 3 = 0$

52. $2x^2 - 5x + 6 = 0$

53. $2a^2 + 4a - 5 = 0$

54. $t(t + 5) = 36$

55. $x^2 + 4x + 9 = 0$

56. $(x - 4)(x - 2) = 80$

57. $\dfrac{3}{x} + \dfrac{2}{x + 3} = 1$

58. $2x^4 - 23x^2 + 56 = 0$

59. $\dfrac{3}{n - 2} = \dfrac{n + 5}{4}$

For Problems 60–70, set up an equation and solve each problem.

60. The wing of an airplane is in the shape of a 30°-60° right triangle. If the side opposite the 30° angle measures 20 feet, find the measure of the other two sides of the wing. Round the answers to the nearest tenth of a foot.

61. An agency is using photo surveillance of a rectangular plot of ground that measures 40 meters by 25 meters. If, during the surveillance, someone is observed moving from one corner of the plot to the corner diagonally opposite, how far has the observed person moved? Round the answer to the nearest tenth of a meter.

62. One leg of an isosceles right triangle measures 4 inches. Find the length of the hypotenuse of the triangle. Express the answer in radical form.

63. Find two numbers whose sum is 6 and whose product is 2.

64. A landscaper agreed to design and plant a flower bed for \$40. It took him three hours less than he anticipated, and therefore he earned \$3 per hour more than he anticipated. How long did he expect it would take to design and plant the flower bed?

65. Andre traveled 270 miles in 1 hour more than it took Sandy to travel 260 miles. Sandy drove 7 miles per hour faster than Andre. How fast did each one travel?

66. The area of a square is numerically equal to twice its perimeter. Find the length of a side of the square.

67. Find two consecutive even whole numbers such that the sum of their squares is 164.

68. The perimeter of a rectangle is 38 inches, and its area is 84 square inches. Find the length and width of the rectangle.

69. It takes Billy 2 hours longer to do a certain job than it takes Reena. They worked together for 2 hours; then Reena left, and Billy finished the job in 1 hour. How long would it take each of them to do the job alone?

70. A company has a rectangular parking lot 40 meters wide and 60 meters long. The company plans to increase the area of the lot by 1100 square meters by adding a strip of equal width to one side and one end. Find the width of the strip to be added.

For Problems 71–78, solve each inequality and indicate the solution set on a number line graph.

71. $x^2 + 3x - 10 > 0$

72. $2x^2 + x - 21 \le 0$

73. $4x^2 - 1 \le 0$

74. $x^2 - 7x + 10 > 0$

75. $\frac{x - 4}{x + 6} \ge 0$

76. $\frac{2x - 1}{x + 1} > 4$

77. $\frac{3x + 1}{x - 4} < 2$

78. $\frac{3x + 1}{x - 1} \le 0$

Chapter 8 Test

1. ______

2. ______

3. ______

4. ______

5. ______

6. ______

7. ______

8. ______

9. ______

10. ______

11. ______

12. ______

13. ______

14. ______

15. ______

16. ______

17. ______

18. ______

19. ______

20. ______

21. ______

22. ______

1. Find the product $(3 - 4i)(5 + 6i)$ and express the result in the standard form of a complex number.

2. Find the quotient $\frac{2 - 3i}{3 + 4i}$ and express the result in the standard form of a complex number.

For Problems 3–15, solve each equation.

3. $x^2 = 7x$

4. $(x - 3)^2 = 16$

5. $x^2 + 3x - 18 = 0$

6. $x^2 - 2x - 1 = 0$

7. $5x^2 - 2x + 1 = 0$

8. $x^2 + 30x = -224$

9. $(3x - 1)^2 + 36 = 0$

10. $(5x - 6)(4x + 7) = 0$

11. $(2x + 1)(3x - 2) = 55$

12. $n(3n - 2) = 40$

13. $x^4 + 12x^2 - 64 = 0$

14. $\frac{3}{x} + \frac{2}{x + 1} = 4$

15. $3x^2 - 2x - 3 = 0$

16. Does the equation $4x^2 + 20x + 25 = 0$ have (a) two nonreal complex solutions, (b) two equal real solutions, or (c) two unequal real solutions?

17. Does the equation $4x^2 - 3x = -5$ have (a) two non-real complex solutions, (b) two equal real solutions, or (c) two unequal real solutions?

For Problems 18–20, solve each inequality and express the solution set using interval notation.

18. $x^2 - 3x - 54 \leq 0$

19. $\frac{3x - 1}{x + 2} > 0$

20. $\frac{x - 2}{x + 6} \geq 3$

For Problems 21–25, set up an equation and solve each problem.

21. A 24-foot ladder leans against a building and makes an angle of 60° with the ground. How far up on the building does the top of the ladder reach? Express your answer to the nearest tenth of a foot.

22. A rectangular plot of ground measures 16 meters by 24 meters. Find, to the nearest meter, the distance from one corner of the plot to the diagonally opposite corner.

23. Amy agreed to clean her brother's room for $36. It took her 1 hour longer than she expected, and therefore she earned $3 per hour less than she anticipated. How long did she expect it would take to clean the room?

23. ________________

24. The perimeter of a rectangle is 41 inches and its area is 91 square inches. Find the length of its shortest side.

24. ________________

25. The sum of two numbers is 6 and their product is 4. Find the larger of the two numbers.

25. ________________

Chapters 1–8 Cumulative Review Problem Set

For Problems 1–4, evaluate each algebraic expression for the given values of the variables.

1. $\dfrac{4a^2b^3}{12a^3b}$ for $a = 5$ and $b = -8$

2. $\dfrac{\frac{1}{x}+\frac{1}{y}}{\frac{1}{x}-\frac{1}{y}}$ for $x = 4$ and $y = 7$

3. $\dfrac{3}{n}+\dfrac{5}{2n}-\dfrac{4}{3n}$ for $n = 25$

4. $2\sqrt{2x+y}-5\sqrt{3x-y}$ for $x = 5$ and $y = 6$

For Problems 5–16, perform the indicated operations and express the answers in simplified form.

5. $(3a^2b)(-2ab)(4ab^3)$

6. $(x+3)(2x^2-x-4)$

7. $\dfrac{6xy^2}{14y}\cdot\dfrac{7x^2y}{8x}$

8. $\dfrac{a^2+6a-40}{a^2-4a}\div\dfrac{2a^2+19a-10}{a^3+a^2}$

9. $\dfrac{3x+4}{6}-\dfrac{5x-1}{9}$

10. $\dfrac{4}{x^2+3x}+\dfrac{5}{x}$

11. $\dfrac{3n^2+n}{n^2+10n+16}\cdot\dfrac{2n^2-8}{3n^3-5n^2-2n}$

12. $\dfrac{3}{5x^2+3x-2}-\dfrac{2}{5x^2-22x+8}$

13. $\dfrac{y^3-7y^2+16y-12}{y-2}$

14. $(4x^3-17x^2+7x+10)\div(4x-5)$

15. $(3\sqrt{2}+2\sqrt{5})(5\sqrt{2}-\sqrt{5})$

16. $(\sqrt{x}-3\sqrt{y})(2\sqrt{x}+4\sqrt{y})$

For Problems 17–24, evaluate each of the numerical expressions.

17. $-\sqrt{\dfrac{9}{64}}$

18. $\sqrt[3]{-\dfrac{8}{27}}$

19. $\sqrt[3]{0.008}$

20. $32^{-\frac{1}{5}}$

21. $3^0+3^{-1}+3^{-2}$

22. $-9^{\frac{3}{2}}$

23. $\left(\dfrac{3}{4}\right)^{-2}$

24. $\dfrac{1}{\left(\frac{2}{3}\right)^{-3}}$

For Problems 25–30, factor each of the algebraic expressions completely.

25. $3x^4+81x$

26. $6x^2+19x-20$

27. $12+13x-14x^2$

28. $9x^4+68x^2-32$

29. $2ax-ay-2bx+by$

30. $27x^3-8y^3$

For Problems 31–54, solve each of the equations.

31. $3(x-2)-2(3x+5)=4(x-1)$

32. $0.06n+0.08(n+50)=25$

33. $4\sqrt{x}+5=x$

34. $\sqrt[3]{n^2-1}=-1$

35. $6x^2-24=0$

36. $a^2+14a+49=0$

37. $3n^2+14n-24=0$

38. $\dfrac{2}{5x-2}=\dfrac{4}{6x+1}$

39. $\sqrt{2x-1}-\sqrt{x+2}=0$

40. $5x-4=\sqrt{5x-4}$

41. $|3x-1|=11$

42. $(3x-2)(4x-1)=0$

43. $(2x+1)(x-2)=7$

44. $\dfrac{5}{6x}-\dfrac{2}{3}=\dfrac{7}{10x}$

45. $\dfrac{3}{y+4}+\dfrac{2y-1}{y^2-16}=\dfrac{-2}{y-4}$

46. $6x^4-23x^2-4=0$

47. $3n^3+3n=0$

48. $n^2-13n-114=0$

49. $12x^2+x-6=0$

50. $x^2-2x+26=0$

51. $(x+2)(x-6)=-15$

52. $(3x-1)(x+4)=0$

53. $x^2+4x+20=0$

54. $2x^2-x-4=0$

For Problems 55–64, solve each inequality and express the solution set using interval notation.

55. $6 - 2x \geq 10$

56. $4(2x - 1) < 3(x + 5)$

57. $\frac{n+1}{4} + \frac{n-2}{12} > \frac{1}{6}$

58. $|2x - 1| < 5$

59. $|3x + 2| > 11$

60. $\frac{1}{2}(3x - 1) - \frac{2}{3}(x + 4) \leq \frac{3}{4}(x - 1)$

61. $x^2 - 2x - 8 \leq 0$

62. $3x^2 + 14x - 5 > 0$

63. $\frac{x+2}{x-7} \geq 0$

64. $\frac{2x-1}{x+3} < 1$

For Problems 65–70, graph the following equations. Label the x and y intercepts on the graph.

65. $2x - y = 4$

66. $x - 3y = 6$

67. $y = \frac{1}{2}x + 3$

68. $y = -3x + 1$

69. $y = -4$

70. $x = 2$

For Problems 71 and 72, find the distance between the two points. Express the answer in simplest radical form.

71. $(3, -1)$ and $(-4, 6)$

72. $(8, 0)$ and $(3, 4)$

For Problems 73–76, write the equation of a line that satisfies the given conditions. Express the answer in standard form.

73. x intercept of 2 and slope of $\frac{3}{5}$

74. Contains the points $(-1, 4)$ and $(0, 3)$

75. Contains the point $(-3, 5)$ and is parallel to the line $4x + 2y = -5$

76. Contains the point $(1, -2)$ and is perpendicular to the line $x - 3y = 3$

For Problems 77 and 78, solve each system of equations.

77. $\begin{pmatrix} y < \frac{1}{2}x + 1 \\ y > -2x + 2 \end{pmatrix}$

78. $\begin{pmatrix} y \leq 3 \\ y \geq x \end{pmatrix}$

For Problems 79–84, solve each system of equations.

79. $\begin{pmatrix} y = 2x + 5 \\ 2x + 3y = 7 \end{pmatrix}$

80. $\begin{pmatrix} x = y - 3 \\ 5x + 2y = 20 \end{pmatrix}$

81. $\begin{pmatrix} 3x + y = 8 \\ 5x - 2y = -16 \end{pmatrix}$

82. $\begin{pmatrix} 2x - 3y = 10 \\ 3x - 5y = 18 \end{pmatrix}$

83. $\begin{pmatrix} x + 2y - z = -1 \\ 2x + y - 2z = 4 \\ 3x + 3y + z = 7 \end{pmatrix}$

84. $\begin{pmatrix} 2x + y + z = 1 \\ x + 2y - z = 8 \\ 3x - y + 2z = -1 \end{pmatrix}$

For Problems 85–93, solve each problem by setting up and solving the appropriate equation or system of equations.

85. How many quarts of 1% fat milk should be mixed with 4% fat milk to obtain 12 quarts of 2% fat milk?

86. The area of a rectangular plot is 120 square feet and its perimeter is 44 feet. Find the dimensions of the rectangle.

87. How many liters of a 60% acid solution must be added to 14 liters of a 10% acid solution to produce a 25% acid solution?

88. A sum of \$2250 is to be divided between two people in the ratio of 2 to 3. How much does each person receive?

89. The length of a picture without its border is 7 inches less than twice its width. If the border is 1 inch wide and its area is 62 square inches, what are the dimensions of the picture alone?

90. Working together, Lolita and Doug can paint a shed in 3 hours and 20 minutes. If Doug can paint the shed by himself in 10 hours, how long would it take Lolita to paint the shed by herself?

91. A jogger who can run an 8-minute mile starts half a mile ahead of a jogger who can run a 6-minute mile. How long will it take the faster jogger to catch the slower jogger?

92. Suppose that \$100 is invested at a certain rate of interest compounded annually for 2 years. If the accumulated value at the end of 2 years is \$114.49, find the rate of interest.

93. A room contains 120 chairs arranged in rows. The number of chairs per row is one less than twice the number of rows. Find the number of chairs per row.

Conic Sections

9

Examples of conic sections—in particular, parabolas and ellipses—can be found in corporate logos throughout the world.

Parabolas, circles, ellipses, and hyperbolas can be formed when a plane intersects a conical surface as shown in Figure 9.1; we often refer to these curves as the **conic sections**. A flashlight produces a "cone of light" that can be cut by the plane of a wall to illustrate the conic sections. Try shining a flashlight against a wall at different angles to produce a circle, an ellipse, a parabola, and one branch of a hyperbola. (You may find it difficult to distinguish between a parabola and a branch of a hyperbola.)

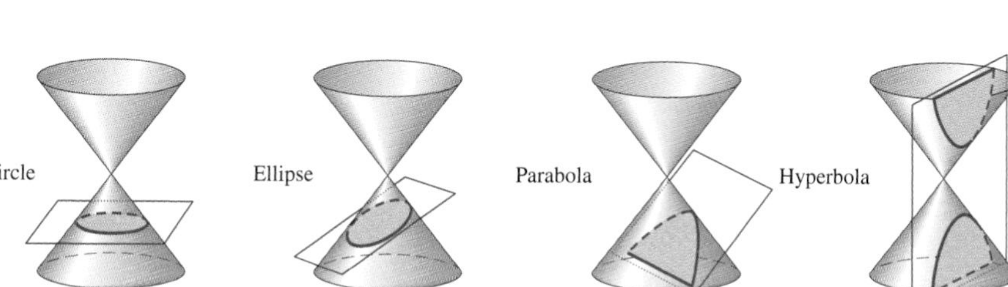

Figure 9.1

Video tutorials for all section learning objectives are available in a variety of delivery modes.

INTERNET PROJECT

Throughout Chapter 9 you will learn the techniques for graphing conic sections. In many of the problems sets you will be asked to check your graphs with a graphing calculator or graphing utility. Conduct an Internet search to find an online graphing calculator/utility; most of them require that you input the equation after it is solved for y. Can you find an online graphing utility that allows you to enter the equation implicitly in a form such as $x^2 + y^2 = 25$?

9.1 Graphing Nonlinear Equations

OBJECTIVE

1 Graph Nonlinear Equations Using Symmetries as an Aid

1 Graph Nonlinear Equations Using Symmetries as an Aid

Equations such as $y = x^2 - 4$, $x = y^2$, $y = \frac{1}{x}$, $x^2y = -2$, and $x = y^3$ are all examples of nonlinear equations. The graphs of these equations are figures other than straight lines that can be determined by plotting a sufficient number of points. Let's plot the points and observe some characteristics of these graphs that we then can use to supplement the point-plotting process.

EXAMPLE 1 Graph $y = x^2 - 4$.

Solution

Let's begin by finding the intercepts. If $x = 0$, then

$$y = 0^2 - 4 = -4$$

The point $(0, -4)$ is on the graph. If $y = 0$, then

$$0 = x^2 - 4$$
$$0 = (x + 2)(x - 2)$$
$$x + 2 = 0 \quad \text{or} \quad x - 2 = 0$$
$$x = -2 \quad \text{or} \quad x = 2$$

The points $(-2, 0)$ and $(2, 0)$ are on the graph. The given equation is in a convenient form for setting up a table of values.

Plotting these points and connecting them with a smooth curve produces Figure 9.2.

x	y	
0	−4	
−2	0	Intercepts
2	0	
1	−3	
−1	−3	
3	5	Other points
−3	5	

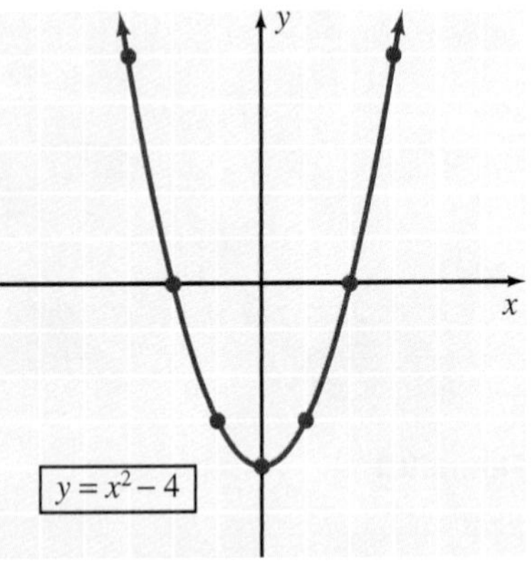

Figure 9.2

▼ PRACTICE YOUR SKILL

Graph $y = x^2 - 3$. ■

The curve in Figure 9.2 is called a parabola; we will study parabolas in more detail in the next section. However, at this time we want to emphasize that the parabola in Figure 9.2 is said to be *symmetric with respect to the y axis*. In other words, the y axis is a line of symmetry. Each half of the curve is a mirror image of the other half through the y axis. Note, in the table of values, that for each ordered pair (x, y), the ordered pair $(-x, y)$ is also a solution. A general test for y-axis symmetry can be stated as follows.

y-Axis Symmetry

The graph of an equation is symmetric with respect to the y axis if replacing x with $-x$ results in an equivalent equation.

The equation $y = x^2 - 4$ exhibits symmetry with respect to the y axis because replacing x with $-x$ produces $y = (-x)^2 - 4 = x^2 - 4$. Let's test some equations for such symmetry. We will replace x with $-x$ and check for an equivalent equation.

Equation	Test for symmetry with respect to the **y** axis	Equivalent equation	Symmetric with respect to the **y** axis
$y = -x^2 + 2$	$y = -(-x)^2 + 2 = -x^2 + 2$	Yes	Yes
$y = 2x^2 + 5$	$y = 2(-x)^2 + 5 = 2x^2 + 5$	Yes	Yes
$y = x^4 + x^2$	$y = (-x)^4 + (-x)^2$ $= x^4 + x^2$	Yes	Yes
$y = x^3 + x^2$	$y = (-x)^3 + (-x)^2$ $= -x^3 + x^2$	No	No
$y = x^2 + 4x + 2$	$y = (-x)^2 + 4(-x) + 2$ $= x^2 - 4x + 2$	No	No

Some equations yield graphs that have x-axis symmetry. In the next example we will see the graph of a parabola that is symmetric with respect to the x axis.

EXAMPLE 2 Graph $x = y^2$.

Solution

First, we see that $(0, 0)$ is on the graph and determines both intercepts. Second, the given equation is in a convenient form for setting up a table of values.

Plotting these points and connecting them with a smooth curve produces Figure 9.3.

x	y	
0	0	Intercepts
1	1	
1	−1	Other points
4	2	
4	−2	

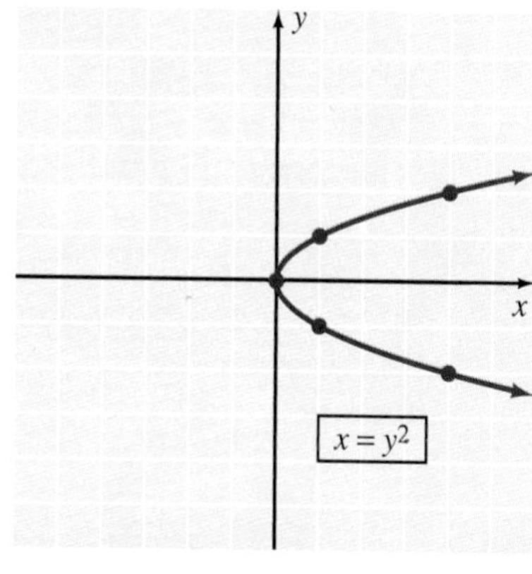

Figure 9.3

▼ PRACTICE YOUR SKILL

Graph $x = y^2 - 4$. ■

The parabola in Figure 9.3 is said to be *symmetric with respect to the x axis*. Each half of the curve is a mirror image of the other half through the x axis. Also note, in the table of values, that for each ordered pair (x, y), the ordered pair $(x, -y)$ is a solution. A general test for x-axis symmetry can be stated as follows.

x-Axis Symmetry

The graph of an equation is symmetric with respect to the x axis if replacing y with $-y$ results in an equivalent equation.

The equation $x = y^2$ exhibits x-axis symmetry because replacing x with $-y$ produces $y = (-y)^2 = y^2$. Let's test some equations for x-axis symmetry. We will replace y with $-y$ and check for an equivalent equation.

Equation	Test for symmetry with respect to the **x** axis	Equivalent equation	Symmetric with respect to the **x** axis
$x = y^2 + 5$	$x = (-y)^2 + 5 = y^2 + 5$	Yes	Yes
$x = -3y^2$	$x = -3(-y)^2 = -3y^2$	Yes	Yes
$x = y^3 + 2$	$x = (-y)^3 + 2 = -y^3 + 2$	No	No
$x = y^2 - 5y + 6$	$x = (-y)^2 - 5(-y) + 6$ $= y^2 + 5y + 6$	No	No

In addition to y-axis and x-axis symmetry, some equations yield graphs that have symmetry with respect to the origin. In the next example we will see a graph that is symmetric with respect to the origin.

EXAMPLE 3

Graph $y = \dfrac{1}{x}$.

Solution

First, let's find the intercepts. Let $x = 0$; then $y = \dfrac{1}{x}$ becomes $y = \dfrac{1}{0}$, and $\dfrac{1}{0}$ is undefined. Thus there is no y intercept. Let $y = 0$; then $y = \dfrac{1}{x}$ becomes $0 = \dfrac{1}{x}$, and there are no values of x that will satisfy this equation. In other words, this graph has no points on either the x axis or the y axis. Second, let's set up a table of values and keep in mind that neither x nor y can equal zero.

In Figure 9.4(a) we plotted the points associated with the solutions from the table. Because the graph does not intersect either axis, it must consist of two branches. Thus connecting the points in the first quadrant with a smooth curve and then connecting the points in the third quadrant with a smooth curve, we obtain the graph shown in Figure 9.4(b).

x	y
$\frac{1}{2}$	2
1	1
2	$\frac{1}{2}$
3	$\frac{1}{3}$
$-\frac{1}{2}$	-2
-1	-1
-2	$-\frac{1}{2}$
-3	$-\frac{1}{3}$

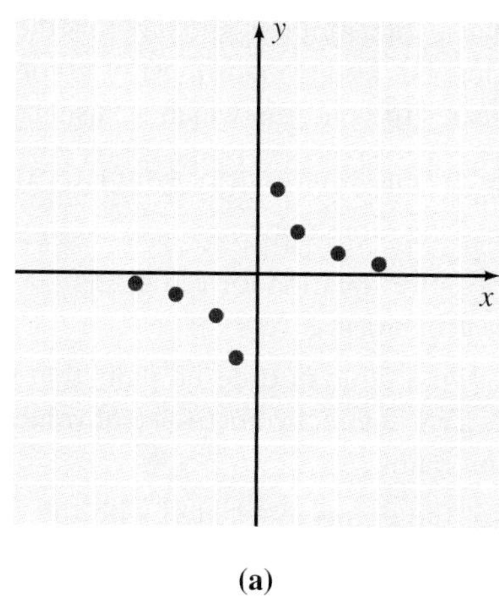

(a)

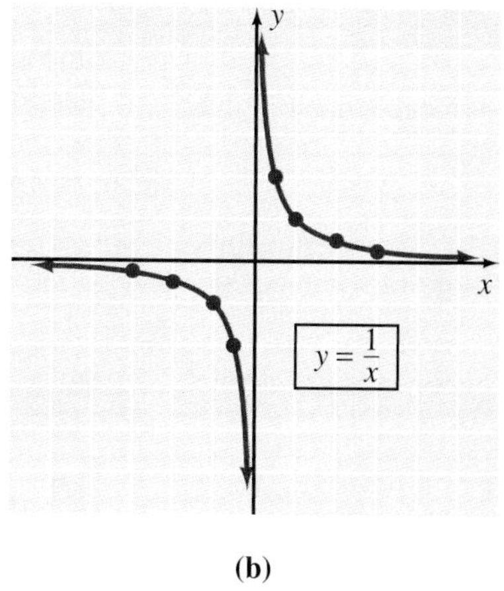

(b)

Figure 9.4

▼ PRACTICE YOUR SKILL

Graph $y = x^3$. ■

The curve in Figure 9.4 is said to be *symmetric with respect to the origin*. Each half of the curve is a mirror image of the other half through the origin. Note, in the table of values, that for each ordered pair (x, y), the ordered pair $(-x, -y)$ is also a solution. A general test for origin symmetry can be stated as follows.

Origin Symmetry

The graph of an equation is symmetric with respect to the origin if replacing x with $-x$ and y with $-y$ results in an equivalent equation.

The equation $y = \frac{1}{x}$ exhibits symmetry with respect to the origin because replacing y with $-y$ and x with $-x$ produces $-y = \frac{1}{-x}$, which is equivalent to $y = \frac{1}{x}$. Let's test some equations for symmetry with respect to the origin. We will replace y with $-y$, replace x with $-x$, and then check for an equivalent equation.

Equation	Test for symmetry with respect to the origin	Equivalent equation	Symmetric with respect to the origin
$y = x^3$	$(-y) = (-x)^3$ $-y = -x^3$ $y = x^3$	Yes	Yes
$x^2 + y^2 = 4$	$(-x)^2 + (-y)^2 = 4$ $x^2 + y^2 = 4$	Yes	Yes
$y = x^2 - 3x + 4$	$(-y) = (-x)^2 - 3(-x) + 4$ $-y = x^2 + 3x + 4$ $y = -x^2 - 3x - 4$	No	No

Let's pause for a moment and pull together the graphing techniques that we have introduced thus far. Following is a list of graphing suggestions. The order of the suggestions indicates the order in which we usually attack a new graphing problem.

1. Determine what type of symmetry the equation exhibits.
2. Find the intercepts.
3. Solve the equation for y in terms of x or for x in terms of y if it is not already in such a form.
4. Set up a table of ordered pairs that satisfy the equation. The type of symmetry will affect your choice of values in the table. (We will illustrate this in a moment.)
5. Plot the points associated with the ordered pairs from the table, and connect them with a smooth curve. Then, if appropriate, reflect this part of the curve according to the symmetry shown by the equation.

EXAMPLE 4

Graph $x^2y = -2$.

Solution

Because replacing x with $-x$ produces $(-x)^2y = -2$ or, equivalently, $x^2y = -2$, the equation exhibits y-axis symmetry. There are no intercepts because neither x nor y can equal 0. Solving the equation for y produces $y = \frac{-2}{x^2}$. The equation exhibits y-axis symmetry, so let's use only positive values for x and then reflect the curve across the y axis.

x	y
1	-2
2	$-\frac{1}{2}$
3	$-\frac{2}{9}$
4	$-\frac{1}{8}$
$\frac{1}{2}$	-8

Let's plot the points determined by the table, connect them with a smooth curve, and reflect this portion of the curve across the y axis. Figure 9.5 is the result of this process.

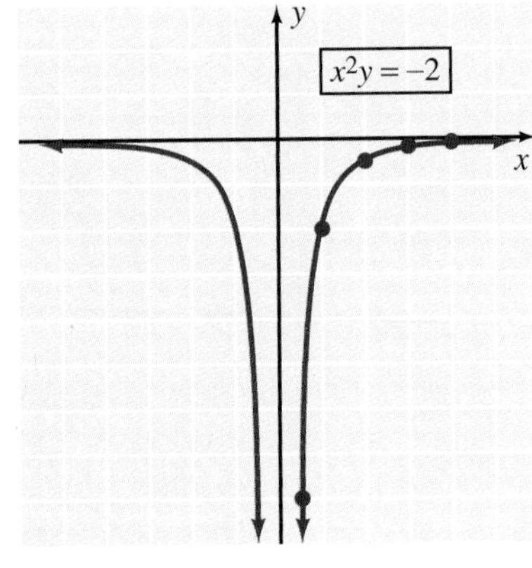

Figure 9.5

▼ PRACTICE YOUR SKILL

Graph $x^2y = 4$. ■

EXAMPLE 5

Graph $x = y^3$.

Solution

Because replacing x with $-x$ and y with $-y$ produces $-x = (-y)^3 = -y^3$, which is equivalent to $x = y^3$, the given equation exhibits origin symmetry. If $x = 0$ then $y = 0$, so the origin is a point of the graph. The given equation is in an easy form for deriving a table of values.

x	y
0	0
8	2
$\frac{1}{8}$	$\frac{1}{2}$
$\frac{27}{64}$	$\frac{3}{4}$

Let's plot the points determined by the table, connect them with a smooth curve, and reflect this portion of the curve through the origin to produce Figure 9.6.

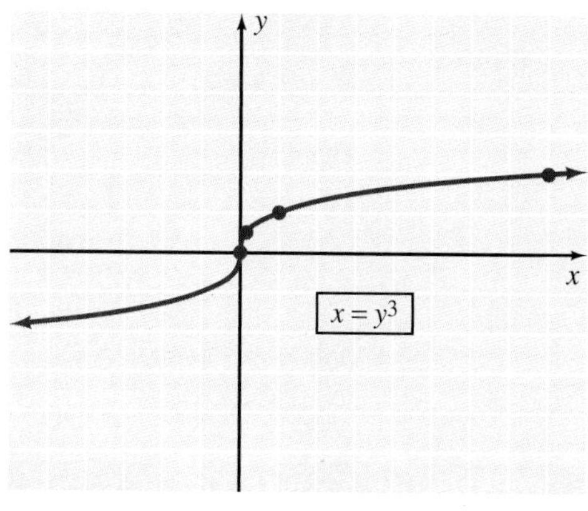

Figure 9.6

▼ PRACTICE YOUR SKILL

Graph $xy = -4$. ■

EXAMPLE 6 Use a graphing utility to obtain a graph of the equation $x = y^3$.

Solution

First, we may need to solve the equation for y in terms of x. (We say we "may need to" because some graphing utilities are capable of graphing two-variable equations without solving for y in terms of x.)

$$y = \sqrt[3]{x} = x^{1/3}$$

Now we can enter the expression $x^{1/3}$ for Y_1 and obtain the graph shown in Figure 9.7.

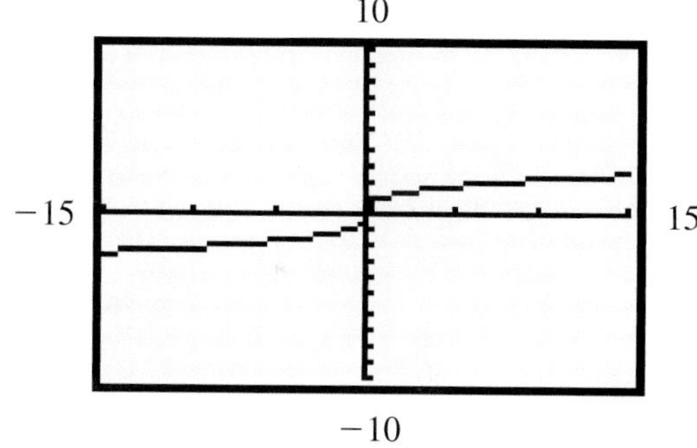

Figure 9.7

▼ PRACTICE YOUR SKILL

Use a graphing utility to obtain a graph of the equation $xy = 2$. ■

As indicated in Figure 9.7, the **viewing rectangle** of a graphing utility is a portion of the xy plane shown on the display of the utility. In this display, the boundaries were set so that $-15 \le x \le 15$ and $-10 \le y \le 10$. These boundaries were set automatically; however, boundaries can be reassigned as necessary, which is an important feature of graphing utilities.

CONCEPT QUIZ

For Problems 1–10, answer true or false.

1. When replacing y with $-y$ in an equation results in an equivalent equation, then the graph of the equation is symmetric with respect to the x axis.
2. If the graph of an equation is symmetric with respect to the x axis, then it cannot be symmetric with respect to the y axis.
3. If, for each ordered pair (x, y) that is a solution of the equation, the ordered pair $(-x, -y)$ is also a solution, then the graph of the equation is symmetric with respect to the origin.

4. The equation $xy^2 = -4$ exhibits y-axis symmetry.
5. The equation $x^2y + 2y = 5$ exhibits y-axis symmetry.
6. The equation $5x^2 - 9y^2 = 36$ exhibits both x-axis and y-axis symmetry.
7. The graph of the equation $x = 0$ is a vertical line.
8. The graph of $y = 3x - 4$ is the same as the graph of $x = 3y - 4$.
9. The graph of $xy = 4$ is the same as the graph of $y = \frac{4}{x}$.
10. The equation $-xy = 5$ exhibits origin symmetry.

Problem Set 9.1

1 Graph Nonlinear Equations Using Symmetries as an Aid

For each of the points in Problems 1–5, determine the points that are symmetric with respect to (a) the x axis, (b) the y axis, and (c) the origin.

1. $(-3, 1)$
2. $(-2, -4)$
3. $(7, -2)$
4. $(0, -4)$
5. $(5, 0)$

For Problems 6–25, determine the type(s) of symmetry (symmetry with respect to the x axis, y axis, and/or origin) exhibited by the graph of each of the following equations. Do not sketch the graph.

6. $x^2 + 2y = 4$
7. $-3x + 2y^2 = -4$
8. $x = -y^2 + 5$
9. $y = 4x^2 + 13$
10. $xy = -6$
11. $2x^2y^2 = 5$
12. $2x^2 + 3y^2 = 9$
13. $x^2 - 2x - y^2 = 4$
14. $y = x^2 - 6x - 4$
15. $y = 2x^2 - 7x - 3$
16. $y = x$
17. $y = 2x$
18. $y = x^4 + 4$
19. $y = x^4 - x^2 + 2$
20. $x^2 + y^2 = 13$
21. $x^2 - y^2 = -6$
22. $y = -4x^2 - 2$
23. $x = -y^2 + 9$
24. $x^2 + y^2 - 4x - 12 = 0$
25. $2x^2 + 3y^2 + 8y + 2 = 0$

For Problems 26–59, graph each of the equations.

26. $y = x + 1$
27. $y = x - 4$
28. $y = 3x - 6$
29. $y = 2x + 4$
30. $y = -2x + 1$
31. $y = -3x - 1$
32. $y = \frac{2}{3}x - 1$
33. $y = -\frac{1}{3}x + 2$
34. $y = \frac{1}{3}x$
35. $y = \frac{1}{2}x$
36. $2x + y = 6$
37. $2x - y = 4$
38. $x + 3y = -3$
39. $x - 2y = 2$
40. $y = x^2 - 1$
41. $y = x^2 + 2$
42. $y = -x^3$
43. $y = x^3$
44. $y = \frac{2}{x^2}$
45. $y = \frac{-1}{x^2}$
46. $y = 2x^2$
47. $y = -3x^2$
48. $xy = -3$
49. $xy = 2$
50. $x^2y = 4$
51. $xy^2 = -4$
52. $y^3 = x^2$
53. $y^2 = x^3$
54. $y = \frac{-2}{x^2 + 1}$
55. $y = \frac{4}{x^2 + 1}$
56. $x = -y^3$
57. $y = x^4$
58. $y = -x^4$
59. $x = -y^3 + 2$

THOUGHTS INTO WORDS

60. How would you convince someone that there are infinitely many ordered pairs of real numbers that satisfy $x + y = 7$?
61. What is the graph of $x = 0$? What is the graph of $y = 0$? Explain your answers.
62. Is a graph symmetric with respect to the origin if it is symmetric with respect to both axes? Defend your answer.
63. Is a graph symmetric with respect to both axes if it is symmetric with respect to the origin? Defend your answer.

GRAPHING CALCULATOR ACTIVITIES

This set of activities is designed to help you get started with your graphing utility by setting different boundaries for the viewing rectangle; you will notice the effect on the graphs produced. These boundaries are usually set by using a menu displayed by a key marked either WINDOW or RANGE. You may need to consult the user's manual for specific key-punching instructions.

64. Graph the equation $y = \frac{1}{x}$ (Example 4) using the following boundaries.

(a) $-15 \leq x \leq 15$ and $-10 \leq y \leq 10$

(b) $-10 \leq x \leq 10$ and $-10 \leq y \leq 10$

(c) $-5 \leq x \leq 5$ and $-5 \leq y \leq 5$

65. Graph the equation $y = \frac{-2}{x^2}$ (Example 5) using the following boundaries.

(a) $-15 \leq x \leq 15$ and $-10 \leq y \leq 10$

(b) $-5 \leq x \leq 5$ and $-10 \leq y \leq 10$

(c) $-5 \leq x \leq 5$ and $-10 \leq y \leq 1$

66. Graph the two equations $y = \pm\sqrt{x}$ (Example 3) on the same set of axes using the following boundaries. (Let $Y_1 = \sqrt{x}$ and $Y_2 = -\sqrt{x}$.)

(a) $-15 \leq x \leq 15$ and $-10 \leq y \leq 10$

(b) $-1 \leq x \leq 15$ and $-10 \leq y \leq 10$

(c) $-1 \leq x \leq 15$ and $-5 \leq y \leq 5$

67. Graph $y = \frac{1}{x}$, $y = \frac{5}{x}$, $y = \frac{10}{x}$, and $y = \frac{20}{x}$ on the same set of axes. (Choose your own boundaries.) What effect does increasing the constant seem to have on the graph?

68. Graph $y = \frac{10}{x}$ and $y = \frac{-10}{x}$ on the same set of axes. What relationship exists between the two graphs?

69. Graph $y = \frac{10}{x^2}$ and $y = \frac{-10}{x^2}$ on the same set of axes. What relationship exists between the two graphs?

Answers to the Concept Quiz

1. True **2.** False **3.** True **4.** False **5.** True **6.** True **7.** True **8.** False **9.** True **10.** True

Answers to the Example Practice Skills

1.

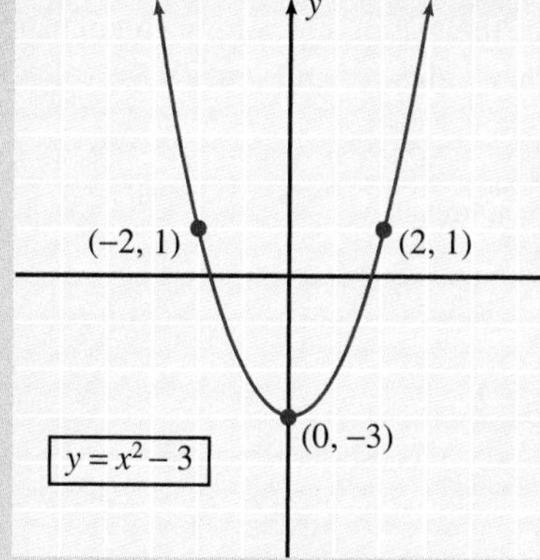

2.

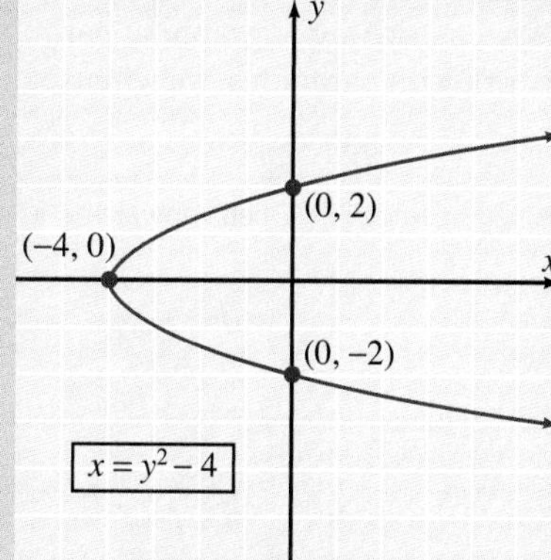

3.

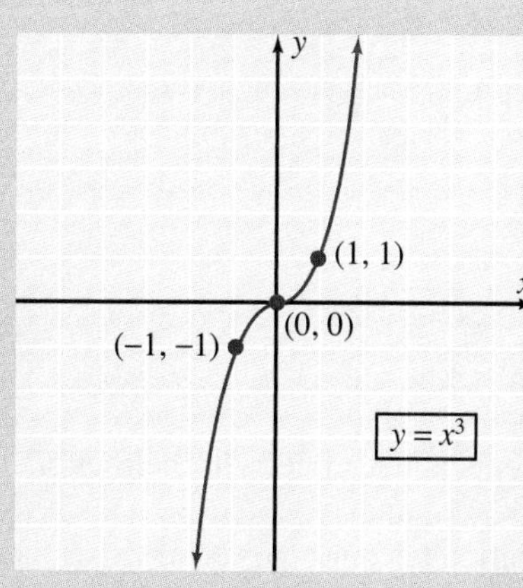

4.

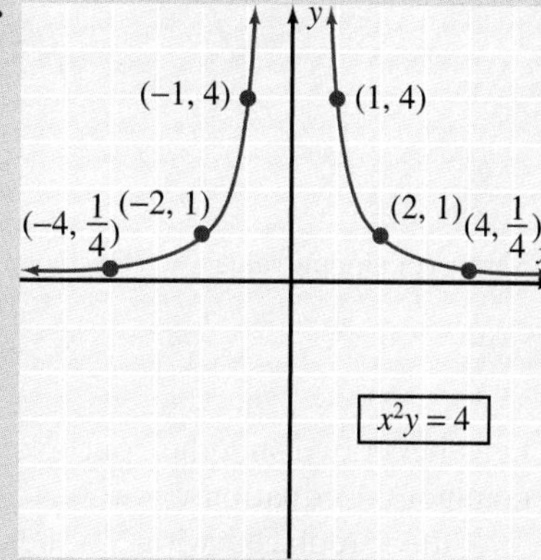

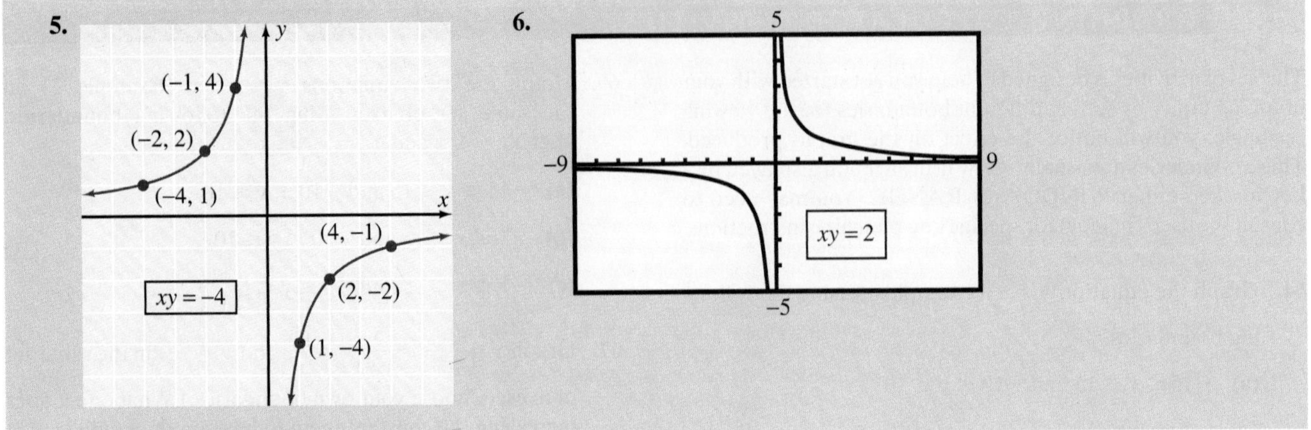

9.2 Graphing Parabolas

OBJECTIVE

1 Graph Parabolas

1 Graph Parabolas

In general, the graph of any equation of the form $y = ax^2 + bx + c$, where a, b, and c are real numbers and $a \neq 0$, is a parabola. At this time we want to develop an easy and systematic way of graphing parabolas without the use of a graphing calculator. As we work with parabolas, we will use the vocabulary indicated in Figure 9.8.

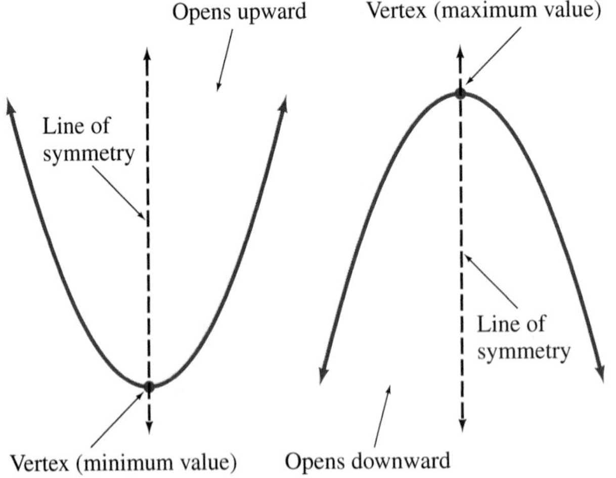

Figure 9.8

Let's begin by using the concepts of intercepts and symmetry to help us sketch the graph of the equation $y = x^2$.

If we replace x with $-x$, the given equation becomes $y = (-x)^2 = x^2$; therefore, we have y-axis symmetry. The origin, (0, 0), is a point of the graph. We can recognize from the equation that 0 is the minimum value of y; hence the point (0, 0) is the vertex of the parabola. Now we can set up a table of values that uses nonnegative values for x. Plot the points determined by the table, connect them with a smooth curve, and reflect that portion of the curve across the y axis to produce Figure 9.9.

x	y
0	0
$\frac{1}{2}$	$\frac{1}{4}$
1	1
2	4
3	9

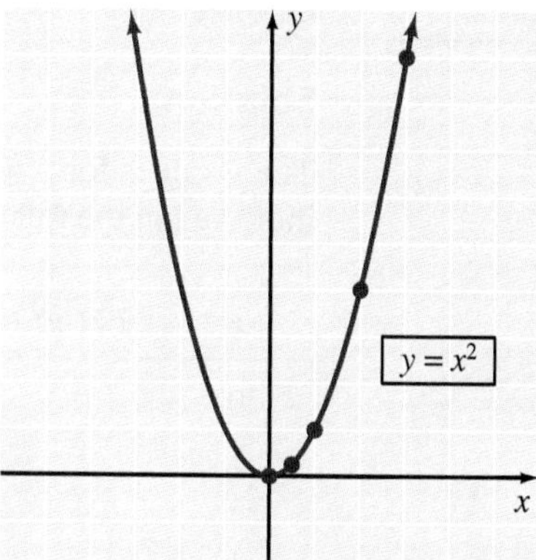

Figure 9.9

To graph parabolas, we need to be able to:

1. Find the vertex.
2. Determine whether the parabola opens upward or downward.
3. Locate two points on opposite sides of the line of symmetry.
4. Compare the parabola to the basic parabola $y = x^2$.

To graph parabolas produced by the various types of equations such as $y = x^2 + k$, $y = ax^2$, $y = (x - h)^2$, and $y = a(x - h)^2 + k$, we can compare these equations to that of the basic parabola, $y = x^2$. First, let's consider some equations of the form $y = x^2 + k$, where k is a constant.

EXAMPLE 1 Graph $y = x^2 + 1$.

Solution

Let's set up a table of values to compare y values for $y = x^2 + 1$ to corresponding y values for $y = x^2$.

x	$y = x^2$	$y = x^2 + 1$
0	0	1
1	1	2
2	4	5
−1	1	2
−2	4	5

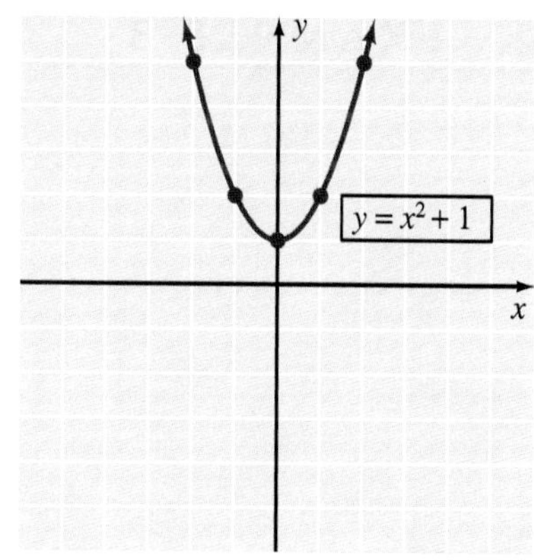

Figure 9.10

It should be evident that y values for $y = x^2 + 1$ are *1 greater than* corresponding y values for $y = x^2$. For example, if $x = 2$, then $y = 4$ for the equation $y = x^2$; but if $x = 2$, then $y = 5$ for the equation $y = x^2 + 1$. Thus the graph of $y = x^2 + 1$ is the same as the graph of $y = x^2$ but moved up 1 unit (Figure 9.10). The vertex will move from (0, 0) to (0, 1).

▼ **PRACTICE YOUR SKILL**

Graph $y = x^2 + 3$. ■

EXAMPLE 2

Graph $y = x^2 - 2$.

Solution

The y values for $y = x^2 - 2$ are 2 *less than* the corresponding y values for $y = x^2$, as indicated in the following table.

x	$y = x^2$	$y = x^2 - 2$
0	0	−2
1	1	−1
2	4	2
−1	1	−1
−2	4	2

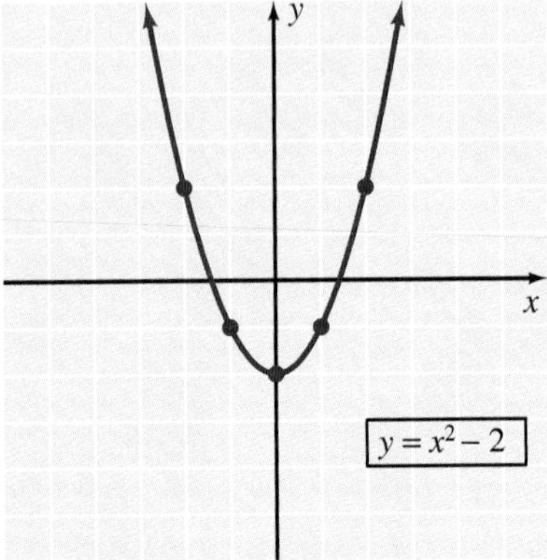

Figure 9.11

Thus the graph of $y = x^2 - 2$ is the same as the graph of $y = x^2$ but moved down 2 units (Figure 9.11). The vertex will move from (0, 0) to (0, −2).

▼ **PRACTICE YOUR SKILL**

Graph $y = x^2 - 4$. ■

In general, the graph of a quadratic equation of the form $y = x^2 + k$ is the same as the graph of $y = x^2$ but moved up or down $|k|$ units, depending on whether k is positive or negative.

Now, let's consider some quadratic equations of the form $y = ax^2$, where a is a nonzero constant.

EXAMPLE 3

Graph $y = 2x^2$.

Solution

Again, let's use a table to make some comparisons of y values.

x	$y = x^2$	$y = 2x^2$
0	0	0
1	1	2
2	4	8
−1	1	2
−2	4	8

Obviously, the y values for $y = 2x^2$ are *twice* the corresponding y values for $y = x^2$. Thus the parabola associated with $y = 2x^2$ has the same vertex (the origin) as the graph of $y = x^2$, but it is narrower (Figure 9.12).

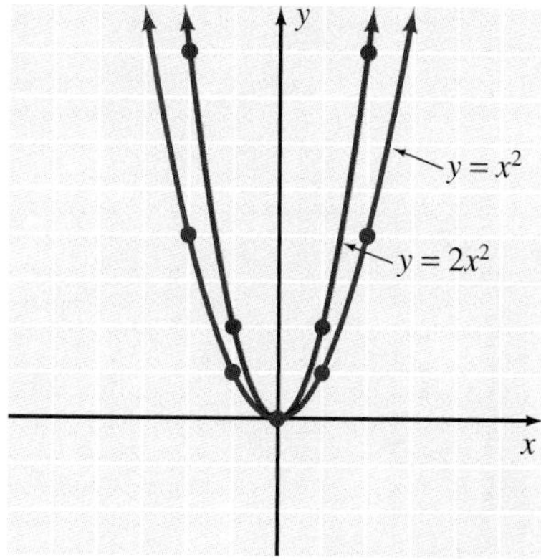

Figure 9.12

▼ PRACTICE YOUR SKILL

Graph $y = 3x^2$. ■

EXAMPLE 4

Graph $y = \frac{1}{2}x^2$.

Solution

The following table indicates some comparisons of y values.

x	$y = x^2$	$y = \frac{1}{2}x^2$
0	0	0
1	1	$\frac{1}{2}$
2	4	2
−1	1	$\frac{1}{2}$
−2	4	2

The y values for $y = \frac{1}{2}x^2$ are one-half of the corresponding y values for $y = x^2$. Therefore, the graph of $y = \frac{1}{2}x^2$ has the same vertex (the origin) as the graph of $y = x^2$ but it is wider (Figure 9.13).

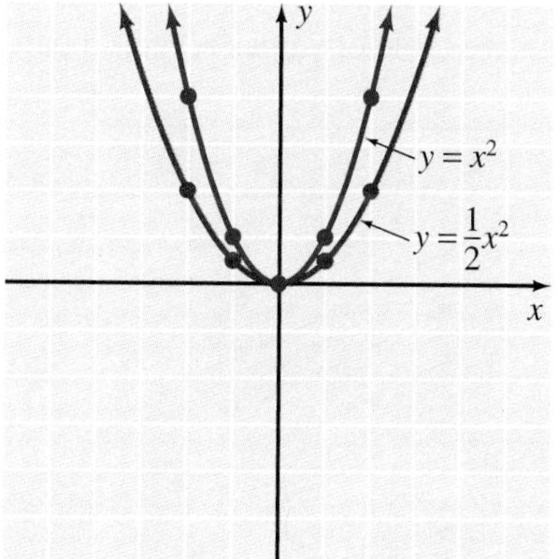

Figure 9.13

▼ PRACTICE YOUR SKILL

Graph $y = \frac{1}{4}x^2$. ■

EXAMPLE 5

Graph $y = -x^2$.

Solution

x	$y = x^2$	$y = -x^2$
0	0	0
1	1	−1
2	4	−4
−1	1	−1
−2	4	−4

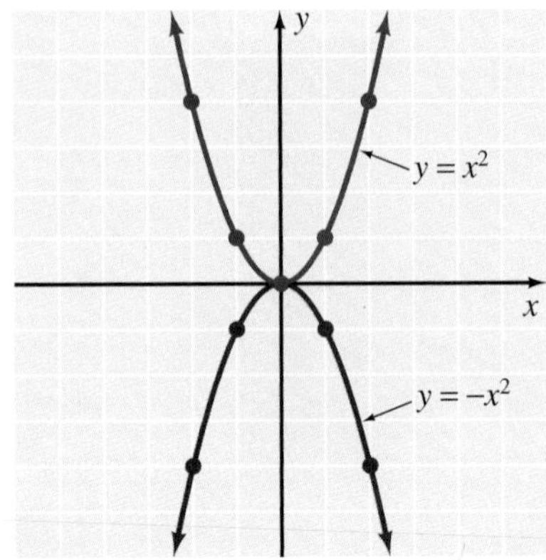

Figure 9.14

The y values for $y = -x^2$ are the opposites of the corresponding y values for $y = x^2$. Thus the graph of $y = -x^2$ has the same vertex (the origin) as the graph of $y = x^2$, but it is a reflection across the x axis of the basic parabola (Figure 9.14).

▼ PRACTICE YOUR SKILL

Graph $y = -x^2 + 2$. ■

> In general, the graph of a quadratic equation of the form $y = ax^2$ has its vertex at the origin and opens upward if a is positive and downward if a is negative. The parabola is narrower than the basic parabola if $|a| > 1$ and wider if $|a| < 1$.

Let's continue our investigation of quadratic equations by considering those of the form $y = (x - h)^2$, where h is a nonzero constant.

EXAMPLE 6

Graph $y = (x - 2)^2$.

Solution

A fairly extensive table of values reveals a pattern.

x	$y = x^2$	$y = (x - 2)^2$
−2	4	16
−1	1	9
0	0	4
1	1	1
2	4	0
3	9	1
4	16	4
5	25	9

Note that $y = (x - 2)^2$ and $y = x^2$ take on the same y values but for different values of x. More specifically, if $y = x^2$ achieves a certain y value at x equals a constant, then $y = (x - 2)^2$ achieves the same y value at x equals the *constant plus 2*. In other words, the graph of $y = (x - 2)^2$ is the same as the graph of $y = x^2$ but moved 2 units to the right (Figure 9.15). The vertex will move from (0, 0) to (2, 0).

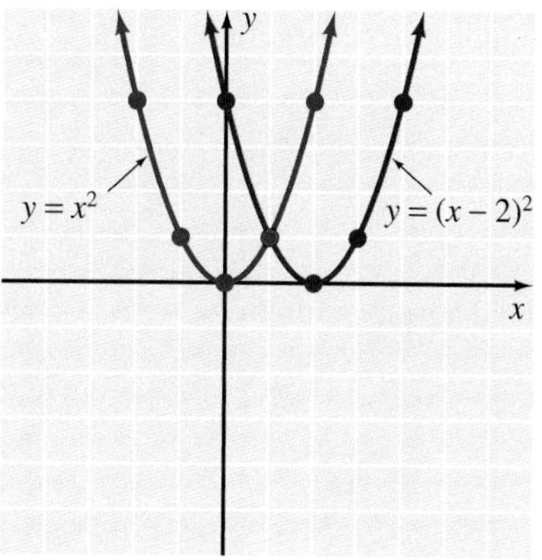

Figure 9.15

▼ PRACTICE YOUR SKILL

Graph $y = (x - 4)^2$. ■

EXAMPLE 7 Graph $y = (x + 3)^2$.

Solution

x	$y = x^2$	$y = (x + 3)^2$
−3	9	0
−2	4	1
−1	1	4
0	0	9
1	1	16
2	4	25
3	9	36

If $y = x^2$ achieves a certain y value at x equals a constant, then $y = (x + 3)^2$ achieves that same y value at x equals that *constant minus 3*. Therefore, the graph of $y = (x + 3)^2$ is the same as the graph of $y = x^2$ but moved 3 units to the left (Figure 9.16). The vertex will move from (0, 0) to (−3, 0).

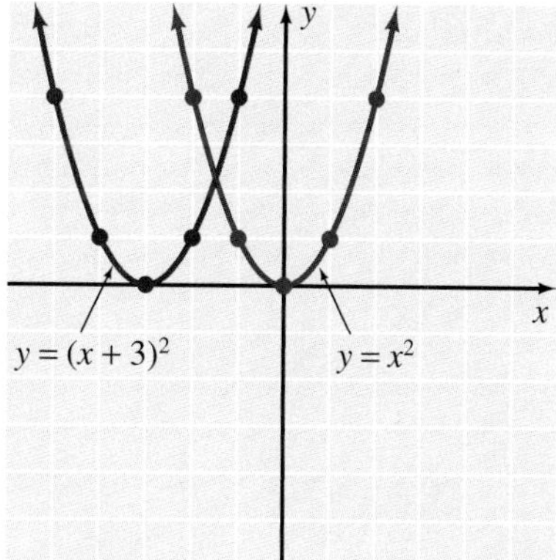

Figure 9.16

▼ PRACTICE YOUR SKILL

Graph $y = (x + 1)^2$. ■

In general, the graph of a quadratic equation of the form $y = (x - h)^2$ is the same as the graph of $y = x^2$ but moved to the right h units if h is positive or moved to the left $|h|$ units if h is negative.

$y = (x - 4)^2$ → Moved to the right 4 units

$y = (x + 2)^2 = (x - (-2))^2$ → Moved to the left 2 units

The following diagram summarizes our work with graphing quadratic equations.

$y = x^2 + (k)$ → Moves the parabola up or down

$y = x^2$ → $y = (a)x^2$ → Affects the width and which way the parabola opens

Basic parabola $y = (x - (h))^2$ → Moves the parabola right or left

Equations of the form $y = x^2 + k$ and $y = ax^2$ are symmetric about the y axis. The next two examples of this section show how we can combine these ideas to graph a quadratic equation of the form $y = a(x - h)^2 + k$.

EXAMPLE 8

Graph $y = 2(x - 3)^2 + 1$.

Solution

$y = 2(x - 3)^2 + 1$

- 2: Narrows the parabola and opens it upward
- -3: Moves the parabola 3 units to the right
- $+1$: Moves the parabola 1 unit up

The vertex will be located at the point (3, 1). In addition to the vertex, two points are located to determine the parabola. The parabola is drawn in Figure 9.17.

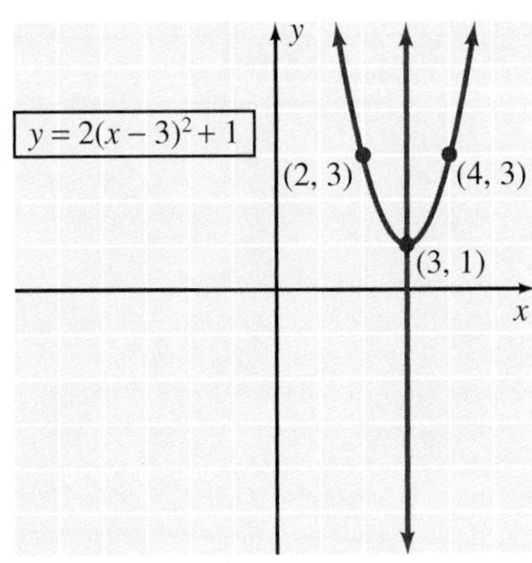

Figure 9.17

▼ PRACTICE YOUR SKILL

Graph $y = 2(x - 1)^2 + 3$. ■

EXAMPLE 9

Graph $y = -\frac{1}{2}(x + 1)^2 - 2$.

Solution

$$y = -\frac{1}{2}(x + 1)^2 - 2$$

- $-\frac{1}{2}$: Widens the parabola and opens it downward
- $+1$: Moves the parabola 1 unit to the left
- -2: Moves the parabola 2 units down

The parabola is drawn in Figure 9.18.

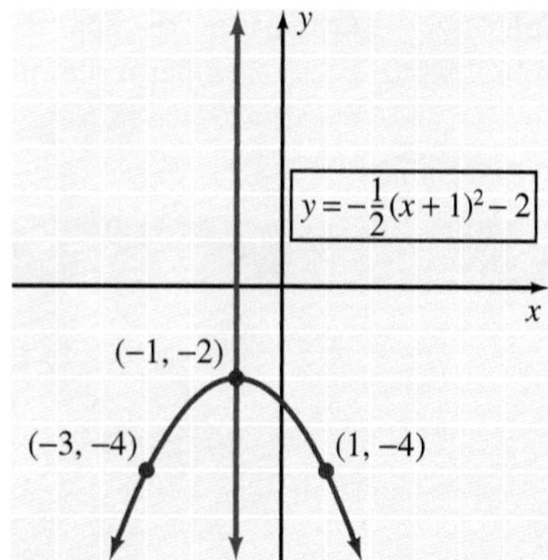

Figure 9.18

▼ PRACTICE YOUR SKILL

Graph $y = -\frac{1}{2}(x - 3)^2 + 4$. ■

Finally, we can use a graphing utility to demonstrate some of the ideas of this section. Let's graph $y = x^2$, $y = -3(x - 7)^2 - 1$, $y = 2(x + 9)^2 + 5$, and $y = -0.2(x + 8)^2 - 3.5$ on the same set of axes, as shown in Figure 9.19. Certainly, Figure 9.19 is consistent with the ideas we presented in this section.

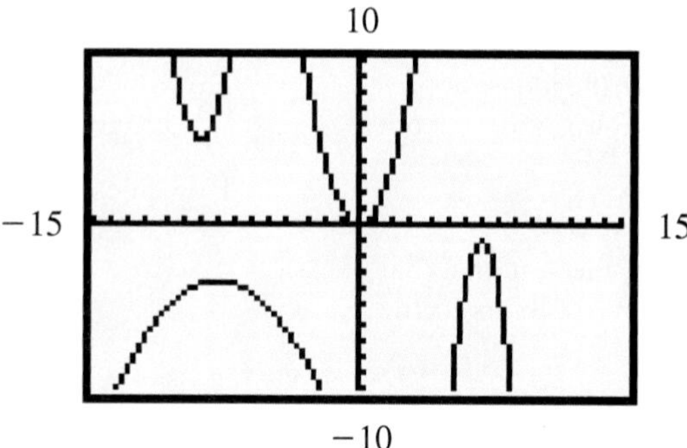

Figure 9.19

CONCEPT QUIZ

For Problems 1–10, answer true or false.

1. The graph of $y = (x - 3)^2$ is the same as the graph of $y = x^2$ but moved 3 units to the right.
2. The graph of $y = x^2 - 4$ is the same as the graph of $y = x^2$ but moved 4 units to the right.
3. The graph of $y = x^2 + 1$ is the same as the graph of $y = x^2$ but moved 1 unit up.
4. The graph of $y = -x^2$ is the same as the graph of $y = x^2$ but is reflected across the y axis.
5. The vertex of the parabola given by the equation $y = (x + 2)^2 - 5$ is located at $(-2, -5)$.
6. The graph of $y = \frac{1}{3}x^2$ is narrower than the graph of $y = x^2$.
7. The graph of $y = -\frac{2}{3}x^2$ is a parabola that opens downward.
8. The graph of $y = x^2 - 9$ is a parabola that intersects the x axis at $(9, 0)$ and $(-9, 0)$.

9. The graph of $y = -(x - 3)^2 - 7$ is a parabola whose vertex is at $(3, -7)$.
10. The graph of $y = -x^2 + 6$ is a parabola that does not intersect the x axis.

Problem Set 9.2

1 Graph Parabolas

For Problems 1–30, graph each parabola.

1. $y = x^2 + 2$
2. $y = x^2 + 3$
3. $y = x^2 - 1$
4. $y = x^2 - 5$
5. $y = 4x^2$
6. $y = 3x^2$
7. $y = -3x^2$
8. $y = -4x^2$
9. $y = \frac{1}{3}x^2$
10. $y = \frac{1}{4}x^2$
11. $y = -\frac{1}{2}x^2$
12. $y = -\frac{2}{3}x^2$
13. $y = (x - 1)^2$
14. $y = (x - 3)^2$
15. $y = (x + 4)^2$
16. $y = (x + 2)^2$
17. $y = 3x^2 + 2$
18. $y = 2x^2 + 3$
19. $y = -2x^2 - 2$
20. $y = \frac{1}{2}x^2 - 2$
21. $y = (x - 1)^2 - 2$
22. $y = (x - 2)^2 + 3$
23. $y = (x + 2)^2 + 1$
24. $y = (x + 1)^2 - 4$
25. $y = 3(x - 2)^2 - 4$
26. $y = 2(x + 3)^2 - 1$
27. $y = -(x + 4)^2 + 1$
28. $y = -(x - 1)^2 + 1$
29. $y = -\frac{1}{2}(x + 1)^2 - 2$
30. $y = -3(x - 4)^2 - 2$

THOUGHTS INTO WORDS

31. Write a few paragraphs that summarize the ideas we presented in this section for someone who was absent from class that day.
32. How would you convince someone that $y = (x + 3)^2$ is the basic parabola moved 3 units to the left but that $y = (x - 3)^2$ is the basic parabola moved 3 units to the right?
33. How does the graph of $-y = x^2$ compare to the graph of $y = x^2$? Explain your answer.
34. How does the graph of $y = 4x^2$ compare to the graph of $y = 2x^2$? Explain your answer.

GRAPHING CALCULATOR ACTIVITIES

35. Use a graphing calculator to check your graphs for Problems 21–30.
36. **(a)** Graph $y = x^2$, $y = 2x^2$, $y = 3x^2$, and $y = 4x^2$ on the same set of axes.

    (b) Graph $y = x^2$, $y = \frac{3}{4}x^2$, $y = \frac{1}{2}x^2$, and $y = \frac{1}{5}x^2$ on the same set of axes.

    (c) Graph $y = x^2$, $y = -x^2$, $y = -3x^2$, and $y = -\frac{1}{4}x^2$ on the same set of axes.
37. **(a)** Graph $y = x^2$, $y = (x - 2)^2$, $y = (x - 3)^2$, and $y = (x - 5)^2$ on the same set of axes.

    (b) Graph $y = x^2$, $y = (x + 1)^2$, $y = (x + 3)^2$, and $y = (x + 6)^2$ on the same set of axes.
38. **(a)** Graph $y = x^2$, $y = (x - 2)^2 + 3$, $y = (x + 4)^2 - 2$, and $y = (x - 6)^2 - 4$ on the same set of axes.

    (b) Graph $y = x^2$, $y = 2(x + 1)^2 + 4$, $y = 3(x - 1)^2 - 3$, and $y = \frac{1}{2}(x - 5)^2 + 2$ on the same set of axes.

    (c) Graph $y = x^2$, $y = -(x - 4)^2 - 3$, $y = -2(x + 3)^2 - 1$, and $y = -\frac{1}{2}(x - 2)^2 + 6$ on the same set of axes.
39. **(a)** Graph $y = x^2 - 12x + 41$ and $y = x^2 + 12x + 41$ on the same set of axes. What relationship seems to exist between the two graphs?

    (b) Graph $y = x^2 - 8x + 22$ and $y = -x^2 + 8x - 22$ on the same set of axes. What relationship seems to exist between the two graphs?

    (c) Graph $y = x^2 + 10x + 29$ and $y = -x^2 + 10x - 29$ on the same set of axes. What relationship seems to exist between the two graphs?

    (d) Summarize your findings for parts (a) through (c).

Answers to the Concept Quiz

1. True **2.** False **3.** True **4.** False **5.** True **6.** False **7.** True **8.** False **9.** True **10.** False

Answers to the Example Practice Skills

1.

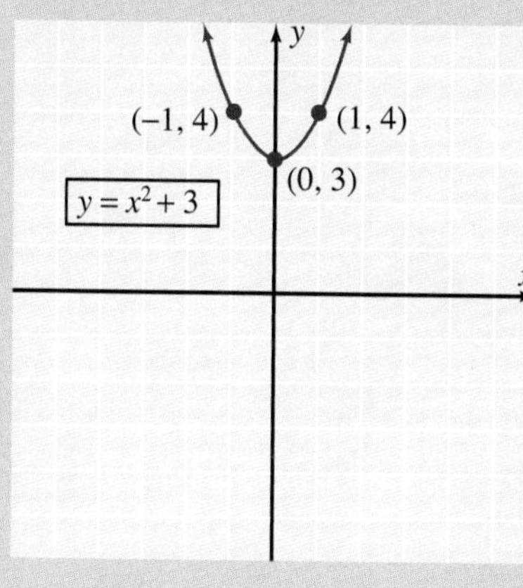

2.

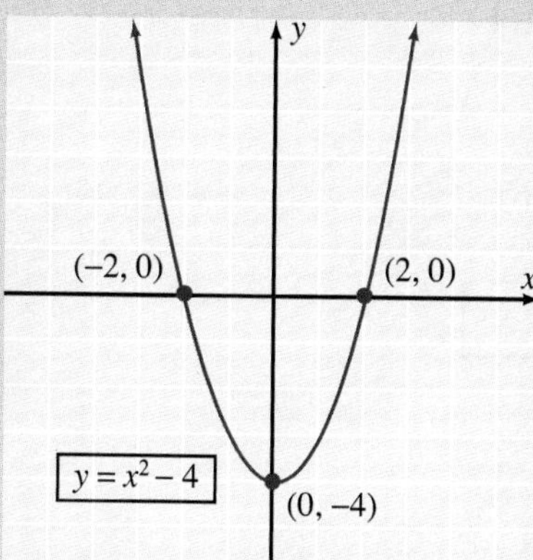

3.

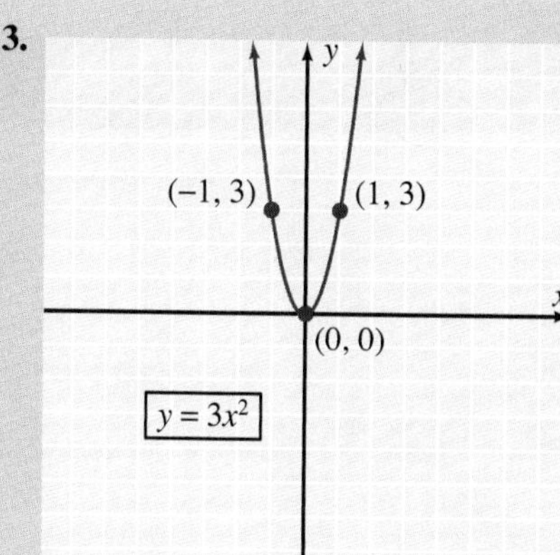

4.

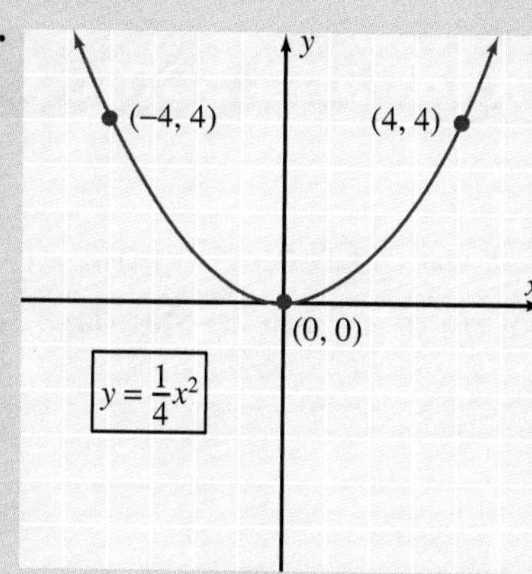

5.

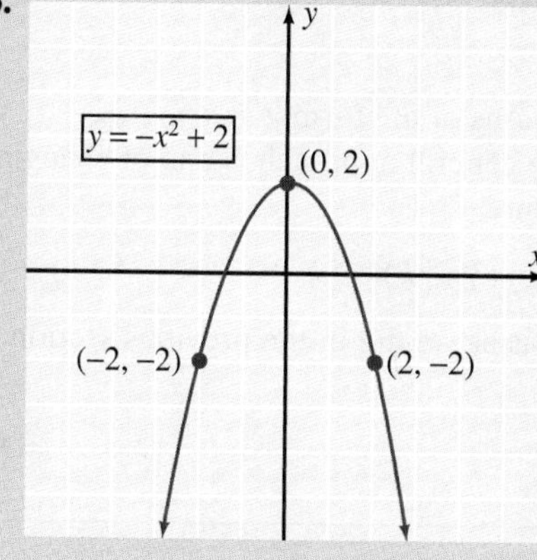

6.

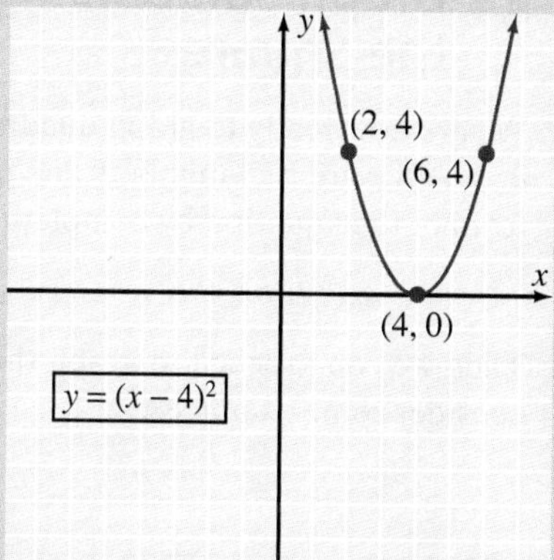

7.

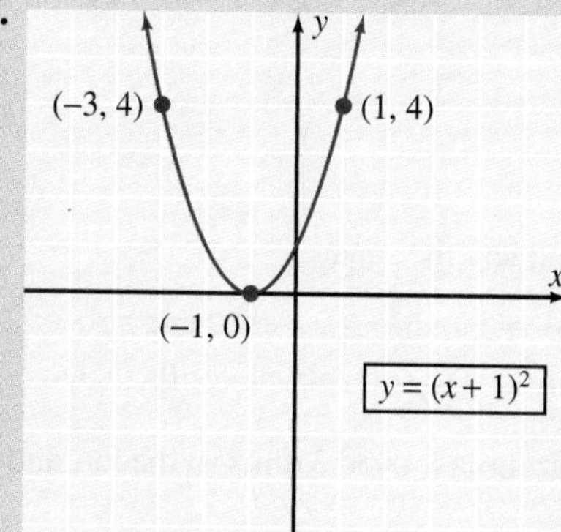

8.

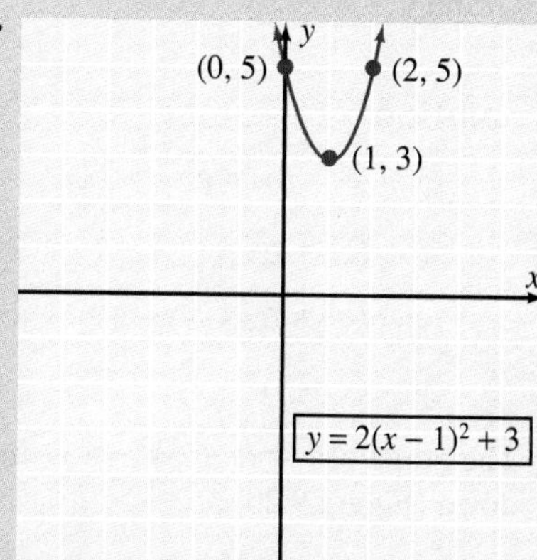

9.

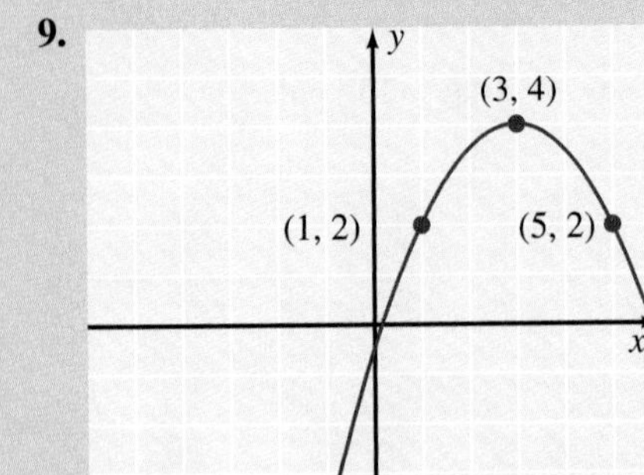

9.3 More Parabolas and Some Circles

OBJECTIVES

1. Graph Parabolas Using Completing the Square
2. Write the Equation of a Circle in Standard Form
3. Graph a Circle

1 Graph Parabolas Using Completing the Square

We are now ready to graph quadratic equations of the form $y = ax^2 + bx + c$, where a, b, and c are real numbers and $a \neq 0$. The general approach is one of changing the form of the equation by completing the square.

$$y = ax^2 + bx + c \longrightarrow y = a(x - h)^2 + k$$

Then we can proceed to graph the parabolas as we did in the previous section. Let's consider some examples.

EXAMPLE 1 Graph $y = x^2 + 6x + 8$.

Solution

$$y = x^2 + 6x + 8$$

$$y = (x^2 + 6x + \_\_) - (\_\_) + 8 \qquad \text{Complete the square}$$

$$y = (x^2 + 6x + 9) - (9) + 8 \qquad \frac{1}{2}(6) = 3 \text{ and } 3^2 = 9\text{. Add 9 and also subtract 9 to compensate for the 9 that was added}$$

$$y = (x + 3)^2 - 1$$

The graph of $y = (x + 3)^2 - 1$ is the basic parabola moved 3 units to the left and 1 unit down (Figure 9.20).

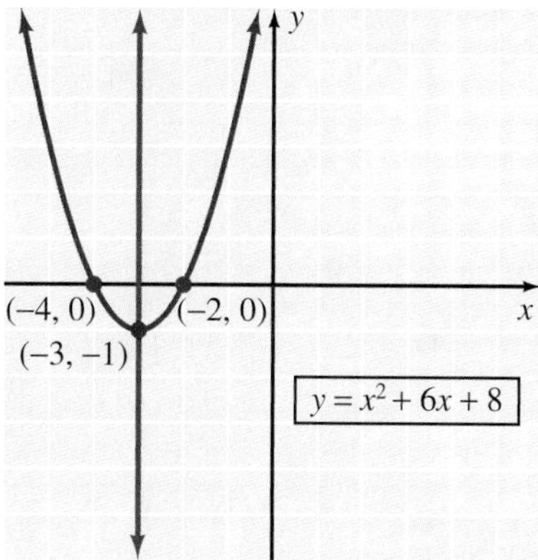

Figure 9.20

▼ PRACTICE YOUR SKILL

Graph $y = x^2 + 4x + 1$. ■

EXAMPLE 2

Graph $y = x^2 - 3x - 1$.

Solution

$$y = x^2 - 3x - 1$$

$$y = (x^2 - 3x + \_\_) - (\_\_) - 1 \quad \text{Complete the square}$$

$$y = \left(x^2 - 3x + \frac{9}{4}\right) - \frac{9}{4} - 1 \quad \frac{1}{2}(-3) = -\frac{3}{2} \text{ and } \left(-\frac{3}{2}\right)^2 = \frac{9}{4}. \text{ Add}$$

$$y = \left(x - \frac{3}{2}\right)^2 - \frac{13}{4} \quad \text{and subtract } \frac{9}{4}$$

The graph of $y = \left(x - \frac{3}{2}\right)^2 - \frac{13}{4}$ is the basic parabola moved $1\frac{1}{2}$ units to the right and $3\frac{1}{4}$ units down (Figure 9.21).

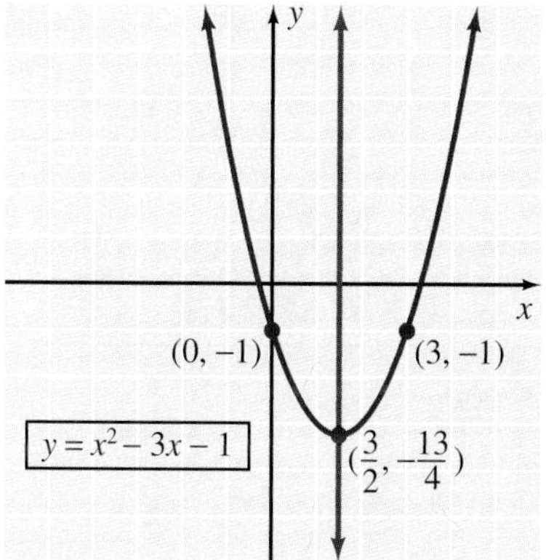

Figure 9.21

▼ PRACTICE YOUR SKILL

Graph $y = x^2 + x + 1$. ■

If the coefficient of x^2 is not 1, then a slight adjustment has to be made before we apply the process of completing the square. The next two examples illustrate this situation.

EXAMPLE 3 Graph $y = 2x^2 + 8x + 9$.

Solution

$y = 2x^2 + 8x + 9$

$y = 2(x^2 + 4x) + 9$ Factor a 2 from the x-variable terms

$y = 2(x^2 + 4x + \_\_) - (2)(\_\_) + 9$ Complete the square. Note that the number being subtracted will be multiplied by a factor of 2

$y = 2(x^2 + 4x + 4) - 2(4) + 9$ $\frac{1}{2}(4) = 2$, and $2^2 = 4$

$y = 2(x^2 + 4x + 4) - 8 + 9$

$y = 2(x + 2)^2 + 1$

See Figure 9.22 for the graph of $y = 2(x + 2)^2 + 1$.

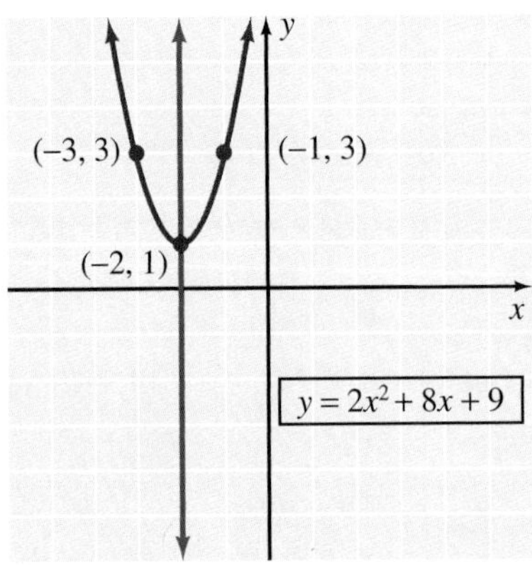

Figure 9.22

▼ **PRACTICE YOUR SKILL**

Graph $y = 2x^2 - 4x + 5$. ■

EXAMPLE 4 Graph $y = -3x^2 + 6x - 5$.

Solution

$y = -3x^2 + 6x - 5$

$y = -3(x^2 - 2x) - 5$ Factor -3 from the x-variable terms

$y = -3(x^2 - 2x + \_\_) - (-3)(\_\_) - 5$ Complete the square. Note that the number being subtracted will be multiplied by a factor of -3

$y = -3(x^2 - 2x + 1) - (-3)(1) - 5$ $\frac{1}{2}(-2) = -1$ and $(-1)^2 = 1$

$y = -3(x^2 - 2x + 1) + 3 - 5$

$y = -3(x - 1)^2 - 2$

The graph of $y = -3(x - 1)^2 - 2$ is shown in Figure 9.23.

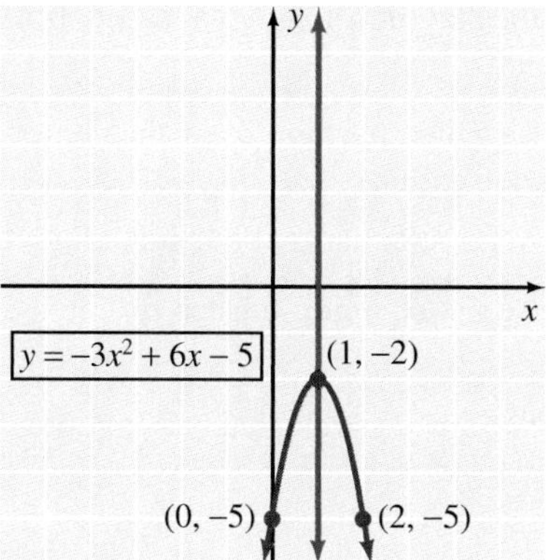

Figure 9.23

▼ **PRACTICE YOUR SKILL**

Graph $y = -2x^2 + 12x - 17$.

2 Write the Equation of a Circle in Standard Form

The distance formula, $d = \sqrt{(x_2 - x_1)^2 + (y_2 - y_1)^2}$ (developed in Section 3.3), when applied to the definition of a circle produces what is known as the **standard equation of a circle**. We start with a precise definition of a circle.

Definition 9.1

A **circle** is the set of all points in a plane equidistant from a given fixed point called the **center.** A line segment determined by the center and any point on the circle is called a **radius.**

Let's consider a circle that has a radius of length r and a center at (h, k) on a coordinate system (Figure 9.24).

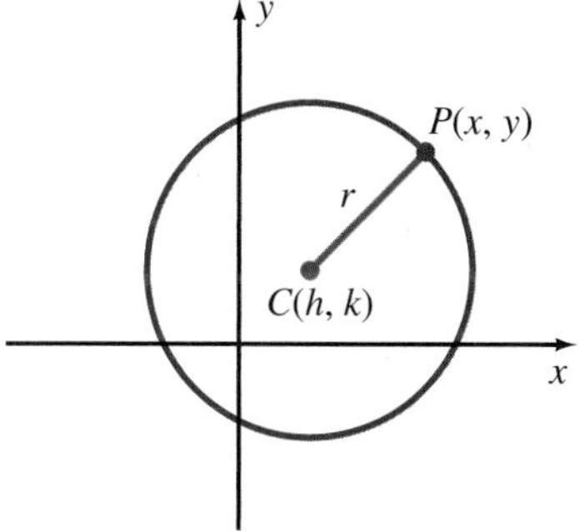

Figure 9.24

By using the distance formula, we can express the length of a radius (denoted by r) for any point $P(x, y)$ on the circle as

$$r = \sqrt{(x - h)^2 + (y - k)^2}$$

Thus squaring both sides of the equation, we obtain the **standard form of the equation of a circle**:

$$(x - h)^2 + (y - k)^2 = r^2$$

We can use the standard form of the equation of a circle to solve two basic kinds of circle problems:

1. Given the coordinates of the center and the length of a radius of a circle, find its equation.
2. Given the equation of a circle, find its center and the length of a radius.

Let's look at some examples of such problems.

EXAMPLE 5

Write the equation of a circle that has its center at $(3, -5)$ and a radius of length 6 units.

Solution

Let's substitute 3 for h, -5 for k, and 6 for r into the standard form $(x - h)^2 + (y - k)^2 = r^2$ to obtain $(x - 3)^2 + (y + 5)^2 = 6^2$, which we can simplify as follows:

$$(x - 3)^2 + (y + 5)^2 = 6^2$$
$$x^2 - 6x + 9 + y^2 + 10y + 25 = 36$$
$$x^2 + y^2 - 6x + 10y - 2 = 0$$

▼ **PRACTICE YOUR SKILL**

Write the equation of a circle that has its center at $(-2, 1)$ and a radius of length 4 units. ■

Note in Example 5 that we simplified the equation to the form $x^2 + y^2 + Dx + Ey + F = 0$, where D, E, and F are integers. This is another form that we commonly use when working with circles.

3 Graph a Circle

EXAMPLE 6

Graph $x^2 + y^2 + 4x - 6y + 9 = 0$.

Solution

This equation is of the form $x^2 + y^2 + Dx + Ey + F = 0$, so its graph is a circle. We can change the given equation into the form $(x - h)^2 + (y - k)^2 = r^2$ by completing the square on x and on y as follows:

$$x^2 + y^2 + 4x - 6y + 9 = 0$$
$$(x^2 + 4x + \_\_\_) + (y^2 - 6y + \_\_\_) = -9$$
$$(x^2 + 4x + 4) + (y^2 - 6y + 9) = -9 + 4 + 9$$

Added 4 to complete the square on x — Added 9 to complete the square on y — Added 4 and 9 to compensate for the 4 and 9 added on the left side

$$(x + 2)^2 + (y - 3)^2 = 4$$
$$(x - (-2))^2 + (y - 3)^2 = 2^2$$

(-2 is h, 3 is k, 2 is r)

The center of the circle is at $(-2, 3)$ and the length of a radius is 2 (Figure 9.25).

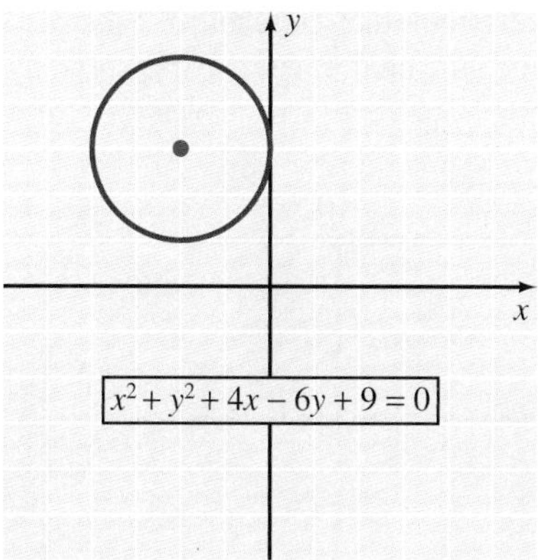

Figure 9.25

▼ PRACTICE YOUR SKILL

Graph $x^2 + y^2 + 2x - 8y + 8 = 0$. ■

As demonstrated by Examples 5 and 6, both forms, $(x - h)^2 + (y - k)^2 = r^2$ and $x^2 + y^2 + Dx + Ey + F = 0$, play an important role when we are solving problems that deal with circles.

Finally, we need to recognize that the standard form of a circle that has its center at the origin is $x^2 + y^2 = r^2$. This is simply the result of letting $h = 0$ and $k = 0$ in the general standard form.

$$(x - h)^2 + (y - k)^2 = r^2$$
$$(x - 0)^2 + (y - 0)^2 = r^2$$
$$x^2 + y^2 = r^2$$

Thus by inspection we can recognize that $x^2 + y^2 = 9$ is a circle with its center at the origin; the length of a radius is 3 units. Likewise, the equation of a circle that has its center at the origin and a radius of length 6 units is $x^2 + y^2 = 36$.

When using a graphing utility to graph a circle, we need to solve the equation for y in terms of x. This will produce two equations that can be graphed on the same set of axes. Furthermore, as with any graph, it may be necessary to change the boundaries on x or y (or both) to obtain a complete graph. If the circle appears oblong, you may want to use a zoom square option so that the graph will appear as a circle. Let's consider an example.

EXAMPLE 7

Use a graphing utility to graph $x^2 - 40x + y^2 + 351 = 0$.

Solution

First, we need to solve for y in terms of x.

$$x^2 - 40x + y^2 + 351 = 0$$
$$y^2 = -x^2 + 40x - 351$$
$$y = \pm\sqrt{-x^2 + 40x - 351}$$

Now we can make the following assignments.

$$Y_1 = \sqrt{-x^2 + 40x - 351}$$
$$Y_2 = -Y_1$$

(Note that we assigned Y_2 in terms of Y_1. By doing this we avoid repetitive key strokes and thus reduce the chance for errors. You may need to consult your user's manual for instructions on how to keystroke $-Y_1$.) Figure 9.26 shows the graph.

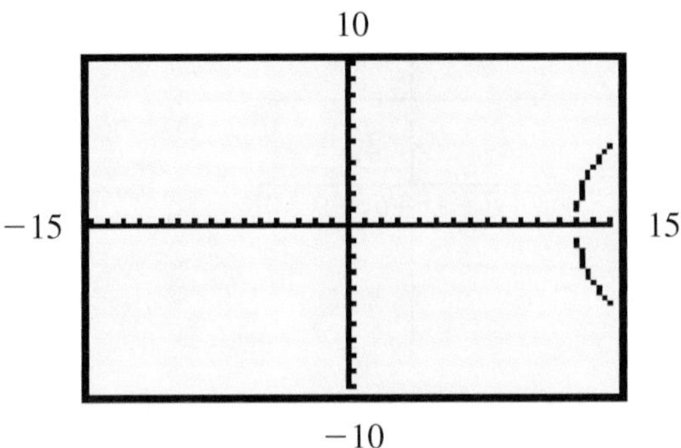

Figure 9.26

Because we know from the original equation that this graph should be a circle, we need to make some adjustments on the boundaries in order to get a complete graph. This can be done by completing the square on the original equation to change its form to $(x - 20)^2 + y^2 = 49$ or simply by a trial-and-error process. By changing the boundaries on x such that $-15 \leq x \leq 30$, we obtain Figure 9.27.

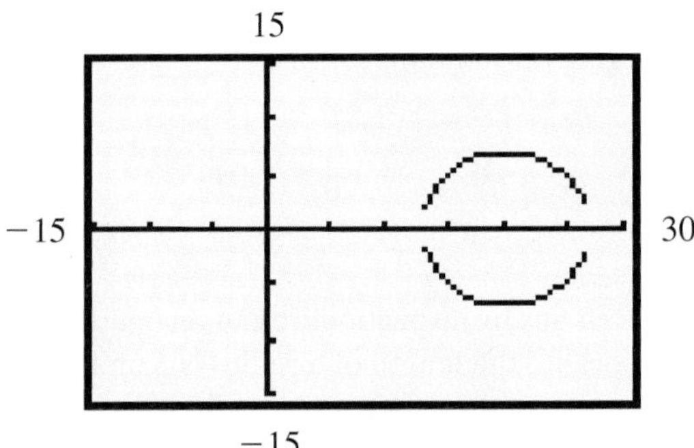

Figure 9.27

▼ **PRACTICE YOUR SKILL**

Use a graphing utility to graph $x^2 + y^2 + 16x - 240 = 0$. ■

CONCEPT QUIZ

For Problems 1–10, answer true or false.

1. Equations of the form $y = ax^2 + bx + c$ can be changed to the form $y = a(x - h)^2 + k$ by completing the square.
2. A circle is the set of points in a plane that are equidistant from a given fixed point.
3. A line segment determined by the center and any point on the circle is called the diameter.
4. The circle $(x + 2)^2 + (y - 5)^2 = 20$ has its center at $(2, -5)$.
5. The circle $(x - 4)^2 + (y + 3)^2 = 10$ has a radius of length 10.
6. The circle $x^2 + y^2 = 16$ has its center at the origin.
7. The graph of $y = -x^2 + 4x - 1$ does not intersect the x axis.
8. The only x intercept of the graph of $y = x^2 - 4x + 4$ is 2.
9. The origin is a point on the circle $x^2 + 4x + y^2 - 2y = 0$.
10. The vertex of the parabola $y = -2x^2 - 8x + 7$ is at $(-2, 15)$.

Problem Set 9.3

1 Graph Parabolas Using Completing the Square

For Problems 1–22, graph each parabola.

1. $y = x^2 - 6x + 13$

2. $y = x^2 - 4x + 7$

3. $y = x^2 + 2x + 6$

4. $y = x^2 + 8x + 14$

5. $y = x^2 - 5x + 3$

6. $y = x^2 + 3x + 1$

7. $y = x^2 + 7x + 14$

8. $y = x^2 - x - 1$

9. $y = 3x^2 - 6x + 5$

10. $y = 2x^2 + 4x + 7$

11. $y = 4x^2 - 24x + 32$

12. $y = 3x^2 + 24x + 49$

13. $y = -2x^2 - 4x - 5$

14. $y = -2x^2 + 8x - 5$

15. $y = -x^2 + 8x - 21$

16. $y = -x^2 - 6x - 7$

17. $y = 2x^2 - x + 2$

18. $y = 2x^2 + 3x + 1$

19. $y = 3x^2 + 2x + 1$

20. $y = 3x^2 - x - 1$

21. $y = -3x^2 - 7x - 2$

22. $y = -2x^2 + x - 2$

2 Write the Equation of a Circle in Standard Form

For Problems 23–34, find the center and the length of a radius of each circle by writing the equation in standard form.

23. $x^2 + y^2 - 2x - 6y - 6 = 0$

24. $x^2 + y^2 + 4x - 12y + 39 = 0$

25. $x^2 + y^2 + 6x + 10y + 18 = 0$

26. $x^2 + y^2 - 10x + 2y + 1 = 0$

27. $x^2 + y^2 = 10$

28. $x^2 + y^2 + 4x + 14y + 50 = 0$

29. $x^2 + y^2 - 16x + 6y + 71 = 0$

30. $x^2 + y^2 = 12$

31. $x^2 + y^2 + 6x - 8y = 0$

32. $x^2 + y^2 - 16x + 30y = 0$

33. $4x^2 + 4y^2 + 4x - 32y + 33 = 0$

34. $9x^2 + 9y^2 - 6x - 12y - 40 = 0$

For Problems 35–44, write the equation of each circle. Express the final equation in the form $x^2 + y^2 + Dx + Ey + F = 0$.

35. Center at (3, 5) and $r = 5$

36. Center at (2, 6) and $r = 7$

37. Center at (−4, 1) and $r = 8$

38. Center at (−3, 7) and $r = 6$

39. Center at (−2, −6) and $r = 3\sqrt{2}$

40. Center at (−4, −5) and $r = 2\sqrt{3}$

41. Center at (0, 0) and $r = 2\sqrt{5}$

42. Center at (0, 0) and $r = \sqrt{7}$

43. Center at (5, −8) and $r = 4\sqrt{6}$

44. Center at (4, −10) and $r = 8\sqrt{2}$

45. Find the equation of the circle that passes through the origin and has its center at (0, 4).

46. Find the equation of the circle that passes through the origin and has its center at (−6, 0).

47. Find the equation of the circle that passes through the origin and has its center at (−4, 3).

48. Find the equation of the circle that passes through the origin and has its center at (8, −15).

3 Graph a Circle

For Problems 49–58, graph each circle.

49. $x^2 + y^2 = 25$

50. $x^2 + y^2 = 36$

51. $(x - 1)^2 + (y + 2)^2 = 9$

52. $(x + 3)^2 + (y - 2)^2 = 1$

53. $x^2 + y^2 + 6x - 2y + 6 = 0$

54. $x^2 + y^2 - 4x - 6y - 12 = 0$

55. $x^2 + y^2 + 4y - 5 = 0$

56. $x^2 + y^2 - 4x + 3 = 0$

57. $x^2 + y^2 + 4x + 4y - 8 = 0$

58. $x^2 + y^2 - 6x + 6y + 2 = 0$

THOUGHTS INTO WORDS

59. What is the graph of $x^2 + y^2 = -4$? Explain your answer.

60. On which axis does the center of the circle $x^2 + y^2 - 8y + 7 = 0$ lie? Defend your answer.

61. Give a step-by-step description of how you would help someone graph the parabola $y = 2x^2 - 12x + 9$.

FURTHER INVESTIGATIONS

62. The points (x, y) and (y, x) are mirror images of each other across the line $y = x$. Therefore, by interchanging x and y in the equation $y = ax^2 + bx + c$, we obtain the equation of its mirror image across the line $y = x$—namely, $x = ay^2 + by + c$. Thus to graph $x = y^2 + 2$, we can first graph $y = x^2 + 2$ and then reflect it across the line $y = x$, as indicated in Figure 9.28.

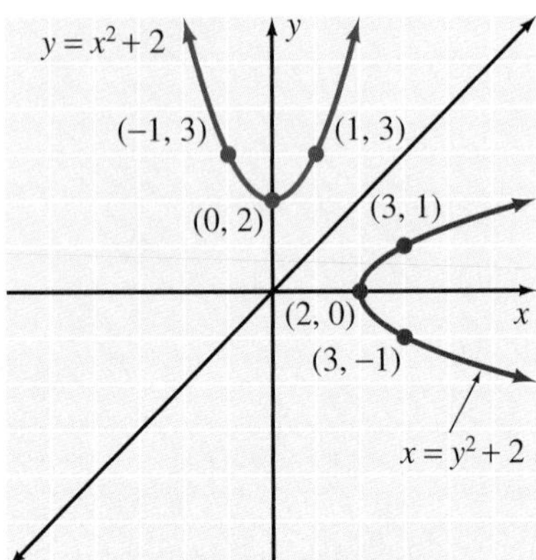

Figure 9.28

Graph each of the following parabolas.

(a) $x = y^2$
(b) $x = -y^2$
(c) $x = y^2 - 1$
(d) $x = -y^2 + 3$
(e) $x = -2y^2$
(f) $x = 3y^2$
(g) $x = y^2 + 4y + 7$
(h) $x = y^2 - 2y - 3$

63. By expanding $(x - h)^2 + (y - k)^2 = r^2$, we obtain $x^2 - 2hx + h^2 + y^2 - 2ky + k^2 - r^2 = 0$. When we compare this result to the form $x^2 + y^2 + Dx + Ey + F = 0$, we see that $D = -2h$, $E = -2k$, and $F = h^2 + k^2 - r^2$. Therefore, the center and length of a radius of a circle can be found by using $h = \frac{D}{-2}$, $k = \frac{E}{-2}$, and $r = \sqrt{h^2 + k^2 - F}$. Use these relationships to find the center and the length of a radius of each of the following circles.

(a) $x^2 + y^2 - 2x - 8y + 8 = 0$
(b) $x^2 + y^2 + 4x - 14y + 49 = 0$
(c) $x^2 + y^2 + 12x + 8y - 12 = 0$
(d) $x^2 + y^2 - 16x + 20y + 115 = 0$
(e) $x^2 + y^2 - 12y - 45 = 0$
(f) $x^2 + y^2 + 14x = 0$

GRAPHING CALCULATOR ACTIVITIES

64. Use a graphing calculator to check your graphs for Problems 1–22.

65. Use a graphing calculator to graph the circles in Problems 23–26. Be sure that your graphs are consistent with the center and the length of a radius that you found when you did the problems.

66. Graph each of the following parabolas and circles. Be sure to set your boundaries so that you get a complete graph.

(a) $x^2 + 24x + y^2 + 135 = 0$
(b) $y = x^2 - 4x + 18$
(c) $x^2 + y^2 - 18y + 56 = 0$
(d) $x^2 + y^2 + 24x + 28y + 336 = 0$
(e) $y = -3x^2 - 24x - 58$
(f) $y = x^2 - 10x + 3$

Answers to the Concept Quiz

1. True **2.** True **3.** False **4.** False **5.** False **6.** True **7.** False **8.** True **9.** True **10.** True

Answers to the Example Practice Skills

1.

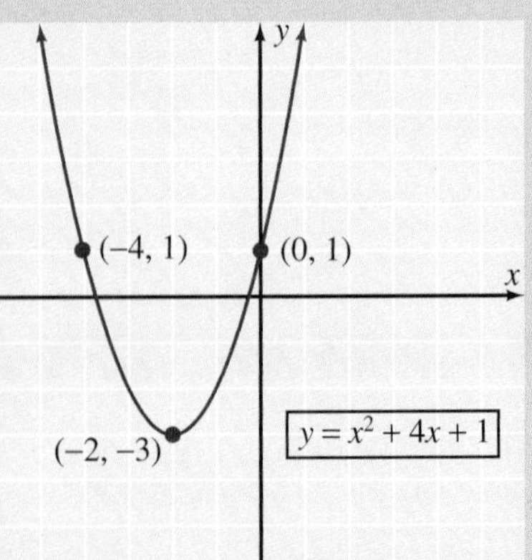

2.

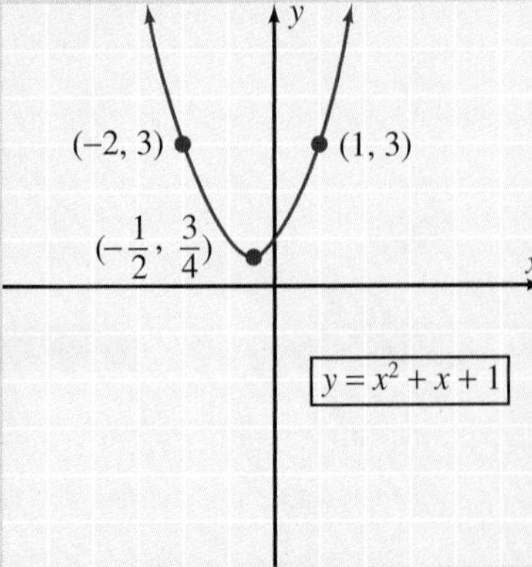

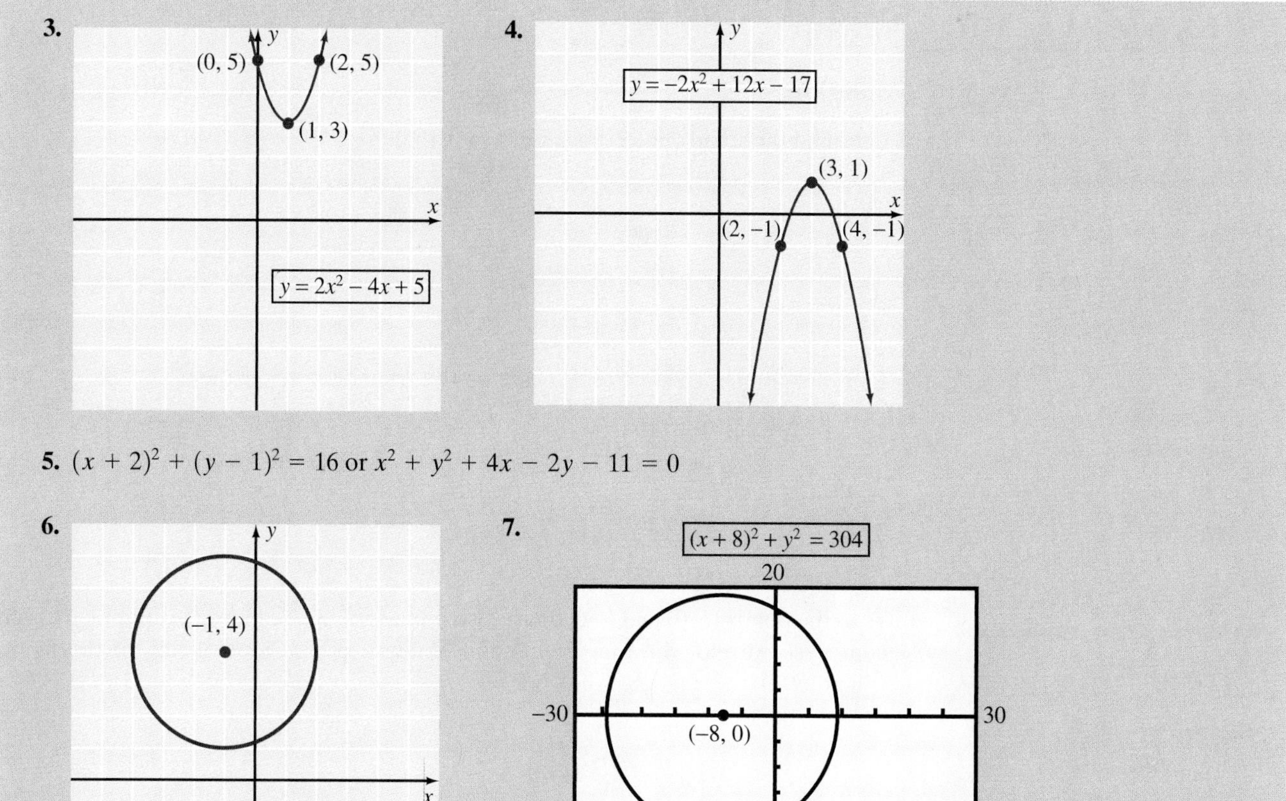

9.4 Graphing Ellipses

OBJECTIVES

1 Graph Ellipses with Centers at the Origin

2 Graph Ellipses with Centers Not at the Origin

1 Graph Ellipses with Centers at the Origin

In the previous section, we found that the graph of the equation $x^2 + y^2 = 36$ is a circle of radius 6 units with its center at the origin. More generally, it is true that any equation of the form $Ax^2 + By^2 = C$, where $A = B$ and where A, B, and C are nonzero constants that have the same sign, is a circle with the center at the origin. For example, $3x^2 + 3y^2 = 12$ is equivalent to $x^2 + y^2 = 4$ (divide both sides of the equation by 3), and thus it is a circle of radius 2 units with its center at the origin.

The general equation $Ax^2 + By^2 = C$ can be used to describe other geometric figures by changing the restrictions on A and B. For example, if A, B, and C are of the same sign but $A \neq B$, then the graph of the equation $Ax^2 + By^2 = C$ is an **ellipse**. Let's consider two examples.

EXAMPLE 1 Graph $4x^2 + 25y^2 = 100$.

Solution

Let's find the x and y intercepts. Let $x = 0$; then

$$\begin{aligned} 4(0)^2 + 25y^2 &= 100 \\ 25y^2 &= 100 \\ y^2 &= 4 \\ y &= \pm 2 \end{aligned}$$

Thus the points $(0, 2)$ and $(0, -2)$ are on the graph. Let $y = 0$; then

$$\begin{aligned} 4x^2 + 25(0)^2 &= 100 \\ 4x^2 &= 100 \\ x^2 &= 25 \\ x &= \pm 5 \end{aligned}$$

Thus the points $(5, 0)$ and $(-5, 0)$ are also on the graph. We know that this figure is an ellipse, so we plot the four points and obtain a pretty good sketch of the figure (Figure 9.29).

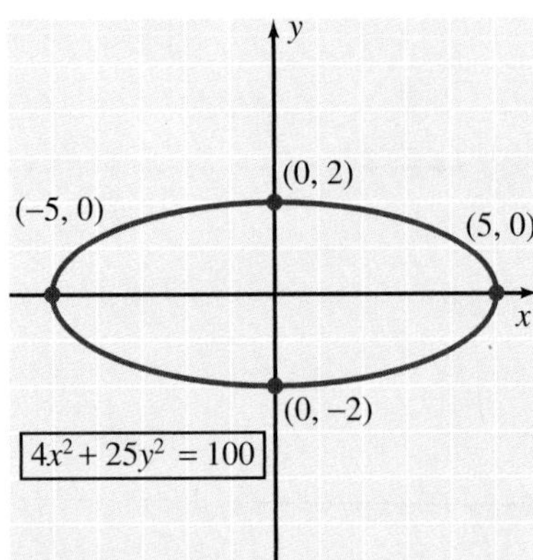

Figure 9.29

▼ **PRACTICE YOUR SKILL**

Graph $9x^2 + 16y^2 = 144$. ■

In Figure 9.29, the line segment with endpoints at $(-5, 0)$ and $(5, 0)$ is called the **major axis** of the ellipse. The shorter line segment, with endpoints at $(0, -2)$ and $(0, 2)$, is called the **minor axis**. Establishing the endpoints of the major and minor axes provides a basis for sketching an ellipse. The point of intersection of the major and minor axes is called the **center** of the ellipse.

EXAMPLE 2 Graph $9x^2 + 4y^2 = 36$.

Solution

Again, let's find the x and y intercepts. Let $x = 0$; then

$$\begin{aligned} 9(0)^2 + 4y^2 &= 36 \\ 4y^2 &= 36 \end{aligned}$$

$$y^2 = 9$$

$$y = \pm 3$$

Thus the points (0, 3) and (0, −3) are on the graph. Let $y = 0$; then

$$9x^2 + 4(0)^2 = 36$$

$$9x^2 = 36$$

$$x^2 = 4$$

$$x = \pm 2$$

Thus the points (2, 0) and (−2, 0) are also on the graph. The ellipse is sketched in Figure 9.30.

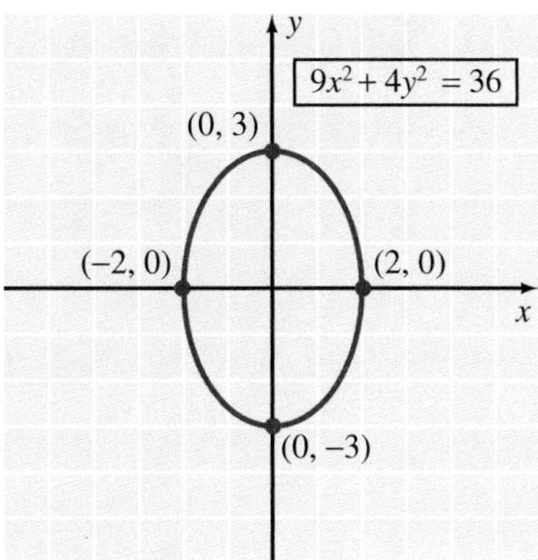

Figure 9.30

▼ **PRACTICE YOUR SKILL**

Graph $25x^2 + 9y^2 = 225$. ■

In Figure 9.30, the major axis has endpoints at (0, −3) and (0, 3), and the minor axis has endpoints at (−2, 0) and (2, 0). The ellipses in Figures 9.29 and 9.30 are symmetric about the x axis and about the y axis. In other words, both the x axis and the y axis serve as **axes of symmetry**.

2 Graph Ellipses with Centers Not at the Origin

Now we turn to some ellipses whose centers are not at the origin but whose major and minor axes are parallel to the x axis and the y axis. We can graph such ellipses in much the same way that we handled circles in Section 9.3. Let's consider two examples to illustrate the procedure.

EXAMPLE 3 Graph $4x^2 + 24x + 9y^2 - 36y + 36 = 0$.

Solution

Let's complete the square on x and y as follows:

$$4x^2 + 24x + 9y^2 - 36y + 36 = 0$$

$$4(x^2 + 6x + \_\_) + 9(y^2 - 4y + \_\_) = -36$$

$$4(x^2 + 6x + 9) + 9(y^2 - 4y + 4) = -36 + 36 + 36$$

$$4(x + 3)^2 + 9(y - 2)^2 = 36$$

$$4(x - (-3))^2 + 9(y - 2)^2 = 36$$

Because 4, 9, and 36 are of the same sign and $4 \neq 9$, the graph is an ellipse. The center of the ellipse is at $(-3, 2)$. We can find the endpoints of the major and minor axes as follows: Use the equation $4(x + 3)^2 + 9(y - 2)^2 = 36$ and let $y = 2$ (the y coordinate of the center).

$$4(x + 3)^2 + 9(2 - 2)^2 = 36$$
$$4(x + 3)^2 = 36$$
$$(x + 3)^2 = 9$$
$$x + 3 = \pm 3$$
$$x + 3 = 3 \quad \text{or} \quad x + 3 = -3$$
$$x = 0 \quad \text{or} \quad x = -6$$

This gives the points $(0, 2)$ and $(-6, 2)$. These are the coordinates of the endpoints of the major axis. Now let $x = -3$ (the x coordinate of the center).

$$4(-3 + 3)^2 + 9(y - 2)^2 = 36$$
$$9(y - 2)^2 = 36$$
$$(y - 2)^2 = 4$$
$$y - 2 = \pm 2$$
$$y - 2 = 2 \quad \text{or} \quad y - 2 = -2$$
$$y = 4 \quad \text{or} \quad y = 0$$

This gives the points $(-3, 4)$ and $(-3, 0)$. These are the coordinates of the endpoints of the minor axis. The ellipse is shown in Figure 9.31.

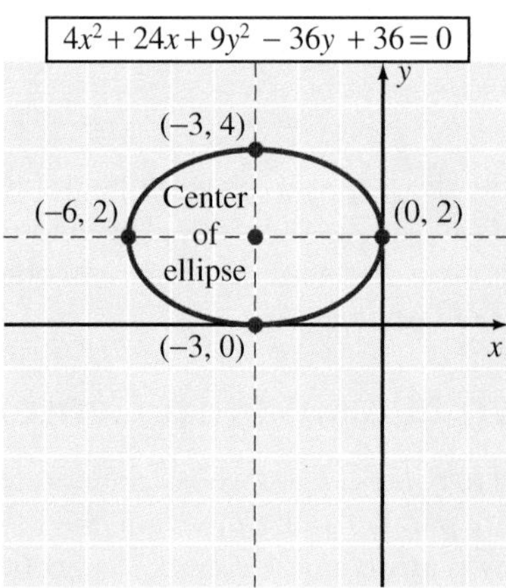

Figure 9.31

▼ PRACTICE YOUR SKILL

Graph $4x^2 + 16x + 9y^2 - 18y - 11 = 0$. ■

EXAMPLE 4

Graph $4x^2 - 16x + y^2 + 6y + 9 = 0$.

Solution

First let's complete the square on x and on y.

$$4x^2 - 16x + y^2 + 6y + 9 = 0$$
$$4(x^2 - 4x + \_\_) + (y^2 + 6y + \_\_) = -9$$

$$4(x^2 - 4x + 4) + (y^2 + 6y + 9) = -9 + 16 + 9$$

$$4(x - 2)^2 + (y + 3)^2 = 16$$

The center of the ellipse is at $(2, -3)$. Now let $x = 2$ (the x coordinate of the center).

$$4(2 - 2)^2 + (y + 3)^2 = 16$$

$$(y + 3)^2 = 16$$

$$y + 3 = \pm 4$$

$$y + 3 = -4 \quad \text{or} \quad y + 3 = 4$$

$$y = -7 \quad \text{or} \quad y = 1$$

This gives the points $(2, -7)$ and $(2, 1)$. These are the coordinates of the endpoints of the major axis. Now let $y = -3$ (the y coordinate of the center).

$$4(x - 2)^2 + (-3 + 3)^2 = 16$$

$$4(x - 2)^2 = 16$$

$$(x - 2)^2 = 4$$

$$x - 2 = \pm 2$$

$$x - 2 = -2 \quad \text{or} \quad x - 2 = 2$$

$$x = 0 \quad \text{or} \quad x = 4$$

This gives the points $(0, -3)$ and $(4, -3)$. These are the coordinates of the endpoints of the minor axis. The ellipse is shown in Figure 9.32.

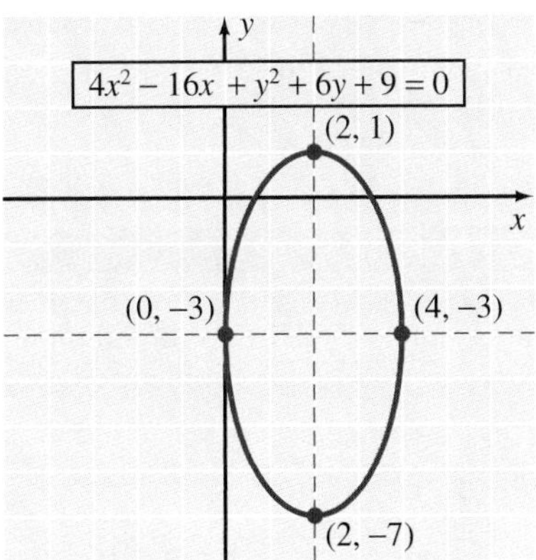

Figure 9.32

▼ PRACTICE YOUR SKILL

Graph $16x^2 - 96x + 9y^2 + 18y + 9 = 0$. ■

CONCEPT QUIZ

For Problems 1–10, answer true or false.

1. The length of the major axis of an ellipse is always greater than the length of the minor axis.
2. The major axis of an ellipse is always parallel to the x axis.
3. The axes of symmetry for an ellipse pass through the center of the ellipse.
4. The ellipse $9(x - 1)^2 + 4(y + 5)^2 = 36$ has its center at $(1, 5)$.
5. The x and y intercepts of the graph of an ellipse centered at the origin and symmetric to both axes are the endpoints of its axes.
6. The endpoints of the major axis of the ellipse $9x^2 + 4y^2 = 36$ are at $(-2, 0)$ and $(2, 0)$.

7. The endpoints of the minor axis of the ellipse $x^2 + 5y^2 = 15$ are at $(0, -\sqrt{3})$ and $(0, \sqrt{3})$.
8. The endpoints of the major axis of the ellipse $3(x - 2)^2 + 5(y + 3)^2 = 12$ are at $(0, -3)$ and $(4, -3)$.
9. The center of the ellipse $7x^2 - 14x + 8y^2 - 32y - 17 = 0$ is at $(1, 2)$.
10. The center of the ellipse $2x^2 + 12x + y^2 + 2 = 0$ is on the y axis.

Problem Set 9.4

1 Graph Ellipses with Centers at the Origin

For Problems 1–16, graph each ellipse.

1. $x^2 + 4y^2 = 36$
2. $x^2 + 4y^2 = 16$
3. $9x^2 + y^2 = 36$
4. $16x^2 + 9y^2 = 144$
5. $4x^2 + 3y^2 = 12$
6. $5x^2 + 4y^2 = 20$
7. $16x^2 + y^2 = 16$
8. $9x^2 + 2y^2 = 18$
9. $25x^2 + 2y^2 = 50$
10. $12x^2 + y^2 = 36$

2 Graph Ellipses with Centers Not at the Origin

11. $4x^2 + 8x + 16y^2 - 64y + 4 = 0$
12. $9x^2 - 36x + 4y^2 - 24y + 36 = 0$
13. $x^2 + 8x + 9y^2 + 36y + 16 = 0$
14. $4x^2 - 24x + y^2 + 4y + 24 = 0$
15. $4x^2 + 9y^2 - 54y + 45 = 0$
16. $x^2 + 2x + 4y^2 - 15 = 0$

THOUGHTS INTO WORDS

17. Is the graph of $x^2 + y^2 = 4$ the same as the graph of $y^2 + x^2 = 4$? Explain your answer.
18. Is the graph of $x^2 + y^2 = 0$ a circle? If so, what is the length of a radius?
19. Is the graph of $4x^2 + 9y^2 = 36$ the same as the graph of $9x^2 + 4y^2 = 36$? Explain your answer.
20. What is the graph of $x^2 + 2y^2 = -16$? Explain your answer.

GRAPHING CALCULATOR ACTIVITIES

21. Use a graphing calculator to graph the ellipses in Examples 1–4 of this section.
22. Use a graphing calculator to check your graphs for Problems 11–16.

Answers to the Concept Quiz

1. True **2.** False **3.** True **4.** False **5.** True **6.** False **7.** True **8.** True **9.** True **10.** False

Answers to the Example Practice Skills

1.

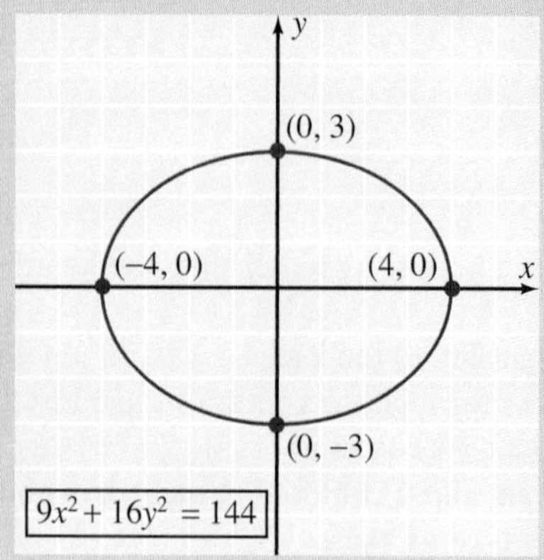

2.

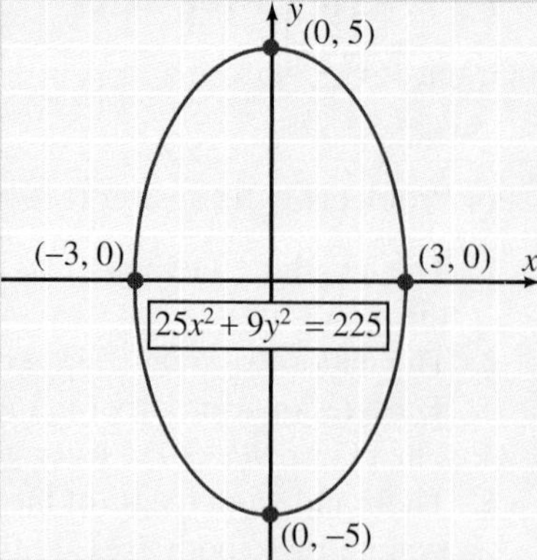

3.

y

$(-2, 3)$

$(-5, 1)$ $(1, 1)$

x

$(-2, -1)$

$4x^2 + 16x + 9y^2 - 18y - 11 = 0$

4.

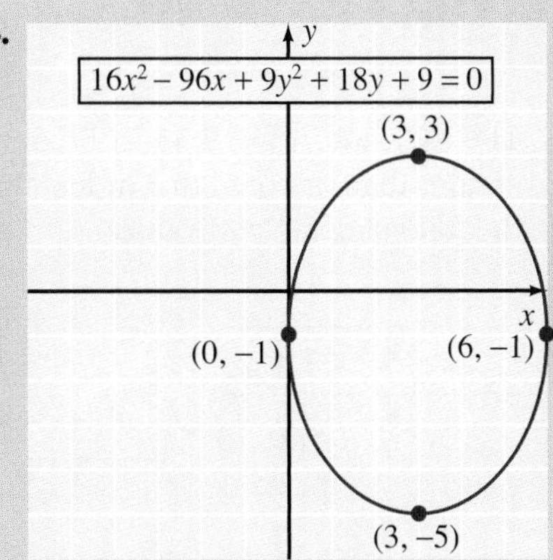

9.5 Graphing Hyperbolas

OBJECTIVES

1. Graph Hyperbolas Symmetric to Both Axes
2. Graph Hyperbolas Not Symmetric to Both Axes

1 Graph Hyperbolas Symmetric to Both Axes

The graph of an equation of the form $Ax^2 + By^2 = C$, where A, B, and C are nonzero real numbers and A and B are of unlike signs, is a **hyperbola**. Let's use some examples to illustrate a procedure for graphing hyperbolas.

EXAMPLE 1

Graph $x^2 - y^2 = 9$.

Solution

If we let $y = 0$, we obtain

$$x^2 - 0^2 = 0$$
$$x^2 = 9$$
$$x = \pm 3$$

Thus the points $(3, 0)$ and $(-3, 0)$ are on the graph. If we let $x = 0$, we obtain

$$0^2 - y^2 = 9$$
$$-y^2 = 9$$
$$y^2 = -9$$

Because $y^2 = -9$ has no real number solutions, there are no points of the y axis on this graph. That is, the graph does not intersect the y axis. Now let's solve the given equation for y so that we have a more convenient form for finding other solutions.

$$x^2 - y^2 = 9$$
$$-y^2 = 9 - x^2$$

$$y^2 = x^2 - 9$$
$$y = \pm\sqrt{x^2 - 9}$$

The radicand, $x^2 - 9$, must be nonnegative, so the values we choose for x must be greater than or equal to 3 or less than or equal to -3. With this in mind, we can form the following table of values.

x	y	
3	0	Intercepts
-3	0	
4	$\pm\sqrt{7}$	Other points
-4	$\pm\sqrt{7}$	
5	± 4	
-5	± 4	

We plot these points and draw the hyperbola as in Figure 9.33. (This graph is also symmetric about both axes.)

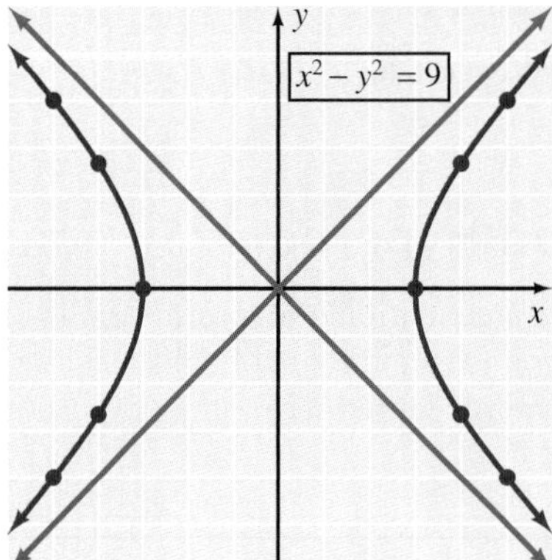

Figure 9.33

▼ PRACTICE YOUR SKILL

Graph $x^2 - y^2 = 25$. ■

Note the blue lines in Figure 9.33; they are called **asymptotes**. Each branch of the hyperbola approaches one of these lines but does not intersect it. Therefore, the ability to sketch the asymptotes of a hyperbola is very helpful when we are graphing the hyperbola. Fortunately, the equations of the asymptotes are easy to determine. They can be found by replacing the constant term in the given equation of the hyperbola with 0 and solving for y. (The reason why this works will become evident in a later course.) So for the hyperbola in Example 3, we obtain

$$x^2 - y^2 = 0$$
$$y^2 = x^2$$
$$y = \pm x$$

Thus the two lines $y = x$ and $y = -x$ are the asymptotes indicated by the blue lines in Figure 9.33.

EXAMPLE 2 Graph $y^2 - 5x^2 = 4$.

Solution

If we let $x = 0$, we obtain

$$y^2 - 5(0)^2 = 4$$
$$y^2 = 4$$
$$y = \pm 2$$

The points (0, 2) and (0, −2) are on the graph. If we let $y = 0$, we obtain

$$0^2 - 5x^2 = 4$$
$$-5x^2 = 4$$
$$x^2 = -\frac{4}{5}$$

Because $x^2 = -\frac{4}{5}$ has no real number solutions, we know that this hyperbola does not intersect the x axis. Solving the given equation for y yields

$$y^2 - 5x^2 = 4$$
$$y^2 = 5x^2 + 4$$
$$y = \pm\sqrt{5x^2 + 4}$$

The table shows some additional solutions for the equation. The equations of the asymptotes are determined as follows:

$$y^2 - 5x^2 = 0$$
$$y^2 = 5x^2$$
$$y = \pm\sqrt{5}x$$

x	y	
0	2	Intercepts
0	−2	
1	±3	Other points
−1	±3	
2	$\pm\sqrt{24}$	
−2	$\pm\sqrt{24}$	

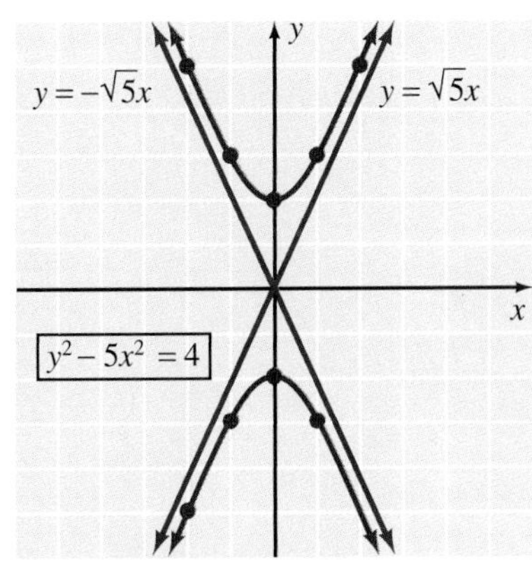

Figure 9.34

Sketch the asymptotes and plot the points shown in the table to determine the hyperbola in Figure 9.34. (Note that this hyperbola is also symmetric about the x axis and the y axis.)

▼ PRACTICE YOUR SKILL

Graph $y^2 - 6x^2 = 9$. ■

EXAMPLE 3 Graph $4x^2 - 9y^2 = 36$.

Solution

If we let $x = 0$, we obtain

$$4(0)^2 - 9y^2 = 36$$
$$-9y^2 = 36$$
$$y^2 = -4$$

Because $y^2 = -4$ has no real number solutions, we know that this hyperbola does not intersect the y axis. If we let $y = 0$, we obtain

$$4x^2 - 9(0)^2 = 36$$
$$4x^2 = 36$$
$$x^2 = 9$$
$$x = \pm 3$$

Thus the points (3, 0) and (−3, 0) are on the graph. Now let's solve the equation for y in terms of x and set up a table of values.

$$4x^2 - 9y^2 = 36$$
$$-9y^2 = 36 - 4x^2$$
$$9y^2 = 4x^2 - 36$$
$$y^2 = \frac{4x^2 - 36}{9}$$
$$y = \pm\frac{\sqrt{4x^2 - 36}}{3}$$

x	y	
3	0	Intercepts
−3	0	
4	$\pm\frac{2\sqrt{7}}{3}$	
−4	$\pm\frac{2\sqrt{7}}{3}$	Other points
5	$\pm\frac{8}{3}$	
−5	$\pm\frac{8}{3}$	

The equations of the asymptotes are found as follows:

$$4x^2 - 9y^2 = 0$$
$$-9y^2 = -4x^2$$
$$9y^2 = 4x^2$$
$$y^2 = \frac{4x^2}{9}$$
$$y = \pm\frac{2}{3}x$$

Sketch the asymptotes and plot the points shown in the table to determine the hyperbola, as shown in Figure 9.35.

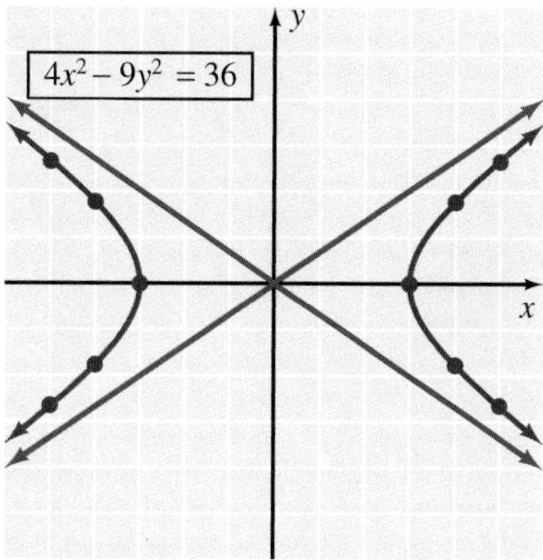

Figure 9.35

▼ **PRACTICE YOUR SKILL**

Graph $9x^2 - 25y^2 = 225$. ■

2 Graph Hyperbolas Not Symmetric to Both Axes

Now let's consider hyperbolas that are not symmetric with respect to the origin but are symmetric with respect to lines parallel to one of the axes—that is, vertical and horizontal lines. Again, let's use examples to illustrate a procedure for graphing such hyperbolas.

EXAMPLE 4

Graph $4x^2 - 8x - y^2 - 4y - 16 = 0$.

Solution

Completing the square on x and y, we obtain

$$4x^2 - 8x - y^2 - 4y - 16 = 0$$

$$4(x^2 - 2x + \_) - (y^2 + 4y + \_) = 16$$

$$4(x^2 - 2x + 1) - (y^2 + 4y + 4) = 16 + 4 - 4$$

$$4(x - 1)^2 - (y + 2)^2 = 16$$

$$4(x - 1)^2 - 1(y - (-2))^2 = 16$$

Because 4 and -1 are of opposite signs, the graph is a hyperbola. The center of the hyperbola is at $(1, -2)$.

Now using the equation $4(x - 1)^2 - (y + 2)^2 = 16$, we can proceed as follows: Let $y = -2$; then

$$4(x - 1)^2 - (-2 + 2)^2 = 16$$

$$4(x - 1)^2 = 16$$

$$(x - 1)^2 = 4$$

$$x - 1 = \pm 2$$

$$x - 1 = 2 \quad \text{or} \quad x - 1 = -2$$

$$x = 3 \quad \text{or} \quad x = -1$$

Thus the hyperbola intersects the horizontal line $y = -2$ at $(3, -2)$ and at $(-1, -2)$. Let $x = 1$; then

$$4(1 - 1)^2 - (y + 2)^2 = 16$$
$$-(y + 2)^2 = 16$$
$$(y + 2)^2 = -16$$

Because $(y + 2)^2 = -16$ has no real number solutions, we know that the hyperbola does not intersect the vertical line $x = 1$. We replace the constant term of $4(x - 1)^2 - (y + 2)^2 = 16$ with 0 and solve for y to produce the equations of the asymptotes as follows:

$$4(x - 1)^2 - (y + 2)^2 = 0$$

The left side can be factored using the pattern of the difference of squares.

$$[2(x - 1) + (y + 2)][2(x - 1) - (y + 2)] = 0$$
$$(2x - 2 + y + 2)(2x - 2 - y - 2) = 0$$
$$(2x + y)(2x - y - 4) = 0$$
$$2x + y = 0 \quad \text{or} \quad 2x - y - 4 = 0$$
$$y = -2x \quad \text{or} \quad 2x - 4 = y$$

Thus the equations of the asymptotes are $y = -2x$ and $y = 2x - 4$. Sketching the asymptotes and plotting the two points $(3, -2)$ and $(-1, -2)$, we can draw the hyperbola as shown in Figure 9.36.

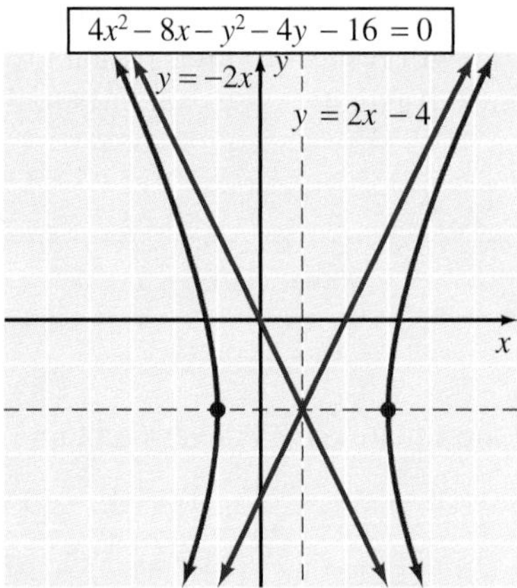

Figure 9.36

▼ **PRACTICE YOUR SKILL**

Graph $4x^2 + 16x - y^2 + 2y - 1 = 0$. ■

EXAMPLE 5

Graph $y^2 - 4y - 4x^2 - 24x - 36 = 0$.

Solution

First let's complete the square on x and on y.

$$y^2 - 4y - 4x^2 - 24x - 36 = 0$$
$$(y^2 - 4y + \_\_) - 4(x^2 + 6x + \_\_) = 36$$
$$(y^2 - 4y + 4) - 4(x^2 + 6x + 9) = 36 + 4 - 36$$
$$(y - 2)^2 - 4(x + 3)^2 = 4$$

The center of the hyperbola is at $(-3, 2)$. Now let $y = 2$.

$$(2 - 2)^2 - 4(x + 3)^2 = 4$$
$$-4(x + 3)^2 = 4$$
$$(x + 3)^2 = -1$$

Because $(x + 3)^2 = -1$ has no real number solutions, the graph does not intersect the line $y = 2$. Now let $x = -3$.

$$(y - 2)^2 - 4(-3 + 3)^2 = 4$$
$$(y - 2)^2 = 4$$
$$y - 2 = \pm 2$$
$$y - 2 = -2 \quad \text{or} \quad y - 2 = 2$$
$$y = 0 \quad \text{or} \quad y = 4$$

Therefore, the hyperbola intersects the line $x = -3$ at $(-3, 0)$ and $(-3, 4)$. Now, to find the equations of the asymptotes, let's replace the constant term of $(y - 2)^2 - 4(x + 3)^2 = 4$ with 0 and solve for y.

$$(y - 2)^2 - 4(x + 3)^2 = 0$$
$$[(y - 2) + 2(x + 3)][(y - 2) - 2(x + 3)] = 0$$
$$(y - 2 + 2x + 6)(y - 2 - 2x - 6) = 0$$
$$(y + 2x + 4)(y - 2x - 8) = 0$$
$$y + 2x + 4 = 0 \quad \text{or} \quad y - 2x - 8 = 0$$
$$y = -2x - 4 \quad \text{or} \quad y = 2x + 8$$

Therefore, the equations of the asymptotes are $y = -2x - 4$ and $y = 2x + 8$. Drawing the asymptotes and plotting the points $(-3, 0)$ and $(-3, 4)$, we can graph the hyperbola as shown in Figure 9.37.

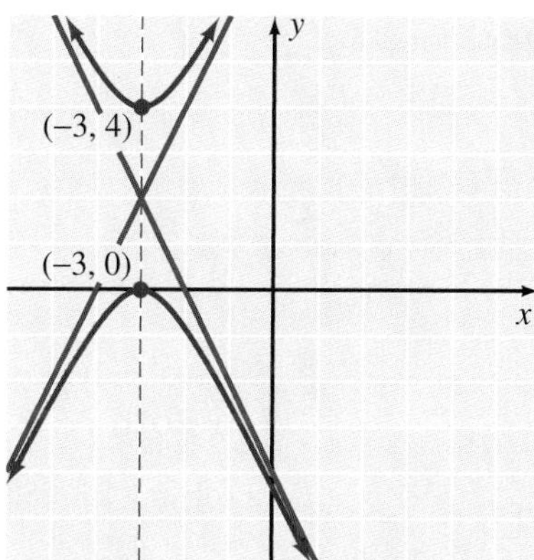

Figure 9.37

▼ PRACTICE YOUR SKILL

Graph $y^2 + 2y - 9x^2 + 36x - 44 = 0$. ■

As a way of summarizing our work with conic sections, let's focus our attention on the continuity pattern used in this chapter. In Sections 9.2 and 9.3, we studied parabolas by considering variations of the basic quadratic equation $y = ax^2 + bx + c$. Also in Section 9.3, we used the definition of a circle to generate a standard form for the equation of a circle. Then, in Sections 9.4 and 9.5, we discussed ellipses and hyperbolas — not from a definition viewpoint but by considering variations of the equations $Ax^2 + By^2 = C$ and $A(x - h)^2 + B(y - k)^2 = C$. In a subsequent mathematics course, parabolas, ellipses, and hyperbolas will be developed from a definition viewpoint. That is, first each concept will be defined and then the definition will be used to generate a standard form of its equation.

CONCEPT QUIZ

For Problems 1–10, answer true or false.

1. The graph of an equation of the form $Ax^2 + By^2 = C$, where A, B, and C are nonzero real numbers, is a hyperbola if A and B are of like sign.
2. The graph of a hyperbola always has two branches.
3. Each branch of the graph of a hyperbola approaches one of the asymptotes but never intersects with the asymptote.
4. To find the equations for the asymptotes, we replace the constant term in the equation of the hyperbola with zero and solve for y.
5. The hyperbola $9(x + 1)^2 - 4(y - 3)^2 = 36$ has its center at $(-1, 3)$.
6. The asymptotes of the graph of a hyperbola intersect at the center of the hyperbola.
7. The equations of the asymptotes for the hyperbola $9x^2 - 4y^2 = 36$ are $y = -\frac{2}{3}x$ and $y = \frac{2}{3}x$.
8. The center of the hyperbola $(y - 2)^2 - 9(x - 6)^2 = 18$ is at $(2, 6)$.
9. The equations of the asymptotes for the hyperbola $(y - 2)^2 - 9(x - 6)^2 = 18$ are $y = 3x - 16$ and $y = -3x + 20$.
10. The center of the hyperbola $x^2 + 6x - y^2 + 4y - 15 = 0$ is at $(-3, 2)$.

Problem Set 9.5

1 Graph Hyperbolas Symmetric to Both Axes

For Problems 1–6, find the intercepts and the equations for the asymptotes.

1. $x^2 - 9y^2 = 16$
2. $16x^2 - y^2 = 25$
3. $y^2 - 9x^2 = 36$
4. $4x^2 - 9y^2 = 16$
5. $25x^2 - 9y^2 = 4$
6. $y^2 - x^2 = 16$

For Problems 7–18, graph each hyperbola.

7. $x^2 - y^2 = 1$
8. $x^2 - y^2 = 4$
9. $y^2 - 4x^2 = 9$
10. $4y^2 - x^2 = 16$
11. $5x^2 - 2y^2 = 20$
12. $9x^2 - 4y^2 = 9$
13. $y^2 - 16x^2 = 4$
14. $y^2 - 9x^2 = 16$
15. $-4x^2 + y^2 = -4$
16. $-9x^2 + y^2 = -36$
17. $25y^2 - 3x^2 = 75$
18. $16y^2 - 5x^2 = 80$

For Problems 19–22, find the equations for the asymptotes.

19. $x^2 + 4x - y^2 - 6y - 30 = 0$
20. $y^2 - 8y - x^2 - 4x + 3 = 0$
21. $9x^2 - 18x - 4y^2 - 24y - 63 = 0$
22. $4x^2 + 24x - y^2 + 4y + 28 = 0$

2 Graph Hyperbolas Not Symmetric to Both Axes

For Problems 23–28, graph each hyperbola.

23. $-4x^2 + 32x + 9y^2 - 18y - 91 = 0$
24. $x^2 - 4x - y^2 + 6y - 14 = 0$
25. $-4x^2 + 24x + 16y^2 + 64y - 36 = 0$
26. $x^2 + 4x - 9y^2 + 54y - 113 = 0$
27. $4x^2 - 24x - 9y^2 = 0$
28. $16y^2 + 64y - x^2 = 0$
29. The graphs of equations of the form $xy = k$, where k is a nonzero constant, are also hyperbolas, sometimes

referred to as rectangular hyperbolas. Graph each of the following.

(a) $xy = 3$ **(b)** $xy = 5$

(c) $xy = -2$ **(d)** $xy = -4$

30. What is the graph of $xy = 0$? Defend your answer.

31. We have graphed various equations of the form $Ax^2 + By^2 = C$, where C is a nonzero constant. Now graph each of the following.

(a) $x^2 + y^2 = 0$ **(b)** $2x^2 + 3y^2 = 0$

(c) $x^2 - y^2 = 0$ **(d)** $4y^2 - x^2 = 0$

THOUGHTS INTO WORDS

32. Explain the concept of an asymptote.

33. Explain how asymptotes can be used to help graph hyperbolas.

34. Are the graphs of $x^2 - y^2 = 0$ and $y^2 - x^2 = 0$ identical? Are the graphs of $x^2 - y^2 = 4$ and $y^2 - x^2 = 4$ identical? Explain your answers.

GRAPHING CALCULATOR ACTIVITIES

35. To graph the hyperbola in Example 1 of this section, we can make the following assignments for the graphing calculator.

$Y_1 = \sqrt{x^2 - 9}$ $Y_2 = -Y_1$

$Y_3 = x$ $Y_4 = -Y_3$

Do this and see if your graph agrees with Figure 9.33. Also graph the asymptotes and hyperbolas for Examples 2 and 3.

36. Use a graphing calculator to check your graphs for Problems 7–18.

37. Use a graphing calculator to check your graphs for Problems 23–28.

38. For each of the following equations, (1) predict the type and location of the graph and (2) use your graphing calculator to check your predictions.

(a) $x^2 + y^2 = 100$ **(b)** $x^2 - y^2 = 100$

(c) $y^2 - x^2 = 100$ **(d)** $y = -x^2 + 9$

(e) $2x^2 + y^2 = 14$ **(f)** $x^2 + 2y^2 = 14$

(g) $x^2 + 2x + y^2 - 4 = 0$ **(h)** $x^2 + y^2 - 4y - 2 = 0$

(i) $y = x^2 + 16$ **(j)** $y^2 = x^2 + 16$

(k) $9x^2 - 4y^2 = 72$ **(l)** $4x^2 - 9y^2 = 72$

(m) $y^2 = -x^2 - 4x + 6$

Answers to the Concept Quiz

1. False **2.** True **3.** True **4.** True **5.** True **6.** True **7.** False **8.** False **9.** True **10.** True

Answers to the Example Practice Skills

1.

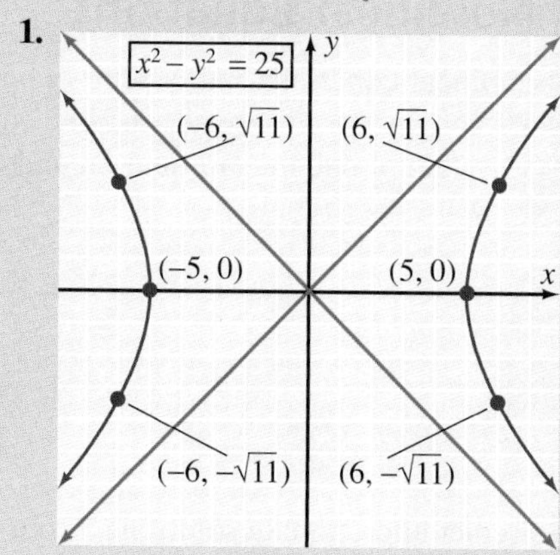

2.

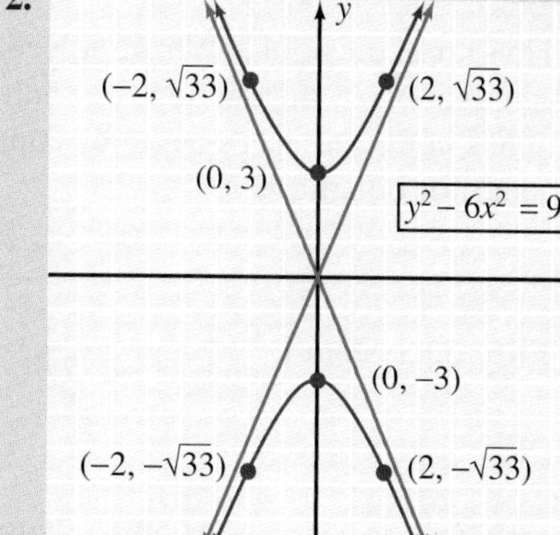

3.

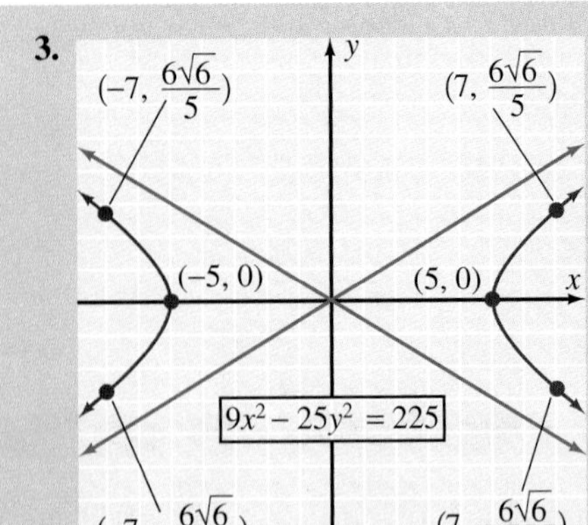

4.

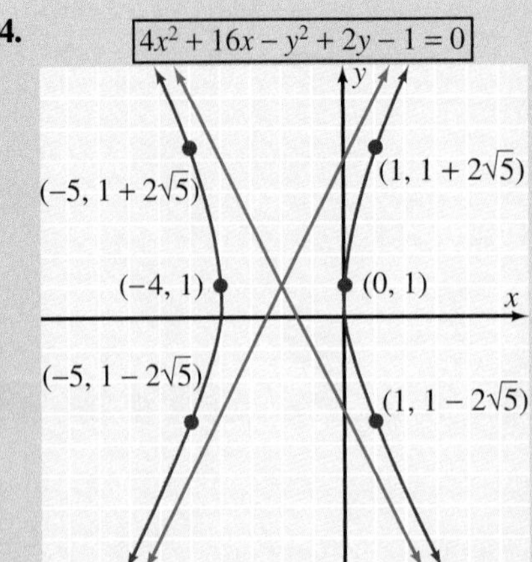

5.

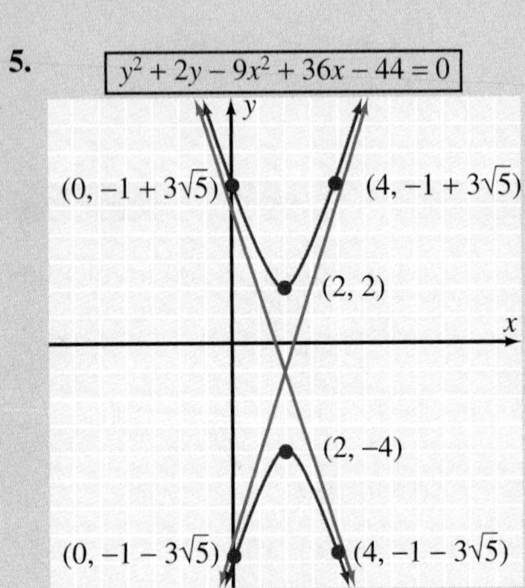

9.6 Systems Involving Nonlinear Equations

OBJECTIVE

1 Solve Systems Involving Nonlinear Equations

1 Solve Systems Involving Nonlinear Equations

Thus far in the book, we have solved systems of linear equations and systems of linear inequalities. In this section, we shall consider some systems where at least one equation is *nonlinear*. Let's begin by considering a system of one linear equation and one quadratic equation.

EXAMPLE 1

Solve the system $\begin{pmatrix} x^2 + y^2 = 17 \\ x + y = 5 \end{pmatrix}$.

Solution

First, let's graph the system so that we can predict approximate solutions. From our previous graphing experiences, we should recognize $x^2 + y^2 = 17$ as a circle and $x + y = 5$ as a straight line (Figure 9.38).

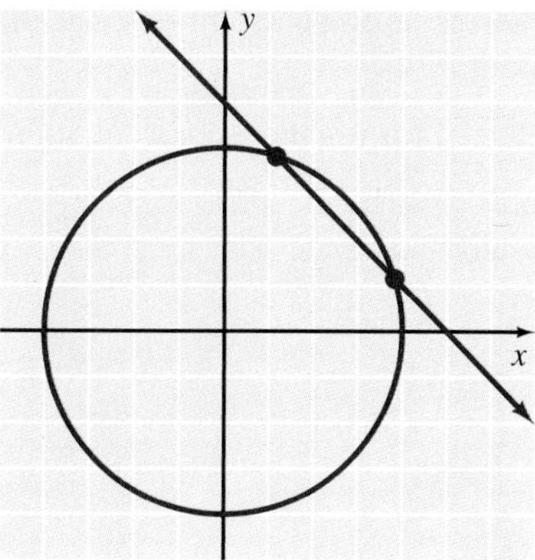

Figure 9.38

The graph indicates that there should be two ordered pairs with positive components (the points of intersection occur in the first quadrant) as solutions for this system. In fact, we could guess that these solutions are (1, 4) and (4, 1) and then verify our guess by checking them in the given equations.

Let's also solve the system analytically using the substitution method as follows: Change the form of $x + y = 5$ to $y = 5 - x$ and substitute $5 - x$ for y in the first equation.

$$x^2 + y^2 = 17$$

$$x^2 + (5 - x)^2 = 17$$

$$x^2 + 25 - 10x + x^2 = 17$$

$$2x^2 - 10x + 8 = 0$$

$$x^2 - 5x + 4 = 0$$

$$(x - 4)(x - 1) = 0$$

$$x - 4 = 0 \quad \text{or} \quad x - 1 = 0$$

$$x = 4 \quad \text{or} \quad x = 1$$

Substitute 4 for x and then 1 for x in the second equation of the system to produce

$$x + y = 5 \qquad x + y = 5$$

$$4 + y = 5 \qquad 1 + y = 5$$

$$y = 1 \qquad y = 4$$

Therefore, the solution set is $\{(1, 4), (4, 1)\}$.

▼ PRACTICE YOUR SKILL

Solve the system $\begin{pmatrix} x^2 + y^2 = 10 \\ x + y = 2 \end{pmatrix}$. ■

EXAMPLE 2 Solve the system $\begin{pmatrix} y = -x^2 + 1 \\ y = x^2 - 2 \end{pmatrix}$.

Solution

Again, let's get an idea of approximate solutions by graphing the system. Both equations produce parabolas, as indicated in Figure 9.39. From the graph, we can predict two nonintegral ordered-pair solutions, one in the third quadrant and the other in the fourth quadrant.

Substitute $-x^2 + 1$ for y in the second equation to obtain

$$y = x^2 - 2$$
$$-x^2 + 1 = x^2 - 2$$
$$3 = 2x^2$$
$$\frac{3}{2} = x^2$$
$$\pm\sqrt{\frac{3}{2}} = x$$
$$\pm\frac{\sqrt{6}}{2} = x$$

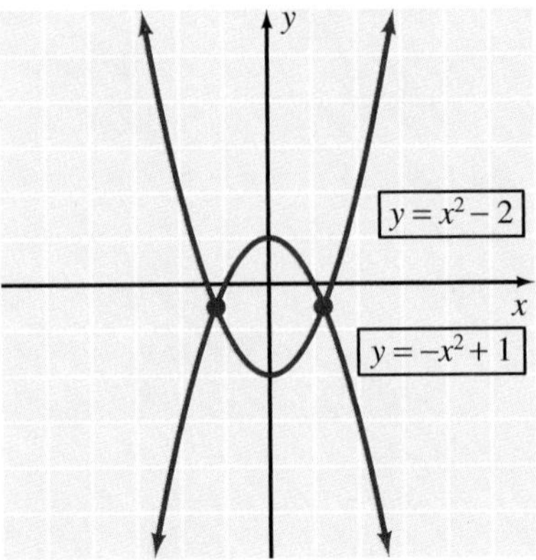

Figure 9.39

Substitute $\frac{\sqrt{6}}{2}$ for x in the second equation to yield

$$y = x^2 - 2$$
$$y = \left(\frac{\sqrt{6}}{2}\right)^2 - 2$$
$$= \frac{6}{4} - 2$$
$$= -\frac{1}{2}$$

Substitute $-\frac{\sqrt{6}}{2}$ for x in the second equation to yield

$$y = x^2 - 2$$
$$y = \left(-\frac{\sqrt{6}}{2}\right)^2 - 2$$
$$= \frac{6}{4} - 2 = -\frac{1}{2}$$

The solution set is $\left\{\left(-\frac{\sqrt{6}}{2}, -\frac{1}{2}\right), \left(\frac{\sqrt{6}}{2}, -\frac{1}{2}\right)\right\}$. Check it!

▼ PRACTICE YOUR SKILL

Solve the system $\begin{pmatrix} y = x^2 + 1 \\ y = -x^2 + 4 \end{pmatrix}$. ■

EXAMPLE 3

Solve the system $\begin{pmatrix} x^2 + y^2 = 13 \\ 9x^2 - 4y^2 = 65 \end{pmatrix}$.

Solution

Let's get an idea of approximate solutions by graphing the system. The graph of $x^2 + y^2 = 13$ is a circle, and the graph of $9x^2 - 4y^2 = 65$ is a hyperbola (Figure 9.40).

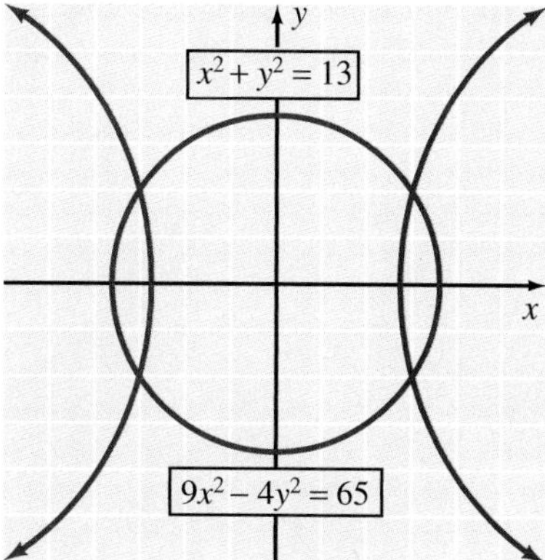

Figure 9.40

The graph indicates that there should be four ordered pairs as solutions. Let's solve the system analytically. This system can be readily solved using the elimination-by-addition method.

$$\begin{pmatrix} x^2 + y^2 = 13 \\ 9x^2 - 4y^2 = 65 \end{pmatrix} \quad \begin{matrix} (1) \\ (2) \end{matrix}$$

Form an equivalent system by multiplying the first equation by 4 and adding the result to the second equation.

$$\begin{pmatrix} x^2 + y^2 = 13 \\ 13x^2 = 117 \end{pmatrix} \quad \begin{matrix} (3) \\ (4) \end{matrix}$$

From equation (4) we can solve for x.

$$13x^2 = 117$$
$$x^2 = 9$$
$$x = \pm 3$$

To find y values, substitute $x = 3$ and $x = -3$ into equation (3).

When $x = 3$

$$x^2 + y^2 = 13$$
$$(3)^2 + y^2 = 13$$
$$9 + y^2 = 13$$
$$y^2 = 4$$
$$y = 2 \text{ or } y = -2$$

When $x = -3$

$$x^2 + y^2 = 13$$
$$(-3)^2 + y^2 = 13$$
$$9 + y^2 = 13$$
$$y^2 = 4$$
$$y = 2 \text{ or } y = -2$$

Therefore the solution set is $\{(3, 2), (3, -2), (-3, 2), (-3, -2)\}$.

▼ **PRACTICE YOUR SKILL**

Solve the system $\begin{pmatrix} 2x^2 + y^2 = 32 \\ x^2 + y^2 = 16 \end{pmatrix}$. ■

EXAMPLE 4

Solve the system $\begin{pmatrix} y = x^2 + 2 \\ 6x - 4y = -5 \end{pmatrix}$.

Solution

From previous graphing experiences, we recognize that $y = x^2 + 2$ is the basic parabola shifted upward 2 units and that $6x - 4y = -5$ is a straight line (see Figure 9.41). Because of the close proximity of the curves, it is difficult to tell whether they intersect. In other words, the graph does not definitely indicate any real number solutions for the system.

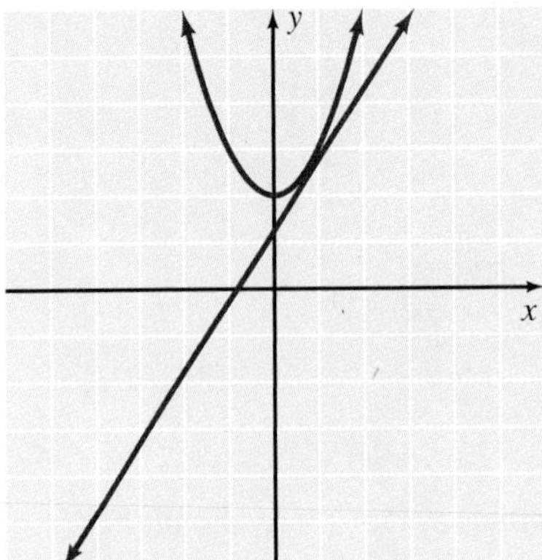

Figure 9.41

Let's solve the system using the substitution method. We can substitute $x^2 + 2$ for y in the second equation, which produces two values for x.

$$6x - 4(x^2 + 2) = -5$$
$$6x - 4x^2 - 8 = -5$$
$$-4x^2 + 6x - 3 = 0$$
$$4x^2 - 6x + 3 = 0$$
$$x = \frac{6 \pm \sqrt{36 - 48}}{8}$$
$$x = \frac{6 \pm \sqrt{-12}}{8}$$
$$x = \frac{6 \pm 2i\sqrt{3}}{8}$$
$$x = \frac{3 \pm i\sqrt{3}}{4}$$

It is now obvious that the system has no real number solutions. That is, the line and the parabola do not intersect in the real number plane. However, there will be two pairs of complex numbers in the solution set. We can substitute $\frac{(3 + i\sqrt{3})}{4}$ for x in the first equation.

$$y = \left(\frac{3 + i\sqrt{3}}{4}\right)^2 + 2$$
$$= \frac{6 + 6i\sqrt{3}}{16} + 2$$
$$= \frac{6 + 6i\sqrt{3} + 32}{16}$$
$$= \frac{38 + 6i\sqrt{3}}{16} = \frac{19 + 3i\sqrt{3}}{8}$$

Likewise, we can substitute $\frac{(3 - i\sqrt{3})}{4}$ for x in the first equation.

$$y = \left(\frac{3 - i\sqrt{3}}{4}\right)^2 + 2$$

$$= \frac{6 - 6i\sqrt{3}}{16} + 2$$

$$= \frac{6 - 6i\sqrt{3} + 32}{16}$$

$$= \frac{38 - 6i\sqrt{3}}{16}$$

$$= \frac{19 - 3i\sqrt{3}}{8}$$

The solution set is $\left\{\left(\frac{3 + i\sqrt{3}}{4}, \frac{19 + 3i\sqrt{3}}{4}\right), \left(\frac{3 - i\sqrt{3}}{4}, \frac{19 - 3i\sqrt{3}}{4}\right)\right\}$.

▼ **PRACTICE YOUR SKILL**

Solve the system $\begin{pmatrix} y = -x^2 + 2 \\ 8x + 4y = 13 \end{pmatrix}$. ■

In Example 4, the use of a graphing utility may not, at first, indicate whether or not the system has any real number solutions. Suppose that we graph the system using a viewing rectangle such that $-15 \leq x \leq 15$ and $-10 \leq y \leq 10$. In Figure 9.42, we cannot tell whether or not the line and parabola intersect. However, if we change the viewing rectangle so that $0 \leq x \leq 2$ and $0 \leq y \leq 4$, as shown in Figure 9.43, then it becomes apparent that the two graphs do not intersect.

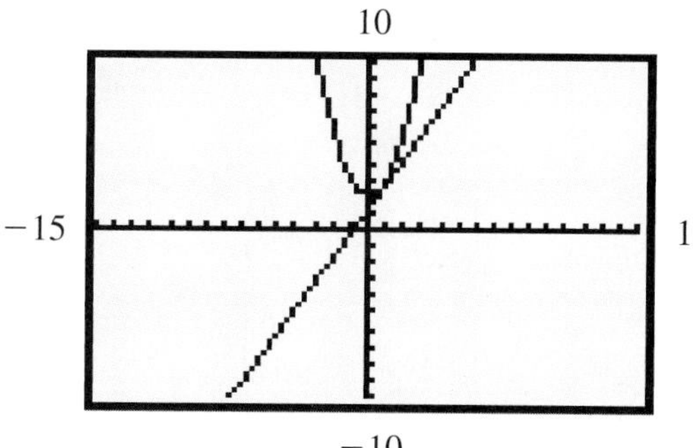

Figure 9.42

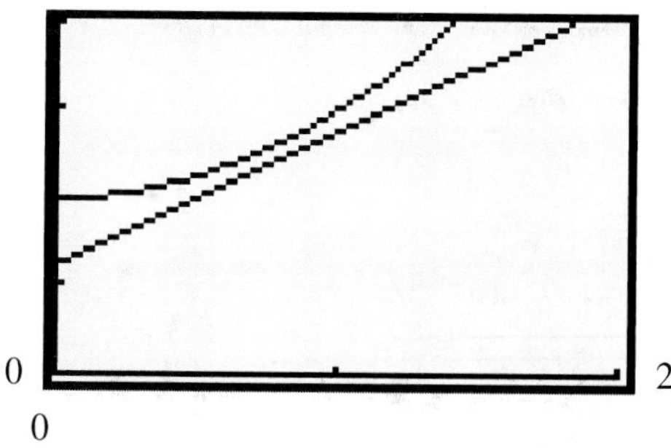

Figure 9.43

CONCEPT QUIZ

For Problems 1–10, answer true or false.

1. Every system of nonlinear equations has a real number solution.
2. If a system of equations has no real number solutions, then the graphs of the equations do not intersect.
3. Every nonlinear system of equations can be solved by substitution.
4. Every nonlinear system of equations can be solved by the elimination method.
5. Graphs of a circle and a line will have one, two, or no points of intersection.
6. Graphs of a circle and an ellipse will have either four points of intersection or no points of intersection.
7. The solution set for the system $\begin{pmatrix} y = x^2 + 1 \\ y = -x^2 + 1 \end{pmatrix}$ is $\{(0, 1)\}$.

8. The solution set for the system $\begin{pmatrix} 3x^2 + 4y^2 = 12 \\ x^2 - y^2 = 9 \end{pmatrix}$ is $\{(4, \sqrt{7})\}$.

9. The solution set for the system $\begin{pmatrix} 2x^2 + y^2 = 8 \\ x^2 + y^2 = 4 \end{pmatrix}$ is the null set.

10. The solution set for the system $\begin{pmatrix} x^2 + y^2 = 16 \\ x^2 + y^2 = 4 \end{pmatrix}$ is the null set.

Problem Set 9.6

1 Solve Systems Involving Nonlinear Equations

For Problems 1–30, (a) graph each system so that approximate real number solutions (if there are any) can be predicted, and (b) solve each system using the substitution method or the elimination-by-addition method.

1. $\begin{pmatrix} y = (x + 2)^2 \\ y = -2x - 4 \end{pmatrix}$

2. $\begin{pmatrix} y = x^2 \\ y = x + 2 \end{pmatrix}$

3. $\begin{pmatrix} x^2 + y^2 = 13 \\ 3x + 2y = 0 \end{pmatrix}$

4. $\begin{pmatrix} x^2 + y^2 = 26 \\ x + y = 6 \end{pmatrix}$

5. $\begin{pmatrix} y = x^2 + 6x + 7 \\ 2x + y = -5 \end{pmatrix}$

6. $\begin{pmatrix} y = x^2 - 4x + 5 \\ -x + y = 1 \end{pmatrix}$

7. $\begin{pmatrix} y = x^2 \\ y = x^2 - 4x + 4 \end{pmatrix}$

8. $\begin{pmatrix} y = -x^2 + 3 \\ y = x^2 + 1 \end{pmatrix}$

9. $\begin{pmatrix} x + y = -8 \\ x^2 - y^2 = 16 \end{pmatrix}$

10. $\begin{pmatrix} x - y = 2 \\ x^2 - y^2 = 16 \end{pmatrix}$

11. $\begin{pmatrix} y = x^2 + 2x - 1 \\ y = x^2 + 4x + 5 \end{pmatrix}$

12. $\begin{pmatrix} 2x^2 + y^2 = 8 \\ x^2 + y^2 = 4 \end{pmatrix}$

13. $\begin{pmatrix} xy = 4 \\ y = x \end{pmatrix}$

14. $\begin{pmatrix} y = x^2 + 2 \\ y = 2x^2 + 1 \end{pmatrix}$

15. $\begin{pmatrix} x^2 + y^2 = 2 \\ x - y = 4 \end{pmatrix}$

16. $\begin{pmatrix} y = -x^2 + 1 \\ x + y = 2 \end{pmatrix}$

17. $\begin{pmatrix} x^2 + y^2 = 5 \\ x + 2y = 5 \end{pmatrix}$

18. $\begin{pmatrix} x^2 + y^2 = 13 \\ 2x + 3y = 13 \end{pmatrix}$

19. $\begin{pmatrix} x^2 + y^2 = 26 \\ x + y = -4 \end{pmatrix}$

20. $\begin{pmatrix} x^2 + y^2 = 10 \\ x + y = -2 \end{pmatrix}$

21. $\begin{pmatrix} 2x + y = -2 \\ y = x^2 + 4x + 7 \end{pmatrix}$

22. $\begin{pmatrix} 2x + y = 0 \\ y = -x^2 + 2x - 4 \end{pmatrix}$

23. $\begin{pmatrix} y = x^2 - 3 \\ x + y = -4 \end{pmatrix}$

24. $\begin{pmatrix} x^2 + y^2 = 3 \\ x - y = -5 \end{pmatrix}$

25. $\begin{pmatrix} x^2 - y^2 = 4 \\ x^2 + y^2 = 4 \end{pmatrix}$

26. $\begin{pmatrix} y = -x^2 + 1 \\ y = x^2 - 2 \end{pmatrix}$

27. $\begin{pmatrix} 2x^2 + y^2 = 11 \\ x^2 - y^2 = 4 \end{pmatrix}$

28. $\begin{pmatrix} 2x^2 - 3y^2 = -1 \\ 2x^2 + 3y^2 = 5 \end{pmatrix}$

29. $\begin{pmatrix} 8y^2 - 9x^2 = 6 \\ 8x^2 - 3y^2 = 5 \end{pmatrix}$

30. $\begin{pmatrix} x^2 - 4y^2 = 16 \\ 2y - x = 2 \end{pmatrix}$

THOUGHTS INTO WORDS

31. What happens if you try to graph the following system?

$$\begin{pmatrix} x^2 + 4y^2 = 16 \\ 2x^2 + 5y^2 = -12 \end{pmatrix}$$

32. Explain how you would solve the following system.

$$\begin{pmatrix} x^2 + y^2 = 9 \\ y^2 = x^2 + 4 \end{pmatrix}$$

GRAPHING CALCULATOR ACTIVITIES

33. Use a graphing calculator to graph the systems in Problems 1–30, and check the reasonableness of your answers.

34. For each of the following systems, (a) use your graphing calculator to show that there are no real number solutions, and (b) solve the system by the substitution method or the elimination-by-addition method to find the complex solutions.

(a) $\begin{pmatrix} y = x^2 + 1 \\ y = -3 \end{pmatrix}$ **(b)** $\begin{pmatrix} y = -x^2 + 1 \\ y = 3 \end{pmatrix}$

(c) $\begin{pmatrix} y = x^2 \\ x - y = 4 \end{pmatrix}$ **(d)** $\begin{pmatrix} y = x^2 + 1 \\ y = -x^2 \end{pmatrix}$

(e) $\begin{pmatrix} x^2 + y^2 = 1 \\ x + y = 2 \end{pmatrix}$ **(f)** $\begin{pmatrix} x^2 + y^2 = 2 \\ x^2 - y^2 = 6 \end{pmatrix}$

35. Graph the system $\begin{pmatrix} y = x^2 + 2 \\ 6x - 4y = -5 \end{pmatrix}$ and use the TRACE and ZOOM features of your calculator to demonstrate clearly that this system has no real number solutions.

Answers to the Concept Quiz

1. False **2.** True **3.** True **4.** False **5.** True **6.** False **7.** True **8.** False **9.** False **10.** True

Answers to the Example Practice Skills

1. $\{(-1, 3), (3, -1)\}$ **2.** $\left\{\left(-\frac{\sqrt{6}}{2}, \frac{5}{2}\right), \left(\frac{\sqrt{6}}{2}, \frac{5}{2}\right)\right\}$ **3.** $\{(-4, 0), (4, 0)\}$

4. $\left\{\left(\frac{2 + i}{2}, \frac{5 - 4i}{4}\right), \left(\frac{2 - i}{2}, \frac{5 + 4i}{4}\right)\right\}$

Chapter 9 Summary

OBJECTIVE	SUMMARY	EXAMPLE	CHAPTER REVIEW PROBLEMS
Graph nonlinear equations using symmetries as an aid. (Sec. 9.1, Obj. 1, p. 476)	The following suggestions are offered for graphing an equation in two variables. **1.** Determine what type of symmetry the equation exhibits. **2.** Find the intercepts. **3.** Solve the equation for y in terms of x or for x in terms of y if it is not already in such a form. **4.** Set up a table of ordered pairs that satisfy the equation. The type of symmetry will affect your choice of values in the table. **5.** Plot the points associated with the ordered pairs from the table, and connect them with a smooth curve. Then, if appropriate, reflect this part of the curve according to the symmetry shown by the equation.	Graph $y = \frac{1}{x^2}$. **Solution** **1.** The graph is symmetric with respect to the y axis because replacing x with $-x$ results in an equivalent equation. The graph is not symmetric with respect to the x axis because replacing y with $-y$ does not result in an equivalent equation. **2.** Zero is excluded from the domain; hence the graph does not intersect the y axis. **3.** x: $\frac{1}{4}$, $\frac{1}{2}$, 2, 4; y: 16, 4, $\frac{1}{4}$, $\frac{1}{16}$ 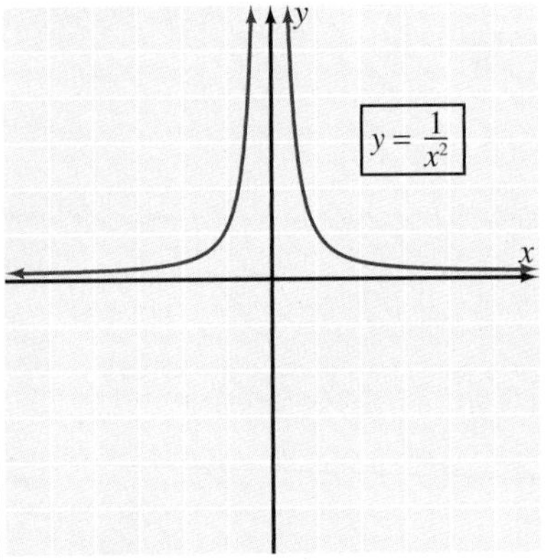 $y = \frac{1}{x^2}$	Problems 1–4
Graph parabolas. (Sec. 9.2, Obj. 1, p. 484)	To graph parabolas, we need to be able to: **1.** Find the vertex. **2.** Determine whether the parabola opens upward or downward. **3.** Locate two points on opposite sides of the line of symmetry. **4.** Compare the parabola to the basic parabola $y = x^2$. The following diagram summarizes the graphing of parabolas. 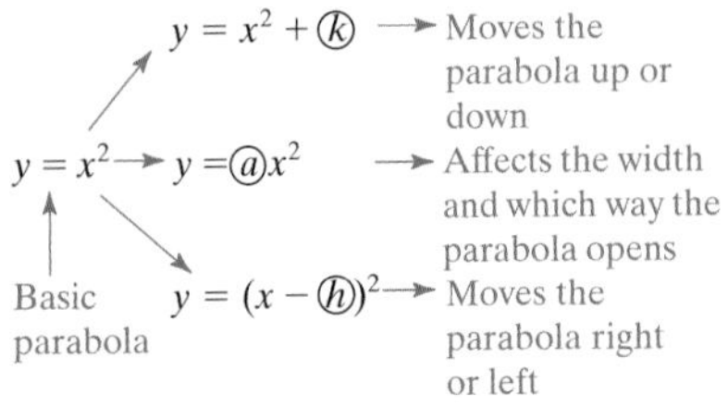 $y = x^2 + (k)$ → Moves the parabola up or down $y = x^2$ → $y = (a)x^2$ → Affects the width and which way the parabola opens $y = (x - (h))^2$ → Moves the parabola right or left Basic parabola	Graph $y = -(x + 2)^2 + 5$. **Solution** The vertex is located at $(-2, 5)$. The parabola opens downward. The points $(-4, 1)$ and $(0, 1)$ are on the parabola. 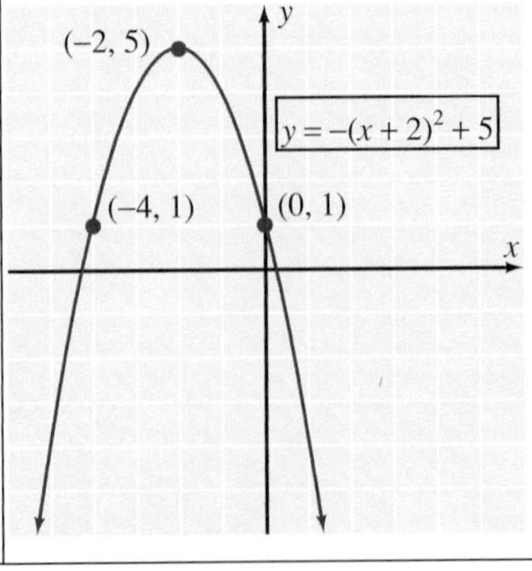 (−2, 5) (−4, 1) (0, 1) $y = -(x + 2)^2 + 5$	Problems 5–8

(continued)

OBJECTIVE	SUMMARY	EXAMPLE	CHAPTER REVIEW PROBLEMS
Graph parabolas using completing the square. (Sec. 9.3, Obj. 1, p. 494)	The general approach to graph equations of the form $y = ax^2 + bx + c$ is to change the form of the equation by completing the square.	Graph $y = x^2 - 6x + 11$. **Solution** $y = x^2 - 6x + 11$ $y = x^2 - 6x + 9 - 9 + 11$ $y = (x - 3)^2 + 2$ The vertex is located at (3, 2). The parabola opens upward. The points (1, 6) and (5, 6) are on the parabola. y (1, 6) (5, 6) (3, 2) x $y = x^2 - 6x + 11$	Problems 9–12
Write the equation of a circle in standard form. (Sec. 9.3, Obj. 2, p. 497)	The standard form of the equation of a circle is $(x - h)^2 + (y - k)^2 = r^2$. We can use the standard form of the equation of a circle to solve two basic kinds of circle problems: **1.** Given the coordinates of the center and the length of a radius of a circle, find its equation. **2.** Given the equation of a circle, find its center and the length of a radius.	Write the equation of a circle that has its center at $(-7, 3)$ and a radius of length 4 units. **Solution** Substitute -7 for h, 3 for k, and 4 for r in $(x - h)^2 + (y - k)^2 = r^2$. Then $(x - (-7))^2 + (y - 3)^2 = 4^2$ $(x + 7)^2 + (y - 3)^2 = 16$	Problems 13–16 *(continued)*

OBJECTIVE	SUMMARY	EXAMPLE	CHAPTER REVIEW PROBLEMS
Graph a circle. (Sec. 9.3, Obj. 3, p. 498)	To graph a circle have the equation in standard form $(x - h)^2 + (y - k)^2 = r^2$. It may be necessary to use completing the square to change the equation to standard form. The center will be at (h, k) and the length of a radius is r.	Graph $x^2 - 4x + y^2 + 2y - 4 = 0$. **Solution** Use completing the square to change the form of the equation. $x^2 - 4x + y^2 + 2y - 4 = 0$ $(x - 2)^2 + (y + 1)^2 = 9$ The center of the circle is at $(2, -1)$ and the length of a radius is 3. 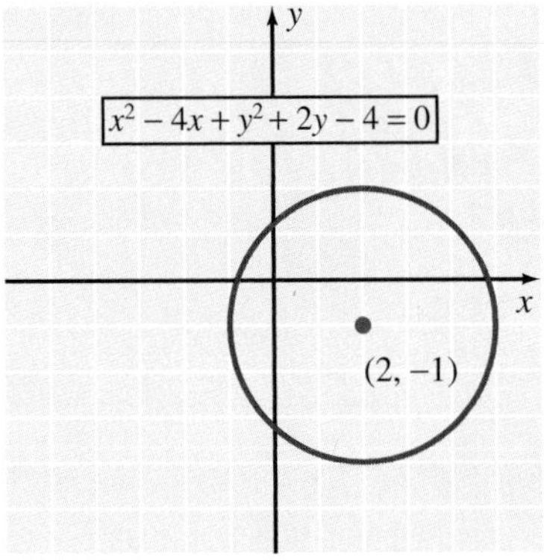 $x^2 - 4x + y^2 + 2y - 4 = 0$ (2, −1)	Problems 17–20
Graph ellipses with centers at the origin. (Sec. 9.4, Obj. 1, p. 503)	The graph of the equation $Ax^2 + By^2 = C$, where A, B, and C are nonzero real numbers of the same sign and $A \neq B$, is an ellipse with the center at $(0, 0)$. The intercepts are the endpoints of the axes of the ellipse. The longer axis is called the major axis and the shorter axis is called the minor axis. The center is at the point of intersection of the major and minor axes.	Graph $4x^2 + y^2 = 16$. **Solution** The coordinates of the intercepts are $(0, 4)$, $(0, -4)$, $(2, 0)$, and $(-2, 0)$. 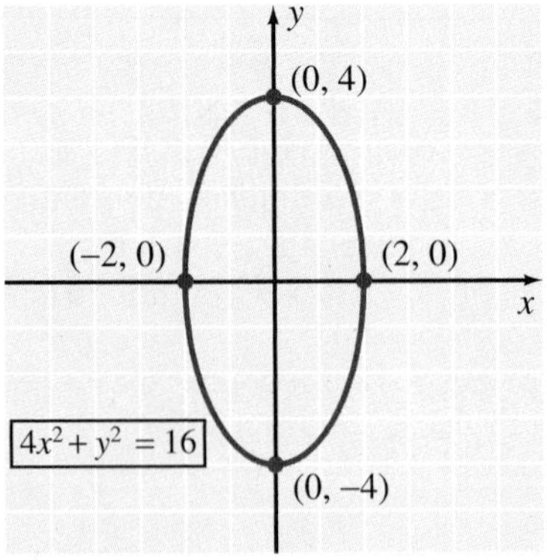 (0, 4) (−2, 0) (2, 0) $4x^2 + y^2 = 16$ (0, −4)	Problems 21–24 *(continued)*

OBJECTIVE	SUMMARY	EXAMPLE	CHAPTER REVIEW PROBLEMS
Graph ellipses with centers not at the origin. (Sec. 9.4, Obj. 2, p. 505)	The standard form of the equation of an ellipse whose center is not at the origin is $A(x-h)^2 + B(y-k)^2 = C$ where A, B, and C are nonzero real numbers of the same sign and $A \neq B$. The center will be at (h, k). Completing the square is often used to change the equation of an ellipse to standard form.	Graph $9x^2 + 36x + 4y^2 - 24y + 36 = 0$ **Solution** Use completing the square to change the equation to the equivalent form $9(x+2)^2 + 4(y-3)^2 = 36$. The center is at $(-2, 3)$. Substitute -2 for x to obtain $(-2, 6)$ and $(-2, 0)$ as the endpoints of the major axis. Substitute 3 for y to obtain $(0, 3)$ and $(-4, 3)$ as the endpoints of the minor axis. 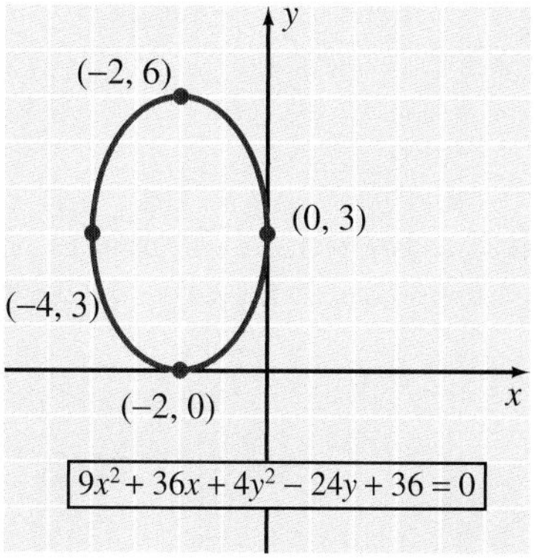 	Problems 25–28
Graph hyperbolas symmetric to both axes. (Sec. 9.5, Obj. 1, p. 509)	The graph of the equation $Ax^2 + By^2 = C$, where A, B, and C are nonzero real numbers and A and B are of unlike signs, is a hyperbola that is symmetric to both axes. Each branch of the hyperbola approaches a line called the asymptote. The equation for the asymptotes can be found by replacing the constant term with zero and solving for y.	Graph $9x^2 - y^2 = 9$. **Solution** If $y = 0$, then $x = \pm 1$. Hence the points $(1, 0)$ and $(-1, 0)$ are on the graph. To find the asymptotes, replace the constant term with zero and solve for y. $9x^2 - y^2 = 0$ $9x^2 = y^2$ $y = \pm 3x$ So the equations of the asymptotes are $y = 3x$ and $y = -3x$. 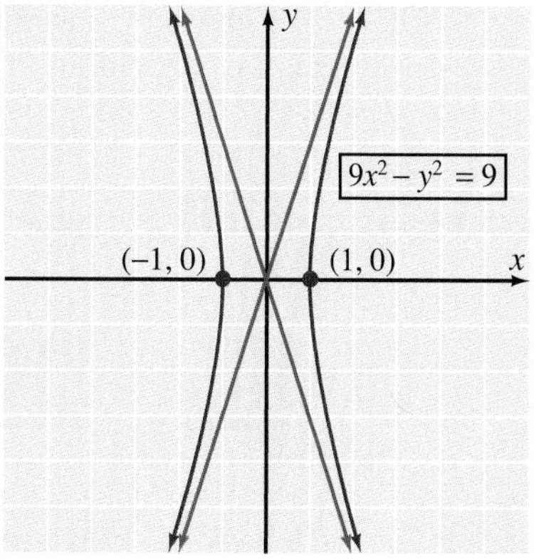 	Problems 29–32

(continued)

OBJECTIVE	SUMMARY	EXAMPLE	CHAPTER REVIEW PROBLEMS
Graph hyperbolas not symmetric to both axes. (Sec. 9.5, Obj. 2, p. 513)	The graph of the equation $A(x-h)^2 + B(y-k)^2 = C$, where A, B, and C are nonzero real numbers and A and B are of unlike signs, is a hyperbola that is not symmetric with respect to both axes.	Graph $4y^2 + 40y - x^2 + 4x + 92 = 0$ **Solution** Complete the square to change the equation to the equivalent form $4(y+5)^2 - (x-2)^2 = 4.$ If $x = 2$, then $y = -4$ or $y = -6$. Hence the points $(2, -4)$ and $(2, -6)$ are on the graph. To find the asymptotes, replace the constant term with zero and solve for y. The equations of the asymptotes are $y = \frac{1}{2}x - 6$ and $y = -\frac{1}{2}x - 4.$	Problems 33–34
Graph conic sections.	Part of the challenge of graphing conic sections is to know the characteristics of the equations that produce each type of conic section.		Problems 35–48
Solve systems involving nonlinear equations. (Sec. 9.6, Obj. 1, p. 518)	Systems that contain at least one nonlinear equation can often be solved by substitution or by the elimination-by-addition method. Graphing the system will often provide a basis for predicting approximate real number solutions if there are any.	Solve the system $\begin{pmatrix} x^2 + 2y^2 = 9 \\ x - 4y = -9 \end{pmatrix}.$ **Solution** Soving the second equation for x yields $x = 4y - 9$. Substitute $4y - 9$ for x in the first equation. $(4y-9)^2 + 2y^2 = 9$ Now solve this equation for y. $16y^2 - 72y + 81 + 2y^2 = 9$ $18y^2 - 72y + 72 = 0$ $18(y^2 - 4y + 4) = 0$ $18(y-2)^2 = 0$ $y = 2$ To find x, substitute 2 for y in the equation $x = 4y - 9$. $x = 4(2) - 9 = -1.$ The solution set is $\{(-1, 2)\}$.	Problems 49–50

Chapter 9 Review Problem Set

For Problems 1–4, graph each of the equations.

1. $xy = 6$

2. $x^2 = y^3$

3. $y = \frac{2}{x^2}$

4. $y = -x^3 + 2$

For Problems 5–12, find the vertex of each parabola and graph.

5. $y = x^2 + 6$

6. $y = -x^2 - 8$

7. $y = (x + 3)^2 - 1$

8. $y = (x + 1)^2$

9. $y = x^2 - 14x + 54$

10. $y = -x^2 + 12x - 44$

11. $y = 3x^2 + 24x + 39$

12. $y = 2x^2 + 8x + 5$

For Problems 13–16, write the equation of the circle satisfying the given conditions. Express your answers in the form $x^2 + y^2 + Dx + Ey + F = 0$.

13. Center at $(0, 0)$ and $r = 6$.

14. Center at $(2, -6)$ and $r = 5$

15. Center at $(-4, -8)$ and $r = 2\sqrt{3}$

16. Center at $(0, 5)$ and passes through the origin.

For Problems 17–20, graph each circle.

17. $x^2 + 14x + y^2 - 8y + 16 = 0$

18. $x^2 + 16x + y^2 + 39 = 0$

19. $x^2 - 12x + y^2 + 16y = 0$

20. $x^2 + y^2 = 24$

For Problems 21–28, graph each ellipse.

21. $16x^2 + y^2 = 64$

22. $16x^2 + 9y^2 = 144$

23. $4x^2 + 25y^2 = 100$

24. $2x^2 + 7y^2 = 28$

25. $x^2 - 4x + 9y^2 + 54y + 76 = 0$

26. $x^2 + 6x + 4y^2 - 16y + 9 = 0$

27. $9x^2 + 72x + 4y^2 - 8y + 112 = 0$

28. $16x^2 - 32x + 9y^2 - 54y - 47 = 0$

For Problems 29–34, graph each hyperbola.

29. $x^2 - 9y^2 = 25$

30. $4x^2 - y^2 = 16$

31. $9y^2 - 25x^2 = 36$

32. $16y^2 - 4x^2 = 17$

33. $25x^2 + 100x - 4y^2 + 24y - 36 = 0$

34. $36y^2 - 288y - x^2 + 2x + 539 = 0$

For Problems 35–48, graph each equation.

35. $9x^2 + y^2 = 81$

36. $9x^2 - y^2 = 81$

37. $y = -2x^2 + 3$

38. $y = 4x^2 - 16x + 19$

39. $x^2 + 4x + y^2 + 8y + 11 = 0$

40. $4x^2 - 8x + y^2 + 8y + 4 = 0$

41. $y^2 + 6y - 4x^2 - 24x - 63 = 0$

42. $y = -2x^2 - 4x - 3$

43. $x^2 - y^2 = -9$

44. $4x^2 + 16y^2 + 96y = 0$

45. $(x - 3)^2 + (y + 1)^2 = 4$

46. $(x + 1)^2 + (y - 2)^2 = 4$

47. $x^2 + y^2 - 6x - 2y + 4 = 0$

48. $x^2 + y^2 - 2y - 8 = 0$

49. Solve the system $\begin{pmatrix} y = x^2 + 2 \\ 4x - y = -7 \end{pmatrix}$.

50. Solve the system $\begin{pmatrix} x^2 + y^2 = 16 \\ y^2 - x^2 = 4 \end{pmatrix}$.

Chapter 9 Test

For Problems 1–4, find the vertex of each parabola.

1. ____________ **1.** $y = -2x^2 + 9$

2. ____________ **2.** $y = -x^2 + 2x + 6$

3. ____________ **3.** $y = 4x^2 + 32x + 62$

4. ____________ **4.** $y = x^2 - 6x + 9$

For Problems 5–7, write the equation of the circle that satisfies the given conditions. Express your answers in the form $x^2 + y^2 + Dx + Ey + F = 0$.

5. ____________ **5.** Center at $(-4, 0)$ and $r = 3\sqrt{5}$

6. ____________ **6.** Center at $(2, 8)$ and $r = 3$

7. ____________ **7.** Center at $(-3, -4)$ and $r = 5$

For Problems 8–10, find the center and the length of a radius of each circle.

8. ____________ **8.** $x^2 + y^2 = 32$

9. ____________ **9.** $x^2 - 12x + y^2 + 8y + 3 = 0$

10. ____________ **10.** $x^2 + 10x + y^2 + 2y - 38 = 0$

11. ____________ **11.** Find the length of the major axis of the ellipse $9x^2 + 2y^2 = 32$.

12. ____________ **12.** Find the length of the minor axis of the ellipse $8x^2 + 3y^2 = 72$.

13. ____________ **13.** Find the length of the major axis of the ellipse $3x^2 - 12x + 5y^2 + 10y - 10 = 0$.

14. ____________ **14.** Find the length of the minor axis of the ellipse $8x^2 - 32x + 5y^2 + 30y + 45 = 0$.

For Problems 15–17, find the equations of the asymptotes for each hyperbola.

15. ____________ **15.** $y^2 - 16x^2 = 36$

16. ____________ **16.** $25x^2 - 16y^2 = 50$

17. ____________ **17.** $x^2 - 2x - 25y^2 - 50y - 54 = 0$

For Problems 18–24, graph each equation.

18. ____________ **18.** $x^2 - 4y^2 = -16$

19. ____________ **19.** $y = x^2 + 4x$

20. ____________ **20.** $x^2 + 2x + y^2 + 8y + 8 = 0$

21. ____________ **21.** $2x^2 + 3y^2 = 12$

22. ____________ **22.** $y = 2x^2 + 12x + 22$

23. ____________ **23.** $9x^2 - y^2 = 9$

24. ____________ **24.** $x^2 - 4x - y^2 + 4y - 9 = 0$

25. ____________ **25.** Solve the system $\begin{pmatrix} 2x - y = 7 \\ 3x^2 + y^2 = 21 \end{pmatrix}$.

Functions

Moodboard/Digital Railroad

The price of goods may be decided by using a function to describe the relationship between the price and the demand. Such a function gives us a means of studying the demand when the price is varied.

A golf pro-shop operator finds that she can sell 30 sets of golf clubs at \$500 per set in a year. Furthermore, she predicts that for each \$25 decrease in price, three additional sets of golf clubs could be sold. At what price should she sell the clubs to maximize gross income? We can use the quadratic function $f(x) = (30 + 3x)(500 - 25x)$ to determine that the clubs should be sold at \$375 per set.

One of the fundamental concepts of mathematics is the concept of a function. Functions are used to unify mathematics and also to apply mathematics to many real-world problems. Functions provide a means of studying quantities that vary with one another—that is, change in one quantity causes a corresponding change in the other.

In this chapter we will (1) introduce the basic ideas that pertain to the function concept, (2) review and extend some concepts from Chapter 9, and (3) discuss some applications of functions.

Video tutorials for all section learning objectives are available in a variety of delivery modes.

INTERNET PROJECT

John Napier, a Scottish mathematician, is considered the inventor of logarithms and Napier's bones. Logarithms can be used to simplify the arithmetic for operations of multiplication and division. Conduct an Internet search to learn about Napier's bones and their use. How do Napier's bones differ from logarithms?

10.1 Relations and Functions

OBJECTIVES

1 Determine If a Relation Is a Function
2 Use Function Notation When Evaluating a Function
3 Specify the Domain of a Function
4 Find the Difference Quotient of a Given Function
5 Apply Function Notation to a Problem

1 Determine If a Relation Is a Function

Mathematically, a function is a special kind of **relation**, so we will begin our discussion with a simple definition of a relation.

> **Definition 10.1**
>
> A **relation** is a set of ordered pairs.

Thus a set of ordered pairs such as $\{(1, 2), (3, 7), (8, 14)\}$ is a relation. The set of all first components of the ordered pairs is the **domain** of the relation, and the set of all second components is the **range** of the relation. The relation $\{(1, 2), (3, 7), (8, 14)\}$ has a domain of $\{1, 3, 8\}$ and a range of $\{2, 7, 14\}$.

The ordered pairs we refer to in Definition 10.1 may be generated by various means, such as a graph or a chart. However, one of the most common ways of generating ordered pairs is by using equations. Because the solution set of an equation in two variables is a set of ordered pairs, such an equation describes a relation. Each of the following equations describes a relation between the variables x and y. We have listed some of the infinitely many ordered pairs (x, y) of each relation.

1. $x^2 + y^2 = 4$: $(1, \sqrt{3}), (1, -\sqrt{3}), (0, 2), (0, -2)$

2. $y^2 = x^3$: $(0, 0), (1, 1), (1, -1), (4, 8), (4, -8)$

3. $y = x + 2$: $(0, 2), (1, 3), (2, 4), (-1, 1), (5, 7)$

4. $y = \dfrac{1}{x - 1}$: $(0, -1), (2, 1), \left(3, \dfrac{1}{2}\right), \left(-1, -\dfrac{1}{2}\right), \left(-2, -\dfrac{1}{3}\right)$

5. $y = x^2$: $(0, 0), (1, 1), (2, 4), (-1, 1), (-2, 4)$

Now we direct your attention to the ordered pairs associated with equations 3, 4, and 5. Note that in each case, no two ordered pairs have the same first component. Such a set of ordered pairs is called a **function**.

Definition 10.2

A **function** is a relation in which no two ordered pairs have the same first component.

Stated another way, Definition 10.2 means that a function is a relation wherein each member of the domain is assigned *one and only one* member of the range. The following table lists the five equations and determines if the generated ordered pairs fit the definition of a function.

Equation	Ordered pairs	Function
1. $x^2 + y^2 = 4$	$(1, \sqrt{3}), (1, -\sqrt{3}), (0, 2), (0, -2)$ *Note:* The ordered pairs $(1, \sqrt{3})$ and $(1, -\sqrt{3})$ have the same first component and different second components.	No
2. $y^2 = x^3$	$(0, 0), (1, 1), (1, -1), (4, 8), (4, -8)$ *Note:* The ordered pairs $(1, 1)$ and $(1, -1)$ have the same first component and different second components.	No
3. $y = x + 2$	$(0, 2), (1, 3), (2, 4), (-1, 1), (5, 7)$	Yes
4. $y = \dfrac{1}{x - 1}$	$(0, -1), (2, 1), \left(3, \frac{1}{2}\right), \left(-1, \frac{1}{2}\right), \left(-2, -\frac{1}{3}\right)$	Yes
5. $y = x^2$	$(0, 0), (1, 1), (2, 4), (-1, 1), (-2, 4)$	Yes

EXAMPLE 1

Determine if the following sets of ordered pairs determine a function. Specify the domain and range.

(a) $\{(1, 3), (2, 5), (3, 7), (4, 8)\}$ **(b)** $\{(2, 1), (2, 3), (2, 5), (2, 7)\}$
(c) $\{(0, -2), (2, -2), (4, 6), (6, 6)\}$

Solution

(a) Domain $= \{1, 2, 3, 4\}$, range $= \{3, 5, 7, 8\}$ Yes, the set of ordered pairs does determine a function. No first component is ever repeated. Therefore every first component has one and only one second component.

(b) Domain $= \{2\}$, range $= \{1, 3, 5, 7\}$ No, the set of ordered pairs does not determine a function. The ordered pairs (2, 1) and (2, 3) have the same first component and different second components.

(c) Domain $= \{0, 2, 4, 6\}$, range $= \{-2, 6\}$ Yes, the set of ordered pairs does determine a function. No first component is ever repeated. Therefore every first component has one and only one second component.

▼ PRACTICE YOUR SKILL

Determine if the following sets of ordered pairs determine a function. Specify the domain and range.

(a) $\{(2, 3), (3, 3), (5, 3), (7, 3)\}$ **(b)** $\{(2, 3), (-2, 3), (2, 5)(-2, 5)\}$
(c) $\{(4, 2), (4, -2), (9, 3), (9, -3)\}$

2 Use Function Notation When Evaluating a Function

The three sets of ordered pairs (listed in the previous table) that generated functions could be named as follows:

$$f = \{(x, y) | y = x + 2\} \quad g = \left\{(x, y) | y = \frac{1}{x - 1}\right\} \quad h = \{(x, y) | y = x^2\}$$

For the first set of ordered pairs, the notation would be read "the function f is the set of ordered pairs (x, y) such that y is equal to $x + 2$."

Note that we named the functions f, g, and h. It is customary to name functions by means of a single letter, and the letters f, g, and h are often used. We would suggest more meaningful choices when functions are used to portray real-world situations. For example, if a problem involves a profit function, then naming the function p or even P would seem natural.

The symbol for a function can be used along with a variable that represents an element in the domain to represent the associated element in the range. For example, suppose that we have a function f specified in terms of the variable x. The symbol $f(x)$, which is read "f of x" or "the value of f at x," represents the element in the range associated with the element x from the domain. The function $f = \{(x, y) \mid y = x + 2\}$ can be written as $f = \{(x, f(x)) \mid f(x) = x + 2\}$ and is usually shortened to read "f is the function determined by the equation $f(x) = x + 2$."

Remark: Be careful with the notation $f(x)$. As we have stated here, it means the value of the function f at x. It does *not* mean f times x.

This **function notation** is very convenient for computing and expressing various values of the function. For example, the value of the function $f(x) = 3x - 5$ at $x = 1$ is

$$f(1) = 3(1) - 5 = -2$$

Likewise, the functional values for $x = 2$, $x = -1$, and $x = 5$ are

$$f(2) = 3(2) - 5 = 1$$
$$f(-1) = 3(-1) - 5 = -8$$
$$f(5) = 3(5) - 5 = 10$$

Thus this function f contains the ordered pairs $(1, -2)$, $(2, 1)$, $(-1, -8)$, $(5, 10)$ and, in general, all ordered pairs of the form $(x, f(x))$, where $f(x) = 3x - 5$ and x is any real number.

It may be helpful for you to picture the concept of a function in terms of a *function machine*, as in Figure 10.1. Each time that a value of x is put into the machine, the equation $f(x) = x + 2$ is used to generate one and only one value for $f(x)$ to be ejected from the machine. For example, if 3 is put into this machine, then $f(3) = 3 + 2 = 5$ and so 5 is ejected. Thus the ordered pair $(3, 5)$ is one element of the function. Now let's look at some examples to illustrate evaluating functions.

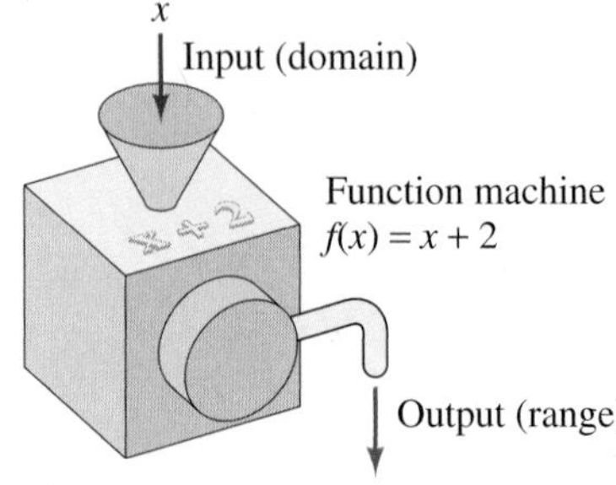

Figure 10.1

EXAMPLE 2

Consider the function $f(x) = x^2$. Evaluate $f(-2), f(0)$, and $f(4)$.

Solution

$$f(-2) = (-2)^2 = 4$$
$$f(0) = (0)^2 = 0$$
$$f(4) = (4)^2 = 16$$

▼ **PRACTICE YOUR SKILL**

Consider the function $f(x) = x^2 + 4$. Evaluate $f(-1), f(0)$, and $f(2)$. ■

EXAMPLE 3

If $f(x) = -2x + 7$ and $g(x) = x^2 - 5x + 6$, find $f(3), f(-4), f(b), f(3c), g(2), g(-1), g(a)$, and $g(a + 4)$.

Solution

$$f(x) = -2x + 7$$
$$f(3) = -2(3) + 7 = 1$$
$$f(-4) = -2(-4) + 7 = 15$$
$$f(b) = -2(b) + 7 = -2b + 7$$
$$f(3c) = -2(3c) + 7 = -6c + 7$$

$$g(x) = x^2 - 5x + 6$$
$$g(2) = (2)^2 - 5(2) + 6 = 0$$
$$g(-1) = (-1)^2 - 5(-1) + 6 = 12$$
$$g(a) = (a)^2 - 5(a) + 6 = a^2 - 5a + 6$$
$$g(a + 4) = (a + 4)^2 - 5(a + 4) + 6$$
$$= a^2 + 8a + 16 - 5a - 20 + 6$$
$$= a^2 + 3a + 2$$

▼ **PRACTICE YOUR SKILL**

If $f(x) = -x + 6$ and $g(x) = x^2 - 9$, find $f(2), f(-5), f(a), f(2b), g(-2), g(1), g(2a)$, and $g(a + 1)$. ■

In Example 3, note that we are working with two different functions in the same problem. Thus different names, f and g, are used.

3 Specify the Domain of a Function

For our purposes in this text, if the domain of a function is not specifically indicated or determined by a real-world application, then we assume the domain to be all **real number** replacements for the variable, which represents an element in the domain that will produce **real number** functional values. Consider the following examples.

EXAMPLE 4

Specify the domain for each of the following:

(a) $f(x) = \dfrac{1}{x - 1}$ **(b)** $f(t) = \dfrac{1}{t^2 - 4}$ **(c)** $f(s) = \sqrt{s - 3}$

Solution

(a) We can replace x with any real number except 1, because 1 makes the denominator zero. Thus the domain, D, is given by

$$D = \{x \mid x \neq 1\} \qquad \text{or} \qquad D: (-\infty, 1) \cup (1, \infty)$$

Here you may consider set builder notation to be easier than interval notation for expressing the domain.

(b) We need to eliminate any value of t that will make the denominator zero, so let's solve the equation $t^2 - 4 = 0$.

$$t^2 - 4 = 0$$
$$t^2 = 4$$
$$t = \pm 2$$

The domain is the set

$$D = \{t \mid t \neq -2 \text{ and } t \neq 2\} \quad \text{or} \quad D: (-\infty, -2) \cup (-2, 2) \cup (2, \infty)$$

When the domain is all real numbers except a few numbers, set builder notation is the more compact notation.

(c) The radicand, $s - 3$, must be nonnegative.

$$s - 3 \geq 0$$
$$s \geq 3$$

The domain is the set

$$D = \{s \mid s \geq 3\} \quad \text{or} \quad D: [3, \infty)$$

▼ **PRACTICE YOUR SKILL**

Specify the domain for each of the following:

(a) $f(x) = \dfrac{x+3}{2x+1}$ **(b)** $g(x) = \dfrac{4}{x^2 - x - 12}$ **(c)** $h(x) = \sqrt{x+4}$ ■

4 Find the Difference Quotient of a Given Function

The quotient $\dfrac{f(a+h) - f(a)}{h}$ is often called a **difference quotient**, and we use it extensively with functions when studying the limit concept in calculus. The next two examples show how we found the difference quotient for two specific functions.

EXAMPLE 5

If $f(x) = 3x - 5$, find $\dfrac{f(a+h) - f(a)}{h}$.

Solution

$$f(a+h) = 3(a+h) - 5$$
$$= 3a + 3h - 5$$

and

$$f(a) = 3a - 5$$

Therefore,

$$f(a+h) - f(a) = (3a + 3h - 5) - (3a - 5)$$
$$= 3a + 3h - 5 - 3a + 5$$
$$= 3h$$

and

$$\frac{f(a+h) - f(a)}{h} = \frac{3h}{h} = 3$$

▼ **PRACTICE YOUR SKILL**

If $f(x) = 4x + 1$, find $\dfrac{f(a + h) - f(a)}{h}$. ■

EXAMPLE 6

If $f(x) = x^2 + 2x - 3$, find $\dfrac{f(a + h) - f(a)}{h}$.

Solution

$$\begin{aligned} f(a + h) &= (a + h)^2 + 2(a + h) - 3 \\ &= a^2 + 2ah + h^2 + 2a + 2h - 3 \end{aligned}$$

and

$$f(a) = a^2 + 2a - 3$$

Therefore,

$$\begin{aligned} f(a + h) - f(a) &= (a^2 + 2ah + h^2 + 2a + 2h - 3) - (a^2 + 2a - 3) \\ &= a^2 + 2ah + h^2 + 2a + 2h - 3 - a^2 - 2a + 3 \\ &= 2ah + h^2 + 2h \end{aligned}$$

and

$$\begin{aligned} \frac{f(a + h) - f(a)}{h} &= \frac{2ah + h^2 + 2h}{h} \\ &= \frac{\cancel{h}(2a + h + 2)}{\cancel{h}} \\ &= 2a + h + 2 \end{aligned}$$

▼ **PRACTICE YOUR SKILL**

If $f(x) = x^2 - 3x + 1$, find $\dfrac{f(a + h) - f(a)}{h}$. ■

5 Apply Function Notation to a Problem

Functions and functional notation provide the basis for describing many real-world relationships. The next example illustrates this point.

EXAMPLE 7

Suppose a factory determines that the overhead for producing a quantity of a certain item is \$500 and that the cost for producing each item is \$25. Express the total expenses as a function of the number of items produced, and compute the expenses for producing 12, 25, 50, 75, and 100 items.

Solution

Let n represent the number of items produced. Then $25n + 500$ represents the total expenses. Let's use E to represent the expense function, so that we have

$$E(n) = 25n + 500, \quad \text{where } n \text{ is a whole number}$$

From this we obtain

$$\begin{aligned} E(12) &= 25(12) + 500 = 800 \\ E(25) &= 25(25) + 500 = 1125 \\ E(50) &= 25(50) + 500 = 1750 \\ E(75) &= 25(75) + 500 = 2375 \\ E(100) &= 25(100) + 500 = 3000 \end{aligned}$$

Thus the total expenses for producing 12, 25, 50, 75, and 100 items are \$800, \$1125, \$1750, \$2375, and \$3000, respectively.

▼ PRACTICE YOUR SKILL

Suppose Colin pays \$29.99 a month and \$0.03 per minute for his cell phone. Express the monthly cost of his cell phone, in dollars, as a function of the number of minutes used. Compute the cost for using 300 minutes, 500 minutes, and 1000 minutes. ■

CONCEPT QUIZ

For Problems 1–10, answer true or false.

1. A function is a special type of relation.
2. The relation {(John, Mary), (Mike, Ada), (Kyle, Jenn), (Mike, Sydney)} is a function.
3. Given $f(x) = 3x + 4$, the notation $f(7)$ means to find the value of f when $x = 7$.
4. The set of all first components of the ordered pairs of a relation is called the range.
5. The domain of a function can never be the set of all real numbers.
6. The domain of the function $f(x) = \dfrac{x}{x-3}$ is the set of all real numbers.
7. The range of the function $f(x) = x + 1$ is the set of all real numbers.
8. If $f(x) = -x^2 - 1$, then $f(2) = -5$.
9. The range of the function $f(x) = \sqrt{x-1}$ is the set of all real numbers greater than or equal to 1.
10. If $f(x) = -x^2 + 3x$, then $f(2a) = 4a^2 + 6a$.

Problem Set 10.1

1 Determine If a Relation Is a Function

For Problems 1–10, specify the domain and the range for each relation. Also state whether or not the relation is a function.

1. $\{(1, 5), (2, 8), (3, 11), (4, 14)\}$
2. $\{(0, 0), (2, 10), (4, 20), (6, 30), (8, 40)\}$
3. $\{(0, 5), (0, -5), (1, 2\sqrt{6}), (1, -2\sqrt{6})\}$
4. $\{(1, 1), (1, 2), (1, -1), (1, -2), (1, 3)\}$
5. $\{(1, 2), (2, 5), (3, 10), (4, 17), (5, 26)\}$
6. $\{(-1, 5), (0, 1), (1, -3), (2, -7)\}$
7. $\{(x, y) \mid 5x - 2y = 6\}$
8. $\{(x, y) \mid y = -3x\}$
9. $\{(x, y) \mid x^2 = y^3\}$
10. $\{(x, y) \mid x^2 - y^2 = 16\}$

2 Use Function Notation When Evaluating a Function

11. If $f(x) = 5x - 2$, find $f(0)$, $f(2)$, $f(-1)$, and $f(-4)$.
12. If $f(x) = -3x - 4$, find $f(-2)$, $f(-1)$, $f(3)$, and $f(5)$.
13. If $f(x) = \dfrac{1}{2}x - \dfrac{3}{4}$, find $f(-2)$, $f(0)$, $f\left(\dfrac{1}{2}\right)$, $f\left(\dfrac{2}{3}\right)$
14. If $g(x) = x^2 + 3x - 1$, find $g(1)$, $g(-1)$, $g(3)$, and $g(-4)$.
15. If $g(x) = 2x^2 - 5x - 7$, find $g(-1)$, $g(2)$, $g(-3)$, and $g(4)$.
16. If $h(x) = -x^2 - 3$, find $h(1)$, $h(-1)$, $h(-3)$, and $h(5)$.
17. If $h(x) = -2x^2 - x + 4$, find $h(-2)$, $h(-3)$, $h(4)$, and $h(5)$.
18. If $f(x) = \sqrt{x-1}$, find $f(1)$, $f(5)$, $f(13)$, and $f(26)$.
19. If $f(x) = \sqrt{2x+1}$, find $f(3)$, $f(4)$, $f(10)$, and $f(12)$.

20. If $f(x) = \frac{3}{x-2}$, find $f(3)$, $f(0)$, $f(-1)$, and $f(-5)$.

21. If $f(x) = \frac{-4}{x+3}$, find $f(1)$, $f(-1)$, $f(3)$, and $f(-6)$.

22. If $f(x) = -2x + 7$, find $f(a)$, $f(a+2)$, and $f(a+h)$.

23. If $f(x) = x^2 - 7x$, find $f(a)$, $f(a-3)$, and $f(a+h)$.

24. If $f(x) = x^2 - 4x + 10$, find $f(-a)$, $f(a-4)$, and $f(a+h)$.

25. If $f(x) = 2x^2 - x - 1$, find $f(-a)$, $f(a+1)$, and $f(a+h)$.

26. If $f(x) = -x^2 + 3x + 5$, find $f(-a)$, $f(a+6)$, and $f(-a+1)$.

27. If $f(x) = -x^2 - 2x - 7$, find $f(-a)$, $f(-a-2)$, and $f(a+7)$.

28. If $f(x) = 2x^2 - 7$ and $g(x) = x^2 + x - 1$, find $f(-2)$, $f(3)$, $g(-4)$, and $g(5)$.

29. If $f(x) = |3x - 2|$ and $g(x) = |x| + 2$, find $f(1)$, $f(-1)$, $g(2)$, and $g(-3)$.

30. If $f(x) = 3|x| - 1$ and $g(x) = -|x| + 1$, find $f(-2)$, $f(3)$, $g(-4)$, and $g(5)$.

3 Specify the Domain of a Function

For Problems 31–56, specify the domain for each of the functions.

31. $f(x) = 7x - 2$

32. $f(x) = x^2 + 1$

33. $f(x) = \frac{1}{x-1}$

34. $f(x) = \frac{-3}{x+4}$

35. $g(x) = \frac{3x}{4x-3}$

36. $g(x) = \frac{5x}{2x+7}$

37. $h(x) = \frac{2}{(x+1)(x-4)}$

38. $h(x) = \frac{-3}{(x-6)(2x+1)}$

39. $f(x) = \frac{14}{x^2 + 3x - 40}$

40. $f(x) = \frac{7}{x^2 - 8x - 20}$

41. $f(x) = \frac{-4}{x^2 + 6x}$

42. $f(x) = \frac{9}{x^2 - 12x}$

43. $f(t) = \frac{4}{t^2 + 9}$

44. $f(t) = \frac{8}{t^2 + 1}$

45. $f(t) = \frac{3t}{t^2 - 4}$

46. $f(t) = \frac{-2t}{t^2 - 25}$

47. $h(x) = \sqrt{x+4}$

48. $h(x) = \sqrt{5x-3}$

49. $f(s) = \sqrt{4s-5}$

50. $f(s) = \sqrt{s-2} + 5$

51. $f(x) = \sqrt{x^2 - 16}$

52. $f(x) = \sqrt{x^2 - 49}$

53. $f(x) = \sqrt{x^2 - 3x - 18}$

54. $f(x) = \sqrt{x^2 + 4x - 32}$

55. $f(x) = \sqrt{1 - x^2}$

56. $f(x) = \sqrt{9 - x^2}$

4 Find the Difference Quotient of a Given Function

For Problems 57–64, find $\frac{f(a+h) - f(a)}{h}$ for each of the given functions.

57. $f(x) = -3x + 6$

58. $f(x) = 5x - 4$

59. $f(x) = -x^2 - 1$

60. $f(x) = x^2 + 5$

61. $f(x) = 2x^2 - x + 8$

62. $f(x) = x^2 - 3x + 7$

63. $f(x) = -4x^2 - 7x - 9$

64. $f(x) = -3x^2 + 4x - 1$

5 Apply Function Notation to a Problem

65. The height of a projectile fired vertically into the air (neglecting air resistance) at an initial velocity of 64 feet per second is a function of the time (t) and is given by the equation

$$h(t) = 64t - 16t^2$$

Compute $h(1)$, $h(2)$, $h(3)$, and $h(4)$.

66. Suppose that the cost function for producing a certain item is given by $C(n) = 3n + 5$, where n represents the number of items produced. Compute $C(150)$, $C(500)$, $C(750)$, and $C(1500)$.

67. A car rental agency charges \$50 per day plus \$0.32 a mile. Therefore, the daily charge for renting a car is a function of the number of miles traveled (m) and can be expressed as $C(m) = 50 + 0.32m$. Compute $C(75)$, $C(150)$, $C(225)$, and $C(650)$.

68. The profit function for selling n items is given by $P(n) = -n^2 + 500n - 61{,}500$. Compute $P(200)$, $P(230)$, $P(250)$, and $P(260)$.

69. The equation $I(r) = 500r$ expresses the amount of simple interest earned by an investment of \$500 for 1 year as a function of the rate of interest (r). Compute $I(0.11)$, $I(0.12)$, $I(0.135)$, and $I(0.15)$.

70. The equation $A(r) = \pi r^2$ expresses the area of a circular region as a function of the length of a radius (r). Use 3.14 as an approximation for π and compute $A(2)$, $A(3)$, $A(12)$, and $A(17)$.

THOUGHTS INTO WORDS

71. Are all functions also relations? Are all relations also functions? Defend your answers.

72. What does it mean to say that the domain of a function may be restricted if the function represents a real-world situation? Give two or three examples of such situations.

73. Does $f(a + b) = f(a) + f(b)$ for all functions? Defend your answer.

74. Are there any functions for which $f(a + b) = f(a) + f(b)$? Defend your answer.

Answers to the Concept Quiz

1. True **2.** False **3.** True **4.** False **5.** False **6.** False **7.** True **8.** True **9.** False **10.** False

Answers to the Example Practice Skills

1. (a) Function, domain = $\{2, 3, 5, 7\}$, range = $\{3\}$ **(b)** Not a function, domain = $\{-2, 2\}$, range = $\{3, 5\}$ **(c)** Not a function, domain = $\{4, 9\}$, range = $\{-3, -2, 2, 3\}$ **2.** $f(-1) = 5, f(0) = 4, f(2) = 8$ **3.** $f(2) = 4,$ $f(-5) = 11, f(a) = -a + 6, f(2b) = -2b + 6; g(-2) = -5, g(1) = -8, g(2a) = 4a^2 - 9, g(a + 1) = a^2 + 2a - 8$ **4. (a)** $D = \left\{x \mid x \neq -\frac{1}{2}\right\}$ or $D = \left(-\infty, -\frac{1}{2}\right) \cup \left(-\frac{1}{2}, \infty\right)$ **(b)** $D = \{x \mid x \neq -3 \text{ and } x \neq 4\}$ or $D = (-\infty, -3) \cup (-3, 4) \cup (4, \infty)$ **(c)** $D = \{x \mid x \geq -4\}$ or $D = [-4, \infty)$ **5.** 4 **6.** $2a + h - 3$ **7.** $C(n) = 29.99 + 0.03\,n$, \$38.99, \$44.99, \$59.99

10.2 Functions: Their Graphs and Applications

OBJECTIVES

1. Graph Linear Functions
2. Apply Linear Functions
3. Graph Quadratic Functions
4. Solve Problems Using Quadratic Functions
5. Graph Functions with a Graphing Utility

1 Graph Linear Functions

In Section 3.1, we made statements such as "The graph of the solution set of the equation $y = x - 1$ (or simply the graph of the equation $y = x - 1$) is a line that contains the points $(0, -1)$ and $(1, 0)$." Because the equation $y = x - 1$ (which can be written as $f(x) = x - 1$) can be used to specify a function, that line we previously referred to is also called the **graph of the function specified by the equation** or simply the **graph of the function**. Generally speaking, the graph of any equation that determines a function is also called the graph of the function. Thus the graphing techniques we discussed earlier will continue to play an important role as we graph functions.

As we use the function concept in our study of mathematics, it is helpful to classify certain types of functions and to become familiar with their equations, characteristics, and graphs. In this section we will discuss two special types of functions—**linear** and **quadratic functions**. These functions are merely an outgrowth of our earlier study of linear and quadratic equations.

Any function that can be written in the form

$$f(x) = ax + b$$

where a and b are real numbers, is called a **linear function**. The following equations are examples of linear functions.

$$f(x) = -3x + 6 \qquad f(x) = 2x + 4 \qquad f(x) = -\frac{1}{2}x - \frac{3}{4}$$

Graphing linear functions is quite easy because the graph of every linear function is a straight line. Therefore, all we need to do is determine two points of the graph and draw the line determined by those two points. You may want to continue using a third point as a check point.

EXAMPLE 1 Graph the function $f(x) = -3x + 6$.

Solution

Because $f(0) = 6$, the point (0, 6) is on the graph. Likewise, because $f(1) = 3$, the point (1, 3) is on the graph. Plot these two points, and draw the line determined by the two points to produce Figure 10.2.

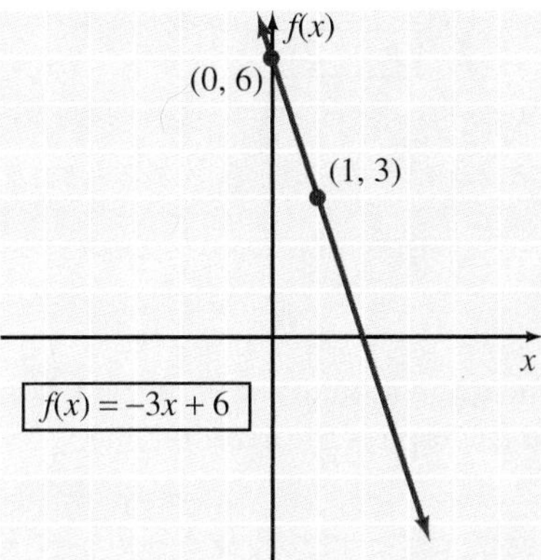

Figure 10.2

▼ PRACTICE YOUR SKILL

Graph the function $f(x) = \frac{1}{2}x + 3$. ■

Remark: Note in Figure 10.2 that we labeled the vertical axis $f(x)$. We could also label it y, because $f(x) = -3x + 6$ and $y = -3x + 6$ mean the same thing. We will continue to use the label $f(x)$ in this chapter to help you adjust to the function notation.

Now let's graph the function $f(x) = x$. The equation $f(x) = x$ can be written as $f(x) = 1x + 0$; thus it is a linear function. Because $f(0) = 0$ and $f(2) = 2$, the points (0, 0) and (2, 2) determine the line in Figure 10.3. The function $f(x) = x$ is often called the **identity function**.

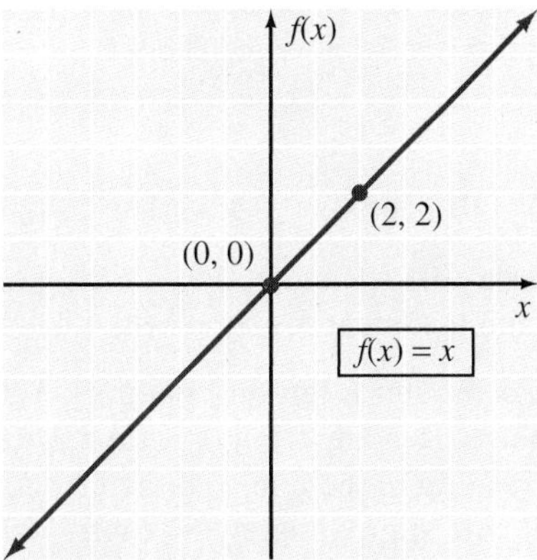

Figure 10.3

As we use function notation to graph functions, it is often helpful to think of the ordinate of every point on the graph as the value of the function at a specific value of x. Geometrically, this functional value is the *directed distance* of the point from the x axis, as illustrated in Figure 10.4 with the function $f(x) = 2x - 4$. For example, consider the graph of the function $f(x) = 2$. The function $f(x) = 2$ means that every functional value is 2, or, geometrically, that every point on the graph is 2 units above the x axis. Thus the graph is the horizontal line shown in Figure 10.5.

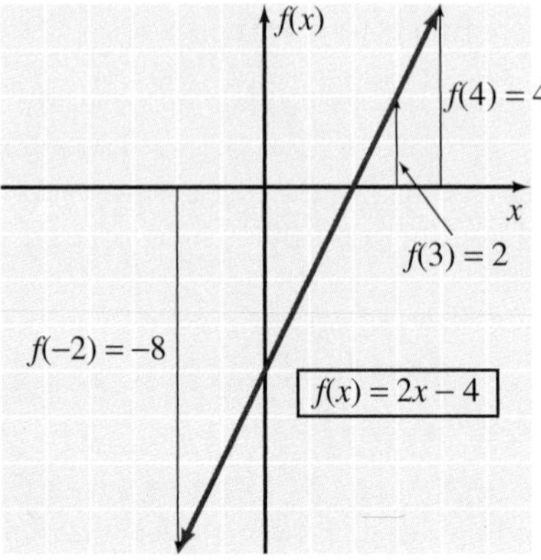

Figure 10.4

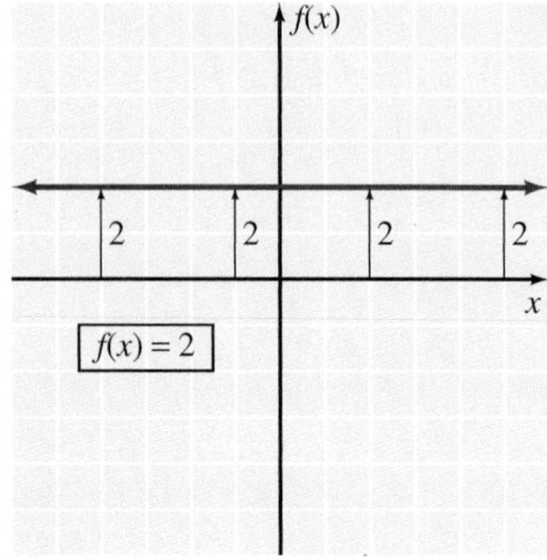

Figure 10.5

Any linear function of the form $f(x) = ax + b$, where $a = 0$, is called a **constant function**, and its graph is a horizontal line.

2 Apply Linear Functions

We worked with some applications of linear equations in Section 3.2. Let's consider some additional applications that use the concept of a linear function to connect mathematics to the real world.

EXAMPLE 2

The cost for operating a desktop computer is given by the function $c(h) = 0.0036h$, where h represents the number of hours that the computer is on.

(a) How much does it cost to operate the computer for 3 hours per night for a 30-day month?

(b) Graph the function $c(h) = 0.0036h$.

(c) Suppose that the computer is accidentally left on for a week while the owner is on vacation. Use the graph from part (b) to approximate the cost of operating the computer for a week. Then use the function to find the exact cost.

Solution

(a) $c(90) = 0.0036(90) = 0.324$. The cost, to the nearest cent, is \$.32.

(b) Because $c(0) = 0$ and $c(100) = 0.36$, we can use the points $(0, 0)$ and $(100, 0.36)$ to graph the linear function $c(h) = 0.0036h$ (Figure 10.6).

(c) If the computer is left on 24 hours per day for a week, then it runs for $24(7) = 168$ hours. Reading from the graph, we can approximate 168 on the horizontal axis, read up to the line, and then read across to the vertical axis. It looks as if it will cost approximately 60 cents. Using $c(h) = 0.0036h$, we obtain exactly $c(168) = 0.0036(168) = 0.6048$.

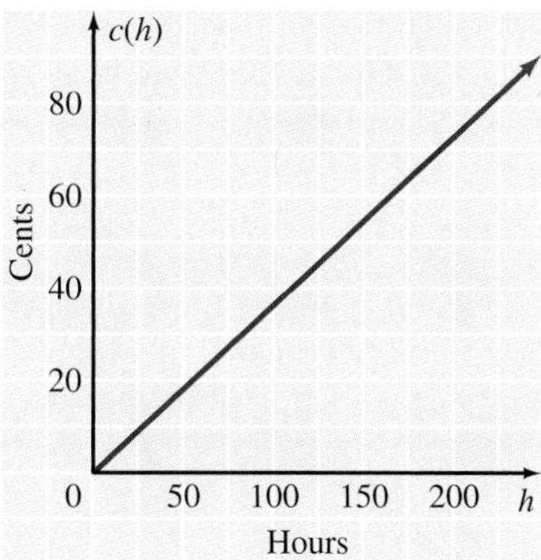

Figure 10.6

▼ PRACTICE YOUR SKILL

The cost of a tutoring session is given by the function $c(m) = 0.5m$, where m represents the number of minutes for the tutoring session.

(a) How much does it cost for three tutoring sessions each lasting 2 hours per night?
(b) Graph the function $c(m) = 0.5m$.
(c) Use the graph from part (b) to approximate the cost for 100 minutes of tutoring. Then use the function to find the exact cost for 100 minutes of tutoring. ■

EXAMPLE 3

The EZ Car Rental charges a fixed amount per day plus an amount per mile for renting a car. For two different day trips, Ed has rented a car from EZ. He paid $70 for 100 miles on one day and $120 for 350 miles on another day. Determine the linear function that the EZ Car Rental uses to determine its daily rental charges.

Solution

The linear function $f(x) = ax + b$, where x represents the number of miles, models this situation. Ed's two day trips can be represented by the ordered pairs (100, 70) and (350, 120). From these two ordered pairs we can determine a, which is the slope of the line.

$$a = \frac{120 - 70}{350 - 100} = \frac{50}{250} = \frac{1}{5} = 0.2$$

Thus $f(x) = ax + b$ becomes $f(x) = 0.2x + b$. Now either ordered pair can be used to determine the value of b. Using (100, 70), we have $f(100) = 70$; therefore,

$$f(100) = 0.2(100) + b = 70$$
$$b = 50$$

The linear function is $f(x) = 0.2x + 50$. In other words, the EZ Car Rental charges a daily fee of $50 plus $.20 per mile.

▼ PRACTICE YOUR SKILL

For legal consultations, the ETF Group charges a fixed amount plus an amount per minute. Morgan had two different consultations with the ETF Group. One consultation lasted 20 minutes and cost $240 and the other consultation lasted 30 minutes and cost $280. Determine the linear function that the ETF Group uses to determine its cost for consultations. ■

EXAMPLE 4

Suppose that Ed (Example 3) also has access to the A-OK Car Rental agency, which charges a daily fee of \$25 plus \$0.30 per mile. Should Ed use the EZ Car Rental from Example 3 or A-OK Car Rental?

Solution

The linear function $g(x) = 0.3x + 25$, where x represents the number of miles, can be used to determine the daily charges of A-OK Car Rental. Let's graph this function and $f(x) = 0.2x + 50$ from Example 4 on the same set of axes (Figure 10.7).

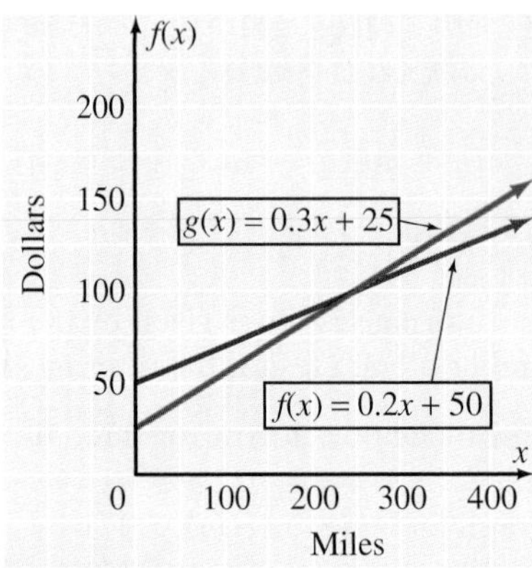

Figure 10.7

Now we see that the two functions have equal values at the point of intersection of the two lines. To find the coordinates of this point, we can set $0.3x + 25$ equal to $0.2x + 50$ and solve for x.

$$0.3x + 25 = 0.2x + 50$$
$$0.1x = 25$$
$$x = 250$$

If $x = 250$, then $0.3(250) + 25 = 100$, and the point of intersection is $(250, 100)$. Again looking at the lines in Figure 10.7, we see that Ed should use A-OK Car Rental for day trips of less than 250 miles, but he should use EZ Car Rental for day trips of more than 250 miles.

▼ PRACTICE YOUR SKILL

Suppose Morgan (Example 3 practice problem) is considering using the Ever Ready Legal Corporation, which charges a consultation fee of \$100 plus \$8 per minute. For what length of consultations should Morgan use the Ever Ready Legal Corporation instead of the ETF Group? ■

3 Graph Quadratic Functions

Any function that can be written in the form

$$f(x) = ax^2 + bx + c$$

where a, b, and c are real numbers with $a \neq 0$, is called a **quadratic function**. The following equations are examples of quadratic functions.

$$f(x) = 3x^2 \qquad f(x) = -2x^2 + 5x \qquad f(x) = 4x^2 - 7x + 1$$

The techniques discussed in Chapter 9 for graphing quadratic equations of the form $y = ax^2 + bx + c$ provide the basis for graphing quadratic functions. Let's review some work we did in Chapter 9 with an example.

EXAMPLE 5

Graph the function $f(x) = 2x^2 - 4x + 5$.

Solution

$$\begin{aligned} f(x) &= 2x^2 - 4x + 5 \\ &= 2(x^2 - 2x + \_\_) + 5 \quad \text{Recall the process of completing the square!} \\ &= 2(x^2 - 2x + 1) + 5 - 2 \\ &= 2(x - 1)^2 + 3 \end{aligned}$$

From this form we can obtain the following information about the parabola.

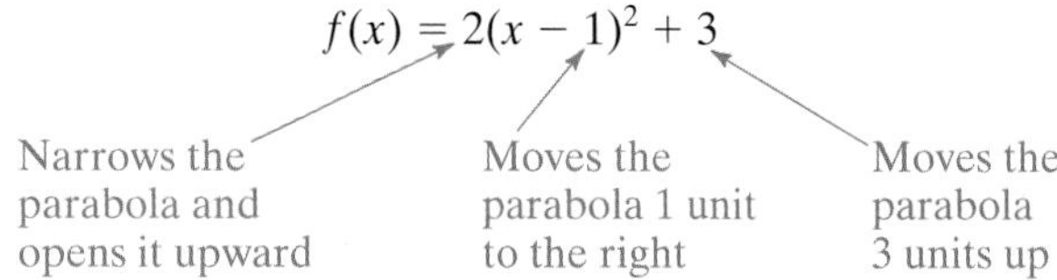

Thus the parabola can be drawn as shown in Figure 10.8.

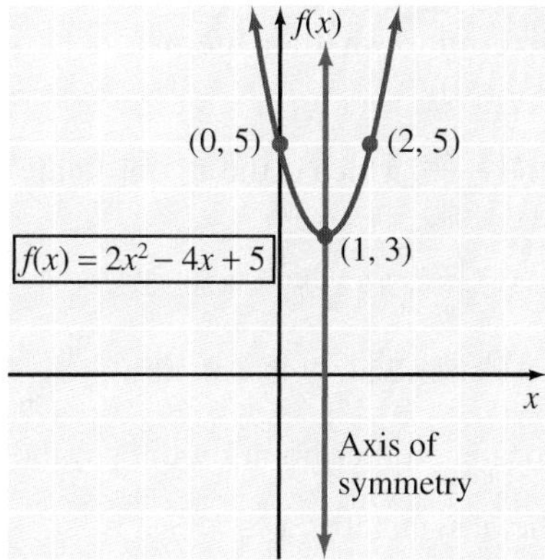

Figure 10.8

▼ PRACTICE YOUR SKILL

Graph the function $f(x) = x^2 - 6x + 7$. ■

In general, if we complete the square on

$$f(x) = ax^2 + bx + c$$

we obtain

$$\begin{aligned} f(x) &= a\left(x^2 + \frac{b}{a}x + \_\_\_\right) + c \\ &= a\left(x^2 + \frac{b}{a}x + \frac{b^2}{4a^2}\right) + c - \frac{b^2}{4a} \\ &= a\left(x + \frac{b}{2a}\right)^2 + \frac{4ac - b^2}{4a} \end{aligned}$$

Therefore, the parabola associated with $f(x) = ax^2 + bx + c$ has its vertex at $\left(-\frac{b}{2a}, \frac{4ac - b^2}{4a}\right)$ and the equation of its axis of symmetry is $x = -\frac{b}{2a}$. These facts are illustrated in Figure 10.9.

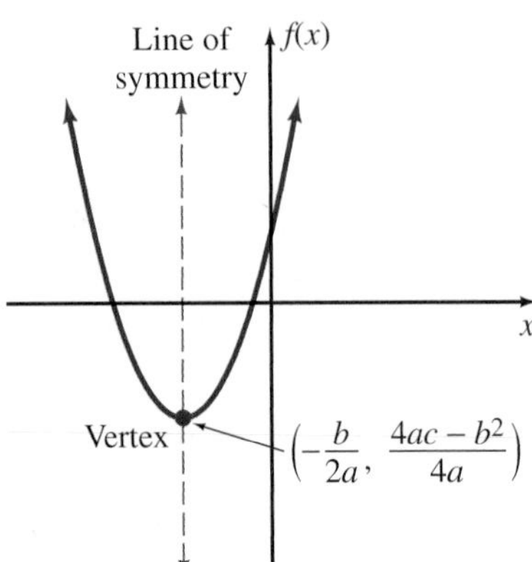

Figure 10.9

By using the information from Figure 10.9, we now have another way of graphing quadratic functions of the form $f(x) = ax^2 + bx + c$, as shown by the following steps.

1. Determine whether the parabola opens upward (if $a > 0$) or downward (if $a < 0$).
2. Find $-\frac{b}{2a}$, which is the x coordinate of the vertex.
3. Find $f\left(-\frac{b}{2a}\right)$, which is the y coordinate of the vertex. $\left(\text{You could also find the } y \text{ coordinate by evaluating } \frac{4ac - b^2}{4a}.\right)$
4. Locate another point on the parabola, and also locate its image across the line of symmetry, $x = -\frac{b}{2a}$.

The three points in Steps 2, 3, and 4 should determine the general shape of the parabola. Let's use these steps in the following two examples.

EXAMPLE 6

Graph $f(x) = 3x^2 - 6x + 5$.

Solution

Step 1 Because $a = 3$, the parabola opens upward.

Step 2 $-\frac{b}{2a} = -\frac{-6}{6} = 1$

Step 3 $f\left(-\frac{b}{2a}\right) = f(1) = 3 - 6 + 5 = 2$. Thus the vertex is at (1, 2).

Step 4 Letting $x = 2$, we obtain $f(2) = 12 - 12 + 5 = 5$. Thus (2, 5) is on the graph and so is its reflection (0, 5) across the line of symmetry $x = 1$.

The three points (1, 2), (2, 5), and (0, 5) are used to graph the parabola in Figure 10.10.

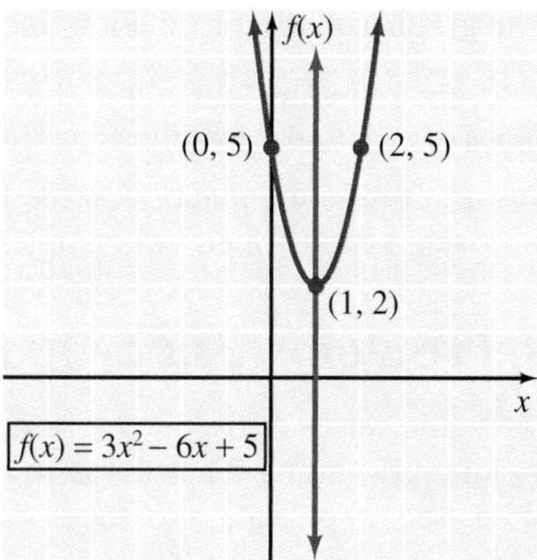

Figure 10.10

▼ **PRACTICE YOUR SKILL**

Graph the function $f(x) = 2x^2 - 8x + 3$. ■

EXAMPLE 7

Graph $f(x) = -x^2 - 4x - 7$.

Solution

Step 1 Because $a = -1$, the parabola opens downward.

Step 2 $-\dfrac{b}{2a} = -\dfrac{-4}{-2} = -2$.

Step 3 $f\left(-\dfrac{b}{2a}\right) = f(-2) = -(-2)^2 - 4(-2) - 7 = -3$. So the vertex is at $(-2, -3)$.

Step 4 Letting $x = 0$, we obtain $f(0) = -7$. Thus $(0, -7)$ is on the graph and so is its reflection $(-4, -7)$ across the line of symmetry $x = -2$.

The three points $(-2, -3)$, $(0, -7)$ and $(-4, -7)$ are used to draw the parabola in Figure 10.11.

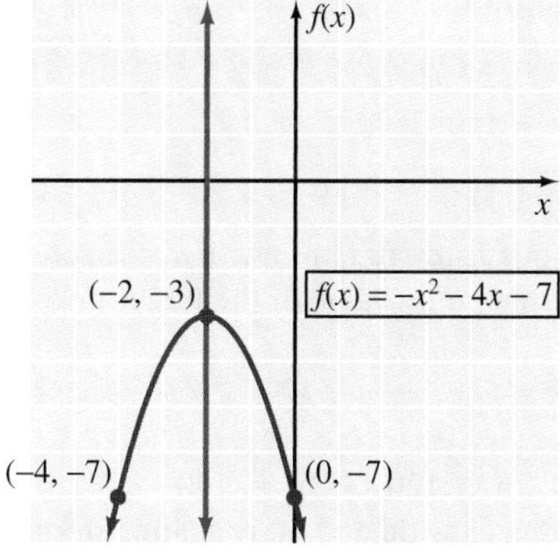

Figure 10.11

▼ **PRACTICE YOUR SKILL**

Graph the function $f(x) = -x^2 + 6x - 2$. ■

In summary, to graph a quadratic function, we have two methods.

1. We can express the function in the form $f(x) = a(x - h)^2 + k$ and use the values of a, h, and k to determine the parabola.
2. We can express the function in the form $f(x) = ax^2 + bx + c$ and use the approach demonstrated in Examples 6 and 7.

4 Solve Problems Using Quadratic Functions

As we have seen, the vertex of the graph of a quadratic function is either the lowest or the highest point on the graph. Thus the term *minimum value* or *maximum value* of a function is often used in applications of the parabola. The x value of the vertex indicates where the minimum or maximum occurs, and $f(x)$ yields the minimum or maximum value of the function. Let's consider some examples that illustrate these ideas.

EXAMPLE 8

Apply Your Skill

A farmer has 120 rods of fencing and wants to enclose a rectangular plot of land that requires fencing on only three sides because it is bounded by a river on one side. Find the length and width of the plot that will maximize the area.

Solution

Let x represent the width; then $120 - 2x$ represents the length, as indicated in Figure 10.12.

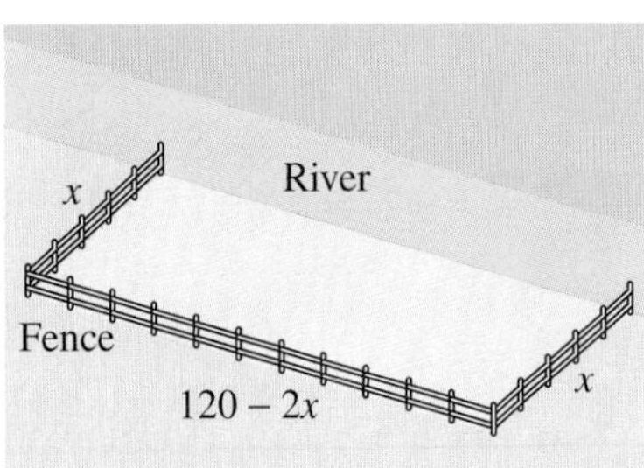

Figure 10.12

The function $A(x) = x(120 - 2x)$ represents the area of the plot in terms of the width x. Because

$$\begin{aligned} A(x) &= x(120 - 2x) \\ &= 120x - 2x^2 \\ &= -2x^2 + 120x \end{aligned}$$

we have a quadratic function with $a = -2$, $b = 120$, and $c = 0$. Therefore, the x value where the maximum value of the function is obtained is

$$-\frac{b}{2a} = -\frac{120}{2(-2)} = 30$$

If $x = 30$, then $120 - 2x = 120 - 2(30) = 60$. Thus the farmer should make the plot 30 rods wide and 60 rods long in order to maximize the area at $(30)(60) = 1800$ square rods.

▼ **PRACTICE YOUR SKILL**

A dog owner has 80 yards of fencing and wants to build a rectangular dog run that requires fencing on all four sides. Find the length and width of the plot that will maximize the area. ■

EXAMPLE 9

Apply Your Skill

Find two numbers whose sum is 30 such that the sum of their squares is a minimum.

Solution

Let x represent one of the numbers; then $30 - x$ represents the other number. By expressing the sum of the squares as a function of x, we obtain

$$f(x) = x^2 + (30 - x)^2$$

which can be simplified to

$$f(x) = x^2 + 900 - 60x + x^2$$
$$= 2x^2 - 60x + 900$$

This is a quadratic function with $a = 2$, $b = -60$, and $c = 900$. Therefore, the x value where the minimum occurs is

$$-\frac{b}{2a} = -\frac{-60}{4} = 15$$

If $x = 15$, then $30 - x = 30 - (15) = 15$. Thus the two numbers should both be 15.

▼ PRACTICE YOUR SKILL

Find two numbers whose difference is 14 such that the sum of the squares is a minimum. ■

EXAMPLE 10

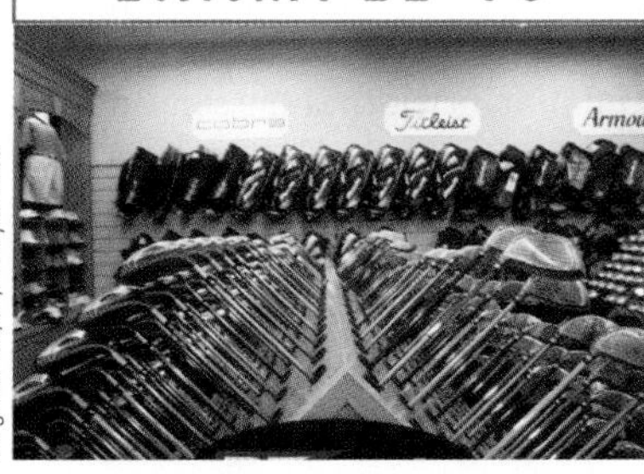

Magnus Rew/dk/Alamy Limited

Apply Your Skill

A golf pro-shop operator finds that she can sell 30 sets of golf clubs at \$500 per set in a year. Furthermore, she predicts that for each \$25 decrease in price, three more sets of golf clubs could be sold. At what price should she sell the clubs to maximize gross income?

Solution

Sometimes, when we are analyzing such a problem, it helps to set up a table.

	Number of sets	×	Price per set	=	Income
(3 additional sets can	30	×	\$500	=	\$15,000
be sold for a \$25	33	×	\$475	=	\$15,675
decrease in price)	36	×	\$450	=	\$16,200

Let x represent the number of \$25 decreases in price. Then we can express the income as a function of x as follows:

$$f(x) = (30 + 3x)(500 - 25x)$$

(30 + 3x): Number of sets; (500 − 25x): Price per set

When we simplify, we obtain

$$f(x) = 15{,}000 - 750x + 1500x - 75x^2$$
$$= -75x^2 + 750x + 15{,}000$$

Completing the square yields

$$\begin{aligned} f(x) &= -75x^2 + 750x + 15{,}000 \\ &= -75(x^2 - 10x + \_\_) + 15{,}000 \\ &= -75(x^2 - 10x + 25) + 15{,}000 + 1875 \\ &= -75(x - 5)^2 + 16{,}875 \end{aligned}$$

From this form we know that the vertex of the parabola is at (5, 16875). Thus 5 decreases of \$25 each—that is, a \$125 reduction in price—will give a maximum income of \$16,875. The golf clubs should be sold at \$375 per set.

▼ PRACTICE YOUR SKILL

A fitness equipment director finds that he can sell 40 treadmills at \$400 apiece in a year. Furthermore, he predicts that for each \$20 decrease in price, five more treadmills could be sold. At what price should he sell the treadmills to maximize the income? ■

5 Graph Functions with a Graphing Utility

What we know about parabolas and the process of completing the square can be helpful when we are using a graphing utility to graph a quadratic function. Consider the following example.

EXAMPLE 11

Use a graphing utility to obtain the graph of the quadratic function

$$f(x) = -x^2 + 37x - 311$$

Solution

First, we know that the parabola opens downward and that its width is the same as that of the basic parabola $f(x) = x^2$. Then we can start the process of completing the square to determine an approximate location of the vertex.

$$\begin{aligned} f(x) &= -x^2 + 37x - 311 \\ &= -(x^2 - 37x + \_\_\_) - 311 \\ &= -\left[x^2 - 37x + \left(\frac{37}{2}\right)^2\right] - 311 + \left(\frac{37}{2}\right)^2 \\ &= -\ [(x^2 - 37x + (18.5)^2] - 311 + 342.25 \\ &= -(x - 18.5)^2 + 31.25 \end{aligned}$$

Thus the vertex is near $x = 18$ and $y = 31$. Therefore, setting the boundaries of the viewing rectangle so that $-2 \le x \le 25$ and $-10 \le y \le 35$, we obtain the graph shown in Figure 10.13.

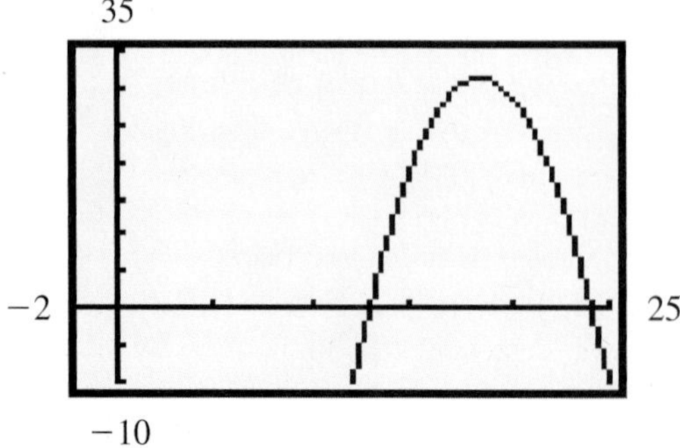

Figure 10.13

▼ **PRACTICE YOUR SKILL**

Use a graphing utility to obtain the graph of the quadratic function $f(x) = x^2 + 36x + 300$. ■

Remark: The graph in Figure 10.13 is sufficient for most purposes because it shows the vertex and the x intercepts of the parabola. Certainly other boundaries could be used that would also give this information.

CONCEPT QUIZ

For Problems 1–10, answer true or false.

1. The function $f(x) = 3x^2 + 4$ is a linear function.
2. The graph of a linear function of the form $f(x) = b$ is a horizontal line.
3. The graph of a quadratic function is a parabola.
4. The vertex of the graph of a quadratic function is either the lowest or highest point on the graph.
5. The axis of symmetry for a parabola passes through the vertex of the parabola.
6. The parabola for the quadratic function $f(x) = -2x^2 + 3x + 7$ opens upward.
7. The linear function $f(x) = 1$ is called the identity function.
8. The parabola $f(x) = x^2 + 6x + 5$ is symmetric with respect to the line $x = 3$.
9. The vertex of the parabola $f(x) = -2x^2 + 4x + 2$ is at $(1, 4)$.
10. The graph of the function $f(x) = -3$ is symmetric with respect to the $f(x)$ axis.

Problem Set 10.2

1 Graph Linear Functions

For Problems 1–8, graph each of the following linear functions.

1. $f(x) = 2x - 4$
2. $f(x) = 3x + 3$
3. $f(x) = -3x$
4. $f(x) = -4x$
5. $f(x) = -x + 3$
6. $f(x) = -2x - 4$
7. $f(x) = -3$
8. $f(x) = 1$

2 Apply Linear Functions

9. The cost for burning a 75-watt bulb is given by the function $c(h) = 0.0045h$, where h represents the number of hours that the bulb burns.
   (a) How much does it cost to burn a 75-watt bulb for 3 hours per night for a 31-day month? Express your answer to the nearest cent.
   (b) Graph the function $c(h) = 0.0045h$.
   (c) Use the graph in part (b) to approximate the cost of burning a 75-watt bulb for 225 hours.
   (d) Use $c(h) = 0.0045h$ to find the exact cost, to the nearest cent, of burning a 75-watt bulb for 225 hours.

10. The Rent-Me Car Rental charges \$35 per day plus \$0.32 per mile to rent a car. Determine a linear function that can be used to calculate daily car rentals. Then use that function to determine the cost of renting a car for a day and driving: 150 miles; 230 miles; 360 miles; 430 miles.

11. The ABC Car Rental uses the function $f(x) = 100$ for any daily use of a car up to and including 200 miles. For driving more than 200 miles per day, ABC uses the function $g(x) = 100 + 0.25(x - 200)$ to determine the charges. How much would ABC charge for daily driving of 150 miles? of 230 miles? of 360 miles? of 430 miles?

12. Suppose that a car-rental agency charges a fixed amount per day plus an amount per mile for renting a car. Heidi rented a car one day and paid \$80 for 200 miles. On another day she rented a car from the same agency and paid \$117.50 for 350 miles. Find the linear function that the agency could use to determine its daily rental charges.

13. A retailer has a number of items that she wants to sell and make a profit of 40% of the cost of each item. The function $s(c) = c + 0.4c = 1.4c$, where c represents the cost of an item, can be used to determine the selling price. Find the selling price of items that cost \$1.50, \$3.25, \$14.80, \$21, and \$24.20.

14. Zack wants to sell five items that cost him \$1.20, \$2.30, \$6.50, \$12, and \$15.60. He wants to make a profit of 60% of the cost. Create a function that you can use to determine the selling price of each item, and then use the function to calculate each selling price.

15. "All Items 20% Off Marked Price" is a sign at a local golf course. Create a function and then use it to determine how much one has to pay for each of the following marked items: a \$9.50 hat, a \$15 umbrella, a \$75 pair of golf shoes, a \$12.50 golf glove, a \$750 set of golf clubs.

16. The linear depreciation method assumes that an item depreciates the same amount each year. Suppose a new piece of machinery costs \$32,500 and it depreciates \$1950 each year for t years.
   (a) Set up a linear function that yields the value of the machinery after t years.
   (b) Find the value of the machinery after 5 years.
   (c) Find the value of the machinery after 8 years.
   (d) Graph the function from part (a).
   (e) Use the graph from part (d) to approximate how many years it takes for the value of the machinery to become zero.
   (f) Use the function to determine how long it takes for the value of the machinery to become zero.

3 Graph Quadratic Functions

For Problems 17–38, graph each quadratic function.

17. $f(x) = -2x^2$ **18.** $f(x) = -4x^2$

19. $f(x) = -(x+1)^2 - 2$ **20.** $f(x) = -(x-2)^2 + 4$

21. $f(x) = x^2 + 2x - 2$ **22.** $f(x) = x^2 - 4x - 1$

23. $f(x) = -x^2 + 6x - 8$ **24.** $f(x) = -x^2 - 8x - 15$

25. $f(x) = 2x^2 - 20x + 52$ **26.** $f(x) = 2x^2 + 12x + 14$

27. $f(x) = -3x^2 + 6x$ **28.** $f(x) = -4x^2 - 8x$

29. $f(x) = x^2 - x + 2$ **30.** $f(x) = x^2 + 3x + 2$

31. $f(x) = 2x^2 + 10x + 11$ **32.** $f(x) = 2x^2 - 10x + 15$

33. $f(x) = -2x^2 - 1$ **34.** $f(x) = -3x^2 + 2$

35. $f(x) = -3x^2 + 12x - 7$

36. $f(x) = -3x^2 - 18x - 23$

37. $f(x) = -2x^2 + 14x - 25$

38. $f(x) = -2x^2 - 10x - 14$

4 Solve Problems Using Quadratic Functions

39. Suppose that the cost function for a particular item is given by the equation $C(x) = 2x^2 - 320x + 12{,}920$, where x represents the number of items. How many items should be produced to minimize the cost?

40. Suppose that the equation $p(x) = -2x^2 + 280x - 1000$, where x represents the number of items sold, describes the profit function for a certain business. How many items should be sold to maximize the profit?

41. Find two numbers whose sum is 30 such that the sum of the square of one number plus ten times the other number is a minimum.

42. The height of a projectile fired vertically into the air (neglecting air resistance) at an initial velocity of 96 feet per second is a function of the time and is given by the equation $f(x) = 96x - 16x^2$, where x represents the time. Find the highest point reached by the projectile.

43. Two hundred and forty meters of fencing is available to enclose a rectangular playground. What should be the dimensions of the playground to maximize the area?

44. Find two numbers whose sum is 50 and whose product is a maximum.

45. A movie rental company has 1000 subscribers, and each pays \$15 per month. On the basis of a survey, company managers feel that for each decrease of \$0.25 on the monthly rate, they could obtain 20 additional subscribers. At what rate will maximum revenue be obtained and how many subscribers will it take at that rate?

46. A restaurant advertises that it will provide beer, pizza, and wings for \$50 per person at a Super Bowl party. It must have a guarantee of 30 people. Furthermore, it will agree that for each person in excess of 30, it will reduce the price per person for all attending by \$0.50. How many people will it take to maximize the restaurant's revenue?

5 Graph Functions with a Graphing Utility

GRAPHING CALCULATOR ACTIVITIES

47. Use a graphing calculator to check your graphs for Problems 25–38.

48. Graph each of the following parabolas, and keep in mind that you may need to change the dimensions of the viewing window to obtain a good picture.
   (a) $f(x) = x^2 - 2x + 12$ **(b)** $f(x) = -x^2 - 4x - 16$
   (c) $f(x) = x^2 + 12x + 44$ **(d)** $f(x) = x^2 - 30x + 229$
   (e) $f(x) = -2x^2 + 8x - 19$

49. Graph each of the following parabolas, and use the TRACE feature to find whole number estimates of the vertex. Then either complete the square or use $\left(-\frac{b}{2a}, \frac{4ac - b^2}{4a}\right)$ to find the vertex.
   (a) $f(x) = x^2 - 6x + 3$
   (b) $f(x) = x^2 - 18x + 66$
   (c) $f(x) = -x^2 + 8x - 3$
   (d) $f(x) = -x^2 + 24x - 129$
   (e) $f(x) = 14x^2 - 7x + 1$
   (f) $f(x) = -0.5x^2 + 5x - 8.5$

50. **(a)** Graph $f(x) = |x|$, $f(x) = 2|x|$, $f(x) = 4|x|$, and $f(x) = \frac{1}{2}|x|$ on the same set of axes.
   (b) Graph $f(x) = |x|$, $f(x) = -|x|$, $f(x) = -3|x|$, and $f(x) = -\frac{1}{2}|x|$ on the same set of axes.

(c) Use your results from parts (a) and (b) to make a conjecture about the graphs of $f(x) = a|x|$, where a is a nonzero real number.

(d) Graph $f(x) = |x|$, $f(x) = |x| + 3$, $f(x) = |x| - 4$, and $f(x) = |x| + 1$ on the same set of axes. Make a conjecture about the graphs of $f(x) = |x| + k$, where k is a nonzero real number.

(e) Graph $f(x) = |x|$, $f(x) = |x - 3|$, $f(x) = |x - 1|$, and $f(x) = |x + 4|$ on the same set of axes. Make a conjecture about the graphs of $f(x) = |x - h|$, where h is a nonzero real number.

(f) On the basis of your results from parts (a) through (e), sketch each of the following graphs. Then use a graphing calculator to check your sketches.

(1) $f(x) = |x - 2| + 3$ **(2)** $f(x) = |x + 1| - 4$

(3) $f(x) = 2|x - 4| - 1$ **(4)** $f(x) = -3|x + 2| + 4$

(5) $f(x) = \frac{1}{2}|x - 3| - 2$

THOUGHTS INTO WORDS

51. Give a step-by-step description of how you would use the ideas of this section to graph $f(x) = -4x^2 + 16x - 13$.

52. Is $f(x) = (3x - 2) - (2x + 1)$ a linear function? Explain your answer.

53. Suppose that Bianca walks at a constant rate of 3 miles per hour. Explain what it means that the distance Bianca walks is a linear function of the time that she walks.

Answers to the Concept Quiz

1. False **2.** True **3.** True **4.** True **5.** True **6.** False **7.** False **8.** False **9.** True **10.** True

Answers to the Example Practice Skills

1. $f(x) = \frac{1}{2}x + 3$ **2.** **(a)** $180.00 **(b)**  **(c)** $50.00

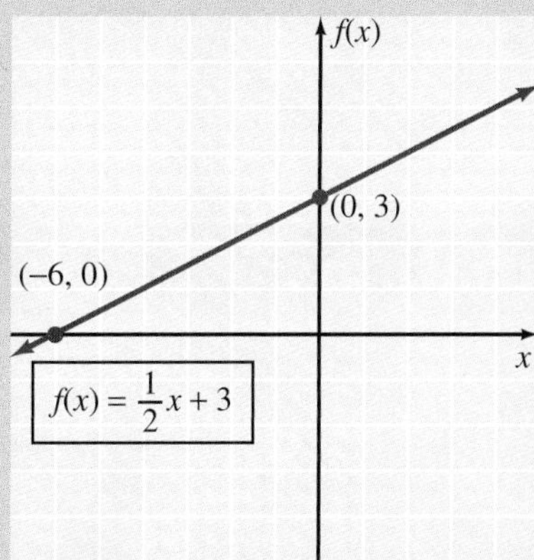

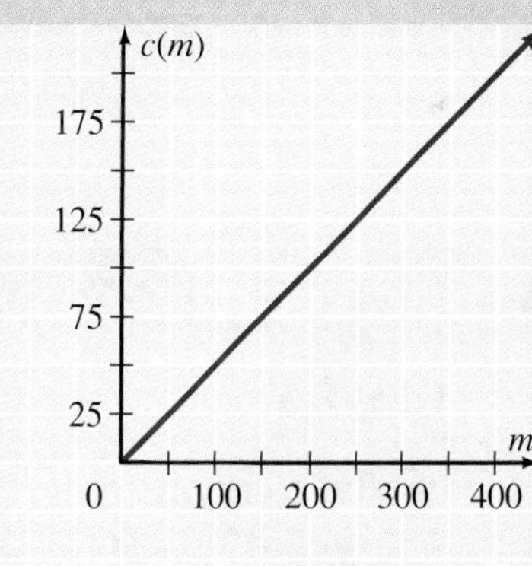

3. $f(x) = 4x + 160$, where x represents the number of minutes **4.** 15 minutes or more

5. $f(x) = x^2 - 6x + 7$ **6.** $f(x) = 2x^2 - 8x + 3$

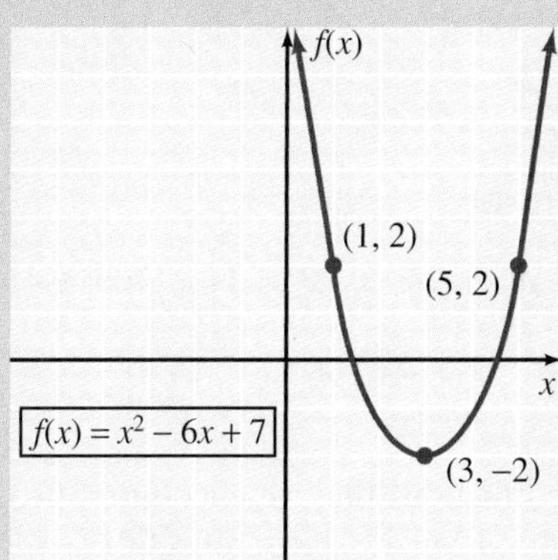

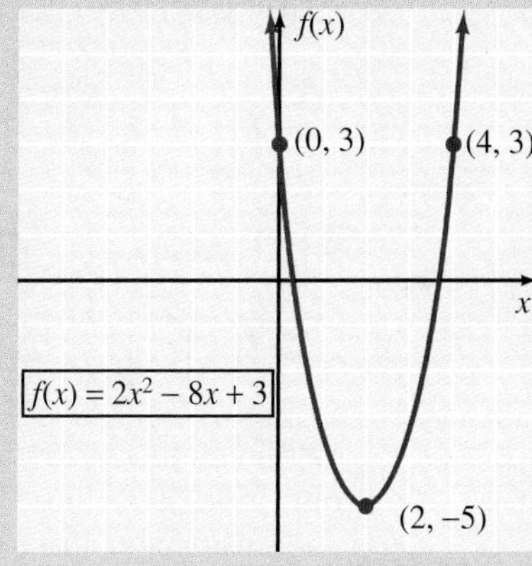

7. $f(x) = -x^2 + 6x - 2$

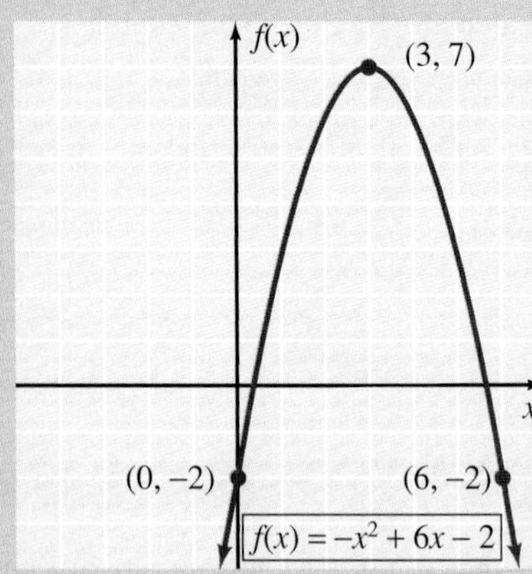

8. Width = 20 yd, length = 20 yd **9.** −7 and 7 **10.** \$280
11. $f(x) = x^2 + 36x + 300$

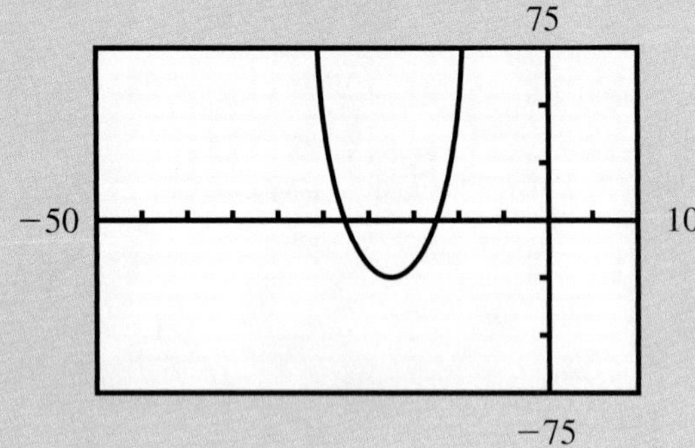

10.3 Graphing Made Easy Via Transformations

OBJECTIVES

1. Know the Basic Graphs of $f(x) = x^2$, $f(x) = x^3$, $f(x) = \frac{1}{x}$, $f(x) = \sqrt{x}$, and $f(x) = |x|$
2. Graph Functions by Using Translations
3. Graph Functions by Using Reflections
4. Graph Functions by Using Vertical Stretching or Shrinking
5. Graph Functions by Using Successive Transformations

1 Know the Basic Graphs of $f(x) = x^2$, $f(x) = x^3$, $f(x) = \frac{1}{x}$, $f(x) = \sqrt{x}$, and $f(x) = |x|$

Within mathematics there are several basic functions that we encounter throughout our work. Many functions that you have to graph will be shifts, reflections, stretching, and shrinking of these basic graphs. The objective of this section is to be able to graph functions by making transformations to the basic graphs.

The five basic functions you will encounter in this section are

$$f(x) = x^2, \qquad f(x) = x^3, \qquad f(x) = \frac{1}{x}, \qquad f(x) = \sqrt{x}, \qquad f(x) = |x|$$

Figures 10.14–10.16 show the graphs of the functions $f(x) = x^2$, $f(x) = x^3$, and $f(x) = \frac{1}{x}$, respectively.

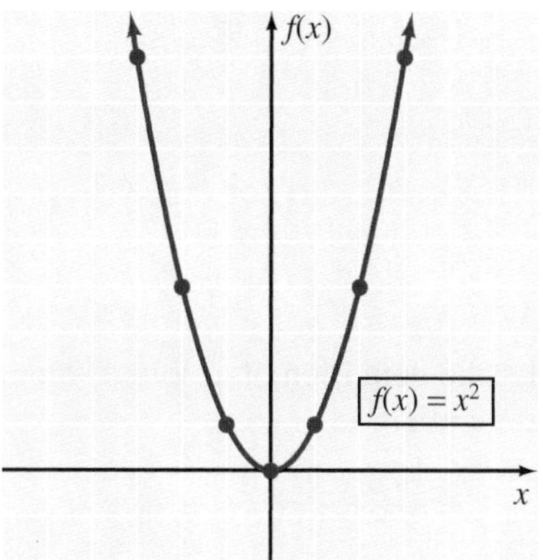

Figure 10.14

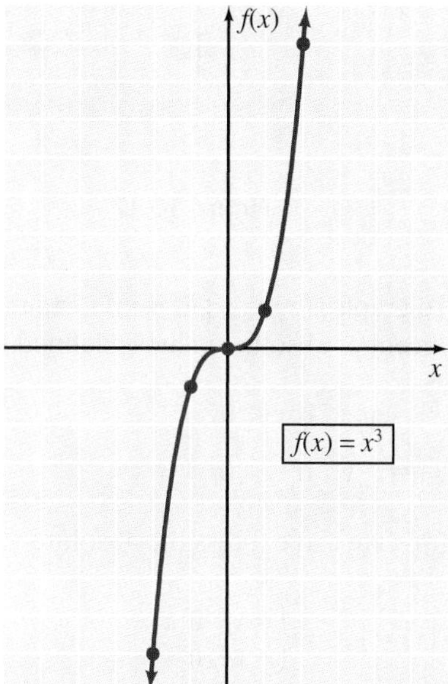

Figure 10.15

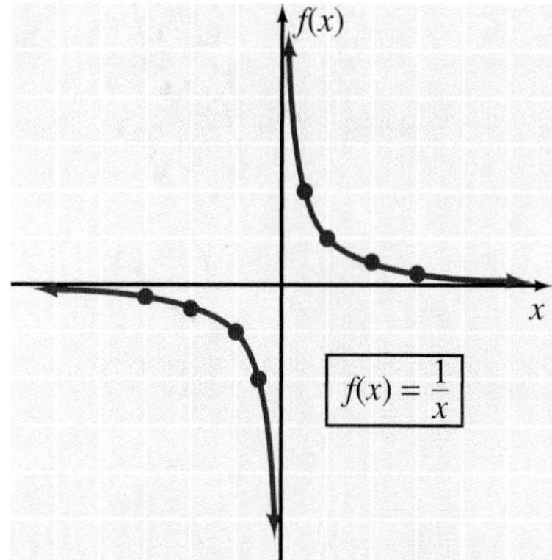

Figure 10.16

To graph a new function—that is, one you are not familiar with—use some of the graphing suggestions we offered in Chapter 3. We will restate those suggestions in terms of function vocabulary and notation. Pay special attention to suggestions 2 and 3, where we have restated the concepts of intercepts and symmetry using function notation.

1. Determine the domain of the function.
2. Determine any types of symmetry that the equation possesses. If $f(-x) = f(x)$, then the function exhibits y-axis symmetry. If $f(-x) = -f(x)$, then the function exhibits origin symmetry. (Note that the definition of a function rules out the possibility that the graph of a function has x-axis symmetry.)
3. Find the y intercept (we are labeling the y axis with $f(x)$) by evaluating $f(0)$. Find the x intercept by finding the value(s) of x such that $f(x) = 0$.
4. Set up a table of ordered pairs that satisfy the equation. The type of symmetry and the domain will affect your choice of values of x in the table.
5. Plot the points associated with the ordered pairs and connect them with a smooth curve. Then, if appropriate, reflect this part of the curve according to any symmetries the graph exhibits.

Let's consider these suggestions as we determine the graphs of $f(x) = \sqrt{x}$ and $f(x) = |x|$.

To graph $f(x) = \sqrt{x}$, let's first determine the domain. The radicand must be nonnegative, so the domain is the set of nonnegative real numbers. Because $x \geq 0$, $f(-x)$ is not a real number; thus there is no symmetry for this graph. We see that $f(0) = 0$, so both intercepts are 0. That is, the origin $(0, 0)$ is a point of the graph. Now let's set up a table of values, keeping in mind that $x \geq 0$. Plotting these points and connecting them with a smooth curve produces Figure 10.17.

x	f(x)
0	0
1	1
4	2
9	3

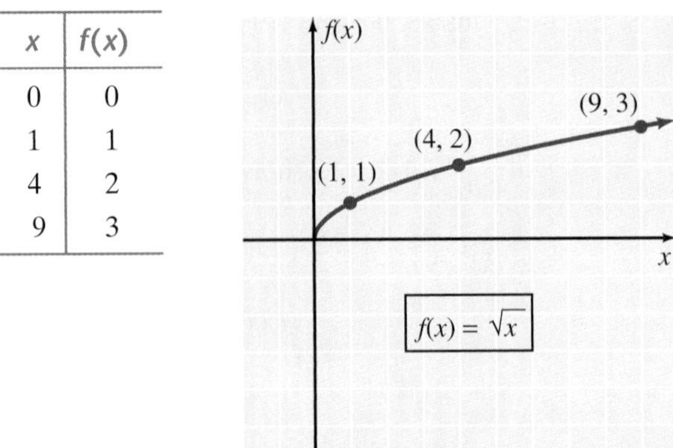

Figure 10.17

To graph $f(x) = |x|$, it is important to consider the definition of absolute value. The concept of absolute value is defined for all real numbers as

$$|x| = x \quad \text{if } x \geq 0$$
$$|x| = -x \quad \text{if } x < 0$$

Therefore, we can express the absolute value function as

$$f(x) = |x| = \begin{cases} x & \text{if } x \geq 0 \\ -x & \text{if } x < 0 \end{cases}$$

The graph of $f(x) = x$ for $x \geq 0$ is the ray in the first quadrant, and the graph of $f(x) = -x$ for $x < 0$ is the half-line in the second quadrant, as indicated in Figure 10.18.

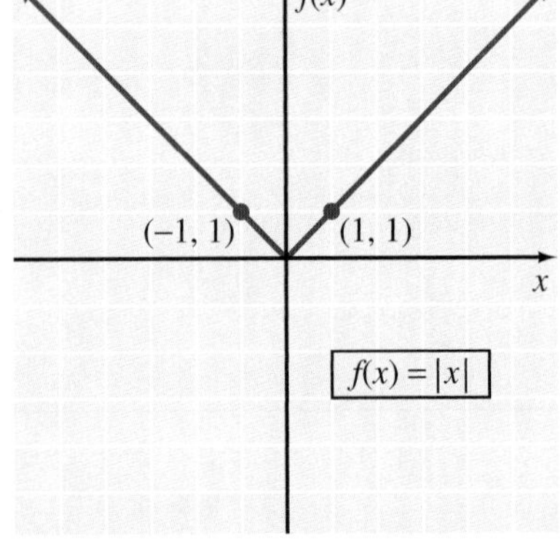

Figure 10.18

Remark: Note that the equation $f(x) = |x|$ does exhibit y-axis symmetry because $f(-x) = |-x| = |x|$. Even though we did not use the symmetry idea to sketch the curve, you should recognize that the symmetry does exist.

2 Graph Functions by Using Translations

From our work in Chapter 9, we know that the graph of $f(x) = x^2 + 3$ is the graph of $f(x) = x^2$ moved up 3 units. Likewise, the graph of $f(x) = x^2 - 2$ is the graph of $f(x) = x^2$ moved down 2 units. Now we will describe in general the concept of **vertical translation**.

Vertical Translation

The graph of $y = f(x) + k$ is the graph of $y = f(x)$ shifted k units upward if $k > 0$ or shifted $|k|$ units downward if $k < 0$.

EXAMPLE 1 Graph **(a)** $f(x) = |x| + 2$ and **(b)** $f(x) = |x| - 3$.

Solution

In Figure 10.19, we obtain the graph of $f(x) = |x| + 2$ by shifting the graph of $f(x) = |x|$ upward 2 units, and we obtain the graph of $f(x) = |x| - 3$ by shifting the graph of $f(x) = |x|$ downward 3 units. (Remember that we can write $f(x) = |x| - 3$ as $f(x) = |x| + (-3)$.)

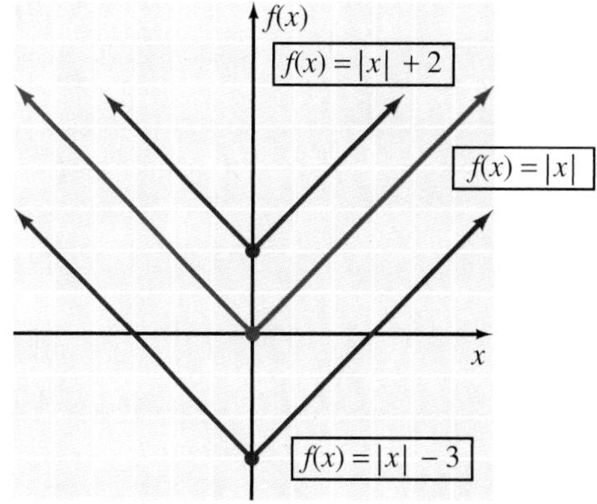

Figure 10.19

▼ **PRACTICE YOUR SKILL**

Graph $f(x) = x^2 + 1$. ■

We also graphed horizontal translations of the basic parabola in Chapter 9. For example, the graph of $f(x) = (x - 4)^2$ is the graph of $f(x) = x^2$ shifted 4 units to the right, and the graph of $f(x) = (x + 5)^2$ is the graph of $f(x) = x^2$ shifted 5 units to the left. We describe the general concept of a **horizontal translation** as follows:

> **Horizontal Translation**
>
> The graph of $y = f(x - h)$ is the graph of $y = f(x)$ shifted h units to the right if $h > 0$ or shifted $|h|$ units to the left if $h < 0$.

EXAMPLE 2 Graph **(a)** $f(x) = (x - 3)^2$ and **(b)** $f(x) = (x + 2)^3$.

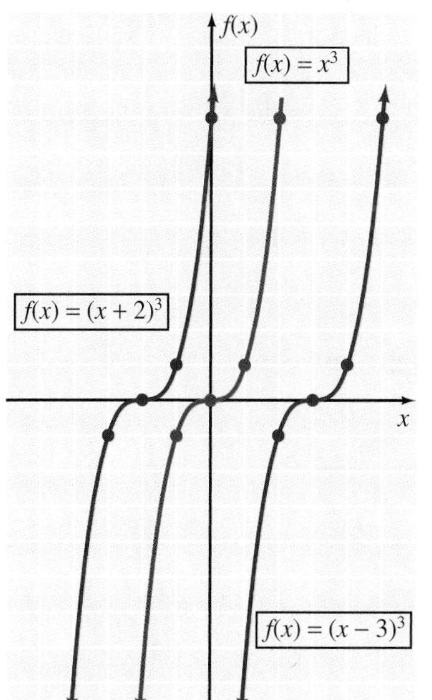

Figure 10.20

Solution

In Figure 10.20, we obtain the graph of $f(x) = (x - 3)^3$ by shifting the graph of $f(x) = x^3$ to the right 3 units. Likewise, we obtain the graph of $f(x) = (x + 2)^3$ by shifting the graph of $f(x) = x^3$ to the left 2 units.

▼ **PRACTICE YOUR SKILL**

Graph $f(x) = |x + 4|$. ■

3 Graph Functions by Using Reflections

From our work in Chapter 9, we know that the graph of $f(x) = -x^2$ is the graph of $f(x) = x^2$ reflected through the x axis. We describe the general concept of an ***x*-axis reflection** as follows.

> ***x*-Axis Reflection**
>
> The graph of $y = -f(x)$ is the graph of $y = f(x)$ reflected through the x axis.

EXAMPLE 3

Graph $f(x) = -\sqrt{x}$.

Solution

In Figure 10.21, we obtain the graph of $f(x) = -\sqrt{x}$ by reflecting the graph of $f(x) = \sqrt{x}$ through the x axis. Reflections are sometimes referred to as **mirror images**. Thus, in Figure 10.21, if we think of the x axis as a mirror, the graphs of $f(x) = \sqrt{x}$ and $f(x) = -\sqrt{x}$ are mirror images of each other.

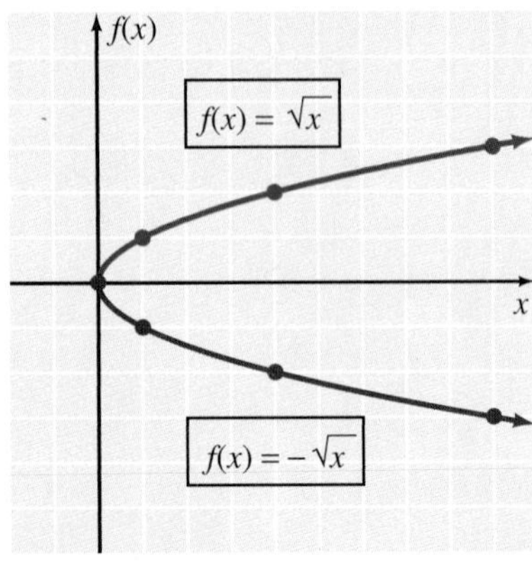

Figure 10.21

▼ PRACTICE YOUR SKILL

Graph $f(x) = -|x|$. ■

In Chapter 9, we did not consider a y-axis reflection of the basic parabola $f(x) = x^2$ because it is symmetric with respect to the y axis. In other words, a y-axis reflection of $f(x) = x^2$ produces the same figure in the same location. At this time we will describe the general concept of a y-axis reflection.

y-Axis Reflection

The graph of $y = f(-x)$ is the graph of $y = f(x)$ reflected through the y axis.

Now suppose that we want to do a y-axis reflection of $f(x) = \sqrt{x}$. Because $f(x) = \sqrt{x}$ is defined for $x \geq 0$, the y-axis reflection $f(x) = \sqrt{-x}$ is defined for $-x \geq 0$, which is equivalent to $x \leq 0$. Figure 10.22 shows the y-axis reflection of $f(x) = \sqrt{x}$.

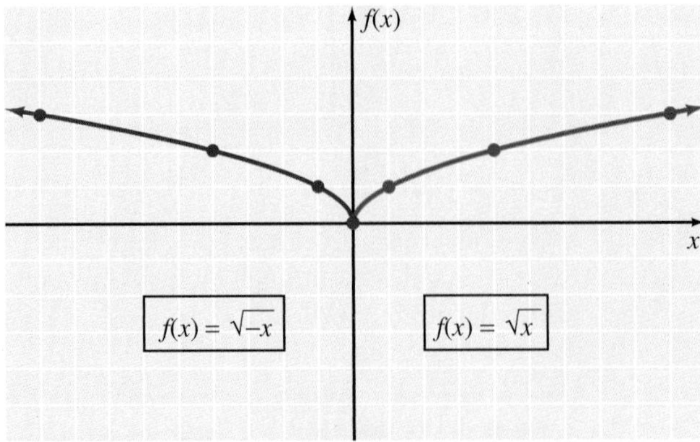

Figure 10.22

4 Graph Functions by Using Vertical Stretching or Shrinking

Translations and reflections are called **rigid transformations** because the basic shape of the curve being transformed is not changed. In other words, only the positions of the graphs are changed. Now we want to consider some transformations that distort the shape of the original figure somewhat.

In Chapter 9, we graphed the equation $y = 2x^2$ by doubling the y coordinates of the ordered pairs that satisfy the equation $y = x^2$. We obtained a parabola with its vertex at the origin, symmetric with respect to the y axis, but narrower than the basic parabola. Likewise, we graphed the equation $y = \frac{1}{2}x^2$ by halving the y coordinates of the ordered pairs that satisfy $y = x^2$. In this case, we obtained a parabola with its vertex at the origin, symmetric with respect to the y axis, but wider than the basic parabola.

We can use the concepts of narrower and wider to describe parabolas, but they cannot be used to describe some other curves accurately. Instead, we use the more general concepts of vertical stretching and shrinking.

Vertical Stretching and Shrinking

The graph of $y = cf(x)$ is obtained from the graph of $y = f(x)$ by multiplying the y coordinates of $y = f(x)$ by c. If $|c| > 1$, the graph is said to be *stretched* by a factor of $|c|$, and if $0 < |c| < 1$, the graph is said to be *shrunk* by a factor of $|c|$.

EXAMPLE 4

Graph **(a)** $f(x) = 2\sqrt{x}$ and **(b)** $f(x) = \frac{1}{2}\sqrt{x}$.

Solution

In Figure 10.23, the graph of $f(x) = 2\sqrt{x}$ is obtained by doubling the y coordinates of points on the graph of $f(x) = \sqrt{x}$. Likewise, in Figure 10.23, the graph of $f(x) = \frac{1}{2}\sqrt{x}$ is obtained by halving the y coordinates of points on the graph of $f(x) = \sqrt{x}$.

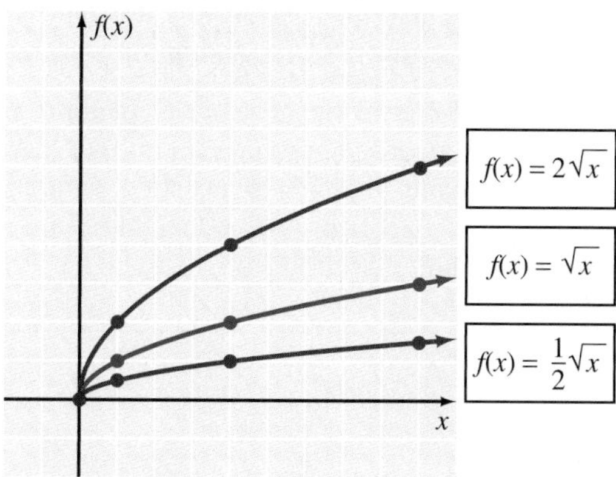

Figure 10.23

▼ **PRACTICE YOUR SKILL**

Graph $f(x) = 2|x|$. ■

5 Graph Functions by Using Successive Transformations

Some curves are the result of performing more than one transformation on a basic curve. Let's consider the graph of a function that involves a stretching, a reflection, a horizontal translation, and a vertical translation of the basic absolute value function.

EXAMPLE 5 Graph $f(x) = -2|x - 3| + 1$.

Solution

This is the basic absolute value curve stretched by a factor of 2, reflected through the x axis, shifted 3 units to the right, and shifted 1 unit upward. To sketch the graph, we locate the point (3, 1) and then determine a point on each of the rays. The graph is shown in Figure 10.24.

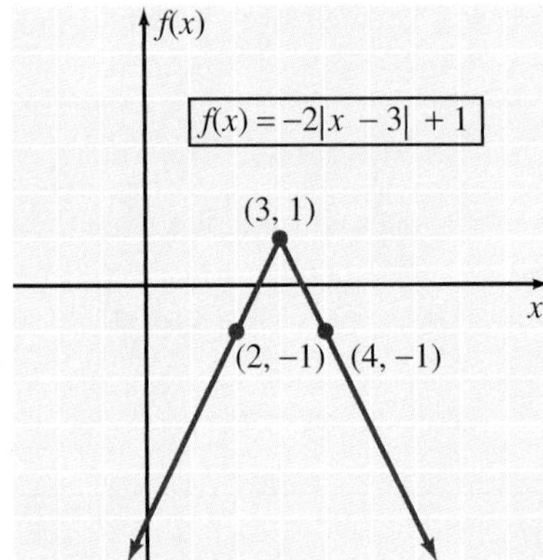

Figure 10.24

▼ **PRACTICE YOUR SKILL**

Graph $f(x) = \frac{1}{2}|x - 1| + 3$. ■

Remark: Note in Example 5 that we did not sketch the original basic curve $f(x) = |x|$ or any of the intermediate transformations. However, it is helpful to mentally picture each transformation. This locates the point (3, 1) and establishes the fact that the two rays point downward. Then a point on each ray determines the final graph.

You also need to realize that changing the order of doing the transformations may produce an incorrect graph. In Example 5, performing the translations first, followed by the stretching and x-axis reflection, would produce an incorrect graph that has its vertex at $(3, -1)$ instead of (3, 1). Unless parentheses indicate otherwise, stretchings, shrinkings, and x-axis reflections should be performed before translations.

EXAMPLE 6 Graph the function $f(x) = \frac{1}{x + 2} + 3$.

Solution

This is the basic curve $f(x) = \frac{1}{x}$ moved 2 units to the left and 3 units upward. Remember that the x axis is a horizontal asymptote and the y axis a vertical asymptote for the curve $f(x) = \frac{1}{x}$. Thus, for this curve, the vertical asymptote is shifted 2 units to the left and its equation is $x = -2$. Likewise, the horizontal asymptote is shifted 3 units upward and its equation is $y = 3$. Therefore, in Figure 10.25 we have drawn the asymptotes as dashed lines and then located a few points to help determine each branch of the curve.

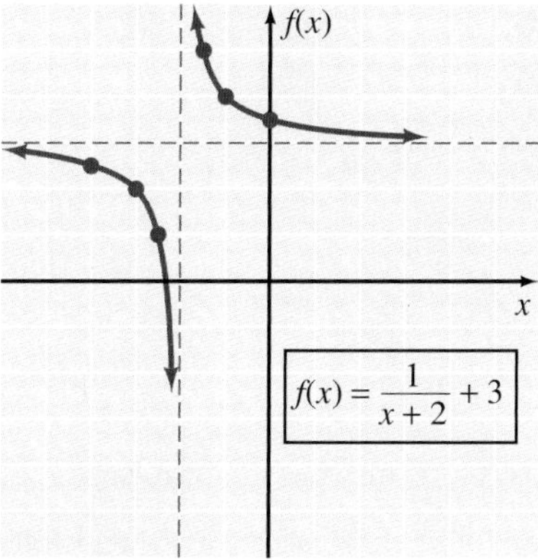

Figure 10.25

▼ PRACTICE YOUR SKILL

Graph $f(x) = \dfrac{1}{x - 3} + 1$. ■

Finally, let's use a graphing utility to give another illustration of the concepts of stretching and shrinking a curve.

EXAMPLE 7

If $f(x) = \sqrt{25 - x^2}$, sketch a graph of $y = 2(f(x))$ and $y = \frac{1}{2}(f(x))$.

Solution

If $y = f(x) = \sqrt{25 - x^2}$, then

$$y = 2(f(x)) = 2\sqrt{25 - x^2} \qquad \text{and} \qquad y = \frac{1}{2}(f(x)) = \frac{1}{2}\sqrt{25 - x^2}$$

Graphing all three of these functions on the same set of axes produces Figure 10.26.

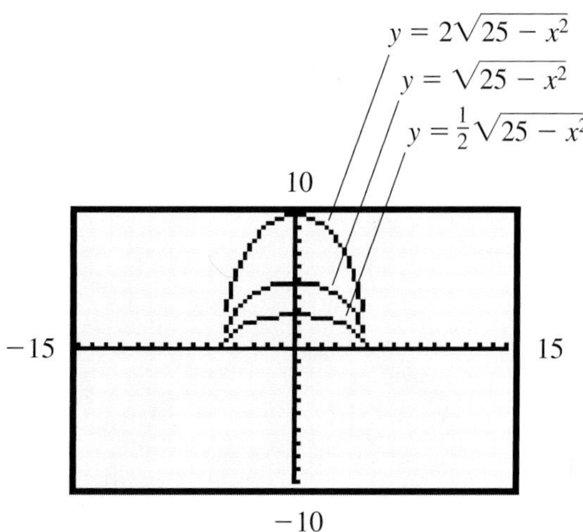

Figure 10.26

▼ **PRACTICE YOUR SKILL**

If $f(x) = -(x-2)^2$, use a graphing utility to graph $y = 3(f(x))$ and $y = \frac{1}{3}(f(x))$. ■

CONCEPT QUIZ

For Problems 1–5, match the function with the description of its graph relative to the graph of $f(x) = \sqrt{x}$.

1. $f(x) = \sqrt{x} + 3$	**A.** Stretched by a factor of three
2. $f(x) = -\sqrt{x}$	**B.** Reflected across the y axis
3. $f(x) = \sqrt{x+3}$	**C.** Shifted up three units
4. $f(x) = \sqrt{-x}$	**D.** Reflected across the x axis
5. $f(x) = 3\sqrt{x}$	**E.** Shifted three units to the left

For Problems 6–10, answer true or false.

6. The graph of $f(x) = \sqrt{x-1}$ is the graph of $f(x) = \sqrt{x}$ shifted 1 unit downward.

7. The graph of $f(x) = |x+2|$ is symmetric with respect to the line $x = -2$.

8. The graph of $f(x) = x^3 - 3$ is the graph of $f(x) = x^3$ shifted 3 units downward.

9. The graph of $f(x) = \frac{1}{x+2}$ intersects the line $x = -2$ at the point $(-2, 0)$.

10. The graph of $f(x) = \frac{1}{x} + 2$ is the graph of $f(x) = \frac{1}{x}$ shifted 2 units upward.

Problem Set 10.3

1 Know the Basic Graphs of $f(x) = x^2$, $f(x) = x^3$, $f(x) = \frac{1}{x}$, $f(x) = \sqrt{x}$, and $f(x) = |x|$

For Problems 1–6, match the function with its graph.

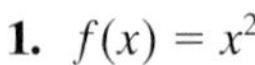

1. $f(x) = x^2$ **2.** $f(x) = x^3$

3. $f(x) = \frac{1}{x}$ **4.** $f(x) = \sqrt{x}$

5. $f(x) = |x|$ **6.** $f(x) = x$

A.

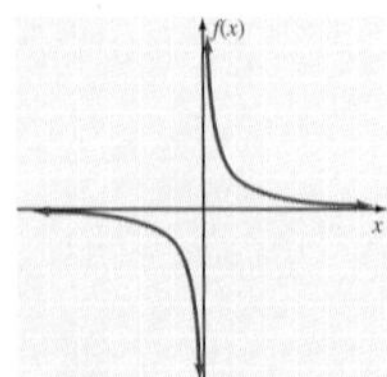

B.

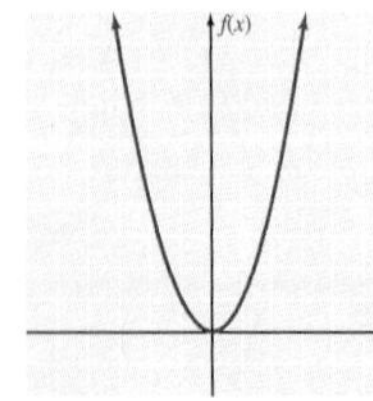

C.

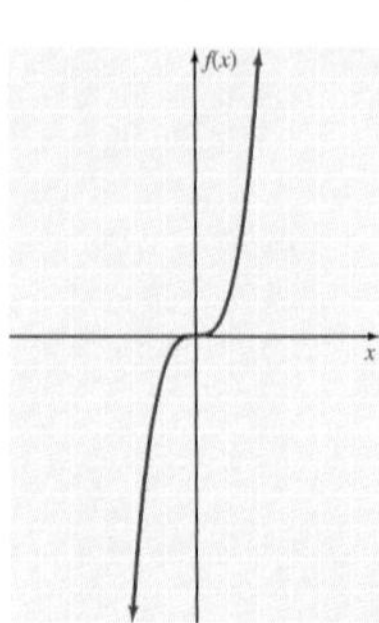

D.

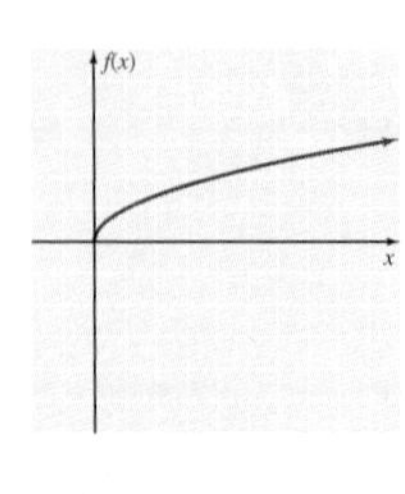

E.

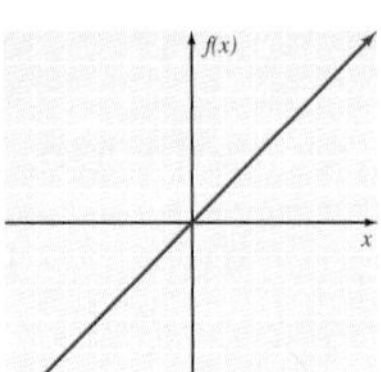

F.

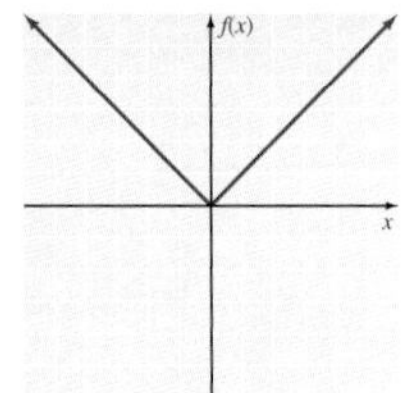

2 Graph Functions by Using Translations

For Problems 7–14, graph each of the functions.

7. $f(x) = x^3 - 2$ **8.** $f(x) = |x| + 3$

9. $f(x) = \sqrt{x} + 3$ **10.** $f(x) = \frac{1}{x} - 2$

11. $f(x) = \sqrt{x+3}$ **12.** $f(x) = \frac{1}{x-2}$

13. $f(x) = |x+1|$ **14.** $f(x) = (x-2)^2$

3 Graph Functions by Using Reflections

For Problems 15–18, graph each of the functions.

15. $f(x) = -x^3$ **16.** $f(x) = -x^2$

17. $f(x) = \sqrt{-x}$ **18.** $f(x) = -\sqrt{x}$

4 Graph Functions by Using Vertical Stretching or Shrinking

For Problems 19–22, graph each of the functions.

19. $f(x) = \frac{1}{2}|x|$

20. $f(x) = \frac{1}{4}x^2$

21. $f(x) = 2x^2$

22. $f(x) = 2\sqrt{x}$

5 Graph Functions by Using Successive Transformations

For Problems 23–50, graph each of the functions.

23. $f(x) = -(x - 4)^2 + 2$

24. $f(x) = -2(x + 3)^2 - 4$

25. $f(x) = |x - 1| + 2$

26. $f(x) = -|x + 2|$

27. $f(x) = -2\sqrt{x}$

28. $f(x) = 2\sqrt{x - 1}$

29. $f(x) = \sqrt{x + 2} - 3$

30. $f(x) = -\sqrt{x + 2} + 2$

31. $f(x) = \frac{2}{x - 1} + 3$

32. $f(x) = \frac{3}{x + 3} - 4$

33. $f(x) = \sqrt{2 - x}$

34. $f(x) = \sqrt{-1 - x}$

35. $f(x) = -3(x - 2)^2 - 1$

36. $f(x) = (x + 5)^2 - 2$

37. $f(x) = 3(x - 2)^3 - 1$

38. $f(x) = -2(x + 1)^3 + 2$

39. $f(x) = 2x^3 + 3$

40. $f(x) = -2x^3 - 1$

41. $f(x) = -2\sqrt{x + 3} + 4$

42. $f(x) = -3\sqrt{x - 1} + 2$

43. $f(x) = \frac{-2}{x + 2} + 2$

44. $f(x) = \frac{-1}{x - 1} - 1$

45. $f(x) = \frac{x - 1}{x}$

46. $f(x) = \frac{x + 2}{x}$

47. $f(x) = -3|x + 4| + 3$

48. $f(x) = -2|x - 3| - 4$

49. $f(x) = 4|x| + 2$

50. $f(x) = -3|x| - 4$

51. Suppose that the graph of $y = f(x)$ with a domain of $-2 \le x \le 2$ is shown in Figure 10.27.

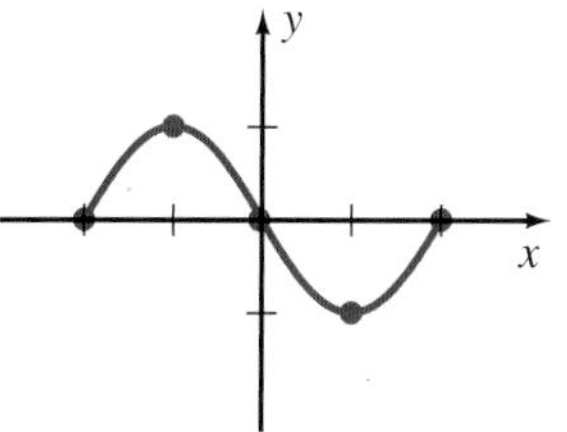

Figure 10.27

Sketch the graph of each of the following transformations of $y = f(x)$.

(a) $y = f(x) + 3$ **(b)** $y = f(x - 2)$

(c) $y = -f(x)$ **(d)** $y = f(x + 3) - 4$

52. Use the definition of absolute value to help you sketch the following graphs.

(a) $f(x) = x + |x|$ **(b)** $f(x) = x - |x|$

(c) $f(x) = |x| - x$ **(d)** $f(x) = \frac{x}{|x|}$ **(e)** $f(x) = \frac{|x|}{x}$

THOUGHTS INTO WORDS

53. Is the graph of $f(x) = x^2 + 2x + 4$ a y-axis reflection of $f(x) = x^2 - 2x + 4$? Defend your answer.

54. Is the graph of $f(x) = x^2 - 4x - 7$ an x-axis reflection of $f(x) = x^2 + 4x + 7$? Defend your answer.

55. Your friend claims that the graph of $f(x) = \frac{2x + 1}{x}$ is the graph of $f(x) = \frac{1}{x}$ shifted 2 units upward. How could you verify whether she is correct?

GRAPHING CALCULATOR ACTIVITIES

56. Use a graphing calculator to check your graphs for Problems 28–43.

57. Use a graphing calculator to check your graphs for Problem 52.

58. For each of the following, answer the question on the basis of your knowledge of transformations, and then use a graphing calculator to check your answer.

(a) Is the graph of $f(x) = 2x^2 + 8x + 13$ a y-axis reflection of $f(x) = 2x^2 - 8x + 13$?

(b) Is the graph of $f(x) = 3x^2 - 12x + 16$ an x-axis reflection of $f(x) = -3x^2 + 12x - 16$?

(c) Is the graph of $f(x) = \sqrt{4 - x}$ a y-axis reflection of $f(x) = \sqrt{x + 4}$?

(d) Is the graph of $f(x) = \sqrt{3 - x}$ a y-axis reflection of $f(x) = \sqrt{x - 3}$?

(e) Is the graph of $f(x) = -x^3 + x + 1$ a y-axis reflection of $f(x) = x^3 - x + 1$?

(f) Is the graph of $f(x) = -(x - 2)^3$ an x-axis reflection of $f(x) = (x - 2)^3$?

(g) Is the graph of $f(x) = -x^3 - x^2 - x + 1$ an x-axis reflection of $f(x) = x^3 + x^2 + x - 1$?

(h) Is the graph of $f(x) = \frac{3x + 1}{x}$ a vertical translation of $f(x) = \frac{1}{x}$ upward 3 units?

(i) Is the graph of $f(x) = 2 + \frac{1}{x}$ a y-axis reflection of $f(x) = \frac{2x - 1}{x}$?

59. Are the graphs of $f(x) = 2\sqrt{x}$ and $g(x) = \sqrt{2x}$ identical? Defend your answer.

60. Are the graphs of $f(x) = \sqrt{x + 4}$ and $g(x) = \sqrt{-x + 4}$ y-axis reflections of each other? Defend your answer.

Answers to the Concept Quiz

1. C **2.** D **3.** E **4.** B **5.** A **6.** False **7.** True **8.** True **9.** False **10.** True

Answers to the Example Practice Skills

1.

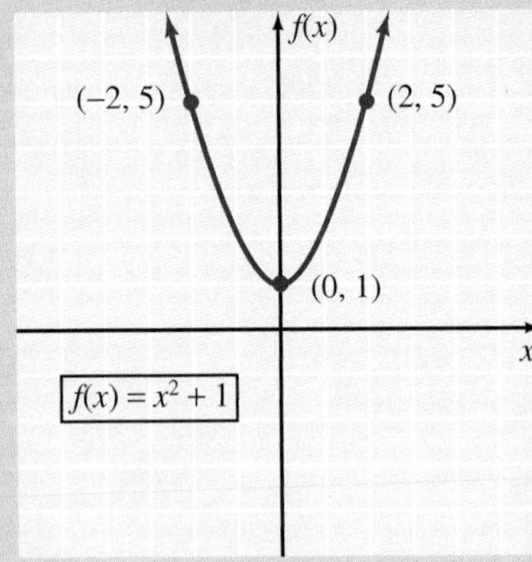

2.

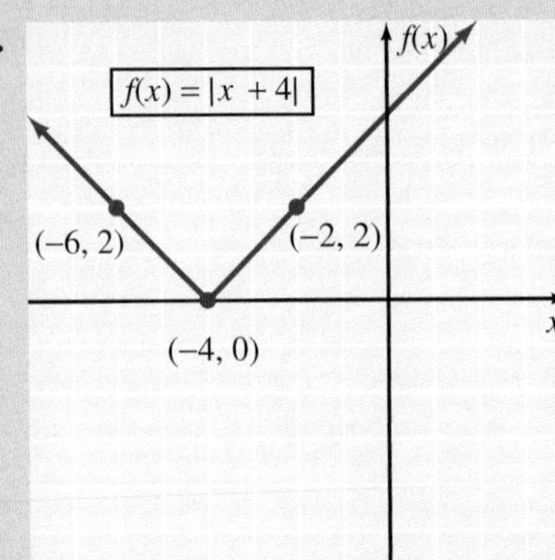

3.

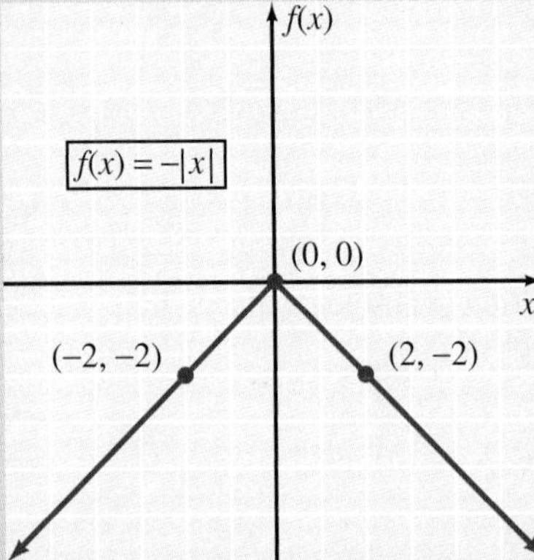

4.

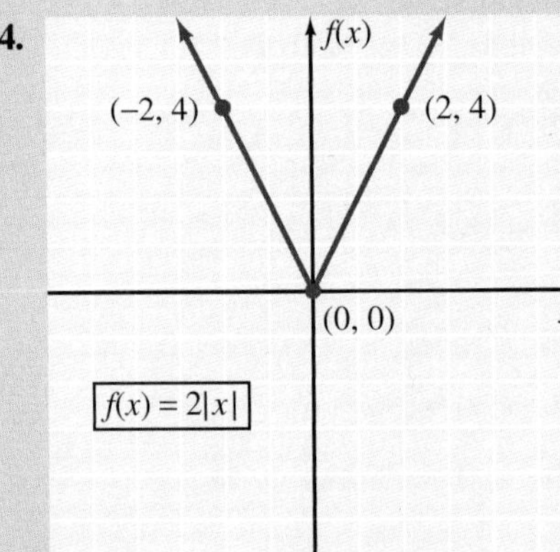

5.

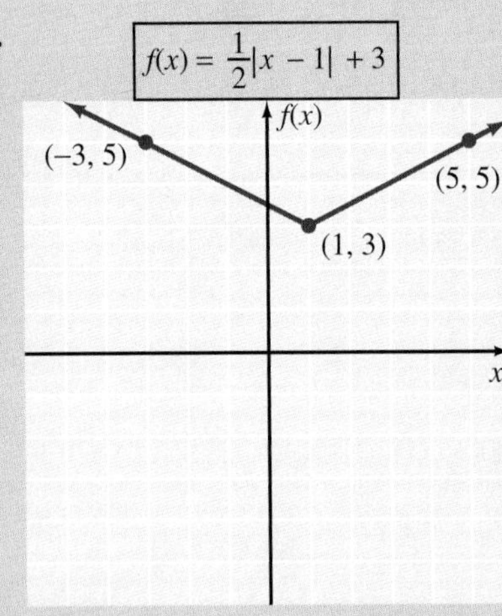

6.

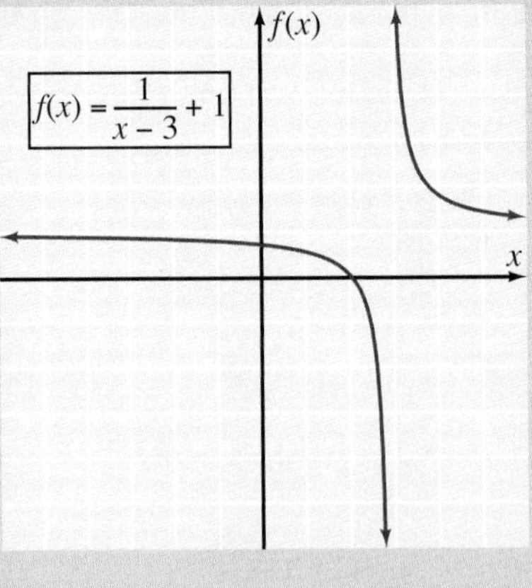

7.

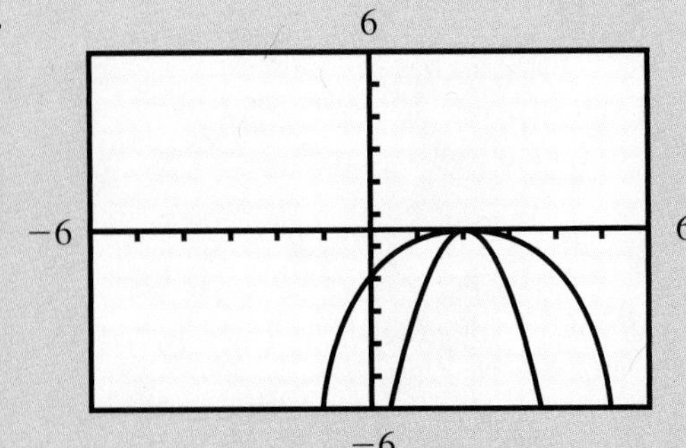

10.4 Composition of Functions

OBJECTIVES

1. Find the Composition of Two Functions and Determine the Domain
2. Determine Functional Values for Composite Functions
3. Graph a Composite Function Using a Graphing Utility

1 Find the Composition of Two Functions and Determine the Domain

The basic operations of addition, subtraction, multiplication, and division can be performed on functions. However, there is an additional operation, called **composition**, that we will use in the next section. Let's start with the definition and an illustration of this operation.

> **Definition 10.3**
>
> The **composition** of functions f and g is defined by
>
> $$(f \circ g)(x) = f(g(x))$$
>
> for all x in the domain of g such that $g(x)$ is in the domain of f.

The left side, $(f \circ g)(x)$, of the equation in Definition 10.3 can be read "the composition of f and g," and the right side, $f(g(x))$, can be read "f of g of x." It may also be helpful for you to picture Definition 10.3 as two function machines hooked together to produce another function (often called a **composite function**) as illustrated in Figure 10.28. Note that what comes out of the function g is substituted into the function f. Thus composition is sometimes called the substitution of functions.

Figure 10.28 also vividly illustrates the fact that $f \circ g$ is defined for all x in the domain of g such that $g(x)$ is in the domain of f. In other words, what comes out of g must be capable of being fed into f. Let's consider some examples.

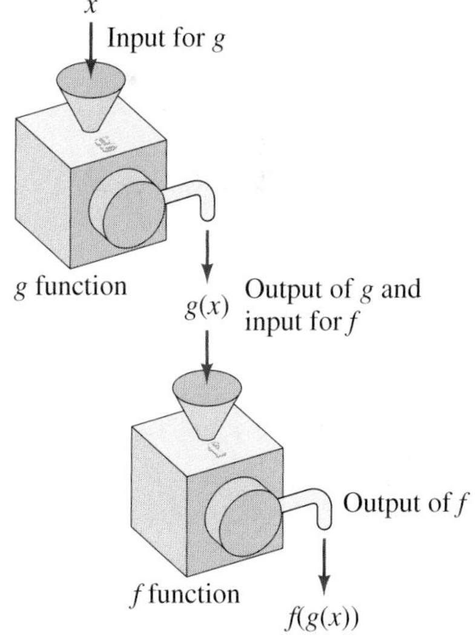

Figure 10.28

EXAMPLE 1

If $f(x) = x^2$ and $g(x) = x - 3$, find $(f \circ g)(x)$ and determine its domain.

Solution

Applying Definition 10.3, we obtain

$$\begin{aligned}(f \circ g)(x) &= f(g(x)) \\ &= f(x - 3) \\ &= (x - 3)^2\end{aligned}$$

Because g and f are both defined for all real numbers, so is $f \circ g$.

▼ PRACTICE YOUR SKILL

If $f(x) = 3x + 7$ and $g(x) = 5x + 2$, find $(f \circ g)(x)$ and determine its domain. ■

EXAMPLE 2

If $f(x) = \sqrt{x}$ and $g(x) = x - 4$, find $(f \circ g)(x)$ and determine its domain.

Solution

Applying Definition 10.3, we obtain

$$\begin{aligned}(f \circ g)(x) &= f(g(x)) \\ &= f(x - 4) \\ &= \sqrt{x - 4}\end{aligned}$$

The domain of g is all real numbers, but the domain of f is only the nonnegative real numbers. Thus $g(x)$, which is $x - 4$, must be nonnegative. Therefore,

$$\begin{aligned}x - 4 &\geq 0 \\ x &\geq 4\end{aligned}$$

and the domain of $f \circ g$ is $D = \{x \mid x \geq 4\}$ or D: $[4, \infty)$.

▼ PRACTICE YOUR SKILL

If $f(x) = \sqrt{x}$ and $g(x) = x + 2$, find $(f \circ g)(x)$ and determine its domain. ■

Definition 10.3, with f and g interchanged, defines the composition of g and f as $(g \circ f)(x) = g(f(x))$.

EXAMPLE 3

If $f(x) = x^2$ and $g(x) = x - 3$, find $(g \circ f)(x)$ and determine its domain.

Solution

$$\begin{aligned}(g \circ f)(x) &= g(f(x)) \\ &= g(x^2) \\ &= x^2 - 3\end{aligned}$$

Because f and g are both defined for all real numbers, the domain of $g \circ f$ is the set of all real numbers.

▼ PRACTICE YOUR SKILL

If $f(x) = 3x + 7$ and $g(x) = 5x + 2$, find $(g \circ f)(x)$ and determine its domain. ■

The results of Examples 1 and 3 demonstrate an important idea: the composition of functions is *not a commutative operation.* In other words, it is not true that $f \circ g = g \circ f$ for all functions f and g. However, as we will see in the next section, there is a special class of functions where $f \circ g = g \circ f$.

EXAMPLE 4

If $f(x) = \dfrac{2}{x-1}$ and $g(x) = \dfrac{1}{x}$, find $(f \circ g)(x)$ and $(g \circ f)(x)$. Determine the domain for each composite function.

Solution

$$
\begin{aligned}
(f \circ g)(x) &= f(g(x)) \\
&= f\left(\frac{1}{x}\right) \\
&= \frac{2}{\frac{1}{x} - 1} = \frac{2}{\frac{1-x}{x}} \\
&= \frac{2x}{1-x}
\end{aligned}
$$

The domain of g is all real numbers except 0, and the domain of f is all real numbers except 1. Because $g(x)$, which is $\frac{1}{x}$, cannot equal 1, we have

$$\frac{1}{x} \neq 1$$

$$x \neq 1$$

Therefore, the domain of $f \circ g$ is $D = \{x \mid x \neq 0 \text{ and } x \neq 1\}$ or $D: (-\infty, 0) \cup (0, 1) \cup (1, \infty)$.

$$
\begin{aligned}
(g \circ f)(x) &= g(f(x)) \\
&= g\left(\frac{2}{x-1}\right) \\
&= \frac{1}{\frac{2}{x-1}} \\
&= \frac{x-1}{2}
\end{aligned}
$$

The domain of f is all real numbers except 1, and the domain of g is all real numbers except 0. Because $f(x)$, which is $\frac{2}{x-1}$, will never equal 0, it follows that the domain of $g \circ f$ is $D = \{x | x \neq 1\}$ or $D: (-\infty, 1) \cup (1, \infty)$.

▼ **PRACTICE YOUR SKILL**

If $f(x) = \dfrac{5}{x+3}$ and $g(x) = \dfrac{1}{x}$, find $(f \circ g)(x)$ and $(g \circ f)(x)$. Determine the domain for each composite function. ■

2 Determine Functional Values for Composite Functions

Composite functions can be evaluated for values of x in the domain. In the next example, the composite function is formed, and then functional values are determined for the composite function.

EXAMPLE 5 If $f(x) = 2x + 3$ and $g(x) = \sqrt{x - 1}$, determine each of the following.

(a) $(f \circ g)(x)$ **(b)** $(g \circ f)(x)$ **(c)** $(f \circ g)(5)$ **(d)** $(g \circ f)(7)$

Solution

(a) $$\begin{aligned}(f \circ g)(x) &= f(g(x)) \\ &= f(\sqrt{x - 1}) \\ &= 2\sqrt{x - 1} + 3\end{aligned}$$

(b) $$\begin{aligned}(g \circ f)(x) &= g(f(x)) \\ &= g(2x + 3) \\ &= \sqrt{2x + 3 - 1} \\ &= \sqrt{2x + 2}\end{aligned}$$

(c) $(f \circ g)(5) = 2\sqrt{5 - 1} + 3 = 7$

(d) $(g \circ f)(7) = \sqrt{2(7) + 2} = 4$

▼ **PRACTICE YOUR SKILL**

If $f(x) = 3x + 4$ and $g(x) = \sqrt{x + 2}$, determine each of the following.

(a) $(f \circ g)(x)$ **(b)** $(g \circ f)(x)$ **(c)** $(f \circ g)(-1)$ **(d)** $(g \circ f)(10)$ ■

3 Graph a Composite Function Using a Graphing Utility

A graphing utility can be used to find the graph of a composite function without actually forming the function algebraically. Let's see how this works.

EXAMPLE 6 If $f(x) = x^3$ and $g(x) = x - 4$, use a graphing utility to obtain the graph of $y = (f \circ g)(x)$ and the graph of $y = (g \circ f)(x)$.

Solution

To find the graph of $y = (f \circ g)(x)$, we can make the following assignments.

$$Y_1 = x - 4$$
$$Y_2 = (Y_1)^3$$

(Note that we have substituted Y_1 for x in $f(x)$ and assigned this expression to Y_2, much the same way as we would algebraically.) Now, by showing only the graph of Y_2, we obtain Figure 10.29.

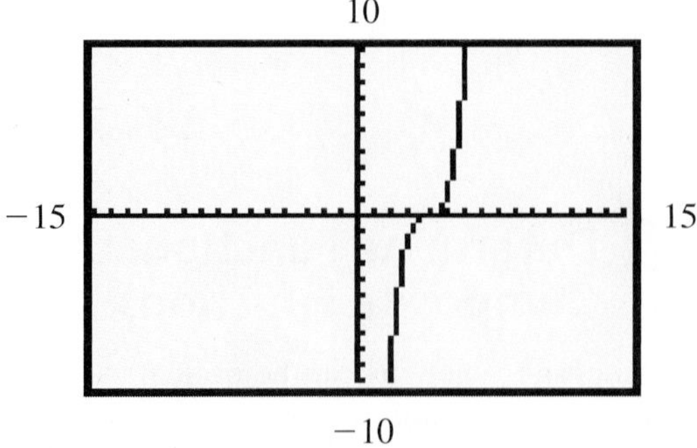

Figure 10.29

To find the graph of $y = (g \circ f)(x)$, we can make the following assignments.

$$Y_1 = x^3$$
$$Y_2 = Y_1 - 4$$

The graph of $y = (g \circ f)(x)$ is the graph of Y_2, as shown in Figure 10.30.

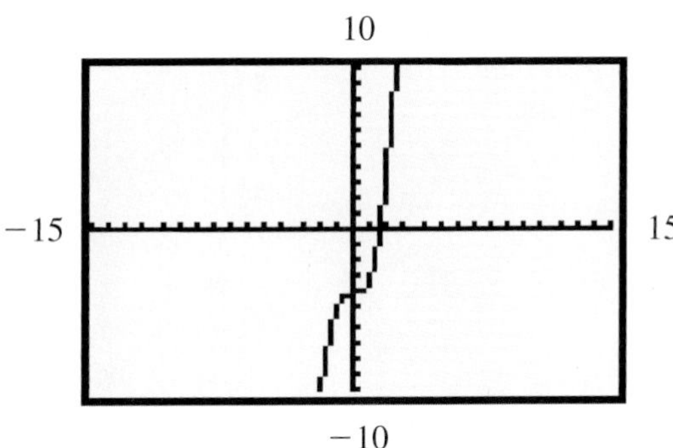

Figure 10.30

▼ PRACTICE YOUR SKILL

If $f(x) = x^2$ and $g(x) = 3x - 1$, use a graphing utility to obtain the graph of $y = (f \circ g)(x)$ and the graph of $y = (g \circ f)(x)$. ■

Take another look at Figures 10.29 and 10.30. Note that in Figure 10.29 the graph of $y = (f \circ g)(x)$ is the basic cubic curve $f(x) = x^3$ shifted 4 units to the right. Likewise, in Figure 10.30 the graph of $y = (g \circ f)(x)$ is the basic cubic curve shifted 4 units downward. These are examples of a more general concept of using composite functions to represent various geometric transformations.

CONCEPT QUIZ

For Problems 1–10, answer true or false.

1. The composition of functions is a commutative operation.
2. To find $(h \circ k)(x)$, the function k will be substituted into the function h.
3. The notation $(f \circ g)(x)$ is read as "the substitution of g and f."
4. The domain for $(f \circ g)(x)$ is always the same as the domain of g.
5. The notation $f(g(x))$ is read "f of g of x."
6. If $f(x) = x + 2$ and $g(x) = 3x - 1$, then $f(g(2)) = 7$.
7. If $f(x) = x + 2$ and $g(x) = 3x - 1$, then $g(f(2)) = 7$.
8. If $f(x) = \sqrt{x - 1}$ and $g(x) = 2x - 3$, then $f(g(1))$ is undefined.
9. If $f(x) = \sqrt{x - 1}$ and $g(x) = 2x - 3$, then $g(f(1))$ is undefined.
10. If $f(x) = -x^2 - x + 2$ and $g(x) = x + 1$, then $f(g(x)) = x^2 + 2x + 1$.

Problem Set 10.4

1 Find the Composition of Two Functions and Determine the Domain

For Problems 1–18, determine $(f \circ g)(x)$ and $(g \circ f)(x)$ for each pair of functions. Also specify the domain of $(f \circ g)(x)$ and $(g \circ f)(x)$.

1. $f(x) = 3x$ and $g(x) = 5x - 1$
2. $f(x) = 4x - 3$ and $g(x) = -2x$
3. $f(x) = -2x + 1$ and $g(x) = 7x + 4$
4. $f(x) = 6x - 5$ and $g(x) = -x + 6$
5. $f(x) = 3x + 2$ and $g(x) = x^2 + 3$
6. $f(x) = -2x + 4$ and $g(x) = 2x^2 - 1$
7. $f(x) = 2x^2 - x + 2$ and $g(x) = -x + 3$
8. $f(x) = 3x^2 - 2x - 4$ and $g(x) = -2x + 1$
9. $f(x) = \dfrac{3}{x}$ and $g(x) = 4x - 9$

10. $f(x) = -\frac{2}{x}$ and $g(x) = -3x + 6$

11. $f(x) = \sqrt{x + 1}$ and $g(x) = 5x + 3$

12. $f(x) = 7x - 2$ and $g(x) = \sqrt{2x - 1}$

13. $f(x) = \frac{1}{x}$ and $g(x) = \frac{1}{x - 4}$

14. $f(x) = \frac{2}{x + 3}$ and $g(x) = -\frac{3}{x}$

15. $f(x) = \sqrt{x}$ and $g(x) = \frac{4}{x}$

16. $f(x) = \frac{2}{x}$ and $g(x) = |x|$

17. $f(x) = \frac{3}{2x}$ and $g(x) = \frac{1}{x + 1}$

18. $f(x) = \frac{4}{x - 2}$ and $g(x) = \frac{3}{4x}$

For Problems 19–26, show that $(f \circ g)(x) = x$ and $(g \circ f)(x) = x$ for each pair of functions.

19. $f(x) = 3x$ and $g(x) = \frac{1}{3}x$

20. $f(x) = -2x$ and $g(x) = -\frac{1}{2}x$

21. $f(x) = 4x + 2$ and $g(x) = \frac{x - 2}{4}$

22. $f(x) = 3x - 7$ and $g(x) = \frac{x + 7}{3}$

23. $f(x) = \frac{1}{2}x + \frac{3}{4}$ and $g(x) = \frac{4x - 3}{2}$

24. $f(x) = \frac{2}{3}x - \frac{1}{5}$ and $g(x) = \frac{3}{2}x + \frac{3}{10}$

25. $f(x) = -\frac{1}{4}x - \frac{1}{2}$ and $g(x) = -4x - 2$

26. $f(x) = -\frac{3}{4}x + \frac{1}{3}$ and $g(x) = -\frac{4}{3}x + \frac{4}{9}$

2 Determine Functional Values for Composite Functions

For Problems 27–38, determine the indicated functional values.

27. If $f(x) = 9x - 2$ and $g(x) = -4x + 6$, find $(f \circ g)(-2)$ and $(g \circ f)(4)$.

28. If $f(x) = -2x - 6$ and $g(x) = 3x + 10$, find $(f \circ g)(5)$ and $(g \circ f)(-3)$.

29. If $f(x) = 4x^2 - 1$ and $g(x) = 4x + 5$, find $(f \circ g)(1)$ and $(g \circ f)(4)$.

30. If $f(x) = -5x + 2$ and $g(x) = -3x^2 + 4$, find $(f \circ g)(-2)$ and $(g \circ f)(-1)$.

31. If $f(x) = \frac{1}{x}$ and $g(x) = \frac{2}{x - 1}$, find $(f \circ g)(2)$ and $(g \circ f)(-1)$.

32. If $f(x) = \frac{2}{x - 1}$ and $g(x) = -\frac{3}{x}$, find $(f \circ g)(1)$ and $(g \circ f)(-1)$.

33. If $f(x) = \frac{1}{x - 2}$ and $g(x) = \frac{4}{x - 1}$, find $(f \circ g)(3)$ and $(g \circ f)(2)$.

34. If $f(x) = \sqrt{x + 6}$ and $g(x) = 3x - 1$, find $(f \circ g)(-2)$ and $(g \circ f)(-2)$.

35. If $f(x) = \sqrt{3x - 2}$ and $g(x) = -x + 4$, find $(f \circ g)(1)$ and $(g \circ f)(6)$.

36. If $f(x) = -5x + 1$ and $g(x) = \sqrt{4x + 1}$, find $(f \circ g)(6)$ and $(g \circ f)(-1)$.

37. If $f(x) = |4x - 5|$ and $g(x) = x^3$, find $(f \circ g)(-2)$ and $(g \circ f)(2)$.

38. If $f(x) = -x^3$ and $g(x) = |2x + 4|$, find $(f \circ g)(-1)$ and $(g \circ f)(-3)$.

GRAPHING CALCULATOR ACTIVITIES

3 Graph a Composite Function Using a Graphing Utility

39. For each of the following, use your graphing calculator to find the graph of $y = (f \circ g)(x)$ and $y = (g \circ f)(x)$. Then algebraically find $(f \circ g)(x)$ and $(g \circ f)(x)$ to see whether your results agree.

(a) $f(x) = x^2$ and $g(x) = x - 3$
(b) $f(x) = x^3$ and $g(x) = x + 4$
(c) $f(x) = x - 2$ and $g(x) = -x^3$
(d) $f(x) = x + 6$ and $g(x) = \sqrt{x}$
(e) $f(x) = \sqrt{x}$ and $g(x) = x - 5$

THOUGHTS INTO WORDS

40. How would you explain the concept of composition of functions to a friend who missed class the day it was discussed?

41. Explain why the composition of functions is not a commutative operation.

Answers to the Concept Quiz

1. False **2.** True **3.** False **4.** False **5.** True **6.** True **7.** False **8.** True **9.** False **10.** False

Answers to the Example Practice Skills

1. $(f \circ g)(x) = 15x + 13, D = \{\text{all reals}\}$ **2.** $(f \circ g)(x) = \sqrt{x+2}, D = \{x|x \geq -2\}$ or $D = [-2, \infty)$

3. $(f \circ g)(x) = 15x + 37, D = \{\text{all reals}\}$ **4.** $(f \circ g)(x) = \dfrac{5x}{3x+1}, D = \left\{x|x \neq -\dfrac{1}{3} \text{ and } x \neq 0\right\}$, or $D = \left(-\infty, -\dfrac{1}{3}\right) \cup \left(-\dfrac{1}{3}, 0\right) \cup (0, \infty)$, $(g \circ f)(x) = \dfrac{x+3}{5}, D = \{x|x \neq -3\}$ or $D = (-\infty, -3) \cup (-3, \infty)$

5. **(a)** $(f \circ g)(x) = 3\sqrt{x+2} + 4$ **(b)** $(g \circ f)(x) = \sqrt{3x+6}$ **(c)** $(f \circ g)(-1) = 7$ **(d)** $(g \circ f)(10) = 6$

6.

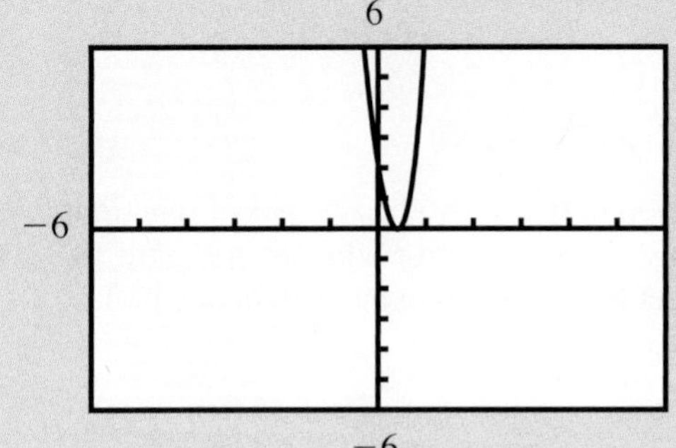

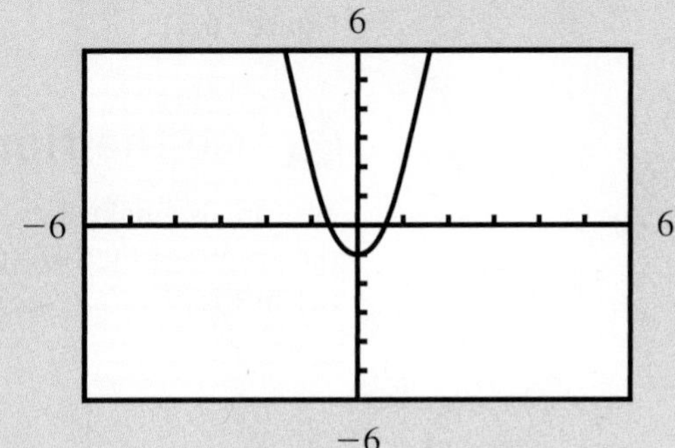

10.5 Inverse Functions

OBJECTIVES

1. Use the Vertical Line Test
2. Use the Horizontal Line Test
3. Find the Inverse Function in Terms of Ordered Pairs
4. Find the Inverse of a Function

1 Use the Vertical Line Test

Graphically, the distinction between a relation and a function can be easily recognized. In Figure 10.31, we sketched four graphs. Which of these are graphs of functions and which are graphs of relations that are not functions? Think in terms of the principle that to each member of the domain there is assigned one and only one member of the range; this is the basis for what is known as the **vertical line test** for functions. Because each value of x produces only one value of $f(x)$, any vertical line drawn through a graph of a function must not intersect the graph in more than one point. Therefore, parts (a) and (c) of Figure 10.31 are graphs of functions, whereas parts (b) and (d) are graphs of relations that are not functions.

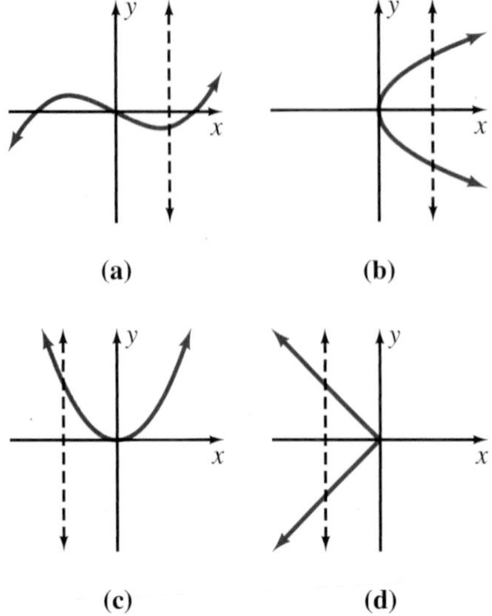

Figure 10.31

2 Use the Horizontal Line Test

We can also make a useful distinction between two basic types of functions. Consider the graphs of the two functions $f(x) = 2x - 1$ and $f(x) = x^2$ in Figure 10.32. In part (a), any *horizontal line* will intersect the graph in no more than one point.

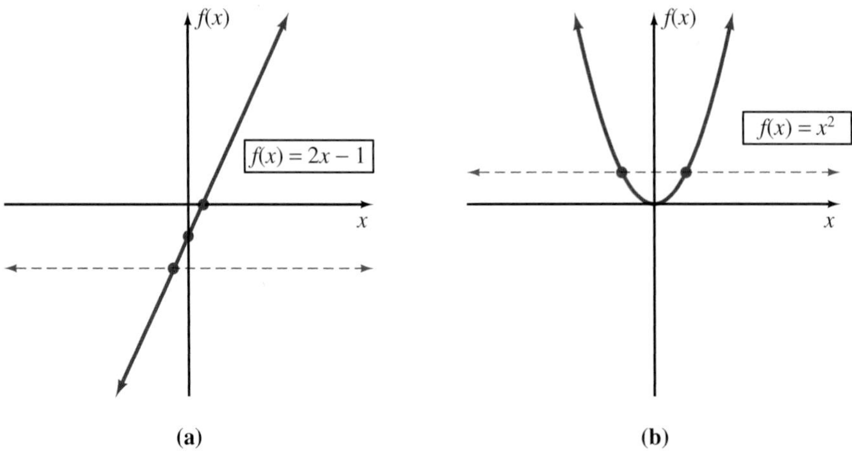

Figure 10.32

Therefore, every value of $f(x)$ has only one value of x associated with it. Any function that has the additional property of having only one value of x associated with each value of $f(x)$ is called a **one-to-one function**. The function $f(x) = x^2$ is not a one-to-one function because the horizontal line in part (b) of Figure 10.32 intersects the parabola in two points.

In terms of ordered pairs, a one-to-one function does not contain any ordered pairs that have the same second component. For example,

$$f = \{(1, 3), (2, 6), (4, 12)\}$$

is a one-to-one function, but

$$g = \{(1, 2), (2, 5), (-2, 5)\}$$

is not a one-to-one function.

3 Find the Inverse Function in Terms of Ordered Pairs

If the components of each ordered pair of a given one-to-one function are interchanged, then the resulting function and the given function are called **inverses** of each other. Thus

$$\{(1, 3), (2, 6), (4, 12)\} \quad \text{and} \quad \{(3, 1), (6, 2), (12, 4)\}$$

are **inverse functions**. The inverse of a function f is denoted by f^{-1} (which is read "f inverse" or "the inverse of f"). If (a, b) is an ordered pair of f, then (b, a) is an ordered pair of f^{-1}. The domain and range of f^{-1} are the range and domain, respectively, of f.

Remark: Do not confuse the -1 in f^{-1} with a negative exponent. The symbol f^{-1} does not mean $\frac{1}{f^1}$ but rather refers to the inverse function of function f.

EXAMPLE 1

For the function $f = \{(1, 4), (6, 2), (8, 5), (9, 7)\}$: **(a)** list the domain and range of the function; **(b)** form the inverse function; and **(c)** list the domain and range of the inverse function.

Solution

The domain is the set of all first components of the ordered pairs and the range is the set of all second components of the ordered pairs.

$$D = \{1, 6, 8, 9\} \quad \text{and} \quad R = \{2, 4, 5, 7\}$$

The inverse function is found by interchanging the components of the ordered pairs.

$$f^{-1} = \{(4, 1), (2, 6), (5, 8), (7, 9)\}$$

The domain for f^{-1} is $D = \{2, 4, 5, 7\}$ and the range for f^{-1} is $R = \{1, 6, 8, 9\}$.

▼ **PRACTICE YOUR SKILL**

For the function $f = \{(0, 3), (1, 4), (5, 6), (7, 9)\}$: **(a)** list the domain and range of the function; **(b)** form the inverse function; and **(c)** list the domain and range of the inverse function. ■

Graphically, two functions that are inverses of each other are mirror images with reference to the line $y = x$. This is due to the fact that ordered pairs (a, b) and (b, a) are mirror images with respect to the line $y = x$, as illustrated in Figure 10.33.

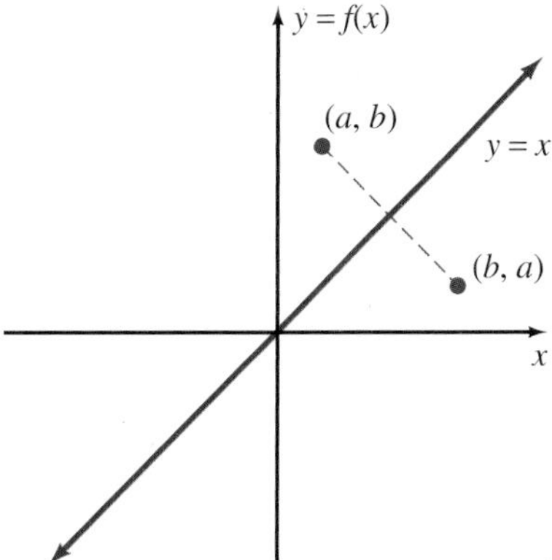

Figure 10.33

Therefore, if we know the graph of a function f, as in Figure 10.34(a), then we can determine the graph of f^{-1} by reflecting f across the line $y = x$ (Figure 10.34b).

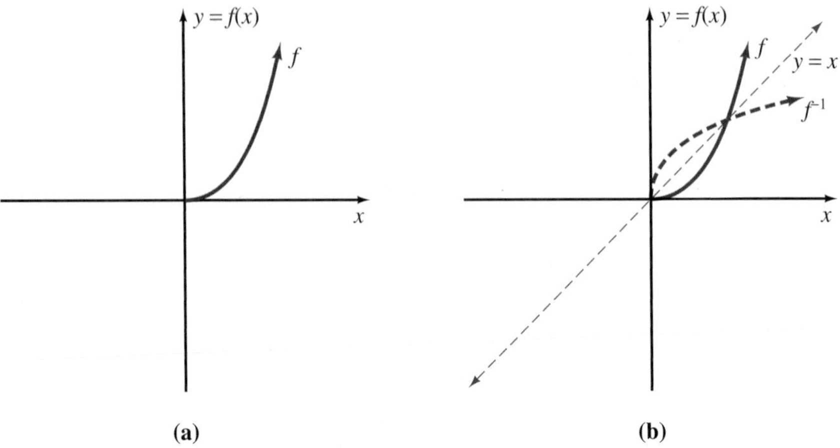

Figure 10.34

Another useful way of viewing inverse functions is in terms of composition. Basically, inverse functions *undo* each other, and this can be more formally stated as follows: If f and g are inverses of each other, then

1. $(f \circ g)(x) = f(g(x)) = x$ for all x in domain of g
2. $(g \circ f)(x) = g(f(x)) = x$ for all x in domain of f

As we will see in a moment, this relationship of inverse functions can be used to verify whether two functions are indeed inverses of each other.

4 Find the Inverse of a Function

The idea of inverse functions undoing each other provides the basis for a rather informal approach to finding the inverse of a function. Consider the function

$$f(x) = 2x + 1$$

To each x this function assigns *twice x plus 1*. To undo this function, we could *subtract 1 and divide by 2*. Thus the inverse should be

$$f^{-1}(x) = \frac{x - 1}{2}$$

Now let's verify that f and f^{-1} are inverses of each other.

$$(f \circ f^{-1})(x) = f(f^{-1}(x)) = f\left(\frac{x - 1}{2}\right) = 2\left(\frac{x - 1}{2}\right) + 1 = x$$

and

$$(f^{-1} \circ f)(x) = f^{-1}(f(x)) = f^{-1}(2x + 1) = \frac{2x + 1 - 1}{2} = x$$

Thus the inverse of f is given by

$$f^{-1}(x) = \frac{x - 1}{2}$$

Let's consider another example of finding an inverse function by the undoing process.

EXAMPLE 2 Find the inverse of $f(x) = 3x - 5$.

Solution

To each x, the function f assigns *three times x minus 5*. To undo this, we can *add 5 and then divide by 3*, so the inverse should be

$$f^{-1}(x) = \frac{x+5}{3}$$

To verify that f and f^{-1} are inverses, we can show that

$$(f \circ f^{-1})(x) = f(f^{-1}(x)) = f\left(\frac{x+5}{3}\right)$$
$$= 3\left(\frac{x+5}{3}\right) - 5 = x$$

and

$$(f^{-1} \circ f)(x) = f^{-1}(f(x)) = f^{-1}(3x - 5)$$
$$= \frac{3x - 5 + 5}{3} = x$$

Thus f and f^{-1} are inverses, and we can write

$$f^{-1}(x) = \frac{x+5}{3}$$

▼ **PRACTICE YOUR SKILL**

Find the inverse of $f(x) = 2x + 7$. ■

This informal approach may not work very well with more complex functions, but it does emphasize how inverse functions are related to each other. A more formal and systematic technique for finding the inverse of a function can be described as follows:

1. Replace the symbol $f(x)$ by y.
2. Interchange x and y.
3. Solve the equation for y in terms of x.
4. Replace y by the symbol $f^{-1}(x)$.

Now let's use two examples to illustrate this technique.

EXAMPLE 3 Find the inverse of $f(x) = -3x + 11$.

Solution

When we replace $f(x)$ by y, the given equation becomes

$$y = -3x + 11$$

Interchanging x and y produces

$$x = -3y + 11$$

Now, solving for y yields

$$x = -3y + 11$$
$$3y = -x + 11$$
$$y = \frac{-x + 11}{3}$$

Finally, replacing y by $f^{-1}(x)$, we can express the inverse function as

$$f^{-1}(x) = \frac{-x + 11}{3}$$

▼ **PRACTICE YOUR SKILL**

Find the inverse of $f(x) = -6x - 5$. ■

EXAMPLE 4

Find the inverse of $f(x) = \frac{3}{2}x - \frac{1}{4}$.

Solution

When we replace $f(x)$ by y, the given equation becomes

$$y = \frac{3}{2}x - \frac{1}{4}$$

Interchanging x and y produces

$$x = \frac{3}{2}y - \frac{1}{4}$$

Now, solving for y yields

$$x = \frac{3}{2}y - \frac{1}{4}$$

$$4(x) = 4\left(\frac{3}{2}y - \frac{1}{4}\right) \quad \text{Multiply both sides by the LCD}$$

$$4x = 4\left(\frac{3}{2}y\right) - 4\left(\frac{1}{4}\right)$$

$$4x = 6y - 1$$

$$4x + 1 = 6y$$

$$\frac{4x + 1}{6} = y$$

$$\frac{2}{3}x + \frac{1}{6} = y$$

Finally, replacing y by $f^{-1}(x)$, we can express the inverse function as

$$f^{-1}(x) = \frac{2}{3}x + \frac{1}{6}$$

▼ **PRACTICE YOUR SKILL**

Find the inverse of $f(x) = \frac{2}{5}x + \frac{1}{3}$. ■

For both Examples 3 and 4, you should be able to show that $(f \circ f^{-1})(x) = x$ and $(f^{-1} \circ f)(x) = x$.

CONCEPT QUIZ

For Problems 1–10, answer true or false.

1. If a horizontal line intersects the graph of a function in exactly two points, then the function is said to be one-to-one.
2. The notation f^{-1} refers to the inverse of function f.
3. The graph of two functions that are inverses of each other are mirror images with reference to the y axis.
4. If $g = \{(1, 3), (5, 9)\}$, then $g^{-1} = \{(3, 1), (9, 5)\}$.

5. If f and g are inverse functions, then the range of f is the domain of g.
6. The functions $f(x) = x + 1$ and $g(x) = x - 1$ are inverse functions.
7. The functions $f(x) = 2x$ and $g(x) = -2x$ are inverse functions.
8. The functions $f(x) = 2x - 7$ and $g(x) = \dfrac{x + 7}{2}$ are inverse functions.
9. The function $f(x) = -x$ has no inverse.
10. The function $f(x) = x^2 + 4$ for x any real number has no inverse.

Problem Set 10.5

1 Use the Vertical Line Test

For Problems 1–8, use the vertical line test to identify each graph as (a) the graph of a function or (b) the graph of a relation that is not a function.

1.

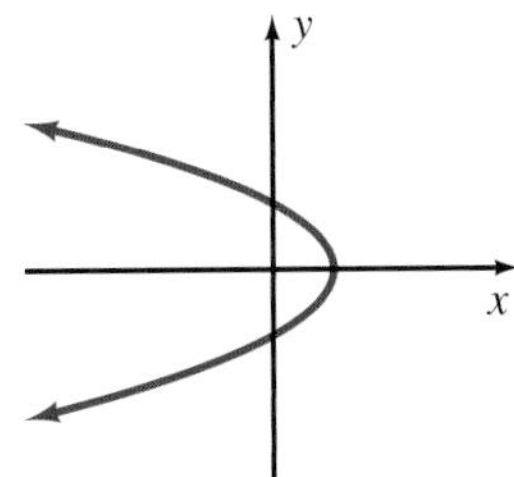

2.

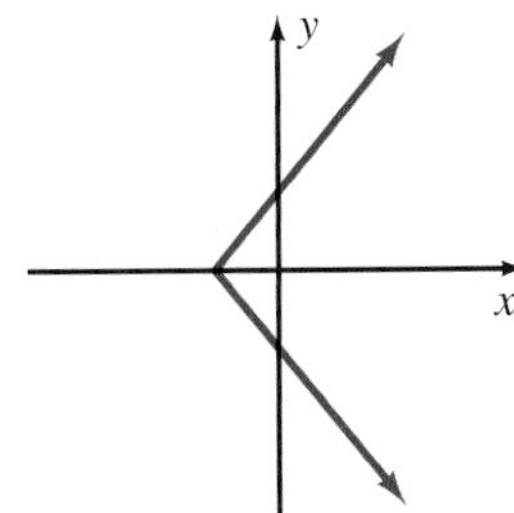

3.

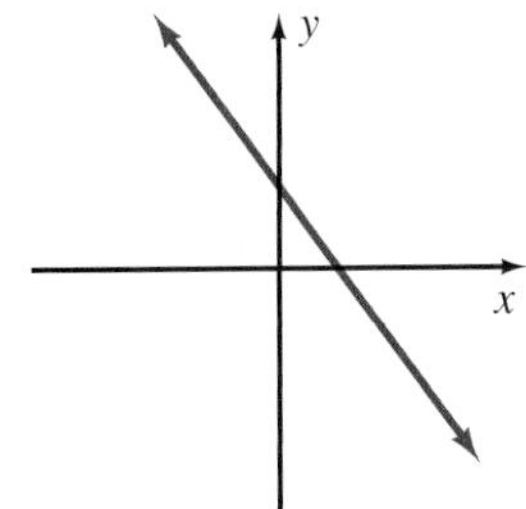

4.

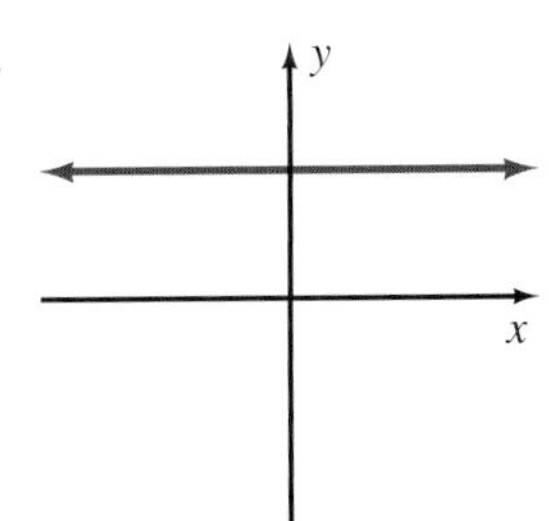

5.

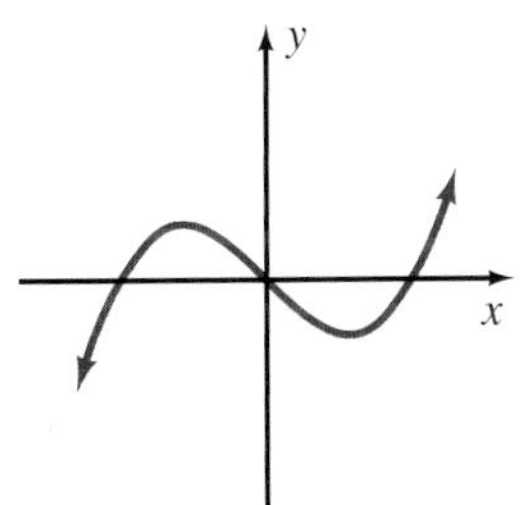

6.

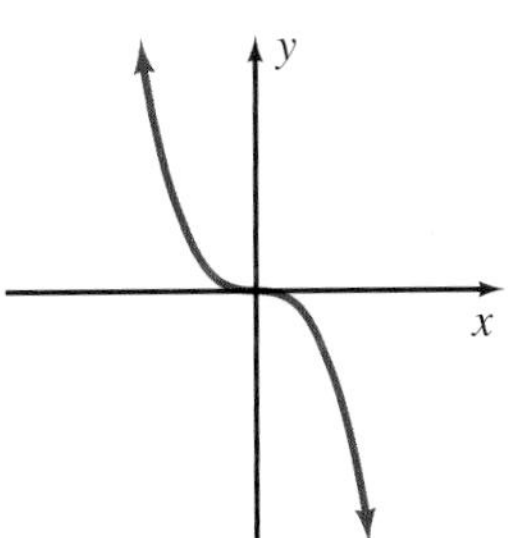

7.

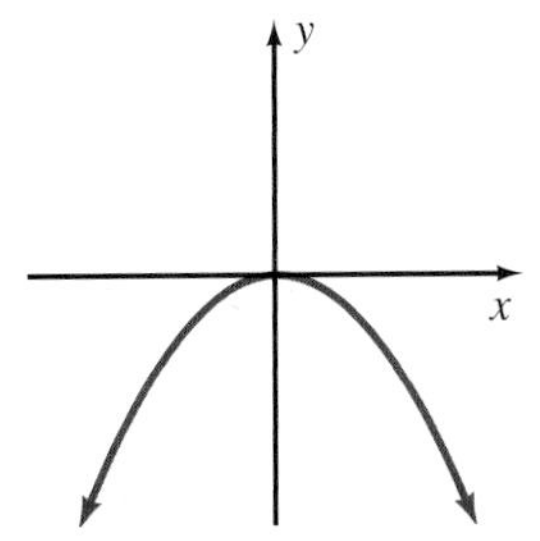

8.

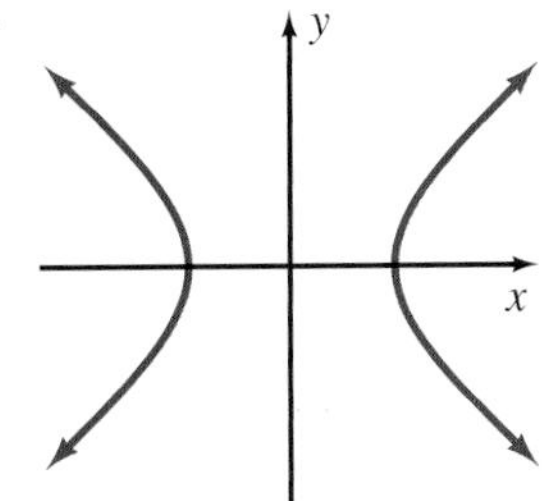

2 Use the Horizontal Line Test

For Problems 9–16, identify each graph as (a) the graph of a one-to-one function or (b) the graph of a function that is not one-to-one. Use the horizontal line test.

9. 10.

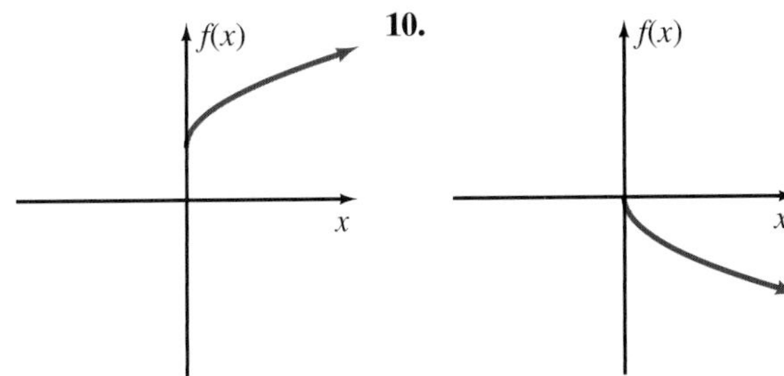

11.

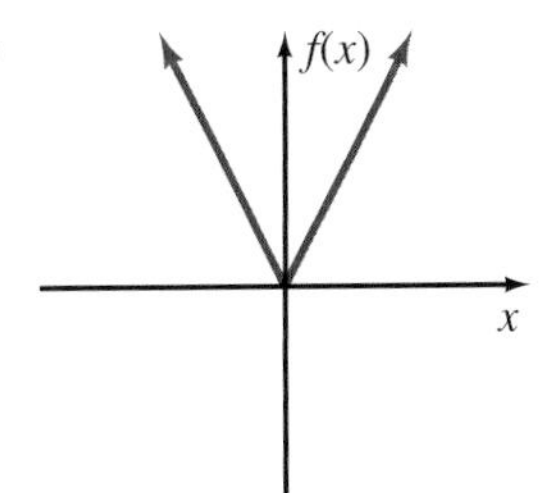

12.

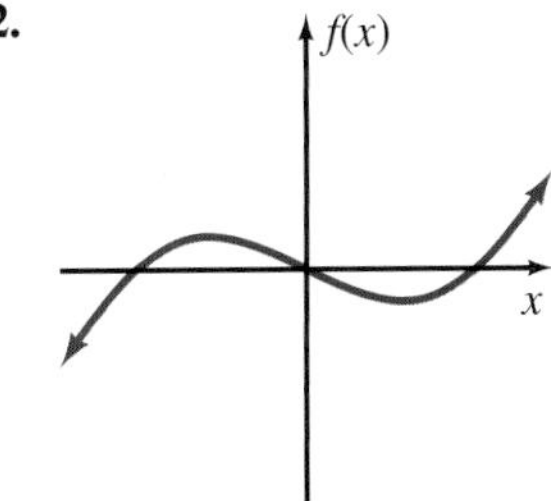

13.

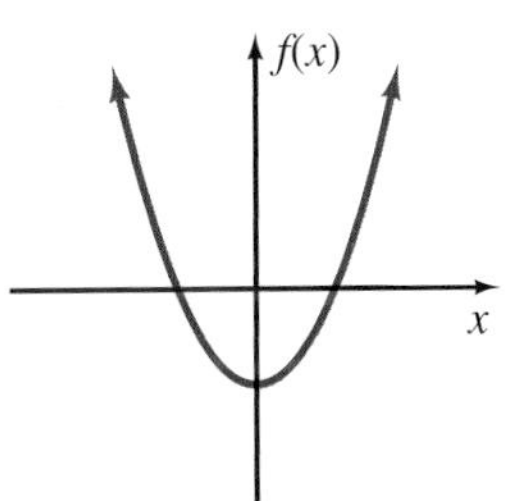

14.

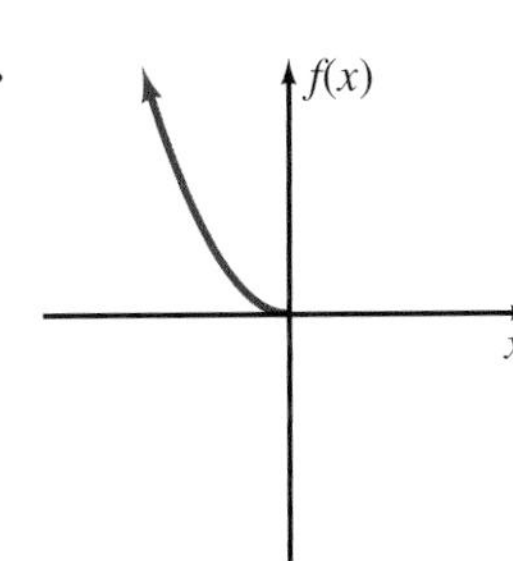

15.

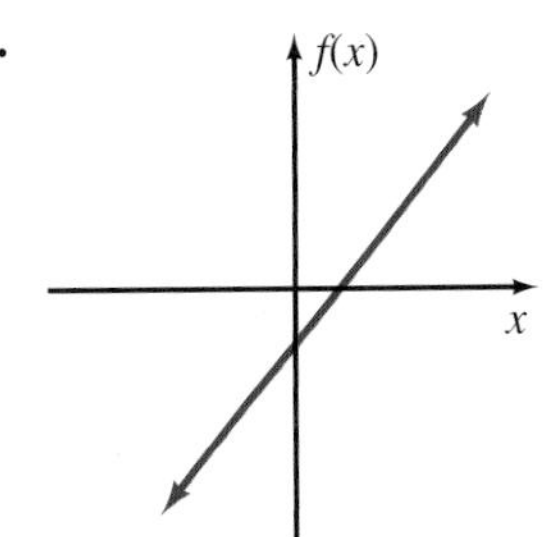

16.

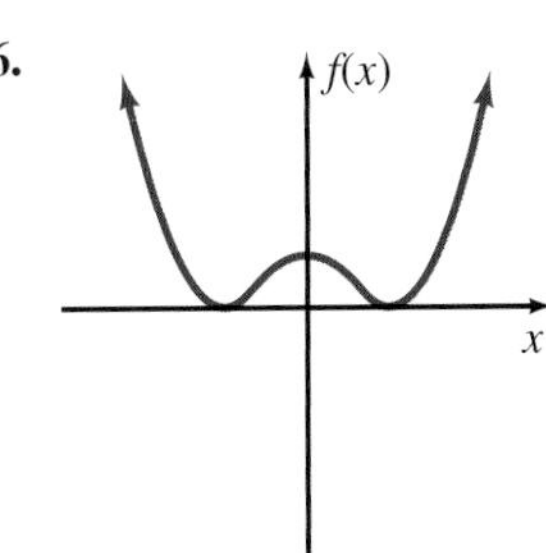

3 Find the Inverse Function in Terms of Ordered Pairs

For Problems 17–20, (a) list the domain and range of the given function, (b) form the inverse function, and (c) list the domain and range of the inverse function.

17. $f = \{(1, 3), (2, 6), (3, 11), (4, 18)\}$

18. $f = \{(0, -4), (1, -3), (4, -2)\}$

19. $f = \{(-2, -1), (-1, 1), (0, 5), (5, 10)\}$

20. $f = \{(-1, 1), (-2, 4), (1, 9), (2, 12)\}$

4 Find the Inverse of a Function

For Problems 21–30, find the inverse of the given function by using the "undoing process," and then verify that $(f \circ f^{-1})(x) = x$ and $(f^{-1} \circ f)(x) = x$.

21. $f(x) = 5x - 4$

22. $f(x) = 7x + 9$

23. $f(x) = -2x + 1$

24. $f(x) = -4x - 3$

25. $f(x) = \frac{4}{5}x$

26. $f(x) = -\frac{2}{3}x$

27. $f(x) = \frac{1}{2}x + 4$

28. $f(x) = \frac{3}{4}x - 2$

29. $f(x) = \frac{1}{3}x - \frac{2}{5}$

30. $f(x) = \frac{2}{5}x + \frac{1}{4}$

For Problems 31–40, find the inverse of the given function by using the process illustrated in Examples 3 and 4 of this section, and then verify that $(f \circ f^{-1})(x) = x$ and $(f^{-1} \circ f)(x) = x$.

31. $f(x) = 9x + 4$

32. $f(x) = 8x - 5$

33. $f(x) = -5x - 4$

34. $f(x) = -6x + 2$

35. $f(x) = -\frac{2}{3}x + 7$

36. $f(x) = -\frac{3}{5}x + 1$

37. $f(x) = \frac{4}{3}x - \frac{1}{4}$

38. $f(x) = \frac{5}{2}x + \frac{2}{7}$

39. $f(x) = -\frac{3}{7}x - \frac{2}{3}$

40. $f(x) = -\frac{3}{5}x + \frac{3}{4}$

For Problems 41–50, (a) find the inverse of the given function, and (b) graph the given function and its inverse on the same set of axes.

41. $f(x) = 4x$

42. $f(x) = \frac{2}{5}x$

43. $f(x) = -\frac{1}{3}x$

44. $f(x) = -6x$

45. $f(x) = 3x - 3$

46. $f(x) = 2x + 2$

47. $f(x) = -2x - 4$

48. $f(x) = -3x + 9$

49. $f(x) = x^2, x \geq 0$

50. $f(x) = x^2 + 2, x \geq 0$

THOUGHTS INTO WORDS

51. Does the function $f(x) = 4$ have an inverse? Explain your answer.

52. Explain why every nonconstant linear function has an inverse.

FURTHER INVESTIGATIONS

53. The composition idea can also be used to find the inverse of a function. For example, to find the inverse of $f(x) = 5x + 3$, we could proceed as follows:

$$f(f^{-1}(x)) = 5(f^{-1}(x)) + 3 \quad \text{and} \quad f(f^{-1}(x)) = x$$

Therefore, equating the two expressions for $f(f^{-1}(x))$, we obtain

$$5(f^{-1}(x)) + 3 = x$$

$$5(f^{-1}(x)) = x - 3$$

$$f^{-1}(x) = \frac{x - 3}{5}$$

Use this approach to find the inverse of each of the following functions.

(a) $f(x) = 2x + 1$

(b) $f(x) = 3x - 2$

(c) $f(x) = -4x + 5$

(d) $f(x) = -x + 1$

(e) $f(x) = 2x$

(f) $f(x) = -5x$

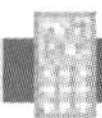

GRAPHING CALCULATOR ACTIVITIES

54. For Problems 31–40, graph the given function, the inverse function that you found, and $f(x) = x$ on the same set of axes. In each case the given function and its inverse should produce graphs that are reflections of each other through the line $f(x) = x$.

55. Let's use a graphing calculator to show that $(f \circ g)(x) = x$ and $(g \circ f)(x) = x$ for two functions that we think are inverses of each other. Consider the functions $f(x) = 3x + 4$ and $g(x) = \frac{x-4}{3}$. We can make the following assignments.

$$f\colon Y_1 = 3x + 4$$

$$g\colon Y_2 = \frac{x-4}{3}$$

$$f \circ g\colon Y_3 = 3Y_2 + 4$$

$$g \circ f\colon Y_4 = \frac{Y_1 - 4}{3}$$

Now we can graph Y_3 and Y_4 and show that they both produce the line $f(x) = x$.

Use this approach to check your answers for Problems 41–50.

56. Use the approach demonstrated in Problem 55 to show that $f(x) = x^2 - 2$ (for $x \geq 0$) and $g(x) = \sqrt{x+2}$ (for $x \geq -2$) are inverses of each other.

Answers to the Concept Quiz

1. False **2.** True **3.** False **4.** True **5.** True **6.** True **7.** False **8.** True **9.** False **10.** True

Answers to the Example Practice Skills

1. (a) $D = \{0, 1, 5, 7\}$ and $R = \{3, 4, 6, 9\}$ **(b)** $f^{-1} = \{(3, 0), (4, 1), (6, 5), (9, 7)\}$ **(c)** $D = \{3, 4, 6, 9\}$ and $R = \{0, 1, 5, 7\}$ **2.** $f^{-1}(x) = \frac{x-7}{2}$ **3.** $f^{-1}(x) = \frac{-x-5}{6}$ **4.** $f^{-1}(x) = \frac{5}{2}x - \frac{5}{6}$

10.6 Direct and Inverse Variations

OBJECTIVES

1 Solve Direct Variation Problems

2 Solve Inverse Variation Problems

3 Solve Variation Problems with Two or More Variables

1 Solve Direct Variation Problems

"The distance a car travels at a fixed rate varies directly as the time." "At a constant temperature, the volume of an enclosed gas varies inversely as the pressure." Such statements illustrate two basic types of functional relationships, called **direct** and **inverse variation**, that are widely used, especially in the physical sciences. These relationships can be expressed by equations that specify functions. The purpose of this section is to investigate these special functions.

The statement "y varies directly as x" means

$$y = kx$$

where k is a nonzero constant called the **constant of variation**. The phrase "y is directly proportional to x" is also used to indicate direct variation; k is then referred to as the **constant of proportionality**.

Remark: Note that the equation $y = kx$ defines a function and could be written as $f(x) = kx$ in function notation. However, in this section it is more convenient to avoid function notation and instead use variables that are meaningful in terms of the physical entities involved in the problem.

Statements that indicate direct variation may also involve powers of x. For example, "y varies directly as the square of x" can be written as

$$y = kx^2$$

In general, "y varies directly as the nth power of x ($n > 0$)" means

$$y = kx^n$$

The three types of problems that deal with direct variation are

1. Translating an English statement into an equation that expresses the direct variation
2. Finding the constant of variation from given values of the variables
3. Finding additional values of the variables once the constant of variation has been determined

Let's consider an example of each of these types of problems.

EXAMPLE 1

Translate the statement "the tension on a spring varies directly as the distance it is stretched" into an equation, and use k as the constant of variation.

Solution

If we let t represent the tension and d the distance, the equation is

$$t = kd$$

▼ **PRACTICE YOUR SKILL**

Translate the statement "the height of a person varies directly with the length of the femur bone" into an equation, and use k as the constant of variation. ■

EXAMPLE 2

If A varies directly as the square of s and if $A = 28$ when $s = 2$, find the constant of variation.

Solution

Because A varies directly as the square of s, we have

$$A = ks^2$$

Substituting $A = 28$ and $s = 2$, we obtain

$$28 = k(2)^2$$

Solving this equation for k yields

$$28 = 4k$$

$$7 = k$$

The constant of variation is 7.

▼ PRACTICE YOUR SKILL

If P varies directly as the square of r and if $P = -108$ when $r = 6$, find the constant of variation. ■

EXAMPLE 3

If y is directly proportional to x and if $y = 6$ when $x = 9$, find the value of y when $x = 24$.

Solution

The statement "y is directly proportional to x" translates into

$$y = kx$$

If we let $y = 6$ and $x = 9$, the constant of variation becomes

$$6 = k(9)$$

$$6 = 9k$$

$$\frac{6}{9} = k$$

$$\frac{2}{3} = k$$

Thus the specific equation is $y = \frac{2}{3}x$. Now, letting $x = 24$, we obtain

$$y = \frac{2}{3}(24) = 16$$

The required value of y is 16.

▼ PRACTICE YOUR SKILL

If m is directly proportional to n and if $m = 60$ when $n = 15$, find the value of m when $n = 18$. ■

2 Solve Inverse Variation Problems

We define the second basic type of variation, called **inverse variation**, as follows: The statement "y varies inversely as x" means

$$y = \frac{k}{x}$$

where k is a nonzero constant; again we refer to it as the constant of variation. The phrase "y is inversely proportional to x" is also used to express inverse variation. As with direct variation, statements that indicate inverse variation may involve powers of x. For example, "y varies inversely as the square of x" can be written as

$$y = \frac{k}{x^2}$$

In general, "y varies inversely as the nth power of x $(n > 0)$" means

$$y = \frac{k}{x^n}$$

The following examples illustrate the three basic kinds of problems we run across that involve inverse variation.

EXAMPLE 4

Translate the statement "the length of a rectangle of a fixed area varies inversely as the width" into an equation that uses k as the constant of variation.

Solution

Let l represent the length and w the width; then the equation is

$$l = \frac{k}{w}$$

▼ PRACTICE YOUR SKILL

Translate the statement "the time it takes to travel a fixed distance varies inversely as the speed" into an equation, and use k as the constant of variation. ■

EXAMPLE 5

If y is inversely proportional to x and if $y = 4$ when $x = 12$, find the constant of variation.

Solution

Because y is inversely proportional to x, we have

$$y = \frac{k}{x}$$

Substituting $y = 4$ and $x = 12$, we obtain

$$4 = \frac{k}{12}$$

Solving this equation for k by multiplying both sides of the equation by 12 yields

$$k = 48$$

The constant of variation is 48.

▼ PRACTICE YOUR SKILL

If r is inversely proportional to t and if $r = 50$ when $t = 3$, find the constant of variation. ■

EXAMPLE 6

Suppose the number of days it takes to complete a construction job varies inversely as the number of people assigned to the job. If it takes 7 people 8 days to do the job, how long would it take 14 people to complete the job?

Solution

Let d represent the number of days and p the number of people. The phrase "number of days . . . varies inversely with the number of people" translates into

$$d = \frac{k}{p}$$

Let $d = 8$ when $p = 7$; then the constant of variation becomes

$$8 = \frac{k}{7}$$

$$k = 56$$

Thus the specific equation is

$$d = \frac{56}{p}$$

Now, let $p = 14$ to obtain

$$d = \frac{56}{14}$$

$$d = 4$$

It should take 14 people 4 days to complete the job.

▼ **PRACTICE YOUR SKILL**

The volume of a gas at a constant temperature varies inversely with the pressure. If the gas occupies 4 liters under a pressure of 30 pounds, what is the volume of the gas under a pressure of 40 pounds? ■

The terms *direct* and *inverse*, as applied to variation, refer to the relative behavior of the variables involved in the equation. That is: in *direct* variation ($y = kx$), an assignment of *increasing* absolute values for x produces increasing absolute values for y; whereas in *inverse* variation $\left(y = \frac{k}{x}\right)$, an assignment of increasing absolute values for x produces *decreasing* absolute values for y.

3 Solve Variation Problems with Two or More Variables

Variation may involve more than two variables. The following table illustrates some variation statements and their equivalent algebraic equations that use k as the constant of variation.

Statements 1, 2, and 3 illustrate the concept of **joint variation**. Statements 4 and 5 show that both direct and inverse variation may occur in the same problem. Statement 6 combines joint variation with inverse variation.

The two final examples of this section illustrate some of these variation situations.

Variation statement	Algebraic equation
1. y varies jointly as x and z.	$y = kxz$
2. y varies jointly as x, z, and w.	$y = kxzw$
3. V varies jointly as h and the square of r.	$V = khr^2$
4. h varies directly as V and inversely as w.	$h = \frac{kV}{w}$
5. y is directly proportional to x and inversely proportional to the square of z.	$y = \frac{kx}{z^2}$
6. y varies jointly as w and z and inversely as x.	$y = \frac{kwz}{x}$

EXAMPLE 7

Suppose that y varies jointly as x and z and inversely as w. If $y = 154$ when $x = 6$, $z = 11$, and $w = 3$, find the constant of variation.

Solution

The statement "y varies jointly as x and z and inversely as w" translates into

$$y = \frac{kxz}{w}$$

Substitute $y = 154$, $x = 6$, $z = 11$, and $w = 3$ to obtain

$$154 = \frac{k(6)(11)}{3}$$

$$154 = 22k$$

$$7 = k$$

The constant of variation is 7.

▼ **PRACTICE YOUR SKILL**

Suppose that q varies jointly as m and n and inversely as p. If $q = 25$ when $m = 10$, $n = 15$, and $p = 3$, find the constant of variation. ■

EXAMPLE 8

The length of a rectangular box with a fixed height varies directly as the volume and inversely as the width. If the length is 12 centimeters when the volume is 960 cubic centimeters and the width is 8 centimeters, find the length when the volume is 700 centimeters and the width is 5 centimeters.

Solution

Use l for length, V for volume, and w for width; then the phrase "length varies directly as the volume and inversely as the width" translates into

$$l = \frac{kV}{w}$$

Substitute $l = 12$, $V = 960$, and $w = 8$. Hence the constant of variation is

$$12 = \frac{k(960)}{8}$$

$$12 = 120k$$

$$\frac{1}{10} = k$$

Thus the specific equation is

$$l = \frac{\frac{1}{10}V}{w} = \frac{V}{10w}$$

Now let $V = 700$ and $w = 5$ to obtain

$$l = \frac{700}{10(5)} = \frac{700}{50} = 14$$

The length is 14 centimeters.

▼ **PRACTICE YOUR SKILL**

The volume of a gas varies directly as the absolute temperature and inversely as the pressure. If the gas occupies 3 liters when the temperature is 300°K and the pressure is 50 pounds, what is the volume of the gas when the temperature is 330°K and the pressure is 50 pounds? ■

CONCEPT QUIZ

For Problems 1–5, answer true or false.

1. In the equation $y = kx$, the k is a quantity that varies as y.
2. The equation $y = kx$ defines a function and could be written in functional notation as $f(x) = kx$.
3. Variation that involves more than two variables is called proportional variation.
4. Every equation of variation will have a constant of variation.
5. In joint variation, both direct and inverse variation may occur in the same problem.

For Problems 6–10, match the statement of variation with its equation.

6. y varies directly as x — **A.** $y = \frac{k}{x}$
7. y varies inversely as x — **B.** $y = kxz$
8. y varies directly as the square of x — **C.** $y = kx^2$
9. y varies directly as the square root of x — **D.** $y = kx$
10. y varies jointly as x and z — **E.** $y = k\sqrt{x}$

Problem Set 10.6

1 Solve Direct Variation Problems

For Problems 1–6, translate each statement of variation into an equation, and use k as the constant of variation.

1. T varies directly as r.
2. W varies directly as x.
3. The area of a circle (A) varies directly as the square of the radius (r).
4. The surface area (S) of a cube varies directly as the square of the length of an edge (e).
5. y varies directly as the cube of x.
6. The volume (V) of a sphere is directly proportional to the cube of its radius (r).

For Problems 7–10, find the constant of variation for each of the stated conditions.

7. y varies directly as x, and $y = 8$ when $x = 12$.
8. y varies directly as x, and $y = 60$ when $x = 24$.
9. y varies directly as the square of x, and $y = -144$ when $x = 6$.
10. y varies directly as the cube of x, and $y = 48$ when $x = -2$.

For Problems 11–16, solve each of the problems.

11. If y is directly proportional to x and if $y = 36$ when $x = 48$, find the value of y when $x = 12$.
12. If y is directly proportional to x and if $y = 42$ when $x = 28$, find the value of y when $x = 38$.
13. The amount of simple interest earned in a year at a fixed interest rate varies directly with the amount of principal. If \$4600 earned \$299 in interest, how much interest will \$8000 earn?
14. The distance that a freely falling body falls varies directly as the square of the time it falls. If a body falls 144 feet in 3 seconds, how far will it fall in 5 seconds?
15. The period (the time required for one complete oscillation) of a simple pendulum varies directly as the square root of its length. If a pendulum 12 feet long has a period of 4 seconds, find the period of a pendulum 3 feet long.
16. The period (the time required for one complete oscillation) of a simple pendulum varies directly as the square root of its length. If a pendulum 9 inches long has a period of 2.4 seconds, find the period of a pendulum 12 inches long. Express your answer to the nearest tenth of a second.

2 Solve Inverse Variation Problems

For Problems 17–20, translate each statement of variation into an equation, and use k as the constant of variation.

17. y varies inversely as the square of x.

18. B varies inversely as w.

19. At a constant temperature, the volume (V) of a gas varies inversely as the pressure (P).

20. The intensity of illumination (I) received from a source of light is inversely proportional to the square of the distance (d) from the source.

For Problems 21–24, find the constant of variation for each of the stated conditions.

21. y varies inversely as x, and $y = -4$ when $x = \frac{1}{2}$.

22. y varies inversely as x, and $y = -6$ when $x = \frac{4}{3}$.

23. r varies inversely as the square of t, and $r = \frac{1}{8}$ when $t = 4$.

24. r varies inversely as the cube of t, and $r = \frac{1}{16}$ when $t = 4$.

For Problems 25–30, solve each of the problems.

25. If y is inversely proportional to x and if $y = \frac{1}{9}$ when $x = 12$, find the value of y when $x = 8$.

26. If y is inversely proportional to x and if $y = \frac{1}{35}$ when $x = 14$, find the value of y when $x = 16$.

27. If y is inversely proportional to the square root of x and if $y = 0.08$ when $x = 225$, find y when $x = 625$.

28. If y is inversely proportional to the square of x and if $y = 64$ when $x = 2$, find y when $x = 4$.

29. The time required for a car to travel a certain distance varies inversely as the rate at which it travels. If it takes 4 hours at 50 miles per hour to travel the distance, how long will it take at 40 miles per hour?

30. The volume of a gas at a constant temperature varies inversely as the pressure. What is the volume of a gas under pressure of 25 pounds if the gas occupies 15 cubic centimeters under a pressure of 20 pounds?

3 Solve Variation Problems with Two or More Variables

For Problems 31–34, translate each statement of variation into an equation, and use k as the constant of variation.

31. C varies directly as g and inversely as the cube of t.

32. V varies jointly as l and w.

33. The volume (V) of a cone varies jointly as its height and the square of its radius.

34. The volume (V) of a gas varies directly as the absolute temperature (T) and inversely as the pressure (P).

For Problems 35–40, find the constant of variation for each of the stated conditions.

35. V varies jointly as B and h, and $V = 96$ when $B = 24$ and $h = 12$.

36. A varies jointly as b and h, and $A = 72$ when $b = 16$ and $h = 9$.

37. y varies directly as x and inversely as z, and $y = 45$ when $x = 18$ and $z = 2$.

38. y varies directly as x and inversely as z, and $y = 24$ when $x = 36$ and $z = 18$.

39. y is directly proportional to x and inversely proportional to the square of z, and $y = 81$ when $x = 36$ and $z = 2$.

40. y is directly proportional to the square of x and inversely proportional to the cube of z, and $y = 4\frac{1}{2}$ when $x = 6$ and $z = 4$.

For Problems 41–48, solve each of the problems.

41. If A varies jointly as b and h and if $A = 60$ when $b = 12$ and $h = 10$, find A when $b = 16$ and $h = 14$.

42. If V varies jointly as B and h and if $V = 51$ when $B = 17$ and $h = 9$, find V when $B = 19$ and $h = 12$.

43. The volume (V) of a gas varies directly as the temperature (T) and inversely as the pressure (P). If $V = 48$ when $T = 320$ and $P = 20$, find V when $T = 280$ and $P = 30$.

44. The simple interest earned by a certain amount of money varies jointly as the rate of interest and the time (in years) that the money is invested. If the money is invested at 12% for 2 years, \$120 is earned. How much is earned if the money is invested at 14% for 3 years?

45. The electrical resistance of a wire varies directly as its length and inversely as the square of its diameter. If the resistance of 200 meters of wire that has a diameter of $\frac{1}{2}$ centimeter is 1.5 ohms, find the resistance of 400 meters of wire with a diameter of $\frac{1}{4}$ centimeter.

46. The volume of a cylinder varies jointly as its altitude and the square of the radius of its base. If the volume of a cylinder is 1386 cubic centimeters when the radius of the base is 7 centimeters and its altitude is 9 centimeters, find the volume of a cylinder that has a base of radius 14 centimeters and the altitude of the cylinder is 5 centimeters.

47. The simple interest earned by a certain amount of money varies jointly as the rate of interest and the time (in years) that the money is invested.

(a) If some money invested at 11% for 2 years earns \$385, how much would the same amount earn at 12% for 1 year?
(b) If some money invested at 12% for 3 years earns \$819, how much would the same amount earn at 14% for 2 years?
(c) If some money invested at 14% for 4 years earns \$1960, how much would the same amount earn at 15% for 2 years?

48. The volume of a cylinder varies jointly as its altitude and the square of the radius of its base. If a cylinder that has a base with a radius of 5 meters and an altitude of 7 meters has a volume of 549.5 cubic meters, find the volume of a cylinder that has a base with a radius of 9 meters and an altitude of 14 meters.

THOUGHTS INTO WORDS

49. How would you explain the difference between direct variation and inverse variation?

50. Suppose that y varies directly as the square of x. Does doubling the value of x also double the value of y? Explain your answer.

51. Suppose that y varies inversely as x. Does doubling the value of x also double the value of y? Explain your answer.

Answers to the Concept Quiz

1. False **2.** True **3.** False **4.** True **5.** True **6.** D **7.** A **8.** C **9.** E **10.** B

Answers to the Example Practice Skills

1. $h = kl$ **2.** $k = -3$ **3.** 72 **4.** $t = \frac{k}{s}$ **5.** $k = 150$ **6.** 3 liters **7.** $\frac{1}{2}$ **8.** 3.3 liters

Chapter 10 Summary

OBJECTIVE	SUMMARY	EXAMPLE	CHAPTER REVIEW PROBLEMS
Determine if a relation is a function. (Sec. 10.1, Obj. 1, p. 534)	A relation is a set of ordered pairs; a function is a relation in which no two ordered pairs have the same first component. The domain of a relation (or function) is the set of all first components, and the range is the set of all second components.	Specify the domain and range of the relation and state whether or not it is a function. $\{(1, 8), (2, 7), (5, 6), (3, 8)\}$ **Solution** $D = \{1, 2, 3, 5\}$, $R = \{6, 7, 8\}$ It is a function.	Problems 1–4
Use function notation when evaluating a function. (Sec. 10.1, Obj. 2, p. 536)	Single letters such as f, g, and h are commonly used to name functions. The symbol $f(x)$ represents the element in the range associated with x from the domain.	If $f(x) = 2x^2 + 3x - 5$, find $f(4)$. **Solution** Substitute 4 for x in the equation. $f(4) = 2(4)^2 + 3(4) - 5$ $f(4) = 32 + 12 - 5$ $f(4) = 39$	Problems 5–6
Specify the domain of a function. (Sec. 10.1, Obj. 3, p. 537)	The domain of a function is the set of all real number replacements for the variable that will produce real number functional values. Replacement values that make a denominator zero or a radical expression undefined are excluded from the domain.	Specify the domain for $f(x) = \dfrac{x+5}{2x-3}$. **Solution** The values that make the denominator zero must be excluded from the domain. To find those values, set the denominator equal to zero and solve. $2x - 3 = 0$ $x = \dfrac{3}{2}$ The domain is the set $\left\{x \mid x \neq \dfrac{3}{2}\right\}$ or $\left(-\infty, \dfrac{3}{2}\right) \cup \left(\dfrac{3}{2}, \infty\right)$.	Problems 7–10 *(continued)*

OBJECTIVE	SUMMARY	EXAMPLE	CHAPTER REVIEW PROBLEMS
Find the difference quotient of a given function. (Sec. 10.1, Obj. 4, p. 538)	The quotient $\frac{f(a+h)-f(a)}{h}$ is called the difference quotient.	If $f(x) = 5x + 7$, find the difference quotient. **Solution** $\frac{f(a+h)-f(a)}{h} = \frac{5(a+h)+7-(5a+7)}{h} = \frac{5a+5h+7-5a-7}{h} = \frac{5h}{h} = 5$	Problems 11–12
Apply function notation to a problem. (Sec. 10.1, Obj. 5, p. 539)	Functions provide the basis for many application problems.	The function $E(d) = 0.693d$ exchanges the value of currency in U.S. dollars (d) for Euros. Compute $E(20)$, $E(100)$, and $E(500)$. **Solution** $E(20) = 0.693(20) = 13.86$ $E(100) = 0.693(100) = 69.30$ $E(500) = 0.693(500) = 346.50$	Problems 13–14
Graph linear functions. (Sec. 10.2, Obj. 1, p. 542)	Any function that can be written in the form $f(x) = ax + b$, where a and b are real numbers, is a linear function. The graph of a linear function is a straight line.	Graph $f(x) = 3x + 1$. **Solution** Because $f(0) = 1$, the point (0, 1) is on the graph. Also $f(1) = 4$, so the point (1, 4) is on the graph. f(x) (1, 4) (0, 1) x f(x) = 3x + 1	Problems 15–18 *(continued)*

OBJECTIVE	SUMMARY	EXAMPLE	CHAPTER REVIEW PROBLEMS
Apply linear functions. (Sec. 10.2, Obj. 2, p. 544)	Linear functions and their graphs can be an aid in problem solving.	The FixItFast computer repair company uses the equation $C(m) = 2m + 15$, where m is the number of minutes for the service call, to determine the charge for a service call. Graph the function and use the graph to approximate the charge for a 25-minute service call. Then use the function to find the exact charge for a 25-minute service call. **Solution** 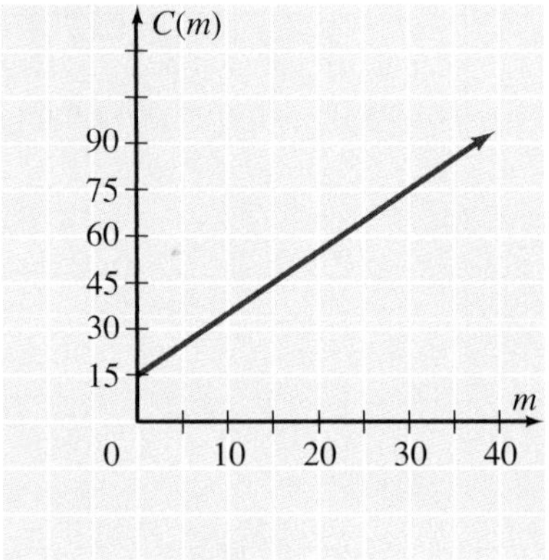 Compare your approximation to the exact charge, $C(25) = 2(25) + 15 = 65$.	Problems 19–20
Graph quadratic functions. (Sec. 10.2, Obj. 3, p. 546)	Any function that can be written in the form $f(x) = ax^2 + bx + c$, where a, b, and c are real numbers and $a \neq 0$, is a quadratic function. The graph of any quadratic function is a parabola, which can be drawn using either of the following methods. 1. Express the function in the form $f(x) = a(x - h)^2 + k$, and use the values of a, h, and k to determine the parabola. 2. Express the function in the form $f(x) = ax^2 + bx + c$, and use the facts that the vertex is at $\left(-\frac{b}{2a}, f\left(-\frac{b}{2a}\right)\right)$ and the axis of symmetry is $x = -\frac{b}{2a}$.	Graph $f(x) = 2x^2 + 8x + 7$. **Solution** $f(x) = 2x^2 + 8x + 7$ $= 2(x^2 + 4x) + 7$ $= 2(x^2 + 4x + 4) - 8 + 7$ $= 2(x + 2)^2 - 1$ 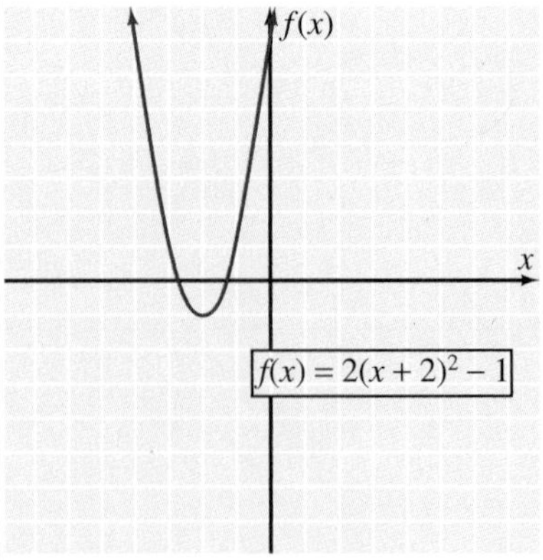 $f(x) = 2(x + 2)^2 - 1$	Problems 21–24 *(continued)*

OBJECTIVE	SUMMARY	EXAMPLE	CHAPTER REVIEW PROBLEMS
Solve problems using quadratic functions. (Sec. 10.2, Obj. 4, p. 556)	We can solve some applications that involve maximum and minimum values with our knowledge of parabolas that are generated by quadratic functions.	Suppose the cost function for producing a particular item is given by the equation $C(x) = 3x^2 - 270x + 15800$, where x represents the number of items. How many items should be produced to minimize the cost? **Solution** The function represents a parabola. The minimum will occur at the vertex, so we want to find the x coordinate of the vertex. $x = -\frac{b}{2a}$ $\quad x = -\frac{-270}{2(3)} = 45$ Therefore, 45 items should be produced to minimize the cost.	Problems 25–28

Know the five basic graphs shown here. In order to shift and reflect these graphs, it is necessary to know their basic shapes.
(Sec. 10.3, Obj. 1, p. 556)

$f(x) = x^2$

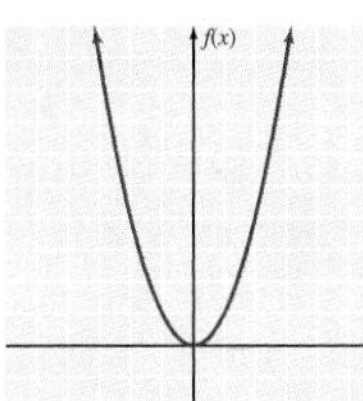

$f(x) = \frac{1}{x}$

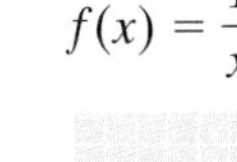

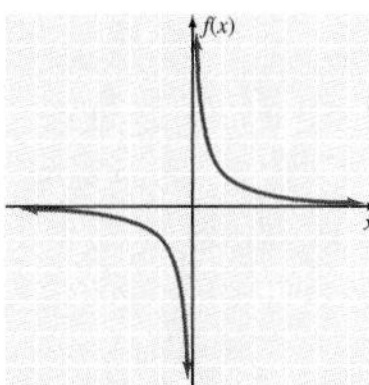

$f(x) = x^3$

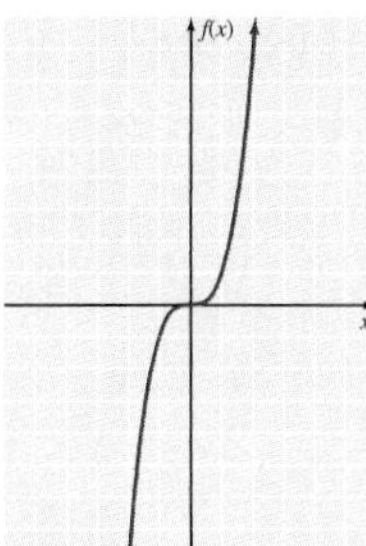

$f(x) = \sqrt{x}$

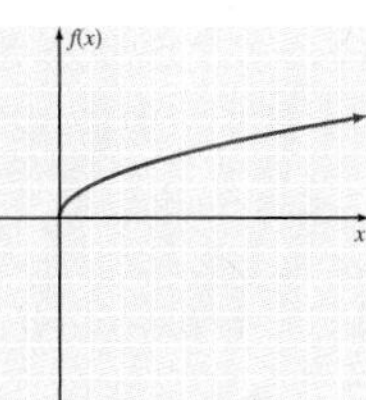

$f(x) = |x|$

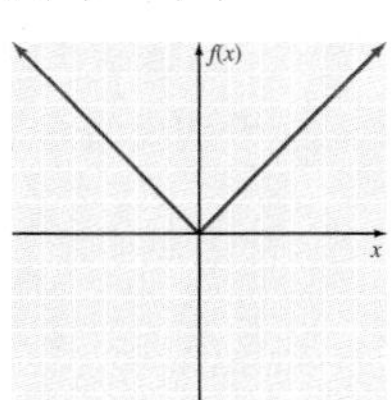

(continued)

OBJECTIVE	SUMMARY	EXAMPLE	CHAPTER REVIEW PROBLEMS
Graph functions by using translations. (Sec. 10.3, Obj. 2, p. 558)	Vertical translation: The graph of $y = f(x) + k$ is the graph of $y = f(x)$ shifted k units upward if k is positive and $\|k\|$ units downward if k is negative. Horizontal translation: The graph of $y = f(x - h)$ is the graph of $y = f(x)$ shifted h units to the right if h is positive and $\|h\|$ units to the left if h is negative.	Graph $f(x) = \|x + 4\|$. **Solution** To fit the form, change the equation to the equivalent form $f(x) = \|x - (-4)\|$. Because h is negative, the graph of $f(x) = \|x\|$ is shifted 4 units to the left. 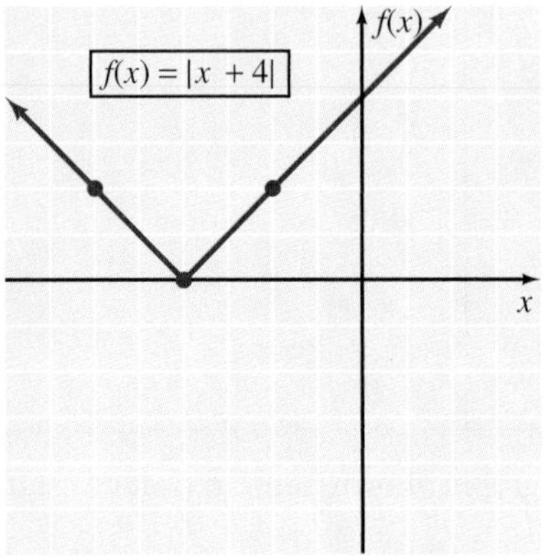 $f(x)$ $f(x) = \|x + 4\|$ x	Problems 29–32
Graph functions by using reflections. (Sec. 10.3, Obj. 3, p. 559)	x-axis reflection: The graph of $y = -f(x)$ is the graph of $y = f(x)$ reflected through the x-axis. y-axis reflection: The graph of $y = f(-x)$ is the graph of $y = f(x)$ reflected through the y-axis.	Graph $f(x) = \sqrt{-x}$. **Solution** The graph of $f(x) = \sqrt{-x}$ is the graph of $f(x) = \sqrt{x}$ reflected through the y axis. 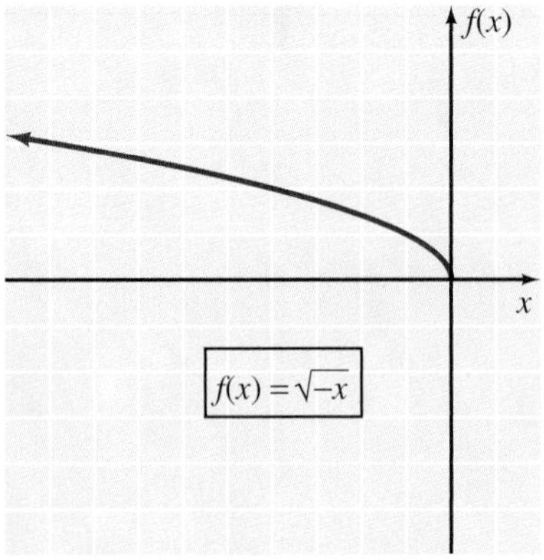 $f(x)$ x $f(x) = \sqrt{-x}$	Problems 33–34 *(continued)*

OBJECTIVE	SUMMARY	EXAMPLE	CHAPTER REVIEW PROBLEMS
Graph functions by using vertical stretching or shrinking. (Sec. 10.3, Obj. 4, p. 561)	Vertical stretching and shrinking: The graph of $y = cf(x)$ is obtained from the graph of $y = f(x)$ by multiplying the y coordinates of $y = f(x)$ by c. If $\|c\| > 1$ then the graph is said to be stretched by a factor of $\|c\|$, and if $0 < \|c\| < 1$ then the graph is said to be shrunk by a factor of $\|c\|$.	Graph $f(x) = \frac{1}{4}x^2$. **Solution** The graph of $f(x) = \frac{1}{4}x^2$ is the graph of $f(x) = x^2$ shrunk by a factor of $\frac{1}{4}$. 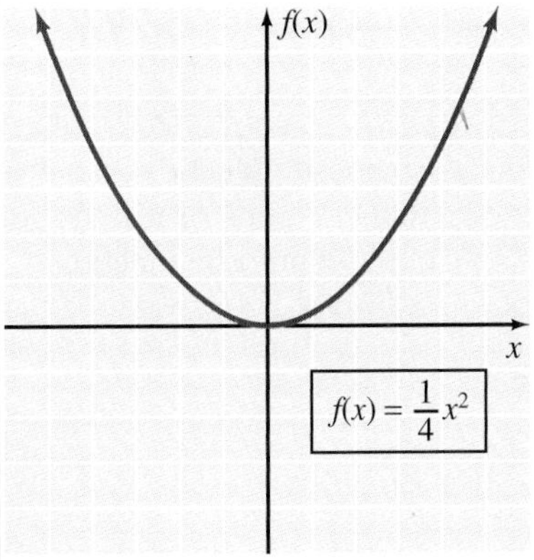 $f(x)$, x, $f(x) = \frac{1}{4}x^2$	Problems 35–36
Graph functions by using successive transformations. (Sec. 10.3, Obj. 5, p. 562)	Some curves are the result of performing more than one transformation on a basic curve. Unless parentheses indicate otherwise, stretchings, shrinkings, and x-axis reflections should be performed before translations.	Graph $f(x) = -2(x + 1)^2 + 3$. **Solution** $f(x) = -2(x + 1)^2 + 3$ Narrows the parabola and opens it downward; Moves the parabola 1 unit to the left; Moves the parabola 3 units up $f(x) = -2(x + 1)^2 + 3$, $f(x)$, x 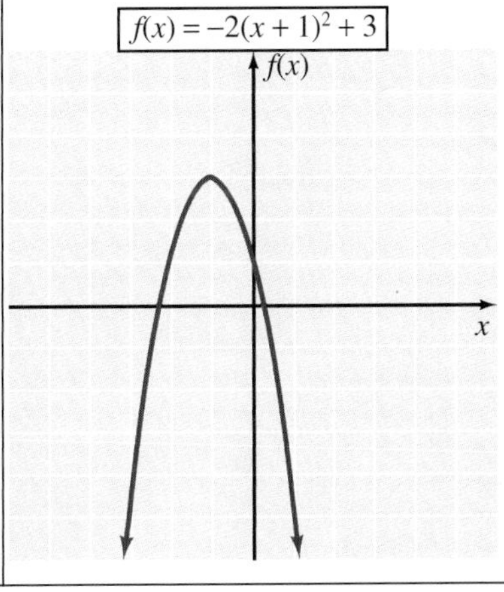	Problems 37–40 *(continued)*

OBJECTIVE	SUMMARY	EXAMPLE	CHAPTER REVIEW PROBLEMS
Find the composition of two functions and determine the domain. (Sec. 10.4, Obj. 1, p. 567)	The composition of two functions f and g is defined by $(f \circ g)(x) = f(g(x))$ for all x in the domain of g such that $g(x)$ is in the domain of f. Remember that the composition of functions is not a commutative operation.	If $f(x) = x + 5$ and $g(x) = x^2 + 4x - 6$, find $(g \circ f)(x)$. **Solution** In the function g, substitute $f(x)$ for x. $(g \circ f)(x)$ $= (f(x))^2 + 4(f(x)) - 6$ $= (x + 5)^2 + 4(x + 5) - 6$ $= x^2 + 10x + 25 + 4x + 20 - 6$ $= x^2 + 14x + 39$	Problems 41–43
Determine functional values for composite functions. (Sec. 10.4, Obj. 2, p. 569)	Composite functions can be evaluated for values of x in the domain of the composite function.	If $f(x) = 3x - 1$ and $g(x) = -x^2 + 9$, find $(f \circ g)(-4)$. **Solution** First, form the composite function $(f \circ g)(x)$: $(f \circ g)(x) = 3(-x^2 + 9) - 1$ $= -3x^2 + 26$ To find $(f \circ g)(-4)$, substitute -4 for x in the composite function. $(f \circ g)(-4)$ $= -3(-4)^2 + 26 = -22$	Problems 44–45
Use the vertical line test. (Sec. 10.5, Obj. 1, p. 573)	The vertical line test is used to determine if a graph is the graph of a function. Any vertical line drawn through the graph of a function must not intersect the graph in more than one point.	Identify the graph as the graph of a function or the graph of a relation that is not a function. 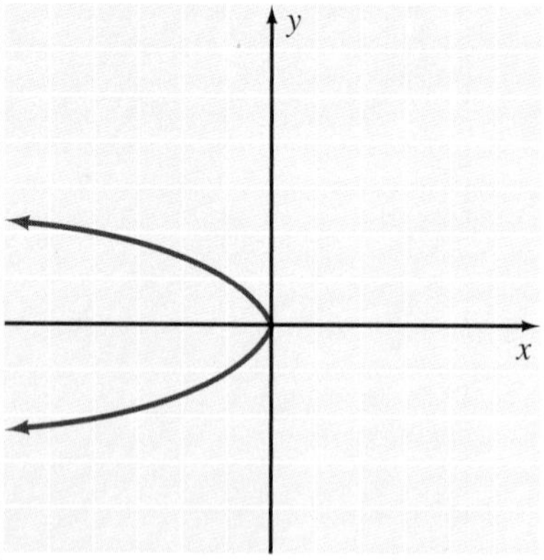 **Solution** It is the graph of a relation that is not a function because a vertical line will intersect the graph in more than one point.	Problems 46–48 *(continued)*

OBJECTIVE	SUMMARY	EXAMPLE	CHAPTER REVIEW PROBLEMS
Use the horizontal line test. (Sec. 10.5, Obj. 2, p. 574)	The horizontal line test is used to determine if a graph is the graph of a one-to-one function. If the graph is the graph of a one-to-one function, then a horizontal line will intersect the graph in only one point.	Identify the graph as the graph of a one-to-one function or the graph of a function that is not one-to-one. 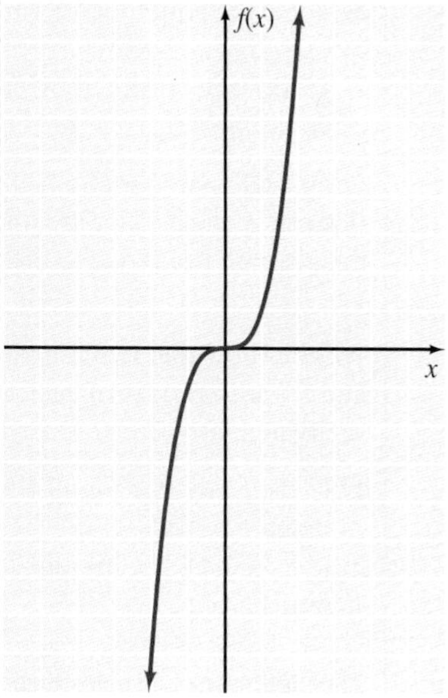 $f(x)$ x **Solution** The graph is the graph of a one-to-one function because a horizontal line intersects the graph in only one point.	Problems 49–51
Find the inverse function in terms of ordered pairs. (Sec. 10.5, Obj. 3, p. 575)	If the components of each ordered pair of a given one-to-one function are interchanged, then the resulting function and the given function are inverses of each other. The inverse of a function f is denoted by f^{-1}.	Given $f = \{(2, 4), (3, 5), (7, 7), (8, 9)\}$, find the inverse function. **Solution** To find the inverse function, interchange the components of the ordered pairs. $f^{-1} = \{(4, 2), (5, 3), (7, 7), (9, 8)\}$	Problems 52–53 *(continued)*

OBJECTIVE	SUMMARY	EXAMPLE	CHAPTER REVIEW PROBLEMS
Find the inverse of a function. (Sec. 10.5, Obj. 4, p. 576)	A technique for finding the inverse of a function is as follows. **1.** Let $y = f(x)$. **2.** Interchange x and y. **3.** Solve the equation for y in terms of x. **4.** $f^{-1}(x)$ is determined by the final equation. Graphically, two functions that are inverses of each other are mirror images with reference to the line $y = x$. We can show that two functions f and f^{-1} are inverses of each other by verifying that **1.** $(f^{-1} \circ f)(x) = x$ for all x in the domain of f **2.** $(f \circ f^{-1})(x) = x$ for all x in the domain of f^{-1}	Find the inverse of the function $f(x) = \frac{2}{5}x - 7$. **Solution** 1. Let $y = \frac{2}{5}x - 7$. 2. Interchange x and y. $x = \frac{2}{5}y - 7$ 3. Solve for y. $x = \frac{2}{5}y - 7$ $5x = 2y - 35$ Multiply both sides by 5. $5x + 35 = 2y$ $y = \frac{5x + 35}{2}$ 4. $f^{-1}(x) = \frac{5x + 35}{2}$	Problems 54–56
Solve direct variation problems. (Sec. 10.6, Obj. 1, p. 581)	The statement "y varies directly as x" means $y = kx$, where k is a nonzero constant called the constant of variation. The phrase "y is directly proportional to x" is also used to indicate direct variation.	The cost of electricity varies directly with the number of kilowatt hours used. If it cost \$127.20 for 1200 kilowatt hours, what will 1500 kilowatt hours cost? **Solution** Let C represent the cost and w represent the number of kilowatt hours. The equation of variation is $C = kw$. Substitute 127.20 for C and 1200 for w and then solve for k. $127.20 = k(1200)$ $k = \frac{127.20}{1200} = 0.106$ Now find the cost when $w = 1500$. $C = 0.106(1500) = 159$ Therefore, 1500 kilowatt hours cost \$159.00.	Problems 57–58 *(continued)*

OBJECTIVE	SUMMARY	EXAMPLE	CHAPTER REVIEW PROBLEMS
Solve inverse variation problems. (Sec. 10.6, Obj. 2, p. 583)	The statement "y varies inversely as x" means $y = \frac{k}{x}$, where k is a nonzero constant called the constant of variation. The phrase "y is inversely proportional to x" is also used to indicate inverse variation.	Suppose the number of hours it takes to conduct a telephone research study varies inversely as the number of people assigned to the job. If it takes 15 people 4 hours to do the study, how long would it take 20 people to complete the study? **Solution** Let T represent the time and n represent the number of people. The equation of variation is $T = \frac{k}{n}$. Substitute 4 for T and 15 for n and then solve for k. $4 = \frac{k}{15}$ $k = 4(15) = 60$ Now find the time when $n = 20$. $T = \frac{60}{20} = 3$ Therefore, it will take 20 people 3 hours to do the study.	Problems 59–60
Solve variation problems with more than two variables. (Sec. 10.6, Obj. 3, p. 585)	Variation may involve more than two variables. The statement "y varies jointly as x and z" means $y = kxz$.	If a pool company is designing a swimming pool in the shape of a rectangular solid where the width of the pool is fixed, then the volume of the pool will vary jointly with the length and depth. If a pool that has a length of 30 feet and a depth of 5 feet has a volume of 1800 cubic feet, find the volume of a pool that has a length of 25 feet and a depth of 4 feet. **Solution** Let V represent the volume, l represent the length, and d represent the depth. The equation of variation is $V = kld$. Substitute 1800 for V, 30 for l, and 5 for d. Solve for k. $1800 = k(30)(5)$ $k = 12$ Now find V when l is 25 feet and d is 4 feet. $V = 12(25)(4) = 1200$ Therefore, the volume is 1200 cubic feet when the length is 25 feet and the depth is 4 feet.	Problems 61–62

Chapter 10 Review Problem Set

For Problems 1–4, determine if the following relations determine a function. Specify the domain and range.

1. $\{(9, 2), (8, 3), (7, 4), (6, 5)\}$

2. $\{(1, 1), (2, 3), (1, 5), (2, 7)\}$

3. $\{(0, -6), (0, -5), (0, -4), (0, -3)\}$

4. $\{(0, 8), (1, 8), (2, 8), (3, 8)\}$

5. If $f(x) = x^2 - 2x - 1$, find $f(2), f(-3)$, and $f(a)$.

6. If $g(x) = \frac{2x - 4}{x + 2}$, find $f(2), f(-1)$, and $f(3a)$.

For Problems 7–10, specify the domain of each function.

7. $f(x) = x^2 - 9$

8. $f(x) = \frac{4}{x - 5}$

9. $f(x) = \frac{3}{x^2 + 4x}$

10. $f(x) = \sqrt{x^2 - 25}$

11. If $f(x) = 6x + 8$, find $\frac{f(a + h) - f(a)}{h}$.

12. If $f(x) = 2x^2 + x - 7$, find $\frac{f(a + h) - f(a)}{h}$.

13. The cost for burning a 100-watt bulb is given by the function $c(h) = 0.006h$, where h represents the number of hours that the bulb burns. How much, to the nearest cent, does it cost to burn a 100-watt bulb for 4 hours per night for a 30-day month?

14. "All Items 30% Off Marked Price" is a sign in a local department store. Form a function and then use it to determine how much one must pay for each of the following marked items: a \$65 pair of shoes, a \$48 pair of slacks, a \$15.50 belt.

For Problems 15–18, graph each of the functions.

15. $f(x) = \frac{1}{2}x + 3$

16. $f(x) = -3x + 2$

17. $f(x) = 4$

18. $f(x) = 2x$

19. A college math placement test has 50 questions. The score is computed by awarding 4 points for each correct answer and subtracting 2 points for each incorrect answer. Determine the linear function that would be used to compute the score. Use the function to determine the score when a student gets 35 questions correct.

20. An outpatient operating room charges each patient a fixed amount per surgery plus an amount per minute for use. One patient was charged \$250 for a 30-minute surgery and another patient was charged \$450 for a 90-minute surgery. Determine the linear function that would be used to compute the charge. Use the function to determine the charge when a patient has a 45-minute surgery.

For Problems 21–23, graph each of the functions.

21. $f(x) = x^2 + 2x + 2$

22. $f(x) = -\frac{1}{2}x^2$

23. $f(x) = -3x^2 + 6x - 2$

24. Find the coordinates of the vertex and the equation of the line of symmetry for each of the following parabolas.

    (a) $f(x) = x^2 + 10x - 3$

    (b) $f(x) = -2x^2 - 14x + 9$

25. Find two numbers whose sum is 40 and whose product is a maximum.

26. Find two numbers whose sum is 50 such that the square of one number plus six times the other number is a minimum.

27. A gardener has 60 yards of fencing and wants to enclose a rectangular garden that requires fencing only on three sides. Find the length and width of the plot that will maximize the area.

28. Suppose that 50 students are able to raise \$250 for a gift when each one contributes \$5. Furthermore, they figure that for each additional student they can find to contribute, the cost per student will decrease by a nickel. How many additional students will they need to maximize the amount of money they will have for a gift?

For Problems 29–40, graph the functions.

29. $f(x) = x^3 - 2$

30. $f(x) = |x| + 4$

31. $f(x) = (x + 1)^2$

32. $f(x) = \frac{1}{x - 4}$

33. $f(x) = \sqrt{-x}$

34. $f(x) = -|x|$

35. $f(x) = \frac{1}{3}|x|$

36. $f(x) = 2\sqrt{x}$

37. $f(x) = -\sqrt{x + 1} - 2$

38. $f(x) = \sqrt{x - 2} - 3$

39. $f(x) = -|x - 2|$

40. $f(x) = -(x + 3)^2 - 2$

For Problems 41–43, determine $(f \circ g)(x)$ and $(g \circ f)(x)$ for each pair of function.

41. If $f(x) = 2x - 3$ and $g(x) = 3x - 4$

42. $f(x) = x - 4$ and $g(x) = x^2 - 2x + 3$

43. $f(x) = x^2 - 5$ and $g(x) = -2x + 5$

44. If $f(x) = x^2 + 3x - 1$ and $g(x) = 4x - 7$, find $(f \circ g)(-2)$.

45. If $f(x) = \frac{1}{2}x - 6$ and $g(x) = 2x + 10$, find $(g \circ f)(-4)$.

For Problems 46–48, use the vertical line test to identify each graph as the graph of a function or the graph of a relation that is not a function.

46. 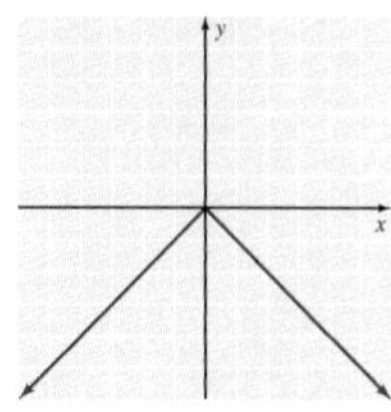

47. 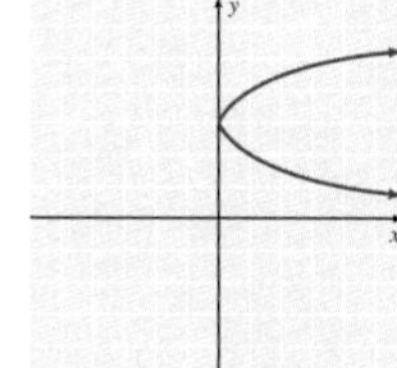

48. 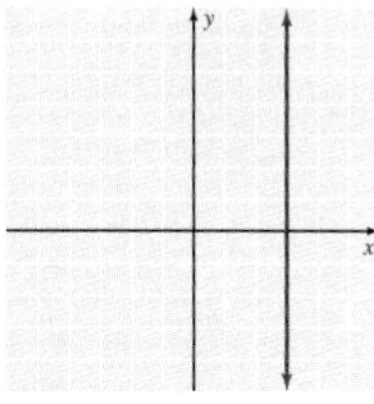

For Problems 49–51, use the horizontal line test to identify each graph as the graph of a one-to-one function or the graph of a function that is not one-to-one.

49. 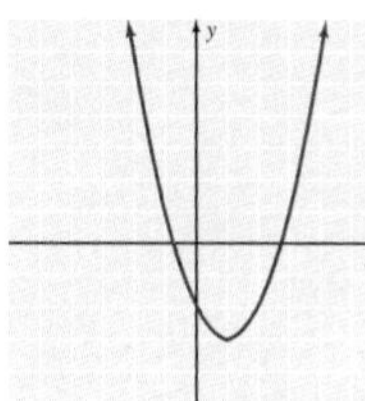

50. 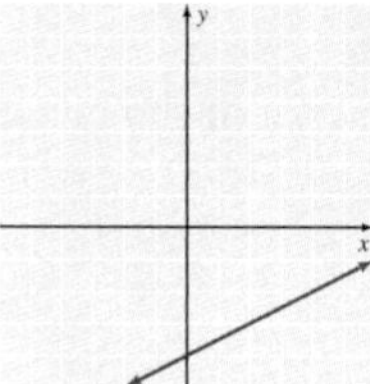

51. 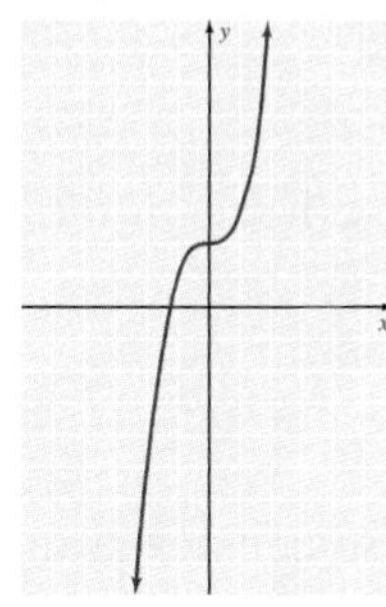

For Problems 52–53, form the inverse function.

52. $f = \{(-1, 4), (0, 5), (1, 7), (2, 9)\}$

53. $f = \{(2, 4), (3, 9), (4, 16), (5, 25)\}$

For Problems 54–56, find the inverse of the given function.

54. $f(x) = 6x - 1$

55. $f(x) = \frac{2}{3}x + 7$

56. $f(x) = -\frac{3}{5}x - \frac{2}{7}$

57. Andrew's paycheck varies directly with the number of hours he works. If he is paid \$475 when he works 38 hours, find his pay when he works 30 hours.

58. The surface area of a cube varies directly as the square of the length of an edge. If the surface area of a cube that has edges 8 inches long is 384 square inches, find the surface area of a cube that has edges 10 inches long.

59. The time it takes to fill an aquarium varies inversely with the square of the hose diameter. If it takes 40 minutes to fill the aquarium when the hose diameter is $\frac{1}{4}$ inch, find the time it takes to fill the aquarium when the hose diameter is $\frac{1}{2}$ inch.

60. The weight of a body above the surface of the earth varies inversely as the square of its distance from the center of the earth. Assume that the radius of the earth is 4000 miles. How much would a man weigh 1000 miles above the earth's surface if he weighs 200 pounds on the surface?

61. If y varies directly as x and inversely as z and if $y = 21$ when $x = 14$ and $z = 6$, find the constant of variation.

62. If y varies jointly as x and the square root of z and if $y = 60$ when $x = 2$ and $z = 9$, find y when $x = 3$ and $z = 16$.

Chapter 10 Test

1. ______

2. ______

3. ______

4. ______

5. ______

6. ______

7. ______

8. ______

9. ______

10. ______

11. ______

12. ______

13. ______

14. ______

15. ______

16. ______

17. ______

18. ______

19. ______

1. Determine the domain of the function $f(x) = \dfrac{-3}{2x^2 + 7x - 4}$.

2. Determine the domain of the function $f(x) = \sqrt{5 - 3x}$.

3. If $f(x) = -\frac{1}{2}x + \frac{1}{3}$, find $f(-3)$.

4. If $f(x) = -x^2 - 6x + 3$, find $f(-2a)$.

5. Find the vertex of the parabola $f(x) = -2x^2 - 24x - 69$.

6. If $f(x) = 3x^2 + 2x - 5$, find $\dfrac{f(a+h) - f(a)}{h}$.

7. If $f(x) = -3x + 4$ and $g(x) = 7x + 2$, find $(f \circ g)(x)$.

8. If $f(x) = 2x + 5$ and $g(x) = 2x^2 - x + 3$, find $(g \circ f)(x)$.

9. If $f(x) = \dfrac{3}{x-2}$ and $g(x) = \dfrac{2}{x}$, find $(f \circ g)(x)$.

For Problems 10–12, find the inverse of the given function.

10. $f(x) = 5x - 9$

11. $f(x) = -3x - 6$

12. $f(x) = \frac{2}{3}x - \frac{3}{5}$

13. If y varies inversely as x and if $y = \frac{1}{2}$ when $x = -8$, find the constant of variation.

14. If y varies jointly as x and z, and if $y = 18$ when $x = 8$ and $z = 9$, find y when $x = 5$ and $z = 12$.

15. Find two numbers whose sum is 60 such that the sum of the square of one number plus twelve times the other number is a minimum.

16. The simple interest earned by a certain amount of money varies jointly as the rate of interest and the time (in years) that the money is invested. If \$140 is earned for a certain amount of money invested at 7% for 5 years, how much is earned if the same amount is invested at 8% for 3 years?

For Problems 17–19, use the concepts of translation and/or reflection to describe how the second curve can be obtained from the first curve.

17. $f(x) = x^3$, $f(x) = (x - 6)^3 - 4$

18. $f(x) = |x|$, $f(x) = -|x| + 8$

19. $f(x) = \sqrt{x}$, $f(x) = -\sqrt{x + 5} + 7$

For Problems 20–25, graph each function.

20. $f(x) = -x - 1$

21. $f(x) = -2x^2 - 12x - 14$

22. $f(x) = 2\sqrt{x} - 2$

23. $f(x) = 3|x - 2| - 1$

24. $f(x) = -\dfrac{1}{x} + 3$

25. $f(x) = \sqrt{-x + 2}$

20. ______

21. ______

22. ______

23. ______

24. ______

25. ______

Chapters 1–10 Cumulative Review Problem Set

For Problems 1–5, evaluate each algebraic expression for the given values of the variables.

1. $-5(x - 1) - 3(2x + 4) + 3(3x - 1)$ for $x = -2$

2. $\dfrac{14a^3b^2}{7a^2b}$ for $a = -1$ and $b = 4$

3. $\dfrac{2}{n} - \dfrac{3}{2n} + \dfrac{5}{3n}$ for $n = 4$

4. $-4\sqrt{2x - y} + 5\sqrt{3x + y}$ for $x = 16$ and $y = 16$

5. $\dfrac{3}{x - 2} - \dfrac{5}{x + 3}$ for $x = 3$

For Problems 6–15, perform the indicated operations and express the answers in simplified form.

6. $(-5\sqrt{6})(3\sqrt{12})$

7. $(2\sqrt{x} - 3)(\sqrt{x} + 4)$

8. $(3\sqrt{2} - \sqrt{6})(\sqrt{2} + 4\sqrt{6})$

9. $(2x - 1)(x^2 + 6x - 4)$

10. $\dfrac{x^2 - x}{x + 5} \cdot \dfrac{x^2 + 5x + 4}{x^4 - x^2}$

11. $\dfrac{16x^2y}{24xy^3} \div \dfrac{9xy}{8x^2y^2}$

12. $\dfrac{x + 3}{10} + \dfrac{2x + 1}{15} - \dfrac{x - 2}{18}$

13. $\dfrac{7}{12ab} - \dfrac{11}{15a^2}$

14. $\dfrac{8}{x^2 - 4x} + \dfrac{2}{x}$

15. $(8x^3 - 6x^2 - 15x + 4) \div (4x - 1)$

For Problems 16–19, simplify each of the complex fractions.

16. $\dfrac{\dfrac{5}{x^2} - \dfrac{3}{x}}{\dfrac{1}{y} + \dfrac{2}{y^2}}$

17. $\dfrac{\dfrac{2}{x} - 3}{\dfrac{3}{y} + 4}$

18. $\dfrac{2 - \dfrac{1}{n + 2}}{3 + \dfrac{4}{n + 3}}$

19. $\dfrac{3a}{2 - \dfrac{1}{a}} - 1$

For Problems 20–25, factor each of the algebraic expressions completely.

20. $20x^2 + 7x - 6$

21. $16x^3 + 54$

22. $4x^4 - 25x^2 + 36$

23. $12x^3 - 52x^2 - 40x$

24. $xy - 6x + 3y - 18$

25. $10 + 9x - 9x^2$

For Problems 26–33, evaluate each of the numerical expressions.

26. $\left(\dfrac{2}{3}\right)^{-4}$

27. $\dfrac{3}{\left(\dfrac{4}{3}\right)^{-1}}$

28. $\sqrt[3]{-\dfrac{27}{64}}$

29. $-\sqrt{0.09}$

30. $(27)^{-\frac{4}{3}}$

31. $4^0 + 4^{-1} + 4^{-2}$

32. $\left(\dfrac{3^{-1}}{2^{-3}}\right)^{-2}$

33. $(2^{-3} - 3^{-2})^{-1}$

For Problems 34–36, find the indicated products and quotients, and express the final answers with positive integral exponents only.

34. $(-3x^{-1}y^2)(4x^{-2}y^{-3})$

35. $\dfrac{48x^{-4}y^2}{6xy}$

36. $\left(\dfrac{27a^{-4}b^{-3}}{-3a^{-1}b^{-4}}\right)^{-1}$

For Problems 37–44, express each radical expression in simplest radical form.

37. $\sqrt{80}$

38. $-2\sqrt{54}$

39. $\sqrt{\dfrac{75}{81}}$

40. $\dfrac{4\sqrt{6}}{3\sqrt{8}}$

41. $\sqrt[3]{56}$

42. $\dfrac{\sqrt[3]{3}}{\sqrt[3]{4}}$

43. $4\sqrt{52x^3y^2}$

44. $\sqrt{\dfrac{2x}{3y}}$

For Problems 45–47, use the distributive property to help simplify each of the following.

45. $-3\sqrt{24} + 6\sqrt{54} - \sqrt{6}$

46. $\dfrac{\sqrt{8}}{3} - \dfrac{3\sqrt{18}}{4} - \dfrac{5\sqrt{50}}{2}$

47. $8\sqrt[3]{3} - 6\sqrt[3]{24} - 4\sqrt[3]{81}$

For Problems 48 and 49, rationalize the denominator and simplify.

48. $\dfrac{\sqrt{3}}{\sqrt{6} - 2\sqrt{2}}$

49. $\dfrac{3\sqrt{5} - \sqrt{3}}{2\sqrt{3} + \sqrt{7}}$

For Problems 50–52, use scientific notation to help perform the indicated operations.

50. $\dfrac{(0.00016)(300)(0.028)}{0.064}$

51. $\dfrac{0.00072}{0.0000024}$

52. $\sqrt{0.00000009}$

For Problems 53–56, find each of the indicated products or quotients, and express the answers in standard form.

53. $(5 - 2i)(4 + 6i)$

54. $(-3 - i)(5 - 2i)$

55. $\frac{5}{4i}$

56. $\frac{-1 + 6i}{7 - 2i}$

57. Find the slope of the line determined by the points $(2, -3)$ and $(-1, 7)$.

58. Find the slope of the line determined by the equation $4x - 7y = 9$.

59. Find the length of the line segment whose endpoints are $(4, 5)$ and $(-2, 1)$.

60. Write the equation of the line that contains the points $(3, -1)$ and $(7, 4)$.

61. Write the equation of the line that is perpendicular to the line $3x - 4y = 6$ and contains the point $(-3, -2)$.

62. Find the center and the length of a radius of the circle $x^2 + 4x + y^2 - 12y + 31 = 0$.

63. Find the coordinates of the vertex of the parabola $y = x^2 + 10x + 21$.

64. Find the length of the major axis of the ellipse $x^2 + 4y^2 = 16$.

For Problems 65–70, graph each of the equations.

65. $-x + 2y = -4$

66. $x^2 + y^2 = 9$

67. $x^2 - y^2 = 9$

68. $x^2 + 2y^2 = 8$

69. $y = -3x$

70. $x^2y = 4$

For Problems 71–76, graph each of the functions.

71. $f(x) = -2x - 4$

72. $f(x) = -2x^2 - 2$

73. $f(x) = x^2 - 2x - 2$

74. $f(x) = \sqrt{x + 1} + 2$

75. $f(x) = 2x^2 + 8x + 9$

76. $f(x) = -|x - 2| + 1$

77. If $f(x) = x - 3$ and $g(x) = 2x^2 - x - 1$, find $(g \circ f)(x)$ and $(f \circ g)(x)$.

78. Find the inverse of $f(x) = 3x - 7$.

79. Find the inverse of $f(x) = -\frac{1}{2}x + \frac{2}{3}$.

80. Find the constant of variation if y varies directly as x, and $y = 2$ when $x = -\frac{2}{3}$.

81. If y is inversely proportional to the square of x, and $y = 4$ when $x = 3$, find y when $x = 6$.

82. The volume of a gas at a constant temperature varies inversely as the pressure. What is the volume of a gas under a pressure of 25 pounds if the gas occupies 15 cubic centimeters under a pressure of 20 pounds?

For Problems 83–103, solve each of the equations.

83. $3(2x - 1) - 2(5x + 1) = 4(3x + 4)$

84. $n + \frac{3n - 1}{9} - 4 = \frac{3n + 1}{3}$

85. $0.92 + 0.9(x - 0.3) = 2x - 5.95$

86. $|4x - 1| = 11$

87. $3x^2 = 7x$

88. $x^3 - 36x = 0$

89. $30x^2 + 13x - 10 = 0$

90. $8x^3 + 12x^2 - 36x = 0$

91. $x^4 + 8x^2 - 9 = 0$

92. $(n + 4)(n - 6) = 11$

93. $2 - \frac{3x}{x - 4} = \frac{14}{x + 7}$

94. $\frac{2n}{6n^2 + 7n - 3} - \frac{n - 3}{3n^2 + 11n - 4} = \frac{5}{2n^2 + 11n + 12}$

95. $\sqrt{3y} - y = -6$

96. $\sqrt{x + 19} - \sqrt{x + 28} = -1$

97. $(3x - 1)^2 = 45$

98. $(2x + 5)^2 = -32$

99. $2x^2 - 3x + 4 = 0$

100. $3n^2 - 6n + 2 = 0$

101. $\frac{5}{n - 3} - \frac{3}{n + 3} = 1$

102. $12x^4 - 19x^2 + 5 = 0$

103. $2x^2 + 5x + 5 = 0$

For Problems 104–113, solve each of the inequalities.

104. $-5(y - 1) + 3 > 3y - 4 - 4y$

105. $0.06x + 0.08(250 - x) \geq 19$

106. $|5x - 2| > 13$

107. $|6x + 2| < 8$

108. $\frac{x - 2}{5} - \frac{3x - 1}{4} \leq \frac{3}{10}$

109. $(x - 2)(x + 4) \leq 0$

110. $(3x - 1)(x - 4) > 0$

111. $x(x + 5) < 24$

112. $\frac{x-3}{x-7} \geq 0$

113. $\frac{2x}{x+3} > 4$

For Problems 114–118, solve each of the systems of equations.

114. $\begin{pmatrix} 4x - 3y = 18 \\ 3x - 2y = 15 \end{pmatrix}$

115. $\begin{pmatrix} y = \frac{2}{5}x - 1 \\ 3x + 5y = 4 \end{pmatrix}$

116. $\begin{pmatrix} \frac{x}{2} - \frac{y}{3} = 1 \\ \frac{2x}{5} + \frac{y}{2} = 2 \end{pmatrix}$

117. $\begin{pmatrix} 4x - y + 3z = -12 \\ 2x + 3y - z = 8 \\ 6x + y + 2z = -8 \end{pmatrix}$

118. $\begin{pmatrix} x - y + 5z = -10 \\ 5x + 2y - 3z = 6 \\ -3x + 2y - z = 12 \end{pmatrix}$

For Problems 119–132, set up an equation, an inequality, or a system of equations to help solve each problem.

119. Find three consecutive odd integers whose sum is 57.

120. Suppose that Eric has a collection of 63 coins consisting of nickels, dimes, and quarters. The number of dimes is 6 more than the number of nickels, and the number of quarters is 1 more than twice the number of nickels. How many coins of each kind are in the collection?

121. One of two supplementary angles is 4° more than one-third of the other angle. Find the measure of each of the angles.

122. If a ring costs a jeweler \$300, at what price should it be sold for the jeweler to make a profit of 50% on the selling price?

123. Last year Beth invested a certain amount of money at 8% and \$300 more than that amount at 9%. Her total yearly interest was \$316. How much did she invest at each rate?

124. Two trains leave the same depot at the same time, one traveling east and the other traveling west. At the end of $4\frac{1}{2}$ hours, they are 639 miles apart. If the rate of the train traveling east is 10 miles per hour greater than that of the other train, find their rates.

125. Suppose that a 10-quart radiator contains a 50% solution of antifreeze. How much needs to be drained out and replaced with pure antifreeze to obtain a 70% antifreeze solution?

126. Sam shot rounds of 70, 73, and 76 on the first three days of a golf tournament. What must he shoot on the fourth day of the tournament to average 72 or less for the four days?

127. The cube of a number equals nine times the same number. Find the number.

128. A strip of uniform width is to be cut off of both sides and both ends of a sheet of paper that is 8 inches by 14 inches in order to reduce the size of the paper to an area of 72 square inches. (See Figure 10.35.) Find the width of the strip.

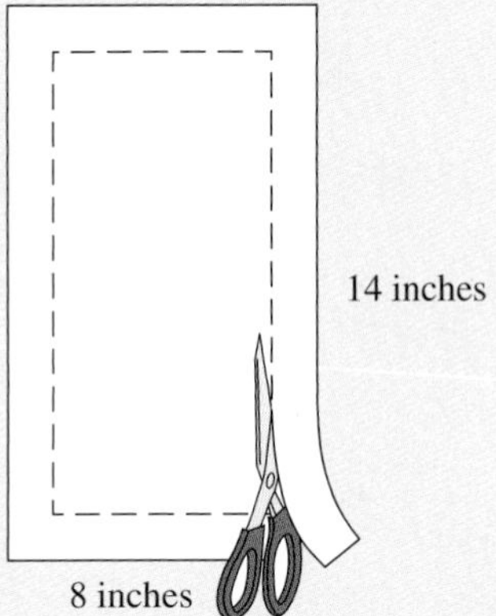

Figure 10.35

129. A sum of \$2450 is to be divided between two people in the ratio of 3 to 4. How much does each person receive?

130. Working together, Crystal and Dean can complete a task in $1\frac{1}{5}$ hours. Dean can do the task by himself in 2 hours. How long would it take Crystal to complete the task by herself?

131. The units digit of a two-digit number is 1 more than twice the tens digit. The sum of the digits is 10. Find the number.

132. The sum of the two smallest angles of a triangle is 40° less than the other angle. The sum of the smallest and largest angles is twice the other angle. Find the measures of the three angles of the triangle.

Exponential and Logarithmic Functions

Romeo Gacad/AFP/Getty Images

■ *Because the Richter number for reporting the intensity of an earthquake is calculated from a logarithm, it is referred to as a logarithmic scale. Logarithmic scales are commonly used in science and mathematics to transform very large numbers to a smaller scale.*

How long will it take \$100 to triple itself if it is invested at 8% interest compounded continuously? We can use the formula $A = Pe^{rt}$ to generate the equation $300 = 100e^{0.08t}$, which can be solved for t by using logarithms. It will take approximately 13.7 years for the money to triple itself.

This chapter will expand the meaning of an exponent and introduce the concept of a logarithm. We will (1) work with some exponential functions, (2) work with some logarithmic functions, and (3) use the concepts of exponent and logarithm to expand our capabilities for solving problems. Your calculator will be a valuable tool throughout this chapter.

Video tutorials for all section learning objectives are available in a variety of delivery modes.

INTERNET PROJECT

Leonardo da Vinci, the prototype for a "Renaissance man," had extraordinary talents and broad interests. He was fascinated with the mathematical relationships found in the human body, and his drawing, *Vitruvian Man*, was a study of its proportions. Da Vinci made a series of observations concerning the human body; for instance, the length of a man's outstretched arms is equal to his height. Conduct an Internet search to find two more of da Vinci's observations on the human body that are examples of direct variation.

11.1 Exponents and Exponential Functions

OBJECTIVES

1. Solve Exponential Equations
2. Graph Exponential Functions

1 Solve Exponential Equations

In Chapter 1, the expression b^n was defined as n factors of b, where n is any positive integer and b is any real number. For example,

$$4^3 = 4 \cdot 4 \cdot 4 = 64$$

$$\left(\frac{1}{2}\right)^4 = \left(\frac{1}{2}\right)\left(\frac{1}{2}\right)\left(\frac{1}{2}\right)\left(\frac{1}{2}\right) = \frac{1}{16}$$

$$(-0.3)^2 = (-0.3)(-0.3) = 0.09$$

In Chapter 7, by defining $b^0 = 1$ and $b^{-n} = \frac{1}{b^n}$, where n is any positive integer and b is any nonzero real number, we extended the concept of an exponent to include all integers. For example,

$$2^{-3} = \frac{1}{2^3} = \frac{1}{2 \cdot 2 \cdot 2} = \frac{1}{8} \qquad \left(\frac{1}{3}\right)^{-2} = \frac{1}{\left(\frac{1}{3}\right)^2} = \frac{1}{\frac{1}{9}} = 9$$

$$(0.4)^{-1} = \frac{1}{(0.4)^1} = \frac{1}{0.4} = 2.5 \qquad (-0.98)^0 = 1$$

In Chapter 7 we also provided for the use of all rational numbers as exponents by defining $b^{\frac{m}{n}} = \sqrt[n]{b^m}$, where n is a positive integer greater than 1 and b is a real-number such that $\sqrt[n]{b}$ exists. For example,

$$8^{\frac{2}{3}} = \sqrt[3]{8^2} = \sqrt[3]{64} = 4$$

$$16^{\frac{1}{4}} = \sqrt[4]{16^1} = 2$$

$$32^{-\frac{1}{5}} = \frac{1}{32^{\frac{1}{5}}} = \frac{1}{\sqrt[5]{32}} = \frac{1}{2}$$

To extend the concept of an exponent formally to include the use of irrational numbers requires some ideas from calculus and is therefore beyond the scope of this text. However, here's a glance at the general idea involved. Consider the number $2^{\sqrt{3}}$. By using the nonterminating and nonrepeating decimal representation 1.73205 . . . for $\sqrt{3}$, form the sequence of numbers $2^1, 2^{1.7}, 2^{1.73}, 2^{1.732}, 2^{1.7320}, 2^{1.73205}, \ldots$. It would seem reasonable that each successive power gets closer to $2^{\sqrt{3}}$. This is precisely what happens when b^n, where n is irrational, is properly defined by using the concept of a limit.

From now on, then, we can use any real number as an exponent, and the basic properties stated in Chapter 7 can be extended to include all real numbers as exponents. Let's restate those properties at this time with the restriction that the bases a and b are to be positive numbers (to avoid expressions such as $(-4)^{\frac{1}{2}}$, which do not represent real numbers).

Property 11.1

If a and b are positive real numbers and m and n are any real numbers, then

1. $b^n \cdot b^m = b^{n+m}$ — Product of two powers
2. $(b^n)^m = b^{mn}$ — Power of a power
3. $(ab)^n = a^n b^n$ — Power of a product
4. $\left(\frac{a}{b}\right)^n = \frac{a^n}{b^n}$ — Power of a quotient
5. $\frac{b^n}{b^m} = b^{n-m}$ — Quotient of two powers

Another property that can be used to solve certain types of equations involving exponents can be stated as follows.

Property 11.2

If $b > 0$, $b \neq 1$, and m and n are real numbers, then

$$b^n = b^m \quad \text{if and only if } n = m$$

The following examples illustrate the use of Property 11.2.

EXAMPLE 1

Solve $2^x = 32$.

Solution

$$2^x = 32$$

$$2^x = 2^5 \qquad 32 = 2^5$$

$$x = 5 \qquad \text{Property 11.2}$$

The solution set is $\{5\}$.

▼ PRACTICE YOUR SKILL

Solve $3^x = 81$. ■

EXAMPLE 2

Solve $3^{2x} = \frac{1}{9}$.

Solution

$$3^{2x} = \frac{1}{9} = \frac{1}{3^2}$$

$$3^{2x} = 3^{-2}$$

$$2x = -2 \qquad \text{Property 11.2}$$

$$x = -1$$

The solution set is $\{-1\}$.

▼ PRACTICE YOUR SKILL

Solve $2^{3x} = \frac{1}{8}$. ■

EXAMPLE 3

Solve $\left(\frac{1}{5}\right)^{x-2} = \frac{1}{125}$.

Solution

$$\left(\frac{1}{5}\right)^{x-2} = \frac{1}{125} = \left(\frac{1}{5}\right)^3$$

$$x - 2 = 3 \qquad \text{Property 11.2}$$

$$x = 5$$

The solution set is {5}.

▼ PRACTICE YOUR SKILL

Solve $\left(\frac{1}{4}\right)^{x+2} = \frac{1}{64}$. ■

EXAMPLE 4

Solve $8^x = 32$.

Solution

$$8^x = 32$$

$$(2^3)^x = 2^5 \qquad 8 = 2^3$$

$$2^{3x} = 2^5$$

$$3x = 5 \qquad \text{Property 11.2}$$

$$x = \frac{5}{3}$$

The solution set is $\left\{\frac{5}{3}\right\}$.

▼ PRACTICE YOUR SKILL

Solve $9^x = 27$. ■

EXAMPLE 5

Solve $(3^{x+1})(9^{x-2}) = 27$.

Solution

$$(3^{x+1})(9^{x-2}) = 27$$

$$(3^{x+1})(3^2)^{x-2} = 3^3$$

$$(3^{x+1})(3^{2x-4}) = 3^3$$

$$3^{3x-3} = 3^3$$

$$3x - 3 = 3 \qquad \text{Property 11.2}$$

$$3x = 6$$

$$x = 2$$

The solution set is {2}.

▼ **PRACTICE YOUR SKILL**

Solve $(2^{x+3})(4^{x-1}) = 16$. ■

2 Graph Exponential Functions

If b is any positive number, then the expression b^x designates exactly one real number for every real value of x. Thus the equation $f(x) = b^x$ defines a function whose domain is the set of real numbers. Furthermore, if we impose the additional restriction $b \neq 1$, then any equation of the form $f(x) = b^x$ describes a one-to-one function and is called an **exponential function**. This leads to the following definition.

Definition 11.1

If $b > 0$ and $b \neq 1$, then the function f defined by

$$f(x) = b^x$$

where x is any real number, is called the **exponential function with base b.**

Remark: The function $f(x) = 1^x$ is a constant function whose graph is a horizontal line, and therefore it is not a one-to-one function. Remember from Chapter 10 that one-to-one functions have inverses; this becomes a key issue in a later section.

Now let's consider graphing some exponential functions.

EXAMPLE 6

Graph the function $f(x) = 2^x$.

Solution

First, let's set up a table of values.

x	$f(x) = 2^x$
-2	$\frac{1}{4}$
-1	$\frac{1}{2}$
0	1
1	2
2	4
3	8

Plot these points and connect them with a smooth curve to produce Figure 11.1.

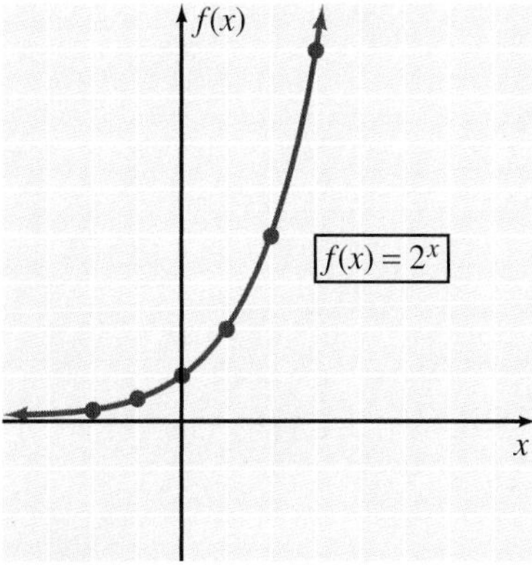

Figure 11.1

▼ **PRACTICE YOUR SKILL**

Graph the function $f(x) = 3^x$. ■

EXAMPLE 7 Graph $f(x) = \left(\frac{1}{2}\right)^x$.

Solution

Again, let's set up a table of values. Plot these points and connect them with a smooth curve to produce Figure 11.2.

x	$f(x) = \left(\frac{1}{2}\right)^x$
−2	4
−1	2
0	1
1	$\frac{1}{2}$
2	$\frac{1}{4}$
3	$\frac{1}{8}$

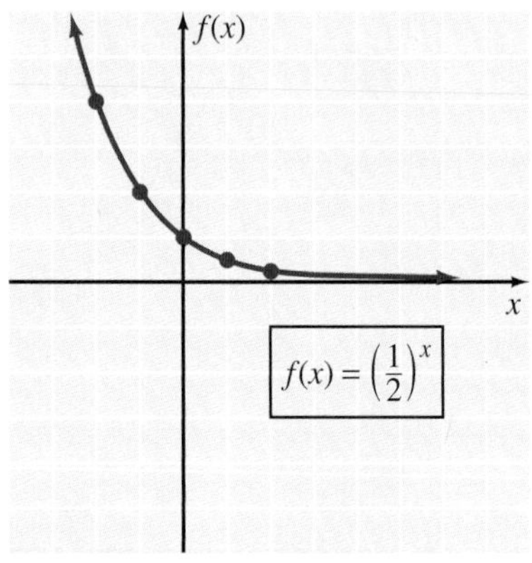

Figure 11.2

▼ **PRACTICE YOUR SKILL**

Graph the function $f(x) = \left(\frac{1}{3}\right)^x$. ■

In the tables for Examples 6 and 7 we chose integral values for x to keep the computation simple. However, with the use of a calculator, we could easily acquire functional values by using nonintegral exponents. Consider the following additional values for each of the tables.

$f(x) = 2^x$	
$f(0.5) \approx 1.41$	$f(-0.5) \approx 0.71$
$f(1.7) \approx 3.25$	$f(-2.6) \approx 0.16$

$f(x) = \left(\frac{1}{2}\right)^x$	
$f(0.7) \approx 0.62$	$f(-0.8) \approx 1.74$
$f(2.3) \approx 0.20$	$f(-2.1) \approx 4.29$

Use your calculator to check these results. Also, it would be worthwhile for you to go back and see that the points determined do fit the graphs in Figures 11.1 and 11.2.

The graphs in Figures 11.1 and 11.2 illustrate a general behavior pattern of exponential functions. That is, if $b > 1$, then the graph of $f(x) = b^x$ *goes up to the right* and the function is called an **increasing function**. If $0 < b < 1$, then the graph of $f(x) = b^x$ *goes down to the right* and the function is called a **decreasing function**. These facts are illustrated in Figure 11.3. Note that because $b^0 = 1$ for any $b > 0$, all graphs of $f(x) = b^x$ contain the point (0, 1).

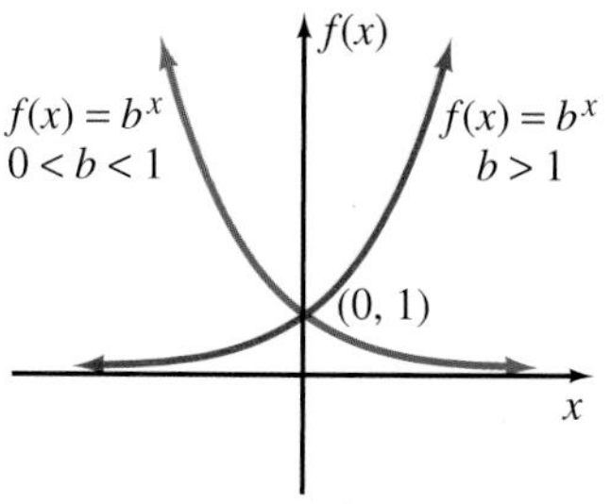

Figure 11.3

As you graph exponential functions, don't forget your previous graphing experiences.

1. The graph of $f(x) = 2^x - 4$ is the graph of $f(x) = 2^x$ moved down 4 units.
2. The graph of $f(x) = 2^{x+3}$ is the graph of $f(x) = 2^x$ moved 3 units to the left.
3. The graph of $f(x) = -2^x$ is the graph of $f(x) = 2^x$ reflected across the x axis.

We used a graphing calculator to graph these four functions on the same set of axes, as shown in Figure 11.4.

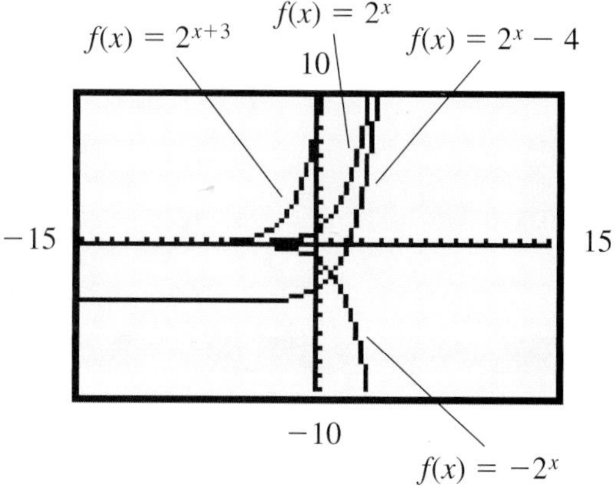

Figure 11.4

CONCEPT QUIZ

For Problems 1–10, answer true or false.

1. If $2^{x+1} = 2^{3x}$, then $x + 1 = 3x$.
2. The numerical expression 9^x is equivalent to 3^{2x}.
3. For the exponential function $f(x) = b^x$, the base b can be any positive number.
4. All the graphs of $f(x) = b^x$ for all positive values of b pass through the point (0, 1).
5. The graphs of $f(x) = 3^x$ and $f(x) = \left(\frac{1}{3}\right)^x$ are reflections of each other across the y axis.
6. The solution set for $(8^x)(4^{x-2}) = 16$ is {1}.
7. The graph of $f(x) = 2^x - 4$ is the graph of $f(x) = 2^x$ shifted 4 units to the left.
8. The graph of $f(x) = 3^{-x}$ is the graph of $f(x) = 3^x$ reflected across the y axis.
9. The function $f(x) = 1^x$ is a one-to-one function.
10. The function $f(x) = -3^x$ is a decreasing function.

Problem Set 11.1

1 Solve Exponential Equations

For Problems 1–34, solve each equation.

1. $3^x = 27$
2. $2^x = 64$
3. $2^{2x} = 16$
4. $3^{2x} = 81$
5. $\left(\frac{1}{4}\right)^x = \frac{1}{256}$
6. $\left(\frac{1}{2}\right)^x = \frac{1}{128}$
7. $5^{x+2} = 125$
8. $4^{x-3} = 16$
9. $3^{-x} = \frac{1}{243}$
10. $5^{-x} = \frac{1}{25}$
11. $6^{3x-1} = 36$
12. $2^{2x+3} = 32$
13. $4^x = 8$
14. $16^x = 64$
15. $8^{2x} = 32$
16. $9^{3x} = 27$
17. $\left(\frac{1}{2}\right)^{2x} = 64$
18. $\left(\frac{1}{3}\right)^{5x} = 243$
19. $\left(\frac{3}{4}\right)^x = \frac{64}{27}$
20. $\left(\frac{2}{3}\right)^x = \frac{9}{4}$
21. $9^{4x-2} = \frac{1}{81}$
22. $8^{3x+2} = \frac{1}{16}$
23. $6^{2x} + 3 = 39$
24. $5^{2x} - 2 = 123$
25. $10^x = 0.1$
26. $10^x = 0.0001$
27. $32^x = \frac{1}{4}$
28. $9^x = \frac{1}{27}$
29. $(2^{x+1})(2^x) = 64$
30. $(2^{2x-1})(2^{x+2}) = 32$
31. $(27)(3^x) = 9^x$
32. $(3^x)(3^{5x}) = 81$
33. $(4^x)(16^{3x-1}) = 8$
34. $(8^{2x})(4^{2x-1}) = 16$

2 Graph Exponential Functions

For Problems 35–52, graph each exponential function.

35. $f(x) = 4^x$
36. $f(x) = 3^x$
37. $f(x) = 6^x$
38. $f(x) = 5^x$
39. $f(x) = \left(\frac{1}{3}\right)^x$
40. $f(x) = \left(\frac{1}{4}\right)^x$
41. $f(x) = \left(\frac{3}{4}\right)^x$
42. $f(x) = \left(\frac{2}{3}\right)^x$
43. $f(x) = 2^{x-2}$
44. $f(x) = 2^{x+1}$
45. $f(x) = 3^{-x}$
46. $f(x) = 2^{-x}$
47. $f(x) = 3^{2x}$
48. $f(x) = 2^{2x}$
49. $f(x) = 3^x - 2$
50. $f(x) = 2^x + 1$
51. $f(x) = 2^{-x-2}$
52. $f(x) = 3^{-x+1}$

THOUGHTS INTO WORDS

53. Explain how you would solve the equation

$$(2^{x+1})(8^{2x-3}) = 64.$$

54. Why is the base of an exponential function restricted to positive numbers not including 1?

55. Explain how you would graph the function

$$f(x) = -\left(\frac{1}{3}\right)^x.$$

GRAPHING CALCULATOR ACTIVITIES

56. Use a graphing calculator to check your graphs for Problems 35–52.

57. Graph $f(x) = 2^x$. Where should the graphs of $f(x) = 2^{x-5}$, $f(x) = 2^{x-7}$, and $f(x) = 2^{x+5}$ be located? Graph all three functions on the same set of axes with $f(x) = 2^x$.

58. Graph $f(x) = 3^x$. Where should the graphs of $f(x) = 3^x + 2$, $f(x) = 3^x - 3$, and $f(x) = 3^x - 7$ be located? Graph all three functions on the same set of axes with $f(x) = 3^x$.

59. Graph $f(x) = \left(\frac{1}{2}\right)^x$. Where should the graphs of $f(x) = -\left(\frac{1}{2}\right)x$, $f(x) = \left(\frac{1}{2}\right)^{-x}$, and $f(x) = -\left(\frac{1}{2}\right)^{-x}$ be located? Graph all three functions on the same set of axes with $f(x) = \left(\frac{1}{2}\right)^x$.

60. Graph $f(x) = (1.5)^x$, $f(x) = (5.5)^x$, $f(x) = (0.3)^x$, and $f(x) = (0.7)^x$ on the same set of axes. Are these graphs consistent with Figure 11.3?

61. What is the solution for $3^x = 5$? Do you agree that it is between 1 and 2 because $3^1 = 3$ and $3^2 = 9$? Now graph $f(x) = 3^x - 5$ and use the ZOOM and TRACE features of your graphing calculator to find an approximation, to the nearest hundredth, for the x intercept. You should get an answer of 1.46, to the nearest hundredth. Do you see that this is an approximation for the solution of $3^x = 5$? Try it; raise 3 to the 1.46 power.

Find an approximate solution, to the nearest hundredth, for each of the following equations by graphing the appropriate function and finding the x intercept.

(a) $2^x = 19$ **(b)** $3^x = 50$

(c) $4^x = 47$ **(d)** $5^x = 120$

(e) $2^x = 1500$ **(f)** $3^{x-1} = 34$

Answers to the Concept Quiz

1. True **2.** True **3.** False **4.** True **5.** True **6.** False **7.** False **8.** True **9.** False **10.** True

Answers to the Example Practice Skills

1. {4} **2.** {−1} **3.** {1} **4.** $\left\{\frac{3}{2}\right\}$ **5.** {1}

6.

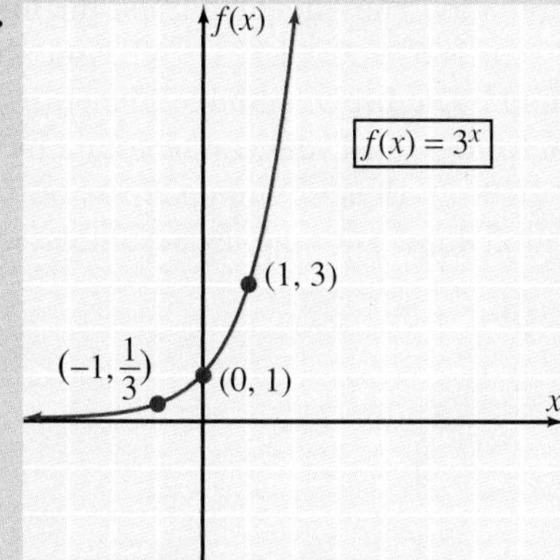

7.

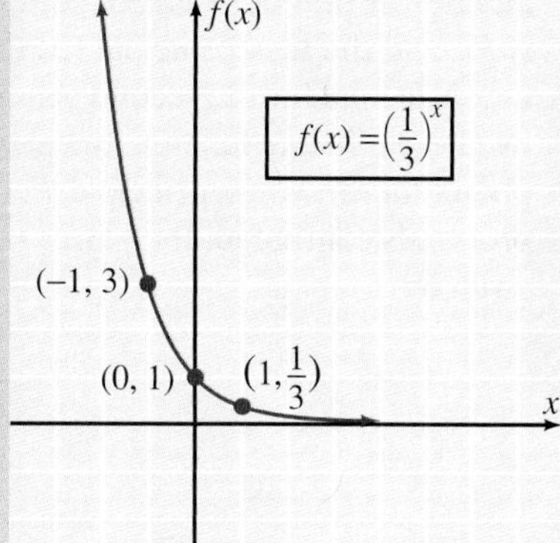

11.2 Applications of Exponential Functions

OBJECTIVES

1. Solve Exponential Growth and Compound Interest Problems
2. Solve Exponential Decay Problems
3. Solve Growth Problems Involving the Number e

1 Solve Exponential Growth and Compound Interest Problems

Equations that describe exponential functions can represent many real-world situations that exhibit growth or decay. For example, suppose that an economist predicts an annual inflation rate of 5% for the next 10 years. This means an item that presently costs \$8 will cost $8(105\%) = 8(1.05) = \$8.40$ a year from now. The same item will cost $(105\%)[8(105\%)] = 8(1.05)^2 = \$8.82$ in 2 years. In general, the equation

$$P = P_0(1.05)^t$$

yields the predicted price P of an item in t years at the annual inflation rate of 5%, where that item presently costs P_0. By using this equation, we can look at some future prices based on the prediction of a 5% inflation rate.

A \$1.00 jar of mustard will cost $\$1.00(1.05)^3 = \$1.15$ in 3 years.

A \$2.69 bag of potato chips will cost $\$2.69(1.05)^5 = \$3.43$ in 5 years.

A \$6.69 can of coffee will cost $\$6.69(1.05)^7 = \$9.41$ in 7 years.

Compound interest provides another illustration of exponential growth. Suppose that \$500 (called the **principal**) is invested at an interest rate of 8% **compounded annually**. The interest earned the first year is $\$500(0.08) = \$40$, and this amount is added to the original \$500 to form a new principal of \$540 for the second year. The interest earned during the second year is $\$540(0.08) = \$43.20$, and this amount is added to \$540 to form a new principal of \$583.20 for the third year. Each year a new principal is formed by reinvesting the interest earned that year.

In general, suppose that a sum of money P (called the principal) is invested at an interest rate of r percent compounded annually. The interest earned the first year is Pr, and the new principal for the second year is $P + Pr$ or $P(1 + r)$. Note that the new principal for the second year can be found by multiplying the original principal P by $(1 + r)$. In like fashion, the new principal for the third year can be found by multiplying the previous principal $P(1 + r)$ by $(1 + r)$, thus obtaining $P(1 + r)^2$. If this process is continued, then after t years the total amount of money accumulated, A, is given by

$$A = P(1 + r)^t$$

Consider the following examples of investments made at a certain rate of interest compounded annually.

1. \$750 invested for 5 years at 9% compounded annually produces

   $$A = \$750(1.09)^5 = \$1153.97$$

2. \$1000 invested for 10 years at 7% compounded annually produces

   $$A = \$1000(1.07)^{10} = \$1967.15$$

3. \$5000 invested for 20 years at 6% compounded annually produces

   $$A = \$5000(1.06)^{20} = \$16{,}035.68$$

If we invest money at a certain rate of interest to be compounded more than once a year, then we can adjust the basic formula, $A = P(1 + r)^t$, according to the number of compounding periods in the year. For example, for **compounding semiannually**, the formula becomes $A = P\left(1 + \frac{r}{2}\right)^{2t}$ and for **compounding quarterly**, the formula becomes $A = P\left(1 + \frac{r}{4}\right)^{4t}$. In general, we have the following formula, where n represents the number of compounding periods in a year.

$$A = P\left(1 + \frac{r}{n}\right)^{nt}$$

The following example should clarify the use of this formula.

EXAMPLE 1

Apply Your Skill

Find the amount produced by the given conditions.

(a) \$750 invested for 5 years at 9% compounded semiannually

(b) \$1000 invested for 10 years at 7% compounded quarterly

(c) \$5000 invested for 20 years at 6% compounded monthly

Solution

(a) \$750 invested for 5 years at 9% compounded semiannually produces

$$A = \$750\left(1 + \frac{0.09}{2}\right)^{2(5)} = \$750(1.045)^{10} = \$1164.73$$

(b) \$1000 invested for 10 years at 7% compounded quarterly produces

$$A = \$1000\left(1 + \frac{0.07}{4}\right)^{4(10)} = \$1000(1.0175)^{40} = \$2001.60$$

(c) \$5000 invested for 20 years at 6% compounded monthly produces

$$A = \$5000\left(1 + \frac{0.06}{12}\right)^{12(20)} = \$5000(1.005)^{240} = \$16{,}551.02$$

▼ PRACTICE YOUR SKILL

Find the amount produced by the given conditions.

(a) \$2500 invested for 4 years at 8% compounded quarterly

(b) \$3000 invested for 5 years at 6% compounded monthly ■

You may find it interesting to compare these results with those obtained earlier for compounding annually.

2 Solve Exponential Decay Problems

Suppose it is estimated that the value of a car depreciates 15% per year for the first 5 years. Thus a car that costs \$19,500 will be worth 19,500(100% − 15%) = 19,500(85%) = 19,500(0.85) = \$16,575 in 1 year. In 2 years the value of the car will have depreciated to $19{,}500(0.85)^2$ = \$14,089 (nearest dollar). The equation

$$V = V_0(0.85)^t$$

yields the value V of a car in t years at the annual depreciation rate of 15%, where the car initially cost V_0. By using this equation, we can estimate some car values, to the nearest dollar, as follows:

A \$17,000 car will be worth $\$17{,}000(0.85)^5$ = \$7543 in 5 years.

A \$25,000 car will be worth $\$25{,}000(0.85)^4$ = \$13,050 in 4 years.

A \$40,000 car will be worth $\$40{,}000(0.85)^3$ = \$24,565 in 3 years.

Another example of exponential decay is associated with radioactive substances. We can describe the rate of decay exponentially, on the basis of the half-life of a substance. The *half-life* of a radioactive substance is the amount of time that it takes for one-half of an initial amount of the substance to disappear as the result of decay. For example, suppose that we have 200 grams of a certain substance that has a half-life of 5 days. After 5 days, $200\left(\frac{1}{2}\right) = 100$ grams remain. After 10 days, $200\left(\frac{1}{2}\right)^2 = 50$ grams remain. After 15 days, $200\left(\frac{1}{2}\right)^3 = 25$ grams remain. In general, after t days, $200\left(\frac{1}{2}\right)^{\frac{t}{5}}$ grams remain.

This discussion leads us to the following half-life formula. Suppose there is an initial amount, Q_0, of a radioactive substance with a half-life of h. The amount of substance remaining, Q, after a time period of t is given by the formula

$$Q = Q_0\left(\frac{1}{2}\right)^{\frac{t}{h}}$$

The units of measure for t and h must be the same.

EXAMPLE 2

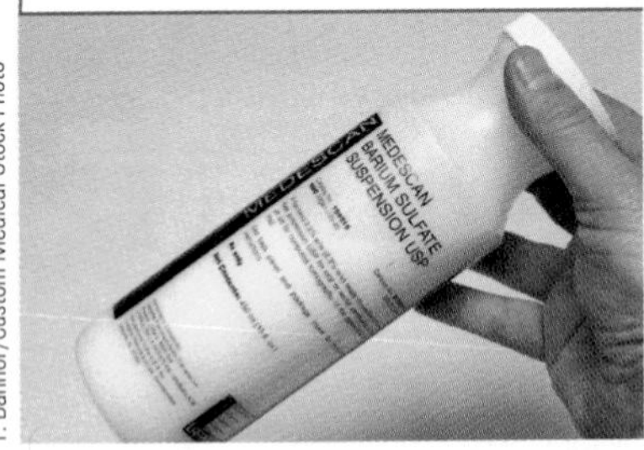

T. Bannor/Custom Medical Stock Photo

Apply Your Skill

Barium-140 has a half-life of 13 days. If there are 500 milligrams of barium initially, how many milligrams remain after 26 days? After 100 days?

Solution

When we use $Q_0 = 500$ and $h = 13$, the half-life formula becomes

$$Q = 500\left(\frac{1}{2}\right)^{\frac{t}{13}}$$

If $t = 26$, then

$$\begin{aligned} Q &= 500\left(\frac{1}{2}\right)^{\frac{26}{13}} \\ &= 500\left(\frac{1}{2}\right)^2 \\ &= 500\left(\frac{1}{4}\right) \\ &= 125 \end{aligned}$$

Thus 125 milligrams remain after 26 days. If $t = 100$, then

$$\begin{aligned} Q &= 500\left(\frac{1}{2}\right)^{\frac{100}{13}} \\ &= 500(0.5)^{\frac{100}{13}} \\ &= 2.4, \quad \text{to the nearest tenth of a milligram} \end{aligned}$$

Thus approximately 2.4 milligrams remain after 100 days.

▼ PRACTICE YOUR SKILL

A radioactive isotope has a half-life of 12 hours. If there are 20 milligrams of the isotope initially, how many milligrams remain after 36 hours? ■

Remark: The solution to Example 2 clearly demonstrates one facet of the role of the calculator in the application of mathematics. We solved the first part of the problem easily without the calculator, but the calculator certainly was helpful for the second part of the problem.

3 Solve Growth Problems Involving the Number *e*

An interesting situation occurs if we consider the compound interest formula for $P = \$1$, $r = 100\%$, and $t = 1$ year. The formula becomes $A = 1\left(1 + \frac{1}{n}\right)^n$. The accompanying table shows some values, rounded to eight decimal places, of $\left(1 + \frac{1}{n}\right)^n$ for different values of n.

n	$\left(1 + \frac{1}{n}\right)^n$
1	2.00000000
10	2.59374246
100	2.70481383
1000	2.71692393
10,000	2.71814593
100,000	2.71826824
1,000,000	2.71828047
10,000,000	2.71828169
100,000,000	2.71828181
1,000,000,000	2.71828183

The table suggests that as n increases, the value of $\left(1 + \frac{1}{n}\right)^n$ gets closer and closer to some fixed number. This does happen, and the fixed number is called e. To five decimal places, $e = 2.71828$.

The function defined by the equation $f(x) = e^x$ is the **natural exponential function**. It has a great many real-world applications; some we will look at in a moment. First, however, let's get a picture of the natural exponential function. Because $2 < e < 3$, the graph of $f(x) = e^x$ must fall between the graphs of $f(x) = 2^x$ and $f(x) = 3^x$. To be more specific, let's use our calculator to determine a table of values. Use the $\boxed{e^x}$ key, and round the results to the nearest tenth to obtain the following table. Plot the points determined by this table, and connect them with a smooth curve to produce Figure 11.5.

x	$f(x) = e^x$
0	1.0
1	2.7
2	7.4
−1	0.4
−2	0.1

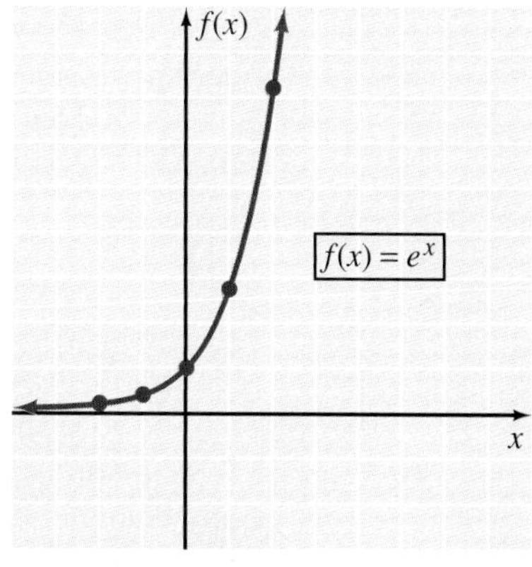

Figure 11.5

Let's return to the concept of compound interest. If the number of compounding periods in a year is increased indefinitely, we arrive at the concept of **compounding continuously**. Mathematically, this can be accomplished by applying the limit concept to the expression $P\left(1 + \frac{r}{n}\right)^{nt}$. We will not show the details here, but the following result is obtained. The formula

$$A = Pe^{rt}$$

yields the accumulated value, A, of a sum of money, P, that has been invested for t years at a rate of r percent compounded continuously. The following example shows the use of the formula.

EXAMPLE 3

Apply Your Skill

Find the amount produced by the given conditions.

(a) \$750 invested for 5 years at 9% compounded continuously

(b) \$1000 invested for 10 years at 7% compounded continuously

(c) \$5000 invested for 20 years at 6% compounded continuously

Solution

(a) \$750 invested for 5 years at 9% compounded continuously produces

$$A = \$750e^{(0.09)(5)} = 750e^{0.45} = \$1176.23$$

(b) \$1000 invested for 10 years at 7% compounded continuously produces

$$A = \$1000e^{(0.07)(10)} = 1000e^{0.7} = \$2013.75$$

(c) \$5000 invested for 20 years at 6% compounded continuously produces

$$A = \$5000e^{(0.06)(20)} = 5000e^{1.2} = \$16{,}600.58$$

▼ PRACTICE YOUR SKILL

Find the amount produced by the given conditions.

(a) \$2500 invested for 4 years at 8% compounded continuously

(b) \$3000 invested for 5 years at 6% compounded continuously ■

Again you may find it interesting to compare these results with those you obtained earlier when using a different number of compounding periods.

The ideas behind compounding continuously carry over to other growth situations. The law of exponential growth,

$$Q(t) = Q_0e^{kt}$$

is used as a mathematical model for numerous growth-and-decay applications. In this equation, $Q(t)$ represents the quantity of a given substance at any time t; Q_0 is the initial amount of the substance (when $t = 0$); and k is a constant that depends on the particular application. If $k < 0$, then $Q(t)$ decreases as t increases, and we refer to the model as the **law of decay**. Let's consider some growth-and-decay applications.

EXAMPLE 4

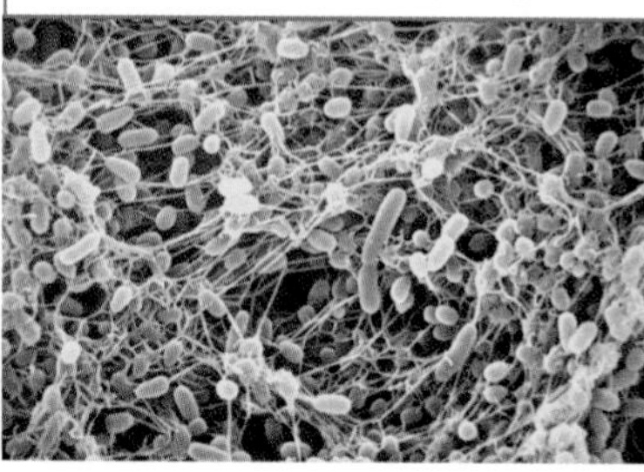

Manfred Kage/Peter Arnold, Inc./Alamy Limited

Apply Your Skill

Suppose that in a certain culture, the equation $Q(t) = 15{,}000e^{0.3t}$ expresses the number of bacteria present as a function of the time t, where t is expressed in hours. Find (a) the initial number of bacteria and (b) the number of bacteria after 3 hours.

Solution

(a) The initial number of bacteria is produced when $t = 0$.

$$Q(0) = 15{,}000e^{0.3(0)}$$
$$= 15{,}000e^{0}$$
$$= 15{,}000 \qquad e^0 = 1$$

(b) $Q(3) = 15{,}000e^{0.3(3)}$

$$= 15{,}000e^{0.9}$$
$$= 36{,}894, \quad \text{to the nearest whole number}$$

Thus there should be approximately 36,894 bacteria present after 3 hours.

▼ PRACTICE YOUR SKILL

Suppose that in a certain culture, the equation $Q(t) = 28{,}000e^{0.5t}$ expresses the number of bacteria present as a function of the time t, where t is expressed in hours. Find (a) the initial number of bacteria, and (b) the number of bacteria, to the nearest whole number, after 1 hour. ■

EXAMPLE 5

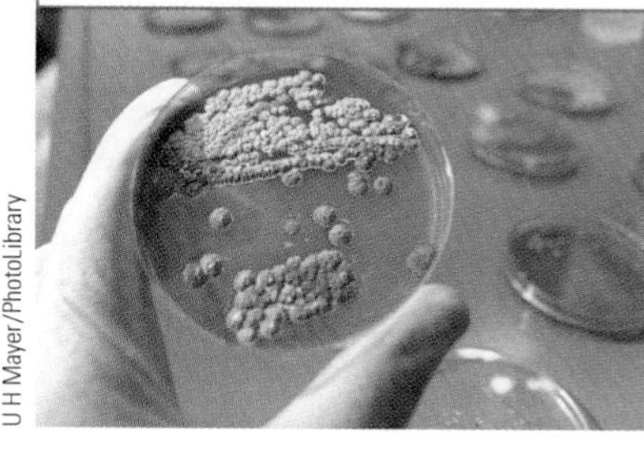

U H Mayer/PhotoLibrary

Apply Your Skill

Suppose the number of bacteria present in a certain culture after t minutes is given by the equation $Q(t) = Q_0e^{0.05t}$, where Q_0 represents the initial number of bacteria. If 5000 bacteria are present after 20 minutes, how many bacteria were present initially?

Solution

If 5000 bacteria are present after 20 minutes, then $Q(20) = 5000$.

$$5000 = Q_0e^{0.05(20)}$$
$$5000 = Q_oe^{1}$$
$$\frac{5000}{e} = Q_0$$
$$1839 = Q_0, \quad \text{to the nearest whole number}$$

Thus there were approximately 1839 bacteria present initially.

▼ PRACTICE YOUR SKILL

Suppose the number of bacteria present in a certain culture after t minutes is given by the equation $Q(t) = Q_0\,e^{0.04t}$, where Q_0 represents the initial number of bacteria. If 2500 bacteria are present after 35 minutes, how many bacteria, to the nearest whole number, were present initially? ■

EXAMPLE 6

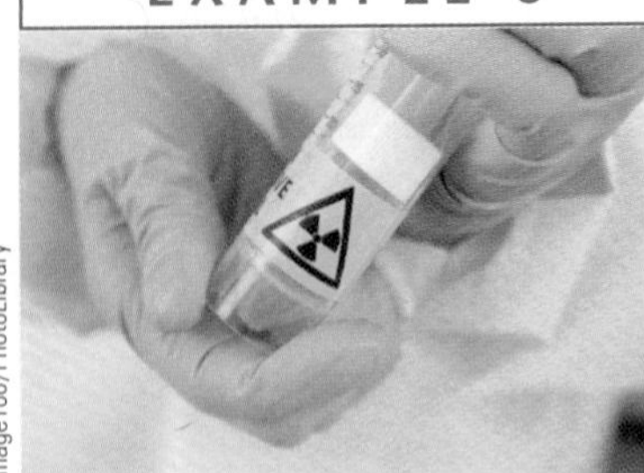
Image100/PhotoLibrary

Apply Your Skill

The number of grams Q of a certain radioactive substance present after t seconds is given by $Q(t) = 200e^{-0.3t}$. How many grams remain after 7 seconds?

Solution

We need to evaluate $Q(7)$.

$$Q(7) = 200e^{-0.3(7)}$$
$$= 200e^{-2.1}$$
$$= 24, \quad \text{to the nearest whole number}$$

Therefore, approximately 24 grams remain after 7 seconds.

▼ **PRACTICE YOUR SKILL**

The number of grams Q of a certain radioactive substance present after t seconds is given by $Q(t) = 500e^{-0.08t}$. How many grams remain after 12 seconds? ■

Finally, let's use a graphing calculator to graph a special exponential function.

EXAMPLE 7

Graph the function

$$y = \frac{1}{\sqrt{2\pi}}e^{-x^2/2}$$

and find its maximum value.

Solution

If $x = 0$, then $y = \frac{1}{\sqrt{2\pi}}e^0 = \frac{1}{\sqrt{2\pi}} \approx 0.4$. Let's set the boundaries of the viewing rectangle so that $-5 \le x \le 5$ and $0 \le y \le 1$ with a y scale of 0.1; the graph of the function is shown in Figure 11.6. From the graph, we see that the maximum value of the function occurs at $x = 0$, which we have already determined to be approximately 0.4.

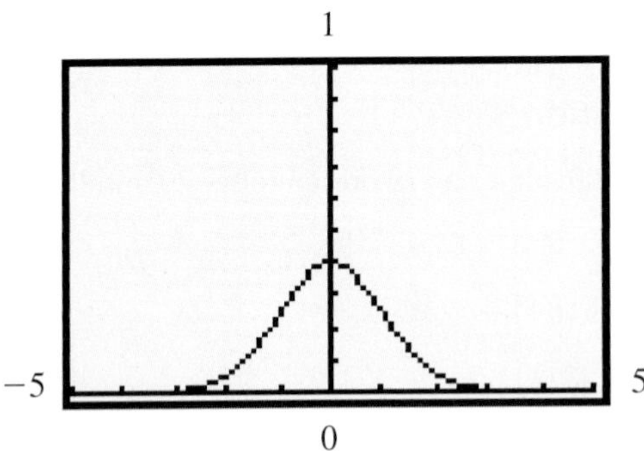

Figure 11.6

Remark: The curve in Figure 11.6 is called the standard normal distribution curve. You may want to ask your instructor to explain what it means to assign grades on the basis of the normal distribution curve.

CONCEPT QUIZ

For Problems 1–5, match each type of problem with its formula.

1. Compound continuously
2. Exponential growth or decay
3. Interest compounded annually

4. Compound interest
5. Half-life

A. $A = P\left(1 + \frac{r}{n}\right)^{nt}$ **B.** $Q = Q_0\left(\frac{1}{2}\right)^{\frac{t}{h}}$ **C.** $A = P(1 + r)^t$

D. $Q(t) = Q_0 e^{kt}$ **E.** $A = Pe^{rt}$

For Problems 6–10, answer true or false.

6. \$500 invested for 2 years at 7% compounded semiannually produces \$573.76.
7. \$500 invested for 2 years at 7% compounded continuously produces \$571.14.
8. The graph of $f(x) = e^{x-5}$ is the graph of $f(x) = e^x$ shifted 5 units to the right.
9. The graph of $f(x) = e^x - 5$ is the graph of $f(x) = e^x$ shifted 5 units downward.
10. The graph of $f(x) = -e^x$ is the graph of $f(x) = e^x$ reflected across the x axis.

Problem Set 11.2

1 Solve Exponential Growth and Compound Interest Problems

1. Assuming that the rate of inflation is 4% per year, the equation $P = P_0(1.04)^t$ yields the predicted price P, in t years, of an item that presently costs P_0. Find the predicted price of each of the following items for the indicated years ahead.

(a) \$1.38 can of soup in 3 years

(b) \$3.43 container of cocoa mix in 5 years

(c) \$1.99 jar of coffee creamer in 4 years

(d) \$1.54 can of beans and bacon in 10 years

(e) \$18,000 car in 5 years (nearest dollar)

(f) \$180,000 house in 8 years (nearest dollar)

(g) \$500 TV set in 7 years (nearest dollar)

2. Suppose it is estimated that the value of a car depreciates 30% per year for the first 5 years. The equation $A = P_0(0.7)^t$ yields the value (A) of a car after t years if the original price is P_0. Find the value (to the nearest dollar) of each of the following cars after the indicated time.

(a) \$16,500 car after 4 years

(b) \$22,000 car after 2 years

(c) \$27,000 car after 5 years

(d) \$40,000 car after 3 years

For Problems 3–14, use the formula $A = P\left(1 + \frac{r}{n}\right)^{nt}$ to find the total amount of money accumulated at the end of the indicated time period for each of the following investments.

3. \$200 for 6 years at 6% compounded annually

4. \$250 for 5 years at 7% compounded annually

5. \$500 for 7 years at 8% compounded semiannually

6. \$750 for 8 years at 8% compounded semiannually

7. \$800 for 9 years at 9% compounded quarterly

8. \$1200 for 10 years at 10% compounded quarterly

9. \$1500 for 5 years at 12% compounded monthly

10. \$2000 for 10 years at 9% compounded monthly

11. \$5000 for 15 years at 8.5% compounded annually

12. \$7500 for 20 years at 9.5% compounded semiannually

13. \$8000 for 10 years at 10.5% compounded quarterly

14. \$10,000 for 25 years at 9.25% compounded monthly

For Problems 15–23, use the formula $A = Pe^{rt}$ to find the total amount of money accumulated at the end of the indicated time period by compounding continuously.

15. \$400 for 5 years at 7%

16. \$500 for 7 years at 6%

17. \$750 for 8 years at 8%

18. \$1000 for 10 years at 9%

19. \$2000 for 15 years at 10%

20. \$5000 for 20 years at 11%

21. \$7500 for 10 years at 8.5%

22. \$10,000 for 25 years at 9.25%

23. \$15,000 for 10 years at 7.75%

24. Complete the following chart, which illustrates what happens to \$1000 invested at various rates of interest for different lengths of time but always compounded continuously. Round your answers to the nearest dollar.

\$1000 compounded continuously				
	8%	10%	12%	14%
5 years				
10 years				
15 years				
20 years				
25 years				

25. Complete the following chart, which illustrates what happens to $1000 invested at 12% for different lengths of time and different numbers of compounding periods. Round all of your answers to the nearest dollar.

$1000 at 12%			
Compounded annually			
1 year	5 years	10 years	20 years
Compounded semiannually			
1 year	5 years	10 years	20 years
Compounded quarterly			
1 year	5 years	10 years	20 years
Compounded monthly			
1 year	5 years	10 years	20 years
Compounded continuously			
1 year	5 years	10 years	20 years

26. Complete the following chart, which illustrates what happens to $1000 in 10 years for different rates of interest and different numbers of compounding periods. Round your answers to the nearest dollar.

$1000 for 10 years			
Compounded annually			
8%	10%	12%	14%
Compounded semiannually			
8%	10%	12%	14%
Compounded quarterly			
8%	10%	12%	14%
Compounded monthly			
8%	10%	12%	14%
Compounded continuously			
8%	10%	12%	14%

27. Suppose that Nora invested $500 at 8.25% compounded annually for 5 years and Patti invested $500 at 8% compounded quarterly for 5 years. At the end of 5 years, who will have the most money and by how much?

28. Two years ago, Daniel invested some money at 8% interest compounded annually. Today it is worth $758.16. How much did he invest two years ago?

29. What rate of interest (to the nearest hundredth of a percent) is needed so that an investment of $2500 will yield $3000 in 2 years if the money is compounded annually?

2 Solve Exponential Decay Problems

30. Suppose that a certain radioactive substance has a half-life of 20 years. If there are presently 2500 milligrams of the substance then how much, to the nearest milligram, will remain after 40 years? After 50 years?

31. Strontium-90 has a half-life of 29 years. If there are 400 grams of strontium initially then how much, to the nearest gram, will remain after 87 years? After 100 years?

32. The half-life of radium is approximately 1600 years. If the present amount of radium in a certain location is 500 grams, how much will remain after 800 years? Express your answer to the nearest gram.

For Problems 33–38, graph each of the exponential functions.

33. $f(x) = e^x + 1$

34. $f(x) = e^x - 2$

35. $f(x) = 2e^x$

36. $f(x) = -e^x$

37. $f(x) = e^{2x}$

38. $f(x) = e^{-x}$

3 Solve Growth Problems Involving the Number *e*

For Problems 39–44, express your answers to the nearest whole number.

39. Suppose that in a certain culture, the equation $Q(t) = 1000e^{0.4t}$ expresses the number of bacteria present as a function of the time t, where t is expressed in hours. How many bacteria are present at the end of 2 hours? 3 hours? 5 hours?

40. The number of bacteria present at a given time under certain conditions is given by the equation $Q = 5000e^{0.05t}$, where t is expressed in minutes. How many bacteria are present at the end of 10 minutes? 30 minutes? 1 hour?

41. The number of bacteria present in a certain culture after t hours is given by the equation $Q = Q_0e^{0.3t}$, where Q_0 represents the initial number of bacteria. If 6640 bacteria are present after 4 hours, how many bacteria were present initially?

42. The number of grams Q of a certain radioactive substance present after t seconds is given by the equation $Q = 1500e^{-0.4t}$. How many grams remain after 5 seconds? 10 seconds? 20 seconds?

43. Suppose that the present population of a city is 75,000. Using the equation $P(t) = 75{,}000e^{0.01t}$ to estimate future growth, estimate the population:

(a) 10 years from now **(b)** 15 years from now

(c) 25 years from now

44. Suppose that the present population of a city is 150,000. Use the equation $P(t) = 150{,}000e^{0.032t}$ to estimate future growth. Estimate the population:

(a) 10 years from now **(b)** 20 years from now

(c) 30 years from now

45. The atmospheric pressure, measured in pounds per square inch, is a function of the altitude above sea level. The equation $P(a) = 14.7e^{-0.21a}$, where a is the altitude measured in miles, can be used to approximate atmospheric pressure. Find the atmospheric pressure at each of the following locations. Express each answer to the nearest tenth of a pound per square inch.

(a) Mount McKinley in Alaska—altitude of 3.85 miles

(b) Denver, Colorado—the "mile-high" city (5280 feet = 1 mile)

(c) Asheville, North Carolina—altitude of 1985 feet

(d) Phoenix, Arizona—altitude of 1090 feet

THOUGHTS INTO WORDS

46. Explain the difference between simple interest and compound interest.

47. Would it be better to invest \$5000 at 6.25% interest compounded annually for 5 years or to invest \$5000 at 6% interest compounded continuously for 5 years? Defend your answer.

GRAPHING CALCULATOR ACTIVITIES

48. Use a graphing calculator to check your graphs for Problems 33–38.

49. Graph $f(x) = 2^x$, $f(x) = e^x$, and $f(x) = 3^x$ on the same set of axes. Are these graphs consistent with the discussion prior to Figure 11.5?

50. Graph $f(x) = e^x$. Where should the graphs of $f(x) = e^{x-4}$, $f(x) = e^{x-6}$, and $f(x) = e^{x+5}$ be located? Graph all three functions on the same set of axes with $f(x) = e^x$.

51. Graph $f(x) = e^x$. Now predict the graphs for $f(x) = -e^x$, $f(x) = e^{-x}$, and $f(x) = -e^{-x}$. Graph all three functions on the same set of axes with $f(x) = e^x$.

52. How do you think the graphs of $f(x) = e^x$, $f(x) = e^{2x}$, and $f(x) = 2e^x$ will compare? Graph them on the same set of axes to see whether you were correct.

Answers to the Concept Quiz

1. E **2.** D **3.** C **4.** A **5.** B **6.** True **7.** False **8.** True **9.** True **10.** True

Answers to the Example Practice Skills

1. (a) \$3431.96 **(b)** \$4046.55 **2.** 2.5 mg **3. (a)** \$3442.82 **(b)** \$4049.58 **4. (a)** 28,000 **(b)** 46,164 **5.** 616 **6.** 191 g

11.3 Logarithms

OBJECTIVES

1. Change Equations between Exponential and Logarithmic Form
2. Evaluate a Logarithmic Expression
3. Solve Logarithmic Equations by Switching to Exponential Form
4. Use the Properties of Logarithms
5. Solve Logarithmic Equations

1 Change Equations between Exponential and Logarithmic Form

In Sections 11.1 and 11.2: (1) we learned about exponential expressions of the form b^n, where b is any positive real number and n is any real number; (2) we used exponential expressions of the form b^n to define exponential functions; and (3) we used exponential functions to help solve problems. In the next three sections we will follow the same basic pattern with respect to a new concept, that of a **logarithm**. Let's begin with the following definition.

Definition 11.2

If r is any positive real number, then the unique exponent t such that $b^t = r$ is called the **logarithm of r with base b** and is denoted by $\log_b r$.

According to Definition 11.2, the logarithm of 8 base 2 is the exponent t such that $2^t = 8$; thus we can write $\log_2 8 = 3$. Likewise, we can write $\log_{10} 100 = 2$ because $10^2 = 100$. In general, we can remember Definition 11.2 in terms of the statement

$$\log_b r = t \quad \text{is equivalent to } b^t = r$$

Thus we can easily switch back and forth between exponential and logarithmic forms of equations, as the next examples illustrate.

$$\log_3 81 = 4 \quad \text{is equivalent to } 3^4 = 81$$
$$\log_{10} 100 = 2 \quad \text{is equivalent to } 10^2 = 100$$
$$\log_{10} 0.001 = -3 \quad \text{is equivalent to } 10^{-3} = 0.001$$
$$\log_2 128 = 7 \quad \text{is equivalent to } 2^7 = 128$$
$$\log_m n = p \quad \text{is equivalent to } m^p = n$$
$$2^4 = 16 \quad \text{is equivalent to } \log_2 16 = 4$$
$$5^2 = 25 \quad \text{is equivalent to } \log_5 25 = 2$$
$$\left(\frac{1}{2}\right)^4 = \frac{1}{16} \quad \text{is equivalent to } \log_{\frac{1}{2}}\left(\frac{1}{16}\right) = 4$$
$$10^{-2} = 0.01 \quad \text{is equivalent to } \log_{10} 0.01 = -2$$
$$a^b = c \quad \text{is equivalent to } \log_a c = b$$

EXAMPLE 1

Change the form of the equation $\log_b 9 = x$ to exponential form.

Solution

$\log_b 9 = x$ is equivalent to $b^x = 9$.

▼ PRACTICE YOUR SKILL

Change the form of each equation to exponential form.

(a) $\log_5 25 = 2$ **(b)** $\log_6\left(\frac{1}{216}\right) = -3$ **(c)** $\log_{10} 10{,}000 = 4$

(d) $\log_a b = c$ ■

EXAMPLE 2

Change the form of the equation $7^2 = 49$ to logarithmic form.

Solution

$7^2 = 49$ is equivalent to $\log_7 49 = 2$.

▼ PRACTICE YOUR SKILL

Change the form of each equation to logarithmic form.

(a) $2^4 = 16$ **(b)** $4^{-1} = \frac{1}{4}$ **(c)** $10^3 = 1000$ **(d)** $x^y = z$ ■

2 Evaluate a Logarithmic Expression

We can conveniently calculate some logarithms by changing to exponential form, as in the next examples.

EXAMPLE 3

Evaluate $\log_4 64$.

Solution

Let $\log_4 64 = x$. Then, by switching to exponential form, we have $4^x = 64$, which we can solve as we did back in Section 11.1.

$$4^x = 64$$
$$4^x = 4^3$$
$$x = 3$$

Therefore, we can write $\log_4 64 = 3$.

▼ PRACTICE YOUR SKILL

Evaluate $\log_2 128$. ■

EXAMPLE 4

Evaluate $\log_{10} 0.1$.

Solution

Let $\log_{10} 0.1 = x$. Then, by switching to exponential form, we have $10^x = 0.1$, which can be solved as follows:

$$10^x = 0.1$$
$$10^x = \frac{1}{10}$$
$$10^x = 10^{-1}$$
$$x = -1$$

Thus we obtain $\log_{10} 0.1 = -1$.

▼ PRACTICE YOUR SKILL

Evaluate $\log_{10} 0.001$. ■

3 Solve Logarithmic Equations by Switching to Exponential Form

The link between logarithms and exponents also provides the basis for solving some equations that involve logarithms, as the next two examples illustrate.

EXAMPLE 5 Solve $\log_8 x = \frac{2}{3}$.

Solution

$$\log_8 x = \frac{2}{3}$$

$$8^{\frac{2}{3}} = x \quad \text{By switching to exponential form}$$

$$\sqrt[3]{8^2} = x$$

$$(\sqrt[3]{8})^2 = x$$

$$4 = x$$

The solution set is {4}.

▼ **PRACTICE YOUR SKILL**

Solve $\log_9 x = \frac{5}{2}$. ■

EXAMPLE 6 Solve $\log_b 1000 = 3$.

Solution

$$\log_b 1000 = 3$$

$$b^3 = 1000$$

$$b = 10$$

The solution set is {10}.

▼ **PRACTICE YOUR SKILL**

Solve $\log_b 121 = 2$. ■

4 Use the Properties of Logarithms

There are some properties of logarithms that are a direct consequence of Definition 11.2 and our knowledge of exponents. For example, writing the exponential equations $b^1 = b$ and $b^0 = 1$ in logarithmic form yields the following property.

Property 11.3

For $b > 0$ and $b \neq 1$,

1. $\log_b b = 1$ **2.** $\log_b 1 = 0$

Thus we can write

$$\log_{10} 10 = 1$$

$$\log_2 2 = 1$$

$$\log_{10} 1 = 0$$

$$\log_5 1 = 0$$

By Definition 11.2, $\log_b r$ is the exponent t such that $b^t = r$. Therefore, raising b to the $\log_b r$ power must produce r. We state this fact in Property 11.4.

Property 11.4

For $b > 0$, $b \neq 1$, and $r > 0$,

$$b^{\log_b r} = r$$

The following examples illustrate Property 11.4.

$$10^{\log_{10} 19} = 19$$

$$2^{\log_2 14} = 14$$

$$e^{\log_e 5} = 5$$

Because a logarithm is by definition an exponent, it would seem reasonable to predict that there are some properties of logarithms that correspond to the basic exponential properties. This is an accurate prediction; these properties provide a basis for computational work with logarithms. Let's state the first of these properties and show how it can be verified by using our knowledge of exponents.

Property 11.5

For positive real numbers b, r, and s, where $b \neq 1$,

$$\log_b rs = \log_b r + \log_b s$$

To verify Property 11.5 we can proceed as follows. Let $m = \log_b r$ and $n = \log_b s$. Change each of these equations to exponential form.

$m = \log_b r$ becomes $r = b^m$

$n = \log_b s$ becomes $s = b^n$

Thus the product rs becomes

$$rs = b^m \cdot b^n = b^{m+n}$$

Now, by changing $rs = b^{m+n}$ back to logarithmic form, we obtain

$$\log_b rs = m + n$$

Replacing m with $\log_b r$ and n with $\log_b s$ yields

$$\log_b rs = \log_b r + \log_b s$$

The following three examples demonstrate a use of Property 11.5.

EXAMPLE 7

If $\log_2 5 = 2.3222$ and $\log_2 3 = 1.5850$, evaluate $\log_2 15$.

Solution

Because $15 = 5 \cdot 3$, we can apply Property 11.5.

$$\begin{aligned}\log_2 15 &= \log_2(5 \cdot 3)\\ &= \log_2 5 + \log_2 3\\ &= 2.3222 + 1.5850\\ &= 3.9072\end{aligned}$$

▼ PRACTICE YOUR SKILL

If $\log_3 8 = 1.8928$ and $\log_3 5 = 1.4650$, evaluate $\log_3 40$. ■

EXAMPLE 8

If $\log_{10} 178 = 2.2504$ and $\log_{10} 89 = 1.9494$, evaluate $\log_{10}(178 \cdot 89)$.

Solution

$$\begin{aligned}\log_{10}(178 \cdot 89) &= \log_{10} 178 + \log_{10} 89\\ &= 2.2504 + 1.9494\\ &= 4.1998\end{aligned}$$

▼ PRACTICE YOUR SKILL

If $\log_6 45 = 2.1245$ and $\log_6 92 = 2.5237$, evaluate $\log_6(45 \cdot 92)$. ■

EXAMPLE 9

If $\log_3 8 = 1.8928$, evaluate $\log_3 72$.

Solution

$$\begin{aligned}\log_3 72 &= \log_3(9 \cdot 8)\\ &= \log_3 9 + \log_3 8\\ &= 2 + 1.8928 \qquad \log_3 9 = 2 \text{ because } 3^2 = 9\\ &= 3.8928\end{aligned}$$

▼ PRACTICE YOUR SKILL

If $\log_2 7 = 2.8074$, evaluate $\log_2 56$. ■

Because $\frac{b^m}{b^n} = b^{m-n}$, we would expect a corresponding property pertaining to logarithms. There is such a property, Property 11.6.

Property 11.6

For positive numbers b, r, and s, where $b \neq 1$,

$$\log_b\left(\frac{r}{s}\right) = \log_b r - \log_b s$$

This property can be verified by using an approach similar to the one we used to verify Property 11.5. We leave it for you to do as an exercise in the next problem set.

We can use Property 11.6 to change a division problem into a subtraction problem, as in the next two examples.

EXAMPLE 10

If $\log_5 36 = 2.2265$ and $\log_5 4 = 0.8614$, evaluate $\log_5 9$.

Solution

Because $9 = \frac{36}{4}$, we can use Property 11.6.

$$\begin{aligned}\log_5 9 &= \log_5\left(\frac{36}{4}\right)\\ &= \log_5 36 - \log_5 4\end{aligned}$$

$$= 2.2265 - 0.8614$$
$$= 1.3651$$

▼ **PRACTICE YOUR SKILL**

If $\log_3 24 = 2.8928$ and $\log_3 2 = 0.6309$, evaluate $\log_3 12$. ■

EXAMPLE 11

Evaluate $\log_{10}\left(\frac{379}{86}\right)$ given that $\log_{10} 379 = 2.5786$ and $\log_{10} 86 = 1.9345$.

Solution

$$\log_{10}\left(\frac{379}{86}\right) = \log_{10} 379 - \log_{10} 86$$
$$= 2.5786 - 1.9345$$
$$= 0.6441$$

▼ **PRACTICE YOUR SKILL**

Evaluate $\log_{10}\left(\frac{436}{72}\right)$ given that $\log_{10} 436 = 2.6395$ and $\log_{10} 72 = 1.8573$. ■

The next property of logarithms provides the basis for evaluating expressions such as $3^{\sqrt{2}}$, $(\sqrt{5})^{\frac{2}{3}}$, and $(0.076)^{\frac{2}{3}}$. We cite the property, consider a basis for its justification, and offer illustrations of its use.

Property 11.7

If r is a positive real number, b is a positive real number other than 1, and p is any real number, then

$$\log_b r^p = p(\log_b r)$$

As you might expect, the exponential property $(b^n)^m = b^{mn}$ plays an important role in the verification of Property 11.7. This is an exercise for you in the next problem set. Let's look at some uses of Property 11.7.

EXAMPLE 12

Evaluate $\log_2 22^{\frac{1}{3}}$ given that $\log_2 22 = 4.4598$.

Solution

$$\log_2 22^{\frac{1}{3}} = \frac{1}{3}\log_2 22 \qquad \text{Property 11.7}$$
$$= \frac{1}{3}(4.4598)$$
$$= 1.4866$$

▼ **PRACTICE YOUR SKILL**

Evaluate $\log_4 28^{\frac{1}{2}}$ given that $\log_4 28 = 2.4037$. ■

Together, the properties of logarithms enable us to change the forms of various logarithmic expressions. For example, an expression such as $\log_b\sqrt{\frac{xy}{z}}$ can be rewritten in terms of sums and differences of simpler logarithmic quantities as follows:

$$\log_b\sqrt{\frac{xy}{z}} = \log_b\left(\frac{xy}{z}\right)^{\frac{1}{2}}$$

$$= \frac{1}{2}\log_b\left(\frac{xy}{z}\right) \qquad \text{Property 11.7}$$

$$= \frac{1}{2}(\log_b xy - \log_b z) \qquad \text{Property 11.6}$$

$$= \frac{1}{2}(\log_b x + \log_b y - \log_b z) \qquad \text{Property 11.5}$$

EXAMPLE 13

Write each expression as the sums or differences of simpler logarithmic quantities. Assume that all variables represent positive real numbers.

(a) $\log_b xy^2$ **(b)** $\log_b\frac{\sqrt{x}}{y}$ **(c)** $\log_b\frac{x^3}{yz}$

Solution

(a) $\log_b xy^2 = \log_b x + \log_b y^2 = \log_b x + 2\log_b y$

(b) $\log_b\frac{\sqrt{x}}{y} = \log_b\sqrt{x} - \log_b y = \frac{1}{2}\log_b x - \log_b y$

(c) $\log_b\frac{x^3}{yz} = \log_b x^3 - \log_b yz = 3\log_b x - (\log_b y + \log_b z)$

$$= 3\log_b x - \log_b y - \log_b z$$

▼ PRACTICE YOUR SKILL

Write each expression as the sums or differences of simpler logarithmic quantities. Assume that all variables represent positive real numbers.

(a) $\log_b x^3\sqrt{y}$ **(b)** $\log_b\frac{\sqrt[3]{x}}{y^2}$ **(c)** $\log_b\frac{\sqrt[4]{x}}{y^2z}$ ■

5 Solve Logarithmic Equations

Sometimes we need to change from indicated sums or differences of logarithmic quantities to indicated products or quotients. This is especially helpful when solving certain kinds of equations that involve logarithms. Note in these next two examples how we can use the properties—along with the process of changing from logarithmic form to exponential form—to solve some equations.

EXAMPLE 14

Solve $\log_{10} x + \log_{10}(x + 9) = 1$.

Solution

$$\log_{10} x + \log_{10}(x + 9) = 1$$

$$\log_{10}[x(x + 9)] = 1 \qquad \text{Property 11.5}$$

$$10^1 = x(x + 9) \qquad \text{Change to exponential form}$$

$$10 = x^2 + 9x$$
$$0 = x^2 + 9x - 10$$
$$0 = (x + 10)(x - 1)$$
$$x + 10 = 0 \quad \text{or} \quad x - 1 = 0$$
$$x = -10 \quad \text{or} \quad x = 1$$

Because logarithms are defined only for positive numbers, x and $x + 9$ must be positive. Therefore, the solution of -10 must be discarded. The solution set is $\{1\}$.

▼ **PRACTICE YOUR SKILL**

Solve $\log_{10} x + \log_{10}(x + 3) = 1$. ■

EXAMPLE 15 Solve $\log_5(x + 4) - \log_5 x = 2$.

Solution

$$\log_5(x + 4) - \log_5 x = 2$$
$$\log_5\left(\frac{x + 4}{x}\right) = 2 \qquad \text{Property 11.6}$$
$$5^2 = \frac{x + 4}{x} \qquad \text{Change to exponential form}$$
$$25 = \frac{x + 4}{x}$$
$$25x = x + 4$$
$$24x = 4$$
$$x = \frac{4}{24} = \frac{1}{6}$$

The solution set is $\left\{\frac{1}{6}\right\}$.

▼ **PRACTICE YOUR SKILL**

Solve $2 = \log_3(x + 6) - \log_3 x$. ■

Because logarithms are defined only for positive numbers, we should realize that some logarithmic equations may not have any solutions. (The solution set is the null set.) It is also possible that a logarithmic equation has a negative solution, as the next example illustrates.

EXAMPLE 16 Solve $\log_2 3 + \log_2(x + 4) = 3$.

Solution

$$\log_2 3 + \log_2(x + 4) = 3$$
$$\log_2 3(x + 4) = 3 \qquad \text{Property 11.5}$$
$$3(x + 4) = 2^3 \qquad \text{Change to exponential form}$$
$$3x + 12 = 8$$

$$3x = -4$$

$$x = -\frac{4}{3}$$

The only restriction is that $x + 4 > 0$ or $x > -4$. Therefore, the solution set is $\left\{-\frac{4}{3}\right\}$. Perhaps you should check this answer.

▼ **PRACTICE YOUR SKILL**

Solve $\log_3 2 + \log_3(x + 6) = 2$. ■

CONCEPT QUIZ

For Problems 1–10, answer true or false.

1. The $\log_m n = q$ is equivalent to $m^q = n$.
2. The $\log_7 7$ equals 0.
3. A logarithm is by definition an exponent.
4. The $\log_5 9^2$ is equivalent to $2 \log_5 9$.
5. For the expression $\log_3 9$, the base of the logarithm is 9.
6. The expression $\log_2 x - \log_2 y + \log_2 z$ is equivalent to $\log_2 xyz$.
7. $\log_4 4 + \log_4 1 = 1$.
8. $\log_2 8 - \log_3 9 + \log_4\left(\frac{1}{16}\right) = -1$.
9. The solution set for $\log_{10} x + \log_{10}(x - 3) = 1$ is $\{10\}$.
10. The solution for $\log_6 x + \log_6(x + 5) = 2$ is $\{4\}$.

Problem Set 11.3

1 Change Equations between Exponential and Logarithmic Form

For Problems 1–10, write each of the following in logarithmic form. For example, $2^3 = 8$ becomes $\log_2 8 = 3$ in logarithmic form.

1. $2^7 = 128$ **2.** $3^3 = 27$

3. $5^3 = 125$ **4.** $2^6 = 64$

5. $10^3 = 1000$ **6.** $10^1 = 10$

7. $2^{-2} = \left(\frac{1}{4}\right)$ **8.** $3^{-4} = \left(\frac{1}{81}\right)$

9. $10^{-1} = 0.1$ **10.** $10^{-2} = 0.01$

For Problems 11–20, write each of the following in exponential form. For example, $\log_2 8 = 3$ becomes $2^3 = 8$ in exponential form.

11. $\log_3 81 = 4$ **12.** $\log_2 256 = 8$

13. $\log_4 64 = 3$ **14.** $\log_5 25 = 2$

15. $\log_{10} 10{,}000 = 4$ **16.** $\log_{10} 100{,}000 = 5$

17. $\log_2\left(\frac{1}{16}\right) = -4$ **18.** $\log_5\left(\frac{1}{125}\right) = -3$

19. $\log_{10} 0.001 = -3$ **20.** $\log_{10} 0.000001 = -6$

2 Evaluate a Logarithmic Expression

For Problems 21–40, evaluate each expression.

21. $\log_2 16$ **22.** $\log_3 9$

23. $\log_3 81$ **24.** $\log_2 512$

25. $\log_6 216$ **26.** $\log_4 256$

27. $\log_7 \sqrt{7}$ **28.** $\log_2 \sqrt[3]{2}$

29. $\log_{10} 1$ **30.** $\log_{10} 10$

31. $\log_{10} 0.1$ **32.** $\log_{10} 0.0001$

33. $10^{\log_{10} 5}$ **34.** $10^{\log_{10} 14}$

35. $\log_2\left(\frac{1}{32}\right)$ **36.** $\log_5\left(\frac{1}{25}\right)$

37. $\log_5(\log_2 32)$ **38.** $\log_2(\log_4 16)$

39. $\log_{10}(\log_7 7)$ **40.** $\log_2(\log_5 5)$

3 Solve Logarithmic Equations by Switching to Exponential Form

For Problems 41–50, solve each equation.

41. $\log_7 x = 2$ **42.** $\log_2 x = 5$

43. $\log_8 x = \frac{4}{3}$ **44.** $\log_{16} x = \frac{3}{2}$

45. $\log_9 x = \frac{3}{2}$

46. $\log_8 x = -\frac{2}{3}$

47. $\log_4 x = -\frac{3}{2}$

48. $\log_9 x = -\frac{5}{2}$

49. $\log_x 2 = \frac{1}{2}$

50. $\log_x 3 = \frac{1}{2}$

4 Use the Properties of Logarithms

For Problems 51–59, you are given that $\log_2 5 = 2.3219$ and $\log_2 7 = 2.8074$. Evaluate each expression using Properties 11.5–11.7.

51. $\log_2 35$

52. $\log_2\left(\frac{7}{5}\right)$

53. $\log_2 125$

54. $\log_2 49$

55. $\log_2\sqrt{7}$

56. $\log_2 \sqrt[3]{5}$

57. $\log_2 175$

58. $\log_2 56$

59. $\log_2 80$

For Problems 60–68, you are given that $\log_8 5 = 0.7740$ and $\log_8 11 = 1.1531$. Evaluate each expression using Properties 11.5–11.7.

60. $\log_8 55$

61. $\log_8\left(\frac{5}{11}\right)$

62. $\log_8 25$

63. $\log_8 \sqrt{11}$

64. $\log_8(5)^{\frac{2}{3}}$

65. $\log_8 88$

66. $\log_8 320$

67. $\log_8\left(\frac{25}{11}\right)$

68. $\log_8\left(\frac{121}{25}\right)$

For Problems 69–80, express each of the following as sums or differences of simpler logarithmic quantities. Assume that all variables represent positive real numbers. For example,

$$\log_b \frac{x^3}{y^2} = \log_b x^3 - \log_b y^2$$
$$= 3\log_b x - 2\log_b y$$

69. $\log_b xyz$

70. $\log_b 5x$

71. $\log_b\left(\frac{y}{z}\right)$

72. $\log_b\left(\frac{x^2}{y}\right)$

73. $\log_b y^3z^4$

74. $\log_b x^2y^3$

75. $\log_b\left(\frac{x^{\frac{1}{2}}y^{\frac{1}{3}}}{z^4}\right)$

76. $\log_b x^{\frac{2}{3}}y^{\frac{3}{4}}$

77. $\log_b\sqrt[3]{x^2z}$

78. $\log_b\sqrt{xy}$

79. $\log_b\left(x\sqrt{\frac{x}{y}}\right)$

80. $\log_b\sqrt{\frac{x}{y}}$

5 Solve Logarithmic Equations

For Problems 81–97, solve each of the equations.

81. $\log_3 x + \log_3 4 = 2$

82. $\log_7 5 + \log_7 x = 1$

83. $\log_{10} x + \log_{10}(x - 21) = 2$

84. $\log_{10} x + \log_{10}(x - 3) = 1$

85. $\log_2 x + \log_2(x - 3) = 2$

86. $\log_3 x + \log_3(x - 2) = 1$

87. $\log_{10}(2x - 1) - \log_{10}(x - 2) = 1$

88. $\log_{10}(9x - 2) = 1 + \log_{10}(x - 4)$

89. $\log_5(3x - 2) = 1 + \log_5(x - 4)$

90. $\log_6 x + \log_6(x + 5) = 2$

91. $\log_8(x + 7) + \log_8 x = 1$

92. $\log_6(x + 1) + \log_6(x - 4) = 2$

93. $\log_2 5 + \log_2(x + 6) = 3$

94. $\log_2(x - 1) - \log_2(x + 3) = 2$

95. $\log_5 x = \log_5(x + 2) + 1$

96. $\log_3(x + 3) + \log_3(x + 5) = 1$

97. $\log_2(x + 2) = 1 - \log_2(x + 3)$

98. Verify Property 11.6.

99. Verify Property 11.7.

THOUGHTS INTO WORDS

100. Explain, without using Property 11.4, why $4^{\log_4 9}$ equals 9.

101. How would you explain the concept of a logarithm to someone who had just completed an elementary algebra course?

102. In the next section we will show that the logarithmic function $f(x) = \log_2 x$ is the inverse of the exponential function $f(x) = 2^x$. From that information, how could you sketch a graph of $f(x) = \log_2 x$?

Answers to the Concept Quiz

1. True **2.** False **3.** True **4.** True **5.** False **6.** False **7.** True **8.** True **9.** False **10.** True

Answers to the Example Practice Skills

1. (a) $5^2 = 25$ **(b)** $6^{-3} = \frac{1}{216}$ **(c)** $10^4 = 10{,}000$ **(d)** $a^c = b$ **2. (a)** $\log_2 16 = 4$ **(b)** $\log_4\left(\frac{1}{4}\right) = -1$ **(c)** $\log_{10} 1000 = 3$ **(d)** $\log_x z = y$ **3.** 7 **4.** −3 **5.** {243} **6.** {11} **7.** 3.3578 **8.** 4.6482 **9.** 5.8074 **10.** 2.2619 **11.** 0.7822 **12.** 1.2019 **13. (a)** $3\log_b x + \frac{1}{2}\log_b y$ **(b)** $\frac{1}{3}\log_b x - 2\log_b y$ **(c)** $\frac{1}{4}\log_b x - 2\log_b y - \log_b z$ **14.** {2} **15.** $\left\{\frac{3}{4}\right\}$ **16.** $\left\{-\frac{3}{2}\right\}$

11.4 Logarithmic Functions

OBJECTIVES

1. Evaluate and Solve Equations for Common Logarithms
2. Evaluate and Solve Equations for Natural Logarithms
3. Graph Logarithmic Functions

1 Evaluate and Solve Equations for Common Logarithms

The properties of logarithms we discussed in Section 11.3 are true for any valid base. For example, because the Hindu-Arabic numeration system that we use is a base-10 system, logarithms to base 10 have historically been used for computational purposes. Base-10 logarithms are called **common logarithms.**

Originally, common logarithms were developed to assist in complicated numerical calculations that involved products, quotients, and powers of real numbers. Today they are seldom used for that purpose because the calculator and computer can much more effectively handle the messy computational problems. However, common logarithms do still occur in applications; they are deserving of our attention.

As we know from earlier work, the definition of a logarithm provides the basis for evaluating $\log_{10} x$ for values of x that are integral powers of 10. Consider the following examples.

$\log_{10} 1000 = 3$ because $10^3 = 1000$

$\log_{10} 100 = 2$ because $10^2 = 100$

$\log_{10} 10 = 1$ because $10^1 = 10$

$\log_{10} 1 = 0$ because $10^0 = 1$

$\log_{10} 0.1 = -1$ because $10^{-1} = \frac{1}{10} = 0.1$

$\log_{10} 0.01 = -2$ because $10^{-2} = \frac{1}{10^2} = 0.01$

$\log_{10} 0.001 = -3$ because $10^{-3} = \frac{1}{10^3} = 0.001$

When working with base-10 logarithms, it is customary to omit writing the numeral 10 to designate the base. Thus the expression $\log_{10} x$ is written as $\log x$, and a statement such as $\log_{10} 1000 = 3$ becomes $\log 1000 = 3$. We will follow this practice from now on in this chapter, but don't forget that the base is understood to be 10.

$$\log_{10} x = \log x$$

To find the common logarithm of a positive number that is not an integral power of 10, we can use an appropriately equipped calculator. Using a calculator equipped with a common logarithm function (ordinarily a key labeled log is used), we obtained the following results, rounded to four decimal places.

$\log 1.75 = 0.2430$

$\log 23.8 = 1.3766$

$\log 134 = 2.1271$ (Be sure that you can use a calculator and obtain these results.)

$\log 0.192 = -0.7167$

$\log 0.0246 = -1.6091$

In order to use logarithms to solve problems, we sometimes need to be able to determine a number when only the logarithm of the number is known. That is, we may need to determine x when $\log x$ is known. Let's consider an example.

EXAMPLE 1

Find x if $\log x = 0.2430$.

Solution

If $\log x = 0.2430$, then changing to exponential form yields $10^{0.2430} = x$. Therefore, using the 10^x key, we can find x.

$x = 10^{0.2430} \approx 1.749846689$

Thus $x = 1.7498$, rounded to five significant digits.

▼ **PRACTICE YOUR SKILL**

Find x if $\log x = 2.4568$ rounded to five significant digits. ■

Be sure that you can use your calculator to obtain the following results. We have rounded the values for x to five significant digits.

If $\log x = 0.7629$, then $x = 10^{0.7629} = 5.7930$.

If $\log x = 1.4825$, then $x = 10^{1.4825} = 30.374$.

If $\log x = 4.0214$, then $x = 10^{4.0214} = 10{,}505$.

If $\log x = -1.5162$, then $x = 10^{-1.5162} = 0.030465$.

If $\log x = -3.8921$, then $x = 10^{-3.8921} = 0.00012820$.

2 Evaluate and Solve Equations for Natural Logarithms

In many practical applications of logarithms, the number e (remember that $e \approx 2.71828$) is used as a base. Logarithms with a base of e are called **natural logarithms**, and the symbol $\ln x$ is commonly used instead of $\log_e x$.

$$\log_e x = \ln x$$

Natural logarithms can be found with an appropriately equipped calculator. Using a calculator with a natural logarithm function (ordinarily a key labeled [ln x]), we can obtain the following results, rounded to four decimal places.

$\ln 3.21 = 1.1663$

$\ln 47.28 = 3.8561$

$\ln 842 = 6.7358$

$\ln 0.21 = -1.5606$

$\ln 0.0046 = -5.3817$

$\ln 10 = 2.3026$

Be sure that you can use your calculator to obtain these results. Keep in mind the significance of a statement such as $\ln 3.21 = 1.1663$. By changing to exponential form, we are claiming that e raised to the 1.1663 power is approximately 3.21. Using a calculator, we obtain $e^{1.1663} = 3.210093293$.

Let's do a few more problems and find x when given $\ln x$. Be sure that you agree with these results.

If $\ln x = 2.4156$, then $x = e^{2.4156} = 11.196$.

If $\ln x = 0.9847$, then $x = e^{0.9847} = 2.6770$.

If $\ln x = 4.1482$, then $x = e^{4.1482} = 63.320$.

If $\ln x = -1.7654$, then $x = e^{-1.7654} = 0.17112$.

3 Graph Logarithmic Functions

We can now use the concept of a logarithm to define a new function.

Definition 11.3

If $b > 0$ and $b \neq 1$, then the function f defined by

$$f(x) = \log_b x$$

where x is any positive real number, is called the **logarithmic function with base b.**

We can obtain the graph of a specific logarithmic function in various ways. For example, we can change the equation $y = \log_2 x$ to the exponential equation $2^y = x$, where we can determine a table of values. The next set of exercises asks you to graph some logarithmic functions with this approach.

We can obtain the graph of a logarithmic function by setting up a table of values directly from the logarithmic equation. Example 1 illustrates this approach.

EXAMPLE 2 Graph $f(x) = \log_2 x$.

Solution

Let's choose some values for x where the corresponding values for $\log_2 x$ are easily determined. (Remember that logarithms are defined only for the positive real numbers.) Plot these points and connect them with a smooth curve to produce Figure 11.7.

x	$f(x)$
$\frac{1}{8}$	-3
$\frac{1}{4}$	-2
$\frac{1}{2}$	-1
1	0
2	1
4	2
8	3

$\log_2 \frac{1}{8} = -3$

because $2^{-3} = \frac{1}{2^3} = \frac{1}{8}$

$\log_2 1 = 0$
because $2^0 = 1$

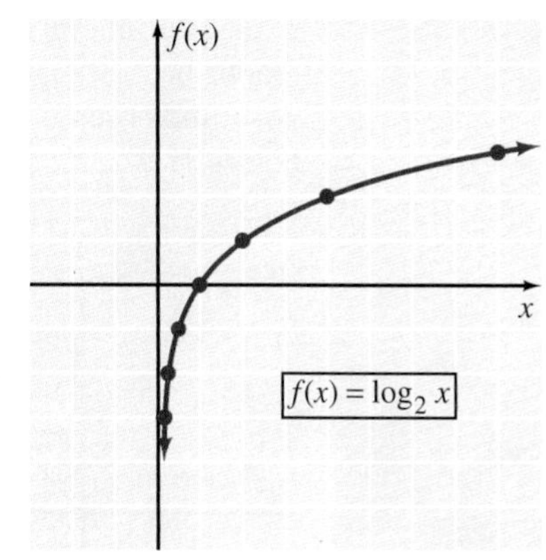

Figure 11.7

▼ PRACTICE YOUR SKILL

Graph $f(x) = \log_3 x$. ■

Suppose that we consider the following two functions f and g.

$f(x) = b^x$ Domain: all real numbers
Range: positive real numbers

$g(x) = \log_b x$ Domain: positive real numbers
Range: all real numbers

Furthermore, suppose that we consider the composition of f and g and the composition of g and f.

$$(f \circ g)(x) = f(g(x)) = f(\log_b x) = b^{\log_b x} = x$$

$$(g \circ f)(x) = g(f(x)) = g(b^x) = \log_b b^x = x \log_b b = x(1) = x$$

Therefore, because the domain of f is the range of g, the range of f is the domain of g, $f(g(x)) = x$, and $g(f(x)) = x$, the two functions f and g are inverses of each other.

Remember also from Chapter 10 that the graphs of a function and its inverse are reflections of each other through the line $y = x$. Thus the graph of a logarithmic function can also be determined by reflecting the graph of its inverse exponential function through the line $y = x$. We see this in Figure 11.8, where the graph of $y = 2^x$ has been reflected across the line $y = x$ to produce the graph of $y = \log_2 x$.

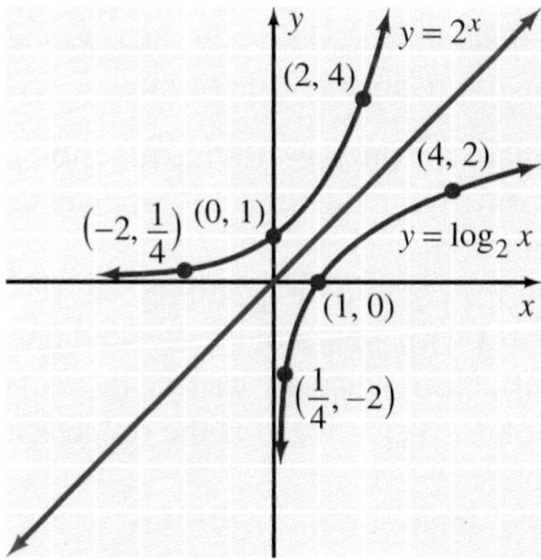

Figure 11.8

The general behavior patterns of exponential functions were illustrated by two graphs back in Figure 11.3. We can now reflect each of those graphs through the line $y = x$ and observe the general behavior patterns of logarithmic functions, as shown in Figure 11.9.

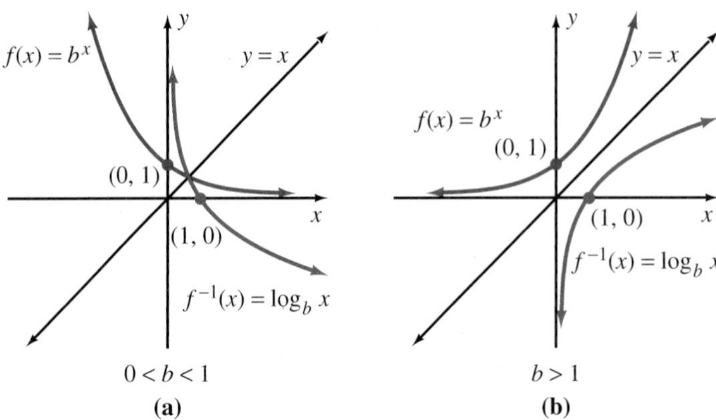

Figure 11.9

Finally, when graphing logarithmic functions, don't forget about variations of the basic curves.

1. The graph of $f(x) = 3 + \log_2 x$ is the graph of $f(x) = \log_2 x$ moved up 3 units. (Because $\log_2 x + 3$ is apt to be confused with $\log_2(x + 3)$, we commonly write $3 + \log_2 x$.)
2. The graph of $f(x) = \log_2(x - 4)$ is the graph of $f(x) = \log_2 x$ moved 4 units to the right.
3. The graph of $f(x) = -\log_2 x$ is the graph of $f(x) = \log_2 x$ reflected across the x axis.

The **common logarithmic function** is defined by the equation $f(x) = \log x$. It should now be a simple matter to set up a table of values and sketch the function. We will have you do this in the next set of exercises. Remember that $f(x) = 10^x$ and $g(x) = \log x$ are inverses of each other. Therefore, we could also get the graph of $g(x) = \log x$ by reflecting the exponential curve $f(x) = 10^x$ across the line $y = x$.

The **natural logarithmic function** is defined by the equation $f(x) = \ln x$. It is the inverse of the natural exponential function $f(x) = e^x$. Thus one way to graph $f(x) = \ln x$ is to reflect the graph of $f(x) = e^x$ across the line $y = x$. We will have you do this in the next set of problems.

CONCEPT QUIZ

For Problems 1–10, answer true or false.

1. The domain for the logarithmic function $f(x) = \log_b x$ is all real numbers.
2. Every logarithmic function has an inverse function that is an exponential function.
3. The base for common logarithms is 2.
4. Logarithms with a base of e are called empirical logarithms.
5. The symbol $\ln x$ is usually written instead of $\log_e x$.
6. The graph of $f(x) = \log_4 x$ is the reflection with reference to the line $y = x$ of the graph $f(x) = 4^x$.
7. If $\ln x = 1$, then $x = e$.

8. If $\log x = 1$, then $x = 10$.
9. The graph of $f(x) = \ln x$ is the graph of $g(x) = e^x$ reflected across the x axis.
10. The domain of the function $f(x) = \ln x$ is the set of real numbers.

Problem Set 11.4

1 Evaluate and Solve Equations for Common Logarithms

For Problems 1–10, use a calculator to find each common logarithm. Express answers to four decimal places.

1. log 7.24
2. log 2.05
3. log 52.23
4. log 825.8
5. log 3214.1
6. log 14,189
7. log 0.729
8. log 0.04376
9. log 0.00034
10. log 0.000069

For Problems 11–20, use your calculator to find x when given $\log x$. Express answers to five significant digits.

11. $\log x = 2.6143$
12. $\log x = 1.5263$
13. $\log x = 4.9547$
14. $\log x = 3.9335$
15. $\log x = 1.9006$
16. $\log x = 0.5517$
17. $\log x = -1.3148$
18. $\log x = -0.1452$
19. $\log x = -2.1928$
20. $\log x = -2.6542$

2 Evaluate and Solve Equations for Natural Logarithms

For Problems 21–30, use your calculator to find each natural logarithm. Express answers to four decimal places.

21. ln 5
22. ln 18
23. ln 32.6
24. ln 79.5
25. ln 430
26. ln 371.8
27. ln 0.46
28. ln 0.524
29. ln 0.0314
30. ln 0.008142

For Problems 31–40, use your calculator to find x when given $\ln x$. Express answers to five significant digits.

31. $\ln x = 0.4721$
32. $\ln x = 0.9413$
33. $\ln x = 1.1425$
34. $\ln x = 2.7619$
35. $\ln x = 4.6873$
36. $\ln x = 3.0259$
37. $\ln x = -0.7284$
38. $\ln x = -1.6246$
39. $\ln x = -3.3244$
40. $\ln x = -2.3745$

3 Graph Logarithmic Functions

41. **(a)** Complete the following table and then graph $f(x) = \log x$. (Express the values for $\log x$ to the nearest tenth.)

x	0.1	0.5	1	2	4	8	10
$\log x$							

(b) Complete the following table and express values for 10^x to the nearest tenth.

x	−1	−0.3	0	0.3	0.6	0.9	1
10^x							

Then graph $f(x) = 10^x$ and reflect it across the line $y = x$ to produce the graph for $f(x) = \log x$.

42. **(a)** Complete the following table and then graph $f(x) = \ln x$. (Express the values for $\ln x$ to the nearest tenth.)

x	0.1	0.5	1	2	4	8	10
$\ln x$							

(b) Complete the following table and express values for e^x to the nearest tenth.

x	−2.3	−0.7	0	0.7	1.4	2.1	2.3
e^x							

Then graph $f(x) = e^x$ and reflect it across the line $y = x$ to produce the graph for $f(x) = \ln x$.

43. Graph $y = \log_{\frac{1}{2}} x$ by graphing $\left(\frac{1}{2}\right)^y = x$.
44. Graph $y = \log_2 x$ by graphing $2^y = x$.
45. Graph $f(x) = \log_3 x$ by reflecting the graph of $g(x) = 3^x$ across the line $y = x$.
46. Graph $f(x) = \log_4 x$ by reflecting the graph of $g(x) = 4^x$ across the line $y = x$.

For Problems 47–53, graph each of the functions. Remember that the graph of $f(x) = \log_2 x$ is given in Figure 11.7.

47. $f(x) = 3 + \log_2 x$
48. $f(x) = -2 + \log_2 x$
49. $f(x) = \log_2(x + 3)$
50. $f(x) = \log_2(x - 2)$

51. $f(x) = \log_2 2x$

52. $f(x) = -\log_2 x$

53. $f(x) = 2 \log_2 x$

For Problems 54–61, perform the following calculations and express answers to the nearest hundredth. (These calculations are in preparation for our work in the next section.)

54. $\dfrac{\log 7}{\log 3}$

55. $\dfrac{\ln 2}{\ln 7}$

56. $\dfrac{2 \ln 3}{\ln 8}$

57. $\dfrac{\ln 5}{2 \ln 3}$

58. $\dfrac{\ln 3}{0.04}$

59. $\dfrac{\ln 2}{0.03}$

60. $\dfrac{\log 2}{5 \log 1.02}$

61. $\dfrac{\log 5}{3 \log 1.07}$

THOUGHTS INTO WORDS

62. Why is the number 1 excluded from being a base of a logarithmic function?

63. How do we know that $\log_2 6$ is between 2 and 3?

GRAPHING CALCULATOR ACTIVITIES

64. Graph $f(x) = x$, $f(x) = e^x$, and $f(x) = \ln x$ on the same set of axes.

65. Graph $f(x) = x$, $f(x) = 10^x$, and $f(x) = \log x$ on the same set of axes.

66. Graph $f(x) = \ln x$. How should the graphs of $f(x) = 2 \ln x$, $f(x) = 4 \ln x$, and $f(x) = 6 \ln x$ compare to the graph of $f(x) = \ln x$? Graph the three functions on the same set of axes with $f(x) = \ln x$.

67. Graph $f(x) = \log x$. Now predict the graphs for $f(x) = 2 + \log x$, $f(x) = -2 + \log x$, and $f(x) = -6 + \log x$. Graph the three functions on the same set of axes with $f(x) = \log x$.

68. Graph $\ln x$. Now predict the graphs for $f(x) = \ln (x - 2)$, $f(x) = \ln (x - 6)$, and $f(x) = \ln (x + 4)$. Graph the three functions on the same set of axes with $f(x) = \ln x$.

69. For each of the following, (a) predict the general shape and location of the graph, and (b) use your graphing calculator to graph the function and thus check your prediction.

(a) $f(x) = \log x + \ln x$

(b) $f(x) = \log x - \ln x$

(c) $f(x) = \ln x - \log x$

(d) $f(x) = \ln x^2$

Answers to the Concept Quiz

1. False **2.** True **3.** False **4.** False **5.** True **6.** True **7.** True **8.** True **9.** False **10.** False

Answers to the Example Practice Skills

1. 286.29 **2.**

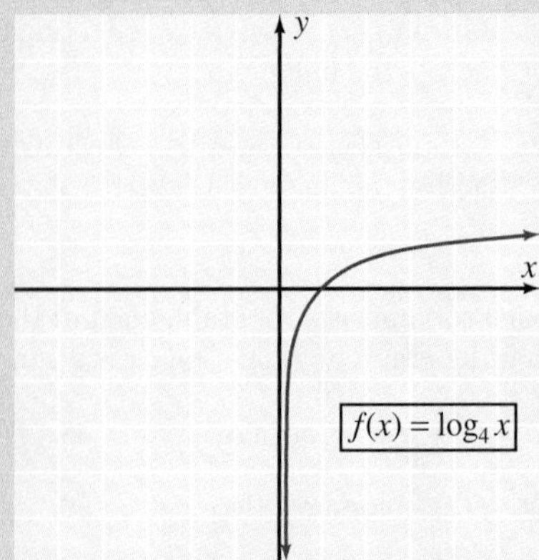

11.5 Exponential Equations, Logarithmic Equations, and Problem Solving

OBJECTIVES

1. Solve Exponential Equations
2. Solve Logarithmic Equations
3. Use Logarithms to Solve Problems
4. Solve Problems Involving Richter Numbers
5. Evaluate Logarithms Using the Change-of-Base Formula

1 Solve Exponential Equations

In Section 11.1 we solved exponential equations such as $3^x = 81$ when we expressed both sides of the equation as a power of 3 and then applied the property "if $b^n = b^m$, then $n = m$." However, if we try to use this same approach with an equation such as $3^x = 5$, we face the difficulty of expressing 5 as a power of 3. We can solve this type of problem by using the properties of logarithms and the following property of equality.

Property 11.8

If $x > 0$, $y > 0$, and $b \neq 1$, then

$$x = y \quad \text{if and only if } \log_b x = \log_b y$$

Property 11.8 is stated in terms of any valid base b; however, for most applications we use either common logarithms (base 10) or natural logarithms (base e). Let's consider some examples.

EXAMPLE 1

Solve $3^x = 5$ to the nearest hundredth.

Solution

By using common logarithms, we can proceed as follows:

$$3^x = 5$$

$$\log 3^x = \log 5 \qquad \text{Property 11.8}$$

$$x \log 3 = \log 5 \qquad \log r^p = p \log r$$

$$x = \frac{\log 5}{\log 3}$$

$$x = 1.46, \qquad \text{to the nearest hundredth}$$

✔ **Check**

Because $3^{1.46} \approx 4.972754647$, we say that, to the nearest hundredth, the solution set for $3^x = 5$ is $\{1.46\}$.

▼ **PRACTICE YOUR SKILL**

Solve $4^x = 24$ to the nearest hundredth. ■

EXAMPLE 2

Solve $e^{x+1} = 5$ to the nearest hundredth.

Solution

The base e is used in the exponential expression, so let's use natural logarithms to help solve this equation.

$$e^{x+1} = 5$$

$$\ln e^{x+1} = \ln 5 \quad \text{Property 11.8}$$

$$(x + 1)\ln e = \ln 5 \quad \ln r^p = p \ln r$$

$$(x + 1)(1) = \ln 5 \quad \ln e = 1$$

$$x = \ln 5 - 1$$

$$x \approx 0.609437912$$

$$x = 0.61, \quad \text{to the nearest hundredth.}$$

The solution set is {0.61}. Check it!

▼ PRACTICE YOUR SKILL

Solve $e^{x-2} = 20$ to the nearest hundredth. ■

2 Solve Logarithmic Equations

In Example 14 of Section 11.3 we solved the logarithmic equation

$$\log_{10} x + \log_{10}(x + 9) = 1$$

by simplifying the left side of the equation to $\log_{10} [x(x + 9)]$ and then changing the equation to exponential form to complete the solution. Now, using Property 11.8, we can solve such a logarithmic equation another way and also expand our equation-solving capabilities. Let's consider some examples.

EXAMPLE 3

Solve $\log x + \log(x - 15) = 2$.

Solution

Because $\log 100 = 2$, the given equation becomes

$$\log x + \log(x - 15) = \log 100$$

Now simplify the left side, apply Property 11.8, and proceed as follows:

$$\log(x)(x - 15) = \log 100$$

$$x(x - 15) = 100$$

$$x^2 - 15x - 100 = 0$$

$$(x - 20)(x + 5) = 0$$

$$x - 20 = 0 \quad \text{or} \quad x + 5 = 0$$

$$x = 20 \quad \text{or} \quad x = -5$$

The domain of a logarithmic function must contain only positive numbers, so x and $x - 15$ must be positive in this problem. Therefore, we discard the solution -5; the solution set is {20}.

▼ PRACTICE YOUR SKILL

Solve $\log x + \log(x + 21) = 2$. ■

EXAMPLE 4

Solve $\ln(x + 2) = \ln(x - 4) + \ln 3$.

Solution

$$\ln(x + 2) = \ln(x - 4) + \ln 3$$

$$\ln(x + 2) = \ln[3(x - 4)]$$

$$x + 2 = 3(x - 4)$$

$$x + 2 = 3x - 12$$

$$14 = 2x$$

$$7 = x$$

The solution set is $\{7\}$.

▼ PRACTICE YOUR SKILL

Solve $\ln(x + 22) = \ln(x - 2) + \ln 4$. ■

3 Use Logarithms to Solve Problems

In Section 11.2 we used the compound interest formula

$$A = P\left(1 + \frac{r}{n}\right)^{nt}$$

to determine the amount of money (A) accumulated at the end of t years if P dollars is invested at r rate of interest compounded n times per year. Now let's use this formula to solve other types of problems that deal with compound interest.

EXAMPLE 5

Apply Your Skill

How long will \$500 take to double itself if it is invested at 12% interest compounded quarterly?

Solution

To "double itself" means that the \$500 will grow into \$1000. Thus

$$1000 = 500\left(1 + \frac{0.12}{4}\right)^{4t}$$

$$= 500(1 + 0.03)^{4t}$$

$$= 500(1.03)^{4t}$$

Multiply both sides of $1000 = 500(1.03)^{4t}$ by $\frac{1}{500}$ to yield

$$2 = (1.03)^{4t}$$

Therefore,

$$\log 2 = \log(1.03)^{4t} \quad \text{Property 11.8}$$

$$\log 2 = 4t \log 1.03 \quad \log r^p = p \log r$$

Solve for t to obtain

$$\log 2 = 4t \log 1.03$$

$$\frac{\log 2}{\log 1.03} = 4t$$

$$\frac{\log 2}{4 \log 1.03} = t \quad \text{Multiply both sides by } \frac{1}{4}$$

$$t \approx 5.862443063$$

$$t = 5.9, \quad \text{to the nearest tenth}$$

Therefore, we are claiming that \$500 invested at 12% interest compounded quarterly will double itself in approximately 5.9 years.

✔ **Check**

\$500 invested at 12% compounded quarterly for 5.9 years will produce

$$A = \$500\left(1 + \frac{0.12}{4}\right)^{4(5.9)}$$

$$= \$500(1.03)^{23.6}$$

$$= \$1004.45$$

▼ **PRACTICE YOUR SKILL**

How long will it take \$2000 to double itself if it is invested at 6% interest compounded monthly? ■

In Section 11.2, we also used the formula $A = Pe^{rt}$ when money was to be compounded continuously. At this time, with the help of natural logarithms, we can extend our use of this formula.

EXAMPLE 6 Apply Your Skill

How long will it take \$100 to triple itself if it is invested at 8% interest compounded continuously?

Solution

To "triple itself" means that the \$100 will grow into \$300. Thus, using the formula for interest that is compounded continuously, we can proceed as follows:

$$A = Pe^{rt}$$

$$\$300 = \$100e^{(0.08)t}$$

$$3 = e^{0.08t}$$

$$\ln 3 = \ln e^{0.08t} \quad \text{Property 11.8}$$

$$\ln 3 = 0.08t \ln e \quad \ln r^p = p \ln r$$

$$\ln 3 = 0.08t \quad \ln e = 1$$

$$\frac{\ln 3}{0.08} = t$$

$$t \approx 13.73265361$$

$$t = 13.7, \quad \text{to the nearest tenth}$$

Therefore, \$100 invested at 8% interest compounded continuously will triple itself in approximately 13.7 years.

✔ Check

\$100 invested at 8% compounded continuously for 13.7 years produces

$$\begin{aligned} A &= Pe^{rt} \\ &= \$100e^{0.08(13.7)} \\ &= \$100e^{1.096} \\ &\approx \$299.22 \end{aligned}$$

▼ **PRACTICE YOUR SKILL**

How long will it take \$500 to triple itself if it is invested at 6% interest compounded continuously? ■

EXAMPLE 7

Sheer Photo, Inc./Photodisc/Getty Images

Apply Your Skill

For a certain virus, the equation $Q(t) = Q_0e^{0.35t}$, where t is the time in hours and Q_0 is the initial number of virus particles, yields the number of virus particles as a function of time. How long will it take 200 particles to increase to 2000 particles?

Solution

Substitute 200 for Q_0 and 2000 for $Q(t)$ in the equation and proceed to solve for t.

$$2000 = 200e^{0.35t}$$

$$10 = e^{0.35t} \qquad \text{Take the natural logarithm of each side}$$

$$\ln 10 = \ln e^{0.35t}$$

$$\ln 10 = 0.35t \ln e \qquad \text{Property 11.7}$$

$$\ln 10 = 0.35t \qquad \text{Property 11.3}$$

$$t = \frac{\ln 10}{0.35}$$

$$t \approx 6.578814551$$

$$t = 6.6, \quad \text{to the nearest tenth}$$

Therefore, it takes approximately 6.6 hours for 200 virus particles to grow to 2000 particles.

▼ **PRACTICE YOUR SKILL**

The number of grams of a certain radioactive isotope present after t hours is given by the equation $Q(t) = Q_0e^{-0.4t}$, where Q_0 represents the initial number of grams. How long will it take 2000 grams to be reduced to 1000 grams? ■

4 Solve Problems Involving Richter Numbers

Seismologists use the Richter scale to measure and report the magnitude of earthquakes. The equation

$$R = \log \frac{I}{I_0} \qquad R \text{ is called a Richter number}$$

compares the intensity I of an earthquake to a minimum or reference intensity I_0. The reference intensity is the smallest earth movement that can be recorded on a seismograph. Suppose that the intensity of an earthquake was determined to be 50,000 times the reference intensity. In this case, $I = 50{,}000I_0$, and the Richter number is calculated as follows:

$$R = \log \frac{50{,}000I_0}{I_0}$$

$$R = \log 50{,}000$$

$$R \approx 4.698970004$$

Thus a Richter number of 4.7 would be reported. Let's consider two more examples that involve Richter numbers.

EXAMPLE 8

Paul Sakuma/AP Photos

Apply Your Skill

An earthquake in San Francisco in 1989 was reported to have a Richter number of 6.9. How did its intensity compare to the reference intensity?

Solution

$$6.9 = \log \frac{I}{I_0}$$

$$10^{6.9} = \frac{I}{I_0}$$

$$I = (10^{6.9})(I_0)$$

$$I \approx 7{,}943{,}282I_0$$

Thus its intensity was nearly 8 million times the reference intensity.

▼ **PRACTICE YOUR SKILL**

An earthquake in Indonesia in 2004 had a Richter number of 5.8. How did its intensity compare to the reference intensity? ■

EXAMPLE 9

Bernard Bisson/Sygma/CORBIS

Apply Your Skill

An earthquake in Iran in 1990 had a Richter number of 7.7. Compare the intensity level of that earthquake to the one in San Francisco (Example 8).

Solution

From Example 8 we have $I = (10^{6.9})(I_0)$ for the earthquake in San Francisco. Using a Richter number of 7.7, we obtain $I = (10^{7.7})(I_0)$ for the earthquake in Iran. Therefore, by comparison,

$$\frac{(10^{7.7})(I_0)}{(10^{6.9})(I_0)} = 10^{7.7-6.9} = 10^{0.8} \approx 6.3$$

The earthquake in Iran was about 6 times as intense as the one in San Francisco.

▼ **PRACTICE YOUR SKILL**

An earthquake on May 12, 2008, in eastern Sichuan China had a Richter number of 7.9. Compare its intensity to the earthquake in San Francisco (Example 8). ■

5 Evaluate Logarithms Using the Change-of-Base Formula

Now let's use either common or natural logarithms to evaluate logarithms that have bases other than 10 or e. Consider the following example.

EXAMPLE 10 Apply Your Skill

Evaluate $\log_3 41$.

Solution

Let $x = \log_3 41$. Change to exponential form to obtain

$$3^x = 41$$

Now we can apply Property 11.8 and proceed as follows:

$$\log 3^x = \log 41$$

$$x \log 3 = \log 41$$

$$x = \frac{\log 41}{\log 3}$$

$$x = \frac{1.6128}{0.4771} = 3.3804, \quad \text{rounded to four decimal places}$$

▼ PRACTICE YOUR SKILL

Evaluate $\log_5 38$. ■

Using the method of Example 10 to evaluate $\log_a r$ produces the following formula, which we often refer to as the **change-of-base** formula for logarithms.

Property 11.9

If a, b, and r are positive numbers with $a \neq 1$ and $b \neq 1$, then

$$\log_a r = \frac{\log_b r}{\log_b a}$$

By using Property 11.9, we can easily determine a relationship between logarithms of different bases. For example, suppose that in Property 11.9 we let $a = 10$ and $b = e$. Then

$$\log_a r = \frac{\log_b r}{\log_b a}$$

becomes

$$\log_{10} r = \frac{\log_e r}{\log_e 10}$$

which can be written as

$$\log_e r = (\log_e 10)(\log_{10} r)$$

Because $\log_e 10 = 2.3026$, rounded to four decimal places, we have

$$\log_e r = (2.3026)(\log_{10} r)$$

Thus the natural logarithm of any positive number is approximately equal to the common logarithm of the number times 2.3026.

Now we can use a graphing utility to graph logarithmic functions such as $f(x) = \log_2 x$. Using the change-of-base formula, we can express this function as $f(x) = \dfrac{\log x}{\log 2}$ or as $f(x) = \dfrac{\ln x}{\ln 2}$. The graph of $f(x) = \log_2 x$ is shown in Figure 11.10.

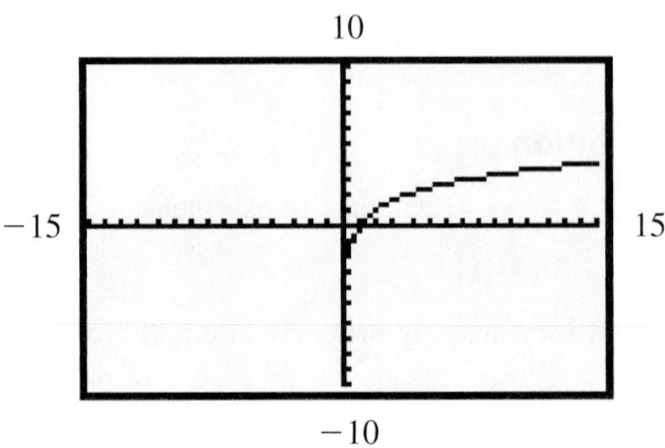

Figure 11.10

CONCEPT QUIZ

For Problems 1–10, answer true or false.

1. The equation $5^{2x+1} = 70$ can be solved by taking the logarithm of both sides of the equation.
2. All solutions of the equation $\log_b x + \log_b(x + 2) = 3$ must be positive numbers.
3. The Richter number compares the intensity of an earthquake to the logarithm of 100,000.
4. If the difference in Richter numbers for two earthquakes is 2, then it can be said that one earthquake was 2 times more intense than the other earthquake.
5. The expression $\log_3 7$ is equivalent to $\dfrac{\ln 7}{\ln 3}$.
6. The expression $\log_3 7$ is equivalent to $\dfrac{\log 7}{\log 3}$.
7. The solution set for $\ln(3x - 4) - \ln(x + 1) = \ln 2$ is $\{8\}$.
8. The value of $\log_5 47$ can be determined by solving the equation $x = \log_5 47$.
9. \$750 will double in value if it is invested at 8% interest compounded quarterly for approximately 8.75 years.
10. \$300 will triple in value if it is invested at 6% interest compounded continuously for approximately 12.2 years.

Problem Set 11.5

1 Solve Exponential Equations

For Problems 1–14, solve each exponential equation and express solutions to the nearest hundredth.

1. $3^x = 32$
2. $2^x = 40$
3. $4^x = 21$
4. $5^x = 73$
5. $3^{x-2} = 11$
6. $2^{x+1} = 7$
7. $5^{3x+1} = 9$
8. $7^{2x-1} = 35$
9. $e^x = 5.4$
10. $e^x = 45$
11. $e^{x-2} = 13.1$
12. $e^{x-1} = 8.2$
13. $3e^x = 35.1$
14. $4e^x - 2 = 26$

2 Solve Logarithmic Equations

For Problems 15–22, solve each logarithmic equation.

15. $\log x + \log(x + 21) = 2$
16. $\log x + \log(x + 3) = 1$
17. $\log(3x - 1) = 1 + \log(5x - 2)$

18. $\log(2x - 1) - \log(x - 3) = 1$

19. $\log(x + 2) - \log(2x + 1) = \log x$

20. $\log(x + 1) - \log(x + 2) = \log\frac{1}{x}$

21. $\ln(2t + 5) = \ln 3 + \ln(t - 1)$

22. $\ln(3t - 4) - \ln(t + 1) = \ln 2$

3 Use Logarithms to Solve Problems

For Problems 23–32, solve each problem and express answers to the nearest tenth.

23. How long will it take \$750 to be worth \$1000 if it is invested at 12% interest compounded quarterly?

24. How long will it take \$1000 to double itself if it is invested at 9% interest compounded semiannually?

25. How long will it take \$2000 to double itself if it is invested at 13% interest compounded continuously?

26. How long will it take \$500 to triple itself if it is invested at 9% interest compounded continuously?

27. For a certain strain of bacteria, the number present after t hours is given by the equation $Q = Q_0e^{0.34t}$, where Q_0 represents the initial number of bacteria. How long will it take 400 bacteria to increase to 4000 bacteria?

28. A piece of machinery valued at \$30,000 depreciates at a rate of 10% yearly. How long will it take until it has a value of \$15,000?

29. The number of grams of a certain radioactive substance present after t hours is given by the equation $Q = Q_0e^{-0.45t}$, where Q_0 represents the initial number of grams. How long will it take 2500 grams to be reduced to 1250 grams?

30. For a certain culture the equation $Q(t) = Q_0e^{0.4t}$, where Q_0 is an initial number of bacteria and t is time measured in hours, yields the number of bacteria as a function of time. How long will it take 500 bacteria to increase to 2000?

31. Suppose that the equation $P(t) = P_0e^{0.02t}$, where P_0 represents an initial population and t is the time in years, is used to predict population growth. How long does this equation predict it would take a city of 50,000 to double its population?

32. The equation $P(a) = 14.7e^{-0.21a}$, where a is the altitude above sea level measured in miles, yields the atmospheric pressure in pounds per square inch. If the atmospheric pressure at Cheyenne, Wyoming, is approximately 11.53 pounds per square inch, find that city's altitude above sea level. Express your answer to the nearest hundred feet.

4 Solve Problems Involving Richter Numbers

Solve each of Problems 33–36.

33. An earthquake in Los Angeles in 1971 had an intensity of approximately 5 million times the reference intensity. What was the Richter number associated with that earthquake?

34. An earthquake in San Francisco in 1906 was reported to have a Richter number of 8.3. How did its intensity compare to the reference intensity?

35. Calculate how many times more intense an earthquake with a Richter number of 7.3 is than an earthquake with a Richter number of 6.4.

36. Calculate how many times more intense an earthquake with a Richter number of 8.9 is than an earthquake with a Richter number of 6.2.

5 Evaluate Logarithms Using the Change-of-Base Formula

For Problems 37–46, approximate each of the following logarithms to three decimal places.

37. $\log_2 23$

38. $\log_3 32$

39. $\log_6 0.214$

40. $\log_5 1.4$

41. $\log_7 421$

42. $\log_8 514$

43. $\log_9 0.0017$

44. $\log_4 0.00013$

45. $\log_3 720$

46. $\log_2 896$

THOUGHTS INTO WORDS

47. Explain how to determine $\log_7 46$ without using Property 11.9.

48. Explain the concept of a Richter number.

49. Explain how you would solve the equation $2^x = 64$ and also how you would solve the equation $2^x = 53$.

50. How do logarithms with a base of 9 compare to logarithms with a base of 3? Explain how you reached your conclusion.

GRAPHING CALCULATOR ACTIVITIES

51. Graph $f(x) = x$, $f(x) = 2^x$, and $f(x) = \log_2 x$ on the same set of axes.

52. Graph $f(x) = x$, $f(x) = (0.5)^x$, and $f(x) = \log_{0.5} x$ on the same set of axes.

53. Graph $f(x) = \log_2 x$. Now predict the graphs for $f(x) = \log_3 x$, $f(x) = \log_4 x$, and $f(x) = \log_8 x$. Graph these three functions on the same set of axes with $f(x) = \log_2 x$.

54. Graph $f(x) = \log_5 x$. Now predict the graphs for $f(x) = 2 \log_5 x$, $f(x) = -4 \log_5 x$, and $f(x) = \log_5(x + 4)$. Graph these three functions on the same set of axes with $f(x) = \log_5 x$.

Answers to the Concept Quiz

1. True **2.** True **3.** False **4.** False **5.** True **6.** True **7.** False **8.** True **9.** True **10.** False

Answers to the Example Practice Skills

1. {2.29} **2.** {5.00} **3.** {4} **4.** {10} **5.** Approximately 11.6 yr **6.** Approximately 18.3 yr **7.** Approximately 1.7 hr **8.** 630,957 times the reference intensity **9.** 10 times as intense **10.** 2.2602, rounded to four decimal places

Chapter 11 Summary

<table>
<tr><th>OBJECTIVE</th><th>SUMMARY</th><th>EXAMPLE</th><th>CHAPTER REVIEW PROBLEMS</th></tr>
<tr><td>Graph exponential functions.
(Sec. 11.2, Obj. 2, p. 611)</td><td>A function defined by an equation of the form $f(x) = b^x$, where $b > 0$ and $b \neq 0$, is called an exponential function with base b.</td><td>Graph $f(x) = 4^x$.
Solution
Set up a table of values.
<table><tr><td>x</td><td>-2</td><td>-1</td><td>0</td><td>1</td><td>2</td></tr><tr><td>$f(x)$</td><td>$\frac{1}{16}$</td><td>$\frac{1}{4}$</td><td>1</td><td>4</td><td>16</td></tr></table>
y
$f(x) = 4^x$</td><td>Problems 1–2</td></tr>
<tr><td>Solve exponential equations.
(Sec. 11.1, Obj. 1, p. 608)</td><td>Some exponential equations can be solved by applying Property 11.2: If $b > 0$, $b \neq 1$, and m and n are real numbers, then $b^n = b^m$ if and only if $n = m$.

This property can be applied only for equations that can be written in the form where the bases are equal.</td><td>Solve $4^{3x} = 32$.
Solution
Rewrite each side of the equation as a power of 2. Then apply property 11.2 to solve the equation.
$4^{3x} = 32$
$(2^2)^{3x} = 2^5$
$2^{6x} = 2^5$
$6x = 5$
$x = \frac{5}{6}$
The solution set is $\left\{\frac{5}{6}\right\}$.</td><td>Problems 3–6</td></tr>
<tr><td>Change equations between exponential and logarithmic form.
(Sec. 11.3, Obj. 1, p. 625)</td><td>If r is any positive real number, then the unique exponent t such that $b^t = r$ is called the logarithm of r with base b and is denoted by $\log_b r$.

The solutions to many problems are based on being able to convert between the exponential and logarithmic form of equations.

$\log_b r = t$ is equivalent to $b^t = r$</td><td>1. Write $6^2 = 36$ in logarithmic form.
2. Write $\log_3 81 = 4$ in exponential form.
Solution
1. $6^2 = 36$ is equivalent to $\log_6 36 = 2$.
2. $\log_3 81 = 4$ is equivalent to $3^4 = 81$.</td><td>Problems 7–14

(continued)</td></tr>
</table>

OBJECTIVE	SUMMARY	EXAMPLE	CHAPTER REVIEW PROBLEMS
Evaluate a logarithmic expression. (Sec. 11.3, Obj. 2, p. 627)	Some logarithmic expressions can be evaluated by switching to exponential form and then applying Property 11.2.	Evaluate $\log_2\left(\frac{1}{8}\right)$. **Solution** Let $\log_2\left(\frac{1}{8}\right) = x$. Switching to exponential form gives $2^x = \frac{1}{8}$. $2^x = \frac{1}{8}$ $2^x = 2^{-3}$ $x = -3$ Therefore, $\log_2\left(\frac{1}{8}\right) = -3$.	Problems 15–22
Evaluate a common logarithm. (Sec. 11.4, Obj. 1, p. 636)	Base-10 logarithms are called common logarithms. When working with base-10 logarithms, it is customary to omit writing the numeral 10 to designate the base; hence $\log_{10} x$ is written as $\log x$. A calculator with a logarithm function key, typically labeled log, is used to evaluate a common logarithm.	Use a calculator to find log 245. Express the answer to four decimal places. **Solution** Follow the instructions for your calculator. $\log 245 = 2.3892$	Problems 23–24
Evaluate a natural logarithm. (Sec. 11.4, Obj. 2, p. 637)	In many practical applications of logarithms, the number e (remember that $e \approx 2.71828$) is used as a base. Logarithms with a base of e are called natural logarithms and are written as $\ln x$.	Use a calculator to find ln 486. Express the answer to four decimal places. **Solution** Follow the instructions for your calculator. $\ln 486 = 6.1860$	Problems 25–26
Evaluate logarithms using the change-of-base formula. (Sec. 11.5, Obj. 5, p. 649)	To evaluate logarithms other than base-10 or base-e, a change-of-base formula, $\log_a r = \frac{\log_b r}{\log_b a}$, is used. When applying the formula, you can choose base-10 or base-e for b.	Find $\log_8 724$. Round the answer to the nearest hundredth. **Solution** $\log_8 724 = \frac{\log 724}{\log 8} = 3.17$	Problems 27–30 *(continued)*

<table>
<tr><th>OBJECTIVE</th><th>SUMMARY</th><th>EXAMPLE</th><th>CHAPTER REVIEW PROBLEMS</th></tr>
<tr>
<td>Graph logarithmic functions.
(Sec. 11.4, Obj. 3, p. 638)</td>
<td>A function defined by an equation of the form $f(x) = \log_b x$, where $b > 0$ and $b \neq 1$, is called a logarithmic function. The equation $y = \log_b x$ is equivalent to $x = b^y$. The two functions $f(x) = b^x$ and $g(x) = \log_b x$ are inverses of each other.</td>
<td>Graph $f(x) = \log_4 x$.

Solution
Change $y = \log_4 x$ to $4^y = x$ and determine a table of values.

<table>
<tr><td>x</td><td>$\frac{1}{16}$</td><td>$\frac{1}{4}$</td><td>1</td><td>4</td><td>16</td></tr>
<tr><td>y</td><td>−2</td><td>−1</td><td>0</td><td>1</td><td>2 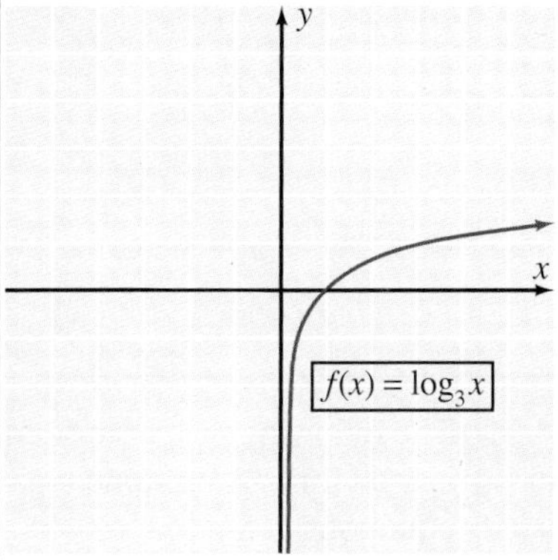</td></tr>
</table>
</td>
<td>Problems 31–32</td>
</tr>
<tr>
<td>Use the properties of logarithms.
(Sec. 11.3, Obj. 4, p. 628)</td>
<td>The following properties of logarithms are derived from the definition of a logarithm and the properties of exponents. For positive real numbers b, r, and s, where $b \neq 1$, we have:

1. $\log_b b = 1$
2. $\log_b 1 = 0$
3. $b^{\log_b r} = r$
4. $\log_b rs = \log_b r + \log_b s$
5. $\log_b\left(\frac{r}{s}\right) = \log_b r - \log_b s$
6. $\log_b r^p = p \log_b r$ where p is any real number</td>
<td>Express $\log_b \frac{\sqrt{x}}{y^2 z}$ as indicated sums or differences of simpler logarithmic quantities.

Solution
$\log_b \frac{\sqrt{x}}{y^2 z} = \log_b \sqrt{x} - \log_b(y^2 z)$
$= \frac{1}{2}\log_b x - [\log_b y^2 + \log_b z]$
$= \frac{1}{2}\log_b x - [2\log_b y + \log_b z]$
$= \frac{1}{2}\log_b x - 2\log_b y - \log_b z$</td>
<td>Problems 33–34</td>
</tr>
<tr>
<td>Solve logarithmic and exponential equations.</td>
<td colspan="2">The properties of equality and the properties of exponents and logarithms combine to help us solve a variety of logarithmic and exponential equations. The techniques for the various types of equations are given as follows. Note the differences in the technique depending on the form of the equation.</td>
<td>(*continued*)</td>
</tr>
</table>

OBJECTIVE	SUMMARY	EXAMPLE	CHAPTER REVIEW PROBLEMS
Solve logarithmic equations by switching to exponential form. (Sec. 11.3, Obj. 3, p. 627; Sec. 11.3, Obj. 5, p. 632)	This technique is used when a single logarithmic expression is equal to a constant, such as $\log_4(3x + 5) = 2$. You may have to apply properties of logarithms to rewrite a sum or difference of logarithms as a single logarithm.	Solve $\log_3 x + \log_3(x + 8) = 2$. **Solution** Rewrite the left-hand side of the equation as a single logarithm. $\log_3 x(x + 8) = 2$ Switch to exponential form and solve the equation. $x(x + 8) = 3^2$ $x^2 + 8x = 9$ $x^2 + 8x - 9 = 0$ $(x + 9)(x - 1) = 0$ $x = -9$ or $x = 1$ *Always check your answer.* Because -9 would make the expression $\log_3 x$ undefined, the solution set is $\{1\}$.	Problems 35–44
Solve logarithmic equations of the form $\log_b x = \log_b y$ (Sec. 11.5, Obj. 2, p. 644)	The following property of equality is used for solving some logarithmic equations: If $x > 0$, and $y > 0$, and $b \neq 1$, then $x = y$ if and only if $\log_b x = \log_b y$. One application of this property is for solving equations where a single logarithm equals another single logarithm.	Solve: $\ln(9x + 1) - \ln(x + 4) = \ln 4$. **Solution** Write the left-hand side of the equation as a single logarithm. $\ln\frac{9x + 1}{x + 4} = \ln 4$ Now apply the property of equality: if $\log_b x = \log_b y$, then $x = y$. $\frac{9x + 1}{x + 4} = 4$ $9x + 1 = 4(x + 4)$ $9x + 1 = 4x + 16$ $5x = 15$ $x = 3$ The answer does check so the solution set is $\{3\}$.	Problems 45–46
Solve exponential equations by taking the logarithm of each side. (Sec. 11.5, Obj. 1, p. 643)	Some exponential equations can be solved by applying the property of equality in the form "If $x = y$ then $\log_b x = \log_b y$, where $x > 0$, and $y > 0$, and $b \neq 1$." If the base of the exponential equation is e then use natural logarithms, and if the base is 10 then use common logarithms.	Solve $6^{3x-2} = 45$. Express the solution to the nearest hundredth. **Solution** Take the logarithm of both sides of the equation. $\log 6^{3x-2} = \log 45$ $(3x - 2)\log 6 = \log 45$ $3x(\log 6) - 2 \log 6 = \log 45$ $3x(\log 6) = \log 45 + 2 \log 6$ $x = \frac{\log 45 + 2 \log 6}{3 \log 6}$ $x = 1.37$ The solution set is $\{1.37\}$.	Problems 47–50

(continued)

OBJECTIVE	SUMMARY	EXAMPLE	CHAPTER REVIEW PROBLEMS
Solve exponential growth problems. (Sec. 11.2, Obj. 1, p. 615)	In general, the equation $P = P_0(1.06)^t$ yields the predicted price P of an item in t years at the annual inflation rate of 6%, where that item presently costs P_0.	Assuming that the rate of inflation is 3% per year, find the price of a \$20.00 haircut in 5 years. **Solution** Use the formula $P = P_0(1.03)^t$ and substitute \$20.00 for P_0 and 5 for t. $P = 20.00(1.03)^5 \approx 23.19$ The haircut would cost about \$23.19.	Problems 51–52
Solve compound interest problems. (Sec. 11.2, Obj. 1, p. 615; Sec. 11.5, Obj. 3, p. 645)	A general formula for any principal, P, invested for any number of years (t) at a rate of r percent compounded n times per year is $A = P\left(1 + \frac{r}{n}\right)^{nt}$ where A represents the total amount of money accumulated at the end of t years.	Find the total amount of money accumulated at the end of 8 years when \$4000 is invested at 6% compounded quarterly. **Solution** $A = 4000\left(1 + \frac{0.06}{4}\right)^{4(8)}$ $A = 4000(1.015)^{32}$ $A = 6441.30$	Problems 53–56
Solve continuous compounding problems. (Sec. 11.2, Obj. 3, p. 619)	The value of $\left(1 + \frac{1}{n}\right)^n$, as n gets infinitely large, approaches the number e, where $e = 2.71828$ to five decimal places. The formula $A = Pe^{rt}$ yields the accumulated value, A, of a sum of money, P, that has been invested for t years at a rate of r percent compounded continuously.	If \$8300 is invested at 6.5% interest compounded continuously, how much will accumulate in 10 years? **Solution** $A = 8300e^{0.065(10)} = 8300e^{0.65}$ $= 15{,}898.99$ The accumulated amount will be \$15,898.99.	Problems 57–58
Solve exponential decay half-life problems. (Sec. 11.2, Obj. 2, p. 617)	For radioactive substances, the rate of decay is exponential and can be described in terms of the half-life of a substance, which is the amount of time that it takes for one-half of an initial amount of the substance to disappear as the result of decay. Suppose there is an initial amount, Q_0, of a radioactive substance with a half-life of h. The amount of substance remaining, Q, after a time period of t is given by the formula $Q = Q_0\left(\frac{1}{2}\right)^{\frac{t}{h}}$. The units of measure for t and h must be the same.	Molybdenum-99 has a half-life of 66 hours. If there are 200 milligrams of molybdenum-99 initially, how many milligrams remain after 16 hours? **Solution** $Q = 200\left(\frac{1}{2}\right)^{\frac{16}{66}}$ $Q = 169.1$, to the nearest tenth of a milligram Thus approximately 169.1 milligrams remain after 16 hours.	Problems 59–60 *(continued)*

OBJECTIVE	SUMMARY	EXAMPLE	CHAPTER REVIEW PROBLEMS
Solve growth problems involving the number e. (Sec. 11.2, Obj. 3, p. 619)	The law of exponential growth $Q(t) = Q_0e^{kt}$ is used as a mathematical model for growth-and-decay problems. In this equation, $Q(t)$ represents the quantity of a given substance at any time t; Q_0 is the initial amount of the substance (when $t = 0$); and k is a constant that depends on the particular application. If $k < 0$, then $Q(t)$ decreases as t increases, and we refer to the model as the law of decay.	The number of bacteria present in a certain culture after t hours is given by the equation $Q(t) = Q_0e^{0.42t}$, where Q_0 represents the initial number of bacteria. How long will it take 100 bacteria to increase to 500 bacteria? **Solution** Substitute 100 for Q_0 and 500 for $Q(t)$ into $Q(t) = Q_0e^{0.42t}$. $500 = 100e^{0.42t}$ $5 = e^{0.42t}$ $\ln 5 = \ln e^{0.42t}$ $\ln 5 = 0.42t$ $t = \frac{\ln 5}{0.42} = 3.83$, to the nearest hundredth It will take approximately 3.83 hours for 100 bacteria to increase to 500 bacteria.	Problems 61–62
Solve problems involving Richter numbers. (Sec. 11.5, Obj. 4, p. 647)	Seismologists use the Richter scale to measure and report the magnitude of earthquakes. The equation $R = \log\frac{I}{I_0}$ compares the intensity I of an earthquake to a minimum or reference intensity I_0.	An earthquake in the Sandwich Islands in 2006 was about 5,011,872 times as intense as the reference intensity. Find the Richter number for that earthquake. **Solution** $R = \log\frac{I}{I_0}$ $= \log\frac{5{,}011{,}872\ I_0}{I_0} = 6.7$ The Richter number for the earthquake is 6.7.	Problems 63–64

Chapter 11 Review Problem Set

For Problems 1 and 2, graph each of the functions.

1. $f(x) = 2^x - 3$

2. $f(x) = -3^x$

For Problems 3–6, solve each equation without using your calculator.

3. $2^x = \frac{1}{16}$

4. $16^x = \frac{1}{8}$

5. $3^x - 4 = 23$

6. $4^{3x+1} = 32$

For Problems 7–10, write each of the following in logarithmic form.

7. $8^2 = 64$

8. $a^b = c$

9. $3^{-2} = \left(\frac{1}{9}\right)$

10. $10^{-1} = \left(\frac{1}{10}\right)$

For Problems 11–14, write each of the following in exponential form.

11. $\log_5 125 = 3$

12. $\log_x y = z$

13. $\log_2\left(\frac{1}{8}\right) = -3$

14. $\log_{10} 10{,}000 = 4$

For Problems 15–22, evaluate each expression without using a calculator.

15. $\log_2 128$

16. $\log_4 64$

17. $\log 10{,}000$

18. $\log 0.001$

19. $\ln e^2$

20. $5^{\log_5 13}$

21. $\log(\log_3 3)$

22. $\log_2\left(\frac{1}{4}\right)$

For Problems 23–26, use your calculator to find each logarithm. Express answers to four decimal places.

23. $\log 73.14$

24. $\log 0.00235$

25. $\ln 0.014$

26. $\ln 114.2$

For Problems 27–30, approximate a value for each logarithm to three decimal places.

27. $\log_3 97$

28. $\log_8 200$

29. $\log_5 0.0065$

30. $\log_2 0.0036$

For Problems 31 and 32, graph each of the functions.

31. $f(x) = -1 + \log_2 x$

32. $f(x) = \log_2(x + 1)$

33. Express $\log_b \frac{x\sqrt{y}}{z^3}$ as the sum or difference of simpler logarithmic quantities. Assume that all variables represent positive real numbers.

34. Express $\log_b \frac{b}{cd}$ as the sum or difference of simpler logarithmic quantities. Assume that all variables represent positive real numbers.

For Problems 35–38, use your calculator to find x when given $\log x$ or $\ln x$. Express the answer to five significant figures.

35. $\ln x = 0.1724$

36. $\log x = 3.4215$

37. $\log x = -1.8765$

38. $\ln x = -2.5614$

For Problems 39–46, solve each equation without using your calculator.

39. $\log_8 x = \frac{2}{3}$

40. $\log_x 3 = \frac{1}{2}$

41. $\log_5 2 + \log_5(3x + 1) = 1$

42. $\log_2 x + \log_2(x - 4) = 5$

43. $\log_2(x + 5) - \log_2 x = 1$

44. $\log_3 x - \log_3(x - 5) = 2$

45. $\ln x + \ln(x + 2) = \ln 35$

46. $\log(2x) - \log(x + 3) = \log\left(\frac{5}{6}\right)$

For Problems 47–50, use your calculator to help solve each equation. Express solutions to the nearest hundredth.

47. $3^x = 42$

48. $2e^x = 14$

49. $2^{x+1} = 79$

50. $e^{x-2} = 37$

51. Assuming the rate of inflation is 3% per year, the equation $P = P_0(1.03)^t$ yields the predicted price P, in t years, of an item that presently costs P_0. Find the predicted price of each of the following items for the indicated years ahead.

(a) \$829 cruise in 4 years

(b) \$280 suit in 3 years

(c) \$85 bicycle in 5 years

52. Suppose it is estimated that the value of a car depreciates 25% per year for the first 5 years. The equation $A = P_0(0.75)^t$ yields the value (A) of a car after t years if the original price is P_0. Find the value (to the nearest dollar) of each of the following cars after the indicated time.

(a) \$25,000 car after 4 years

(b) \$35,000 car after 3 years

(c) \$20,000 car after 1 year

53. Suppose that \$800 is invested at 9% interest compounded monthly. How much money has accumulated at the end of 15 years?

54. If \$2500 is invested at 10% interest compounded quarterly, how much money has accumulated at the end of 12 years?

55. How long will it take \$100 to double itself if it is invested at 8% interest compounded semiannually?

56. How long will it take \$1000 to be worth \$3500 if it is invested at 10.5% interest compounded quarterly?

57. If \$3500 is invested at 8% interest compounded continuously, how much money will accumulate in 6 years?

58. How long will it take \$5000 to be worth \$20,000 if it is invested at 7% interest compounded continuously?

59. Suppose that a certain radioactive substance has a half-life of 40 days. If there are presently 750 grams of the substance, how much (to the nearest gram) will remain after 100 days?

60. Suppose that a certain radioactive substance has a half-life of 20 hours. If there are presently 500 grams, how long will it take for only 100 grams to remain?

61. Suppose that the present population of a city is 50,000. Use the equation $P = P_0 e^{0.02t}$, where P_0 represents an initial population, to estimate future populations. Estimate the population of that city in 10 years, 15 years, and 20 years.

62. The number of bacteria present in a certain culture after t hours is given by the equation $Q = Q_0 e^{0.29t}$, where Q_0 represents the initial number of bacteria. How long will it take 500 bacteria to increase to 2000 bacteria?

63. An earthquake in Mexico City in 1985 had an intensity level about 125,000,000 times the reference intensity. Find the Richter number for that earthquake.

64. Calculate how many times more intense an earthquake with a Richter number of 7.8 is than an earthquake with a Richter number of 7.1.

Chapter 11 Test

For Problems 1–4, evaluate each expression.

1. $\log_3 \sqrt{3}$ 1. ______
2. $\log_2(\log_2 4)$ 2. ______
3. $-2 + \ln e^3$ 3. ______
4. $\log_2(0.5)$ 4. ______

For Problems 5–10, solve each equation.

5. $4^x = \frac{1}{64}$ 5. ______
6. $9^x = \frac{1}{27}$ 6. ______
7. $2^{3x-1} = 128$ 7. ______
8. $\log_9 x = \frac{5}{2}$ 8. ______
9. $\log x + \log(x + 48) = 2$ 9. ______
10. $\ln x = \ln 2 + \ln(3x - 1)$ 10. ______

For Problems 11–14, given that $\log_3 4 = 1.2619$ and $\log_3 5 = 1.4650$, evaluate each of the following.

11. $\log_3 100$ 11. ______
12. $\log_3 1.25$ 12. ______
13. $\log_3 \sqrt{5}$ 13. ______
14. $\log_3(16 \cdot 25)$ 14. ______
15. Solve $e^x = 176$ to the nearest hundredth. 15. ______
16. Solve $2^{x-2} = 314$ to the nearest hundredth. 16. ______
17. Determine $\log_5 632$ to four decimal places. 17. ______
18. Express $3 \log_b x + 2 \log_b y - \log_b z$ as a single logarithm with a coefficient of 1. 18. ______
19. If \$3500 is invested at 7.5% interest compounded quarterly, how much money has accumulated at the end of 8 years? 19. ______
20. How long will it take \$5000 to be worth \$12,500 if it is invested at 7% compounded annually? Express your answer to the nearest tenth of a year. 20. ______
21. The number of bacteria present in a certain culture after t hours is given by $Q(t) = Q_0e^{0.23t}$, where Q_0 represents the initial number of bacteria. How long will it take 400 bacteria to increase to 2400 bacteria? Express your answer to the nearest tenth of an hour. 21. ______
22. Suppose that a certain radioactive substance has a half-life of 50 years. If there are presently 7500 grams of the substance, how much will remain after 32 years? Express your answer to the nearest gram. 22. ______

For Problems 23–25, graph each of the functions.

23. $f(x) = e^x - 2$ 23. ______
24. $f(x) = -3^{-x}$ 24. ______
25. $f(x) = \log_2(x - 2)$ 25. ______

Appendices

A Prime Numbers and Operations with Fractions

This appendix reviews the operations with rational numbers in common fraction form. Throughout this section, we will speak of "multiplying fractions." Be aware that this phrase means multiplying rational numbers in common fraction form. A strong foundation here will simplify your later work in rational expressions. Because prime numbers and prime factorization play an important role in the operations with fractions, let's begin by considering two special kinds of whole numbers, prime numbers and composite numbers.

Definition A.1

A **prime number** is a whole number greater than 1 that has no factors (divisors) other than itself and 1. Whole numbers greater than 1 that are not prime numbers are called **composite numbers**.

The prime numbers less than 50 are 2, 3, 5, 7, 11, 13, 17, 19, 23, 29, 31, 37, 41, 43, and 47. Note that each of these has no factors other than itself and 1. We can express every composite number as the indicated product of prime numbers. Consider the following examples:

$$4 = 2 \cdot 2 \qquad 6 = 2 \cdot 3 \qquad 8 = 2 \cdot 2 \cdot 2 \qquad 10 = 2 \cdot 5 \qquad 12 = 2 \cdot 2 \cdot 3$$

In each case we express a composite number as the indicated product of prime numbers. The indicated product form is called the prime-factored form of the number. There are various procedures to find the prime factors of a given composite number. For our purposes, the simplest technique is to factor the given composite number into any two easily recognized factors and then continue to factor each of these until we obtain only prime factors. Consider these examples:

$$18 = 2 \cdot 9 = 2 \cdot 3 \cdot 3 \qquad 27 = 3 \cdot 9 = 3 \cdot 3 \cdot 3$$

$$24 = 4 \cdot 6 = 2 \cdot 2 \cdot 2 \cdot 3 \qquad 150 = 10 \cdot 15 = 2 \cdot 5 \cdot 3 \cdot 5$$

It does not matter which two factors we choose first. For example, we might start by expressing 18 as $3 \cdot 6$ and then factor 6 into $2 \cdot 3$, which produces a final result of $18 = 3 \cdot 2 \cdot 3$. Either way, 18 contains two prime factors of 3 and one prime factor of 2. The order in which we write the prime factors is not important.

Least Common Multiple

It is sometimes necessary to determine the smallest common nonzero multiple of two or more whole numbers. We call this nonzero number the **least common multiple**. In our work with fractions, there will be problems where it will be necessary to find the least common multiple of some numbers, usually the denominators of fractions. So let's review the concepts of multiples. We know that 35 is a multiple of 5 because $5 \cdot 7 = 35$. The set of all whole numbers that are multiples of 5 consists of 0, 5, 10, 15, 20, 25, and so on. In other words, 5 times each successive whole number ($5 \cdot 0 = 0$, $5 \cdot 1 = 5$, $5 \cdot 2 = 10$, $5 \cdot 3 = 15$, etc.) produces the multiples of 5. In a like manner, the set of multiples of 4 consists of 0, 4, 8, 12, 16, and so on. We can illustrate the concept of least common multiple and find the least common multiple of 5 and 4 by using a simple listing of the multiples of 5 and the multiples of 4.

Multiples of 5 are 0, 5, 10, 15, 20, 25, 30, 35, 40, 45, . . .

Multiples of 4 are 0, 4, 8, 12, 16, 20, 24, 28, 32, 36, 40, 44, 48, . . .

The nonzero numbers in common on the lists are 20 and 40. The least of these, 20, is the least common multiple. Stated another way, 20 is the smallest nonzero whole number that is divisible by both 4 and 5.

Often, from your knowledge of arithmetic, you will be able to determine the least common multiple by inspection. For instance, the least common multiple of 6 and 8 is 24. Therefore, 24 is the smallest nonzero whole number that is divisible by both 6 and 8. If we cannot determine the least common multiple by inspection, then using the prime-factorized form of composite numbers is helpful. The procedure is as follows.

Step 1 Express each number as a product of prime factors.

Step 2 The least common multiple contains each different prime factor as many times as the most times it appears in any one of the factorizations from Step 1.

The following examples illustrate this technique for finding the least common multiple of two or more numbers.

EXAMPLE 1

Find the least common multiple of 24 and 36.

Solution

Let's first express each number as a product of prime factors.

$24 = 2 \cdot 2 \cdot 2 \cdot 3$

$36 = 2 \cdot 2 \cdot 3 \cdot 3$

The prime factor 2 occurs the most times (three times) in the factorization of 24. Because the factorization of 24 contains three 2s, the least common multiple must have three 2s. The prime factor 3 occurs the most times (two times) in the factorization of 36. Because the factorization of 36 contains two 3s, the least common multiple must have two 3s. The least common multiple of 24 and 36 is therefore $2 \cdot 2 \cdot 2 \cdot 3 \cdot 3 = 72$. ■

EXAMPLE 2

Find the least common multiple of 48 and 84.

Solution

$48 = 2 \cdot 2 \cdot 2 \cdot 2 \cdot 3$

$84 = 2 \cdot 2 \cdot 3 \cdot 7$

We need four 2s in the least common multiple because of the four 2s in 48. We need one 3 because of the 3 in each of the numbers, and we need one 7 because of the 7 in 84. The least common multiple of 48 and 84 is $2 \cdot 2 \cdot 2 \cdot 2 \cdot 3 \cdot 7 = 336$. ■

EXAMPLE 3

Find the least common multiple of 12, 18, and 28.

Solution

$28 = 2 \cdot 2 \cdot 7$

$18 = 2 \cdot 3 \cdot 3$

$12 = 2 \cdot 2 \cdot 3$

The least common multiple is $2 \cdot 2 \cdot 3 \cdot 3 \cdot 7 = 252$. ■

EXAMPLE 4

Find the least common multiple of 8 and 9.

Solution

$$9 = 3 \cdot 3$$

$$8 = 2 \cdot 2 \cdot 2$$

The least common multiple is $2 \cdot 2 \cdot 2 \cdot 3 \cdot 3 = 72$. ■

Multiplying Fractions

We can define the multiplication of fractions in common fractional form as follows.

Multiplying Fractions

If a, b, c, and d are integers with b and d not equal to zero, then $\frac{a}{b} \cdot \frac{c}{d} = \frac{a \cdot c}{b \cdot d}$.

To multiply fractions in common fractional form, we simply multiply numerators and multiply denominators. The following examples illustrate the multiplying of fractions.

$$\frac{1}{3} \cdot \frac{2}{5} = \frac{1 \cdot 2}{3 \cdot 5} = \frac{2}{15}$$

$$\frac{3}{4} \cdot \frac{5}{7} = \frac{3 \cdot 5}{4 \cdot 7} = \frac{15}{28}$$

$$\frac{3}{5} \cdot \frac{5}{3} = \frac{15}{15} = 1$$

The last of these examples is a very special case. If the product of two numbers is 1, then the numbers are said to be reciprocals of each other.

Before we proceed too far with multiplying fractions, we need to learn about reducing fractions. The following property is applied throughout our work with fractions. We call this property the fundamental property of fractions.

Fundamental Property of Fractions

If b and k are nonzero integers and if a is any integer, then $\frac{a \cdot k}{b \cdot k} = \frac{a}{b}$.

The fundamental property of fractions provides the basis for what is often called reducing fractions to lowest terms, or expressing fractions in simplest or reduced form. Let's apply the property to a few examples.

EXAMPLE 5

Reduce $\frac{12}{18}$ to lowest terms.

Solution

$$\frac{12}{18} = \frac{2 \cdot \cancel{6}}{3 \cdot \cancel{6}} = \frac{2}{3}$$ A common factor of 6 has been divided out of both numerator and denominator ■

EXAMPLE 6 Change $\frac{14}{35}$ to simplest form.

Solution

$$\frac{14}{35} = \frac{2 \cdot \cancel{7}}{5 \cdot \cancel{7}} = \frac{2}{5}$$ A common factor of 7 has been divided out of both numerator and denominator ■

EXAMPLE 7 Reduce $\frac{72}{90}$.

Solution

$$\frac{72}{90} = \frac{\cancel{2} \cdot 2 \cdot 2 \cdot \cancel{3} \cdot \cancel{3}}{\cancel{2} \cdot \cancel{3} \cdot \cancel{3} \cdot 5} = \frac{4}{5}$$ The prime-factored forms of the numerator and denominator may be used to find common factors ■

We are now ready to consider multiplication problems with the understanding that the final answer should be expressed in reduced form. Study the following examples carefully; we use different methods to simplify the problems.

EXAMPLE 8 Multiply $\left(\frac{9}{4}\right)\left(\frac{14}{15}\right)$.

Solution

$$\left(\frac{9}{4}\right)\left(\frac{14}{15}\right) = \frac{\cancel{3} \cdot 3 \cdot \cancel{2} \cdot 7}{\cancel{2} \cdot 2 \cdot \cancel{3} \cdot 5} = \frac{21}{10}$$ ■

EXAMPLE 9 Find the product of $\frac{8}{9}$ and $\frac{18}{24}$.

Solution

$$\frac{\overset{1}{\cancel{8}}}{\underset{1}{\cancel{9}}} \cdot \frac{\overset{2}{\cancel{18}}}{\underset{3}{\cancel{24}}} = \frac{2}{3}$$ A common factor of 8 has been divided out of 8 and 24, and a common factor of 9 has been divided out of 9 and 18 ■

Dividing Fractions

The next example motivates a definition for division of rational numbers in fractional form:

$$\frac{\frac{3}{4}}{\frac{2}{3}} = \left(\frac{\frac{3}{4}}{\frac{2}{3}}\right)\left(\frac{\frac{3}{2}}{\frac{3}{2}}\right) = \frac{\left(\frac{3}{4}\right)\left(\frac{3}{2}\right)}{1} = \left(\frac{3}{4}\right)\left(\frac{3}{2}\right) = \frac{9}{8}$$

Note that $\left(\frac{\frac{3}{2}}{\frac{3}{2}}\right)$ is a form of 1 and that $\frac{3}{2}$ is the reciprocal of $\frac{2}{3}$. In other words, $\frac{3}{4}$ divided by $\frac{2}{3}$ is equivalent to $\frac{3}{4}$ times $\frac{3}{2}$. The following definition for division should now seem reasonable.

Division of Fractions

If b, c, and d are nonzero integers and if a is any integer, then $\frac{a}{b} \div \frac{c}{d} = \frac{a}{b} \cdot \frac{d}{c}$.

Note that to divide $\frac{a}{b}$ by $\frac{c}{d}$, we multiply $\frac{a}{b}$ by the reciprocal of $\frac{c}{d}$, which is $\frac{d}{c}$. The next examples demonstrate the important steps of a division problem.

$$\frac{2}{3} \div \frac{1}{2} = \frac{2}{3} \cdot \frac{2}{1} = \frac{4}{3}$$

$$\frac{5}{6} \div \frac{3}{4} = \frac{5}{6} \cdot \frac{4}{3} = \frac{5 \cdot 4}{6 \cdot 3} = \frac{5 \cdot \cancel{2} \cdot 2}{\cancel{2} \cdot 3 \cdot 3} = \frac{10}{9}$$

$$\frac{\frac{6}{7}}{2} = \frac{\overset{3}{\cancel{6}}}{7} \cdot \frac{1}{\underset{1}{\cancel{2}}} = \frac{3}{7}$$

Adding and Subtracting Fractions

Suppose that it is one-fifth of a mile between your dorm and the union and two-fifths of a mile between the union and the library along a straight line, as indicated in Figure A.1. The total distance between your dorm and the library is three-fifths of a mile, and we write $\frac{1}{5} + \frac{2}{5} = \frac{3}{5}$.

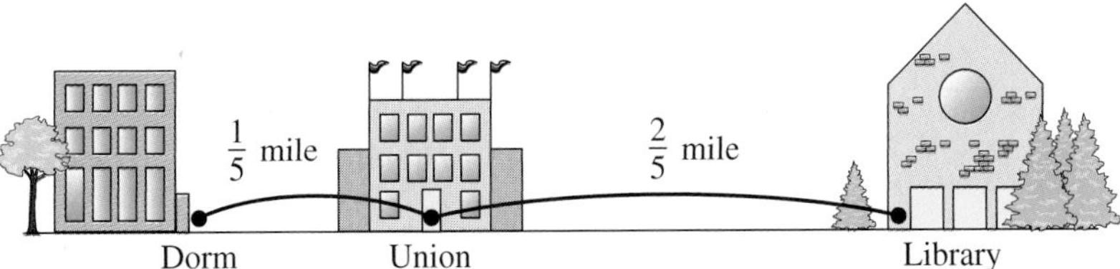

Figure A.1

A pizza is cut into seven equal pieces and you eat two of the pieces (see Figure A.2). How much of the pizza remains? We represent the whole pizza by $\frac{7}{7}$ and conclude that $\frac{7}{7} - \frac{2}{7} = \frac{5}{7}$ of the pizza remains.

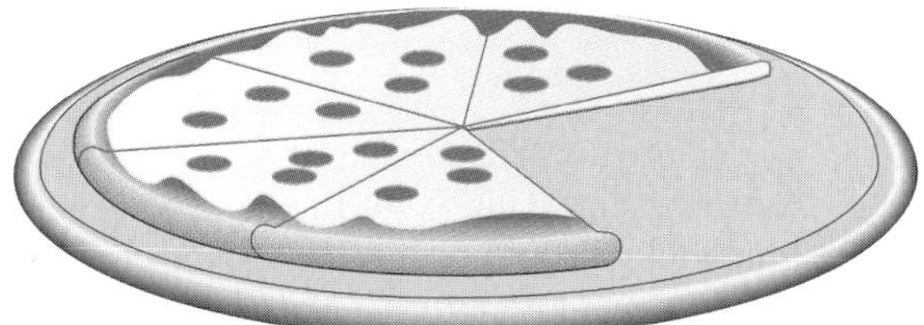

Figure A.2

These examples motivate the following definition for addition and subtraction of rational numbers in $\frac{a}{b}$ form.

Addition and Subtraction of Fractions

If a, b, and c are integers and if b is not zero, then

$$\frac{a}{b} + \frac{c}{b} = \frac{a + c}{b} \quad \text{Addition}$$

$$\frac{a}{b} - \frac{c}{b} = \frac{a - c}{b} \quad \text{Subtraction}$$

We say that fractions with common denominators can be added or subtracted by adding or subtracting the numerators and placing the results over the common denominator. Consider the following examples:

$$\frac{3}{7} + \frac{2}{7} = \frac{3 + 2}{7} = \frac{5}{7}$$

$$\frac{7}{8} - \frac{2}{8} = \frac{7 - 2}{8} = \frac{5}{8}$$

$$\frac{5}{6} - \frac{1}{6} = \frac{5 - 1}{6} = \frac{4}{6} = \frac{2}{3} \quad \text{We agree to reduce the final answer}$$

How do we add or subtract if the fractions do not have a common denominator? We use the fundamental principle of fractions, $\frac{a \cdot k}{b \cdot k} = \frac{a}{b}$, to obtain equivalent fractions that have a common denominator. **Equivalent fractions** are fractions that name the same number. Consider the next example, which shows the details.

EXAMPLE 10 Add $\frac{1}{4} + \frac{2}{5}$.

Solution

$$\frac{1}{4} = \frac{1 \cdot 5}{4 \cdot 5} = \frac{5}{20} \quad \frac{1}{4} \text{ and } \frac{5}{20} \text{ are equivalent fractions}$$

$$\frac{2}{5} = \frac{2 \cdot 4}{5 \cdot 4} = \frac{8}{20} \quad \frac{2}{5} \text{ and } \frac{8}{20} \text{ are equivalent fractions}$$

$$\frac{5}{20} + \frac{8}{20} = \frac{13}{20}$$

■

Note that in Example 10 we chose 20 as the common denominator, and 20 is the least common multiple of the original denominators 4 and 5. (Recall that the least common multiple is the smallest nonzero whole number divisible by the given numbers.) In general, we use the least common multiple of the denominators of the fractions to be added or subtracted as a **least common denominator** (LCD).

Recall that the least common multiple may be found either by inspection or by using prime factorization forms of the numbers. Consider some examples involving these procedures.

EXAMPLE 11 Subtract $\frac{5}{8} - \frac{7}{12}$.

Solution

By inspection, the LCD is 24.

$$\frac{5}{8} - \frac{7}{12} = \frac{5 \cdot 3}{8 \cdot 3} - \frac{7 \cdot 2}{12 \cdot 2} = \frac{15}{24} - \frac{14}{24} = \frac{1}{24}$$ ■

If the LCD is not obvious by inspection, then we can use the technique of prime factorization to find the least common multiple.

EXAMPLE 12 Add $\frac{5}{18} + \frac{7}{24}$.

Solution

If we cannot find the LCD by inspection, then we can use the prime-factorized forms.

$$\left.\begin{array}{l} 18 = 2 \cdot 3 \cdot 3 \\ 24 = 2 \cdot 2 \cdot 2 \cdot 3 \end{array}\right\} \longrightarrow \text{LCD} = 2 \cdot 2 \cdot 2 \cdot 3 \cdot 3 = 72$$

$$\frac{5}{18} + \frac{7}{24} = \frac{5 \cdot 4}{18 \cdot 4} + \frac{7 \cdot 3}{24 \cdot 3} = \frac{20}{72} + \frac{21}{72} = \frac{41}{72}$$ ■

EXAMPLE 13 Marcey put $\frac{5}{8}$ pound of chemicals in the spa to adjust the water quality. Michael, not realizing Marcey had already put in chemicals, put $\frac{3}{14}$ pound of chemicals in the spa. The chemical manufacturer states that you should never add more than 1 pound of chemicals. Have Marcey and Michael together put in more than 1 pound of chemicals?

Solution

Add $\frac{5}{8} + \frac{3}{14}$.

$$\left.\begin{array}{l} 8 = 2 \cdot 2 \cdot 2 \\ 14 = 2 \cdot 7 \end{array}\right\} \longrightarrow \text{LCD} = 2 \cdot 2 \cdot 2 \cdot 7 = 56$$

$$\frac{5}{8} + \frac{3}{14} = \frac{5 \cdot 7}{8 \cdot 7} + \frac{3 \cdot 4}{14 \cdot 4} = \frac{35}{56} + \frac{12}{56} = \frac{47}{56}$$

No, Marcey and Michael have not added more than 1 pound of chemicals. ■

Simplifying Numerical Expressions

We now consider simplifying numerical expressions that contain fractions. In agreement with the order of operations, first multiplications and divisions are done as they appear from left to right, and then additions and subtractions are performed as they appear from left to right. In these next examples, we show only the major steps. Be sure you can fill in all the details.

EXAMPLE 14 Simplify $\frac{3}{4} + \frac{2}{3} \cdot \frac{3}{5} - \frac{1}{2} \cdot \frac{1}{5}$.

Solution

$$\frac{3}{4} + \frac{2}{3} \cdot \frac{3}{5} - \frac{1}{2} \cdot \frac{1}{5} = \frac{3}{4} + \frac{2}{5} - \frac{1}{10}$$

$$= \frac{15}{20} + \frac{8}{20} - \frac{2}{20} = \frac{15 + 8 - 2}{20} = \frac{21}{20}$$

■

EXAMPLE 15 Simplify $\frac{5}{8}\left(\frac{1}{2} + \frac{1}{3}\right)$.

Solution

$$\frac{5}{8}\left(\frac{1}{2} + \frac{1}{3}\right) = \frac{5}{8}\left(\frac{3}{6} + \frac{2}{6}\right) = \frac{5}{8}\left(\frac{5}{6}\right) = \frac{25}{48}$$

■

Appendix A Exercises

For Problems 1–12, factor each composite number into a product of prime numbers; for example, $18 = 2 \cdot 3 \cdot 3$.

1. 26
2. 16
3. 36
4. 80
5. 49
6. 92
7. 56
8. 144
9. 120
10. 84
11. 135
12. 98

For Problems 13–24, find the least common multiple of the given numbers.

13. 6 and 8
14. 8 and 12
15. 12 and 16
16. 9 and 12
17. 28 and 35
18. 42 and 66
19. 49 and 56
20. 18 and 24
21. 8, 12, and 28
22. 6, 10, and 12
23. 9, 15, and 18
24. 8, 14, and 24

For Problems 25–30, reduce each fraction to lowest terms.

25. $\frac{8}{12}$
26. $\frac{12}{16}$
27. $\frac{16}{24}$
28. $\frac{18}{32}$
29. $\frac{15}{9}$
30. $\frac{48}{36}$

For Problems 31–36, multiply or divide as indicated, and express answers in reduced form.

31. $\frac{3}{4} \cdot \frac{5}{7}$
32. $\frac{4}{5} \cdot \frac{3}{11}$
33. $\frac{2}{7} \div \frac{3}{5}$
34. $\frac{5}{6} \div \frac{11}{13}$
35. $\frac{3}{8} \cdot \frac{12}{15}$
36. $\frac{4}{9} \cdot \frac{3}{2}$

37. A certain recipe calls for $\frac{3}{4}$ cup of milk. To make half of the recipe, how much milk is needed?

38. John is adding a diesel fuel additive to his fuel tank, which is half full. The directions say to add $\frac{1}{3}$ of the bottle to a full fuel tank. What portion of the bottle should he add to the fuel tank?

39. Mark shares a computer with his roommates. He has partitioned the hard drive in such a way that he gets $\frac{1}{3}$ of the disk space. His part of the hard drive is currently $\frac{2}{3}$ full. What portion of the computer's hard drive space is he currently taking up?

40. Angelina teaches $\frac{2}{3}$ of the deaf children in her local school. Her local school educates $\frac{1}{2}$ of the deaf children in the school district. What portion of the school district's deaf children is Angelina teaching?

For Problems 41–57, add or subtract as indicated and express answers in lowest terms.

41. $\frac{2}{7}+\frac{3}{7}$

42. $\frac{3}{11}+\frac{5}{11}$

43. $\frac{7}{9}-\frac{2}{9}$

44. $\frac{11}{13}-\frac{6}{13}$

45. $\frac{3}{4}+\frac{9}{4}$

46. $\frac{5}{6}+\frac{7}{6}$

47. $\frac{11}{12}-\frac{3}{12}$

48. $\frac{13}{16}-\frac{7}{16}$

49. $\frac{5}{24}+\frac{11}{24}$

50. $\frac{7}{36}+\frac{13}{36}$

51. $\frac{1}{3}+\frac{1}{5}$

52. $\frac{1}{6}+\frac{1}{8}$

53. $\frac{15}{16}-\frac{3}{8}$

54. $\frac{13}{12}-\frac{1}{6}$

55. $\frac{7}{10}+\frac{8}{15}$

56. $\frac{7}{12}+\frac{5}{8}$

57. $\frac{11}{24}+\frac{5}{32}$

58. Alicia and her brother Jeff shared a pizza Alicia ate $\frac{1}{8}$ of the pizza while Jeff ate $\frac{2}{3}$ of the pizza. How much of the pizza has been eaten?

59. Rosa has $\frac{1}{3}$ pound of blueberries, $\frac{1}{4}$ pound of strawberries, and $\frac{1}{2}$ pound of raspberries. If she combines these for a fruit salad, how many pounds of these berries will be in the salad?

60. A chemist has $\frac{11}{16}$ of an ounce of dirt residue to perform crime lab tests. He needs $\frac{3}{8}$ of an ounce to perform a test for iron content. How much of the dirt residue will be left for the chemist to use in other testing?

For Problems 61–68, simplify each numerical expression, expressing answers in reduced form.

61. $\frac{1}{4}-\frac{3}{8}+\frac{5}{12}-\frac{1}{24}$

62. $\frac{3}{4}+\frac{2}{3}-\frac{1}{6}+\frac{5}{12}$

63. $\frac{5}{6}+\frac{2}{3}\cdot\frac{3}{4}-\frac{1}{4}\cdot\frac{2}{5}$

64. $\frac{2}{3}+\frac{1}{2}\cdot\frac{2}{5}-\frac{1}{3}\cdot\frac{1}{5}$

65. $\frac{3}{4}\cdot\frac{6}{9}-\frac{5}{6}\cdot\frac{8}{10}+\frac{2}{3}\cdot\frac{6}{8}$

66. $\frac{3}{5}\cdot\frac{5}{7}+\frac{2}{3}\cdot\frac{3}{5}-\frac{1}{7}\cdot\frac{2}{5}$

67. $\frac{7}{13}\left(\frac{2}{3}-\frac{1}{6}\right)$

68. $48\left(\frac{5}{12}-\frac{1}{6}+\frac{3}{8}\right)$

69. Blake Scott leaves $\frac{1}{4}$ of his estate to the Boy Scouts, $\frac{2}{5}$ to the local cancer fund, and the rest to his church. What fractional part of the estate does the church receive?

70. Franco has $\frac{7}{8}$ of an ounce of gold. He wants to give $\frac{3}{16}$ of an ounce to his friend Julie. He plans to divide the remaining amount of his gold in half to make two rings. How much gold will he have for each ring?

B Matrix Approach to Solving Systems

The primary objective of Appendices B, C, and D is to introduce a variety of techniques for solving systems of linear equations. The techniques we have discussed thus far lend themselves to "small" systems. As the number of equations and variables increases, the systems become more difficult to solve and require other techniques. In these appendices we will continue to work with small systems for the sake of convenience, but you will learn some techniques that can be extended to larger systems. This appendix introduces a matrix approach to solving systems.

A **matrix** is simply an array of numbers arranged in horizontal rows and vertical columns. For example, the matrix

$$\text{2 rows}\rightarrow\begin{bmatrix} 2 & 1 & -4 \\ 5 & -7 & 6 \end{bmatrix}$$

$\uparrow\ \uparrow\ \uparrow$ 3 columns

has 2 rows and 3 columns, which we refer to as a 2×3 (read "two-by-three") matrix. Some additional examples of matrices (*matrices* is the plural of *matrix*) are as follows:

$$\underset{3\times 2}{\begin{bmatrix} 3 & 2 \\ -1 & 4 \\ 5 & 7 \end{bmatrix}} \qquad \underset{2\times 2}{\begin{bmatrix} 4 & 1 \\ 0 & -5 \end{bmatrix}} \qquad \underset{1\times 4}{\begin{bmatrix} 1 & 2 & 6 & 8 \end{bmatrix}} \qquad \underset{5\times 1}{\begin{bmatrix} 3 \\ 7 \\ 10 \\ 2 \\ -4 \end{bmatrix}}$$

In general, a matrix of m rows and n columns is called a matrix of dimension $m \times n$. With every system of linear equations we can associate a matrix that consists of the coefficients and constant terms. For example, with the system

$$\begin{pmatrix} x - 3y = -17 \\ 2x + 7y = 31 \end{pmatrix}$$

we can associate the matrix

$$\left[\begin{array}{cc:c} 1 & -3 & -17 \\ 2 & 7 & 31 \end{array}\right]$$

which is called the **augmented matrix** of the system. The dashed line separates the coefficients from the constant terms; technically the dashed line is not necessary.

Because augmented matrices represent systems of equations, we can operate with them as we do with systems of equations. Our previous work with systems of equations was based on the following properties.

1. Any two equations of a system may be interchanged.

   EXAMPLE When we interchange the two equations, the system $\begin{pmatrix} 2x - 5y = 9 \\ x + 3y = 4 \end{pmatrix}$ is equivalent to the system $\begin{pmatrix} x + 3y = 4 \\ 2x - 5y = 9 \end{pmatrix}$.

2. Any equation of the system may be multiplied by a nonzero constant.

   EXAMPLE When we multiply the top equation by -2, the system $\begin{pmatrix} x + 3y = 4 \\ 2x - 5y = 9 \end{pmatrix}$ is equivalent to the system $\begin{pmatrix} -2x - 6y = -8 \\ 2x - 5y = 9 \end{pmatrix}$.

3. Any equation of the system can be replaced by adding a nonzero multiple of another equation to that equation.

   EXAMPLE When we add -2 times the first equation to the second equation, the system $\begin{pmatrix} x + 3y = 4 \\ 2x - 5y = 9 \end{pmatrix}$ is equivalent to the system $\begin{pmatrix} x + 3y = 4 \\ -11y = 1 \end{pmatrix}$.

Each of the properties geared to solving a system of equations produces a corresponding property of the augmented matrix of the system. For example, exchanging two equations of a system corresponds to exchanging two rows of the augmented matrix that represents the system. We usually refer to these properties as elementary row operations, and we can state them as follows.

Elementary Row Operations

1. Any two rows of an augmented matrix can be interchanged.
2. Any row can be multiplied by a nonzero constant.
3. Any row of the augmented matrix can be replaced by adding a nonzero multiple of another row to that row.

Using the elementary row operations on an augmented matrix provides a basis for solving systems of linear equations. Study the following examples very carefully; keep in mind that the general scheme, called **Gaussian elimination**, is one of using elementary row operations on a matrix to continue replacing a system of equations with an equivalent system until a system is obtained where the solutions are easily determined. We will use a format similar to the one used in Section 4.3, except that here we will represent systems of equations by matrices.

EXAMPLE 1

Solve the system $\begin{pmatrix} x - 3y = -17 \\ 2x + 7y = 31 \end{pmatrix}$.

Solution

The augmented matrix of the system is

$$\left[\begin{array}{cc|c} 1 & -3 & -17 \\ 2 & 7 & 31 \end{array}\right]$$

We can multiply row one by -2 and add this result to row two to produce a new row two.

$$\left[\begin{array}{cc|c} 1 & -3 & -17 \\ 0 & 13 & 65 \end{array}\right]$$

This matrix represents the system

$$\begin{pmatrix} x - 3y = -17 \\ 13y = 65 \end{pmatrix}$$

From the last equation we can determine the value of y.

$$13y = 65$$
$$y = 5$$

Now we can substitute 5 for y in the equation $x - 3y = -17$.

$$x - 3(5) = -17$$
$$x - 15 = -17$$
$$x = -2$$

The solution set is $\{(-2, 5)\}$. ■

EXAMPLE 2

Solve the system $\begin{pmatrix} 3x + 2y = 3 \\ 30x - 6y = 17 \end{pmatrix}$.

Solution

The augmented matrix of the system is

$$\left[\begin{array}{cc|c} 3 & 2 & 3 \\ 30 & -6 & 17 \end{array}\right]$$

We can multiply row one by -10 and add this result to row two to produce a new row two.

$$\left[\begin{array}{cc|c} 3 & 2 & 3 \\ 0 & -26 & -13 \end{array}\right]$$

This matrix represents the system

$$\begin{pmatrix} 3x + 2y = 3 \\ -26y = -13 \end{pmatrix}$$

From the last equation we can determine the value of y.

$$-26y = -13$$

$$y = \frac{-13}{-26} = \frac{1}{2}$$

Now we can substitute $\frac{1}{2}$ for y in the equation $3x + 2y = 3$.

$$3x + 2\left(\frac{1}{2}\right) = 3$$

$$3x + 1 = 3$$

$$3x = 2$$

$$x = \frac{2}{3}$$

The solution set is $\left\{\left(\frac{2}{3}, \frac{1}{2}\right)\right\}$. ■

EXAMPLE 3

Solve the system $\begin{pmatrix} 2x - 3y - z = -2 \\ x - 2y + 3z = 9 \\ 3x + y - 5z = -8 \end{pmatrix}$.

Solution

The augmented matrix of the system is

$$\left[\begin{array}{ccc|c} 2 & -3 & -1 & -2 \\ 1 & -2 & 3 & 9 \\ 3 & 1 & -5 & -8 \end{array}\right]$$

Let's begin by interchanging the top two rows.

$$\left[\begin{array}{ccc|c} 1 & -2 & 3 & 9 \\ 2 & -3 & -1 & -2 \\ 3 & 1 & -5 & -8 \end{array}\right]$$

Now we can multiply row one by -2 and add this result to row two to produce a new row two. Also, we can multiply row one by -3 and add this result to row three to produce a new row three.

$$\left[\begin{array}{ccc|c} 1 & -2 & 3 & 9 \\ 0 & 1 & -7 & -20 \\ 0 & 7 & -14 & -35 \end{array}\right]$$

Now we can multiply row two by -7 and add this result to row three to produce a new row three.

$$\left[\begin{array}{ccc|c} 1 & -2 & 3 & 9 \\ 0 & 1 & -7 & -20 \\ 0 & 0 & 35 & 105 \end{array}\right]$$

This last matrix represents the system $\begin{pmatrix} x - 2y + 3z = 9 \\ y - 7z = -20 \\ 35z = 105 \end{pmatrix}$, which is said to be in **triangular form**. We can use the third equation to determine the value of z.

$$35z = 105$$

$$z = 3$$

Now we can substitute 3 for z in the second equation.

$$y - 7z = -20$$
$$y - 7(3) = -20$$
$$y - 21 = -20$$
$$y = 1$$

Finally, we can substitute 3 for z and 1 for y in the first equation.

$$x - 2y + 3z = 9$$
$$x - 2(1) + 3(3) = 9$$
$$x - 2 + 9 = 9$$
$$x + 7 = 9$$
$$x = 2$$

The solution set is $\{(2, 1, 3)\}$. ■

Appendix B Exercises

Solve each of the following systems and use matrices as we did in the examples of this section.

1. $\begin{pmatrix} x - 2y = 14 \\ 4x + 5y = 4 \end{pmatrix}$

2. $\begin{pmatrix} x + 5y = -3 \\ 3x - 2y = -26 \end{pmatrix}$

3. $\begin{pmatrix} 3x + 7y = -40 \\ x + 4y = -20 \end{pmatrix}$

4. $\begin{pmatrix} 7x - 9y = 53 \\ x - 3y = 11 \end{pmatrix}$

5. $\begin{pmatrix} x - 3y = 4 \\ 4x - 5y = 3 \end{pmatrix}$

6. $\begin{pmatrix} x + 3y = 7 \\ 2x - 4y = 9 \end{pmatrix}$

7. $\begin{pmatrix} 6x + 7y = -15 \\ 4x - 9y = 31 \end{pmatrix}$

8. $\begin{pmatrix} 5x - 3y = -16 \\ 6x + 5y = -2 \end{pmatrix}$

9. $\begin{pmatrix} x + 3y - 4z = 5 \\ -2x - 5y + z = 9 \\ 7x - y - z = -2 \end{pmatrix}$

10. $\begin{pmatrix} x - y + 5z = -2 \\ -3x + 2y + z = 17 \\ 4x - 5y - 3z = -36 \end{pmatrix}$

11. $\begin{pmatrix} x - 2y - 3z = -11 \\ 2x - 3y + z = 7 \\ -3x - 5y + 7z = 14 \end{pmatrix}$

12. $\begin{pmatrix} x + y + 3z = -8 \\ 3x + 2y - 5z = 19 \\ 5x - y - 4z = 23 \end{pmatrix}$

13. $\begin{pmatrix} y + 3z = -3 \\ 2x - 5z = 18 \\ 3x - y + 2z = 5 \end{pmatrix}$

14. $\begin{pmatrix} x - z = -1 \\ -2x + y + 3z = 4 \\ 3x - 4y = 31 \end{pmatrix}$

15. $\begin{pmatrix} -x - 5y + 2z = -5 \\ 3x + 14y - z = 13 \\ 4x - 3y + 5z = -26 \end{pmatrix}$

16. $\begin{pmatrix} -x - 3y + 4z = -3 \\ 3x + 8y - z = 27 \\ 5x - y + 2z = -5 \end{pmatrix}$

17. $\begin{pmatrix} x + 2y - z = -5 \\ 3x + 4y + 2z = -8 \\ -2x - y + 5z = 10 \end{pmatrix}$

18. $\begin{pmatrix} x - 3y + 2z = 0 \\ 2x - 4y - 3z = 19 \\ -3x - y + z = -11 \end{pmatrix}$

19. $\begin{pmatrix} -3x + 2y - z = 12 \\ 5x + 2y - 3z = 6 \\ x - y + 5z = -10 \end{pmatrix}$

20. $\begin{pmatrix} -2x - 3y + 5z = 15 \\ 4x - y + 2z = -4 \\ x + y - 3z = -7 \end{pmatrix}$

21. $\begin{pmatrix} -2x + 5y - z = -1 \\ 4x + y - 5z = 23 \\ x - 2y + 3z = -7 \end{pmatrix}$

22. $\begin{pmatrix} 2x + 5y + z = 1 \\ x + 2y - 3z = -13 \\ 3x - y - 2z = -4 \end{pmatrix}$

THOUGHTS INTO WORDS

23. What is a matrix? What is an augmented matrix of a system of linear equations?

24. Describe how to use matrices to solve the system $\begin{pmatrix} x - 2y = 5 \\ 2x + 7y = 9 \end{pmatrix}$.

FURTHER INVESTIGATIONS

25. Solve the system $\begin{pmatrix} x - 3y - 2z + w = -3 \\ -2x + 7y + z - 2w = -1 \\ 3x - 7y - 3z + 3w = -5 \\ 5x + y + 4z - 2w = 18 \end{pmatrix}$.

26. Solve the system $\begin{pmatrix} x - 2y + 2z - w = -2 \\ -3x + 5y - z - 3w = 2 \\ 2x + 3y + 3z + 5w = -9 \\ 4x - y - z - 2w = 8 \end{pmatrix}$.

27. Suppose that the augmented matrix of a system of three equations in three variables can be changed to the following matrix.

$$\left[\begin{array}{ccc|c} 1 & 1 & -2 & 4 \\ 0 & -5 & 11 & -13 \\ 0 & 0 & 0 & -9 \end{array}\right]$$

What can be said about the solution set of the system?

28. Suppose that the augmented matrix of a system of three linear equations in three variables can be changed to the following matrix.

$$\left[\begin{array}{ccc|c} 1 & 0 & 1 & 1 \\ 0 & 1 & -1 & 0 \\ 0 & 0 & 0 & 0 \end{array}\right]$$

What can be said about the solution set of the system?

GRAPHING CALCULATOR ACTIVITIES

29. If your graphing calculator has the capability to manipulate matrices, this is a good time to become familiar with those operations. You may need to refer to your calculator manual for the specific instructions. To begin the familiarization process, load your calculator with the three augmented matrices in Examples 1, 2, and 3 of this section. Then, for each one, carry out the row operations as described in the text.

C Determinants

A **square matrix** is one that has the same number of rows as columns. Associated with each square matrix that has real number entries is a real number called the **determinant** of the matrix. For a 2×2 matrix

$$\begin{bmatrix} a_1 & b_1 \\ a_2 & b_2 \end{bmatrix}$$

the determinant is written as

$$\begin{vmatrix} a_1 & b_1 \\ a_2 & b_2 \end{vmatrix}$$

and defined by

$$\begin{vmatrix} a_1 & b_1 \\ a_2 & b_2 \end{vmatrix} = a_1b_2 - a_2b_1 \tag{1}$$

Note that a determinant is simply a number and that the determinant notation used on the left side of equation (1) is a way of expressing the number on the right side.

EXAMPLE 1

Find the determinant of the matrix $\begin{bmatrix} 3 & -2 \\ 5 & 8 \end{bmatrix}$.

Solution

In this case, $a_1 = 3$, $b_1 = -2$, $a_2 = 5$, and $b_2 = 8$. Thus we have

$$\begin{vmatrix} 3 & -2 \\ 5 & 8 \end{vmatrix} = 3(8) - 5(-2) = 24 + 10 = 34$$

■

Finding the determinant of a square matrix is commonly called *evaluating the determinant*, and the matrix notation is sometimes omitted.

EXAMPLE 2

Evaluate $\begin{vmatrix} -3 & 5 \\ 1 & 2 \end{vmatrix}$.

Solution

$$\begin{vmatrix} -3 & 5 \\ 1 & 2 \end{vmatrix} = -3(2) - 1(5) = -11$$

■

Cramer's Rule

Determinants provide the basis for another method of solving linear systems. Consider the system

$$\begin{pmatrix} a_1x + b_1y = c_1 \\ a_2x + b_2y = c_2 \end{pmatrix} \quad \begin{matrix} (1) \\ (2) \end{matrix}$$

We shall solve this system by using the elimination method; observe that our solutions can be conveniently written in determinant form. To solve for x, we can multiply equation (1) by b_2 and equation (2) by $-b_1$ and then add.

$$\begin{aligned} a_1b_2x + b_1b_2y &= c_1b_2 \\ -a_2b_1x - b_1b_2y &= -c_2b_1 \\ \hline a_1b_2x - a_2b_1x &= c_1b_2 - c_2b_1 \\ (a_1b_2 - a_2b_1)x &= c_1b_2 - c_2b_1 \\ x &= \frac{c_1b_2 - c_2b_1}{a_1b_2 - a_2b_1} \quad \text{If } a_1b_2 - a_2b_1 \neq 0 \end{aligned}$$

To solve for y, we can multiply equation (1) by $-a_2$ and equation (2) by a_1 and add.

$$\begin{aligned} -a_1a_2x - a_2b_1y &= -a_2c_1 \\ a_1a_2x + a_1b_2y &= a_1c_2 \\ \hline a_1b_2y - a_2b_1y &= a_1c_2 - a_2c_1 \\ (a_1b_2 - a_2b_1)y &= a_1c_2 - a_2c_1 \\ y &= \frac{a_1c_2 - a_2c_1}{a_1b_2 - a_2b_1} \quad \text{If } a_1b_2 - a_2b_1 \neq 0 \end{aligned}$$

We can express the solutions for x and y in determinant form as follows:

$$x = \frac{c_1b_2 - c_2b_1}{a_1b_2 - a_2b_1} = \frac{\begin{vmatrix} c_1 & b_1 \\ c_2 & b_2 \end{vmatrix}}{\begin{vmatrix} a_1 & b_1 \\ a_2 & b_2 \end{vmatrix}} \qquad y = \frac{a_1c_2 - a_2c_1}{a_1b_2 - a_2b_1} = \frac{\begin{vmatrix} a_1 & c_1 \\ a_2 & c_2 \end{vmatrix}}{\begin{vmatrix} a_1 & b_1 \\ a_2 & b_2 \end{vmatrix}}$$

For convenience, we shall denote the three determinants in the solution as

$$\begin{vmatrix} a_1 & b_1 \\ a_2 & b_2 \end{vmatrix} = D \qquad \begin{vmatrix} c_1 & b_1 \\ c_2 & b_2 \end{vmatrix} = D_x \qquad \begin{vmatrix} a_1 & c_1 \\ a_2 & c_2 \end{vmatrix} = D_y$$

Note that the elements of D are the coefficients of the variables in the given system. In D_x, we obtain the elements by replacing the coefficients of x with the respective constants. In D_y, we replace the coefficients of y with the respective constants. This method of using determinants to solve a system of two linear equations in two variables is called **Cramer's rule**. We state it as follows.

Cramer's Rule

Given the system

$$\begin{pmatrix} a_1x + b_1y = c_1 \\ a_2x + b_2y = c_2 \end{pmatrix} \quad \text{with } a_1b_2 - a_2b_1 \neq 0$$

then

$$x = \frac{\begin{vmatrix} c_1 & b_1 \\ c_2 & b_2 \end{vmatrix}}{\begin{vmatrix} a_1 & b_1 \\ a_2 & b_2 \end{vmatrix}} = \frac{D_x}{D} \quad \text{and} \quad y = \frac{\begin{vmatrix} a_1 & c_1 \\ a_2 & c_2 \end{vmatrix}}{\begin{vmatrix} a_1 & b_1 \\ a_2 & b_2 \end{vmatrix}} = \frac{D_y}{D}$$

Let's use Cramer's rule to solve some systems.

EXAMPLE 3

Solve the system $\begin{pmatrix} x + 2y = 11 \\ 2x - y = 2 \end{pmatrix}$.

Solution

Let's find D, D_x, and D_y.

$$D = \begin{vmatrix} 1 & 2 \\ 2 & -1 \end{vmatrix} = -1 - 4 = -5$$

$$D_x = \begin{vmatrix} 11 & 2 \\ 2 & -1 \end{vmatrix} = -11 - 4 = -15$$

$$D_y = \begin{vmatrix} 1 & 11 \\ 2 & 2 \end{vmatrix} = 2 - 22 = -20$$

Thus we have

$$x = \frac{D_x}{D} = \frac{-15}{-5} = 3$$

$$y = \frac{D_y}{D} = \frac{-20}{-5} = 4$$

The solution set is $\{(3, 4)\}$, which we can verify, as always, by substituting back into the original equations. ■

Remark: Note that Cramer's rule has the restriction $a_1b_2 - a_2b_1 \neq 0$; that is, $D \neq 0$. Thus it is a good idea to find D first. Then if $D = 0$, Cramer's rule does not apply, and you must use one of the other methods to determine whether the solution set is empty or has infinitely many solutions.

EXAMPLE 4

Solve the system $\begin{pmatrix} 2x - 3y = -8 \\ 3x + 5y = 7 \end{pmatrix}$.

Solution

$$D = \begin{vmatrix} 2 & -3 \\ 3 & 5 \end{vmatrix} = 10 - (-9) = 19$$

$$D_x = \begin{vmatrix} -8 & -3 \\ 7 & 5 \end{vmatrix} = -40 - (-21) = -19$$

$$D_y = \begin{vmatrix} 2 & -8 \\ 3 & 7 \end{vmatrix} = 14 - (-24) = 38$$

Thus we obtain

$$x = \frac{D_x}{D} = \frac{-19}{19} = -1 \quad \text{and} \quad y = \frac{D_y}{D} = \frac{38}{19} = 2$$

The solution set is $\{(-1, 2)\}$. ■

EXAMPLE 5

Solve the system $\begin{pmatrix} y = -2x - 2 \\ 4x - 5y = 17 \end{pmatrix}$.

Solution

First, we must change the form of the first equation so that the system fits the form given in Cramer's rule. The equation $y = -2x - 2$ can be written as $2x + y = -2$. The system now becomes

$$\begin{pmatrix} 2x + y = -2 \\ 4x - 5y = 17 \end{pmatrix}$$

and we can proceed as before.

$$D = \begin{vmatrix} 2 & 1 \\ 4 & -5 \end{vmatrix} = -10 - 4 = -14$$

$$D_x = \begin{vmatrix} -2 & 1 \\ 17 & -5 \end{vmatrix} = 10 - 17 = -7$$

$$D_y = \begin{vmatrix} 2 & -2 \\ 4 & 17 \end{vmatrix} = 34 - (-8) = 42$$

Thus the solutions are

$$x = \frac{D_x}{D} = \frac{-7}{-14} = \frac{1}{2} \quad \text{and} \quad y = \frac{D_y}{D} = \frac{42}{-14} = -3.$$

The solution set is $\left\{\left(\frac{1}{2}, -3\right)\right\}$. ■

Appendix C Exercises

Evaluate each of the following determinants.

1. $\begin{vmatrix} 6 & 2 \\ 4 & 3 \end{vmatrix}$

2. $\begin{vmatrix} 7 & 6 \\ 2 & 5 \end{vmatrix}$

3. $\begin{vmatrix} 4 & 7 \\ 8 & 2 \end{vmatrix}$

4. $\begin{vmatrix} 3 & 9 \\ 6 & 4 \end{vmatrix}$

5. $\begin{vmatrix} -3 & 2 \\ 7 & 5 \end{vmatrix}$

6. $\begin{vmatrix} 5 & 1 \\ 8 & -4 \end{vmatrix}$

7. $\begin{vmatrix} 8 & -3 \\ 6 & 4 \end{vmatrix}$

8. $\begin{vmatrix} 5 & 9 \\ -3 & 6 \end{vmatrix}$

9. $\begin{vmatrix} -3 & 2 \\ 5 & -6 \end{vmatrix}$

10. $\begin{vmatrix} -2 & 4 \\ 9 & -7 \end{vmatrix}$

11. $\begin{vmatrix} 3 & -3 \\ -6 & 8 \end{vmatrix}$

12. $\begin{vmatrix} 6 & -5 \\ -8 & 12 \end{vmatrix}$

13. $\begin{vmatrix} -7 & -2 \\ -2 & 4 \end{vmatrix}$

14. $\begin{vmatrix} 6 & -1 \\ -8 & -3 \end{vmatrix}$

15. $\begin{vmatrix} -2 & -3 \\ -4 & -5 \end{vmatrix}$

16. $\begin{vmatrix} -9 & -7 \\ -6 & -4 \end{vmatrix}$

17. $\begin{vmatrix} \frac{1}{4} & -2 \\ \frac{3}{2} & 8 \end{vmatrix}$

18. $\begin{vmatrix} -\frac{2}{3} & 10 \\ -\frac{1}{2} & 6 \end{vmatrix}$

19. $\begin{vmatrix} \frac{3}{2} & -\frac{1}{2} \\ \frac{1}{2} & -\frac{2}{5} \end{vmatrix}$

20. $\begin{vmatrix} -\frac{1}{4} & \frac{1}{3} \\ \frac{3}{2} & \frac{2}{3} \end{vmatrix}$

Use Cramer's rule to find the solution set for each of the following systems.

21. $\begin{pmatrix} 2x + y = 14 \\ 3x - y = 1 \end{pmatrix}$

22. $\begin{pmatrix} 4x - y = 11 \\ 2x + 3y = 23 \end{pmatrix}$

23. $\begin{pmatrix} -x + 3y = 17 \\ 4x - 5y = -33 \end{pmatrix}$

24. $\begin{pmatrix} 5x + 2y = -15 \\ 7x - 3y = 37 \end{pmatrix}$

25. $\begin{pmatrix} 9x + 5y = -8 \\ 7x - 4y = -22 \end{pmatrix}$

26. $\begin{pmatrix} 8x - 11y = 3 \\ -x + 4y = -3 \end{pmatrix}$

27. $\begin{pmatrix} x + 5y = 4 \\ 3x + 15y = -1 \end{pmatrix}$

28. $\begin{pmatrix} 4x - 7y = 0 \\ 7x + 2y = 0 \end{pmatrix}$

29. $\begin{pmatrix} 6x - y = 0 \\ 5x + 4y = 29 \end{pmatrix}$

30. $\begin{pmatrix} 3x - 4y = 2 \\ 9x - 12y = 6 \end{pmatrix}$

31. $\begin{pmatrix} -4x + 3y = 3 \\ 4x - 6y = -5 \end{pmatrix}$

32. $\begin{pmatrix} x - 2y = -1 \\ x = -6y + 5 \end{pmatrix}$

33. $\begin{pmatrix} 6x - 5y = 1 \\ 4x + 7y = 2 \end{pmatrix}$

34. $\begin{pmatrix} y = 3x + 5 \\ y = 6x + 6 \end{pmatrix}$

35. $\begin{pmatrix} 7x + 2y = -1 \\ y = -x + 2 \end{pmatrix}$

36. $\begin{pmatrix} 9x - y = -2 \\ y = 4 - 8x \end{pmatrix}$

37. $\begin{pmatrix} -\frac{2}{3}x + \frac{1}{2}y = -7 \\ \frac{1}{3}x - \frac{3}{2}y = 6 \end{pmatrix}$

38. $\begin{pmatrix} \frac{1}{2}x + \frac{2}{3}y = -6 \\ \frac{1}{4}x - \frac{1}{3}y = -1 \end{pmatrix}$

39. $\begin{pmatrix} x + \frac{2}{3}y = -6 \\ -\frac{1}{4}x + 3y = -8 \end{pmatrix}$

40. $\begin{pmatrix} 3x - \frac{1}{2}y = 6 \\ -2x + \frac{1}{3}y = -4 \end{pmatrix}$

THOUGHTS INTO WORDS

41. Explain the difference between a matrix and a determinant.

42. Give a step-by-step description of how you would solve the system $\begin{pmatrix} 3x - 2y = 7 \\ 5x + 9y = 14 \end{pmatrix}$ using determinants.

FURTHER INVESTIGATIONS

43. Verify each of the following. The variables represent real numbers.

(a) $\begin{vmatrix} a & b \\ a & b \end{vmatrix} = 0$

(b) $\begin{vmatrix} a & a \\ b & b \end{vmatrix} = 0$

(c) $\begin{vmatrix} a & b \\ c & d \end{vmatrix} = -\begin{vmatrix} b & a \\ d & c \end{vmatrix}$

(d) $\begin{vmatrix} a & b \\ c & d \end{vmatrix} = -\begin{vmatrix} c & d \\ a & b \end{vmatrix}$

(e) $k\begin{vmatrix} a & b \\ c & d \end{vmatrix} = \begin{vmatrix} ka & b \\ kc & d \end{vmatrix}$

(f) $k\begin{vmatrix} a & b \\ c & d \end{vmatrix} = \begin{vmatrix} ka & kb \\ c & d \end{vmatrix}$

GRAPHING CALCULATOR ACTIVITIES

44. Use the determinant function of your graphing calculator to check your answers for Problems 1–16.

45. Make up two or three examples for each part of Problem 43, and evaluate the determinants using your graphing calculator.

D 3 × 3 Determinants and Systems of Three Linear Equations in Three Variables

This appendix will extend the concept of a determinant to include 3 × 3 determinants and then extend the use of determinants to solve systems of three linear equations in three variables.

For a 3 × 3 matrix

$$\begin{bmatrix} a_1 & b_1 & c_1 \\ a_2 & b_2 & c_2 \\ a_3 & b_3 & c_3 \end{bmatrix}$$

the determinant is written as

$$\begin{vmatrix} a_1 & b_1 & c_1 \\ a_2 & b_2 & c_2 \\ a_3 & b_3 & c_3 \end{vmatrix}$$

and defined by

$$\begin{vmatrix} a_1 & b_1 & c_1 \\ a_2 & b_2 & c_2 \\ a_3 & b_3 & c_3 \end{vmatrix} = a_1b_2c_3 + b_1c_2a_3 + c_1a_2b_3 - a_3b_2c_1 - b_3c_2a_1 - c_3a_2b_1 \tag{1}$$

It is evident that the definition given by equation (1) is a bit complicated to be very useful in practice. Fortunately, there is a method called **expansion of a determinant by minors** that we can use to calculate such a determinant.

The **minor** of an element in a determinant is the determinant that remains after deleting the row and column in which the element appears. For example, consider the determinant of equation (1).

The minor of a_1 is $\begin{vmatrix} b_2 & c_2 \\ b_3 & c_3 \end{vmatrix}$.

The minor of a_2 is $\begin{vmatrix} b_1 & c_1 \\ b_3 & c_3 \end{vmatrix}$.

The minor of a_3 is $\begin{vmatrix} b_1 & c_1 \\ b_2 & c_2 \end{vmatrix}$.

Now let's consider the terms, in pairs, of the right side of equation (1) and show the tie-in with minors.

$$\begin{aligned} a_1b_2c_3 - b_3c_2a_1 &= a_1(b_2c_3 - b_3c_2) \\ &= a_1\begin{vmatrix} b_2 & c_2 \\ b_3 & c_3 \end{vmatrix} \\ c_1a_2b_3 - c_3a_2b_1 &= -(c_3a_2b_1 - c_1a_2b_3) \\ &= -a_2(b_1c_3 - b_3c_1) \\ &= -a_2\begin{vmatrix} b_1 & c_1 \\ b_3 & c_3 \end{vmatrix} \\ b_1c_2a_3 - a_3b_2c_1 &= a_3(b_1c_2 - b_2c_1) \\ &= a_3\begin{vmatrix} b_1 & c_1 \\ b_2 & c_2 \end{vmatrix} \end{aligned}$$

Therefore, we have

$$\begin{vmatrix} a_1 & b_1 & c_1 \\ a_2 & b_2 & c_2 \\ a_3 & b_3 & c_3 \end{vmatrix} = a_1\begin{vmatrix} b_2 & c_2 \\ b_3 & c_3 \end{vmatrix} - a_2\begin{vmatrix} b_1 & c_1 \\ b_3 & c_3 \end{vmatrix} + a_3\begin{vmatrix} b_1 & c_1 \\ b_2 & c_2 \end{vmatrix}$$

and this is called the **expansion of the determinant by minors about the first column**.

EXAMPLE 1

Evaluate $\begin{vmatrix} 1 & 2 & -1 \\ 3 & 1 & -2 \\ 2 & 4 & 3 \end{vmatrix}$ by expanding by minors about the first column.

Solution

$$\begin{vmatrix} 1 & 2 & -1 \\ 3 & 1 & -2 \\ 2 & 4 & 3 \end{vmatrix} = 1\begin{vmatrix} 1 & -2 \\ 4 & 3 \end{vmatrix} - 3\begin{vmatrix} 2 & -1 \\ 4 & 3 \end{vmatrix} + 2\begin{vmatrix} 2 & -1 \\ 1 & -2 \end{vmatrix}$$

$$= 1[3 - (-8)] - 3[6 - (-4)] + 2[(-4) - (-1)]$$

$$= 1(11) - 3(10) + 2(-3) = -25$$

■

It is possible to expand a determinant by minors about any row or any column. To help determine the signs of the terms in the expansions, the following *sign array* is very useful.

$$\begin{matrix} + & - & + \\ - & + & - \\ + & - & + \end{matrix}$$

For example, let's expand the determinant in Example 1 by minors about the second row. The second row in the sign array is $- + -$. Therefore,

$$\begin{vmatrix} 1 & 2 & -1 \\ 3 & 1 & -2 \\ 2 & 4 & 3 \end{vmatrix} = -3\begin{vmatrix} 2 & -1 \\ 4 & 3 \end{vmatrix} + 1\begin{vmatrix} 1 & -1 \\ 2 & 3 \end{vmatrix} - (-2)\begin{vmatrix} 1 & 2 \\ 2 & 4 \end{vmatrix}$$

$$= -3[6 - (-4)] + 1[3 - (-2)] + 2(4 - 4)$$

$$= -3(10) + 1(5) + 2(0)$$

$$= -25$$

Your decision as to which row or column to use for expanding a particular determinant by minors may depend on the numbers involved in the determinant. A row or column with one or more zeros is frequently a good choice, as the next example illustrates.

EXAMPLE 2

Evaluate $\begin{vmatrix} 3 & -1 & 4 \\ 5 & 2 & 0 \\ -2 & 6 & 0 \end{vmatrix}$.

Solution

Because the third column has two zeros, we shall expand about it.

$$\begin{vmatrix} 3 & -1 & 4 \\ 5 & 2 & 0 \\ -2 & 6 & 0 \end{vmatrix} = 4\begin{vmatrix} 5 & 2 \\ -2 & 6 \end{vmatrix} - 0\begin{vmatrix} 3 & -1 \\ -2 & 6 \end{vmatrix} + 0\begin{vmatrix} 3 & -1 \\ 5 & 2 \end{vmatrix}$$

$$= 4[30 - (-4)] - 0 + 0 = 136$$

(Note that because of the zeros, there is no need to evaluate the last two minors.)

■

Remark 1: The expansion-by-minors method can be extended to determinants of size 4×4, 5×5, and so on. However, it should be obvious that it becomes increasingly tedious with bigger determinants. Fortunately, the computer handles the calculation of such determinants with a different technique.

Remark 2: There is another method for evaluating 3×3 determinants. This method is demonstrated in Problem 32 of the next problem set. If you choose to use that method, keep in mind that it works *only* for 3×3 determinants.

Without showing all of the details, we will simply state that Cramer's rule also applies to solving systems of three linear equations in three variables. It can be restated as follows.

Cramer's Rule

Given the system

$$\begin{pmatrix} a_1x + b_1y + c_1z = d_1 \\ a_2x + b_2y + c_2z = d_2 \\ a_3x + b_3y + c_3z = d_3 \end{pmatrix}$$

with

$$D = \begin{vmatrix} a_1 & b_1 & c_1 \\ a_2 & b_2 & c_2 \\ a_3 & b_3 & c_3 \end{vmatrix} \neq 0 \qquad D_x = \begin{vmatrix} d_1 & b_1 & c_1 \\ d_2 & b_2 & c_2 \\ d_3 & b_3 & c_3 \end{vmatrix}$$

$$D_y = \begin{vmatrix} a_1 & d_1 & c_1 \\ a_2 & d_2 & c_2 \\ a_3 & d_3 & c_3 \end{vmatrix} \qquad D_z = \begin{vmatrix} a_1 & b_1 & d_1 \\ a_2 & b_2 & d_2 \\ a_3 & b_3 & d_3 \end{vmatrix}$$

then $x = \dfrac{D_x}{D}$, $y = \dfrac{D_y}{D}$, and $z = \dfrac{D_z}{D}$.

Note that the elements of D are the coefficients of the variables in the given system. Then D_x, D_y, and D_z are formed by replacing the elements in the x, y, and z columns, respectively, by the constants of the system d_1, d_2, and d_3. Again, note the restriction $D \neq 0$. As before, if $D = 0$ then Cramer's rule does not apply, and you can use the elimination method to determine whether the system has no solution or infinitely many solutions.

EXAMPLE 3

Use Cramer's rule to solve the system $\begin{pmatrix} x - 2y + z = -4 \\ 2x + y - z = 5 \\ 3x + 2y + 4z = 3 \end{pmatrix}$.

Solution

To find D, let's expand about row 1.

$$D = \begin{vmatrix} 1 & -2 & 1 \\ 2 & 1 & -1 \\ 3 & 2 & 4 \end{vmatrix} = 1\begin{vmatrix} 1 & -1 \\ 2 & 4 \end{vmatrix} - (-2)\begin{vmatrix} 2 & -1 \\ 3 & 4 \end{vmatrix} + 1\begin{vmatrix} 2 & 1 \\ 3 & 2 \end{vmatrix}$$

$$= 1[4 - (-2)] + 2[8 - (-3)] + 1(4 - 3)$$

$$= 1(6) + 2(11) + 1(1) = 29$$

To find D_x, let's expand about column 3.

$$D_x = \begin{vmatrix} -4 & -2 & 1 \\ 5 & 1 & -1 \\ 3 & 2 & 4 \end{vmatrix} = 1\begin{vmatrix} 5 & 1 \\ 3 & 2 \end{vmatrix} - (-1)\begin{vmatrix} -4 & -2 \\ 3 & 2 \end{vmatrix} + 4\begin{vmatrix} -4 & -2 \\ 5 & 1 \end{vmatrix}$$

$$= 1(10 - 3) + 1[-8 - (-6)] + 4[-4 - (-10)]$$

$$= 1(7) + 1(-2) + 4(6)$$

$$= 29$$

To find D_y, let's expand about row 1.

$$D_y = \begin{vmatrix} 1 & -4 & 1 \\ 2 & 5 & -1 \\ 3 & 3 & 4 \end{vmatrix} = 1\begin{vmatrix} 5 & -1 \\ 3 & 4 \end{vmatrix} - (-4)\begin{vmatrix} 2 & -1 \\ 3 & 4 \end{vmatrix} + 1\begin{vmatrix} 2 & 5 \\ 3 & 3 \end{vmatrix}$$

$$= 1[20 - (-3)] + 4[8 - (-3)] + 1(6 - 15)$$

$$= 1(23) + 4(11) + 1(-9)$$

$$= 58$$

To find D_z, let's expand about column 1.

$$D_z = \begin{vmatrix} 1 & -2 & -4 \\ 2 & 1 & 5 \\ 3 & 2 & 3 \end{vmatrix} = 1\begin{vmatrix} 1 & 5 \\ 2 & 3 \end{vmatrix} - 2\begin{vmatrix} -2 & -4 \\ 2 & 3 \end{vmatrix} + 3\begin{vmatrix} -2 & -4 \\ 1 & 5 \end{vmatrix}$$

$$= 1(3 - 10) - 2[-6 - (-8)] + 3[-10 - (-4)]$$

$$= 1(-7) - 2(2) + 3(-6)$$

$$= -29$$

Thus

$$x = \frac{D_x}{D} = \frac{29}{29} = 1$$

$$y = \frac{D_y}{D} = \frac{58}{29} = 2$$

$$z = \frac{D_z}{D} = \frac{-29}{29} = -1$$

The solution set is $\{(1, 2, -1)\}$. (Be sure to check it!) ■

EXAMPLE 4

Use Cramer's rule to solve the system $\begin{pmatrix} 2x - y + 3z = -17 \\ 3y + z = 5 \\ x - 2y - z = -3 \end{pmatrix}$.

Solution

To find D, let's expand about column 1.

$$D = \begin{vmatrix} 2 & -1 & 3 \\ 0 & 3 & 1 \\ 1 & -2 & -1 \end{vmatrix} = 2\begin{vmatrix} 3 & 1 \\ -2 & -1 \end{vmatrix} - 0\begin{vmatrix} -1 & 3 \\ -2 & -1 \end{vmatrix} + 1\begin{vmatrix} -1 & 3 \\ 3 & 1 \end{vmatrix}$$

$$= 2[-3 - (-2)] - 0 + 1(-1 - 9)$$

$$= 2(-1) - 0 - 10 = -12$$

To find D_x, let's expand about column 3.

$$D_x = \begin{vmatrix} -17 & -1 & 3 \\ 5 & 3 & 1 \\ -3 & -2 & -1 \end{vmatrix} = 3\begin{vmatrix} 5 & 3 \\ -3 & -2 \end{vmatrix} - 1\begin{vmatrix} -17 & -1 \\ -3 & -2 \end{vmatrix} + (-1)\begin{vmatrix} -17 & -1 \\ 5 & 3 \end{vmatrix}$$

$$= 3[-10 - (-9)] - 1(34 - 3) - 1[-51 - (-5)]$$

$$= 3(-1) - 1(31) - 1(-46) = 12$$

To find D_y, let's expand about column 1.

$$D_y = \begin{vmatrix} 2 & -17 & 3 \\ 0 & 5 & 1 \\ 1 & -3 & -1 \end{vmatrix} = 2\begin{vmatrix} 5 & 1 \\ -3 & -1 \end{vmatrix} - 0\begin{vmatrix} -17 & 3 \\ -3 & -1 \end{vmatrix} + 1\begin{vmatrix} -17 & 3 \\ 5 & 1 \end{vmatrix}$$

$$= 2[-5 - (-3)] - 0 + 1(-17 - 15)$$

$$= 2(-2) - 0 + 1(-32) = -36$$

To find D_z, let's expand about column 1.

$$D_z = \begin{vmatrix} 2 & -1 & -17 \\ 0 & 3 & 5 \\ 1 & -2 & -3 \end{vmatrix} = 2\begin{vmatrix} 3 & 5 \\ -2 & -3 \end{vmatrix} - 0\begin{vmatrix} -1 & -17 \\ -2 & -3 \end{vmatrix} + 1\begin{vmatrix} -1 & -17 \\ 3 & 5 \end{vmatrix}$$

$$= 2[-9 - (-10)] - 0 + 1[-5 - (-51)]$$

$$= 2(1) - 0 + 1(46) = 48$$

Thus

$$x = \frac{D_x}{D} = \frac{12}{-12} = -1 \qquad y = \frac{D_y}{D} = \frac{-36}{-12} = 3$$

$$z = \frac{D_z}{D} = \frac{48}{-12} = -4.$$

The solution set is $\{(-1, 3, -4)\}$. ■

Appendix D Exercises

For Problems 1–10, use expansion by minors to evaluate each determinant.

1. $\begin{vmatrix} 2 & 7 & 5 \\ 1 & -1 & 1 \\ -4 & 3 & 2 \end{vmatrix}$

2. $\begin{vmatrix} 2 & 4 & 1 \\ -1 & 5 & 1 \\ -3 & 6 & 2 \end{vmatrix}$

3. $\begin{vmatrix} 3 & -2 & 1 \\ 2 & 1 & 4 \\ -1 & 3 & 5 \end{vmatrix}$

4. $\begin{vmatrix} 1 & -1 & 2 \\ 2 & 1 & 3 \\ -1 & -2 & 1 \end{vmatrix}$

5. $\begin{vmatrix} -3 & -2 & 1 \\ 5 & 0 & 6 \\ 2 & 1 & -4 \end{vmatrix}$

6. $\begin{vmatrix} -5 & 1 & -1 \\ 3 & 4 & 2 \\ 0 & 2 & -3 \end{vmatrix}$

7. $\begin{vmatrix} 3 & -4 & -2 \\ 5 & -2 & 1 \\ 1 & 0 & 0 \end{vmatrix}$

8. $\begin{vmatrix} -6 & 5 & 3 \\ 2 & 0 & -1 \\ 4 & 0 & 7 \end{vmatrix}$

9. $\begin{vmatrix} 4 & -2 & 7 \\ 1 & -1 & 6 \\ 3 & 5 & -2 \end{vmatrix}$

10. $\begin{vmatrix} -5 & 2 & 6 \\ 1 & -1 & 3 \\ 4 & -2 & -4 \end{vmatrix}$

For Problems 11–30, use Cramer's rule to find the solution set of each system.

11. $\begin{pmatrix} 2x - y + 3z = -10 \\ x + 2y - 3z = 2 \\ 3x - 2y + 5z = -16 \end{pmatrix}$

12. $\begin{pmatrix} -x + y - z = 1 \\ 2x + 3y - 4z = 10 \\ -3x - y + z = -5 \end{pmatrix}$

13. $\begin{pmatrix} x - y + 2z = -8 \\ 2x + 3y - 4z = 18 \\ -x + 2y - z = 7 \end{pmatrix}$

14. $\begin{pmatrix} x - 2y + z = 3 \\ 3x + 2y + z = -3 \\ 2x - 3y - 3z = -5 \end{pmatrix}$

15. $\begin{pmatrix} 3x - 2y - 3z = -5 \\ x + 2y + 3z = -3 \\ -x + 4y - 6z = 8 \end{pmatrix}$

16. $\begin{pmatrix} 2x - 3y + 3z = -3 \\ -2x + 5y - 3z = 5 \\ 3x - y + 6z = -1 \end{pmatrix}$

17. $\begin{pmatrix} -x + y + z = -1 \\ x - 2y + 5z = -4 \\ 3x + 4y - 6z = -1 \end{pmatrix}$

18. $\begin{pmatrix} x - 2y + 3z = 1 \\ 2x + y + z = 4 \\ 4x - 3y + 7z = 6 \end{pmatrix}$

19. $\begin{pmatrix} x - y + 2z = 4 \\ 3x - 2y + 4z = 6 \\ 2x - 2y + 4z = -1 \end{pmatrix}$

20. $\begin{pmatrix} -x - 2y + z = 8 \\ 3x + y - z = 5 \\ 5x - y + 4z = 33 \end{pmatrix}$

21. $\begin{pmatrix} 2x - y + 3z = -5 \\ 3x + 4y - 2z = -25 \\ -x + z = 6 \end{pmatrix}$

22. $\begin{pmatrix} 3x - 2y + z = 11 \\ 5x + 3y = 17 \\ x + y - 2z = 6 \end{pmatrix}$

23. $\begin{pmatrix} 2y - z = 10 \\ 3x + 4y = 6 \\ x - y + z = -9 \end{pmatrix}$

24. $\begin{pmatrix} 6x - 5y + 2z = 7 \\ 2x + 3y - 4z = -21 \\ 2y + 3z = 10 \end{pmatrix}$

25. $\begin{pmatrix} -2x + 5y - 3z = -1 \\ 2x - 7y + 3z = 1 \\ 4x - y - 6z = -6 \end{pmatrix}$

26. $\begin{pmatrix} 7x - 2y + 3z = -4 \\ 5x + 2y - 3z = 4 \\ -3x - 6y + 12z = -13 \end{pmatrix}$

27. $\begin{pmatrix} -x - y + 5z = 4 \\ x + y - 7z = -6 \\ 2x + 3y + 4z = 13 \end{pmatrix}$

28. $\begin{pmatrix} x + 7y - z = -1 \\ -x - 9y + z = 3 \\ 3x + 4y - 6z = 5 \end{pmatrix}$

29. $\begin{pmatrix} 5x - y + 2z = 10 \\ 7x + 2y - 2z = -4 \\ -3x - y + 4z = 1 \end{pmatrix}$

30. $\begin{pmatrix} 4x - y - 3z = -12 \\ 5x + y + 6z = 4 \\ 6x - y - 3z = -14 \end{pmatrix}$

31. **(a)** Show that $\begin{vmatrix} 2 & 1 & 2 \\ 4 & -1 & -2 \\ 6 & 3 & 1 \end{vmatrix} = 2\begin{vmatrix} 1 & 1 & 2 \\ 2 & -1 & -2 \\ 3 & 3 & 1 \end{vmatrix}$.

Make a conjecture about the result of factoring a common factor from each element of a column in a determinant.

(b) Use your conjecture from part (a) to help evaluate the following determinant.

$$\begin{vmatrix} 2 & 4 & -1 \\ -3 & -4 & -2 \\ 5 & 4 & 3 \end{vmatrix}$$

32. We can describe another technique for evaluating 3×3 determinants as follows: First, let's write the given determinant with its first two columns repeated on the right.

$$\begin{vmatrix} a_1 & b_1 & c_1 \\ a_2 & b_2 & c_2 \\ a_3 & b_3 & c_3 \end{vmatrix} \begin{matrix} a_1 & b_1 \\ a_2 & b_2 \\ a_3 & b_3 \end{matrix}$$

Then we can add the three products shown with + and subtract the three products shown with −.

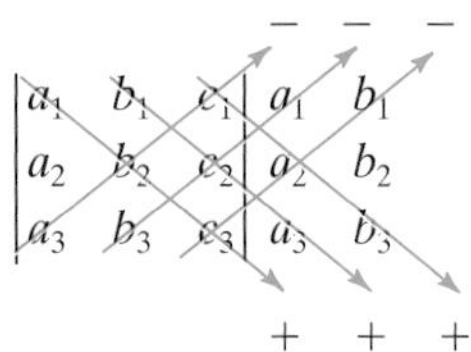

(a) Be sure that the previous description will produce equation (1) on page 525.

(b) Use this technique to do Problems 1–10.

THOUGHTS INTO WORDS

33. How would you explain the process of evaluating 3×3 determinants to a friend who missed class the day it was discussed?

34. Explain how to use determinants to solve the system

$$\begin{pmatrix} x - 2y + z = 1 \\ 2x - y - z = 5 \\ 5x + 3y + 4z = -6 \end{pmatrix}$$

FURTHER INVESTIGATIONS

35. Evaluate the following determinant by expanding about the second column.

$$\begin{vmatrix} a & e & a \\ b & f & b \\ c & g & c \end{vmatrix}$$

Make a conjecture about determinants that contain two identical columns.

36. Show that $\begin{vmatrix} 1 & -1 & 2 \\ 2 & 3 & -1 \\ -1 & 2 & 4 \end{vmatrix} = -\begin{vmatrix} -1 & 1 & 2 \\ 3 & 2 & -1 \\ 2 & -1 & 4 \end{vmatrix}$.

Make a conjecture about the result of interchanging two columns of a determinant.

Practice Your Skills Solutions

CHAPTER 1

Section 1.1

1. First do the division.

$15 + 45 \div 5 \cdot 3 = 15 + 9 \cdot 3$

Next do the multiplication.

$15 + 9 \cdot 3 = 15 + 27$

Then do the addition.

$15 + 27 = 42$

2. $$\begin{aligned} 5 \cdot 6 \div 3 \cdot 2 &= 30 \div 3 \cdot 2 \\ &= 10 \cdot 2 \\ &= 20 \end{aligned}$$

3. $$\begin{aligned} 2 \cdot 5 + 15 \div 5 - 2 &= 10 + 3 - 2 \\ &= 13 - 2 \\ &= 11 \end{aligned}$$

4. $$\begin{aligned} (18 - 6)(2 + 3) &= (12)(5) \\ &= 60 \end{aligned}$$

5. $$\begin{aligned} (3 \cdot 4 - 5 \cdot 2)(3 \cdot 6 - 2 \cdot 4) &= (12 - 10)(18 - 8) \\ &= (2)(10) \\ &= 20 \end{aligned}$$

6. $$\begin{aligned} 1 + 3[2(8 - 3)] &= 1 + 3[2(5)] \\ &= 1 + 3(10) \\ &= 1 + 30 \\ &= 31 \end{aligned}$$

7. $$\begin{aligned} \frac{12 \cdot 6 \div 3 - 3}{7 \cdot 5 - 4 \cdot 8} &= \frac{72 \div 3 - 3}{35 - 32} \\ &= \frac{24 - 3}{3} \\ &= \frac{21}{3} \\ &= 7 \end{aligned}$$

Section 1.2

1. (a) $$\begin{aligned} -8.42 + 10.75 &= (|10.75| - |8.42)|) \\ &= 10.75 - 8.42 \\ &= 2.33 \end{aligned}$$

(b) $$\begin{aligned} \left(-\frac{2}{3}\right) + \left(-\frac{1}{4}\right) &= -\left(\left|-\frac{2}{3}\right| + \left|-\frac{1}{4}\right|\right) \\ &= -\left(\frac{2}{3} + \frac{1}{4}\right) \\ &= -\left(\frac{8}{12} + \frac{3}{12}\right) \\ &= -\frac{11}{12} \end{aligned}$$

(c) $$\begin{aligned} 145 + (-213) &= -(|-213| - |145|) \\ &= -(213 - 145) \\ &= -68 \end{aligned}$$

2. (a) $$\begin{aligned} -2 - 9 &= -2 + (-9) \\ &= -11 \end{aligned}$$

(b) $$\begin{aligned} 6 - (-10) &= 6 + 10 \\ &= 16 \end{aligned}$$

(c) $$\begin{aligned} -3.2 - (-7.2) &= -3.2 + 7.2 \\ &= 4 \end{aligned}$$

(d) $$\begin{aligned} \frac{3}{4} - \left(-\frac{1}{2}\right) &= \frac{3}{4} + \frac{1}{2} \\ &= \frac{3}{4} + \frac{2}{4} \\ &= \frac{5}{4} \end{aligned}$$

3. $$\begin{aligned} &4 - 10 - 3 + 12 - 2 - 8 \\ &= 4 + (-10) + (-3) + 12 + (-2) + (-8) \\ &= -7 \end{aligned}$$

4. $$\begin{aligned} -1\frac{3}{4} - \left(-\frac{3}{8}\right) + \frac{1}{2} &= -1\frac{3}{4} + \frac{3}{8} + \frac{1}{2} \\ &= -\frac{7}{4} + \frac{3}{8} + \frac{1}{2} \\ &= -\frac{14}{8} + \frac{3}{8} + \frac{4}{8} \\ &= -\frac{7}{8} \end{aligned}$$

5. $$\begin{aligned} 8 - 4 - 3 + 6 - 1 &= 8 + (-4) + (-3) + 6 + (-1) \\ &= 6 \end{aligned}$$

6. $$\begin{aligned} &\left(\frac{1}{3} - \frac{1}{2}\right) - \left(\frac{2}{5} - \frac{1}{10}\right) \\ &= \left[\frac{2}{6} + \left(-\frac{3}{6}\right)\right] - \left[\frac{4}{10} + \left(-\frac{1}{10}\right)\right] \\ &= \left(-\frac{1}{6}\right) - \left(\frac{3}{10}\right) \end{aligned}$$

$$= \left(-\frac{5}{30}\right) + \left(-\frac{9}{30}\right)$$
$$= -\frac{14}{30}$$
$$= -\frac{7}{15}$$

7. **(a)** $(-12)(3) = -36$

(b) $(-1)(-9) = 9$

(c) $\left(\frac{5}{8}\right)\left(-\frac{2}{5}\right) = -\frac{5 \cdot 2}{8 \cdot 5}$
$$= -\frac{1}{4}$$

8. **(a)** $-\frac{24}{3} = -8$

(b) $\frac{-4.8}{-0.8} = 6$

(c) $\frac{0}{-7} = 0$

9. $5\frac{3}{4} - \left(-1\frac{1}{4}\right) - (-2)\left(-\frac{1}{2}\right) = 5\frac{3}{4} + 1\frac{1}{4} + 2\left(-\frac{1}{2}\right)$
$$= \frac{23}{4} + \frac{5}{4} + (-1)$$
$$= \frac{28}{4} + (-1)$$
$$= 7 + (-1)$$
$$= 6$$

10. $12 - 8 \div (-2) + 6(-4) = 12 - (-4) + (-24)$
$$= 12 + 4 + (-24)$$
$$= -8$$

11. $-6.8 - 3[8 - (-2.1)(-5)] = -6.8 - 3[8 - (10.5)]$
$$= -6.8 - 3(-2.5)$$
$$= -6.8 - (-7.5)$$
$$= -6.8 + 7.5$$
$$= 0.7$$

12. $[-2(-4) - 3(6)][4(-3) + 2(-1)]$
$$= [8 - 18][-12 + (-2)]$$
$$= (-10)(-14)$$
$$= 140$$

13. $10(5) + 3(3) + 8(-1) + 4(-2) + 5(-3)$
$$= 50 + 9 + (-8) + (-8) + (-15)$$
$$= 59 + (-16) + (-15)$$
$$= 28$$

Section 1.3

1. $25 + [(-25) + 119] = [25 + (-25)] + 119$
$$= 0 + 119$$
$$= 119$$

2. $4[(-25)(-57)] = [(4)(-25)](-57)$
$$= (-100)(-57)$$
$$= 5700$$

3. $22 + (-14) + (-42) + 12 + (-11) + 15$

22	−14	49
12	−42	+ −67
+ 15	+ −11	−18
49	−67	

4. $-20(-5 + 150) = (-20)(-5) + (-20)(150)$
$$= 100 + (-3000)$$
$$= -2900$$

5. $-15(47 - 44) = -15(3)$
$$= -45$$

6. $1.4(-5) + 1.4(15) = 1.4(-5 + 15)$
$$= 1.4(10)$$
$$= 14$$

7. $5(-3)^2 + 6(-2)^2 = 5(9) + 6(4)$
$$= 45 + 24$$
$$= 69$$

8. $(2 - 6)^2 = (-4)^2$
$$= 16$$

9. $[-5(-3) - 2(6)]^3 = [15 - 12]^3$
$$= (3)^3$$
$$= 27$$

10. $2 + 8\left(\frac{1}{2}\right) - 3\left(\frac{1}{2}\right)^2 + 4\left(\frac{1}{2}\right)^3$
$$= 2 + 4 - 3\left(\frac{1}{4}\right) + 4\left(\frac{1}{8}\right)$$
$$= 6 - \frac{3}{4} + \frac{1}{2}$$
$$= \frac{24}{4} - \frac{3}{4} + \frac{2}{4}$$
$$= \frac{23}{4}$$

Section 1.4

1. **(a)** $3(x - 4) + 5(x + 2)$
$$= 3(x) + 3(-4) + 5(x) + 5(2)$$
$$= 3x - 12 + 5x + 10$$
$$= 8x - 2$$

(b) $-3(b + 8) - 5(b - 1)$
$$= -3(b) - 3(8) - 5(b) - 5(-1)$$
$$= -3b - 24 - 5b + 5$$
$$= -3b - 5b - 24 + 5$$
$$= -8b - 19$$

(c) $-2(a + b) - (a + b)$
$= -2(a) - 2(b) - 1(a) - 1(b)$
$= -2a - 2b - a - b$
$= -2a - a - 2b - b$
$= -3a - 3b$

2. $-4(3x + 7y) + 2(-5x + 3y)$
$= -4(3x) - 4(7y) + 2(-5x) + 2(3y)$
$= -12x - 28y - 10x + 6y$
$= -12x - 10x - 28y + 6y$
$= -22x - 22y$

3. $5a - 3b = 5(-8) - 3(4)$ when $a = -8, b = 4$
$= -40 - 12$
$= -52$

4. $x^2 + 3xy - y^2 = (-1)^2 + 3(-1)(3) - (3)^2$
when $x = -1, y = 3$
$= 1 - 9 - 9$
$= -17$

5. $(x + y)^3 = [(-6) + (2)]^3$ when $x = -6, y = 2$
$= (-4)^3$
$= -64$

6. $(2a + 5b)(a - 2b) = [2(-3) + 5(-1)][(-3) - 2(-1)]$
when $a = -3, b = -1$
$= (-6 - 5)(-3 + 2)$
$= (-11)(-1) = 11$

7. $4a - 3b - 6a + 2b = -2a - b$
$= -2\left(-\frac{3}{4}\right) - \left(\frac{1}{3}\right)$
when $a = -\frac{3}{4}, b = \frac{1}{3}$
$= \frac{3}{2} - \frac{1}{3}$
$= \frac{9}{6} - \frac{2}{6} = \frac{7}{6}$

8. $-4(2y + 3) + 5(3y - 1) = -8y - 12 + 15y - 5$
$= 7y - 17$ when $y = -3.1$
$= 7(-3.1) - 17$
$= -21.7 - 17 = -38.7$

9. $3(x^2 + 2) - 2(x^2 - 1) - 5(x^2 - 3)$
$= 3x^2 + 6 - 2x^2 + 2 - 5x^2 + 15$
$= -4x^2 + 23$
$= -4(4)^2 + 23$ when $x = 4$
$= -4(16) + 23$
$= -64 + 23 = -41$

10. The total number of cars painted is the product of the rate per hour and the number of hours. So $8h$ parts will be painted in h hours.

11. Each quarter is worth 25 cents, and each dime is worth 10 cents. Michelle has $(25q + 10d)$ cents.

12. Find the cost per pound by dividing the total cost by the number of pounds. The cost per pound is $\frac{d}{25}$.

13. Because 1 yard equals 3 feet, the distance in feet is $(3y)$ feet.

14. Because 1 meter equals 1000 millimeters and 1 centimeter equals 10 millimeters, the height in millimeters of the corn plant is $(1000m + 10c)$.

15. The perimeter of the rectangle is $l + w + l + w$, which simplifies to $2l + 2w$. Because 1 inch equals $\frac{1}{12}$ feet, the perimeter in feet is
$\frac{1}{12}(2l + 2w) = \frac{2l}{12} + \frac{2w}{12} = \frac{l}{6} + \frac{w}{6} = \frac{l + w}{6}$

CHAPTER 2

Section 2.1

1. $-4x + 3 = -41$
$-4x + 3 - 3 = -41 - 3$
$-4x = -44$
$\frac{-4x}{-4} = \frac{-44}{-4}$
$x = 11$

The solution set is $\{11\}$.

2. $15 = -2x + 38$
$15 + (-38) = -2x + 38 + (-38)$
$-23 = -2x$
$-\frac{1}{2}(-23) = -\frac{1}{2}(-2x)$
$\frac{23}{2} = x$

The solution set is $\left\{\frac{23}{2}\right\}$.

3. $3y + 4 = 8y - 26$
$3y + 4 + (-8y) = 8y - 26 + (-8y)$
$-5y + 4 = -26$
$-5y + 4 + (-4) = -26 + (-4)$
$-5y = -30$

$$-\frac{1}{5}(-5y) = -\frac{1}{5}(-30)$$
$$y = 6$$

The solution set is {6}.

4. $5(x - 4) + 3(x + 7) = 2(x - 1)$
$$5x - 20 + 3x + 21 = 2x - 2$$
$$8x + 1 = 2x - 2$$
$$8x + 1 + (-2x) = 2x - 2 + (-2x)$$
$$6x + 1 = -2$$
$$6x + 1 + (-1) = -2 + (-1)$$
$$6x = -3$$
$$\frac{1}{6}(6x) = \frac{1}{6}(-3)$$
$$x = -\frac{1}{2}$$

The solution set is $\left\{-\frac{1}{2}\right\}$.

5. Let n represent the number.
$$2n + 43 = -19$$
$$2n + 43 + (-43) = -19 + (-43)$$
$$2n = -62$$
$$\frac{1}{2}(2n) = \frac{1}{2}(-62)$$
$$n = -31$$

The number is −31.

6. Let x represent the first integer. Then $x + 1$ represents the second integer, and $x + 2$ represents the third integer.
$$x + (x + 1) = 14 + (x + 2)$$
$$2x + 1 = 14 + x + 2$$
$$2x + 1 = 16 + x$$
$$2x + 1 + (-x) = 16 + x + (-x)$$
$$x + 1 = 16$$
$$x + 1 + (-1) = 16 + (-1)$$
$$x = 15$$

The integers are 15, 16, and 17.

7. Let x represent the number of cell phone minutes.
$$\begin{pmatrix}\text{Cell phone}\\ \text{bill}\end{pmatrix} = \begin{pmatrix}\text{Service}\\ \text{charge}\end{pmatrix} + (\text{Taxes}) + \begin{pmatrix}\text{Cost of}\\ \text{calls}\end{pmatrix}$$
$$89 = 49 + 21 + 0.05x$$
$$89 = 70 + 0.05x$$
$$89 + (-70) = 70 + (-70) + 0.05x$$
$$19 = 0.05x$$
$$\frac{19}{0.05} = x$$
$$380 = x$$

There were 380 cell phone minutes used.

Section 2.2

1. $\frac{2}{5}a - \frac{3}{2} = \frac{1}{3}$
$$30\left(\frac{2}{5}a - \frac{3}{2}\right) = 30\left(\frac{1}{3}\right)$$
$$30\left(\frac{2}{5}a\right) - 30\left(\frac{3}{2}\right) = 10$$
$$12a - 45 = 10$$
$$12a = 55$$
$$\frac{12a}{12} = \frac{55}{12}$$
$$a = \frac{55}{12}$$

The solution set is $\left\{\frac{55}{12}\right\}$.

2. $\frac{y}{6} + \frac{y}{4} = 8$
$$12\left(\frac{y}{6} + \frac{y}{4}\right) = 12(8)$$
$$12\left(\frac{y}{6}\right) + 12\left(\frac{y}{4}\right) = 96$$
$$2y + 3y = 96$$
$$5y = 96$$
$$\frac{5y}{5} = \frac{96}{5}$$
$$y = \frac{96}{5}$$

The solution set is $\left\{\frac{96}{5}\right\}$.

3. $\frac{y + 4}{4} + \frac{y - 1}{3} = \frac{5}{2}$
$$12\left(\frac{y + 4}{4} + \frac{y - 1}{3}\right) = 12\left(\frac{5}{2}\right)$$
$$12\left(\frac{y + 4}{4}\right) + 12\left(\frac{y - 1}{3}\right) = 6(5)$$
$$3(y + 4) + 4(y - 1) = 30$$
$$3y + 12 + 4y - 4 = 30$$
$$7y + 8 = 30$$
$$7y = 22$$
$$\frac{7y}{7} = \frac{22}{7}$$
$$y = \frac{22}{7}$$

The solution set is $\left\{\frac{22}{7}\right\}$.

4.
$$\frac{2a+5}{3} - \frac{a-6}{2} = 1$$
$$6\left(\frac{2a+5}{3} - \frac{a-6}{2}\right) = 6(1)$$
$$6\left(\frac{2a+5}{3}\right) - 6\left(\frac{a-6}{2}\right) = 6$$
$$4a + 10 - 3a + 18 = 6$$
$$a + 28 = 6$$
$$a = -22$$

The solution set is $\{-22\}$.

5. Let c represent the number.
$$\frac{3}{4}c + \frac{1}{3}c = 2 + c$$
$$12\left(\frac{3}{4}c + \frac{1}{3}c\right) = 12(2 + c)$$
$$12\left(\frac{3}{4}c\right) + 12\left(\frac{1}{3}c\right) = 24 + 12c$$
$$9c + 4c = 24 + 12c$$
$$13c = 24 + 12c$$
$$c = 24$$

The number is 24.

6. Let x represent the length of the field. Then $\frac{3}{4}x - 20$ represents the width of the field.

Use $P = 2l + 2w$ to form the equation.
$$1080 = 2x + 2\left(\frac{3}{4}x - 20\right)$$
$$1080 = 2x + \frac{3}{2}x - 40$$
$$2(1080) = 2\left(2x + \frac{3}{2}x - 40\right)$$
$$2160 = 4x + 3x - 80$$
$$2160 = 7x - 80$$
$$2240 = 7x$$
$$320 = x$$

If $x = 320$, then $\frac{3}{4}x - 20 = \frac{3}{4}(320) - 20 = 220$.

The length is 320 feet, and the width is 220 feet.

7. Let a represent the measure of one angle. Then $4 + \frac{3}{5}a$ represents the other angle.

Because the angles are supplementary, the sum of their measures is 180°.
$$a + \left(4 + \frac{3}{5}a\right) = 180$$
$$5\left[a + \left(4 + \frac{3}{5}a\right)\right] = 5(180)$$
$$5a + 5\left(4 + \frac{3}{5}a\right) = 900$$
$$5a + 20 + 3a = 900$$
$$8a + 20 = 900$$
$$8a = 880$$
$$a = 110$$

If $a = 110$, then $4 + \frac{3}{5}a = 4 + \frac{3}{5}(110) = 70$.

The measures of the angles are 110° and 70°.

8.

	Present age	**Age in 4 years**
Kay	z	$z + 4$
Raymond	$z - 6$	$z - 6 + 4 = z - 2$

$$z - 2 = \frac{5}{8}(z + 4)$$
$$8(z - 2) = 8\left[\frac{5}{8}(z + 4)\right]$$
$$8z - 16 = 5(z + 4)$$
$$8z - 16 = 5z + 20$$
$$3z - 16 = 20$$
$$3z = 36$$
$$z = 12$$

If $z = 12$, then $z - 6 = 12 - 6 = 6$. Kay's present age is 12, and Raymond's age is 6.

Section 2.3

1.
$$0.14a - 0.8 = 0.07a + 3.4$$
$$100(0.14a - 0.8) = 100(0.07a + 3.4)$$
$$100(0.14a) - 100(0.8) = 100(0.07a) + 100(3.4)$$
$$14a - 80 = 7a + 340$$
$$7a - 80 = 340$$
$$7a = 420$$
$$a = 60$$

The solution set is $\{60\}$.

2.
$$0.4y + 1.1y = 3.15$$
$$1.5y = 3.15$$
$$\frac{1.5y}{1.5} = \frac{3.15}{1.5}$$
$$y = 2.1$$

The solution set is $\{2.1\}$.

3.
$$x = 4.5 + 0.25x$$
$$x + (-0.25x) = 4.5 + 0.25x + (-0.25x)$$
$$0.75x = 4.5$$
$$x = \frac{4.5}{0.75}$$
$$x = 6$$

The solution set is {6}.

4.
$$0.06n + 0.05(3000 - n) = 167$$
$$100[0.06n + 0.05(3000 - n)] = 100(167)$$
$$100(0.06n) + 100[0.05(3000 - n)] = 16{,}700$$
$$6n + 5(3000 - n) = 16{,}700$$
$$6n + 15{,}000 - 5n = 16{,}700$$
$$n + 15{,}000 = 16{,}700$$
$$n = 1700$$

The solution set is {1700}.

5. Let c represent the original price of the jeans. Use the following formula:

$$\begin{pmatrix}\text{Original}\\\text{price}\end{pmatrix} - (\text{Discount}) = \begin{pmatrix}\text{Discount}\\\text{sale price}\end{pmatrix}$$
$$c - (40\%)c = 45$$
$$c - 0.40c = 45$$
$$0.60c = 45$$
$$c = \frac{45}{0.60}$$
$$c = 75$$

The original price of the jeans was $75.00.

6. Let d represent the discount sale price. Use the following formula:

$$\begin{pmatrix}\text{Original}\\\text{price}\end{pmatrix} - (\text{Discount}) = \begin{pmatrix}\text{Discount}\\\text{sale price}\end{pmatrix}$$
$$980 - (12\%)980 = d$$
$$980 - (0.12)980 = d$$
$$980 - 117.60 = d$$
$$862.40 = d$$

The discount sale price is $862.40.

7. Let s represent the selling price. Use the following formula:

$$\begin{pmatrix}\text{Selling}\\\text{price}\end{pmatrix} = (\text{Cost}) + (\text{Profit})$$
$$s = 400 + (40\%)(400)$$
$$s = 400 + (0.40)(400)$$
$$s = 400 + 160$$
$$s = 560$$

Heather should list the art work for $560.

8. Let c represent the selling price. Use the following formula:

$$\begin{pmatrix}\text{Selling}\\\text{price}\end{pmatrix} = (\text{Cost}) + (\text{Profit})$$
$$c = 54 + (60\%)c$$
$$c = 54 + (0.60)c$$
$$0.40c = 54$$
$$c = \frac{54}{0.40}$$
$$c = 135$$

The selling price should be $135.

9. Let x represent the rate of profit. Use the following formula:

$$\begin{pmatrix}\text{Selling}\\\text{price}\end{pmatrix} = (\text{Cost}) + (\text{Profit})$$
$$300 = 200 + (x)(200)$$
$$300 = 200 + 200x$$
$$100 = 200x$$
$$\frac{100}{200} = x$$
$$\frac{50}{100} = x$$

The rate of profit is 50%.

10. Let d represent the number of dimes. Then $20 - d$ represents the number of quarters.

$$\begin{pmatrix}\text{Money from}\\\text{dimes}\end{pmatrix} + \begin{pmatrix}\text{Money from}\\\text{quarters}\end{pmatrix} = \begin{pmatrix}\text{Total amount}\\\text{of money}\end{pmatrix}$$
$$0.10d + 0.25(20 - d) = 3.95$$
$$100[0.10d + 0.25(20 - d)] = 100(3.95)$$
$$100(0.10d) + 100[0.25(20 - d)] = 395$$
$$10d + 25(20 - d) = 395$$
$$10d + 500 - 25d = 395$$
$$-15d + 500 = 395$$
$$-15d = -105$$
$$d = 7$$

If $d = 7$, then $20 - d = 20 - 7 = 13$. Lane has 7 dimes and 13 quarters.

11. Let a represent the amount invested at 7%. Then $10{,}000 - a$ represents the amount invested at 5%.

$$\begin{pmatrix}\text{Money from}\\\text{dimes}\end{pmatrix} + \begin{pmatrix}\text{Money from}\\\text{quarters}\end{pmatrix} = \begin{pmatrix}\text{Total amount}\\\text{of money}\end{pmatrix}$$
$$(7\%)(a) + (5\%)(10{,}000 - a) = 630$$
$$0.07a + 0.05(10{,}000 - a) = 630$$
$$100[0.07a + 0.05(10{,}000 - a)] = 100(630)$$
$$100(0.07a) + 100[0.05(10{,}000 - a)] = 63{,}000$$
$$7a + 5(10{,}000 - a) = 63{,}000$$

$$7a + 50{,}000 - 5a = 63{,}000$$
$$2a + 50{,}000 = 63{,}000$$
$$2a = 13{,}000$$
$$a = 6500$$

If $a = 6500$, then $10{,}000 - a = 10{,}000 - 6500 = 3500$. So, \$6500 was invested at 7%, and \$3500 was invested at 5%.

Section 2.4

1. Substitute \$2500 for P, 6% for r, and 3 for t.

$$i = Prt$$
$$i = (2500)(6\%)(3)$$
$$i = (2500)(0.06)(3)$$
$$i = 450$$

The interest earned is \$450.

2. Substitute \$1000 for P, 5% for r, and \$1800 for A.

$$A = P + Prt$$
$$1800 = 1000 + 1000(5\%)(t)$$
$$1800 = 1000 + 1000(0.05)(t)$$
$$800 = 1000(0.05)(t)$$
$$800 = 50t$$
$$16 = t$$

It will take 16 years to accumulate \$1800.

3. $V = \frac{1}{3}Bh$

$$3(V) = 3\left(\frac{1}{3}Bh\right)$$
$$3V = Bh$$
$$\frac{3V}{h} = \frac{Bh}{h}$$
$$\frac{3V}{h} = B$$

4. $S = 4lw + 2lh$

$$S - 4lw = 2lh$$
$$\frac{S - 4lw}{2l} = \frac{2lh}{2l}$$
$$\frac{S - 4lw}{2l} = h$$

5. $S = ad + an$

$$S = a(d + n)$$
$$\frac{S}{d + n} = \frac{a(d + n)}{(d + n)}$$
$$\frac{S}{d + n} = a$$

6. $P = 2(l + w)$

$$P = 2l + 2w$$
$$P - 2l = 2w$$
$$\frac{P - 2l}{2} = \frac{2w}{2}$$
$$\frac{P - 2l}{2} = w$$

7. $3x - 4y = 5$

$$-4y = -3x + 5$$
$$-\frac{1}{4}(-4y) = -\frac{1}{4}(-3x + 5)$$
$$y = \frac{3x}{4} - \frac{5}{4} = \frac{3x - 5}{4}$$

8. $\frac{x}{c} + \frac{y}{d} = 1$

$$cd\left(\frac{x}{c} + \frac{y}{d}\right) = cd(1)$$
$$cd\left(\frac{x}{c}\right) + cd\left(\frac{y}{d}\right) = cd$$
$$dx + cy = cd$$
$$dx = cd - cy$$
$$dx = c(d - y)$$
$$\frac{dx}{d} = \frac{c(d - y)}{d}$$
$$x = \frac{c(d - y)}{d}$$

9. For \$3000 to grow to \$4200, it must earn \$1200 in interest. So we let t represent the number of years it will take \$3000 to earn \$1200.

$$i = Prt$$
$$1200 = (3000)(5\%)(t)$$
$$1200 = (3000)(0.05)(t)$$
$$1200 = 150t$$
$$\frac{1200}{150} = t$$
$$8 = t$$

It will take 8 years.

10. Let t represent Franco's time. Then $t + 1$ represents Brittany's time. Use the chart to organize the problem.

Rate · Time = Distance

	Rate	Time	Distance
Brittany	8	$t + 1$	$8(t + 1)$
Franco	12	t	$12t$

$$\begin{pmatrix}\text{Brittany's}\\ \text{distance}\end{pmatrix} = \begin{pmatrix}\text{Franco's}\\ \text{distance}\end{pmatrix}$$

$$\begin{aligned} 8(t + 1) &= 12t \\ 8t + 8 &= 12t \\ 8 &= 4t \\ 2 &= t \end{aligned}$$

It will take 2 hours.

11. Let r represent the rate of the northbound truck. Then $r - 6$ represents the rate of the southbound truck. Both trucks travel for $1\frac{1}{2}$ hours. Use the chart to organize the problem.

Rate · Time = Distance

	Rate	**Time**	**Distance**
Northbound	r	$1\frac{1}{2}$	$\frac{3}{2}r$
Southbound	$r - 6$	$1\frac{1}{2}$	$\frac{3}{2}(r - 6)$

The sum of the distances is 159 miles.

$$\begin{aligned} \frac{3}{2}r + \frac{3}{2}(r - 6) &= 159 \\ \frac{3}{2}r + \frac{3}{2}r - 9 &= 159 \\ \frac{3}{2}r + \frac{3}{2}r &= 168 \\ \frac{6}{2}r &= 168 \\ 3r &= 168 \\ r &= 56 \end{aligned}$$

The northbound truck travels at 56 mph, and the southbound truck travels at 50 mph.

12. Let x represent the ounces of 6% solution. Then $16 - x$ represents the ounces of 14% solution.

$$\begin{pmatrix}\text{Medicine in}\\ \text{6\% solution}\end{pmatrix} + \begin{pmatrix}\text{Medicine in}\\ \text{14\% solution}\end{pmatrix} = \begin{pmatrix}\text{Medicine in}\\ \text{10\% solution}\end{pmatrix}$$

$$\begin{aligned} (6\%)(x) + (14\%)(16 - x) &= (10\%)(16) \\ (0.06)(x) + (0.14)(16 - x) &= (0.10)(16) \\ 0.06x + 2.24 - 0.14x &= 1.6 \\ -0.08x + 2.24 &= 1.6 \\ -0.08x &= -0.64 \\ x &= 8 \end{aligned}$$

If $x = 8$, then $16 - x = 16 - 8 = 8$. There are 8 ounces of 6% cough medicine and 8 ounces of 14% cough medicine.

13. Let a represent the number of quarts of pure antifreeze.

$$\begin{pmatrix}\text{Antifreeze in}\\ \text{30\% solution}\end{pmatrix} + \begin{pmatrix}\text{Pure antifreeze}\\ \text{added}\end{pmatrix} = \begin{pmatrix}\text{Antifreeze in}\\ \text{40\% solution}\end{pmatrix}$$

$$\begin{aligned} (30\%)(12) + (100\%)(a) &= (40\%)(12 + a) \\ (0.30)(12) + 1a &= (0.40)(12 + a) \\ 3.6 + a &= 4.8 + 0.40a \\ 3.6 + 0.60a &= 4.8 \\ 0.60a &= 1.2 \\ a &= \frac{1.2}{0.60} \\ a &= 2 \end{aligned}$$

The amount of pure antifreeze added is 2 quarts.

Section 2.5

1. (a) $x < 4$

$(-\infty, 4)$

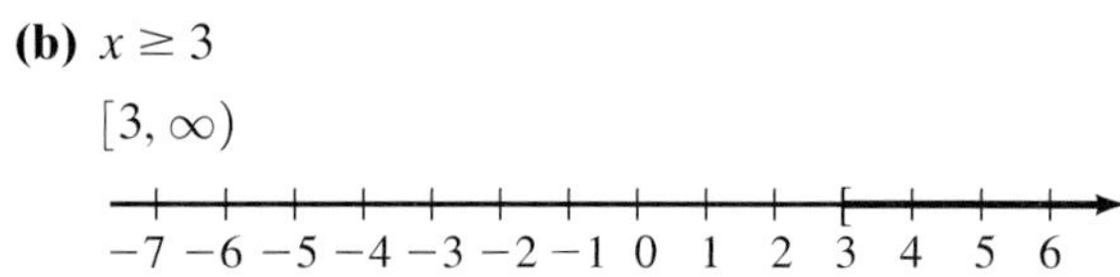

(b) $x \geq 3$

$[3, \infty)$

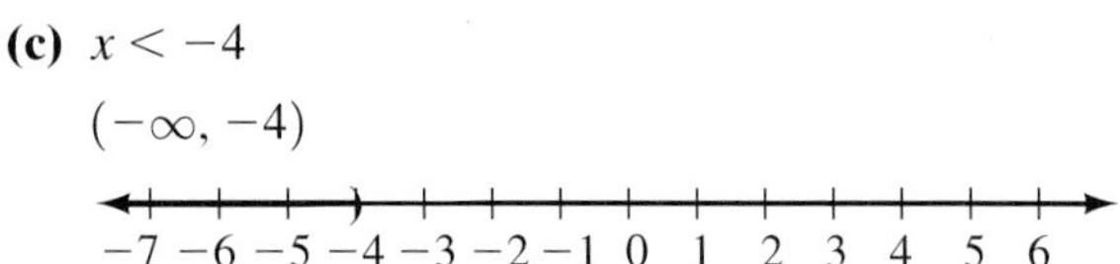

(c) $x < -4$

$(-\infty, -4)$

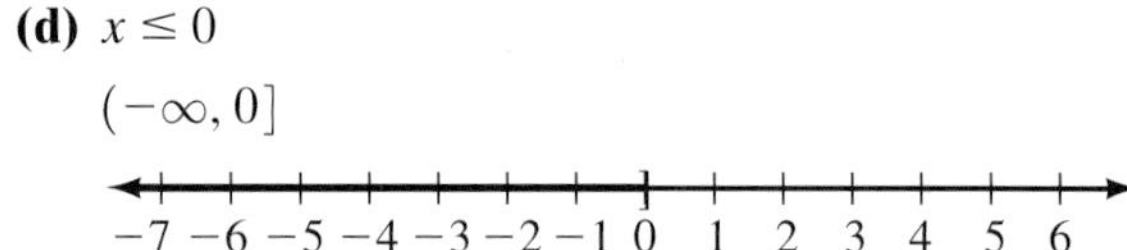

(d) $x \leq 0$

$(-\infty, 0]$

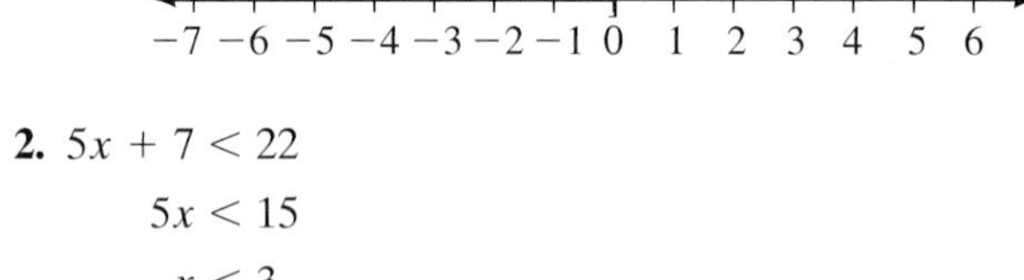

2. $5x + 7 < 22$

$$\begin{aligned} 5x &< 15 \\ x &< 3 \end{aligned}$$

The solution set is $(-\infty, 3)$.

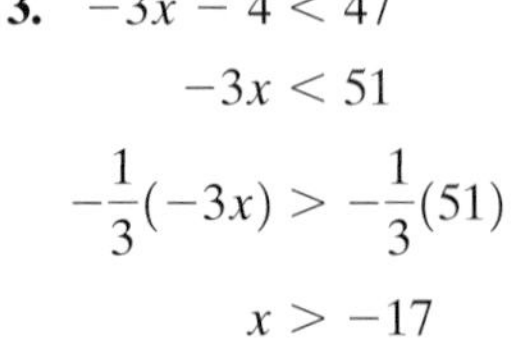

3. $-3x - 4 < 47$

$$\begin{aligned} -3x &< 51 \\ -\frac{1}{3}(-3x) &> -\frac{1}{3}(51) \\ x &> -17 \end{aligned}$$

The solution set is $(-17, \infty)$.

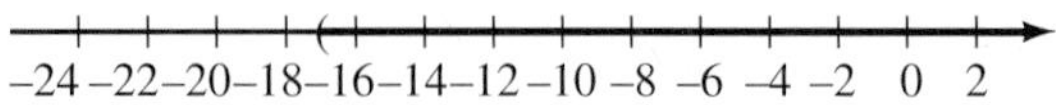

4. $-x + 4x + 8 \geq 6x - 5 - 2x$

$$\begin{aligned} 3x + 8 &\geq 4x - 5 \\ -x + 8 &\geq -5 \\ -x &\geq -13 \\ (-1)(-x) &\leq (-1)(-13) \\ x &\leq 13 \end{aligned}$$

The solution set is $(-\infty, 13]$.

–4 –2 0 2 4 6 8 10 12 14 16 18 20 22

5. $-4(x + 3) \geq 28$

$$\begin{aligned} -4x - 12 &\geq 28 \\ -4x &\geq 40 \\ -\frac{1}{4}(-4x) &\leq -\frac{1}{4}(40) \\ x &\leq -10 \end{aligned}$$

The solution set is $(-\infty, -10]$.

–20 –18 –16 –14 –12 –10 –8 –6 –4 –2 0 2 4 6

6. $2(x - 1) \geq 5(x + 3)$

$$\begin{aligned} 2x - 2 &\geq 5x + 15 \\ -3x - 2 &\geq 15 \\ -3x &\geq 17 \\ -\frac{1}{3}(-3x) &\leq -\frac{1}{3}(17) \\ x &\leq -\frac{17}{3} \end{aligned}$$

The solution set is $\left(-\infty, -\frac{17}{3}\right]$.

7. $2(3x - 4) - 5(x + 1) < 3(2x + 5)$

$$\begin{aligned} 6x - 8 - 5x - 5 &< 6x + 15 \\ x - 13 &< 6x + 15 \\ -5x - 13 &< 15 \\ -5x &< 28 \\ -\frac{1}{5}(-5x) &> -\frac{1}{5}(28) \\ x &> -\frac{28}{5} \end{aligned}$$

The solution set is $\left(-\frac{28}{5}, \infty\right)$.

Section 2.6

1. $\frac{2}{5}x - \frac{1}{3}x > -\frac{4}{5}$

$$\begin{aligned} 15\left(\frac{2}{5}x - \frac{1}{3}x\right) &> 15\left(-\frac{4}{5}\right) \\ 15\left(\frac{2}{5}x\right) - 15\left(\frac{1}{3}x\right) &> 3(-4) \\ 3(2x) - 5(x) &> -12 \\ 6x - 5x &> -12 \\ x &> -12 \end{aligned}$$

The solution set is $(-12, \infty)$.

2. $\frac{x - 1}{2} + \frac{x + 4}{5} \leq 3$

$$\begin{aligned} 10\left(\frac{x - 1}{2} + \frac{x + 4}{2}\right) &\leq 10(3) \\ 10\left(\frac{x - 1}{2}\right) + 10\left(\frac{x + 4}{2}\right) &\leq 30 \\ 5(x - 1) + 2(x + 4) &\leq 30 \\ 5x - 5 + 2x + 8 &\leq 30 \\ 7x + 3 &\leq 30 \\ 7x &\leq 27 \\ x &\leq \frac{27}{7} \end{aligned}$$

The solution set is $\left(-\infty, \frac{27}{7}\right]$.

3. $\frac{y}{3} - \frac{y - 2}{5} > \frac{y + 1}{15} - 1$

$$\begin{aligned} 15\left(\frac{y}{3} - \frac{y - 2}{5}\right) &> 15\left(\frac{y + 1}{15} - 1\right) \\ 15\left(\frac{y}{3}\right) - 15\left(\frac{y - 2}{5}\right) &> 15\left(\frac{y + 1}{15}\right) - 15(1) \\ 5y - 3(y - 2) &> y + 1 - 15 \\ 5y - 3y + 6 &> y - 14 \\ 2y + 6 &> y - 14 \\ y + 6 &> -14 \\ y &> -20 \end{aligned}$$

The solution set is $(-20, \infty)$.

4. $0.12x + 0.6 \leq 0.48$

$$\begin{aligned} 100(0.12x + 0.6) &\leq 100(0.48) \\ 100(0.12x) + 100(0.6) &\leq 100(0.48) \\ 12x + 60 &\leq 48 \\ 12x &\leq -12 \\ x &\leq -1 \end{aligned}$$

The solution set is $(-\infty, -1]$.

5. $0.05x + 0.07(x + 500) \geq 287$

$$\begin{aligned} 100[0.05x + 0.07(x + 500)] &\geq 100(287) \\ 100(0.05x) + 100[0.07(x + 500)] &\geq 28{,}700 \\ 5x + 7(x + 500) &\geq 28{,}700 \\ 5x + 7x + 3500 &\geq 28{,}700 \\ 12x &\geq 25{,}200 \\ x &\geq 2100 \end{aligned}$$

The solution set is $[2100, \infty)$.

6. $x \geq 1$ and $x \leq 6$

The solution set is $[1, 6]$.

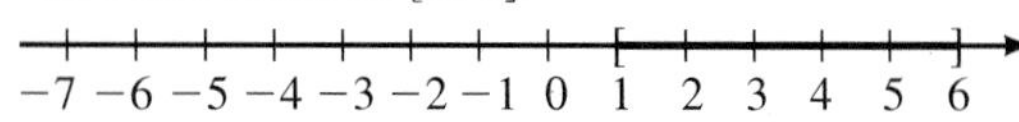

7. $2x - 4 > -6$ and $3x + 5 > 14$

$2x > -2$ and $3x > 9$

$x > -1$ and $x > 3$

The solution set is $(3, \infty)$.

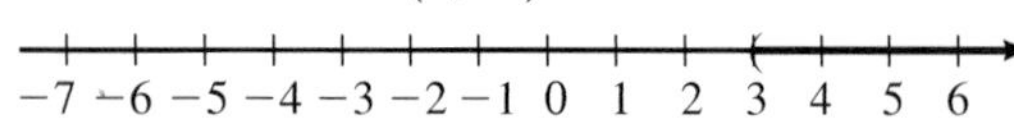

8. $x \leq 0$ or $x \geq 5$

The solution set is $(-\infty, 0] \cup [5, \infty)$.

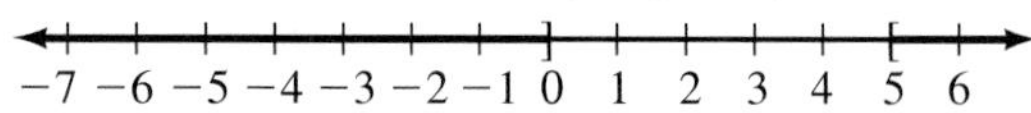

9. $3x - 1 < 5$ or $2x + 5 > 15$

$3x < 6$ or $2x > 10$

$x < 2$ or $x > 5$

The solution set is $(-\infty, 2) \cup (5, \infty)$.

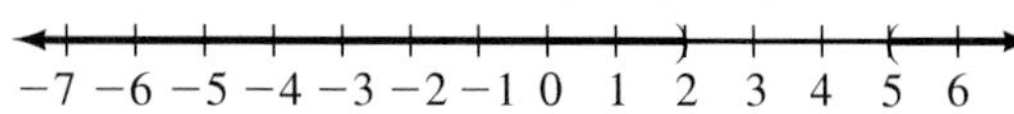

10. Let s represent his score on the fifth exam.

$$\frac{86 + 75 + 71 + 80 + s}{5} \geq 80$$

$$\frac{312 + s}{5} \geq 80$$

$$5\left(\frac{312 + s}{5}\right) \geq 5(80)$$

$$312 + s \geq 5(80)$$

$$312 + s \geq 400$$

$$s \geq 88$$

He must earn an 88 or better on the fifth exam.

11. Let r represent the unknown rate of interest.

$$\begin{pmatrix}\text{Interest from} \\ 6\% \text{ investment}\end{pmatrix} + \begin{pmatrix}\text{Interest from} \\ r\% \text{ investment}\end{pmatrix} > 335$$

$$(6\%)(1500) + (r)(3500) > 335$$

$$(0.06)(1500) + r(3500) > 335$$

$$90 + 3500r > 335$$

$$3500r > 245$$

$$r > \frac{245}{3500}$$

$$r > \frac{7}{100}$$

She must invest \$3500 at a rate greater than 7%.

12. Use $F = \frac{9}{5}C + 32$ to solve.

$$50 \leq \frac{9}{5}C + 32 \leq 86$$

$$50 \leq \frac{9}{5}C + 32 \leq 86$$

$$18 \leq \frac{9}{5}C \leq 54$$

$$\frac{5}{9}(18) \leq \left(\frac{5}{9}\right)\left(\frac{9}{5}C\right) \leq \frac{5}{9}(54)$$

$$10 \leq C \leq 30$$

The range must be between 10°C and 30°C, inclusive.

Section 2.7

1. $|x| = 7$

$x = -7$ or $x = 7$

The solution set is $\{-7, 7\}$.

2. $|x - 9| = 4$

$x - 9 = -4$ or $x - 9 = 4$

$x = 5$ or $x = 13$

The solution set is $\{5, 13\}$.

3. $|4x - 6| = 10$

$4x - 6 = -10$ or $4x - 6 = 10$

$4x = -4$ or $4x = 16$

$x = -1$ or $x = 4$

The solution set is $\{-1, 4\}$.

4. $|x - 1| + 7 = 15$

$|x - 1| = 8$

$x - 1 = -8$ or $x - 1 = 8$

$x = -7$ or $x = 9$

The solution set is $\{-7, 9\}$.

5. $|x| < 4$

$x > -4$ and $x < 4$

The solution set is $(-4, 4)$.

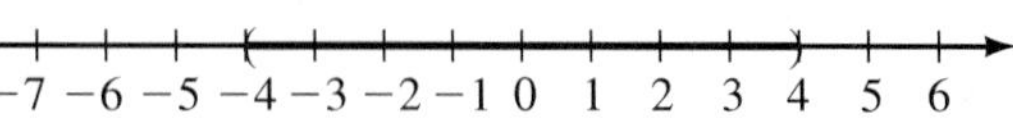

6. $|x - 2| \leq 1$

$x - 2 \geq -1$ and $x - 2 \leq 1$

$x \geq 1$ and $x \leq 3$

The solution set is $[1, 3]$.

−7 −6 −5 −4 −3 −2 −1 0 1 2 3 4 5 6

7. $|2x - 3| < 9$

$-9 < 2x - 3 < 9$

$-9 + 3 < 2x - 3 + 3 < 9 + 3$

$-6 < 2x < 12$

$-\frac{6}{2} < \frac{2x}{2} < \frac{12}{2}$

$-3 < x < 6$

The solution set is $(-3, 6)$.

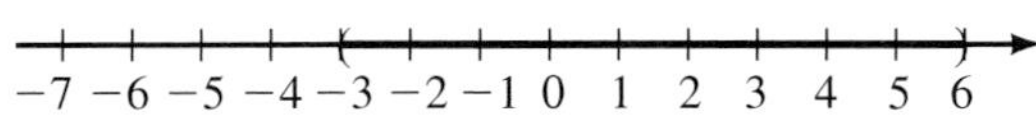

8. $|x| > 4$

$x < -4$ or $x > 4$

The solution set is $(-\infty, -4) \cup (4, \infty)$.

9. $|x + 2| \geq 4$

$x + 2 \leq -4$ or $x + 2 \geq 4$

$x \leq -6$ or $x \geq 2$

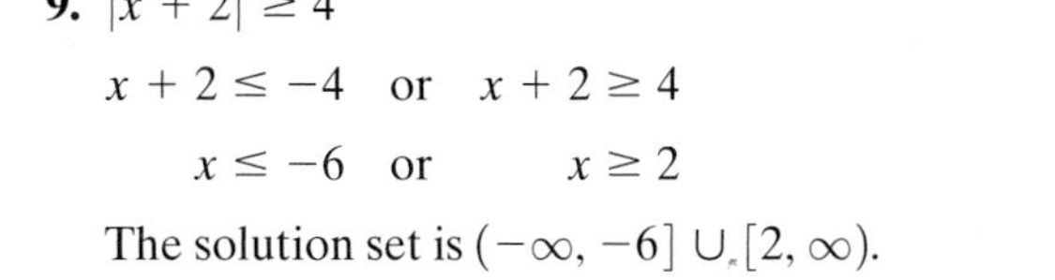

The solution set is $(-\infty, -6] \cup [2, \infty)$.

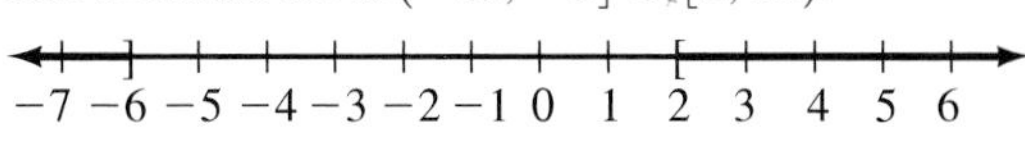

10. $|2x - 5| - 3 > 1$

$|2x - 5| > 4$

$2x - 5 < -4$ or $2x - 5 > 4$

$2x < 1$ or $2x > 9$

$x < \frac{1}{2}$ or $x > \frac{9}{2}$

The solution set is $\left(-\infty, \frac{1}{2}\right) \cup \left(\frac{9}{2}, \infty\right)$.

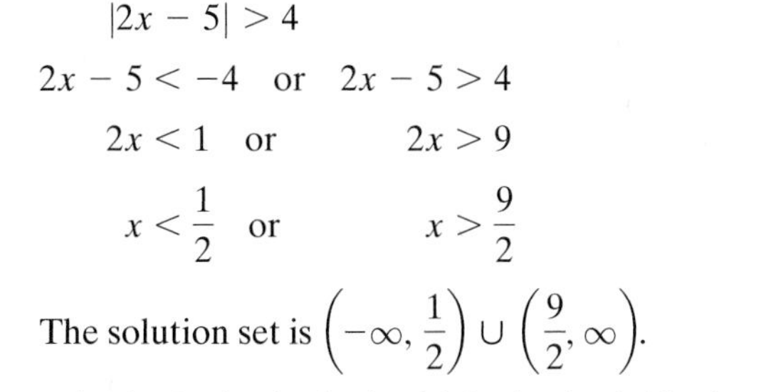

CHAPTER 3

Section 3.1

1.

x	$y = -2x + 4$	Ordered pairs
−4	12	(−4, 12)
−2	8	(−2, 8)
0	4	(0, 4)
1	2	(1, 2)
3	−2	(3, −2)

2.

x	$y = 2x - 2$	Ordered pairs
−3	−8	(−3, −8)
−1	−4	(−1, −4)
0	−2	(0, −2)
2	2	(2, 2)
4	6	(4, 6)
6	10	(6, 10)

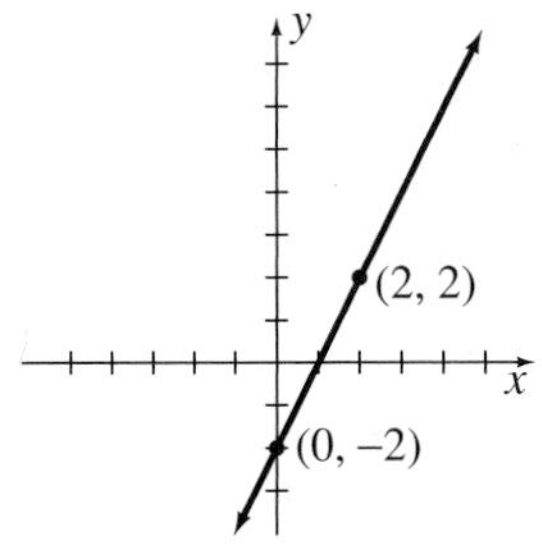

3. Graph $3x - y = 6$.

To find the y intercept, let $x = 0$.

If $x = 0$, then

$3(0) - y = 6$

$-y = 6$

$y = -6$

To find the x intercept, let $y = 0$.

If $y = 0$, then

$3x - (0) = 6$

$3x = 6$

$x = 2$

To find a check point, let $x = 4$.

If $x = 4$, then

$3(4) - y = 6$

$12 - y = 6$

$-y = -6$

$y = 6$

Three solutions are $(0, -6)$, $(2, 0)$, and $(4, 6)$.

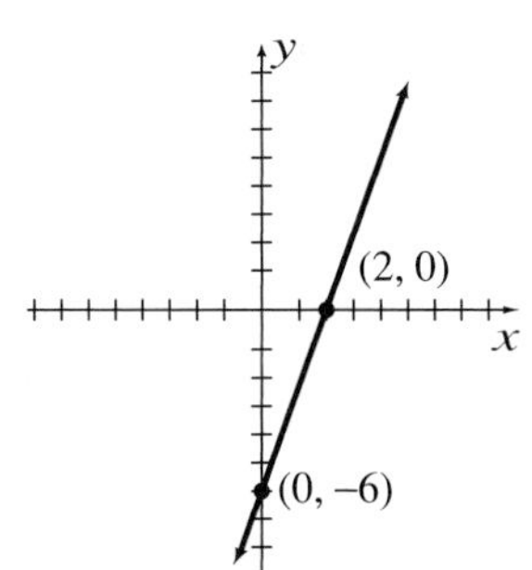

4. Graph $x + 2y = 3$.

Complete the chart with intercepts and check point.

x	y
0	$\frac{3}{2}$
3	0
5	−1

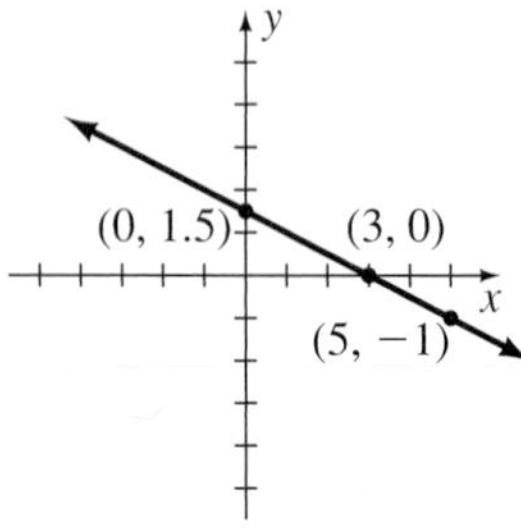

5. Graph $y = -3x$.

Complete the chart with intercept, additional point, and check point.

x	y
0	0
1	−3
−1	3

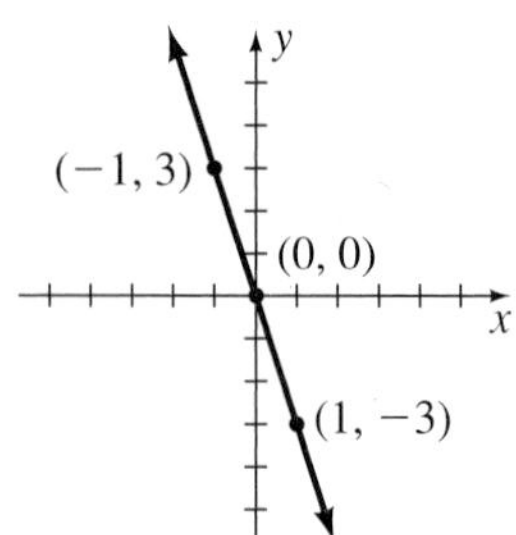

6. Graph $x = -3$.

For any value of y, $x = -3$.

$(-3, 0)$, $(-3, -2)$, $(-3, 4)$

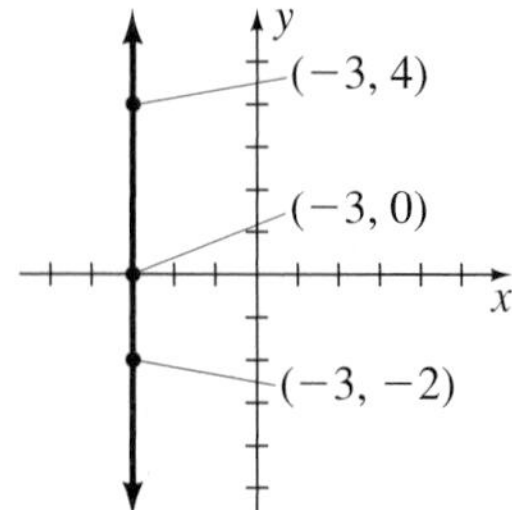

7. Graph $y = 4$.

For any value of x, $y = 4$.

$(0, 4)$, $(-2, 4)$, $(2, 4)$

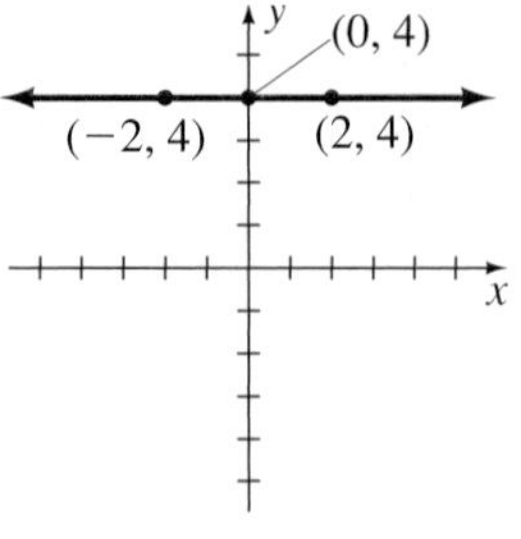

8. To use a graphing utility, solve for y.

$$3.4x - 2.5y = 6.8$$
$$-2.5y = -3.4x + 6.8$$
$$\frac{-2.5y}{-2.5} = \frac{-3.4x + 6.8}{-2.5}$$
$$y = \frac{-3.4x + 6.8}{-2.5} = \frac{-3.4}{-2.5}x + \frac{6.8}{-2.5}$$
$$y = 1.36x - 2.72$$

Graph $y = 1.36x - 2.72$.

x	y
0	−2.72
1	−1.36
−1	−4.08

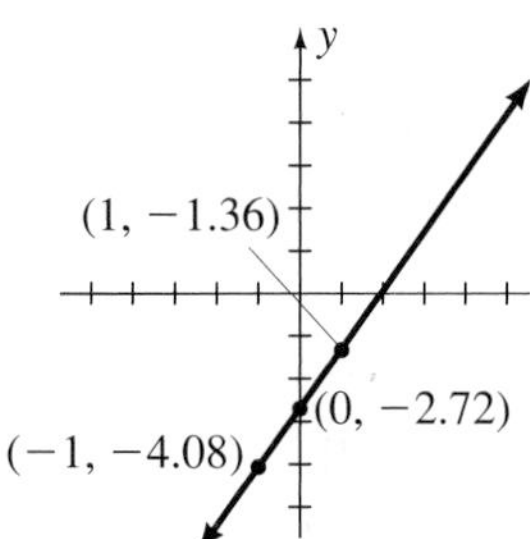

Section 3.2

1. Graph $3x + y < 3$.

Graph $3x + y = 3$ as a dashed line. Choose $(0, 0)$ as a test point, so $3x + y < 3$ becomes $3(0) + 0 < 3$, which is a true statement. Since the test point satisfies the inequality, the graph is the half-plane that does contain the test point.

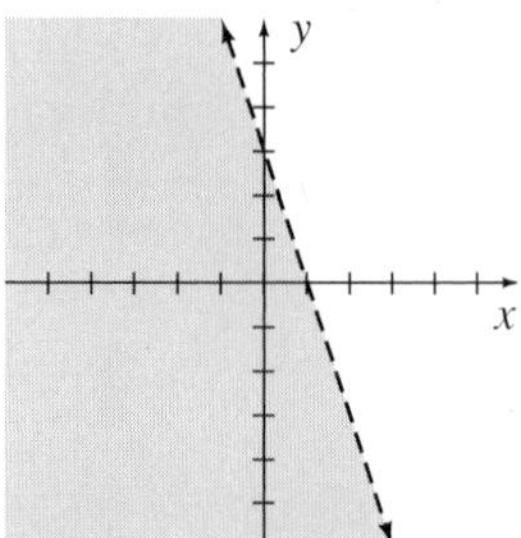

2. Graph $x - 4y \geq 4$.

Graph $x - 4y = 4$ as a solid line. Choose (0, 0) as a test point, so $x - 4y \geq 4$ becomes $0 - 4(0) \geq 4$, which is a false statement. Since the test point does not satisfy the inequality, the graph is the line and the half-plane that does not contain the test point.

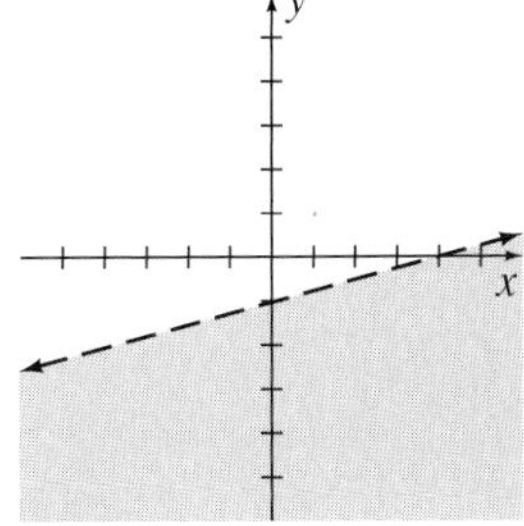

3. Graph $y > 2x$.

Graph $y = 2x$ as a dashed line. Choose a test point other than the origin since the line passes through (0, 0). Use (4, 0) as the test point, so $y > 2x$ becomes $0 > 2(4)$, which is a false statement. Since the test point does not satisfy the inequality, the graph is the half-plane that does not contain the test point.

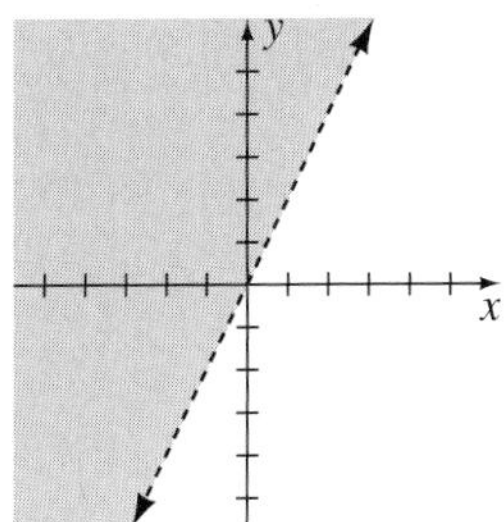

Section 3.3

1. $A(3, 6)$; $B(3, -2)$

The length of $\overline{AB}$ is $|-2 - 6| = |-8| = 8$ units.

2. Find the distance between $A(-4, 1)$ and $B(8, 6)$. Let d represent the length of $\overline{AB}$.

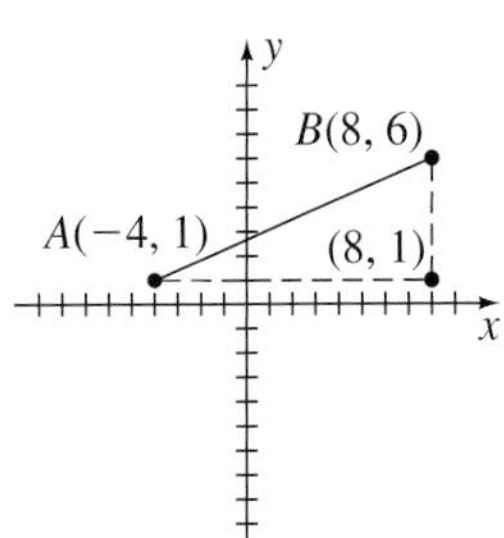

$\overline{AC} = |-4 - 8| = |-12| = 12$

$\overline{BC} = |1 - 6| = |-5| = 5$

Use the Pythagorean theorem.

$d^2 = (12)^2 + (5)^2$

$d^2 = 144 + 25$

$d^2 = 169$

$d = \pm\sqrt{169} = \pm 13$

Distance is nonnegative, so discard $d = -13$. The length of $\overline{AB}$ is 13 units.

3. Find the distance between $A(-1, 3)$ and $B(4, 5)$.

Use $d = \sqrt{(x_2 - x_1)^2 + (y_2 - y_1)^2}$.

Let $(-1, 3)$ be P_1 and let $(4, 5)$ be P_2.

$d = \sqrt{[4 - (-1)]^2 + (5 - 3)^2}$

$d = \sqrt{(4 + 1)^2 + (2)^2}$

$d = \sqrt{(5)^2 + (2)^2}$

$d = \sqrt{25 + 4}$

$d = \sqrt{29}$

4. Verify that $(-2, -2)$, $(7, -1)$, and $(2, 3)$ are vertices of an isosceles triangle.

Let d_1 = the distance between $(-2, -2)$ and $(7, -1)$.

Let d_2 = the distance between $(2, 3)$ and $(7, -1)$.

Let d_3 = the distance between $(-2, -2)$ and $(2, 3)$.

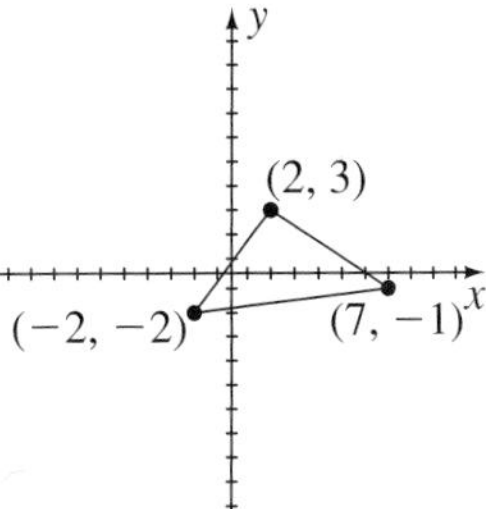

$d_1 = \sqrt{[7 - (-2)]^2 + [-1 - (-2)]^2}$

$d_1 = \sqrt{(7 + 2)^2 + (-1 + 2)^2}$

$d_1 = \sqrt{(9)^2 + (1)^2}$

$d_1 = \sqrt{81 + 1}$

$d_1 = \sqrt{82}$

$d_2 = \sqrt{[3 - (-1)]^2 + (2 - 7)^2}$

$d_2 = \sqrt{(3 + 1)^2 + (-5)^2}$

$d_2 = \sqrt{(4)^2 + (-5)^2}$

$d_2 = \sqrt{16 + 25}$

$d_2 = \sqrt{41}$

$d_3 = \sqrt{(-2 - 2)^2 + (-2 - 3)^2}$

$d_3 = \sqrt{(-4)^2 + (-5)^2}$

$d_3 = \sqrt{16 + 25}$

$d_3 = \sqrt{41}$

$d_1 = \sqrt{82}$

$d_2 = d_3 = \sqrt{41}$

Therefore, the vertices form an isosceles triangle.

5. **(a)** Let $(-4, 2)$ be P_1 and let $(2, 5)$ be P_2.

$$m = \frac{y_2 - y_1}{x_2 - x_1} = \frac{5 - 2}{2 - (-4)} = \frac{3}{6} = \frac{1}{2}$$

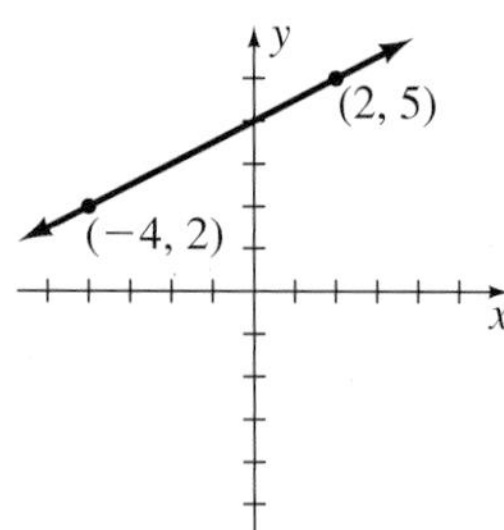

(b) Let $(-3, 4)$ be P_1 and let $(1, 4)$ be P_2.

$$m = \frac{y_2 - y_1}{x_2 - x_1} = \frac{4 - 4}{1 - (-3)} = \frac{0}{4} = 0$$

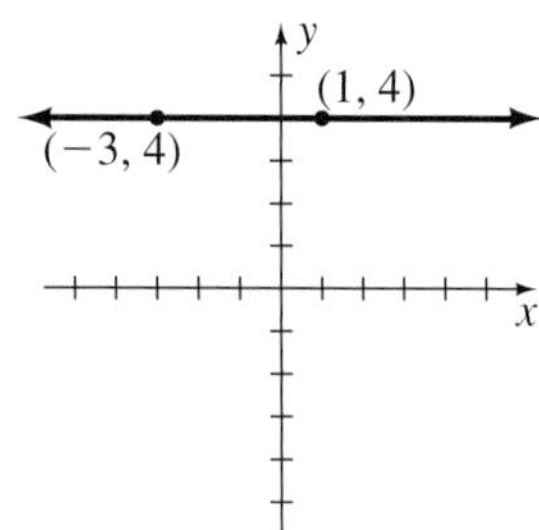

(c) Let $(3, -2)$ be P_1 and let $(0, 2)$ be P_2.

$$m = \frac{y_2 - y_1}{x_2 - x_1} = \frac{2 - (-2)}{0 - 3} = \frac{4}{-3} = -\frac{4}{3}$$

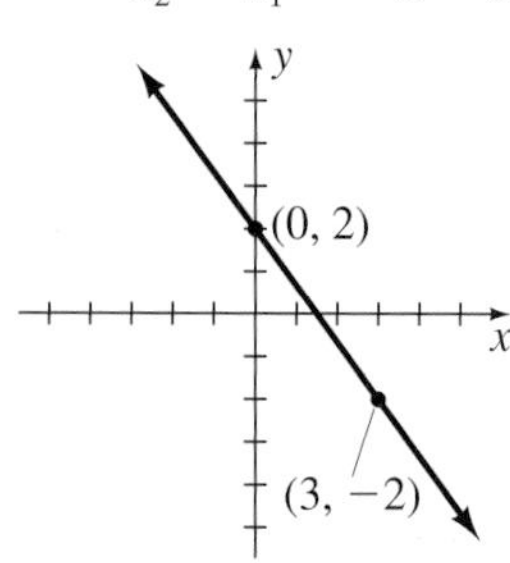

6. Graph the line that passes through $(-3, 2)$ and has a slope of $\frac{2}{5}$.

$$\text{Slope} = \frac{\text{Vertical change}}{\text{Horizontal change}} = \frac{2}{5}.$$

Start at $(-3, 2)$. Move up two units and right five units. Because $\frac{2}{5} = \frac{-2}{-5}$, we can locate another point by moving down two units and left five units.

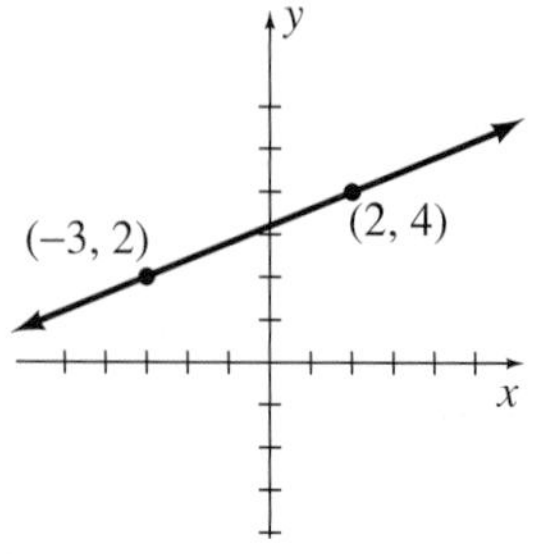

7. Graph the line that passes through $(2, 0)$ and has a slope of $-\frac{1}{3}$.

$$\text{Slope} = \frac{\text{Vertical change}}{\text{Horizontal change}} = \frac{-1}{3}.$$

Start at $(2, 0)$. Move down one unit and right three units. Because $\frac{-1}{3} = \frac{1}{-3}$, we can locate another point by moving up one unit and left three units.

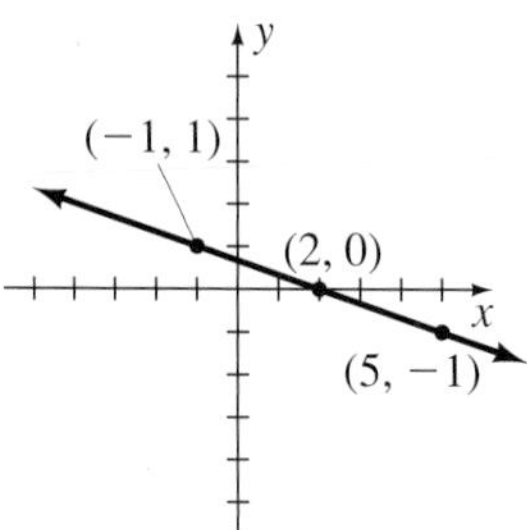

8. A 2.5% grade means a slope of $\frac{2.5}{100}$. Let y represent the unknown vertical distance. Solve the following proportion.

$$\frac{2.5}{100} = \frac{y}{2000}$$
$$(2000)(2.5) = 100y$$
$$5000 = 100y$$
$$50 = y$$

The vertical distance is 50 feet.

Section 3.4

1. The slope determined by $(3, -1)$ and any point on the line, (x, y), is $\frac{3}{4}$.

$$\frac{y - (-1)}{x - 3} = \frac{3}{4}$$
$$\frac{y + 1}{x - 3} = \frac{3}{4}$$

Cross multiply.

$$4(y + 1) = 3(x - 3)$$
$$4y + 4 = 3x - 9$$
$$-3x + 4y + 4 = -9$$
$$-3x + 4y = -13$$
$$3x - 4y = 13$$

2. The line contains $(2, 5)$ and $(-4, 10)$. Find the slope.

$$m = \frac{y_2 - y_1}{x_2 - x_1} = \frac{10 - 5}{-4 - 2} = \frac{5}{-6} = -\frac{5}{6}$$
$$\frac{y - 5}{x - 2} = \frac{-5}{6}$$

Cross multiply.

$$6(y - 5) = -5(x - 2)$$
$$6y - 30 = -5x + 10$$
$$5x + 6y - 30 = 10$$
$$5x + 6y = 40$$

3. A y intercept of -4 means that the point $(0, -4)$ is on the line. The slope is $\frac{3}{2}$.

$$\frac{y - (-4)}{x - 0} = \frac{3}{2}$$
$$\frac{y + 4}{x} = \frac{3}{2}$$

Cross multiply.

$$3x = 2(y + 4)$$
$$3x = 2y + 8$$
$$3x - 2y = 8$$

4. Determine the equation by substituting $\frac{4}{3}$ for m and $(-2, 5)$ for (x_1, y_1) in the point-slope form.

$$y - y_1 = m(x - x_1)$$
$$y - 5 = \frac{4}{3}[x - (-2)]$$
$$y - 5 = \frac{4}{3}(x + 2)$$
$$3(y - 5) = 4(x + 2)$$
$$3y - 15 = 4x + 8$$
$$-15 = 4x - 3y + 8$$
$$-23 = 4x - 3y$$
$$4x - 3y = -23$$

5. Because $m = -3$ and the y intercept is 8, use $y = mx + b$ to write the equation of the line.

$$y = mx + b$$
$$y = -3x + 8$$
$$3x + y = 8$$

6. To find the slope, solve for y.

$$4x - 5y = 10$$
$$-5y = -4x + 10$$
$$\frac{-5y}{-5} = \frac{-4x}{-5} + \frac{10}{-5}$$
$$y = \frac{4}{5}x - 2$$

The slope is $\frac{4}{5}$.

7. Graph $y = \frac{1}{4}x + 2$.

The slope is $\frac{1}{4}$. The y intercept is 2, which gives the point $(0, 2)$. Because the slope is $\frac{1}{4}$, move four units right and one unit up from $(0, 2)$ to locate $(4, 3)$. Because $\frac{1}{4} = \frac{-1}{-4}$, determine a third point by moving four units left and one unit down from $(0, 2)$ to locate $(-4, 1)$.

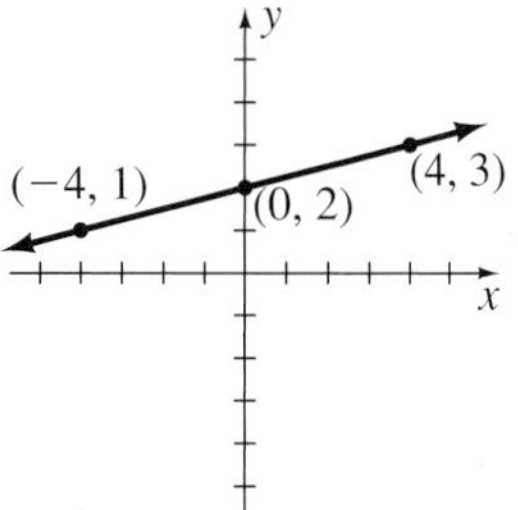

8. (a) Write each equation in slope-intercept form.

$$x - 3y = 2$$
$$-3y = -x + 2$$
$$\frac{-3y}{-3} = \frac{-x}{-3} + \frac{2}{-3}$$
$$y = \frac{1}{3}x - \frac{2}{3}$$
$$2x - 6y = 7$$
$$-6y = -2x + 7$$
$$\frac{-6y}{-6} = \frac{-2x}{-6} + \frac{7}{-6}$$
$$y = \frac{1}{3}x - \frac{7}{6}$$

Both lines have a slope of $\frac{1}{3}$ and different y intercepts. Therefore, the two lines are parallel.

(b) Write each equation in slope-intercept form.

$$2x - 5y = 3$$
$$-5y = -2x + 3$$
$$\frac{-5y}{-5} = \frac{-2x}{-5} + \frac{3}{-5}$$
$$y = \frac{2}{5}x - \frac{3}{5}$$
$$5x + 2y = 8$$
$$2y = -5x + 8$$
$$\frac{2y}{2} = \frac{-5x}{2} + \frac{8}{2}$$
$$y = -\frac{5}{2}x + 4$$

Because $\frac{2}{5}$ and $-\frac{5}{2}$ are negative reciprocals, the two lines are perpendicular.

9. The line contains $(-2, 7)$ and is parallel to the line determined by $3x + y = 4$.

Find the slope of $3x + y = 4$.

$$3x + y = 4$$
$$y = -3x + 4$$

The slope is -3.

Use the point-slope form.

$$y_2 - y_1 = m(x_2 - x_1)$$
$$y - 7 = -3[x - (-2)]$$
$$y - 7 = -3(x + 2)$$
$$y - 7 = -3x - 6$$
$$3x + y - 7 = -6$$
$$3x + y = 1$$

10. The line contains $(3, -1)$ and is perpendicular to the line determined by $5x + 2y = -10$.

Find the slope of $5x + 2y = -10$.

$$5x + 2y = -10$$
$$2y = -5x - 10$$
$$\frac{2y}{2} = \frac{-5x}{2} - \frac{10}{2}$$
$$y = -\frac{5}{2}x - 5$$

The slope is $-\frac{5}{2}$. The negative reciprocal is $\frac{2}{5}$.

Use the point-slope form.

$$y_2 - y_1 = m(x_2 - x_1)$$
$$y - (-1) = \frac{2}{5}(x - 3)$$
$$y + 1 = \frac{2}{5}(x - 3)$$
$$5(y + 1) = 2(x - 3)$$
$$5y + 5 = 2x - 6$$
$$-2x + 5y + 5 = -6$$
$$-2x + 5y = -11$$
$$2x - 5y = 11$$

CHAPTER 4

Section 4.1

1. Solve the system $\begin{pmatrix} y = -x - 2 \\ x + 2y = -4 \end{pmatrix}$.

$y = -x - 2$

x	y
0	−2
−2	0
−3	1

$x + 2y = -4$

x	y
0	−2
−4	0
2	−3

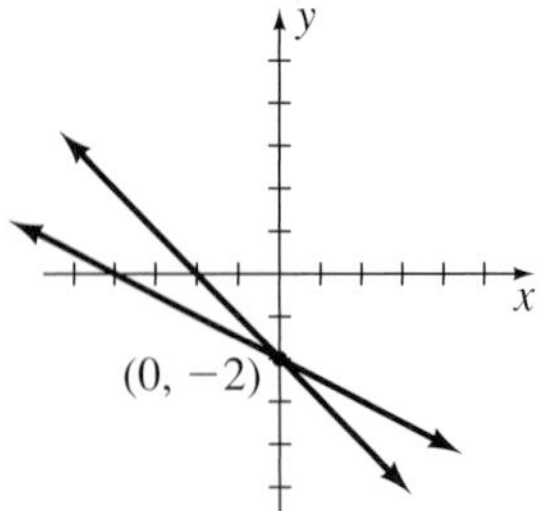

The solution set is $\{(0, -2)\}$.

2. Solve the system $\begin{pmatrix} 4x + 2y = -2 \\ 2x + y = -1 \end{pmatrix}$.

$4x + 2y = -2$

x	y
0	−1
$-\frac{1}{2}$	0
−2	3

$2x + y = -1$

x	y
0	−1
$-\frac{1}{2}$	0
−2	3

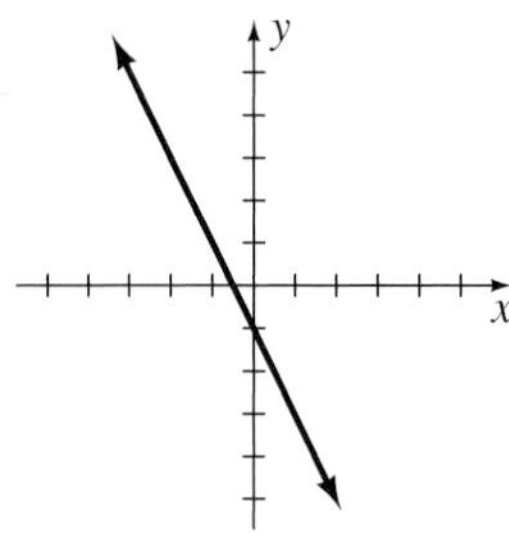

The solution set is $\{(x, y) | 2x + y = -1\}$.

3. Solve the system $\begin{pmatrix} x + y = 5 \\ y = -x + 2 \end{pmatrix}$.

$x + y = 5$

x	y
0	5
5	0
2	3

$y = -x + 2$

x	y
0	2
2	0
3	−1

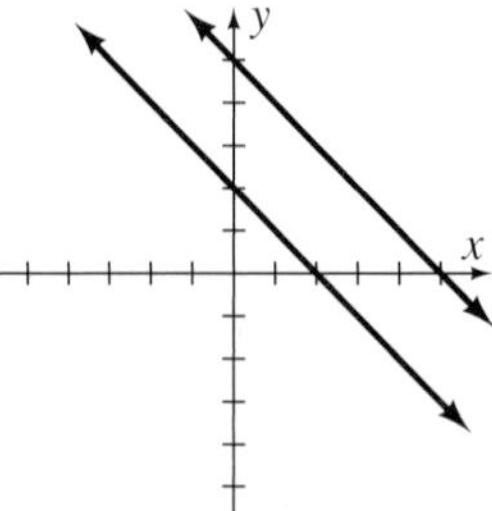

The solution set is $\varnothing$.

4. Solve the system $\begin{pmatrix} 3x + y > -3 \\ 2x - y \geq 0 \end{pmatrix}$.

Graph $3x + y = -3$ as a dashed line.
Graph $2x - y = 0$ as a solid line.

$3x + y = -3$

x	y
0	−3
−1	0
−3	6

$2x - y = 0$

x	y
0	0
1	2
−1	−2

The graph of $3x + y > -3$ consists of all points above the line $3x + y = -3$. The graph of $2x - y \geq 0$ consists of all points on or below the line $2x - y = 0$. The graph of the system is indicated by all the points in the shaded region on or below $2x - y = 0$ and above $3x + y = -3$.

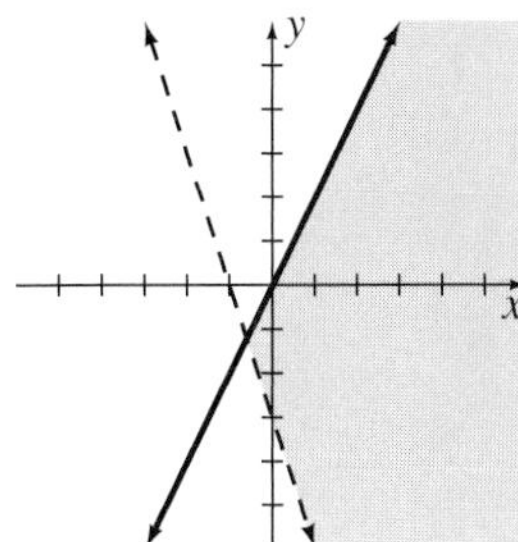

5. Solve the system $\begin{pmatrix} y < -\frac{4}{3}x + 5 \\ y < x - 2 \end{pmatrix}$.

Graph the following lines as dashed lines.

$y = -\frac{4}{3}x + 5$

x	y
0	5
$\frac{15}{4}$	0
3	1

$y = x - 2$

x	y
0	−2
2	0
3	1

The graph of $y < -\frac{4}{3}x + 5$ consists of all points below the line $y = -\frac{4}{3}x + 5$. The graph of $y < x - 2$ consists of all points below the line $y = x - 2$.

The graph of the system is indicated by all the points in the shaded region.

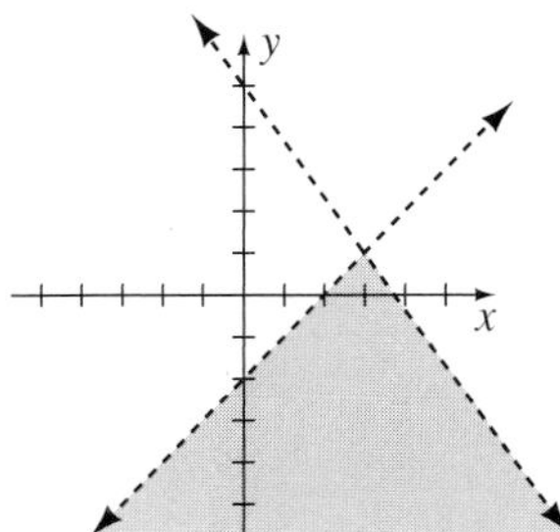

6. Solve the system $\begin{pmatrix} y < 4 \\ x > -1 \end{pmatrix}$.

Graph the following lines as dashed lines.

$y = 4$

x	y
0	4
−4	4

$x = -1$

x	y
−1	−2
−1	4

The graph of $y < 4$ consists of all points below the line $y = 4$. The graph of $x > -1$ consists of all points to the right of the line $x = -1$.

The graph of the system is indicated by all the points in the shaded region below the line $y = 4$ and to the right of the line $x = -1$.

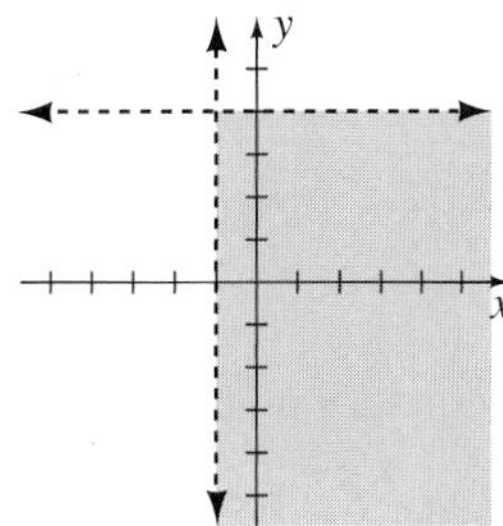

7. Solve the system $\begin{pmatrix} 2.35x + 4.16y = 10.13 \\ 5.18x - 1.17y = -8.69 \end{pmatrix}$.

Solve each equation for y in terms of x.

$$2.35x + 4.16y = 10.13$$
$$4.16y = -2.35x + 10.13$$
$$y = \frac{-2.35x + 10.13}{4.16}$$
$$5.18x - 1.17y = -8.69$$
$$-1.17y = -5.18x - 8.69$$
$$y = \frac{5.18x + 8.69}{1.17}$$

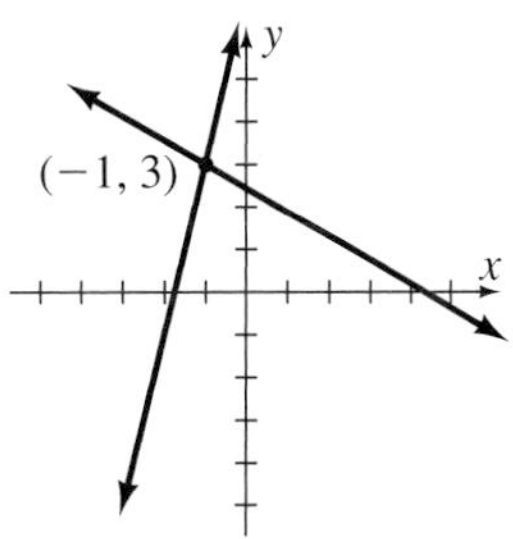

The solution set is $\{(-1, 3)\}$.

Section 4.2

1. Solve the system $\begin{pmatrix} 3x + y = 5 \\ y = x - 11 \end{pmatrix}$.

Substitute $x - 11$ for y in the first equation.

$$3x + y = 5$$
$$3x + (x - 11) = 5$$

$$4x - 11 = 5$$
$$4x = 16$$
$$x = 4$$

Substitute 4 for x in the second equation to find y.

$$y = x - 11$$
$$y = 4 - 11$$
$$y = -7$$

The solution set is $\{(4, -7)\}$.

2. Solve the system $\begin{pmatrix} x = 2y - 13 \\ 2x + 3y = 30 \end{pmatrix}$.

Substitute $2y - 13$ for x in the second equation.

$$2x + 3y = 30$$
$$2(2y - 13) + 3y = 30$$
$$4y - 26 + 3y = 30$$
$$7y - 26 = 30$$
$$7y = 56$$
$$y = 8$$

Substitute 8 for y in the first equation to find x.

$$x = 2y - 13$$
$$x = 2(8) - 13$$
$$x = 16 - 13 = 3$$

The solution set is $\{(3, 8)\}$.

3. Solve the system $\begin{pmatrix} x + 3y = 5 \\ 2x - 6y = -8 \end{pmatrix}$.

Solve the first equation for x.

$$x + 3y = 5$$
$$x = 5 - 3y$$

Substitute $5 - 3y$ for x in the second original equation to find y.

$$2(5 - 3y) - 6y = -8$$
$$10 - 6y - 6y = -8$$
$$10 - 12y = -8$$
$$-12y = -18$$
$$y = \frac{-18}{-12} = \frac{3}{2}$$

Substitute $\frac{3}{2}$ for y in the first original equation to find x.

$$x + 3y = 5$$
$$x + 3\left(\frac{3}{2}\right) = 5$$
$$x + \frac{9}{2} = \frac{10}{2}$$
$$x = \frac{1}{2}$$

The solution set is $\left\{\left(\frac{1}{2}, \frac{3}{2}\right)\right\}$.

4. Solve the system $\begin{pmatrix} 3x - 4y = 7 \\ 2x + 3y = 16 \end{pmatrix}$.

Solve the second equation for x.

$$2x + 3y = 16$$
$$2x = 16 - 3y$$
$$x = \frac{16 - 3y}{2}$$

Substitute $\frac{16 - 3y}{2}$ for x in the first original equation to find y.

$$3x - 4y = 7$$
$$3\left(\frac{16 - 3y}{2}\right) - 4y = 7$$
$$\frac{48 - 9y}{2} - 4y = 7$$
$$2\left(\frac{48 - 9y}{2} - 4y\right) = 2(7)$$
$$48 - 9y - 8y = 14$$
$$48 - 17y = 14$$
$$-17y = -34$$
$$y = 2$$

Substitute 2 for y in the second original equation to find x.

$$2x + 3y = 16$$
$$2x + 3(2) = 16$$
$$2x + 6 = 16$$
$$2x = 10$$
$$x = 5$$

The solution set is $\{(5, 2)\}$.

5. Let x represent the amount invested at 6% and let y represent the amount invested at 7%.

The total interest earned was \$484.

$$0.06x + 0.07y = 484$$

The amount invested at 7% was \$600 more than the amount invested at 6%.

$$y = x + 600$$

Solve the system $\begin{pmatrix} 0.06x + 0.07y = 484 \\ y = x + 600 \end{pmatrix}$.

Substitute $x + 600$ for y in the first original equation.

$$0.06x + 0.07y = 484$$
$$0.06x + 0.07(x + 600) = 484$$
$$0.06x + 0.07x + 42 = 484$$
$$0.13x + 42 = 484$$
$$0.13x = 442$$
$$x = 3400$$

Substitute 3400 for x in the second original equation.

$y = x + 600$

$y = 3400 + 600$

$y = 4000$

Chris invested \$3400 at 6% and \$4000 at 7%.

6. Let w represent the width of the rectangle and l represent the length of the rectangle.

The length is 9 more than the width.

$l = 9 + w$

The perimeter is 50 inches.
Use $P = 2l + 2w$.

$50 = 2l + 2w$

Use substitution to solve. Substitute $9 + w$ for l in the second original equation.

$50 = 2(9 + w) + 2w$

$50 = 18 + 2w + 2w$

$50 = 18 + 4w$

$32 = 4w$

$8 = w$

Substitute 8 for w into the first original equation.

$l = 9 + w$

$l = 9 + 8$

$l = 17$

The length is 17 inches, and the width is 8 inches.

Section 4.3

1. Solve the system $\begin{pmatrix} 3x + 5y = 14 & (1) \\ -3x + 4y = 22 & (2) \end{pmatrix}$.

Multiply equation (1) by 1 and add to equation (2) to replace equation (2).

$\begin{pmatrix} 3x + 5y = 14 & (3) \\ 9y = 36 & (4) \end{pmatrix}$

From equation (4) we can find y.

$9y = 36$

$y = 4$

Substitute 4 for y in equation (3).

$3x + 5(4) = 14$

$3x + 20 = 14$

$3x = -6$

$x = -2$

The solution set is $\{(-2, 4)\}$.

2. Solve the system $\begin{pmatrix} 2x + y = -4 & (1) \\ 5x + 3y = -9 & (2) \end{pmatrix}$.

Multiply equation (1) by -3 and add to equation (2) to replace equation (2).

$\begin{pmatrix} 2x + y = -4 & (3) \\ -x = 3 & (4) \end{pmatrix}$

From equation (4) we can find x.

$-x = 3$

$x = -3$

Substitute -3 for x in equation (3).

$2(-3) + y = -4$

$-6 + y = -4$

$y = 2$

The solution set is $\{(-3, 2)\}$.

3. Solve the system $\begin{pmatrix} 3x - 5y = 23 & (1) \\ 5x + 4y = 26 & (2) \end{pmatrix}$.

Multiply equation (2) by 5 to form equation (4).

$\begin{pmatrix} 3x - 5y = 23 & (3) \\ 25x + 20y = 130 & (4) \end{pmatrix}$

Multiply equation (3) by 4 and add to equation (4) to replace equation (4).

$\begin{pmatrix} 3x - 5y = 23 & (5) \\ 37x = 222 & (6) \end{pmatrix}$

From equation (6) we can find x.

$37x = 222$

$x = 6$

Substitute 6 for x in equation (5).

$3(6) - 5y = 23$

$18 - 5y = 23$

$-5y = 5$

$y = -1$

The solution set is $\{(6, -1)\}$.

4. Solve the system $\begin{pmatrix} 6x + 5y = 5 & (1) \\ 5x - 2y = 6 & (2) \end{pmatrix}$.

Multiply equation (2) by 5 to form equation (4).

$\begin{pmatrix} 6x + 5y = 5 & (3) \\ 25x - 10y = 30 & (4) \end{pmatrix}$

Multiply equation (3) by 2 and add to equation (4) to replace equation (4).

$\begin{pmatrix} 6x + 5y = 5 & (5) \\ 37x = 40 & (6) \end{pmatrix}$

From equation (6) we can find x.

$37x = 40$

$x = \frac{40}{37}$

Substitute $\frac{40}{37}$ for x in equation (5).

$6\left(\frac{40}{37}\right) + 5y = 5$

$\frac{240}{37} + 5y = 5$

$5y = 5 - \frac{240}{37}$

$5y = \frac{185}{37} - \frac{240}{37}$

$5y = \frac{-55}{37}$

$$\frac{1}{5}(5y) = \left(\frac{-55}{37}\right)\left(\frac{1}{5}\right)$$

$$y = -\frac{11}{37}$$

The solution set is $\left\{\left(\frac{40}{37}, -\frac{11}{37}\right)\right\}$.

5. Solve the system $\begin{pmatrix} 2x + 6y = -2 & (1) \\ 4x + 9y = -3 & (2) \end{pmatrix}$.

Multiply equation (1) by -2 and add to equation (2) to replace equation (2).

$$\begin{pmatrix} 2x + 6y = -2 & (3) \\ -3y = 1 & (4) \end{pmatrix}$$

From equation (4) we can find y.

$$-3y = 1$$

$$y = \frac{1}{-3} = -\frac{1}{3}$$

Substitute $-\frac{1}{3}$ for y in equation (3).

$$2x + 6y = -2$$

$$2x + 6\left(-\frac{1}{3}\right) = -2$$

$$2x - 2 = -2$$

$$2x = 0$$

$$x = 0$$

The solution set is $\left\{\left(0, -\frac{1}{3}\right)\right\}$.

6. Solve the system $\begin{pmatrix} x = 3y + 14 \\ 2x + y = 7 \end{pmatrix}$.

Substitute $3y + 14$ for x in the second equation.

$$2x + y = 7$$

$$2(3y + 14) + y = 7$$

$$6y + 28 + y = 7$$

$$7y + 28 = 7$$

$$7y = -21$$

$$y = -3$$

Substitute -3 for y in the first equation.

$$x = 3y + 14$$

$$x = 3(-3) + 14$$

$$x = -9 + 14$$

$$x = 5$$

The solution set is $\{(5, -3)\}$.

7. Solve the system $\begin{pmatrix} \frac{x+1}{5} + \frac{y+1}{2} = 2 & (1) \\ \frac{x+6}{2} + \frac{y-3}{3} = \frac{5}{2} & (2) \end{pmatrix}$.

Simplify equation (1) by multiplying both sides by 10 to form equation (3).

$$2(x + 1) + 5(y + 1) = 20$$

$$2x + 2 + 5y + 5 = 20$$

$$2x + 5y + 7 = 20$$

$$2x + 5y = 13$$

Simplify equation (2) by multiplying both sides by 6 to form equation (4).

$$3(x + 6) + 2(y - 3) = 15$$

$$3x + 18 + 2y - 6 = 15$$

$$3x + 2y + 12 = 15$$

$$3x + 2y = 3$$

Solve the system $\begin{pmatrix} 2x + 5y = 13 & (3) \\ 3x + 2y = 3 & (4) \end{pmatrix}$.

Multiply equation (3) by -3 to form equation (5).

$$\begin{pmatrix} -6x - 15y = -39 & (5) \\ 3x + 2y = 3 & (6) \end{pmatrix}$$

Multiply equation (6) by 2 and add to equation (5) to replace equation (6).

$$\begin{pmatrix} -6x - 15y = -39 & (7) \\ -11y = -33 & (8) \end{pmatrix}$$

From equation (8) we can find y.

$$-11y = -33$$

$$y = 3$$

Substitute 3 for y in equation (3).

$$2x + 5y = 13$$

$$2x + 5(3) = 13$$

$$2x + 15 = 13$$

$$2x = -2$$

$$x = -1$$

The solution set is $\{(-1, 3)\}$.

8. Solve the system $\begin{pmatrix} y = -2x + 1 \\ 4x + 2y = 3 \end{pmatrix}$.

Substitute $-2x + 1$ for y in the second equation.

$$4x + 2y = 3$$

$$4x + 2(-2x + 1) = 3$$

$$4x - 4x + 2 = 3$$

$$0 + 2 = 3$$

$$2 = 3$$

Because $2 = 3$ is a false statement, there is no solution. The solution set is $\varnothing$.

9. Solve the system $\begin{pmatrix} 4x - 6y = 2 & (1) \\ 2x - 3y = 1 & (2) \end{pmatrix}$.

Multiply equation (2) by -2 and add to equation (1) to replace equation (2).

$$\begin{pmatrix} 4x - 6y = 2 & (3) \\ 0 + 0 = 0 & (4) \end{pmatrix}$$

The true numerical statement, $0 + 0 = 0$, implies that the system has infinitely many solutions. The solution set is $\{(x, y) | 2x - 3y = 1\}$.

Section 4.4

1. Solve the system $\begin{pmatrix} 2x + 2y + z = 5 & (1) \\ 3y + z = -9 & (2) \\ 2z = 6 & (3) \end{pmatrix}$.

From equation (3) find z.

$2z = 6$

$z = 3$

Substitute 3 for z in equation (2).

$3y + z = -9$

$3y + (3) = -9$

$3y = -12$

$y = -4$

Substitute 3 for z and -4 for y in equation (1).

$2x + 2y + z = 5$

$2x + 2(-4) + (3) = 5$

$2x - 8 + 3 = 5$

$2x - 5 = 5$

$2x = 10$

$x = 5$

The solution set is $\{(5, -4, 3)\}$.

2. Solve the system $\begin{pmatrix} 3x - 2y + 4z = -16 & (1) \\ 5y + 2z = 13 & (2) \\ y - 4z = 7 & (3) \end{pmatrix}$.

Multiply equation (2) by 2 and add to equation (3) to replace equation (3).

$\begin{pmatrix} 3x - 2y + 4z = -16 & (4) \\ 5y + 2z = 13 & (5) \\ 11y = 33 & (6) \end{pmatrix}$

From equation (3) find y.

$11y = 33$

$y = 3$

Substitute 3 for y in equation (2) to find z.

$5y + 2z = 13$

$5(3) + 2z = 13$

$15 + 2z = 13$

$2z = -2$

$z = -1$

Substitute 3 for y and -1 for z in equation (1) to find x.

$3x - 2y + 4z = -16$

$3x - 2(3) + 4(-1) = -16$

$3x - 6 - 4 = -16$

$3x - 10 = -16$

$3x = -6$

$x = -2$

The solution set is $\{(-2, 3, -1)\}$.

3. Solve the system $\begin{pmatrix} x - 3y + 2z = 3 & (1) \\ 2x + 7y - 3z = -3 & (2) \\ 3x - 2y + 2z = 3 & (3) \end{pmatrix}$.

Multiply equation (1) by -2 and add to equation (2) to replace equation (2). Also multiply equation (1) by -3 and add to equation (3) to replace equation (3).

$\begin{pmatrix} x - 3y + 2z = 3 & (1) \\ 13y - 7z = -9 & (4) \\ 7y - 4z = -6 & (5) \end{pmatrix}$

Multiply equation (5) by -7 to form equation (6).

$\begin{pmatrix} x - 3y + 2z = 3 & (1) \\ 13y - 7z = -9 & (4) \\ -49y + 28z = 42 & (6) \end{pmatrix}$

Multiply equation (4) by 4 and add to equation (6) to replace equation (6).

$\begin{pmatrix} x - 3y + 2z = 3 & (1) \\ 13y - 7z = -9 & (4) \\ 3y = 6 & (7) \end{pmatrix}$

From equation (7) find y.

$3y = 6$

$y = 2$

Substitute 2 for y in equation (4) to find z.

$13y - 7z = -9$

$13(2) - 7z = -9$

$26 - 7x = -9$

$-7z = -35$

$z = 5$

Substitute 2 for y and 5 for z in equation (1) to find x.

$x - 3y + 2z = 3$

$x - 3(2) + 2(5) = 3$

$x - 6 + 10 = 3$

$x + 4 = 3$

$x = -1$

The solution set is $\{(-1, 2, 5)\}$.

4. Solve the system $\begin{pmatrix} 3x + y + 2z = 10 & (1) \\ 4x + 2y + z = 11 & (2) \\ 2x - 3y + 3z = 12 & (3) \end{pmatrix}$.

Multiply equation (1) by -2 and add to equation (2) to replace equation (2). Also multiply equation (1) by 3 and add to equation (3) to replace equation (3).

$\begin{pmatrix} 3x + y + 2z = 10 & (1) \\ -2x \quad - 3z = -9 & (4) \\ 11x \quad + 9z = 42 & (5) \end{pmatrix}$

Multiply equation (4) by 3 and add to equation (5) to replace equation (5).

$\begin{pmatrix} 3x + y + 2z = 10 & (1) \\ -2x \quad - 3z = -9 & (4) \\ 5x \quad = 15 & (6) \end{pmatrix}$

From equation (6) find x.

$5x = 15$

$x = 3$

Substitute 3 for x in equation (4) to find z.

$$\begin{aligned} -2x - 3z &= -9 \\ -2(3) - 3z &= -9 \\ -6 - 3z &= -9 \\ -3x &= -3 \\ z &= 1 \end{aligned}$$

Substitute 3 for x and 1 for z in equation (1) to find y.

$$\begin{aligned} 3x + y + 2z &= 10 \\ 3(3) + y + 2(1) &= 10 \\ 9 + y + 2 &= 10 \\ y + 11 &= 10 \\ y &= -1 \end{aligned}$$

The solution set is $\{(3, -1, 1)\}$.

5. Solve the system $\left(\begin{array}{rl} x + 2y + 5z = 8 & (1) \\ 3x - 5y + 2z = 12 & (2) \\ 2x + 4y + 10z = 14 & (3) \end{array}\right)$.

Multiply equation (1) by -3 and add to equation (2) to replace equation (2). Also multiply equation (1) by -2 and add to equation (3) to replace equation (3).

$$\left(\begin{array}{rl} x + 2y + 5z = 8 & (1) \\ -11y - 13z = -12 & (4) \\ 0 = -2 & (5) \end{array}\right).$$

The false statement, $0 = -2$, indicates that the system is inconsistent. Therefore, the solution set is $\varnothing$.

6. Solve the system $\left(\begin{array}{rl} 3x + y + 2z = 3 & (1) \\ x + 2y + 5z = 8 & (2) \\ -2x + y + 3z = 5 & (3) \end{array}\right)$.

Multiply equation (1) by -2 and add to equation (2) to replace equation (2). Also multiply equation (1) by -1 and add to equation (3) to replace equation (3).

$$\left(\begin{array}{rl} 3x + y + 2z = 3 & (4) \\ -5x \quad\;\; + z = 2 & (5) \\ -5x \quad\;\; + z = 2 & (6) \end{array}\right)$$

Multiply equation (5) by -1 and add to equation (6) to replace equation (6).

$$\left(\begin{array}{rl} 3x + y + 2z = 3 & (7) \\ -5x \quad\;\; + z = 2 & (8) \\ 0 + 0 = 0 & (9) \end{array}\right)$$

The true numerical statement, $0 + 0 = 0$, indicates that the system has infinitely many solutions.

7. Let x represent the amount invested at 6%, let y represent the amount invested at 8%, and let z represent the amount invested at 9%. The total yearly interest is \$2340.

$0.06x + 0.08y + 0.09z = 2340.$

The total invested amount is \$30,000.

$x + y + z = 30{,}000$

The sum of the amounts at 6% and 8% equals \$10,000 more than the amount invested at 9%.

$x + y = 10{,}000 + z$

Solve this system

$$\left(\begin{array}{rl} 0.06x + 0.08y + 0.09z = 2340 & (1) \\ x + y + z = 30{,}000 & (2) \\ x + y = 10{,}000 + z & (3) \end{array}\right)$$

Multiply equation (1) by 100 to form equation (4) and simplify equation (3) to form equation (5).

$$\left(\begin{array}{rl} 6x + 8y + 9z = 234{,}000 & (4) \\ x + y + z = 30{,}000 & (2) \\ x + y - z = 10{,}000 & (5) \end{array}\right)$$

Multiply equation (5) by 9 and then add to equation (4) to form equation (6). Leave equation (2) alone. Add equation (2) to equation (5) to form equation (7).

$$\left(\begin{array}{rl} 15x + 17y \quad\;\; = 324{,}000 & (6) \\ x + y + z = 30{,}000 & (2) \\ 2x + 2y \quad\;\; = 40{,}000 & (7) \end{array}\right)$$

Multiply equation (7) by $\frac{1}{2}$ to form equation (8).

$$\left(\begin{array}{rl} 15x + 17y \quad\;\; = 324{,}000 & (6) \\ x + y + z = 30{,}000 & (2) \\ x + y \quad\;\; = 20{,}000 & (8) \end{array}\right)$$

Multiply equation (8) by -15 and then add to equation (6) to form equation (9).

$$\left(\begin{array}{rl} 2y = 24{,}000 & (9) \\ x + y + z = 30{,}000 & (2) \\ x + y = 20{,}000 & (8) \end{array}\right)$$

Solve equation (9) for y.

$$\begin{aligned} 2y &= 24{,}000 \\ y &= 12{,}000 \end{aligned}$$

Substitute 12,000 for y in equation (8) to find x.

$$\begin{aligned} x + y &= 20{,}000 \\ x + 12{,}000 &= 20{,}000 \\ x &= 8000 \end{aligned}$$

Substitute 12,000 for y, 8000 for x in equation (2) to find z.

$$\begin{aligned} x + y + z &= 30{,}000 \\ 8000 + 12{,}000 + z &= 30{,}000 \\ 20{,}000 + z &= 30{,}000 \\ z &= 10{,}000 \end{aligned}$$

At 6%, \$8000 is invested; at 8%, \$12,000 is invested; and at 9%, \$10,000 is invested.

CHAPTER 5

Section 5.1

1. $(3x^2 - 7x + 3) + (5x^2 + 11x - 7)$
$= (3x^2 + 5x^2) + (-7x + 11x) + (3 - 7)$
$= 8x^2 + 4x - 4$

2. $(7x + 2) + (2x + 6) + (6x - 1)$
$= (7x + 2x + 6x) + (2 + 6 - 1)$
$= 15x + 7$

3. $(5x^2y - 2xy^2) + (-10x^2y - 4xy^2) + (-2x^2y + 7xy^2)$
$= (5x^2y - 10x^2y - 2x^2y)$
$+ (-2xy^2 - 4xy^2 + 7xy^2)$
$= -7x^2y + xy^2$

4. $(5x^2 - 6x + 8) - (2x^2 + 4x - 3)$
$= (5x^2 - 6x + 8) + (-2x^2 - 4x + 3)$
$= (5x^2 - 2x^2) + (-6x - 4x) + (8 + 3)$
$= 3x^2 - 10x + 11$

5. $(2y^2 + 10) - (-5y^2 + 3y - 6)$
$= (2y^2 + 10) + (5y^2 - 3y + 6)$
$= (2y^2 + 5y^2) + (-3y) + (10 + 6)$
$= 7y^2 - 3y + 16$

6. To subtract add the opposite.

$$\begin{array}{rcr} 7x^2 - 3xy + y^2 & \Rightarrow & 7x^2 - 3xy + y^2 \\ -\quad x^2 + 4xy + 6y^2 & \Rightarrow & -x^2 - 4xy - 6y^2 \\ \hline & & 6x^2 - 7xy - 5y^2 \end{array}$$

7. $(3x - 5) + (4x - 6) - (3x + 4)$
$= 1(3x - 5) + 1(4x - 6) - 1(3x + 4)$
$= 3x - 5 + 4x - 6 - 3x - 4$
$= (3x + 4x - 3x) + (-5 - 6 - 4)$
$= 4x - 15$

8. $(8a^2 - 3b) - (a^2 + 5) + (-6b - 8)$
$= 8a^2 - 3b - a^2 - 5 - 6b - 8$
$= 7a^2 - 9b - 13$

9. $(6a^2 - 2a - 3) - (8a^2 + 4a - 5)$
$= 6a^2 - 2a - 3 - 8a^2 - 4a + 5$
$= -2a^2 - 6a + 2$

10. $10x + [4x - (x + 5)]$
$10x + [4x - x - 5]$
$10x + [3x - 5]$
$10x + 3x - 5$
$13x - 5$

11. Rectangular solid has six sides.
Area of front $= 8x$
Area of back $= 8x$
Area of top $= 8(5) = 40$
Area of bottom $= 8(5) = 40$
Area of left side $= 5x$
Area of right side $= 5x$
Total surface area $= 26x + 80$
When height is 4, $x = 4$: $26(4) + 80 = 184$ square units
When height is 6, $x = 6$: $26(6) + 80 = 236$ square units
When height is 12, $x = 12$: $26(12) + 80 =$ 392 square units

Section 5.2

1. $(5xy^2)(2x^4y^2) = 5 \cdot 2 \cdot x \cdot x^4 \cdot y^2 \cdot y^2$
$= 10x^{1+4}y^{2+2}$
$= 10x^5y^4$

2. $(-2a^2b^3)(6a^4b^5) = -2 \cdot 6 \cdot a^2 \cdot a^4 \cdot b^3 \cdot b^5$
$= -12a^{2+4}b^{3+5}$
$= -12a^6b^8$

3. $\left(\frac{2}{3}x^2y\right)\left(\frac{1}{5}x^3y^5\right) = \frac{2}{3} \cdot \frac{1}{5} \cdot x^2 \cdot x^3 \cdot y \cdot y^5$
$= \frac{2}{15}x^5y^6$

4. $(-3a^2b)(-a^3b^2) = (-3)(-1)(a^2)(a^3)(b)(b^2)$
$= 3a^{2+3}b^{1+2}$
$= 3a^5b^3$

5. $(5x^3y)(2xy^3)(3x^2) = 5 \cdot 2 \cdot 3 \cdot x^3 \cdot x \cdot x^2 \cdot y \cdot y^3$
$= 30x^{3+1+2}y^{1+3}$
$= 30x^6y^4$

6. $(x^3y^4)^3 = (x^3)^3(y^4)^3 = x^9y^{12}$

7. $(2a^3)^4 = (2)^4(a^3)^4 = 16a^{12}$

8. $(-3x^2y^5)^3 = (-3)^3(x^2)^3(y^5)^3 = -27x^6y^{15}$

9. (a) $\frac{72a^4b}{-8ab} = -9a^{4-1}(1) = -9a^3$

(b) $\frac{-5m^5n^4}{m^3n^2} = -5m^{5-3}n^{4-2} = -5m^2n^2$

(c) $\frac{-16x^3}{2x^2} = -8x^{3-2} = -8x$

Section 5.3

1. $4x^3(3x^2 + 2x - 5) = 4x^3(3x^2) + 4x^3(2x) + 4x^3(-5)$
$= 12x^5 + 8x^4 - 20x^3$

2. $-4ab^2(2a^2 + ab^2 - 3ab + b^2)$
$= -4ab^2(2a^2) + (-4ab^2)(ab^2) + (-4ab^2)(-3ab)$
$+ (-4ab^2)(b^2)$
$= -8a^3b^2 - 4a^2b^4 + 12a^2b^3 - 4ab^4$

3. $(a + 3)(b + 4) = a(b + 4) + 3(b + 4)$
$= ab + 4a + 3b + 12$

4. $(a - 4)(a + b + 5) = a(a + b + 5) - 4(a + b + 5)$
$= a^2 + ab + 5a - 4a - 4b - 20$
$= a^2 + ab + a - 4b - 20$

5. $(a + 8)(a + 4) = a(a + 4) + 8(a + 4)$
$= a^2 + 4a + 8a + 32$
$= a^2 + 12a + 32$

6. $(a - 3)(a^2 + 2a - 5)$
$= a(a^2 + 2a - 5) - 3(a^2 + 2a - 5)$
$= a^3 + 2a^2 - 5a - 3a^2 - 6a + 15$
$= a^3 - a^2 - 11a + 15$

7. $(4a - b)(3a^2 - ab - b^2)$
$= 4a(3a^2 - ab - b^2) - b(3a^2 - ab - b^2)$
$= 12a^3 - 4a^2b - 4ab^2 - 3a^2b + ab^2 + b^3$
$= 12a^3 - 7a^2b - 3ab^2 + b^3$

8.

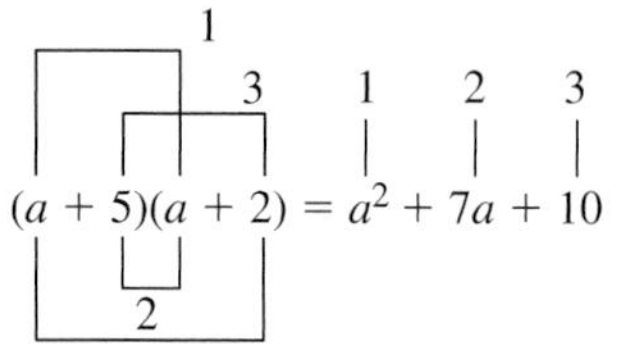

$(a + 5)(a + 2) = a^2 + 7a + 10$

$(a + 5)(a + 2) = a^2 + 7a + 10$

9.

$(5a - 1)(3a + 2) = 15a^2 + 7a - 2$

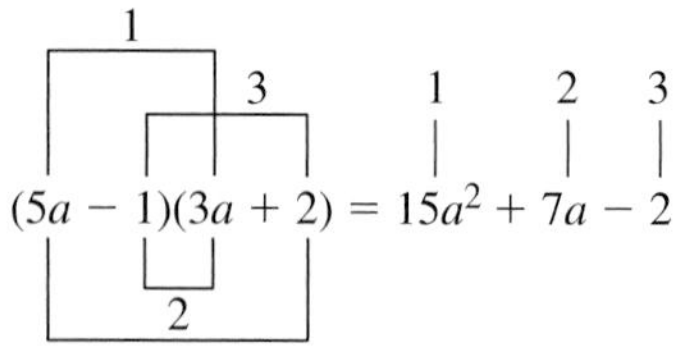

$(5a - 1)(3a + 2) = 15a^2 + 7a - 2$

10. (a) $(x + 3)^2 = (x)^2 + 2(x)(3) + (3)^2$
$= x^2 + 6x + 9$

(b) $(3x + y)^2 = (3x)^2 + 2(3x)(y) + (y)^2$
$= 9x^2 + 6xy + y^2$

(c) $(3a + 5b)^2 = (3a)^2 + 2(3a)(5b) + (5b)^2$
$= 9a^2 + 30ab + 25b^2$

11. (a) $(x - 5)^2 = (x)^2 - 2(x)(5) + (5)^2$
$= x^2 - 10x + 25$

(b) $(x - 2y)^2 = (x)^2 - 2(x)(2) + (2y)^2$
$= x^2 - 4xy + 4y^2$

(c) $(2a - 3b)^2 = (2a)^2 - 2(2a)(3b) + (3b)^2$
$= 4a^2 - 12ab + 9b^2$

12. (a) $(x + 6)(x - 6) = (x)^2 - (6)^2$
$= x^2 - 36$

(b) $(x + 9y)(x - 9y) = (x)^2 - (9y)^2$
$= x^2 - 81y^2$

(c) $(7a - 5b)(7a + 5b) = (7a)^2 - (5b)^2$
$= 49a^2 - 25b^2$

13. $(a + b)^3 = a^3 + 3a^2b + 3ab^2 + b^3$
$(x + 5)^3 = (x)^3 + 3(x)^2(5) + 3(x)(5)^2 + (5)^3$
$= x^3 + 15x^2 + 75x + 125$

14. $(a - b)^3 = a^3 - 3a^2b + 3ab^2 - b^3$
$(4x - 3y)^3 = (4x)^3 - 3(4x)^2(3y)$
$+ 3(4x)(3y)^2 - (3y)^3$
$= 64x^3 - 144x^2y + 108xy^2 - 27y^3$

15. Let x represent the length of each side of the cut-out square.

Let $8 - 2x$ represent the length of the box. Let $8 - 2x$ represent the width of the box. Then x also represents the height of the box.

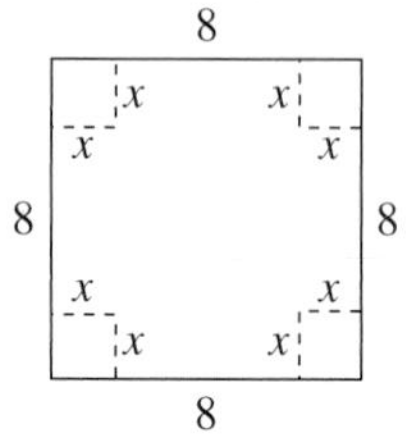

Volume = Length · Width · Height
$V = LWH$
$V = (8 - 2x)(8 - 2x)(x)$
$V = (64 - 16x - 16x + 4x^2)(x)$
$V = (64 - 32x + 4x^2)(x)$
$V = 64x - 32x^2 + 4x^3$

Area of original square $= (8)(8) = 64$
Area of four corners $= 4(x^2) = 4x^2$

(Surface area) = (Area of original piece) − (Area of four corners)

(Surface area) $= 64 - 4x^2$

Section 5.4

1. (a) No, there are coefficients that are not integers.

(b) No, the polynomials inside the parentheses can be factored further.

(c) Yes

2. (a) The common factor is $5a^2$.
$10a^2 - 15a^3 + 35a^4 = 5a^2(2 - 3a + 7a^2)$

(b) The common factor is $2m$.
$2mn - 8m^3 = 2m(n - 4m^2)$

3. (a) $6(xy + 8) + z(xy + 8) = (xy + 8)(6 + z)$

(b) $x^2(x + y) - y^3(x + y) = (x + y)(x^2 - y^3)$

(c) $x(2x + y) + y(2x + y) + z(2x + y)$
$= (2x + y)(x + y + z)$

4. (a) $4x^3y + 8xy + 3x^2 + 6$
$= 4xy(x^2 + 2) + 3(x^2 + 2)$
$= (x^2 + 2)(4xy + 3)$

(b) $7y^2 - 14y + 5y - 10$
$= 7y(y - 2) + 5(y - 2)$
$= (y - 2)(7y + 5)$

(c) $2x^2 + 3xy - 4xy - 6y^2$
$= x(2x + 3y) - 2y(2x + 3y)$
$= (2x + 3y)(x - 2y)$

5. $y^2 - 4y = 0$
$y(y - 4) = 0$
$y = 0$ or $y - 4 = 0$
$y = 0$ or $y = 4$
The solution set is $\{0, 4\}$.

6. $x^2 = -12x$
$x^2 + 12x = 0$
$x(x + 12) = 0$
$x = 0$ or $x + 12 = 0$
$x = 0$ or $x = -12$
The solution set is $\{-12, 0\}$.

7. $7y^2 + 2y = 0$
$y(7y + 2) = 0$
$y = 0$ or $7y + 2 = 0$
$y = 0$ or $7y = -2$
$y = 0$ or $y = -\frac{2}{7}$
The solution set is $\left\{-\frac{2}{7}, 0\right\}$.

8. $8cy^2 - dy = 0$
$y(8cy - d) = 0$
$y = 0$ or $8cy - d = 0$
$y = 0$ or $8cy = d$
$y = 0$ or $y = \frac{d}{8c}$
The solution set is $\left\{0, \frac{d}{8c}\right\}$.

9. Let s represent the length of the side of the square.
Use these formulas:
$A = s^2$: Area of a square
$P = 4s$: Perimeter of a square
$$\begin{pmatrix}\text{Area of}\\ \text{square}\end{pmatrix} = 2 \cdot \begin{pmatrix}\text{Perimeter}\\ \text{of square}\end{pmatrix}$$
$s^2 = 2(4s)$
$s^2 = 8s$
$s^2 - 8s = 0$
$s(s - 8) = 0$
$s = 0$ or $s - 8 = 0$
$s = 0$ or $s = 8$
Discard $s = 0$, because length cannot be 0. The length is 8 units.

10. Let y represent the length of the side of the cube. There are six sides of the cube, each side a square.
Use this formula:
$A = s^2$: Area of a square
$A = (y)^2 = y^2$
Because there are six sides each with area of y^2, the total surface area of the cube is the sum of the six areas, $6y^2$.
To find volume, use this formula:
$V = LWH$
$V = (y)(y)(y) = y^3$
The volume of the cube is to be equal to the surface area of the cube.
$y^3 = 6y^2$
$y^3 - 6y^2 = 0$
$y^2(y - 6) = 0$
$y = 0$ or $y = 0$ or $y - 6 = 0$
$y = 0$ or $y = 0$ or $y = 6$
Discard $y = 0$, because length cannot be 0. The length of the side of the cube is 6 units.

Section 5.5

1. (a) $m^2 - 36 = (m)^2 - (6)^2$
$= (m + 6)(m - 6)$

(b) $9y^2 - 49 = (3y)^2 - (7)^2$
$= (3y + 7)(3y - 7)$

(c) $64 - 25b^2 = (8)^2 - (5b)^2$
$= (8 + 5b)(8 - 5b)$.

2. $256a^4 - b^4 = (16a^2)^2 - (b^2)^2$
$= (16a^2 + b^2)(16a^2 - b^2)$
$= (16a^2 + b^2)(4a - b)(4a + b)$

3. $(2x + y)^2 - 9 = (2x + y)^2 - (3)^2$
$= (2x + y + 3)(2x + y - 3)$

4. $18a^2 - 50 = 2(9a^2 - 25)$
$= 2(3a + 5)(3a - 5)$

5. $x^3 + 8y^3 = (x)^3 + (2y)^3$
$= (x + 2y)(x^2 - 2xy + 4y^2)$

6. $m^2 = 81$
$m^2 - 81 = 0$
$(m - 9)(m + 9) = 0$
$m - 9 = 0$ or $m + 9 = 0$
$m = 9$ or $m = -9$
The solution set is $\{-9, 9\}$.

7.
$$25a^2 = 36$$
$$25a^2 - 36 = 0$$
$$(5a - 6)(5a + 6) = 0$$
$$5a - 6 = 0 \quad \text{or} \quad 5a + 6 = 0$$
$$5a = 6 \quad \text{or} \quad 5a = -6$$
$$a = \frac{6}{5} \quad \text{or} \quad a = -\frac{6}{5}$$
The solution set is $\left\{-\frac{6}{5}, \frac{6}{5}\right\}$.

8.
$$3 - 12x^2 = 0$$
$$3(1 - 4x^2) = 0$$
$$(1 - 4x^2) = 0$$
$$(1 - 2x)(1 + 2x) = 0$$
$$1 - 2x = 0 \quad \text{or} \quad 1 + 2x = 0$$
$$1 = 2x \quad \text{or} \quad 1 = -2x$$
$$\frac{1}{2} = x \quad \text{or} \quad -\frac{1}{2} = x$$
The solution set is $\left\{-\frac{1}{2}, \frac{1}{2}\right\}$.

9.
$$x^4 - 81 = 0$$
$$(x^2 + 9)(x^2 - 9) = 0$$
$$(x^2 + 9)(x - 3)(x + 3) = 0$$
$$x^2 + 9 = 0 \quad \text{or} \quad x - 3 = 0 \quad \text{or} \quad x + 3 = 0$$
$$x^2 = -9 \quad \text{or} \quad x = 3 \quad \text{or} \quad x = -3$$
No real solution or $x = 3$ or $x = -3$
The solution set is $\{-3, 3\}$.

10.
$$y^3 - 16y = 0$$
$$y(y^2 - 16) = 0$$
$$y(y - 4)(y + 4) = 0$$
$$y = 0 \quad \text{or} \quad y - 4 = 0 \quad \text{or} \quad y + 4 = 0$$
$$y = 0 \quad \text{or} \quad y = 4 \quad \text{or} \quad y = -4$$
The solution set is $\{-4, 0, 4\}$.

11. Let y represent the length of the side of the smaller square. Then $2y$ represents the length of the side of the larger square.
The sum of the areas of the squares is 125 square feet.
$$\begin{pmatrix}\text{Area of} \\ \text{smaller square}\end{pmatrix} + \begin{pmatrix}\text{Area of} \\ \text{larger square}\end{pmatrix} = 125$$
$$y^2 + (2y)^2 = 125$$
$$y^2 + 4y^2 = 125$$
$$5y^2 = 125$$
$$\frac{5y^2}{5} = \frac{125}{5}$$
$$y^2 = 25$$
$$y^2 - 25 = 0$$
$$(y + 5)(y - 5) = 0$$
$$y + 5 = 0 \quad \text{or} \quad y - 5 = 0$$
$$y = -5 \quad \text{or} \quad y = 5$$
Discard $y = -5$ because length cannot be negative. The length of the side of the smaller square is 5 inches, and the length of the side of the larger square is $2(5) = 10$ inches.

Section 5.6

1. $y^2 + 11y + 24$
We need two integers whose sum is 11 and whose product is 24.

Factors of 24	Sum of factors
$1(24) = 24$	$1 + 24 = 25$
$2(12) = 24$	$2 + 12 = 14$
$3(8) = 24$	$3 + 8 = 11$

The two integers are 3 and 8.
$y^2 + 11y + 24 = (y + 3)(y + 8)$

2. $a^2 - 18a + 32$
We need two integers whose sum is -18 and whose product is 32.

Factors of 32	Sum of factors
$(-1)(-32) = 32$	$-1 + (-32) = -33$
$(-2)(-16) = 32$	$-2 + (-16) = -18$

The two integers are -2 and -16.
$a^2 - 18a + 32 = (a - 2)(a - 16)$

3. $y^2 - 8y - 20$
We need two integers whose sum is -8 and whose product is -20.

Factors of -20	Sum of factors
$(-1)(20) = -20$	$-1 + 20 = 19$
$(-2)(10) = -20$	$-2 + 10 = 8$
$2(-10) = -20$	$2 + (-10) = -8$

The two integers are 2 and -10.
$y^2 - 8y - 20 = (y + 2)(y - 10)$

4. $m^2 + 8m + 24$
We need two integers whose sum is 8 and whose product is 24.

Factors of 24	Sum of factors
$1(24) = 24$	$1 + 24 = 25$
$2(12) = 24$	$2 + 12 = 14$
$3(8) = 24$	$3 + 8 = 11$
$4(6) = 24$	$4 + 6 = 10$
$(-1)(-24) = 24$	$-1 + (-24) = -25$

We have exhausted all possible factors of 24 and none has a sum of 8. Therefore, $m^2 + 8m + 24$ is not factorable.

5. $a^2 + a - 30$
We need two integers whose sum is 1 and whose product is -30.
The integers are 6 and -5.
$a^2 + a - 30 = (a + 6)(a - 5)$

6. $y^2 - 6y - 216$

We need two integers whose sum is -6 and whose product is -216.

Prime factor 216.

$216 = 2 \cdot 2 \cdot 2 \cdot 3 \cdot 3 \cdot 3$

Using different combinations of factors, we find 12 and -18.

$y^2 - 6y - 216 = (y + 12)(y - 18)$

7. $3m^2 + 11m + 10$

The binomial will have the form $(3m + \_)(m + \_)$. The possible factors of 10 are $1 \cdot 10$ and $2 \cdot 5$, which will give the following possibilities:

$(3m + 1)(m + 10)$ $(3m + 2)(m + 5)$

$(3m + 10)(m + 1)$ $(3m + 5)(m + 2)$

By checking the middle term in these products, we find that

$3m^2 + 11m + 10 = (3m + 5)(m + 2)$

8. $14a^2 - 37a + 5$

We find that $14a^2$ can be written as $14a \cdot a$ or $7a \cdot 2a$.

Because the last term is positive and the middle term is negative, we have two possible forms:

$(14a - \_)(a - \_)$ or $(7a - \_)(2a - \_)$

The factors of 5 are 5 and 1, so we have the following possibilities:

$(14a - 1)(a - 5)$ $(7a - 1)(2a - 5)$

$(14a - 5)(a - 1)$ $(7a - 5)(2a - 1)$

By checking the middle term in these products, we find that

$14a^2 - 37a + 5 = (7a - 1)(2a - 5)$

9. $6b^2 + 10b + 3$

We find that $6b^2$ can be written as $6b \cdot b$ or $3b \cdot 2b$.

Because the last term is positive and the middle term is positive, we have two possible forms:

$(6b + \_)(b + \_)$ or $(3b + \_)(2b + \_)$

The factors of 3 are 3 and 1, so we have the following possibilities:

$(6b + 3)(b + 1)$ $(3b + 1)(2b + 3)$

$(6b + 1)(b + 3)$ $(3b + 3)(2b + 1)$

Because none of the possibilities gives the middle term, $10b$, then $6b^2 + 10b + 3$ is not factorable.

10. $4a^2 - 3a - 10$

First, multiply the coefficient of the first term by the last term.

$(4)(-10) = -40$

We need two integers whose sum is -3 and whose product is -40.

Factors of -40	Sum of factors
$(-1)(40) = -40$	$-1 + 40 = 39$
$(-2)(20) = -40$	$-2 + 20 = 18$
$(-4)(10) = -40$	$-4 + 10 = 6$
$(-5)(8) = -40$	$-5 + 8 = 3$
$(5)(-8) = -40$	$5 + (-8) = -3$

The two integers are 5 and -8.

Rewrite $4a^2 - 3a - 10$ as $4a^2 - 8a + 5a - 10$.

$$4a^2 - 8a + 5a - 10 = 4a(a - 2) + 5(a - 2)$$
$$= (a - 2)(4a + 5)$$

11. $12a^2 + a - 6$

First, multiply the coefficient of the first term by the last term.

$(12)(-6) = -72$

We need two integers whose sum is 1 and whose product is -72.

Factors of -72	Sum of factors
$(-1)(72) = -72$	$-1 + 72 = 72$
$(-2)(36) = -72$	$-2 + 36 = 36$
$(-3)(24) = -72$	$-3 + 24 = 21$
$(-4)(18) = -72$	$-4 + 18 = 14$
$(-6)(12) = -72$	$-6 + 12 = 6$
$(-8)(9) = -72$	$-8 + 9 = 1$

The two integers are 9 and -8.

Rewrite $12a^2 + a - 6$ as $12a^2 - 8a + 9a - 6$.

$$12a^2 - 8a + 9a - 6 = 4a(3a - 2) + 3(3a - 2)$$
$$= (3a - 2)(4a + 3)$$

12. **(a)** $a^2 + 22a + 121 = (a + 11)^2$

(b) $25x^2 - 60xy + 36y^2 = (5x - 6y)^2$

Section 5.7

1. $y^2 - 5y - 24 = 0$

$(y - 8)(y + 3) = 0$

$y - 8 = 0$ or $y + 3 = 0$

$y = 8$ or $y = -3$

The solution set is $\{-3, 8\}$.

2. $6a^2 + a - 5 = 0$

$(6a - 5)(a + 1) = 0$

$6a - 5 = 0$ or $a + 1 = 0$

$6a = 5$ or $a = -1$

$a = \frac{5}{6}$ or $a = -1$

The solution set is $\left\{-1, \frac{5}{6}\right\}$.

3. $3m^2 - 6m - 24 = 0$

$3(m^2 - 2m - 8) = 0$

$3(m - 4)(m + 2) = 0$

$(m - 4)(m + 2) = 0$

$m - 4 = 0$ or $m + 2 = 0$

$m = 4$ or $m = -2$

The solution set is $\{-2, 4\}$.

4. $9y^2 + 48y + 64 = 0$

$(3y + 8)(3y + 8) = 0$

$3y + 8 = 0$ or $3y + 8 = 0$

$3y = -8$ or $3y = -8$

$y = -\frac{8}{3}$ or $y = -\frac{8}{3}$

The solution set is $\left\{-\frac{8}{3}\right\}$.

5. $x(2x - 1) = 10$

$2x^2 - x = 10$

$2x^2 - x - 10 = 0$

$(2x - 5)(x + 2) = 0$

$2x - 5 = 0$ or $x + 2 = 0$

$2x = 5$ or $x = -2$

$x = \frac{5}{2}$ or $x = -2$

The solution set is $\left\{-2, \frac{5}{2}\right\}$.

6. $(x + 7)(x - 5) = 13$

$x^2 + 2x - 35 = 13$

$x^2 + 2x - 48 = 0$

$(x + 8)(x - 6) = 0$

$x + 8 = 0$ or $x - 6 = 0$

$x = -8$ or $x = 6$

The solution set is $\{-8, 6\}$.

7. Let r represent the number of rows. Then $2r + 2$ represents the number of columns.

$$\binom{\text{Number}}{\text{of rows}} \cdot \binom{\text{Number}}{\text{of columns}} = 60$$

$r(2r + 2) = 60$

$2r^2 + 2r = 60$

$2r^2 + 2r - 60 = 0$

$2(r^2 + r - 30) = 0$

$(r^2 + r - 30) = 0$

$(r + 6)(r - 5) = 0$

$r + 6 = 0$ or $r - 5 = 0$

$r = -6$ or $r = 5$

Discard $r = -6$. If $r = 5$, then $2r + 2 = 12$. There are 5 rows and 12 columns.

8. Let x represent the uniform width to be cut (cropped). Let $7 - 2x$ represent the cropped length. Let $5 - 2x$ represent the cropped width.

x x x x 5 7

Area = Length · Width

$A = LW$

$15 = (7 - 2x)(5 - 2x)$

$15 = 35 - 24x + 4x^2$

$0 = 4x^2 - 24x + 20$

$0 = 4(x^2 - 6x + 5)$

$0 = x^2 - 6x + 5$

$0 = (x - 5)(x - 1)$

$x - 5 = 0$ or $x - 1 = 0$

$x = 5$ or $x = 1$

Discard $x = 5$ since the width of the photo itself is 5 inches. Therefore, the length of the amount to be cropped is 1 inch.

9. Let x represent the length of a leg of a right triangle and let $x + 1$ represent the length of the other leg of the right triangle. Also let $x + 2$ represent the length of the hypotenuse of the right triangle.

x $x + 2$ $x + 1$

Use the Pythagorean theorem.

$a^2 + b^2 = c^2$

$(x)^2 + (x + 1)^2 = (x + 2)^2$

$x^2 + x^2 + 2x + 1 = x^2 + 4x + 4$

$2x^2 + 2x + 1 = x^2 + 4x + 4$

$x^2 - 2x - 3 = 0$

$(x - 3)(x + 1) = 0$

$x - 3 = 0$ or $x + 1 = 0$

$x = 3$ or $x = -1$

Discard $x = -1$ since length must be positive.
If $x = 3$, then $x + 1 = 4$ and $x + 2 = 5$. The lengths of the three sides of the right triangle are 3 inches, 4 inches, and 5 inches.

CHAPTER 6

Section 6.1

1. $\dfrac{12}{20} = \dfrac{4 \cdot 3}{4 \cdot 5} = \dfrac{3}{5}$

2. $\dfrac{45}{65} = \dfrac{\overset{9}{\cancel{45}}}{\underset{13}{\cancel{65}}} = \dfrac{9}{13}$

3. $\dfrac{-20}{68} = -\dfrac{20}{68} = -\dfrac{4 \cdot 5}{4 \cdot 17} = -\dfrac{5}{17}$

4. $\dfrac{84}{-120} = -\dfrac{84}{120}$

$= -\dfrac{\cancel{2} \cdot \cancel{2} \cdot \cancel{3} \cdot 7}{\cancel{2} \cdot \cancel{2} \cdot 2 \cdot \cancel{3} \cdot 5}$

$= -\dfrac{7}{10}$

5. $\dfrac{18ab}{4a} = \dfrac{\cancel{2} \cdot 3 \cdot 3 \cdot \cancel{a} \cdot b}{\cancel{2} \cdot 2 \cdot \cancel{a}} = \dfrac{9b}{2}$

6. $\dfrac{-6}{24ab^3} = -\dfrac{\overset{1}{\cancel{6}}}{\underset{4}{\cancel{24}}ab^3} = -\dfrac{1}{4ab^3}$

7. $\dfrac{-42x^3y^4}{-60x^5y^4} = \dfrac{\cancel{6} \cdot 7 \cdot \cancel{x^3} \cdot \cancel{y^4}}{\cancel{6} \cdot 10 \cdot \underset{x^2}{\cancel{x^5}} \cdot \cancel{y^4}} = \dfrac{7}{10x^2}$

8. $\dfrac{y^2 - 5y}{y^2 - 25} = \dfrac{y\cancel{(y-5)}}{\cancel{(y-5)}(y+5)} = \dfrac{y}{y+5}$

9. $\dfrac{9x^2 + 24x + 16}{3x + 4} = \dfrac{\cancel{(3x+4)}(3x+4)}{1\cancel{(3x+4)}}$

$\dfrac{3x+4}{1} = 3x + 4$

10. $\dfrac{2x^2 - 7x - 15}{4x^2 + 12x + 9} = \dfrac{\cancel{(2x+3)}(x-5)}{\cancel{(2x+3)}(2x+3)}$

$= \dfrac{x-5}{2x+3}$

11. $\dfrac{3x^3y - 12xy}{x^3 + 3x^2 - 10x} = \dfrac{3xy(x^2-4)}{x(x^2+3x-10)}$

$= \dfrac{3xy(x+2)(x-2)}{x(x+5)(x-2)}$

$= \dfrac{3\cancel{x}y(x+2)\cancel{(x-2)}}{\cancel{x}(x+5)\cancel{(x-2)}}$

$= \dfrac{3y(x+2)}{x+5}$

12. $\dfrac{x^2 - 9}{6x - 2x^2} = \dfrac{(x+3)(x-3)}{2x(3-x)}$

$= \left(\dfrac{x+3}{2x}\right)(-1)$

$= -\dfrac{x+3}{2x}$ or $\dfrac{-x-3}{2x}$

Section 6.2

1. $\dfrac{m}{n^2 - 16} \cdot \dfrac{n-4}{m^3} = \dfrac{\overset{1}{\cancel{m}}\cancel{(n-4)}}{\underset{m^2}{\cancel{m^3}}(n+4)\cancel{(n-4)}}$

$= \dfrac{1}{m^2(n+4)}$

2. $\dfrac{y^2 + 2y}{y+3} \cdot \dfrac{y^2 - 4y - 5}{y^5 - 3y^4 - 10y^3}$

$= \dfrac{y(y+2)}{y+3} \cdot \dfrac{(y-5)(y+1)}{y^3(y^2 - 3y - 10)}$

$= \dfrac{\cancel{y}\cancel{(y+2)}\cancel{(y-5)}(y+1)}{\underset{y^2}{\cancel{y^3}}(y+3)\cancel{(y-5)}\cancel{(y+2)}}$

$= \dfrac{y+1}{y^2(y+3)}$

3. $\dfrac{12x^2 - x - 1}{x^2 + 2x - 8} \cdot \dfrac{3x^2 - 4x - 4}{12x^2 + 11x + 2}$

$= \dfrac{(4x+1)(3x-1)}{(x+4)(x-2)} \cdot \dfrac{(3x+2)(x-2)}{(4x+1)(3x+2)}$

$= \dfrac{\cancel{(4x+1)}(3x-1)\cancel{(3x+2)}\cancel{(x-2)}}{(x+4)\cancel{(x-2)}\cancel{(4x+1)}\cancel{(3x+2)}}$

$= \dfrac{3x-1}{x+4}$

4. $\dfrac{20xy^3}{15x^2y} \div \dfrac{4y^5}{12x^2y} = \dfrac{20xy^3}{15x^2y} \cdot \dfrac{12x^2y}{4y^5}$

$= \dfrac{\overset{5}{\cancel{20}} \cdot \overset{4}{\cancel{12}} \cdot \overset{x}{\cancel{x^3}} \cdot \cancel{y^4}}{\underset{5}{\cancel{15}} \cdot \underset{1}{\cancel{4}} \cdot \cancel{x^2} \cdot \underset{y^2}{\cancel{y^6}}}$

$= \dfrac{20x}{5y^2} = \dfrac{4x}{y^2}$

5. $\dfrac{2y^2 + 18}{2y + 12} \div \dfrac{y^4 - 81}{y^2 + 3y - 18}$

$= \dfrac{2y^2 + 18}{2y + 12} \cdot \dfrac{y^2 + 3y - 18}{y^4 - 81}$

$= \dfrac{2(y^2+9)}{2(y+6)} \cdot \dfrac{(y+6)(y-3)}{(y^2+9)(y+3)(y-3)}$

$= \dfrac{\overset{1}{\cancel{2}}\cancel{(y^2+9)}\cancel{(y+6)}\cancel{(y-3)}}{\cancel{2}\cancel{(y^2+9)}\cancel{(y+6)}(y+3)\cancel{(y-3)}}$

$= \dfrac{1}{y+3}$

6. $\dfrac{10y^4 - y^3 - 2y^2}{25y^2 + 20y + 4} \div (2y - 1)$

$= \dfrac{10y^4 - y^3 - 2y^2}{25y^2 + 20y + 4} \cdot \dfrac{1}{2y-1}$

$= \dfrac{y^2(10y^2 - y - 2)}{(5y+2)(5y+2)} \cdot \dfrac{1}{2y-1}$

$$= \frac{y^2(5y + 2)(2y - 1)}{(5y + 2)(5y + 2)} \cdot \frac{1}{2y - 1}$$

$$= \frac{y^2\cancel{(5y + 2)}\cancel{(2y - 1)}}{\cancel{(5y + 2)}(5y + 2)\cancel{(2y - 1)}}$$

$$= \frac{y^2}{5y + 2}$$

7. $\dfrac{3x + 6}{2x^2 + 5x + 3} \cdot \dfrac{x^2 + 4}{x^2 + x - 2} \div \dfrac{3xy}{2x^2 + x - 3}$

$$= \frac{3x + 6}{2x^2 + 5x + 3} \cdot \frac{x^2 + 4}{x^2 + x - 2} \cdot \frac{2x^2 + x - 3}{3xy}$$

$$= \frac{3(x + 2)}{(2x + 3)(x + 1)} \cdot \frac{x^2 + 4}{(x + 2)(x + 1)} \cdot \frac{(2x + 3)(x - 1)}{3xy}$$

$$= \frac{\cancel{3}\cancel{(x + 2)}(x^2 + 4)\cancel{(2x + 3)}\cancel{(x - 1)}}{\cancel{3}xy\cancel{(2x + 3)}(x + 1)\cancel{(x + 2)}\cancel{(x - 1)}}$$

$$= \frac{x^2 + 4}{xy(x + 1)}$$

Section 6.3

1. The LCD is 30.

$$\frac{13}{15} - \frac{3}{10} = \frac{13}{15}\left(\frac{2}{2}\right) - \frac{3}{10}\left(\frac{3}{3}\right)$$

$$= \frac{26}{30} - \frac{9}{30}$$

$$= \frac{26 - 9}{30}$$

$$= \frac{17}{30}$$

2. The LCD is 24.

$$\frac{3}{8} + \frac{1}{6} - \frac{3}{4} = \frac{3}{8}\left(\frac{3}{3}\right) + \frac{1}{6}\left(\frac{4}{4}\right) - \frac{3}{4}\left(\frac{6}{6}\right)$$

$$= \frac{9}{24} + \frac{4}{24} - \frac{18}{24}$$

$$= \frac{9 + 4 - 18}{24}$$

$$= \frac{-5}{24}$$

$$= -\frac{5}{24}$$

3. Find the LCD of 20 and 24.

$20 = 2 \cdot 2 \cdot 5$

$24 = 2 \cdot 2 \cdot 2 \cdot 3$

$\text{LCD} = 2 \cdot 2 \cdot 2 \cdot 3 \cdot 5 = 120$

$$\frac{7}{24} + \frac{3}{20} = \frac{7}{24}\left(\frac{5}{5}\right) + \frac{3}{20}\left(\frac{6}{6}\right)$$

$$= \frac{35}{120} + \frac{18}{120}$$

$$= \frac{53}{120}$$

4. $\dfrac{x - 4}{3} - \dfrac{x - 2}{9} = \left(\dfrac{x - 4}{3}\right)\left(\dfrac{3}{3}\right) - \dfrac{x - 2}{9}$

$$= \frac{3(x - 4)}{9} - \frac{x - 2}{9}$$

$$= \frac{3(x - 4) - (x - 2)}{9}$$

$$= \frac{3x - 12 - x + 2}{9}$$

$$= \frac{2x - 10}{9}$$

5. $\text{LCD} = 2 \cdot 2 \cdot 3 \cdot 3 = 36.$

$$\frac{y - 5}{12} + \frac{y + 3}{18} - \frac{y - 5}{9}$$

$$= \left(\frac{y - 5}{12}\right)\left(\frac{3}{3}\right) + \left(\frac{y + 3}{18}\right)\left(\frac{2}{2}\right) - \left(\frac{y - 5}{9}\right)\left(\frac{4}{4}\right)$$

$$= \frac{3(y - 5)}{36} + \frac{2(y + 3)}{36} - \frac{4(y - 5)}{36}$$

$$= \frac{3(y - 5) + 2(y + 3) - 4(y - 5)}{36}$$

$$= \frac{3y - 15 + 2y + 6 - 4y + 20}{36}$$

$$= \frac{y + 11}{36}$$

6. $\text{LCD} = 2 \cdot 2 \cdot 3 \cdot 3 = 36.$

$$\frac{y + 2}{4} + \frac{3y - 1}{6} - \frac{y - 4}{18}$$

$$= \left(\frac{y + 2}{4}\right)\left(\frac{9}{9}\right) + \left(\frac{3y - 1}{6}\right)\left(\frac{6}{6}\right) - \left(\frac{y - 4}{18}\right)\left(\frac{2}{2}\right)$$

$$= \frac{9(y + 2)}{36} + \frac{6(3y - 1)}{36} - \frac{2(y - 4)}{36}$$

$$= \frac{9(y + 2) + 6(3y - 1) - 2(y - 4)}{36}$$

$$= \frac{9y + 18 + 18y - 6 - 2y + 8}{36}$$

$$= \frac{25y + 20}{36}$$

7. $\text{LCD} = 15ab.$

$$\frac{7}{3a} + \frac{4}{5b} = \frac{7}{3a}\left(\frac{5b}{5b}\right) + \left(\frac{4}{5b}\right)\left(\frac{3a}{3a}\right)$$

$$= \frac{35b}{15ab} + \frac{12a}{15ab}$$

$$= \frac{35b + 12a}{15ab}$$

8. LCD $= 30xy^2$.

$$\frac{5}{6xy} - \frac{3}{10y^2} = \frac{5}{6xy}\left(\frac{5y}{5y}\right) - \left(\frac{3}{10y^2}\right)\left(\frac{3x}{3x}\right)$$
$$= \frac{25y}{30xy^2} - \frac{9x}{30xy^2}$$
$$= \frac{25y - 9x}{30xy^2}$$

9. LCD $= y(y + 1)$.

$$\frac{8}{y} + \frac{2y}{y+1} = \frac{8}{y}\left(\frac{y+1}{y+1}\right) + \left(\frac{2y}{y+1}\right)\left(\frac{y}{y}\right)$$
$$= \frac{8(y+1)}{y(y+1)} + \frac{2y^2}{y(y+1)}$$
$$= \frac{8y+8}{y(y+1)} + \frac{2y^2}{y(y+1)}$$
$$= \frac{8y + 8 + 2y^2}{y(y+1)}$$
$$= \frac{2y^2 + 8y + 8}{y(y+1)}$$
$$= \frac{2(y^2 + 4y + 4)}{y(y+1)}$$
$$= \frac{2(y+2)^2}{y(y+1)}$$

10. $$\frac{4y}{y+5} - 7 = \frac{4y}{y+5} - 7\left(\frac{y+5}{y+5}\right)$$
$$= \frac{4y}{y+5} - 7\left(\frac{y+5}{y+5}\right)$$
$$= \frac{4y}{y+5} - \frac{7(y+5)}{y+5}$$
$$= \frac{4y - 7(y+5)}{y+5}$$
$$= \frac{4y - 7y - 35}{y+5}$$
$$= \frac{-3y - 35}{y+5}$$

Section 6.4

1. LCD $= y(y - 3)$.

$$\frac{8}{y^2 - 3y} + \frac{2}{y} = \frac{8}{y(y-3)} + \frac{2}{y}$$
$$= \frac{8}{y(y-3)} + \left(\frac{2}{y}\right)\left(\frac{y-3}{y-3}\right)$$
$$= \frac{8}{y(y-3)} + \frac{2(y-3)}{y(y-3)}$$
$$= \frac{8 + 2(y-3)}{y(y-3)}$$
$$= \frac{8 + 2y - 6}{y(y-3)}$$
$$= \frac{2y + 2}{y(y-3)}$$
$$= \frac{2(y+1)}{y(y-3)}$$

2. $$\frac{2x}{x^2 - 16} - \frac{3}{x-4} = \frac{2x}{(x+4)(x-4)} - \frac{3}{x-4}$$

The LCD $= (x + 4)(x - 4)$.

$$= \frac{2x}{(x+4)(x-4)} - \left(\frac{3}{x-4}\right)\left(\frac{x+4}{x+4}\right)$$
$$= \frac{2x}{(x+4)(x-4)} - \frac{3(x+4)}{(x-4)(x+4)}$$
$$= \frac{2x - 3(x+4)}{(x+4)(x-4)}$$
$$= \frac{2x - 3x - 12}{(x+4)(x-4)}$$
$$= \frac{-x - 12}{(x+4)(x-4)}$$

3. $$\frac{x}{x^2 + 6x + 8} + \frac{6}{x^2 + x - 12}$$
$$= \frac{x}{(x+4)(x+2)} + \frac{6}{(x+4)(x-3)}$$

LCD $= (x + 4)(x + 2)(x - 3)$.

$$\frac{x}{(x+4)(x+2)}\left(\frac{x-3}{x-3}\right) + \frac{6}{(x+4)(x-3)}\left(\frac{x+2}{x+2}\right)$$
$$= \frac{x(x-3)}{(x+4)(x+2)(x-3)}$$
$$+ \frac{6(x+2)}{(x+4)(x+2)(x-3)}$$
$$= \frac{x(x-3) + 6(x+2)}{(x+4)(x+2)(x-3)}$$
$$= \frac{x^2 - 3x + 6x + 12}{(x+4)(x+2)(x-3)}$$
$$= \frac{x^2 + 3x + 12}{(x+4)(x+2)(x-3)}$$

4. $$\frac{3}{y-2} - \frac{y}{y^2 - 4} + \frac{y^2}{y^4 - 16}$$
$$= \frac{3}{y-2} - \frac{y}{(y+2)(y-2)}$$
$$+ \frac{y^2}{(y^2+4)(y+2)(y-2)}$$

LCD $= (y + 2)(y - 2)(y^2 + 4)$.

$$\frac{3}{y-2}\left[\frac{(y+2)(y^2+4)}{(y+2)(y^2+4)}\right] - \frac{y}{(y+2)(y-2)}\left(\frac{y^2+4}{y^2+4}\right)$$
$$+ \frac{y^2}{(y^2+4)(y+2)(y-2)}$$
$$= \frac{3(y+2)(y^2+4)}{(y-2)(y+2)(y^2+4)}$$

$$- \frac{y(y^2+4)}{(y+2)(y-2)(y^2+4)}$$

$$+ \frac{y^2}{(y^2+4)(y+2)(y-2)}$$

$$= \frac{3(y+2)(y^2+4) - y(y^2+4) + y^2}{(y-2)(y+2)(y^2+4)}$$

$$= \frac{3(y^3+4y+2y^2+8) - y^3 - 4y + y^2}{(y-2)(y+2)(y^2+4)}$$

$$= \frac{3y^3+12y+6y^2+24 - y^3 - 4y + y^2}{(y-2)(y+2)(y^2+4)}$$

$$= \frac{2y^3+7y^2+8y+24}{(y-2)(y+2)(y^2+4)}$$

5. $\dfrac{\frac{b}{3}}{\frac{6}{ab}} = \dfrac{b}{3} \div \dfrac{6}{ab} = \dfrac{b}{3} \cdot \dfrac{ab}{6} = \dfrac{ab^2}{18}$

6. $\dfrac{\frac{2}{3}+\frac{1}{6}}{\frac{3}{4}-\frac{1}{8}}$ LCM $= 2 \cdot 2 \cdot 2 \cdot 3 = 24.$

$$\frac{\frac{2}{3}+\frac{1}{6}}{\frac{3}{4}-\frac{1}{8}} = \frac{24}{24} \cdot \frac{\left(\frac{2}{3}+\frac{1}{6}\right)}{\left(\frac{3}{4}-\frac{1}{8}\right)}$$

$$= \frac{24\left(\frac{2}{3}+\frac{1}{6}\right)}{24\left(\frac{3}{4}-\frac{1}{8}\right)}$$

$$= \frac{24\left(\frac{2}{3}\right)+24\left(\frac{1}{6}\right)}{24\left(\frac{3}{4}\right)-24\left(\frac{1}{8}\right)}$$

$$= \frac{16+4}{18-3} = \frac{20}{15} = \frac{4 \cdot \cancel{5}}{3 \cdot \cancel{5}} = \frac{4}{3}$$

7. $\dfrac{\frac{4}{a^2}+\frac{3}{b^2}}{\frac{2}{a}-\frac{5}{b}}$ LCM $= a^2b^2.$

$$\frac{\frac{4}{a^2}+\frac{3}{b^2}}{\frac{2}{a}-\frac{5}{b}} = \frac{a^2b^2}{a^2b^2} \cdot \frac{\left(\frac{4}{a^2}+\frac{3}{b^2}\right)}{\left(\frac{2}{a}-\frac{5}{b}\right)}$$

$$= \frac{a^2b^2\left(\frac{4}{a^2}+\frac{3}{b^2}\right)}{a^2b^2\left(\frac{2}{a}-\frac{5}{b}\right)}$$

$$= \frac{a^2b^2\left(\frac{4}{a^2}\right)+a^2b^2\left(\frac{3}{b^2}\right)}{a^2b^2\left(\frac{2}{a}\right)-a^2b^2\left(\frac{5}{b}\right)}$$

$$= \frac{\frac{4a^2b^2}{a^2}+\frac{3a^2b^2}{b^2}}{\frac{2a^2b^2}{a}-\frac{5a^2b^2}{b}}$$

$$= \frac{4b^2+3a^2}{2ab^2-5a^2b}$$

$$= \frac{4b^2+3a^2}{ab(2b-5a)}$$

8. $\dfrac{\frac{1}{a}+\frac{1}{b}}{5}$ LCM $= ab.$

$$\frac{\frac{1}{a}+\frac{1}{b}}{5} = \frac{ab}{ab} \cdot \frac{\left(\frac{1}{a}+\frac{1}{b}\right)}{5}$$

$$= \frac{ab\left(\frac{1}{a}+\frac{1}{b}\right)}{ab(5)}$$

$$= \frac{ab\left(\frac{1}{a}\right)+ab\left(\frac{1}{b}\right)}{5ab}$$

$$= \frac{b+a}{5ab}$$

9. $\dfrac{-4}{\frac{2}{a}-\frac{5}{b}}$ LCM $= ab.$

$$\frac{-4}{\frac{2}{a}-\frac{5}{b}} = \frac{ab}{ab} \cdot \frac{(-4)}{\frac{2}{a}-\frac{5}{b}}$$

$$= \frac{-4ab}{ab\left(\frac{2}{a}-\frac{5}{b}\right)}$$

$$= \frac{-4ab}{ab\left(\frac{2}{a}\right)-ab\left(\frac{5}{b}\right)}$$

$$= \frac{-4ab}{2b-5a}$$

10. $2 - \dfrac{x}{1+\frac{3}{x}}$

$$\left(2 - \frac{x}{1+\frac{3}{x}}\right) \cdot \left(\frac{x}{x}\right) = \frac{2x}{x} - \frac{x^2}{x+3}$$

$$= \frac{2x}{x}\left(\frac{x+3}{x+3}\right) - \frac{x^2}{x+3}\left(\frac{x}{x}\right)$$

$$= \frac{2x(x+3)}{x(x+3)} - \frac{x^3}{x(x+3)}$$

$$= \frac{2x(x+3) - x^3}{x(x+3)}$$

$$= \frac{2x^2 + 6x - x^3}{x(x+3)}$$

$$= \frac{\cancel{x}(2x + 6 - x^2)}{\cancel{x}(x+3)}$$

$$= \frac{2x + 6 - x^2}{x+3}$$

Section 6.5

1.

$$\begin{array}{r|l} & 2x + 1 \\ \hline x + 3 & 2x^2 + 7x + 3 \\ & \underline{2x^2 + 6x} \\ & x + 3 \\ & \underline{x + 3} \\ & 0 \end{array}$$

$$\frac{2x^2 + 7x + 3}{x + 3} = 2x + 1$$

2.

$$\begin{array}{r|l} & 3x + 2 \\ \hline x - 4 & 3x^2 - 10x - 4 \\ & \underline{3x^2 - 12x} \\ & 2x - 4 \\ & \underline{2x - 8} \\ & 4 \end{array}$$

$$\frac{3x^2 - 10x - 4}{x - 4} = 3x + 2 + \frac{4}{x - 4}$$

3.

$$\begin{array}{r|l} & y^2 - 3y + 9 \\ \hline y + 3 & y^3 + 0y^2 + 0y + 27 \\ & \underline{y^3 + 3y^2} \\ & -3y^2 + 0y + 27 \\ & \underline{-3y^2 - 9y} \\ & 9y + 27 \\ & \underline{9y + 27} \\ & 0 \end{array}$$

$$\frac{y^3 + 27}{y + 3} = y^2 - 3y + 9$$

4.

$$\begin{array}{r|l} & x - 4 \\ \hline x^2 + 3x & x^3 - x^2 - 9x + 2 \\ & \underline{x^3 + 3x^2} \\ & -4x^2 - 9x + 2 \\ & \underline{-4x^2 - 12x} \\ & 3x + 2 \end{array}$$

$$\frac{x^3 - x^2 - 9x + 2}{x^2 + 3x} = x - 4 + \frac{3x + 2}{x^2 + 3x}$$

5. $(3x^4 - 7x^3 + 6x^2 - 3x - 10) \div (x - 2)$

$$\begin{array}{r|rrrrr} 2 & 3 & -7 & 6 & -3 & -10 \\ & & 6 & -2 & 8 & 10 \\ \hline & 3 & -1 & 4 & 5 & 0 \end{array}$$

Quotient: $3x^3 - x^2 + 4x + 5$

Section 6.6

1. $\dfrac{x-2}{2} + \dfrac{x-2}{4} = \dfrac{x+3}{8}$

Multiply both sides by 8.

$$8\left(\frac{x-2}{2} + \frac{x-2}{4}\right) = 8\left(\frac{x+3}{8}\right)$$
$$4(x-2) + 2(x-2) = x + 3$$
$$4x - 8 + 2x - 4 = x + 3$$
$$6x - 12 = x + 3$$
$$5x - 12 = 3$$
$$5x = 15$$
$$x = 3$$

The solution set is $\{3\}$.

2. $\dfrac{5}{n} + \dfrac{1}{3} = \dfrac{7}{n}$ $\quad n \neq 0$

Multiply both sides by $3n$.

$$3n\left(\frac{5}{n} + \frac{1}{3}\right) = 3n\left(\frac{7}{n}\right)$$
$$15 + n = 21$$
$$n = 6$$

The solution set is $\{6\}$.

3. $\dfrac{30 - x}{x} = 3 + \dfrac{2}{x}$ $\quad x \neq 0$

Multiply both sides by x.

$$x\left(\frac{30 - x}{x}\right) = x\left(3 + \frac{2}{x}\right)$$
$$30 - x = 3x + 2$$
$$30 - 4x = 2$$
$$-4x = -28$$
$$x = 7$$

The solution set is $\{7\}$.

4. $\dfrac{3}{x-2} = \dfrac{6}{x+2}$ $\quad x \neq 2, x \neq -2$

Multiply both sides by $(x + 2)(x - 2)$.

$$(x+2)(x-2)\left(\frac{3}{x-2}\right) = (x+2)(x-2)\left(\frac{6}{x+2}\right)$$
$$3(x + 2) = 6(x - 2)$$
$$3x + 6 = 6x - 12$$
$$-3x + 6 = -12$$

$$-3x = -18$$
$$x = 6$$

The solution set is {6}.

5. $\dfrac{x}{x-4} + \dfrac{1}{2} = \dfrac{4}{x-4} \qquad x \neq 4$

Multiply both sides by $2(x - 4)$.

$$2(x-4)\left(\frac{x}{x-4} + \frac{1}{2}\right) = 2(x-4)\left(\frac{4}{x-4}\right)$$
$$2x + x - 4 = 8$$
$$3x - 4 = 8$$
$$3x = 12$$
$$x = 4$$

Because $x \neq 4$, there is no solution. The solution set is $\varnothing$.

6. $\dfrac{7}{x+3} = \dfrac{2}{x-8} \qquad x \neq 8, x \neq -3$

Cross multiply.

$$7(x - 8) = 2(x + 3)$$
$$7x - 56 = 2x + 6$$
$$5x - 56 = 6$$
$$5x = 62$$
$$x = \frac{62}{5}$$

The solution set is $\left\{\dfrac{62}{5}\right\}$.

7. $\dfrac{y}{5} = \dfrac{3}{y-2} \qquad y \neq 2$

Cross multiply.

$$y(y - 2) = 3(5)$$
$$y^2 - 2y = 15$$
$$y^2 - 2y - 15 = 0$$
$$(y - 5)(y + 3) = 0$$
$$y - 5 = 0 \quad \text{or} \quad y + 3 = 0$$
$$y = 5 \quad \text{or} \quad y = -3$$

The solution set is $\{-3, 5\}$.

8. Let x represent the actual number of millimeters between the blood vessels.

$\dfrac{\text{Centimeters}}{\text{Millimeters}}: \quad \dfrac{2}{1.5} = \dfrac{8}{x}$

$$2x = 8(1.5)$$
$$2x = 12$$
$$x = 6$$

The actual number of millimeters between blood vessels is 6 mm.

9. Let x represent the amount of money one person receives. Then $2000 - x$ represents the amount for the other person.

$$\frac{x}{2000 - x} = \frac{1}{4}$$

Cross multiply.

$$4x = 2000 - x$$
$$5x = 2000$$
$$x = 400$$

If $x = 400$, then $2000 - x = 1600$. One person receives \$400 and the other person receives \$1600.

Section 6.7

1. $\dfrac{x}{3x+6} + \dfrac{8}{x^2-4} = \dfrac{1}{3} \qquad x \neq 2, x \neq -2$

$$\frac{x}{3(x+2)} + \frac{8}{(x+2)(x-2)} = \frac{1}{3}$$

Multiply both sides by $3(x + 2)(x - 2)$.

$$3(x+2)(x-2)\left[\frac{x}{3(x+2)} + \frac{8}{(x+2)(x-2)}\right]$$
$$= 3(x+2)(x-2)\left(\frac{1}{3}\right)$$
$$x(x - 2) + 3(8) = (x + 2)(x - 2)$$
$$x^2 - 2x + 24 = x^2 - 4$$
$$-2x + 24 = -4$$
$$-2x = -28$$
$$x = 14$$

The solution set is {14}.

2. $\dfrac{5}{x-2} - \dfrac{2}{3x+1} = \dfrac{x+1}{3x^2-5x-2}$

$x \neq 2, x \neq -\dfrac{1}{3}$

$$\frac{5}{x-2} - \frac{2}{3x+1} = \frac{x+1}{(3x+1)(x-2)}$$

Multiply both sides by $(3x + 1)(x - 2)$.

$$(3x+1)(x-2)\left[\frac{5}{x-2} - \frac{2}{3x+1}\right] = x + 1$$
$$5(3x + 1) - 2(x - 2) = x + 1$$
$$15x + 5 - 2x + 4 = x + 1$$
$$13x + 9 = x + 1$$
$$12x + 9 = 1$$
$$12x = -8$$
$$x = \frac{-8}{12} = -\frac{2}{3}$$

The solution set is $\left\{-\dfrac{2}{3}\right\}$.

3. $2 - \frac{6}{x+3} = \frac{18}{x^2+3x} \qquad x \neq 0, x \neq -3$

$2 - \frac{6}{x+3} = \frac{18}{x(x+3)}$

Multiply both sides by $x(x + 3)$.

$$x(x+3)\left(2 - \frac{6}{x+3}\right) = x(x+3)\left[\frac{18}{x(x+3)}\right]$$

$$\begin{aligned} 2x(x+3) - 6x &= 18 \\ 2x^2 + 6x - 6x &= 18 \\ 2x^2 &= 18 \\ 2x^2 - 18 &= 0 \\ x^2 - 9 &= 0 \\ (x+3)(x-3) &= 0 \end{aligned}$$

$x + 3 = 0 \quad \text{or} \quad x - 3 = 0$

$x = -3 \quad \text{or} \quad x = 3$

Discard $x = -3$, since -3 is a restricted value. The solution set is $\{3\}$.

4. Solve $S = \frac{a}{1-r}$ for r.

$\frac{S}{1} = \frac{a}{1-r}$

Cross multiply.

$$\begin{aligned} S(1-r) &= 1a \\ S - Sr &= a \\ -Sr &= a - S \\ Sr &= -a + S \\ Sr &= S - a \\ r &= \frac{S-a}{S} \end{aligned}$$

5. Let r represent the rate of the car. Then $r + 480$ represents the rate of the plane.

The information is recorded in the following table:

	Distance	**Rate**	**Time** $\left(t = \frac{d}{r}\right)$
Plane	2220	$r + 480$	$\frac{2200}{r+480}$
Car	280	r	$\frac{280}{r}$

$\begin{pmatrix}\text{Time for}\\ \text{plane}\end{pmatrix} = \begin{pmatrix}\text{Time for}\\ \text{car}\end{pmatrix}$

$\frac{2200}{r+480} = \frac{280}{r} \qquad r \neq 0$

Cross multiply.

$$\begin{aligned} 2200r &= 280(r + 480) \\ 2200r &= 280r + 134{,}400 \\ 1920r &= 134{,}400 \\ r &= 70 \end{aligned}$$

If $r = 70$, then $r + 480 = 550$. The rate of the car is 70 mph, and the rate of the plane is 550 mph.

6. Let r represent the rate of the tour bus. Then $r + 15$ represents the rate of the car.

The information is recorded in the following table:

	Distance	**Rate**	**Time** $\left(t = \frac{d}{r}\right)$
Tour bus	220	r	$\frac{220}{r}$
Car	210	$r + 15$	$\frac{210}{r+15}$

$\begin{pmatrix}\text{Time for}\\ \text{tour bus}\end{pmatrix} = 1 + \begin{pmatrix}\text{Time}\\ \text{for car}\end{pmatrix}$

$\frac{220}{r} = 1 + \frac{210}{r+15} \qquad r \neq 0$

Multiply both sides by $r(r + 15)$.

$$r(r+15)\left(\frac{220}{r}\right) = r(r+15)\left(1 + \frac{210}{r+15}\right)$$

$$\begin{aligned} 220(r+15) &= r(r+15) + 210r \\ 220r + 3300 &= r^2 + 15r + 210r \\ 220r + 3300 &= r^2 + 225r \\ 0 &= r^2 + 5r - 3300 \\ 0 &= (r+60)(r-55) \end{aligned}$$

$r + 60 = 0 \quad \text{or} \quad r - 55 = 0$

$r = -60 \quad \text{or} \quad r = 55$

Discard $r = -60$. If $r = 55$, then $r + 15 = 70$. The tour bus travels 4 hours at 55 mph, and the car travels 3 hours at 70 mph.

7. Maria's rate is $\frac{1}{120}$ of the job per minute. Stephanie's rate is $\frac{1}{200}$ of the job per minute.

Let m represent the number of minutes they work together. Then $\frac{1}{m}$ represents the rate they work together.

$\begin{pmatrix}\text{Maria's}\\ \text{rate}\end{pmatrix} + \begin{pmatrix}\text{Stephanie's}\\ \text{rate}\end{pmatrix} = \begin{pmatrix}\text{Combined}\\ \text{rate}\end{pmatrix}$

$$\frac{1}{120} + \frac{1}{200} = \frac{1}{m} \qquad m \neq 0$$

$$600m\left(\frac{1}{120} + \frac{1}{200}\right) = 600m\left(\frac{1}{m}\right)$$

$$\begin{aligned} 5m + 3m &= 600 \\ 8m &= 600 \end{aligned}$$

$$m = 75$$

It will take Maria and Stephanie 75 minutes to clean the house together.

8. Let h represent the number of hours it takes Mayra to print and fold by herself. Then $\frac{1}{h}$ represents the rate Mayra works. Since Josh completes the job in 4 hours, his rate is $\frac{1}{4}$.

	Time	Rate
Mayra	h	$\frac{1}{h}$
Josh	4	$\frac{1}{4}$
Together	$2\frac{2}{3} = \frac{8}{3}$	$\frac{1}{\frac{8}{3}} = \frac{3}{8}$

$$\begin{pmatrix}\text{Mayra's}\\ \text{rate}\end{pmatrix} + \begin{pmatrix}\text{Josh's}\\ \text{rate}\end{pmatrix} = \begin{pmatrix}\text{Combined}\\ \text{rate}\end{pmatrix}$$

$$\frac{1}{h} + \frac{1}{4} = \frac{3}{8}$$

$$8h\left(\frac{1}{h} + \frac{1}{4}\right) = 8h\left(\frac{3}{8}\right) \qquad h \neq 0$$

$$8 + 2h = 3h$$

$$8 = h$$

It will take Mayra 8 hours to complete the job herself.

9. Let h represent the number of hours Franco can do the job himself. Then $\frac{1}{h}$ represents Franco's rate. John worked for 5 hours (2 hours before Franco and 3 hours after). John's working rate is $\frac{1}{8}$, because he can do the job himself in 8 hours.

$$\begin{pmatrix}\text{Fractional part}\\ \text{of job done}\end{pmatrix} = \begin{pmatrix}\text{Working}\\ \text{rate}\end{pmatrix} \times (\text{Time})$$

$$\begin{pmatrix}\text{John's fractional}\\ \text{part of job done}\end{pmatrix} = \frac{1}{8}(5) = \frac{5}{8}$$

$$\begin{pmatrix}\text{Franco's fractional}\\ \text{part of job done}\end{pmatrix} = \frac{1}{h}(3) = \frac{3}{h}$$

Use the following chart to organize information:

	Time to do entire job	Working rate	Time worked	Fractional part of the job done
John	8	$\frac{1}{8}$	5	$\frac{5}{8}$
Franco	h	$\frac{1}{h}$	3	$\frac{3}{h}$

$$\begin{pmatrix}\text{John's fractional}\\ \text{part of job}\end{pmatrix} + \begin{pmatrix}\text{Franco's fractional}\\ \text{part of job}\end{pmatrix} = \begin{pmatrix}\text{Total job}\\ \text{done}\end{pmatrix}$$

$$\frac{5}{8} + \frac{3}{h} = 1$$

$$8h\left(\frac{5}{8} + \frac{3}{h}\right) = 8h(1) \qquad h \neq 0$$

$$5h + 24 = 8h$$

$$24 = 3h$$

$$8 = h$$

It will take Franco 8 hours to detail the boat by himself.

CHAPTER 7

Section 7.1

1. (a) $6^{-2} \cdot 6^5 = 6^{-2+5} = 6^3 = 216$

(b) $(2^{-2})^{-2} = 2^{(-2)(-2)} = 2^4 = 16$

(c) $(2^{-2} \cdot 3^{-1})^{-1} = (2^{-2})^{-1} \cdot (3^{-1})^{-1} = 2^2 \cdot 3^1 = 4 \cdot 3 = 12$

(d) $\left(\frac{3^{-3}}{2^{-2}}\right)^{-1} = \frac{(3^{-3})^{-1}}{(2^{-2})^{-1}} = \frac{3^3}{2^2} = \frac{27}{4}$

(e) $\frac{4^{-3}}{4^{-6}} = 4^{-3-(-6)} = 4^3 = 64$

2. (a) $y^4 \cdot y^{-1} = y^{4-1} = y^3$

(b) $(x^3)^{-2} = x^{-6} = \frac{1}{x^6}$

(c) $(a^{-2}b^3)^{-3} = (a^{-2})^{-3} \cdot (b^3)^{-3} = a^{(-2)(-3)} \cdot b^{3(-3)} = a^6b^{-9} = \frac{a^6}{b^9}$

(d) $\left(\frac{x^2}{y^3}\right)^{-3} = \frac{(x^2)^{-3}}{(y^3)^{-3}} = \frac{x^{-6}}{y^{-9}} = \frac{y^9}{x^6}$

(e) $\frac{y^{-2}}{y^{-5}} = y^{-2-(-5)} = y^3$

3. (a) $(2xy^{-3})(5x^{-2}y^5) = 10x^{1-2}y^{-3+5} = 10x^{-1}y^2 = \frac{10y^2}{x}$

(b) $\dfrac{8x^2y}{-2x^{-3}y^4} = -4x^{2-(-3)}y^{1-4}$

$= -4x^5y^{-3} = \dfrac{-4x^5}{y^3}$

(c) $\left(\dfrac{12a^{-2}b^3}{3a^3b^{-4}}\right)^{-1} = (4a^{-2-3}b^{3-(-4)})^{-1}$

$= (4a^{-5}b^7)^{-1}$

$= (4)^{-1}(a^{-5})^{-1}(b^7)^{-1}$

$= 4^{-1}a^5b^{-7}$

$= \dfrac{a^5}{4b^7}$

4. $3^{-2} + 4^{-1} = \dfrac{1}{3^2} + \dfrac{1}{4^1} = \dfrac{1}{9} + \dfrac{1}{4}.$

$= \dfrac{1}{9}\left(\dfrac{4}{4}\right) + \dfrac{1}{4}\left(\dfrac{9}{9}\right)$

$= \dfrac{4}{36} + \dfrac{9}{36} = \dfrac{13}{36}$

5. $(2^{-3} - 3^{-1}) = \left(\dfrac{1}{2^3} - \dfrac{1}{3^1}\right)^{-1}$

$= \left(\dfrac{1}{8} - \dfrac{1}{3}\right)^{-1} = \left(\dfrac{3}{24} - \dfrac{8}{24}\right)^{-1}$

$= \left(\dfrac{-5}{24}\right)^{-1} = \dfrac{1}{\left(\dfrac{-5}{24}\right)} = -\dfrac{24}{5}$

6. $x^{-3} + y^{-2} = \dfrac{1}{x^3} + \dfrac{1}{y^2}$

$= \dfrac{1}{x^3}\left(\dfrac{y^2}{y^2}\right) + \dfrac{1}{y^2}\left(\dfrac{x^3}{x^3}\right)$

$= \dfrac{y^2}{x^3y^2} + \dfrac{x^3}{x^3y^2} = \dfrac{y^2 + x^3}{x^3y^2}$

Section 7.2

1. (a) $\sqrt{20} = \sqrt{4 \cdot 5} = \sqrt{4}\sqrt{5} = 2\sqrt{5}$

(b) $\sqrt{18} = \sqrt{9 \cdot 2} = \sqrt{9}\sqrt{2} = 3\sqrt{2}$

(c) $\sqrt[3]{32} = \sqrt[3]{8 \cdot 4} = \sqrt[3]{8}\sqrt[3]{4} = 2\sqrt[3]{4}$

(d) $\sqrt[3]{128} = \sqrt[3]{64 \cdot 2} = \sqrt[3]{64}\sqrt[3]{2} = 4\sqrt[3]{2}$

2. (a) $\sqrt{48} = \sqrt{2 \cdot 2 \cdot 2 \cdot 2 \cdot 3}$

$= \sqrt{2^4}\sqrt{3} = 2^2\sqrt{3} = 4\sqrt{3}$

(b) $5\sqrt{32} = 5\sqrt{2 \cdot 2 \cdot 2 \cdot 2 \cdot 2}$

$= 5\sqrt{2^4}\sqrt{2} = 5(2^2)\sqrt{2}$

$= 5(4)\sqrt{2} = 20\sqrt{2}$

(c) $\sqrt[3]{375} = \sqrt[3]{3 \cdot 5 \cdot 5 \cdot 5} = \sqrt[3]{5^3}\sqrt[3]{3} = 5\sqrt[3]{3}$

3. $\sqrt{\dfrac{3}{5}} = \dfrac{\sqrt{3}}{\sqrt{5}} = \dfrac{\sqrt{3}}{\sqrt{5}} \cdot \dfrac{\sqrt{5}}{\sqrt{5}} = \dfrac{\sqrt{15}}{5}$

4. $\dfrac{\sqrt{7}}{\sqrt{12}} = \dfrac{\sqrt{7}}{\sqrt{4}\sqrt{3}} = \dfrac{\sqrt{7}}{2\sqrt{3}} \cdot \dfrac{\sqrt{3}}{\sqrt{3}} = \dfrac{\sqrt{21}}{2(3)} = \dfrac{\sqrt{21}}{6}$

5. (a) $\dfrac{5\sqrt{2}}{3\sqrt{5}} = \dfrac{5\sqrt{2}}{3\sqrt{5}} \cdot \dfrac{\sqrt{5}}{\sqrt{5}} = \dfrac{5\sqrt{10}}{3(5)} = \dfrac{\sqrt{10}}{3}$

(b) $\dfrac{2\sqrt{5}}{7\sqrt{12}} = \dfrac{2\sqrt{5}}{7\sqrt{4}\sqrt{3}} = \dfrac{2\sqrt{5}}{7(2)\sqrt{3}}$

$= \dfrac{\sqrt{5}}{7\sqrt{3}} \cdot \dfrac{\sqrt{3}}{\sqrt{3}} = \dfrac{\sqrt{15}}{7(3)} = \dfrac{\sqrt{15}}{21}$

(c) $\sqrt[3]{\dfrac{3}{4}} = \dfrac{\sqrt[3]{3}}{\sqrt[3]{4}} \cdot \dfrac{\sqrt[3]{2}}{\sqrt[3]{2}} = \dfrac{\sqrt[3]{6}}{\sqrt[3]{8}} = \dfrac{\sqrt[3]{6}}{2}$

(d) $\dfrac{\sqrt[3]{7}}{\sqrt[3]{32}} = \dfrac{\sqrt[3]{7}}{\sqrt[3]{8}\sqrt[3]{4}} = \dfrac{\sqrt[3]{7}}{2\sqrt[3]{4}} \cdot \dfrac{\sqrt[3]{2}}{\sqrt[3]{2}}$

$= \dfrac{\sqrt[3]{14}}{2\sqrt[3]{8}} = \dfrac{\sqrt[3]{14}}{2(2)} = \dfrac{\sqrt[3]{14}}{4}$

6. Use the formula: $S = \sqrt{30Df}$.

Substitute 0.25 for f and 290 for D to find the rate.

$S = \sqrt{30Df}$

$= \sqrt{30(290)(0.25)}$

$= \sqrt{2175}$

≈ 46.6369

To the nearest whole number, the rate is 47 mph.

7. Use the formula: $T = 2\pi\sqrt{\dfrac{L}{32}}$.

Substitute 4.5 for L to find time.

$T = 2\pi\sqrt{\dfrac{L}{32}}$

$= 2\pi\sqrt{\dfrac{4.5}{32}}$

$= 2\pi\sqrt{0.140625}$

$= 2\pi(0.375)$

$= 2.356$

The time to the nearest second is 2.4 seconds.

8. Use the following formulas:

$s = \dfrac{a + b + c}{2}$

$K = \sqrt{s(s - a)(s - b)(s - c)}$

Substitute 14, 18, and 20, respectively, for a, b, and c.

$s = \dfrac{14 + 18 + 20}{2} = \dfrac{52}{2} = 26$

$K = \sqrt{s(s - a)(s - b)(s - c)}$

$= \sqrt{26(26 - 14)(26 - 18)(26 - 20)}$

$= \sqrt{26(12)(8)(6)}$

$= \sqrt{14{,}976} = 122.376$

The area of the triangle to the nearest tenth is 122.4 square inches.

Section 7.3

1. $5\sqrt{12} - 3\sqrt{75} + 6\sqrt{27}$
$= 5\sqrt{4}\sqrt{3} - 3\sqrt{25}\sqrt{3} + 6\sqrt{9}\sqrt{3}$
$= 5(2)\sqrt{3} - 3(5)\sqrt{3} + 6(3)\sqrt{3}$
$= 10\sqrt{3} - 15\sqrt{3} + 18\sqrt{3}$
$= (10 - 15 + 18)\sqrt{3} = 13\sqrt{3}$

2. $\frac{1}{5}\sqrt{8} + \frac{1}{3}\sqrt{18} = \frac{1}{5}\sqrt{4}\sqrt{2} + \frac{1}{3}\sqrt{9}\sqrt{2}$
$= \frac{1}{5}(2)\sqrt{2} + \frac{1}{3}(3)\sqrt{2} = \frac{2}{5}\sqrt{2} + \sqrt{2}$
$= \left(\frac{2}{5} + 1\right)\sqrt{2} = \left(\frac{2}{5} + \frac{5}{5}\right)\sqrt{2}$
$= \frac{7}{5}\sqrt{2}$

3. $\sqrt[3]{2} - 4\sqrt[3]{54} + 3\sqrt[3]{16}$
$= \sqrt[3]{2} - 4\sqrt[3]{27}\sqrt[3]{2} + 3\sqrt[3]{8}\sqrt[3]{2}$
$= \sqrt[3]{2} - 4(3)\sqrt[3]{2} + 3(2)\sqrt[3]{2}$
$= \sqrt[3]{2} - 12\sqrt[3]{2} + 6\sqrt[3]{2}$
$= (1 - 12 + 6)\sqrt[3]{2} = -5\sqrt[3]{2}$

4. (a) $\sqrt{98y^3} = \sqrt{49y^2}\sqrt{2y} = 7y\sqrt{2y}$

(b) $\sqrt{28a^5b^3} = \sqrt{4a^4b^2}\sqrt{7ab} = 2a^2b\sqrt{7ab}$

(c) $\sqrt{240x^3y^4} = \sqrt{2 \cdot 2 \cdot 2 \cdot 2 \cdot 5 \cdot 3 \cdot x^3 \cdot y^4}$
$= \sqrt{16 \cdot 15 \cdot x^3 \cdot y^4}$
$= \sqrt{16x^2y^4}\sqrt{15x} = 4xy^2\sqrt{15x}$

(d) $\sqrt[3]{54a^2b^5} = \sqrt[3]{27b^3}\sqrt[3]{2a^2b^2}$
$= 3b\sqrt[3]{2a^2b^2}$

5. (a) $\sqrt{\frac{3a}{5b}} = \frac{\sqrt{3a}}{\sqrt{5b}} = \frac{\sqrt{3a}}{\sqrt{5b}} \cdot \frac{\sqrt{5b}}{\sqrt{5b}} = \frac{\sqrt{15ab}}{5b}$

(b) $\frac{\sqrt{7}}{\sqrt{18x^3}} = \frac{\sqrt{7}}{\sqrt{9x^2}\sqrt{2x}} \cdot \frac{\sqrt{2x}}{\sqrt{2x}}$
$= \frac{\sqrt{14x}}{\sqrt{36x^4}} = \frac{\sqrt{14x}}{\sqrt{6x^2}}$

(c) $\frac{\sqrt{75a^3}}{\sqrt{8b}} = \frac{\sqrt{25a^2}\sqrt{3a}}{\sqrt{4}\sqrt{2b}} = \frac{5a\sqrt{3a}}{2\sqrt{2b}}$
$= \frac{5a\sqrt{3a}}{2\sqrt{2b}} \cdot \frac{\sqrt{2b}}{\sqrt{2b}} = \frac{5a\sqrt{6ab}}{2(2b)} = \frac{5a\sqrt{6ab}}{4b}$

(d) $\frac{2}{\sqrt[3]{3y^2}} = \frac{2}{\sqrt[3]{3y^2}} \cdot \frac{\sqrt[3]{9y}}{\sqrt[3]{9y}} = \frac{2\sqrt[3]{9y}}{\sqrt[3]{27y^3}} = \frac{2\sqrt[3]{9y}}{3y}$

(e) $\frac{\sqrt[3]{81a^2}}{\sqrt[3]{2b}} = \frac{\sqrt[3]{27}\sqrt[3]{3a^2}}{\sqrt[3]{2b}} = \frac{3\sqrt[3]{3a^2}}{\sqrt[3]{2b}} \cdot \frac{\sqrt[3]{4b^2}}{\sqrt[3]{4b^2}}$
$= \frac{3\sqrt[3]{12a^2b^2}}{\sqrt[3]{8b^3}} = \frac{3\sqrt[3]{12a^2b^2}}{2b}$

Section 7.4

1. (a) $(2\sqrt{6})(2\sqrt{5}) = 2 \cdot 2 \cdot \sqrt{6} \cdot \sqrt{5} = 4\sqrt{30}$

(b) $(5\sqrt{18})(2\sqrt{2}) = 5 \cdot 2 \cdot \sqrt{18} \cdot \sqrt{2}$
$= 10\sqrt{36} = 10(6) = 60$

(c) $(2\sqrt{21})(4\sqrt{3}) = 2 \cdot 4 \cdot \sqrt{21} \cdot \sqrt{3}$
$= 8\sqrt{63} = 8\sqrt{9}\sqrt{7} = 8 \cdot 3 \cdot \sqrt{7}$
$= 24\sqrt{7}$

(d) $(4\sqrt[3]{6})(5\sqrt[3]{9}) = 4 \cdot 5 \cdot \sqrt[3]{54}$
$= 20\sqrt[3]{27}\sqrt[3]{2} = 20(3)\sqrt[3]{2} = 60\sqrt[3]{2}$

2. (a) $\sqrt{2}(\sqrt{10} + \sqrt{8}) = \sqrt{2}\sqrt{10} + \sqrt{2}\sqrt{8}$
$= \sqrt{20} + \sqrt{16}$
$= \sqrt{4}\sqrt{5} + 4 = 2\sqrt{5} + 4$

(b) $4\sqrt{3}(5\sqrt{2} - 3\sqrt{6}) = (4\sqrt{3})(5\sqrt{2}) - (4\sqrt{3})(3\sqrt{6})$
$= 20\sqrt{6} - 12\sqrt{18}$
$= 20\sqrt{6} - 12\sqrt{9}\sqrt{2}$
$= 20\sqrt{6} - 36\sqrt{2}$

(c) $\sqrt{2a}(\sqrt{14a} + \sqrt{12ab}) = (\sqrt{2a})(\sqrt{14a})$
$+ (\sqrt{2a})(\sqrt{12ab})$
$= \sqrt{28a^2} + \sqrt{24a^2b}$
$= \sqrt{4a^2}\sqrt{7} + \sqrt{4a^2}\sqrt{6b}$
$= 2a\sqrt{7} + 2a\sqrt{6b}$

(d) $\sqrt[3]{3}(4\sqrt[3]{9} + \sqrt[3]{81}) = (\sqrt[3]{3})(4\sqrt[3]{9}) + (\sqrt[3]{3})(\sqrt[3]{81})$
$= 4\sqrt[3]{27} + \sqrt[3]{243}$
$= 4(3) + \sqrt[3]{27}\sqrt[3]{9}$
$= 12 + 3\sqrt[3]{9}$

3. (a) $(\sqrt{6} + \sqrt{3})(\sqrt{5} + \sqrt{14})$
$= \sqrt{6}(\sqrt{5} + \sqrt{14}) + \sqrt{3}(\sqrt{5} + \sqrt{14})$
$= \sqrt{6}\sqrt{5} + \sqrt{6}\sqrt{14} + \sqrt{3}\sqrt{5} + \sqrt{3}\sqrt{14}$
$= \sqrt{30} + \sqrt{84} + \sqrt{15} + \sqrt{42}$
$= \sqrt{30} + 2\sqrt{21} + \sqrt{15} + \sqrt{42}$

(b) $(5\sqrt{2} + \sqrt{3})(4\sqrt{2} - 2\sqrt{3})$
$= 5\sqrt{2}(4\sqrt{2} - 2\sqrt{3}) + \sqrt{3}(4\sqrt{2} - 2\sqrt{3})$
$= (5\sqrt{2})(4\sqrt{2}) - (5\sqrt{2})(2\sqrt{3}) + (\sqrt{3})(4\sqrt{2})$
$- (\sqrt{3})(2\sqrt{3})$
$= 20\sqrt{4} - 10\sqrt{6} + 4\sqrt{6} - 2\sqrt{9}$
$= 40 - 6\sqrt{6} - 6 = 34 - 6\sqrt{6}$

(c) $(\sqrt{5} + \sqrt{3})(\sqrt{5} - \sqrt{3})$
$= \sqrt{5}(\sqrt{5} - \sqrt{3}) + \sqrt{3}(\sqrt{5} - \sqrt{3})$
$= (\sqrt{5})(\sqrt{5}) - (\sqrt{5})(\sqrt{3}) + (\sqrt{3})(\sqrt{5})$
$- (\sqrt{3})(\sqrt{3})$
$= 5 - \sqrt{15} + \sqrt{15} - 3$
$= 5 - 3 = 2$

(d) $(\sqrt{a}-\sqrt{b})(\sqrt{a}+\sqrt{b})$

$$= \sqrt{a}(\sqrt{a}+\sqrt{b}) - (\sqrt{b})(\sqrt{a}+\sqrt{b})$$
$$= (\sqrt{a})(\sqrt{a}) + (\sqrt{a})(\sqrt{b}) - (\sqrt{b})(\sqrt{a}) - (\sqrt{b})(\sqrt{b})$$
$$= a + \sqrt{ab} - \sqrt{ab} - b$$
$$= a - b$$

4. $$\frac{5}{\sqrt{7}+\sqrt{3}} = \frac{5}{\sqrt{7}+\sqrt{3}} \cdot \frac{\sqrt{7}-\sqrt{3}}{\sqrt{7}-\sqrt{3}} = \frac{5(\sqrt{7}-\sqrt{3})}{(\sqrt{7}+\sqrt{3})(\sqrt{7}-\sqrt{3})} = \frac{5(\sqrt{7}-\sqrt{3})}{7-3} = \frac{5(\sqrt{7}-\sqrt{3})}{4}$$

5. (a) $$\frac{\sqrt{2}}{\sqrt{6}-4} = \frac{\sqrt{2}}{\sqrt{6}-4} \cdot \frac{\sqrt{6}+4}{\sqrt{6}+4} = \frac{\sqrt{2}(\sqrt{6}+4)}{(\sqrt{6}-4)(\sqrt{6}+4)} = \frac{\sqrt{12}+4\sqrt{2}}{6-16} = \frac{2\sqrt{3}+4\sqrt{2}}{-10}$$
$$= \frac{\overset{1}{\cancel{2}}(\sqrt{3}+2\sqrt{2})}{-\underset{5}{\cancel{10}}} = -\frac{\sqrt{3}+2\sqrt{2}}{5}$$

(b) $$\frac{4}{3\sqrt{2}-5\sqrt{3}} = \frac{4}{3\sqrt{2}-5\sqrt{3}} \cdot \frac{3\sqrt{2}+5\sqrt{3}}{3\sqrt{2}+5\sqrt{3}} = \frac{4(3\sqrt{2}+5\sqrt{3})}{(3\sqrt{2}-5\sqrt{3})(3\sqrt{2}+5\sqrt{3})} = \frac{12\sqrt{2}+20\sqrt{3}}{18-75} = \frac{12\sqrt{2}+20\sqrt{3}}{-57} = -\frac{12\sqrt{2}+20\sqrt{3}}{57}$$

(c) $$\frac{\sqrt{a}-5}{\sqrt{a}+4} = \frac{\sqrt{a}-5}{\sqrt{a}+4} \cdot \frac{\sqrt{a}-4}{\sqrt{a}-4} = \frac{(\sqrt{a}-5)(\sqrt{a}-4)}{(\sqrt{a}+4)(\sqrt{a}-4)} = \frac{a-4\sqrt{a}-5\sqrt{a}+20}{a-16} = \frac{a-9\sqrt{a}+20}{a-16}$$

(d) $$\frac{\sqrt{a}-2\sqrt{b}}{\sqrt{a}-\sqrt{b}} = \frac{\sqrt{a}-2\sqrt{b}}{\sqrt{a}-\sqrt{b}} \cdot \frac{\sqrt{a}+\sqrt{b}}{\sqrt{a}+\sqrt{b}} = \frac{(\sqrt{a}-2\sqrt{b})(\sqrt{a}+\sqrt{b})}{(\sqrt{a}-\sqrt{b})(\sqrt{a}+\sqrt{b})} = \frac{a+\sqrt{ab}-2\sqrt{ab}-2b}{a-b} = \frac{a-\sqrt{ab}-2b}{a-b}$$

Section 7.5

1.
$$\sqrt{3x+1} = 8$$
$$(\sqrt{3x+1})^2 = (8)^2$$
$$3x+1 = 64$$
$$3x = 63$$
$$x = 21$$

Check

$$\sqrt{3(21)+1} \stackrel{?}{=} 8$$
$$\sqrt{63+1} \stackrel{?}{=} 8$$
$$\sqrt{64} \stackrel{?}{=} 8$$
$$8 = 8$$

The solution set is {21}.

2.
$$\sqrt{5x-1} = -3$$
$$(\sqrt{5x-1})^2 = (-3)^2$$
$$5x-1 = 9$$
$$5x = 10$$
$$x = 2$$

Check

$$\sqrt{5(2)-1} \stackrel{?}{=} -3$$
$$\sqrt{10-1} \stackrel{?}{=} -3$$
$$\sqrt{9} \stackrel{?}{=} -3$$
$$3 \neq -3$$

Because the solution does not check, the original equation has no real number solution. Therefore the solution set is $\varnothing$.

3.
$$\sqrt{2x-6} = x-3$$
$$(\sqrt{2x-6})^2 = (x-3)^2$$
$$2x-6 = x^2-6x+9$$
$$0 = x^2-8x+15$$
$$0 = (x-5)(x-3)$$
$$x-5 = 0 \quad \text{or} \quad x-3 = 0$$
$$x = 5 \quad \text{or} \quad x = 3$$

Check $x = 5$

$\sqrt{2(5) - 6} \stackrel{?}{=} 5 - 3$

$\sqrt{10 - 6} \stackrel{?}{=} 2$

$\sqrt{4} \stackrel{?}{=} 2$

$2 = 2$

Check $x = 3$

$\sqrt{2(3) - 6} \stackrel{?}{=} 3 - 3$

$\sqrt{6 - 6} \stackrel{?}{=} 0$

$\sqrt{0} \stackrel{?}{=} 0$

$0 = 0$

The solution set is $\{3, 5\}$.

4. $\sqrt{x} + 12 = x$

$\sqrt{x} = x - 12$

$(\sqrt{x})^2 = (x - 12)^2$

$x = x^2 - 24x + 144$

$0 = x^2 - 25x + 144$

$0 = (x - 16)(x - 9)$

$x - 16 = 0$ or $x - 9 = 0$

$x = 16$ or $x = 9$

Check $x = 16$

$\sqrt{16} + 12 \stackrel{?}{=} 16$

$4 + 12 \stackrel{?}{=} 16$

$16 = 16$

Check $x = 9$

$\sqrt{9} + 12 \stackrel{?}{=} 9$

$3 + 12 \stackrel{?}{=} 9$

$15 \neq 9$

The solution set is $\{16\}$.

5. $\sqrt[3]{y^2 + 2} = 3$

$(\sqrt[3]{y^2 + 2})^3 = (3)^3$

$y^2 + 2 = 27$

$y^2 - 25 = 0$

$(y + 5)(y - 5) = 0$

$y + 5 = 0$ or $y - 5 = 0$

$y = -5$ or $y = 5$

Check $y = -5$

$\sqrt[3]{(-5)^2 + 2} \stackrel{?}{=} 3$

$\sqrt[3]{25 + 2} \stackrel{?}{=} 3$

$\sqrt[3]{27} \stackrel{?}{=} 3$

$3 = 3$

Check $y = 5$

$\sqrt[3]{(5)^2 + 2} \stackrel{?}{=} 3$

$\sqrt[3]{25 + 2} \stackrel{?}{=} 3$

$\sqrt[3]{27} \stackrel{?}{=} 3$

$3 = 3$

The solution set is $\{-5, 5\}$.

6. $\sqrt{x + 8} = 1 + \sqrt{x + 1}$

$(\sqrt{x + 8})^2 = (1 + \sqrt{x + 1})^2$

$x + 8 = 1 + 2\sqrt{x + 1} + x + 1$

$x + 8 = x + 2 + 2\sqrt{x + 1}$

$6 = 2\sqrt{x + 1}$

$3 = \sqrt{x + 1}$

$(3)^2 = (\sqrt{x + 1})^2$

$9 = x + 1$

$8 = x$

Check

$\sqrt{8 + 8} \stackrel{?}{=} 1 + \sqrt{8 + 1}$

$\sqrt{16} \stackrel{?}{=} 1 + \sqrt{9}$

$4 = 1 + 3$

The solution set is $\{8\}$.

7. Use $D = \dfrac{S^2}{30f}$.

Substitute 0.45 for f and 70 for S.

$D = \dfrac{(70)^2}{30(0.45)} = \dfrac{4900}{13.5} = 362.96$

The car will skid approximately 363 feet.

Section 7.6

1. (a) $36^{\frac{1}{2}} = \sqrt{36} = 6$

(b) $81^{\frac{3}{4}} = (\sqrt[4]{81})^3 = (3)^3 = 27$

(c) $(9)^{-\frac{3}{2}} = \dfrac{1}{(9)^{\frac{3}{2}}} = \dfrac{1}{(\sqrt{9})^3} = \dfrac{1}{(3)^3} = \dfrac{1}{27}$

(d) $(-125)^{\frac{2}{3}} = (\sqrt[3]{-125})^2 = (-5)^2 = 25$

(e) $-16^{\frac{1}{2}} = -\sqrt{16} = -4$

2. (a) $y^{\frac{5}{6}} = \sqrt[6]{y^5}$

(b) $4a^{\frac{3}{5}} = 4\sqrt[5]{a^3}$

(c) $a^{\frac{3}{5}}b^{\frac{4}{5}} = \sqrt[5]{a^3b^4}$

(d) $(a - b)^{\frac{3}{4}} = \sqrt[4]{(a - b)^3}$

3. (a) $\sqrt{ab} = (ab)^{\frac{1}{2}} = a^{\frac{1}{2}}b^{\frac{1}{2}}$

(b) $\sqrt[5]{a^2b^4} = (a^2b^4)^{\frac{1}{5}} = a^{\frac{2}{5}}b^{\frac{4}{5}}$

(c) $7\sqrt[3]{x} = 7x^{\frac{1}{3}}$

(d) $\sqrt[5]{(a - b)^2} = (a - b)^{\frac{2}{5}}$

4. (a) $(6a^{\frac{3}{4}})(3a^{\frac{5}{8}}) = 6 \cdot 3 \cdot a^{\frac{3}{4}} \cdot a^{\frac{5}{8}}$
$= 18a^{\frac{3}{4}+\frac{5}{8}} = 18a^{\frac{6}{8}+\frac{5}{8}} = 18a^{\frac{11}{8}}$

(b) $(8x^{\frac{1}{5}}y^{\frac{1}{4}})^2 = (8)^2 \cdot (x^{\frac{1}{5}})^2 \cdot (y^{\frac{1}{4}})^2$
$= 64x^{\frac{2}{5}}y^{\frac{1}{2}}$

(c) $\dfrac{15a^{\frac{1}{4}}}{5a^{\frac{2}{3}}} = 3a^{\frac{1}{4}-\frac{2}{3}} = 3a^{\frac{3}{12}-\frac{8}{12}} = 3a^{-\frac{5}{12}} = \dfrac{3}{a^{\frac{5}{12}}}$

(d) $\left(\dfrac{3a^{\frac{1}{3}}}{4b^{\frac{2}{5}}}\right)^2 = \dfrac{(3a^{\frac{1}{3}})^2}{(4b^{\frac{2}{5}})^2} = \dfrac{3^2 \cdot (a^{\frac{1}{3}})^2}{4^2 \cdot (b^{\frac{2}{5}})^2} = \dfrac{9a^{\frac{2}{3}}}{16b^{\frac{4}{5}}}$

5. (a) $\sqrt[3]{5}\sqrt[4]{5} = 5^{\frac{1}{3}} \cdot 5^{\frac{1}{4}} = 5^{\frac{1}{3}+\frac{1}{4}} = 5^{\frac{4}{12}+\frac{3}{12}}$
$= 5^{\frac{7}{12}} = \sqrt[12]{5^7}$

(b) $\dfrac{\sqrt[3]{6}}{\sqrt[4]{6}} = \dfrac{6^{\frac{1}{3}}}{6^{\frac{1}{4}}} = 6^{\frac{1}{3}-\frac{1}{4}} = 6^{\frac{4}{12}-\frac{3}{12}}$
$= 6^{\frac{1}{12}} = \sqrt[12]{6}$

(c) $\dfrac{\sqrt[3]{9}}{\sqrt[4]{3}} = \dfrac{9^{\frac{1}{3}}}{3^{\frac{1}{4}}} = \dfrac{(3^2)^{\frac{1}{3}}}{3^{\frac{1}{4}}} = \dfrac{3^{\frac{2}{3}}}{3^{\frac{1}{4}}}$
$= 3^{\frac{2}{3}-\frac{1}{4}} = 3^{\frac{8}{12}-\frac{3}{12}}$
$= 3^{\frac{5}{12}} = \sqrt[12]{3^5}$

Section 7.7

1. (a) $(0.000031)(3000) = (3.1)(10^{-5})(3)(10^3)$
$= (3.1)(3)(10^{-5})(10^3)$
$= (9.3)(10^{-2})$
$= 0.093$

(b) $\dfrac{4{,}500{,}000}{0.15} = \dfrac{(4.5)(10^6)}{(1.5)(10^{-1})}$
$= (3)(10^7)$
$= 30{,}000{,}000$

(c) $\dfrac{(0.00000036)(5400)}{(270{,}000)(0.00012)}$
$= \dfrac{(3.6)(10^{-7})(5.4)(10^3)}{(2.7)(10^5)(1.2)(10^{-4})}$
$= \dfrac{(\overset{3}{\cancel{3.6}})(\overset{2}{\cancel{5.4}})(10^{-4})}{(\underset{1}{\cancel{2.7}})(\underset{1}{\cancel{1.2}})(10^{1})}$
$= (6)(10^{-5}) = 0.00006$

(d) $\sqrt{0.00000016} = \sqrt{(16)(10^{-8})}$
$= [(16)(10^{-8})]^{\frac{1}{2}}$
$= (16)^{\frac{1}{2}}[(10^{-8})]^{\frac{1}{2}}$
$= (4)(10^{-4})$
$= 0.0004$

2. Size of virus: $d = 0.0000001$ m
Size of *E. coli*: $d = 0.000002$ m
Let x represent how many times larger the *E. coli* is.

$\left(\begin{matrix}\text{Size of}\\ E.\ coli\text{ cell}\end{matrix}\right) = x \cdot \left(\begin{matrix}\text{Size of}\\ \text{virus cell}\end{matrix}\right)$

$0.000002 = x(0.0000001)$

$\dfrac{0.000002}{0.0000001} = x$

$\dfrac{(2)(10^{-6})}{(1)(10^{-7})} = x$

$(2)(10^{-6-(-7)}) = x$

$(2)(10^1) = x$

$20 = x$

The *E. coli* cell is 20 times larger than the virus cell.

CHAPTER 8

Section 8.1

1. (a) $(6 + 2i) + (4 + 5i) = (6 + 4) + (2 + 5)i$
$= 10 + 7i$

(b) $(5 - 3i) + (-9 - 2i) = (5 - 9) + (-3 - 2)i$
$= -4 - 5i$

(c) $\left(\dfrac{1}{4} + \dfrac{2}{3}i\right) + \left(\dfrac{1}{2} + \dfrac{3}{4}i\right) = \left(\dfrac{1}{4} + \dfrac{1}{2}\right) + \left(\dfrac{2}{3} + \dfrac{3}{4}\right)i$
$= \left(\dfrac{1}{4} + \dfrac{2}{4}\right) + \left(\dfrac{8}{12} + \dfrac{9}{12}\right)i$
$= \dfrac{3}{4} + \dfrac{17}{12}i$

2. (a) $\sqrt{-25} = i\sqrt{25} = 5i$

(b) $\sqrt{-13} = i\sqrt{13}$

(c) $\sqrt{-18} = i\sqrt{18} = i\sqrt{9}\sqrt{2} = 3i\sqrt{2}$

3. (a) $\sqrt{-3}\sqrt{-12} = (i\sqrt{3})(i\sqrt{12}) = i^2\sqrt{36}$
$= (-1)(6) = -6$

(b) $\sqrt{-10}\sqrt{-2} = (i\sqrt{10})(i\sqrt{2}) = i^2\sqrt{20}$
$= (-1)\sqrt{4}\sqrt{5} = -2\sqrt{5}$

(c) $\dfrac{\sqrt{-98}}{\sqrt{-2}} = \dfrac{i\sqrt{98}}{i\sqrt{2}} = \dfrac{\sqrt{98}}{\sqrt{2}} = \sqrt{49} = 7$

(d) $\dfrac{\sqrt{-108}}{\sqrt{3}} = \dfrac{i\sqrt{108}}{\sqrt{3}} = i\sqrt{\dfrac{108}{3}} = i\sqrt{36} = 6i$

4. (a) $(4 + 3i)(6 + i) = 4(6 + i) + 3i(6 + i)$
$= 24 + 4i + 18i + 3i^2$
$= 24 + 22i + 3(-1)$
$= 21 + 22i$

(b) $(-2 + 5i)(1 - 3i) = -2(1 - 3i) + 5i(1 - 3i)$
$= -2 + 6i + 5i - 15i^2$
$= -2 + 11i + (-15)(-1)$
$= -2 + 11i + 15$
$= 13 + 11i$

(c) $(6 - 2i)^2 = (6 - 2i)(6 - 2i)$
$= 6(6 - 2i) - 2i(6 - 2i)$
$= 36 - 12i - 12i + 4i^2$
$= 36 - 24i + 4(-1)$
$= 32 - 24i$

(d) $(5 + 2i)(5 - 2i) = 5(5 - 2i) + 2i(5 - 2i)$
$= 25 - 10i + 10i - 4i^2$
$= 25 - 4(-1)$
$= 25 + 4 = 29$

5. (a) $\dfrac{3 + 5i}{6 + i} = \dfrac{(3 + 5i)(6 - i)}{(6 + i)(6 - i)}$
$= \dfrac{18 - 3i + 30i - 5i^2}{36 - i^2}$
$= \dfrac{18 + 27i - 5(-1)}{36 - (-1)}$
$= \dfrac{18 + 27i + 5}{37} = \dfrac{23 + 27i}{37}$

(b) $\dfrac{5 - 8i}{3i} = \dfrac{(5 - 8i)(i)}{(3i)(i)} = \dfrac{5i - 8i^2}{3i^2}$
$= \dfrac{5i - 8(-1)}{3(-1)} = \dfrac{5i + 8}{-3}$
$= \dfrac{-8 - 5i}{3} = -\dfrac{8}{3} - \dfrac{5}{3}i$

Section 8.2

1. $2x^2 - 7x - 15 = 0$
$(2x + 3)(x - 5) = 0$
$2x + 3 = 0$ or $x - 5 = 0$
$2x = -3$ or $x = 5$
$x = -\dfrac{3}{2}$ or $x = 5$

The solution set is $\left\{-\dfrac{3}{2}, 5\right\}$.

2. $\sqrt{x} = x - 6$
$(\sqrt{x})^2 = (x - 6)^2$
$x = x^2 - 12x + 36$
$0 = x^2 - 13x + 36$
$0 = (x - 9)(x - 4)$
$x - 9 = 0$ or $x - 4 = 0$
$x = 9$ or $x = 4$

Check $x = 9$.
$\sqrt{x} = x - 6$
$\sqrt{9} \overset{?}{=} 9 - 6$
$3 = 3$

Check $x = 4$.
$\sqrt{x} = x - 6$
$\sqrt{4} \overset{?}{=} 4 - 6$
$2 \neq -2$

The solution set is $\{9\}$.

3. $y^2 = 75$
$y = \pm\sqrt{75}$
$y = \pm 5\sqrt{3}$
The solution set is $\{\pm 5\sqrt{3}\}$.

4. $a^2 = -36$
$a = \pm\sqrt{-36}$
$y = \pm 6i$
The solution set is $\{\pm 6i\}$.

5. $5x^2 = 48$
$x^2 = \dfrac{48}{5}$
$x = \pm\sqrt{\dfrac{48}{5}} = \pm\dfrac{\sqrt{48}}{\sqrt{5}} = \pm\dfrac{\sqrt{16}\sqrt{3}}{\sqrt{5}}$
$x = \pm\dfrac{4\sqrt{3}}{\sqrt{5}} \cdot \dfrac{\sqrt{5}}{\sqrt{5}} = \pm\dfrac{4\sqrt{15}}{5}$

The solution set is $\left\{\pm\dfrac{4\sqrt{15}}{5}\right\}$.

6. $(4x - 3)^2 = 81$
$4x - 3 = \pm\sqrt{81}$
$4x - 3 = \pm 9$
$4x - 3 = -9$ or $4x - 3 = 9$
$4x = -6$ or $4x = 12$
$x = -\dfrac{6}{4} = -\dfrac{3}{2}$ or $x = 3$

The solution set is $\left\{-\dfrac{3}{2}, 3\right\}$.

7. $(x + 8)^2 = -15$
$x + 8 = \pm\sqrt{-15}$
$x + 8 = \pm i\sqrt{15}$
$x = -8 \pm i\sqrt{15}$
The solution set is $\{-8 \pm i\sqrt{15}\}$.

8. $2(5x + 1)^2 + 6 = 62$
$2(5x + 1)^2 = 56$
$(5x + 1)^2 = 28$
$5x + 1 = \pm\sqrt{28}$

$$5x + 1 = \pm 2\sqrt{7}$$
$$5x = -1 \pm 2\sqrt{7}$$
$$x = \frac{-1 \pm 2\sqrt{7}}{5}$$

The solution set is $\left\{\frac{-1 \pm 2\sqrt{7}}{5}\right\}$.

9. Let h represent the height of the cell tower. Use the Pythagorean theorem to solve $a^2 + b^2 = c^2$.

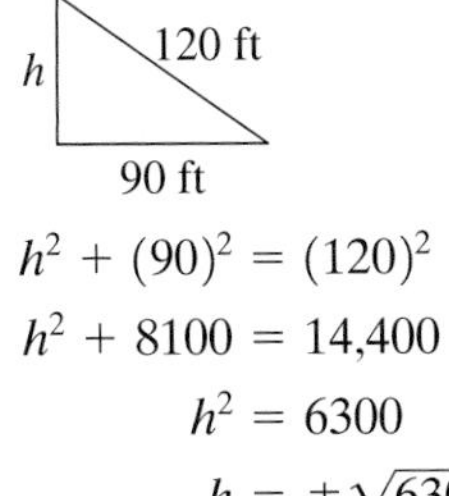

$$h^2 + (90)^2 = (120)^2$$
$$h^2 + 8100 = 14{,}400$$
$$h^2 = 6300$$
$$h = \pm\sqrt{6300}$$
$$h = \pm 79.37$$

Discard the negative value, $h = -79.37$. To the nearest tenth, the cell tower is approximately 79.4 feet high.

10. Let x represent the height of each leg of the isosceles right triangle. Use the Pythagorean theorem to solve.

8 in. x x

$$(x)^2 + (x)^2 = (8)^2$$
$$2x^2 = 64$$
$$x^2 = 32$$
$$x = \pm\sqrt{32} = \pm 4\sqrt{2}$$

Discard the negative value, $x = -4\sqrt{2}$. The length of each leg is $4\sqrt{2}$ inches.

11. Let h represent the height at which the ladder reaches the building. The side opposite the 30° angle is one-half the length of the hypotenuse, which is $\frac{1}{2}(16) = 8$ feet. Use the Pythagorean theorem to solve.

16 ft 30° h 60° 8 ft

$$h^2 + (8)^2 = (16)^2$$
$$h^2 + 64 = 256$$
$$h^2 = 192$$
$$h = \pm\sqrt{192}$$
$$h \approx \pm 13.86$$

Discard the negative value, $h = -13.86$. To the nearest tenth, the height of the ladder is 13.9 feet, so the ladder does reach the window.

Section 8.3

1. $y^2 + 6y - 4 = 0$

$$y^2 + 6y = 4$$
$$\frac{1}{2}(6) = 3 \quad \text{and} \quad 3^2 = 9$$
$$y^2 + 6y + 9 = 4 + 9$$
$$(y + 3)^2 = 13$$
$$y + 3 = \pm\sqrt{13}$$
$$y = -3 \pm \sqrt{13}$$

The solution set is $\{-3 \pm \sqrt{13}\}$.

2. $y(y + 4) = -10$

$$y^2 + 4y = -10$$
$$\frac{1}{2}(4) = 2 \quad \text{and} \quad 2^2 = 4$$
$$y^2 + 4y + 4 = -10 + 4$$
$$(y + 2)^2 = -6$$
$$y + 2 = \pm\sqrt{-6}$$
$$y + 2 = \pm i\sqrt{6}$$
$$y = -2 \pm i\sqrt{6}$$

The solution set is $\{-2 \pm i\sqrt{6}\}$.

3. $y^2 + y + 3 = 0$

$$y^2 + y = -3$$
$$\frac{1}{2}(1) = \frac{1}{2} \quad \text{and} \quad \left(\frac{1}{2}\right)^2 = \frac{1}{4}$$
$$y^2 + y + \frac{1}{4} = -3 + \frac{1}{4}$$
$$\left(y + \frac{1}{2}\right)^2 = -\frac{12}{4} + \frac{1}{4}$$
$$\left(y + \frac{1}{2}\right)^2 = -\frac{11}{4}$$
$$y + \frac{1}{2} = \pm\sqrt{-\frac{11}{4}} = \pm i\frac{\sqrt{11}}{\sqrt{4}}$$
$$y + \frac{1}{2} = \pm i\frac{\sqrt{11}}{2}$$
$$y = -\frac{1}{2} \pm i\frac{\sqrt{11}}{2} = \frac{-1 \pm i\sqrt{11}}{2}$$

The solution set is $\left\{\frac{-1 \pm i\sqrt{11}}{2}\right\}$.

4. $3y^2 + 24y - 7 = 0$

$$3y^2 + 24y = 7$$
$$y^2 + 8y = \frac{7}{3}$$
$$\frac{1}{2}(8) = 4 \quad \text{and} \quad (4)^2 = 16$$
$$y^2 + 8y + 16 = \frac{7}{3} + 16$$

$$(y+4)^2 = \frac{7}{3} + \frac{48}{3}$$

$$(y+4)^2 = \frac{55}{3}$$

$$y+4 = \pm\sqrt{\frac{55}{3}} = \pm\frac{\sqrt{55}}{\sqrt{3}}$$

$$y+4 = \pm\frac{\sqrt{55}}{\sqrt{3}} \cdot \frac{\sqrt{3}}{\sqrt{3}}$$

$$y+4 = \pm\frac{\sqrt{165}}{3}$$

$$y = -4 \pm \frac{\sqrt{165}}{3} = \frac{-12}{3} \pm \frac{\sqrt{165}}{3}$$

$$y = \frac{-12 \pm \sqrt{165}}{3}$$

The solution set is $\left\{\frac{-12 \pm \sqrt{165}}{3}\right\}$.

5. $y^2 - 6y - 72 = 0$

$$y^2 - 6y = 72$$

$$y^2 - 6y + 9 = 72 + 9$$

$$(y-3)^2 = 81$$

$$y - 3 = \pm\sqrt{81}$$

$$y - 3 = \pm 9$$

$y - 3 = -9$ or $y - 3 = 9$

$y = -6$ or $y = 12$

The solution set is $\{-6, 12\}$.

Section 8.4

1. $x^2 + 7x + 5 = 0$

Use the quadratic formula.

$a = 1 \quad b = 7 \quad c = 5$

$$x = \frac{-b \pm \sqrt{b^2 - 4ac}}{2a}$$

$$= \frac{-(7) \pm \sqrt{(7)^2 - 4(1)(5)}}{2(1)}$$

$$= \frac{-7 \pm \sqrt{49 - 20}}{2}$$

$$= \frac{-7 \pm \sqrt{29}}{2}$$

The solution set is $\left\{\frac{-7 \pm \sqrt{29}}{2}\right\}$.

2. $x^2 - 6x - 4 = 0$

Use the quadratic formula.

$a = 1 \quad b = -6 \quad c = -4$

$$x = \frac{-b \pm \sqrt{b^2 - 4ac}}{2a}$$

$$= \frac{-(-6) \pm \sqrt{(-6)^2 - 4(1)(-4)}}{2(1)}$$

$$= \frac{6 \pm \sqrt{36 + 16}}{2}$$

$$= \frac{6 \pm \sqrt{52}}{2} = \frac{6 \pm \sqrt{4}\sqrt{13}}{2}$$

$$= \frac{6 \pm 2\sqrt{13}}{2} = \frac{2(3 \pm \sqrt{13})}{2}$$

$$= 3 \pm \sqrt{13}$$

The solution set is $\{3 \pm \sqrt{13}\}$.

3. $x^2 - 2x + 8 = 0$

Use the quadratic formula.

$a = 1 \quad b = -2 \quad c = 8$

$$x = \frac{-b \pm \sqrt{b^2 - 4ac}}{2a}$$

$$= \frac{-(-2) \pm \sqrt{(-2)^2 - 4(1)(8)}}{2(1)}$$

$$= \frac{2 \pm \sqrt{4 - 32}}{2} = \frac{2 \pm \sqrt{-28}}{2}$$

$$= \frac{2 \pm i\sqrt{4}\sqrt{7}}{2} = \frac{2 \pm 2i\sqrt{7}}{2}$$

$$= \frac{2(1 \pm i\sqrt{7})}{2} = \frac{2(1 \pm i\sqrt{7})}{2}$$

The solution set is $\{1 \pm i\sqrt{7}\}$.

4. $5x^2 + 3x - 4 = 0$

Use the quadratic formula.

$a = 5 \quad b = 3 \quad c = -4$

$$x = \frac{-b \pm \sqrt{b^2 - 4ac}}{2a}$$

$$= \frac{-(3) \pm \sqrt{(3)^2 - 4(5)(-4)}}{2(5)}$$

$$= \frac{-3 \pm \sqrt{9 + 80}}{10}$$

$$= \frac{-3 \pm \sqrt{89}}{10}$$

The solution set is $\left\{\frac{-3 \pm \sqrt{89}}{10}\right\}$.

5. $n(2n + 5) = 12$

$$2n^2 + 5n = 12$$

$$2n^2 + 5n - 12 = 0$$

Use the quadratic formula.

$a = 2 \quad b = 5 \quad c = -12$

$$x = \frac{-b \pm \sqrt{b^2 - 4ac}}{2a}$$

$$= \frac{-(5) \pm \sqrt{(5)^2 - 4(2)(-12)}}{2(2)}$$

$$= \frac{-5 \pm \sqrt{25 + 96}}{4} = \frac{-5 \pm \sqrt{121}}{4}$$

$$= \frac{-5 \pm 11}{4}$$

$$= \frac{-5 - 11}{4} \quad \text{or} \quad x = \frac{-5 + 11}{4}$$

$$= \frac{-16}{4} \quad \text{or} \quad x = \frac{6}{4}$$

$$= -4 \quad \text{or} \quad x = \frac{3}{2}$$

The solution set is $\left\{-4, \frac{3}{2}\right\}$.

6. $4x^2 + 12x + 9 = 0$

$a = 4 \qquad b = 12 \qquad c = 9$

$$b^2 - 4ac = (12)^2 - 4(4)(9)$$
$$= 144 - 144 = 0$$

Because $b^2 - 4ac = 0$, the equation has one real solution with multiplicity of two.

Section 8.5

1. $3x^2 - 5x + 1 = 0$

Use the quadratic formula.

$a = 3 \qquad b = -5 \qquad c = 1$

$$x = \frac{-b \pm \sqrt{b^2 - 4ac}}{2a}$$

$$= \frac{-(-5) \pm \sqrt{(-5)^2 - 4(3)(1)}}{2(3)}$$

$$= \frac{5 \pm \sqrt{25 - 12}}{6}$$

$$= \frac{5 \pm \sqrt{13}}{6}$$

The solution set is $\left\{\frac{5 \pm \sqrt{13}}{6}\right\}$.

2.

$$\frac{6}{x} + \frac{3}{x + 4} = 1$$

$$x(x + 4)\left(\frac{6}{x} + \frac{3}{x + 4}\right) = x(x + 4)(1)$$

$$x(x + 4)\left(\frac{6}{x}\right) + x(x + 4)\left(\frac{3}{x + 4}\right) = x^2 + 4x$$

$$6(x + 4) + 3x = x^2 + 4x$$

$$6x + 24 + 3x = x^2 + 4x$$

$$9x + 24 = x^2 + 4x$$

$$0 = x^2 - 5x - 24$$

$$0 = (x - 8)(x + 3)$$

$$x - 8 = 0 \quad \text{or} \quad x + 3 = 0$$

$$x = 8 \quad \text{or} \quad x = -3$$

The solution set is $\{-3, 8\}$.

3. $y^2 + 28y + 192 = 0$

$$y^2 + 28y = -192$$

$\frac{1}{2}(28) = 14 \quad \text{and} \quad 14^2 = 196$

$$y^2 + 28y + 196 = -192 + 196$$

$$(y + 14)^2 = 4$$

$$y + 14 = \pm\sqrt{4}$$

$$y + 14 = \pm 2$$

$$y + 14 = -2 \quad \text{or} \quad y + 14 = 2$$

$$y = -16 \quad \text{or} \quad y = -12$$

The solution set is $\{-16, -12\}$.

4. $y^4 - 17y^2 - 60 = 0$

$$(y^2 + 3)(y^2 - 20) = 0$$

$$y^2 + 3 = 0 \quad \text{or} \quad y^2 - 20 = 0$$

$$y^2 = -3 \quad \text{or} \quad y^2 = 20$$

$$y = \pm\sqrt{-3} \qquad y = \pm\sqrt{20}$$

$$y = \pm i\sqrt{3} \qquad y = \pm 2\sqrt{5}$$

The solution set is $\{\pm 2\sqrt{5}, \pm i\sqrt{3}\}$.

5. Let x represent the length of the original digital image and let $3x$ represent the width of the original digital image.

Then $x - 2$ represents the length of the cropped picture, and $3x - 2$ represents the width of the cropped picture.

$$\left(\begin{matrix}\text{Length of}\\ \text{cropped picture}\end{matrix}\right) \cdot \left(\begin{matrix}\text{Width of}\\ \text{cropped picture}\end{matrix}\right) = 64$$

$$(3x - 2)(x - 2) = 64$$

$$3x^2 - 6x - 2x + 4 = 64$$

$$3x^2 - 8x - 60 = 0$$

$$(3x + 10)(x - 6) = 0$$

$$3x + 10 = 0 \quad \text{or} \quad x - 6 = 0$$

$$3x = -10 \quad \text{or} \quad x = 6$$

$$x = -\frac{10}{3} \quad \text{or} \quad x = 6$$

Discard the negative solution. If $x = 6$, then $x - 2 = 4$ and $3x - 2 = 16$. The dimensions of the cropped image are 4 centimeters and 16 centimeters.

6. Let r represent the rate of interest. Use the formula: $A = P(1 + r)^t$. Substitute 3000 for P, 3500 for A, and 2 for t.

$$A = P(1 + r)^t$$

$$3500 = 3000(1 + r)^2$$

$$1.16666 = (1 + r)^2$$

$$\pm\sqrt{1.16666} = 1 + r$$

$$-1 \pm 1.080 = r$$

$$-1 - 1.080 = r \quad \text{or} \quad -1 + 1.080 = r$$
$$-2.080 = r \quad \text{or} \quad 0.080 = r$$

Discard the negative solution. If $r = 0.080$, then $r = 8.0\%$. The rate of interest is 8.0%.

7. Let r represent Pete's rate when riding his bicycle. Then $r - 6$ represents his walking rate.

Use the formula:

Rate · Time = Distance

Organize the information with the following chart.

	Rate	Time	Distance
Cycling	r	$\frac{15}{r}$	15
Walking	$r - 6$	$\frac{5}{r-6}$	5

$$\left(\begin{matrix}\text{Time}\\ \text{cycling}\end{matrix}\right) + \left(\begin{matrix}\text{Time}\\ \text{walking}\end{matrix}\right) = 2\frac{3}{4}$$
$$\frac{15}{r} + \frac{5}{r-6} = \frac{11}{4}$$
$$4r(r-6)\left(\frac{15}{r} + \frac{5}{r-6}\right) = 4r(r-6)\left(\frac{11}{4}\right)$$
$$4r(r-6)\left(\frac{15}{r}\right) + 4r(r-6)\left(\frac{5}{r-6}\right) = 11r(r-6)$$
$$60(r-6) + 20r = 11r^2 - 66r$$
$$60r - 360 + 20r = 11r^2 - 66r$$
$$80r - 360 = 11r^2 - 66r$$
$$0 = 11r^2 - 146r + 360$$
$$0 = (11r - 36)(r - 10)$$
$$11r - 36 = 0 \quad \text{or} \quad r - 10 = 0$$
$$11r = 36 \quad \text{or} \quad r = 10$$
$$r = \frac{36}{11} \quad \text{or} \quad r = 10$$

Discard $r = \frac{36}{11}$ since $r - 6$ becomes negative. Pete bicycles at 10 mph.

8. Let x represent the anticipated time to proofread the term paper and let $x - \frac{1}{2}$ represent the actual number of hours to proofread the term paper.

Use the formula:

Hours · Rate = Pay

Organize the information with the following chart.

	Hours	Rate	Pay
Actual	$x - \frac{1}{2}$	$\frac{24}{x - \frac{1}{2}}$	24
Anticipated	x	$\frac{24}{x}$	24

$$\left(\begin{matrix}\text{Actual}\\ \text{rate}\end{matrix}\right) = \left(\begin{matrix}\text{Anticipated}\\ \text{rate}\end{matrix}\right) + 4.00$$
$$\frac{24}{x - 0.5} = \frac{24}{x} + 4$$
$$x(x-0.5)\left(\frac{24}{x-0.5}\right) = x(x-0.5)\left(\frac{24}{x} + 4\right)$$
$$24x = x(x-0.5)\left(\frac{24}{x}\right) + x(x-0.5)(4)$$
$$24x = 24(x - 0.5) + 4x(x - 0.5)$$
$$24x = 24x - 12 + 4x^2 - 2x$$
$$0 = 4x^2 - 2x - 12$$
$$0 = 2x^2 - x - 6$$
$$0 = (2x + 3)(x - 2)$$
$$2x + 3 = 0 \quad \text{or} \quad x - 2 = 0$$
$$2x = -3 \quad \text{or} \quad x = 2$$
$$x = -\frac{3}{2} \quad \text{or} \quad x = 2$$

Discard the negative solution. The tutor expected to proofread for 2 hours.

Section 8.6

1.
$$y^2 - y - 30 > 0$$
$$y^2 - y - 30 = 0$$
$$(y - 6)(y + 5) = 0$$
$$y - 6 = 0 \quad \text{or} \quad y + 5 = 0$$
$$y = 6 \quad \text{or} \quad y = -5$$

Test point	−6	0	7
$y - 6$:	Negative	Negative	Positive
$y + 5$:	Negative	Positive	Positive
Product:	Positive	Negative	Positive

(Boundaries: −5 and 6)

The solution set is $(-\infty, -5) \cup (6, \infty)$.

−7 −6 −5 −4 −3 −2 −1 0 1 2 3 4 5 6

2.
$$y^2 - 7y + 10 \le 0$$
$$y^2 - 7y + 10 = 0$$
$$(y - 5)(y - 2) = 0$$
$$y - 5 = 0 \quad \text{or} \quad y - 2 = 0$$
$$y = 5 \quad \text{or} \quad y = 2$$

Critical values: 2, 5

Test point	0	3	6
$y - 5$:	Negative	Negative	Positive
$y - 2$:	Negative	Positive	Positive
Product:	Positive	Negative	Positive

The solution set is $[2, 5]$.

−7 −6 −5 −4 −3 −2 −1 2 3 4 5 6

3. $\dfrac{x-4}{x+3} > 0 \qquad x \neq -3$

$x - 4 = 0 \quad \text{or} \quad x + 3 = 0$

$x = 4 \quad \text{or} \quad x = -3$

Critical values: −3, 4

Test point	−4	0	5
$x - 4$:	Negative	Negative	Positive
$x + 3$:	Negative	Positive	Positive
Quotient:	Positive	Negative	Positive

The solution set is $(-\infty, -3) \cup (4, \infty)$.

−7 −6 −5 −4 −3 −2 −1 0 1 2 3 4 5 6

4. $\dfrac{x-6}{x-2} \leq 0 \qquad x \neq 2$

$x - 6 = 0 \quad \text{or} \quad x - 2 = 0$

$x = 6 \quad \text{or} \quad x = 2$

Critical values: 2, 6

Test point	0	4	7
$x - 6$:	Negative	Negative	Positive
$x - 2$:	Negative	Positive	Positive
Quotient:	Positive	Negative	Positive

The solution set is $(2, 6]$.

5. $\dfrac{5x}{x+3} \geq 2 \qquad x \neq -3$

$$\frac{5x}{x+3} - 2 \geq 0$$

$$\frac{5x}{x+3} - \frac{2(x+3)}{(x+3)} \geq 0$$

$$\frac{5x - 2(x+3)}{(x+3)} \geq 0$$

$$\frac{5x - 2x - 6}{(x+3)} \geq 0$$

$$\frac{3x-6}{x+3} \geq 0$$

$$\frac{3x-6}{x+3} = 0$$

$3x - 6 = 0 \quad \text{or} \quad x + 3 = 0$

$3x = 6 \quad \text{or} \quad x = -3$

$x = 2 \quad \text{or} \quad x = -3$

Critical values: −3, 2

Test point	−4	0	3
$3x - 6$:	Negative	Negative	Positive
$x + 3$:	Negative	Positive	Positive
Quotient:	Positive	Negative	Positive

The solution set is $(-\infty, -3) \cup [2, \infty)$.

CHAPTER 9

Section 9.1

1. Graph $y = x^2 - 3$.

If $x = 0$, then $y = (0)^2 - 3 = -3$. So $(0, -3)$ is on the graph. If $y = 0$, then

$0 = x^2 - 3$

$-x^2 = -3$

$x^2 = 3$

$x = \pm\sqrt{3}$

Intercepts:

x	y
0	−3
$\sqrt{3}$	0
$-\sqrt{3}$	0

Other points:

x	y
2	1
−2	1
3	6
−3	6

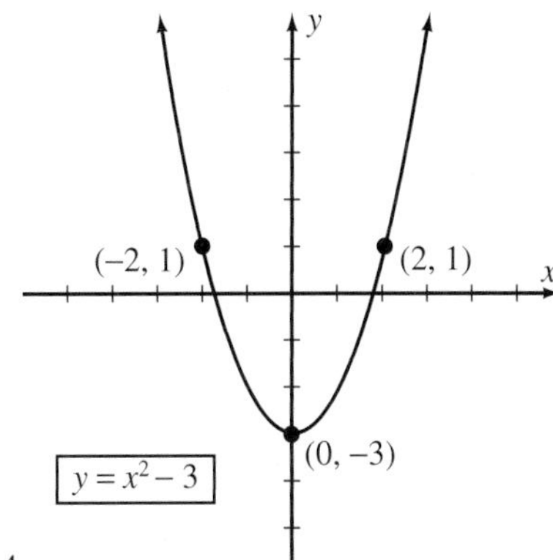

2. Graph $x = y^2 - 4$.

If $y = 0$, then $x = (0)^2 - 4 = -4$. So $(0, -4)$ is on the graph. If $x = 0$, then

$0 = y^2 - 4$

$-y^2 = -4$

$y^2 = 4$

$y = \pm 2$

Intercepts:

x	y
−4	0
0	2
0	−2

Other points:

x	y
−3	1
−3	−1
5	3
5	−3

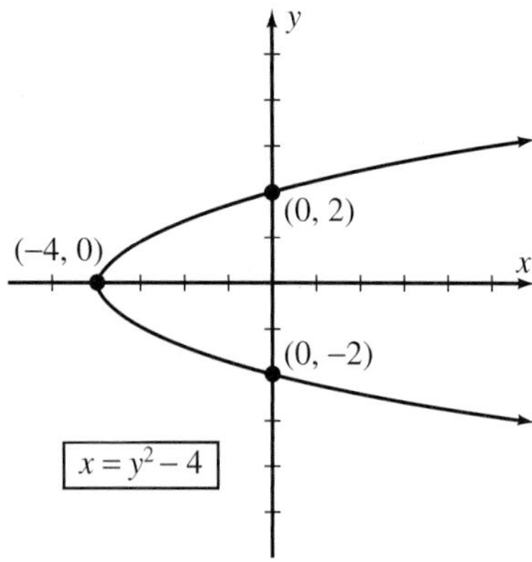

3. Graph $y = x^3$.

Points:

x	y
0	0
1	1
−1	−1
2	8
−2	−8

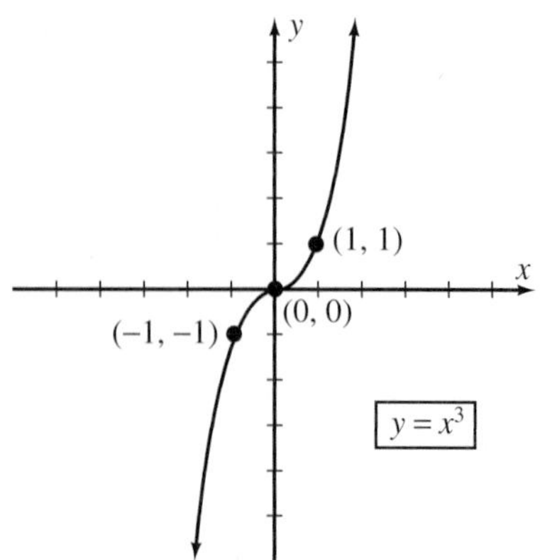

4. Graph $x^2y = 4$.

Points:

x	y
1	4
−1	4
2	1
−2	1
4	$\frac{1}{4}$
−4	$\frac{1}{4}$

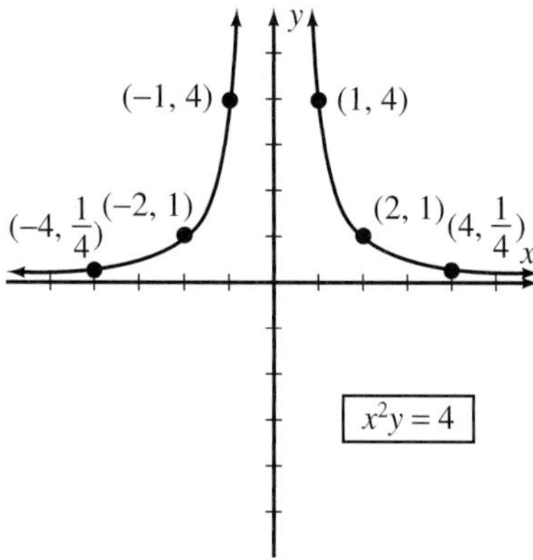

5. Graph $xy = -4$.

Points:

x	y
1	−4
−1	4
2	−2
−2	2
4	−1
−4	1

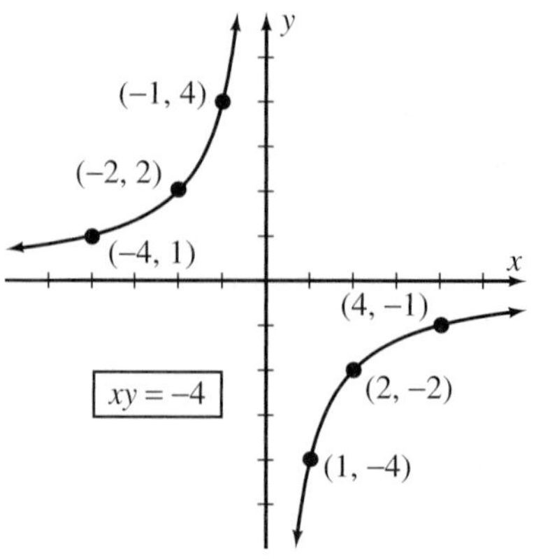

6. Graph $xy = 2$.

Points:

x	y
1	2
−1	−2
2	1
−2	−1
4	$\frac{1}{2}$
−4	$-\frac{1}{2}$

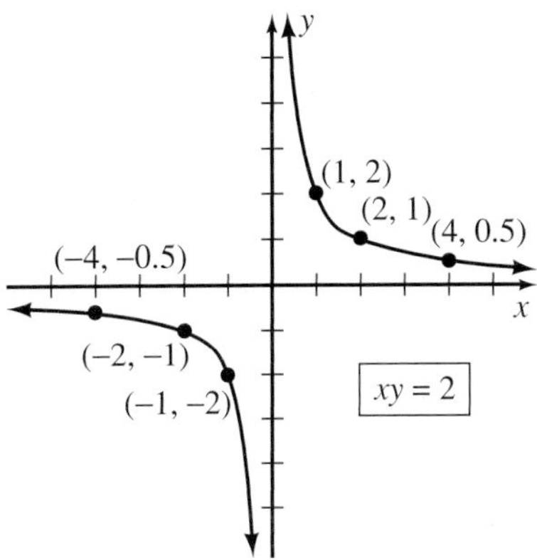

Section 9.2

1. Graph $y = x^2 + 3$.

x	$y = x^2 + 3$
0	3
1	4
2	7
−1	4
−2	7

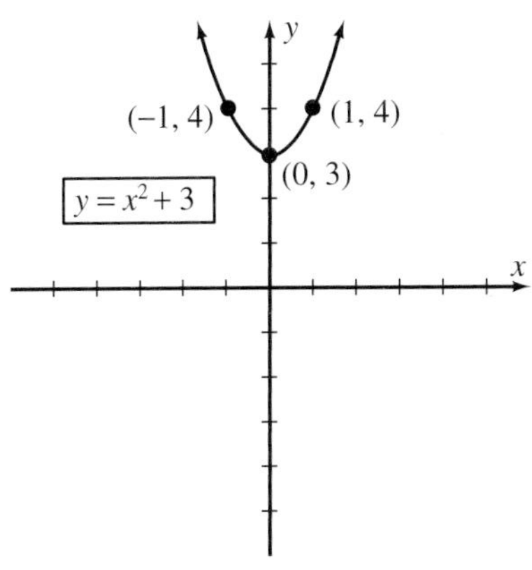

2. Graph $y = x^2 - 4$.

x	$y = x^2 - 4$
0	−4
1	−3
2	0
−1	−3
−2	0
3	5
−3	5

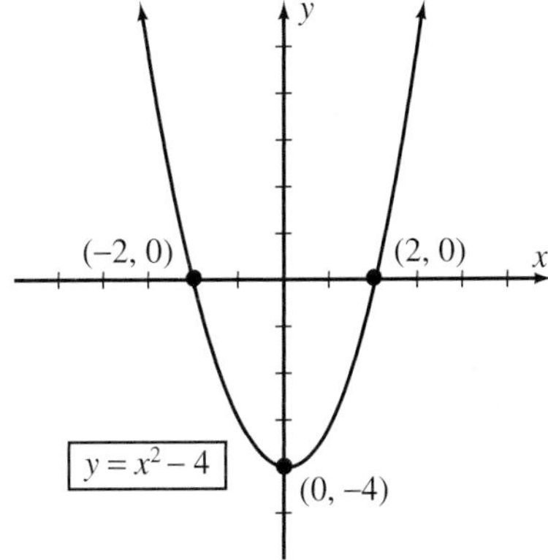

3. Graph $y = 3x^2$.

x	$y = 3x^2$
0	0
1	3
2	12
−1	3
−2	12

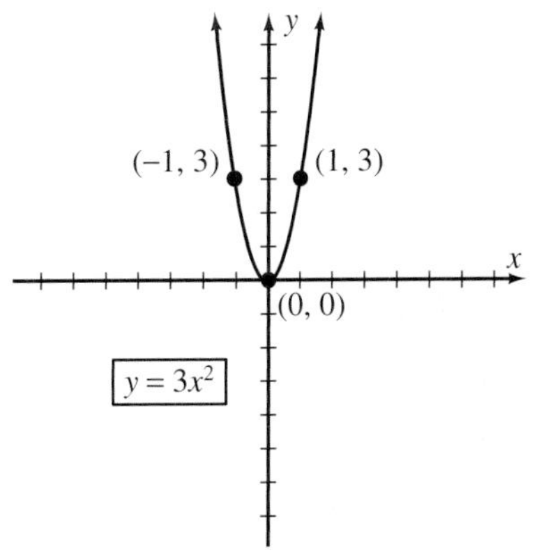

4. Graph $y = \frac{1}{4}x^2$.

x	$y = \frac{1}{4}x^2$
0	0
2	1
−2	1
4	4
−4	4

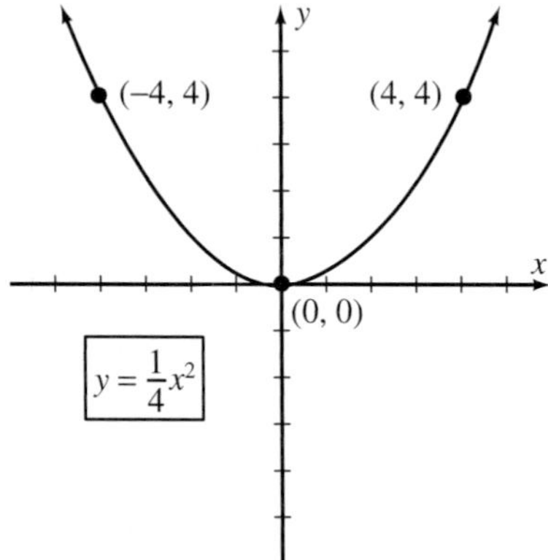

5. Graph $y = -x^2 + 2$.

x	$y = -x^2 + 2$
0	2
1	1
2	−2
−1	1
−2	−2
3	−7
−3	−7

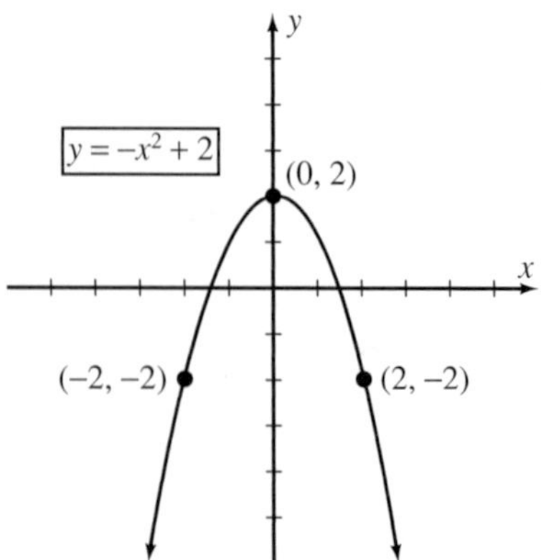

6. Graph $y = (x - 4)^2$.

x	$y = (x - 4)^2$
1	9
2	4
3	1
4	0
5	1
6	4
7	9

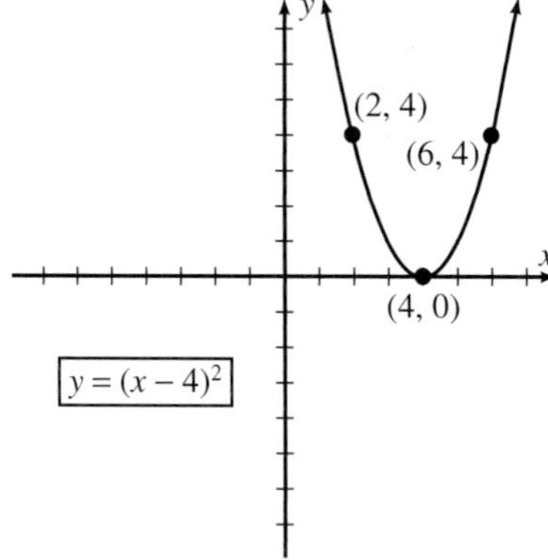

7. Graph $y = (x + 1)^2$.

x	$y = (x + 1)^2$
−4	9
−3	4
−2	1
−1	0
0	1
1	4
2	9

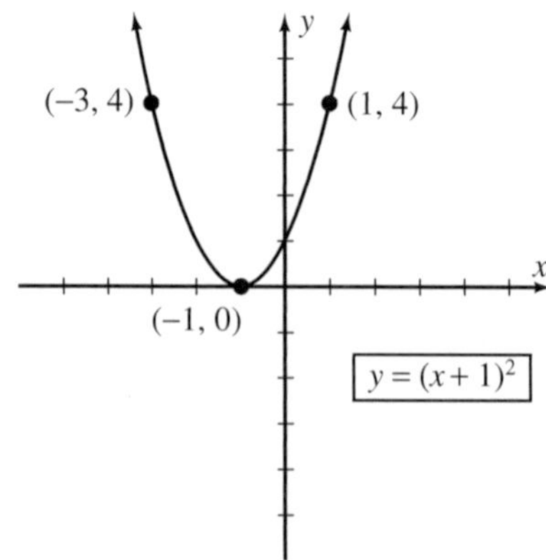

8. Graph $y = 2(x - 1)^2 + 3$.

The 2 narrows the parabola and opens it upward.
The 1 shifts the parabola one unit to the right.
The 3 shifts the parabola three units up.

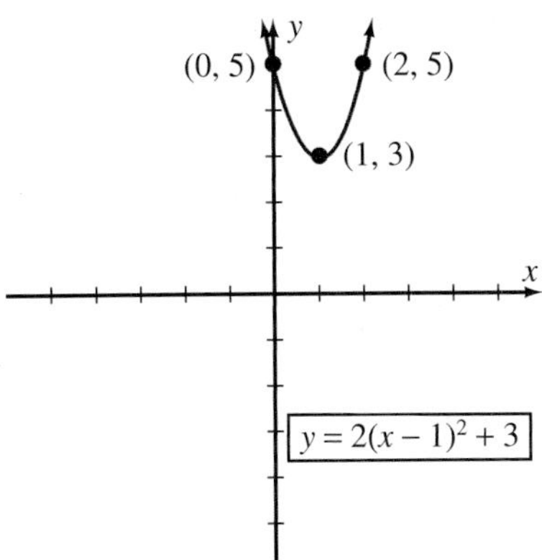

9. Graph $y = -\frac{1}{2}(x - 3)^2 + 4$.

The $-\frac{1}{2}$ widens the parabola and opens it downward.

The 3 shifts the parabola three units to the right.
The 4 shifts the parabola four units up.

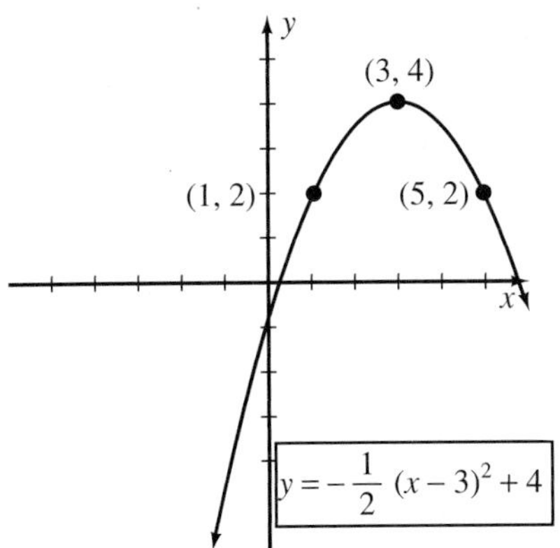

Section 9.3

1. Graph $y = x^2 + 4x + 1$.

$y = x^2 + 4x + 1$

$y = (x^2 + 4x + 4) + 1 - 4$

$y = (x + 2)^2 - 3$

The graph is the basic parabola, $y = x^2$, moved two units to the left and three units down.

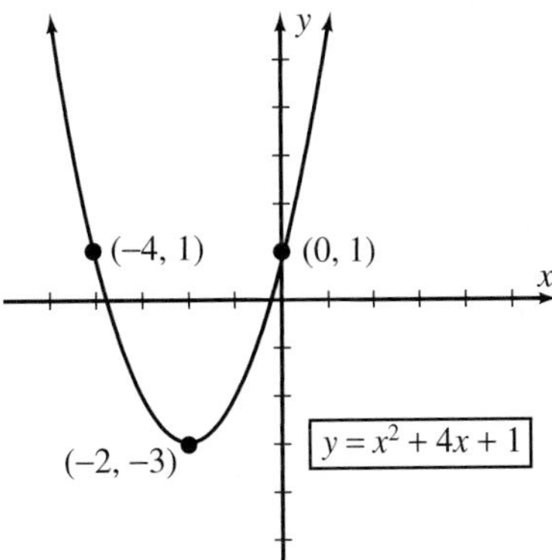

2. Graph $y = x^2 + x + 1$.

$y = x^2 + x + 1$

$y = \left(x^2 + x + \frac{1}{4}\right) - \frac{1}{4} + 1$

$y = \left(x + \frac{1}{2}\right)^2 - \frac{1}{4} + \frac{4}{4}$

$y = \left(x + \frac{1}{2}\right)^2 + \frac{3}{4}$

The graph is the basic parabola, $y = x^2$, moved one-half unit to the left and three-fourths unit up.

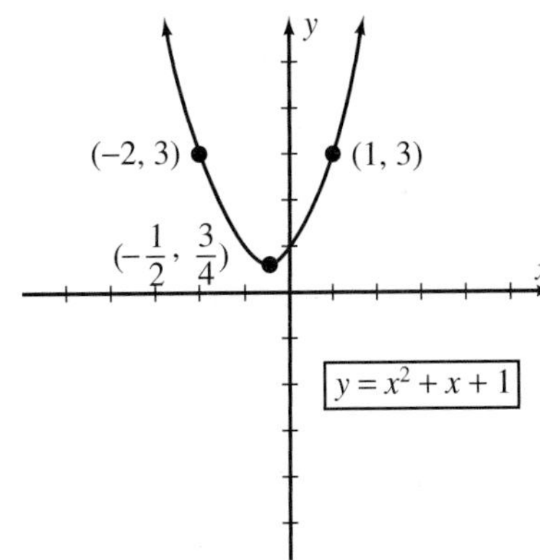

3. Graph $y = 2x^2 - 4x + 5$.

$y = 2x^2 - 4x + 5$

$y = 2(x^2 - 2x + 1) - 2 + 5$

$y = 2(x - 1)^2 + 3$

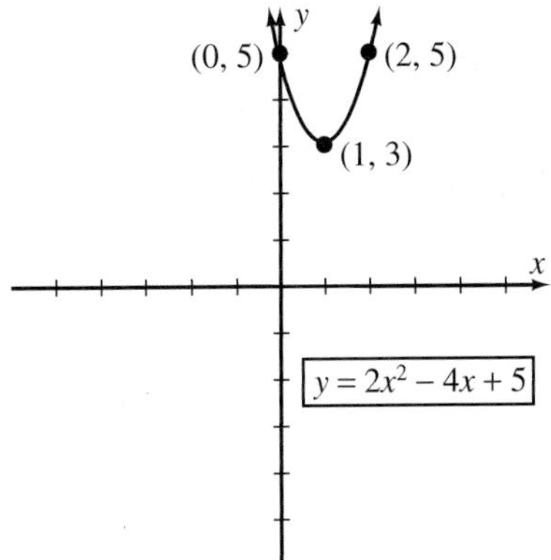

4. Graph $y = -2x^2 + 12x - 17$.

$y = -2x^2 + 12x - 17$

$y = -2(x^2 - 6x + 9) - 17 + 18$

$y = -2(x - 3)^2 + 1$

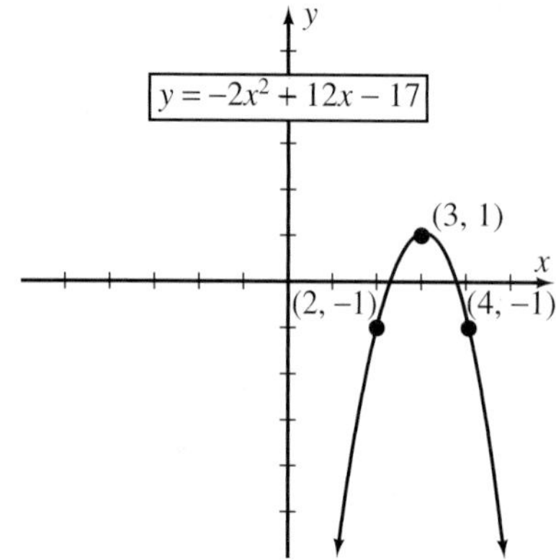

5. Standard form: $(x - h)^2 + (y - k)^2 = r^2$

Because the center is at $(-2, 1)$, we have $h = -2$ and $k = 1$. Substitute the values for h, k, and r into the standard form and simplify.

$$[x - (-2)]^2 + (y - 1)^2 = (4)^2$$
$$(x + 2)^2 + (y - 1)^2 = 16$$
$$x^2 + 4x + 4 + y^2 - 2y + 1 = 16$$
$$x^2 + y^2 + 4x - 2y + 5 = 16$$
$$x^2 + y^2 + 4x - 2y - 11 = 0$$

6. Graph $x^2 + y^2 + 2x - 8y + 8 = 0$.

$$x^2 + y^2 + 2x - 8y + 8 = 0$$
$$x^2 + 2x + y^2 - 8y = -8$$
$$(x^2 + 2x + 1) + (y^2 - 8x + 16) = -8 + 1 + 16$$
$$(x + 1)^2 + (y - 4)^2 = 9$$

The center is at $(-1, 4)$, and $r = \sqrt{9} = 3$.

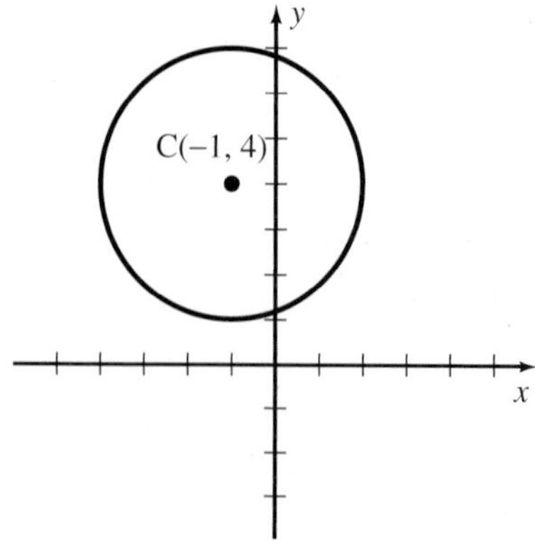

7. Graph $x^2 + y^2 + 16x - 240 = 0$.

$$x^2 + 16x + y^2 - 240 = 0$$
$$x^2 + 16x + y^2 = 240$$
$$(x^2 + 16x + 64) + y^2 = 240 + 64$$
$$(x + 8)^2 + (y - 0)^2 = 304$$

The center is at $(-8, 0)$, and $r = \sqrt{304} \approx 17.4$.

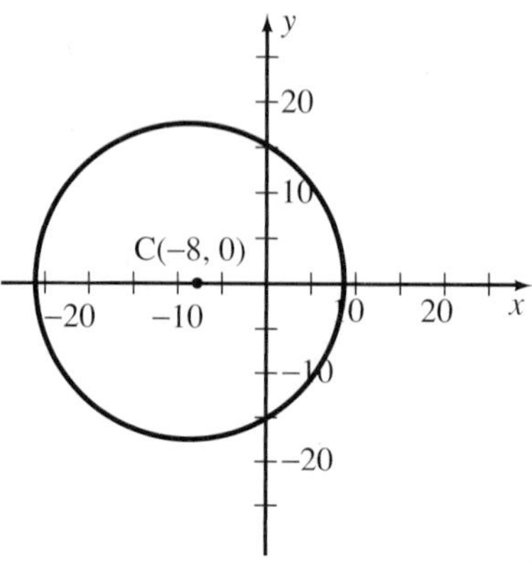

Section 9.4

1. Graph $9x^2 + 16y^2 = 144$.

If $x = 0$, then

$$9x^2 + 16y^2 = 144$$
$$9(0)^2 + 16y^2 = 144$$
$$16y^2 = 144$$
$$y^2 = 9$$
$$y = \pm 3$$

So $(0, -3)$ and $(0, 3)$ are on the graph.

If $y = 0$, then

$$9x^2 + 16y^2 = 144$$
$$9x^2 + 16(0)^2 = 144$$
$$9x^2 = 144$$
$$x^2 = 16$$
$$x = \pm 4$$

So $(-4, 0)$ and $(4, 0)$ are on the graph.

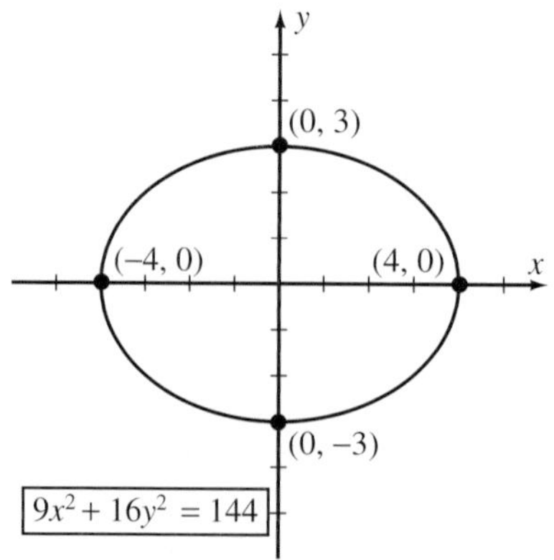

2. Graph $25x^2 + 9y^2 = 225$.

If $x = 0$, then

$$25x^2 + 9y^2 = 225$$

$$25(0)^2 + 9y^2 = 225$$
$$9y^2 = 225$$
$$y^2 = 25$$
$$y = \pm 5$$

So $(0, -5)$ and $(0, 5)$ are on the graph.

If $y = 0$, then

$$25x^2 + 9y^2 = 225$$
$$25x^2 + 9(0)^2 = 225$$
$$25x^2 = 225$$
$$x^2 = 9$$
$$x = \pm 3$$

So $(-3, 0)$ and $(3, 0)$ are on the graph.

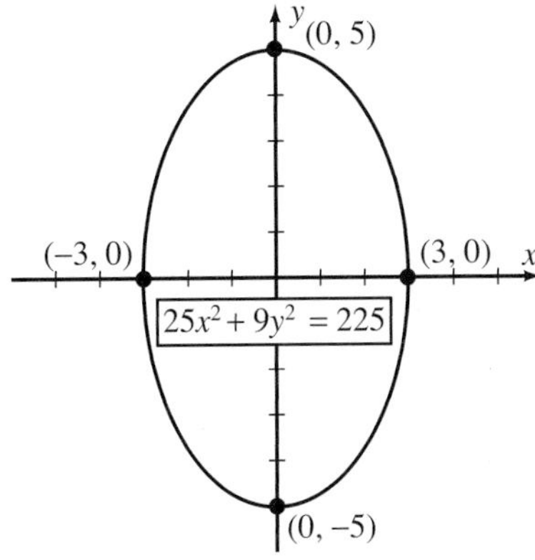

3. Graph $4x^2 + 16x + 9y^2 - 18y - 11 = 0$.

$$4x^2 + 16x + 9y^2 - 18y - 11 = 0$$
$$4(x^2 + 4x + 4) + 9(y^2 - 2y + 1) = 11 + 16 + 9$$
$$4(x + 2)^2 + 9(y - 1)^2 = 36$$

The center is at $(-2, 1)$.

Let $x = -2$. Then

$$4(-2 + 2)^2 + 9(y - 1)^2 = 36$$
$$4(0)^2 + 9(y - 1)^2 = 36$$
$$9(y - 1)^2 = 36$$
$$(y - 1)^2 = 4$$
$$y - 1 = \pm 2$$
$$y - 1 = -2 \quad \text{or} \quad y - 1 = 2$$
$$y = -1 \quad \text{or} \quad y = 3$$

So $(-2, -1)$ and $(-2, 3)$ are on the graph.

Let $y = 1$. Then

$$4(x + 2)^2 + 9(1 - 1)^2 = 36$$
$$4(x + 2)^2 + 9(0)^2 = 36$$
$$4(x + 2)^2 = 36$$
$$(x + 2)^2 = 9$$
$$x + 2 = \pm 3$$
$$x + 2 = -3 \quad \text{or} \quad x + 2 = 3$$
$$x = -5 \quad \text{or} \quad x = 1$$

So $(-5, 1)$ and $(1, 1)$ are on the graph.

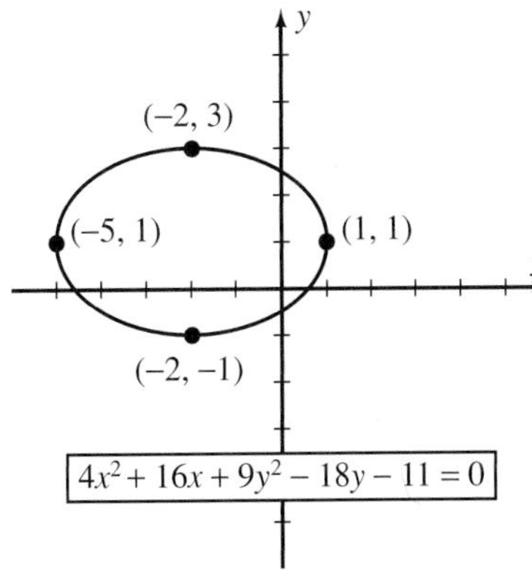

4. Graph $16x^2 - 96x + 9y^2 + 18y + 9 = 0$.

$$16x^2 - 96x + 9y^2 + 18y + 9 = 0$$
$$16(x^2 - 6x + 9) + 9(y^2 + 2y + 1) = -9 + 144 + 9$$
$$16(x - 3)^2 + 9(y + 1)^2 = 144$$

The center is at $(3, -1)$.

Let $x = 3$. Then

$$16(3 - 3)^2 + 9(y + 1)^2 = 144$$
$$16(0)^2 + 9(y + 1)^2 = 144$$
$$9(y + 1)^2 = 144$$
$$(y + 1)^2 = 16$$
$$y + 1 = \pm 4$$
$$y + 1 = -4 \quad \text{or} \quad y + 1 = 4$$
$$y = -5 \quad \text{or} \quad y = 3$$

So $(3, -5)$ and $(3, 3)$ are on the graph.

Let $y = -1$. Then

$$16(x - 3)^2 + 9(y + 1)^2 = 144$$
$$16(x - 3)^2 + 9(0)^2 = 144$$
$$16(x - 3)^2 = 144$$
$$(x - 3)^2 = 9$$
$$x - 3 = \pm 3$$
$$x - 3 = -3 \quad \text{or} \quad x - 3 = 3$$
$$x = 0 \quad \text{or} \quad x = 6$$

So $(0, -1)$ and $(6, -1)$ are on the graph.

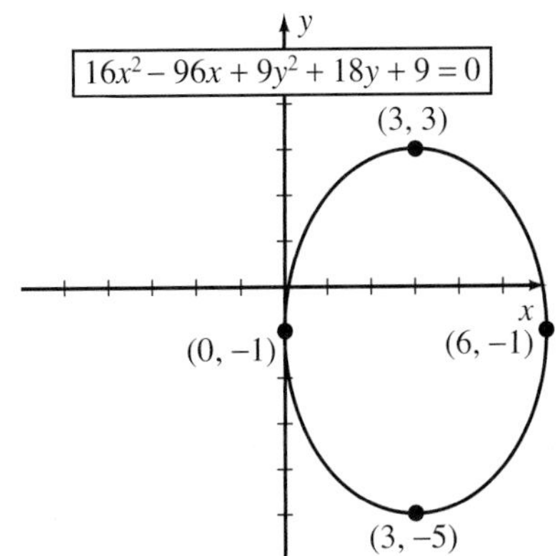

Section 9.5

1. Graph $x^2 - y^2 = 25$.

Let $y = 0$. Then

$x^2 - (0)^2 = 25$

$x^2 = 25$

$x = \pm 5$

So $(-5, 0)$ and $(5, 0)$ are on the graph.

Let $x = 0$. Then

$(0)^2 - y^2 = 25$

$-y^2 = 25$

$y^2 = -25$

Because $y^2 = -25$ has no real number solutions, there are no points on the y axis.

To find other solutions, solve $x^2 - y^2 = 25$ for y.

$x^2 - y^2 = 25$

$-y^2 = 25 - x^2$

$y^2 = x^2 - 25$

$y = \pm\sqrt{x^2 - 25}$

Make a chart:

x	y
6	$\sqrt{11} \approx 3.3$
6	$-\sqrt{11} \approx -3.3$
-6	$\sqrt{11} \approx 3.3$
-6	$-\sqrt{11} \approx -3.3$
7	$\pm\sqrt{24} \approx \pm 4.9$
-7	$\pm\sqrt{24} \approx \pm 4.9$

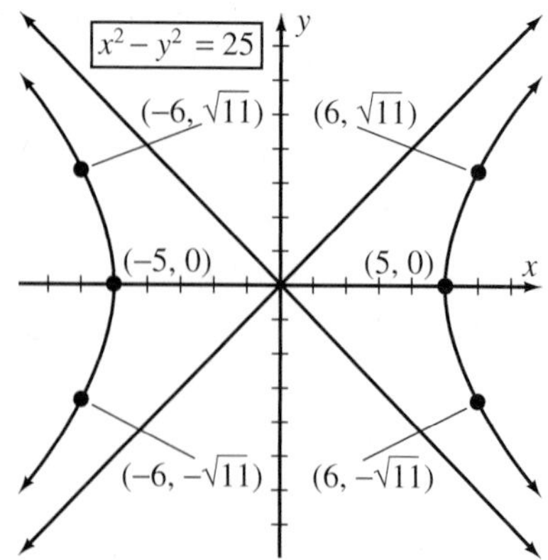

2. Graph $y^2 - 6x^2 = 9$.

Let $x = 0$. Then

$y^2 - 6(0)^2 = 9$

$y^2 = 9$

$y = \pm 3$

So $(0, -3)$ and $(0, 3)$ are on the graph.

Let $y = 0$. Then

$(0)^2 - 6x^2 = 9$

$-6x^2 = 9$

$x^2 = -\frac{9}{6}$

Because $x^2 = -\frac{9}{6}$ has no real number solutions, there are no points on the x axis.

To find the asymptotes, solve $y^2 - 6x^2 = 0$ for y.

$y^2 - 6x^2 = 0$

$y^2 = 6x^2$

$y = \pm\sqrt{6x^2}$

$y = \pm\sqrt{6}x$

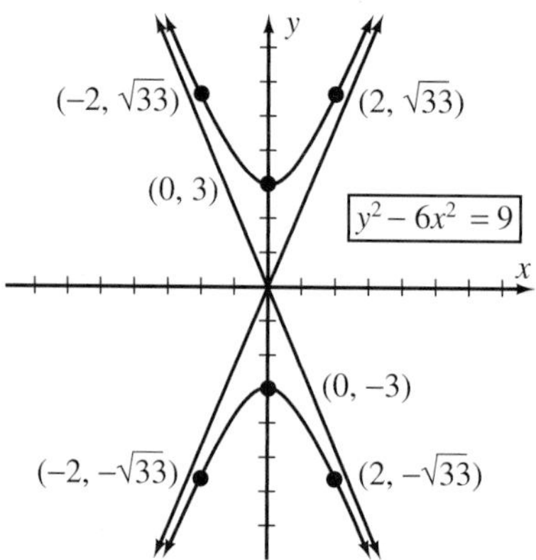

3. Graph $9x^2 - 25y^2 = 225$.

Let $y = 0$. Then

$9x^2 - 25(0)^2 = 225$

$9x^2 = 225$

$x^2 = 25$

$x = \pm 5$

So $(-5, 0)$ and $(5, 0)$ are on the graph.

Let $x = 0$. Then

$9(0)^2 - 25y^2 = 225$

$-25y^2 = 225$

$y^2 = -9$

Because $y^2 = -9$ has no real number solutions, there are no points on the y axis.

To find the asymptotes, solve $9x^2 - 25y^2 = 0$ for y.

$9x^2 - 25y^2 = 0$

$-25y^2 = -9x^2$

$$25y^2 = 9x^2$$
$$y^2 = \frac{9}{25}x^2$$
$$y = \pm\sqrt{\frac{9}{25}x^2}$$
$$y = \pm\frac{3}{5}x$$

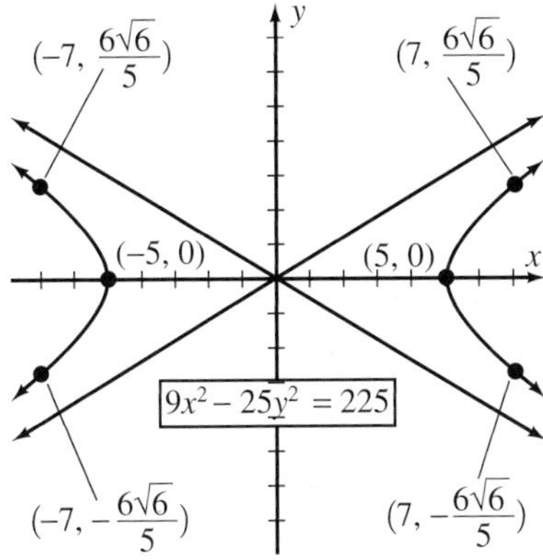

4. Graph $4x^2 + 16x - y^2 + 2y - 1 = 0$.

$$4x^2 + 16x - y^2 + 2y - 1 = 0$$
$$4(x^2 + 4x + 4) - (y^2 - 2y + 1) = 1 + 16 - 1$$
$$4(x + 2)^2 - (y - 1)^2 = 16$$

Let $y = 1$. Then

$$4(x + 2)^2 - (1 - 1)^2 = 16$$
$$4(x + 2)^2 - (0)^2 = 16$$
$$4(x + 2)^2 = 16$$
$$(x + 2)^2 = 4$$
$$x + 2 = \pm 2$$
$$x + 2 = -2 \quad \text{or} \quad x + 2 = 2$$
$$x = -4 \quad \text{or} \quad x = 0$$

So $(-4, 1)$ and $(0, 1)$ are on the graph.

Let $x = -2$. Then

$$4(-2 + 2)^2 - (y - 1)^2 = 16$$
$$4(0)^2 - (y - 1)^2 = 16$$
$$-(y - 1)^2 = 16$$
$$(y - 1)^2 = -16$$

Because $(y - 1)^2 = -16$ has no real number solutions, there are no points on the line $x = -2$.

To find the asymptotes, solve $4(x + 2)^2 - (y - 1)^2 = 0$ for y.

$$4(x + 2)^2 - (y - 1)^2 = 0$$

Factor the difference of two squares.

$$[2(x + 2) + (y - 1)][2(x + 2) - (y - 1)] = 0$$
$$(2x + 4 + y - 1)(2x + 4 - y + 1) = 0$$
$$(2x + y + 3)(2x - y + 5) = 0$$
$$2x + y + 3 = 0 \quad \text{or} \quad 2x - y + 5 = 0$$
$$y = -2x - 3 \quad \text{or} \quad 2x + 5 = y$$

The asymptotes are $y = -2x - 3$ and $y = 2x + 5$.

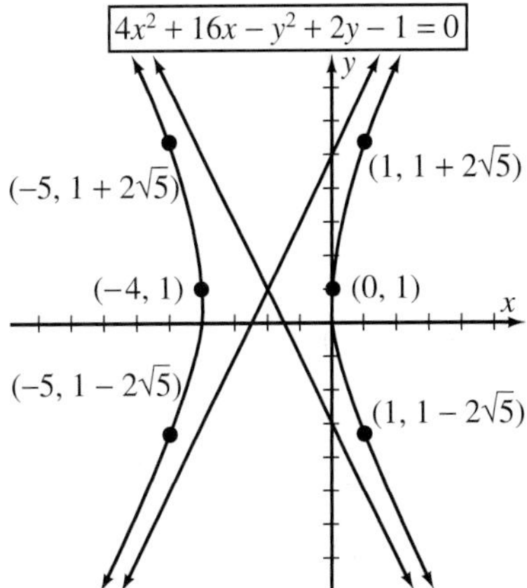

5. Graph $y^2 + 2y - 9x^2 + 36x - 44 = 0$.

$$y^2 + 2y - 9x^2 + 36x - 44 = 0$$
$$(y^2 + 2y + 1) - 9(x^2 - 4x + 4) = 44 + 1 - 36$$
$$(y + 1)^2 - 9(x - 2)^2 = 9$$

Let $y = -1$. Then

$$(-1 + 1)^2 - 9(x - 2)^2 = 9$$
$$(0)^2 - 9(x - 2)^2 = 9$$
$$-9(x - 2)^2 = 9$$
$$(x - 2)^2 = -1$$

Because $(x - 2)^2 = -1$ has no real number solutions, there are no points on the line $y = -1$.

Let $x = 2$. Then

$$(y + 1)^2 - 9(2 - 2)^2 = 9$$
$$(y + 1)^2 - 9(0)^2 = 9$$
$$(y + 1)^2 = 9$$
$$y + 1 = \pm 3$$
$$y + 1 = -3 \quad \text{or} \quad y + 1 = 3$$
$$y = -4 \quad \text{or} \quad y = 2$$

So $(2, -4)$ and $(2, 2)$ are on the graph.

To find the asymptotes, solve $(y + 1)^2 - 9(x - 2)^2 = 0$ for y.

$$(y + 1)^2 - 9(x - 2)^2 = 0$$

Factor the difference of two squares.

$$[y + 1 - 3(x - 2)][y + 1 + 3(x - 2)] = 0$$

$$(y + 1 - 3x + 6)(y + 1 + 3x - 6) = 0$$
$$(y - 3x + 7)(y + 3x - 5) = 0$$
$$y - 3x + 7 = 0 \quad \text{or} \quad y + 3x - 5 = 0$$
$$y = 3x - 7 \quad \text{or} \quad y = -3x + 5$$

The asymptotes are $y = 3x - 7$ and $y = -3x + 5$.

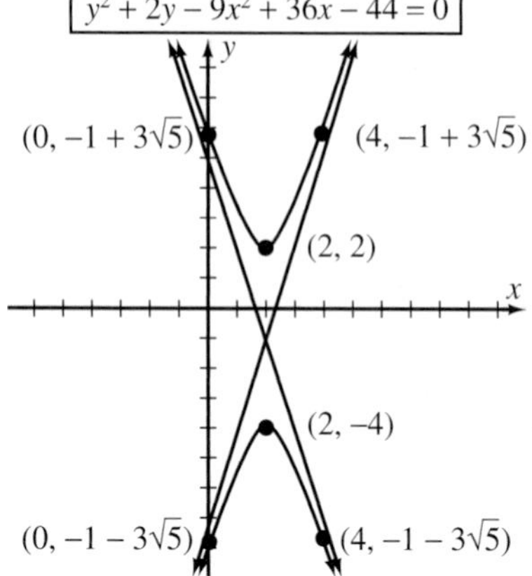

Section 9.6

1. Solve the system $\begin{pmatrix} x^2 + y^2 = 10 \\ x + y = 2 \end{pmatrix}$.

The graph of $x^2 + y^2 = 10$ is a circle, and the graph of $x + y = 2$ is a line. There appear to be two real solutions.

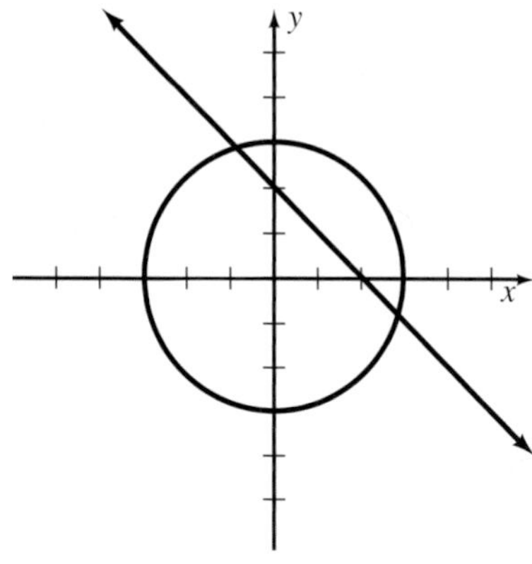

Solving $x + y = 2$ for y gives $y = 2 - x$. Substitute $2 - x$ for y in the first equation and solve for x.

$$x^2 + y^2 = 10$$
$$x^2 + (2 - x)^2 = 10$$
$$x^2 + 4 - 4x + x^2 = 10$$
$$2x^2 - 4x - 6 = 0$$
$$x^2 - 2x - 3 = 0$$
$$(x - 3)(x + 1) = 0$$
$$x - 3 = 0 \quad \text{or} \quad x + 1 = 0$$
$$x = 3 \quad \text{or} \quad x = -1$$

If $x = 3$, then $y = 2 - (3) = -1$.
If $x = -1$, then $y = 2 - (-1) = 3$.
The solution set is $\{(3, -1), (-1, 3)\}$.

2. Solve the system $\begin{pmatrix} y = x^2 + 1 \\ y = -x^2 + 4 \end{pmatrix}$.

Both equations are the graphs of parabolas. There appear to be two real solutions.

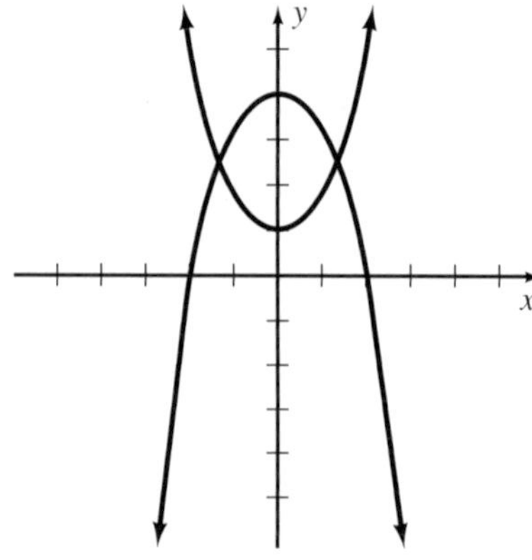

Substitute $x^2 + 1$ for y in the second equation.

$$y = -x^2 + 4$$
$$x^2 + 1 = -x^2 + 4$$
$$2x^2 = 3$$
$$x^2 = \frac{3}{2}$$
$$x = \pm\sqrt{\frac{3}{2}} = \pm\frac{\sqrt{3}}{\sqrt{2}} \cdot \frac{\sqrt{2}}{\sqrt{2}}$$
$$= \pm\frac{\sqrt{6}}{2}$$

Substitute $-\frac{\sqrt{6}}{2}$ for x in the second equation.

$$y = -x^2 + 4$$
$$= -\left(\frac{\sqrt{6}}{2}\right)^2 + 4$$
$$= -\frac{6}{4} + 4 = -\frac{3}{2} + \frac{8}{2}$$
$$= \frac{5}{2}$$

Similarly, when $x = \frac{\sqrt{6}}{2}$, $y = \frac{5}{2}$.

The solution set is $\left\{\left(-\frac{\sqrt{6}}{2}, \frac{5}{2}\right), \left(\frac{\sqrt{6}}{2}, \frac{5}{2}\right)\right\}$.

3. Solve the system $\begin{pmatrix} 2x^2 + y^2 = 32 \\ x^2 + y^2 = 16 \end{pmatrix}$.

The graph of $2x^2 + y^2 = 32$ is an ellipse, and the graph of $x^2 + y^2 = 16$ is a circle. There appear to be two real solutions.

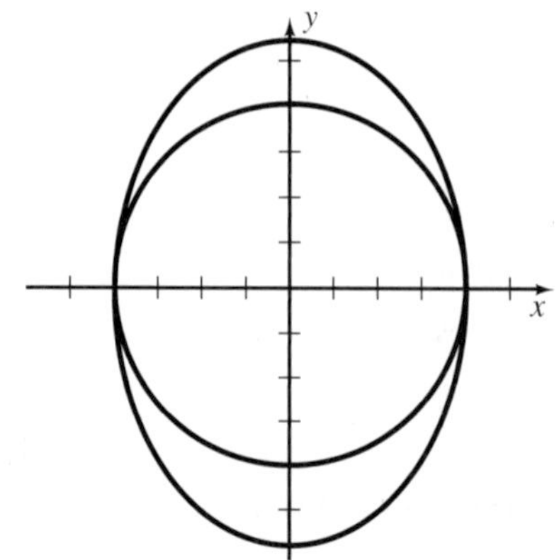

Solve the system using the elimination-by-addition method. Form an equivalent system by multiplying the second equation by -1 and adding the result to the first equation.

$$\begin{pmatrix} 2x^2 + y^2 = 32 \\ x \quad\quad = 16 \end{pmatrix}$$

Solve $x^2 = 16$ for x.

$x^2 = 16$

$x = \pm 4$

To find y, substitute $x = -4$ and $x = 4$ in the second equation.

$$(-4)^2 + y^2 = 16$$
$$16 + y^2 = 16$$
$$y^2 = 0$$
$$y = 0$$

Similarly, when $x = 4$, $y = 0$.

The solution set is $\{(-4, 0), (4, 0)\}$.

4. Solve the system $\begin{pmatrix} y = -x^2 + 2 \\ 8x + 4y = 13 \end{pmatrix}$.

The graph of $y = -x^2 + 2$ is a parabola, and the graph of $8x + 4y = 13$ is a line. It appears that there are no real solutions.

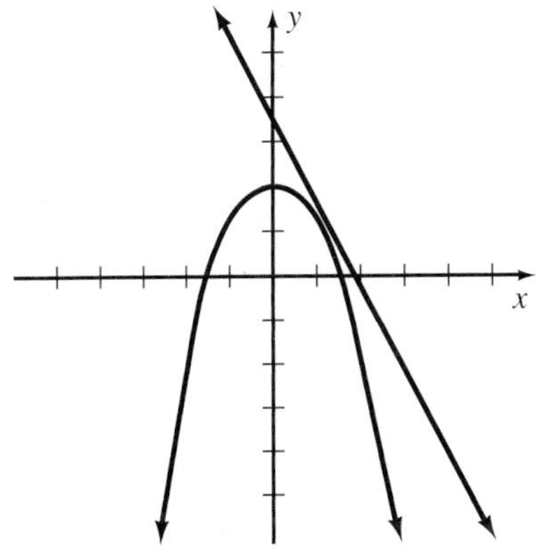

Substitute $-x^2 + 2$ for y in the second equation.

$$8x + 4y = 13$$
$$8x + 4(-x^2 + 2) = 13$$
$$8x - 4x^2 + 8 = 13$$
$$-4x^2 + 8x - 5 = 0$$
$$4x^2 - 8x + 5 = 0$$

Use the quadratic formula to solve.

$a = 4, b = -8, c = 5$

$$x = \frac{-(-8) \pm \sqrt{(-8)^2 - 4(4)(5)}}{2(4)}$$
$$= \frac{8 \pm \sqrt{64 - 80}}{8} = \frac{8 \pm \sqrt{-16}}{8}$$
$$= \frac{8 \pm 4i}{8} = \frac{4(2 \pm i)}{8}$$
$$= \frac{2 \pm i}{2}$$

Substitute $\frac{2 - i}{2}$ for x in the first equation to find y.

$$y = -x^2 + 2$$
$$= -\left(\frac{2 - i}{2}\right)^2 + 2$$
$$= -\left(\frac{4 - 4i + i^2}{4}\right) + 2$$
$$= \frac{-4 + 4i - i^2}{4} + 2$$
$$= \frac{-4 + 4i - (-1)}{4} + \frac{8}{4}$$
$$= \frac{-4 + 4i + 1 + 8}{4}$$
$$= \frac{5 + 4i}{4}$$

Similarly, when $x = \frac{2 + i}{2}$, $y = \frac{5 - 4i}{4}$.

The solution set is

$$\left\{\left(\frac{2 - i}{2}, \frac{5 + 4i}{4}\right), \left(\frac{2 + i}{2}, \frac{5 - 4i}{4}\right)\right\}.$$

CHAPTER 10

Section 10.1

1. (a) D = $\{2, 3, 5, 7\}$, R = $\{3\}$.

Yes, the set of ordered pairs does determine a function. No first component is repeated. Every first component has one and only one second component.

(b) D = $\{-2, 2\}$, R = $\{3, 5\}$.

No, the set of ordered pairs does not determine a function. The ordered pairs $(2, 3)$ and $(2, 5)$ have the same first component and different second components.

(c) D = $\{4, 9\}$, R = $\{-3, -2, 2, 3\}$.

No, the set of ordered pairs does not determine a function. The ordered pairs $(4, 2)$ and $(4, -2)$ have the same first component and different second components.

2. $f(x) = x^2 + 4$

$$f(-1) = (-1)^2 + 4$$
$$= 1 + 4 = 5$$
$$f(0) = (0)^2 + 4$$
$$= 4$$

$$f(2) = (2)^2 + 4 = 4 + 4 = 8$$

3.
$$\begin{aligned} f(x) &= -x + 6 \\ f(2) &= -(2) + 6 \\ &= -2 + 6 = 4 \\ f(-5) &= -(-5) + 6 \\ &= 5 + 6 = 11 \\ f(a) &= -(a) + 6 \\ &= -a + 6 \\ f(2b) &= -(2b) + 6 \\ &= -2b + 6 \\ g(x) &= x^2 - 9 \\ g(-2) &= (-2)^2 - 9 \\ &= 4 - 9 = -5 \\ g(1) &= (1)^2 - 9 \\ &= 1 - 9 = -8 \\ g(2a) &= (2a)^2 - 9 \\ &= 4a^2 - 9 \\ g(a+1) &= (a+1)^2 - 9 \\ &= a^2 + 2a + 1 - 9 \\ &= a^2 + 2a - 8 \end{aligned}$$

4. (a) $f(x) = \dfrac{x+3}{2x+1}$.

Eliminate any value of x that would make the denominator 0. So solve

$$\begin{aligned} 2x + 1 &= 0 \\ 2x &= -1 \\ x &= -\frac{1}{2} \end{aligned}$$

$D = \left\{x \mid x \neq -\frac{1}{2}\right\}$ or D: $\left(-\infty, -\frac{1}{2}\right) \cup \left(-\frac{1}{2}, \infty\right)$.

(b) $g(x) = \dfrac{4}{x^2 - x - 12}$.

Eliminate any value of x that would make the denominator 0. So solve

$$x^2 - x - 12 = 0$$
$$(x-4)(x+3) = 0$$
$$x - 4 = 0 \quad \text{or} \quad x + 3 = 0$$
$$x = 4 \quad \text{or} \quad x = -3$$

$D = \{x \mid x \neq -3 \text{ and } x \neq 4\}$ or
D: $(-\infty, -3) \cup (-3, 4) \cup (4, \infty)$.

(c) $h(x) = \sqrt{x+4}$.

The radicand $x + 4$ must be nonnegative. So

$$\begin{aligned} x + 4 &\geq 0 \\ x &\geq -4 \end{aligned}$$

$D = \{x \mid x \geq -4\}$ or D: $[-4, \infty)$.

5.
$$\begin{aligned} f(x) &= 4x + 1 \\ f(a+h) &= 4(a+h) + 1 \\ &= 4a + 4h + 1 \\ f(a) &= 4(a) + 1 \\ &= 4a + 1 \\ f(a+h) - f(a) &= 4a + 4h + 1 - (4a + 1) \\ &= 4a + 4h + 1 - 4a - 1 \\ &= 4h \\ \frac{f(a+h) - f(a)}{h} &= \frac{4h}{h} = 4 \end{aligned}$$

6.
$$\begin{aligned} f(x) &= x^2 - 3x + 1 \\ f(a+h) &= (a+h)^2 - 3(a+h) + 1 \\ &= a^2 + 2ah + h^2 - 3a - 3h + 1 \\ f(a) &= a^2 - 3a + 1 \\ f(a+h) - f(a) &= a^2 + 2ah + h^2 - 3a - 3h + 1 \\ &\quad - (a^2 - 3a + 1) \\ &= a^2 + 2ah + h^2 - 3a - 3h + 1 \\ &\quad - a^2 + 3a - 1 \\ &= 2ah + h^2 - 3h \\ \frac{f(a+h) - f(a)}{h} &= \frac{2ah + h^2 - 3h}{h} \\ &= \frac{h(2a + h - 3)}{h} \\ &= 2a + h - 3 \end{aligned}$$

7. Let x represent the number of cell phone minutes. Then $0.03x + 29.99$ represents the monthly cell phone cost.

$$\begin{aligned} C(x) &= 0.03x + 29.99 \\ C(300) &= 0.03(300) + 29.99 \\ &= 9 + 29.99 = 38.99 \\ C(500) &= 0.03(500) + 29.99 \\ &= 15 + 29.99 = 44.99 \\ C(1000) &= 0.03(1000) + 29.99 \\ &= 30 + 29.99 = 59.99 \end{aligned}$$

The cost for using 300, 500, and 1000 minutes is \$38.99, \$44.99, and \$59.99, respectively.

Section 10.2

1. Graph $f(x) = \frac{1}{2}x + 3$.

Because $f(0) = 3$ and $f(-6) = 0$, plot $(0, -3)$ and $(-6, 0)$ to determine the graph of the line.

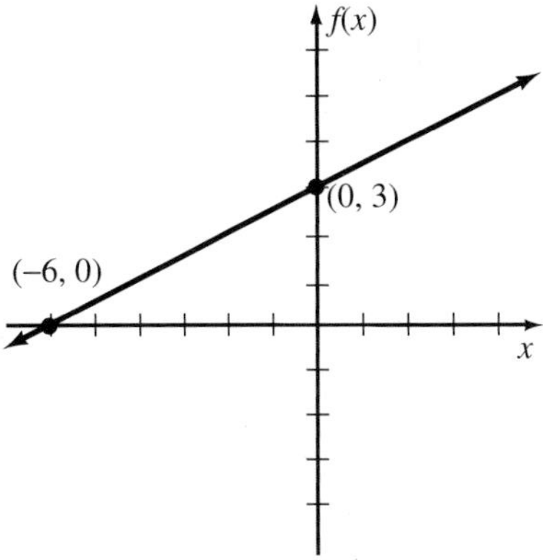

2. (a) $c(m) = 0.5m$.

Three sessions at 2 hours (120 minutes) is $3(120) = 360$ minutes.

$c(360) = 0.5(360) = 180$

The cost for tutoring is \$180.

(b) Graph $c(m) = 0.5m$.

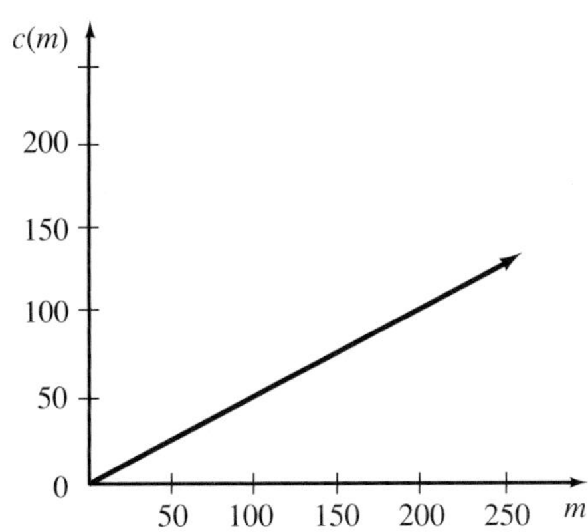

(c) $c(100) \approx \$50$

$c(100) = 0.5(100) = \$50$

3. Model the situation using the linear function $f(x) = ax + b$, where x represents the number of minutes of consultation.

Morgan's two consultations can be represented by the ordered pairs (20, 240) and (30, 280). Use these to determine a, the slope.

$$a = \frac{280 - 240}{30 - 20} = \frac{40}{10} = 4$$

$f(x) = 4x + b$. Use either ordered pair to find b.

$240 = 4(20) + b$

$240 = 80 + b$

$160 = b$

The linear function is $f(x) = 4x + 160$, where x represents the number of minutes.

4. ETF Group charges according to the function $f(x) = 4x + 160$ and Ever Ready Legal Corporation charges according to the function $g(x) = 8x + 100$.

To find equal values, set $f(x) = g(x)$.

$4x + 160 = 8x + 100$

$-4x + 160 = 100$

$-4x = -60$

$x = 15$

If $x = 15$, then $g(15) = 8(15) + 100 = 220$.

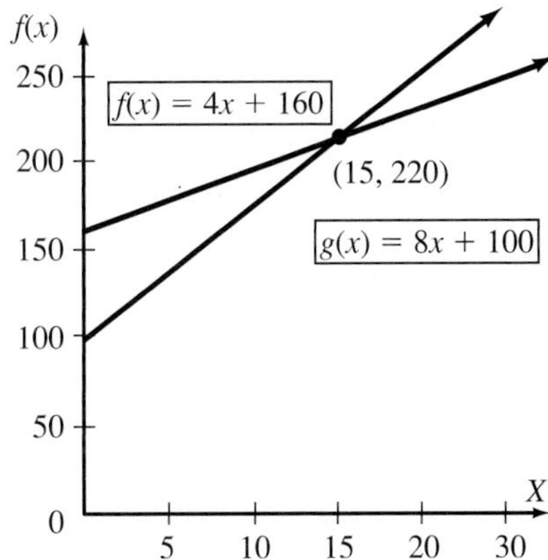

The point of intersection is (15, 220). Looking at the lines, Morgan should use Ever Ready Legal for consultations of 15 minutes or less and ETF Group for consultations of 15 minutes or more.

5. Graph $f(x) = x^2 - 6x + 7$.

$f(x) = x^2 - 6x + 7$

$f(x) = (x^2 - 6x + 9) + 7 - 9$

$= (x - 3)^2 - 2$

From the graph $f(x) = x^2$, the parabola is moved three units right and two units downward.

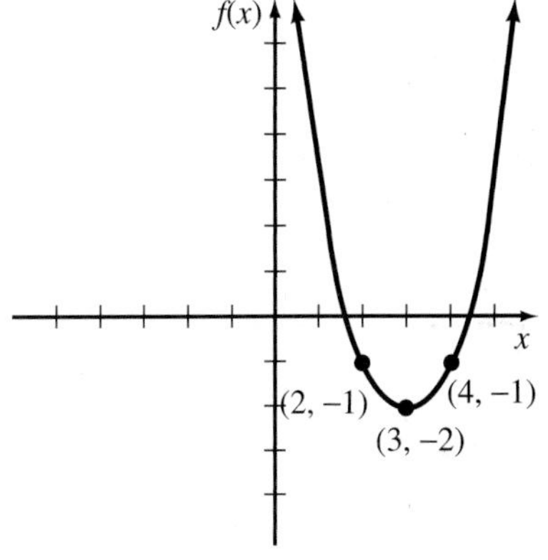

6. Graph $f(x) = 2x^2 - 8x + 3$.

Because $a = 2$, the parabola opens upward.

$$-\frac{b}{2a} = -\frac{-8}{2(2)} = \frac{8}{4} = 2$$

$$f\left(-\frac{b}{2a}\right) = f(2) = 2(2)^2 - 8(2) + 3$$

$$= 8 - 16 - 3 = -5$$

The vertex is at $(2, -5)$.

Let $x = 3$, so $f(3) = 2(3)^2 - 8(3) + 3 = -3$. Therefore $(3, -3)$ is on the graph. And so across the line of symmetry, $x = 2$, the point $(1, -3)$ is also on the graph.

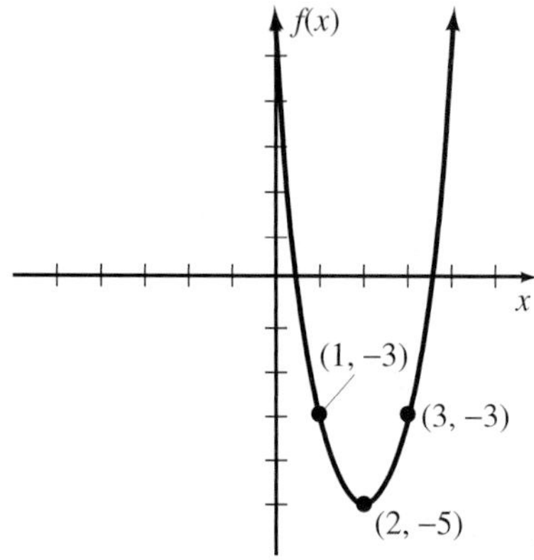

7. Graph $f(x) = -x^2 + 6x - 2$.

Because $a = -1$, the parabola opens downward.

$$-\frac{b}{2a} = -\frac{6}{2(-1)} = -\frac{6}{-2} = 3$$

$$f\left(-\frac{b}{2a}\right) = f(3) = -(3)^2 + 6(3) - 2$$

$$= -9 + 18 - 2 = 7$$

The vertex is at $(3, 7)$.

Let $x = 2$, so $f(2) = -(2)^2 + 6(2) - 2 = 6$. Therefore, $(2, 6)$ is on the graph. And so is its reflection $(4, 6)$ across the line of symmetry, $x = 3$.

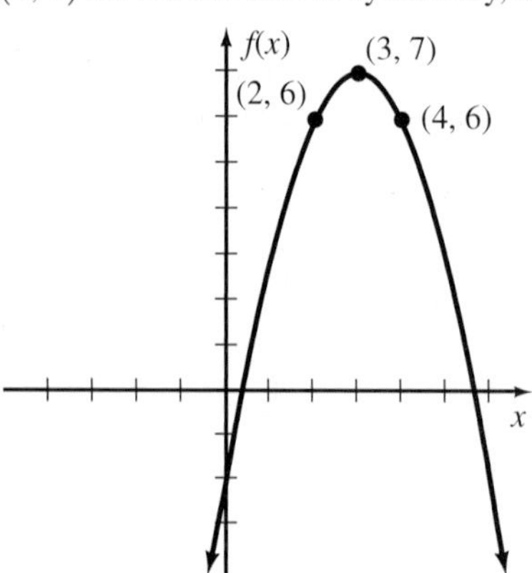

8. Let x represent the width of the dog run. Then $\frac{80 - 2x}{2} = 40 - x$ represents the length.

The function $A(x) = x(40 - x)$ represents the area in terms of x.

$$A(x) = x(40 - x)$$
$$= 40x - x^2$$
$$= -x^2 + 40x$$

In this quadratic function, $a = -1, b = 40, c = 0$, so the x value for the maximum value of the function is

$$-\frac{b}{2a} = -\frac{40}{2(-1)} = \frac{-40}{-2} = 20$$

If $x = 20$, then $40 - x = 20$. The length of the plot is 20 yards, and the width of the plot is 20 yards.

9. Let x represent one number. Then x^2 represents its square. Let $x - 14$ represent the other number. Then $(x - 14)^2$ represents its square.

$$f(x) = x^2 + (x - 14)^2$$
$$f(x) = x^2 + x^2 - 28x + 196$$
$$f(x) = 2x^2 - 28x + 196$$

Because $a = 2, b = -28$, and $c = 196$,

$$-\frac{b}{2a} = -\frac{-28}{2(2)} = -\frac{-28}{4} = 7$$

If $x = 7$, then $x - 14 = -7$. The two numbers are 7 and -7.

10. Use a table to analyze the problem.

$$\begin{pmatrix}\text{Number of}\\ \text{treadmills}\end{pmatrix} \times \begin{pmatrix}\text{Price per}\\ \text{treadmill}\end{pmatrix} = \text{Income}$$

	Number of treadmills	Price per treadmill	Income
Five additional units can be sold for each $20 increase	40	400	16,000
	45	380	17,100
	50	360	18,000

Let x represent the number of \$20 decreases. Then express the income as a function of x.

$$f(x) = (40 + 5x)(400 - 20x)$$

Simplify.

$$f(x) = 16{,}000 - 800x + 2000x - 100x^2$$
$$f(x) = 16{,}000 + 1200x - 100x^2$$

Complete the square.

$$f(x) = -100x^2 + 1200x + 16{,}000$$
$$f(x) = -100(x^2 - 12x + 36) + 16{,}000 + 3600$$
$$f(x) = -100(x - 6)^2 + 196{,}000$$

From this we can determine that the vertex is at (6, 19600).

So six decreases of \$20 each, a \$120 reduction in price, will give a minimum income of \$19600. Therefore, the price of the treadmill is $400 - 20x = 400 - 20(6) = \$280$.

11. Graph $f(x) = x^2 + 36x + 300$.

The parabola opens upward. Complete the square.

$f(x) = x^2 + 36x + 300$

$f(x) = (x^2 + 36x + 324) + 300 - 324$

$f(x) = (x + 18)^2 - 24$

The vertex is at $(-18, -24)$.

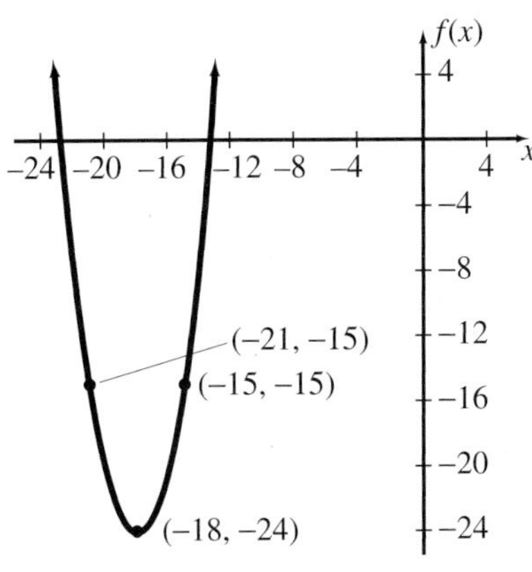

Section 10.3

1. Graph $f(x) = x^2 + 1$.

The graph of $f(x) = x^2 + 1$ is the graph of $f(x) = x^2$ moved up one unit.

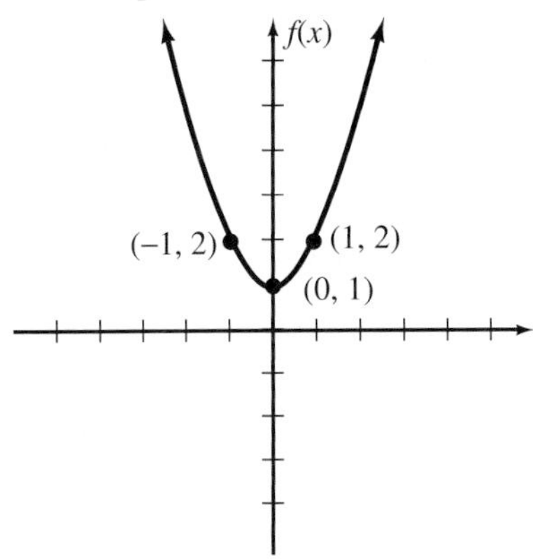

2. Graph $f(x) = |x + 4|$.

The graph of $f(x) = |x + 4|$ is the graph of $f(x) = |x|$ shifted four units to the left.

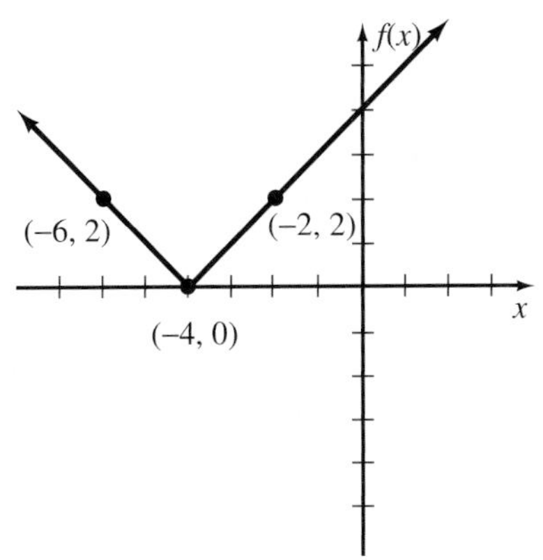

3. Graph $f(x) = -|x|$.

The graph of $f(x) = -|x|$ is the graph of $f(x) = |x|$ reflected through the x axis.

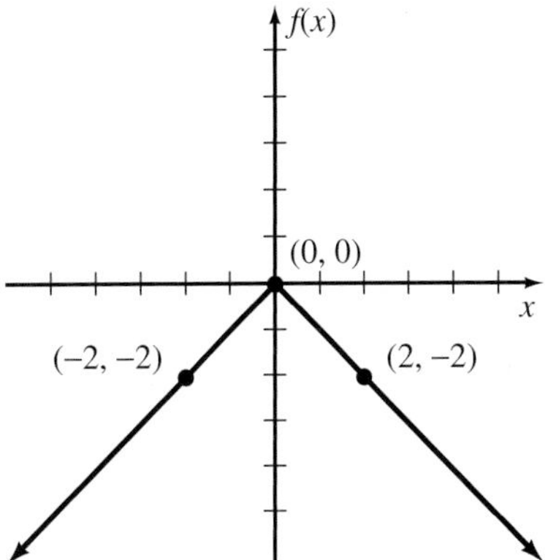

4. Graph $f(x) = 2|x|$.

The graph of $f(x) = 2|x|$ is the graph of $f(x) = |x|$ stretched by a factor of 2.

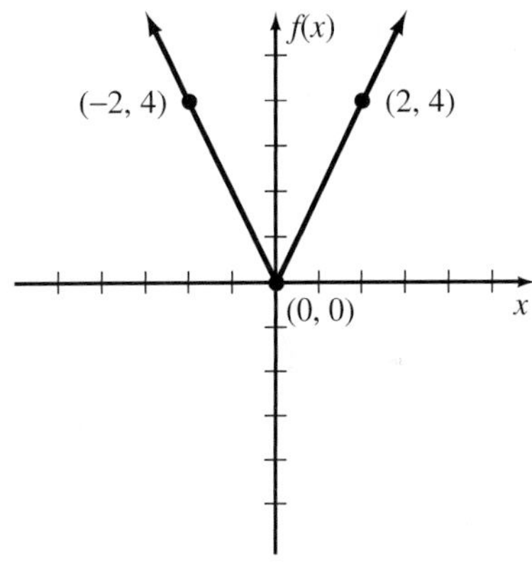

5. Graph $f(x) = \frac{1}{2}|x - 1| + 3$.

The graph of $f(x) = \frac{1}{2}|x - 1| + 3$ is the graph of $f(x) = |x|$ shrunk by a factor of $\frac{1}{2}$, moved one unit to the right, and shifted three units upward.

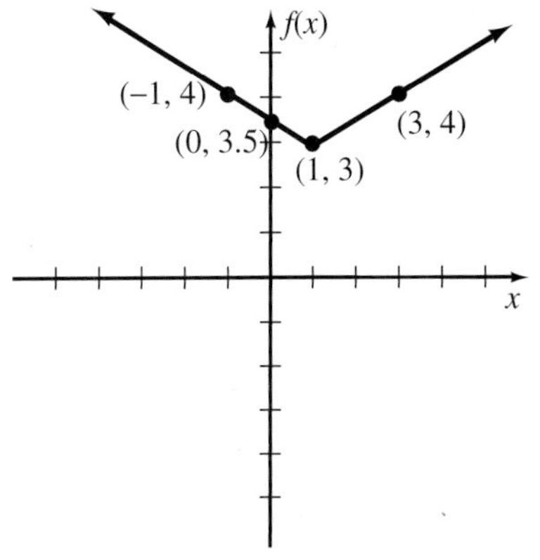

6. Graph $f(x) = \dfrac{1}{x-3} + 1$.

The graph of $f(x) = \dfrac{1}{x-3} + 1$ is the graph of $f(x) = \dfrac{1}{x}$ moved three units right and one unit upward. The vertical asymptote is shifted three units right to $x = 3$, and the horizontal asymptote is shifted one unit upward to $y = 1$.

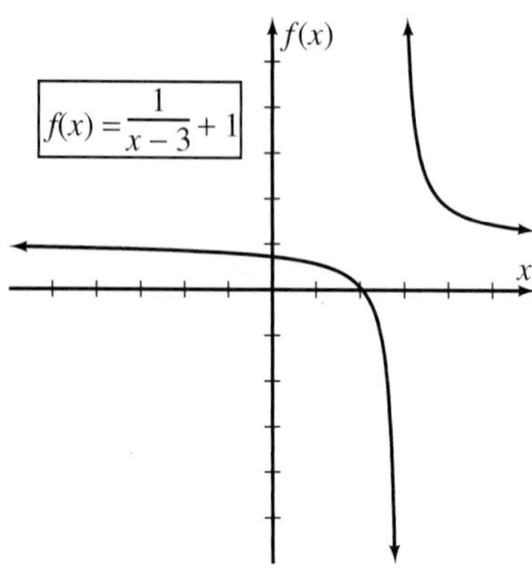

7. Graph $f(x) = -(x-2)^2$.

x	$y = f(x)$
0	-4
1	-1
2	0
3	-1
4	-4

$y = 3f(x) = -3(x-2)^2$

x	$y = 3f(x)$
0	-12
1	-3
2	0
3	-3
4	-12

$y = \frac{1}{3}f(x) = -\frac{1}{3}(x-2)^2$

x	$y = \frac{1}{3}f(x)$
0	$-\frac{4}{3}$
1	$-\frac{1}{3}$
2	0
3	$-\frac{1}{3}$
4	$-\frac{4}{3}$

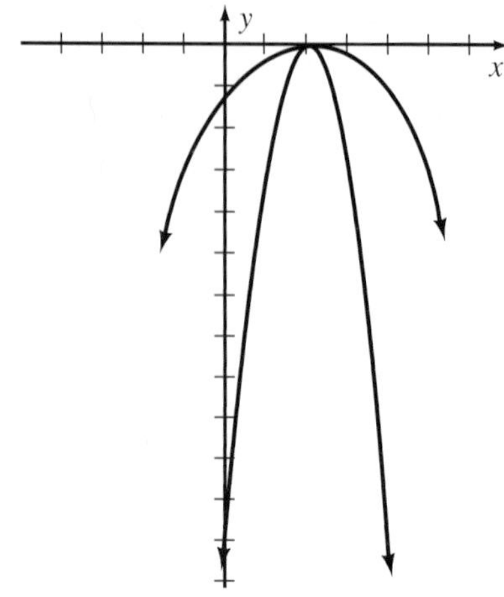

Section 10.4

1. $f(x) = 3x + 7$, $g(x) = 5x + 2$.

$$\begin{aligned}(f \circ g)(x) &= f(g(x)) \\ &= f(5x+2) \\ &= 3(5x+2)+7 \\ &= 15x+6+7 \\ &= 15x+13\end{aligned}$$

Because g and f are both defined for all real numbers, so is $f \circ g$. Therefore, D = {all reals}.

2. $f(x) = \sqrt{x}$, $g(x) = x + 2$.

$$\begin{aligned}(f \circ g)(x) &= f(g(x)) \\ &= f(x+2) \\ &= \sqrt{x+2}\end{aligned}$$

The domain of g is all real numbers, but the domain of f is only the nonnegative real numbers. So $g(x)$, which is $x + 2$, must be nonnegative. So

$$\begin{aligned}x + 2 &\geq 0 \\ x &\geq -2\end{aligned}$$

and the domain of $f \circ g$ is D = $\{x \mid x \geq -2\}$ or D: $[-2, \infty)$.

3. $f(x) = 3x + 7$, $g(x) = 5x + 2$.

$$\begin{aligned}(g \circ f)(x) &= g(f(x)) \\ &= g(3x+7) \\ &= 5(3x+7)+2 \\ &= 15x+35+2 \\ &= 15x+37\end{aligned}$$

Because g and f are both defined for all real numbers, so is $g \circ f$. Therefore, D = {all reals}.

4. $f(x) = \dfrac{5}{x+3}$, $g(x) = \dfrac{1}{x}$.

$$\begin{aligned}(f \circ g)(x) &= f\left(\frac{1}{x}\right) \\ &= \frac{5}{\frac{1}{x}+3} \\ &= \frac{5}{\frac{1}{x}+\frac{3x}{x}} = \frac{5}{\frac{1+3x}{x}}\end{aligned}$$

$$= \frac{5x}{1 + 3x}$$

The domain of g is all real numbers except 0, and the domain of f is all real numbers except -3. Because $g(x)$, which is $\frac{1}{x}$, cannot equal -3,

$$\frac{1}{x} \neq -3$$

$$1 \neq -3x$$

$$-\frac{1}{3} \neq x$$

Therefore, the domain of $f \circ g$ is D = $\left\{x|x \neq -\frac{1}{3} \text{ and } x \neq 0\right\}$ or D: $\left(-\infty, -\frac{1}{3}\right) \cup \left(-\frac{1}{3}, 0\right) \cup (0, \infty)$.

$$(g \circ f)(x) = g\left(\frac{5}{x+3}\right)$$

$$= \frac{1}{\frac{5}{x+3}}$$

$$= \frac{x+3}{5}$$

The domain of f is all real numbers except -3. The domain of g is all real numbers except 0. Because $f(x)$, which is $\frac{5}{x+3}$, will never equal 0, the domain of $g \circ f$ is D = $\{x|x \neq -3\}$ or D: $(-\infty, -3) \cup (-3, \infty)$.

5. $f(x) = 3x + 4$, $g(x) = \sqrt{x+2}$.

(a) $(f \circ g)(x) = f(g(x))$
$= f(\sqrt{x+2})$
$= 3\sqrt{x+2} + 4$

(b) $(g \circ f)(x) = g(f(x))$
$= g(3x + 4)$
$= \sqrt{(3x+4) + 2}$
$= \sqrt{3x+6}$

(c) $(f \circ g)(-1) = 3\sqrt{(-1) + 2} + 4$
$= 3\sqrt{1} + 4$
$= 3 + 4 = 7$

(d) $(g \circ f)(10) = \sqrt{3(10) + 6}$
$= \sqrt{36}$
$= 6$

6. $f(x) = x^2$, $g(x) = 3x - 1$.

$(f \circ g)(x) = f(g(x))$
$= f(3x - 1)$
$= (3x - 1)^2$
$= 9x^2 - 6x + 1$

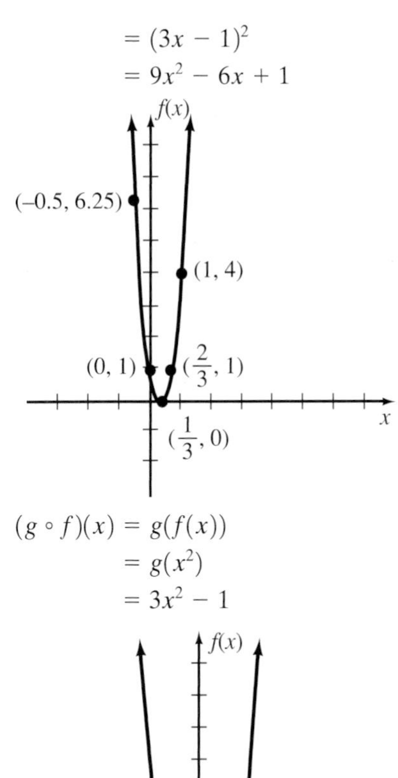

$(g \circ f)(x) = g(f(x))$
$= g(x^2)$
$= 3x^2 - 1$

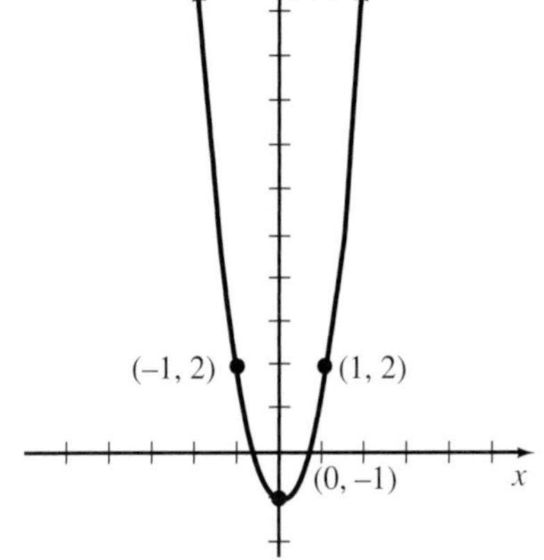

Section 10.5

1. $f(x) = \{(0, 3), (1, 4), (5, 6), (7, 9)\}$.

(a) D = $\{0, 1, 5, 7\}$, R = $\{3, 4, 6, 9\}$.

(b) $f^{-1} = \{(3, 0), (4, 1), (6, 5), (9, 7)\}$.

(c) D = $\{3, 4, 6, 9\}$, R = $\{0, 1, 5, 7\}$.

2. $f(x) = 2x + 7$.

To find $f^{-1}(x)$, subtract 7 and divide by 2.

$$f^{-1}(x) = \frac{x-7}{2}$$

$(f \circ f^{-1})(x) = f(f^{-1}(x))$
$= f\left(\frac{x-7}{2}\right)$
$= 2\left(\frac{x-7}{2}\right) + 7$
$= x - 7 + 7 = x$

$(f^{-1} \circ f)(x) = f^{-1}(f(x))$
$= f(2x + 7)$
$= \frac{2x + 7 - 7}{2}$

$$= \frac{2x}{2} = x$$

f and f^{-1} are inverses, so $f^{-1}(x) = \frac{x-7}{2}$.

3. $f(x) = -6x - 5$, $y = -6x - 5$.
Interchange x and y. Then solve for y.

$$x = -6y - 5$$
$$x + 5 = -6y$$
$$\frac{x+5}{-6} = y$$
$$\frac{-x-5}{6} = y$$

So $f^{-1}(x) = \frac{-x-5}{6}$.

4. $f(x) = \frac{2}{5}x + \frac{1}{3}$, $y = \frac{2}{5}x + \frac{1}{3}$.
Interchange x and y. Then solve for y.

$$x = \frac{2}{5}y + \frac{1}{3}$$
$$15(x) = 15\left(\frac{2}{5}y + \frac{1}{3}\right)$$
$$15x = 15\left(\frac{2}{5}y\right) + 15\left(\frac{1}{3}\right)$$
$$15x = 6y + 5$$
$$15x - 5 = 6y$$
$$\frac{15x-5}{6} = \frac{6y}{6}$$
$$\frac{15x}{6} - \frac{5}{6} = y$$
$$\frac{5x}{2} - \frac{5}{6} = y$$

So $f^{-1}(x) = \frac{5x}{2} - \frac{5}{6}$.

Section 10.6

1. Let h represent the height of a person and let l represent the length of the femur bone. Then $h = kl$.

2. "P varies directly as the square of r" gives

$$P = kr^2$$
$$-108 = k(6)^2$$
$$-108 = 36k$$
$$-3 = k$$

3. "m is directly proportional to n" gives

$$m = kn$$
$$60 = k(15)$$
$$4 = k$$
$$m = 4n$$
$$m = 4(18)$$
$$m = 72$$

4. Let t represent time and s represent speed. Then $t = \frac{k}{s}$.

5. "r is inversely proportional to t" gives

$$r = \frac{k}{t}$$
$$50 = \frac{k}{3}$$
$$150 = k$$

6. Let V represent volume of gas and let p represent pressure. Then

$$V = \frac{k}{p}$$
$$4 = \frac{k}{30}$$
$$120 = k$$
$$V = \frac{120}{p}$$
$$40 = \frac{120}{p}$$
$$(p)40 = (p)\left(\frac{120}{p}\right)$$
$$40p = 120$$
$$p = 3$$

The volume of gas is 3 liters.

7. "q varies jointly as m and n and inversely as p" gives

$$q = \frac{kmn}{p}$$

Substitute values for q, m, n, and p to find the value of the constant of variation, k.

$$25 = \frac{k(10)(15)}{3}$$
$$75 = 150k$$
$$\frac{75}{150} = k$$
$$k = \frac{1}{2}$$

The constant of variation is $\frac{1}{2}$.

8. Let V represent the volume of the gas, let t represent the temperature, and let p represent the pressure. Then

$$V = \frac{tk}{p}$$
$$3 = \frac{300k}{50}$$
$$3 = 6k$$
$$\frac{1}{2} = k$$
$$V = \frac{330\left(\frac{1}{2}\right)}{50}$$
$$V = \frac{165}{50} = 3.3$$

The volume of the gas is 3.3 liters.

CHAPTER 11

Section 11.1

1. $3^x = 81$

$3^x = 3^4$

$x = 4$

The solution set is $\{4\}$.

2. $2^{3x} = \frac{1}{8}$

$2^{3x} = \frac{1}{8} = \frac{1}{2^3}$

$2^{3x} = 2^{-3}$

$3x = -3$

$x = -1$

The solution set is $\{-1\}$.

3. $\left(\frac{1}{4}\right)^{x+2} = \frac{1}{64}$

$\left(\frac{1}{4}\right)^{x+2} = \frac{1}{64} = \frac{1^3}{4^3}$

$\left(\frac{1}{4}\right)^{x+2} = \left(\frac{1}{4}\right)^3$

$x + 2 = 3$

$x = 1$

The solution set is $\{1\}$.

4. $9^x = 27$

$(3^2)^x = 3^3$

$3^{2x} = 3^3$

$2x = 3$

$x = \frac{3}{2}$

The solution set is $\left\{\frac{3}{2}\right\}$.

5. $(2^{x+3})(4^{x-1}) = 16$

$(2^{x+3})(2^2)^{x-1} = 16$

$(2^{x+3})(2^{2x-2}) = 16$

$2^{x+3+2x-2} = 2^4$

$2^{3x+1} = 2^4$

$3x + 1 = 4$

$3x = 3$

$x = 1$

The solution set is $\{1\}$.

6. Graph $f(x) = 3^x$.

x	$f(x)$
-2	$\frac{1}{9}$
-1	$\frac{1}{3}$
0	1
1	3
2	9

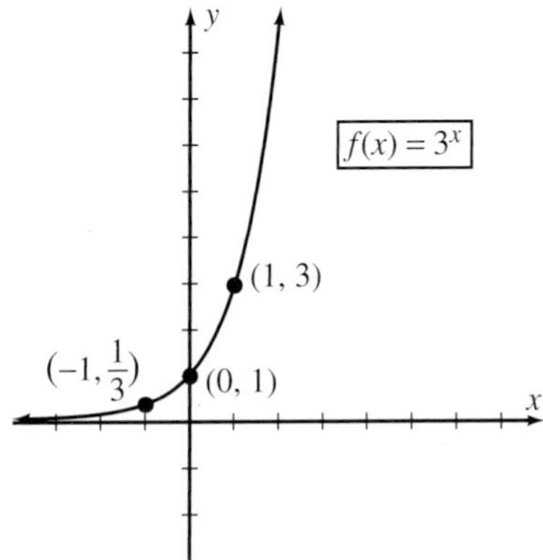

7. Graph $f(x) = \left(\frac{1}{3}\right)^x$.

x	$f(x)$
-2	9
-1	3
0	1
1	$\frac{1}{3}$
2	$\frac{1}{9}$

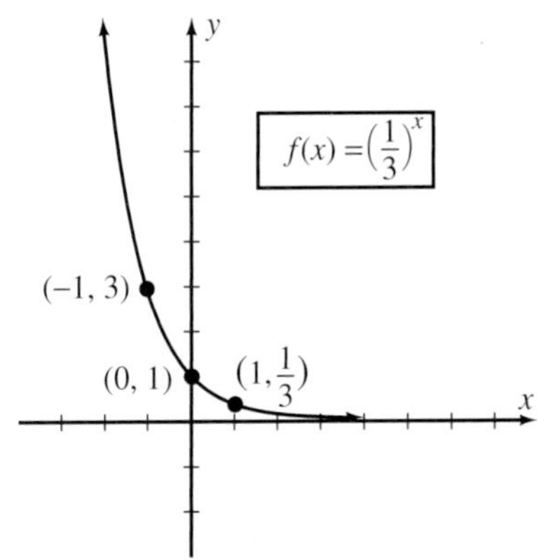

Section 11.2

1. (a) $A = P\left(1 + \frac{r}{n}\right)^{nt}$

$A = 2500\left(1 + \frac{0.08}{4}\right)^{4(4)}$

$A = 2500(1 + 0.02)^{16}$

$A = 2500(1.02)^{16}$

$A = \$3431.96$

(b) $A = P\left(1 + \frac{r}{n}\right)^{nt}$

$A = 3000\left(1 + \frac{0.06}{12}\right)^{12(5)}$

$A = 3000(1 + 0.005)^{60}$

$A = 3000(1.005)^{60}$

$A = \$4046.55$

2. $Q = Q_0\left(\frac{1}{2}\right)^{\frac{t}{n}}$

$Q = 20\left(\frac{1}{2}\right)^{\frac{36}{12}}$

$Q = 20\left(\frac{1}{2}\right)^{3}$

$Q = 20\left(\frac{1}{8}\right)$

$Q = 2.5\,\text{mg}$

3. (a) $A = Pe^{rt}$

$A = 2500e^{(0.08)4}$

$A = 2500e^{0.32}$

$A = \$3442.82$

(b) $A = Pe^{rt}$

$A = 3000e^{(0.06)5}$

$A = 3000e^{0.3}$

$A = \$4049.58$

4. $Q(t) = 28{,}000e^{0.5t}$

(a) The initial number of bacteria is produced when $t = 0$.

$Q(0) = 28{,}000e^{0.5(0)}$

$Q(0) = 28{,}000e^{0}$

$Q(0) = 28{,}000$

(b) $Q(1) = 28{,}000e^{0.5(1)}$

$Q(1) = 28{,}000e^{0.5}$

$Q(1) = 41{,}164$

Initially, there are 28,000 bacteria, and after one hour there should be approximately 46,164 bacteria.

5. $Q(t) = Q_0e^{0.04t}$

$Q(35) = Q_0e^{0.04(35)}$

$2500 = Q_0e^{1.4}$

$\frac{2500}{e^{1.4}} = Q_0$

$616 = Q_0$

There were approximately 616 bacteria present initially.

6. $Q(t) = 500e^{-0.08t}$

$Q(12) = 500e^{-0.08(12)}$

$Q(12) = 500e^{-0.96}$

$Q(12) = 191$

Approximately 191 grams remain after 12 seconds.

Section 11.3

1. (a) $\log_5 25 = 2$

$5^2 = 25$

(b) $\log_6\left(\frac{1}{216}\right) = -3$

$6^{-3} = \frac{1}{216}$

(c) $\log_{10} 10{,}000 = 4$

$10^4 = 10{,}000$

(d) $\log_a b = c$

$a^c = b$

2. (a) $2^4 = 16$

$\log_2 16 = 4$

(b) $4^{-1} = \frac{1}{4}$

$\log_4\left(\frac{1}{4}\right) = -1$

(c) $10^3 = 1000$

$\log_{10} 1000 = 3$

(d) $x^y = z$

$\log_x z = y$

3. Let $\log_2 128 = x$.

$2^x = 128$

$2^x = 2^7$

$x = 7$

$\log_2 128 = 7$

4. Let $\log_{10} 0.001 = x$.

$10^x = 0.001$

$10^x = \frac{1}{1000}$

$$10^x = 10^{-3}$$
$$x = -3$$
$$\log_{10} 0.001 = -3$$

5. $\log_9 x = \frac{5}{2}$.

$$9^{\frac{5}{2}} = x$$
$$(\sqrt{9})^5 = x$$
$$(3)^5 = x$$
$$243 = x$$

The solution set is {243}.

6. $\log_b 121 = 2$

$$b^2 = 121$$
$$b = \pm\sqrt{121}$$
$$b = 11$$

Discard $b = -11$, since b must be positive. The solution set is {11}.

7. $\log_3 8 = 1.8928$, $\log_3 5 = 1.4650$

$$\begin{aligned}\log_3 40 &= \log_3(8 \cdot 5)\\ &= \log_3 8 + \log_3 5\\ &= 1.8928 + 1.4650\\ &= 3.3578\end{aligned}$$

8. $\log_6 45 = 2.1245$, $\log_6 92 = 2.5237$

$$\begin{aligned}\log_6(45 \cdot 92) &= \log_6 45 + \log_6 92\\ &= 2.1245 + 2.5237\\ &= 4.6482\end{aligned}$$

9. $\log_2 7 = 2.8074$

$$\begin{aligned}\log_2 56 &= \log_2(8 \cdot 7)\\ &= \log_2 8 + \log_2 7\\ &= \log_2(2)^3 + \log_2 7\\ &= 3 + 2.8074\\ &= 5.8074\end{aligned}$$

10. $\log_3 24 = 2.8928$, $\log_3 2 = 0.6309$

$$\begin{aligned}\log_3 12 &= \log_3\left(\frac{24}{2}\right)\\ &= \log_3 24 - \log_3 2\\ &= 2.8928 - 0.6309\\ &= 2.2619\end{aligned}$$

11. $\log_{10} 436 = 2.6395$, $\log_{10} 72 = 1.8573$

$$\begin{aligned}\log_3\left(\frac{436}{72}\right) &= \log_{10} 436 - \log_{10} 72\\ &= 2.6395 - 1.8573\\ &= 0.7822\end{aligned}$$

12. $\log_4 28 = 2.4037$

$$\begin{aligned}\log_4 28^{\frac{1}{2}} &= \frac{1}{2}(\log_4 28)\\ &= \frac{1}{2}(2.4037)\\ &= 1.2019\end{aligned}$$

13. (a) $\log_b x^3\sqrt{y} = \log_b x^3 - \log_b\sqrt{y}$

$$\begin{aligned}&= 3\log_b x - \log_b y^{\frac{1}{2}}\\ &= 3\log_b x - \frac{1}{2}\log_b y\end{aligned}$$

(b) $\log_b\left(\frac{\sqrt[3]{x}}{y^2}\right) = \log_b\sqrt[3]{x} - \log_b y^2$

$$\begin{aligned}&= \log_b x^{\frac{1}{3}} - 2\log_b y\\ &= \frac{1}{3}\log_b x - 2\log_b y\end{aligned}$$

(c) $\log_b\left(\frac{\sqrt[4]{x}}{y^2 z}\right) = \log_b\sqrt[4]{x} - \log_b y^2 z$

$$\begin{aligned}&= \log_b x^{\frac{1}{4}} - [\log_b y^2 + \log_b z]\\ &= \frac{1}{4}\log_b x - [2\log_b y + \log_b z]\\ &= \frac{1}{4}\log_b x - 2\log_b y - \log_b z\end{aligned}$$

14. $\log_{10} x + \log_{10}(x + 3) = 1$

$$\log_{10}[x(x + 3)] = 1$$
$$10^1 = x(x + 3)$$
$$10 = x^2 + 3x$$
$$0 = x^2 + 3x - 10$$
$$0 = (x + 5)(x - 2)$$
$$x + 5 = 0 \quad \text{or} \quad x - 2 = 0$$
$$x = -5 \quad \text{or} \quad x = 2$$

If $x = -5$, then x and $x + 3$ would be negative. So discard $x = -5$, since logarithms are only defined for positive numbers. The solution set is {2}.

15. $2 = \log_3(x + 6) - \log_3 x$

$$2 = \log_3\left(\frac{x + 6}{x}\right)$$
$$3^2 = \frac{x + 6}{x}$$
$$x(9) = x\left(\frac{x + 6}{x}\right)$$
$$9x = x + 6$$
$$8x = 6$$
$$x = \frac{6}{8} = \frac{3}{4}$$

The solution set is $\left\{\frac{3}{4}\right\}$.

16. $\log_3 2 + \log_3(x + 6) = 2$

$$\log_3[2(x + 6)] = 2$$
$$3^2 = 2(x + 6)$$
$$9 = 2x + 12$$
$$-3 = 2x$$
$$-\frac{3}{2} = x$$

The only restriction is that $x + 6 > 0$ or $x > -6$.

The solution set is $\left\{-\frac{3}{2}\right\}$.

Section 11.4

1. $\log_{10} x = 2.4568$

$$x = 10^{2.4568}$$
$$x = 286.29$$

2. $f(x) = \log_3 x$ implies $3^y = x$.

x	y
1	0
3	1
9	2
$\frac{1}{3}$	−1
$\frac{1}{9}$	−2

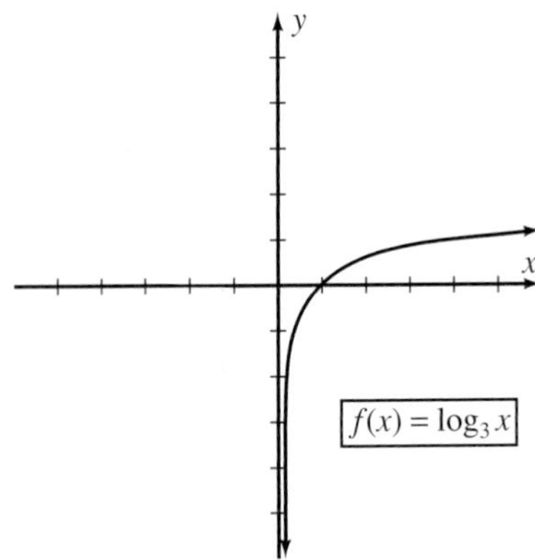

Section 11.5

1. $4^x = 24$

$$\log 4^x = \log 24$$
$$x \log 4 = \log 24$$
$$x = \frac{\log 24}{\log 2}$$
$$x = 2.29 \text{ to the nearest hundredth}$$

The solution set is {2.29}.

2. $e^{x-2} = 20$

$$\ln e^{x-2} = \ln 20$$
$$(x - 2) \ln e = \ln 20$$
$$(x - 2)(1) = \ln 20$$
$$x - 2 = \ln 20$$
$$x = 2 + \ln 20$$
$$x = 5.00 \text{ to the nearest hundredth}$$

The solution set is {5.00}.

3. $\log x + \log(x + 21) = 2$

$$\log[x(x + 21)] = 2$$
$$\log[x(x + 21)] = \log 100 \quad \textit{Hint}: (\log 100 = 2)$$
$$x(x + 21) = 100$$
$$x^2 + 21x = 100$$
$$x^2 + 21x - 100 = 0$$
$$(x - 4)(x + 25) = 0$$
$$x - 4 = 0 \quad \text{or} \quad x + 25 = 0$$
$$x = 4 \quad \text{or} \quad x = -25$$

The domain of a logarithmic function must contain only positive numbers, so x and $x + 21$ must be positive. Therefore, discard $x = -25$. The solution set is {4}.

4. $\ln(x + 22) = \ln(x - 2) + \ln 4$

$$\ln(x + 22) = \ln[4(x - 2)]$$
$$x + 22 = 4(x - 2)$$
$$x + 22 = 4x - 8$$
$$-3x + 22 = -8$$
$$-3x = -30$$
$$x = 10$$

The solution set is {10}.

5. To "double itself," \$2000 will become \$4000.

$$A = P\left(1 + \frac{r}{n}\right)^{nt}$$
$$4000 = 2000\left(1 + \frac{0.06}{12}\right)^{12(t)}$$
$$2 = (1 + 0.005)^{12t}$$
$$2 = (1.005)^{12t}$$
$$\log 2 = \log(1.005)^{12t}$$
$$\log 2 = 12t \log(1.005)$$
$$\frac{\log 2}{\log 1.005} = 12t$$
$$\frac{\log 2}{12 \log 1.005} = t$$
$$11.58 = t$$

It will take approximately 11.6 years.

6. To "triple itself," \$500 will become \$1500.

Use $A = Pe^{rt}$.

$1500 = 500e^{(0.06)t}$

$3 = e^{0.06t}$

$\ln 3 = \ln e^{0.06t}$

$\ln 3 = 0.06t(\ln e)$

$\ln 3 = 0.06t(1)$

$\frac{\ln 3}{0.06} = t$

$18.31 = t$

It will take approximately 18.3 years.

7. Let $Q_0 = 2000$ and $Q(t) = 1000$.

$Q(t) = Q_0e^{-0.4t}$

$1000 = 2000e^{-0.4t}$

$0.5 = e^{-0.4t}$

$\ln 0.5 = \ln e^{-0.4t}$

$\ln 0.5 = -0.4t(\ln e)$

$\ln 0.5 = -0.4t(1)$

$\frac{\ln 0.5}{-0.4} = t$

$1.73 = t$

It will take approximately 1.7 hours.

8. $R = \log\frac{I}{I_0}$

$5.8 = \log\frac{I}{I_0}$

$10^{5.8} = \frac{I}{I_0}$

$(10^{5.8})I_0 = I$

$(630{,}957)I_0 \approx I$

The intensity was 630,957 times the reference intensity.

9. In Example 7, $I = (10^{6.9})I_0$. Using 7.9 as the Richter number for the earthquake in China, we have

$I = (10^{7.9})I_0$

By comparison,

$\frac{(10^{7.9})I_0}{(10^{6.9})I_0} = 10^{7.9-6.9} = 10^1 = 10$

The earthquake in China was 10 times as intense as the one in San Francisco.

10. Let $x = \log_5 38$.

$5^x = 38$

$\log 5^x = \log 38$

$x \log 5 = \log 38$

$x = \frac{\log 38}{\log 5}$

$x = 2.2602$ to four decimal places.

Answers to Odd-Numbered Problems and All Chapter Review, Chapter Test, Cumulative Review, and Appendix Problems

CHAPTER 1

Problem Set 1.1 (page 9)

1. True **3.** False **5.** True **7.** False **9.** True **11.** 0 and 14 **13.** $0, 14, \frac{2}{3}, -\frac{11}{14}, 2.34, 3.2\overline{1}, \frac{55}{8}, -19$, and -2.6 **15.** 0 and 14 **17.** All of them **19.** $\not\subseteq$ **21.** $\subseteq$ **23.** $\not\subseteq$ **25.** $\subseteq$ **27.** $\not\subseteq$ **29.** Real, rational, an integer, and negative **31.** Real, irrational, and negative **33.** $\{1, 2\}$ **35.** $\{0, 1, 2, 3, 4, 5\}$ **37.** $\{\ldots, -1, 0, 1, 2\}$ **39.** $\varnothing$ **41.** $\{0, 1, 2, 3, 4\}$ **43.** -6 **45.** 2 **47.** $3x + 1$ **49.** $5x$ **51.** 26 **53.** 84 **55.** 23 **57.** 65 **59.** 60 **61.** 33 **63.** 1320 **65.** 20 **67.** 119 **69.** 18 **71.** 4 **73.** 31

Problem Set 1.2 (page 20)

1. 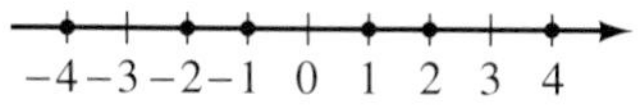

3. (a) 7 **(b)** 0 **(c)** (15) **5.** -7 **7.** -19 **9.** $3\frac{1}{2}$ **11.** -4.8 **13.** $-\frac{13}{12}$ **15.** -22 **17.** -7 **19.** $5\frac{1}{2}$ **21.** -60 **23.** 14.13 **25.** -6.5 **27.** $-\frac{3}{4}$ **29.** $-\frac{13}{9}$ **31.** 108 **33.** -70 **35.** $-\frac{2}{15}$ **37.** -38.88 **39.** $-\frac{3}{5}$ **41.** 14 **43.** -7 **45.** -4 **47.** 0 **49.** Undefined **51.** 0.2 **53.** $-\frac{3}{2}$ **55.** -12 **57.** -24 **59.** $\frac{35}{4}$ **61.** 15 **63.** -17 **65.** $\frac{47}{12}$ **67.** 5 **69.** 0 **71.** 26 **73.** 6 **75.** 25 **77.** 78 **79.** -10 **81.** 5 **83.** -5 **85.** 10.5 **87.** -3.3 **89.** 19.5 **91.** $\frac{3}{4}$ **93.** $\frac{5}{2}$ **97.** 10 over par **99.** Lost \$16.50 **101.** A gain of \$0.88 **103.** No; they made it 49.1 pounds lighter

Problem Set 1.3 (page 29)

1. Associative property of addition
3. Commutative property of addition
5. Additive inverse property
7. Multiplication property of negative one
9. Commutative property of multiplication
11. Distributive property
13. Associative property of multiplication
15. 18 **17.** 2 **19.** -1300 **21.** 1700 **23.** -47 **25.** 3200 **27.** -19 **29.** -41 **31.** -17 **33.** -39 **35.** 24 **37.** 20 **39.** 55 **41.** 16 **43.** 49 **45.** -216 **47.** -14 **49.** -8 **51.** $\frac{3}{16}$ **53.** $-\frac{10}{9}$ **57.** 2187 **59.** -2048 **61.** $-15{,}625$ **63.** 3.9525416

Problem Set 1.4 (page 39)

1. $4x$ **3.** $-a^2$ **5.** $-6n$ **7.** $-5x + 2y$ **9.** $6a^2 + 5b^2$ **11.** $21x - 13$ **13.** $-2a^2b - ab^2$ **15.** $8x + 21$ **17.** $-5a + 2$ **19.** $-5n^2 + 11$ **21.** $-7x^2 + 32$ **23.** $22x - 3$ **25.** $-14x - 7$ **27.** $-10n^2 + 4$ **29.** $4x - 30y$ **31.** $-13x - 31$ **33.** $-21x - 9$ **35.** -17 **37.** 12 **39.** 4 **41.** 3 **43.** -38 **45.** -14 **47.** 64 **49.** 104 **51.** 5 **53.** 4 **55.** $-\frac{22}{3}$ **57.** $\frac{29}{4}$ **59.** 221.6 **61.** 1092.4 **63.** 1420.5 **65.** $n + 12$ **67.** $n - 5$ **69.** $50n$ **71.** $\frac{1}{2}n - 4$ **73.** $\frac{n}{8}$ **75.** $2n - 9$ **77.** $10(n - 6)$ **79.** $n + 20$ **81.** $2t - 3$ **83.** $n + 47$ **85.** $8y$ **87.** 25 cm **89.** $\frac{c}{25}$ **91.** $n + 2$ **93.** $\frac{c}{5}$ **95.** $12d$ **97.** $3y + f$ **99.** $5280m$

Chapter 1 Review Problem Set (page 44)

1. (a) 67 **(b)** $0, -8$, and 67 **(c)** 0 and 67 **(d)** $0, \frac{3}{4}, -\frac{5}{6}, \frac{25}{3}, -8, 0.34, 0.2\overline{3}, 67$, and $\frac{9}{7}$ **(e)** $\sqrt{2}$ and $-\sqrt{3}$
2. Associative property of addition
3. Substitution property of equality
4. Multiplication property of negative one
5. Distributive property
6. Associative property of multiplication
7. Commutative property of addition
8. Distributive property
9. Multiplicative inverse property
10. Symmetric property of equality
11. 6.2 **12.** $\frac{7}{3}$ **13.** $\sqrt{15}$ **14.** 8 **15.** $-6\frac{1}{2}$ **16.** $-6\frac{1}{6}$ **17.** -8 **18.** -15 **19.** 20 **20.** 49 **21.** -56 **22.** 8 **23.** -24 **24.** 6 **25.** 4 **26.** 100 **27.** $-4a^2 - 5b^2$ **28.** $3x - 2$ **29.** ab^2 **30.** $-\frac{7}{3}x^2y$ **31.** $10n^2 - 17$ **32.** $-13a + 4$ **33.** $-2n + 2$ **34.** $-7x - 29y$ **35.** $-7a - 9$

36. $-9x^2 + 7$ **37.** $-6\frac{1}{2}$ **38.** $-\frac{5}{16}$ **39.** -55 **40.** 144
41. -16 **42.** -44 **43.** 19.4 **44.** 59.6 **45.** $-\frac{59}{3}$ **46.** $\frac{9}{2}$
47. $4 + 2n$ **48.** $3n - 50$ **49.** $\frac{2}{3}n - 6$ **50.** $10(n - 14)$
51. $5n - 8$ **52.** $\frac{n}{n - 3}$ **53.** $5(n + 2) - 3$ **54.** $\frac{3}{4}(n + 12)$
55. $37 - n$ **56.** $\frac{w}{60}$ **57.** $2y - 7$ **58.** $n + 3$
59. $p + 5n + 25q$ **60.** $\frac{i}{48}$ **61.** $24f + 72y$ **62.** $10d$
63. $12f + i$ **64.** $25 - c$ **65.** 1 min **66.** Loss of \$0.03
67. 0.2 oz **68.** 32 lb

Chapter 1 Test (page 46)
1. Symmetric property **2.** Distributive property **3.** -3
4. -23 **5.** $-\frac{23}{6}$ **6.** 11 **7.** 8 **8.** -94 **9.** -4 **10.** 960
11. -32 **12.** $-x^2 - 8x - 2$ **13.** $-19n - 20$ **14.** 27
15. $\frac{11}{16}$ **16.** $\frac{2}{3}$ **17.** 77 **18.** -22.5 **19.** 93 **20.** -5
21. $6n - 30$ **22.** $3n + 28$ or $3(n + 8) + 4$ **23.** $\frac{72}{n}$
24. $5n + 10d + 25q$ **25.** $6x + 2y$

CHAPTER 2

Problem Set 2.1 (page 56)
1. $\{4\}$ **3.** $\{-3\}$ **5.** $\{-14\}$ **7.** $\{6\}$ **9.** $\left\{\frac{19}{3}\right\}$ **11.** $\{1\}$
13. $\left\{-\frac{10}{3}\right\}$ **15.** $\{4\}$ **17.** $\left\{-\frac{13}{3}\right\}$ **19.** $\{3\}$ **21.** $\{8\}$
23. $\{-9\}$ **25.** $\{-3\}$ **27.** $\{0\}$ **29.** $\left\{-\frac{7}{2}\right\}$ **31.** $\{-2\}$
33. $\left\{-\frac{5}{3}\right\}$ **35.** $\left\{\frac{33}{2}\right\}$ **37.** $\{-35\}$ **39.** $\left\{\frac{1}{2}\right\}$ **41.** $\left\{\frac{1}{6}\right\}$
43. $\{5\}$ **45.** $\{-1\}$ **47.** $\left\{-\frac{21}{16}\right\}$ **49.** $\left\{\frac{12}{7}\right\}$ **51.** 14
53. 13, 14, and 15 **55.** 9, 11, and 13 **57.** 14 and 81
59. \$11 per hour **61.** 30 pennies, 50 nickels, and 70 dimes
63. \$300 **65.** 20 three-bedroom, 70 two-bedroom, and 140 one-bedroom **73. (a)** $\varnothing$ **(c)** $\{0\}$ **(e)** $\varnothing$

Problem Set 2.2 (page 64)
1. $\{12\}$ **3.** $\left\{-\frac{3}{5}\right\}$ **5.** $\{3\}$ **7.** $\{-2\}$ **9.** $\{-36\}$ **11.** $\left\{\frac{20}{9}\right\}$
13. $\{3\}$ **15.** $\{3\}$ **17.** $\{-2\}$ **19.** $\left\{\frac{8}{5}\right\}$ **21.** $\{-3\}$
23. $\left\{\frac{48}{17}\right\}$ **25.** $\left\{\frac{103}{6}\right\}$ **27.** $\{3\}$ **29.** $\left\{\frac{40}{3}\right\}$ **31.** $\left\{-\frac{20}{7}\right\}$
33. $\left\{\frac{24}{5}\right\}$ **35.** $\{-10\}$ **37.** $\left\{-\frac{25}{4}\right\}$ **39.** $\{0\}$ **41.** 18
43. 16 in. long and 5 in. wide **45.** 14, 15, and 16
47. 8 ft **49.** Angie is 22 and her mother is 42.
51. Sydney is 18 and Marcus is 36. **53.** 80, 90, and 94
55. 48° and 132° **57.** 78°

Problem Set 2.3 (page 73)
1. $\{20\}$ **3.** $\{50\}$ **5.** $\{40\}$ **7.** $\{12\}$ **9.** $\{6\}$ **11.** $\{400\}$
13. $\{400\}$ **15.** $\{38\}$ **17.** $\{6\}$ **19.** $\{3000\}$ **21.** $\{3000\}$
23. $\{400\}$ **25.** $\{14\}$ **27.** $\{15\}$ **29.** \$90 **31.** \$54.40
33. \$48 **35.** \$400 **37.** 65% **39.** 62.5% **41.** \$32,500
43. \$3000 at 10% and \$4500 at 11% **45.** \$53,000
47. 8 pennies, 15 nickels, and 18 dimes
49. 15 dimes, 45 quarters, and 10 half-dollars **55.** $\{7.5\}$
57. $\{-4775\}$ **59.** $\{8.7\}$ **61.** $\{17.1\}$ **63.** $\{13.5\}$

Problem Set 2.4 (page 84)
1. \$120 **3.** 3 years **5.** 6% **7.** \$800 **9.** \$1600
11. 8% **13.** \$200 **15.** 6 ft; 14 ft; 10 ft; 20 ft; 7 ft; 2 ft
17. $h = \frac{V}{B}$ **19.** $h = \frac{V}{\pi r^2}$ **21.** $r = \frac{C}{2\pi}$ **23.** $C = \frac{100M}{I}$
25. $C = \frac{5}{9}(F - 32)$ or $C = \frac{5F - 160}{9}$ **27.** $x = \frac{y - b}{m}$
29. $x = \frac{y - y_1 + mx_1}{m}$ **31.** $x = \frac{ab + bc}{b - a}$
33. $x = a + bc$ **35.** $x = \frac{3b - 6a}{2}$ **37.** $x = \frac{5y + 7}{2}$
39. $y = -7x - 4$ **41.** $x = \frac{6y + 4}{3}$ **43.** $x = \frac{cy - ac - b^2}{b}$
45. $y = \frac{x - a + 1}{a - 3}$ **47.** 22 m long and 6 m wide
49. $11\frac{1}{9}$ yr **51.** $11\frac{1}{9}$ yr **53.** 4 hr **55.** 3 hr **57.** 40 miles
59. 15 qt of 30% solution and 5 qt of 70% solution
61. 25 ml **67.** \$596.25 **69.** 1.5 yr **71.** 14.5% **73.** \$1850

Problem Set 2.5 (page 93)
1. $(1, \infty)$

1

3. $[-1, \infty)$

-1

5. $(-\infty, -2)$

-2

7. $(-\infty, 2]$

2

9. $x < 4$ **11.** $x \le -7$ **13.** $x > 8$ **15.** $x \ge -7$
17. $(1, \infty)$

1

19. $(-\infty, -4]$

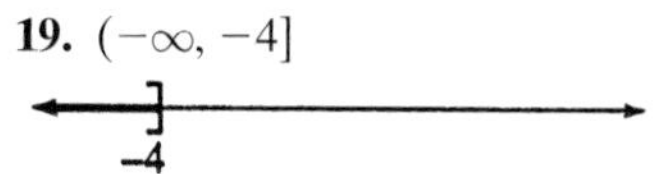

21. $(-\infty, -2]$

23. $(-\infty, 2)$

25. $(-1, \infty)$

27. $[-1, \infty)$

29. $(-2, \infty)$

31. $(-2, \infty)$

33. $(-\infty, -2)$

35. $[-3, \infty)$

37. $(0, \infty)$

39. $[4, \infty)$

41. $\left(\frac{7}{2}, \infty\right)$ **43.** $\left(\frac{12}{5}, \infty\right)$ **45.** $\left(-\infty, -\frac{5}{2}\right]$

47. $\left[\frac{5}{12}, \infty\right)$ **49.** $(-6, \infty)$ **51.** $(-5, \infty)$

53. $\left(-\infty, \frac{5}{3}\right]$ **55.** $(-36, \infty)$ **57.** $\left(-\infty, -\frac{8}{17}\right]$

59. $\left(-\frac{11}{2}, \infty\right)$ **61.** $(23, \infty)$ **63.** $(-\infty, 3)$

65. $\left(-\infty, -\frac{1}{7}\right]$ **67.** $(-22, \infty)$ **69.** $\left(-\infty, \frac{6}{5}\right)$

Problem Set 2.6 (page 102)

1. $(4, \infty)$ **3.** $\left(-\infty, \frac{23}{3}\right)$ **5.** $[5, \infty)$ **7.** $[-9, \infty)$

9. $\left(-\infty, -\frac{37}{3}\right]$ **11.** $\left(-\infty, -\frac{19}{6}\right)$ **13.** $(-\infty, 50]$

15. $(300, \infty)$ **17.** $[4, \infty)$

19. $(-1, 2)$

21. $(-1, 2]$

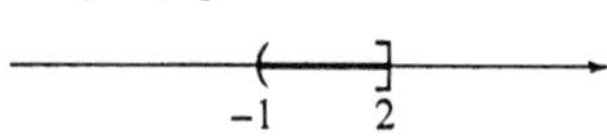

23. $(-\infty, -1) \cup (2, \infty)$

25. $(-\infty, 1] \cup (3, \infty)$

27. $(0, \infty)$

29. $\varnothing$

31. $(-\infty, \infty)$

33. $(-1, \infty)$

35. $(1, 3)$

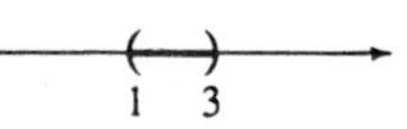

37. $(-\infty, -5) \cup (1, \infty)$

39. $[3, \infty)$

41. $\left(\frac{1}{3}, \frac{2}{5}\right)$

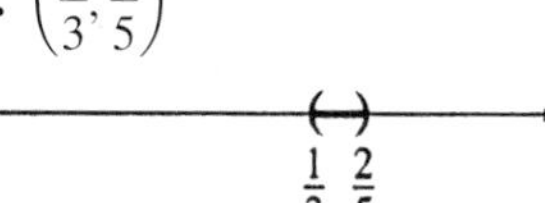

43. $(-\infty, -1) \cup \left(-\frac{1}{3}, \infty\right)$

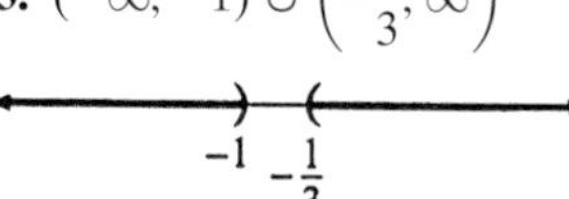

45. $(-2, 2)$ **47.** $[-5, 4]$ **49.** $\left(-\frac{1}{2}, \frac{3}{2}\right)$

51. $\left(-\frac{1}{4}, \frac{11}{4}\right)$ **53.** $[-11, 13]$ **55.** $(-1, 5)$

57. More than 10% **59.** 5 ft 10 in. or more

61. 168 or better **63.** 77 or less **65.** $163°F \le C \le 218°F$

67. $6.3 \le M \le 11.25$

Problem Set 2.7 (page 109)

1. $\{-7, 9\}$ **3.** $\{-1, 5\}$ **5.** $\left\{-5, \frac{7}{3}\right\}$ **7.** $\{-1, 5\}$

9. $\left\{\frac{1}{12}, \frac{7}{12}\right\}$ **11.** $\{0, 3\}$ **13.** $\{-6, 2\}$ **15.** $\left\{\frac{3}{4}\right\}$

17. $(-5, 5)$

−5 5

19. $[-2, 2]$

−2 2

21. $(-\infty, -2) \cup (2, \infty)$

−2 2

23. $(-1, 3)$

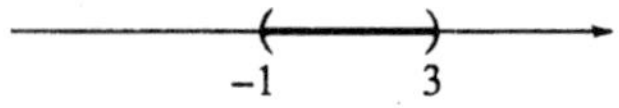

−1 3

25. $[-6, 2]$

−6 2

27. $(-\infty, -3) \cup (-1, \infty)$

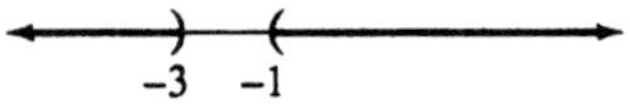

−3 −1

29. $(-\infty, 1] \cup [5, \infty)$

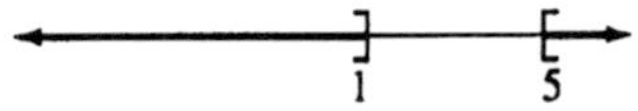

1 5

31. $(-\infty, -4) \cup (8, \infty)$ **33.** $(-8, 2)$ **35.** $[-4, 5]$

37. $\left(-\infty, -\frac{7}{2}\right] \cup \left[\frac{5}{2}, \infty\right)$ **39.** $(-\infty, -2) \cup (6, \infty)$

41. $\left(-\frac{1}{2}, \frac{3}{2}\right)$ **43.** $\left[-5, \frac{7}{5}\right]$ **45.** $[-3, 10]$ **47.** $(-5, 11)$

49. $\left(-\infty, -\frac{3}{2}\right) \cup \left(\frac{1}{2}, \infty\right)$ **51.** $(-\infty, -14] \cup [0, \infty)$

53. $[-2, 3]$ **55.** $\varnothing$ **57.** $(-\infty, \infty)$ **59.** $\left\{\frac{2}{5}\right\}$ **61.** $\varnothing$

63. $\varnothing$ **69.** $\left\{-2, -\frac{4}{3}\right\}$ **71.** $\{-2\}$ **73.** $\{0\}$

Chapter 2 Review Problem Set (page 117)

1. $\{18\}$ **2.** $\{-14\}$ **3.** $\{0\}$ **4.** $\left\{\frac{1}{2}\right\}$ **5.** $\{10\}$ **6.** $\left\{\frac{7}{3}\right\}$

7. $\left\{\frac{28}{17}\right\}$ **8.** $\left\{-\frac{1}{38}\right\}$ **9.** $\left\{\frac{27}{17}\right\}$ **10.** $\left\{-\frac{21}{13}\right\}$

11. $\{50\}$ **12.** $\left\{-\frac{39}{2}\right\}$ **13.** $\{200\}$ **14.** $\{-8\}$

15. The length is 15 m and the width is 7 m **16.** 4, 5, and 6

17. \$10.50 per hour **18.** 20 nickels, 50 dimes, 75 quarters

19. 80° **20.** \$200 invested at 7%, \$300 invested at 8%

21. \$45.60 **22.** \$300.00 **23.** 60% **24.** \$64.00

25. \$8000 **26.** 4.5% **27.** 11 m **28.** −20°

29. $x = \frac{2b + 2}{a}$ **30.** $x = \frac{c}{a - b}$ **31.** $x = \frac{pb - ma}{m - p}$

32. $x = \frac{11 + 7y}{5}$ **33.** $x = \frac{by + b + ac}{c}$ **34.** $s = \frac{A - \pi r^2}{\pi r}$

35. $b_2 = \frac{2A - hb_1}{h}$ **36.** $n = \frac{2S_n}{a_1 + a_2}$ **37.** $R = \frac{R_1 R_2}{R_1 + R_2}$

38. $y = \frac{c - ax}{b}$ **39.** $6\frac{2}{3}$ pints **40.** 55 mph

41. Sonya for $3\frac{1}{4}$ hr, Rita for $4\frac{1}{2}$ hr

42. $6\frac{1}{4}$ cups of orange juice **43.** $[-2, \infty)$ **44.** $(6, \infty)$

45. $(-\infty, -1)$ **46.** $(-\infty, 0]$ **47.** $[-5, \infty)$ **48.** $(4, \infty)$

49. $\left(-\frac{7}{3}, \infty\right)$ **50.** $\left[\frac{17}{2}, \infty\right)$ **51.** $(-\infty, -17)$

52. $\left(-\infty, \frac{1}{3}\right)$ **53.** $\left(\frac{53}{11}, \infty\right)$ **54.** $\left(-\infty, -\frac{15}{4}\right)$

55. $[6, \infty)$ **56.** $(-\infty, 100]$

57. −1 1

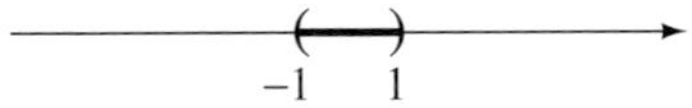

58. −3 2

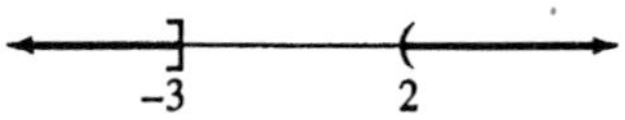

59. 3

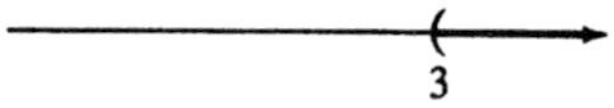

60.

61. −2 1

62. −2 1

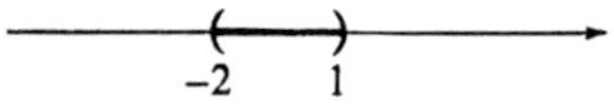

63. $\frac{1}{2}$ 3

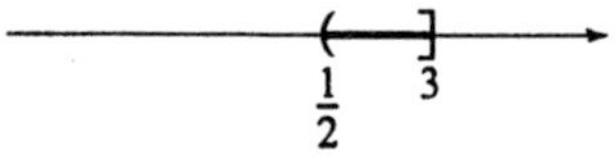

64. $\varnothing$ **65.** 88 or better **66.** More than \$4000

67. $\left\{-\frac{10}{3}, 4\right\}$ **68.** $\left\{-\frac{7}{2}, \frac{1}{2}\right\}$ **69.** $\left\{-\frac{11}{3}, \infty\right\}$

70. $\{-18, 6\}$ **71.** $(-5, 6)$ **72.** $\left(-\infty, -\frac{11}{3}\right) \cup (3, \infty)$

73. $\left(-\infty, -\frac{4}{5}\right] \cup \left[\frac{12}{5}, \infty\right)$ **74.** $[-28, 20]$

Chapter 2 Test (page 119)

1. $\{-3\}$ **2.** $\{5\}$ **3.** $\left\{\frac{1}{2}\right\}$ **4.** $\left\{\frac{16}{5}\right\}$ **5.** $\left\{-\frac{14}{5}\right\}$ **6.** $\{-1\}$

7. $\left\{-\frac{3}{2}, 3\right\}$ **8.** $\{3\}$ **9.** $\left\{\frac{31}{3}\right\}$ **10.** $\{650\}$ **11.** $y = \frac{8x - 24}{9}$

12. $h = \frac{S - 2\pi r^2}{2\pi r}$ **13.** $(-2, \infty)$ **14.** $[-4, \infty)$

15. $(-\infty, -35]$ **16.** $(-\infty, 10)$ **17.** $(3, \infty)$ **18.** $(-\infty, 200]$

19. $\left(-1, \frac{7}{3}\right)$ **20.** $\left(-\infty, -\frac{11}{4}\right] \cup \left[\frac{1}{4}, \infty\right)$ **21.** \$72

22. 19 cm **23.** $\frac{2}{3}$ of a cup **24.** 97 or better **25.** 70°

CHAPTER 3

Problem Set 3.1 (page 134)

1. (2, 4), (−1, −5) **3.** (−2, 10), (3, 0)

5.

x	−2	−1	0	4
y	5	4	3	−1

7.

x	−2	0	2	4
y	−10	−6	−2	2

9.

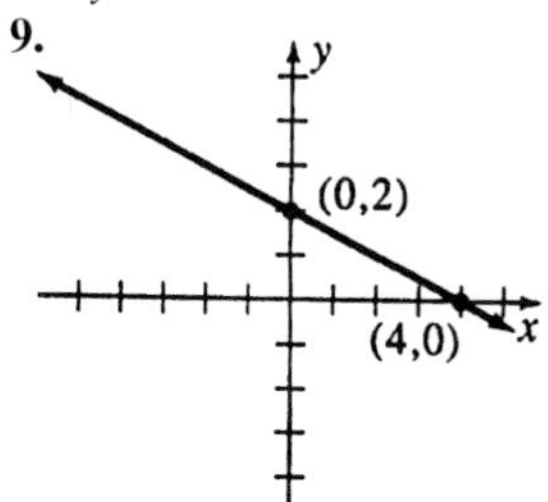

11.

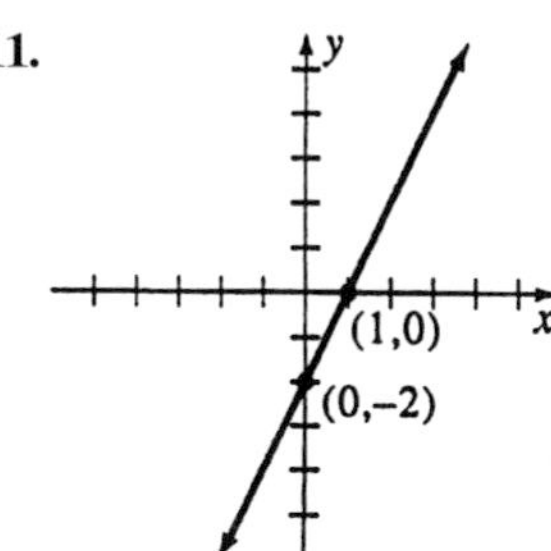

13.

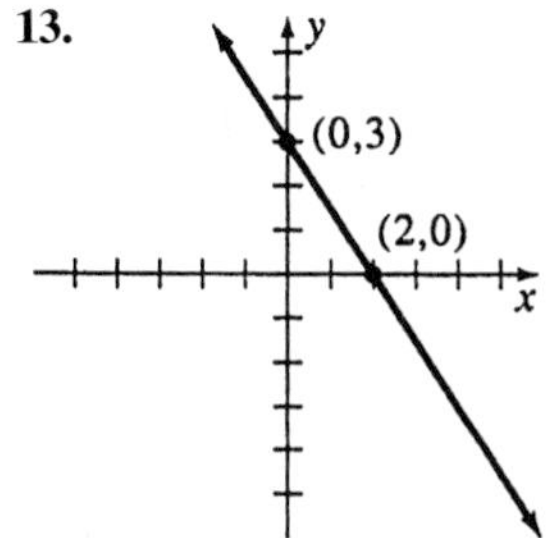

15.

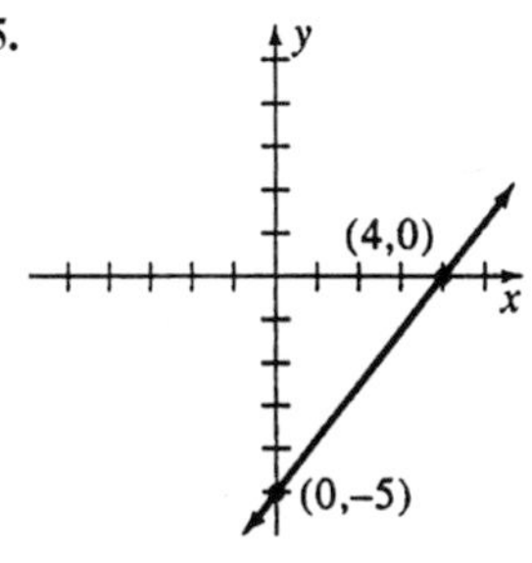

17.

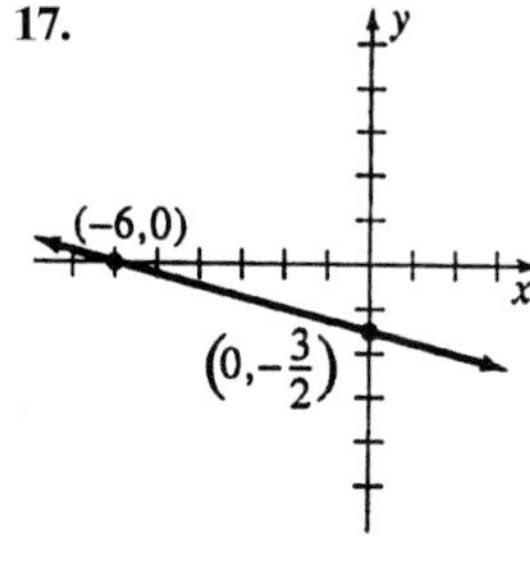

19.

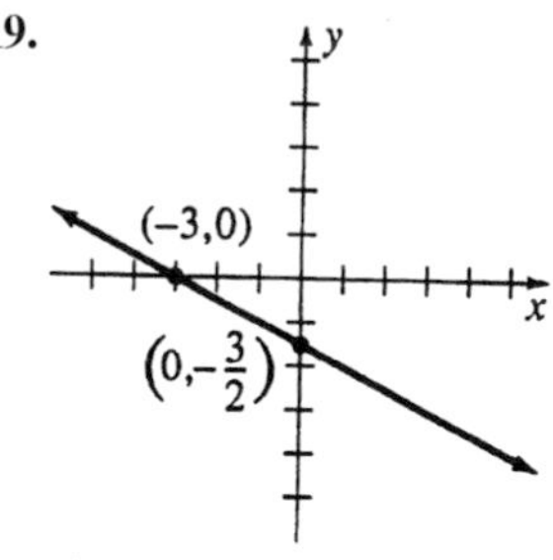

21.

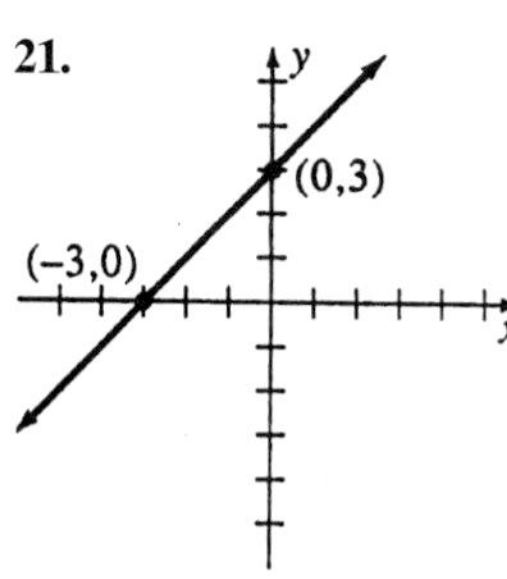

23.

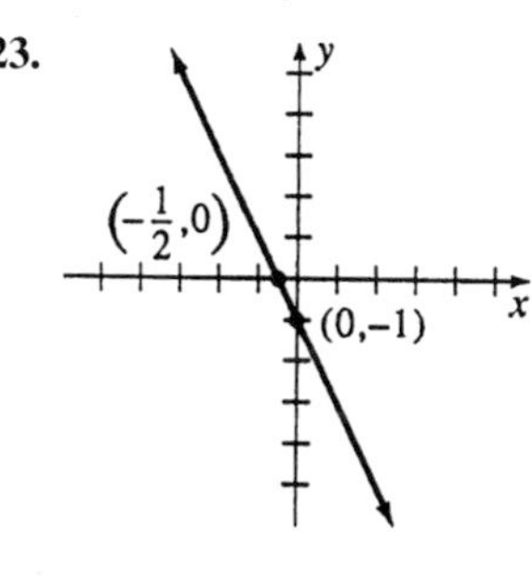

25.

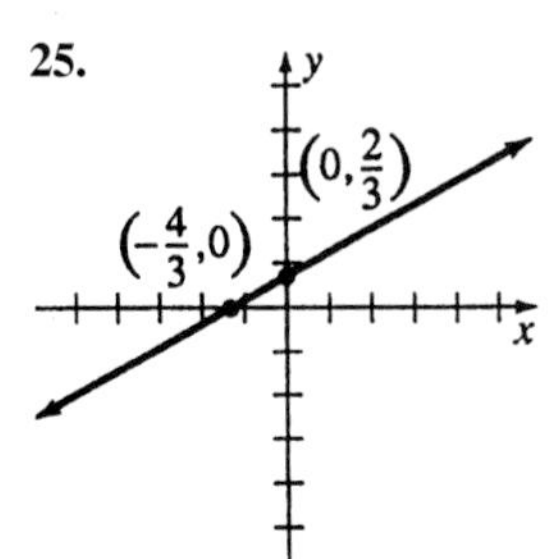

27.

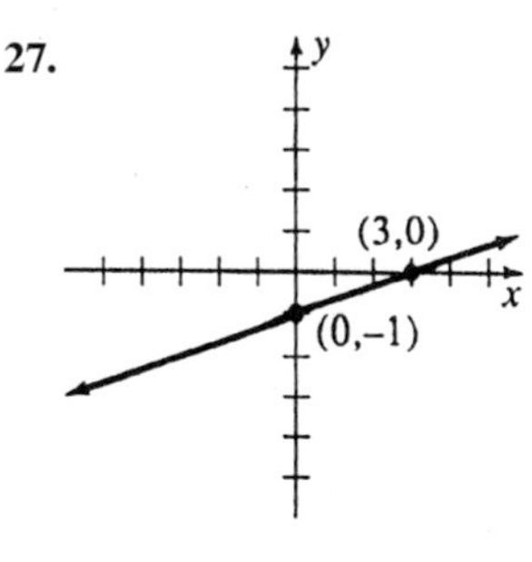

29. **31.**

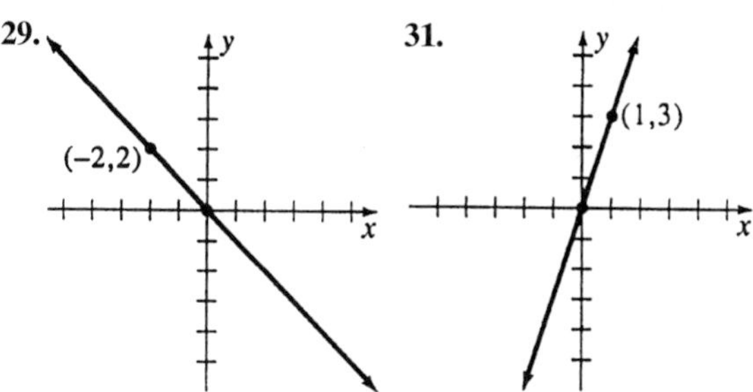

33. **35.**

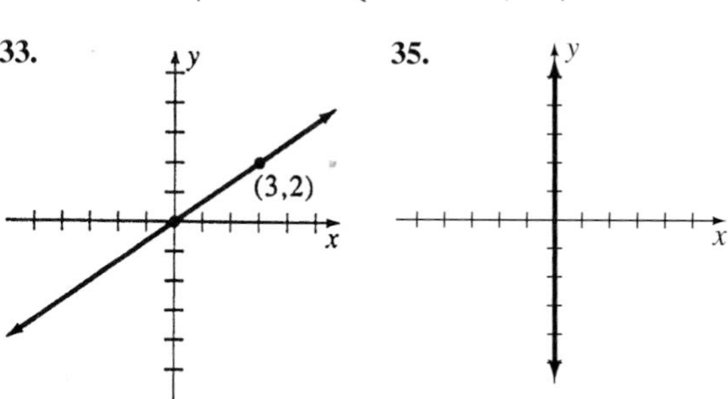

37. **39.**

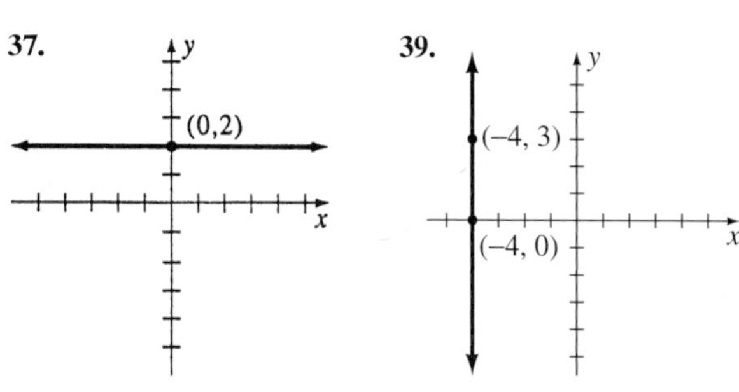

41. (a)

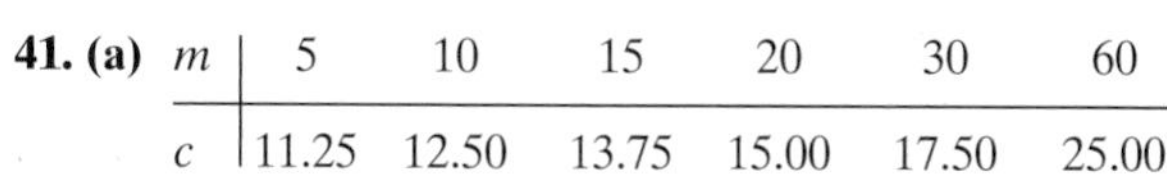

m	5	10	15	20	30	60
c	11.25	12.50	13.75	15.00	17.50	25.00

(b)

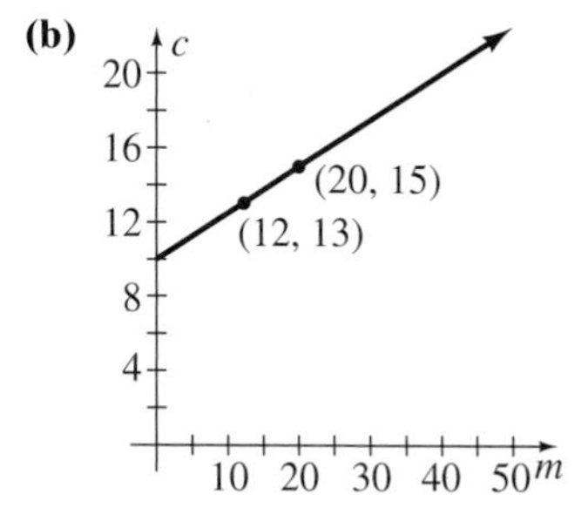

(d) 16.25, 20.00, 21.25

43. (a)

h	6.0	6.5	7.0	8.0	8.5	9.0	10.0
G	120	135	150	180	195	210	240

(b)

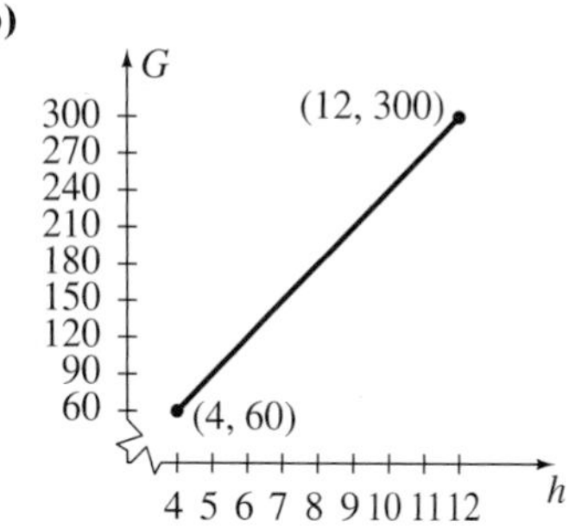

45.

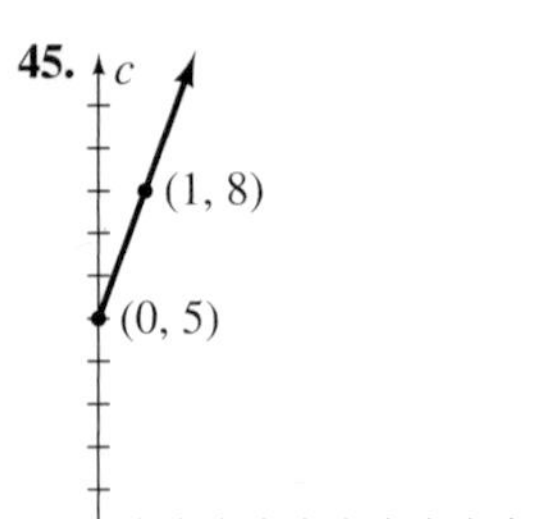

47.

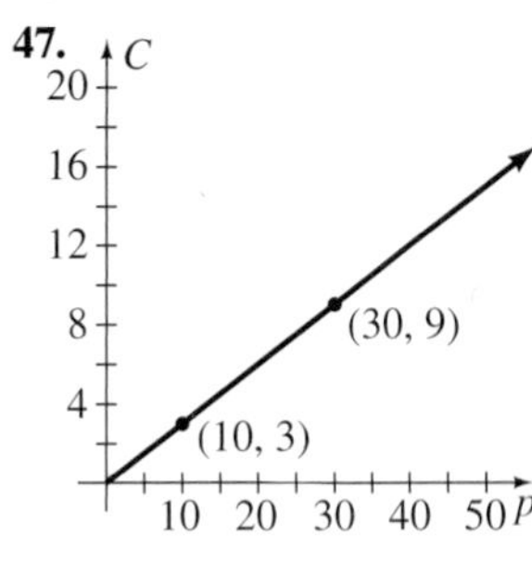

53.

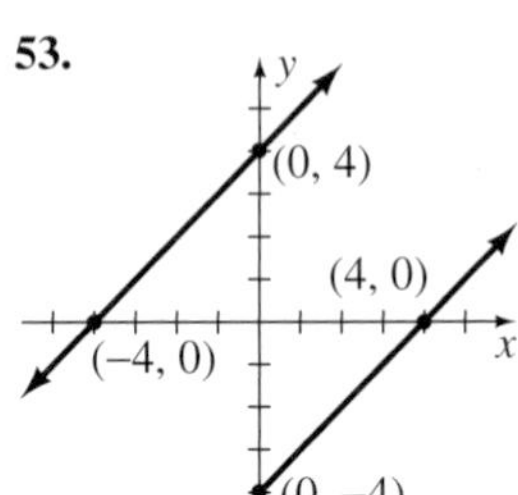

55.

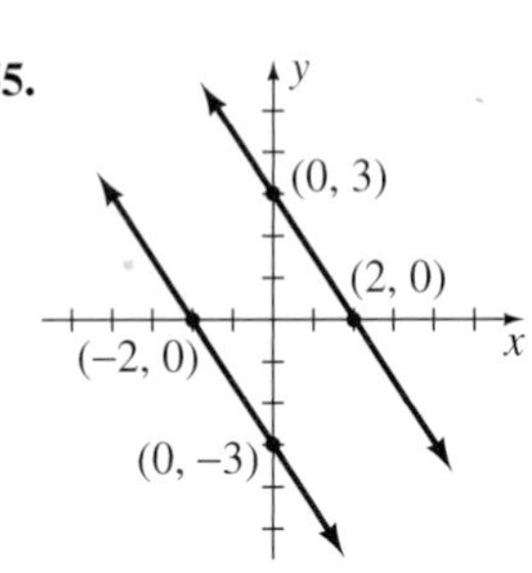

Problem Set 3.2 (page 142)

1.

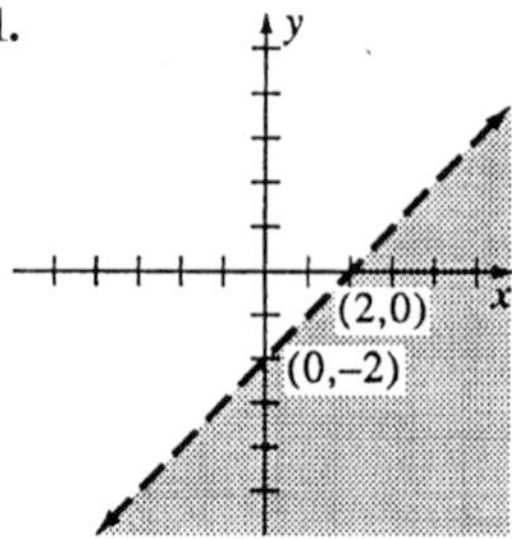

3.

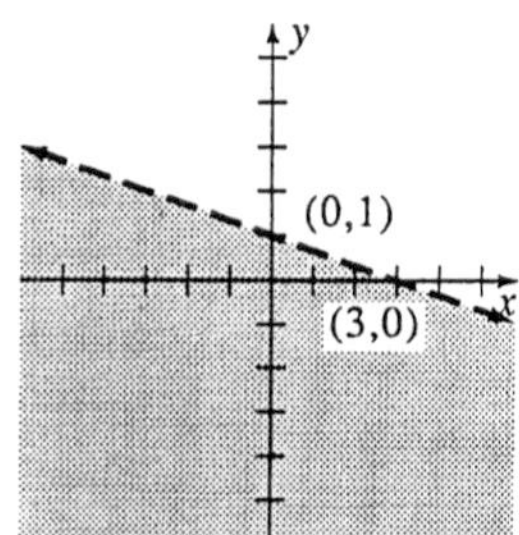

5.

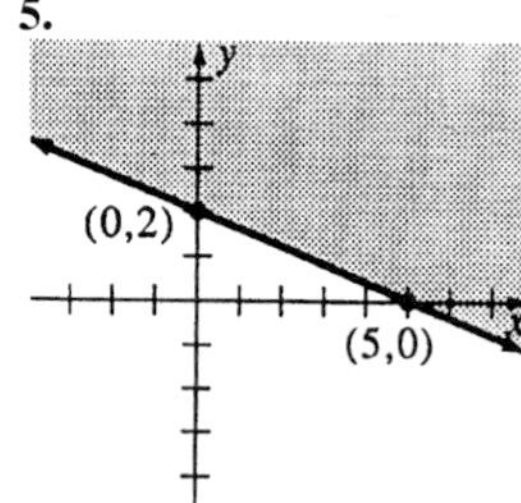

7.

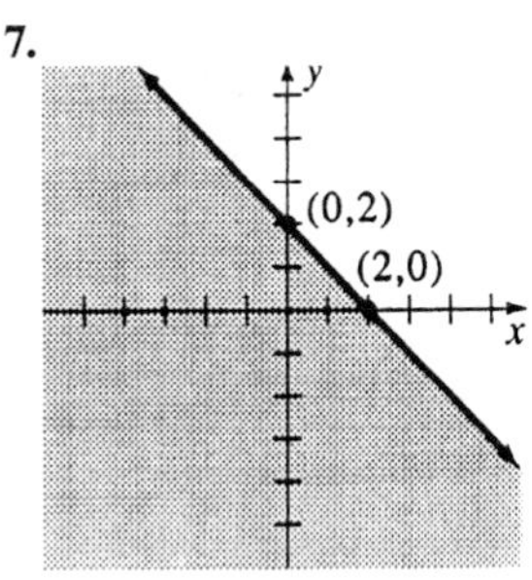

9.

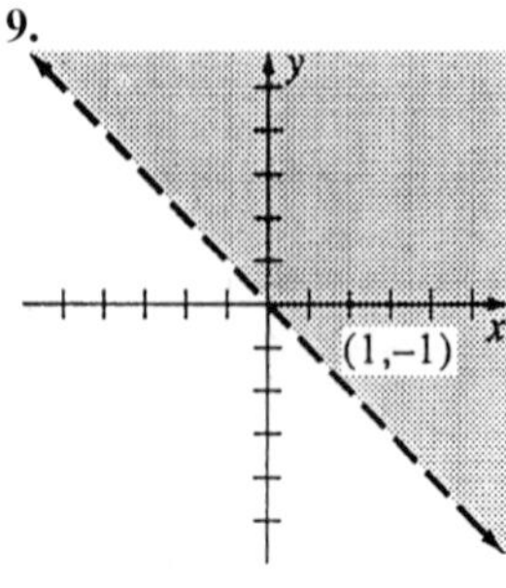

11.

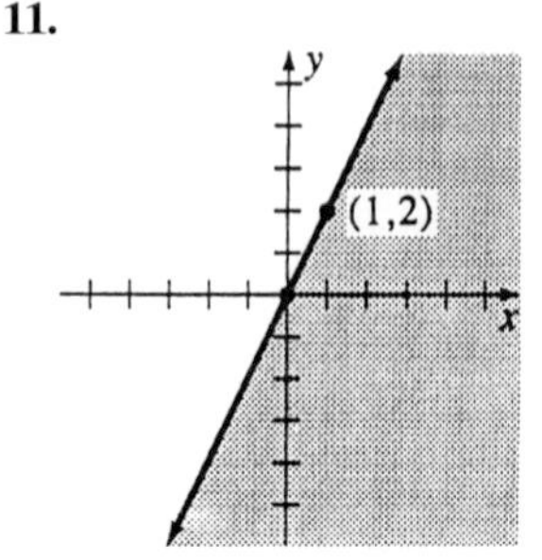

13.

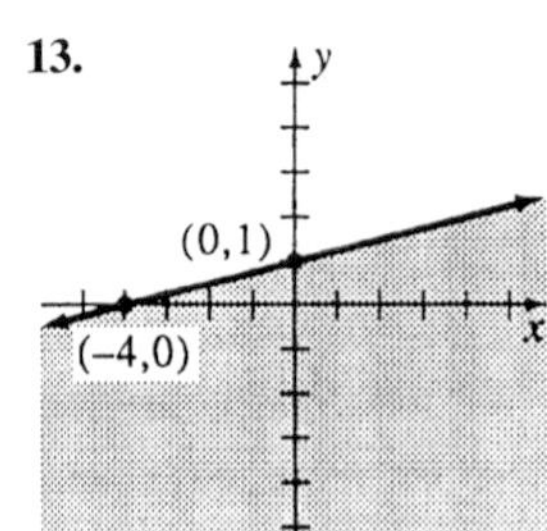

15.

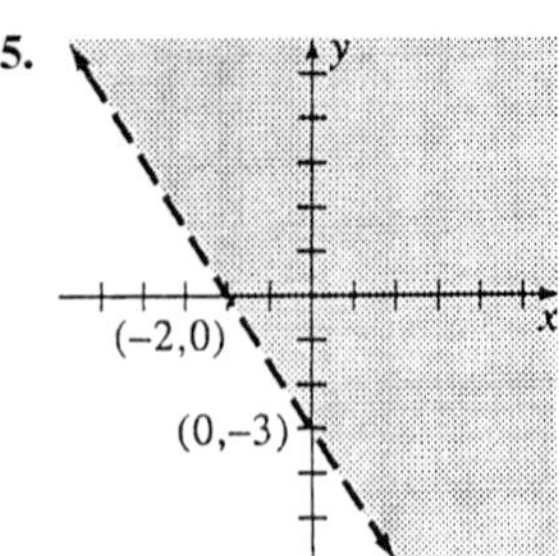

17.

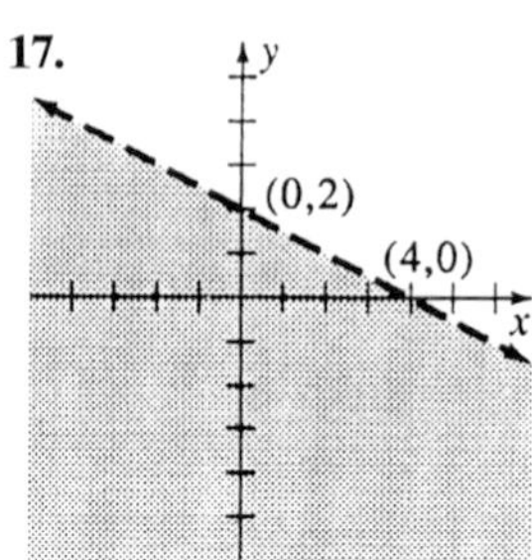

19.

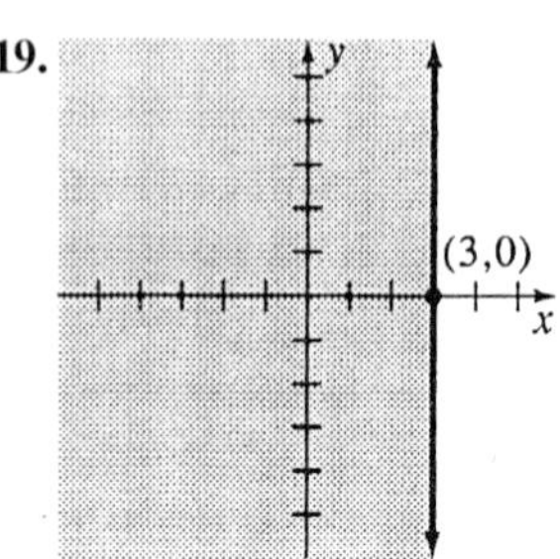

21.

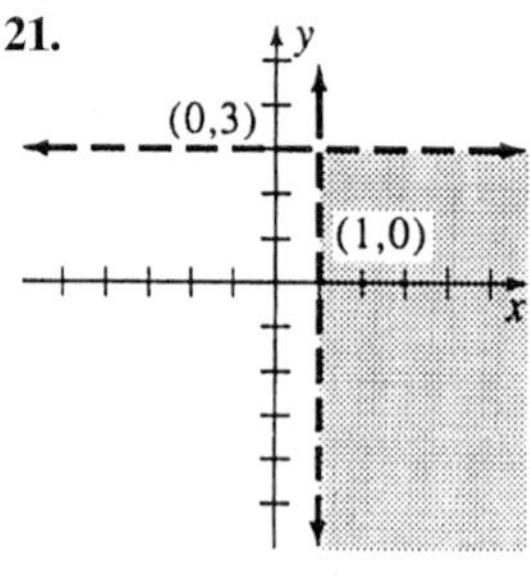

23.

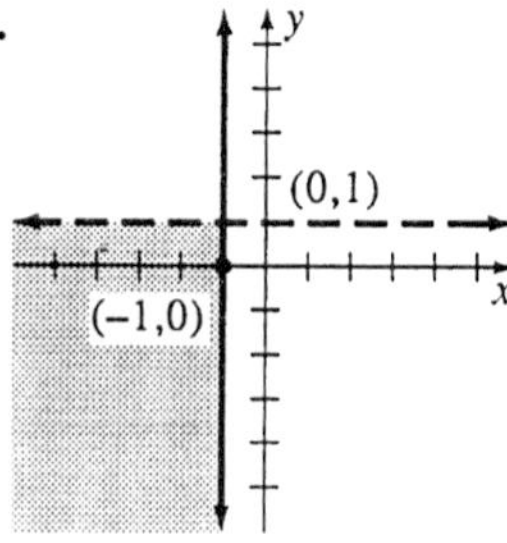

27.

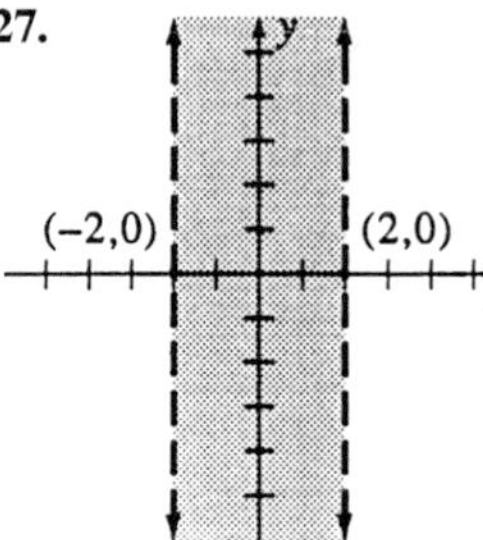

29.

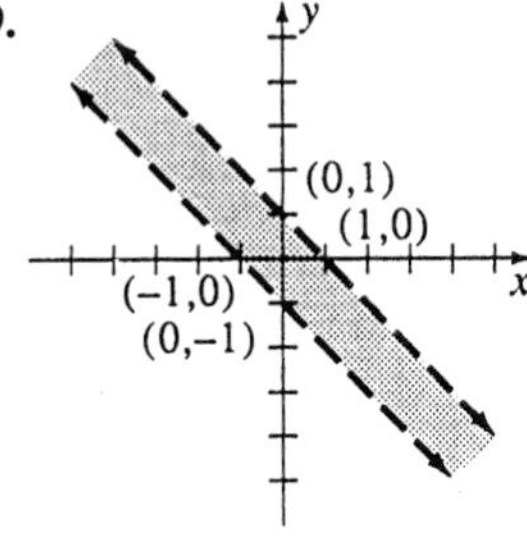

Problem Set 3.3 (page 152)

1. 15 **3.** $\sqrt{13}$ **5.** $\sqrt{97}$ **7.** $\sqrt{34}$ **9.** 6 **11.** $\sqrt{73}$
13. The lengths of the sides are 5, 12, and 13. Because $5^2 + 12^2 = 13^2$, it is a right triangle.
15. The distances between (1, 2) and (3, 5), between (3, 5) and (5, 8), and between (5, 8) and (7, 11) are all $\sqrt{13}$ units.

17. $\frac{4}{3}$ **19.** $-\frac{7}{3}$ **21.** −2 **23.** $\frac{3}{5}$ **25.** 0 **27.** $\frac{1}{2}$

29. 7 **31.** −2 **33–39.** Answers will vary.

41. $-\frac{2}{3}$ **43.** $\frac{1}{2}$ **45.** $\frac{4}{7}$ **47.** 0 **49.** −5

51.

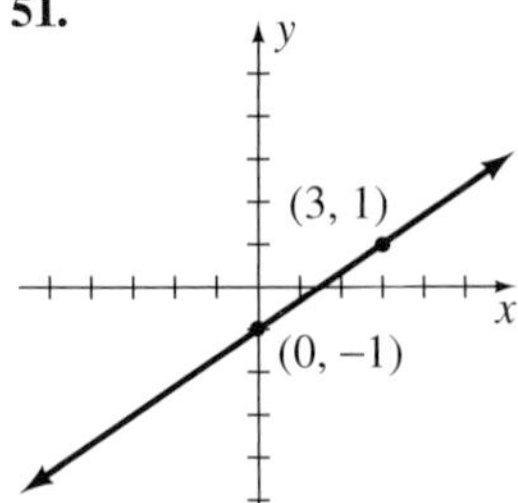

53.

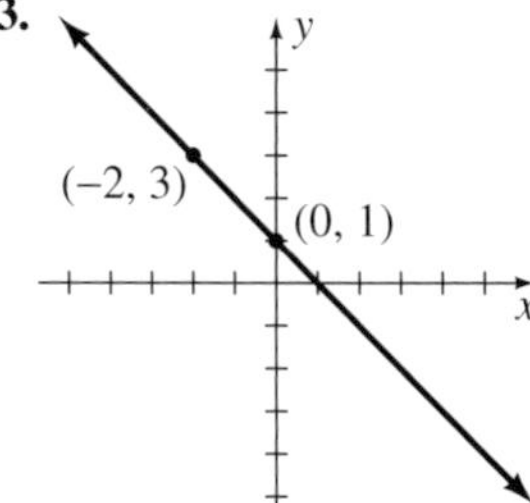

55.

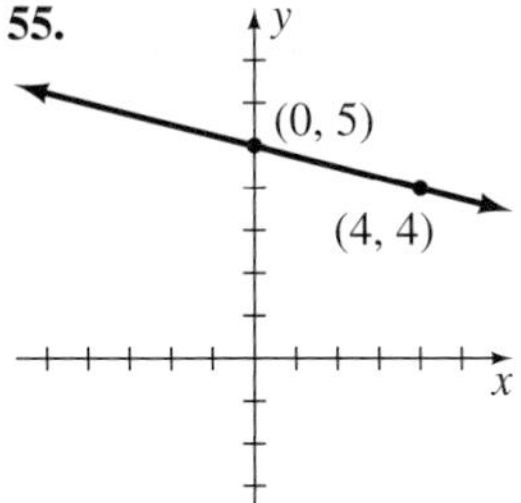

57.

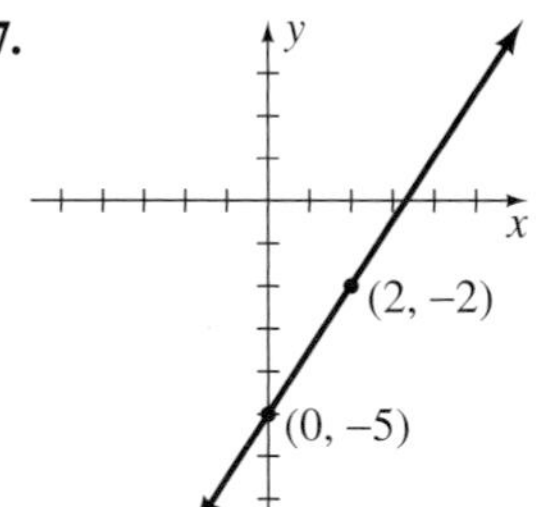

59. 105.6 feet **61.** 8.1% **63.** 19 centimeters

69. **(a)** (3, 5) **(b)** (5, 10) **(c)** (2, 5) **(d)** $\left(-1, \frac{18}{5}\right)$

(e) $\left(\frac{17}{8}, -7\right)$ **(f)** $\left(-\frac{9}{8}, -\frac{15}{2}\right)$

Problem Set 3.4 (page 165)

1. $x - 2y = -7$ **3.** $3x - y = -10$ **5.** $3x + 4y = -15$
7. $5x - 4y = 28$ **9.** $5x + 8y = -15$ **11.** $x - y = 1$
13. $5x - 2y = -4$ **15.** $x + 7y = 11$ **17.** $x + 2y = -9$
19. $7x - 5y = 0$ **21.** $2x - y = 4$ **23.** $y = \frac{1}{1000}x + 2$
25. $y = \frac{9}{5}x + 32$ **27.** $y = 30x - 60$ **29.** $y = \frac{3}{7}x + 4$
31. $y = 2x - 3$ **33.** $y = -\frac{2}{5}x + 1$ **35.** $y = 0(x) - 4$
37. $5x - 2y = -23$ **39.** $2x + y = 18$ **41.** $x + 3y = 5$
43. $m = -3$ and $b = 7$ **45.** $m = -\frac{3}{2}$ and $b = \frac{9}{2}$
47. $m = \frac{1}{5}$ and $b = -\frac{12}{5}$

49.

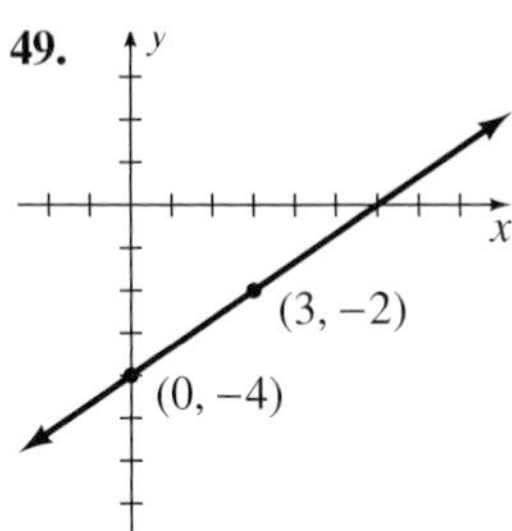

51.

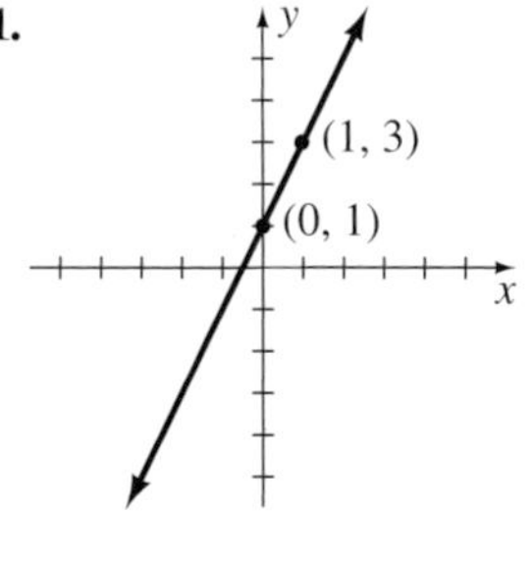

53.

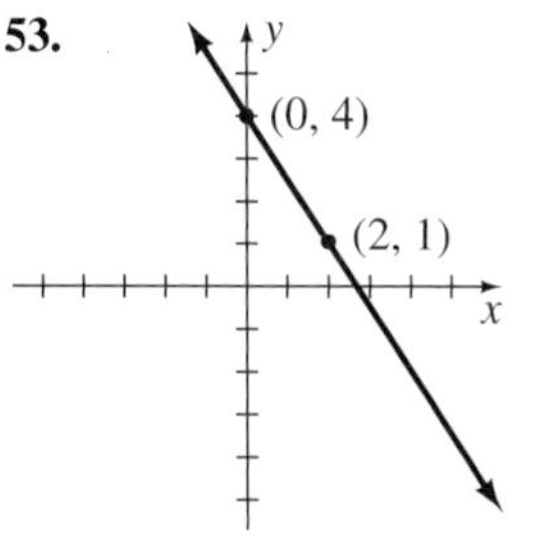

55.

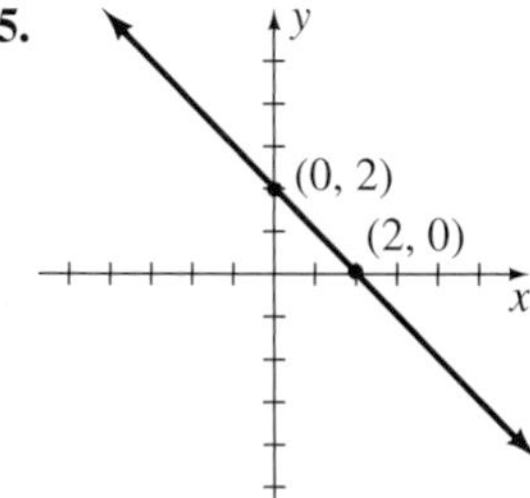

57.

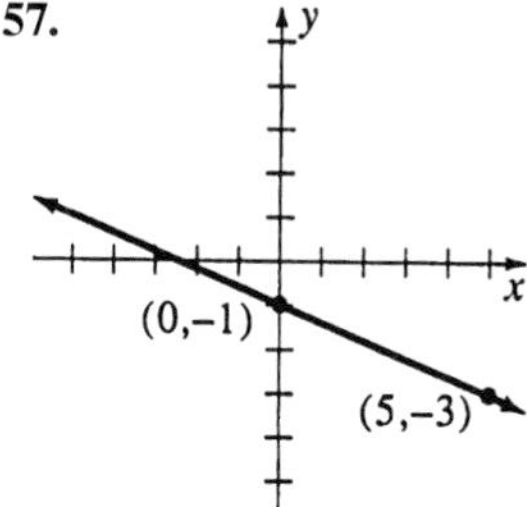

59.

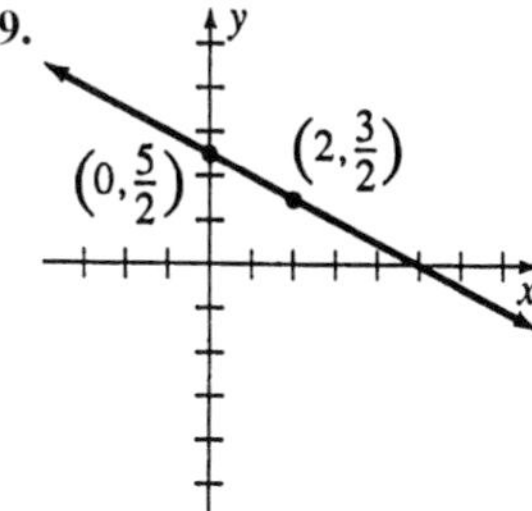

61.

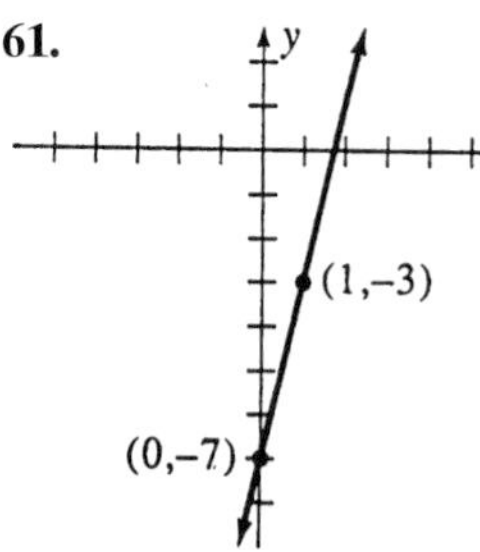

63.

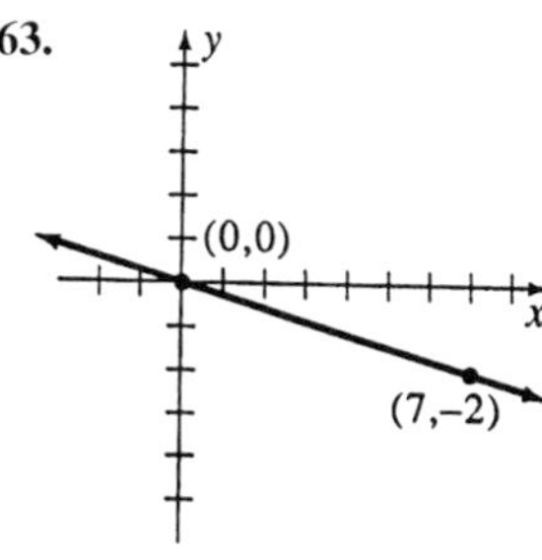

65.

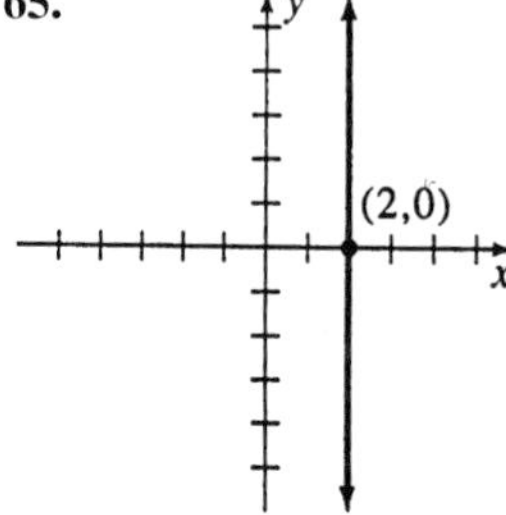

67. $x + 0(y) = 2$ **69.** $0(x) + y = 6$ **71.** $x + 5y = 16$
73. $4x - 7y = 0$ **75.** $x + 2y = 5$ **77.** $3x + 2y = 0$
85. **(a)** $2x - y = 1$ **(b)** $5x - 6y = 29$ **(c)** $x + y = 2$
(d) $3x - 2y = 18$

Chapter 3 Review Problem Set (page 174)

1. (1, 2), (−1, 10) **2.** (0, 2)

3. (2, 3), (−2, 9) **4.** (−3, 0)

5.

x	−1	0	1	4
y	−7	−5	−3	3

6.

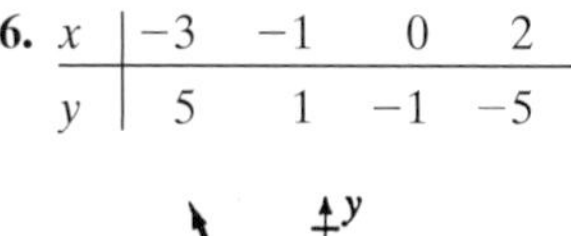

x	−3	−1	0	2
y	5	1	−1	−5

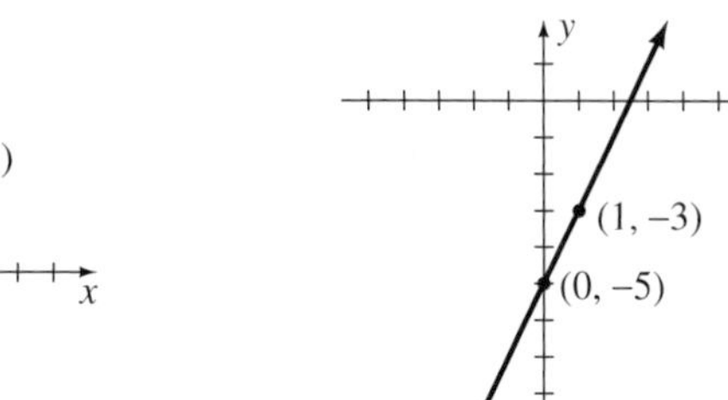

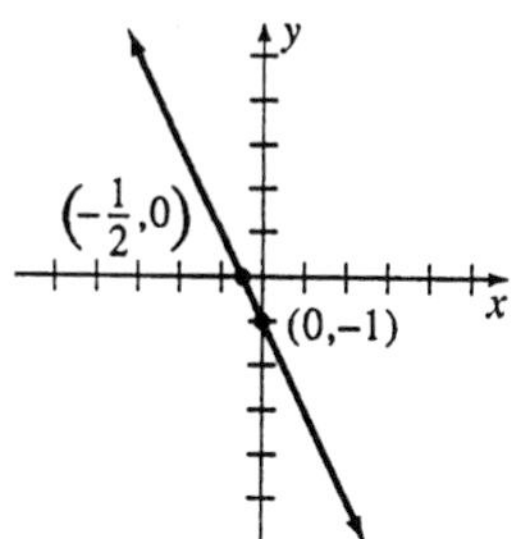

7.

x	-2	0	2	4
y	-5	-2	1	4

8.

x	-3	0	3
y	-3	-1	1

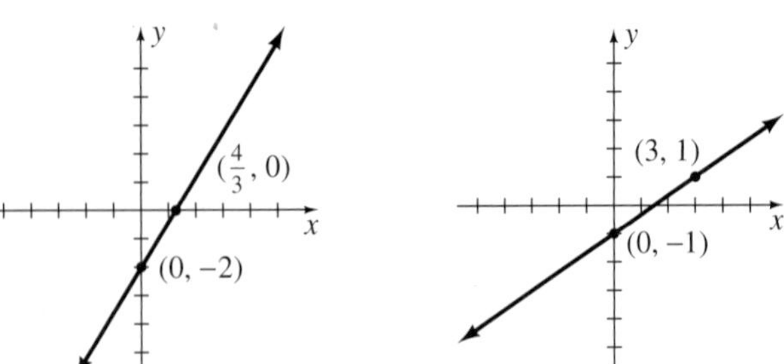

9.

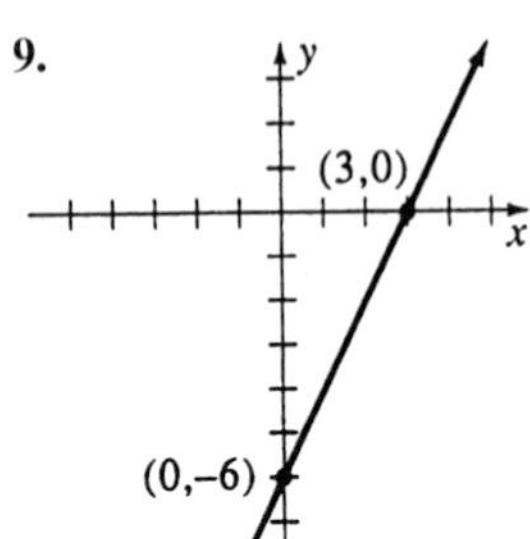

10.

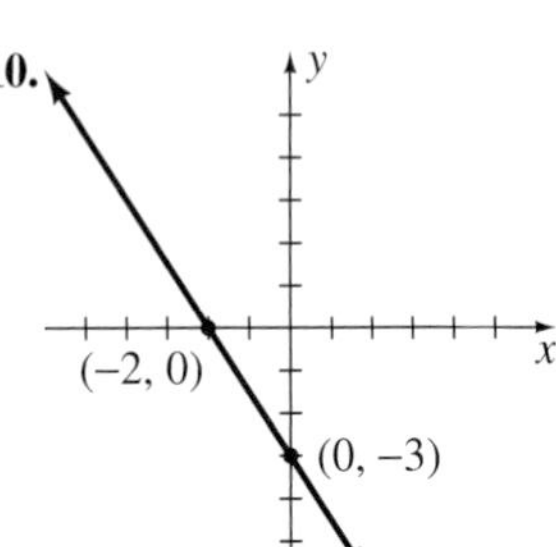

11.

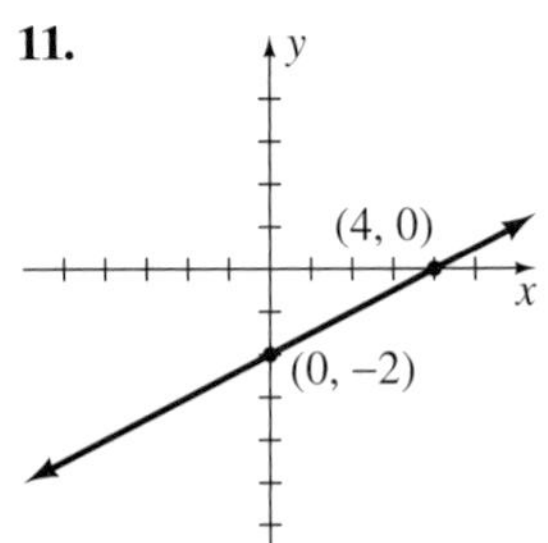

12.

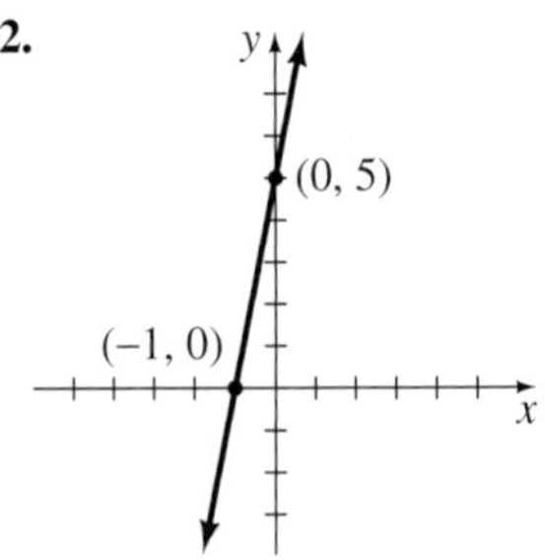

13.

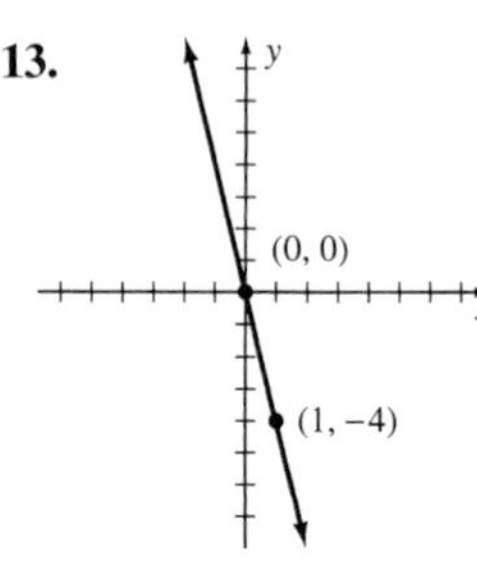

14.

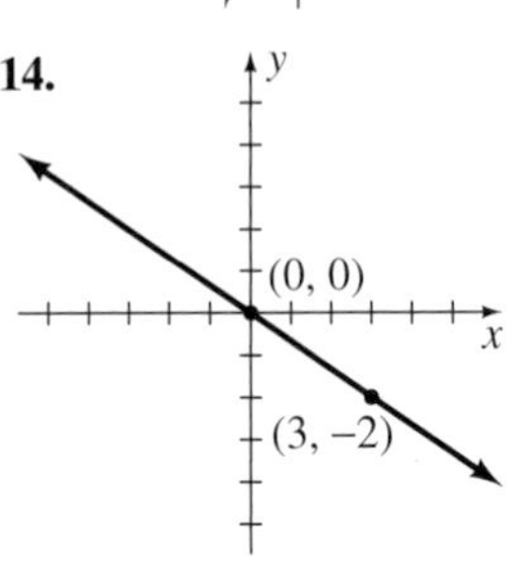

15.

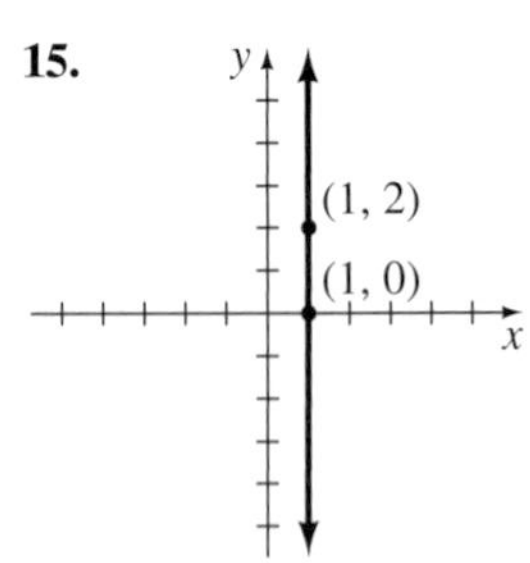

16.

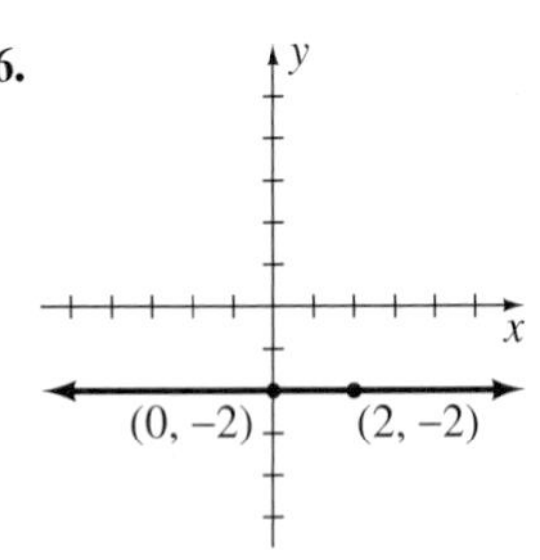

17.

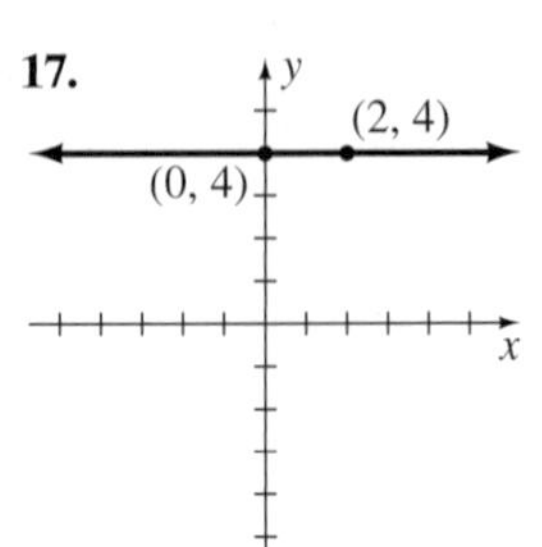

18.

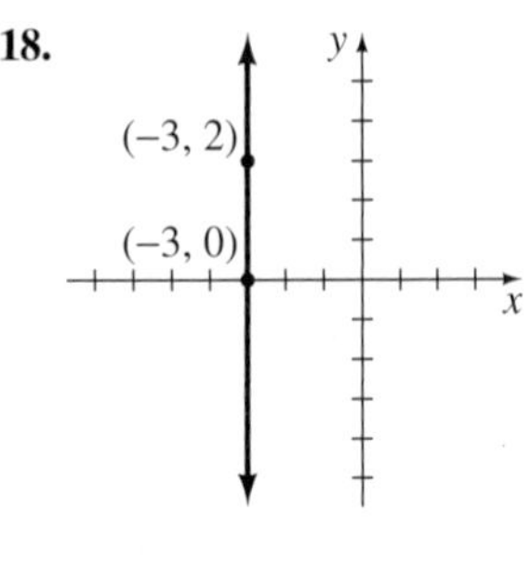

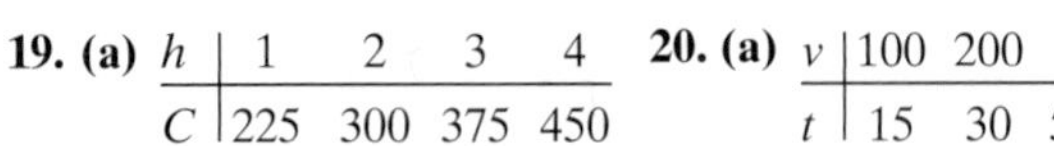

19. (a)

h	1	2	3	4
C	225	300	375	450

20. (a)

v	100	200	350	400
t	15	30	52.50	60

(b)

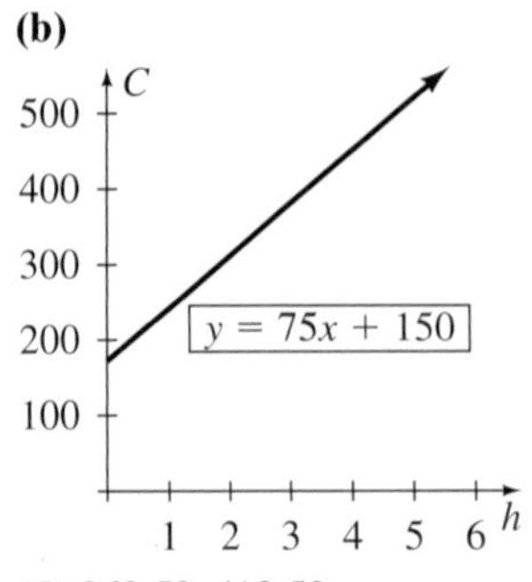

(b)

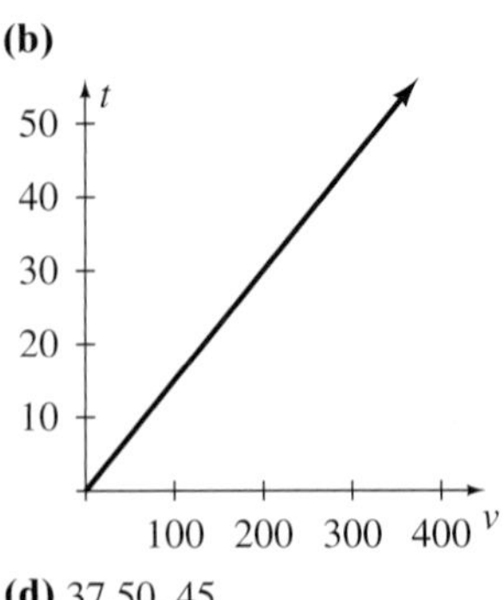

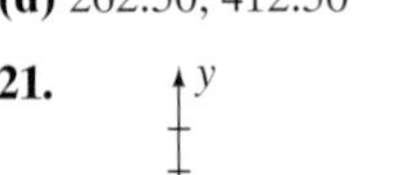

(d) 262.50, 412.50

(d) 37.50, 45

21.

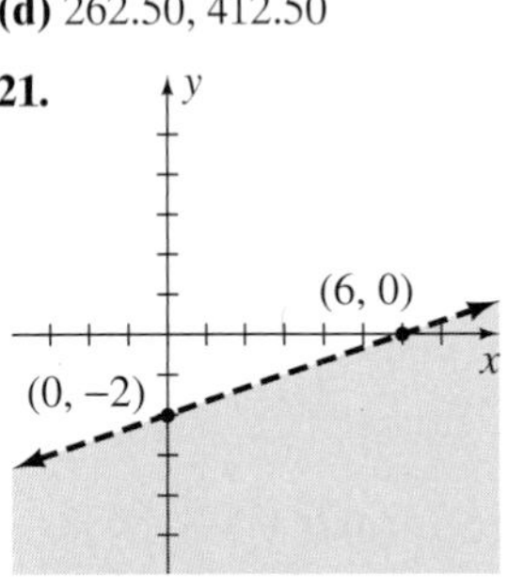

22.

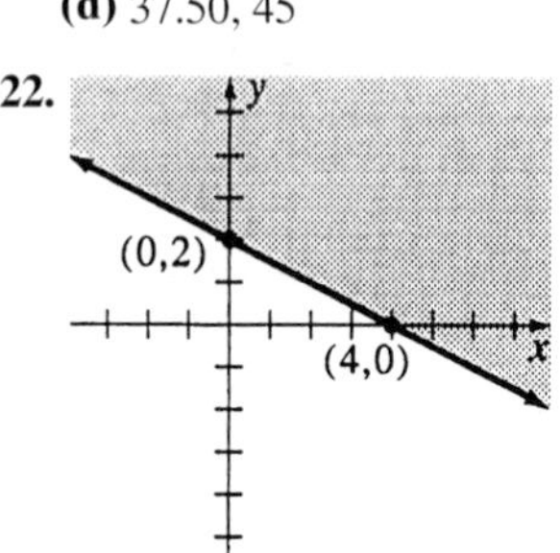

23.

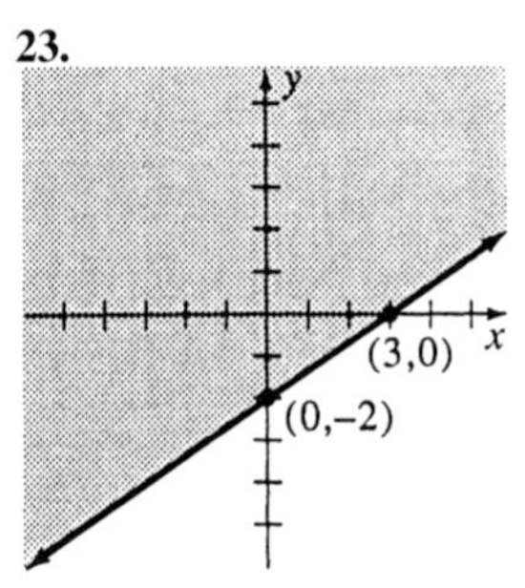

24.

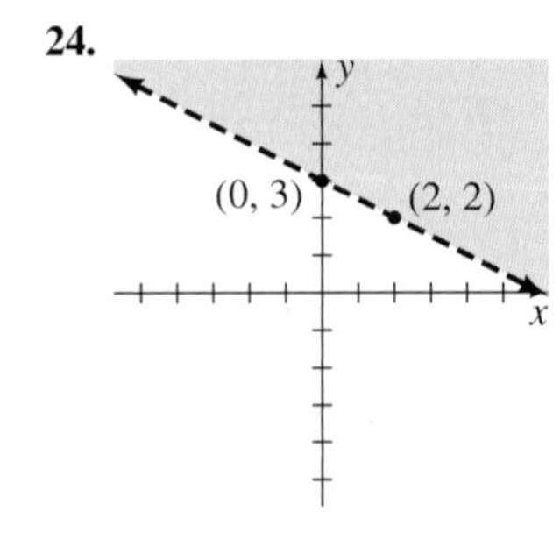

25.

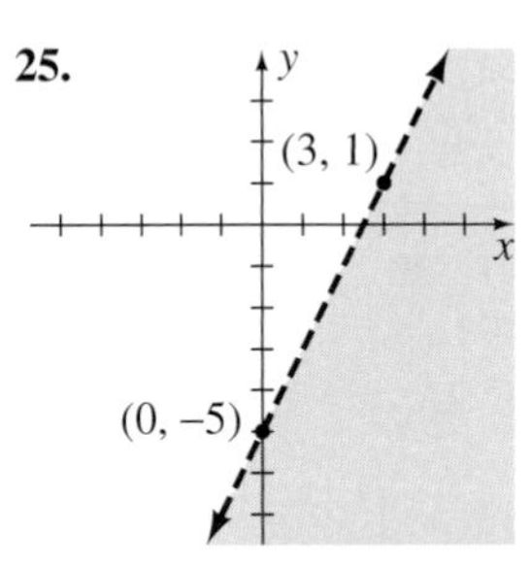

26.

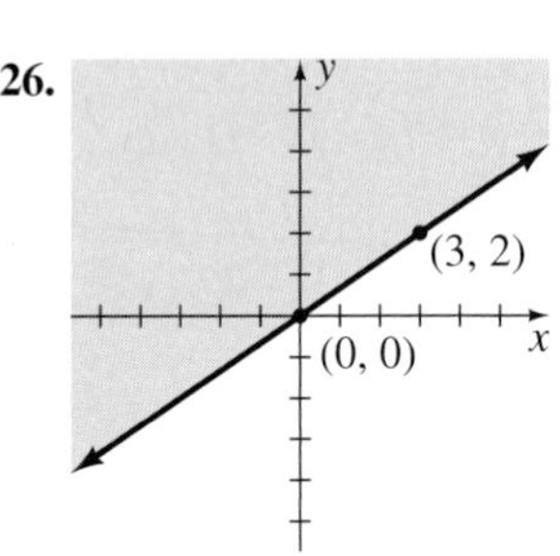

27. (a) $\sqrt{53}$ **(b)** $\sqrt{58}$ **28.** 5, 10, and $\sqrt{97}$

29. The distances between $(-3, -1)$ and $(1, 2)$ and between $(1, 2)$ and $(5, 5)$ are 5.

30. (a) $\frac{6}{5}$ **(b)** $-\frac{2}{3}$ **31.** 5 **32.** -1

33.

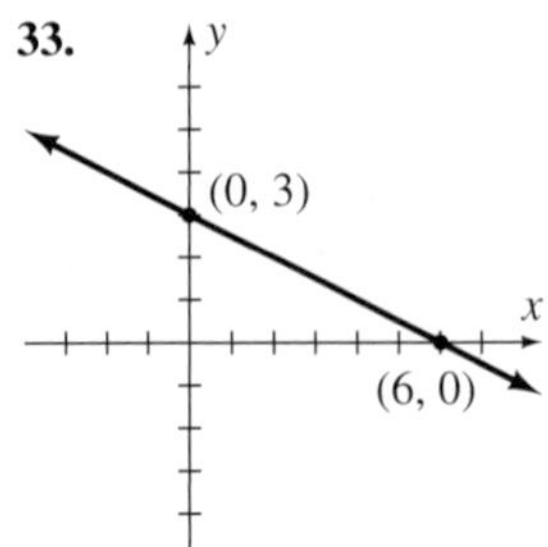

34.

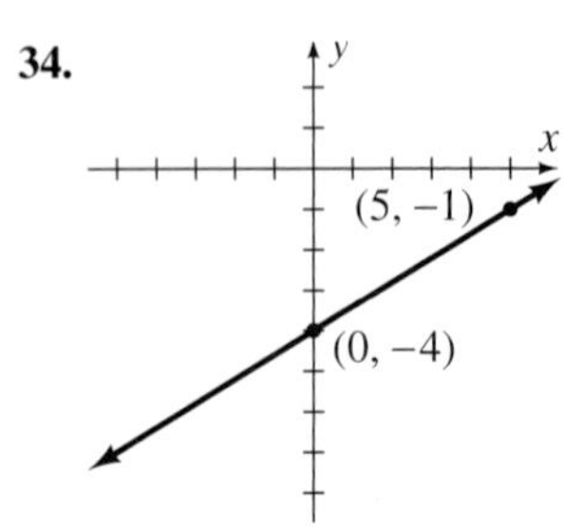

35.

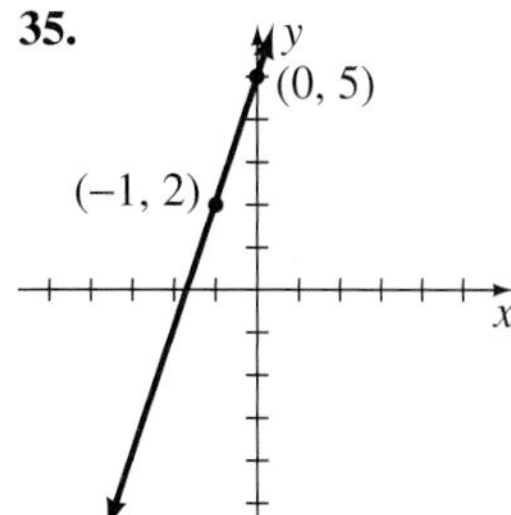

36.

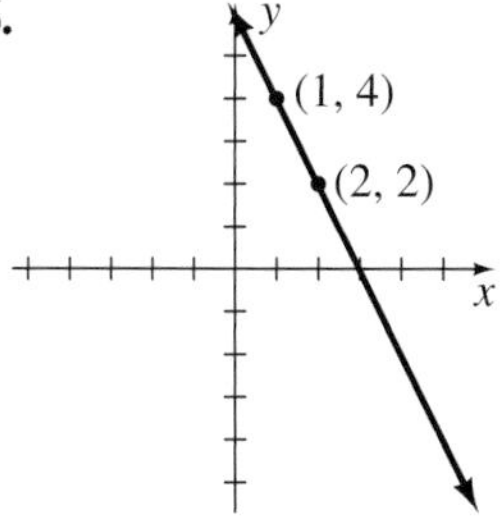

37. 316.8 ft **38.** 8 in. **39. (a)** $m = -4$ **(b)** $m = \frac{2}{7}$
40. $m = -\frac{5}{3}$ **41.** $m = -\frac{4}{5}$ **42.** $3x + 7y = 28$
43. $2x - 3y = 16$ **44.** $x + y = -2$ **45.** $7x + 4y = 1$
46. $x - y = -4$ **47.** $x - 2y = -8$ **48.** $2x - 3y = 14$
49. $4x + y = -29$ **50.** $y = \frac{3}{200}x - 600$
51. $y = \frac{1}{5}x - 20$ **52.** $y = 8x$ **53.** $y = 300x - 150$

Chapter 3 Test (page 177)

1. $(-1, 4), (-3, 0), (10, 26)$ **2.** $-\frac{6}{5}$ **3.** $\frac{3}{7}$ **4.** $\sqrt{58}$ **5.** $\frac{7}{2}$
6. $\frac{9}{4}$ **7.** 480 ft **8.** 6.7% **9.** 43 cm **10.** -2 **11.** $-\frac{2}{3}$

12.

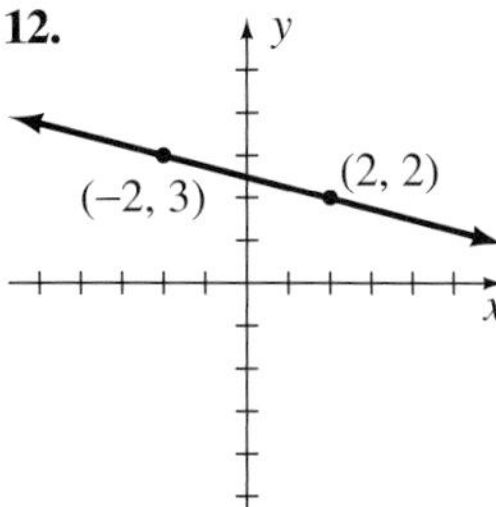

13.

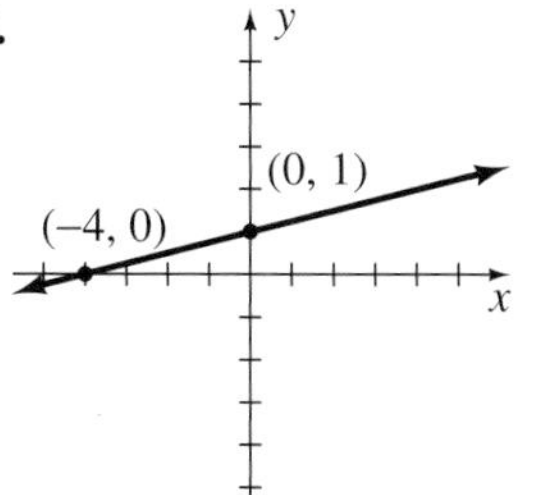

14.

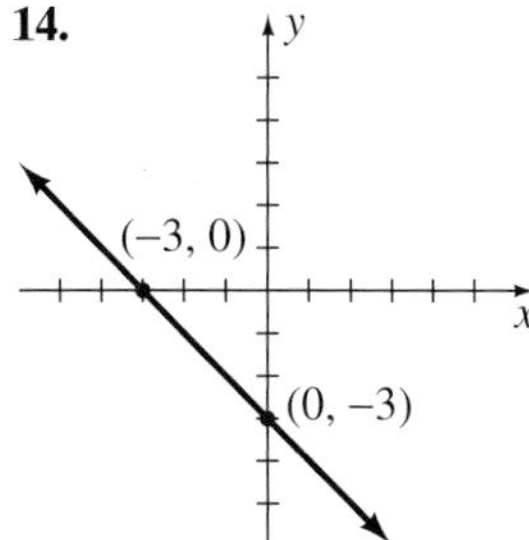

15.

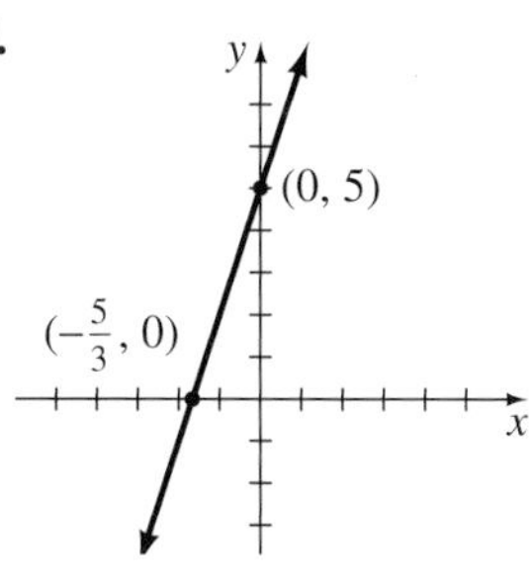

16.

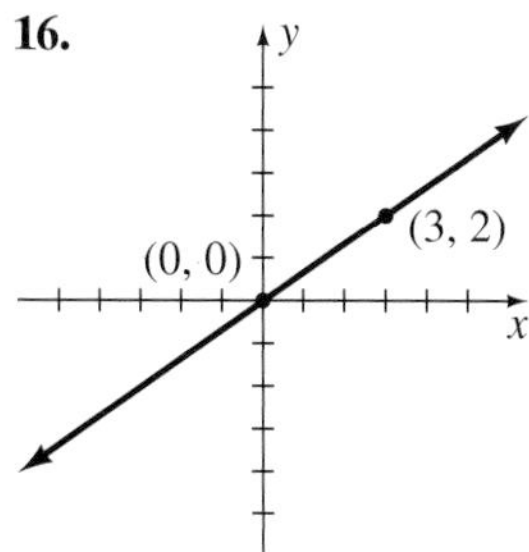

17.

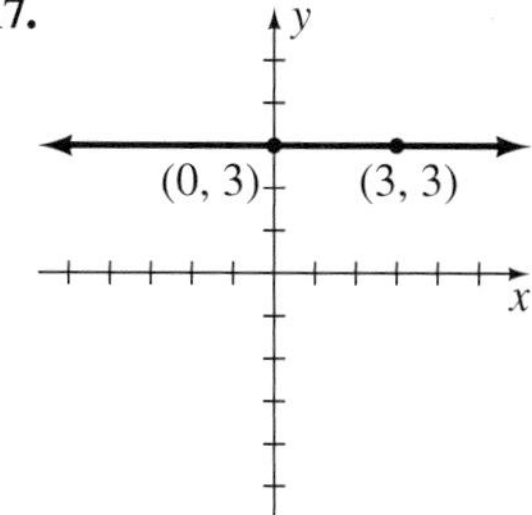

18.

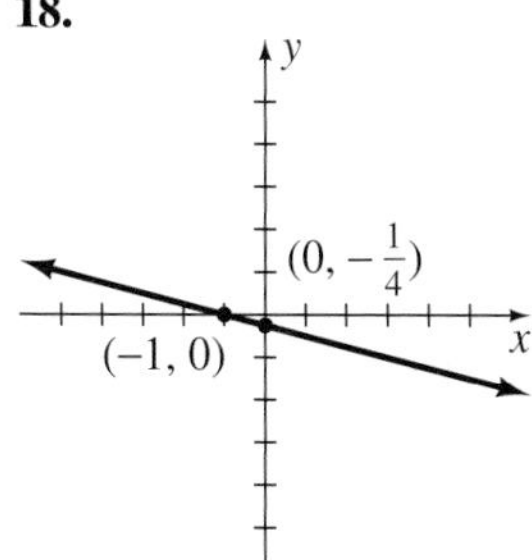

19.

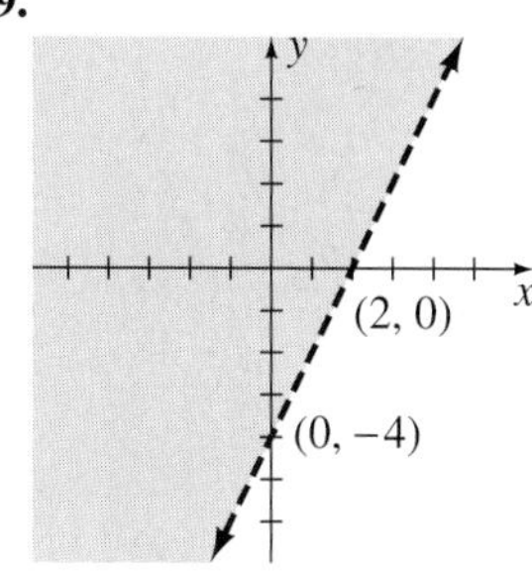

20.

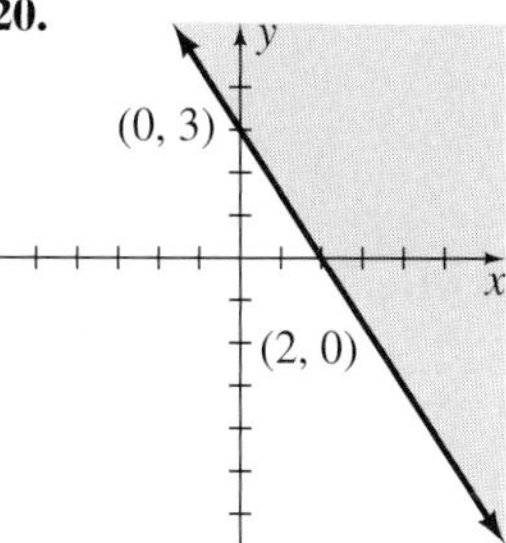

21. $3x + 2y = 2$ **22.** $y = -\frac{1}{6}x + \frac{4}{3}$
23. $5x + 2y = -18$
24. $6x + y = 31$
25. $y = \frac{1}{50}x + 20$ or $y = 0.02x + 20$

CHAPTER 4

Problem Set 4.1 (page 186)

1. $\{(3, 2)\}$ **3.** $\{(-2, 1)\}$ **5.** Dependent **7.** $\{(4, -3)\}$
9. Inconsistent **11.** $\{(2, 1)\}$ **13.** $\{(-3, -1)\}$ **15.** $\{(2, 4)\}$

17.

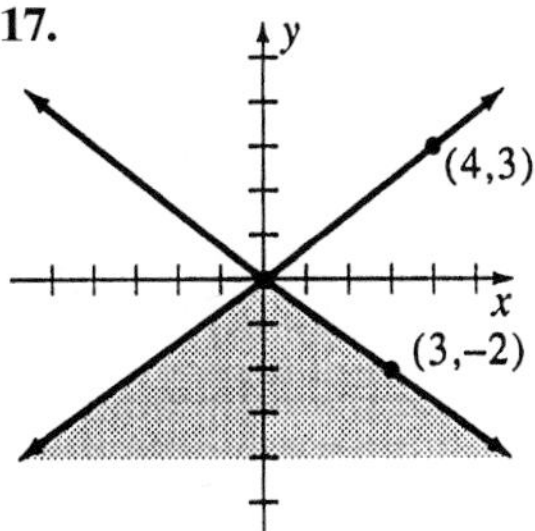

19.

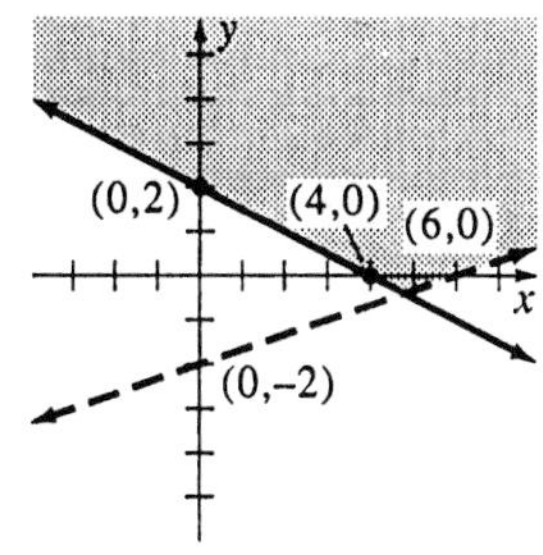

21.

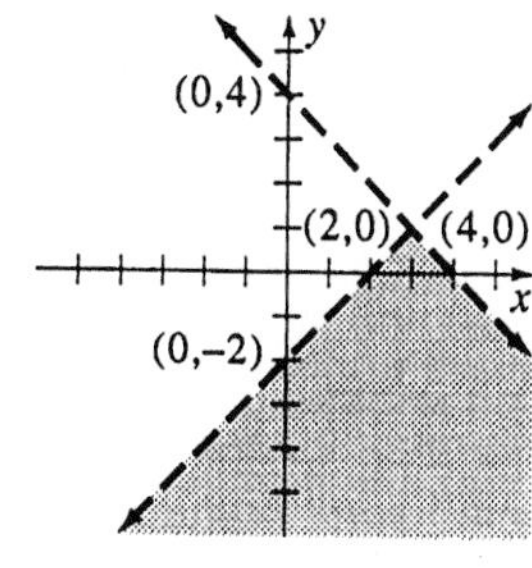

23.

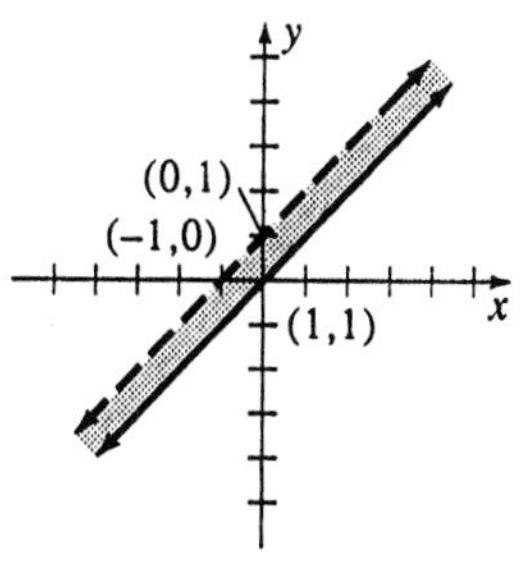

25.

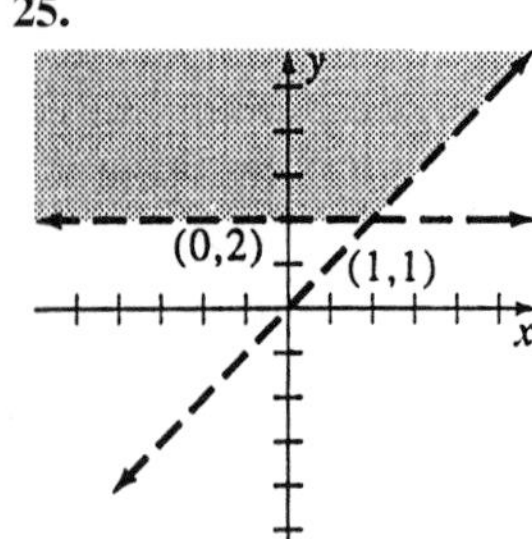

27.

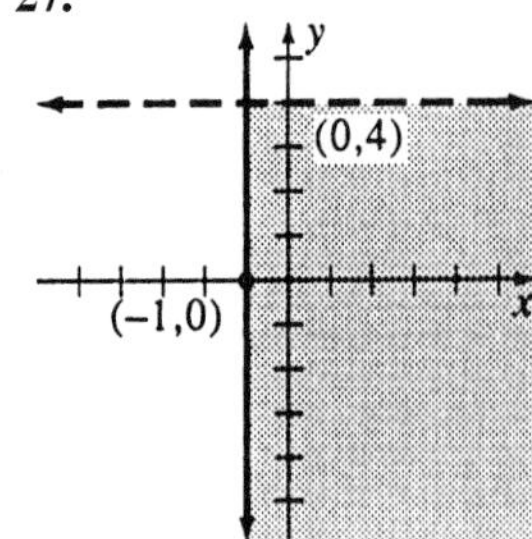

29.

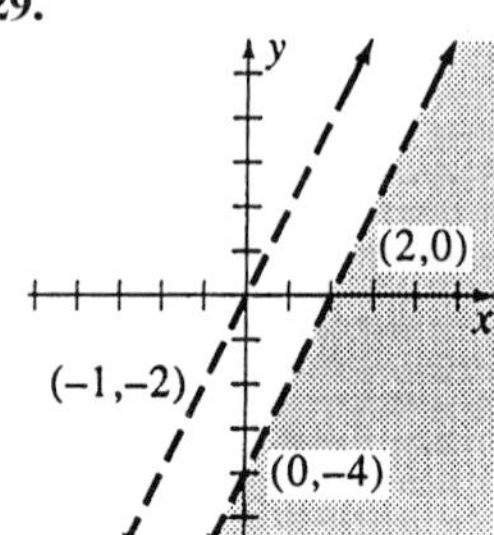

31.

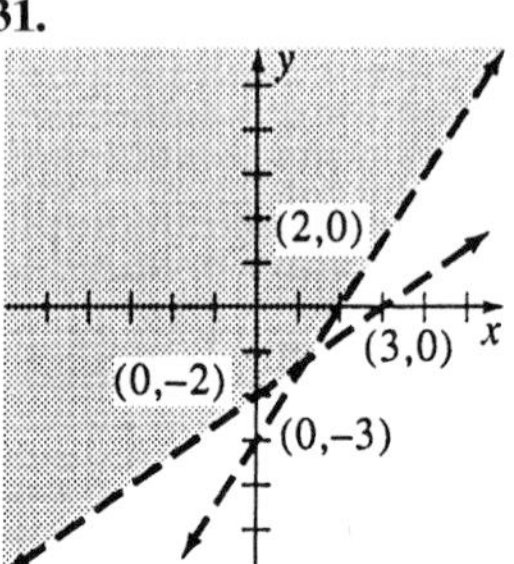

Problem Set 4.2 (page 192)

1. $\{(8, 12)\}$ **3.** $\{(-4, -6)\}$ **5.** $\{(-9, 3)\}$ **7.** $\{(1, 3)\}$

9. $\left\{\left(5, \frac{3}{2}\right)\right\}$ **11.** $\left\{\left(\frac{9}{5}, -\frac{7}{25}\right)\right\}$ **13.** $\{(-2, -4)\}$ **15.** $\{(5, 2)\}$

17. $\{(-4, -8)\}$ **19.** $\left\{\left(\frac{11}{20}, \frac{7}{20}\right)\right\}$ **21.** $\{(-1, 5)\}$

23. $\left\{\left(-\frac{3}{4}, -\frac{6}{5}\right)\right\}$ **25.** $\left\{\left(\frac{5}{27}, -\frac{26}{27}\right)\right\}$

27. \$2000 at 7% and \$8000 at 8% **29.** 34 and 97
31. 20°, 70° **33.** 42 females **35.** 20 in. by 27 in.
37. 60 five-dollar bills and 40 ten-dollar bills
39. 2500 student tickets and 500 nonstudent tickets

Problem Set 4.3 (page 201)

1. $\{(4, -3)\}$ **3.** $\{(-1, -3)\}$ **5.** $\{(-8, 2)\}$ **7.** $\{(-4, 0)\}$

9. $\left\{\left(-\frac{1}{2}, \frac{3}{4}\right)\right\}$ **11.** Inconsistent **13.** $\left\{\left(-\frac{1}{11}, \frac{4}{11}\right)\right\}$

15. Every point on the line is a solution.
17. $\{(4, -9)\}$ **19.** $\{(7, 0)\}$ **21.** $\{(7, 12)\}$

23. $\left\{\left(\frac{7}{11}, \frac{2}{11}\right)\right\}$ **25.** Inconsistent **27.** $\left\{\left(\frac{51}{31}, -\frac{32}{31}\right)\right\}$

29. $\{(-2, -4)\}$ **31.** $\left\{\left(-1, -\frac{14}{3}\right)\right\}$ **33.** $\{(-6, 12)\}$

35. $\{(2, 8)\}$ **37.** $\{(-1, 3)\}$ **39.** $\{(16, -12)\}$

41. $\left\{\left(-\frac{3}{4}, \frac{3}{2}\right)\right\}$ **43.** $\{(5, -5)\}$

45. 5 gal of 10% solution and 15 gal of 20% solution
47. \$1 for a tennis ball and \$2 for a golf ball
49. 40 double rooms and 15 single rooms

51. 50 cheese pizzas and 35 supreme pizzas

53. $\frac{3}{4}$ **55.** 18 cm by 24 cm

61. **(a)** $\{(4, 6)\}$ **(b)** $\{(3, 5)\}$ **(c)** $\{(2, -3)\}$

(d) $\left\{\left(\frac{1}{2}, \frac{1}{3}\right)\right\}$ **(e)** $\left\{\left(\frac{1}{4}, -\frac{2}{3}\right)\right\}$ **(f)** $\{(-4, -5)\}$

Problem Set 4.4 (page 211)

1. $\{(-2, 5, 2)\}$ **3.** $\{(4, -1, -2)\}$ **5.** $\{(-1, 3, 5)\}$

7. Infinitely many solutions **9.** $\varnothing$

11. $\left\{\left(-2, \frac{3}{2}, 1\right)\right\}$ **13.** $\left\{\left(\frac{1}{3}, -\frac{1}{2}, 1\right)\right\}$

15. $\left\{\left(\frac{2}{3}, -4, \frac{3}{4}\right)\right\}$ **17.** $\{(-2, 4, 0)\}$

19. $\left\{\left(\frac{1}{2}, \frac{1}{3}, \frac{1}{6}\right)\right\}$ **21.** 194

23. \$1.22 per bottle of catsup, \$1.77 per jar of peanut butter, and \$1.80 per jar of pickles
25. −2, 6, and 16 **27.** 40°, 60°, and 80°
29. \$500 at 12%, \$1000 at 13%, and \$1500 at 14%

Chapter 4 Review Problem Set (page 217)

1. $\{(2, 3)\}$ **2.** $\{(-3, 1)\}$ **3** $\varnothing$

4.

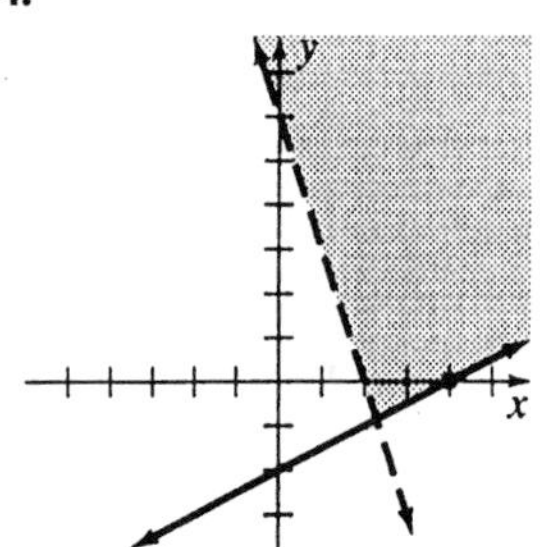

5.

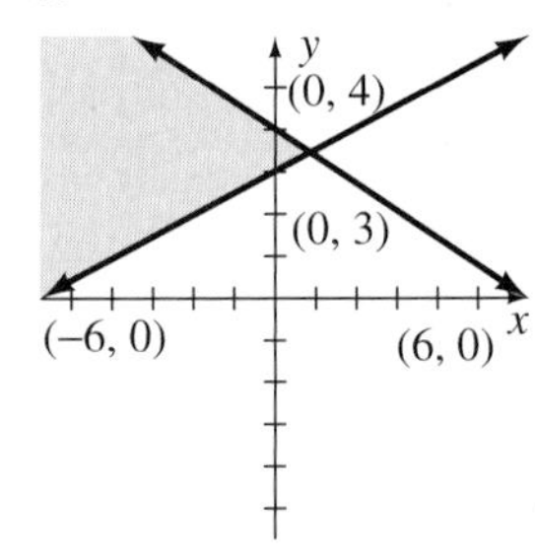

6.

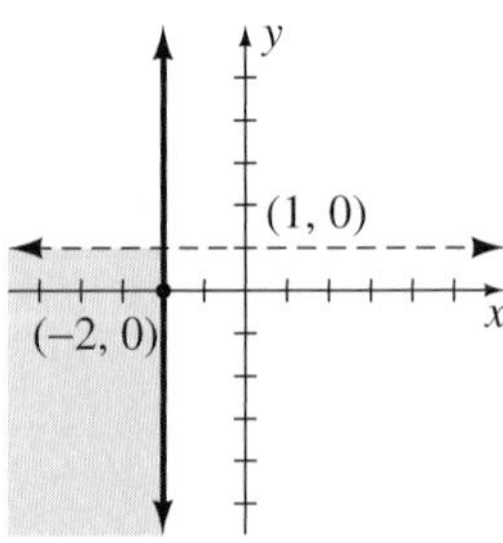

7. $\{(2, 6)\}$ **8.** $\{(-3, 7)\}$ **9.** $\{(-9, 8)\}$ **10.** $\left\{\left(\frac{89}{23}, -\frac{12}{23}\right)\right\}$

11. $\{(4, -7)\}$ **12.** $\{(-2, 8)\}$ **13.** $\{(0, -3)\}$ **14.** $\{(6, 7)\}$
15. $\{(1, -6)\}$ **16.** $\{(-4, 0)\}$ **17.** $\{(-12, 18)\}$ **18.** $\{(24, 8)\}$

19. $\left\{\left(\frac{2}{3}, -\frac{3}{4}\right)\right\}$ **20.** $\left\{\left(-\frac{1}{2}, \frac{3}{5}\right)\right\}$ **21.** $\varnothing$

22. $\{(x, y) | y = -4x + 4\}$
23. \$9 per lb for cashews and \$5 per lb for Spanish peanuts
24. \$1.50 for a carton of pop and \$2.25 for a pound of candy
25. The fixed fee is \$2, and the additional fee is \$.10 per pound.
26. $6\frac{2}{3}$ qt of the 1% and $3\frac{1}{3}$ qt of the 4%
27. 7 cm by 21 cm **28.** \$1200 at 1% and \$3000 at 1.5%
29. 5 five-dollar bills and 25 one-dollar bills
30. 30 review problems and 80 new material problems
31. $\{(2, -4, -6)\}$ **32.** $\{(-1, 5, -7)\}$ **33.** $\{(2, -3, 1)\}$
34. $\{(2, -1, -2)\}$ **35.** $\left\{\left(-\frac{1}{3}, -1, 4\right)\right\}$ **36.** $\{(0, -2, -4)\}$
37. 6 in., 12 in., 15 in.
38. \$2100 on Bank of US, \$1600 on Community Bank, and \$2700 on First National
39. 40°, 60°, 80°
40. 24 five-dollar bills, 30 ten-dollar bills, 10 twenty-dollar bills

Chapter 4 Test (page 219)

1. II **2.** I and IV **3.** III **4.** I **5.** $\{(-2, 3)\}$
6. $\{(x, y) | x - 3y = 6\}$ **7.** $\{(2, 7)\}$ **8.** $\{(-3, 5)\}$
9. $\left\{\left(-\frac{1}{2}, 1\right)\right\}$ **10.** $\{(24, 18)\}$ **11.** $\{(4, -6)\}$ **12.** $\varnothing$
13. $\{(-3, 0)\}$ **14.** $\{(3, 7)\}$ **15.** $x = 7$ **16.** $\{(3, -1, -2)\}$
17. $\{(5, 0, 3)\}$

18.

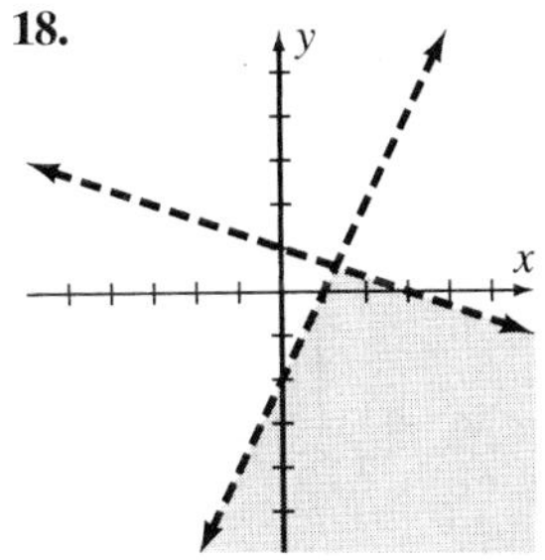

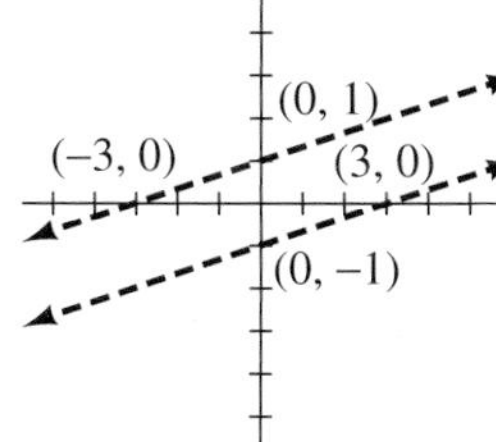

20.

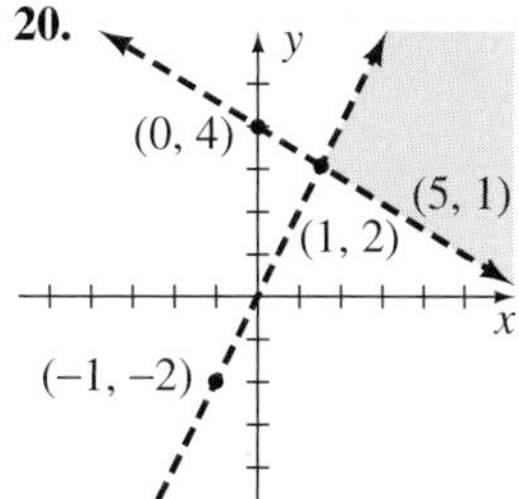

21.

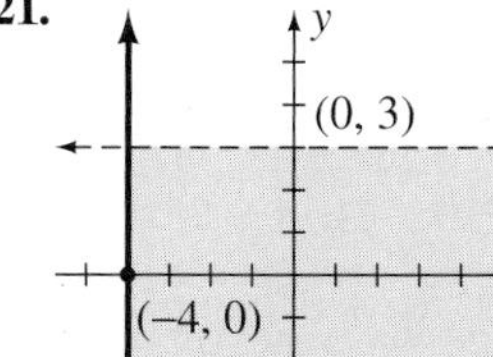

22. 15 in. by 26 in. **23.** \$1400 at 7% and \$2600 at 8%
24. 3 liters **25.** 24 quarters

Cumulative Review Problem Set (page 221)

1. 4 **2.** −19 **3.** 9 **4.** 21 **5.** −78 **6.** −33 **7.** −43
8. −11 **9.** −39 **10.** 57 **11.** $\{-4\}$ **12.** $\{-10\}$ **13.** $\{15\}$
14. $\left\{-\frac{25}{4}\right\}$ **15.** $\left\{-\frac{5}{3}, 3\right\}$ **16.** $\{-1, 5\}$ **17.** $\{400\}$
18. $x = \frac{2y + 6}{5}$ **19.** $y = \frac{12 - 3x}{4}$ **20.** $h = \frac{V - 2\pi r^2}{2\pi r}$
21. $R_1 = \frac{RR_2}{R_2 - R}$ **22.** 10.5% **23.** −15° **24.** $(-22, \infty)$

25. $(23, \infty)$ **26.** $(-\infty, -3) \cup (4, \infty)$ **27.** $\left(-7, \frac{7}{3}\right)$
28. $(300, \infty)$ **29.** $\left(-\infty, \frac{7}{8}\right]$ **30.** $\left[\frac{5}{2}, \infty\right)$ **31.** $\left(-\infty, \frac{32}{31}\right)$
32. $(0, 4), (-2, -4), \left(-\frac{1}{2}, 2\right)$

33.

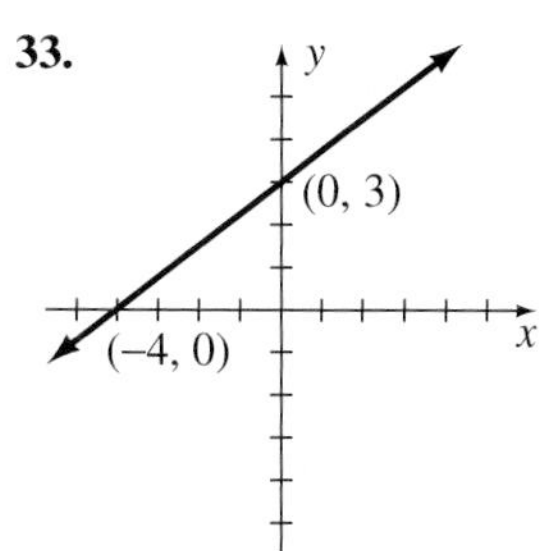

34.

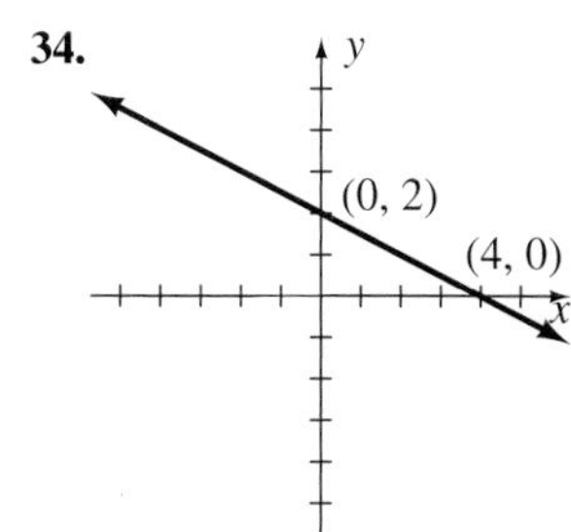

35.

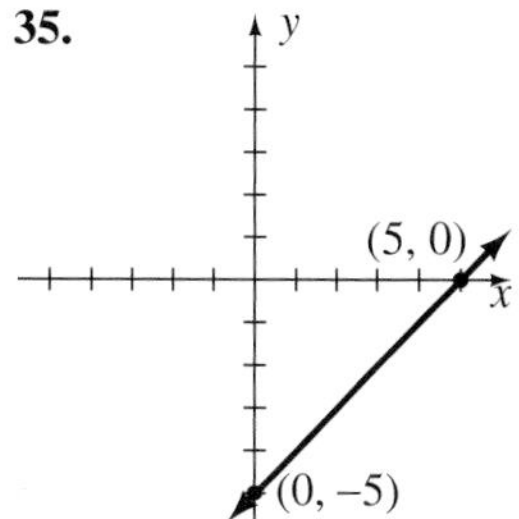

36.

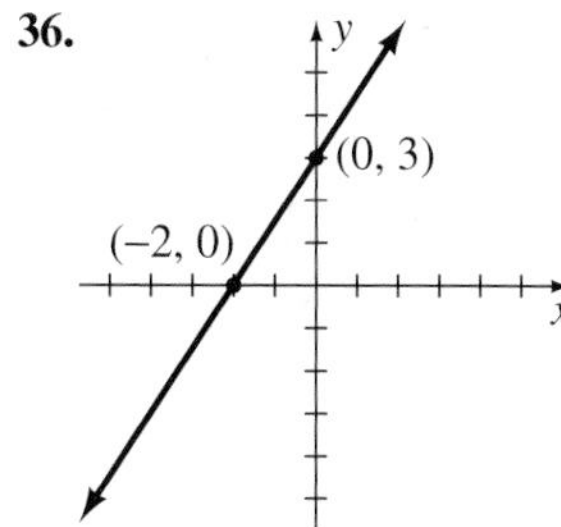

37.

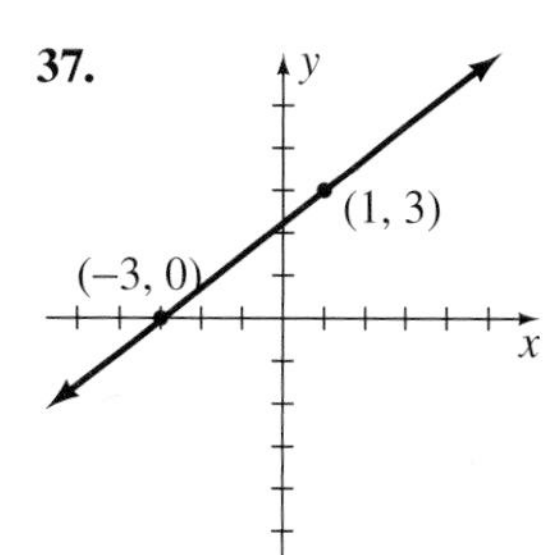

38.

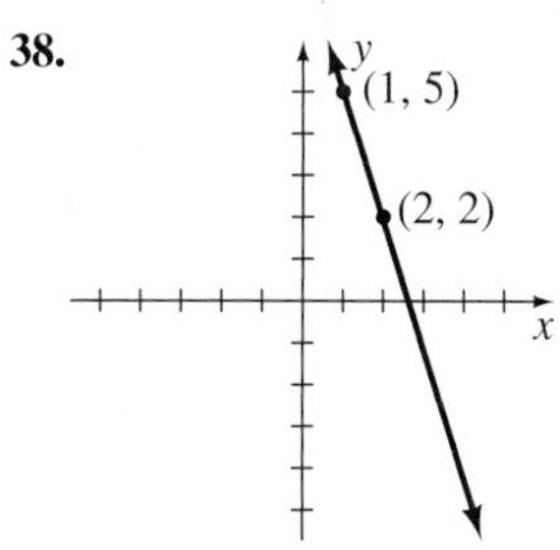

39.

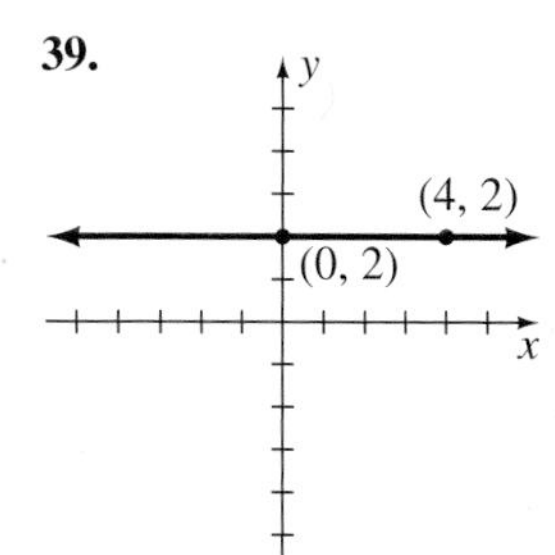

40.

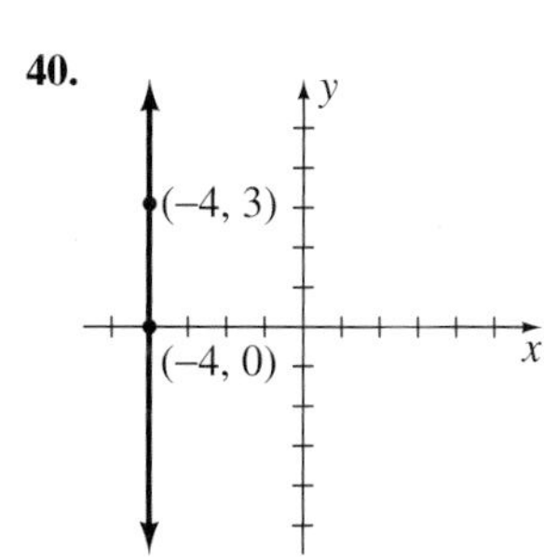

41.

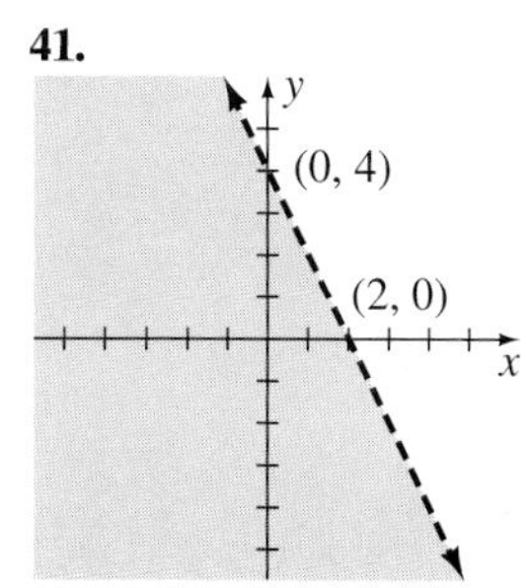

42.

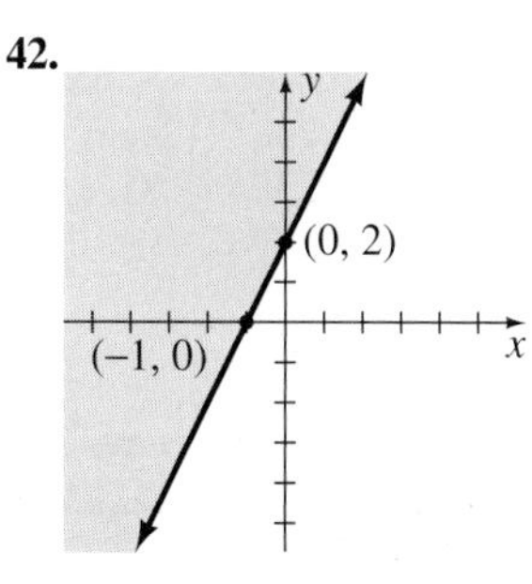

43.

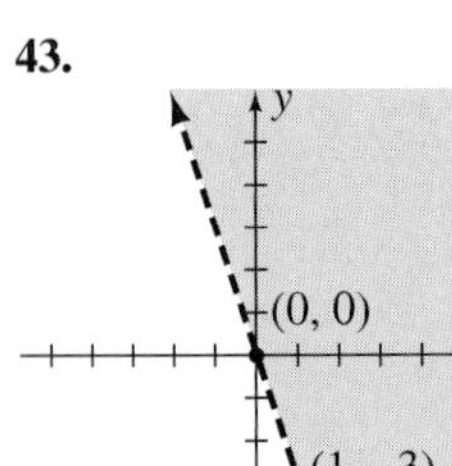

44.

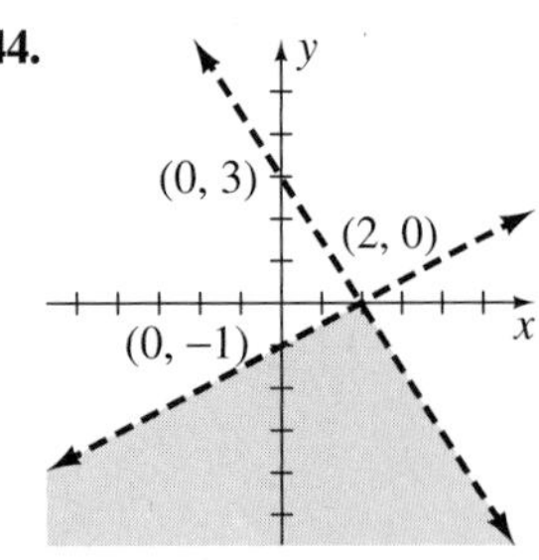

45.

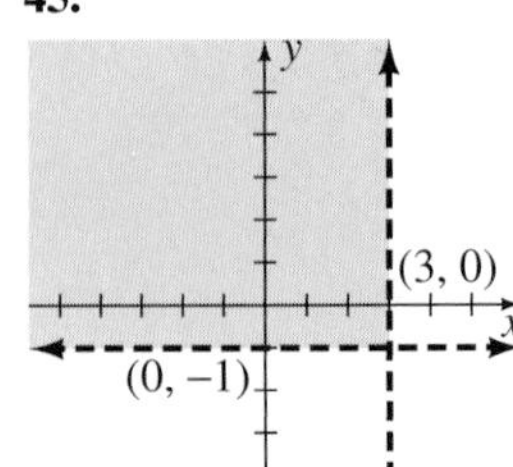

46.

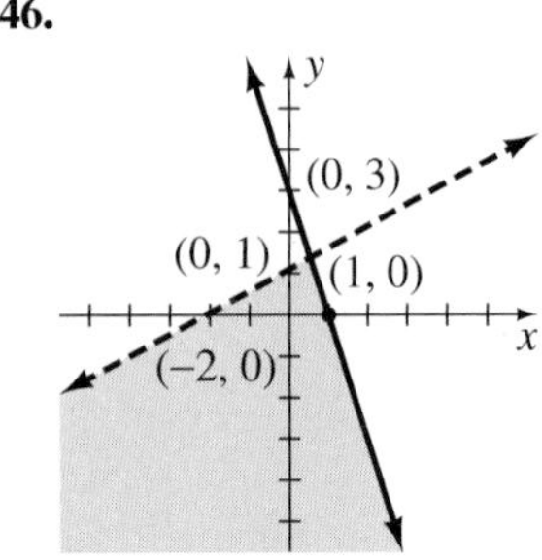

47. 13 **48.** $\sqrt{85}$ **49.** 4 **50.** $\frac{1}{3}$
51. $m = \frac{2}{5}$, y intercept is -2 **52.** $m = -\frac{3}{4}$, y intercept is 0
53. $2x + y = 8$ **54.** $2x - 3y = 6$ **55.** $3x + y = -10$
56. $5x - 2y = 27$ **57.** $y = \frac{1}{7}x - \frac{117}{7}$ **58.** $\{(-1, 3)\}$
59. $\{(-8, 3)\}$ **60.** $\{(4, 8)\}$ **61.** $\{(6, -4)\}$ **62.** $\{(2, 0, -1)\}$
63. 7, 9, 11 **64.** Joey = 12, Mom = 34 **65.** 1hr 40 min
66. 40% **67.** Better than 88
68. 8 nickels, 15 dimes, 25 quarters
69. 118°, 62° **70.** \$400 invested at 8%, \$600 invested at 9%
71. 35 pennies, 40 nickels, 70 dimes **72.** 25 ml of pure acid

CHAPTER 5

Problem Set 5.1 (page 229)

1. 2 **3.** 3 **5.** 2 **7.** 6 **9.** 0 **11.** $10x - 3$
13. $-11t + 5$ **15.** $-x^2 + 2x - 2$ **17.** $17a^2b^2 - 5ab$
19. $-9x + 7$ **21.** $-2x + 6$ **23.** $10a + 7$
25. $4x^2 + 10x + 6$ **27.** $-6a^2 + 12a + 14$
29. $3x^3 + x^2 + 13x - 11$ **31.** $7x + 8$ **33.** $-3x - 16$
35. $2x^2 - 2x - 8$ **37.** $-3x^3 + 5x^2 - 2x + 9$
39. $5x^2 - 4x + 11$ **41.** $-6x^2 + 9x + 7$
43. $-2x^2 + 9x + 4$ **45.** $-10n^2 + n + 9$ **47.** $8x - 2$
49. $8x - 14$ **51.** $-9x^2 - 12x + 4$ **53.** $10x^2 + 13x - 18$
55. $-n^2 - 4n - 4$ **57.** $-x + 6$ **59.** $6x^2 - 4$
61. $-7n^2 + n + 6$ **63.** $t^2 - 4t + 8$ **65.** $4n^2 - n - 12$
67. $-4x - 2y$ **69.** $-x^3 - x^2 + 3x$ **71.** **(a)** $8x + 4$
(c) $12x + 6$ **73.** $8\pi h + 32\pi$ **(a)** 226.1 **(c)** 452.2

Problem Set 5.2 (page 236)

1. $36x^4$ **3.** $-12x^5$ **5.** $4a^3b^4$ **7.** $-3x^3y^2z^6$ **9.** $-30xy^4$
11. $27a^4b^5$ **13.** $-m^3n^3$ **15.** $\frac{3}{10}x^3y^6$ **17.** $-\frac{3}{20}a^3b^4$
19. $-\frac{1}{6}x^3y^4$ **21.** $30x^6$ **23.** $-18x^9$ **25.** $-3x^6y^6$
27. $-24y^9$ **29.** $-56a^4b^2$ **31.** $-18a^3b^3$ **33.** $-10x^7y^7$
35. $50x^5y^2$ **37.** $6x^{3n}$ **39.** a^{5n+3} **41.** x^{4n} **43.** a^{5n+1}
45. $-10x^{2n}$ **47.** $12a^{n+4}$ **49.** $6x^{3n+2}$ **51.** $12x^{n+2}$
53. $27x^3y^6$ **55.** $-32x^{10}y^5$ **57.** $x^{16}y^{20}$ **59.** $a^6b^{12}c^{18}$
61. $64a^{12}b^{18}$ **63.** $81x^2y^8$ **65.** $81a^4b^{12}$ **67.** $-16a^4b^4$
69. $-x^6y^{12}z^{18}$ **71.** $-125a^6b^6c^3$ **73.** $-x^7y^{28}z^{14}$ **75.** $3x^3y^3$
77. $-5x^3y^2$ **79.** $9bc^2$ **81.** $-18xyz^4$ **83.** $-a^2b^3c^2$
85. 9 **87.** $-b^2$ **89.** $-18x^3$ **91.** $22x^2$; $6x^3$
93. $\pi r^2 - 36\pi$

Problem Set 5.3 (page 245)

1. $10x^2y^3 + 6x^3y^4$ **3.** $-12a^3b^3 + 15a^5b$
5. $24a^4b^5 - 16a^4b^6 + 32a^5b^6$ **7.** $-6x^3y^3 - 3x^4y^4 + x^5y^2$
9. $ax + ay + 2bx + 2by$ **11.** $ac + 4ad - 3bc - 12bd$
13. $t^3 - 14t - 15$ **15.** $x^3 + x^2 - 24x + 16$
17. $2x^3 + 9x^2 + 2x - 30$ **19.** $12x^3 - 7x^2 + 25x - 6$
21. $x^4 + 5x^3 + 11x^2 + 11x + 4$
23. $2x^4 - x^3 - 12x^2 + 5x + 4$
25. $x^2 + 16x + 60$ **27.** $y^2 + 6y - 55$
29. $n^2 - 5n - 14$ **31.** $x^2 - 14x + 48$
33. $4x^2 + 33x + 35$ **35.** $14x^2 + 3x - 2$
37. $5 + 3t - 2t^2$ **39.** $18x^2 - 39x - 70$
41. $2x^2 + xy - 15y^2$ **43.** $x^3 - 4x^2 + x + 6$
45. $x^3 - x^2 - 9x + 9$ **47.** $x^{2n} - 16$
49. $x^{2a} + 4x^a - 12$ **51.** $6x^{2n} + x^n - 35$
53. $x^{4a} - 10x^{2a} + 21$ **55.** $4x^{2n} + 20x^n + 25$
57. $x^2 - 12x + 36$ **59.** $t^2 + 18t + 81$
61. $y^2 - 14y + 49$ **63.** $9t^2 + 42t + 49$
65. $49x^2 - 56x + 16$ **67.** $x^2 - 36$
69. $9y^2 - 1$ **71.** $4 - 25x^2$ **73.** $25x^2 - 4a^2$
75. $x^3 + 6x^2 + 12x + 8$ **77.** $x^3 - 12x^2 + 48x - 64$
79. $8x^3 + 36x^2 + 54x + 27$ **81.** $64x^3 - 48x^2 + 12x - 1$
83. $125x^3 + 150x^2 + 60x + 8$ **87.** $2x^2 + 6$
89. $4x^3 - 64x^2 + 256x$; $256 - 4x^2$
93. **(a)** $a^6 + 6a^5b + 15a^4b^2 + 20a^3b^3 + 15a^2b^4 + 6ab^5 + b^6$
(c) $a^8 + 8a^7b + 28a^6b^2 + 56a^5b^3 + 70a^4b^4 + 56a^3b^5 + 28a^2b^6 + 8ab^7 + b^8$

Problem Set 5.4 (page 255)

1. Composite **3.** Prime **5.** Composite **7.** Composite
9. Prime **11.** $2 \cdot 2 \cdot 7$ **13.** $2 \cdot 2 \cdot 11$ **15.** $2 \cdot 2 \cdot 2 \cdot 7$
17. $2 \cdot 2 \cdot 2 \cdot 3 \cdot 3$ **19.** $3 \cdot 29$ **21.** No
23. No **25.** $4y(7y - 1)$ **27.** $5x(4y - 3)$
29. $x^2(7x + 10)$ **31.** $9ab(2a + 3b)$ **33.** $3x^3y^3(4y - 13x)$
35. $4x^2(2x^2 + 3x - 6)$ **37.** $x(5 + 7x + 9x^3)$
39. $5xy^2(3xy + 4 + 7x^2y^2)$ **41.** $(y + 2)(x + 3)$
43. $(2a + b)(3x - 2y)$ **45.** $(x + 2)(x + 5)$
47. $(a + 4)(x + y)$ **49.** $(a - 2b)(x + y)$

51. $(a-b)(3x-y)$ **53.** $(a+1)(2x+y)$
55. $(a-1)(x^2+2)$ **57.** $(a+b)(2c+3d)$
59. $(a+b)(x-y)$ **61.** $(x+9)(x+6)$
63. $(x+4)(2x+1)$ **65.** $\{-7, 0\}$ **67.** $\{0, 1\}$
69. $\{0, 5\}$ **71.** $\left\{-\frac{1}{2}, 0\right\}$ **73.** $\left\{-\frac{7}{3}, 0\right\}$ **75.** $\left\{0, \frac{5}{4}\right\}$
77. $\left\{0, \frac{1}{4}\right\}$ **79.** $\{-12, 0\}$ **81.** $\left\{0, \frac{3a}{5b}\right\}$ **83.** $\left\{-\frac{3a}{2b}, 0\right\}$
85. $\{a, -2b\}$ **87.** 0 or 7 **89.** 6 units **91.** $\frac{4}{\pi}$ units
93. The square is 100 ft by 100 ft, and the rectangle is 50 ft by 100 ft.
95. 6 units **101.** $x^a(2x^a-3)$ **103.** $y^{2m}(y^m+5)$
105. $x^{4a}(2x^{2a}-3x^a+7)$

Problem Set 5.5 (page 263)

1. $(x+1)(x-1)$ **3.** $(4x+5)(4x-5)$
5. $(3x+5y)(3x-5y)$ **7.** $(5xy+6)(5xy-6)$
9. $(2x+y^2)(2x-y^2)$ **11.** $(1+12n)(1-12n)$
13. $(x+2+y)(x+2-y)$ **15.** $(2x+y+1)(2x-y-1)$
17. $(3a+2b+3)(3a-2b-3)$ **19.** $-5(2x+9)$
21. $9(x+2)(x-2)$ **23.** $5(x^2+1)$ **25.** $8(y+2)(y-2)$
27. $ab(a+3)(a-3)$ **29.** Not factorable
31. $(n+3)(n-3)(n^2+9)$ **33.** $3x(x^2+9)$
35. $4xy(x+4y)(x-4y)$ **37.** $6x(1+x)(1-x)$
39. $(1+xy)(1-xy)(1+x^2y^2)$ **41.** $4(x+4y)(x-4y)$
43. $3(x+2)(x-2)(x^2+4)$ **45.** $(a-4)(a^2+4a+16)$
47. $(x+1)(x^2-x+1)$
49. $(3x+4y)(9x^2-12xy+16y^2)$
51. $(1-3a)(1+3a+9a^2)$ **53.** $(xy-1)(x^2y^2+xy+1)$
55. $(x+y)(x-y)(x^2-xy+y^2)(x^2+xy+y^2)$
57. $\{-5, 5\}$ **59.** $\left\{-\frac{7}{3}, \frac{7}{3}\right\}$ **61.** $\{-2, 2\}$ **63.** $\{-1, 0, 1\}$
65. $\{-2, 2\}$ **67.** $\{-3, 3\}$ **69.** $\{0\}$ **71.** -3, 0, or 3
73. 4 cm and 8 cm **75.** 10 m long and 5 m wide
77. 6 in. **79.** 8 yd

Problem Set 5.6 (page 272)

1. $(x+5)(x+4)$ **3.** $(x-4)(x-7)$ **5.** $(a+9)(a-4)$
7. $(y+6)(y+14)$ **9.** $(x-7)(x+2)$ **11.** Not factorable **13.** $(6-x)(1+x)$ **15.** $(x+3y)(x+12y)$
17. $(a-8b)(a+7b)$ **19.** $(x+10)(x+15)$
21. $(n-16)(n-20)$ **23.** $(t+15)(t-12)$
25. $(t^2-3)(t^2-2)$ **27.** $(x+1)(x-1)(x^2-8)$
29. $(x+1)(x-1)(x+4)(x-4)$ **31.** $(3x+1)(5x+6)$
33. $(4x-3)(3x+2)$ **35.** $(a+3)(4a-9)$
37. $(n-4)(3n+5)$ **39.** Not factorable
41. $(2n-7)(5n+3)$ **43.** $(4x-5)(2x+9)$
45. $(1-6x)(6+x)$ **47.** $(5y+9)(4y-1)$
49. $(12n+5)(2n-1)$ **51.** $(5n+3)(n+6)$
53. $(2x^2-1)(5x^2+4)$ **55.** $(3n+1)(3n-1)(2n^2+3)$
57. $(y-8)^2$ **59.** $(2x+3y)^2$ **61.** $2(2y-1)^2$
63. $2(t+2)(t-2)$ **65.** $(4x+5y)(3x-2y)$
67. $3n(2n+5)(3n-1)$ **69.** $(n-12)(n-5)$
71. $(6a-1)^2$ **73.** $6(x^2+9)$ **75.** Not factorable
77. $(x+y-7)(x-y+7)$ **79.** $(1+4x^2)(1+2x)(1-2x)$
81. $(4n+9)(n+4)$ **83.** $n(n+7)(n-7)$
85. $(x-8)(x+1)$ **87.** $3x(x-3)(x^2+3x+9)$
89. $(x^2+3)^2$ **91.** $(x+3)(x-3)(x^2+4)$
93. $(2w-7)(3w+5)$ **95.** Not factorable
97. $2n(n^2+7n-10)$ **99.** $(2x+1)(y+3)$
105. $(x^a+3)(x^a+7)$ **107.** $(2x^a+5)^2$
109. $(5x^n-1)(4x^n+5)$ **111.** $(x-4)(x-2)$
113. $(3x-11)(3x+2)$ **115.** $(3x+4)(5x+9)$

Problem Set 5.7 (page 279)

1. $\{-3, -1\}$ **3.** $\{-12, -6\}$ **5.** $\{4, 9\}$ **7.** $\{-6, 2\}$
9. $\{-1, 5\}$ **11.** $\{-13, -12\}$ **13.** $\left\{-5, \frac{1}{3}\right\}$
15. $\left\{-\frac{7}{2}, -\frac{2}{3}\right\}$ **17.** $\{0, 4\}$ **19.** $\left\{\frac{1}{6}, 2\right\}$ **21.** $\{-6, 0, 6\}$
23. $\{-4, 6\}$ **25.** $\{-4, 4\}$ **27.** $\{-11, 4\}$ **29.** $\{-5, 5\}$
31. $\left\{-\frac{5}{3}, -\frac{3}{5}\right\}$ **33.** $\left\{-\frac{1}{8}, 6\right\}$ **35.** $\left\{\frac{3}{7}, \frac{5}{4}\right\}$ **37.** $\left\{-\frac{2}{7}, \frac{4}{5}\right\}$
39. $\left\{-7, \frac{2}{3}\right\}$ **41.** $\{-20, 18\}$ **43.** $\left\{-2, -\frac{1}{3}, \frac{1}{3}, 2\right\}$
45. $\left\{-\frac{2}{3}, 16\right\}$ **47.** $\left\{-\frac{3}{2}, 1\right\}$ **49.** $\left\{-\frac{5}{2}, -\frac{4}{3}, 0\right\}$
51. $\left\{-1, \frac{5}{3}\right\}$ **53.** $\left\{-\frac{3}{2}, \frac{1}{2}\right\}$ **55.** 8 and 9 or -9 and -8
57. 7 and 15 **59.** 10 in. by 6 in.
61. -7 and -6 or 6 and 7
63. 4 cm by 4 cm and 6 cm by 8 cm
65. 3, 4, and 5 units **67.** 9 in. and 12 in.
69. An altitude of 4 in. and a side 14 in. long

Chapter 5 Review Problem Set (page 286)

1. Third degree **2.** Fourth degree **3.** Sixth degree
4. Fifth degree **5.** $5x-3$ **6.** $3x^2+12x-2$
7. $12x^2-x+5$ **8.** $11x^2+2x+4$ **9.** $2x+y-2$
10. $5x+5y-2$ **11.** $-20x^5y^7$ **12.** $-6a^5b^5$
13. $-8a^7b^3$ **14.** $27x^5y^8$ **15.** $256x^8y^{12}$ **16.** $-8x^6y^9z^3$
17. $-12a^5b^7$ **18.** $6x^{4n}$ **19.** $-13x^2y$ **20.** $2x^3y^3$
21. $-4b^2$ **22.** $4a^3b^3$ **23.** $15a^4-10a^3-5a^2$
24. $-8x^5+6x^4+10x^3$ **25.** $3x^3+7x^2-21x-4$
26. $6x^3-11x^2-7x+2$ **27.** $x^4+x^3-18x^2-x+35$
28. $3x^4+5x^3-21x^2-3x+20$ **29.** $24x^2+2xy-15y^2$
30. $7x^2+19x-36$ **31.** $21+26x-15x^2$
32. x^4+5x^2-24 **33.** $4x^2-12x+9$
34. $25x^2-10x+1$ **35.** $16x^2+24xy+9y^2$
36. $4x^2+20xy+25y^2$ **37.** $4x^2-49$ **38.** $9x^2-1$
39. $x^3-6x^2+12x-8$ **40.** $8x^3+60x^2+150x+125$
41. $2x^2+7x-2$ **42.** $2x^3+2x^2$ **43.** $5ab(2a-b^2-3a^2b)$
44. $xy(3-5xy-15x^2y^2)$ **45.** $(x+4)(a+b)$
46. $(3x-1)(y+7)$ **47.** $(2x+y)(3x^2+z^2)$
48. $(m+5n)(n-4)$ **49.** $(7a-5b)(7a+5b)$
50. $(6x-y)(6x+y)$ **51.** $(5a-2)(25a^2+10a+4)$
52. $(3x+4y)(9x^2-12xy+16y^2)$ **53.** $(x-3)(x-6)$
54. $(x+4)(x+7)$ **55.** $(x-7)(x+3)$ **56.** $(x+8)(x-2)$
57. $(2x+1)(x+4)$ **58.** $(3x-4)(2x-1)$
59. $(3x-2)(4x+1)$ **60.** $(4x+1)(2x-3)$ **61.** $(2x-3y)^2$
62. $(x+8y)^2$ **63.** $(x+7)(x-4)$
64. $2(t+3)(t-3)$ **65.** Not factorable
66. $(4n-1)(3n-1)$ **67.** $x^2(x^2+1)(x+1)(x-1)$
68. $x(x-12)(x+6)$ **69.** $2a^2b(3a+2b-c)$
70. $(x-y+1)(x+y-1)$ **71.** $4(2x^2+3)$

72. $(4x + 7)(3x - 5)$ **73.** $(4n - 5)^2$ **74.** $4n(n - 2)$
75. $3w(w^2 + 6w - 8)$ **76.** $(5x + 2y)(4x - y)$
77. $16a(a - 4)$ **78.** $3x(x + 1)(x - 6)$
79. $(n + 8)(n - 16)$ **80.** $(t + 5)(t - 5)(t^2 + 3)$
81. $(5x - 3)(7x + 2)$ **82.** $(3 - x)(5 - 3x)$
83. $(4n - 3)(16n^2 + 12n + 9)$
84. $2(2x + 5)(4x^2 - 10x + 25)$ **85.** $\{-3, 3\}$
86. $\{-6, 1\}$ **87.** $\left\{\frac{2}{7}\right\}$ **88.** $\left\{-\frac{2}{5}, \frac{1}{3}\right\}$ **89.** $\left\{-\frac{1}{3}, 3\right\}$
90. $\{-3, 0, 3\}$ **91.** $\{-1, 0, 1\}$ **92.** $\{-7, 9\}$ **93.** $\left\{-\frac{4}{7}, \frac{2}{7}\right\}$
94. $\left\{-\frac{4}{5}, \frac{5}{6}\right\}$ **95.** $\{-2, 2\}$ **96.** $\left\{\frac{5}{3}\right\}$ **97.** $\{-8, 6\}$
98. $\left\{-5, \frac{2}{7}\right\}$ **99.** $\{-8, 5\}$ **100.** $\{-12, 1\}$ **101.** $\varnothing$
102. $\left\{-5, \frac{6}{5}\right\}$ **103.** $\{0, 1, 8\}$ **104.** $\left\{-10, \frac{1}{4}\right\}$
105. 8, 9, and 10 or −1, 0, and 1 **106.** −6 and 8
107. 13 and 15 **108.** 12 miles and 16 miles
109. 4 m by 12 m **110.** 9 rows and 16 chairs per row
111. The side is 13 ft long and the altitude is 6 ft.
112. 3 ft **113.** 5 cm by 5 cm and 8 cm by 8 cm **114.** 6 in.

Chapter 5 Test (page 289)

1. $2x - 11$ **2.** $-48x^4y^4$ **3.** $-27x^6y^{12}$
4. $20x^2 + 17x - 63$ **5.** $6n^2 - 13n + 6$
6. $x^3 - 12x^2y + 48xy^2 - 64y^3$ **7.** $2x^3 + 11x^2 - 11x - 30$
8. $-14x^3y$ **9.** $(6x - 5)(x + 4)$ **10.** $3(2x + 1)(2x - 1)$
11. $(4 + t)(16 - 4t + t^2)$ **12.** $2x(3 - 2x)(5 + 4x)$
13. $(x - y)(x + 4)$ **14.** $(3n + 8)(8n - 3)$ **15.** $\{-12, 4\}$
16. $\left\{0, \frac{1}{4}\right\}$ **17.** $\left\{\frac{3}{2}\right\}$ **18.** $\{-4, -1\}$ **19.** $\{-9, 0, 2\}$
20. $\left\{-\frac{3}{7}, \frac{4}{5}\right\}$ **21.** $\left\{-\frac{1}{3}, 2\right\}$ **22.** $\{-2, 2\}$ **23.** 9 in.
24. 15 rows **25.** 8 ft

CHAPTER 6

Problem Set 6.1 (page 297)

1. $\frac{3}{4}$ **3.** $\frac{5}{6}$ **5.** $-\frac{2}{5}$ **7.** $\frac{2}{7}$ **9.** $\frac{2x}{7}$ **11.** $\frac{2a}{5b}$ **13.** $-\frac{y}{4x}$
15. $-\frac{9c}{13d}$ **17.** $\frac{5x^2}{3y^3}$ **19.** $\frac{x - 2}{x}$ **21.** $\frac{3x + 2}{2x - 1}$ **23.** $\frac{a + 5}{a - 9}$
25. $\frac{n - 3}{5n - 1}$ **27.** $\frac{5x^2 + 7}{10x}$ **29.** $\frac{3x + 5}{4x + 1}$ **31.** $\frac{3x}{x^2 + 4x + 16}$
33. $\frac{x + 6}{3x - 1}$ **35.** $\frac{x(2x + 7)}{y(x + 9)}$ **37.** $\frac{y + 4}{5y - 2}$ **39.** $\frac{3x(x - 1)}{x^2 + 1}$
41. $\frac{2(x + 3y)}{3x(3x + y)}$ **43.** $\frac{3n - 2}{7n + 2}$ **45.** $\frac{4 - x}{5 + 3x}$ **47.** $\frac{9x^2 + 3x + 1}{2(x + 2)}$
49. $\frac{-2(x - 1)}{x + 1}$ **51.** $\frac{y + b}{y + c}$ **53.** $\frac{x + 2y}{2x + y}$ **55.** $\frac{x + 1}{x - 6}$
57. $\frac{2s + 5}{3s + 1}$ **59.** −1 **61.** $-n - 7$ **63.** $-\frac{2}{x + 1}$
65. −2 **67.** $-\frac{n + 3}{n + 5}$

Problem Set 6.2 (page 304)

1. $\frac{1}{10}$ **3.** $-\frac{4}{15}$ **5.** $\frac{3}{16}$ **7.** $-\frac{5x^3}{12y^2}$ **9.** $\frac{2a^3}{3b}$ **11.** $\frac{3x^3}{4}$
13. $\frac{3x}{4y}$ **15.** $\frac{3(x^2 + 4)}{5y(x + 8)}$ **17.** $\frac{5(a + 3)}{a(a - 2)}$ **19.** $\frac{3}{2}$ **21.** $\frac{5 + n}{3 - n}$
23. $\frac{x^2 + 1}{x^2 - 10}$ **25.** $\frac{2t^2 + 5}{2(t^2 + 1)(t + 1)}$ **27.** $\frac{t(t + 6)}{4t + 5}$
29. $\frac{n + 3}{n(n - 2)}$ **31.** $-\frac{5}{6}$ **33.** $-\frac{2}{3}$ **35.** $\frac{10}{11}$ **37.** $\frac{25x^3}{108y^2}$
39. $\frac{ac^2}{2b^2}$ **41.** $\frac{3xy}{4(x + 6)}$ **43.** $\frac{5(x - 2y)}{7y}$ **45.** $\frac{6x + 5}{3x + 4}$
47. $\frac{25x^3y^3}{4(x + 1)}$ **49.** $\frac{2(a - 2b)}{a(3a - 2b)}$

Problem Set 6.3 (page 312)

1. $\frac{13}{12}$ **3.** $\frac{11}{40}$ **5.** $\frac{19}{20}$ **7.** $\frac{49}{75}$ **9.** $\frac{17}{30}$ **11.** $-\frac{11}{84}$ **13.** $\frac{2x + 4}{x - 1}$
15. 4 **17.** $\frac{7y - 10}{7y}$ **19.** $\frac{5x + 3}{6}$ **21.** $\frac{12a + 1}{12}$ **23.** $\frac{n + 14}{18}$
25. $-\frac{11}{15}$ **27.** $\frac{3x - 25}{30}$ **29.** $\frac{43}{40x}$ **31.** $\frac{20y - 77x}{28xy}$
33. $\frac{16y + 15x - 12xy}{12xy}$ **35.** $\frac{21 + 22x}{30x^2}$ **37.** $\frac{10n - 21}{7n^2}$
39. $\frac{45 - 6n + 20n^2}{15n^2}$ **41.** $\frac{11x - 10}{6x^2}$ **43.** $\frac{42t + 43}{35t^3}$
45. $\frac{20b^2 - 33a^3}{96a^2b}$ **47.** $\frac{14 - 24y^3 + 45xy}{18xy^3}$ **49.** $\frac{2x^2 + 3x - 3}{x(x - 1)}$
51. $\frac{a^2 - a - 8}{a(a + 4)}$ **53.** $\frac{-41n - 55}{(4n + 5)(3n + 5)}$
55. $\frac{-3x + 17}{(x + 4)(7x - 1)}$ **57.** $\frac{-x + 74}{(3x - 5)(2x + 7)}$
59. $\frac{38x + 13}{(3x - 2)(4x + 5)}$ **61.** $\frac{5x + 5}{2x + 5}$ **63.** $\frac{x + 15}{x - 5}$
65. $\frac{-2x - 4}{2x + 1}$ **67. (a)** −1 **(c)** 0

Problem Set 6.4 (page 321)

1. $\frac{7x + 20}{x(x + 4)}$ **3.** $\frac{-x - 3}{x(x + 7)}$ **5.** $\frac{6x - 5}{(x + 1)(x - 1)}$
7. $\frac{1}{a + 1}$ **9.** $\frac{5n + 15}{4(n + 5)(n - 5)}$ **11.** $\frac{x^2 + 60}{x(x + 6)}$
13. $\frac{11x + 13}{(x + 2)(x + 7)(2x + 1)}$ **15.** $\frac{-3a + 1}{(a - 5)(a + 2)(a + 9)}$
17. $\frac{3a^2 + 14a + 1}{(4a - 3)(2a + 1)(a + 4)}$ **19.** $\frac{3x^2 + 20x - 111}{(x^2 + 3)(x + 7)(x - 3)}$
21. $\frac{x + 6}{(x - 3)^2}$ **23.** $\frac{14x - 4}{(x - 1)(x + 1)^2}$

25. $\frac{-7y-14}{(y+8)(y-2)}$ **27.** $\frac{-2x^2-4x+3}{(x+2)(x-2)}$

29. $\frac{2x^2+14x-19}{(x+10)(x-2)}$ **31.** $\frac{2n+1}{n-6}$

33. $\frac{2x^2-32x+16}{(x+1)(2x-1)(3x-2)}$ **35.** $\frac{1}{(n^2+1)(n+1)}$

37. $\frac{-16x}{(5x-2)(x-1)}$ **39.** $\frac{t+1}{t-2}$ **41.** $\frac{2}{11}$ **43.** $-\frac{7}{27}$

45. $\frac{x}{4}$ **47.** $\frac{3y-2x}{4x-7}$ **49.** $\frac{6ab^2-5a^2}{12b^2+2a^2b}$ **51.** $\frac{2y-3xy}{3x+4xy}$

53. $\frac{3n+14}{5n+19}$ **55.** $\frac{5n-17}{4n-13}$ **57.** $\frac{-x+5y-10}{3y-10}$

59. $\frac{-x+15}{-2x-1}$ **61.** $\frac{3a^2-2a+1}{2a-1}$ **63.** $\frac{-x^2+6x-4}{3x-2}$

Problem Set 6.5 (page 329)

1. $3x^3+6x^2$ **3.** $-6x^4+9x^6$ **5.** $3a^2-5a-8$
7. $-13x^2+17x-28$ **9.** $-3xy+4x^2y-8xy^2$
11. $x-13$ **13.** $x+20$ **15.** $2x+1-\frac{3}{x-1}$
17. $5x-1$ **19.** $3x^2-2x-7$ **21.** x^2+5x-6
23. $4x^2+7x+12+\frac{30}{x-2}$ **25.** x^3-4x^2-5x+3
27. $x^2+5x+25$ **29.** $x^2-x+1+\frac{63}{x+1}$
31. $2x^2-4x+7-\frac{20}{x+2}$ **33.** $4a-4b$
35. $4x+7+\frac{23x-6}{x^2-3x}$ **37.** $8y-9+\frac{8y+5}{y^2+y}$
39. $2x-1$ **41.** $x-3$ **43.** $5a-8+\frac{42a-41}{a^2+3a-4}$
45. $2n^2+3n-4$ **47.** $x^4+x^3+x^2+x+1$
49. x^3-x^2+x-1 **51.** $3x^2+x+1+\frac{7}{x^2-1}$
53. $x-6$ **55.** $x+6, R=14$ **57.** x^2-1
59. x^2-2x-3 **61.** $2x^2-x-6, R=-6$
63. $x^3+7x^2+21x+56, R=167$

Problem Set 6.6 (page 336)

1. $\{2\}$ **3.** $\{-3\}$ **5.** $\{6\}$ **7.** $\left\{-\frac{85}{18}\right\}$ **9.** $\left\{\frac{7}{10}\right\}$ **11.** $\{5\}$

13. $\{58\}$ **15.** $\left\{\frac{1}{4}, 4\right\}$ **17.** $\left\{-\frac{2}{5}, 5\right\}$ **19.** $\{-3, 1\}$ **21.** $\left\{-\frac{5}{2}\right\}$

23. $\varnothing$ **25.** $\left\{-\frac{11}{8}, 2\right\}$ **27.** $\{-29, 0\}$ **29.** $\left\{-2, \frac{23}{8}\right\}$

31. $\left\{3, \frac{7}{2}\right\}$ **33.** $\{-16\}$ **35.** $\left\{-\frac{13}{3}\right\}$ **37.** $\{-51\}$

39. $\left\{-\frac{5}{3}, 4\right\}$ **41.** $\{-9, 3\}$ **43.** $\left\{\frac{11}{23}\right\}$

45. \$750 and \$1000 **47.** 48° and 72° **49.** \$3500
51. \$69 for Tammy and \$51.75 for Laura
53. 14 ft and 6 ft **55.** 690 females and 460 males

Problem Set 6.7 (page 345)

1. $\{-21\}$ **3.** $\{-1, 2\}$ **5.** $\{2\}$ **7.** $\left\{\frac{37}{15}\right\}$ **9.** $\{-1\}$

11. $\{-1\}$ **13.** $\left\{0, \frac{13}{2}\right\}$ **15.** $\left\{-2, \frac{19}{2}\right\}$ **17.** $\{-2\}$

19. $\left\{-\frac{1}{5}\right\}$ **21.** $\varnothing$ **23.** $\left\{\frac{7}{2}\right\}$ **25.** $\{-3\}$ **27.** $\left\{-\frac{7}{9}\right\}$

29. $\left\{-\frac{7}{6}\right\}$ **31.** $x=\frac{18y-4}{15}$ **33.** $y=\frac{-5x+22}{2}$

35. $M=\frac{IC}{100}$ **37.** $R=\frac{ST}{S+T}$ **39.** $y=\frac{bx-x-3b+a}{a-3}$

41. $y=\frac{ab-bx}{a}$ **43.** $y=\frac{-2x-9}{3}$
45. 50 mph for Dave and 54 mph for Kent **47.** 60 min
49. 60 words per minute for Connie and 40 words per minute for Katie
51. Plane B could travel at 400 mph for 5 hr and plane A at 350 mph for 4 hr, or plane B could travel at 250 mph for 8 hr and plane A at 200 mph for 7 hr.
53. 60 min for Nancy and 120 min for Amy **55.** 3 hr
57. 16 mph on the way out and 12 mph on the way back, or 12 mph out and 8 mph back

Chapter 6 Review Problem Set (page 353)

1. $\frac{2y}{3x^2}$ **2.** $\frac{a-3}{a}$ **3.** $\frac{n-5}{n-1}$ **4.** $\frac{x^2+1}{x}$ **5.** $\frac{2x+1}{3}$

6. $\frac{x^2-10}{2x^2+1}$ **7.** $3b$ **8.** $\frac{n(n+5)}{n-1}$ **9.** $\frac{2x}{7y^2}$ **10.** $\frac{x(x-3y)}{x^2+9y^2}$

11. xy^4 **12.** $\frac{3x^2y^5}{4}$ **13.** $\frac{4(x+1)}{x+4}$ **14.** $-\frac{5}{3(x+2)(x+1)}$

15. $\frac{23x-6}{20}$ **16.** $\frac{57-2n}{18n}$ **17.** $\frac{3x^2-2x-14}{x(x+7)}$ **18.** $\frac{2}{x-5}$

19. $\frac{5n-21}{(n-9)(n+4)(n-1)}$ **20.** $\frac{6y-23}{(2y+3)(y-6)}$ **21.** $\frac{3}{22}$

22. $\frac{18y+20x}{48y-9x}$ **23.** $\frac{3x+2}{3x-2}$ **24.** $\frac{x-1}{2x-1}$ **25.** $6x-1$

26. $3x^2-7x+22-\frac{90}{x+4}$ **27.** $3x^3-2x^2-x+2$

28. $2x^3-2x^2+3x-4+\frac{7}{x+1}$ **29.** $\left\{\frac{4}{13}\right\}$ **30.** $\left\{\frac{3}{16}\right\}$

31. $\varnothing$ **32.** $\left\{\frac{2}{7}, \frac{7}{2}\right\}$ **33.** $\left\{-\frac{6}{7}, 3\right\}$ **34.** $\{-17\}$ **35.** $\left\{\frac{3}{4}, \frac{5}{2}\right\}$

36. $\{22\}$ **37.** $\left\{\frac{9}{7}\right\}$ **38.** $\left\{-\frac{5}{4}\right\}$ **39.** $y=\frac{3x+27}{4}$

40. $y=\frac{bx-ab}{a}$ **41.** \$525 and \$875
42. Busboy, \$36; waiter, \$126
43. 20 min for Julio and 30 min for Dan
44. 50 mph and 55 mph or $8\frac{1}{3}$ mph and $13\frac{1}{3}$ mph
45. 9 hr **46.** 80 hr **47.** 13 mph

Chapter 6 Test (page 355)

1. $\frac{13y^2}{24x}$ **2.** $\frac{3x-1}{x(x-6)}$ **3.** $\frac{2n-3}{n+4}$ **4.** $-\frac{2x}{x+1}$ **5.** $\frac{3y^2}{8}$

6. $\frac{a-b}{4(2a+b)}$ **7.** $\frac{x+4}{5x-1}$ **8.** $\frac{13x+7}{12}$ **9.** $\frac{3x}{2}$

10. $\frac{10n - 26}{15n}$ **11.** $\frac{3x^2 + 2x - 12}{x(x - 6)}$ **12.** $\frac{11 - 2x}{x(x - 1)}$
13. $\frac{13n + 46}{(2n + 5)(n - 2)(n + 7)}$ **14.** $3x^2 - 2x - 1$
15. $\frac{18 - 2x}{8 + 9x}$ **16.** $y = \frac{4x + 20}{3}$ **17.** $\{1\}$ **18.** $\left\{\frac{1}{10}\right\}$
19. $\{-35\}$ **20.** $\{-1, 5\}$ **21.** $\left\{\frac{5}{3}\right\}$ **22.** $\left\{-\frac{9}{13}\right\}$
23. $\frac{27}{72}$ **24.** 1 hr **25.** 15 mph

Cumulative Review Problem Set (page 357)

1. 16 **2.** −13 **3.** 104 **4.** $\frac{5}{2}$ **5.** $7a^2 + 11a + 1$
6. $-2x^2 + 9x - 4$ **7.** $-2x^3y^5$ **8.** $16x^2y^6$ **9.** $36a^7b^2$
10. $-24a^6b^4$ **11.** $-18x^4 + 3x^3 - 12x^2$ **12.** $10x^2 + xy - 3y^2$
13. $x^2 + 8xy + 16y^2$ **14.** $a^3 - a^2b - 11ab + 3b^3$
15. $(x - 2)(x - 3)$ **16.** $(2x + 1)(3x - 4)$
17. $2(x - 1)(x - 3)$ **18.** $3(x + 8)(x - 2)$
19. $(3m - 4n)(3m + 4n)$ **20.** $(3a + 2)(9a^2 - 6a + 4)$
21. $-\frac{7y^4}{x^2}$ **22.** $-x$ **23.** $\frac{x(x - 5)}{x - 1}$ **24.** $\frac{x + 7}{x + 3}$ **25.** $\frac{9n - 26}{10}$
26. $\frac{8x - 19}{(x - 2)(x + 3)(x - 3)}$ **27.** $\frac{2y + 3x}{6xy}$ **28.** $\frac{m - n}{mn}$
29. $3x^2 - x + 4$ **30.** $2x^2 + 5x + 3 + \frac{2}{x - 4}$ **31.** $-\frac{3}{2}$

32. **33.**

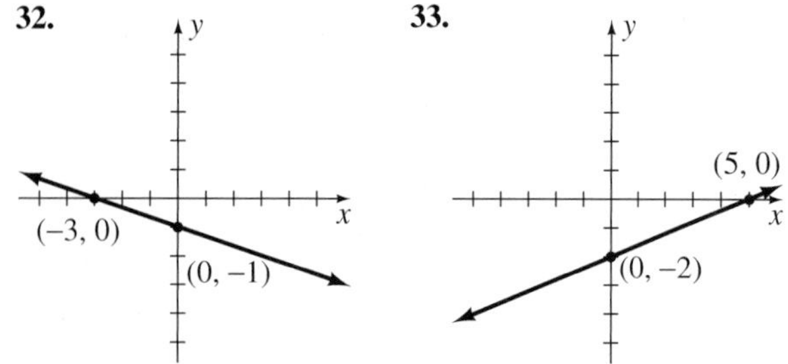

34.

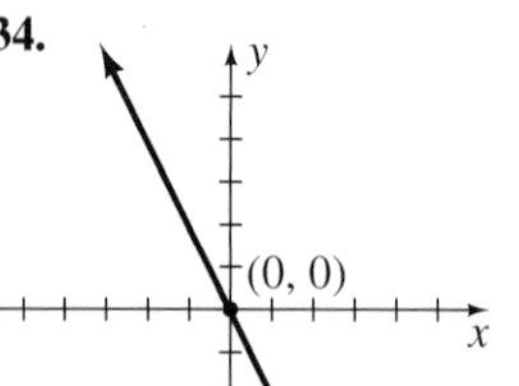

35.

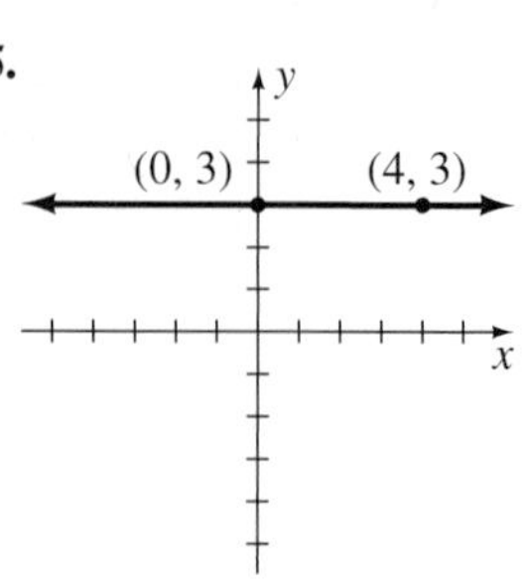

36.

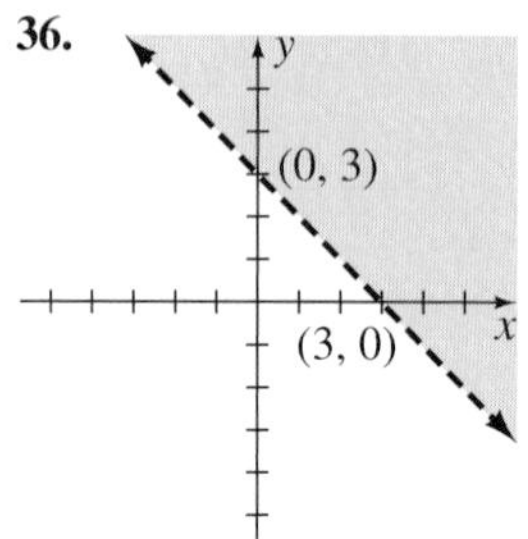

37.

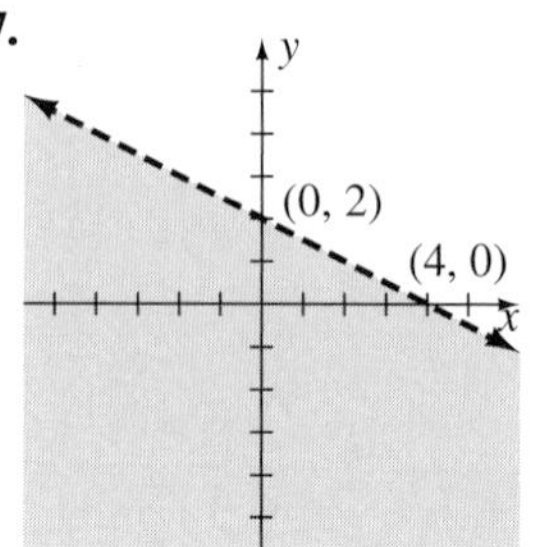

38. $2x - 3y = -14$ **39.** $y = -2x + 4$ **40.** $x - 2y = -9$
41. $\{6\}$ **42.** $\{24\}$ **43.** $\left\{\frac{15}{7}\right\}$ **44.** $\{6\}$ **45.** $\left\{-2, \frac{10}{3}\right\}$
46. $\{-28, 12\}$ **47.** $\{-8, 1\}$ **48.** $\left\{-5, -\frac{3}{2}\right\}$ **49.** $\left\{-\frac{1}{3}, 9\right\}$
50. $\varnothing$ **51.** $\{(-5, 6)\}$ **52.** $\{(0, 8)\}$ **53.** $\{(4, -1)\}$
54. $\{(2, -1, -3)\}$ **55.** $P = \frac{A}{1 + rt}$ **56.** $(-\infty, 2]$
57. $(-4, 2)$ **58.** $\left(-\frac{9}{2}, 3\right)$ **59.** $(-\infty, -13] \cup [7, \infty)$
60. $(-\infty, -10] \cup (2, \infty)$ **61.** \$5.76 **62.** \$690
63. Width is 23 in., length is 38 in.
64. 4 hr **65.** \$13,600 and \$54,400 **66.** 12 min
67. 5 in., 12 in., and 13 quarters
68. $16\frac{2}{3}$ yr **69.** 27 dimes and 13 quarters
70. 12 liters of 10% solution and 18 liters of 40% solution

CHAPTER 7

Problem Set 7.1 (page 366)

1. $\frac{1}{27}$ **3.** $-\frac{1}{100}$ **5.** 81 **7.** −27 **9.** −8 **11.** 1 **13.** $\frac{9}{49}$
15. 16 **17.** $\frac{1}{1000}$ **19.** $\frac{1}{1000}$ **21.** 27 **23.** $\frac{1}{125}$ **25.** $\frac{9}{8}$
27. $\frac{256}{25}$ **29.** $\frac{2}{25}$ **31.** $\frac{81}{4}$ **33.** 81 **35.** $\frac{1}{10{,}000}$ **37.** $\frac{13}{36}$
39. $\frac{1}{2}$ **41.** $\frac{72}{17}$ **43.** $\frac{1}{x^6}$ **45.** $\frac{1}{a^3}$ **47.** $\frac{1}{a^8}$ **49.** $\frac{y^6}{x^2}$ **51.** $\frac{c^8}{a^4b^{12}}$
53. $\frac{y^{12}}{8x^9}$ **55.** $\frac{x^3}{y^{12}}$ **57.** $\frac{4a^4}{9b^2}$ **59.** $\frac{1}{x^2}$ **61.** a^5b^2 **63.** $\frac{6y^3}{x}$
65. $7b^2$ **67.** $\frac{7x}{y^2}$ **69.** $-\frac{12b^3}{a}$ **71.** $\frac{x^5y^5}{5}$ **73.** $\frac{b^{20}}{81}$ **75.** $\frac{x + 1}{x^3}$
77. $\frac{y - x^3}{x^3y}$ **79.** $\frac{3b + 4a^2}{a^2b}$ **81.** $\frac{1 - x^2y}{xy^2}$ **83.** $\frac{2x - 3}{x^2}$

Problem Set 7.2 (page 377)

1. 8 **3.** −10 **5.** 3 **7.** −4 **9.** 3 **11.** $\frac{4}{5}$ **13.** $-\frac{6}{7}$ **15.** $\frac{1}{2}$
17. $\frac{3}{4}$ **19.** 8 **21.** $3\sqrt{3}$ **23.** $4\sqrt{2}$ **25.** $4\sqrt{5}$ **27.** $4\sqrt{10}$
29. $12\sqrt{2}$ **31.** $-12\sqrt{5}$ **33.** $2\sqrt{3}$ **35.** $3\sqrt{6}$ **37.** $-\frac{5}{3}\sqrt{7}$
39. $\frac{\sqrt{19}}{2}$ **41.** $\frac{3\sqrt{3}}{4}$ **43.** $\frac{5\sqrt{3}}{9}$ **45.** $2\sqrt[3]{2}$ **47.** $6\sqrt[3]{3}$
49. $\frac{\sqrt{14}}{7}$ **51.** $\frac{\sqrt{6}}{3}$ **53.** $\frac{\sqrt{15}}{6}$ **55.** $\frac{\sqrt{66}}{12}$ **57.** $\frac{\sqrt{6}}{3}$
59. $\sqrt{5}$ **61.** $\frac{2\sqrt{21}}{7}$ **63.** $-\frac{8\sqrt{15}}{5}$ **65.** $\frac{\sqrt{6}}{4}$ **67.** $-\frac{12}{25}$
69. $\frac{2\sqrt[3]{3}}{3}$ **71.** $\frac{3\sqrt[3]{2}}{2}$ **73.** $\frac{\sqrt[3]{12}}{2}$
75. 42 mph; 49 mph; 65 mph **77.** 107 cm^2 **79.** 140 in.^2
85. **(a)** 1.414 **(c)** 12.490 **(e)** 57.000 **(g)** 0.374 **(i)** 0.930

Problem Set 7.3 (page 383)

1. $13\sqrt{2}$ **3.** $54\sqrt{3}$ **5.** $-30\sqrt{2}$ **7.** $-\sqrt{5}$ **9.** $-21\sqrt{6}$ **11.** $-\frac{7\sqrt{7}}{12}$ **13.** $\frac{37\sqrt{10}}{10}$ **15.** $\frac{41\sqrt{2}}{20}$ **17.** $-9\sqrt[3]{3}$ **19.** $10\sqrt[3]{2}$ **21.** $4\sqrt{2x}$ **23.** $5x\sqrt{3}$ **25.** $2x\sqrt{5y}$ **27.** $8xy^3\sqrt{xy}$ **29.** $3a^2b\sqrt{6b}$ **31.** $3x^3y^4\sqrt{7}$ **33.** $4a\sqrt{10a}$ **35.** $\frac{8y}{3}\sqrt{6xy}$ **37.** $\frac{\sqrt{10xy}}{5y}$ **39.** $\frac{\sqrt{15}}{6x^2}$ **41.** $\frac{5\sqrt{2y}}{6y}$ **43.** $\frac{\sqrt{14xy}}{4y^3}$ **45.** $\frac{3y\sqrt{2xy}}{4x}$ **47.** $\frac{2\sqrt{42ab}}{7b^2}$ **49.** $2\sqrt[3]{3y}$ **51.** $2x\sqrt[3]{2x}$ **53.** $2x^2y^2\sqrt[3]{7y^2}$ **55.** $\frac{\sqrt[3]{21x}}{3x}$ **57.** $\frac{\sqrt[3]{12x^2y}}{4x^2}$ **59.** $\frac{\sqrt[3]{4x^2y^2}}{xy^2}$ **61.** $2\sqrt{2x+3y}$ **63.** $4\sqrt{x+3y}$ **65.** $33\sqrt{x}$ **67.** $-30\sqrt{2x}$ **69.** $7\sqrt{3n}$ **71.** $-40\sqrt{ab}$ **73.** $-7x\sqrt{2x}$ **79.** **(a)** $5|x|\sqrt{5}$ **(b)** $4x^2$ **(c)** $2b\sqrt{2b}$ **(d)** $y^2\sqrt{3y}$ **(e)** $12|x^3|\sqrt{2}$ **(f)** $2m^4\sqrt{7}$ **(g)** $8|c^5|\sqrt{2}$ **(h)** $3d^3\sqrt{2d}$ **(i)** $7|x|$ **(j)** $4n^{10}\sqrt{5}$ **(k)** $9h\sqrt{h}$

Problem Set 7.4 (page 390)

1. $6\sqrt{2}$ **3.** $18\sqrt{2}$ **5.** $-24\sqrt{10}$ **7.** $24\sqrt{6}$ **9.** 120 **11.** 24 **13.** $56\sqrt[3]{3}$ **15.** $\sqrt{6}+\sqrt{10}$ **17.** $6\sqrt{10}-3\sqrt{35}$ **19.** $24\sqrt{3}-60\sqrt{2}$ **21.** $-40-32\sqrt{15}$ **23.** $15\sqrt{2x}+3\sqrt{xy}$ **25.** $5xy-6x\sqrt{y}$ **27.** $2\sqrt{10xy}+2y\sqrt{15y}$ **29.** $-25\sqrt{6}$ **31.** $-25-3\sqrt{3}$ **33.** $23-9\sqrt{5}$ **35.** $6\sqrt{35}+3\sqrt{10}-4\sqrt{21}-2\sqrt{6}$ **37.** $8\sqrt{3}-36\sqrt{2}+6\sqrt{10}-18\sqrt{15}$ **39.** $11+13\sqrt{30}$ **41.** $141-51\sqrt{6}$ **43.** -10 **45.** -8 **47.** $2x-3y$ **49.** $10\sqrt[3]{12}+2\sqrt[3]{18}$ **51.** $12-36\sqrt[3]{2}$ **53.** $\frac{\sqrt{7}-1}{3}$ **55.** $\frac{-3\sqrt{2}-15}{23}$ **57.** $\frac{\sqrt{7}-\sqrt{2}}{5}$ **59.** $\frac{2\sqrt{5}+\sqrt{6}}{7}$ **61.** $\frac{\sqrt{15}-2\sqrt{3}}{2}$ **63.** $\frac{6\sqrt{7}+4\sqrt{6}}{13}$ **65.** $\sqrt{3}-\sqrt{2}$ **67.** $\frac{2\sqrt{x}-8}{x-16}$ **69.** $\frac{x+5\sqrt{x}}{x-25}$ **71.** $\frac{x-8\sqrt{x}+12}{x-36}$ **73.** $\frac{x-2\sqrt{xy}}{x-4y}$ **75.** $\frac{6\sqrt{xy}+9y}{4x-9y}$

Problem Set 7.5 (page 396)

1. {20} **3.** ∅ **5.** $\left\{\frac{25}{4}\right\}$ **7.** $\left\{\frac{4}{9}\right\}$ **9.** {5} **11.** $\left\{\frac{39}{4}\right\}$ **13.** $\left\{\frac{10}{3}\right\}$ **15.** {−1} **17.** ∅ **19.** {1} **21.** $\left\{\frac{3}{2}\right\}$ **23.** {3} **25.** $\left\{\frac{61}{25}\right\}$ **27.** {−3, 3} **29.** {−9, −4} **31.** {0} **33.** {3} **35.** {4} **37.** {−4, −3} **39.** {12} **41.** {25} **43.** {29} **45.** {−15} **47.** $\left\{-\frac{1}{3}\right\}$ **49.** {−3} **51.** {0} **53.** {5} **55.** {2, 6} **57.** 56 ft; 106 ft; 148 ft **59.** 3.2 ft; 5.1 ft; 7.3 ft

Problem Set 7.6 (page 402)

1. 9 **3.** 3 **5.** −2 **7.** −5 **9.** $\frac{1}{6}$ **11.** 3 **13.** 8 **15.** 81 **17.** −1 **19.** −32 **21.** $\frac{81}{16}$ **23.** 4 **25.** $\frac{1}{128}$ **27.** −125 **29.** 625 **31.** $\sqrt[3]{x^4}$ **33.** $3\sqrt{x}$ **35.** $\sqrt[3]{2y}$ **37.** $\sqrt{2x-3y}$ **39.** $\sqrt[3]{(2a-3b)^2}$ **41.** $\sqrt[3]{x^2y}$ **43.** $-3\sqrt[5]{xy^2}$ **45.** $5^{\frac{1}{2}}y^{\frac{1}{2}}$ **47.** $3y^{\frac{1}{2}}$ **49.** $x^{\frac{1}{3}}y^{\frac{2}{3}}$ **51.** $a^{\frac{1}{2}}b^{\frac{3}{4}}$ **53.** $(2x-y)^{\frac{3}{5}}$ **55.** $5xy^{\frac{1}{2}}$ **57.** $-(x+y)^{\frac{1}{3}}$ **59.** $12x^{\frac{13}{20}}$ **61.** $y^{\frac{5}{12}}$ **63.** $\frac{4}{x^{\frac{1}{10}}}$ **65.** $16xy^2$ **67.** $2x^2y$ **69.** $4x^{\frac{4}{15}}$ **71.** $\frac{4}{b^{\frac{5}{12}}}$ **73.** $\frac{36x^{\frac{4}{5}}}{49y^{\frac{4}{3}}}$ **75.** $\frac{y^{\frac{3}{2}}}{x}$ **77.** $4x^{\frac{1}{6}}$ **79.** $\frac{16}{a^{\frac{11}{10}}}$ **81.** $\sqrt[6]{243}$ **83.** $\sqrt[4]{216}$ **85.** $\sqrt[12]{3}$ **87.** $\sqrt{2}$ **89.** $\sqrt[4]{3}$ **93.** **(a)** 12 **(c)** 7 **(e)** 11 **95.** **(a)** 1024 **(c)** 512 **(e)** 49

Problem Set 7.7 (page 408)

1. $(8.9)(10^1)$ **3.** $(4.29)(10^3)$ **5.** $(6.12)(10^6)$ **7.** $(4)(10^7)$ **9.** $(3.764)(10^2)$ **11.** $(3.47)(10^{-1})$ **13.** $(2.14)(10^{-2})$ **15.** $(5)(10^{-5})$ **17.** $(1.94)(10^{-9})$ **19.** 23 **21.** 4190 **23.** 500,000,000 **25.** 31,400,000,000 **27.** 0.43 **29.** 0.000914 **31.** 0.00000005123 **33.** 0.000000074 **35.** 0.77 **37.** 300,000,000,000 **39.** 0.000000004 **41.** 1000 **43.** 1000 **45.** 3000 **47.** 20 **49.** 27,000,000 **51.** $(6.02)(10^{23})$ **53.** 831 **55.** $(3)(10^4)$ dollars **57.** $(1.99)(10^{-26})$ kg **59.** 1833 **63.** **(a)** 7000 **(c)** 120 **(e)** 30 **65.** **(a)** $(4.385)(10^{14})$ **(c)** $(2.322)(10^{17})$ **(e)** $(3.052)(10^{12})$

Chapter 7 Review Problem Set (page 414)

1. $\frac{1}{64}$ **2.** $\frac{9}{4}$ **3.** 3 **4.** 1 **5.** 27 **6.** $\frac{1}{125}$ **7.** $\frac{x^6}{y^8}$ **8.** $\frac{27a^3b^{12}}{8}$ **9.** $\frac{9a^4}{16b^4}$ **10.** $\frac{y^6}{125x^9}$ **11.** $\frac{x^{12}}{9}$ **12.** $\frac{1}{4y^3}$ **13.** $-10x^3$ **14.** $\frac{3b^5}{a^3}$ **15.** $\frac{b^3}{a^5}$ **16.** $\frac{x^4}{y}$ **17.** $-\frac{2}{x^2}$ **18.** $-\frac{2a}{b}$ **19.** $\frac{y+x^2}{x^2y}$ **20.** $\frac{b-2a}{a^2b}$ **21.** $\frac{2y^2+3x}{xy^2}$ **22.** $\frac{y^2+6x}{2xy^2}$ **23.** $3\sqrt{6}$ **24.** $4x\sqrt{3xy}$ **25.** $2\sqrt[3]{7}$ **26.** $3xy^2\sqrt[3]{4xy^2}$ **27.** $\frac{15\sqrt{6}}{4}$ **28.** $2y\sqrt{5xy}$ **29.** $2\sqrt{2}$ **30.** $\frac{\sqrt{15x}}{6x^2}$ **31.** $\frac{\sqrt[3]{6}}{3}$ **32.** $\frac{3\sqrt{5}}{5}$ **33.** $\frac{x\sqrt{21x}}{7}$ **34.** $2\sqrt{x}$ **35.** $\sqrt{5}$ **36.** $5\sqrt[3]{3}$ **37.** $\frac{29\sqrt{6}}{5}$ **38.** $-15\sqrt{3x}$ **39.** $24\sqrt{10}$ **40.** 60 **41.** $2x\sqrt{15y}$ **42.** $-18y^2\sqrt{x}$ **43.** $24\sqrt{3}-6\sqrt{14}$ **44.** $x-2\sqrt{x}-15$ **45.** 17 **46.** $12-8\sqrt{3}$ **47.** $6a-5\sqrt{ab}-4b$ **48.** 70 **49.** $\frac{2(\sqrt{7}+1)}{3}$ **50.** $\frac{2\sqrt{6}-\sqrt{15}}{3}$ **51.** $\frac{3\sqrt{5}-2\sqrt{3}}{11}$ **52.** $\frac{6\sqrt{3}+3\sqrt{5}}{7}$ **53.** $\left\{\frac{19}{7}\right\}$ **54.** {4} **55.** {8} **56.** ∅

57. $\{14\}$ **58.** $\{-10, 1\}$ **59.** $\{2\}$ **60.** $\{8\}$ **61.** 493 ft
62. 4.7 ft **63.** 32 **64.** 1 **65.** $\frac{4}{9}$ **66.** -64 **67.** $\frac{1}{9}$
68. $\frac{1}{4}$ **69.** 27 **70.** 8 **71.** $\sqrt[3]{xy^2}$ **72.** $\sqrt[4]{a^3}$ **73.** $4\sqrt{y}$
74. $\sqrt[3]{(x + 5y)^2}$ **75.** $x^{\frac{3}{5}}y^{\frac{1}{5}}$ **76.** $4^{\frac{1}{3}}a^{\frac{2}{3}}$ **77.** $6y^{\frac{1}{2}}$
78. $(3a + b)^{\frac{5}{3}}$ **79.** $20x^{\frac{7}{10}}$ **80.** $7a^{\frac{5}{12}}$ **81.** $\frac{y^{\frac{4}{3}}}{x}$ **82.** $-\frac{6}{a^{\frac{1}{4}}}$
83. $\frac{1}{x^{\frac{2}{5}}}$ **84.** $-6y^{\frac{5}{12}}$ **85.** $\sqrt[4]{3^3}$ **86.** $3\sqrt[6]{3}$ **87.** $\sqrt[12]{5}$
88. $\sqrt[6]{2^5}$ **89.** $(5.4)(10^8)$ **90.** $(8.4)(10^4)$ **91.** $(3.2)(10^{-8})$
92. $(7.68)(10^{-4})$ **93.** 0.0000014 **94.** 0.000638
95. 41,200,000 **96.** 125,000 **97.** 0.000000006
98. 36,000,000,000 **99.** 6 **100.** 0.15 **101.** 0.000028
102. 0.002 **103.** 0.002 **104.** 8,000,000,000

Chapter 7 Test (page 416)

1. $\frac{1}{32}$ **2.** -32 **3.** $\frac{81}{16}$ **4.** $\frac{1}{4}$ **5.** $3\sqrt{7}$ **6.** $3\sqrt[3]{4}$
7. $2x^2y\sqrt{13y}$ **8.** $\frac{5\sqrt{6}}{6}$ **9.** $\frac{\sqrt{42x}}{12x^2}$ **10.** $72\sqrt{2}$
11. $-5\sqrt{6}$ **12.** $-38\sqrt{2}$ **13.** $\frac{3\sqrt{6} + 3}{10}$ **14.** $\frac{9x^2y^2}{4}$
15. $-\frac{12}{a^{\frac{3}{10}}}$ **16.** $\frac{y^3 + x}{xy^3}$ **17.** $-12x^{\frac{1}{4}}$ **18.** 33 **19.** 600
20. 0.003 **21.** $\left\{\frac{8}{3}\right\}$ **22.** $\{2\}$ **23.** $\{4\}$ **24.** $\{5\}$
25. $\{4, 6\}$

CHAPTER 8

Problem Set 8.1 (page 425)

1. False **3.** True **5.** True **7.** True **9.** $10 + 8i$
11. $-6 + 10i$ **13.** $-2 - 5i$ **15.** $-12 + 5i$ **17.** $-1 - 23i$
19. $-4 - 5i$ **21.** $1 + 3i$ **23.** $\frac{5}{3} - \frac{5}{12}i$ **25.** $-\frac{17}{9} + \frac{23}{30}i$
27. $9i$ **29.** $i\sqrt{14}$ **31.** $\frac{4}{5}i$ **33.** $3i\sqrt{2}$ **35.** $5i\sqrt{3}$
37. $6i\sqrt{7}$ **39.** $-8i\sqrt{5}$ **41.** $36i\sqrt{10}$ **43.** -8
45. $-\sqrt{15}$ **47.** $-3\sqrt{6}$ **49.** $-5\sqrt{3}$ **51.** $-3\sqrt{6}$
53. $4i\sqrt{3}$ **55.** $\frac{5}{2}$ **57.** $2\sqrt{2}$ **59.** $2i$ **61.** $-20 + 0i$
63. $42 + 0i$ **65.** $15 + 6i$ **67.** $-42 + 12i$ **69.** $7 + 22i$
71. $40 - 20i$ **73.** $-3 - 28i$ **75.** $-3 - 15i$
77. $-9 + 40i$ **79.** $-12 + 16i$ **81.** $85 + 0i$
83. $5 + 0i$ **85.** $\frac{3}{5} + \frac{3}{10}i$ **87.** $\frac{5}{17} - \frac{3}{17}i$ **89.** $2 + \frac{2}{3}i$
91. $0 - \frac{2}{7}i$ **93.** $\frac{22}{25} - \frac{4}{25}i$ **95.** $-\frac{18}{41} + \frac{39}{41}i$ **97.** $\frac{9}{2} - \frac{5}{2}i$
99. $\frac{4}{13} - \frac{1}{26}i$ **101. (a)** $-2 - i\sqrt{3}$ **(c)** $\frac{-1 - 3i\sqrt{2}}{2}$
(e) $\frac{10 + 3i\sqrt{5}}{4}$

Problem Set 8.2 (page 433)

1. $\{0, 9\}$ **3.** $\{-3, 0\}$ **5.** $\{-4, 0\}$ **7.** $\left\{0, \frac{9}{5}\right\}$ **9.** $\{-6, 5\}$
11. $\{7, 12\}$ **13.** $\left\{-8, -\frac{3}{2}\right\}$ **15.** $\left\{-\frac{7}{3}, \frac{2}{5}\right\}$ **17.** $\left\{\frac{3}{5}\right\}$
19. $\left\{-\frac{3}{2}, \frac{7}{3}\right\}$ **21.** $\{1, 4\}$ **23.** $\{8\}$ **25.** $\{12\}$ **27.** $\{\pm 1\}$
29. $\{\pm 6i\}$ **31.** $\{\pm\sqrt{14}\}$ **33.** $\{\pm 2\sqrt{7}\}$ **35.** $\{\pm 3\sqrt{2}\}$
37. $\left\{\pm\frac{\sqrt{14}}{2}\right\}$ **39.** $\left\{\pm\frac{2\sqrt{3}}{3}\right\}$ **41.** $\left\{\pm\frac{2i\sqrt{30}}{5}\right\}$
43. $\left\{\pm\frac{\sqrt{6}}{2}\right\}$ **45.** $\{-1, 5\}$ **47.** $\{-8, 2\}$ **49.** $\{-6 \pm 2i\}$
51. $\{1, 2\}$ **53.** $\{4 \pm \sqrt{5}\}$ **55.** $\{-5 \pm 2\sqrt{3}\}$
57. $\left\{\frac{2 \pm 3i\sqrt{3}}{3}\right\}$ **59.** $\{-12, -2\}$ **61.** $\left\{\frac{2 \pm \sqrt{10}}{5}\right\}$
63. $2\sqrt{13}$ cm **65.** $4\sqrt{5}$ in. **67.** 8 yd **69.** $6\sqrt{2}$ in.
71. $a = b = 4\sqrt{2}$ m **73.** $b = 3\sqrt{3}$ in. and $c = 6$ in.
75. $a = 7$ cm and $b = 7\sqrt{3}$ cm
77. $a = \frac{10\sqrt{3}}{3}$ ft and $c = \frac{20\sqrt{3}}{3}$ ft **79.** 17.9 ft
81. 38 m **83.** 53 m **87.** 10.8 cm **89.** $h = s\sqrt{2}$

Problem Set 8.3 (page 440)

1. $\{-6, 10\}$ **3.** $\{4, 10\}$ **5.** $\{-5, 10\}$ **7.** $\{-8, 1\}$
9. $\left\{-\frac{5}{2}, 3\right\}$ **11.** $\left\{-3, \frac{2}{3}\right\}$ **13.** $\{-16, 10\}$
15. $\{-2 \pm \sqrt{6}\}$ **17.** $\{-3 \pm 2\sqrt{3}\}$ **19.** $\{5 \pm \sqrt{26}\}$
21. $\{4 \pm i\}$ **23.** $\{-6 \pm 3\sqrt{3}\}$ **25.** $\{-1 \pm i\sqrt{5}\}$
27. $\left\{\frac{-3 \pm \sqrt{17}}{2}\right\}$ **29.** $\left\{\frac{-5 \pm \sqrt{21}}{2}\right\}$ **31.** $\left\{\frac{7 \pm \sqrt{37}}{2}\right\}$
33. $\left\{\frac{-2 \pm \sqrt{10}}{2}\right\}$ **35.** $\left\{\frac{3 \pm i\sqrt{6}}{3}\right\}$ **37.** $\left\{\frac{-5 \pm \sqrt{37}}{6}\right\}$
39. $\{-12, 4\}$ **41.** $\left\{\frac{4 \pm \sqrt{10}}{2}\right\}$ **43.** $\left\{-\frac{9}{2}, \frac{1}{3}\right\}$
45. $\{-3, 8\}$ **47.** $\{3 \pm 2\sqrt{3}\}$ **49.** $\left\{\frac{3 \pm i\sqrt{3}}{3}\right\}$
51. $\{-20, 12\}$ **53.** $\left\{-1, -\frac{2}{3}\right\}$ **55.** $\left\{\frac{1}{2}, \frac{3}{2}\right\}$
57. $\{-6 \pm 2\sqrt{10}\}$ **59.** $\left\{\frac{-1 \pm \sqrt{3}}{2}\right\}$
61. $\left\{\frac{-b \pm \sqrt{b^2 - 4ac}}{2a}\right\}$ **65.** $x = \frac{a\sqrt{b^2 - y^2}}{b}$
67. $r = \frac{\sqrt{A\pi}}{\pi}$ **69.** $\{2a, 3a\}$ **71.** $\left\{\frac{a}{2}, -\frac{2a}{3}\right\}$
73. $\left\{\frac{2b}{3}\right\}$

Problem Set 8.4 (page 446)

1. $\{-1 \pm \sqrt{2}\}$ **3.** $\left\{\frac{-5 \pm \sqrt{37}}{2}\right\}$ **5.** $\{4 \pm 2\sqrt{5}\}$

7. $\left\{\frac{-5 \pm i\sqrt{7}}{2}\right\}$ **9.** $\{8, 10\}$ **11.** $\left\{\frac{9 \pm \sqrt{61}}{2}\right\}$

13. $\left\{\frac{-1 \pm \sqrt{33}}{4}\right\}$ **15.** $\left\{\frac{-1 \pm i\sqrt{3}}{4}\right\}$ **17.** $\left\{\frac{4 \pm \sqrt{10}}{3}\right\}$

19. $\left\{-1, \frac{5}{2}\right\}$ **21.** $\left\{-5, -\frac{4}{3}\right\}$ **23.** $\left\{\frac{5}{6}\right\}$

25. $\left\{\frac{1 \pm \sqrt{13}}{4}\right\}$ **27.** $\left\{0, \frac{13}{5}\right\}$ **29.** $\left\{\pm\frac{\sqrt{15}}{3}\right\}$

31. $\left\{\frac{-1 \pm \sqrt{73}}{12}\right\}$ **33.** $\{-18, -14\}$ **35.** $\left\{\frac{11}{4}, \frac{10}{3}\right\}$

37. $\left\{\frac{2 \pm i\sqrt{2}}{2}\right\}$ **39.** $\left\{\frac{1 \pm \sqrt{7}}{6}\right\}$

41. Two real solutions; $\{-7, 3\}$ **43.** One real solution; $\left\{\frac{1}{3}\right\}$

45. Two complex solutions; $\left\{\frac{7 \pm i\sqrt{3}}{2}\right\}$

47. Two real solutions; $\left\{-\frac{4}{3}, \frac{1}{5}\right\}$

49. Two real solutions; $\left\{\frac{-2 \pm \sqrt{10}}{3}\right\}$

55. $\{-1.381, 17.381\}$ **57.** $\{-13.426, 3.426\}$
59. $\{-0.347, -8.653\}$ **61.** $\{0.119, 1.681\}$
63. $\{-0.708, 4.708\}$ **65.** $k = 4$ or $k = -4$

Problem Set 8.5 (page 454)

1. $\{2 \pm \sqrt{10}\}$ **3.** $\left\{-9, \frac{4}{3}\right\}$ **5.** $\{9 \pm 3\sqrt{10}\}$

7. $\left\{\frac{3 \pm i\sqrt{23}}{4}\right\}$ **9.** $\{-15, -9\}$ **11.** $\{-8, 1\}$

13. $\left\{\frac{2 \pm i\sqrt{10}}{2}\right\}$ **15.** $\{9 \pm \sqrt{66}\}$ **17.** $\left\{-\frac{5}{4}, \frac{2}{5}\right\}$

19. $\left\{\frac{-1 \pm \sqrt{2}}{2}\right\}$ **21.** $\left\{\frac{3}{4}, 4\right\}$ **23.** $\left\{\frac{11 \pm \sqrt{109}}{2}\right\}$

25. $\left\{\frac{3}{7}, 4\right\}$ **27.** $\left\{\frac{7 \pm \sqrt{129}}{10}\right\}$ **29.** $\left\{-\frac{10}{7}, 3\right\}$

31. $\{1 \pm \sqrt{34}\}$ **33.** $\{\pm\sqrt{6}, \pm 2\sqrt{3}\}$

35. $\left\{\pm 3, \pm\frac{2\sqrt{6}}{3}\right\}$ **37.** $\left\{\pm\frac{i\sqrt{15}}{3}, \pm 2i\right\}$

39. $\left\{\pm\frac{\sqrt{14}}{2}, \pm\frac{2\sqrt{3}}{3}\right\}$ **41.** 8 and 9 **43.** 9 and 12

45. $5 + \sqrt{3}$ and $5 - \sqrt{3}$ **47.** 3 and 6
49. 9 in. and 12 in. **51.** 1 m **53.** 8 in. by 14 in.
55. 20 mph for Lorraine and 25 mph for Charlotte, or 45 mph for Lorraine and 50 mph for Charlotte
57. 55 mph **59.** 6 hr for Tom and 8 hr for Terry
61. 2 hr **63.** 8 students **65.** 50 numbers **67.** 9%

73. $\{9, 36\}$ **75.** $\{1\}$ **77.** $\left\{-\frac{8}{27}, \frac{27}{8}\right\}$ **79.** $\left\{-4, \frac{3}{5}\right\}$

81. $\{4\}$ **83.** $\{\pm 4\sqrt{2}\}$ **85.** $\left\{\frac{3}{2}, \frac{5}{2}\right\}$ **87.** $\{-1, 3\}$

Problem Set 8.6 (page 461)

1. $(-\infty, -2) \cup (1, \infty)$

3. $(-4, -1)$

5. $\left(-\infty, -\frac{7}{3}\right] \cup \left[\frac{1}{2}, \infty\right)$

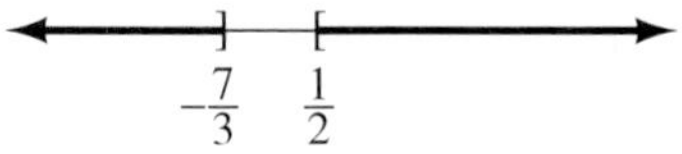

7. $\left[-2, \frac{3}{4}\right]$

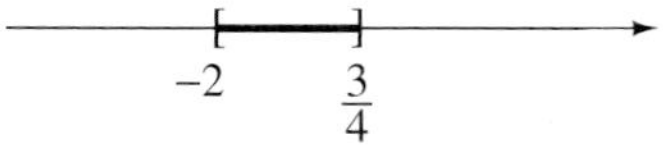

9. $(-1, 1) \cup (3, \infty)$

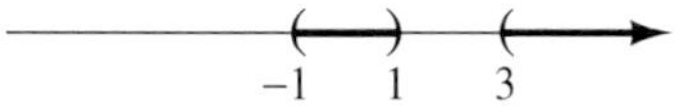

11. $(-\infty, -2] \cup [0, 4]$

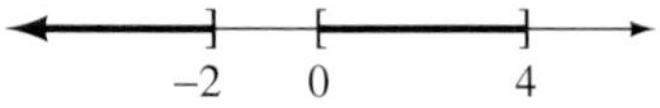

13. $(-7, 5)$ **15.** $(-\infty, 4) \cup (7, \infty)$ **17.** $\left[-5, \frac{2}{3}\right]$

19. $\left(-\infty, -\frac{5}{2}\right] \cup \left[-\frac{1}{4}, \infty\right)$ **21.** $\left(-\infty, -\frac{4}{5}\right) \cup (8, \infty)$

23. $(-\infty, \infty)$ **25.** $\left\{-\frac{5}{2}\right\}$ **27.** $(-1, 3) \cup (3, \infty)$
29. $(-\infty, -2) \cup (2, \infty)$ **31.** $(-6, 6)$ **33.** $(-\infty, \infty)$
35. $(-\infty, 0] \cup [2, \infty)$ **37.** $(-4, 0) \cup (0, \infty)$

39. $(-\infty, -1) \cup (2, \infty)$ **41.** $(-2, 3)$

43. $(-\infty, 0) \cup \left[\frac{1}{2}, \infty\right)$ **45.** $(-\infty, 1) \cup [2, \infty)$ **47.** $(-6, -3)$

49. $(-\infty, 5) \cup [9, \infty)$ **51.** $\left(-\infty, \frac{4}{3}\right) \cup (3, \infty)$

53. $(-4, 6]$ **55.** $(-\infty, 2)$

Chapter 8 Review Problem Set (page 468)

1. $2 - 2i$ **2.** $-3 - i$ **3.** $8 - 8i$ **4.** $-2 + 4i$ **5.** $2i\sqrt{2}$

6. $5i$ **7.** $12i$ **8.** $6i\sqrt{2}$ **9.** $-2\sqrt{3}$ **10.** $6i$ **11.** $\sqrt{7}$
12. $i\sqrt{3}$ **13.** $30 + 15i$ **14.** $86 - 2i$ **15.** $-32 + 4i$

16. $25 + 0i$ **17.** $\frac{9}{20} + \frac{13}{20}i$ **18.** $-\frac{3}{29} + \frac{7}{29}i$ **19.** $2 - \frac{3}{2}i$

20. $-5 - 6i$ **21.** $\{-8, 0\}$ **22.** $\{0, 6\}$ **23.** $\{-4, 7\}$

24. $\left\{-\frac{3}{2}, 1\right\}$ **25.** $\{\pm 3\sqrt{5}\}$ **26.** $3 \pm 3i\sqrt{2}$

27. $\left\{\frac{-3 \pm 2\sqrt{6}}{2}\right\}$ **28.** $\{\pm 3\sqrt{3}\}$ **29.** $\{-9 \pm \sqrt{91}\}$

30. $\{-3 \pm i\sqrt{11}\}$ **31.** $\{5 \pm 2\sqrt{6}\}$ **32.** $\left\{\frac{-5 \pm \sqrt{33}}{2}\right\}$
33. $\{-3 \pm \sqrt{5}\}$ **34.** $\{-2 \pm i\sqrt{2}\}$
35. $\left\{\frac{1 \pm i\sqrt{11}}{3}\right\}$ **36.** $\left\{\frac{1 \pm \sqrt{61}}{10}\right\}$
37. One real solution with a multiplicity of 2
38. Two nonreal complex solutions
39. Two unequal real solutions
40. Two unequal real solutions **41.** $\{0, 17\}$ **42.** $\{-4, 8\}$
43. $\left\{\frac{1 \pm 8i}{2}\right\}$ **44.** $\{-3, 7\}$ **45.** $\{-1 \pm \sqrt{10}\}$
46. $\{3 \pm 5i\}$ **47.** $\{25\}$ **48.** $\left\{-4, \frac{2}{3}\right\}$ **49.** $\{-10, 20\}$
50. $\left\{\frac{-1 \pm \sqrt{61}}{6}\right\}$ **51.** $\left\{\frac{1 \pm i\sqrt{11}}{2}\right\}$
52. $\left\{\frac{5 \pm i\sqrt{23}}{4}\right\}$ **53.** $\left\{\frac{-2 \pm \sqrt{14}}{2}\right\}$ **54.** $\{-9, 4\}$
55. $\{-2 \pm i\sqrt{5}\}$ **56.** $\{-6, 12\}$ **57.** $\{1 \pm \sqrt{10}\}$
58. $\left\{\pm\frac{\sqrt{14}}{2}, \pm 2\sqrt{2}\right\}$ **59.** $\left\{\frac{-3 \pm \sqrt{97}}{2}\right\}$
60. 34.6 ft and 40.0 ft **61.** 47.2 m **62.** $4\sqrt{2}$ in.
63. $3 + \sqrt{7}$ and $3 - \sqrt{7}$ **64.** 8 hr **65.** 45 mph and 52 mph **66.** 8 units **67.** 8 and 10 **68.** 7 in. by 12 in.
69. 4 hr for Reena and 6 hr for Billy **70.** 10 m
71. $(-\infty, -5) \cup (2, \infty)$ **72.** $\left[-\frac{7}{2}, 3\right]$
73. $\left[-\frac{1}{2}, \frac{1}{2}\right]$ **74.** $(-\infty, 2) \cup (5, \infty)$
75. $(-\infty, -6) \cup [4, \infty)$ **76.** $\left(-\frac{5}{2}, -1\right)$
77. $(-9, 4)$ **78.** $\left[-\frac{1}{3}, 1\right)$

Chapter 8 Test (page 470)

1. $39 - 2i$ **2.** $-\frac{6}{25} - \frac{17}{25}i$ **3.** $\{0, 7\}$ **4.** $\{-1, 7\}$
5. $\{-6, 3\}$ **6.** $\{1 - \sqrt{2}, 1 + \sqrt{2}\}$ **7.** $\left\{\frac{1 - 2i}{5}, \frac{1 + 2i}{5}\right\}$
8. $\{-16, -14\}$ **9.** $\left\{\frac{1 - 6i}{3}, \frac{1 + 6i}{3}\right\}$ **10.** $\left\{-\frac{7}{4}, \frac{6}{5}\right\}$
11. $\left\{-3, \frac{19}{6}\right\}$ **12.** $\left\{-\frac{10}{3}, 4\right\}$ **13.** $\{-2, 2, -4i, 4i\}$
14. $\left\{-\frac{3}{4}, 1\right\}$ **15.** $\left\{\frac{1 - \sqrt{10}}{3}, \frac{1 + \sqrt{10}}{3}\right\}$
16. Two equal real solutions **17.** Two nonreal complex solutions **18.** $[-6, 9]$ **19.** $(-\infty, -2) \cup \left(\frac{1}{3}, \infty\right)$
20. $[-10, -6)$ **21.** 20.8 ft **22.** 29 m **23.** 3 hr
24. $6\frac{1}{2}$ in. **25.** $3 + \sqrt{5}$

Cumulative Review Problem Set (page 472)

1. $\frac{64}{15}$ **2.** $\frac{11}{3}$ **3.** $\frac{1}{6}$ **4.** -7 **5.** $-24a^4b^5$
6. $2x^3 + 5x^2 - 7x - 12$ **7.** $\frac{3x^2y^2}{8}$ **8.** $\frac{a(a + 1)}{2a - 1}$
9. $\frac{-x + 14}{18}$ **10.** $\frac{5x + 19}{x(x + 3)}$ **11.** $\frac{2}{n + 8}$
12. $\frac{x - 14}{(5x - 2)(x + 1)(x - 4)}$ **13.** $y^2 - 5y + 6$
14. $x^2 - 3x - 2$ **15.** $20 + 7\sqrt{10}$
16. $2x - 2\sqrt{xy} - 12y$ **17.** $-\frac{3}{8}$ **18.** $-\frac{2}{3}$ **19.** 0.2
20. $\frac{1}{2}$ **21.** $\frac{13}{9}$ **22.** -27 **23.** $\frac{16}{9}$ **24.** $\frac{8}{27}$
25. $3x(x + 3)(x^2 - 3x + 9)$ **26.** $(6x - 5)(x + 4)$
27. $(4 + 7x)(3 - 2x)$ **28.** $(3x + 2)(3x - 2)(x^2 + 8)$
29. $(2x - y)(a - b)$ **30.** $(3x - 2y)(9x^2 + 6xy + 4y^2)$
31. $\left\{-\frac{12}{7}\right\}$ **32.** $\{150\}$ **33.** $\{25\}$ **34.** $\{0\}$ **35.** $\{-2, 2\}$
36. $\{-7\}$ **37.** $\left\{-6, \frac{4}{3}\right\}$ **38.** $\left\{\frac{5}{4}\right\}$ **39.** $\{3\}$ **40.** $\left\{\frac{4}{5}, 1\right\}$
41. $\left\{-\frac{10}{3}, 4\right\}$ **42.** $\left\{\frac{1}{4}, \frac{2}{3}\right\}$ **43.** $\left\{-\frac{3}{2}, 3\right\}$ **44.** $\left\{\frac{1}{5}\right\}$
45. $\left\{\frac{5}{7}\right\}$ **46.** $\{-2, 2\}$ **47.** $\{0\}$ **48.** $\{-6, 19\}$
49. $\left\{-\frac{3}{4}, \frac{2}{3}\right\}$ **50.** $\{1 \pm 5i\}$ **51.** $\{1, 3\}$ **52.** $\left\{-4, \frac{1}{3}\right\}$
53. $\{-2 \pm 4i\}$ **54.** $\left\{\frac{1 \pm \sqrt{33}}{4}\right\}$ **55.** $(-\infty, -2]$
56. $\left(-\infty, \frac{19}{5}\right)$ **57.** $\left(\frac{1}{4}, \infty\right)$ **58.** $(-2, 3)$
59. $\left(-\infty, -\frac{13}{3}\right) \cup (3, \infty)$ **60.** $(-\infty, 29]$ **61.** $[-2, 4]$
62. $(-\infty, -5) \cup \left(\frac{1}{3}, \infty\right)$ **63.** $(-\infty, -2] \cup (7, \infty)$
64. $(-3, 4)$

65.

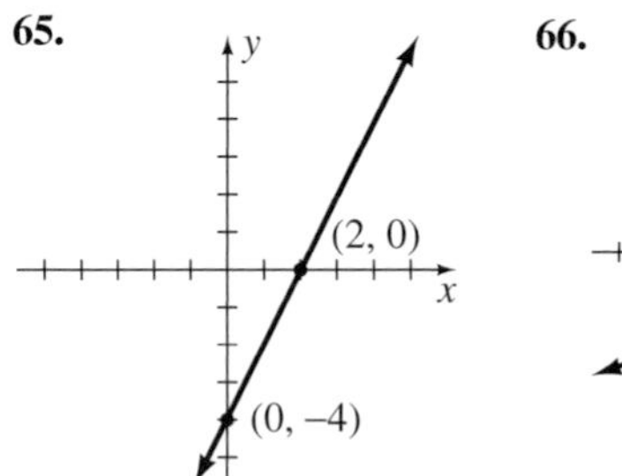

66.

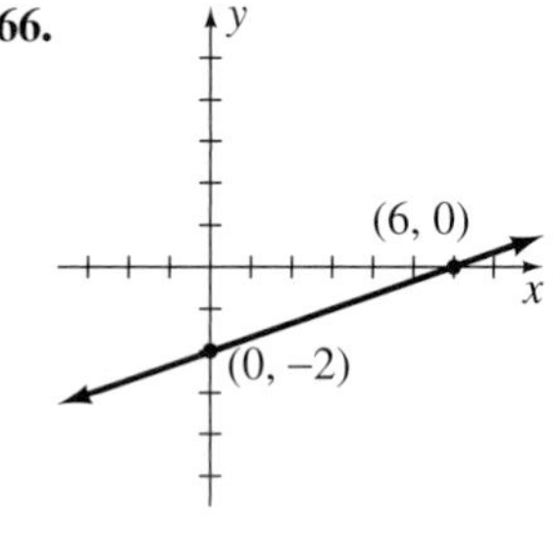

67. **68.**

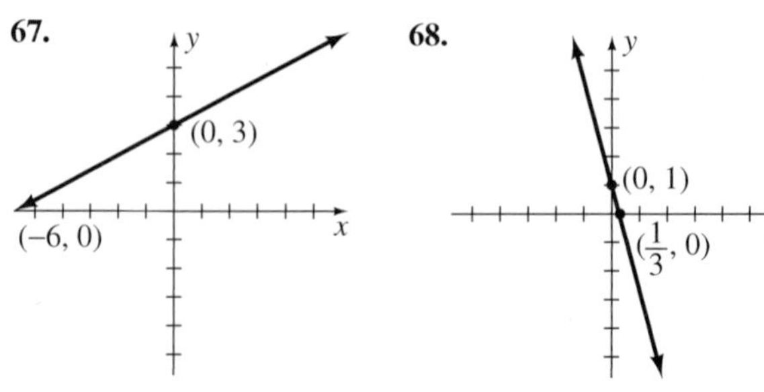

69.

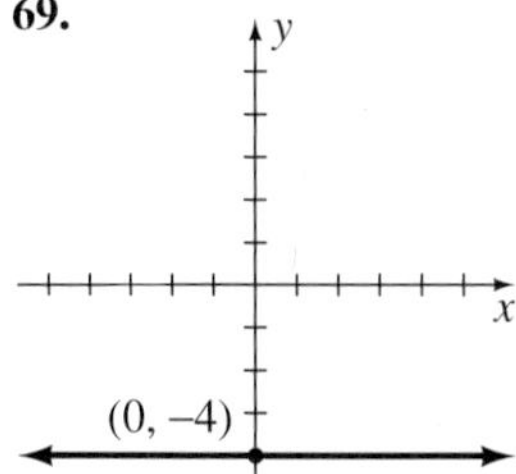

70.

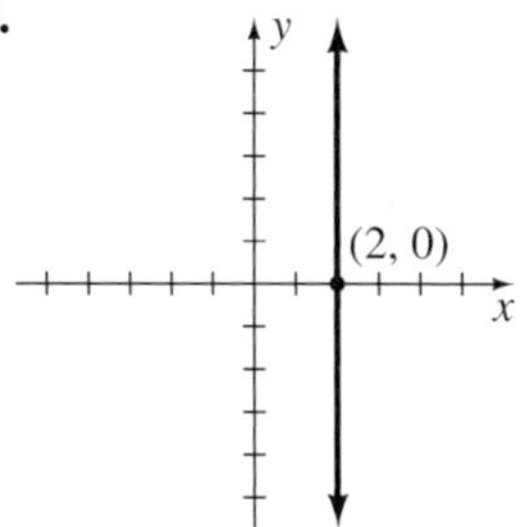

71. $7\sqrt{2}$ 72. $\sqrt{41}$ 73. $3x - 5y = 6$
74. $x + y = 3$ 75. $2x + y = -1$ 76. $3x + y = 1$

77.

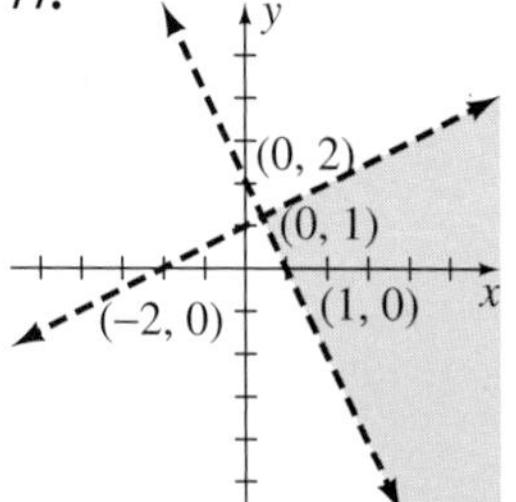

78.

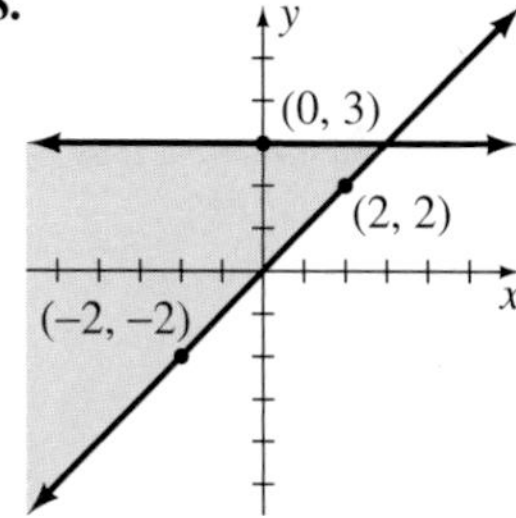

79. $\{(-1, 3)\}$ 80. $\{(2, 5)\}$ 81. $\{(0, 8)\}$ 82. $\{(-4, -6)\}$
83. $\{(4, -2, 1)\}$ 84. $\{(3, 0, -5)\}$
85. 8 qt of 1% and 4 qt of 4% 86. 10 ft by 12 ft
87. 6 liters of 60% acid 88. \$900, \$1350 89. 12 in. by 12 in.
90. 5 hr 91. 12 min 92. 7% 93. 15 chairs per row

CHAPTER 9

Problem Set 9.1 (page 482)

1. $(-3, -1); (3, 1); (3, -1)$ 3. $(7, 2); (-7, -2); (-7, 2)$
5. $(5, 0); (-5, 0); (-5, 0)$ 7. x axis 9. y axis
11. x axis, y axis, and origin 13. x axis 15. None
17. Origin 19. y axis 21. All three 23. x axis
25. y axis

27.

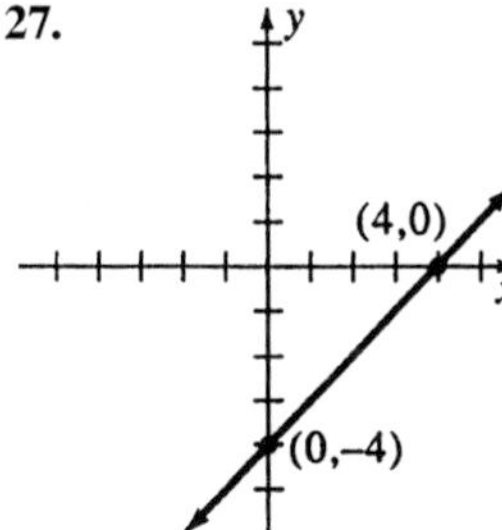

29.

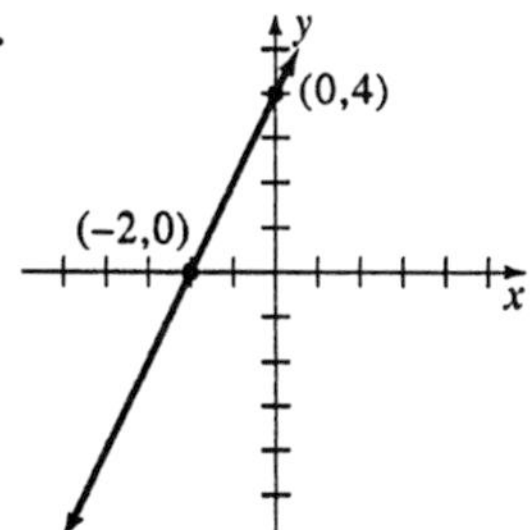

31.

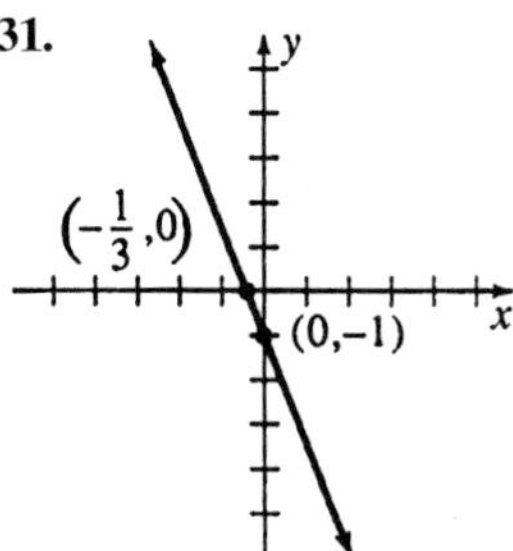

33.

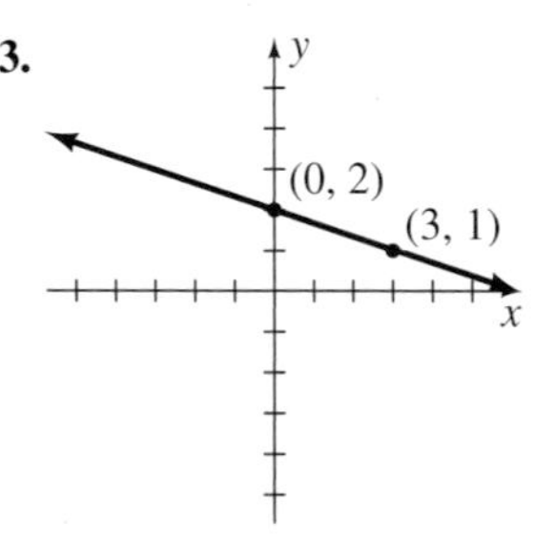

35.

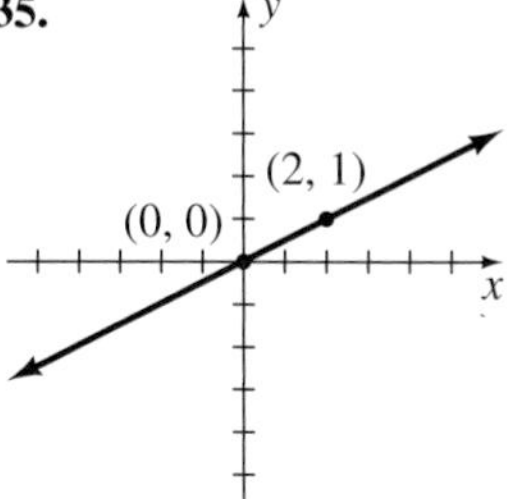

37.

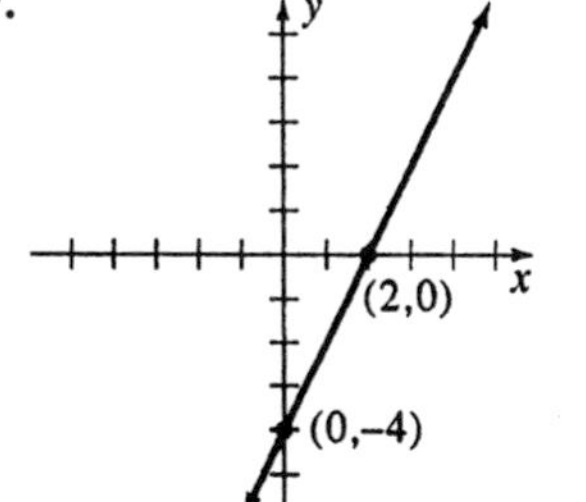

39.

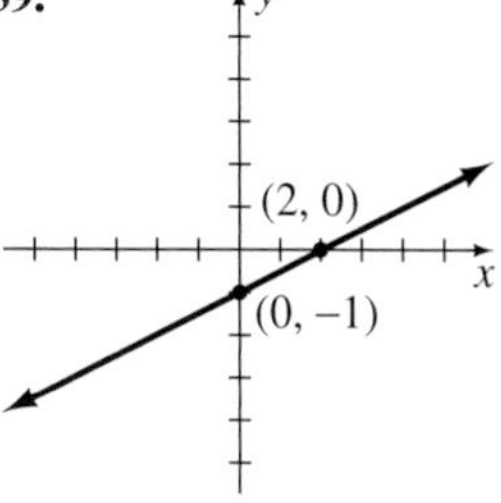

41.

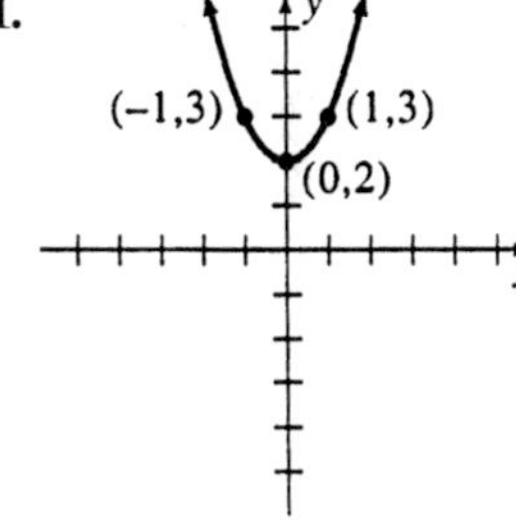

43.

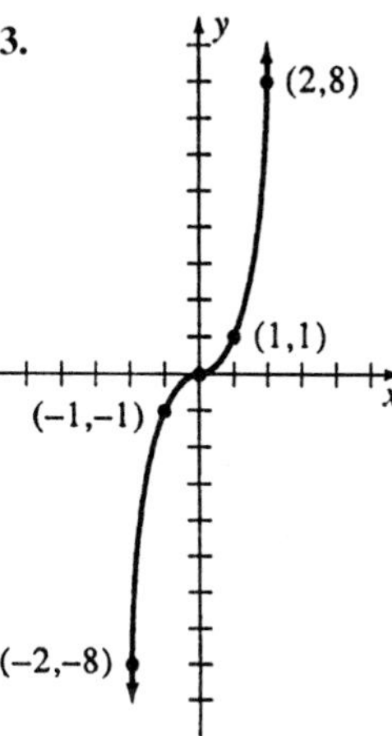

45.

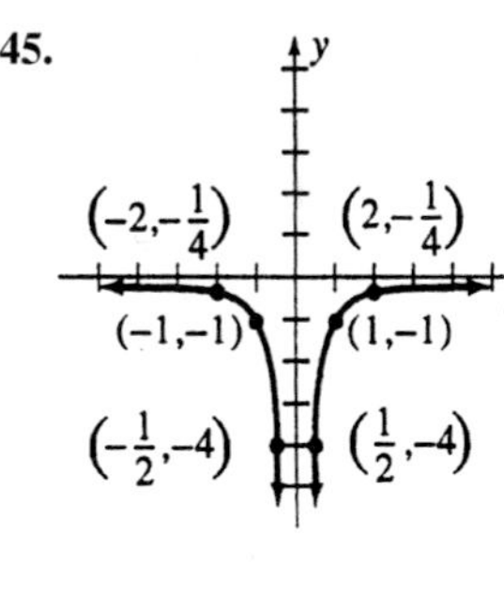

47.

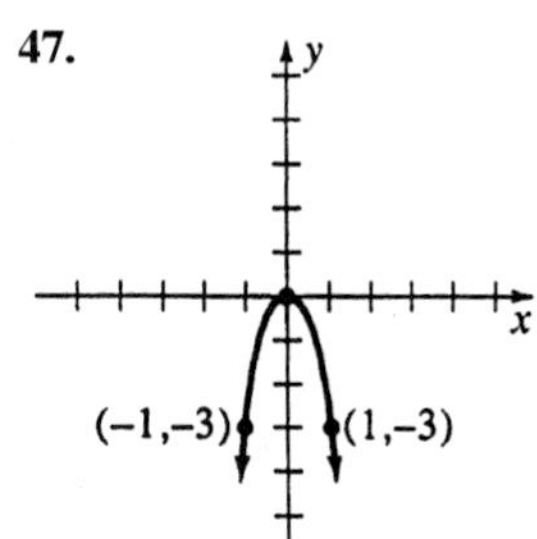

49.

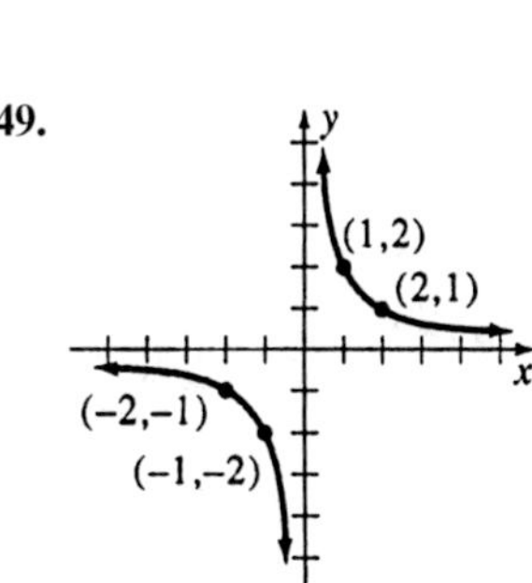

51.

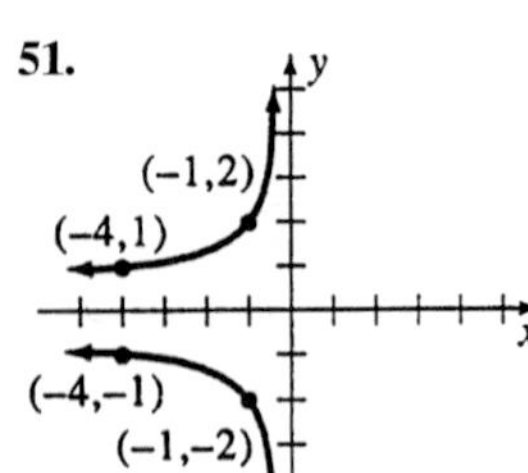

53.

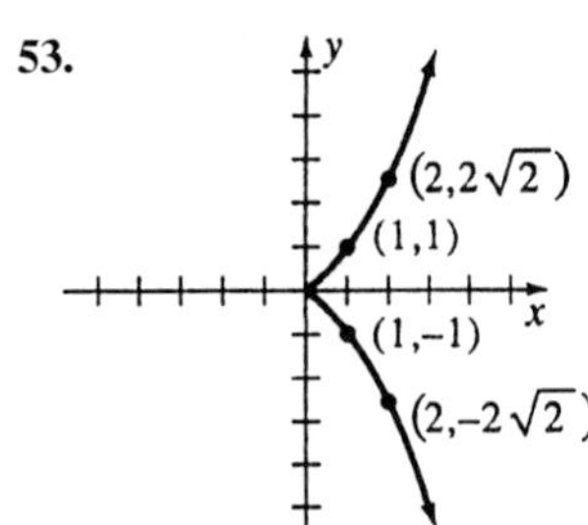

55.

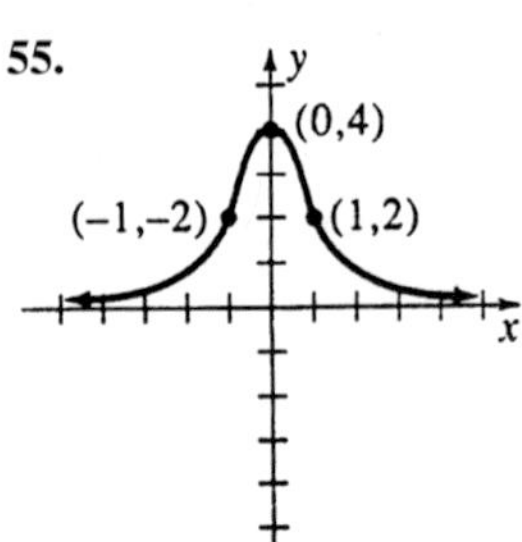

57.

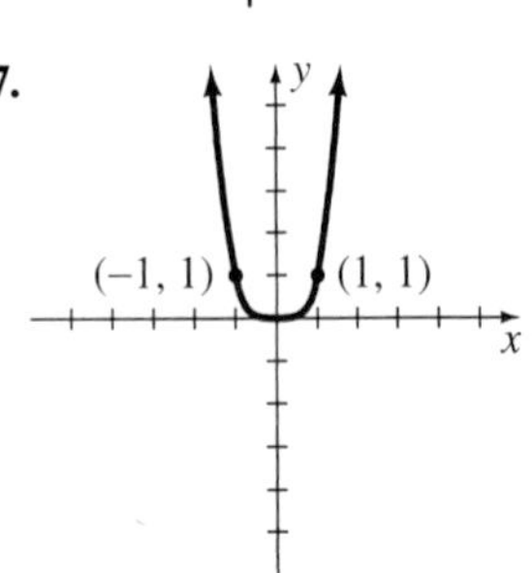

59.

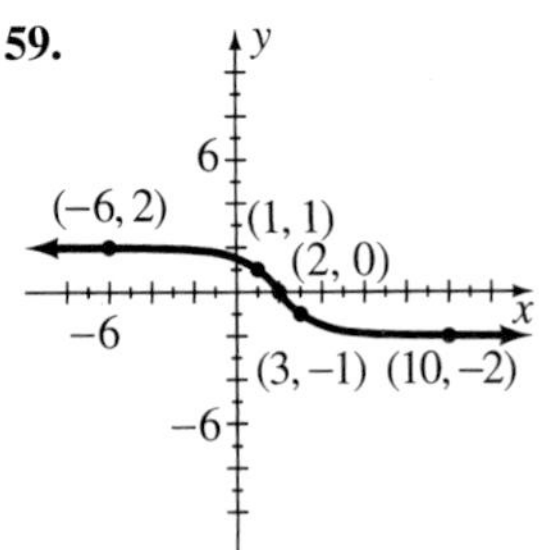

Problem Set 9.2 (page 492)

1.

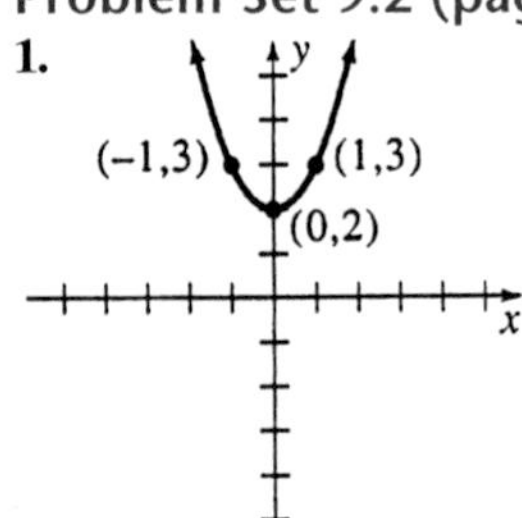

3.

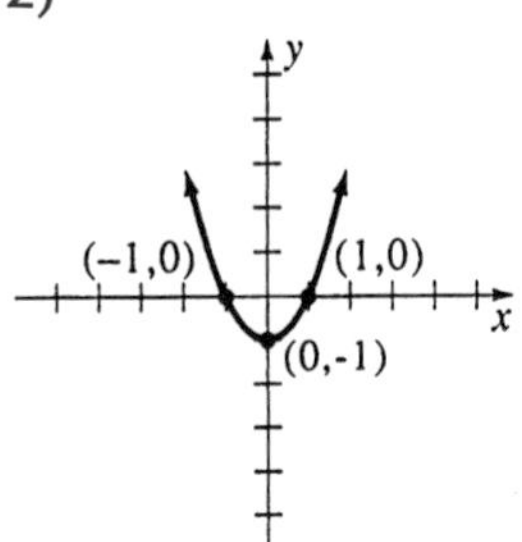

5.

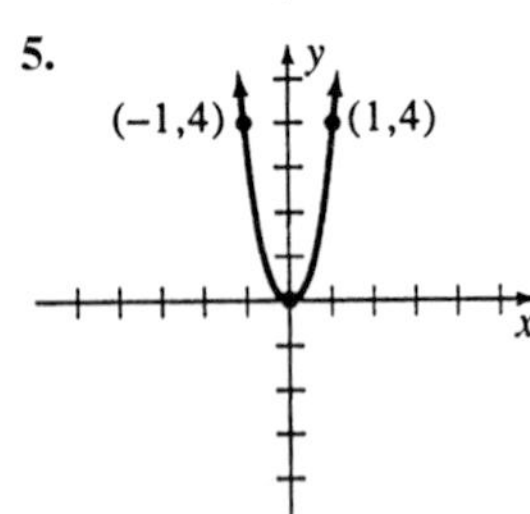

7.

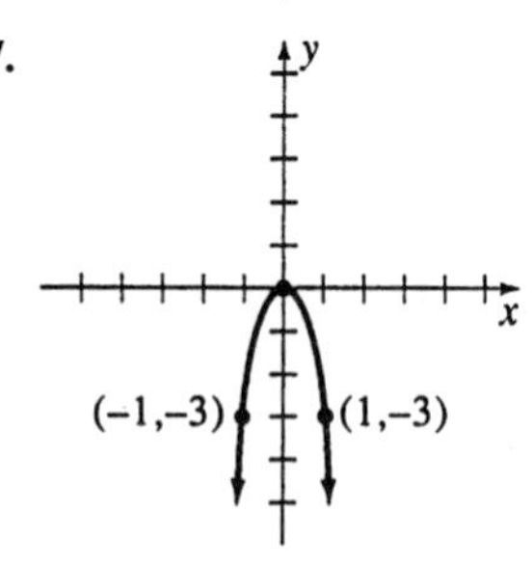

9.

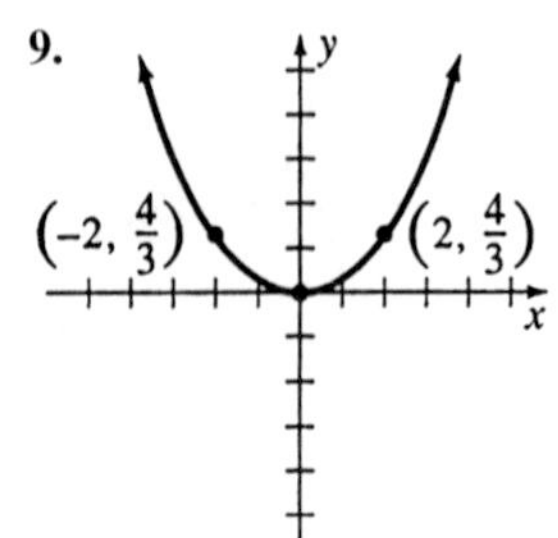

11.

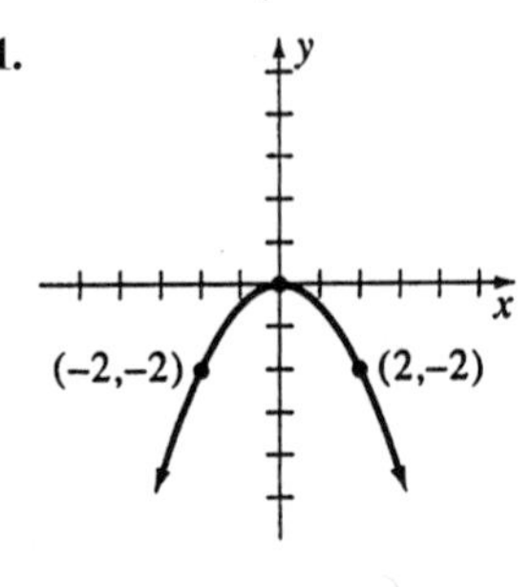

13.

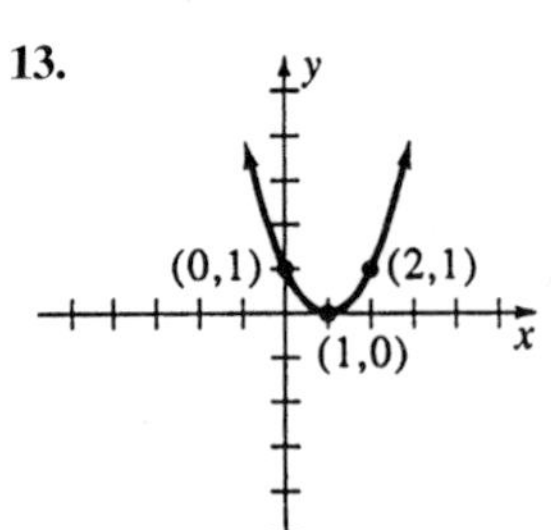

15.

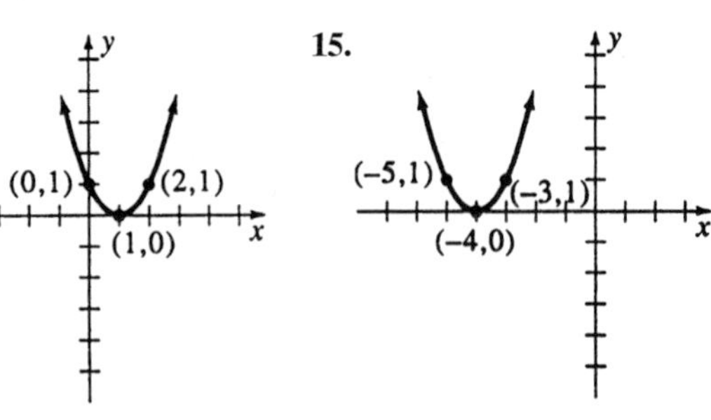

17.

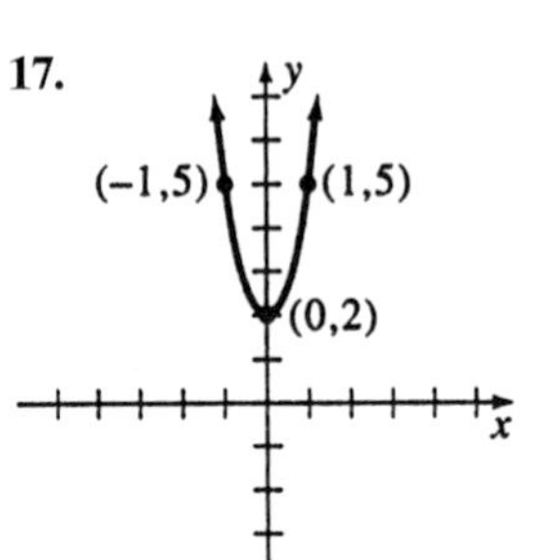

19.

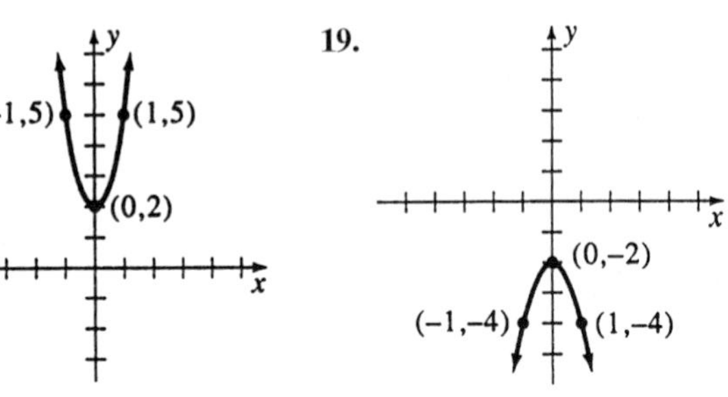

21.

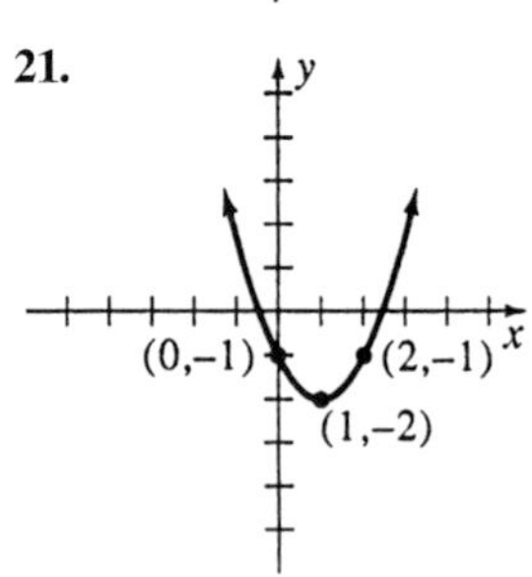

23.

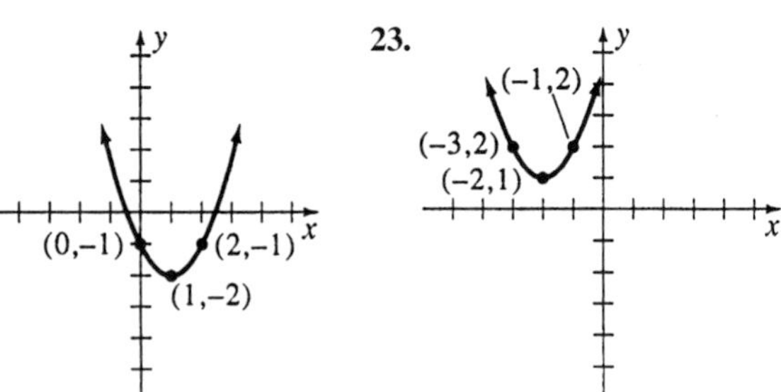

25.

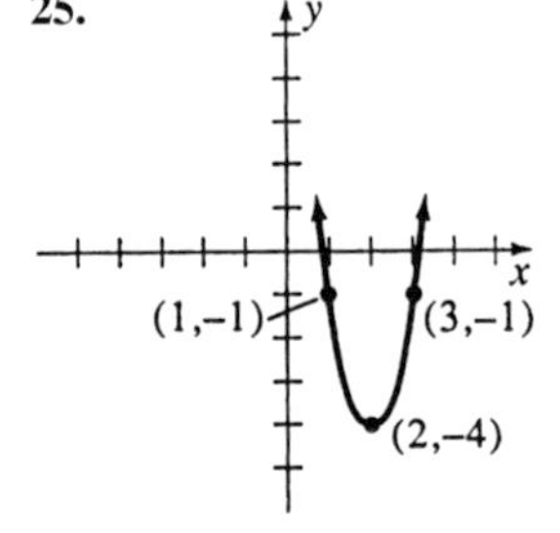

27.

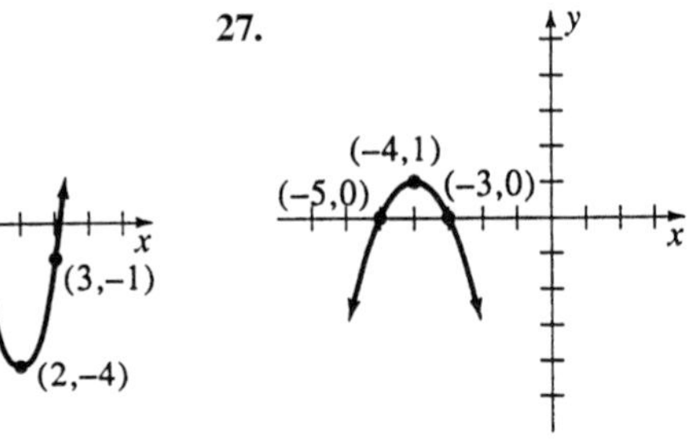

29.

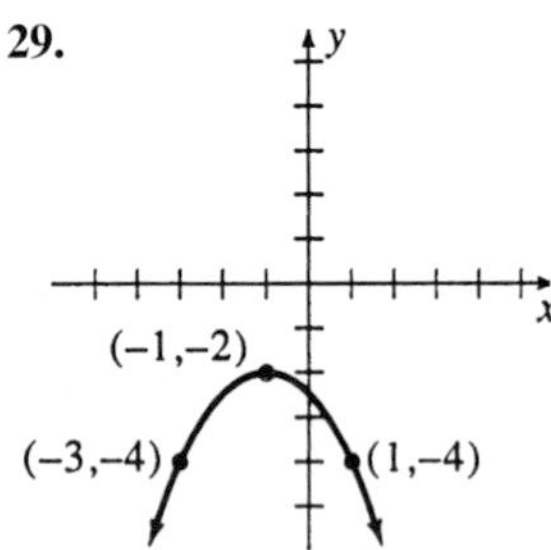

Problem Set 9.3 (page 501)

1.

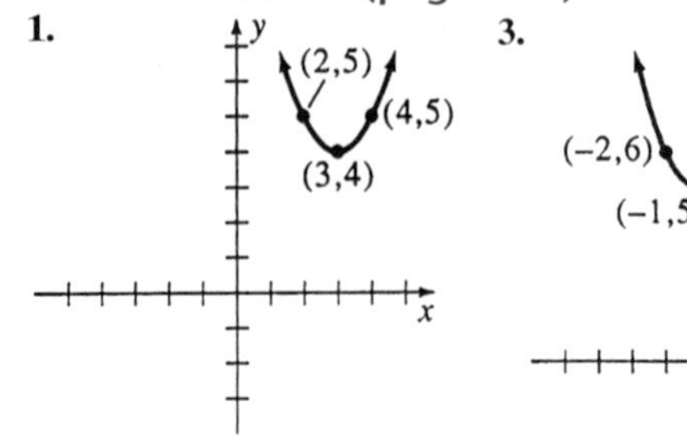

3.

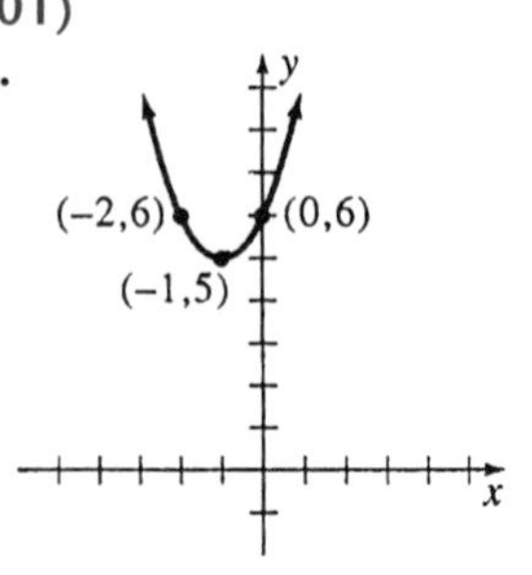

5.
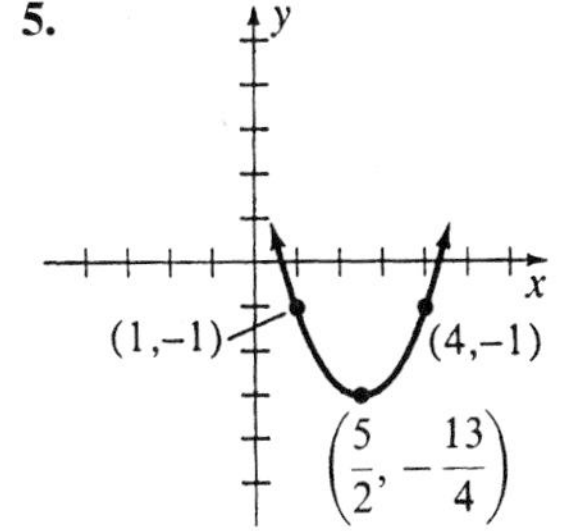

7.
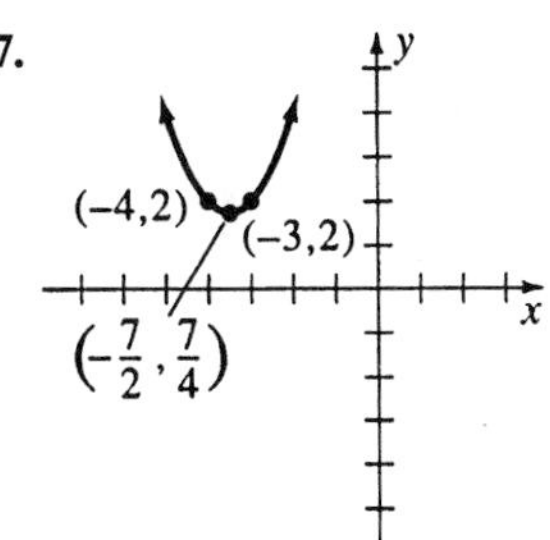

9.
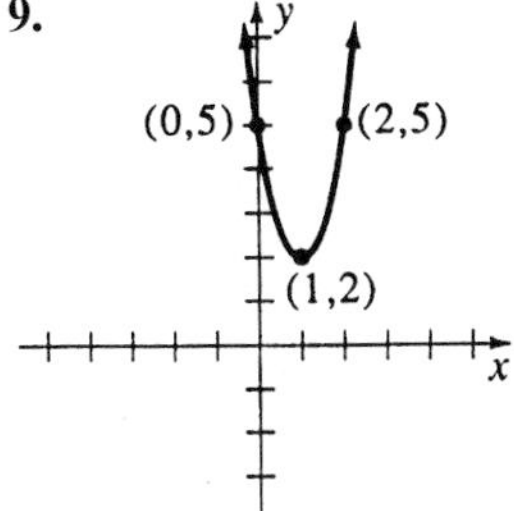

11.
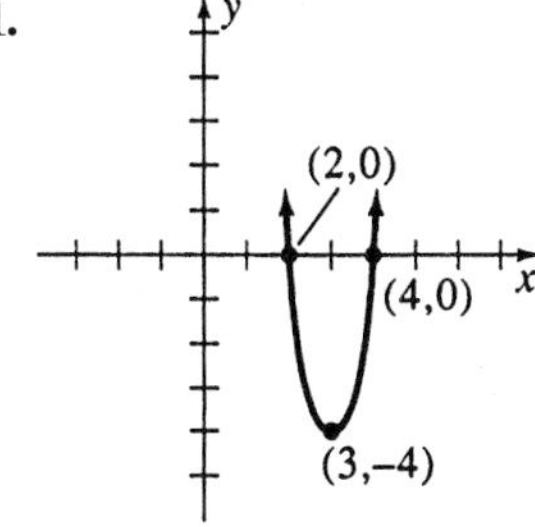

13.
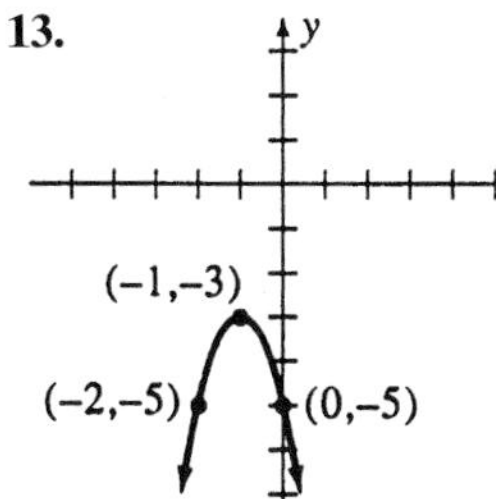

15.
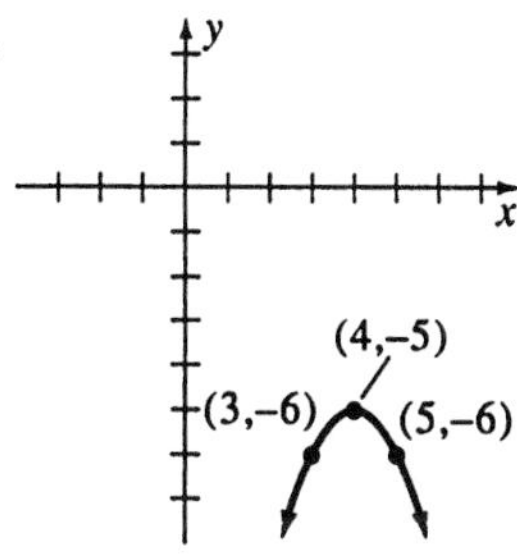

17.
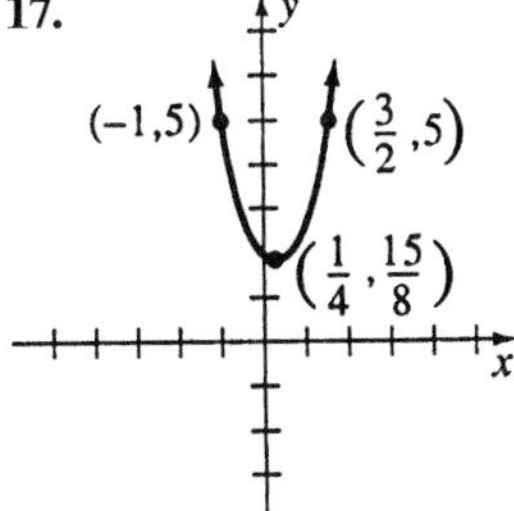

19.
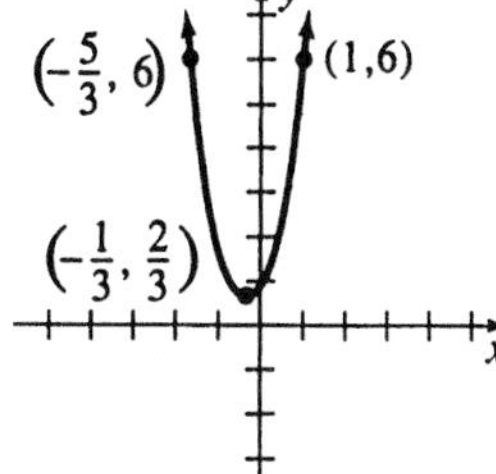

21.
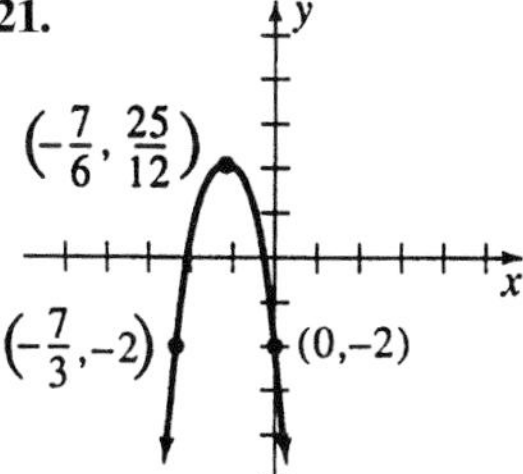

23. $(1, 3), r = 4$

25. $(-3, -5), r = 4$ **27.** $(0, 0), r = \sqrt{10}$
29. $(8, -3), r = \sqrt{2}$ **31.** $(-3, 4), r = 5$
33. $\left(-\frac{1}{2}, 4\right), r = 2\sqrt{2}$

35. $x^2 + y^2 - 6x - 10y + 9 = 0$
37. $x^2 + y^2 + 8x - 2y - 47 = 0$
39. $x^2 + y^2 + 4x + 12y + 22 = 0$ **41.** $x^2 + y^2 - 20 = 0$
43. $x^2 + y^2 - 10x + 16y - 7 = 0$
45. $x^2 + y^2 - 8y = 0$ **47.** $x^2 + y^2 + 8x - 6y = 0$

49.
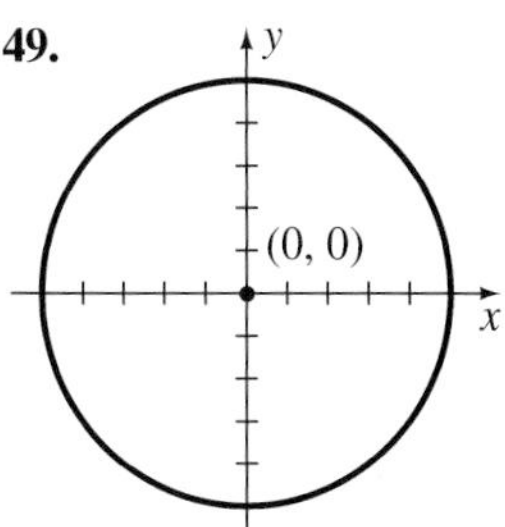

51.
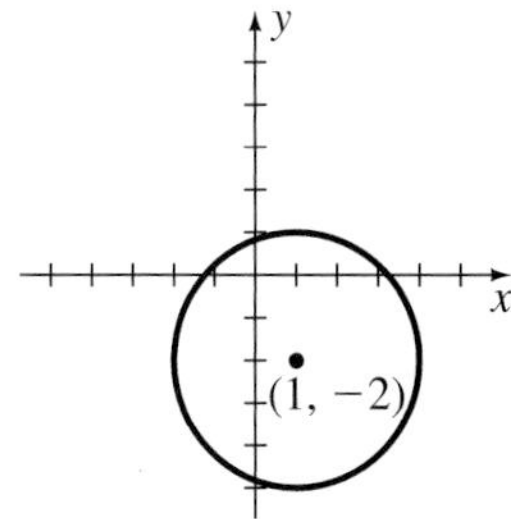

53.
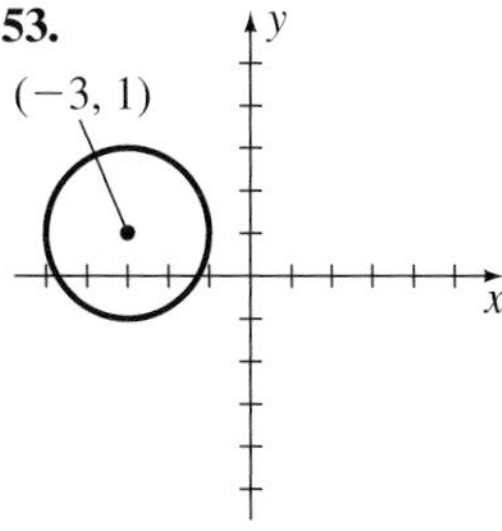

55.
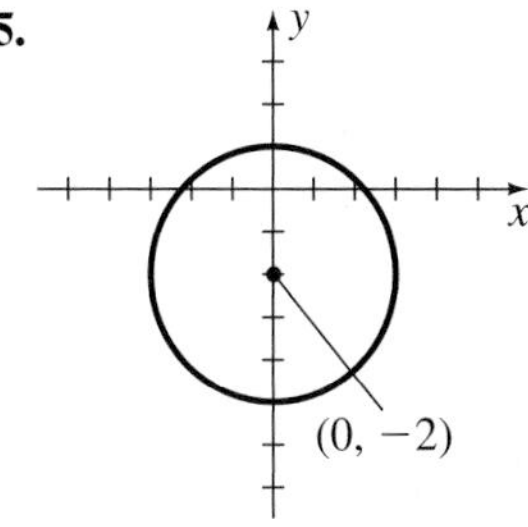

57.
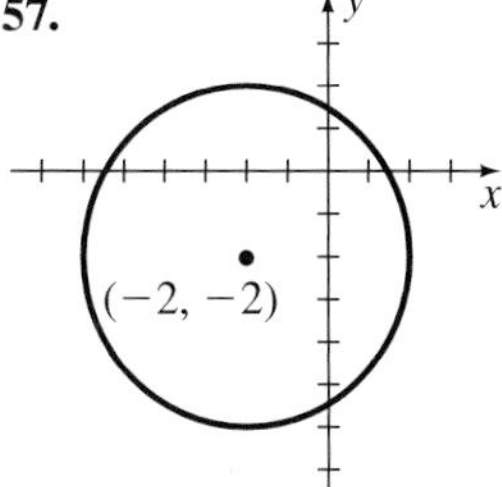

63. (a) $(1, 4), r = 3$ **(c)** $(-6, -4), r = 8$
(e) $(0, 6), r = 9$

Problem Set 9.4 (page 508)

1.
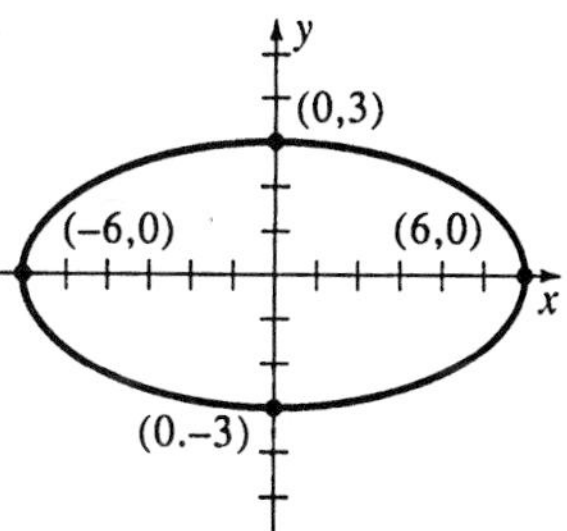

3.
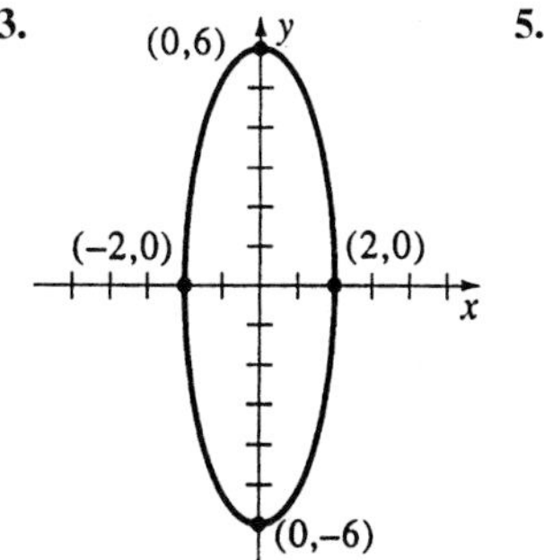

5.
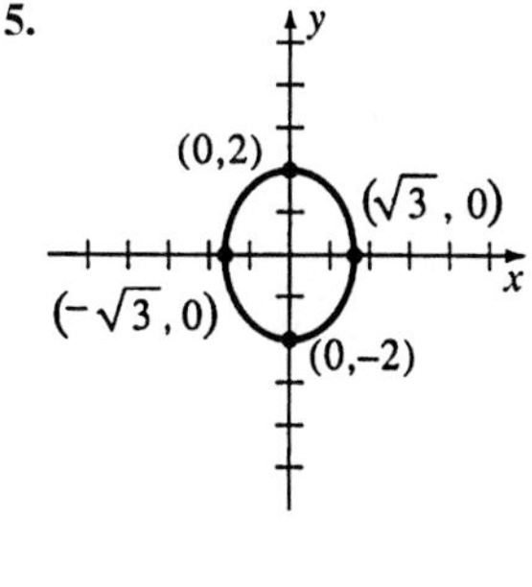

7.

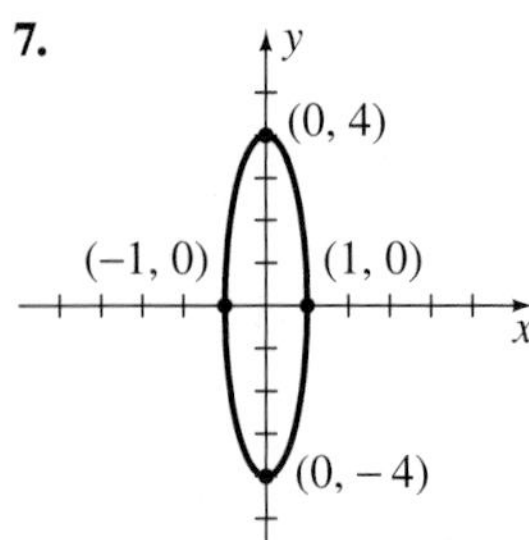

9.

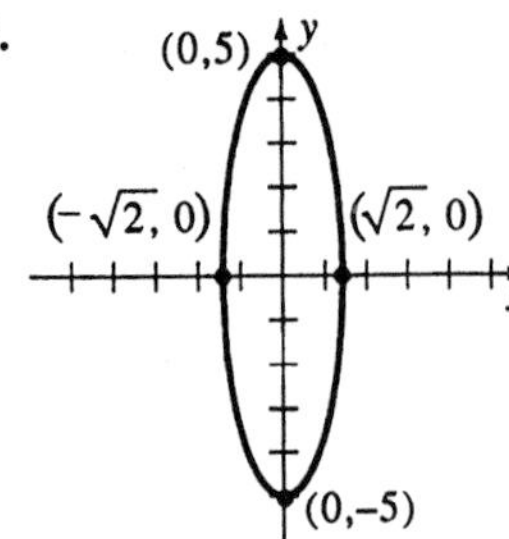

11.

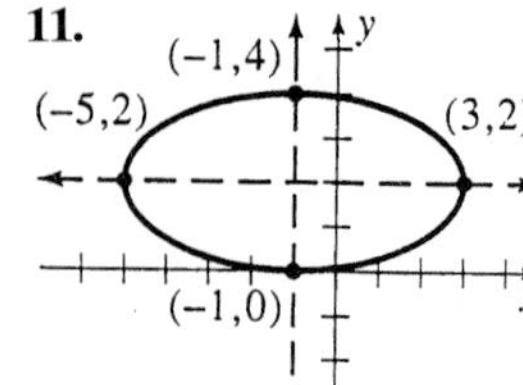

13.

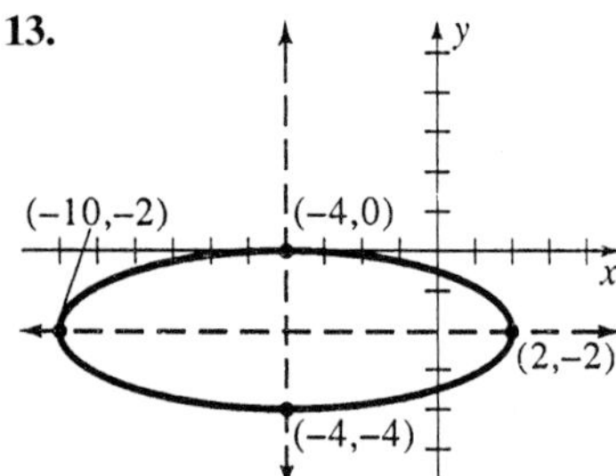

15.

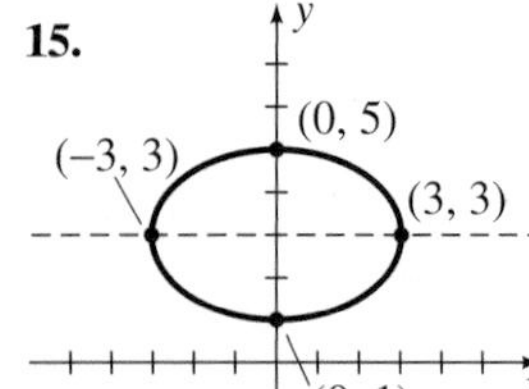

Problem Set 9.5 (page 516)

1. (4, 0), (−4, 0); $y = \frac{1}{3}x$ and $y = -\frac{1}{3}x$

3. (0, 6), (0, −6); $y = 3x$ and $y = -3x$

5. $\left(\frac{2}{5}, 0\right), \left(-\frac{2}{5}, 0\right)$; $y = \frac{5}{3}x$ and $y = -\frac{5}{3}x$

7.

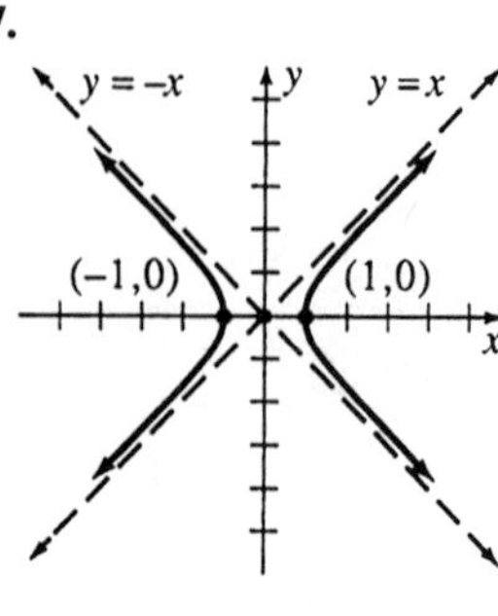

9.

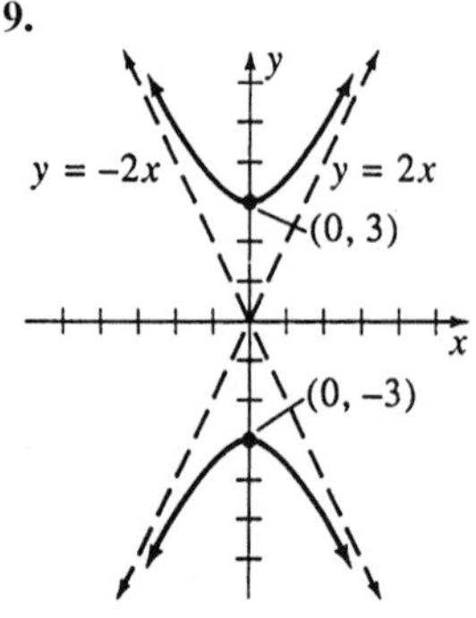

11.

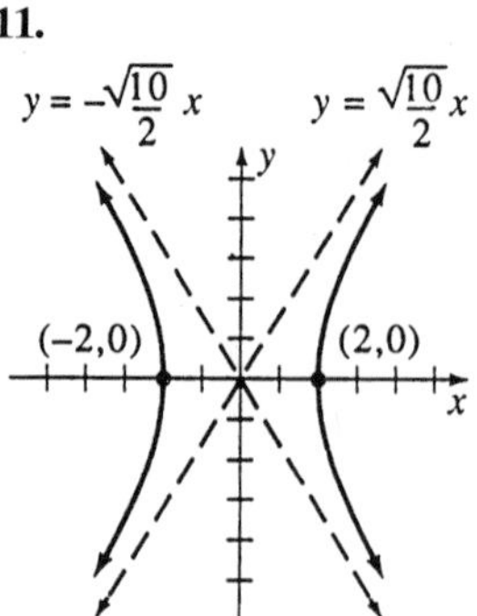

13.

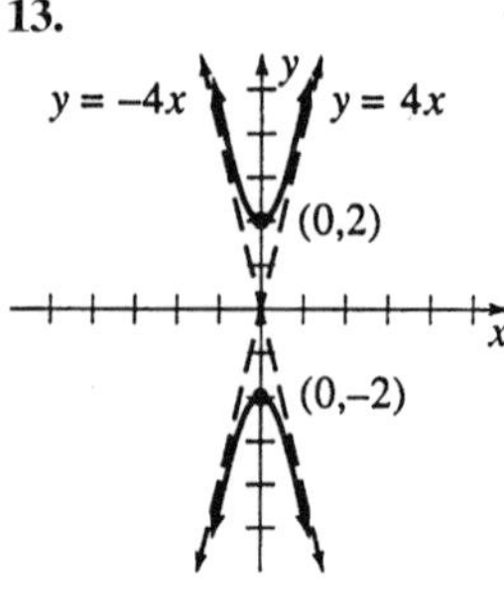

15.

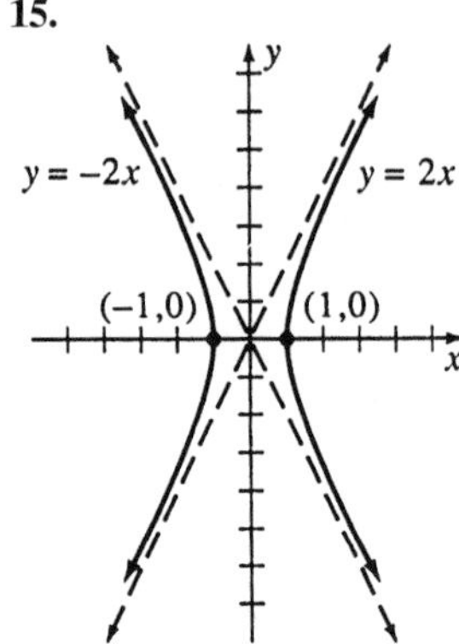

17.

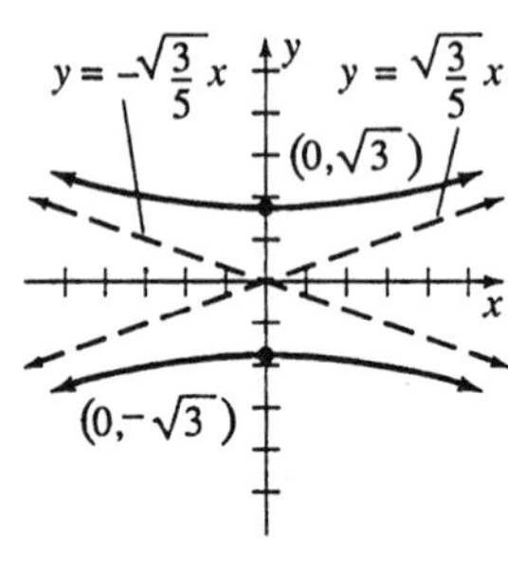

19. $y = -x - 5$ and $y = x - 1$

21. $y = \frac{3}{2}x - \frac{9}{2}$ and $y = -\frac{3}{2}x - \frac{3}{2}$

23.

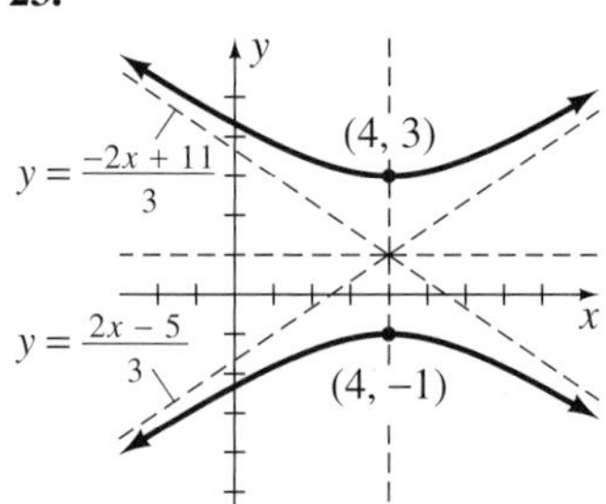

25.

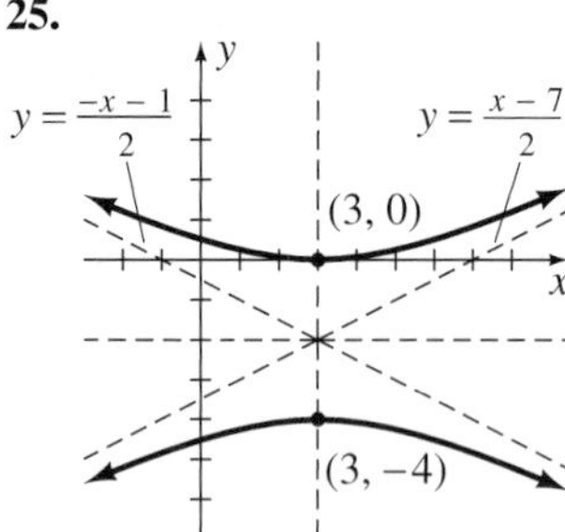

27.

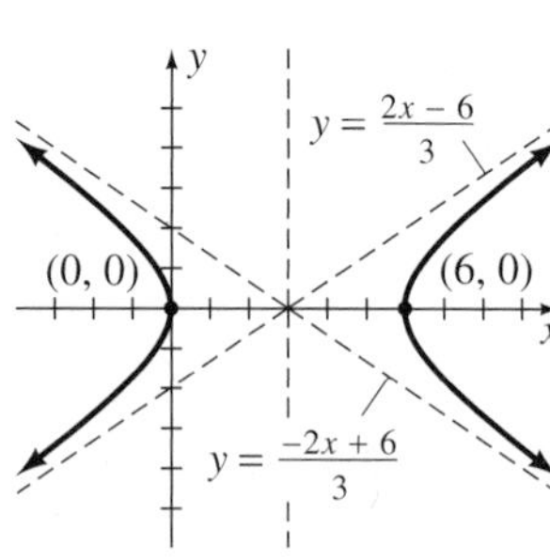

29.

(a)

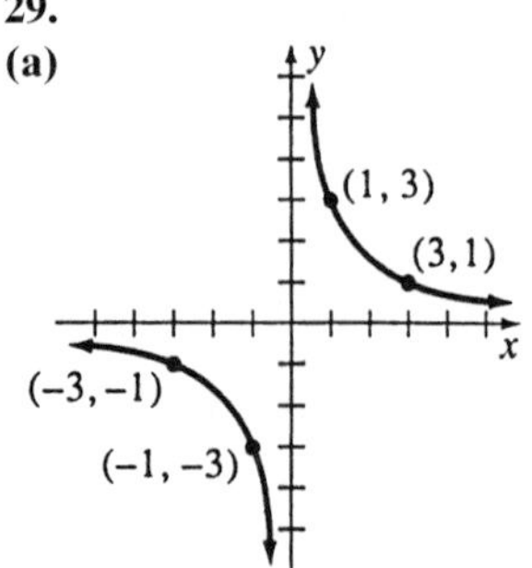

(c)

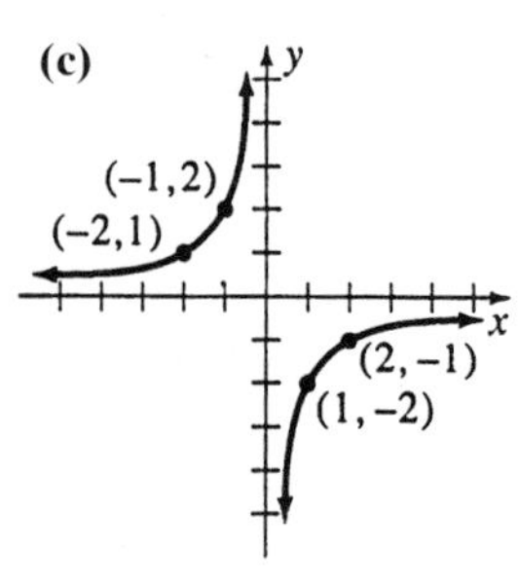

31. **(a)** Origin **(c)**

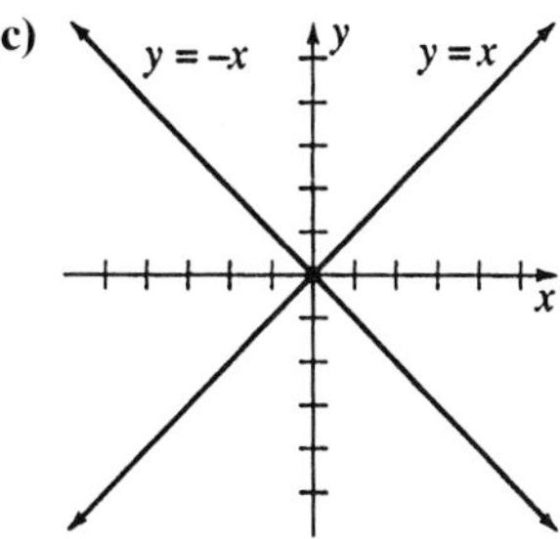

Problem Set 9.6 (page 524)

1. $\{(-2, 0), (-4, 4)\}$ **3.** $\{(2, -3), (-2, 3)\}$
5. $\{(-6, 7), (-2, -1)\}$ **7.** $\{(1, 1)\}$ **9.** $\{(-5, -3)\}$
11. $\{(-3, 2)\}$ **13.** $\{(2, 2), (-2, -2)\}$
15. $\{(2 + i\sqrt{3}, -2 + i\sqrt{3}), (2 - i\sqrt{3}, -2 - i\sqrt{3})\}$
17. $\{(1, 2)\}$ **19.** $\{(1, -5), (-5, 1)\}$ **21.** $\{(-3, 4)\}$
23. $\left\{\left(\frac{-1 + i\sqrt{3}}{2}, \frac{-7 - i\sqrt{3}}{2}\right), \left(\frac{-1 - i\sqrt{3}}{2}, \frac{-7 + i\sqrt{3}}{2}\right)\right\}$
25. $\{(2, 0), (-2, 0)\}$
27. $\{(\sqrt{5}, 1), (\sqrt{5}, -1), (-\sqrt{5}, 1), (-\sqrt{5}, -1)\}$
29. $\{(\sqrt{2}, \sqrt{3}), (\sqrt{2}, -\sqrt{3}), (-\sqrt{2}, \sqrt{3}), (-\sqrt{2}, -\sqrt{3})\}$

Chapter 9 Review Problem Set (page 531)

1.

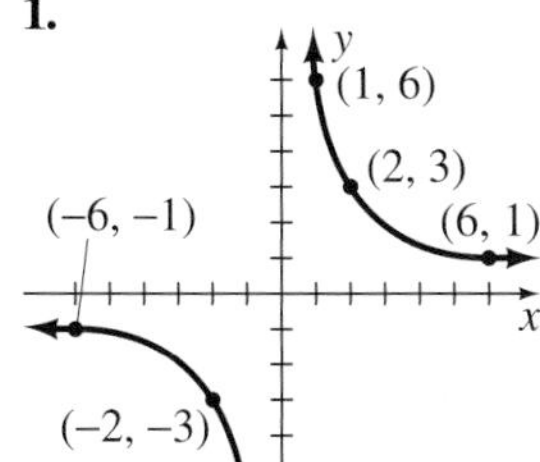

2.

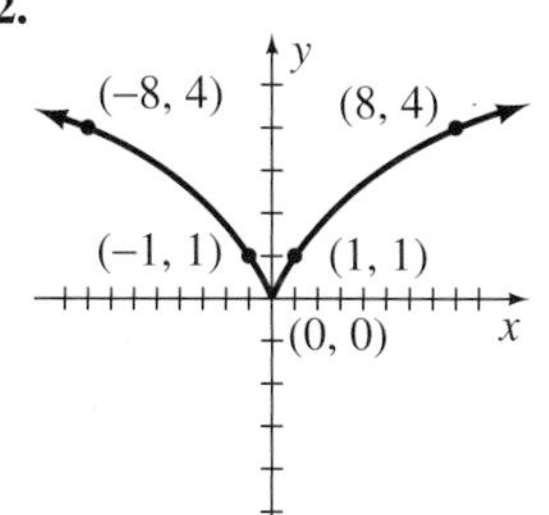

3.

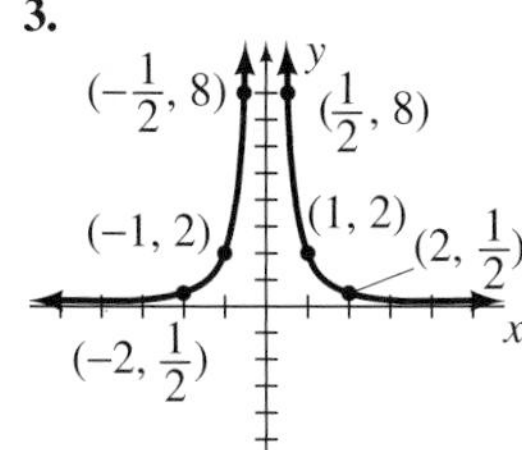

4.

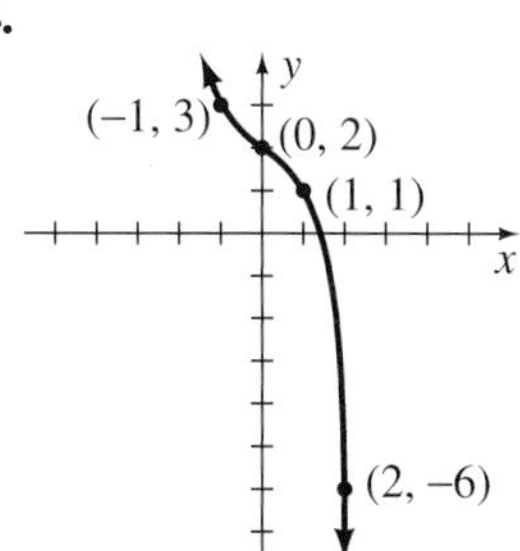

5. $(0, 6)$ **6.** $(0, -8)$ **7.** $(-3, -1)$ **8.** $(-1, 0)$
9. $(7, 5)$ **10.** $(6, -8)$ **11.** $(-4, -9)$ **12.** $(-2, -3)$
13. $x^2 + y^2 - 36 = 0$ **14.** $x^2 + y^2 - 4x + 12y + 15 = 0$
15. $x^2 + y^2 + 8x + 16y + 68 = 0$ **16.** $x^2 + y^2 - 10y = 0$

17.

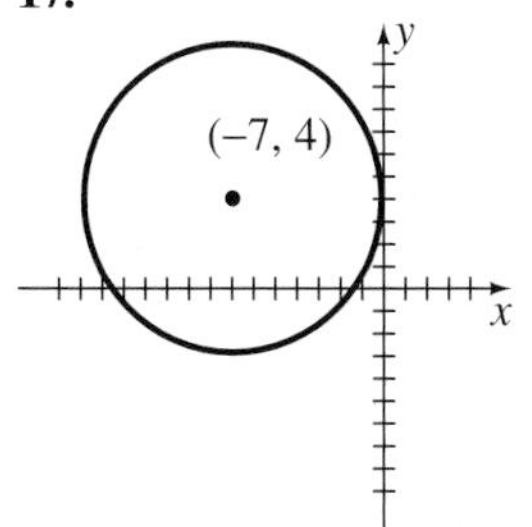

18.

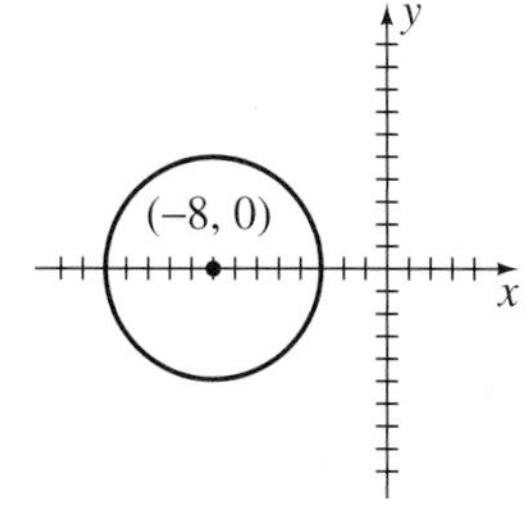

19.

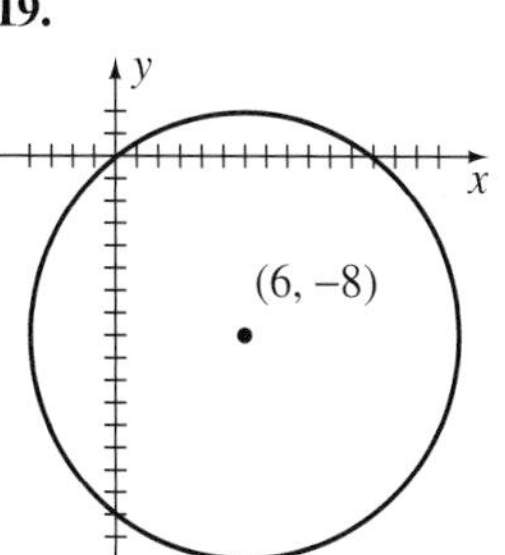

20.

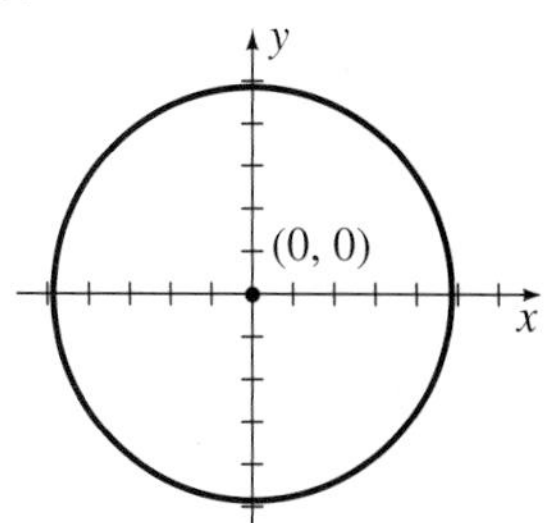

21.

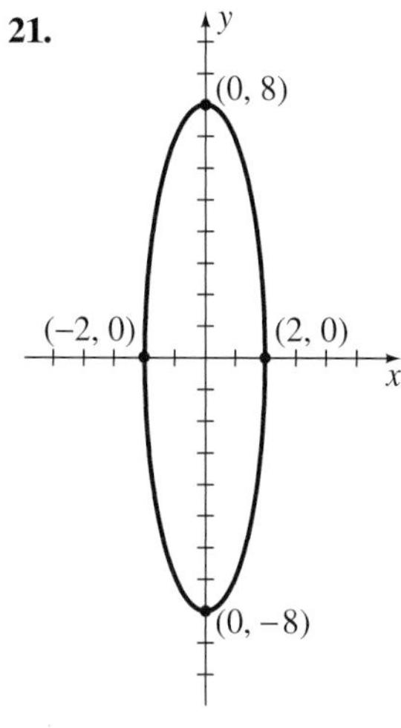

22

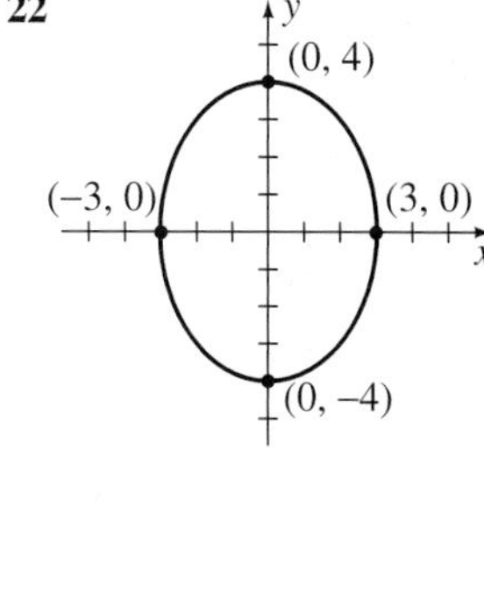

23.

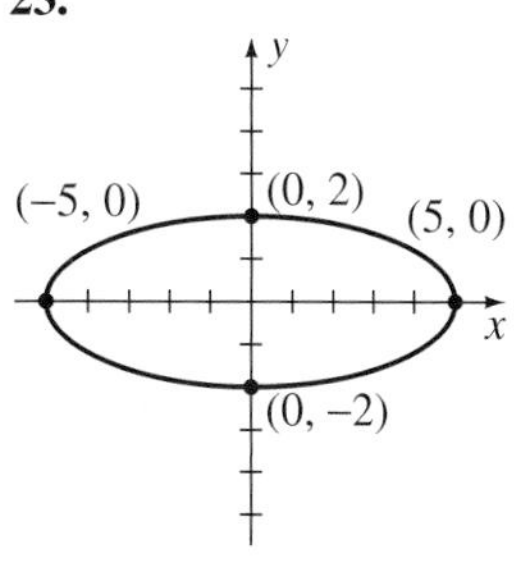

24.

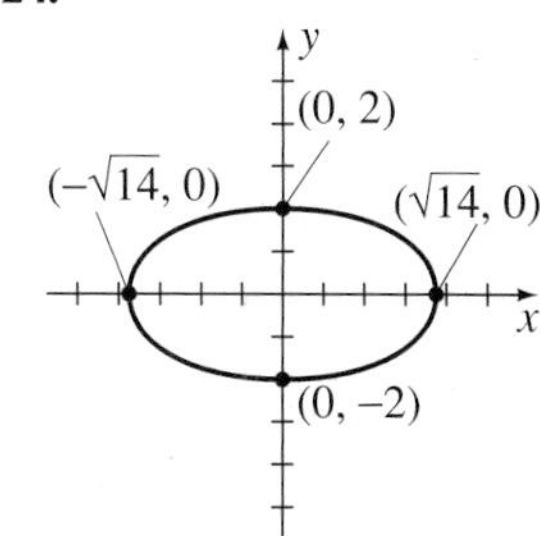

25.

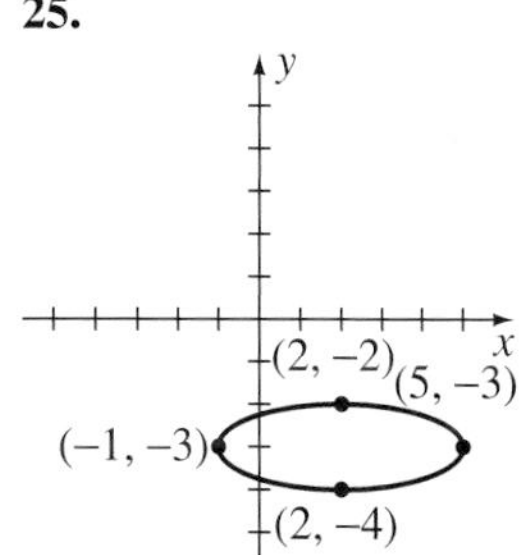

26.

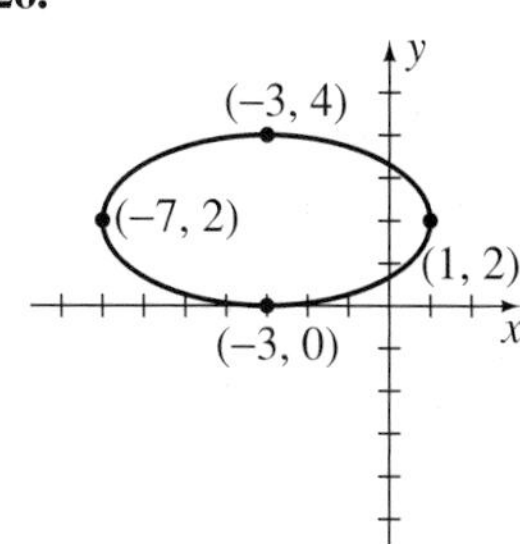

27.

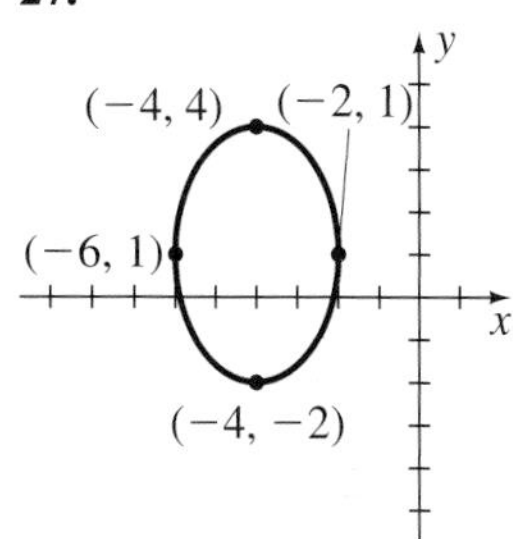

28.

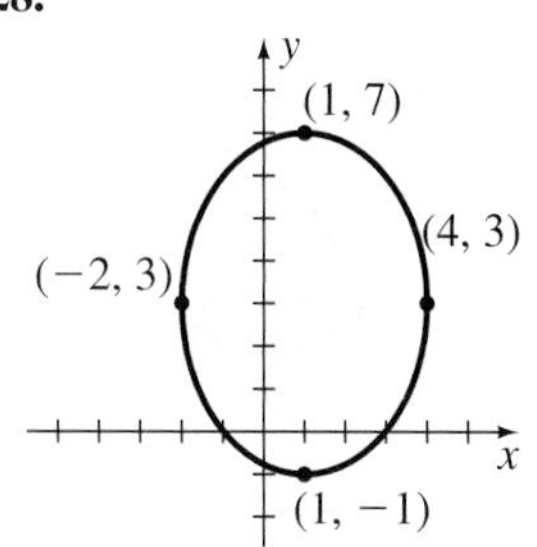

29.

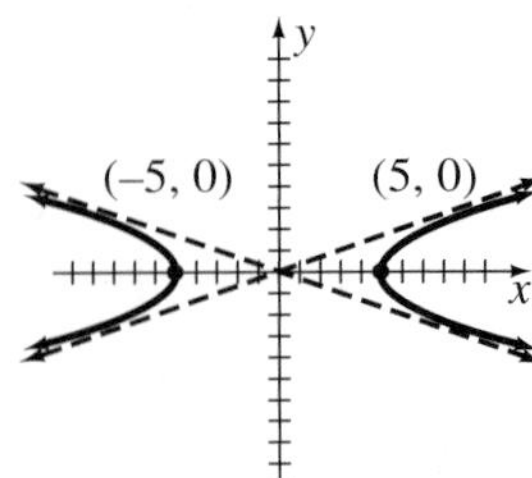

30.

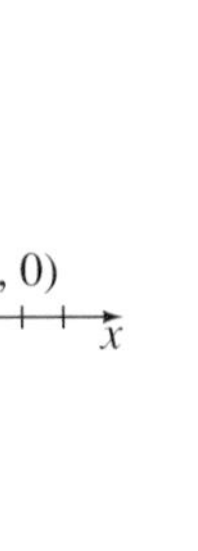

39.

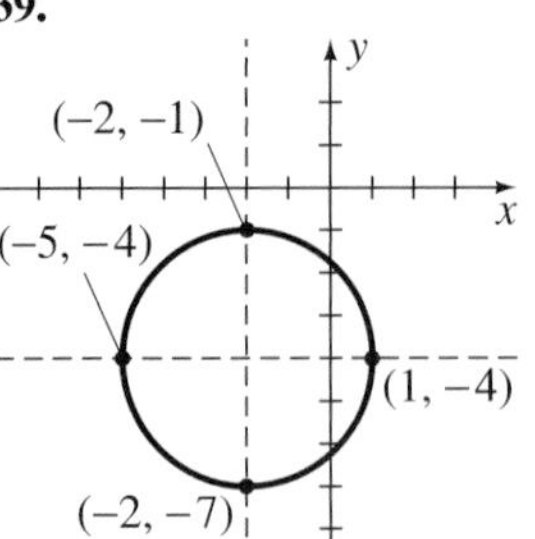

40.

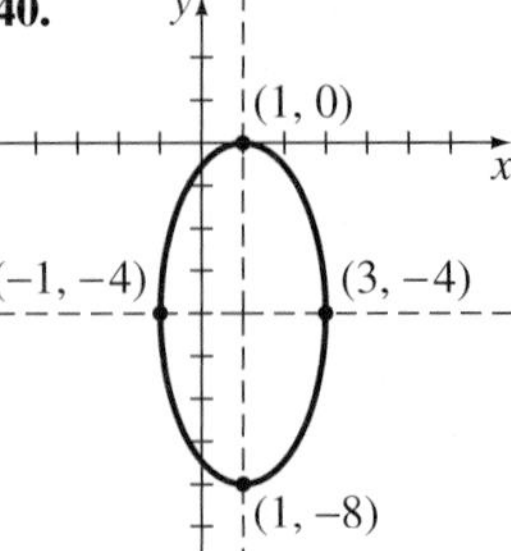

31.

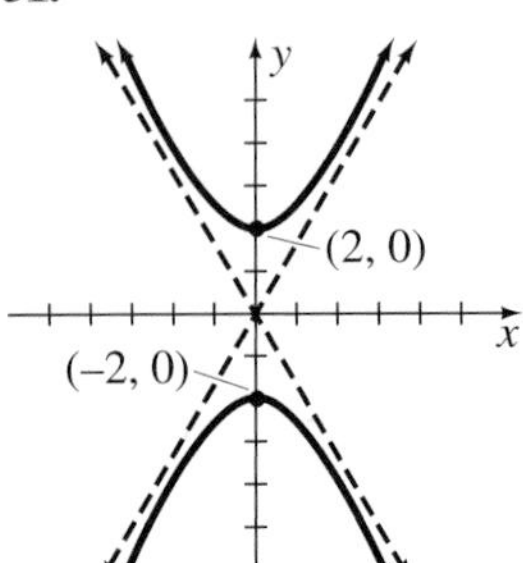

32.

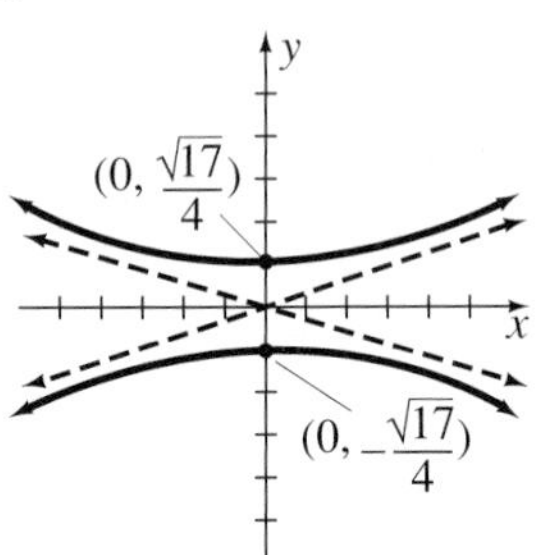

41.

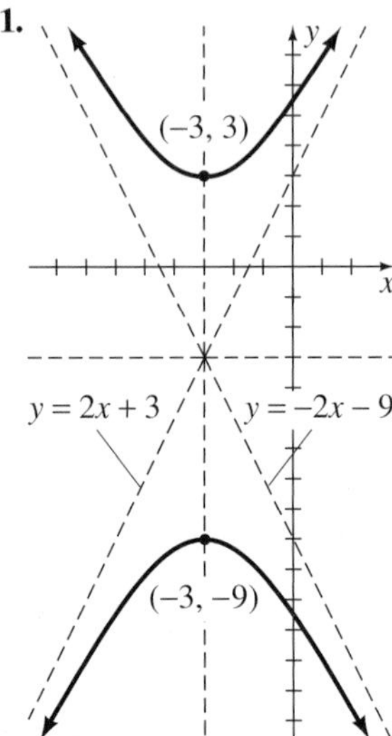

42.

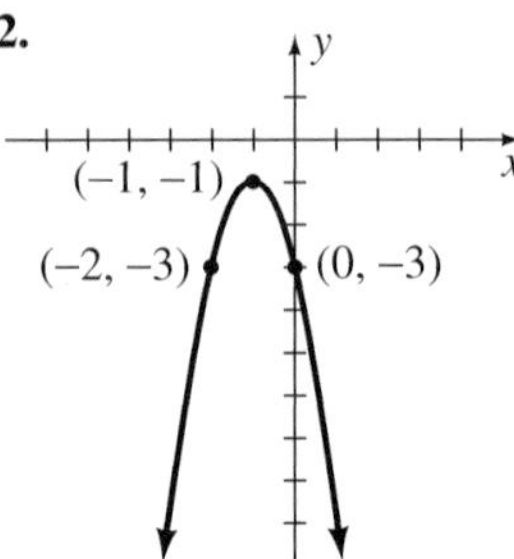

33.

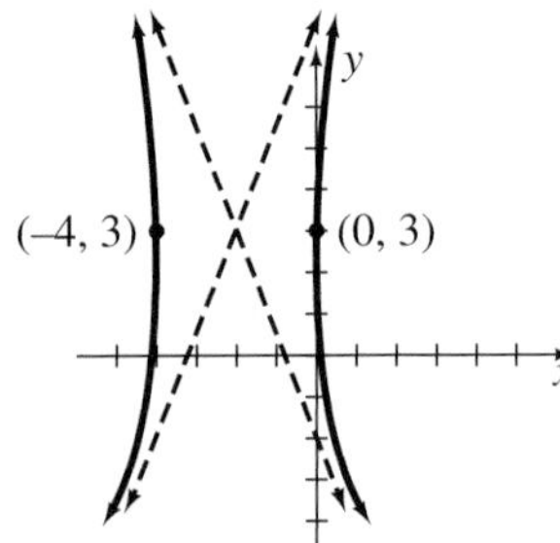

34.

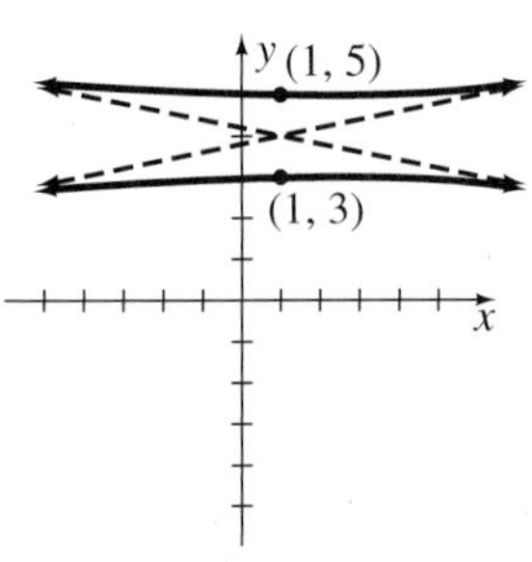

43.

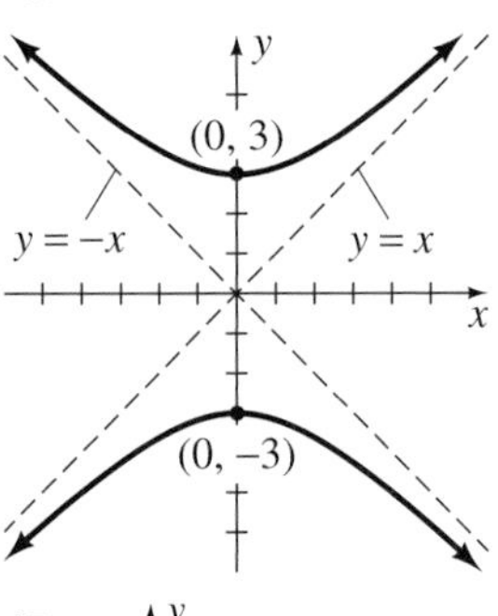

44.

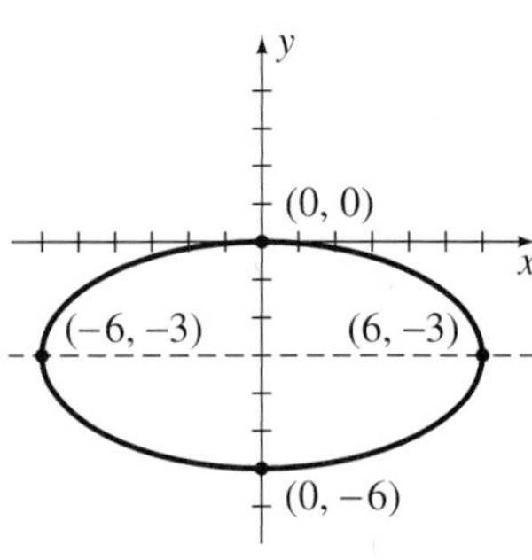

35.

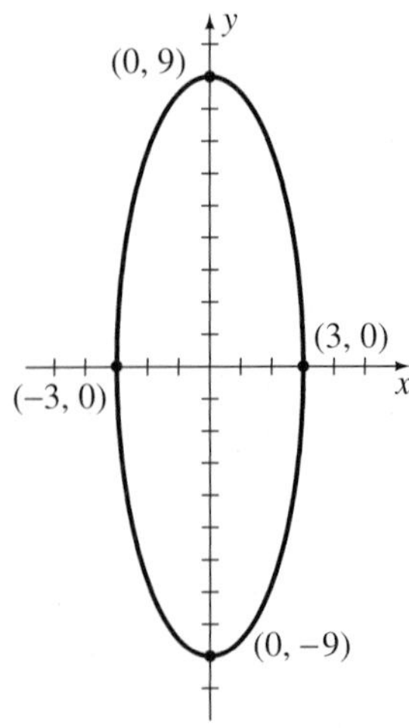

36.

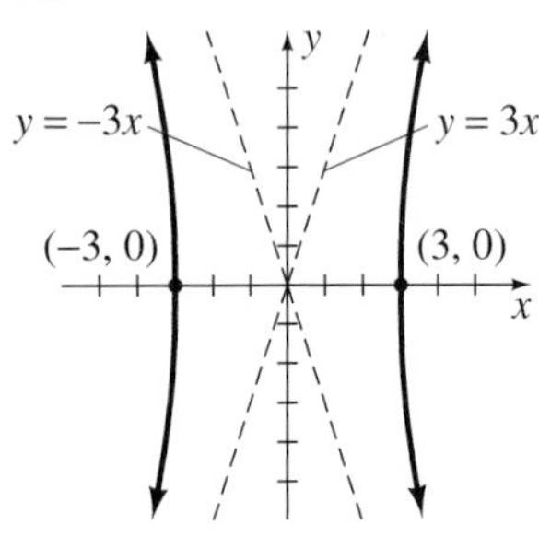

45.

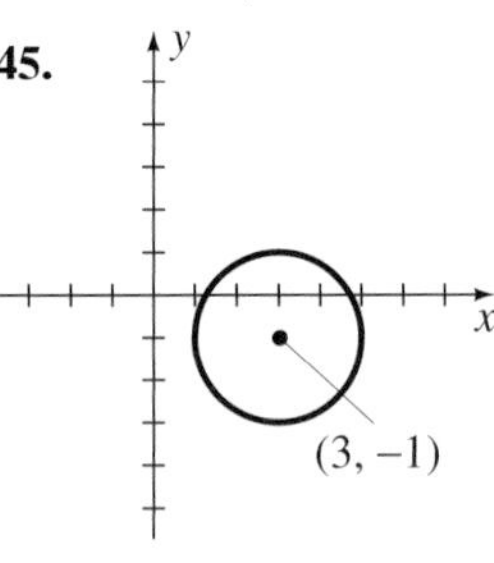

46.

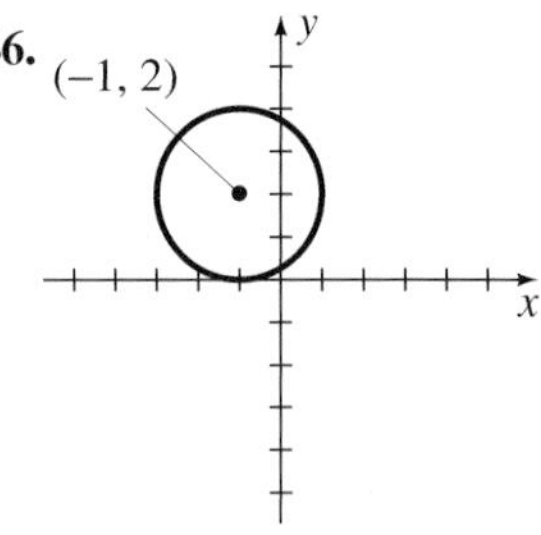

37.

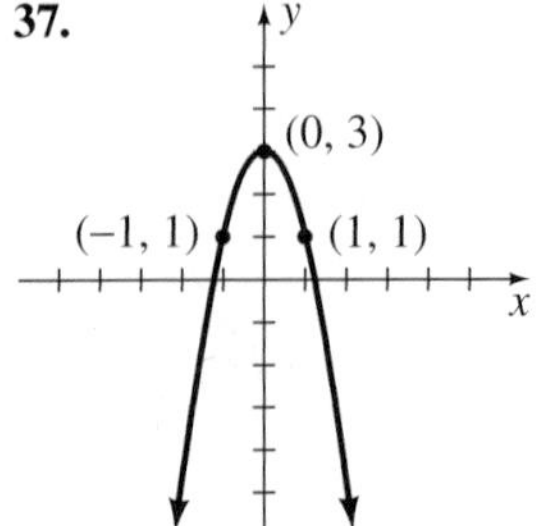

38.

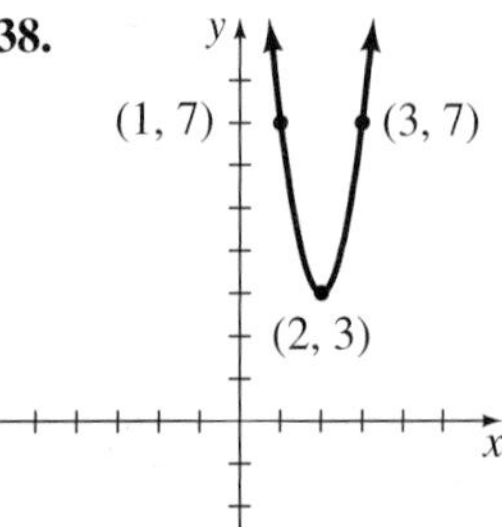

47.

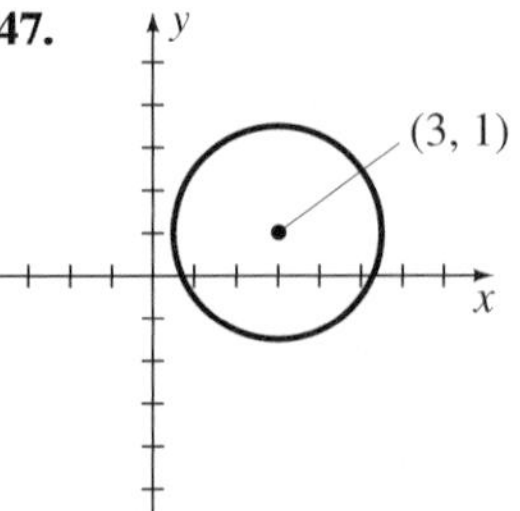

48.

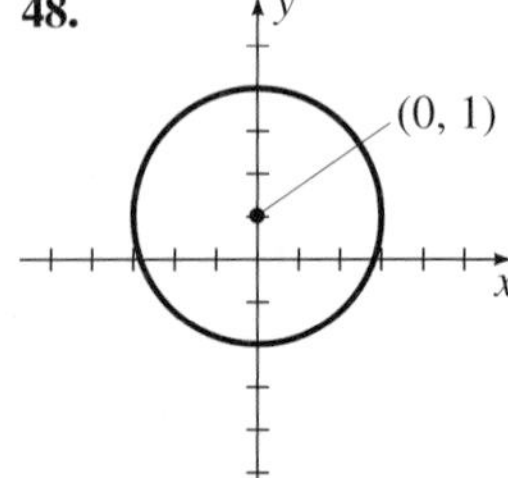

49. $\{(-1, 3), (5, 27)\}$

50. $\{(\sqrt{6}, \sqrt{10}), (\sqrt{6}, -\sqrt{10}), (-\sqrt{6}, \sqrt{10}), (-\sqrt{6}, -\sqrt{10})\}$

Chapter 9 Test (page 532)

1. (0, 9) **2.** (1, 7) **3.** (−4, −2)
4. (3, 0) **5.** $x^2 + 8x + y^2 - 29 = 0$
6. $x^2 + y^2 - 4x - 16y + 59 = 0$
7. $x^2 + y^2 + 6x + 8y = 0$
8. (0, 0) and $r = 4\sqrt{2}$ **9.** (6, −4) and $r = 7$
10. (−5, −1) and $r = 8$ **11.** 8 units **12.** 6 units
13. 6 units **14.** 4 units **15.** $y = \pm 4x$ **16.** $y = \pm\frac{5}{4}x$
17. $y = \frac{1}{5}x - \frac{6}{5}$ and $y = -\frac{1}{5}x - \frac{4}{5}$

18.

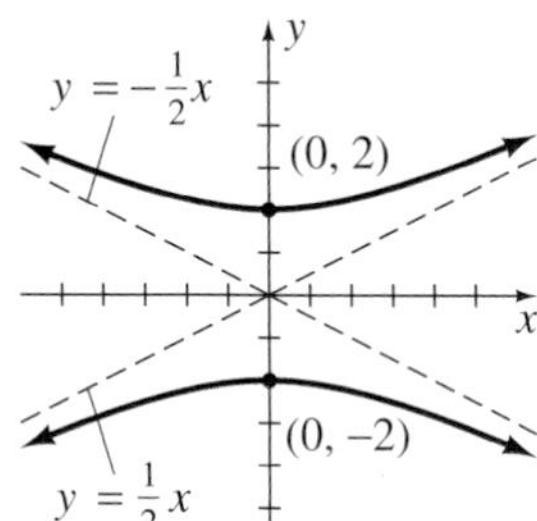

19.

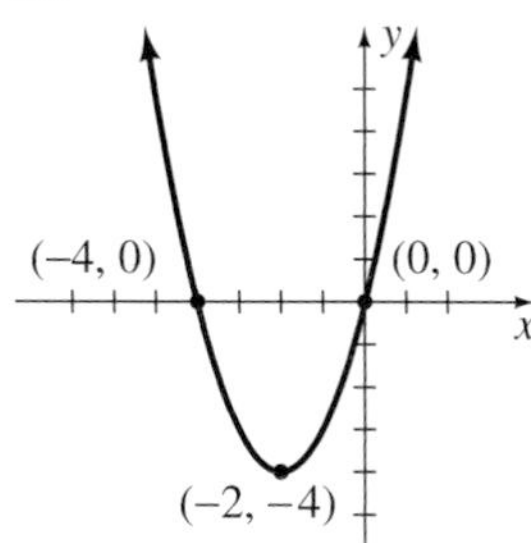

20.

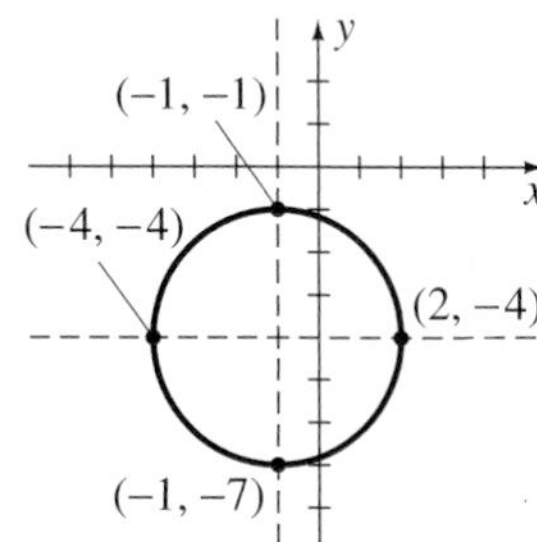

21.

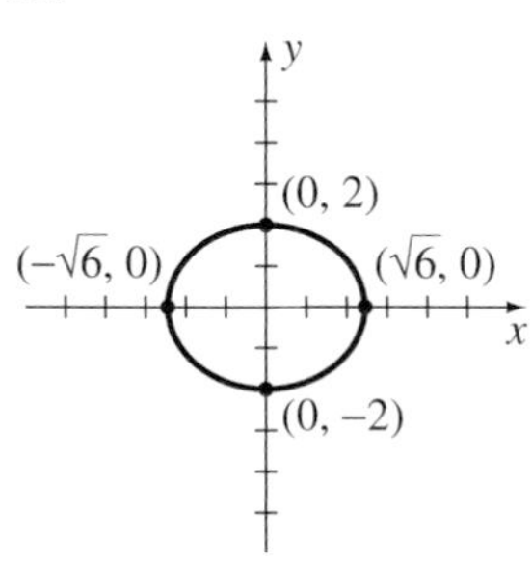

22.

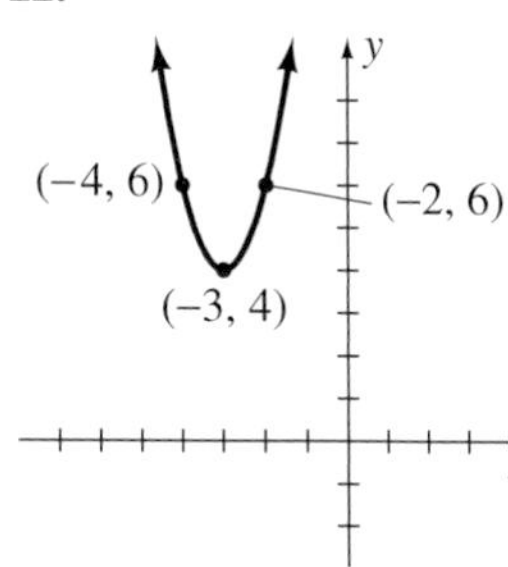

23.

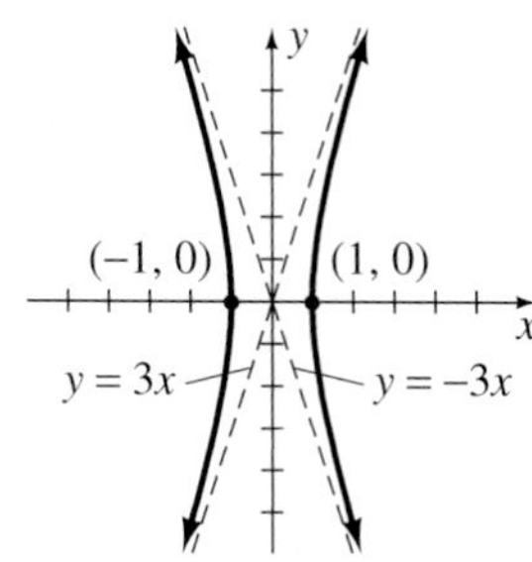

24.

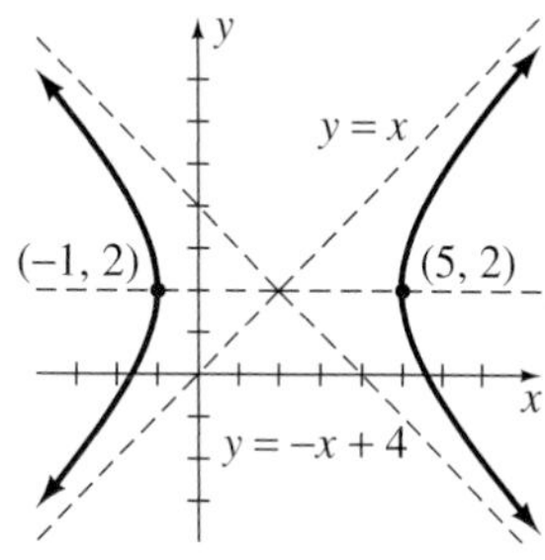

25. {(2, −3)}

CHAPTER 10

Problem Set 10.1 (page 540)

1. $D = \{1, 2, 3, 4\}$, $R = \{5, 8, 11, 14\}$; it is a function.
3. $D = \{0, 1\}$, $R = \{-2\sqrt{6}, -5, 5, 2\sqrt{6}\}$; it is not a function.
5. $D = \{1, 2, 3, 4, 5\}$, $R = \{2, 5, 10, 17, 26\}$; it is a function.
7. $D = \{$all reals$\}$ or D: $(-\infty, \infty)$, $R = \{$all reals$\}$ or R: $(-\infty, \infty)$; yes
9. $D = \{$all reals$\}$ or D: $(-\infty, \infty)$, $R = \{y \mid y \geq 0\}$ or R: $[0, \infty)$; yes
11. $f(0) = -2, f(2) = 8, f(-1) = -7, f(-4) = -22$
13. $f(-2) = -\frac{7}{4}, f(0) = -\frac{3}{4}, f\left(\frac{1}{2}\right) = -\frac{1}{2}, f\left(\frac{2}{3}\right) = -\frac{5}{12}$
15. $g(-1) = 0; g(2) = -9; g(-3) = 26; g(4) = 5$
17. $h(-2) = -2; h(-3) = -11; h(4) = -32; h(5) = -51$
19. $f(3) = \sqrt{7}; f(4) = 3; f(10) = \sqrt{21}; f(12) = 5$
21. $f(1) = -1; f(-1) = -2; f(3) = -\frac{2}{3}; f(-6) = \frac{4}{3}$
23. $f(a) = a^2 - 7a, f(a - 3) = a^2 - 13a + 30, f(a + h) = a^2 + 2ah + h^2 - 7a - 7h$
25. $f(-a) = 2a^2 + a - 1, f(a + 1) = 2a^2 + 3a, f(a + h) = 2a^2 + 4ah + 2h^2 - a - h - 1$
27. $f(-a) = -a^2 + 2a - 7, f(-a - 2) = -a^2 - 2a - 7, f(a + 7) = -a^2 - 16a - 70$
29. $f(1) = 1; f(-1) = 5, g(2) = 4, g(-3) = 5$
31. {all reals} or $(-\infty, \infty)$
33. $\{x \mid x \neq 1\}$ or $(-\infty, 1) \cup (1, \infty)$
35. $\left\{x \mid x \neq \frac{3}{4}\right\}$ or $\left(-\infty, \frac{3}{4}\right) \cup \left(\frac{3}{4}, \infty\right)$
37. $\{x \mid x \neq -1, x \neq 4\}$ or $(-\infty, -1) \cup (-1, 4) \cup (4, \infty)$
39. $\{x \mid x \neq -8, x \neq 5\}$ or $(-\infty, -8) \cup (-8, 5) \cup (5, \infty)$
41. $\{x \mid x \neq -6, x \neq 0\}$ or $(-\infty, -6) \cup (-6, 0) \cup (0, \infty)$
43. {all reals} or $(-\infty, \infty)$
45. $\{t \mid t \neq -2, t \neq 2\}$ or $(-\infty, -2) \cup (-2, 2) \cup (2, \infty)$
47. $\{x \mid x \geq -4\}$ or $[-4, \infty)$ **49.** $\left\{s \mid s \geq \frac{5}{4}\right\}$ or $\left[\frac{5}{4}, \infty\right)$
51. $\{x \mid x \leq -4$ or $x \geq 4\}$ or $(-\infty, -4] \cup [4, \infty)$
53. $\{x \mid x \leq -3$ or $x \geq 6\}$ or $(-\infty, -3] \cup [6, \infty)$
55. $\{x \mid -1 \leq x \leq 1\}$ or $[-1, 1]$
57. −3 **59.** $-2a - h$ **61.** $4a - 1 + 2h$
63. $-8a - 7 - 4h$
65. $h(1) = 48; h(2) = 64; h(3) = 48; h(4) = 0$
67. $C(75) = \$74; C(150) = \$98; C(225) = \$122; C(650) = \$258$
69. $I(0.11) = 55; I(0.12) = 60; I(0.135) = 67.5; I(0.15) = 75$

Problem Set 10.2 (page 553)

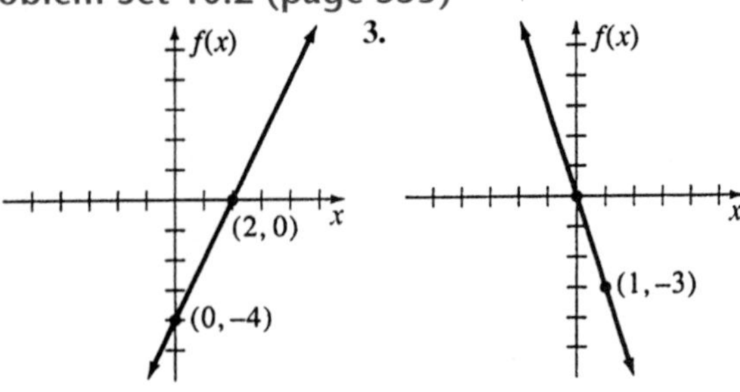

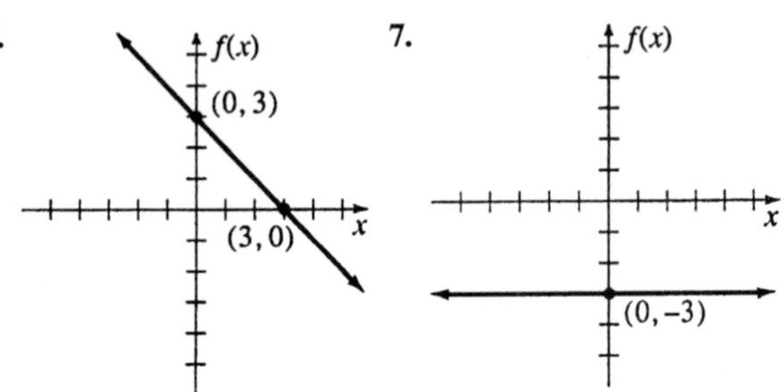

9. (a) \$.42 **(c)** Answers will vary.
11. \$100.00; \$107.50; \$140.00; \$157.50
13. \$2.10; \$4.55; \$20.72; \$29.40; \$33.88
15. $f(p) = 0.8p$; \$7.60; \$12; \$60; \$10; \$600

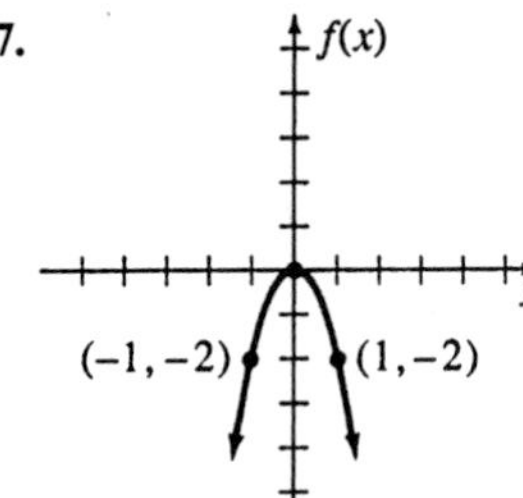

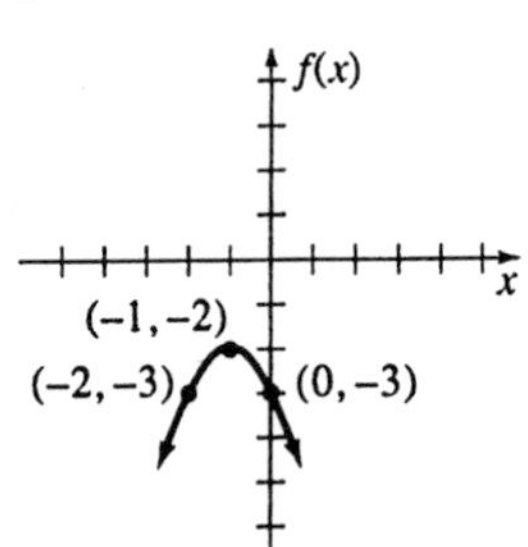

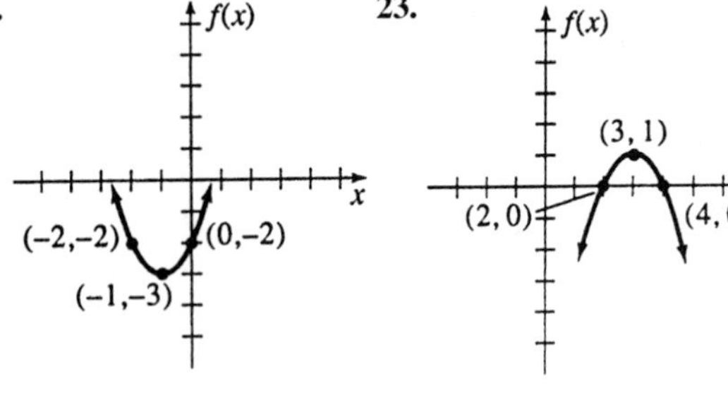

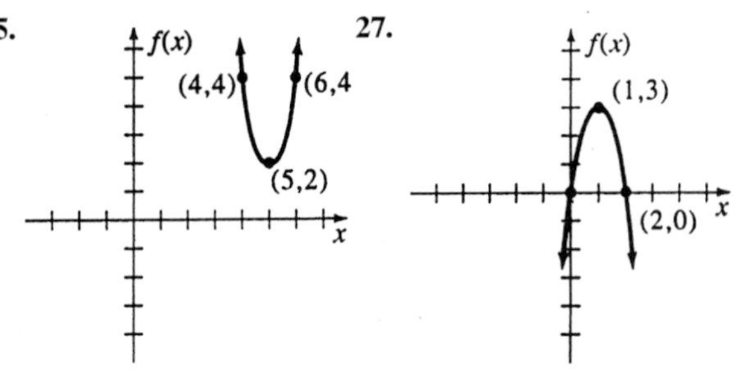

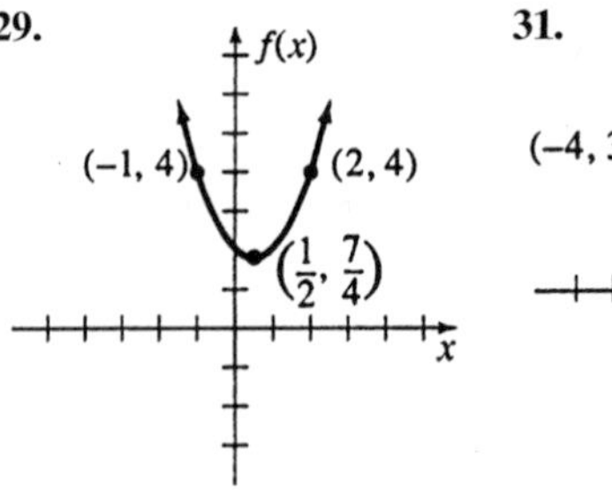

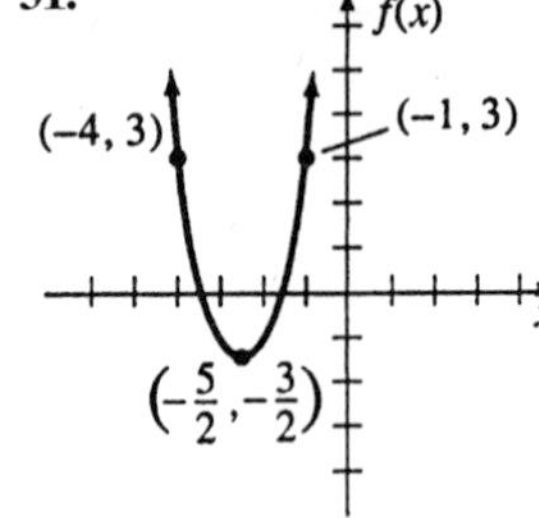

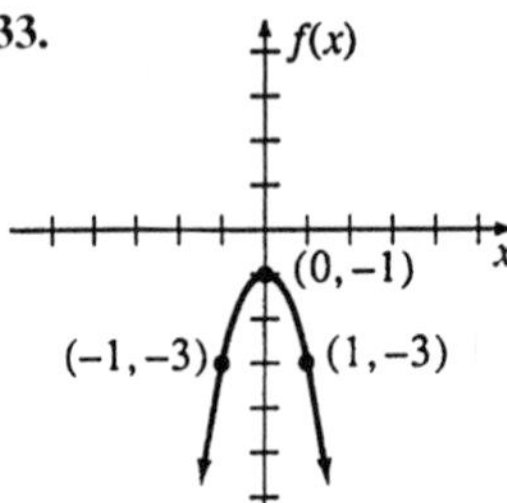

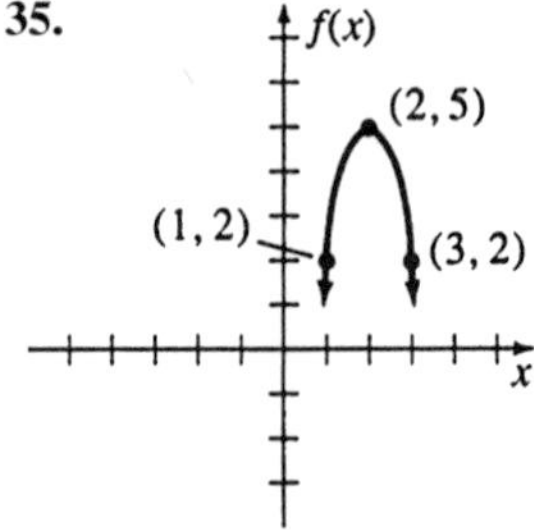

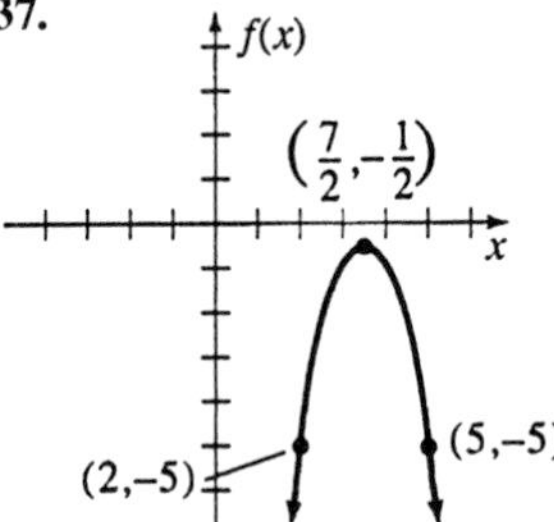

39. 80 items **41.** 5 and 25 **43.** 60 m by 60 m
45. 1100 subscribers at \$13.75 per month

Problem Set 10.3 (page 564)

1. B **3.** A **5.** F

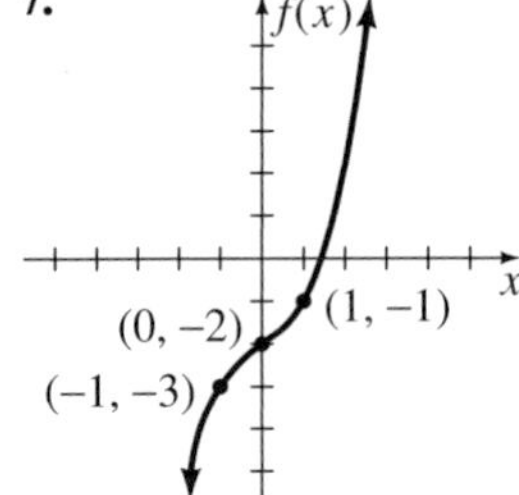

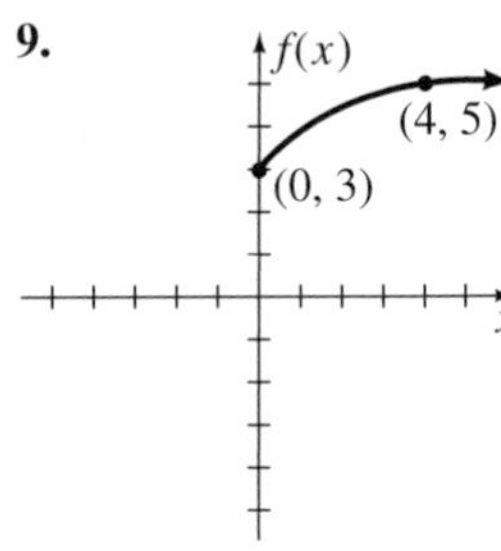

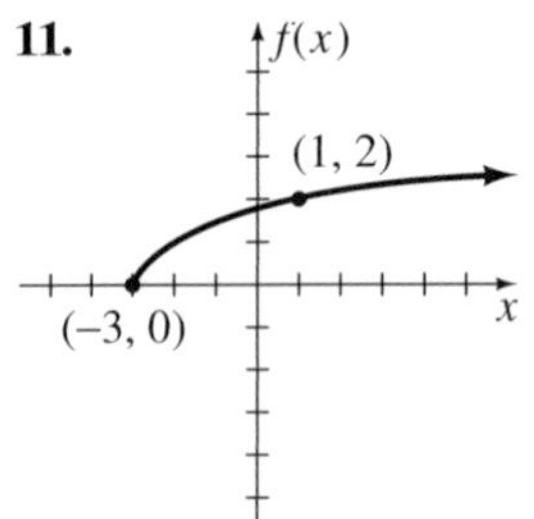

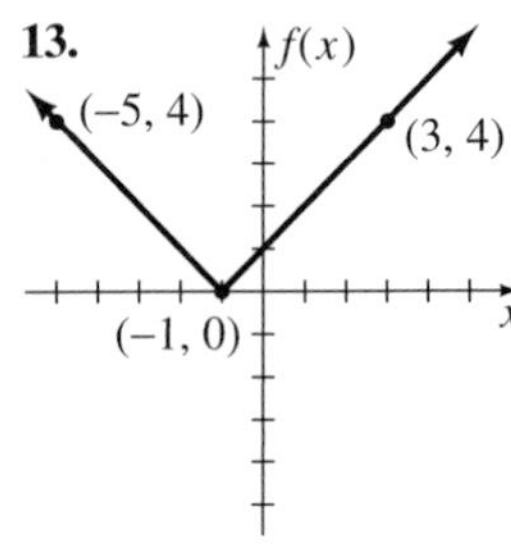

15. (−2, 8) (0, 0) (1, −1) (2, −8)

17. (−4, 2) (0, 0)

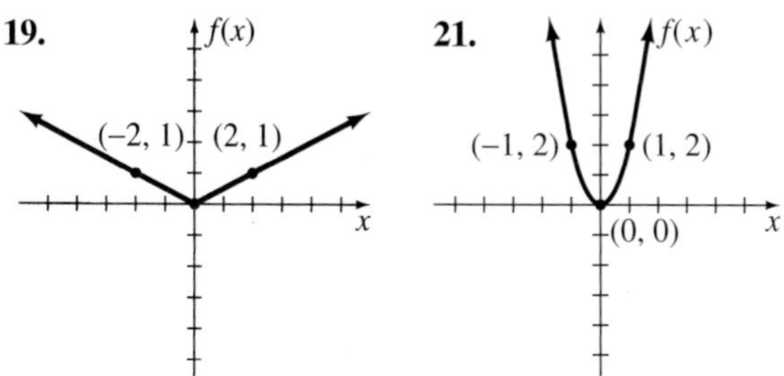

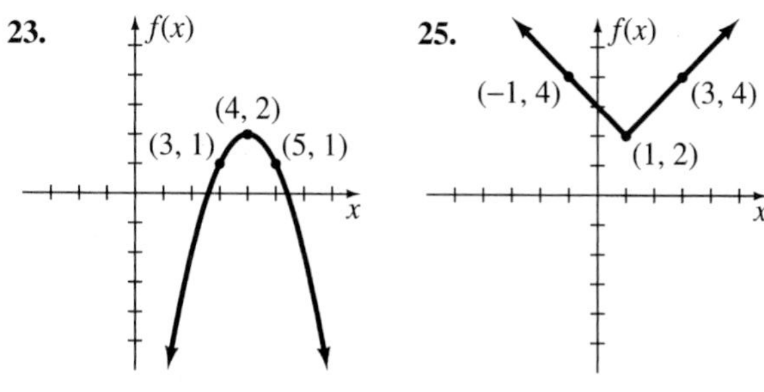

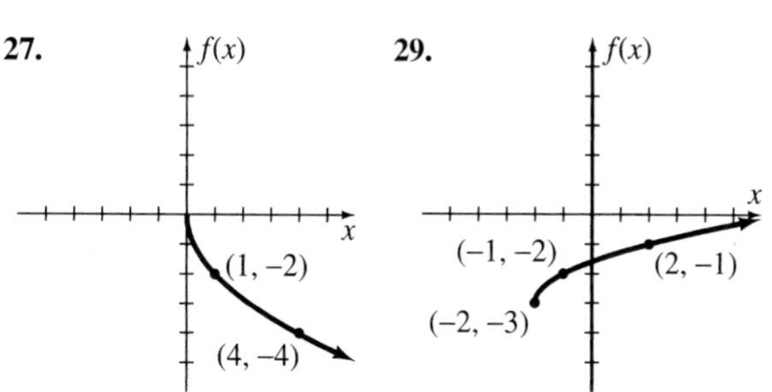

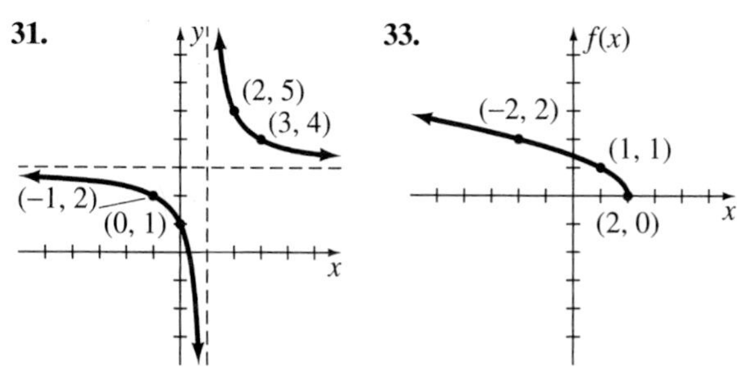

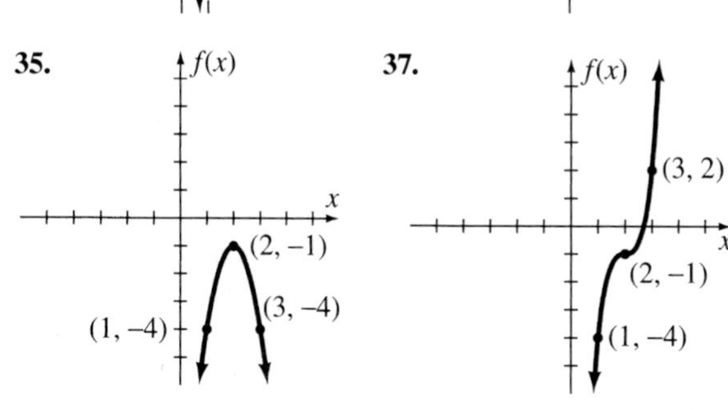

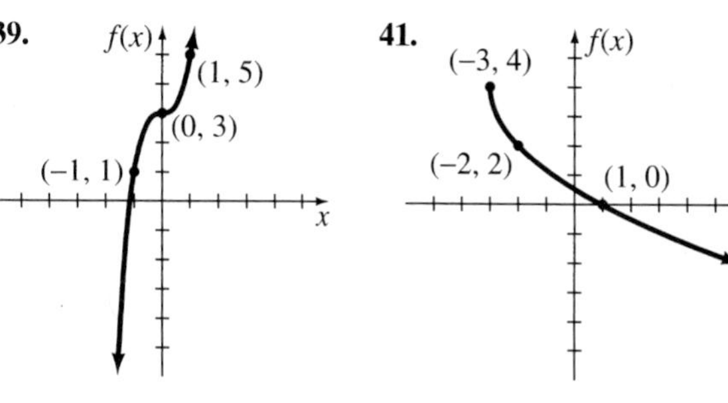

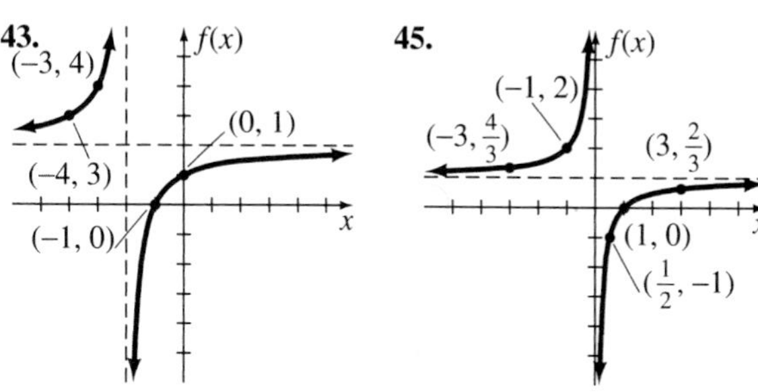

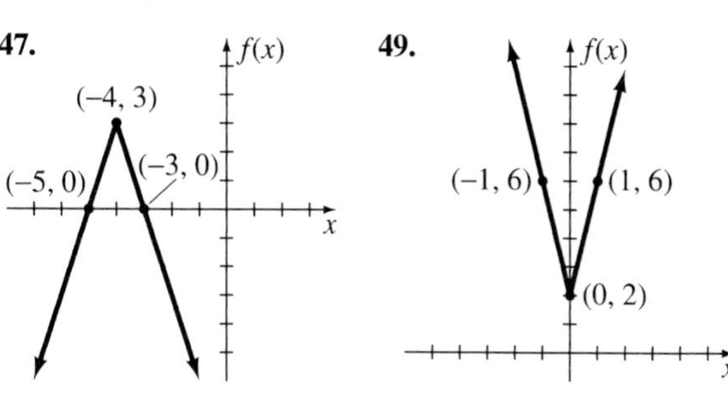

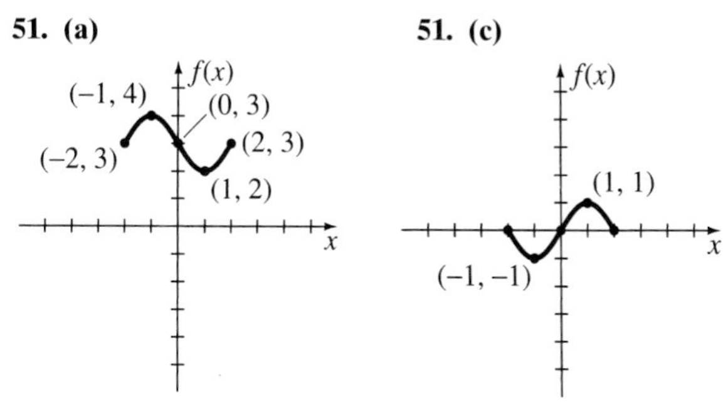

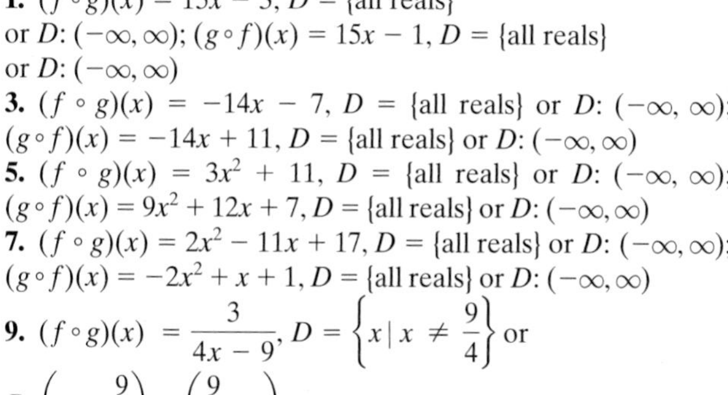

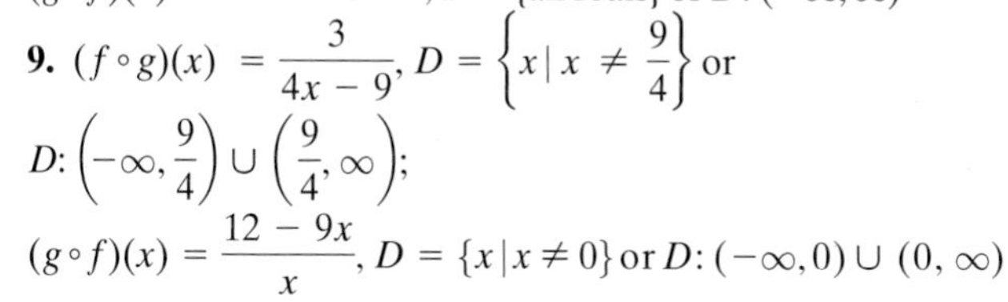

Problem Set 10.4 (page 571)

1. $(f \circ g)(x) = 15x - 3$, $D = \{\text{all reals}\}$ or $D: (-\infty, \infty)$; $(g \circ f)(x) = 15x - 1$, $D = \{\text{all reals}\}$ or $D: (-\infty, \infty)$

3. $(f \circ g)(x) = -14x - 7$, $D = \{\text{all reals}\}$ or $D: (-\infty, \infty)$; $(g \circ f)(x) = -14x + 11$, $D = \{\text{all reals}\}$ or $D: (-\infty, \infty)$

5. $(f \circ g)(x) = 3x^2 + 11$, $D = \{\text{all reals}\}$ or $D: (-\infty, \infty)$; $(g \circ f)(x) = 9x^2 + 12x + 7$, $D = \{\text{all reals}\}$ or $D: (-\infty, \infty)$

7. $(f \circ g)(x) = 2x^2 - 11x + 17$, $D = \{\text{all reals}\}$ or $D: (-\infty, \infty)$; $(g \circ f)(x) = -2x^2 + x + 1$, $D = \{\text{all reals}\}$ or $D: (-\infty, \infty)$

9. $(f \circ g)(x) = \dfrac{3}{4x - 9}$, $D = \left\{x \mid x \neq \dfrac{9}{4}\right\}$ or $D: \left(-\infty, \dfrac{9}{4}\right) \cup \left(\dfrac{9}{4}, \infty\right)$; $(g \circ f)(x) = \dfrac{12 - 9x}{x}$, $D = \{x \mid x \neq 0\}$ or $D: (-\infty, 0) \cup (0, \infty)$

11. $(f \circ g)(x) = \sqrt{5x + 4}, D = \left\{x \mid x \geq -\frac{4}{5}\right\}$ or $D: \left[-\frac{4}{5}, \infty\right)$;
$(g \circ f)(x) = 5\sqrt{x + 1} + 3, D = \{x \mid x \geq -1\}$ or $D: [-1, \infty)$
13. $(f \circ g)(x) = x - 4, D = \{x \mid x \neq 4\}$ or $D: (-\infty, 4) \cup (4, \infty)$;
$(g \circ f)(x) = \frac{x}{1 - 4x}, D = \left\{x \mid x \neq 0 \text{ and } x \neq \frac{1}{4}\right\}$ or
$D: (-\infty, 0) \cup \left(0, \frac{1}{4}\right) \cup \left(\frac{1}{4}, \infty\right)$
15. $(f \circ g)(x) = \frac{2\sqrt{x}}{x}, D = \{x \mid x > 0\}$ or $D: (0, \infty)$;
$(g \circ f)(x) = \frac{4\sqrt{x}}{x}, D = \{x \mid x > 0\}$ or $D: (0, \infty)$
17. $(f \circ g)(x) = \frac{3x + 3}{2}, D = \{x \mid x \neq -1\}$ or $D: (-\infty, -1) \cup (-1, \infty)$;
$(g \circ f)(x) = \frac{2x}{2x + 3}, D = \left\{x \mid x \neq 0 \text{ and } x \neq -\frac{3}{2}\right\}$ or
$D: \left(-\infty, -\frac{3}{2}\right) \cup \left(-\frac{3}{2}, 0\right) \cup (0, \infty)$
27. 124 and −130 **29.** 323 and 257 **31.** $\frac{1}{2}$ and −1
33. Undefined and undefined **35.** $\sqrt{7}$ and 0
37. 37 and 27

Problem Set 10.5 (page 579)

1. Not a function **3.** Function **5.** Function
7. Function **9.** One-to-one function
11. Not a one-to-one function
13. Not a one-to-one function **15.** One-to-one function
17. Domain of f: {1, 2, 3, 4}
Range of f: {3, 6, 11, 18}
f^{-1}: {(3, 1), (6, 2), (11, 3), (18, 4)}
Domain of f^{-1}: {3, 6, 11, 18}
Range of f^{-1}: {1, 2, 3, 4}
19. Domain of f: {−2, −1, 0, 5}
Range of f: {−1, 1, 5, 10}
f^{-1}: {(−1, −2), (1, −1), (5, 0), (10, 5)}
Domain of f^{-1}: {−1, 1, 5, 10}
Range of f^{-1}: {−2, −1, 0, 5}
21. $f^{-1}(x) = \frac{x + 4}{5}$ **23.** $f^{-1}(x) = \frac{1 - x}{2}$
25. $f^{-1}(x) = \frac{5}{4}x$ **27.** $f^{-1}(x) = 2x - 8$
29. $f^{-1}(x) = \frac{15x + 6}{5}$ **31.** $f^{-1}(x) = \frac{x - 4}{9}$
33. $f^{-1}(x) = \frac{-x - 4}{5}$ **35.** $f^{-1}(x) = \frac{-3x + 21}{2}$
37. $f^{-1}(x) = \frac{3}{4}x + \frac{3}{16}$ **39.** $f^{-1}(x) = -\frac{7}{3}x - \frac{14}{9}$
41. $f^{-1}(x) = \frac{1}{4}x$ **43.** $f^{-1}(x) = -3x$
45. $f^{-1}(x) = \frac{x + 3}{3}$ **47.** $f^{-1}(x) = \frac{-x - 4}{2}$
49. $f^{-1}(x) = \sqrt{x}, x \geq 0$ **53. (a)** $f^{-1}(x) = \frac{x - 1}{2}$
(c) $f^{-1}(x) = \frac{-x + 5}{4}$ **(e)** $f^{-1}(x) = \frac{1}{2}x$

Problem Set 10.6 (page 587)

1. $T = kr$ **3.** $A = kr^2$ **5.** $y = kx^3$ **7.** $\frac{2}{3}$ **9.** −4 **11.** 9
13. \$520 **15.** 2 sec **17.** $y = \frac{k}{x^2}$ **19.** $V = \frac{k}{P}$ **21.** −2
23. 2 **25.** $\frac{1}{6}$ **27.** 0.048 **29.** 5 hr **31.** 12 cm^3
33. $C = \frac{kg}{t^3}$ **35.** $\frac{1}{3}$ **37.** 5 **39.** 9 **41.** 112 **43.** 28
45. 12 ohms **47. (a)** \$210 **(b)** \$637 **(c)** \$1050

Chapter 10 Review Problem Set (page 600)

1. Yes; $D = \{-3, -1, 7, 9\}, R = \{-3, -1, 7, 9\}$
2. No; $D = \{1, 2\}, R = \{1, 3, 5, 7\}$
3. No; $D = \{0\}, R = \{-6, -5, -4, -3\}$
4. Yes; $D = \{0, 1, 2, 3\}, R = \{8\}$
5. $f(2) = -1; f(-3) = 14; f(a) = a^2 - 2a - 1$
6. $f(2) = 0; f(-1) = -6; f(3a) = \frac{6a - 4}{3a + 2}$
7. $D = \{\text{all reals}\}$
8. $D = \{x \mid x \neq 5\}$ or $D = (-\infty, 5) \cup (5, \infty)$
9. $D = \{x \mid x \neq 0, x \neq 4\}$ or $D = (-\infty, -4) \cup (-4, 0) \cup (0, \infty)$
10. $D = \{x \mid x \leq -5 \text{ or } x \geq 5\}$ or $D = (-\infty, -5] \cup [5, \infty)$
11. 6 **12.** $4a + 1 + 2h$ **13.** \$0.72
14. $f(x) = 0.7x$; \$45.50; \$33.60; \$10.85

15. **16.**

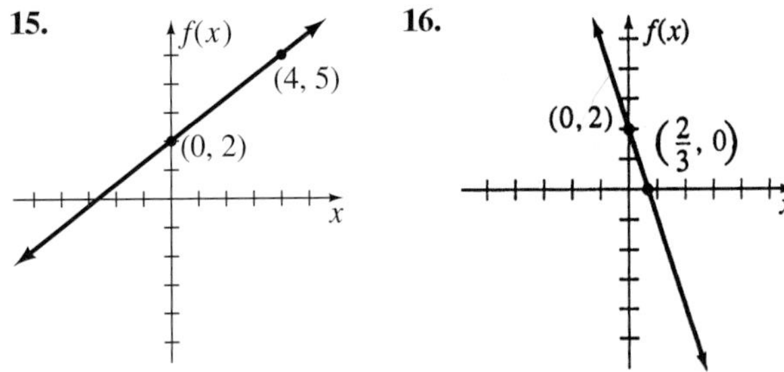

17. **18.**

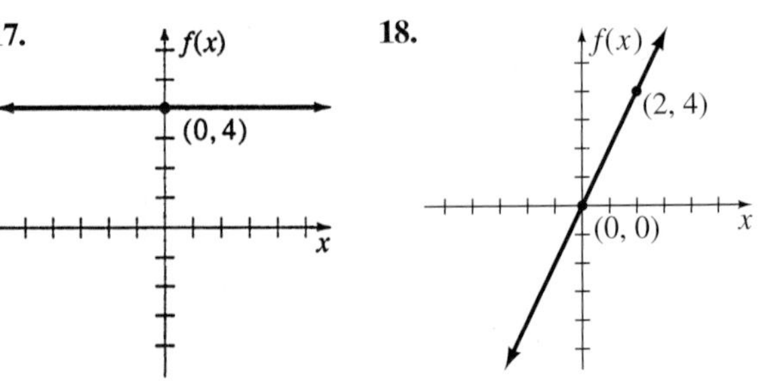

19. $f(x) = 6x - 100$; 110 **20.** $f(x) = \frac{10}{3}x + 150$; \$300

21.

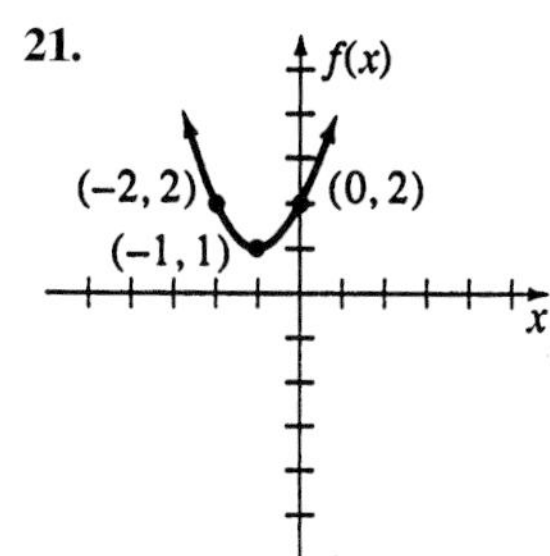

22.

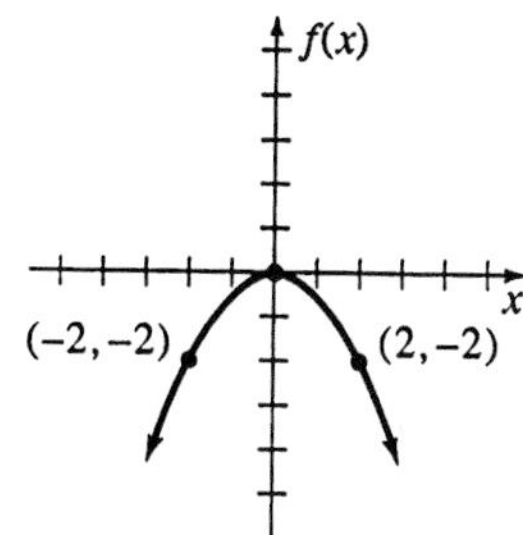

23.

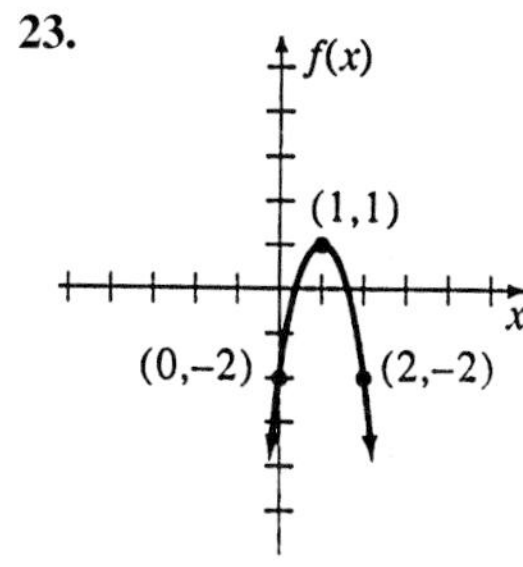

24. **(a)** $(-5, -28)$; $x = -5$ **25.** 20 and 20 **26.** 3, 47
27. Width, 15 yd; length, 30 yd **28.** 25 students

29.

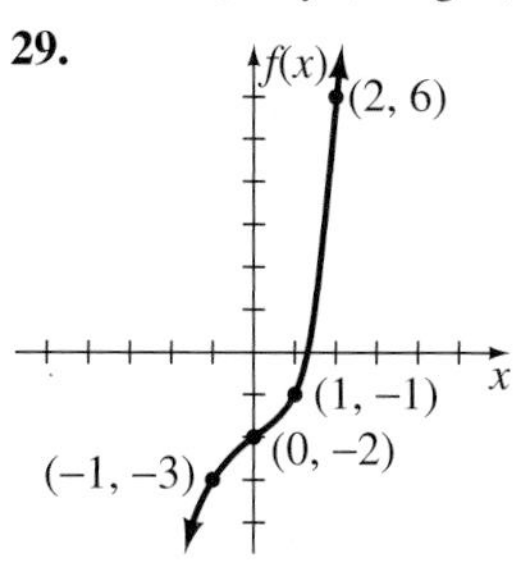

30.

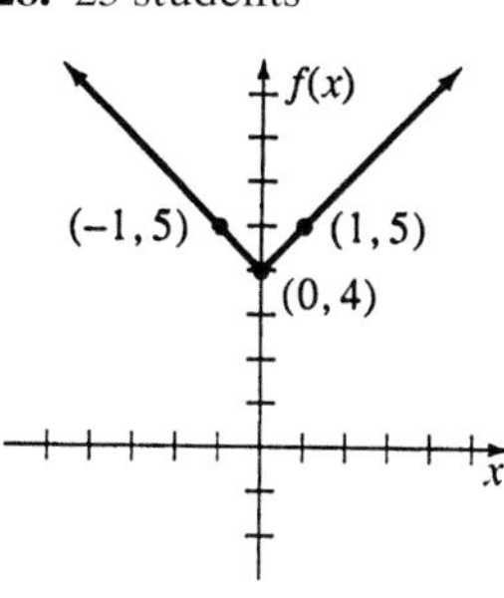

31.

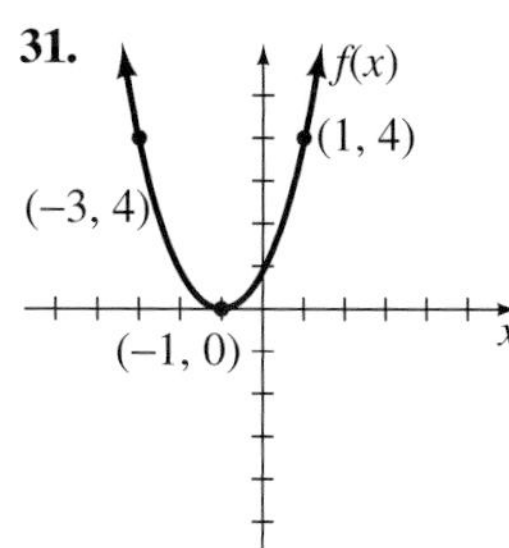

32.

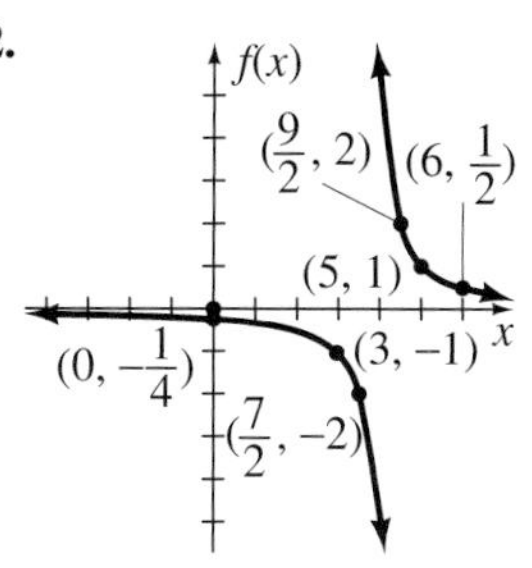

33.

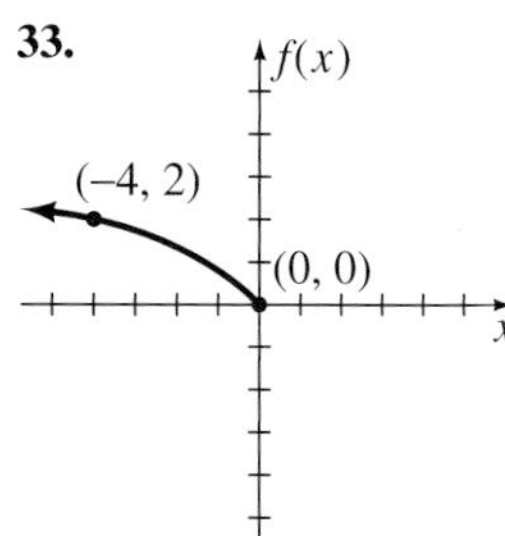

34.

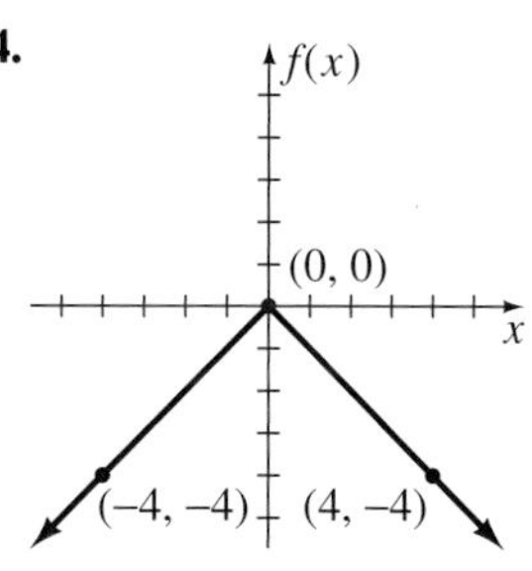

35.

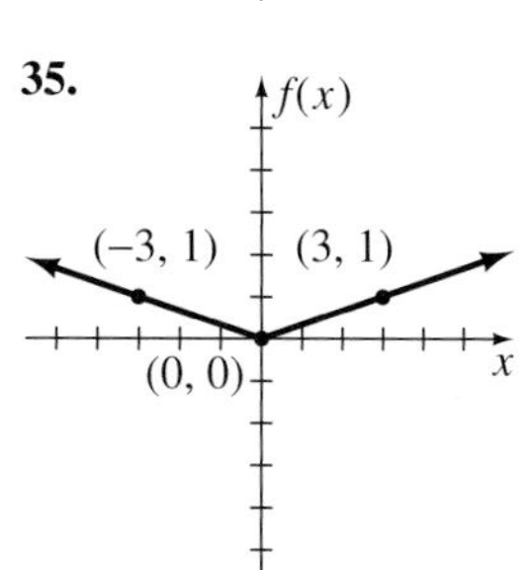

36.

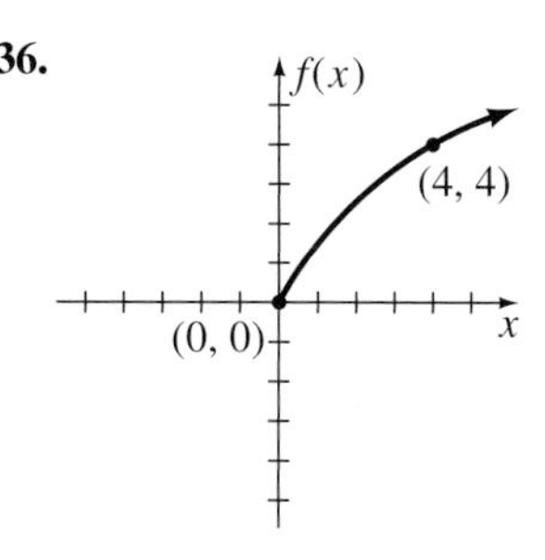

37.

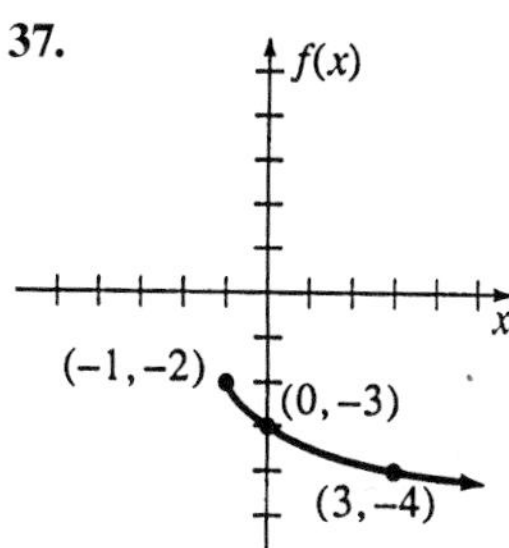

38.

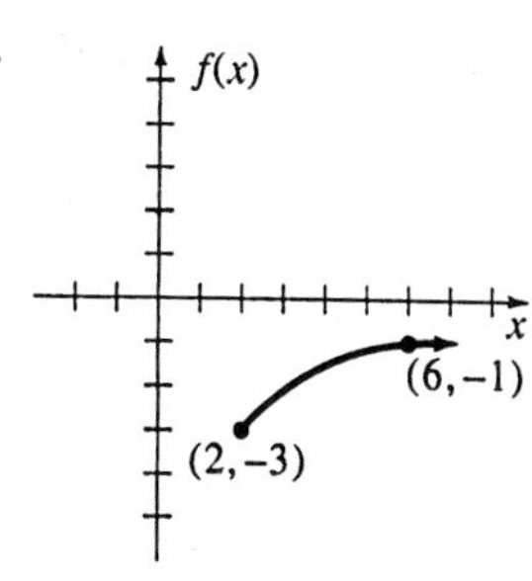

39.

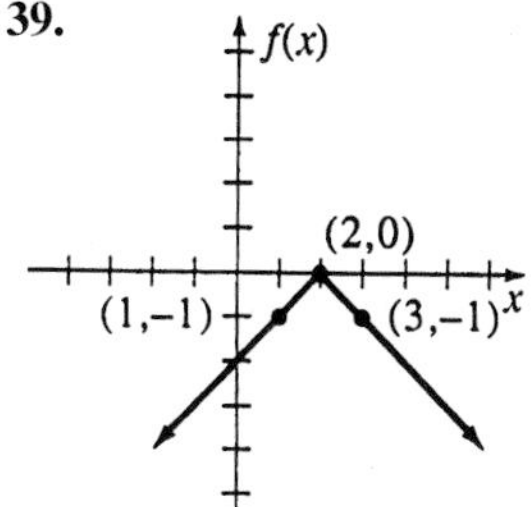

40.

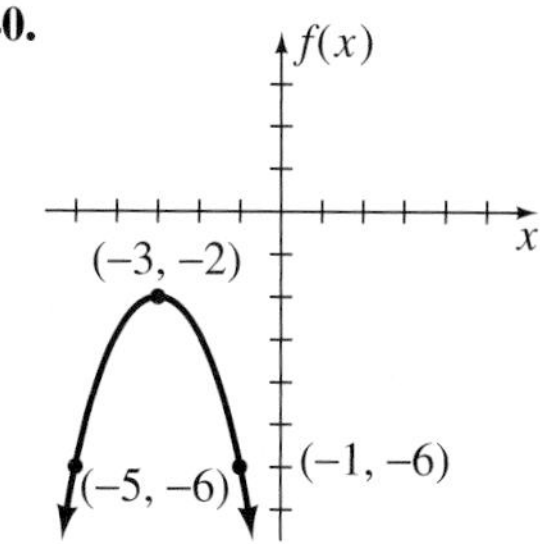

41. $(f \circ g)(x) = 6x - 11$; $(g \circ f)(x) = 6x - 13$
42. $(f \circ g)(x) = x^2 - 2x - 1$; $(g \circ f)(x) = x^2 - 10x + 27$
43. $(f \circ g)(x) = 2x^2 - 20x + 20$; $(g \circ f)(x) = -2x^2 + 5$
44. 179 **45.** $x - 1$ **46.** Function **47.** Not a function
48. Not a function **49.** Not one-to-one **50.** One-to-one
51. One-to-one **52.** $f^{-1} = \{(4, -1), (5, 0), (7, 1), (9, 2)\}$
53. $f^{-1} = \{(4, 2), (9, 3), (16, 4), (25, 5)\}$
54. $f^{-1}(x) = \frac{x + 1}{6}$ **55.** $f^{-1}(x) = \frac{3x - 21}{2}$
56. $f^{-1}(x) = \frac{-35x - 10}{21}$ **57.** \$375 **58.** 600 in.^{2}
59. 10 min **60.** 128 lb **61.** $k = 9$ **62.** $y = 120$

Chapter 10 Test (page 602)

1. $D = \left\{x \mid x \neq -4 \text{ and } x \neq \frac{1}{2}\right\}$ or
$D: (-\infty, -4) \cup \left(-4, \frac{1}{2}\right) \cup \left(\frac{1}{2}, \infty\right)$
2. $D = \left\{x \mid x \leq \frac{5}{3}\right\}$ or $D: \left(-\infty, \frac{5}{3}\right]$ **3.** $\frac{11}{6}$
4. $-4a^2 + 48a + 3$ **5.** $(-6, 3)$ **6.** $6a + 3h + 2$
7. $(f \circ g)(x) = -21x - 2$ **8.** $(g \circ f)(x) = 8x^2 + 38x + 48$
9. $(f \circ g)(x) = \frac{3x}{2 - 2x}$ **10.** $f^{-1}(x) = \frac{x + 9}{5}$
11. $f^{-1}(x) = \frac{-x - 6}{3}$ **12.** $f^{-1}(x) = \frac{15x + 9}{10}$
13. -4 **14.** 15 **15.** 6 and 54 **16.** \$96
17. The graph of $f(x) = (x - 6)^3 - 4$ is the graph of $f(x) = x^3$ translated 6 units to the right and 4 units downward.
18. The graph of $f(x) = -|x| + 8$ is the graph of $f(x) = |x|$ reflected across the x axis and translated 8 units upward.
19. The graph of $f(x) = -\sqrt{x + 5} + 7$ is the graph of $f(x) = \sqrt{x}$ reflected across the x axis and translated 5 units to the left and 7 units upward.

20.

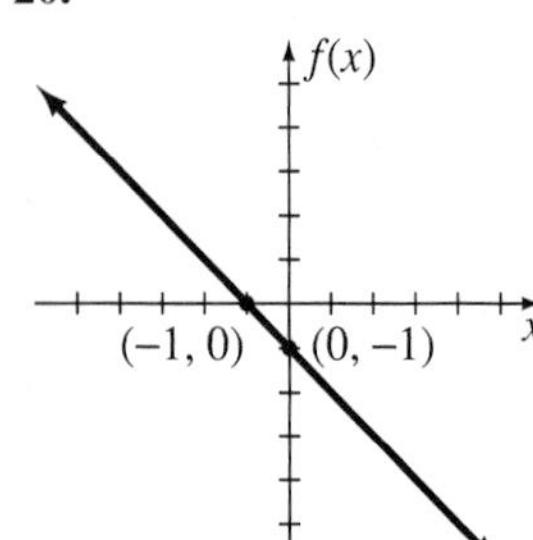

21.

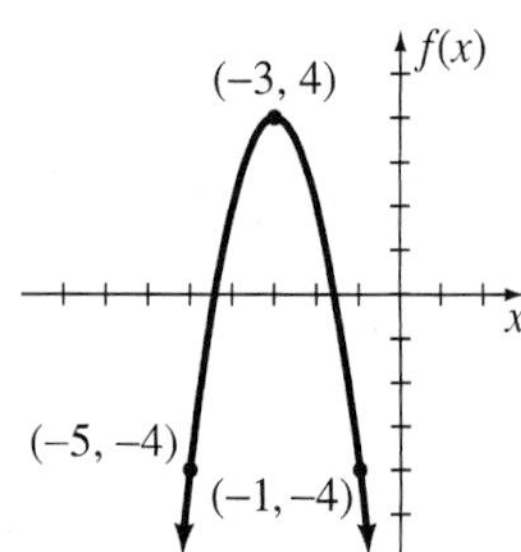

22.

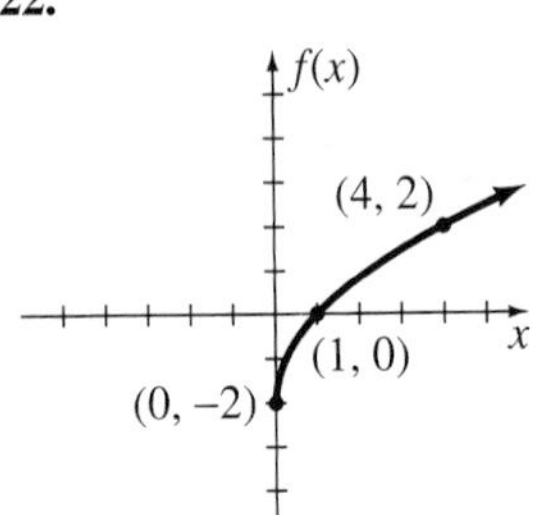

23.

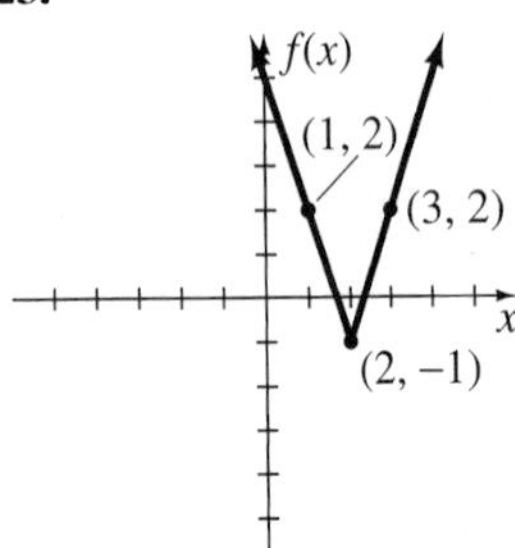

24.

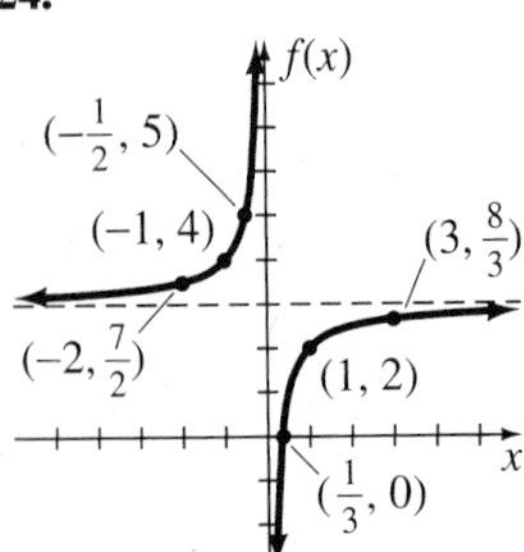

25.

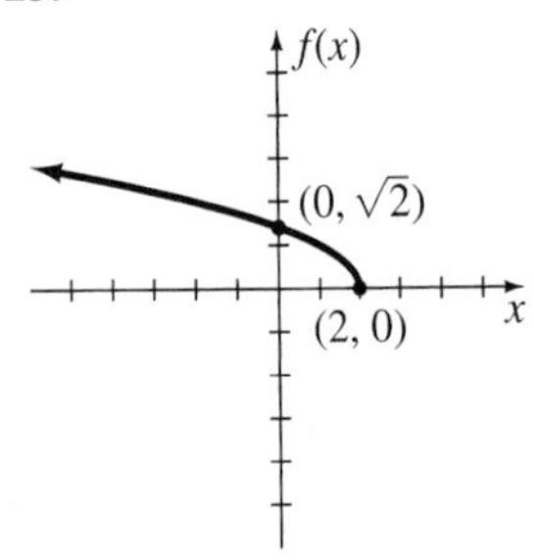

Cumulative Review Problem Set (page 604)

1. -6 **2.** -8 **3.** $\frac{13}{24}$ **4.** 24 **5.** $\frac{13}{6}$ **6.** $-90\sqrt{2}$
7. $2x + 5\sqrt{x} - 12$ **8.** $-18 + 22\sqrt{3}$
9. $2x^3 + 11x^2 - 14x + 4$ **10.** $\frac{x+4}{x(x+5)}$ **11.** $\frac{16x^2}{27y}$
12. $\frac{16x+43}{90}$ **13.** $\frac{35a-44b}{60a^2b}$ **14.** $\frac{2}{x-4}$
15. $2x^2 - x - 4$ **16.** $\frac{5y^2-3xy^2}{x^2y+2x^2}$ **17.** $\frac{2y-3xy}{3x+4xy}$
18. $\frac{(2n+3)(n+3)}{(n+2)(3n+13)}$ **19.** $\frac{3a^2-2a+1}{2a-1}$
20. $(5x-2)(4x+3)$ **21.** $2(2x+3)(4x^2-6x+9)$
22. $(2x+3)(2x-3)(x+2)(x-2)$
23. $4x(3x+2)(x-5)$ **24.** $(y-6)(x+3)$
25. $(5-3x)(2+3x)$ **26.** $\frac{81}{16}$ **27.** 4 **28.** $-\frac{3}{4}$ **29.** -0.3
30. $\frac{1}{81}$ **31.** $\frac{21}{16}$ **32.** $\frac{9}{64}$ **33.** 72 **34.** $-\frac{12}{x^3y}$ **35.** $\frac{8y}{x^5}$
36. $-\frac{a^3}{9b}$ **37.** $4\sqrt{5}$ **38.** $-6\sqrt{6}$ **39.** $\frac{5\sqrt{3}}{9}$ **40.** $\frac{2\sqrt{3}}{3}$
41. $2\sqrt[3]{7}$ **42.** $\frac{\sqrt[3]{6}}{2}$ **43.** $8xy\sqrt{13x}$ **44.** $\frac{\sqrt{6xy}}{3y}$
45. $11\sqrt{6}$ **46.** $-\frac{169\sqrt{2}}{12}$ **47.** $-16\sqrt[3]{3}$
48. $\frac{-3\sqrt{2}-2\sqrt{6}}{2}$ **49.** $\frac{6\sqrt{15}-3\sqrt{35}-6+\sqrt{21}}{5}$
50. 0.021 **51.** 300 **52.** 0.0003 **53.** $32 + 22i$
54. $-17 + i$ **55.** $0 - \frac{5}{4}i$ **56.** $-\frac{19}{53} + \frac{40}{53}i$
57. $-\frac{10}{3}$ **58.** $\frac{4}{7}$ **59.** $2\sqrt{13}$ **60.** $5x - 4y = 19$
61. $4x + 3y = -18$ **62.** $(-2, 6)$ and $r = 3$
63. $(-5, -4)$ **64.** 8 units

65.

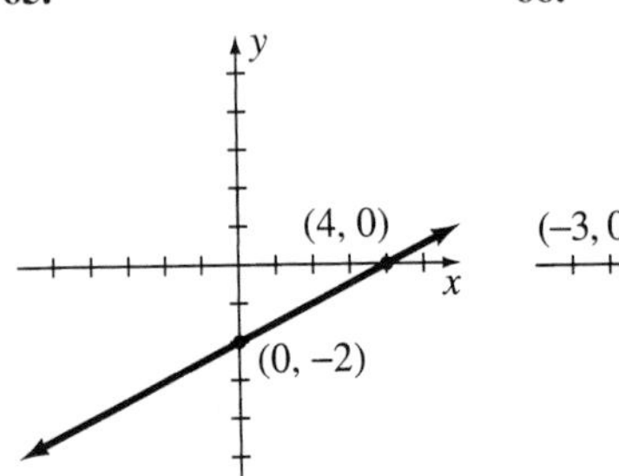

66.

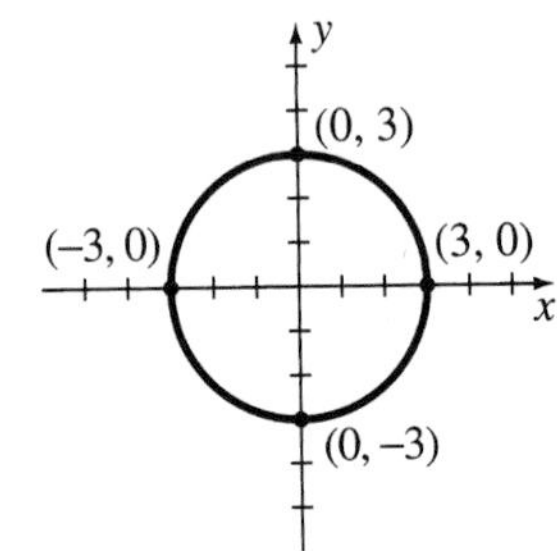

67.

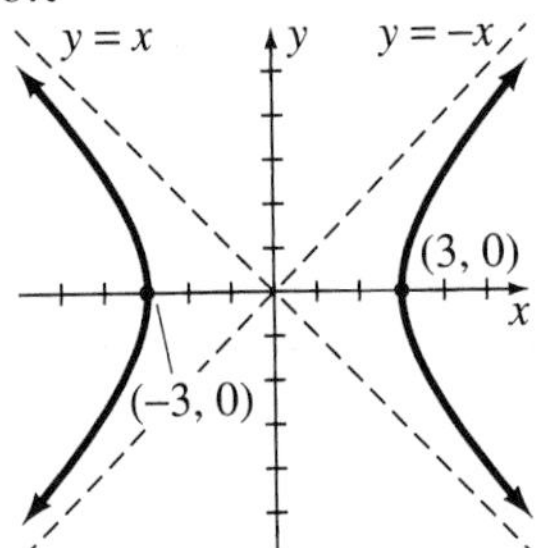

68.

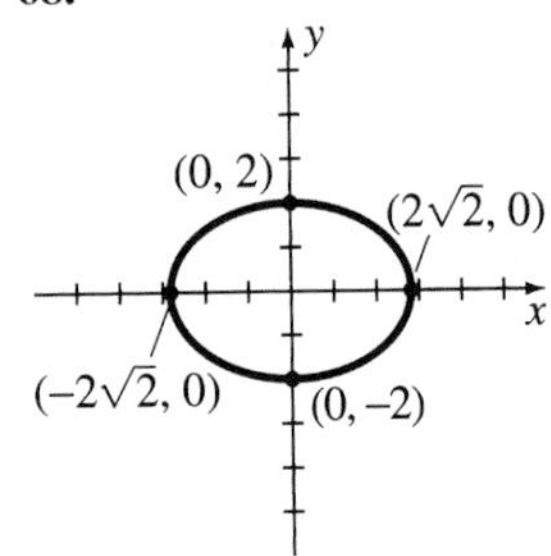

69.

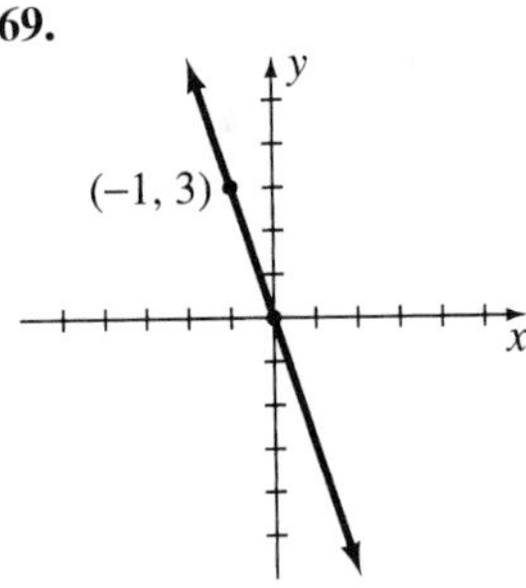

70.

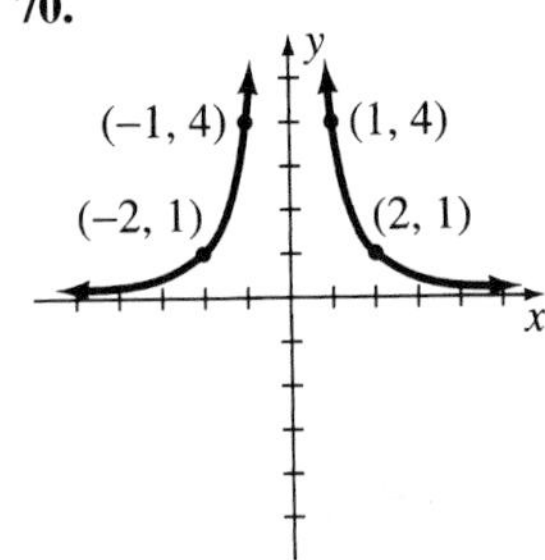

71.

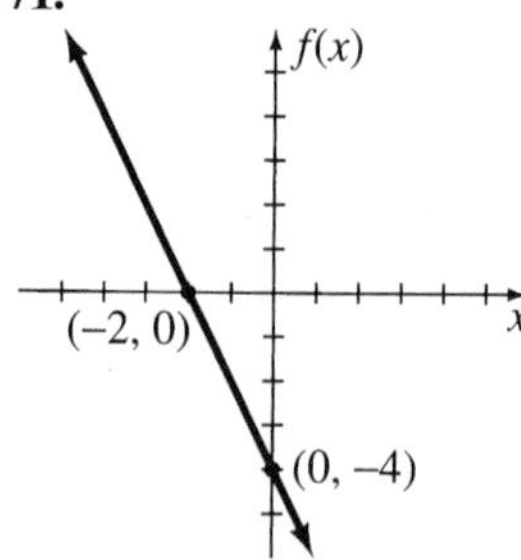

72.

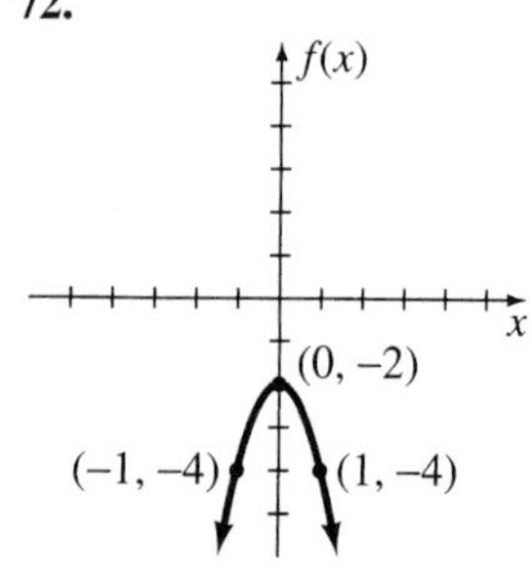

73.

74.

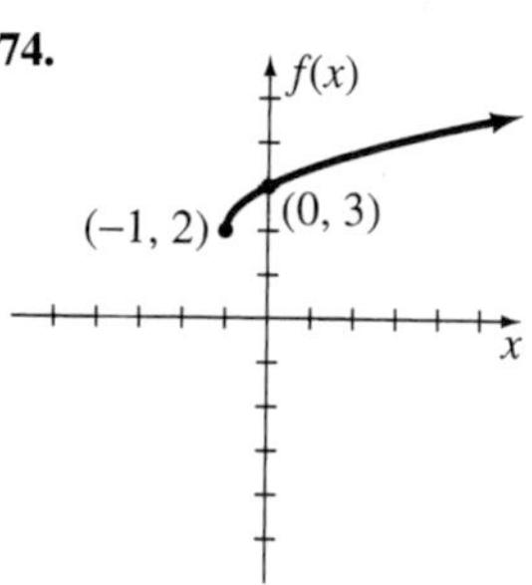

75.

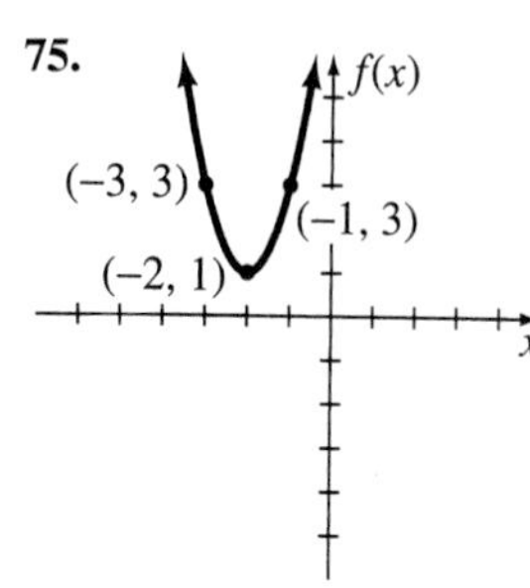

76.

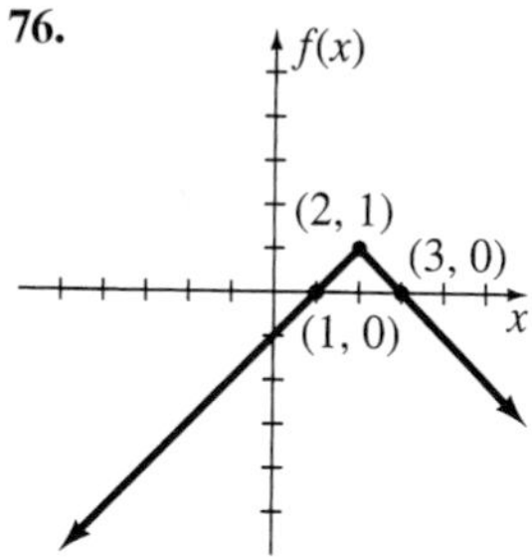

77. $(g \circ f)(x) = 2x^2 - 13x + 20$

$(f \circ g)(x) = 2x^2 - x - 4$ **78.** $f^{-1}(x) = \dfrac{x+7}{3}$

79. $f^{-1}(x) = -2x + \dfrac{4}{3}$ **80.** $k = -3$ **81.** $y = 1$

82. 12 cm^3 **83.** $\left\{-\dfrac{21}{16}\right\}$ **84.** $\left\{\dfrac{40}{3}\right\}$ **85.** $\{6\}$

86. $\left\{-\dfrac{5}{2}, 3\right\}$ **87.** $\left\{0, \dfrac{7}{3}\right\}$ **88.** $\{-6, 0, 6\}$ **89.** $\left\{-\dfrac{5}{6}, \dfrac{2}{5}\right\}$

90. $\left\{-3, 0, \dfrac{3}{2}\right\}$ **91.** $\{\pm 1, \pm 3i\}$ **92.** $\{-5, 7\}$ **93.** $\{-29, 0\}$

94. $\left\{\dfrac{7}{2}\right\}$ **95.** $\{12\}$ **96.** $\{-3\}$ **97.** $\left\{\dfrac{1 \pm 3\sqrt{5}}{3}\right\}$

98. $\left\{\dfrac{-5 \pm 4i\sqrt{2}}{2}\right\}$ **99.** $\left\{\dfrac{3 \pm i\sqrt{23}}{4}\right\}$

100. $\left\{\dfrac{3 \pm \sqrt{3}}{3}\right\}$ **101.** $\{1 \pm \sqrt{34}\}$

102. $\left\{\pm\dfrac{\sqrt{5}}{2}, \pm\dfrac{\sqrt{3}}{3}\right\}$ **103.** $\left\{\dfrac{-5 \pm i\sqrt{15}}{4}\right\}$

104. $(-\infty, 3)$ **105.** $(-\infty, 50]$ **106.** $\left(-\infty, -\dfrac{11}{5}\right) \cup (3, \infty)$

107. $\left(-\dfrac{5}{3}, 1\right)$ **108.** $\left[-\dfrac{9}{11}, \infty\right)$ **109.** $[-4, 2]$

110. $\left(-\infty, \dfrac{1}{3}\right) \cup (4, \infty)$ **111.** $(-8, 3)$

112. $(-\infty, 3] \cup (7, \infty)$ **113.** $(-6, -3)$ **114.** $\{(9, 6)\}$

115. $\left\{\left(\dfrac{9}{5}, -\dfrac{7}{25}\right)\right\}$ **116.** $\left\{\left(\dfrac{70}{23}, \dfrac{36}{23}\right)\right\}$ **117.** $\{(-2, 4, 0)\}$

118. $\{(-1, 4, -1)\}$ **119.** 17, 19, and 21
120. 14 nickels, 20 dimes, and 29 quarters
121. 48° and 132° **122.** \$600
123. \$1700 at 8% and \$2000 at 9%
124. 66 mph and 76 mph **125.** 4 qt **126.** 69 or less
127. −3, 0, or 3 **128.** 1 in. **129.** \$1050 and \$1400
130. 3 hr **131.** 37 **132.** 10°, 60°, and 110°

CHAPTER 11

Problem Set 11.1 (page 614)

1. $\{3\}$ **3.** $\{2\}$ **5.** $\{4\}$ **7.** $\{1\}$ **9.** $\{5\}$ **11.** $\{1\}$ **13.** $\left\{\dfrac{3}{2}\right\}$

15. $\left\{\dfrac{5}{6}\right\}$ **17.** $\{-3\}$ **19.** $\{-3\}$ **21.** $\{0\}$ **23.** $\{1\}$

25. $\{-1\}$ **27.** $\left\{-\dfrac{2}{5}\right\}$ **29.** $\left\{\dfrac{5}{2}\right\}$ **31.** $\{3\}$ **33.** $\left\{\dfrac{1}{2}\right\}$

35.

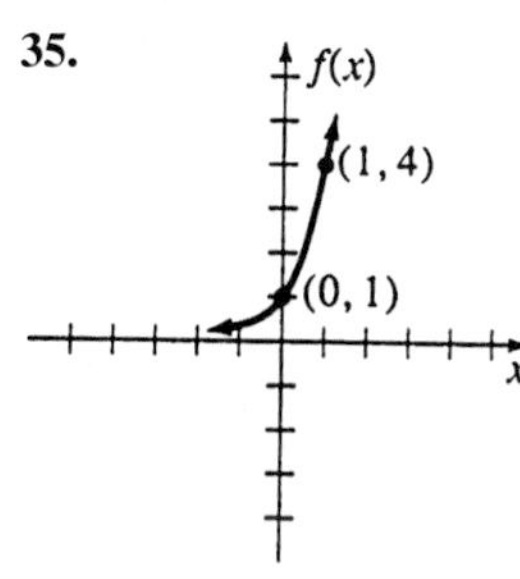

37.

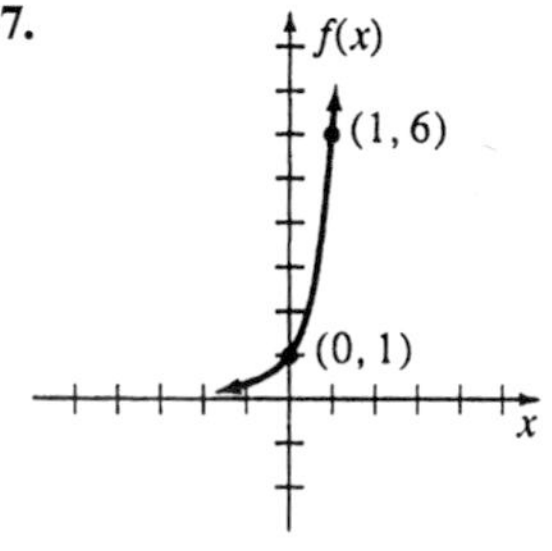

39. **41.**

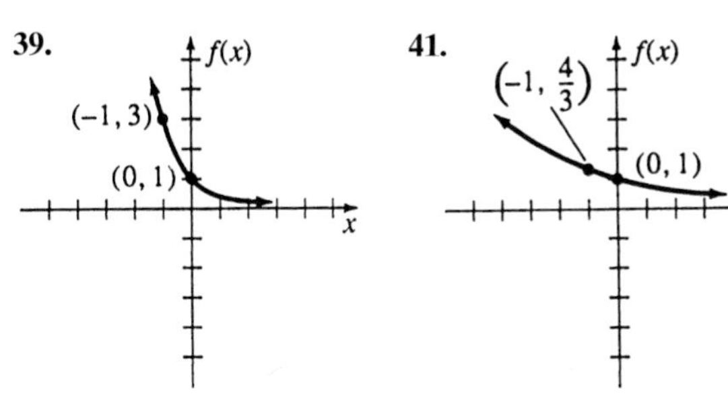

43. **45.**

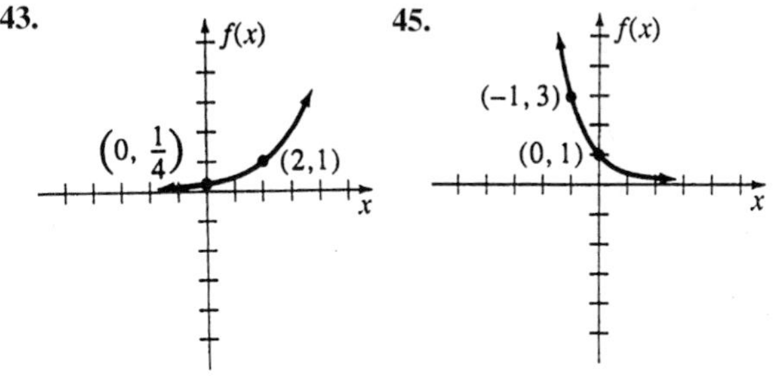

47. **49.**

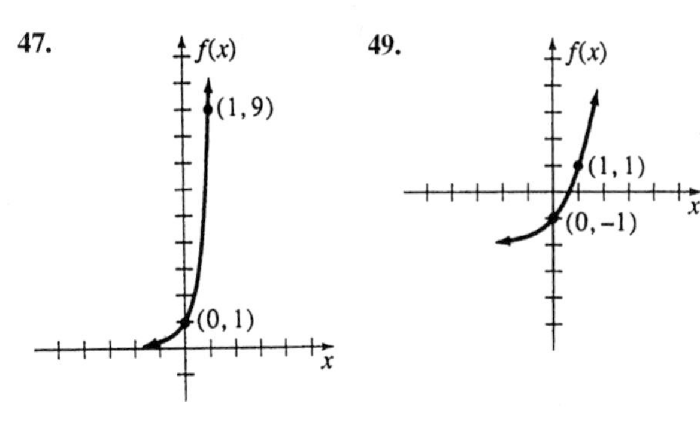

51.

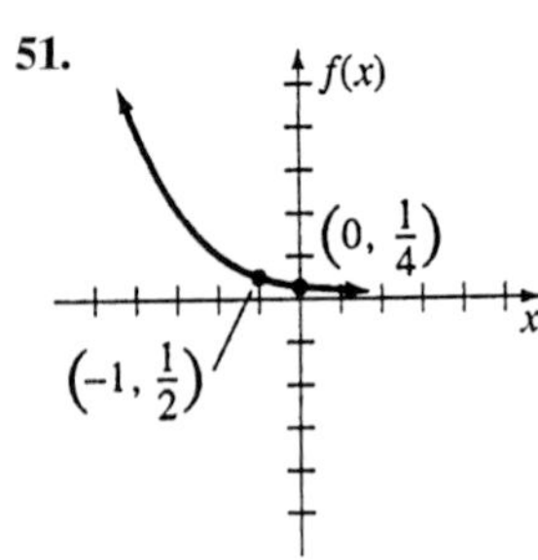

Problem Set 11.2 (page 623)

1. **(a)** \$1.55 **(b)** \$4.17 **(c)** \$2.33 **(d)** \$2.27 **(e)** \$21,900 **(f)** \$246,342 **(g)** \$658 **3.** \$283.70 **5.** \$865.84 **7.** \$1782.25 **9.** \$2725.05 **11.** \$16,998.71 **13.** \$22,553.65 **15.** \$567.63 **17.** \$1422.36 **19.** \$8963.38 **21.** \$17,547.35 **23.** \$32,558.88

25.

	1 yr	5 yr	10 yr	20 yr
Compounded annually	\$1120	1762	3106	9,646
Compounded semiannually	1124	1791	3207	10,286
Compounded quarterly	1126	1806	3262	10,641
Compounded monthly	1127	1817	3300	10,893
Compounded continuously	1127	1822	3320	11,023

27. Nora will have \$.24 more. **29.** 9.54%
31. 50 grams; 37 grams

33. **35.**

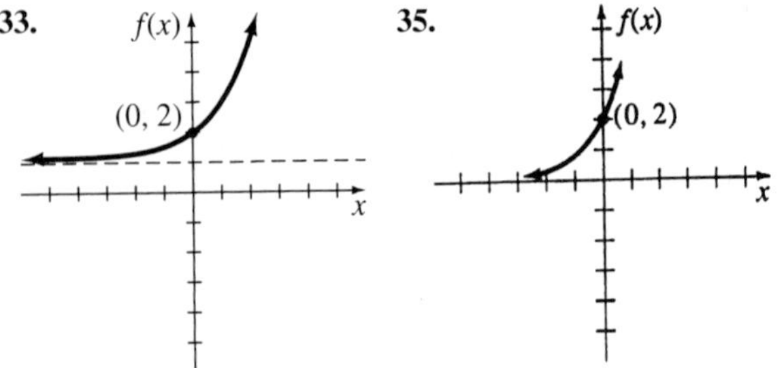

37.

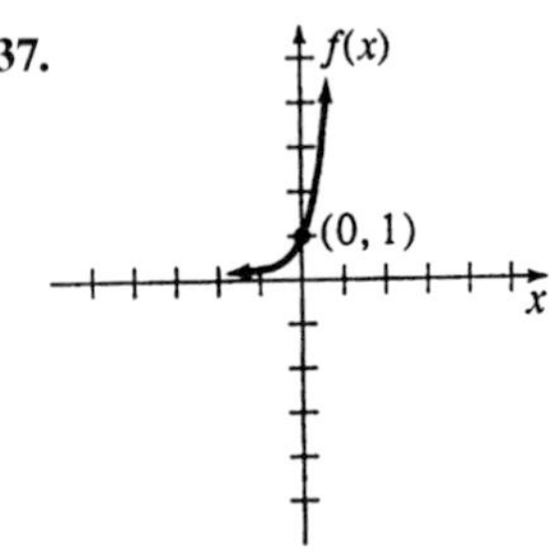

39. 2226; 3320; 7389 **41.** 2000 **43.** **(a)** 82,888 **(c)** 96,302 **45.** **(a)** 6.5 lb/in.^{2} **(c)** 13.6 lb/in.^{2}

Problem Set 11.3 (page 634)

1. $\log_2 128 = 7$ **3.** $\log_5 125 = 3$ **5.** $\log_{10} 1000 = 3$
7. $\log_2\left(\frac{1}{4}\right) = -2$ **9.** $\log_{10} 0.1 = -1$ **11.** $3^4 = 81$
13. $4^3 = 64$ **15.** $10^4 = 10{,}000$ **17.** $2^{-4} = \frac{1}{16}$
19. $10^{-3} = 0.001$ **21.** 4 **23.** 4 **25.** 3 **27.** $\frac{1}{2}$ **29.** 0
31. −1 **33.** 5 **35.** −5 **37.** 1 **39.** 0 **41.** {49}
43. {16} **45.** {27} **47.** $\left\{\frac{1}{8}\right\}$ **49.** {4} **51.** 5.1293
53. 6.9657 **55.** 1.4037 **57.** 7.4512 **59.** 6.3219
61. −0.3791 **63.** 0.5766 **65.** 2.1531 **67.** 0.3949
69. $\log_b x + \log_b y + \log_b z$ **71.** $\log_b y - \log_b z$
73. $3 \log_b y + 4 \log_b z$ **75.** $\frac{1}{2}\log_b x + \frac{1}{3}\log_b y - 4\log_b z$
77. $\frac{2}{3}\log_b x + \frac{1}{3}\log_b z$ **79.** $\frac{3}{2}\log_b x - \frac{1}{2}\log_b y$
81. $\left\{\frac{9}{4}\right\}$ **83.** {25} **85.** {4} **87.** $\left\{\frac{19}{8}\right\}$ **89.** {9}
91. {1} **93.** $\left\{-\frac{22}{5}\right\}$ **95.** ∅ **97.** {−1}

Problem Set 11.4 (page 641)

1. 0.8597 **3.** 1.7179 **5.** 3.5071 **7.** −0.1373 **9.** −3.4685
11. 411.43 **13.** 90,095 **15.** 79.543 **17.** 0.048440
19. 0.0064150 **21.** 1.6094 **23.** 3.4843 **25.** 6.0638
27. −0.7765 **29.** −3.4609 **31.** 1.6034
33. 3.1346 **35.** 108.56 **37.** 0.48268 **39.** 0.035994

41. **43.**

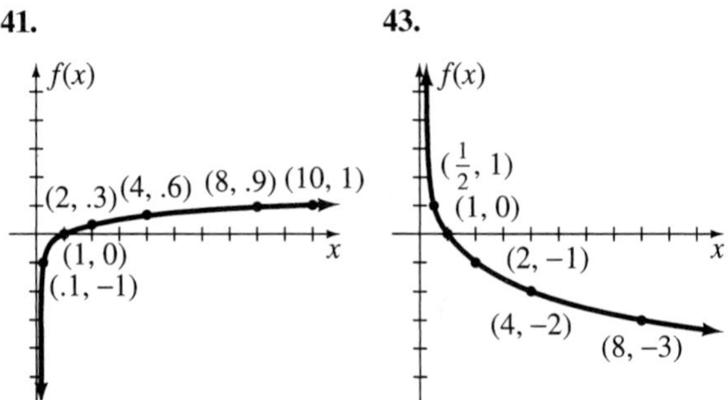

45. **47.**

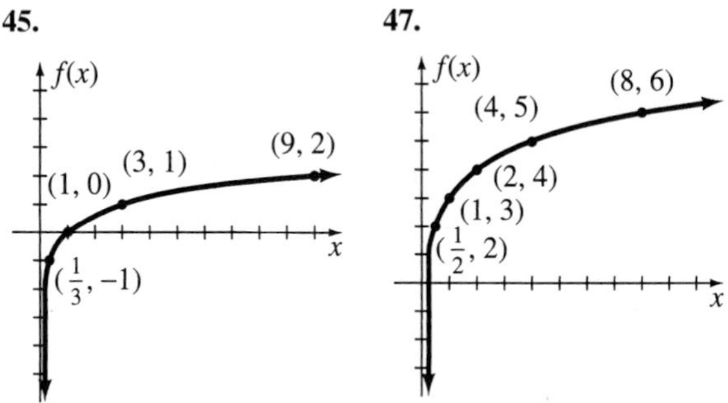

49. **51.**

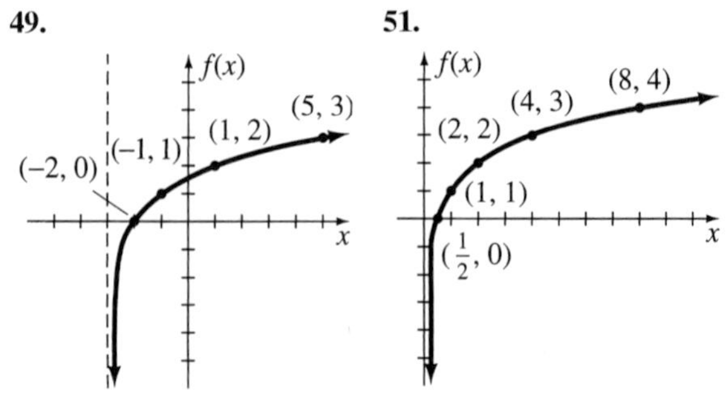

53.

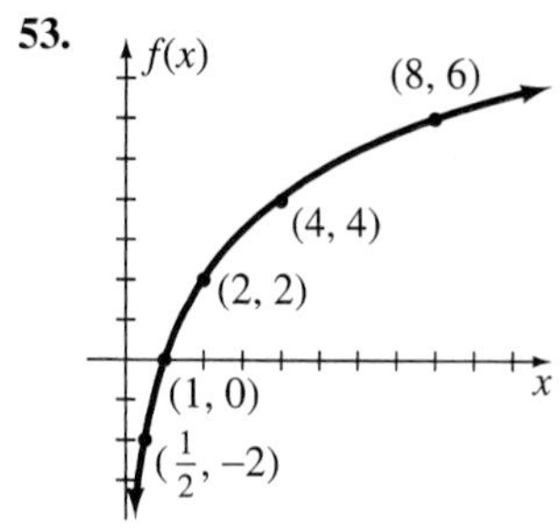

55. 0.36 **57.** 0.73 **59.** 23.10 **61.** 7.93

Problem Set 11.5 (page 650)

1. {3.15} **3.** {2.20} **5.** {4.18} **7.** {0.12} **9.** {1.69}
11. {4.57} **13.** {2.46} **15.** {4} **17.** $\left\{\frac{19}{47}\right\}$
19. {1} **21.** {8} **23.** 2.4 years **25.** 5.3 years
27. 6.8 hours **29.** 1.5 hours **31.** 34.7 years **33.** 6.7
35. Approximately 8 times **37.** 4.524 **39.** −0.860
41. 3.105 **43.** −2.902 **45.** 5.989

Chapter 11 Review Problem Set (page 658)

1. **2.**

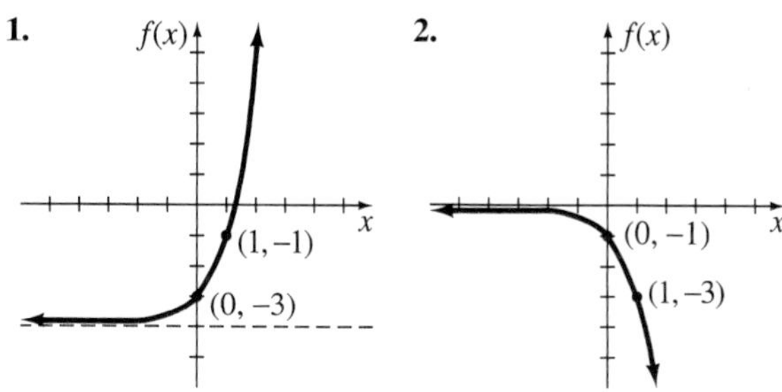

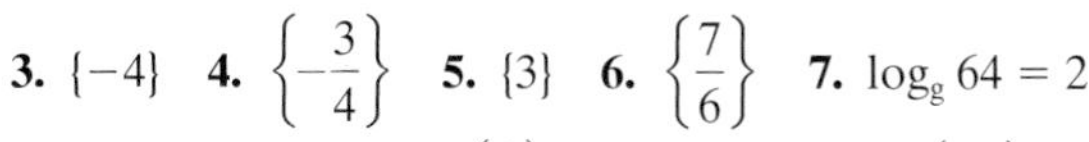

3. {−4} **4.** $\left\{-\frac{3}{4}\right\}$ **5.** {3} **6.** $\left\{\frac{7}{6}\right\}$ **7.** $\log_8 64 = 2$

8. $\log_a c = b$ **9.** $\log_3\left(\frac{1}{9}\right) = -2$ **10.** $\log_{10}\left(\frac{1}{10}\right) = -1$

11. $5^3 = 123$ **11.** $x^2 = y$ **13.** $2^{-3} = \frac{1}{8}$ **14.** $10^4 = 10{,}000$

15. 7 **17.** 4 **18.** −3 **19.** 2 **20.** 13 **21.** 0 **22.** −2
23. 1.8642 **24.** −2.6289 **25.** −4.2687 **26.** 4.7380
27. 4.164 **28.** 2.548 **29.** −3.129 **30.** −8.118

31. **32.**

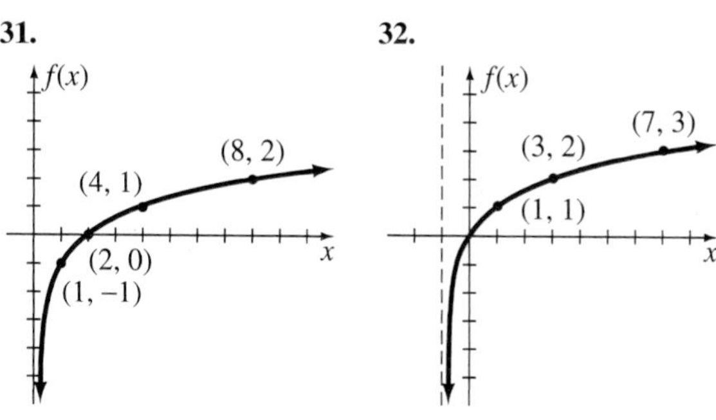

33. $\log_b x + \frac{1}{2}\log_b y - 3\log_b z$ **34.** $1 - \log_b c - \log_b d$
35. {1.1882} **36.** {2639.4} **37.** {0.0132289}
39. {4} **40.** {9} **41.** $\left\{\frac{1}{2}\right\}$ **42.** {8} **43.** {5}
44. $\left\{\frac{45}{8}\right\}$ **45.** {5} **46.** $\left\{\frac{15}{7}\right\}$ **47.** {3.40}
48. {1.95} **49.** {5.30} **50.** {5.61}
51. **(a)** \$933.05 **(b)** \$305.96 **(c)** \$98.54
52. **(a)** \$7910 **(b)** \$14,765 **(c)** \$15,000 **53.** \$3070.43
54. \$8178.72 55. Approximately 8.8 yr
56. Approximately 12.1 yr **57.** \$5656.26
58. Approximately 19.8 yr **59.** 133 g
60. 46.4 hr **61.** 61,070; 67,493; 74,591
62. Approximately 4.8 yr **63.** 81
64. Approximately 5 times

Chapter 11 Test (page 661)

1. $\frac{1}{2}$ **2.** 1 **3.** 1 **4.** −1 **5.** {−3} **6.** $\left\{-\frac{3}{2}\right\}$ **7.** $\left\{\frac{8}{3}\right\}$
8. {243} **9.** {2} **10.** $\left\{\frac{2}{5}\right\}$ **11.** 4.1919 **12.** 0.2031
13. 0.7325 **14.** 5.4538 **15.** {5.17} **16.** {10.29}
17. 4.0069 **18.** $\log_b\left(\frac{x^3y^2}{z}\right)$ **19.** \$6342.08
20. 13.5 years **21.** 7.8 hours **22.** 4813 grams

23. **24.**

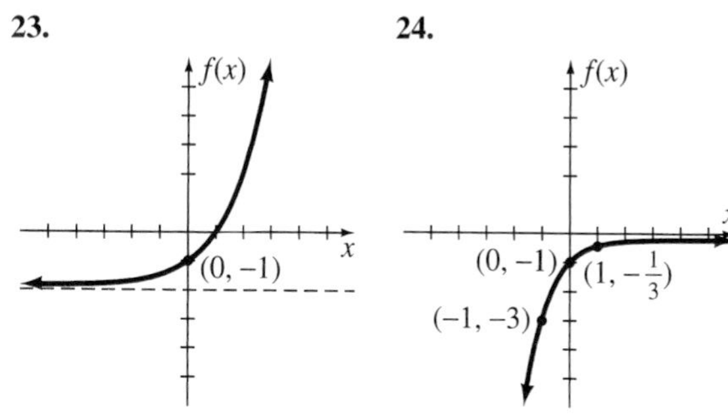

25.

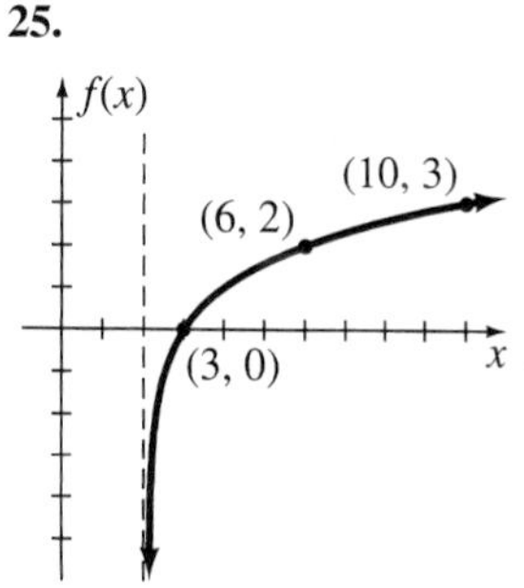

APPENDICES

Appendix A (page 670)

1. $2 \cdot 13$ **2.** $2 \cdot 2 \cdot 2 \cdot 2 \cdot 2$ **3.** $2 \cdot 2 \cdot 3 \cdot 3$
4. $2 \cdot 2 \cdot 2 \cdot 2 \cdot 5$ **5.** $7 \cdot 7$ **6.** $2 \cdot 2 \cdot 23$
7. $2 \cdot 2 \cdot 2 \cdot 7$ **8.** $2 \cdot 2 \cdot 2 \cdot 2 \cdot 3 \cdot 3$
9. $2 \cdot 2 \cdot 2 \cdot 3 \cdot 5$ **10.** $2 \cdot 2 \cdot 3 \cdot 7$ **11.** $3 \cdot 3 \cdot 3 \cdot 5$
12. $2 \cdot 7 \cdot 7$ **13.** 24 **14.** 24 **15.** 48 **16.** 36 **17.** 140
18. 462 **19.** 392 **20.** 72 **21.** 168 **22.** 60 **23.** 90
24. 168 **25.** $\frac{2}{3}$ **26.** $\frac{3}{4}$ **27.** $\frac{2}{3}$ **28.** $\frac{9}{16}$ **29.** $\frac{5}{3}$ **30.** $\frac{4}{3}$
31. $\frac{15}{28}$ **32.** $\frac{12}{55}$ **33.** $\frac{10}{21}$ **34.** $\frac{65}{66}$ **35.** $\frac{3}{10}$ **36.** $\frac{2}{3}$
37. $\frac{3}{8}$ cup **38.** $\frac{1}{6}$ of the bottle **39.** $\frac{2}{9}$ of the disk space
40. $\frac{1}{3}$ **41.** $\frac{5}{7}$ **42.** $\frac{8}{11}$ **43.** $\frac{5}{9}$ **44.** $\frac{5}{13}$ **45.** 3 **46.** 2
47. $\frac{2}{3}$ **48.** $\frac{3}{8}$ **49.** $\frac{2}{3}$ **50.** $\frac{5}{9}$ **51.** $\frac{8}{15}$ **52.** $\frac{7}{24}$ **53.** $\frac{9}{16}$
54. $\frac{11}{12}$ **55.** $\frac{37}{30}$ **56.** $\frac{29}{24}$ **57.** $\frac{59}{96}$ **58.** $\frac{19}{24}$ **59.** $\frac{13}{12}$
60. $\frac{5}{16}$ **61.** $\frac{1}{4}$ **62.** $\frac{5}{3}$ **63.** $\frac{37}{30}$ **64.** $\frac{4}{5}$ **65.** $\frac{1}{3}$ **66.** $\frac{27}{35}$
67. $\frac{7}{26}$ **68.** 30 **69.** $\frac{7}{20}$ **70.** $\frac{11}{32}$

Appendix B (page 675)

1. $\{(6, -4)\}$ **3.** $\{(-4, -4)\}$ **5.** $\left\{\left(-\frac{11}{7}, -\frac{13}{7}\right)\right\}$ **7.** $\{(1, -3)\}$
9. $\{(-1, -2, -3)\}$ **11.** $\{(3, 1, 4)\}$ **13.** $\{(4, 3, -2)\}$
15. $\{(-5, 2, 0)\}$ **17.** $\{(-2, -1, 1)\}$ **19.** $\{(-1, 4, -1)\}$
21. $\{(2, 0, -3)\}$ **25.** $\{(1, -1, 2, -3)\}$ **27.** $\varnothing$

Appendix C (page 679)

1. 10 **3.** −48 **5.** −29 **7.** 50 **9.** 8 **11.** 6 **13.** −32
15. −2 **17.** 5 **19.** $-\frac{7}{20}$ **21.** $\{(3, 8)\}$ **23.** $\{(-2, 5)\}$
25. $\{(-2, 2)\}$ **27.** $\varnothing$ **29.** $\{(1, 6)\}$ **31.** $\left\{\left(-\frac{1}{4}, \frac{2}{3}\right)\right\}$
33. $\left\{\left(\frac{17}{62}, \frac{4}{31}\right)\right\}$ **35.** $\{(-1, 3)\}$ **37.** $\{(9, -2)\}$
39. $\{(-4, -3)\}$

Appendix D (page 685)

1. −57 **3.** 14 **5.** −41 **7.** −8 **9.** −96
11. $\{(-3, 1, -1)\}$ **13.** $\{(0, 2, -3)\}$ **15.** $\left\{\left(-2, \frac{1}{2}, -\frac{2}{3}\right)\right\}$
17. $\{(-1, -1, -1)\}$ **19.** $\varnothing$ **21.** $\{(-5, -2, 1)\}$
23. $\{(-2, 3, -4)\}$ **25.** $\left\{\left(-\frac{1}{2}, 0, \frac{2}{3}\right)\right\}$ **27.** $\{(-6, 7, 1)\}$
29. $\left\{\left(1, -6, -\frac{1}{2}\right)\right\}$ **33.** 0 **35. (b)** −20

Index

Properties of Absolute Value

$|a| \geq 0$

$|a| = |-a|$

$|a - b| = |b - a|$

$|a^2| = |a|^2 = a^2$

Multiplication Patterns

$(a + b)^2 = a^2 + 2ab + b^2$

$(a - b)^2 = a^2 - 2ab + b^2$

$(a + b)(a - b) = a^2 - b^2$

$(a + b)^3 = a^3 + 3a^2b + 3ab^2 + b^3$

$(a - b)^3 = a^3 - 3a^2b + 3ab^2 - b^3$

Properties of Exponents

$b^n \cdot b^m = b^{n+m}$ $\qquad \dfrac{b^n}{b^m} = b^{n-m}$

$(b^n)^m = b^{mn}$

$(ab)^n = a^n b^n$

$\left(\dfrac{a}{b}\right)^n = \dfrac{a^n}{b^n}$

Interval Notation	Set Notation
(a, ∞)	$\{x \mid x > a\}$
$(-\infty, b)$	$\{x \mid x < b\}$
(a, b)	$\{x \mid a < x < b\}$
$[a, \infty)$	$\{x \mid x \geq a\}$
$(-\infty, b]$	$\{x \mid x \leq b\}$
$(a, b]$	$\{x \mid a < x \leq b\}$
$[a, b)$	$\{x \mid a \leq x < b\}$
$[a, b]$	$\{x \mid a \leq x \leq b\}$

Properties of Logarithms

$\log_b b = 1$

$\log_b 1 = 0$

$\log_b rs = \log_b r + \log_b s$

$\log_b\left(\dfrac{r}{s}\right) = \log_b r - \log_b s$

$\log_b r^p = p(\log_b r)$

Factoring Patterns

$a^2 - b^2 = (a + b)(a - b)$

$a^3 - b^3 = (a - b)(a^2 + ab + b^2)$

$a^3 + b^3 = (a + b)(a^2 - ab + b^2)$

Equations Determining Functions

Linear function: $f(x) = ax + b$

Quadratic function: $f(x) = ax^2 + bx + c$

Polynomial function: $f(x) = a_n x^n + a_{n-1} x^{n-1} + \cdots + a_1 x + a_0$, where n is a whole number

Rational function: $f(x) = \dfrac{g(x)}{h(x)}$, where g and h are polynomials; $h(x) \neq 0$

Exponential function: $f(x) = b^x$, where $b > 0$ and $b \neq 1$

Logarithmic function: $f(x) = \log_b x$, where $b > 0$ and $b \neq 1$